# Antenna Azimuth Position Control System



## Layout

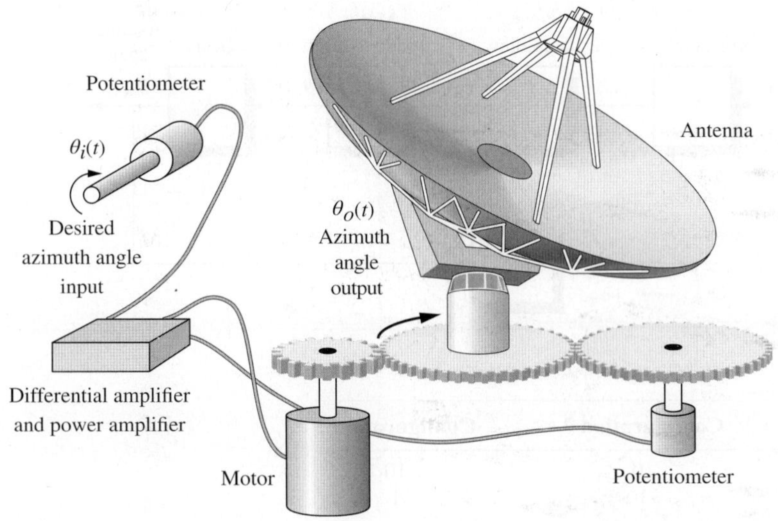


## Schematic

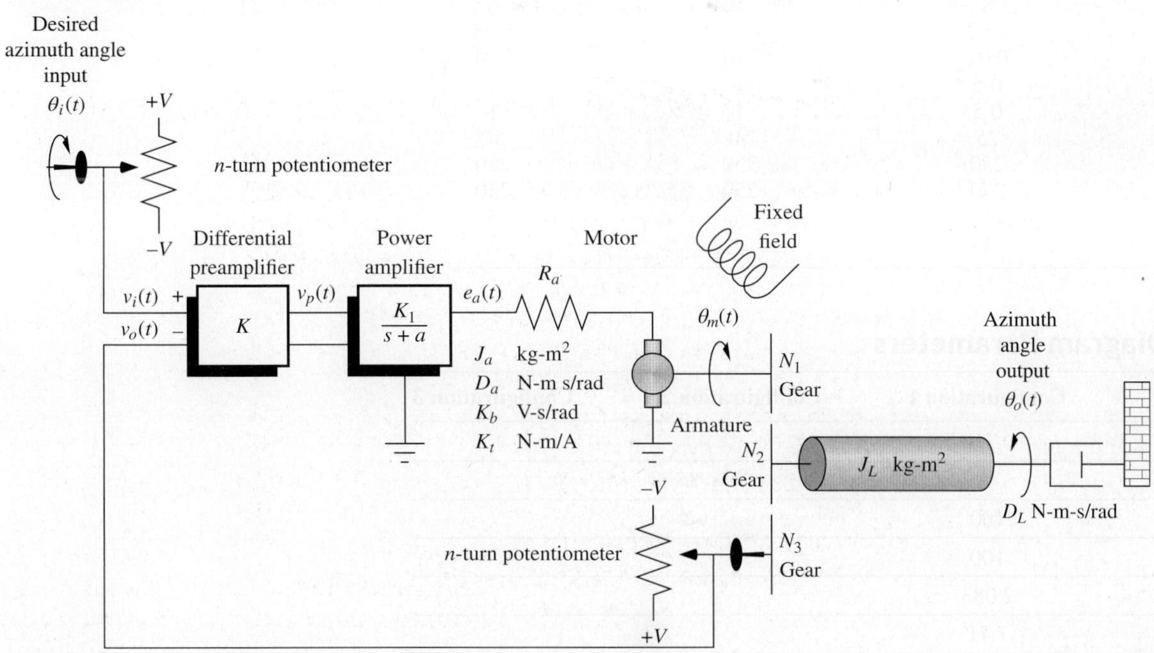

## Block Diagram

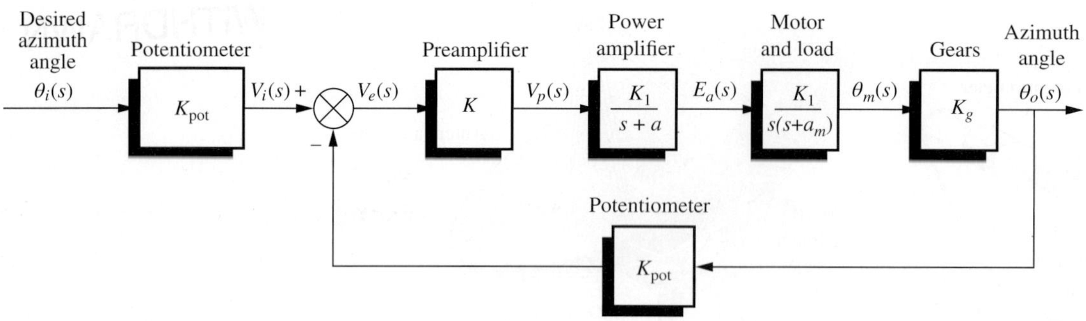


## Schematic Parameters

| Parameter | Configuration 1 | Configuration 2 | Configuration 3 |
|---|---|---|---|
| $V$ | 10 | 10 | 10 |
| $n$ | 10 | 1 | 1 |
| $K$ | — | — | — |
| $K_1$ | 100 | 150 | 100 |
| $a$ | 100 | 150 | 100 |
| $R_a$ | 8 | 5 | 5 |
| $J_a$ | 0.02 | 0.05 | 0.05 |
| $D_a$ | 0.01 | 0.01 | 0.01 |
| $K_b$ | 0.5 | 1 | 1 |
| $K_t$ | 0.5 | 1 | 1 |
| $N_1$ | 25 | 50 | 50 |
| $N_2$ | 250 | 250 | 250 |
| $N_3$ | 250 | 250 | 250 |
| $J_L$ | 1 | 5 | 5 |
| $D_L$ | 1 | 3 | 3 |

## Block Diagram Parameters

| Parameter | Configuration 1 | Configuration 2 | Configuration 3 |
|---|---|---|---|
| $K_{pot}$ | 0.318 | | |
| $K$ | — | | |
| $K_1$ | 100 | | |
| $a$ | 100 | | |
| $K_m$ | 2.083 | | |
| $a_m$ | 1.71 | | |
| $K_g$ | 0.1 | | |

Note: reader may fill in Configuration 2 and Configuration 3 columns after completing the antenna control Case Study challenge problems in Chapters 2 and 10, respectively.

# NISE'S CONTROL SYSTEMS ENGINEERING

**Global Edition**

**Norman S. Nise**
*California State Polytechnic University, Pomona*

WILEY

Cover image: Digital_Art / Shutterstock

Founded in 1807, John Wiley & Sons, Inc. has been a valued source of knowledge and understanding for more than 200 years, helping people around the world meet their needs and fulfill their aspirations. Our company is built on a foundation of principles that include responsibility to the communities we serve and where we live and work. In 2008, we launched a Corporate Citizenship Initiative, a global effort to address the environmental, social, economic, and ethical challenges we face in our business. Among the issues we are addressing are carbon impact, paper specifications and procurement, ethical conduct within our business and among our vendors, and community and charitable support. For more information, please visit our website: www.wiley.com/go/citizenship.



ISBN: 978-1-119-38297-3

Printed in Singapore by Markono Print Media Pte Ltd

10 9 8 7 6 5 4 3 2 1

# Contents

Online location is www.wiley.com/college/nise

# Preface

This book introduces students to the theory and practice of control systems engineering. The text emphasizes the practical application of the subject to the analysis and design of feedback systems.

The study of control systems engineering is essential for students pursuing degrees in electrical, mechanical, aerospace, biomedical, or chemical engineering. Control systems are found in a broad range of applications within these disciplines, from aircraft and spacecraft to robots and process control systems.

*Control Systems Engineering* is suitable for upper-division college and university engineering students and for those who wish to master the subject matter through self-study. The student using this text should have completed typical lower-division courses in physics and mathematics through differential equations. Other required background material, including Laplace transforms and linear algebra, is incorporated in the text, either within chapter discussions or separately in the appendixes or on the book's Companion Web site. This review material can be omitted without loss of continuity if the student does not require it.

## Key Features

The key features of this Global edition are:

- Standardized chapter organization
- Qualitative and quantitative explanations
- **Examples, Skill-Assessment Exercises,** and **Case Studies** throughout the text
- **Cyber Exploration Laboratory and Hardware Interface Laboratory**
- Abundant illustrations
- Numerous end-of-chapter problems
- Emphasis on design
- Flexible coverage
- Emphasis on computer-aided analysis and design including MATLAB®[1] and LabVIEW®[2]
- Labels identifying major topics

Let us look at each feature in more detail.

[1] MATLAB is a registered trademark of The MathWorks, Inc.

[2] LabVIEW is a registered trademark of National Instruments Corporation.

## Standardized Chapter Organization

Each chapter begins with a list of chapter learning outcomes, followed by a list of case study learning outcomes that relate to specific student performance in solving a practical case study problem, such as an antenna azimuth position control system.

Topics are then divided into clearly numbered and labeled sections containing explanations, examples, and, where appropriate, skill-assessment exercises with answers. These numbered sections are followed by one or more case studies, as will be outlined in a few paragraphs. Each chapter ends with a brief summary, several review questions requiring short answers, a set of homework problems, and experiments.

## Qualitative and Quantitative Explanations

Explanations are clear and complete and, where appropriate, include a brief review of required background material. Topics build upon and support one another in a logical fashion. Groundwork for new concepts and terminology is carefully laid to avoid overwhelming the student and to facilitate self-study.

Although quantitative solutions are obviously important, a qualitative or intuitive understanding of problems and methods of solution is vital to producing the insight required to develop sound designs. Therefore, whenever possible, new concepts are discussed from a qualitative perspective before quantitative analysis and design are addressed. For example, in Chapter 8 the student can simply look at the root locus and describe qualitatively the changes in transient response that will occur as a system parameter, such as gain, is varied. This ability is developed with the help of a few simple equations from Chapter 4.

## Examples, Skill-Assessment Exercises, and Case Studies

Explanations are clearly illustrated by means of numerous numbered and labeled **Examples** throughout the text. Where appropriate, sections conclude with **Skill-Assessment Exercises**. These are computation drills, most with answers that test comprehension and provide immediate feedback. Complete solutions can be found within the Student and Instructor Resources.

Broader examples in the form of **Case Studies** can be found after the last numbered section of every chapter, with the exception of Chapter 1. These case studies are practical application problems that demonstrate the concepts introduced in the chapter. Each case study concludes with a "Challenge" problem that students may work in order to test their understanding of the material.

One of the case studies, an antenna azimuth position control system, is carried throughout the book. The purpose is to illustrate the application of new material in each chapter to the same physical system, thus highlighting the continuity of the design process. Another, more challenging case study, involving an Unmannered Free-Swimming Submersible Vehicle, is developed over the course of five chapters.

## Cyber Exploration Laboratory and Hardware Interface Laboratory

Computer experiments using MATLAB, Simulink®[3] and the Control System Toolbox are found at the end of the Problems sections under the sub-heading **Cyber Exploration Laboratory**. The experiments allow the reader to verify the concepts covered in the chapter via simulation. The reader also can change parameters and perform "what if" exploration to gain insight into the effect of parameter and configuration changes. The experiments are written with stated Objectives, Minimum Required Software Packages, as well as Prelab, Lab, and Postlab tasks and questions. Thus, the experiments may be used for a laboratory course that accompanies the class. Cover sheets for these experiments are available within Resources.

---

[3] Simulink is a registered trademark of The MathWorks, Inc.

Subsequent to the Cyber Exploration Laboratory experiments, and new to this edition, are Hardware Interface Laboratory experiments in some chapters. These experiments use National Instruments' myDAQ to interface your computer to actual hardware to test control system principles in the real world.

## Abundant Illustrations

The ability to visualize concepts and processes is critical to the student's understanding. For this reason, approximately 800 photos, diagrams, graphs, and tables appear throughout the book to illustrate the topics under discussion.

## Numerous End-of-Chapter Problems

Each chapter ends with a variety of homework problems that allow students to test their understanding of the material presented in the chapter. Problems vary in degree of difficulty and complexity, and most chapters include several practical, real-life problems to help maintain students' motivation. Also, the homework problems contain progressive analysis and design problems that use the same practical systems to demonstrate the concepts of each chapter.

## Emphasis on Design

This textbook places a heavy emphasis on design. Chapters 8, 9, 11, 12, and 13 focus primarily on design. But. even in chapters that emphasize analysis, simple design examples are included wherever possible.

Design D

Throughout the book, design examples involving physical systems are identified. End-of-chapter problems that involve the design of physical systems are included under the separate heading **Design Problems**. Design Problems also can be found in chapters covering design, under the heading **Progressive Analysis and Design Problems**. In these examples and problems, a desired response is specified, and the student must evaluate certain system parameters, such as gain, or specify a system configuration along with parameter values. In addition, the text includes numerous design examples and problems that involve purely mathematical systems.

Because visualization is so vital to understanding design, this text carefully relates indirect design specifications to more familiar ones. For example, the less familiar and indirect phase margin is carefully related to the more direct and familiar percent overshoot before being used as a design specification.

For each general type of design problem introduced in the text, a methodology for solving the problem is presented—in many cases in the form of a step-by-step procedure, beginning with a statement of design objectives. Example problems serve to demonstrate the methodology by following the procedure, making simplifying assumptions, and presenting the results of the design in tables or plots that compare the performance of the original system to that of the improved system. This comparison also serves as a check on the simplifying assumptions.

Transient response design topics are covered comprehensively in the text. They include:

- Design via gain adjustment using the root locus
- Design of compensation and controllers via the root locus
- Design via gain adjustment using sinusoidal frequency response methods
- Design of compensation via sinusoidal frequency response methods
- Design of controllers in state space using pole-placement techniques

- Design of observers in state-space using pole-placement techniques
- Design of digital control systems via gain adjustment on the root locus
- Design of digital control system compensation via $s$-plane design and the Tustin transformation

Steady-state error design is covered comprehensively in this textbook and includes:

- Gain adjustment
- Design of compensation via the root locus
- Design of compensation via sinusoidal frequency response methods
- Design of integral control in state space

Finally, the design of gain to yield stability is covered from the following perspectives:

- Routh-Hurwitz criterion
- Root locus
- Nyquist criterion
- Bode plots

## Flexible Coverage

The material in this book can be adapted for a one-quarter or a one-semester course. The organization is flexible, allowing the instructor to select the material that best suits the requirements and time constraints of the class.

State Space
SS

Throughout the book, state-space methods are presented along with the classical approach. Chapters and sections (as well as examples, exercises, review questions, and problems) that cover state space are marked by the State Space label and can be omitted without any loss of continuity. Those wishing to add a basic introduction to state-space modeling can include Chapter 3 in the syllabus.

In a one-semester course, the discussions of state-space analysis in Chapters 4, 5, 6 and 7, as well as state-space design in Chapter 12, can be covered along with the classical approach. Another option is to teach state space separately by gathering the appropriate chapters and sections marked with the **State Space** label into a single unit that follows the classical approach. In a one-quarter course, Chapter 13, Digital Control Systems, could be eliminated.

## Emphasis on Computer-Aided Analysis and Design

Control systems problems, particularly analysis and design problems using the root locus, can be tedious, since their solution involves trial and error. To solve these problems, students should be given access to computers or programmable calculators configured with appropriate software. In this edition, MATLAB and LabVIEW continue to be integrated into the text as an optional feature.

Many problems in this text can be solved with either a computer or a hand-held programmable calculator. For example, students can use the programmable calculator to (1) determine whether a point on the $s$-plane is also on the root locus, (2) find magnitude and phase frequency response data for Nyquist and Bode diagrams, and (3) convert between the following representations of a second-order system:

- Pole location in polar coordinates
- Pole location in Cartesian coordinates
- Characteristic polynomial
- Natural frequency and damping ratio
- Settling time and percent overshoot
- Peak time and percent overshoot
- Settling time and peak time

Handheld calculators have the advantage of easy accessibility for homework and exams. Please consult Appendix H for a discussion of computational aids that can be adapted to handheld calculators.

Personal computers are better suited for more computation-intensive applications, such as plotting time responses, root loci, and frequency response curves, as well as finding state-transition matrices. These computers also give the student a real-world environment in which to analyze and design control systems. Those not using MATLAB or LabVIEW can write their own programs or use other programs, such as Program CC. Please consult Appendix H for a discussion of computational aids that can be adapted for use on computers that do not have MATLAB or LabVIEW installed.

Without access to computers or programmable calculators, students cannot obtain meaningful analysis and design results and the learning experience will be limited.

### Label Identifying Major Topics

Several label identify coverage and optional material. The icons are summarized as follows:

MATLAB ML

The MATLAB label identify MATLAB discussions, examples, exercises, and problems. MATLAB coverage is provided as an enhancement and is not required to use the text.

Simulink SL

The Simulink label identify Simulink discussions, examples, exercises, and problems. Simulink coverage is provided as an enhancement and is not required to use the text.

GUI Tool GUIT

The GUI Tool label identify MATLAB GUI Tools discussions, examples, exercises, and problems. The discussion of the tools, which includes the LTI Viewer, the Simulink LTI Viewer, and the SISO Design Tool, is provided as an enhancement and is not required to use the text.

Symbolic Math SM

The Symbolic Math label identify Symbolic Math Toolbox discussions, examples, exercises, and problems. Symbolic Math Toolbox coverage is provided as an enhancement and is not required to use the text.

LabVIEW LV

The LabVIEW label identify LabVIEW discussions, examples, exercises, and problems. LabVIEW is provided as an enhancement and is not required to use the text.

State Space SS

The State Space label highlight state-space discussions, examples, exercises, and problems. State-space material is optional and can be omitted without loss of continuity.

Design D

The Design label clearly identify design problems involving physical systems.

## New to This Edition

The following list describes the key changes in this global edition:

### End-of-chapter problems

Many end-of-chapter problems have been revised. Also, an additional Progressive Analysis and Design Problem has been added at the end of the chapter problems. The new progressive problem analyzes and designs a solar energy parabolic trough collector.

### MATLAB

The use of MATLAB for computer-aided analysis and design continues to be integrated into discussions and problems as an optional feature in the seventh edition. The MATLAB tutorial has been updated to MATLAB Version 8.3 (R2014a), the Control System Toolbox Version 9.7, and the Symbolic Math Toolbox Version 6.0

In addition, MATLAB code continues to be incorporated in the chapters in the form of sidebar boxes entitled TryIt.

### Simulink

The use of Simulink to show the effects of nonlinearities upon the time response of open-loop and closed-loop systems appears again in this seventh edition. We also continue to use Simulink to demonstrate how to simulate digital systems. Finally, the Simulink tutorial has been updated to Simulink 8.3 and a new section has been added: **Using Simulink for Control System Design**.

### LabVIEW

LabVIEW continues to be integrated in problems and experiments. LabVIEW has been updated to LabVIEW 2013.

### Hardware Interface Laboratory

New to this edition are experiments that use National Instruments' myDAQ to control an actual motor. A tutorial to familiarize the reader with myDAQ is added to Appendix D LabVIEW Tutorial.

## Student and Instructor Resources in Learning Space

For the Student:

- All M-files used in the MATLAB, Simulink, GUI Tools, and Symbolic Math Toolbox tutorials, as well as the TryIt exercises
- Copies of the Cyber Exploration Laboratory experiments for use as experiment cover sheets
- Solutions to the Skill-Assessment Exercises in the text
- LabVIEW VIs used in Appendix D
- All files required to perform Hardware Interface Laboratory experiments using National Instruments myDAQ

For the Instructor;

- PowerPoint®[4] files containing the figures from the textbook
- Solutions to end-of-chapter problem sets
- Simulations, developed by JustAsk, for inclusion in lecture presentations

## Book Organization by Chapter

Many times it is helpful to understand an author's reasoning behind the organization of the course material. The following paragraphs hopefully shed light on this topic.

The primary goal of Chapter 1 is to motivate students. In this chapter, students learn about the many applications of control systems in everyday life and about the advantages of study and a career in this field. Control systems engineering design objectives, such as transient response, steady-state error, and stability, are introduced, as is the path to obtaining these objectives. New and unfamiliar terms also are included in the Glossary.

Many students have trouble with an early step in the analysis and design sequence: transforming a physical system into a schematic. This step requires many simplifying assumptions based on experience the typical college student does not yet possess. Identifying some of these assumptions in Chapter 1 helps to fill the experience gap.

Chapters 2, 3, and 5 address the representation of physical systems. Chapters 2 and 3 cover modeling of open-loop systems, using frequency response techniques and state-space techniques, respectively. Chapter 5 discusses the representation and reduction of systems formed of interconnected open-loop subsystems. Only a representative sample of physical

[4] PowerPoint is a registered trademark of Microsoft Corporation.

systems can be covered in a textbook of this length. Electrical, mechanical (both translational and rotational), and electromechanical systems are used as examples of physical systems that are modeled, analyzed, and designed. Linearization of a nonlinear system—one technique used by the engineer to simplify a system in order to represent it mathematically—is also introduced.

Chapter 4 provides an introduction to system analysis, that is, finding and describing the output response of a system. It may seem more logical to reverse the order of Chapters 4 and 5, to present the material in Chapter 4 along with other chapters covering analysis. However, many years of teaching control systems have taught me that the sooner students see an application of the study of system representation, the higher their motivation levels remain.

Chapters 6, 7, 8, and 9 return to control systems analysis and design with the study of stability (Chapter 6), steady-state errors (Chapter 7), and transient response of higher-order systems using root locus techniques (Chapter 8). Chapter 9 covers design of compensators and controllers using the root locus.

Chapters 10 and 11 focus on sinusoidal frequency analysis and design. Chapter 10, like Chapter 8, covers basic concepts for stability, transient response, and steady-state-error analysis. However, Nyquist and Bode methods are used in place of root locus. Chapter 11, like Chapter 9, covers the design of compensators, but from the point of view of sinusoidal frequency techniques rather than root locus.

An introduction to state-space design and digital control systems analysis and design completes the text in Chapters 12 and 13, respectively. Although these chapters can be used as an introduction for students who will be continuing their study of control systems engineering, they are useful by themselves and as a supplement to the discussion of analysis and design in the previous chapters. The subject matter cannot be given a comprehensive treatment in two chapters, but the emphasis is clearly outlined and logically linked to the rest of the book.

## Acknowledgments

The author would like to acknowledge the contributions of faculty and students, both at California State Polytechnic University, Pomona, and across the country, whose suggestions through all editions have made a positive impact on this new edition.

I am deeply indebted to my colleagues, Drs. Elhami T. Ibrahim and Salomon Oldak at California State Polytechnic University, Pomona for authoring the creative new problems you will find at the end of every chapter. The new progressive problem, solar energy parabolic trough collector, that is at the end of every chapter is the creation of Dr Oldak. In addition, I am grateful to Dr. Oldak for creating the National Instruments myDAQ experiments, which you will find in the Hardware Interface Laboratory section of many chapters. Finally, I want to thank Dr. Norali Pernalete, also of California State Polytechnic University, Pomona, for her contribution of the LabVIEW experiments and problems continuing in this edition.

I would like to express my appreciation to those who participated in reviews of this edition. They are: James R. McCusker, Wentworth Institute of Technology, and Ji-chul Ryu, Northern Illinois University.

The author would like to thank John Wiley & Sons, Inc. and its staff for once again providing professional support for this project through all phases of its development. Specifically, the following are due recognition for their contributions: Don Fowley, Vice President and Publisher, who gave full corporate support to the project; Dan Sayre, Executive Editor, with whom I worked closely and who provided guidance and leadership throughout the development of this edition; Francesca Baratta and Jessica Knecht, Editorial Assistants who were always there to answer my questions and respond to my concerns in a professional manner; and Anna Melhorn, Senior Production Editor, who turned the manuscript into the final product you are holding in your hands. Other contributors who worked hard behind the scenes are: Wendy Lai, Designer; Harry Nolan, Director of Design;

Jennifer Welter, Senior Product Designer; Kevin Holm, Learning Space Content Manager; and Wendy Ashenberg, Content Editor.

My sincere appreciation is extended to Erik Luther of National Instruments Corporation and Paul Gilbert, Michel Levis, and Tom Lee of Quanser for conceiving, coordinating, and developing the Virtual Experiments that I am sure will enhance your understanding of control systems. Others from National Instruments who contributed to the successful publication of this book are Margaret Barrett and Kathy Brown.

Finally, last but certainly not least, I want to express my appreciation to my wife, Ellen, for her support in ways too numerous to mention during the writing of all editions. Specifically, though, thanks to her proofing pages for this edition, you, the reader, hopefully will find comprehension rather than apprehension in the pages that follow.

Norman S. Nise

# 1

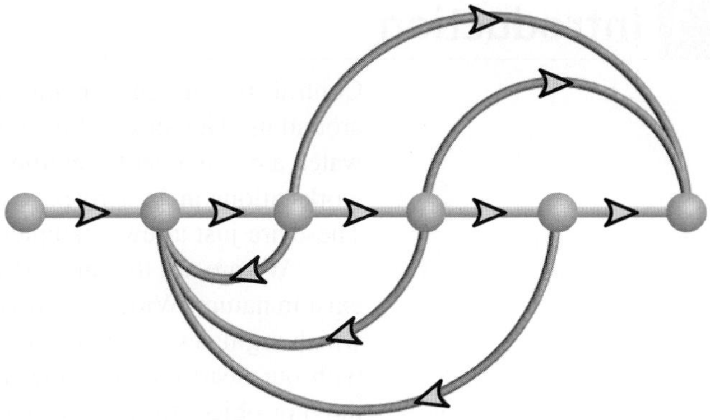

# Introduction

## Chapter Learning Outcomes

After completing this chapter, the student will be able to:

- Define a control system and describe some applications (Section 1.1)
- Describe historical developments leading to modern day control theory (Section 1.2)
- Describe the basic features and configurations of control systems (Section 1.3)
- Describe control systems analysis and design objectives (Section 1.4)
- Describe a control system's design process (Sections 1.5–1.6)
- Describe the benefit from studying control systems (Section 1.7)

## ▶ Case Study Learning Outcomes

- You will be introduced to a running case study—an antenna azimuth position control system—that will serve to illustrate the principles in each subsequent chapter. In this chapter, the system is used to demonstrate qualitatively how a control system works as well as to define performance criteria that are the basis for control systems analysis and design.

# 1.1 Introduction

Control systems are an integral part of modern society. Numerous applications are all around us: The rockets fire, and the space shuttle lifts off to earth orbit; in splashing cooling water, a metallic part is automatically machined; a self-guided vehicle delivering material to workstations in an aerospace assembly plant glides along the floor seeking its destination. These are just a few examples of the automatically controlled systems that we can create.

We are not the only creators of automatically controlled systems; these systems also exist in nature. Within our own bodies are numerous control systems, such as the pancreas, which regulates our blood sugar. In time of "fight or flight," our adrenaline increases along with our heart rate, causing more oxygen to be delivered to our cells. Our eyes follow a moving object to keep it in view; our hands grasp the object and place it precisely at a predetermined location.

Even the nonphysical world appears to be automatically regulated. Models have been suggested showing automatic control of student performance. The input to the model is the student's available study time, and the output is the grade. The model can be used to predict the time required for the grade to rise if a sudden increase in study time is available. Using this model, you can determine whether increased study is worth the effort during the last week of the term.

## Control System Definition

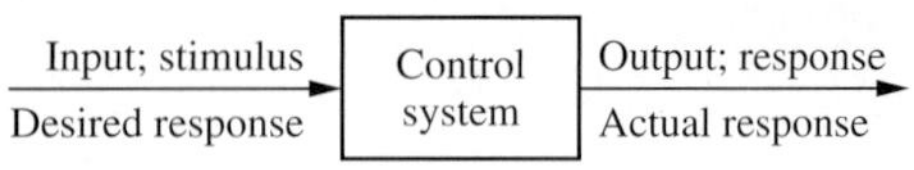


**FIGURE 1.1** Simplified description of a control system

A control system consists of *subsystems* and *processes* (or *plants*) assembled for the purpose of obtaining a desired *output* with desired *performance*, given a specified *input*. Figure 1.1 shows a control system in its simplest form, where the input represents a desired output.

For example, consider an elevator. When the fourth-floor button is pressed on the first floor, the elevator rises to the fourth floor with a speed and floor-leveling accuracy designed for passenger comfort. The push of the fourth-floor button is an *input* that represents our desired *output*, shown as a step function in Figure 1.2. The *performance* of the elevator can be seen from the elevator response curve in the figure.

Two major measures of performance are apparent: (1) the transient response and (2) the steady-state error. In our example, passenger comfort and passenger patience are dependent upon the transient response. If this response is too fast, passenger comfort is sacrificed; if too slow, passenger patience is sacrificed. The steady-state error is another important performance specification since passenger safety and convenience would be sacrificed if the elevator did not level properly.

## Advantages of Control Systems

With control systems we can move large equipment with precision that would otherwise be impossible. We can point huge antennas toward the farthest reaches of the universe to pick up faint radio signals; controlling these antennas by hand would be impossible. Because of control systems, elevators carry us quickly to our destination, automatically

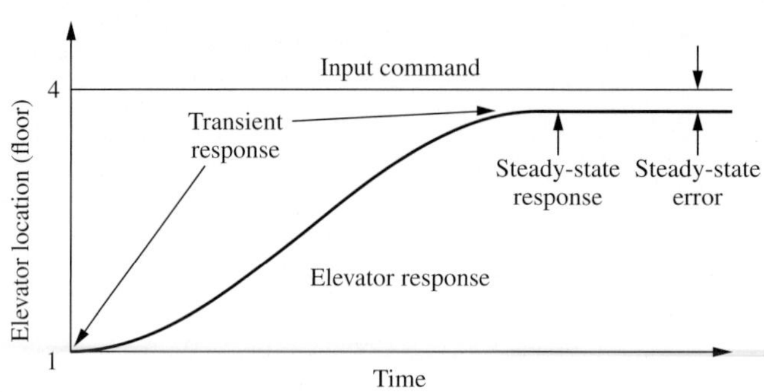


**FIGURE 1.2** Elevator response

(*a*) (*b*)

Bettman/Corbis. Courtesy of United Techonolgies Otis Elevator.

Art on File/Corbis

**FIGURE 1.3** **a.** Early elevators were controlled by hand ropes or an elevator operator. Here a rope is cut to demonstrate the safety brake, an innovation in early elevators; **b.** One of two modern Duo-lift elevators makes its way up the Grande Arche in Paris. Two elevators are driven by one motor, with each car acting as a counterbalance to the other. Today, elevators are fully automatic, using control systems to regulate position and velocity.

stopping at the right floor (Figure 1.3). We alone could not provide the power required for the load and the speed; motors provide the power, and control systems regulate the position and speed.

We build control systems for four primary reasons:

1. Power amplification
2. Remote control
3. Convenience of input form
4. Compensation for disturbances

For example, a radar antenna, positioned by the low-power rotation of a knob at the input, requires a large amount of power for its output rotation. A control system can produce the needed power amplification, or power *gain*.

Robots designed by control system principles can compensate for human disabilities. Control systems are also useful in remote or dangerous locations. For example, a remote-controlled robot arm can be used to pick up material in a radioactive environment. Figure 1.4 shows a robot arm designed to work in contaminated environments.

Control systems can also be used to provide convenience by changing the form of the input. For example, in a temperature control system, the input is a *position* on a thermostat. The output is *heat*. Thus, a convenient position input yields a desired thermal output.

Another advantage of a control system is the ability to compensate for disturbances. Typically, we control such variables as temperature in thermal systems, position and velocity in mechanical systems, and voltage, current, or frequency in electrical systems. The system must be able to yield the correct output even with a disturbance. For example, consider an antenna system that points in a commanded direction. If wind forces the antenna from its commanded position, or if noise enters internally, the system must be able to detect the disturbance and correct the antenna's position.

Hank Mogan/Science Source

**FIGURE 1.4** *Rover* was built to work in contaminated areas at Three Mile Island in Middleton, Pennsylvania, where a nuclear accident occurred in 1979. The remote-controlled robot's long arm can be seen at the front of the vehicle.

Obviously, the system's input will not change to make the correction. Consequently, the system itself must measure the amount that the disturbance has repositioned the antenna and then return the antenna to the position commanded by the input.

## 1.2 A History of Control Systems

Feedback control systems are older than humanity. Numerous biological control systems were built into the earliest inhabitants of our planet. Let us now look at a brief history of human-designed control systems.[1]

### Liquid-Level Control

The Greeks began engineering feedback systems around 300 B.C. A water clock invented by Ktesibios operated by having water trickle into a measuring container at a constant rate. The level of water in the measuring container could be used to tell time. For water to trickle at a constant rate, the supply tank had to be kept at a constant level. This was accomplished using a float valve similar to the water-level control in today's flush toilets.

Soon after Ktesibios, the idea of liquid-level control was applied to an oil lamp by Philon of Byzantium. The lamp consisted of two oil containers configured vertically. The lower pan was open at the top and was the fuel supply for the flame. The closed upper bowl was the fuel reservoir for the pan below. The containers were interconnected by two capillary tubes and another tube, called a *vertical riser*, which was inserted into the oil in the lower pan just below the surface. As the oil burned, the base of the vertical riser was exposed to air, which forced oil in the reservoir above to flow through the capillary tubes and into the pan. The transfer of fuel from the upper reservoir to the pan stopped when the previous oil level in the pan was reestablished, thus blocking the air from entering the vertical riser. Hence, the system kept the liquid level in the lower container constant.

### Steam Pressure and Temperature Controls

Regulation of steam pressure began around 1681 with Denis Papin's invention of the safety valve. The concept was further elaborated on by weighting the valve top. If the upward pressure from the boiler exceeded the weight, steam was released, and the pressure decreased. If it did not exceed the weight, the valve did not open, and the pressure inside the boiler increased. Thus, the weight on the valve top set the internal pressure of the boiler.

Also in the seventeenth century, Cornelis Drebbel in Holland invented a purely mechanical temperature control system for hatching eggs. The device used a vial of alcohol and mercury with a floater inserted in it. The floater was connected to a damper that controlled a flame. A portion of the vial was inserted into the incubator to sense the heat generated by the fire. As the heat increased, the alcohol and mercury expanded, raising the floater, closing the damper, and reducing the flame. Lower temperature caused the float to descend, opening the damper and increasing the flame.

### Speed Control

In 1745, speed control was applied to a windmill by Edmund Lee. Increasing winds pitched the blades farther back, so that less area was available. As the wind decreased, more blade area was available. William Cubitt improved on the idea in 1809 by dividing the windmill sail into movable louvers.

Also in the eighteenth century, James Watt invented the flyball speed governor to control the speed of steam engines. In this device, two spinning flyballs rise as rotational speed increases. A steam valve connected to the flyball mechanism closes with the ascending flyballs and opens with the descending flyballs, thus regulating the speed.

[1] See (*Bennett, 1979*) and (*Mayr, 1970*) for definitive works on the history of control systems.

## Stability, Stabilization, and Steering

Control systems theory as we know it today began to crystallize in the latter half of the nineteenth century. In 1868, James Clerk Maxwell published the stability criterion for a third-order system based on the coefficients of the differential equation. In 1874, Edward John Routh, using a suggestion from William Kingdon Clifford that was ignored earlier by Maxwell, was able to extend the stability criterion to fifth-order systems. In 1877, the topic for the Adams Prize was "The Criterion of Dynamical Stability." In response, Routh submitted a paper entitled *A Treatise on the Stability of a Given State of Motion* and won the prize. This paper contains what is now known as the Routh-Hurwitz criterion for stability, which we will study in Chapter 6. Alexandr Michailovich Lyapunov also contributed to the development and formulation of today's theories and practice of control system stability. A student of P. L. Chebyshev at the University of St. Petersburg in Russia, Lyapunov extended the work of Routh to nonlinear systems in his 1892 doctoral thesis, entitled *The General Problem of Stability of Motion*.

During the second half of the 1800s, the development of control systems focused on the steering and stabilizing of ships. In 1874, Henry Bessemer, using a gyro to sense a ship's motion and applying power generated by the ship's hydraulic system, moved the ship's saloon to keep it stable (whether this made a difference to the patrons is doubtful). Other efforts were made to stabilize platforms for guns as well as to stabilize entire ships, using pendulums to sense the motion.

## Twentieth-Century Developments

It was not until the early 1900s that automatic steering of ships was achieved. In 1922, the Sperry Gyroscope Company installed an automatic steering system that used the elements of compensation and adaptive control to improve performance. However, much of the general theory used today to improve the performance of automatic control systems is attributed to Nicholas Minorsky, a Russian born in 1885. It was his theoretical development applied to the automatic steering of ships that led to what we call today proportional-plus-integral-plus-derivative (PID), or three-mode, controllers, which we will study in Chapters 9 and 11.

In the late 1920s and early 1930s, H. W. Bode and H. Nyquist at Bell Telephone Laboratories developed the analysis of feedback amplifiers. These contributions evolved into sinusoidal frequency analysis and design techniques currently used for feedback control system, and are presented in Chapters 10 and 11.

In 1948, Walter R. Evans, working in the aircraft industry, developed a graphical technique to plot the roots of a characteristic equation of a feedback system whose parameters changed over a particular range of values. This technique, now known as the root locus, takes its place with the work of Bode and Nyquist in forming the foundation of linear control systems analysis and design theory. We will study root locus in Chapters 8, 9, and 13.

## Contemporary Applications

Today, control systems find widespread application in the guidance, navigation, and control of missiles and spacecraft, as well as planes and ships at sea. For example, modern ships use a combination of electrical, mechanical, and hydraulic components to develop rudder commands in response to desired heading commands. The rudder commands, in turn, result in a rudder angle that steers the ship.

We find control systems throughout the process control industry, regulating liquid levels in tanks, chemical concentrations in vats, as well as the thickness of fabricated material. For example, consider a thickness control system for a steel plate finishing mill. Steel enters the finishing mill and passes through rollers. In the finishing mill, X-rays measure the actual thickness and compare it to the desired thickness. Any difference is adjusted by a screw-down position control that changes the roll gap at the rollers through which the steel passes. This change in roll gap regulates the thickness.

Modern developments have seen widespread use of the digital computer as part of control systems. For example, computers in control systems are for industrial robots, spacecraft, and the process control industry. It is hard to visualize a modern control system that does not use a digital computer.

Although recently retired, the space shuttle provides an excellent example of the use of control systems because it contained numerous control systems operated by an onboard computer on a time-shared basis. Without control systems, it would be impossible to guide the shuttle to and from earth's orbit or to adjust the orbit itself and support life on board. Navigation functions programmed into the shuttle's computers used data from the shuttle's hardware to estimate vehicle position and velocity. This information was fed to the guidance equations that calculated commands for the shuttle's flight control systems, which steered the spacecraft. In space, the flight control system gimbaled (rotated) the orbital maneuvering system (OMS) engines into a position that provided thrust in the commanded direction to steer the spacecraft. Within the earth's atmosphere, the shuttle was steered by commands sent from the flight control system to the aerosurfaces, such as the elevons.

Within this large control system represented by navigation, guidance, and control were numerous subsystems to control the vehicle's functions. For example, the elevons required a control system to ensure that their position was indeed that which was commanded, since disturbances such as wind could rotate the elevons away from the commanded position. Similarly, in space, the gimbaling of the orbital maneuvering engines required a similar control system to ensure that the rotating engine can accomplish its function with speed and accuracy. Control systems were also used to control and stabilize the vehicle during its descent from orbit. Numerous small jets that compose the reaction control system (RCS) were used initially in the exoatmosphere, where the aerosurfaces are ineffective. Control was passed to the aerosurfaces as the orbiter descended into the atmosphere.

Inside the shuttle, numerous control systems were required for power and life support. For example, the orbiter had three fuel-cell power plants that converted hydrogen and oxygen (reactants) into electricity and water for use by the crew. The fuel cells involved the use of control systems to regulate temperature and pressure. The reactant tanks were kept at constant pressure as the quantity of reactant diminishes. Sensors in the tanks sent signals to the control systems to turn heaters on or off to keep the tank pressure constant (*Rockwell International, 1984*).

Control systems are not limited to science and industry. For example, a home heating system is a simple control system consisting of a thermostat containing a bimetallic material that expands or contracts with changing temperature. This expansion or contraction moves a vial of mercury that acts as a switch, turning the heater on or off. The amount of expansion or contraction required to move the mercury switch is determined by the temperature setting.

Home entertainment systems also have built-in control systems. For example, in an optical disk recording system microscopic pits representing the information are burned into the disc by a laser during the recording process. During playback, a reflected laser beam focused on the pits changes intensity. The light intensity changes are converted to an electrical signal and processed as sound or picture. A control system keeps the laser beam positioned on the pits, which are cut as concentric circles.

There are countless other examples of control systems, from the everyday to the extraordinary. As you begin your study of control systems engineering, you will become more aware of the wide variety of applications.

## 1.3 System Configurations

In this section, we discuss two major configurations of control systems: open loop and closed loop. We can consider these configurations to be the internal architecture of the total system shown in Figure 1.1. Finally, we show how a digital computer forms part of a control system's configuration.

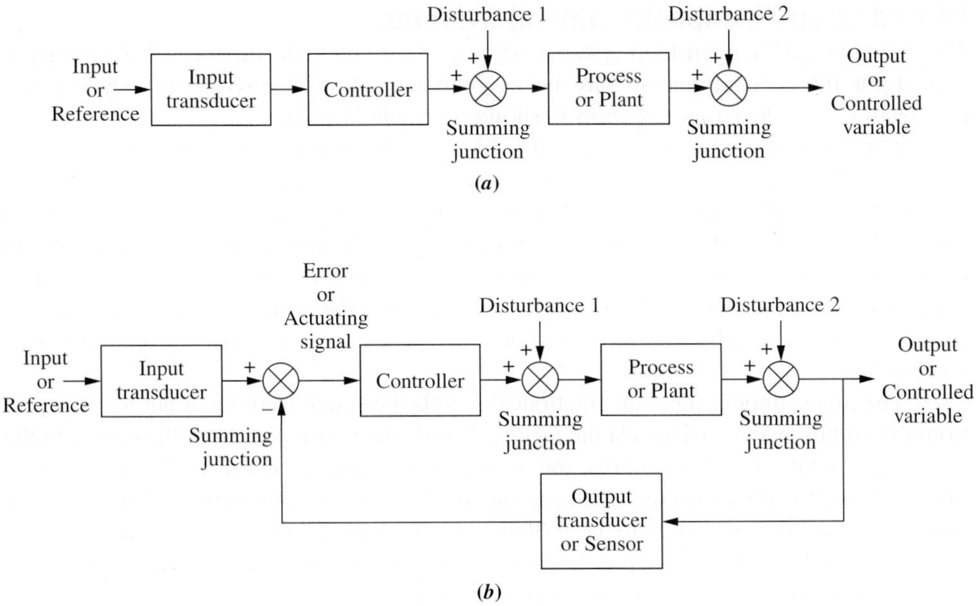


**FIGURE 1.5** Block diagrams of control systems: **a.** open-loop system; **b.** closed-loop system

## Open-Loop Systems

A generic *open-loop system* is shown in Figure 1.5(*a*). It starts with a subsystem called an *input transducer*, which converts the form of the input to that used by the *controller*. The controller drives a *process* or a *plant*. The input is sometimes called the *reference*, while the output can be called the *controlled variable*. Other signals, such as *disturbances*, are shown added to the controller and process outputs via *summing junctions*, which yield the algebraic sum of their input signals using associated signs. For example, the plant can be a furnace or air conditioning system, where the output variable is temperature. The controller in a heating system consists of fuel valves and the electrical system that operates the valves.

The distinguishing characteristic of an open-loop system is that it cannot compensate for any disturbances that add to the controller's driving signal (Disturbance 1 in Figure 1.5(*a*)). For example, if the controller is an electronic amplifier and Disturbance 1 is noise, then any additive amplifier noise at the first summing junction will also drive the process, corrupting the output with the effect of the noise. The output of an open-loop system is corrupted not only by signals that add to the controller's commands but also by disturbances at the output (Disturbance 2 in Figure 1.5(*a*)). The system cannot correct for these disturbances, either.

Open-loop systems, then, do not correct for disturbances and are simply commanded by the input. For example, toasters are open-loop systems, as anyone with burnt toast can attest. The controlled variable (output) of a toaster is the color of the toast. The device is designed with the assumption that the toast will be darker the longer it is subjected to heat. The toaster does not measure the color of the toast; it does not correct for the fact that the toast is rye, white, or sourdough, nor does it correct for the fact that toast comes in different thicknesses.

Other examples of open-loop systems are mechanical systems consisting of a mass, spring, and damper with a constant force positioning the mass. The greater the force, the greater the displacement. Again, the system position will change with a disturbance, such as an additional force, and the system will not detect or correct for the disturbance. Or, assume that you calculate the amount of time you need to study for an examination that covers three chapters in order to get an A. If the professor adds a fourth chapter—a disturbance—you are an open-loop system if you do not detect the disturbance and add study time to that previously calculated. The result of this oversight would be a lower grade than you expected.

## Closed-Loop (Feedback Control) Systems

The disadvantages of open-loop systems, namely sensitivity to disturbances and inability to correct for these disturbances, may be overcome in *closed-loop systems*. The generic architecture of a closed-loop system is shown in Figure 1.5(*b*).

The input transducer converts the form of the input to the form used by the controller. An *output transducer*, or *sensor*, measures the output response and converts it into the form used by the controller. For example, if the controller uses electrical signals to operate the valves of a temperature control system, the input position and the output temperature are converted to electrical signals. The input position can be converted to a voltage by a *potentiometer*, a variable resistor, and the output temperature can be converted to a voltage by a *thermistor*, a device whose electrical resistance changes with temperature.

The first summing junction algebraically adds the signal from the input to the signal from the output, which arrives via the *feedback path*, the return path from the output to the summing junction. In Figure 1.5(*b*), the output signal is subtracted from the input signal. The result is generally called the *actuating signal*. However, in systems where both the input and output transducers have *unity gain* (that is, the transducer amplifies its input by 1), the actuating signal's value is equal to the actual difference between the input and the output. Under this condition, the actuating signal is called the *error*.

The closed-loop system compensates for disturbances by measuring the output response, feeding that measurement back through a feedback path, and comparing that response to the input at the summing junction. If there is any difference between the two responses, the system drives the plant, via the actuating signal, to make a correction. If there is no difference, the system does not drive the plant, since the plant's response is already the desired response.

Closed-loop systems, then, have the obvious advantage of greater accuracy than open-loop systems. They are less sensitive to noise, disturbances, and changes in the environment. Transient response and steady-state error can be controlled more conveniently and with greater flexibility in closed-loop systems, often by a simple adjustment of gain (amplification) in the loop and sometimes by redesigning the controller. We refer to the redesign as *compensating* the system and to the resulting hardware as a *compensator*. On the other hand, closed-loop systems are more complex and expensive than open-loop systems. A standard, open-loop toaster serves as an example: It is simple and inexpensive. A closed-loop toaster oven is more complex and more expensive since it has to measure both color (through light reflectivity) and humidity inside the toaster oven. Thus, the control systems engineer must consider the trade-off between the simplicity and low cost of an open-loop system and the accuracy and higher cost of a closed-loop system.

In summary, systems that perform the previously described measurement and correction are called closed-loop, or feedback control, systems. Systems that do not have this property of measurement and correction are called open-loop systems.

## Computer-Controlled Systems

In many modern systems, the controller (or compensator) is a digital computer. The advantage of using a computer is that many loops can be controlled or compensated by the same computer through time sharing. Furthermore, any adjustments of the compensator parameters required to yield a desired response can be made by changes in software rather than hardware. The computer can also perform supervisory functions, such as scheduling many required applications. For example, the space shuttle main engine (SSME) controller, which contained two digital computers, alone controlled numerous engine functions. It monitored engine sensors that provided pressures, temperatures, flow rates, turbopump speed, valve positions, and engine servo valve actuator positions. The controller further provided closed-loop control of thrust and propellant mixture ratio, sensor excitation, valve actuators, spark igniters, as well as other functions (*Rockwell International, 1984*).

## 1.4 Analysis and Design Objectives

In Section 1.1 we briefly alluded to some control system performance specifications, such as transient response and steady-state error. We now expand upon the topic of performance and place it in perspective as we define our analysis and design objectives.

*Analysis* is the process by which a system's performance is determined. For example, we evaluate its transient response and steady-state error to determine if they meet the desired specifications. *Design* is the process by which a system's performance is created or changed. For example, if a system's transient response and steady-state error are analyzed and found not to meet the specifications, then we change parameters or add additional components to meet the specifications.

A control system is *dynamic:* It responds to an input by undergoing a transient response before reaching a steady-state response that generally resembles the input. We have already identified these two responses and cited a position control system (an elevator) as an example. In this section, we discuss three major objectives of systems analysis and design: producing the desired transient response, reducing steady-state error, and achieving stability. We also address some other design concerns, such as cost and the sensitivity of system performance to changes in parameters.

### Transient Response

Transient response is important. In the case of an elevator, a slow transient response makes passengers impatient, whereas an excessively rapid response makes them uncomfortable. If the elevator oscillates about the arrival floor for more than a second, a disconcerting feeling can result. Transient response is also important for structural reasons: Too fast a transient response could cause permanent physical damage. In a computer, transient response contributes to the time required to read from or write to the computer's disk storage (see Figure 1.6). Since reading and writing cannot take place until the head stops, the speed of the read/write head's movement from one track on the disk to another influences the overall speed of the computer.

In this book, we establish quantitative definitions for transient response. We then analyze the system for its *existing* transient response. Finally, we adjust parameters or design components to yield a *desired* transient response—our first analysis and design objective.

Donald Swartz/iStockphoto

**FIGURE 1.6** Computer hard disk drive, showing disks and read/write head

### Steady-State Response

Another analysis and design goal focuses on the steady-state response. As we have seen, this response resembles the input and is usually what remains after the transients have decayed to zero. For example, this response may be an elevator stopped near the fourth floor or the head of a disk drive finally stopped at the correct track. We are concerned about the accuracy of the steady-state response. An elevator must be level enough with the floor for the passengers to exit, and a read/write head not positioned over the commanded track results in computer errors. An antenna tracking a satellite must keep the satellite well within its beamwidth in order not to lose track. In this text we define steady-state errors quantitatively, analyze a system's steady-state error, and then design corrective action to reduce the steady-state error—our second analysis and design objective.

## Stability

Discussion of transient response and steady-state error is moot if the system does not have *stability*. In order to explain stability, we start from the fact that the total response of a system is the sum of the *natural response* and the *forced response*. When you studied linear differential equations, you probably referred to these responses as the *homogeneous* and the *particular solutions*, respectively. Natural response describes the way the system dissipates or acquires energy. The form or nature of this response is dependent only on the system, not the input. On the other hand, the form or nature of the forced response is dependent on the input. Thus, for a *linear* system, we can write

$$\text{Total response} = \text{Natural response} + \text{Forced response} \qquad (1.1)^2$$

For a control system to be useful, the natural response must (1) eventually approach zero, thus leaving only the forced response, or (2) oscillate. In some systems, however, the natural response grows without bound rather than diminish to zero or oscillate. Eventually, the natural response is so much greater than the forced response that the system is no longer controlled. This condition, called *instability*, could lead to self-destruction of the physical device if limit stops are not part of the design. For example, the elevator would crash through the floor or exit through the ceiling; an aircraft would go into an uncontrollable roll; or an antenna commanded to point to a target would rotate, line up with the target, but then begin to oscillate about the target with *growing* oscillations and *increasing* velocity until the motor or amplifiers reached their output limits or until the antenna was damaged structurally. A time plot of an unstable system would show a transient response that grows without bound and without any evidence of a steady-state response.

Control systems must be designed to be stable. That is, their natural response must decay to zero as time approaches infinity, or oscillate. In many systems the transient response you see on a time response plot can be directly related to the natural response. Thus, if the natural response decays to zero as time approaches infinity, the transient response will also die out, leaving only the forced response. If the system is stable, the proper transient response and steady-state error characteristics can be designed. Stability is our third analysis and design objective.

## Other Considerations

The three main objectives of control system analysis and design have already been enumerated. However, other important considerations must be taken into account. For example, factors affecting hardware selection, such as motor sizing to fulfill power requirements and choice of sensors for accuracy, must be considered early in the design.

Finances are another consideration. Control system designers cannot create designs without considering their economic impact. Such considerations as budget allocations and competitive pricing must guide the engineer. For example, if your product is one of a kind, you may be able to create a design that uses more expensive components without appreciably increasing total cost. However, if your design will be used for many copies, slight increases in cost per copy can translate into many more dollars for your company to propose during contract bidding and to outlay before sales.

[2] You may be confused by the words *transient* vs. *natural*, and *steady-state* vs. *forced*. If you look at Figure 1.2, you can see the transient and steady-state portions of the total response as indicated. The transient response is the sum of the natural and forced responses, while the natural response is large. If we plotted the natural response by itself, we would get a curve that is different from the transient portion of Figure 1.2. The steady-state response of Figure 1.2 is also the sum of the natural and forced responses, but the natural response is small. Thus, the transient and steady-state responses are what you actually see on the plot; the natural and forced responses are the underlying mathematical components of those responses.

Another consideration is *robust* design. System parameters considered constant during the design for transient response, steady-state errors, and stability change over time when the actual system is built. Thus, the performance of the system also changes over time and will not be consistent with your design. Unfortunately, the relationship between parameter changes and their effect on performance is not linear. In some cases, even in the same system, changes in parameter values can lead to small or large changes in performance, depending on the system's nominal operating point and the type of design used. Thus, the engineer wants to create a robust design so that the system will not be sensitive to parameter changes. We discuss the concept of system sensitivity to parameter changes in Chapters 7 and 8. This concept, then, can be used to test a design for robustness.

## Case Study Introduction to a Case Study

Now that our objectives are stated, how do we meet them? In this section we will look at an example of a feedback control system. The system introduced here will be used in subsequent chapters as a running case study to demonstrate the objectives of those chapters. A colored background like this will identify the case study section at the end of each chapter. Section 1.5, which follows this first case study, explores the design process that will help us build our system.

Peter Menzel

**FIGURE 1.7** The search for extraterrestrial life is being carried out with radio antennas like the one pictured here. A radio antenna is an example of a system with position controls

### Antenna Azimuth: An Introduction to Position Control Systems

A position control system converts a position input command to a position output response. Position control systems find widespread applications in antennas, robot arms, and computer disk drives. The radio telescope antenna in Figure 1.7 is one example of a system that uses position control systems. In this section, we will look in detail at an antenna azimuth position control system that could be used to position a radio telescope antenna. We will see how the system works and how we can effect changes in its performance. The discussion here will be on a qualitative level, with the objective of getting an intuitive feeling for the systems with which we will be dealing.

An antenna azimuth position control system is shown in Figure 1.8(*a*), with a more detailed layout and schematic in Figures 1.8(*b*) and 1.8(*c*), respectively. Figure 1.8(*d*) shows *a functional block diagram* of the system. The functions are shown above the blocks, and the required hardware is indicated inside the blocks. Parts of Figure 1.8 are repeated on the front endpapers for future reference.

The purpose of this system is to have the azimuth angle output of the antenna, $\theta_o(t)$, follow the input angle of the potentiometer, $\theta_i(t)$. Let us look at Figure 1.8(*d*) and describe how this system works. The input command is an angular displacement. The potentiometer converts the angular displacement into a voltage. Similarly, the output angular displacement is converted to a voltage by the potentiometer in the feedback path. The signal and power amplifiers boost the difference between the input and output voltages. This amplified actuating signal drives the plant.

The system normally operates to drive the error to zero. When the input and output match, the error will be zero, and the motor will not turn. Thus, the motor is driven only when the output and the input do not match. The greater the difference between the input and the output, the larger the motor input voltage, and the faster the motor will turn.

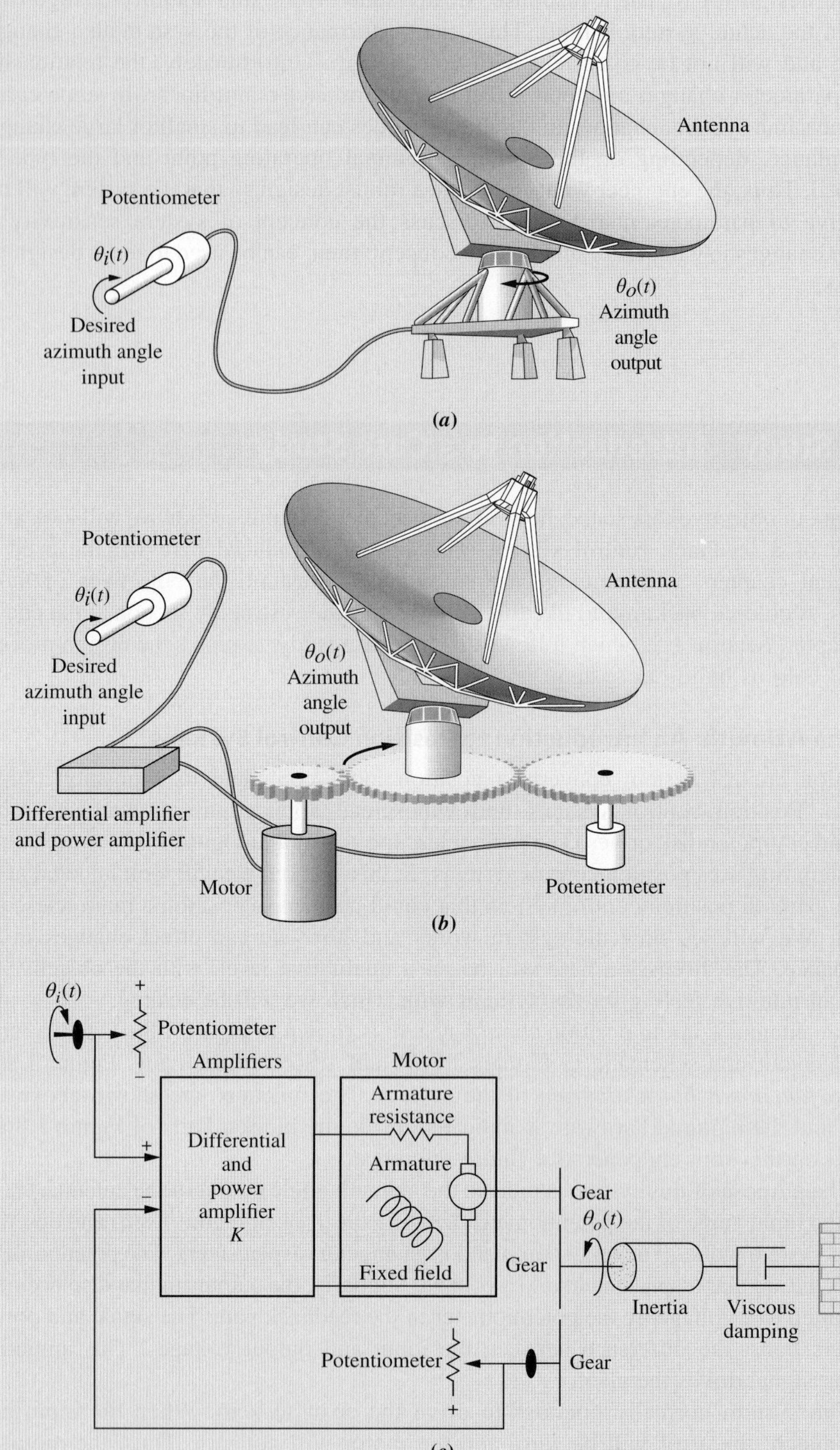


**FIGURE 1.8** Antenna azimuth position control system: **a.** system concept; **b.** detailed layout; **c.** schematic (*figure continues*)

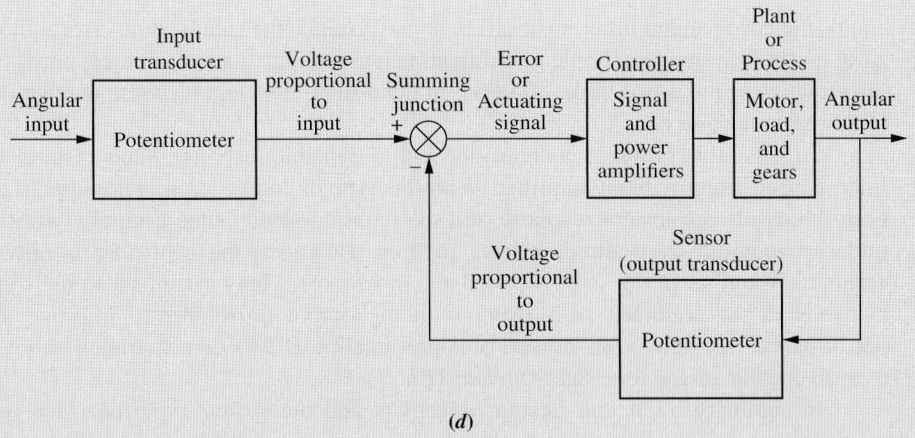


**FIGURE 1.8** (*Continued*) **d.** functional block diagram

If we increase the gain of the signal amplifier, will there be an increase in the steady-state value of the output? If the gain is increased, then for a given actuating signal, the motor will be driven harder. However, the motor will still stop when the actuating signal reaches zero, that is, when the output matches the input. The difference in the response, however, will be in the transients. Since the motor is driven harder, it turns faster toward its final position. Also, because of the increased speed, increased momentum could cause the motor to overshoot the final value and be forced by the system to return to the commanded position. Thus, the possibility exists for a transient response that consists of *damped oscillations* (that is, a sinusoidal response whose amplitude diminishes with time) about the steady-state value if the gain is high. The responses for low gain and high gain are shown in Figure 1.9.

We have discussed the transient response of the position control system. Let us now direct our attention to the steady-state position to see how closely the output matches the input after the transients disappear.

We define steady-state error as the difference between the input and the output after the transients have effectively disappeared. The definition holds equally well for step, ramp, and other types of inputs. Typically, the steady-state error decreases with an increase in gain and increases with a decrease in gain. Figure 1.9 shows zero error in the steady-state response; that is, after the transients have disappeared, the output position

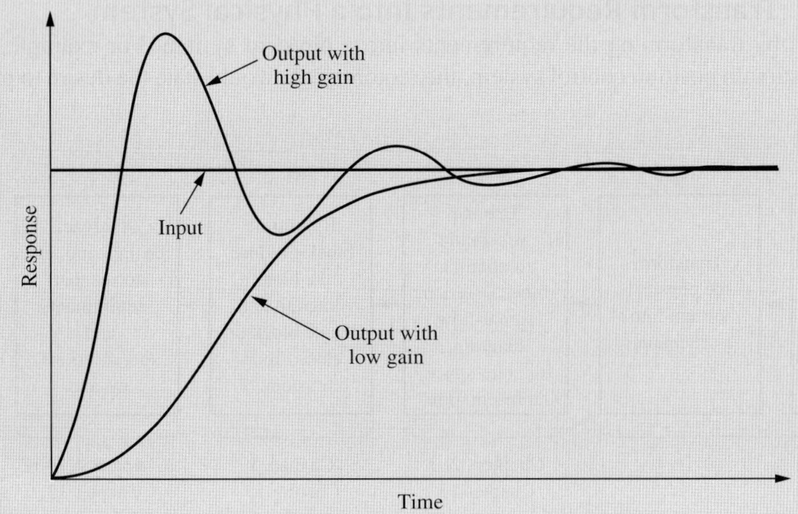


**FIGURE 1.9** Response of a position control system, showing effect of high and low controller gain on the output response

equals the commanded input position. In some systems, the steady-state error will not be zero; for these systems, a simple gain adjustment to regulate the transient response is either not effective or leads to a trade-off between the desired transient response and the desired steady-state accuracy.

To solve this problem, a controller with a dynamic response, such as an electrical filter, is used along with an amplifier. With this type of controller, it is possible to design both the required transient response and the required steady-state accuracy without the trade-off required by a simple setting of gain. However, the controller is now more complex. The filter in this case is called a compensator. Many systems also use dynamic elements in the feedback path along with the output transducer to improve system performance. An animation PowerPoint presentation (PPT) demonstrating this system is available for instructors. See *Antenna (Ch. 1)*.

In summary, then, our design objectives and the system's performance revolve around the transient response, the steady-state error, and stability. Gain adjustments can affect performance and sometimes lead to trade-offs between the performance criteria. Compensators can often be designed to achieve performance specifications without the need for trade-offs. Now that we have stated our objectives and some of the methods available to meet those objectives, we describe the orderly progression that leads us to the final system design.

## 1.5 The Design Process

In this section, we establish an orderly sequence for the design of feedback control systems that will be followed as we progress through the rest of the book. Figure 1.10 shows the described process as well as the chapters in which the steps are discussed.

The antenna azimuth position control system discussed in the last section is representative of control systems that must be analyzed and designed. Inherent in Figure 1.10 is feedback and communication during each phase. For example, if testing (Step 6) shows that requirements have not been met, the system must be redesigned and retested. Sometimes requirements are conflicting and the design cannot be attained. In these cases, the requirements have to be respecified and the design process repeated. Let us now elaborate on each block of Figure 1.10.

### Step 1: Transform Requirements Into a Physical System

We begin by transforming the requirements into a physical system. For example, in the antenna azimuth position control system, the requirements would state the desire to position

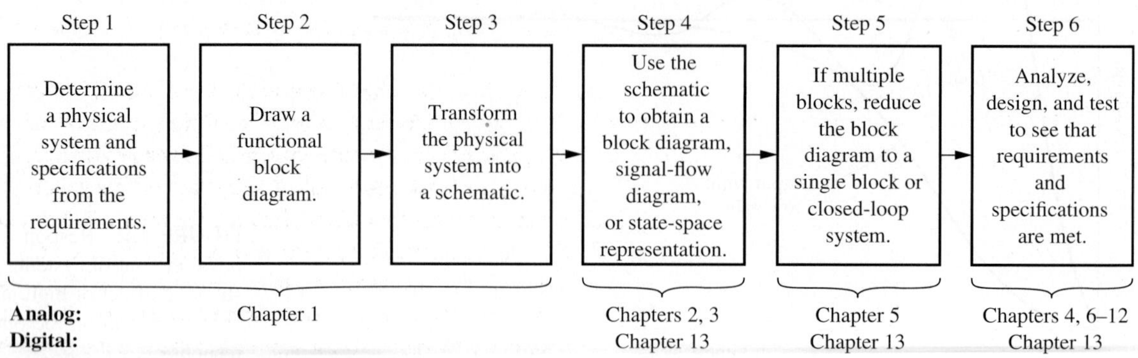


**FIGURE 1.10** The control system design process

the antenna from a remote location and describe such features as weight and physical dimensions. Using the requirements, design specifications, such as desired transient response and steady-state accuracy, are determined. Perhaps an overall concept, such as Figure 1.8(*a*), would result.

## Step 2: Draw a Functional Block Diagram

The designer now translates a qualitative description of the system into a functional block diagram that describes the component parts of the system (that is, function and/or hardware) and shows their interconnection. Figure 1.8(*d*) is an example of a functional block diagram for the antenna azimuth position control system. It indicates functions such as input transducer and controller, as well as possible hardware descriptions such as amplifiers and motors. At this point the designer may produce a detailed layout of the system, such as that shown in Figure 1.8(*b*), from which the next phase of the analysis and design sequence, developing a schematic diagram, can be launched.

## Step 3: Create a Schematic

As we have seen, position control systems consist of electrical, mechanical, and electromechanical components. After producing the description of a physical system, the control systems engineer transforms the physical system into a schematic diagram. The control system designer can begin with the physical description, as contained in Figure 1.8(*a*), to derive a schematic. The engineer must make approximations about the system and neglect certain phenomena, or else the schematic will be unwieldy, making it difficult to extract a useful mathematical model during the next phase of the analysis and design sequence. The designer starts with a simple schematic representation and, at subsequent phases of the analysis and design sequence, checks the assumptions made about the physical system through analysis and computer simulation. If the schematic is too simple and does not adequately account for observed behavior, the control systems engineer adds phenomena to the schematic that were previously assumed negligible. A schematic diagram for the antenna azimuth position control system is shown in Figure 1.8(*c*).

When we draw the potentiometers, we make our first simplifying assumption by neglecting their friction or inertia. These mechanical characteristics yield a dynamic, rather than an instantaneous, response in the output voltage. We assume that these mechanical effects are negligible and that the voltage across a potentiometer changes instantaneously as the potentiometer shaft turns.

A differential amplifier and a power amplifier are used as the controller to yield gain and power amplification, respectively, to drive the motor. Again, we assume that the dynamics of the amplifiers are rapid compared to the response time of the motor; thus, we model them as a pure gain, $K$.

A dc motor and equivalent load produce the output angular displacement. The speed of the motor is proportional to the voltage applied to the motor's *armature circuit*. Both inductance and resistance are part of the armature circuit. In showing just the armature resistance in Figure 1.8(*c*), we assume the effect of the armature inductance is negligible for a dc motor.

The designer makes further assumptions about the load. The load consists of a rotating mass and bearing friction. Thus, the model consists of *inertia* and *viscous damping* whose resistive torque increases with speed, as in an automobile's shock absorber or a screen door damper.

The decisions made in developing the schematic stem from knowledge of the physical system, the physical laws governing the system's behavior, and *practical experience*. These decisions are not easy; however, as you acquire more design experience, you will gain the insight required for this difficult task.

## Step 4: Develop a Mathematical Model (Block Diagram)

Once the schematic is drawn, the designer uses physical laws, such as Kirchhoff's laws for electrical networks and Newton's law for mechanical systems, along with simplifying assumptions, to model the system mathematically. These laws are

| | |
|---|---|
| ***Kirchhoff's voltage law*** | The sum of voltages around a closed path equals zero. |
| ***Kirchhoff's current law*** | The sum of electric currents flowing from a node equals zero. |
| ***Newton's laws*** | The sum of forces on a body equals zero;[3] the sum of moments on a body equals zero. |

Kirchhoff's and Newton's laws lead to mathematical models that describe the relationship between the input and output of dynamic systems. One such model is the *linear, time-invariant differential equation*, Eq. (1.2):

$$\frac{d^m c(t)}{dt^n} + d_{n-1}\frac{d^{m-1}c(t)}{dt^{n-1}} + \cdots + d_0 c(t) = b_m \frac{d^m r(t)}{dt^m} + b_{m-1}\frac{d^{m-1}r(t)}{dt^{m-1}} + \cdots + b_0 r(t) \tag{1.2}$$

(1.2)[4]

Many systems can be approximately described by this equation, which relates the output, $c(t)$, to the input, $r(t)$, by way of the system parameters, $a_i$ and $b_j$. We assume the reader is familiar with differential equations. Problems and a bibliography are provided at the end of the chapter for you to review this subject.

Simplifying assumptions made in the process of obtaining a mathematical model usually leads to a low-order form of Eq. (1.2). Without the assumptions the system model could be of high order or described with nonlinear, time-varying, or partial differential equations. These equations complicate the design process and reduce the designer's insight. Of course, all assumptions must be checked and all simplifications justified through analysis or testing. If the assumptions for simplification cannot be justified, then the model cannot be simplified. We examine some of these simplifying assumptions in Chapter 2.

In addition to the differential equation, the *transfer function* is another way of mathematically modeling a system. The model is derived from the linear, time-invariant differential equation using what we call the *Laplace transform*. Although the transfer function can be used only for linear systems, it yields more intuitive information than the differential equation. We will be able to change system parameters and rapidly sense the effect of these changes on the system response. The transfer function is also useful in modeling the interconnection of subsystems by forming a block diagram similar to Figure 1.8(*d*) but with a mathematical function inside each block.

Still another model is the *state-space representation*. One advantage of state-space methods is that they can also be used for systems that cannot be described by linear differential equations. Further, state-space methods are used to model systems for simulation on the digital computer. Basically, this representation turns an $n$th-order differential equation into $n$ simultaneous first-order differential equations. Let this description suffice for now; we describe this approach in more detail in Chapter 3.

[3] Alternately, $\sum$ forces $= Ma$. In this text the force, $Ma$, will be brought to the left-hand side of the equation to yield $\sum$ forces $= 0$ (D'Alembert's principle). We can then have a consistent analogy between force and voltage, and Kirchhoff's and Newton's laws (that is, $\sum$ forces $= 0$; $\sum$ voltages $= 0$).

[4] The right-hand side of Eq. (1.2) indicates differentiation of the input, $r(t)$. In physical systems, differentiation of the input introduces noise. In Chapters 3 and 5 we show implementations and interpretations of Eq. (1.2) that do not require differentiation of the input.

Finally, we should mention that to produce the mathematical model for a system, we require knowledge of the parameter values, such as equivalent resistance, inductance, mass, and damping, which is often not easy to obtain. Analysis, measurements, or specifications from vendors are sources that the control systems engineer may use to obtain the parameters.

## Step 5: Reduce the Block Diagram

Subsystem models are interconnected to form block diagrams of larger systems, as in Figure 1.8(*d*), where each block has a mathematical description. Notice that many signals, such as proportional voltages and error, are internal to the system. There are also two signals—angular input and angular output—that are external to the system. In order to evaluate system response in this example, we need to reduce this large system's block diagram to a single block with a mathematical description that represents the system from its input to its output, as shown in Figure 1.11. Once the block diagram is reduced, we are ready to analyze and design the system.

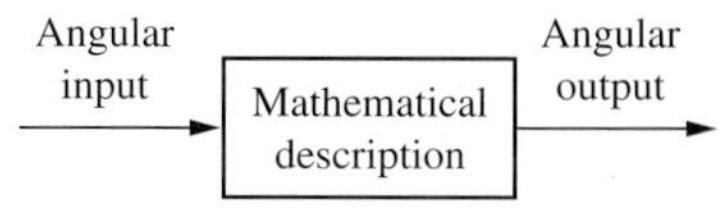


**FIGURE 1.11** Equivalent block diagram for the antenna azimuth position control system

## Step 6: Analyze and Design

The next phase of the process, following block diagram reduction, is analysis and design. If you are interested only in the performance of an individual subsystem, you can skip the block diagram reduction and move immediately into analysis and design. In this phase, the engineer analyzes the system to see if the response specifications and performance requirements can be met by simple adjustments of system parameters. If specifications cannot be met, the designer then designs additional hardware in order to effect a desired performance.

Test input signals are used, both analytically and during testing, to verify the design. It is neither necessarily practical nor illuminating to choose complicated input signals to analyze a system's performance. Thus, the engineer usually selects standard test inputs. These inputs are impulses, steps, ramps, parabolas, and sinusoids, as shown in Table 1.1.

An *impulse* is infinite at $t = 0$ and zero elsewhere. The area under the unit impulse is 1. An approximation of this type of waveform is used to place initial energy into a system so that the response due to that initial energy is only the transient response of a system. From this response the designer can derive a mathematical model of the system.

A *step* input represents a *constant command*, such as position, velocity, or acceleration. Typically, the step input command is of the same form as the output. For example, if the system's output is position, as it is for the antenna azimuth position control system, the step input represents a desired position, and the output represents the actual position. If the system's output is velocity, as is the spindle speed for a video disc player, the step input represents a constant desired speed, and the output represents the actual speed. The designer uses step inputs because both the transient response and the steady-state response are clearly visible and can be evaluated.

The *ramp* input represents a *linearly increasing command*. For example, if the system's output is position, the input ramp represents a linearly increasing position, such as that found when tracking a satellite moving across the sky at constant speed. If the system's output is velocity, the input ramp represents a linearly increasing velocity. The response to an input ramp test signal yields additional information about the steady-state error. The previous discussion can be extended to *parabolic* inputs, which are also used to evaluate a system's steady-state error.

*Sinusoidal* inputs can also be used to test a physical system to arrive at a mathematical model. We discuss the use of this waveform in detail in Chapters 10 and 11.

**TABLE 1.1** Test waveforms used in control systems

| Input | Function | Description | Sketch | Use |
|---|---|---|---|---|
| Impulse | $\delta(t)$ | $\delta(t) = \infty$ for $0- < t < 0+$<br>$= 0$ elsewhere<br>$\int_{0-}^{0+} \delta(t)dt = 1$ | 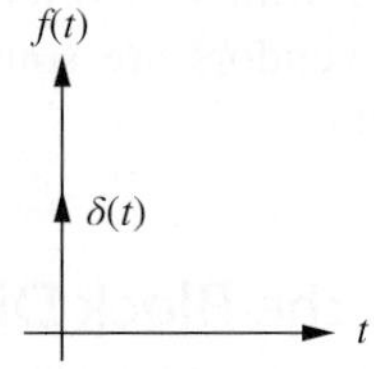 | Transient response<br>Modeling |
| Step | $u(t)$ | $u(t) = 1$ for $t > 0$<br>$= 0$ for $t < 0$ | 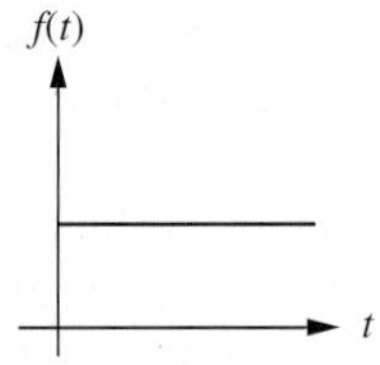 | Transient response<br>Steady-state error |
| Ramp | $tu(t)$ | $tu(t) = t$ for $t \geq 0$<br>$= 0$ elsewhere | 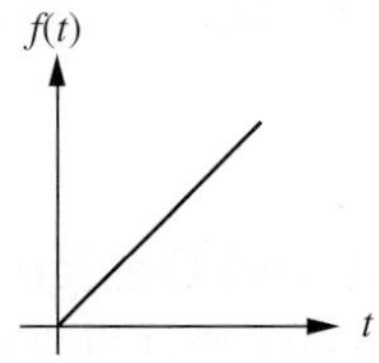 | Steady-state error |
| Parabola | $\frac{1}{2}t^2u(t)$ | $\frac{1}{2}t^2u(t) = \frac{1}{2}t^2$ for $t \geq 0$<br>$= 0$ elsewhere | 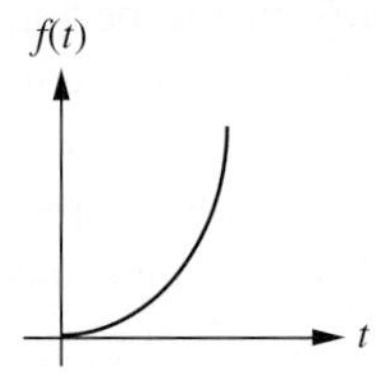 | Steady-state error |
| Sinusoid | $\sin \omega t$ | | 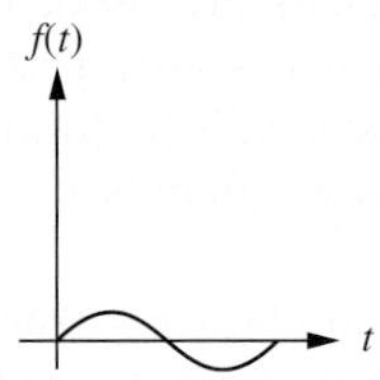 | Transient response<br>Modeling<br>Steady-state error |

We conclude that one of the basic analysis and design requirements is to evaluate the time response of a system for a given input. Throughout the book you will learn numerous methods for accomplishing this goal.

The control systems engineer must take into consideration other characteristics about feedback control systems. For example, control system behavior is altered by fluctuations in component values or system parameters. These variations can be caused by temperature, pressure, or other environmental changes. Systems must be built so that expected fluctuations do not degrade performance beyond specified bounds. A *sensitivity* analysis can yield the percentage of change in a specification as a function of a change in a system parameter. One of the designer's goals, then, is to build a system with minimum sensitivity over an expected range of environmental changes.

In this section we looked at some control systems analysis and design considerations. We saw that the designer is concerned about transient response, steady-state error, stability, and sensitivity. The text pointed out that although the basis of evaluating system performance is the differential equation, other methods, such as transfer functions and state space, will be used. The advantages of these new techniques over differential equations will become apparent as we discuss them in later chapters.

## 1.6 Computer-Aided Design

Now that we have discussed the analysis and design sequence, let us discuss the use of the computer as a computational tool in this sequence. The computer plays an important role in the design of modern control systems. In the past, control system design was labor intensive. Many of the tools we use today were implemented through hand calculations or, at best, using plastic graphical aid tools. The process was slow, and the results not always accurate. Large mainframe computers were then used to simulate the designs.

Today we are fortunate to have computers and software that remove the drudgery from the task. At our own desktop computers, we can perform analysis, design, and simulation with one program. With the ability to simulate a design rapidly, we can easily make changes and immediately test a new design. We can play what-if games and try alternate solutions to see if they yield better results, such as reduced sensitivity to parameter changes. We can include nonlinearities and other effects and test our models for accuracy.

### MATLAB

The computer is an integral part of modern control system design, and many computational tools are available for your use. In this book we use MATLAB and the MATLAB Control System Toolbox, which expands MATLAB to include control system–specific commands. In addition, presented are several MATLAB enhancements that give added functionality to MATLAB and the Control Systems Toolbox. Included are (1) Simulink, which uses a graphical user interface (GUI); (2) the LTI Viewer, which permits measurements to be made directly from time and frequency response curves; (3) the SISO Design Tool, a convenient and intuitive analysis and design tool; and (4) the Symbolic Math Toolbox, which saves labor when making symbolic calculations required in control system analysis and design. Some of these enhancements may require additional software available from The MathWorks, Inc.

MATLAB is presented as an alternate method of solving control system problems. You are encouraged to solve problems first by hand and then by MATLAB so that insight is not lost through mechanized use of computer programs. To this end, many examples throughout the book are solved by hand, followed by suggested use of MATLAB.

As an enticement to begin using MATLAB, simple program statements that you can try are suggested throughout the chapters at appropriate locations. Throughout the book, MATLAB references direct you to the proper program in the proper appendix and tell you what you will learn. Selected end-of-chapter problems and Case Study Challenges to be solved using MATLAB have also been marked. The following list itemizes the specific components of MATLAB used in this book and the appendix in which a description can be found:

MATLAB/Control System Toolbox tutorials and code are found in Appendix B and identified in the text.

MATLAB
**ML**

Simulink tutorials and diagrams are found in Appendix C and identified in the text.

Simulink
**SL**

MATLAB GUI tools, tutorials, and examples are in Appendix E and identified in the text. These tools consist of the LTI Viewer and the SISO Design Tool.

GUI Tool
**GUIT**

Symbolic Math
SM

Symbolic Math Toolbox tutorials and code are found in Appendix F and identified in the text.

MATLAB code itself is not platform specific. The same code runs on PCs and workstations that support MATLAB. Although there are differences in installing and managing MATLAB files, we do not address them in this book. Also, there are many more commands in MATLAB and the MATLAB toolboxes than are covered in the appendixes. Please explore the bibliographies at the end of the applicable appendixes to find out more about MATLAB file management and MATLAB instructions not covered in this textbook.

### LabVIEW

LabVIEW is a programming environment presented as an alternative to MATLAB. This graphical alternative produces front panels of virtual instruments on your computer that are pictorial reproductions of hardware instruments, such as waveform generators or oscilloscopes. Underlying the front panels are block diagrams. The blocks contain underlying code for the controls and indicators on the front panel. Thus, a knowledge of coding is not required. Also, parameters can be easily passed or viewed from the front panel.

LabVIEW
LV

A LabVIEW tutorial is in Appendix D and all LabVIEW material is identified in the text.

You are encouraged to use computational aids throughout this book. Those not using MATLAB or LabVIEW should consult Appendix H for a discussion of other alternatives. Now that we have introduced control systems to you and established a need for computational aids to perform analysis and design, we will conclude with a discussion of your career as a control systems engineer and look at the opportunities and challenges that await you.

## 1.7 The Control Systems Engineer

Control systems engineering is an exciting field in which to apply your engineering talents, because it cuts across numerous disciplines and numerous functions within those disciplines. The control engineer can be found at the top level of large projects, engaged at the conceptual phase in determining or implementing overall system requirements. These requirements include total system performance specifications, subsystem functions, and the interconnection of these functions, including interface requirements, hardware and software design, and test plans and procedures.

Many engineers are engaged in only one area, such as circuit design or software development. However, as a control systems engineer, you may find yourself working in a broad arena and interacting with people from numerous branches of engineering and the sciences. For example, if you are working on a biological system, you will need to interact with colleagues in the biological sciences, mechanical engineering, electrical engineering, and computer engineering, not to mention mathematics and physics. You will be working with these engineers at all levels of project development from concept through design and, finally, testing. At the design level, the control systems engineer can be performing hardware selection, design, and interface, including total subsystem design to meet specified requirements. The control engineer can be working with sensors and motors as well as electronic, pneumatic, and hydraulic circuits.

The space shuttle provides another example of the diversity required of the systems engineer. In the previous section, we showed that the space shuttle's control systems cut across many branches of science: orbital mechanics and propulsion, aerodynamics, electrical engineering, and mechanical engineering. Whether or not you work in the space program, as a control systems engineer you will apply broad-based knowledge to the solution of engineering control problems. You will have the opportunity to expand your engineering horizons beyond your university curriculum.

You are now aware of future opportunities. But for now, what advantages does this course offer to a student of control systems (other than the fact that you need it to graduate)? Engineering curricula tend to emphasize *bottom-up* design. That is, you start from the components, develop circuits, and then assemble a product. In *top-down* design, a high-level picture of the requirements is first formulated; then the functions and hardware required to implement the system are determined. You will be able to take a top-down systems approach as a result of this course.

A major reason for not teaching top-down design throughout the curriculum is the high level of mathematics initially required for the systems approach. For example, control systems theory, which requires differential equations, could not be taught as a lower-division course. However, while progressing through bottom-up design courses, it is difficult to see how such design fits logically into the large picture of the product development cycle.

After completing this control systems course, you will be able to stand back and see how your previous studies fit into the large picture. Your amplifier course or vibrations course will take on new meaning as you begin to see the role design work plays as part of product development. For example, as engineers, we want to describe the physical world mathematically so that we can create systems that will benefit humanity. You will find that you have indeed acquired, through your previous courses, the ability to model physical systems mathematically, although at the time you might not have understood where in the product development cycle the modeling fits. This course will clarify the analysis and design procedures and show you how the knowledge you acquired fits into the total picture of system design.

Understanding control systems enables students from all branches of engineering to speak a common language and develop an appreciation and working knowledge of the other branches. You will find that there really is not much difference among the branches of engineering as far as the goals and applications are concerned. As you study control systems, you will see this commonality.

## Summary

Control systems contribute to every aspect of modern society. In our homes we find them in everything from toasters to heating systems to DVD players. Control systems also have widespread applications in science and industry, from steering ships and planes to guiding missiles. Control systems also exist naturally; our bodies contain numerous control systems. Even economic and psychological system representations have been proposed based on control system theory. Control systems are used where power gain, remote control, or conversion of the form of the input is required.

A control system has an *input*, a *process*, and an *output*. Control systems can be *open loop* or *closed loop*. Open-loop systems do not monitor or correct the output for disturbances; however, they are simpler and less expensive than closed-loop systems. Closed-loop systems monitor the output and compare it to the input. If an error is detected, the system corrects the output and hence corrects the effects of disturbances.

Control systems analysis and design focuses on three primary objectives:

**1.** Producing the desired transient response

**2.** Reducing steady-state errors

**3.** Achieving stability

A system must be stable in order to produce the proper transient and steady-state response. Transient response is important because it affects the speed of the system and influences human patience and comfort, not to mention mechanical stress. Steady-state response determines the accuracy of the control system; it governs how closely the output matches the desired response.

The design of a control system follows these steps:

**Step 1** Determine a physical system and specifications from requirements.
**Step 2** Draw a functional block diagram.
**Step 3** Represent the physical system as a schematic.
**Step 4** Use the schematic to obtain a mathematical model, such as a block diagram.
**Step 5** Reduce the block diagram.
**Step 6** Analyze and design the system to meet specified requirements and specifications that include stability, transient response, and steady-state performance.

In the next chapter we continue through the analysis and design sequence and learn how to use the schematic to obtain a mathematical model.

## Review Questions

1. Name three applications for feedback control systems.
2. Name three reasons for using feedback control systems and at least one reason for not using them.
3. Give three examples of open-loop systems.
4. Functionally, how do closed-loop systems differ from open-loop systems?
5. State one condition under which the error signal of a feedback control system would not be the difference between the input and the output.
6. If the error signal is not the difference between input and output, by what general name can we describe the error signal?
7. Name two advantages of having a computer in the loop.
8. Name the three major design criteria for control systems.
9. Name the two parts of a system's response.
10. Physically, what happens to a system that is unstable?
11. Instability is attributable to what part of the total response?
12. Describe a typical control system analysis task.
13. Describe a typical control system design task.
14. Adjustments of the forward path gain can cause changes in the transient response. True or false?
15. Name three approaches to the mathematical modeling of control systems.
16. Briefly describe each of your answers to Question 15.

## Problems

1. A temperature control system operates by sensing the difference between the thermostat setting and the actual temperature and then opening a fuel valve an amount proportional to this difference. Draw a functional closed-loop block diagram similar to Figure 1.8(*d*) identifying the input and output transducers, the controller, and the plant. Further, identify the input and output signals of all subsystems previously described. [Section 1.4: Introduction to a Case Study]

2. An aircraft's attitude varies in roll, pitch, and yaw as defined in Figure P1.1. Draw a functional block diagram for a closed-loop system that stabilizes the roll as follows: The system measures the actual roll angle with a gyro and compares the actual roll angle with the desired roll angle.

The ailerons respond to the roll-angle error by undergoing an angular deflection. The aircraft responds to this angular deflection, producing a roll angle rate. Identify the input and output transducers, the controller, and the plant. Further, identify the nature of each signal. [Section 1.4: Introduction to a Case Study]

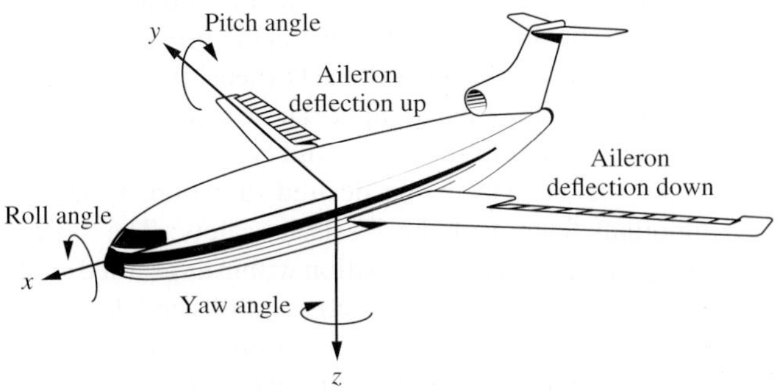


**FIGURE P1.1** Aircraft attitude defined

3. A variable resistor, called a *potentiometer*, is shown in Figure P1.2. The resistance is varied by moving a wiper arm along a fixed resistance. The resistance from *A* to *C* is fixed, but the resistance from *B* to *C* varies with the position of the wiper arm. If it takes 10 turns to move the wiper arm from *A* to *C*, draw a block diagram of the potentiometer showing the input variable, the output variable, and (inside the block) the gain, which is a constant and is the amount by which the input is multiplied to obtain the output. An animation PowerPoint presentation (PPT) demonstrating this system is available for instructors. See *Potentiometer*. [Section 1.4: Introduction to a Case Study]

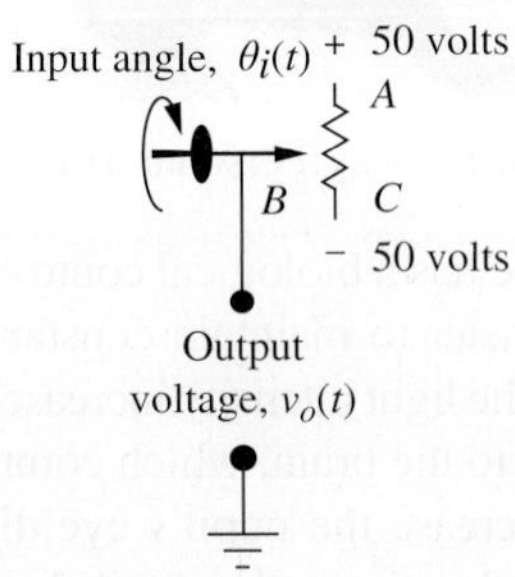


**FIGURE P1.2** Potentiometer

4. Many processes operate on rolled material that moves from a supply reel to a take-up reel. Typically, these systems, called *winders*, control the material so that it travels at a constant velocity. Besides velocity, complex winders also control tension, compensate for roll inertia while accelerating or decelerating, and regulate acceleration due to sudden changes. A winder is shown in Figure P1.3. The force transducer measures tension; the winder pulls against the nip rolls, which provide an opposing force; and the bridle provides slip. In order to compensate for changes in speed, the material is looped around a *dancer*. The loop prevents rapid changes from causing excessive slack or damaging the material. If the dancer position is sensed by a potentiometer or other device, speed variations due to buildup on the take-up reel or other causes can be controlled by comparing the potentiometer voltage to the commanded speed. The system then corrects the speed and resets the dancer to the desired position (*Ayers, 1988*). Draw a functional block diagram for the speed control system, showing each component and signal. [Section 1.4: Introduction to a Case Study]

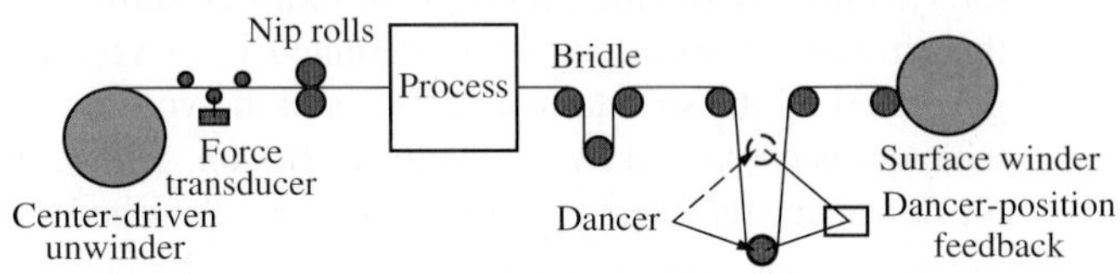


**FIGURE P1.3** Winder

5. In a nuclear power generating plant, heat from a reactor is used to generate steam for turbines. The rate of the fission reaction determines the amount of heat generated, and this rate is controlled by rods inserted into the radioactive core. The rods regulate the flow of neutrons. If the rods are lowered into the core, the rate of fission will diminish; if the rods are raised, the fission rate will increase. By automatically controlling the position of the rods, the amount of heat generated by the reactor can be regulated. Draw a functional block diagram for the nuclear reactor control system shown in Figure P1.4. Show all blocks and signals. [Section 1.4: Introduction to a Case Study]

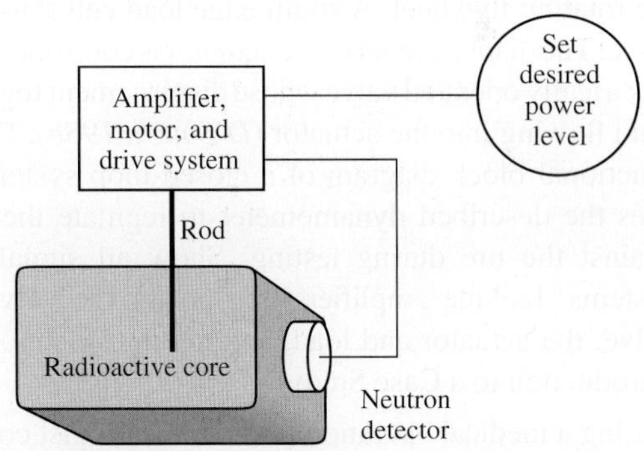


**FIGURE P1.4** Control of a nuclear reactor

6. A university wants to establish a control system model that represents the student population as an output, with the desired student population as an input. The administration determines the rate of admissions by comparing the current and desired student populations.

The admissions office then uses this rate to admit students. Draw a functional block diagram showing the administration and the admissions office as blocks of the system. Also show the following signals: the desired student population, the actual student population, the desired student rate as determined by the administration, the actual student rate as generated by the admissions office, the dropout rate, and the net rate of influx. [Section 1.4: Introduction to a Case Study]

7. We can build a control system that will automatically adjust a motorcycle's radio volume as the noise generated by the motorcycle changes. The noise generated by the motorcycle increases with speed. As the noise increases, the system increases the volume of the radio. Assume that the amount of noise can be represented by a voltage generated by the speedometer cable, and the volume of the radio is controlled by a dc voltage (*Hogan, 1988*). If the dc voltage represents the desired volume disturbed by the motorcycle noise, draw the functional block diagram of the automatic volume control system, showing the input transducer, the volume control circuit, and the speed transducer as blocks. Also, show the following signals: the desired volume as an input, the actual volume as an output, and voltages representing speed, desired volume, and actual volume. An animation PowerPoint presentation (PPT) demonstrating this system is available for instructors. See *Motorcycle*. [Section 1.4: Introduction to a Case Study]

8. A dynamometer is a device used to measure torque and speed and to vary the load on rotating devices. The dynamometer operates as follows to control the amount of torque: A hydraulic actuator attached to the axle presses a tire against a rotating flywheel. The greater the displacement of the actuator, the greater the force that is applied to the rotating flywheel. A strain gage load cell senses the force. The displacement of the actuator is controlled by an electrically operated valve whose displacement regulates fluid flowing into the actuator (*D'Souza, 1988*). Draw a functional block diagram of a closed-loop system that uses the described dynamometer to regulate the force against the tire during testing. Show all signals and systems. Include amplifiers that power the valve, the valve, the actuator and load, and the tire. [Section 1.4: Introduction to a Case Study]

9. During a medical operation an anesthesiologist controls the depth of unconsciousness by controlling the concentration of isoflurane in a vaporized mixture with oxygen and nitrous oxide. The depth of anesthesia is measured by the patient's blood pressure. The anesthesiologist also regulates ventilation, fluid balance, and the administration of other drugs. In order to free the anesthesiologist to devote more time to the latter tasks, and in the interest of the patient's safety, we wish to automate the depth of anesthesia by automating the control of isoflurane concentration. Draw a functional block diagram of the system showing pertinent signals and subsystems (*Meier, 1992*). [Section 1.4: Introduction to a Case Study]

10. A Segway®[5] Personal Transporter (PT) (Figure P1.5) is a two-wheeled vehicle in which the human operator stands vertically on a platform. As the driver leans left, right, forward, or backward, a set of sensitive gyroscopic sensors sense the desired input. These signals are fed to a computer that amplifies them and commands motors to propel the vehicle in the desired direction. One very important feature of the PT is its safety: The system will maintain its vertical position within a specified angle despite road disturbances, such as uphills and downhills or even if the operator over-leans in any direction. Draw a functional block diagram of the PT system that keeps the system in a vertical position. Indicate the input and output signals, intermediate signals, and main subsystems. (http://segway.com)

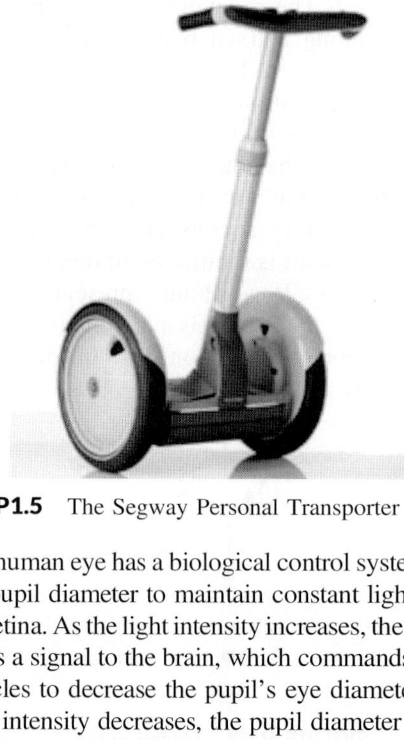

D.A. Winstein/Custom Medial Stock photo

**FIGURE P1.5** The Segway Personal Transporter (PT)

11. The human eye has a biological control system that varies the pupil diameter to maintain constant light intensity to the retina. As the light intensity increases, the optical nerve sends a signal to the brain, which commands internal eye muscles to decrease the pupil's eye diameter. When the light intensity decreases, the pupil diameter increases.
   a. Draw a functional block diagram of the light-pupil system indicating the input, output, and intermediate signals; the sensor; the controller; and the actuator. [Section 1.4: Introduction to a Case Study]
   b. Under normal conditions the incident light will be larger than the pupil. If the incident light is smaller

[5] Segway is a registered trademark of Segway, Inc. in the United States and/or other countries.

than the diameter of the pupil, the feedback path is broken (*Bechhoefer, 2005*). Modify your block diagram from Part **a**, to show where the loop is broken. What will happen if the narrow beam of light varies in intensity, such as in a sinusoidal fashion?

**c.** It has been found (*Bechhoefer, 2005*) that it takes the pupil about 300 milliseconds to react to a change in the incident light. If light shines off center to the retina, describe the response of the pupil with delay present and then without delay present.

**12.** Some skillful drivers can drive and balance a four-wheeled vehicle on two wheels. To verify that a control system can also drive a car in this fashion, a prototype using an RC (remote-controlled) car is equipped with a feedback control system (*Arndt, 2011*). In a simplified system model, the roll angle at which the car balances was calculated a priori and found to be 52.3°. This value was used as the desired input. The desired input is compared with the actual roll angle and the difference is fed to a controller that feeds a servomotor indicating the desired wheel steering angle that controls the vehicle's roll angle on two wheels. The car's actual roll angle is measured using a hinged linkage that rolls along the ground next to the vehicle and is connected to a potentiometer. Draw a block diagram indicating the system functions. Draw blocks for the system controller, the steering servo, and the car dynamics. Indicate in the diagram the following signals: the desired roll angle, the steering wheel angle, and the actual car roll angle.

**13.** Your bathtub at home is a control system that keeps the water level constant. A constant flow from the tap yields a constant water level, because the flow rate through the drain increases as the water level increases, and decreases as the water level decreases. After equilibrium has been reached, the level can be controlled by controlling the input flow rate. A low input flow rate yields a lower level, while a higher input flow rate yields a higher level. [Section 1.4: Introduction to a Case Study]

**a.** Sketch a control system that uses this principle to precisely control the fluid level in a tank. Show the intake and drain valves, the tank, any sensors and transducers, and the interconnection of all components.

**b.** Draw a functional block diagram of the system, identifying the input and output signals of each block.

**14.** Tactile feedback is an important component in the learning of motor skills such as dancing, sports, and physical rehabilitation. A suit with white dots recognized by a vision system to determine arm joint positions with millimetric precision was developed. This suit is worn by both teacher and student to provide position information. (*Lieberman, 2007*). If there is a difference between the teacher's positions and that of the student, vibrational feedback is provided to the student through eight strategically placed vibrotactile actuators in the wrist and arm. This placement takes advantage of a sensory effect known as *cutaneous rabbit* that tricks the subject to feel uniformly spaced stimuli in places where the actuators are not present. These stimuli help the student adjust to correct the motion. In summary, the system consists of an instructor and a student having their movements followed by the vision system. Their movements are fed into a computer that finds the differences between their joint positions and provides proportional vibrational strength feedback to the student. Draw a block diagram describing the system design.

**15.** In the Case Study of Section 1.4, an antenna azimuth angle is controlled, and its corresponding block diagram is shown in Figure 1.8(*d*) in the text. There, the sensor used to measure the antenna's azimuth angle is a potentiometer.

**a.** Modify the block diagram if the sensor used to measure the antenna's angle is an accelerometer.

**b.** Modify the block diagram if the sensor used to measure the antenna's angle is a gyroscope.

**16.** In humans, hormone levels, alertness, and core body temperature are synchronized through a 24-hour circadian cycle. Daytime alertness is at its best when sleep/wake cycles are in sync with the circadian cycle. Thus alertness can be easily affected with a distributed work schedule, such as the one to which astronauts are subjected. It has been shown that the human circadian cycle can be delayed or advanced through light stimulus. To ensure optimal alertness, a system is designed to track astronauts' circadian cycles and increase the quality of sleep during missions. Core body temperature can be used as an indicator of the circadian cycle. A computer model with optimum circadian body temperature variations can be compared to an astronaut's body temperatures. Whenever a difference is detected, the astronaut is subjected to a light stimulus to advance or delay the astronaut's circadian cycle (*Mott, 2003*). Draw a functional block diagram of the system. Indicate the input and output signals, intermediate signals, and main subsystems.

**17.** Oil drilling rigs are used for drilling holes for identification of oil or natural gas sources and for extraction. An oil drilling system can be thought of as a drill inside a straw, which is placed inside a glass. The straw assembly represents the drill string, the drill surrounded by fluid, and the glass represents the volume, the annulus, around the drill string through which slurry and eventually oil will flow as the drilling progresses.

Assume that we want to control the drill pressure output, $P_d(t)$, with a reference voltage input, $V_d(t)$. A control loop model (*Zhao, 2007*) consists of a drill-pressure controller, drill motor subsystem, pulley subsystem, and drill stick subsystem. The output signal of the latter, the drill pressure, $P_d(t)$, is measured using a transducer, which transmits a negative feedback voltage signal, $V_b(t)$, to the drill pressure controller. That signal is compared at the input of the controller to the reference voltage, $V_r(t)$, Based on the error, $e(t) = V_r(t) - V_b(t)$, the drill pressure controller sets the desired drill speed, $\omega_d$, which is the input to the drill motor subsystem whose output is the actual drill speed, $\omega_a$, which is the input to the pulley subsystem. The output of the pulley system drives the drill stick subsystem. The drill stick subsystem may be severely affected by environmental conditions, which may be represented as disturbances acting between the pulley and stick subsystems.

Draw a functional block diagram for the above system, showing its major components as well as all signals.

**18.** Given the electric network shown in Figure P1.6. [Review]

**a.** Write the differential equation for the network if $v(t) = u(t)$, a unit step.

**b.** Solve the differential equation for the current, $i(t)$, if there is no initial energy in the network.

**c.** Make a plot of your solution if $R/L = 1$.

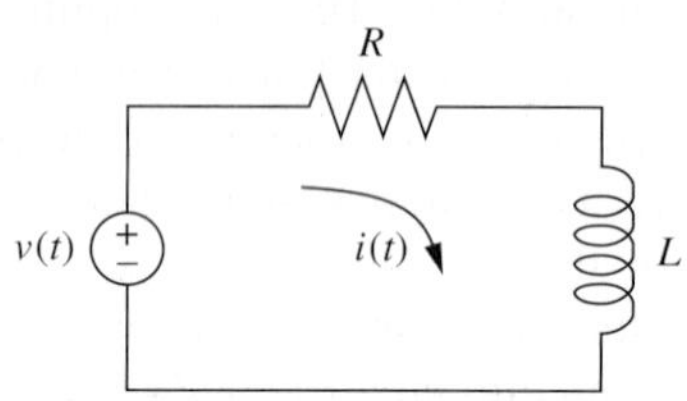


**FIGURE P1.6** *RL* network

**19.** Repeat Problem 18 using the network shown in Figure P1.7. Assume $R = 2\ \Omega$, $L = 1$ H, and $1/LC = 25$. [Review]

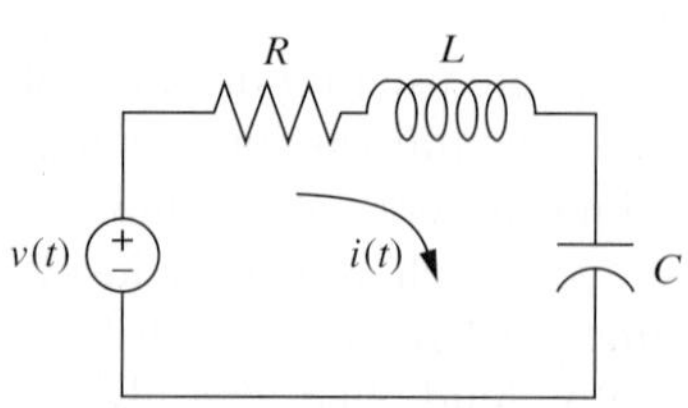


**FIGURE P1.7** *RLC* network

**20.** Solve the following differential equations using classical methods and the given initial conditions: [Review]

**a.** $\dfrac{d^2x}{dt^2} + 2\dfrac{dx}{dt} + 2x = \sin 2t$

$x(0) = 2;\ \dfrac{dx}{dt}(0) = -3$

**b.** $\dfrac{d^2x}{dt^2} + 2\dfrac{dx}{dt} + x = 5e^{-2t} + t$

$x(0) = 2;\ \dfrac{dx}{dt}(0) = 1$

**c.** $\dfrac{d^2x}{dt^2} + 4x = t^2$

$x(0) = 1;\ \dfrac{dx}{dt}(0) = 2$

**21.** Solve the following differential equations using classical methods. Assume zero initial conditions. [Review]

**a.** $\dfrac{dx}{dt} + 7x = 5\cos 2t$

**b.** $\dfrac{d^2x}{dt^2} + 6\dfrac{dx}{dt} + 8x = 5\sin 3t$

**c.** $\dfrac{d^2x}{dt^2} + 8\dfrac{dx}{dt} + 25x = 10u(t)$

## PROGRESSIVE ANALYSIS AND DESIGN PROBLEMS

**22. Contol of HIV/AIDS.** As of 2012, the number of people living worldwide with Human Immunodeficiency Virus/Acquired Immune Deficiency Syndrome (HIV/AIDS) was estimated at 35 million, with 2.3 million new infections per year and 1.6 million deaths due to the disease (*UNAIDS, 2013*). Currently there is no known cure for the disease, and HIV cannot be completely eliminated in an infected individual. Drug combinations can be used to maintain the virus numbers at low levels, which helps prevent AIDS from developing. A common treatment for HIV is the administration of two types of drugs: reverse transcriptase inhibitors (RTIs) and protease inhibitors (PIs). The amount in which each of these drugs is administered is varied according to the amount of HIV viruses in the body (*Craig, 2004*). Draw a block diagram of a feedback system designed to control the amount of HIV viruses in an infected person. The plant input variables are the amount of RTIs and PIs dispensed. Show blocks representing the controller, the system under control, and the transducers. Label the corresponding variables at the input and output of every block.

23. **Hybrid vehicle.** The use of hybrid cars is becoming increasingly popular. A hybrid electric vehicle (HEV) combines electric machine(s) with an internal combustion engine (ICE), making it possible (along with other fuel consumption–reducing measures, such as stopping the ICE at traffic lights) to use smaller and more efficient gasoline engines. Thus, the efficiency advantages of the electric drivetrain are obtained, while the energy needed to power the electric motor is stored in the onboard fuel tank and not in a large and heavy battery pack.

There are various ways to arrange the flow of power in a hybrid car. In a serial HEV (Figure P1.8), the ICE is not connected to the drive shaft. It drives only the generator, which charges the batteries and/or supplies power to the electric motor(s) through an inverter or a converter.

The HEVs sold today are primarily of the parallel or split-power variety. If the combustion engine can turn the drive wheels as well as the generator, then the vehicle is referred to as a *parallel* hybrid, because both an electric motor and the ICE can drive the vehicle. A parallel hybrid car (Figure P1.9) includes a relatively small battery pack (electrical storage) to put out extra power to the electric motor when fast acceleration is needed. See (*Bosch, 5th ed., 2000*), (*Bosch, 7th ed., 2007*), (*Edelson, 2008*), (*Anderson, 2009*) for more detailed information about HEV.

As shown in Figure P1.10, split-power hybrid cars utilize a combination of series and parallel drives (*Bosch, 5th ed., 2007*). These cars use a planetary gear (3) as a split-power transmission to allow some of the ICE power to be applied mechanically to the drive. The other part is converted into electrical energy through the alternator (7) and the inverter (5) to feed the electric motor (downstream of the transmission) and/or

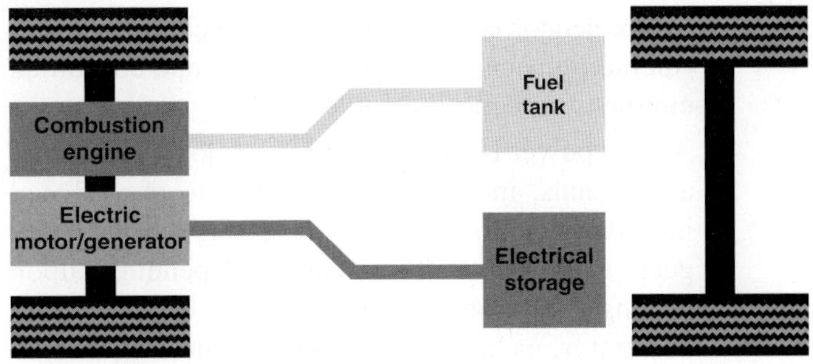


**FIGURE P1.9** Parallel hybrid drive[7]

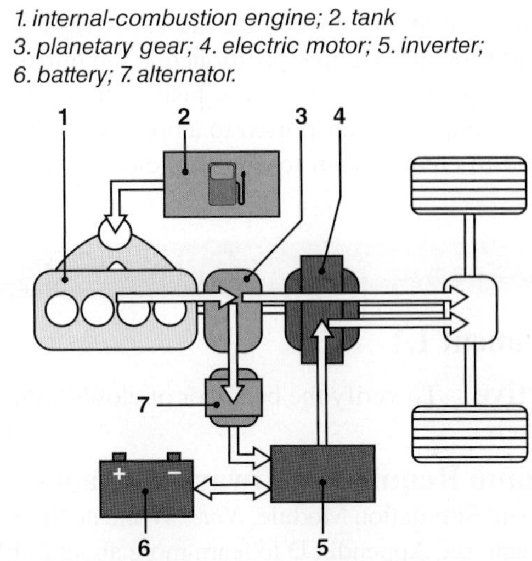


**FIGURE P1.10** Split-power hybrid electric vehicle[8]

to charge the high-voltage battery (6). Depending upon driving conditions, the ICE, the electric motor, or both propel the vehicle.

Draw a functional block dagram for the cruise (speed) control system of:

**a.** A serial hybrid vehicle, showing its major components, including the speed sensor, electronic control unit (ECU), inverter, electric motor, and vehicle dynamics; as well as all signals, including the desired vehicle speed, actual speed, control command (ECU output), controlled voltage (inverter output), the motive force (provided by the electric motor), and running resistive force;[9]

**b.** A parallel hybrid vehicle, showing its major components, which should include also a block that represents

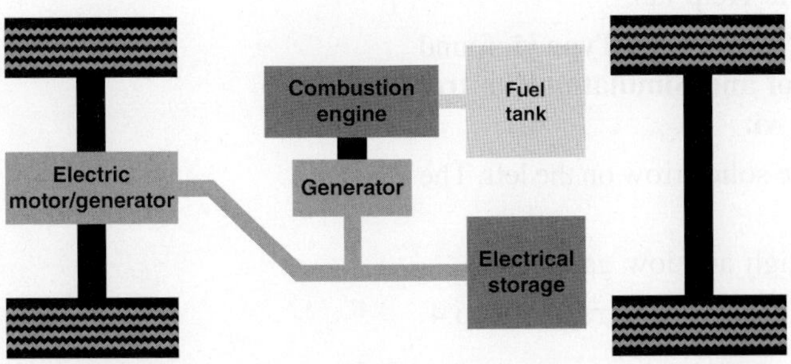


**FIGURE P1.8** Serial hybrid-electric vehicle[6]

[6] Mark Looper, www.Altfuels.org. Alternative Drivetrains, July 2005, http://www.altfuels.org/backgrnd/altdrive.html. Last accessed 10/13/2009.

[7] Mark Looper, www.Altfuels.org. Alternative Drivetrains, July 2005, http://www.altfuels.org/backgrnd/altdrive.html. Last accessed 10/13/2009.

[8] Robert Bosch GmbH, *Bosch Automotive Handbook*, 7th ed. John Wiley & Sons, Ltd., UK, 2007.

[9] These include the aerodynamic drag, rolling resistance, and climbing resistance. The aerodynamic drag is a function of car speed, whereas the other two are proportional to car weight.

the accelerator, engine, and motor, as well as the signals (including accelerator displacement and combined engine/motor motive force);

**c.** A split-power HEV, showing its major components and signals, including, in addition to those listed in Parts **a** and **b**, a block representing the planetary gear and its control, which, depending upon driving conditions, would allow the ICE, the electric motor, or both to propel the vehicle, that is, to provide the necessary total motive force.

**24.** **Parabolic trough collector.** A set of parabolic mirrors can be used to concentrate the sun's rays to heat a fluid flowing in a pipe positioned at the mirrors' focal points (*Camacho, 2012*). The heated fluid, such as oil, for example, is transported to a pressurized tank to be used to create steam to generate electricity or power an industrial process. Since the solar energy varies with time of day, time of year, cloudiness, humidity, etc., a control system has to be developed in order to maintain the fluid temperature constant. The temperature is mainly controlled by varying the amount of fluid flow through the pipes, but possibly also with a solar tracking mechanism that tilts the mirrors at appropriate angles.

Assuming fixed mirror angles, draw the functional block diagram of a system to maintain the fluid temperature a constant. The desired and actual fluid temperature difference is fed to a controller followed by an amplifier and signal conditioning circuit that varies the speed of a fluid circulating pump. Label the blocks and links of your diagram, indicating all the inputs to the system, including external disturbances such as solar variations, cloudiness, humidity, etc.

# Cyber Exploration Laboratory

## Experiment 1.1

**Objective** To verify the behavior of closed-loop systems as described in the Chapter 1 Case Study.

**Minimum Required Software Packages** LabVIEW and the LabVIEW Control Design and Simulation Module. *Note*: While no knowledge of LabVIEW is required for this experiment, see Appendix D to learn more about LabVIEW, which will be pursued in more detail in later chapters.

### Prelab

1. From the discussion in the Case Study, describe the effect of the gain of a closed-loop system upon transient response.
2. From the discussion in the Case Study about steady-state error, sketch a graph of a step input superimposed with a step response output and show the steady-state error. Assume any transient response. Repeat for a ramp input and ramp output response. Describe the effect of gain upon the steady-state error.

### Lab

1. Launch LabVIEW and open **Find Examples . . .** under the **Help** tab.
2. In the **NI Example Finder** window, open **CDEx Effect of Controller Type.vi,** found by navigating to it thourgh **Toolkits and Modules/Control and Simulation/Control Design/Time Analysis/CDEx Effect of Controller Type vi.**
3. On the tool bar click the circulating arrows located next to the solid arrow on the left. The program is now running.
4. Move the slider **Controller Gain** and note the effect of high and low gains.
5. Change the controller by clicking the arrows for **Controller Type** and repeat Step 4.

### Postlab

1. Correlate the responses found in the experiment with those described in your Prelab. Explore other examples provided in the LabVIEW example folders.

# Bibliography


Alternative Drivetrains, July 2005. Available at www.altfuels.org/backgrnd/altdrive.html. Accessed October 13, 2009.

Anderson, S. Field Guide: Hybrid Electric Powertrains, Part 4 of 5. *Automotive Design & Production*. Gardner Publication, Inc. Available at http://www.autofieldguide.com/articles/020904.html. Accessed October 13, 2009.

Arndt, D., Bobrow, J. E., Peters, S., Iagnemma, K., and Dubowsky, S. Two-Wheel Self-Balancing of a Four-wheel Vehicle. *IEEE Control Systems Magazine*, April 2011, pp. 20–37.

Bahill, A. T. *Bioengineering: Biomedical, Medical, and Clinical Engineering*. Prentice Hall, Englewood Cliffs, NJ, 1981.

Bechhoefer, J. Feedback for Physicists: A Tutorial Essay on Control. *To Appear in Review of Modern Physics*, July 2005, pp. 42–45. Also available at http://www.sfu.ca/chaos/papers/2005/rmp_reprint05.pdf.

Bennett, S. *A History of Control Engineering, 1800–1930*. Peter Peregrinus, Stevenage, UK, 1979.

Bode, H. W. *Network Analysis and Feedback Amplifier Design*. Van Nostrand, Princeton, NJ, 1945.

Bosch, R. GmbH. *Automotive Electrics and Automotive Electronics*, 5th ed. John Wiley & Sons Ltd., UK, 2007.

Bosch, R. GmbH. *Bosch Automotive Handbook*, 7th ed. John Wiley & Sons Ltd., UK, 2007.

Camacho, E. F., Berenguel, M., Rubio, F. R., and Martinez, D. *Control of Solar Energy Systems*, Springer-Verlag, London, 2012.

Cannon, R. H., Jr. *Dynamics of Physical Systems*. McGraw-Hill, New York, 1967.

Craig, I. K., Xia, X., and Venter, J. W. Introducing HIV/AIDS Education into the Electrical Engineering Curriculum at the University of Pretoria. *IEEE Transactions on Education*, vol. 47, no. 1, February 2004, pp. 65–73.

D'Azzo, J. J., and Houpis, C. H. *Feedback Control System Analysis and Synthesis*, 2d ed. McGraw-Hill, New York, 1966.

Doebelin, E. O. *Measurement Systems Application and Design*, 4th ed. McGraw-Hill, New York, 1990.

Dorf, R. C. *Modern Control Systems*, 5th ed. Addison-Wesley, Reading, MA, 1989.

D'Souza, A. F. *Design of Control Systems*. Prentice Hall, Upper Saddle River, NJ, 1988.

Edelson, J., et al. Facing the Challenges of the Current Hybrid Electric Drivetrain. *SMMA Technical Conference of the Motor and Motion Association*. Fall 2008. Available at www.ChorusCars.com.

Franklin, G. F., Powell, J. D., and Emami-Naeini, A. *Feedback Control of Dynamic Systems*. Addison-Wesley, Reading, MA, 1986.

Gurkaynak, Y., Li, Z., and Khaligh, A. A Novel Grid-tied, Solar Powered Residential Home with Plug-in Hybrid Electric Vehicle (PHEV) Loads. *IEEE Vehicle Power and Propulsion Conference* 2009, pp. 813–816.

Heller, H. C., Crawshaw, L. I., and Hammel, H. T. The Thermostat of Vertebrate Animals. *Scientific American*, August 1978, pp. 102–113.

Hogan, B. J. As Motorcycle's Speed Changes, Circuit Adjusts Radio's Volume. *Design News*, 18, August 1988, pp. 118–119.

Hostetter, G. H., Savant, C. J., Jr., and Stefani, R. T. *Design of Feedback Control Systems*, 2d ed. Saunders College Publishing, New York, 1989.

Jenkins, H. E., Kurfess, T. R., and Ludwick, S. J. Determination of a Dynamic Grinding Model. *Journal of Dynamic Systems, Measurements, and Control*, vol. 119, June 1997, pp. 289–293.

Klapper, J., and Frankle, J. T. *Phase-Locked and Frequency-Feedback Systems*. Academic Press, New York, 1972.

Lieberman, J., and Breazeal, C. Development of a Wearable Vibrotactile Feedback Suit for Accelerated Human Motor Learning. *2007 IEEE Int. Conf. on Robotics and Automation*. Roma, Italy, April 2007.

Martin, R. H., Jr. *Elementary Differential Equations with Boundary Value Problems*. McGraw-Hill, New York, 1984.

Mayr, O. The Origins of Feedback Control. *Scientific American*, October 1970, pp. 110–118.

Mayr, O. *The Origins of Feedback Control*. MIT Press, Cambridge, MA, 1970.

Meier, R., Nieuwland, J., Zbinden, A. M., and Hacisalihzade, S. S. Fuzzy Logic Control of Blood Pressure during Anesthesia. *IEEE Control Systems*, December 1992, pp. 12–17.

Mott, C., et al. *Modifying the Human Circadian Pacemaker Using Model-Based Predictive Control*. Proceedings of the American Control Conference. Denver, CO, June 2003, pp. 453–458.

Muñoz-Mansilla, R., Aranda, J., Diaz, J. M., Chaos, D., and Reinoso, A. J. Applications of QFT Robust Control Techniques to Marine Systems. *9th IEEE International Conference on Control and Automation*, December 19–21, 2011, pp. 378–385.

Novosad, J. P. *Systems, Modeling, and Decision Making*. Kendall/Hunt, Dubuque, IA, 1982.

Nyquist, H. Regeneration Theory. *Bell System Technical Journal*, January 1932.

Ogata, K. *Modern Control Engineering*, 2d ed. Prentice Hall, Upper Saddle River, NJ, 1990.

Overbye, D. The Big Ear. *Omni,* December 1990, pp. 41–48. Figure caption source for Figure 1.7.

Rockwell International. *Space Shuttle Transportation System*, 1984 (press information).

Shaw, D. A., and Turnbull, G. A. Modern Thickness Control for a Generation III Hot Strip Mill. *The International Steel Rolling Conference—The Science & Technology of Flat Rolling*, vol. 1. Association Technique de la Siderurgie Francaise, Deauville, France, 1–3, June 1987.

UNAIDS. *GLOBAL REPORT: UNAIDS report on the global AIDS epidemic 2013*. Joint United Nations Programme on HIV/AIDS (UNAIDS), 2013, p. 4.

United Technologies Otis Elevator Co. *The World of Otis*. United Technologies Otis Elevator Co., p. 2, 1991. Figure caption source for Figure 1.3(b).

United Technologies Otis Elevator Co. *Tell Me About Elevators*. United Technologies Otis Elevator Co., pp. 20–25, 1991. Figure caption source for Figure 1.3(b).

Vaughan, N. D., and Gamble, J. B. The Modeling and Simulation of a Proportional Solenoid Valve. *Journal of Dynamic Systems, Measurements, and Control*, vol. 118, March 1996, pp. 120–125.

Zhao, Q., Wang, F., Wang, W., and Deng, H. Adaptive Fuzzy Control Technology for Automatic Oil Drilling System. *Proceedings of the IEEE International Conference on Automation and Logistics*, China, August 18–21, 2007.

Zhou, J., Nygaard, G., Godhaven, J., Bretholtz, Ø., and Verfing, E. H. Adaptive Observer for Kick Detection and Switched Control for Bottomhole Pressure Regulation and Kick Attenuation during Managed Pressure Drilling. *2010 American Control Conference*. Baltimore, MD, USA, June 30–July 02, 2010.

# 2

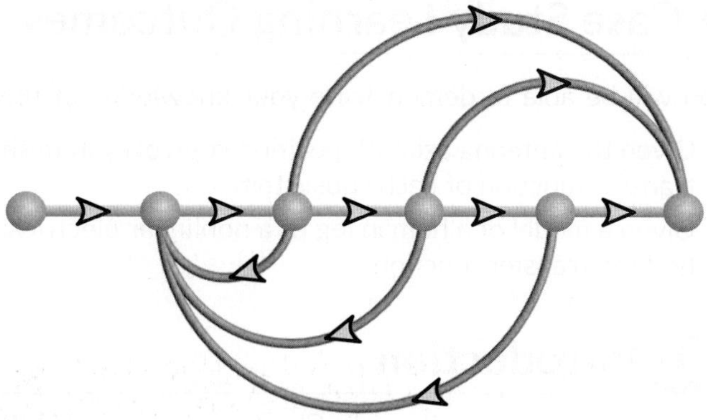

# Modeling in the Frequency Domain

## Chapter Learning Outcomes

After completing this chapter, the student will be able to:

- Find the Laplace transform of time functions and the inverse Laplace transform (Sections 2.1–2.2)
- Find the transfer function from a differential equation and solve the differential equation using the transfer function (Section 2.3)
- Find the transfer function for linear, time-invariant electrical networks (Section 2.4)
- Find the transfer function for linear, time-invariant translational mechanical systems (Section 2.5)
- Find the transfer function for linear, time-invariant rotational mechanical systems (Section 2.6)
- Find the transfer functions for gear systems with loss and for gear systems with no loss (Section 2.7)
- Find the transfer function for linear, time-invariant electromechanical systems (Section 2.8)
- Produce analogous electrical and mechanical circuits (Section 2.9)
- Linearize a nonlinear system in order to find the transfer function (Sections 2.10–2.11)

## ▶ Case Study Learning Outcomes

You will be able to demonstrate your knowledge of the chapter objectives with case studies as follows:

- Given the antenna azimuth position control system shown on the front endpapers, you will be able to find the transfer function of each subsystem.
- Given a model of a human leg or a nonlinear electrical circuit, you will be able to linearize the model and then find the transfer function.

# 2.1 Introduction

In Chapter 1 we discussed the analysis and design sequence that included obtaining the system's schematic and demonstrated this step for a position control system. To obtain a schematic, the control systems engineer must often make many simplifying assumptions in order to keep the ensuing model manageable and still approximate physical reality.

The next step is to develop mathematical models from schematics of physical systems. We will discuss two methods: (1) transfer functions in the frequency domain and (2) state equations in the time domain. These topics are covered in this chapter and in Chapter 3, respectively. As we proceed, we will notice that in every case the first step in developing a mathematical model is to apply the fundamental physical laws of science and engineering. For example, when we model electrical networks, Ohm's law and Kirchhoff's laws, which are basic laws of electric networks, will be applied initially. We will sum voltages in a loop or sum currents at a node. When we study mechanical systems, we will use Newton's laws as the fundamental guiding principles. Here we will sum forces or torques. From these equations we will obtain the relationship between the system's output and input.

In Chapter 1 we saw that a differential equation can describe the relationship between the input and output of a system. The form of the differential equation and its coefficients are a formulation or description of the system. Although the differential equation relates the system to its input and output, it is not a satisfying representation from a system perspective. Looking at Eq. (1.2), a general, $n$th-order, linear, time-invariant differential equation, we see that the system parameters, which are the coefficients, appear throughout the equation. In addition, the output, $c(t)$, and the input, $r(t)$, also appear throughout the equation.

We would prefer a mathematical representation such as that shown in Figure 2.1(*a*), where the input, output, and system are distinct and separate parts. Also, we would like to represent conveniently the interconnection of several subsystems. For example, we would like to represent *cascaded* interconnections, as shown in Figure 2.1(*b*), where a mathematical function, called a transfer function, is inside each block, and block functions

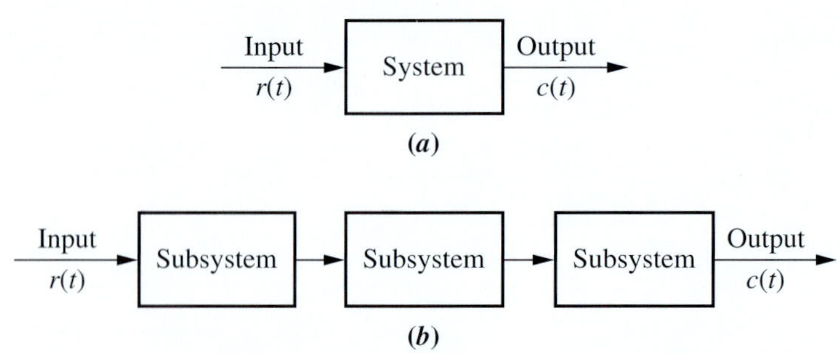


Note: The input, $r(t)$, stands for *reference input*.
The output, $c(t)$, stands for *controlled variable*.

**FIGURE 2.1** **a.** Block diagram representation of a system; **b.** block diagram representation of an interconnection of subsystems

can easily be combined to yield Figure 2.1(*a*) for ease of analysis and design. This convenience cannot be obtained with the differential equation.

## 2.2 Laplace Transform Review

A system represented by a differential equation is difficult to model as a block diagram. Thus, we now lay the groundwork for the Laplace transform, with which we can represent the input, output, and system as separate entities. Further, their interrelationship will be simply algebraic. Let us first define the Laplace transform and then show how it simplifies the representation of physical systems (*Nilsson, 1996*).

The Laplace transform is defined as

$$\mathscr{L}[f(t)] = F(s) = \int_{0-}^{\infty} f(t)e^{-st}dt \tag{2.1}$$

where $s = \sigma + j\omega$, a complex variable. Thus, knowing $f(t)$ and that the integral in Eq. (2.1) exists, we can find a function, $F(s)$, that is called the *Laplace transform* of $f(t)$.[1]

The notation for the lower limit means that even if $f(t)$ is discontinuous at $t = 0$, we can start the integration prior to the discontinuity as long as the integral converges. Thus, we can find the Laplace transform of impulse functions. This property has distinct advantages when applying the Laplace transform to the solution of differential equations where the initial conditions are discontinuous at $t = 0$. Using differential equations, we have to solve for the initial conditions after the discontinuity knowing the initial conditions before the discontinuity. Using the Laplace transform we need only know the initial conditions before the discontinuity. See *Kailath (1980)* for a more detailed discussion.

The inverse Laplace transform, which allows us to find $f(t)$ given $F(s)$, is

$$\mathscr{L}^{-1}[F(s)] = \frac{1}{2\pi j}\int_{\sigma-j\infty}^{\sigma+j\infty} F(s)e^{st}ds = f(t)u(t) \tag{2.2}$$

where

$$\begin{aligned} u(t) &= 1 \quad t > 0 \\ &= 0 \quad t < 0 \end{aligned}$$

is the unit step function. Multiplication of $f(t)$ by $u(t)$ yields a time function that is zero for $t < 0$.

Using Eq. (2.1), it is possible to derive a table relating $f(t)$ to $F(s)$ for specific cases. Table 2.1 shows the results for a representative sample of functions. If we use the tables, we do not have to use Eq. (2.2), which requires complex integration, to find $f(t)$ given $F(s)$.

[1] The Laplace transform exists if the integral of Eq. (2.1) converges. The integral will converge if $\int_{0-}^{\infty}|f(t)|e^{-\sigma_1 t}dt < \infty$. If $|f(t)| < Me^{\sigma_2 t}, 0 < t < \infty$, the integral will converge if $\infty > \sigma_1 > \sigma_2$. We call $\sigma_2$ the *abscissa of convergence*, and it is the smallest value of $\sigma$, where $s = \sigma + j\omega$, for which the integral exists.

**TABLE 2.1** Laplace transform table

| Item no. | $f(t)$ | $F(s)$ |
|---|---|---|
| 1. | $\delta(t)$ | $1$ |
| 2. | $u(t)$ | $\frac{1}{s}$ |
| 3. | $tu(t)$ | $\frac{1}{s^2}$ |
| 4. | $t^n u(t)$ | $\frac{n!}{s^{n+1}}$ |
| 5. | $e^{-at}u(t)$ | $\frac{1}{s+a}$ |
| 6. | $\sin \omega t u(t)$ | $\frac{\omega}{s^2+\omega^2}$ |
| 7. | $\cos \omega t u(t)$ | $\frac{s}{s^2+\omega^2}$ |

In the following example we demonstrate the use of Eq. (2.1) to find the Laplace transform of a time function.

## Example 2.1 Laplace Transform of a Time Function

**PROBLEM:** Find the Laplace transform of $f(t) = Ae^{-at}u(t)$.

**SOLUTION:** Since the time function does not contain an impulse function, we can replace the lower limit of Eq. (2.1) with 0. Hence,

$$F(s) = \int_0^\infty f(t)e^{-st}\,dt = \int_0^\infty Ae^{-at}e^{-st}\,dt = A\int_0^\infty e^{-(s+a)t}\,dt$$
$$= -\frac{A}{s+a}e^{-(s+a)t}\Big|_{t=0}^{\infty} = \frac{A}{s+a} \tag{2.3}$$

In addition to the Laplace transform table, Table 2.1, we can use Laplace transform theorems, listed in Table 2.2, to assist in transforming between $f(t)$ and $F(s)$. In the next example, we demonstrate the use of the Laplace transform theorems shown in Table 2.2 to find $f(t)$ given $F(s)$.

## Example 2.2 Inverse Laplace Transform

**PROBLEM:** Find the inverse Laplace transform of $F_1(s) = 1/(s+3)^2$.

**SOLUTION:** For this example we make use of the frequency shift theorem, Item 4 of Table 2.2, and the Laplace transform of $f(t) = tu(t)$, Item 3 of Table 2.1. If the inverse transform of $F(s) = 1/s^2$ is $tu(t)$, the inverse transform of $F(s+a) = 1/(s+a)^2$ is $e^{-at}tu(t)$. Hence, $f_1(t) = e^{-3t}tu(t)$.

**TABLE 2.2** Laplace transform theorems

| Item no. | Theorem | | Name |
|---|---|---|---|
| 1. | $\mathscr{L}[f(t)] = F(s)$ | $= \int_{0-}^{\infty} f(t)e^{-st}dt$ | Definition |
| 2. | $\mathscr{L}[kf(t)]$ | $= kF(s)$ | Linearity theorem |
| 3. | $\mathscr{L}[f_1(t) + f_2(t)]$ | $= F_1(s) + F_2(s)$ | Linearity theorem |
| 4. | $\mathscr{L}[e^{-at}f(t)]$ | $= F(s+a)$ | Frequency shift theorem |
| 5. | $\mathscr{L}[f(t-T)]$ | $= e^{-sT}F(s)$ | Time shift theorem |
| 6. | $\mathscr{L}[f(at)]$ | $= \frac{1}{a}F\left(\frac{s}{a}\right)$ | Scaling theorem |
| 7. | $\mathscr{L}\left[\frac{df}{dt}\right]$ | $= sF(s) - f(0-)$ | Differentiation theorem |
| 8. | $\mathscr{L}\left[\frac{d^2f}{dt^2}\right]$ | $= s^2F(s) - sf(0-) - f'(0-)$ | Differentiation theorem |
| 9. | $\mathscr{L}\left[\frac{d^nf}{dt^n}\right]$ | $= s^nF(s) - \sum_{k=1}^{n} s^{n-k}f^{k-1}(0-)$ | Differentiation theorem |
| 10. | $\mathscr{L}\left[\int_{0-}^{t} f(\tau)d\tau\right]$ | $= \frac{F(s)}{s}$ | Integration theorem |
| 11. | $f(\infty)$ | $= \lim_{s\to 0} sF(s)$ | Final value theorem[1] |
| 12. | $f(0+)$ | $= \lim_{s\to\infty} sF(s)$ | Initial value theorem[2] |

[1]For this theorem to yield correct finite results, all roots of the denominator of $F(s)$ must have negative real parts, and no more than one can be at the origin.
[2]For this theorem to be valid, $f(t)$ must be continuous or have a step discontinuity at $t = 0$ (that is, no impulses or their derivatives at $t = 0$).

## Partial-Fraction Expansion

To find the inverse Laplace transform of a complicated function, we can convert the function to a sum of simpler terms for which we know the Laplace transform of each term. The result is called a *partial-fraction expansion*. If $F_1(s) = N(s)/D(s)$, where the order of $N(s)$ is less than the order of $D(s)$, then a partial-fraction expansion can be made. If the order of $N(s)$ is greater than or equal to the order of $D(s)$, then $N(s)$ must be divided by $D(s)$ successively until the result has a remainder whose numerator is of order less than its denominator. For example, if

$$F_1(s) = \frac{s^3 + 2s^2 + 6s + 7}{s^2 + s + 5} \tag{2.4}$$

we must perform the indicated division until we obtain a remainder whose numerator is of order less than its denominator. Hence,

$$F_1(s) = s + 1 + \frac{2}{s^2 + s + 5} \tag{2.5}$$

Taking the inverse Laplace transform, using Item 1 of Table 2.1, along with the differentiation theorem (Item 7) and the linearity theorem (Item 3 of Table 2.2), we obtain

$$f_1(t) = \frac{d\delta(t)}{dt} + \delta(t) + \mathscr{L}^{-1}\left[\frac{2}{s^2 + s + 5}\right] \tag{2.6}$$

Using partial-fraction expansion, we will be able to expand functions like $F(s) = 2/(s^2 + s + 5)$ into a sum of terms and then find the inverse Laplace transform for each term. We will now consider three cases and show for each case how an $F(s)$ can be expanded into partial fractions.

### Case 1. Roots of the Denominator of *F*(*s*) Are Real and Distinct

An example of an $F(s)$ with real and distinct roots in the denominator is

$$F(s) = \frac{2}{(s+1)(s+2)} \tag{2.7}$$

The roots of the denominator are distinct, since each factor is raised only to unity power. We can write the partial-fraction expansion as a sum of terms where each factor of the original denominator forms the denominator of each term, and constants, called *residues*, form the numerators. Hence,

$$F(s) = \frac{2}{(s+1)(s+2)} = \frac{K_1}{(s+1)} + \frac{K_2}{(s+2)} \tag{2.8}$$

To find $K_1$, we first multiply Eq. (2.8) by $(s+1)$, which isolates $K_1$. Thus,

$$\frac{2}{(s+2)} = K_1 + \frac{(s+1)K_2}{(s+2)} \tag{2.9}$$

Letting $s$ approach $-1$ eliminates the last term and yields $K_1 = 2$. Similarly, $K_2$ can be found by multiplying Eq. (2.8) by $(s+2)$ and then letting $s$ approach $-2$; hence, $K_2 = -2$.

Each component part of Eq. (2.8) is an $F(s)$ in Table 2.1. Hence, $f(t)$ is the sum of the inverse Laplace transform of each term, or

$$f(t) = (2e^{-t} - 2e^{-2t})u(t) \tag{2.10}$$

In general, then, given an $F(s)$ whose denominator has real and distinct roots, a partial-fraction expansion,

$$\begin{aligned} F(s) = \frac{N(s)}{D(s)} &= \frac{N(s)}{(s+p_1)(s+p_2)\cdots(s+p_m)\cdots(s+p_n)} \\ &= \frac{K_1}{(s+p_1)} + \frac{K_2}{(s+p_2)} + \cdots + \frac{K_m}{(s+p_m)} + \cdots + \frac{K_n}{(s+p_n)} \end{aligned} \tag{2.11}$$

can be made if the order of $N(s)$ is less than the order of $D(s)$. To evaluate each residue, $K_i$, we multiply Eq. (2.11) by the denominator of the corresponding partial fraction. Thus, if we want to find $K_m$, we multiply Eq. (2.11) by $(s+p_m)$ and get

$$\begin{aligned} (s+p_m)F(s) &= \frac{(s+p_m)N(s)}{(s+p_1)(s+p_2)\cdots(s+p_m)\cdots(s+p_n)} \\ &= (s+p_m)\frac{K_1}{(s+p_1)} + (s+p_m)\frac{K_2}{(s+p_2)} + \cdots + K_m + \cdots \\ &\quad + (s+p_m)\frac{K_n}{(s+p_n)} \end{aligned} \tag{2.12}$$

If we let $s$ approach $-p_m$, all terms on the right-hand side of Eq. (2.12) go to zero except the term $K_m$, leaving

$$\left.\frac{\cancel{(s+p_m)}N(s)}{(s+p_1)(s+p_2)\cdots\cancel{(s+p_m)}\cdots(s+p_n)}\right|_{s\to -p_m} = K_m \tag{2.13}$$

The following example demonstrates the use of the partial-fraction expansion to solve a differential equation. We will see that the Laplace transform reduces the task of finding the solution to simple algebra.

## Example 2.3 Laplace Transform Solution of a Differential Equation

**PROBLEM:** Given the following differential equation, solve for $y(t)$ if all initial conditions are zero. Use the Laplace transform.

$$\frac{d^2y}{dt^2} + 12\frac{dy}{dt} + 32y = 32u(t) \tag{2.14}$$

**SOLUTION:** Substitute the corresponding $F(s)$ for each term in Eq. (2.14), using Item 2 in Table 2.1, Items 7 and 8 in Table 2.2, and the initial conditions of $y(t)$ and $dy(t)/dt$ given by $y(0-) = 0$ and $\dot{y}(0-) = 0$, respectively. Hence, the Laplace transform of Eq. (2.14) is

$$s^2Y(s) + 12sY(s) + 32Y(s) = \frac{32}{s} \tag{2.15}$$

Solving for the response, $Y(s)$, yields

$$Y(s) = \frac{32}{s(s^2 + 12s + 32)} = \frac{32}{s(s+4)(s+8)} \tag{2.16}$$

To solve for $y(t)$, we notice that Eq. (2.16) does not match any of the terms in Table 2.1. Thus, we form the partial-fraction expansion of the right-hand term and match each of the resulting terms with $F(s)$ in Table 2.1. Therefore,

$$Y(s) = \frac{32}{s(s+4)(s+8)} = \frac{K_1}{s} + \frac{K_2}{(s+4)} + \frac{K_3}{(s+8)} \tag{2.17}$$

where, from Eq. (2.13),

$$K_1 = \left.\frac{32}{(s+4)(s+8)}\right|_{s\to 0} = 1 \tag{2.18a}$$

$$K_2 = \left.\frac{32}{s(s+8)}\right|_{s\to -4} = -2 \tag{2.18b}$$

$$K_3 = \left.\frac{32}{s(s+4)}\right|_{s\to -8} = 1 \tag{2.18c}$$

Hence,

$$Y(s) = \frac{1}{s} - \frac{2}{(s+4)} + \frac{1}{(s+8)} \tag{2.19}$$

Since each of the three component parts of Eq. (2.19) is represented as an $F(s)$ in Table 2.1, $y(t)$ is the sum of the inverse Laplace transforms of each term. Hence,

$$y(t) = (1 - 2e^{-4t} + e^{-8t})u(t) \tag{2.20}$$

MATLAB **ML**

`Students who are using MATLAB should now run ch2p1 through ch2p8 in Appendix B. This is your first MATLAB exercise. You will learn how to use MATLAB to (1) represent polynomials, (2) find roots of polynomials, (3) multiply polynomials, and (4) find partial-fraction expansions. Finally, Example 2.3 will be solved using MATLAB.`

The $u(t)$ in Eq. (2.20) shows that the response is zero until $t = 0$. Unless otherwise specified, all inputs to systems in the text will not start until $t = 0$. Thus, output responses will also be zero until $t = 0$. For convenience, we will leave off the $u(t)$ notation from now on. Accordingly, we write the output response as

$$y(t) = 1 - 2e^{-4t} + e^{-8t} \tag{2.21}$$

### Case 2. Roots of the Denominator of $F(s)$ Are Real and Repeated

An example of an $F(s)$ with real and repeated roots in the denominator is

$$F(s) = \frac{2}{(s+1)(s+2)^2} \tag{2.22}$$

**TryIt 2.1**

Use the following MATLAB and Control System Toolbox statement to form the linear, time-invariant (LTI) transfer function of Eq. (2.22).

```
F=zpk([],[-1 -2 -2],2)
```

The roots of $(s+2)^2$ in the denominator are repeated, since the factor is raised to an integer power higher than 1. In this case, the denominator root at $-2$ is a *multiple root of multiplicity* 2.

We can write the partial-fraction expansion as a sum of terms, where each factor of the denominator forms the denominator of each term. In addition, each multiple root generates additional terms consisting of denominator factors of reduced multiplicity. For example, if

$$F(s) = \frac{2}{(s+1)(s+2)^2} = \frac{K_1}{(s+1)} + \frac{K_2}{(s+2)^2} + \frac{K_3}{(s+2)} \tag{2.23}$$

then $K_1 = 2$, which can be found as previously described. $K_2$ can be isolated by multiplying Eq. (2.23) by $(s+2)^2$, yielding

$$\frac{2}{s+1} = (s+2)^2 \frac{K_1}{(s+1)} + K_2 + (s+2)K_3 \tag{2.24}$$

Letting $s$ approach $-2$, $K_2 = -2$. To find $K_3$ we see that if we differentiate Eq. (2.24) with respect to $s$,

$$\frac{-2}{(s+1)^2} = \frac{(s+2)s}{(s+1)^2} K_1 + K_3 \tag{2.25}$$

$K_3$ is isolated and can be found if we let $s$ approach $-2$. Hence, $K_3 = -2$.

Each component part of Eq. (2.23) is an $F(s)$ in Table 2.1; hence, $f(t)$ is the sum of the inverse Laplace transform of each term, or

$$f(t) = 2e^{-t} - 2te^{-2t} - 2e^{-2t} \tag{2.26}$$

**TryIt 2.2**

Use the following MATLAB statements to help you get Eq. (2.26).

```
numf=2;
denf=poly([-1-2-2]);
[K,p,k]=residue...
  (numf,denf)
```

If the denominator root is of higher multiplicity than 2, successive differentiation would isolate each residue in the expansion of the multiple root.

In general, then, given an $F(s)$ whose denominator has real and repeated roots, a partial-fraction expansion

$$\begin{aligned} F(s) &= \frac{N(s)}{D(s)} \\ &= \frac{N(s)}{(s+p_1)^r (s+p_2)\cdots(s+p_n)} \\ &= \frac{K_1}{(s+p_1)^r} + \frac{K_2}{(s+p_1)^{r-1}} + \cdots + \frac{K_r}{(s+p_1)} + \frac{K_{r+1}}{(s+p_2)} + \cdots + \frac{K_n}{(s+p_n)} \end{aligned} \tag{2.27}$$

can be made if the order of $N(s)$ is less than the order of $D(s)$ and the repeated roots are of multiplicity $r$ at $-p_1$. To find $K_1$ through $K_r$ for the roots of multiplicity greater than unity, first multiply Eq. (2.27) by $(s+p_1)^r$ getting $F_1(s)$, which is

$$\begin{aligned} F_1(s) &= (s+p_1)^r F(s) \\ &= \frac{(s+p_1)^r N(s)}{(s+p_1)^r (s+p_2)\cdots(s+p_n)} \\ &= K_1 + (s+p_1)K_2 + (s+p_1)^2 K_3 + \cdots + (s+p_1)^{r-1} K_r \\ &\quad + \frac{K_{r+1}(s+p_1)^r}{(s+p_2)} + \cdots + \frac{K_n(s+p_1)^r}{(s+p_n)} \end{aligned} \tag{2.28}$$

Immediately, we can solve for $K_1$ if we let $s$ approach $-p_1$. We can solve for $K_2$ if we differentiate Eq. (2.28) with respect to $s$ and then let $s$ approach $-p_1$. Subsequent differentiation will allow us to find $K_3$ through $K_r$. The general expression for $K_1$ through $K_r$ for the multiple roots is

$$K_i = \frac{1}{(i-1)!} \frac{d^{i-1} F_1(s)}{ds^{i-1}} \bigg|_{s \to -p_1} \qquad i = 1, 2, \ldots, r; \quad 0! = 1 \tag{2.29}$$

### Case 3. Roots of the Denominator of *F(s)* Are Complex or Imaginary

An example of $F(s)$ with complex roots in the denominator is

$$F(s) = \frac{3}{s(s^2+2s+5)} \tag{2.30}$$

> **TryIt 2.3**
>
> Use the following MATLAB and Control System Toolbox statement to form the LTI transfer function of Eq. (2.30).
>
> ```
> F=tf([3],[1 2 5 0])
> ```

This function can be expanded in the following form:

$$\frac{3}{s(s^2+2s+5)} = \frac{K_1}{s} + \frac{K_2 s + K_3}{s^2+2s+5} \tag{2.31}$$

$K_1$ is found in the usual way to be $\frac{3}{5}$. $K_2$ and $K_3$ can be found by first multiplying Eq. (2.31) by the lowest common denominator, $s(s^2+2s+5)$, and clearing the fractions. After simplification with $K_1 = \frac{3}{5}$, we obtain

$$3 = \left(K_2 + \frac{3}{5}\right)s^2 + \left(K_3 + \frac{6}{5}\right)s + 3 \tag{2.32}$$

Balancing coefficients, $\left(K_2 + \frac{3}{5}\right) = 0$ and $\left(K_3 + \frac{6}{5}\right) = 0$. Hence $K_2 = -\frac{3}{5}$ and $K_3 = -\frac{6}{5}$. Thus,

$$F(s) = \frac{3}{s(s^2+2s+5)} = \frac{3/5}{s} - \frac{3}{5}\frac{s+2}{s^2+2s+5} \tag{2.33}$$

The last term can be shown to be the sum of the Laplace transforms of an exponentially damped sine and cosine. Using Item 7 in Table 2.1 and Items 2 and 4 in Table 2.2, we get

$$\mathscr{L}[Ae^{-at}\cos\omega t] = \frac{A(s+a)}{(s+a)^2+\omega^2} \tag{2.34}$$

Similarly,

$$\mathscr{L}[Be^{-at}\sin \omega t] = \frac{B\omega}{(s+a)^2+\omega^2} \tag{2.35}$$

Adding Eqs. (2.34) and (2.35), we get

$$\mathscr{L}[Ae^{-at}\cos \omega t + Be^{-at}\sin \omega t] = \frac{A(s+a)+B\omega}{(s+a)^2+\omega^2} \tag{2.36}$$

We now convert the last term of Eq. (2.33) to the form suggested by Eq. (2.36) by completing the squares in the denominator and adjusting terms in the numerator without changing its value. Hence,

$$F(s) = \frac{3/5}{s} - \frac{3}{5}\frac{(s+1)+(1/2)(2)}{(s+1)^2+2^2} \tag{2.37}$$

**TryIt 2.4**

Use the following MATLAB and Symbolic Math Toolbox statements to get Eq. (2.38) from Eq. (2.30).

```
syms s
f=ilaplace...
  (3/(s*(s^2+2*s+5)));
pretty(f)
```

Comparing Eq. (2.37) to Table 2.1 and Eq. (2.36), we find

$$f(t) = \frac{3}{5} - \frac{3}{5}e^{-t}\left(\cos 2t + \frac{1}{2}\sin 2t\right) \tag{2.38}$$

In order to visualize the solution, an alternate form of *f*(*t*), obtained by trigonometric identities, is preferable. Using the amplitudes of the cos and sin terms, we factor out $\sqrt{1^2+(1/2)^2}$ from the term in parentheses and obtain

$$f(t) = \frac{3}{5} - \frac{3}{5}\sqrt{1^2+(1/2)^2}e^{-t}\left(\frac{1}{\sqrt{1^2+(1/2)^2}}\cos 2t + \frac{1/2}{\sqrt{1^2+(1/2)^2}}\sin 2t\right) \tag{2.39}$$

Letting $1/\sqrt{1^2+(1/2)^2} = \cos\phi$ and $(1/2)/\sqrt{1^2+(1/2)^2} = \sin\phi$,

$$f(t) = \frac{3}{5} - \frac{3}{5}\sqrt{1^2+(1/2)^2}e^{-t}(\cos\phi\cos 2t + \sin\phi\sin 2t) \tag{2.40}$$

or

$$f(t) = 0.6 - 0.671e^{-t}\cos(2t-\phi) \tag{2.41}$$

where $\phi = \arctan 0.5 = 26.57°$. Thus, *f*(*t*) is a constant plus an exponentially damped sinusoid.

In general, then, given an *F*(*s*) whose denominator has complex or purely imaginary roots, a partial-fraction expansion,

$$F(s) = \frac{N(s)}{D(s)} = \frac{N(s)}{(s+p_1)(s^2+as+b)\cdots}$$
$$= \frac{K_1}{(s+p_1)} + \frac{(K_2s+K_3)}{(s^2+as+b)} + \cdots \tag{2.42}$$

can be made if the order of $N(s)$ is less than the order of $D(s)$ $p_1$ is real, and $(s^2 + as + b)$ has complex or purely imaginary roots. The complex or imaginary roots are expanded with $(K_2 s + K_3)$ terms in the numerator rather than just simply $K_i$, as in the case of real roots. The $K_i$'s in Eq. (2.42) are found through balancing the coefficients of the equation after clearing fractions. After completing the squares on $(s^2 + as + b)$ and adjusting the numerator, $(K_2 s + K_3)/(s^2 + as + b)$ can be put into the form shown on the right-hand side of Eq. (2.36).

Finally, the case of purely imaginary roots arises if $a = 0$ in Eq. (2.42). The calculations are the same.

Another method that follows the technique used for the partial-fraction expansion of $F(s)$ with real roots in the denominator can be used for complex and imaginary roots. However, the residues of the complex and imaginary roots are themselves complex conjugates. Then, after taking the inverse Laplace transform, the resulting terms can be identified as

$$\frac{e^{j\theta} + e^{-j\theta}}{2} = \cos\theta \tag{2.43}$$

and

$$\frac{e^{j\theta} - e^{-j\theta}}{2j} = \sin\theta \tag{2.44}$$

For example, the previous $F(s)$ can also be expanded in partial fractions as

$$\begin{aligned} F(s) = \frac{3}{s(s^2+2s+5)} &= \frac{3}{s(s+1+j2)(s+1-j2)} \\ &= \frac{K_1}{s} + \frac{K_2}{s+1+j2} + \frac{K_3}{s+1-j2} \end{aligned} \tag{2.45}$$

Finding $K_2$,

$$K_2 = \left.\frac{3}{s(s+1-j2)}\right|_{s\to -1-j2} = -\frac{3}{20}(2+j1) \tag{2.46}$$

Similarly, $K_3$ is found to be the complex conjugate of $K_2$, and $K_1$ is found as previously described. Hence,

$$F(s) = \frac{3/5}{s} - \frac{3}{20}\left(\frac{2+j1}{s+1+j2} + \frac{2-j1}{s+1-j2}\right) \tag{2.47}$$

**TryIt 2.5**

Use the following MATLAB statements to help you get Eq. (2.47).

```
numf=3
denf=[1 2 5 0]
[K,p,k]=residue...
  (numf, denf)
```

from which

$$\begin{aligned} f(t) &= \frac{3}{5} - \frac{3}{20}\left[(2+j1)e^{-(1+j2)t} + (2-j1)e^{-(1-j2)t}\right] \\ &= \frac{3}{5} - \frac{3}{20}e^{-t}\left[4\left(\frac{e^{j2t}+e^{-j2t}}{2}\right) + 2\left(\frac{e^{j2t}+e^{-j2t}}{2j}\right)\right] \end{aligned} \tag{2.48}$$

Using Eqs. (2.43) and (2.44), we get

$$f(t) = \frac{3}{5} - \frac{3}{5}e^{-t}\left(\cos 2t + \frac{1}{2}\sin 2t\right) = 0.6 - 0.671e^{-t}\cos(2t - \phi) \tag{2.49}$$

where $\phi = \arctan 0.5 = 26.57°$.

Symbolic Math
SM

Students who are performing the MATLAB exercises and want to explore the added capability of MATLAB's Symbolic Math Toolbox should now run ch2sp1 and ch2sp2 in Appendix F at www.wiley.com/college/nise. You will learn how to construct symbolic objects and then find the inverse Laplace and Laplace transforms of frequency and time functions, respectively. The examples in Case 2 and Case 3 in this section will be solved using the Symbolic Math Toolbox.

## Skill-Assessment Exercise 2.1

**PROBLEM:** Find the Laplace transform of $f(t) = te^{-5t}$.

**ANSWER:** $F(s) = 1/(s+5)^2$

The complete solution is at www.wiley.com/college/nise.

## Skill-Assessment Exercise 2.2

**PROBLEM:** Find the inverse Laplace transform of $F(s) = 10/[s(s+2)(s+3)^2]$.

**ANSWER:** $f(t) = \frac{5}{9} - 5e^{-2t} + \frac{10}{3}te^{-3t} + \frac{40}{9}e^{-3t}$

The complete solution is at www.wiley.com/college/nise.

## 2.3 The Transfer Function

In the previous section we defined the Laplace transform and its inverse. We presented the idea of the partial-fraction expansion and applied the concepts to the solution of differential equations. We are now ready to formulate the system representation shown in Figure 2.1 by establishing a viable definition for a function that algebraically relates a system's output to its input. This function will allow separation of the input, system, and output into three separate and distinct parts, unlike the differential equation. The function will also allow us to *algebraically* combine mathematical representations of subsystems to yield a total system representation.

Let us begin by writing a general $n$th-order, linear, time-invariant differential equation,

$$a_n\frac{d^nc(t)}{dt^n} + a_{n-1}\frac{d^{n-1}c(t)}{dt^{n-1}} + \cdots + a_0c(t) = b_m\frac{d^mr(t)}{dt^m} + b_{m-1}\frac{d^{m-1}r(t)}{dt^{m-1}} + \cdots + b_0r(t) \tag{2.50}$$

where $c(t)$ is the output, $r(t)$ is the input, and the $a_i$'s, $b_i$'s, and the form of the differential equation represent the system. Taking the Laplace transform of both sides,

$$\begin{aligned} &a_ns^nC(s) + a_{n-1}s^{n-1}C(s) + \cdots + a_0C(s) + \text{initial condition terms involving } c(t) \\ &= b_ms^mR(s) + b_{m-1}s^{m-1}R(s) + \cdots + b_0R(s) + \text{initial condition terms involving } r(t) \end{aligned} \tag{2.51}$$

Equation (2.51) is a purely algebraic expression. If we assume that *all initial conditions are zero*, Eq. (2.51) reduces to

$$(a_n s^n + a_{n-1} s^{n-1} + \cdots + a_0)C(s) = (b_m s^m + b_{m-1} s^{m-1} + \cdots + b_0)R(s) \tag{2.52}$$

Now form the ratio of the output transform, $C(s)$, divided by the input transform, $R(s)$:

$$\frac{C(s)}{R(s)} = G(s) = \frac{(b_m s^m + b_{m-1} s^{m-1} + \cdots + b_0)}{(a_n s^n + a_{n-1} s^{n-1} + \cdots + a_0)} \tag{2.53}$$

Notice that Eq. (2.53) separates the output, $C(s)$, the input, $R(s)$, and the system, which is the ratio of polynomials in $s$ on the right. We call this ratio, $G(s)$, the *transfer function* and evaluate it with *zero initial conditions*.

The transfer function can be represented as a block diagram, as shown in Figure 2.2, with the input on the left, the output on the right, and the system transfer function inside the block. Notice that the denominator of the transfer function is identical to the characteristic polynomial of the differential equation. Also, we can find the output, $C(s)$ by using

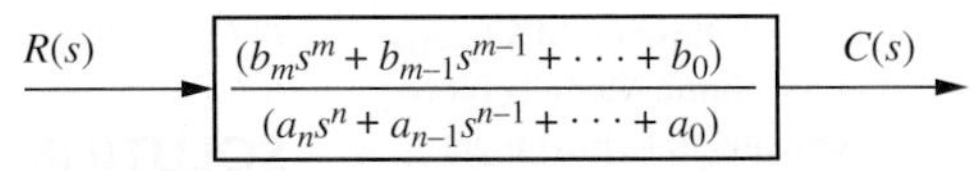


**FIGURE 2.2** Block diagram of a transfer function

$$C(s) = R(s)G(s) \tag{2.54}$$

Let us apply the concept of a transfer function to an example and then use the result to find the response of the system.

## Example 2.4 Transfer Function for a Differential Equation

**PROBLEM:** Find the transfer function represented by

$$\frac{dc(t)}{dt} + 2c(t) = r(t) \tag{2.55}$$

**SOLUTION:** Taking the Laplace transform of both sides, assuming zero initial conditions, we have

$$sC(s) + 2C(s) = R(s) \tag{2.56}$$

The transfer function, $G(s)$, is

$$G(s) = \frac{C(s)}{R(s)} = \frac{1}{s+2} \tag{2.57}$$

MATLAB

**ML**

Students who are using MATLAB should now run ch2p9 through ch2p12 in Appendix B. You will learn how to use MATLAB to create transfer functions with numerators and denominators in polynomial or factored form. You will also learn how to convert between polynomial and factored forms. Finally, you will learn how to use MATLAB to plot time functions.

Symbolic Math

**SM**

Students who are performing the MATLAB exercises and want to explore the added capability of MATLAB's Symbolic Math Toolbox should now run ch2sp3 in Appendix F at www.wiley.com/college/nise. You will learn how to use the Symbolic Math Toolbox to simplify the input of complicated transfer functions as well as improve readability. You will learn how to enter a symbolic transfer function and convert it to a linear, time-invariant (LTI) object as presented in Appendix B, ch2p9.

## Example 2.5 System Response from the Transfer Function

**TryIt 2.6**

Use the following MATLAB and Symbolic Math Toolbox statements to help you get Eq. (2.60).

```
syms s
C=1/(s*(s+2))
C=ilaplace(C)
```

**TryIt 2.7**

Use the following MATLAB statements to plot Eq. (2.60) for $t$ from 0 to 1 sat intervals of 0.01 s.

```
t=0:0.01:3; plot...
(t,(1/2-1/2*exp(-2*t)))
```

**PROBLEM:** Use the result of Example 2.4 to find the response, $c(t)$ to an input, $r(t) = u(t)$, a unit step, assuming zero initial conditions.

**SOLUTION:** To solve the problem, we use Eq. (2.54), where $G(s) = 1/(s+2)$ as found in Example 2.4. Since $r(t) = u(t)$, $R(s) = 1/s$, from Table 2.1. Since the initial conditions are zero,

$$C(s) = R(s)G(s) = \frac{1}{s(s+2)} \tag{2.58}$$

Expanding by partial fractions, we get

$$C(s) = \frac{1/2}{s} - \frac{1/2}{s+2} \tag{2.59}$$

Finally, taking the inverse Laplace transform of each term yields

$$c(t) = \frac{1}{2} - \frac{1}{2}e^{-2t} \tag{2.60}$$

## Skill-Assessment Exercise 2.3

**PROBLEM:** Find the transfer function, $G(s) = C(s)/R(s)$, corresponding to the differential equation $\frac{d^3c}{dt^3} + 3\frac{d^2c}{dt^2} + 7\frac{dc}{dt} + 5c = \frac{d^2r}{dt^2} + 4\frac{dr}{dt} + 3r$.

**ANSWER:** $G(s) = \dfrac{C(s)}{R(s)} = \dfrac{s^2 + 4s + 3}{s^3 + 3s^2 + 7s + 5}$

The complete solution is at www.wiley.com/college/nise.

## Skill-Assessment Exercise 2.4

**PROBLEM:** Find the differential equation corresponding to the transfer function,

$$G(s) = \frac{2s+1}{s^2 + 6s + 2}$$

**ANSWER:** $\dfrac{d^2c}{dt^2} + 6\dfrac{dc}{dt} + 2c = 2\dfrac{dr}{dt} + r$

The complete solution is at www.wiley.com/college/nise.

## Skill-Assessment Exercise 2.5

**PROBLEM:** Find the ramp response for a system whose transfer function is

$$G(s) = \frac{s}{(s+4)(s+8)}$$

**ANSWER:** $c(t) = \frac{1}{32} - \frac{1}{16}e^{-4t} + \frac{1}{32}e^{-8t}$

The complete solution is at www.wiley.com/college/nise.

In general, a physical system that can be represented by a linear, time-invariant differential equation can be modeled as a transfer function. The rest of this chapter will be devoted to the task of modeling individual subsystems. We will learn how to represent electrical networks, translational mechanical systems, rotational mechanical systems, and electromechanical systems as transfer functions. As the need arises, the reader can consult the Bibliography at the end of the chapter for discussions of other types of systems, such as pneumatic, hydraulic, and heat-transfer systems (*Cannon, 1967*).

# 2.4 Electrical Network Transfer Functions

In this section, we formally apply the transfer function to the mathematical modeling of electric circuits including passive networks and operational amplifier circuits. Subsequent sections cover mechanical and electromechanical systems.

Equivalent circuits for the electric networks that we work with first consist of three passive linear components: resistors, capacitors, and inductors.[2] Table 2.3 summarizes the components and the relationships between voltage and current and between voltage and charge under zero initial conditions.

We now combine electrical components into circuits, decide on the input and output, and find the transfer function. Our guiding principles are Kirchhoff's laws. We sum voltages around loops or sum currents at nodes, depending on which technique involves the least

**TABLE 2.3** Voltage-current, voltage-charge, and impedance relationships for capacitors, resistors, and inductors

| Component | Voltage-current | Current-voltage | Voltage-charge | Impedance $Z(s) = V(s)/I(s)$ | Admittance $Y(s) = I(s)/V(s)$ |
|---|---|---|---|---|---|
| 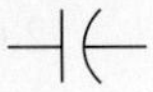 Capacitor | $v(t) = \frac{1}{C}\int_0^1 i(\tau)d\tau$ | $i(t) = C\frac{dv(t)}{dt}$ | $v(t) = \frac{1}{C}q(t)$ | $\frac{1}{Cs}$ | $Cs$ |
|  Resistor | $v(t) = Ri(t)$ | $i(t) = \frac{1}{R}v(t)$ | $v(t) = R\frac{dq(t)}{dt}$ | $R$ | $\frac{1}{R} = G$ |
|  Inductor | $v(t) = L\frac{di(t)}{dt}$ | $i(t) = \frac{1}{L}\int_0^1 v(\tau)d\tau$ | $v(t) = L\frac{d^2q(t)}{dt^2}$ | $Ls$ | $\frac{1}{Ls}$ |

Note: The following set of symbols and units is used throughout this book: $v(t)$ – V (volts), $i(t)$ – A (amps), $q(t)$ – Q (coulombs), $C$ – F (farads), $R$ – Ω (ohms), $G$ – Ω (mhos), $L$ – $H$ (henries).

[2] *Passive* means that there is no internal source of energy.

effort in algebraic manipulation, and then equate the result to zero. From these relationships we can write the differential equations for the circuit. Then we can take the Laplace transforms of the differential equations and finally solve for the transfer function.

## Simple Circuits via Mesh Analysis

Transfer functions can be obtained using Kirchhoff's voltage law and summing voltages around loops or meshes.[3] We call this method *loop* or *mesh analysis* and demonstrate it in the following example.

### Example 2.6 Transfer Function—Single Loop via the Differential Equation

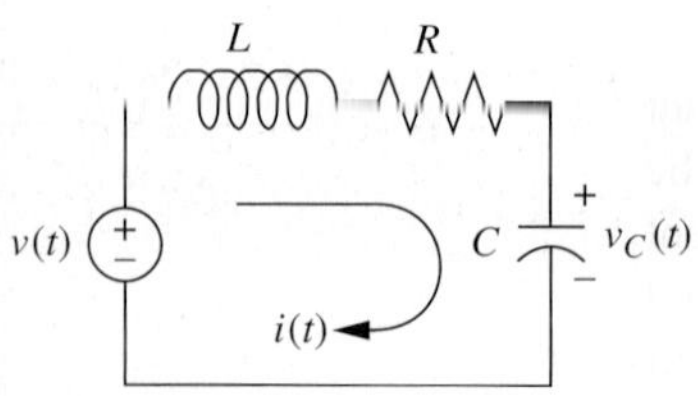


**FIGURE 2.3** *RLC* network

**PROBLEM:** Find the transfer function relating the capacitor voltage, $V_C(s)$, to the input voltage, $V(s)$ in Figure 2.3.

**SOLUTION:** In any problem, the designer must first decide what the input and output should be. In this network, several variables could have been chosen to be the output—for example, the inductor voltage, the capacitor voltage, the resistor voltage, or the current. The problem statement, however, is clear in this case: We are to treat the capacitor voltage as the output and the applied voltage as the input.

Summing the voltages around the loop, assuming zero initial conditions, yields the integro-differential equation for this network as

$$L\frac{di(t)}{dt} + Ri(t) + \frac{1}{C}\int_0^t i(\tau)d\tau = v(t) \tag{2.61}$$

Changing variables from current to charge using $i(t) = dq(t)/dt$ yields

$$L\frac{d^2q(t)}{dt^2} + R\frac{dq(t)}{dt} + \frac{1}{C}q(t) = v(t) \tag{2.62}$$

From the voltage-charge relationship for a capacitor in Table 2.3,

$$q(t) = Cv_C(t) \tag{2.63}$$

Substituting Eq. (2.63) into Eq. (2.62) yields

$$LC\frac{d^2v_C(t)}{dt^2} + RC\frac{dv_C(t)}{dt} + v_C(t) = v(t) \tag{2.64}$$

Taking the Laplace transform assuming zero initial conditions, rearranging terms, and simplifying yields

$$(LCs^2 + RCs + 1)V_C(s) = V(s) \tag{2.65}$$

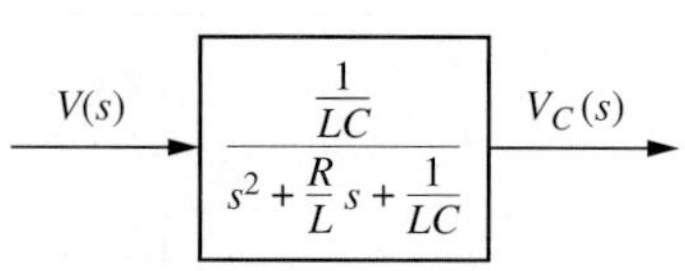


**FIGURE 2.4** Block diagram of series *RLC* electrical network

Solving for the transfer function, $V_C(s)/V(s)$, we obtain

$$\frac{V_C(s)}{V(s)} = \frac{1/LC}{s^2 + \frac{R}{L}s + \frac{1}{LC}} \tag{2.66}$$

as shown in Figure 2.4.

[3] A particular loop that resembles the spaces in a screen or fence is called a *mesh*.

Let us now develop a technique for simplifying the solution for future problems. First, take the Laplace transform of the equations in the voltage-current column of Table 2.3 assuming zero initial conditions.

For the capacitor,

$$V(s) = \frac{1}{Cs} I(s) \tag{2.67}$$

For the resistor,

$$V(s) = RI(s) \tag{2.68}$$

For the inductor,

$$V(s) = LsI(s) \tag{2.69}$$

Now define the following transfer function:

$$\frac{V(s)}{I(s)} = Z(s) \tag{2.70}$$

Notice that this function is similar to the definition of resistance, that is, the ratio of voltage to current. But, unlike resistance, this function is applicable to capacitors and inductors and carries information on the dynamic behavior of the component, since it represents an equivalent differential equation. We call this particular transfer function *impedance*. The impedance for each of the electrical elements is shown in Table 2.3.

Let us now demonstrate how the concept of impedance simplifies the solution for the transfer function. The Laplace transform of Eq. (2.61), assuming zero initial conditions, is

$$\left(Ls + R + \frac{1}{Cs}\right) I(s) = V(s) \tag{2.71}$$

Notice that Eq. (2.71), which is in the form

$$[\text{Sum of impedances}]\, I(s) = [\text{Sum of applied voltages}] \tag{2.72}$$

suggests the series circuit shown in Figure 2.5. Also notice that the circuit of Figure 2.5 could have been obtained immediately from the circuit of Figure 2.3 simply by replacing each element with its impedance. We call this altered circuit the *transformed circuit*. Finally, notice that the transformed circuit leads immediately to Eq. (2.71) if we add impedances in series as we add resistors in series. Thus, rather than writing the differential equation first and then taking the Laplace transform, we can draw the transformed circuit and obtain the Laplace transform of the differential equation simply by applying Kirchhoff's voltage law to the transformed circuit. We summarize the steps as follows:

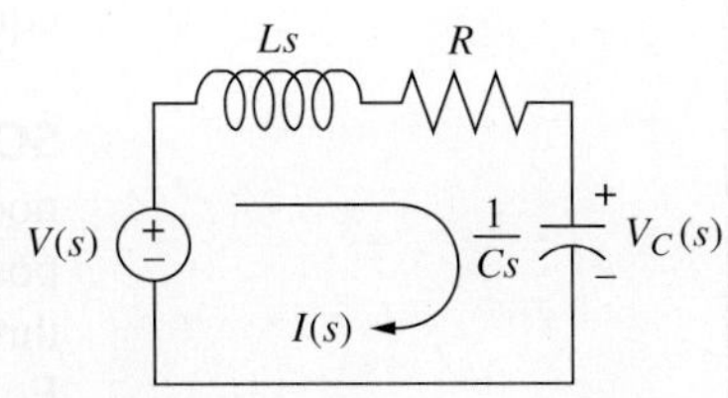


**FIGURE 2.5** Laplace-transformed network

1. Redraw the original network showing all time variables, such as $v(t)$, $i(t)$, and $v_C(t)$, as Laplace transforms $V(s)$, $I(s)$, and $V_C(s)$, respectively.
2. Replace the component values with their impedance values. This replacement is similar to the case of dc circuits, where we represent resistors with their resistance values.

We now redo Example 2.6 using the transform methods just described and bypass the writing of the differential equation.

## Example 2.7 Transfer Function—Single Loop via Transform Methods

**PROBLEM:** Repeat Example 2.6 using mesh analysis and transform methods without writing a differential equation.

**SOLUTION:** Using Figure 2.5 and writing a mesh equation using the impedances as we would use resistor values in a purely resistive circuit, we obtain

$$\left(Ls + R + \frac{1}{Cs}\right)I(s) = V(s) \tag{2.73}$$

Solving for $I(s)/V(s)$,

$$\frac{I(s)}{V(s)} = \frac{1}{Ls + R + \frac{1}{Cs}} \tag{2.74}$$

But the voltage across the capacitor, $V_C(s)$, is the product of the current and the impedance of the capacitor. Thus,

$$V_C(s) = I(s)\frac{1}{Cs} \tag{2.75}$$

Solving Eq. (2.75) for $I(s)$ substituting $I(s)$ into Eq. (2.74), and simplifying yields the same result as Eq. (2.66).

### Simple Circuits via Nodal Analysis

Transfer functions also can be obtained using Kirchhoff's current law and summing currents flowing from nodes. We call this method *nodal analysis*. We now demonstrate this principle by redoing Example 2.6 using Kirchhoff's current law and the transform methods just described to bypass writing the differential equation.

## Example 2.8 Transfer Function—Single Node via Transform Methods

**PROBLEM:** Repeat Example 2.6 using nodal analysis and without writing a differential equation.

**SOLUTION:** The transfer function can be obtained by summing currents flowing out of the node whose voltage is $V_C(s)$ in Figure 2.5. We assume that currents leaving the node are positive and currents entering the node are negative. The currents consist of the current through the capacitor and the current flowing through the series resistor and inductor. From Eq. (2.70), each $I(s) = V(s)/Z(s)$. Hence,

$$\frac{V_C(s)}{1/Cs} + \frac{V_C(s) - V(s)}{R + Ls} = 0 \tag{2.76}$$

where $V_C(s)/(1/Cs)$ is the current flowing out of the node through the capacitor, and $[V_C(s) - V(s)]/(R + Ls)$ is the current flowing out of the node through the series resistor and inductor. Solving Eq. (2.76) for the transfer function, $V_C(s)/V(s)$, we arrive at the same result as Eq. (2.66).

## Simple Circuits via Voltage Division

Example 2.6 can be solved directly by using voltage division on the transformed network. We now demonstrate this technique.

### Example 2.9 Transfer Function—Single Loop via Voltage Division

**PROBLEM:** Repeat Example 2.6 using voltage division and the transformed circuit.

**SOLUTION:** The voltage across the capacitor is some proportion of the input voltage, namely the impedance of the capacitor divided by the sum of the impedances. Thus,

$$V_C(s) = \frac{1/Cs}{\left(Ls + R + \dfrac{1}{Cs}\right)} V(s) \tag{2.77}$$

Solving for the transfer function, $V_C(s)/V(s)$, yields the same result as Eq. (2.66).

Review Examples 2.6 through 2.9. Which method do you think is easiest for this circuit?

The previous example involves a simple, single-loop electrical network. Many electrical networks consist of multiple loops and nodes, and for these circuits we must write and solve simultaneous differential equations in order to find the transfer function, or solve for the output.

## Complex Circuits via Mesh Analysis

To solve complex electrical networks—those with multiple loops and nodes—using mesh analysis, we can perform the following steps:

1. Replace passive element values with their impedances.
2. Replace all sources and time variables with their Laplace transform.
3. Assume a transform current and a current direction in each mesh.
4. Write Kirchhoff's voltage law around each mesh.
5. Solve the simultaneous equations for the output.
6. Form the transfer function.

Let us look at an example.

### Example 2.10 Transfer Function—Multiple Loops

**PROBLEM:** Given the network of Figure 2.6($a$), find the transfer function, $I_2(s)/V(s)$.

**SOLUTION:** The first step in the solution is to convert the network into Laplace transforms for impedances and circuit variables, assuming zero initial conditions. The result is shown in Figure 2.6($b$). The circuit with which we are dealing requires two simultaneous equations to solve for the transfer function. These equations can be found by summing voltages

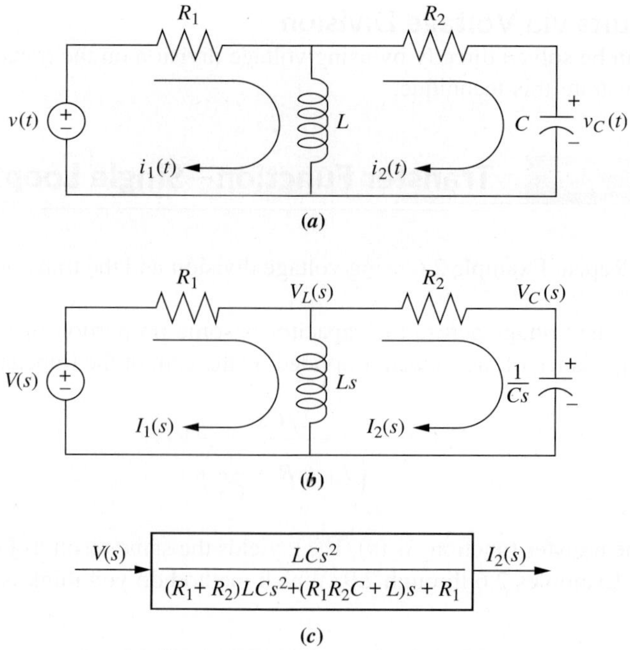


**FIGURE 2.6** **a.** Two-loop electrical network; **b.** transformed two-loop electrical network; **c.** block diagram

around each mesh through which the assumed currents, $I_1(s)$ and $I_2(s)$, flow. Around Mesh 1, where $I_1(s)$ flows,

$$R_1I_1(s) + LsI_1(s) - LsI_2(s) = V(s) \tag{2.78}$$

Around Mesh 2, where $I_2(s)$ flows,

$$LsI_2(s) + R_2I_2(s) + \frac{1}{Cs}I_2(s) - LsI_1(s) = 0 \tag{2.79}$$

Combining terms, Eqs. (2.78) and (2.79) become simultaneous equations in $I_1(s)$ and $I_2(s)$:

$$(R_1 + Ls)I_1(s) \qquad - LsI_2(s) = V(s) \tag{2.80a}$$

$$-LsI_1(s) + \left(Ls + R_2 + \frac{1}{Cs}\right)I_2(s) = 0 \tag{2.80b}$$

We can use Cramer's rule (or any other method for solving simultaneous equations) to solve Eq. (2.80) for $I_2(s)$.[4] Hence,

$$I_2(s) = \frac{\begin{vmatrix} (R_1 + Ls) & V(s) \\ -Ls & 0 \end{vmatrix}}{\Delta} = \frac{LsV(s)}{\Delta} \tag{2.81}$$

where

$$\Delta = \begin{vmatrix} (R_1 + Ls) & -Ls \\ -Ls & \left(Ls + R_2 + \frac{1}{Cs}\right) \end{vmatrix}$$

[4] See Appendix G (Section G.4) at www.wiley.com/college/nise for Cramer's rule.

Forming the transfer function, $G(s)$, yields

$$G(s) = \frac{I_2(s)}{V(s)} = \frac{Ls}{\Delta} = \frac{LCs^2}{(R_1 + R_2)LCs^2 + (R_1R_2C + L)s + R_1} \quad (2.82)$$

as shown in Figure 2.6(*c*).

We have succeeded in modeling a physical network as a transfer function: The network of Figure 2.6(*a*) is now modeled as the transfer function of Figure 2.6(*c*). Before leaving the example, we notice a pattern first illustrated by Eq. (2.72). The form that Eqs. (2.80) take is

$$\begin{bmatrix}\text{Sum of}\\ \text{impedances}\\ \text{around Mesh 1}\end{bmatrix} I_1(s) - \begin{bmatrix}\text{Sum of}\\ \text{impedances}\\ \text{common to the}\\ \text{two meshes}\end{bmatrix} I_2(s) = \begin{bmatrix}\text{Sum of applied}\\ \text{voltages around}\\ \text{Mesh 1}\end{bmatrix} \quad (2.83a)$$

$$-\begin{bmatrix}\text{Sum of}\\ \text{impedances}\\ \text{common to the}\\ \text{two meshes}\end{bmatrix} I_1(s) + \begin{bmatrix}\text{Sum of}\\ \text{impedances}\\ \text{around Mesh 2}\end{bmatrix} I_2(s) = \begin{bmatrix}\text{Sum of applied}\\ \text{voltages around}\\ \text{Mesh 2}\end{bmatrix} \quad (2.83b)$$

Recognizing the form will help us write such equations rapidly; for example, mechanical equations of motion (covered in Sections 2.5 and 2.6) have the same form.

Symbolic Math

SM

`Students who are performing the MATLAB exercises and want to explore the added capability of MATLAB's Symbolic Math Toolbox should now run ch2sp4 in Appendix F at www.wiley.com/college/nise, where Example 2.10 is solved. You will learn how to use the Symbolic Math Toolbox to solve simultaneous equations using Cramer's rule. Specifically, the Symbolic Math Toolbox will be used to solve for the transfer function in Eq. (2.82) using Eq. (2.80).`

## Complex Circuits via Nodal Analysis

Often, the easiest way to find the transfer function is to use nodal analysis rather than mesh analysis. The number of simultaneous differential equations that must be written is equal to the number of nodes whose voltage is unknown. In the previous example we wrote simultaneous mesh equations using Kirchhoff's voltage law. For multiple nodes we use Kirchhoff's current law and sum currents flowing from each node. Again, as a convention, currents flowing from the node are assumed to be positive, and currents flowing into the node are assumed to be negative.

Before progressing to an example, let us first define *admittance*, $Y(s)$, as the reciprocal of impedance, or

$$Y(s) = \frac{1}{Z(s)} = \frac{I(s)}{V(s)} \quad (2.84)$$

When writing nodal equations, it can be more convenient to represent circuit elements by their admittance. Admittances for the basic electrical components are shown in Table 2.3. Let us look at an example.

## Example 2.11 Transfer Function—Multiple Nodes

**PROBLEM:** Find the transfer function, $V_C(s)/V(s)$, for the circuit in Figure 2.6(*b*). Use nodal analysis.

**SOLUTION:** For this problem, we sum currents at the nodes rather than sum voltages around the meshes. From Figure 2.6(*b*) the sum of currents flowing from the nodes marked $V_L(s)$ and $V_C(s)$ are, respectively,

$$\frac{V_L(s) - V(s)}{R_1} + \frac{V_L(s)}{Ls} + \frac{V_L(s) - V_C(s)}{R_2} = 0 \tag{2.85a}$$

$$CsV_C(s) + \frac{V_C(s) - V_L(s)}{R_2} = 0 \tag{2.85b}$$

Rearranging and expressing the resistances as conductances,[5] $G_1 = 1/R_1$ and $G_2 = 1/R_2$, we obtain,

$$\left(G_1 + G_2 + \frac{1}{Ls}\right)V_L(s) \qquad - G_2V_C(s) = V(s)G_1 \tag{2.86a}$$

$$-G_2V_L(s) + (G_2 + Cs)V_C(s) = 0 \tag{2.86b}$$

Solving for the transfer function, $V_C(s)/V(s)$, yields Eq. (2.87) as shown in Figure 2.7.

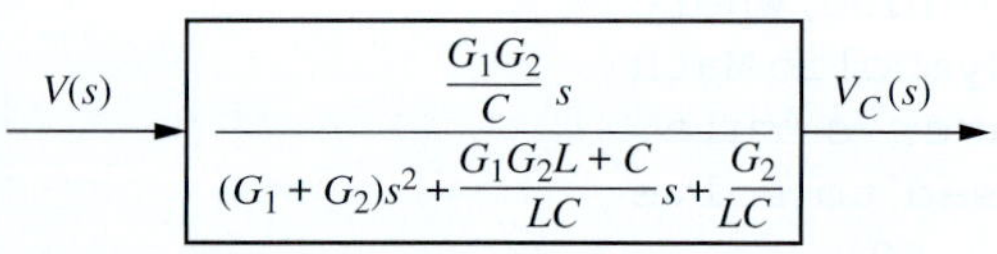


**FIGURE 2.7** Block diagram of the network of Figure 2.6

$$\frac{V_C(s)}{V(s)} = \frac{\dfrac{G_1G_2}{C}s}{(G_1 + G_2)s^2 + \dfrac{G_1G_2L + C}{LC}s + \dfrac{G_2}{LC}} \tag{2.87}$$

Another way to write node equations is to replace voltage sources by current sources. A voltage source presents a constant voltage to any load; conversely, a current source delivers a constant current to any load. Practically, a current source can be constructed from a voltage source by placing a large resistance in series with the voltage source. Thus, variations in the load do not appreciably change the current because the current is determined approximately by the large series resistor and the voltage source. Theoretically, we rely on *Norton's theorem*, which states that a voltage source, $V(s)$, in series with an impedance, $Z_s(s)$, can be replaced by a current source, $I(s) = V(s)/Z_s(s)$, in parallel with $Z_s(s)$.

In order to handle multiple-node electrical networks, we can perform the following steps:

1. Replace passive element values with their admittances.
2. Replace all sources and time variables with their Laplace transform.
3. Replace transformed voltage sources with transformed current sources.

[5] In general, admittance is complex. The real part is called conductance and the imaginary part is called susceptance. But when we take the reciprocal of resistance to obtain the admittance, a purely real quantity results. The reciprocal of resistance is called conductance.

**4.** Write Kirchhoff's current law at each node.

**5.** Solve the simultaneous equations for the output.

**6.** Form the transfer function.

Let us look at an example.

## Example 2.12 Transfer Function—Multiple Nodes with Current Sources

**PROBLEM:** For the network of Figure 2.6, find the transfer function, $V_C(s)/V(s)$, using nodal analysis and a transformed circuit with current sources.

**SOLUTION:** Convert all impedances to admittances and all voltage sources in series with an impedance to current sources in parallel with an admittance using Norton's theorem.

Redrawing Figure 2.6(*b*) to reflect the changes, we obtain Figure 2.8, where $G_1 = 1/R_1$, $G_2 = 1/R_2$, and the node voltages—the voltages across the inductor and the capacitor—have been identified as $V_L(s)$ and $V_C(s)$, respectively. Using the general relationship, $I(s) = Y(s)V(s)$, and summing currents at the node $V_L(s)$,

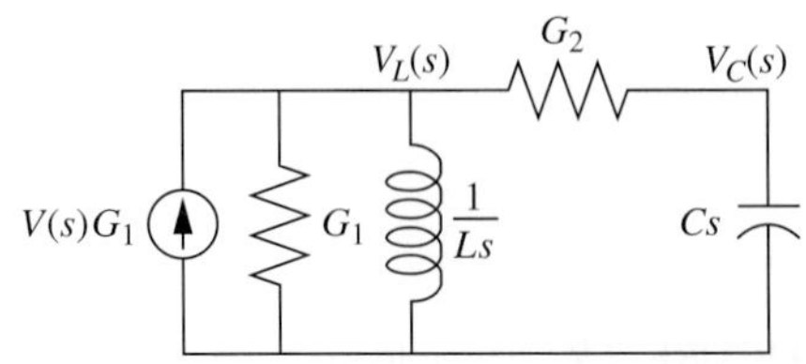


**FIGURE 2.8** Transformed network ready for nodal analysis

$$G_1V_L(s) + \frac{1}{Ls}V_L(s) + G_2[V_L(s) - V_C(s)] = V(s)G_1 \tag{2.88}$$

Summing the currents at the node $V_C(s)$ yields

$$C_sV_C(s) + G_2[V_C(s) - V_L(s)] = 0 \tag{2.89}$$

Combining terms, Eqs. (2.88) and (2.89) become simultaneous equations in $V_C(s)$ and $V_L(s)$, which are identical to Eq. (2.86) and lead to the same solution as Eq. (2.87).

An advantage of drawing this circuit lies in the form of Eq. (2.86) and its direct relationship to Figure 2.8, namely

$$\begin{bmatrix}\text{Sum of admittances}\\\text{connected to Node 1}\end{bmatrix}V_L(s) - \begin{bmatrix}\text{Sum of admittances}\\\text{common to the two}\\\text{nodes}\end{bmatrix}V_C(s) = \begin{bmatrix}\text{Sum of applied}\\\text{currents at Node 1}\end{bmatrix} \tag{2.90a}$$

$$-\begin{bmatrix}\text{Sum of admittances}\\\text{common to the two}\\\text{nodes}\end{bmatrix}V_L(s) + \begin{bmatrix}\text{Sum of admittances}\\\text{connected to Node 2}\end{bmatrix}V_C(s) = \begin{bmatrix}\text{Sum of applied}\\\text{currents at Node 2}\end{bmatrix} \tag{2.90b}$$

## A Problem-Solving Technique

In all of the previous examples, we have seen a repeating pattern in the equations that we can use to our advantage. If we recognize this pattern, we need not write the equations component by component; we can sum impedances around a mesh in the case of mesh equations or sum admittances at a node in the case of node equations. Let us now look at a three-loop electrical network and write the mesh equations by inspection to demonstrate the process.

## Example 2.13 Mesh Equations via Inspection

**PROBLEM:** Write, but do not solve, the mesh equations for the network shown in Figure 2.9.

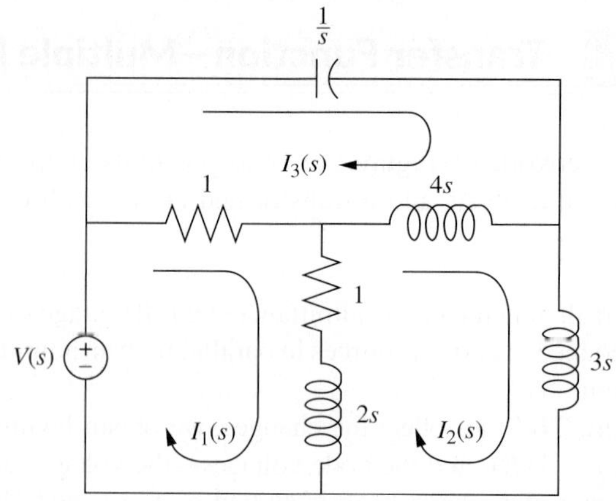


**FIGURE 2.9** Three-loop electrical network

**SOLUTION:** Each of the previous problems has illustrated that the mesh equations and nodal equations have a predictable form. We use that knowledge to solve this three-loop problem. The equation for Mesh 1 will have the following form:

$$\begin{bmatrix}\text{Sum of}\\ \text{impedances}\\ \text{around Mesh 1}\end{bmatrix} I_1(s) - \begin{bmatrix}\text{Sum of}\\ \text{impedances}\\ \text{common to}\\ \text{Mesh 1 and}\\ \text{Mesh 2}\end{bmatrix} I_2(s)$$
$$- \begin{bmatrix}\text{Sum of}\\ \text{impedances}\\ \text{common to}\\ \text{Mesh 1 and}\\ \text{Mesh 3}\end{bmatrix} I_3(s) = \begin{bmatrix}\text{Sum of applied}\\ \text{voltages around}\\ \text{Mesh 1}\end{bmatrix} \tag{2.91}$$

Similarly, Meshes 2 and 3, respectively, are

$$- \begin{bmatrix}\text{Sum of}\\ \text{impedances}\\ \text{common to}\\ \text{Mesh 1 and}\\ \text{Mesh 2}\end{bmatrix} I_1(s) + \begin{bmatrix}\text{Sum of}\\ \text{impedances}\\ \text{around Mesh 2}\end{bmatrix} I_2(s) - \begin{bmatrix}\text{Sum of}\\ \text{impedances}\\ \text{common to}\\ \text{Mesh 2 and}\\ \text{Mesh 3}\end{bmatrix} I_3(s) = \begin{bmatrix}\text{Sum of appied}\\ \text{voltages around}\\ \text{Mesh 2}\end{bmatrix} \tag{2.92}$$

and

$$-\begin{bmatrix}\text{Sum of}\\\text{impedances}\\\text{common to}\\\text{Mesh 1 and}\\\text{Mesh 3}\end{bmatrix}I_1(s)-\begin{bmatrix}\text{Sum of}\\\text{impedances}\\\text{common to}\\\text{Mesh 2 and}\\\text{Mesh 3}\end{bmatrix}I_2(s)+\begin{bmatrix}\text{Sum of}\\\text{impedances}\\\text{around Mesh 3}\end{bmatrix}I_3(s)=\begin{bmatrix}\text{Sum of applied}\\\text{voltages around}\\\text{Mesh 3}\end{bmatrix}\tag{2.93}$$

**TryIt 2.8**

Use the following MATLAB and Symbolic Math Toolbox statements to help you solve for the electrical currents in Eq. (2.94).

```
syms s I1 I2 I3 V
A=[(2*s + 2) -(2*s +
1)...
 -1
 -(2*s+1)(9*s+1)...
 -4*s
 -1 -4*s...
 (4*s+1+1/s)];
B=[I1;I2;I3];
C=[V;0;0];
B=inv(A)*C;
pretty(B)
```

Substituting the values from Figure 2.9 into Eqs. (2.91) through (2.93) yields

$$+(2s+2)I_1(s)-(2s+1)I_2(s)\qquad -I_3(s)=V(s)\tag{2.94a}$$

$$-(2s+1)I_1(s)+(9s+1)I_2(s)\qquad -4sI_3(s)=0\tag{2.94b}$$

$$-I_1(s)\qquad -4sI_2(s)+\left(4s+1+\frac{1}{s}\right)I_3(s)=0\tag{2.94c}$$

which can be solved simultaneously for any desired transfer function, for example, $I_3(s)/V(s)$.

Passive electrical circuits were the topic of discussion up to this point. We now discuss a class of active circuits that can be used to implement transfer functions. These are circuits built around an operational amplifier.

## Operational Amplifiers

An *operational amplifier*, pictured in Figure 2.10($a$), is an electronic amplifier used as a basic building block to implement transfer functions. It has the following characteristics:

1. Differential input, $v_2(t)-v_1(t)$
2. High input impedance, $Z_i=\infty$ (ideal)
3. Low output impedance, $Z_o=0$ (ideal)
4. High constant gain amplification, $A=\infty$ (ideal)

The output, $v_o(t)$, is given by

$$v_o(t)=A(v_2(t)-v_1(t))\tag{2.95}$$

## Inverting Operational Amplifier

If $v_2(t)$ is grounded, the amplifier is called an *inverting operational amplifier*, as shown in Figure 2.10($b$). For the inverting operational amplifier, we have

$$v_o(t)=-Av_1(t)\tag{2.96}$$

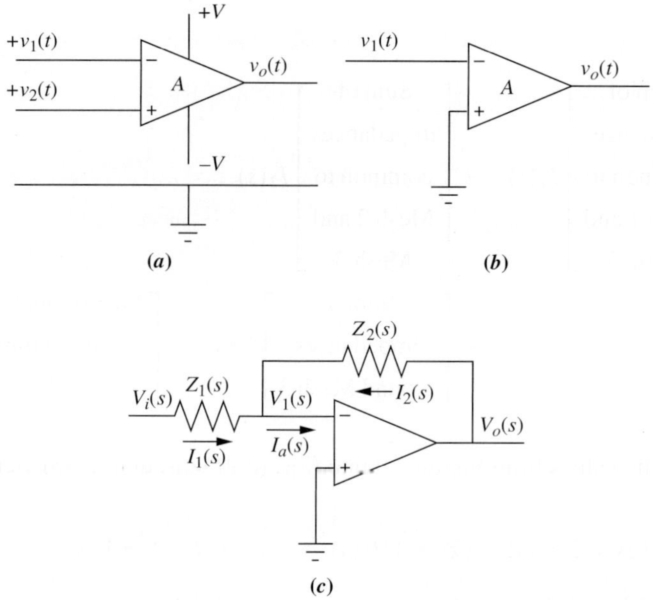


**FIGURE 2.10** **a.** Operational amplifier; **b.** schematic for an inverting operational amplifier; **c.** inverting operational amplifier configured for transfer function realization. Typically, the amplifier gain, $A$, is omitted.

If two impedances are connected to the inverting operational amplifier as shown in Figure 2.10(*c*), we can derive an interesting result if the amplifier has the characteristics mentioned in the beginning of this subsection. If the input impedance to the amplifier is high, then by Kirchhoff's current law $I_a(s) = 0$ and $I_1(s) = -I_2(s)$. Also, since the gain $A$ is large, $v_1(t) \approx 0$. Thus, $I_1(s) = V_i(s)/Z_1(s)$, and $-I_2(s) = -V_o(s)/Z_2(s)$. Equating the two currents, $V_o(s)/Z_2(s) = -V_i(s)/Z_1(s)$, or the transfer function of the inverting operational amplifier configured as shown in Figure 2.10(*c*) is

$$\frac{V_o(s)}{V_i(s)} = -\frac{Z_2(s)}{Z_1(s)} \tag{2.97}$$

## Example 2.14 Transfer Function—Inverting Operational Amplifier Circuit

**PROBLEM:** Find the transfer function, $V_o(s)/V_i(s)$, for the circuit given in Figure 2.11.

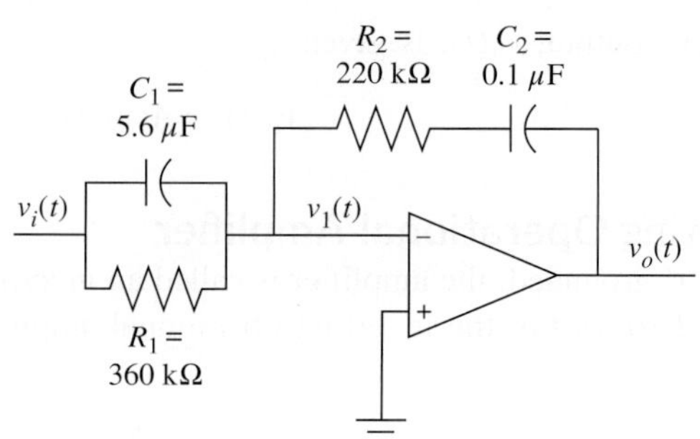


**FIGURE 2.11** Inverting operational amplifier circuit for Example 2.14

**SOLUTION:** The transfer function of the operational amplifier circuit is given by Eq. (2.97). Since the admittances of parallel components add, $Z_1(s)$ is the reciprocal of the sum of the admittances, or

$$Z_1(s) = \frac{1}{C_1 s + \frac{1}{R_1}} = \frac{1}{5.6 + 10^{-6}s + \frac{1}{360 \times 10^3}} = \frac{360 \times 10^3}{2.016s + 1} \tag{2.98}$$

For $Z_2(s)$ the impedances add, or

$$Z_2(s) = R_2 + \frac{1}{C_2 s} = 220 \times 10^3 + \frac{10^7}{s} \tag{2.99}$$

Substituting Eqs. (2.98) and (2.99) into Eq. (2.97) and simplifying, we get

$$\frac{V_o(s)}{V_i(s)} = -1.232\frac{s^2 + 45.95s + 22.55}{s} \tag{2.100}$$

The resulting circuit is called a PID controller and can be used to improve the performance of a control system. We explore this possibility further in Chapter 9.

## Noninverting Operational Amplifier

Another circuit that can be analyzed for its transfer function is the noninverting operational amplifier circuit shown in Figure 2.12. We now derive the transfer function. We see that

$$V_o(s) = A(V_i(s) - V_1(s)) \tag{2.101}$$

But, using voltage division,

$$V_1(s) = \frac{Z_1(s)}{Z_1(s) + Z_2(s)} V_o(s) \tag{2.102}$$

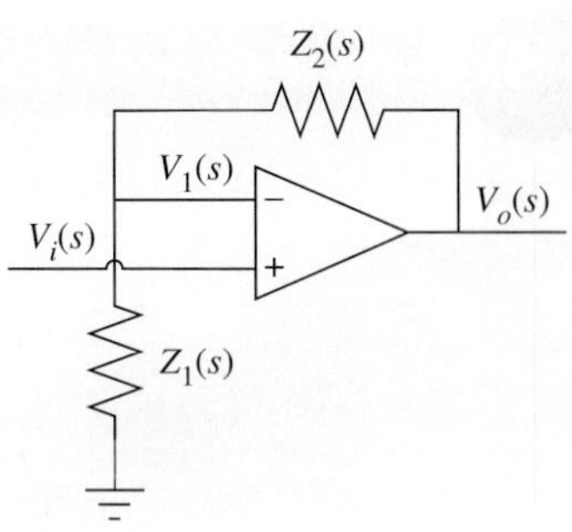


**FIGURE 2.12** General noninverting operational amplifier circuit

Substituting Eq. (2.102) into Eq. (2.101), rearranging, and simplifying, we obtain

$$\frac{V_o(s)}{V_i(s)} = \frac{A}{1 + AZ_1(s)/(Z_1(s) + Z_2(s))} \tag{2.103}$$

For large $A$, we disregard unity in the denominator and Eq. (2.103) becomes

$$\boxed{\frac{V_o(s)}{V_i(s)} = \frac{Z_1(s) + Z_2(s)}{Z_1(s)}} \tag{2.104}$$

Let us now look at an example.

## Example 2.15 Transfer Function—Noninverting Operational Amplifier Circuit

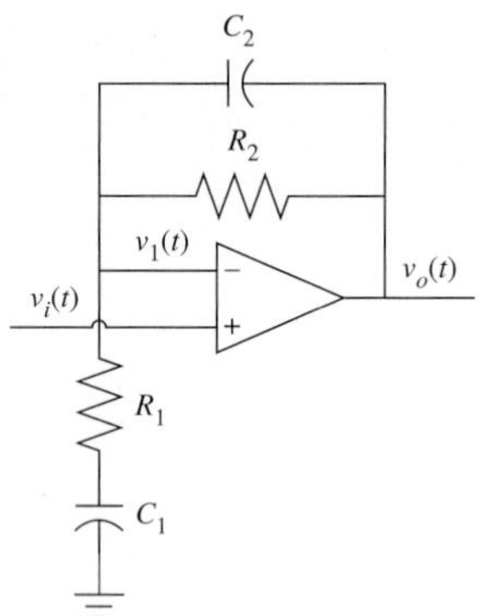


**FIGURE 2.13** Noninverting operational amplifier circuit for Example 2.15

**PROBLEM:** Find the transfer function, $V_o(s)/V_i(s)$, for the circuit given in Figure 2.13.

**SOLUTION:** We find each of the impedance functions, $Z_1(s)$ and $Z_2(s)$, and then substitute them into Eq. (2.104). Thus,

$$Z_1(s) = R_1 + \frac{1}{C_1 s} \tag{2.105}$$

and

$$Z_2(s) = \frac{R_2(1/C_2 s)}{R_2 + (1/C_2 s)} \tag{2.106}$$

Substituting Eqs. (2.105) and (2.106) into Eq. (2.104) yields

$$\frac{V_o(s)}{V_i(s)} = \frac{C_2C_1R_2R_1s^2 + (C_2R_2 + C_1R_2 + C_1R_1)s + 1}{C_2C_1R_2R_1s^2 + (C_2R_2 + C_1R_1)s + 1} \tag{2.107}$$

## Skill-Assessment Exercise 2.6

**PROBLEM:** Find the transfer function, $G(s) = V_L(s)/V(s)$, for the circuit given in Figure 2.14. Solve the problem two ways—mesh analysis and nodal analysis. Show that the two methods yield the same result.

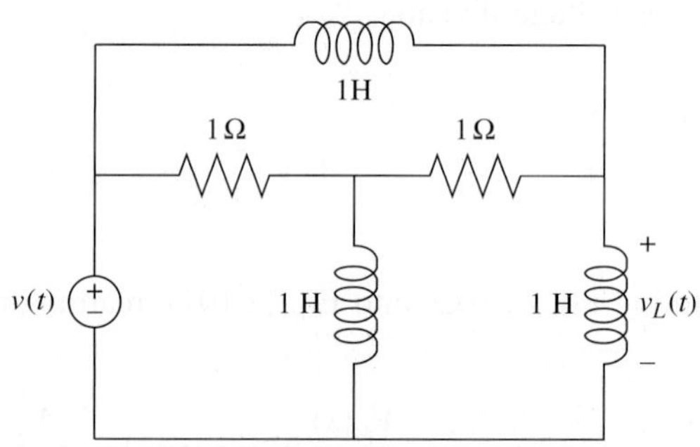


**FIGURE 2.14** Electric circuit for Skill-Assessment Exercise 2.6

**ANSWER:** $V_L(s)/V(s) = (s^2 + 2s + 1)/(s^2 + 5s + 2)$

The complete solution is at www.wiley.com/college/nise.

**Skill-Assessment Exercise 2.7**

**PROBLEM:** If $Z_1(s)$ is the impedance of a 10 $\mu$F capacitor and $Z_2(s)$ is the impedance of a 100 k$\Omega$ resistor, find the transfer function, $G(s) = V_o(s)/V_i(s)$, if these components are used with (a) an inverting operational amplifier and (b) a noninverting amplifier as shown in Figures 2.10(*c*) and 2.12, respectively.

**ANSWER:** $G(s) = -s$ for an inverting operational amplifier; $G(s) = s+1$ for a noninverting operational amplifier.

The complete solution is at www.wiley.com/college/nise.

In this section, we found transfer functions for multiple-loop and multiple-node electrical networks, as well as operational amplifier circuits. We developed mesh and nodal equations, noted their form, and wrote them by inspection. In the next section we begin our work with mechanical systems. We will see that many of the concepts applied to electrical networks can also be applied to mechanical systems via analogies—from basic concepts to writing the describing equations by inspection. This revelation will give you the confidence to move beyond this textbook and study systems not covered here, such as hydraulic or pneumatic systems.

## 2.5 Translational Mechanical System Transfer Functions

We have shown that electrical networks can be modeled by a transfer function, $G(s)$, that algebraically relates the Laplace transform of the output to the Laplace transform of the input. Now we will do the same for mechanical systems. In this section we concentrate on translational mechanical systems. In the next section we extend the concepts to rotational mechanical systems. Notice that the end product, shown in Figure 2.2, will be mathematically indistinguishable from an electrical network. Hence, an electrical network can be interfaced to a mechanical system by cascading their transfer functions, provided that one system is not loaded by the other.[6]

Mechanical systems parallel electrical networks to such an extent that there are analogies between electrical and mechanical components and variables. Mechanical systems, like electrical networks, have three passive, linear components. Two of them, the spring and the mass, are energy-storage elements; one of them, the viscous damper, dissipates energy. The two energy-storage elements are analogous to the two electrical energy-storage elements, the inductor and capacitor. The energy dissipator is analogous to electrical resistance. Let us take a look at these mechanical elements, which are shown in Table 2.4. In the table, $K, f_v$, and $M$ are called *spring constant*, *coefficient of viscous friction*, and *mass*, respectively.

We now create analogies between electrical and mechanical systems by comparing Tables 2.3 and 2.4. Comparing the force-velocity column of Table 2.4 to the voltage-current column of Table 2.3, we see that mechanical force is analogous to electrical voltage and mechanical velocity is analogous to electrical current. Comparing the force-displacement column of Table 2.4 with the voltage-charge column of Table 2.3 leads to the analogy between the mechanical displacement and electrical charge. We also see that the spring is

[6] The concept of loading is explained further in Chapter 5.

**TABLE 2.4** Force-velocity, force-displacement, and impedance translational relationships for springs, viscous dampers, and mass

| Component | Force-velocity | Force-displacement | Impedence $Z_M(s) = F(s)/X(s)$ |
|---|---|---|---|
| Spring 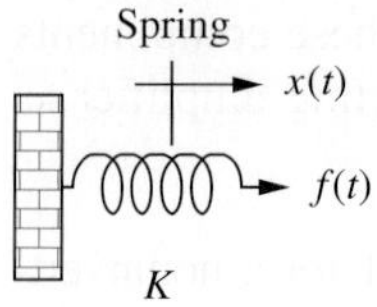  | $f(t) = K\int_0^t v(\tau)d\tau$ | $f(t) = Kx(t)$ | $K$ |
| Viscous damper 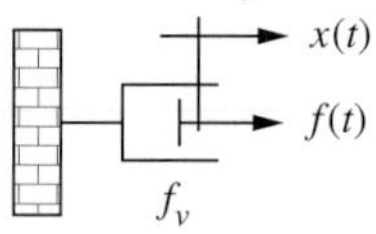  | $f(t) = f_v v(t)$ | $f(t) = f_v \frac{dx(t)}{dt}$ | $f_v s$ |
| Mass 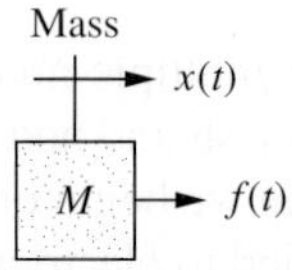  | $f(t) = M\frac{dv(t)}{dt}$ | $f(t) = M\frac{d^2x(t)}{dt^2}$ | $Ms^2$ |

Note: The following set of symbols and units is used throughout this book: $f(t)$ = N (newtons), $x(t)$ = m (meters), $v(t)$ = m/s (meters/second), $K$ = N/m (newtons/meter), $f_v$ = N-s/m (newton-seconds/meter), $M$ = kg (kilograms = newton-seconds$^2$/meter).

analogous to the capacitor, the viscous damper is analogous to the resistor, and the mass is analogous to the inductor. Thus, summing forces written in terms of velocity is analogous to summing voltages written in terms of current, and the resulting mechanical differential equations are analogous to mesh equations. If the forces are written in terms of displacement, the resulting mechanical equations resemble, but are not analogous to, the mesh equations. We, however, will use this model for mechanical systems so that we can write equations directly in terms of displacement.

Another analogy can be drawn by comparing the force-velocity column of Table 2.4 to the current-voltage column of Table 2.3 in reverse order. Here the analogy is between force and current and between velocity and voltage. Also, the spring is analogous to the inductor, the viscous damper is analogous to the resistor, and the mass is analogous to the capacitor. Thus, summing forces written in terms of velocity is analogous to summing currents written in terms of voltage and the resulting mechanical differential equations are analogous to nodal equations. We will discuss these analogies in more detail in Section 2.9.

We are now ready to find transfer functions for translational mechanical systems. Our first example, shown in Figure 2.15($a$), is similar to the simple $RLC$ network of Example 2.6 (see Figure 2.3). The mechanical system requires just one differential equation, called the *equation of motion*, to describe it. We will begin by assuming a positive direction of motion, for example, to the right. This assumed positive direction of motion is similar to assuming a current direction in an electrical loop. Using our assumed direction of positive motion, we first draw a free-body diagram, placing on the body all forces that act on the body either in the direction of motion or opposite to it. Next we use Newton's law to form a differential equation of motion by summing the forces and setting the sum equal to zero. Finally, assuming zero initial conditions, we take the Laplace transform of the differential equation, separate the variables, and arrive at the transfer function. An example follows.

## Example 2.16 Transfer Function—One Equation of Motion

**PROBLEM:** Find the transfer function, $X(s)/F(s)$, for the system of Figure 2.15($a$).

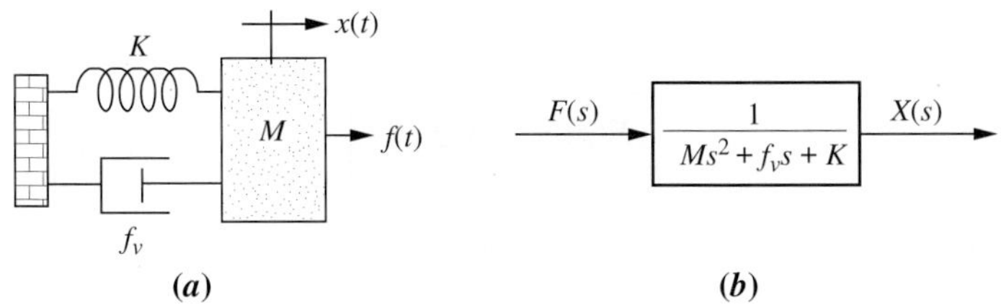


**FIGURE 2.15** **a.** Mass, spring, and damper system; **b.** block diagram

**SOLUTION:** Begin the solution by drawing the free-body diagram shown in Figure 2.16($a$). Place on the mass all forces felt by the mass. We assume the mass is traveling toward the right. Thus, only the applied force points to the right; all other forces impede the motion and act to oppose it. Hence, the spring, viscous damper, and the force due to acceleration point to the left.

We now write the differential equation of motion using Newton's law to sum to zero all of the forces shown on the mass in Figure 2.16($a$):

$$M\frac{d^2x(t)}{dt^2}+f_v\frac{dx(t)}{dt}+Kx(t)=f(t) \tag{2.108}$$

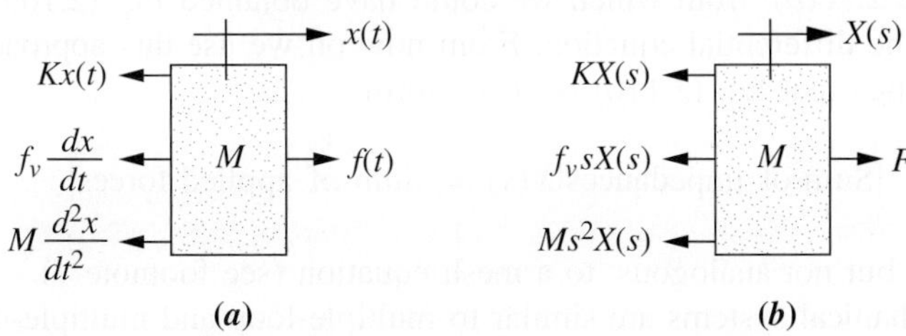


**FIGURE 2.16** **a.** Free-body diagram of mass, spring, and damper system; **b.** transformed free-body diagram

Taking the Laplace transform, assuming zero initial conditions,

$$Ms^2X(s)+f_vsX(s)+KX(s)=F(s) \tag{2.109}$$

or

$$(Ms^2+f_vs+K)X(s)=F(s) \tag{2.110}$$

Solving for the transfer function yields

$$G(s)=\frac{X(s)}{F(s)}=\frac{1}{Ms^2+f_vs+K} \tag{2.111}$$

which is represented in Figure 2.15($b$).

Now can we parallel our work with electrical networks by circumventing the writing of differential equations and by defining impedances for mechanical components? If so, we can apply to mechanical systems the problem-solving techniques learned in the previous section. Taking the Laplace transform of the force-displacement column in Table 2.4, we obtain for the spring,

$$F(s)=KX(s) \tag{2.112}$$

for the viscous damper,

$$F(s) = f_v sX(s) \tag{2.113}$$

and for the mass,

$$F(s) = Ms^2X(s) \tag{2.114}$$

If we define impedance for mechanical components as

$$Z_M(s) = \frac{F(s)}{X(s)} \tag{2.115}$$

and apply the definition to Eqs. (2.112) through (2.114), we arrive at the impedances of each component as summarized in Table 2.4 (*Raven, 1995*).[7]

Replacing each force in Figure 2.16(*a*) by its Laplace transform, which is in the format

$$F(s) = Z_M(s)X(s) \tag{2.116}$$

we obtain Figure 2.16(*b*), from which we could have obtained Eq. (2.109) immediately without writing the differential equation. From now on we use this approach.

Finally, notice that Eq. (2.110) is of the form

$$[\text{Sum of impedances}]X(s) = [\text{Sum of applied forces}] \tag{2.117}$$

which is similar, but not analogous, to a mesh equation (see footnote 7).

Many mechanical systems are similar to multiple-loop and multiple-node electrical networks, where more than one simultaneous differential equation is required to describe the system. In mechanical systems, the number of equations of motion required is equal to the number of *linearly independent* motions. Linear independence implies that a point of motion in a system can still move if all other points of motion are held still. Another name for the number of linearly independent motions is the number of *degrees of freedom*. This discussion is not meant to imply that these motions are not coupled to one another; in general, they are. For example, in a two-loop electrical network, each loop current depends on the other loop current, but if we open-circuit just one of the loops, the other current can still exist if there is a voltage source in that loop. Similarly, in a mechanical system with two degrees of freedom, one point of motion can be held still while the other point of motion moves under the influence of an applied force.

In order to work such a problem, we draw the free-body diagram for each point of motion and then use superposition. For each free-body diagram we begin by holding all other points of motion still and finding the forces acting on the body due only to its own motion. Then we hold the body still and activate the other points of motion one at a time, placing on the original body the forces created by the adjacent motion.

Using Newton's law, we sum the forces on each body and set the sum to zero. The result is a system of simultaneous equations of motion. As Laplace transforms, these

[7] Notice that the impedance column of Table 2.4 is not a direct analogy to the impedance column of Tabel 2.3, since the denominator of Eq. (2.115) is displacement. A direct analogy could be derived by defining mechanical impedance in terms of velocity as $F(s)/V(s)$. We chose Eq. (2.115) as a convenient definition for writing the equations of motion in terms of displacement, rather than velocity. The alternative, however, is available.

equations are then solved for the output variable of interest in terms of the input variable from which the transfer function is evaluated. Example 2.17 demonstrates this problem-solving technique.

## Example 2.17 Transfer Function—Two Degrees of Freedom

**PROBLEM:** Find the transfer function, $X_2(s)/F(s)$, for the system of Figure 2.17($a$).

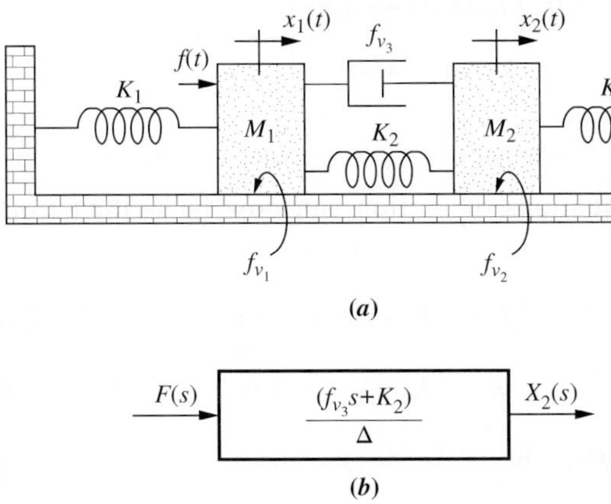


**FIGURE 2.17** **a.** Two-degrees-of-freedom translational mechanical system;[8] **b.** block diagram

**SOLUTION:** The system has two degrees of freedom, since each mass can be moved in the horizontal direction while the other is held still. Thus, two simultaneous equations of motion will be required to describe the system. The two equations come from free-body diagrams of each mass. Superposition is used to draw the free-body diagrams. For example, the forces on $M_1$ are due to (1) its own motion and (2) the motion of $M_2$ transmitted to $M_1$ through the system. We will consider these two sources separately.

If we hold $M_2$ still and move $M_1$ to the right, we see the forces shown in Figure 2.18($a$). If we hold $M_1$ still and move $M_2$ to the right, we see the forces shown in Figure 2.18($b$). The total force on $M_1$ is the superposition, or sum, of the forces just discussed. This result is shown in Figure 2.18($c$). For $M_2$, we proceed in a similar fashion: First we move $M_2$ to the right while holding $M_1$ still; then we move $M_1$ to the right and hold $M_2$ still. For each case we evaluate the forces on $M_2$. The results appear in Figure 2.19.

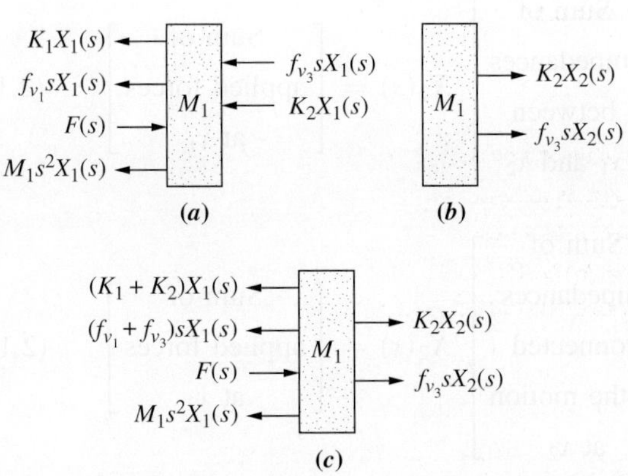


**FIGURE 2.18** **a.** Forces on $M_1$ due only to motion of $M_1$; **b.** forces on $M_1$ due only to motion of $M_2$; **c.** all forces on $M_1$

[8] Friction shown here and throughout the book, unless otherwise indicated, is viscous friction. Thus, $f_{v1}$ and $f_{v2}$ are not Coulomb friction, but arise because of a viscous interface.

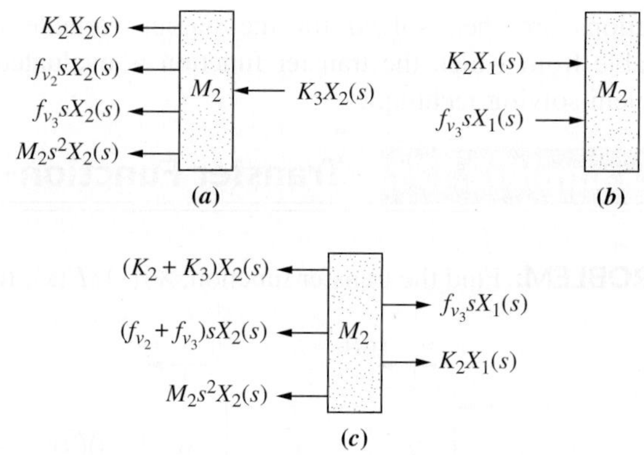


**FIGURE 2.19** **a.** Forces on $M_2$ due only to motion of $M_2$; **b.** forces on $M_2$ due only to motion of $M_1$; **c.** all forces on $M_2$

The Laplace transform of the equations of motion can now be written from Figures 2.18(*c*) and 2.19(*c*) as

$$[M_1s^2(F_{v_1}+f_{v_3})s+(K_1+K_2)]X_1(s)-(f_{v_3}s+K_2)X_2(s)=F(s) \tag{2.118a}$$

$$-(f_{v_3}s+K_2)X_1(s)+[M_2s^2+(f_{v_2}+f_{v_3})s+(K_2+K_3)]X_2(s)=0 \tag{2.118b}$$

From this, the transfer function, $X_2(s)/F(s)$, is

$$\frac{X_2(s)}{F(s)}=G(s)=\frac{(f_{v_3}s+K_2)}{\Delta} \tag{2.119}$$

as shown in Figure 2.17(*b*) where

$$\Delta=\begin{vmatrix}[M_1s^2+(f_{v_1}+f_{v_3})s+(K_1+K_2)] & -(f_{v_3}s+K_2)\\ -(f_{v_3}s+K_2) & [M_2s^2+(f_{v_2}+f_{v_3})s+(K_2+K_3)]\end{vmatrix}$$

Notice again, in Eq. (2.118), that the form of the equations is similar to electrical mesh equations:

$$\begin{bmatrix}\text{Sum of}\\ \text{impedances}\\ \text{connected}\\ \text{to the motion}\\ \text{at } x_1\end{bmatrix}X_1(s)-\begin{bmatrix}\text{Sum of}\\ \text{impedances}\\ \text{between}\\ x_1 \text{ and } x_2\end{bmatrix}X_2(s)=\begin{bmatrix}\text{Sum of}\\ \text{applied forces}\\ \text{at } x_1\end{bmatrix} \tag{2.120a}$$

$$-\begin{bmatrix}\text{Sum of}\\ \text{impedances}\\ \text{between}\\ x_1 \text{ and } x_2\end{bmatrix}X_1(s)+\begin{bmatrix}\text{Sum of}\\ \text{impedances}\\ \text{connected}\\ \text{to the motion}\\ \text{at } x_2\end{bmatrix}X_2(s)=\begin{bmatrix}\text{Sum of}\\ \text{applied forces}\\ \text{at } x_2\end{bmatrix} \tag{2.120b}$$

The pattern shown in Eq. (2.120) should now be familiar to us. Let us use the concept to write the equations of motion of a three-degrees-of-freedom mechanical network by inspection, without drawing the free-body diagram.

## Example 2.18 Equations of Motion by Inspection

**PROBLEM:** Write, but do not solve, the equations of motion for the mechanical network of Figure 2.20.

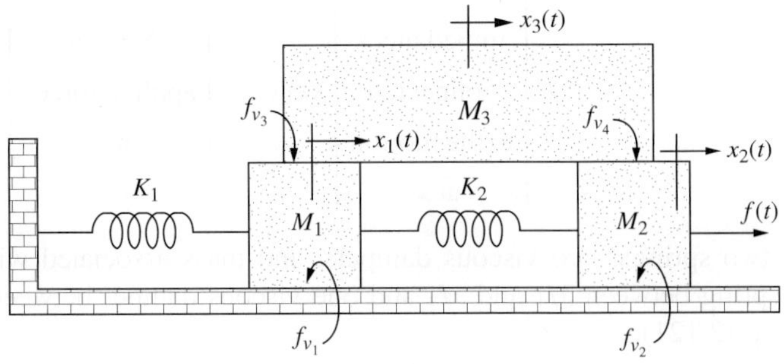


**FIGURE 2.20** Three-degrees-of-freedom translational mechanical system

**SOLUTION:** The system has three degrees of freedom, since each of the three masses can be moved independently while the others are held still. The form of the equations will be similar to electrical mesh equations. For $M_1$,

$$\left[\begin{array}{c}\text{Sum of}\\ \text{impedances}\\ \text{connected}\\ \text{to the motion}\\ \text{at } x_1\end{array}\right] X_1(s) - \left[\begin{array}{c}\text{Sum of}\\ \text{impedances}\\ \text{between}\\ x_1 \text{ and } x_2\end{array}\right] X_2(s)$$

$$- \left[\begin{array}{c}\text{Sum of}\\ \text{impedances}\\ \text{between}\\ x_1 \text{ and } x_3\end{array}\right] X_3(s) = \left[\begin{array}{c}\text{Sum of}\\ \text{applied forces}\\ \text{at } x_1\end{array}\right] \tag{2.121}$$

Similarly, for $M_2$ and $M_3$, respectively,

$$- \left[\begin{array}{c}\text{Sum of}\\ \text{impedances}\\ \text{between}\\ x_1 \text{ and } x_2\end{array}\right] X_1(s) + \left[\begin{array}{c}\text{Sum of}\\ \text{impedances}\\ \text{connected}\\ \text{to the motion}\\ \text{at } x_2\end{array}\right] X_2(s)$$

$$- \left[\begin{array}{c}\text{Sum of}\\ \text{impedances}\\ \text{between}\\ x_2 \text{ and } x_3\end{array}\right] X_3(s) = \left[\begin{array}{c}\text{Sum of}\\ \text{applied forces}\\ \text{at } x_2\end{array}\right] \tag{2.122}$$

$$-\begin{bmatrix}\text{Sum of}\\ \text{impedances}\\ \text{between}\\ x_1 \text{ and } x_3\end{bmatrix}X_1(s)-\begin{bmatrix}\text{Sum of}\\ \text{impedances}\\ \text{between}\\ x_2 \text{ and } x_3\end{bmatrix}X_2(s)$$
$$+\begin{bmatrix}\text{Sum of}\\ \text{impedances}\\ \text{connected}\\ \text{to the motion}\\ \text{at } x_3\end{bmatrix}X_3(s)=\begin{bmatrix}\text{Sum of}\\ \text{applied forces}\\ \text{at } x_3\end{bmatrix} \tag{2.123}$$

$M_1$ has two springs, two viscous dampers, and mass associated with its motion. There is one spring between $M_1$ and $M_2$ and one viscous damper between $M_1$ and $M_3$. Thus, using Eq. (2.121),

$$[M_1s^2+(f_{v_1}+f_{v_3})s+(K_1+K_2)]X_1(s)-K_2X_2(s)-f_{v_3}sX_3(s)=0 \tag{2.124}$$

Similarly, using Eq. (2.122) for $M_2$,

$$-K_2X_1(s)+[M_2s^2+(f_{v_2}+f_{v_4})s+K_2]X_2(s)-f_{v_4}sX_3(s)=F(s) \tag{2.125}$$

and using Eq. (2.123) for $M_3$,

$$-f_{v_3}sX_1(s)-f_{v_4}sX_2(s)+[M_3s^2+(f_{v_3}+f_{v_4})s]X_3(s)=0 \tag{2.126}$$

Equations (2.124) through (2.126) are the equations of motion. We can solve them for any displacement, $X_1(s)$, $X_2(s)$, or $X_3(s)$, or transfer function.

## Skill-Assessment Exercise 2.8

**PROBLEM:** Find the transfer function, $G(s)=X_2(s)/F(s)$, for the translational mechanical system shown in Figure 2.21.

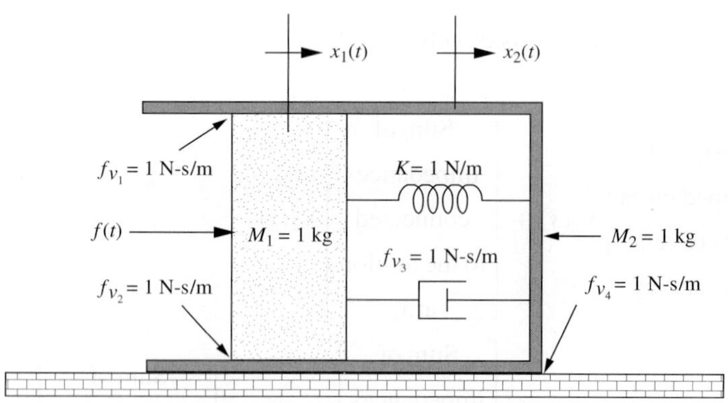


**FIGURE 2.21** Translational mechanical system for Skill-Assessment Exercise 2.8

**ANSWER:** $G(s)=\dfrac{3s+1}{s(s^3+7s^2+5s+1)}$

The complete solution is at www.wiley.com/college/nise.

## 2.6 Rotational Mechanical System Transfer Functions

Having covered electrical and translational mechanical systems, we now move on to consider rotational mechanical systems. Rotational mechanical systems are handled the same way as translational mechanical systems, except that torque replaces force and angular displacement replaces translational displacement. The mechanical components for rotational systems are the same as those for translational systems, except that the components undergo rotation instead of translation. Table 2.5 shows the components along with the relationships between torque and angular velocity, as well as angular displacement. Notice that the symbols for the components look the same as translational symbols, but they are undergoing rotation and not translation.

Also notice that the term associated with the mass is replaced by inertia. The values of $K$, $D$, and $J$ are called *spring constant*, *coefficient of viscous friction*, and *moment of inertia*, respectively. The impedances of the mechanical components are also summarized in the last column of Table 2.5. The values can be found by taking the Laplace transform, assuming zero initial conditions, of the torque-angular displacement column of Table 2.5.

The concept of degrees of freedom carries over to rotational systems, except that we test a point of motion by *rotating* it while holding still all other points of motion. The number of points of motion that can be rotated while all others are held still equals the number of equations of motion required to describe the system.

Writing the equations of motion for rotational systems is similar to writing them for translational systems; the only difference is that the free-body diagram consists of torques rather than forces. We obtain these torques using superposition. First, we rotate a body while holding all other points still and place on its free-body diagram all torques due to the body's own motion. Then, holding the body still, we rotate adjacent points of motion one at a time and add the torques due to the adjacent motion to the free-body diagram. The process is repeated for each point of motion. For each free-body diagram, these torques are summed and set equal to zero to form the equations of motion.

**TABLE 2.5** Torque-angular velocity, torque-angular displacement, and impedance rotational relationships for springs, viscous dampers, and inertia

| Component | Torque-angular velocity | Torque-angular displacement | Impedence $Z_M(s) = T(s)/\theta(s)$ |
|---|---|---|---|
| 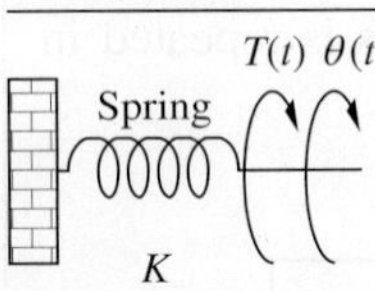 Spring, $K$, $T(t)$ $\theta(t)$ | $T(t) = K\int_0^t \omega(\tau)d\tau$ | $T(t) = K\theta(t)$ | $K$ |
| 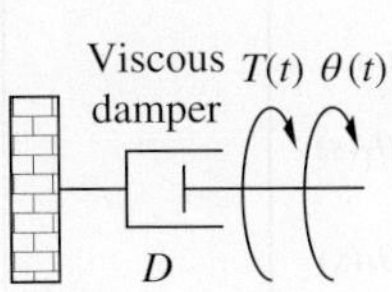 Viscous damper, $D$, $T(t)$ $\theta(t)$ | $T(t) = D\omega(t)$ | $T(t) = D\dfrac{d\theta(t)}{dt}$ | $Ds$ |
| 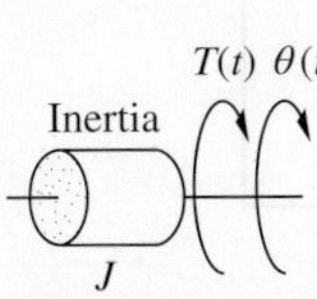 Inertia, $J$, $T(t)$ $\theta(t)$ | $T(t) = J\dfrac{d\omega(t)}{dt}$ | $T(t) = J\dfrac{d^2\theta(t)}{dt^2}$ | $Js^2$ |

Note: The following set of symbols and units is used throughout this book: $T(t)$ – N-m (newton-meters), $\theta(t)$ – rad (radians), $\omega(t)$ – rad/s (radians/second), $K$ – N-m/rad (newton-meters/radian), $D$ – N-m-s/rad (newton-meters-seconds/radian). $J$ – kg-m$^2$ (kilograms-meters$^2$ – newton-meters-seconds$^2$/radian).

Two examples will demonstrate the solution of rotational systems. The first one uses free-body diagrams; the second uses the concept of impedances to write the equations of motion by inspection.

## Example 2.19 Transfer Function—Two Equations of Motion

**PROBLEM:** Find the transfer function, $\theta_2(s)/T(s)$, for the rotational system shown in Figure 2.22(*a*). The rod is supported by bearings at either end and is undergoing torsion. A torque is applied at the left, and the displacement is measured at the right.

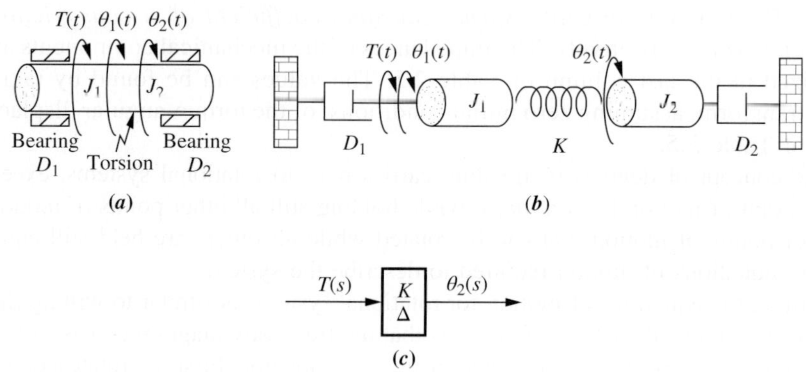


**FIGURE 2.22** **a.** Physical system; **b.** schematic; **c.** block diagram

**SOLUTION:** First, obtain the schematic from the physical system. Even though torsion occurs throughout the rod in Figure 2.22(*a*),[9] we approximate the system by assuming that the torsion acts like a spring concentrated at one particular point in the rod, with an inertia $J_1$ to the left and an inertia $J_2$ to the right.[10] We also assume that the damping inside the flexible shaft is negligible. The schematic is shown in Figure 2.22(*b*). There are two degrees of freedom, since each inertia can be rotated while the other is held still. Hence, it will take two simultaneous equations to solve the system.

Next, draw a free-body diagram of $J_1$, using superposition. Figure 2.23(*a*) shows the torques on $J_1$ if $J_2$ is held still and $J_1$ rotated. Figure 2.23(*b*) shows the torques on $J_1$ if $J_1$ is held still and $J_2$ rotated. Finally, the sum of Figures 2.23(*a*) and 2.23(*b*) is shown in Figure 2.23(*c*), the final free-body diagram for $J_1$. The same process is repeated in Figure 2.24 for $J_2$.

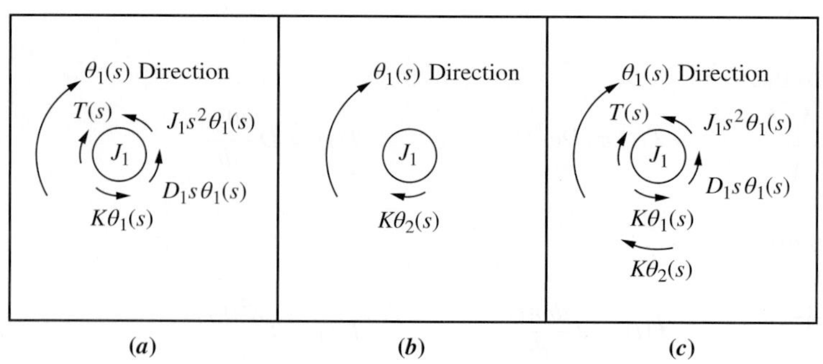


**FIGURE 2.23** **a.** Torques on $J_1$ due only to the motion of $J_1$; **b.** torques on $J_1$ due only to the motion of $J_2$; **c.** final free-body diagram for $J_1$

[9] In this case the parameter is referred to as a *distributed* parameter.
[10] The parameter is now referred to as a *lumped* parameter.

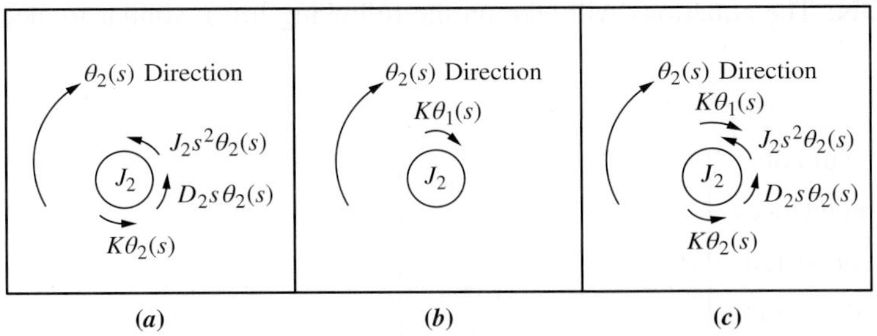


**FIGURE 2.24** **a.** Torques on $J_2$ due only to the motion of $J_2$; **b.** torques on $J_2$ due only to the motion of $J_1$; **c.** final free-body diagram for $J_2$

Summing torques, respectively, from Figures 2.23(*c*) and 2.24(*c*) we obtain the equations of motion,

$$(J_1s^2 + D_1s + K)\theta_1(s) \qquad - K\theta_2(s) = T(s) \tag{2.127a}$$

$$-K\theta_1(s) + (J_2s^2 + D_2s + K)\theta_2(s) = 0 \tag{2.127b}$$

from which the required transfer function is found to be

$$\frac{\theta_2(s)}{T(s)} = \frac{K}{\Delta} \tag{2.128}$$

as shown in Figure 2.22(*c*), where

$$\Delta = \begin{vmatrix} (J_1s^2 + D_1s + K) & -K \\ -K & (J_2s^2 + D_2s + K) \end{vmatrix}$$

Notice that Eqs. (2.127) have that now well-known form

$$\begin{bmatrix} \text{Sum of} \\ \text{impedances} \\ \text{connected} \\ \text{to the motion} \\ \text{at } \theta_1 \end{bmatrix} \theta_1(s) - \begin{bmatrix} \text{Sum of} \\ \text{impedances} \\ \text{between} \\ \theta_1 \text{ and } \theta_2 \end{bmatrix} \theta_2(s) = \begin{bmatrix} \text{Sum of} \\ \text{applied torques} \\ \text{at } \theta_1 \end{bmatrix} \tag{2.129a}$$

$$-\begin{bmatrix} \text{Sum of} \\ \text{impedances} \\ \text{between} \\ \theta_1 \text{ and } \theta_2 \end{bmatrix} \theta_1(s) + \begin{bmatrix} \text{Sum of} \\ \text{impedances} \\ \text{connected} \\ \text{to the motion} \\ \text{at } \theta_2 \end{bmatrix} \theta_2(s) = \begin{bmatrix} \text{Sum of} \\ \text{applied torques} \\ \text{at } \theta_2 \end{bmatrix} \tag{2.129b}$$

**TryIt 2.9**

Use the following MATLAB and Symbolic Math Toolbox statements to help you get Eq. (2.128).

```
syms s J1 D1 K T J2 D2...
  theta1 theta2
A=[(J1*s^2+D1*s+K)-K
  -K(J2*s^2+D2*s+K)];
B=[theta1
     theta2];
C=[T
     0];
B=inv(A)*C;
theta2 = B(2);
'theta2'
pretty (theta2)
```

## Example 2.20 Equations of Motion by Inspection

**PROBLEM:** Write, but do not solve, the Laplace transform of the equations of motion for the system shown in Figure 2.25.

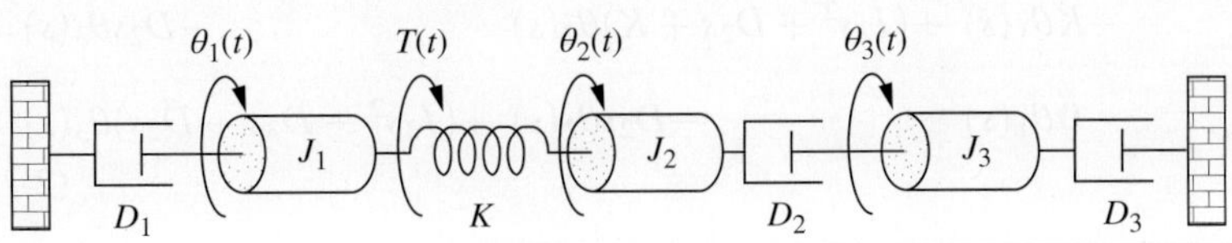


**FIGURE 2.25** Three-degrees-of-freedom rotational system

**SOLUTION:** The equations will take on the following form, similar to electrical mesh equations:

$$\begin{bmatrix}\text{Sum of}\\ \text{impedances}\\ \text{connected}\\ \text{to the motion}\\ \text{at } \theta_1\end{bmatrix}\theta_1(s) - \begin{bmatrix}\text{Sum of}\\ \text{impedances}\\ \text{between}\\ \theta_1 \text{ and } \theta_2\end{bmatrix}\theta_2(s) - \begin{bmatrix}\text{Sum of}\\ \text{impedances}\\ \text{between}\\ \theta_1 \text{ and } \theta_3\end{bmatrix}\theta_3(s) = \begin{bmatrix}\text{Sum of}\\ \text{applied torques}\\ \text{at } \theta_1\end{bmatrix} \tag{2.130a}$$

$$-\begin{bmatrix}\text{Sum of}\\ \text{impedances}\\ \text{between}\\ \theta_1 \text{ and } \theta_2\end{bmatrix}\theta_1(s) + \begin{bmatrix}\text{Sum of}\\ \text{impedances}\\ \text{connected}\\ \text{to the motion}\\ \text{at } \theta_2\end{bmatrix}\theta_2(s) - \begin{bmatrix}\text{Sum of}\\ \text{impedances}\\ \text{between}\\ \theta_2 \text{ and } \theta_3\end{bmatrix}\theta_3(s) = \begin{bmatrix}\text{Sum of}\\ \text{applied torques}\\ \text{at } \theta_2\end{bmatrix} \tag{2.130b}$$

$$-\begin{bmatrix}\text{Sum of}\\ \text{impedances}\\ \text{between}\\ \theta_1 \text{ and } \theta_3\end{bmatrix}\theta_1(s) - \begin{bmatrix}\text{Sum of}\\ \text{impedances}\\ \text{between}\\ \theta_2 \text{ and } \theta_3\end{bmatrix}\theta_2(s) + \begin{bmatrix}\text{Sum of}\\ \text{impedances}\\ \text{connected}\\ \text{to the motion}\\ \text{at } \theta_3\end{bmatrix}\theta_3(s) = \begin{bmatrix}\text{Sum of}\\ \text{applied torques}\\ \text{at } \theta_3\end{bmatrix} \tag{2.130c}$$

Hence,

$$\begin{aligned}
(J_1s^2 + D_1s + K)\theta_1(s) \quad & -K\theta_2(s) \quad & -0\theta_3(s) &= T(s)\\
-K\theta_1(s) \quad & +(J_2s^2 + D_2s + K)\theta_2(s) \quad & -D_2s\theta_3(s) &= 0\\
-0\theta_1(s) \quad & -D_2s\theta_2(s) \quad & +(J_3s^2 + D_3s + D_2s)\theta_3(s) &= 0
\end{aligned} \tag{2.131a,b,c}$$

**Skill-Assessment Exercise 2.9**

**PROBLEM:** Find the transfer function, $G(s) = \theta_2(s)/T(s)$, for the rotational mechanical system shown in Figure 2.26.

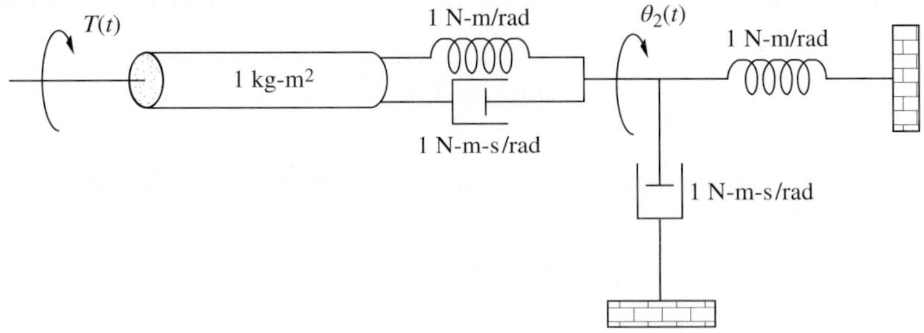


**FIGURE 2.26** Rotational mechanical system for Skill-Assessment Exercise 2.9

**ANSWER:** $G(s) = \dfrac{1}{2s^2 + s + 1}$

The complete solution is at www.wiley.com/college/nise.

# 2.7 Transfer Functions for Systems with Gears

Now that we are able to find the transfer function for rotational systems, we realize that these systems, especially those driven by motors, are rarely seen without associated gear trains driving the load. This section covers this important topic.

Gears provide mechanical advantage to rotational systems. Anyone who has ridden a 10-speed bicycle knows the effect of gearing. Going uphill, you shift to provide more torque and less speed. On the straightaway, you shift to obtain more speed and less torque. Thus, gears allow you to match the drive system and the load—a trade-off between speed and torque.

For many applications, gears exhibit *backlash*, which occurs because of the loose fit between two meshed gears. The drive gear rotates through a small angle before making contact with the meshed gear. The result is that the angular rotation of the output gear does not occur until a small angular rotation of the input gear has occurred. In this section, we idealize the behavior of gears and assume that there is no backlash.

The linearized interaction between two gears is depicted in Figure 2.27. An input gear with radius $r_1$ and $N_1$ teeth is rotated through angle $\theta_1(t)$ due to a torque, $T_1(t)$. An output gear with radius $r_2$ and $N_2$ teeth responds by rotating through angle $\theta_2(t)$ and delivering a torque, $T_2(t)$. Let us now find the relationship between the rotation of Gear 1, $\theta_1(t)$, and Gear 2, $\theta_2(t)$.

From Figure 2.27, as the gears turn, the distance traveled along each gear's circumference is the same. Thus,

$$r_1\theta_1 = r_2\theta_2 \tag{2.132}$$

or

$$\boxed{\frac{\theta_2}{\theta_1} = \frac{r_1}{r_2} = \frac{N_1}{N_2}} \tag{2.133}$$

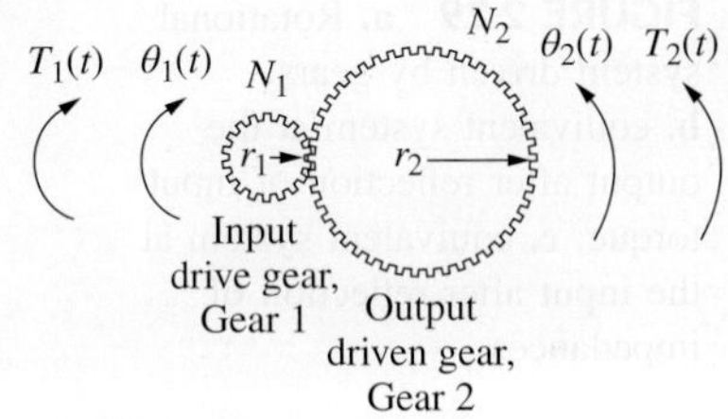


**FIGURE 2.27** A gear system

since the ratio of the number of teeth along the circumference is in the same proportion as the ratio of the radii. We conclude that the ratio of the angular displacement of the gears is inversely proportional to the ratio of the number of teeth.

What is the relationship between the input torque, $T_1$, and the delivered torque, $T_2$? If we assume the gears are *lossless*, that is, they do not absorb or store energy, the energy into Gear 1 equals the energy out of Gear 2.[11] Since the translational energy of force times displacement becomes the rotational energy of torque times angular displacement,

$$T_1\theta_1 = T_2\theta_2 \tag{2.134}$$

Solving Eq. (2.134) for the ratio of the torques and using Eq. (2.133), we get

$$\frac{T_2}{T_1} = \frac{\theta_1}{\theta_2} = \frac{N_2}{N_1} \tag{2.135}$$

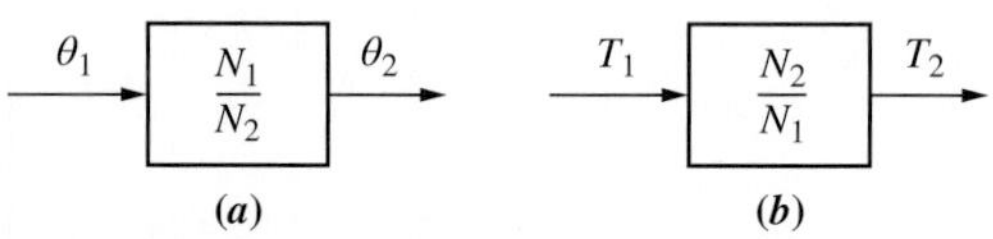


**FIGURE 2.28** Transfer functions for **a.** angular displacement in lossless gears and **b.** torque in lossless gears

Thus, the torques are directly proportional to the ratio of the number of teeth. All results are summarized in Figure 2.28.

Let us see what happens to mechanical impedances that are driven by gears. Figure 2.29(*a*) shows gears driving a rotational inertia, spring, and viscous damper. For clarity, the gears are shown by an end-on view. We want to represent Figure 2.29(*a*) as an equivalent system at $\theta_1$ without the gears. In other words, can the mechanical impedances be reflected from the output to the input, thereby eliminating the gears?

From Figure 2.28(*b*), $T_1$ can be reflected to the output by multiplying by $N_2/N_1$. The result is shown in Figure 2.29(*b*), from which we write the equation of motion as

$$(Js^2 + Ds + K)\,\theta_2(s) = T_1(s)\frac{N_2}{N_1} \tag{2.136}$$

Now convert $\theta_2(s)$ into an equivalent $\theta_1(s)$, so that Eq. (2.136) will look as if it were written at the input. Using Figure 2.28(*a*) to obtain $\theta_2(s)$ in terms of $\theta_1(s)$, we get

$$(Js^2 + Ds + K)\frac{N_1}{N_2}\theta_1(s) = T_1(s)\frac{N_2}{N_1} \tag{2.137}$$

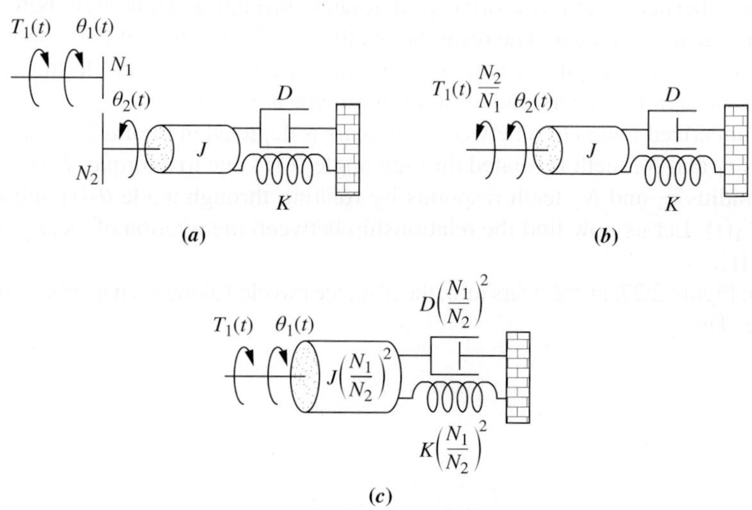


**FIGURE 2.29** **a.** Rotational system driven by gears; **b.** equivalent system at the output after reflection of input torque; **c.** equivalent system at the input after reflection of impedances

[11] This is equivalent to saying that the gears have negligible inertia and damping.

After simplification,

$$\left[J\left(\frac{N_1}{N_2}\right)^2 s^2 + D\left(\frac{N_1}{N_2}\right)^2 s + K\left(\frac{N_1}{N_2}\right)^2\right]\theta_1(s) = T_1(s) \quad (2.138)$$

which suggests the equivalent system at the input and without gears shown in Figure 2.29(*c*). Thus, the load can be thought of as having been reflected from the output to the input.

Generalizing the results, we can make the following statement: *Rotational mechanical impedances can be reflected through gear trains by multiplying the mechanical impedance by the ratio*

$$\left(\frac{\text{Number of teeth of gear on } \textit{destination} \text{ shaft}}{\text{Number of teeth of gear on } \textit{source} \text{ shaft}}\right)^2$$

where the impedance to be reflected is attached to the source shaft and is being reflected to the destination shaft. The next example demonstrates the application of the concept of reflected impedances as we find the transfer function of a rotational mechanical system with gears.

## Example 2.21 Transfer Function—System with Lossless Gears

**PROBLEM:** Find the transfer function, $\theta_2(s)/T_1(s)$, for the system of Figure 2.30(*a*).

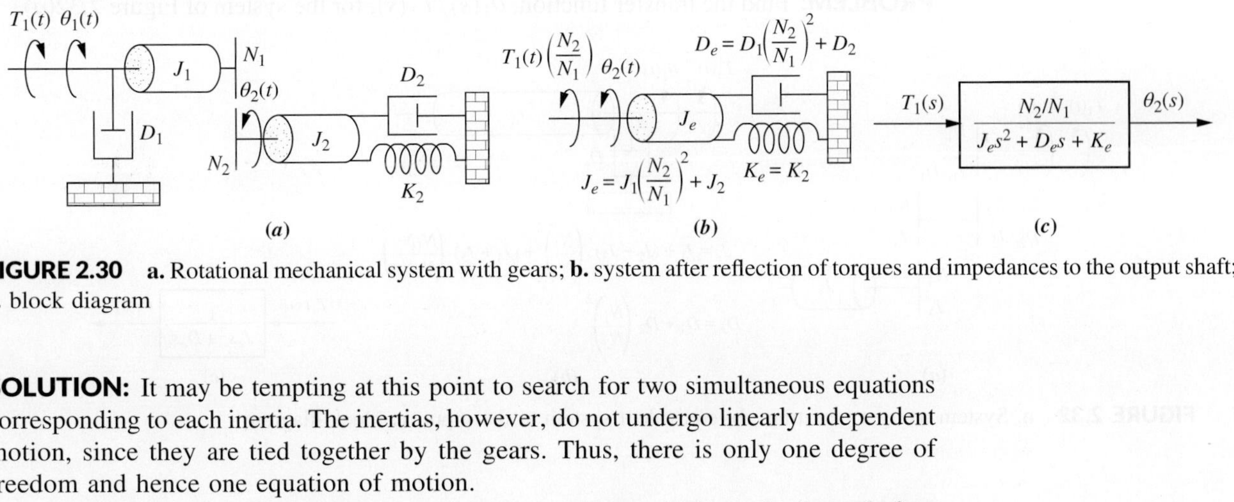


**FIGURE 2.30** **a.** Rotational mechanical system with gears; **b.** system after reflection of torques and impedances to the output shaft; **c.** block diagram

**SOLUTION:** It may be tempting at this point to search for two simultaneous equations corresponding to each inertia. The inertias, however, do not undergo linearly independent motion, since they are tied together by the gears. Thus, there is only one degree of freedom and hence one equation of motion.

Let us first reflect the impedances ($J_1$ and $D_1$) and torque ($T_1$) on the input shaft to the output as shown in Figure 2.30(*b*), where the impedances are reflected by $(N_2/N_1)^2$ and the torque is reflected by $(N_2/N_1)$. The equation of motion can now be written as

$$(J_e s^2 + D_e s + K_e)\theta_2(s) = T_1(s)\frac{N_2}{N_1} \quad (2.139)$$

where

$$J_e = J_1\left(\frac{N_2}{N_1}\right)^2 + J_2;\; D_e = D_1\left(\frac{N_2}{N_1}\right)^2 + D_2;\; K_e = K_2$$

Solving for $\theta_2(s)/T_1(s)$, the transfer function is found to be

$$G(s) = \frac{\theta_2(s)}{T_1(s)} = \frac{N_2/N_1}{J_e s^2 + D_e s + K_e} \tag{2.140}$$

as shown in Figure 2.30($c$).

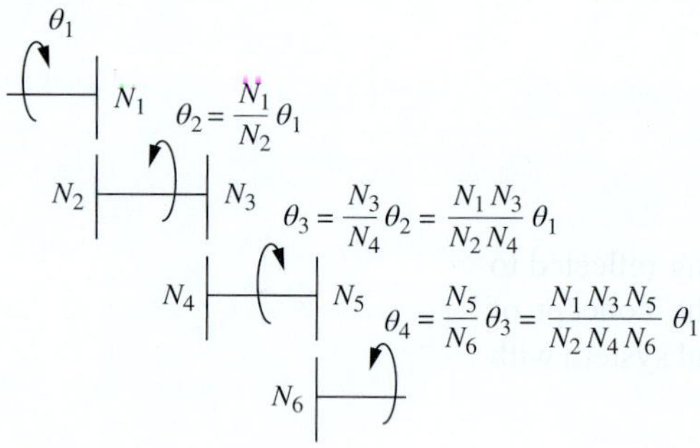


**FIGURE 2.31** Gear train

In order to eliminate gears with large radii, a *gear train* is used to implement large gear ratios by cascading smaller gear ratios. A schematic diagram of a gear train is shown in Figure 2.31. Next to each rotation, the angular displacement relative to $\theta_1$ has been calculated. From Figure 2.31,

$$\theta_4 = \frac{N_1 N_3 N_5}{N_2 N_4 N_6}\theta_1 \tag{2.141}$$

For gear trains, we conclude that the equivalent gear ratio is the product of the individual gear ratios. We now apply this result to solve for the transfer function of a system that does not have lossless gears.

## Example 2.22 Transfer Function—Gears with Loss

**PROBLEM:** Find the transfer function, $\theta_1(s)/T_1(s)$, for the system of Figure 2.32($a$).

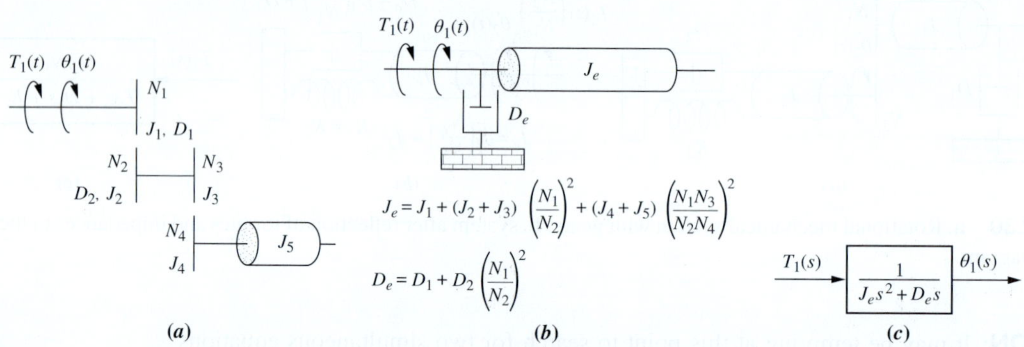


**FIGURE 2.32** **a.** System using a gear train; **b.** equivalent system at the input; **c.** block diagram

**SOLUTION:** This system, which uses a gear train, does not have lossless gears. All of the gears have inertia, and for some shafts there is viscous friction. To solve the problem, we want to reflect all of the impedances to the input shaft, $\theta_1$. The gear ratio is not the same for all impedances. For example, $D_2$ is reflected only through one gear ratio as $D_2(N_1/N_2)^2$, whereas $J_4$ plus $J_5$ is reflected through two gear ratios as $(J_4 + J_5)[(N_3/N_4)(N_1/N_2)]^2$. The result of reflecting all impedances to $\theta_1$ is shown in Figure 2.32(*b*), from which the equation of motion is

$$(J_e s^2 + D_e s)\theta_1(s) = T_1(s) \tag{2.142}$$

where

$$J_e = J_1 + (J_2 + J_3)\left(\frac{N_1}{N_2}\right)^2 + (J_4 + J_5)\left(\frac{N_1 N_3}{N_2 N_4}\right)^2$$

and

$$D_e = D_1 + D_2\left(\frac{N_1}{N_2}\right)^2$$

From Eq. (2.142), the transfer function is

$$G(s) = \frac{\theta_1(s)}{T_1(s)} = \frac{1}{J_e s^2 + D_e s} \tag{2.143}$$

as shown in Figure 2.32(*c*).

## Skill-Assessment Exercise 2.10

**PROBLEM:** Find the transfer function, $G(s) = \theta_2(s)/T(s)$, for the rotational mechanical system with gears shown in Figure 2.33.

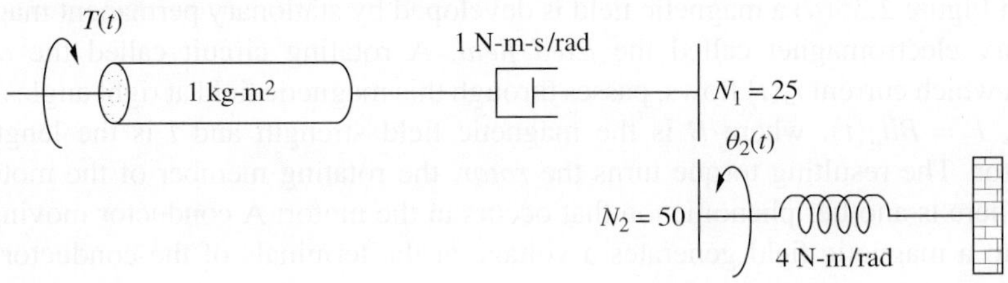


**FIGURE 2.33** Rotational mechanical system with gears for Skill-Assessment Exercise 2.10

**ANSWER:** $G(s) = \dfrac{1/2}{s^2 + s + 1}$

The complete solution is at www.wiley.com/college/nise.

# 2.8 Electromechanical System Transfer Functions

In the last section we talked about rotational systems with gears, which completed our discussion of purely mechanical systems. Now, we move to systems that are hybrids of electrical and mechanical variables, the *electromechanical systems*. We have seen one application of an electromechanical system in Chapter 1, the antenna azimuth position

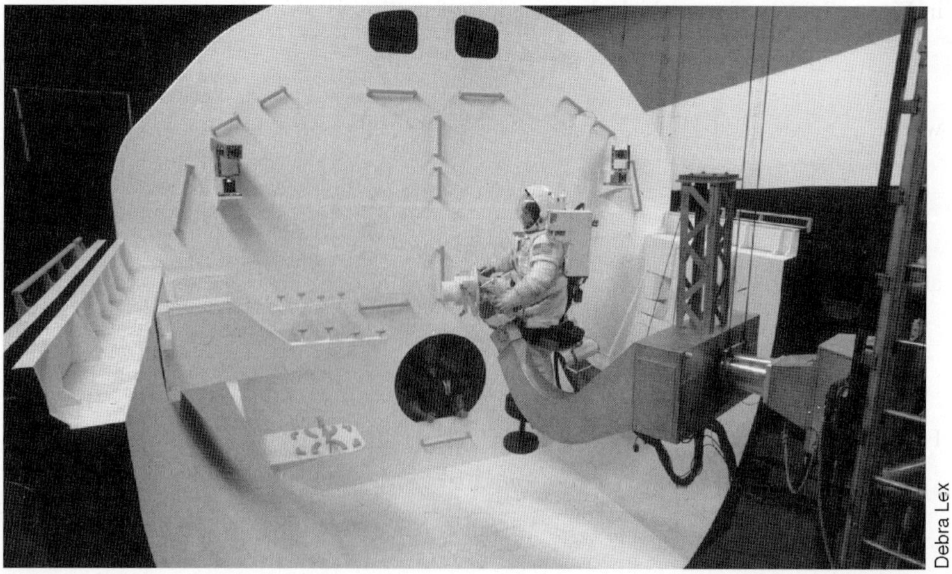


**FIGURE 2.34** NASA flight simulator robot arm with electromechanical control system components

control system. Other applications for systems with electromechanical components are robot controls, sun and star trackers, and computer tape and disk-drive position controls. An example of a control system that uses electromechanical components is shown in Figure 2.34.

A motor is an electromechanical component that yields a displacement output for a voltage input, that is, a mechanical output generated by an electrical input. We will derive the transfer function for one particular kind of electromechanical system, the armature-controlled dc servomotor (*Mablekos, 1980*). The motor's schematic is shown in Figure 2.35(*a*), and the transfer function we will derive appears in Figure 2.35(*b*).

In Figure 2.35(*a*) a magnetic field is developed by stationary permanent magnets or a stationary electromagnet called the *fixed field*. A rotating circuit called the *armature*, through which current $i_a(t)$ flows, passes through this magnetic field at right angles and feels a force, $F = Bli_a(t)$, where $B$ is the magnetic field strength and $l$ is the length of the conductor. The resulting torque turns the *rotor*, the rotating member of the motor.

There is another phenomenon that occurs in the motor: A conductor moving at right angles to a magnetic field generates a voltage at the terminals of the conductor equal to

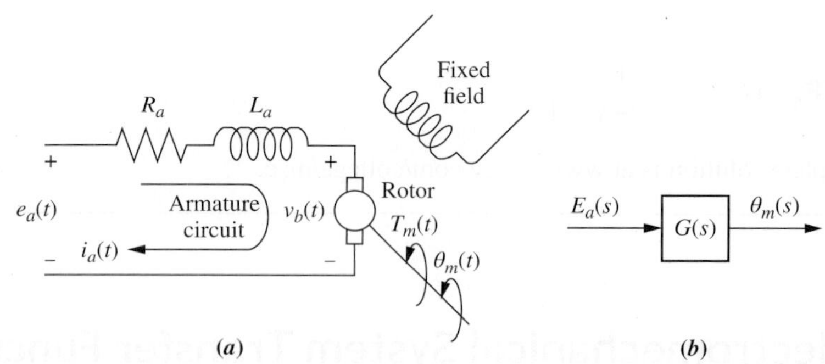


**FIGURE 2.35** DC motor: **a.** schematic;[12] **b.** block diagram

[12] See Appendix I at www.wiley.com/college/nise for a derivation of this schematic and its parameters.

$e = Blv$, where $e$ is the voltage and $v$ is the velocity of the conductor normal to the magnetic field. Since the current-carrying armature is rotating in a magnetic field, its voltage is proportional to speed. Thus,

$$v_b(t) = K_b \frac{d\theta_m(t)}{dt} \tag{2.144}$$

We call $v_b(t)$ the *back electromotive force* (*back emf*); $K_b$ is a constant of proportionality called the back emf constant; and $d\theta_m(t)/dt = \omega_m(t)$ is the angular velocity of the motor. Taking the Laplace transform, we get

$$V_b(s) = K_b s\theta_m(s) \tag{2.145}$$

The relationship between the armature current, $i_a(t)$, the applied armature voltage, $e_a(t)$, and the back emf, $v_b(t)$, is found by writing a loop equation around the Laplace transformed armature circuit (see Figure 3.5($a$)):

$$R_a I_a(s) + L_a s I_a(s) + V_b(s) = E_a(s) \tag{2.146}$$

The torque developed by the motor is proportional to the armature current; thus,

$$T_m(s) = K_t I_a(s) \tag{2.147}$$

where $T_m$ is the torque developed by the motor, and $K_t$ is a constant of proportionality, called the motor torque constant, which depends on the motor and magnetic field characteristics. In a consistent set of units, the value of $K_t$ is equal to the value of $K_b$. Rearranging Eq. (2.147) yields

$$I_a(s) = \frac{1}{K_t} T_m(s) \tag{2.148}$$

To find the transfer function of the motor, we first substitute Eqs. (2.145) and (2.148) into (2.146), yielding

$$\frac{(R_a + L_a s)T_m(s)}{K_t} + K_b s\theta_m(s) = E_a(s) \tag{2.149}$$

Now we must find $T_m(s)$ in terms of $\theta_m(s)$ if we are to separate the input and output variables and obtain the transfer function, $\theta_m(s)/E_a(s)$.

Figure 2.36 shows a typical equivalent mechanical loading on a motor. $J_m$ is the equivalent inertia at the armature and includes both the armature inertia and, as we will see later, the load inertia reflected to the armature. $D_m$ is the equivalent viscous damping at the armature and includes both the armature viscous damping and, as we will see later, the load viscous damping reflected to the armature. From Figure 2.36,

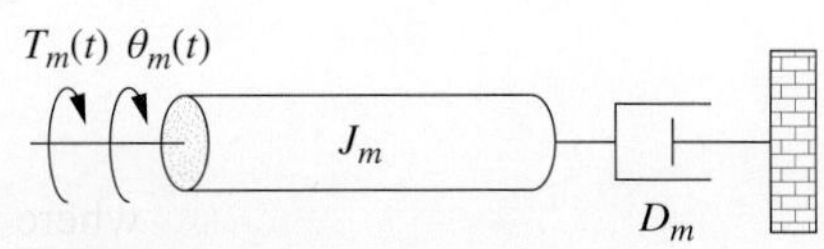


**FIGURE 2.36** Typical equivalent mechanical loading on a motor

$$T_m(s) = (J_m s^2 + D_m s)\theta_m(s) \tag{2.150}$$

Substituting Eq. (2.150) into Eq. (2.149) yields

$$\frac{(R_a + L_a s)(J_m s^2 + D_m s)\theta_m(s)}{K_t} + K_b s\theta_m(s) = E_a(s) \tag{2.151}$$

If we assume that the armature inductance, $L_a$, is small compared to the armature resistance, $R_a$, which is usual for a dc motor, Eq. (2.151) becomes

$$\left[\frac{R_a}{K_t}(J_m s + D_m) + K_b\right] s\theta_m(s) = E_a(s) \tag{2.152}$$

After simplification, the desired transfer function, $\theta_m(s)/E_a(s)$, is found to be

$$\frac{\theta_m(s)}{E_a(s)} = \frac{K_t/(R_a J_m)}{s\left[s + \frac{1}{J_m}\left(D_m + \frac{K_t K_b}{R_a}\right)\right]} \tag{2.153}$$ [13]

Even though the form of Eq. (2.153) is relatively simple, namely

$$\frac{\theta_m(s)}{E_a(s)} = \frac{K}{s(s+\alpha)} \tag{2.154}$$

the reader may be concerned about how to evaluate the constants.

Let us first discuss the mechanical constants, $J_m$ and $D_m$. Consider Figure 2.37, which shows a motor with inertia $J_a$ and damping $D_a$ at the armature driving a load consisting of inertia $J_L$ and damping $D_L$. Assuming that all inertia and damping values shown are known, $J_L$ and $D_L$ can be reflected back to the armature as some equivalent inertia and damping to be added to $J_a$ and $D_a$, respectively. Thus, the equivalent inertia, $J_m$, and equivalent damping, $D_m$, at the armature are

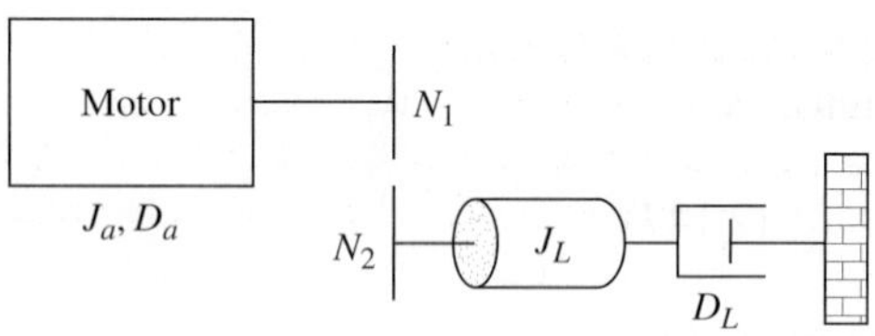


**FIGURE 2.37** DC motor driving a rotational mechanical load

$$J_m = J_a + J_L\left(\frac{N_1}{N_2}\right)^2; \quad D_m = D_a + D_L\left(\frac{N_1}{N_2}\right)^2 \tag{2.155}$$ [14]

Now that we have evaluated the mechanical constants, $J_m$ and $D_m$, what about the electrical constants in the transfer function of Eq. (2.153)? We will show that these constants can be obtained through a *dynamometer* test of the motor, where a dynamometer measures the torque and speed of a motor under the condition of a constant applied voltage. Let us first develop the relationships that dictate the use of a dynamometer.

Substituting Eqs. (2.145) and (2.148) into Eq. (2.146), with $L_a = 0$, yields

$$\frac{R_a}{K_t}T_m(s) + K_b s\theta_m(s) = E_a(s) \tag{2.156}$$

Taking the inverse Laplace transform, we get

$$\frac{R_a}{K_t}T_m(t) + K_b\omega_m(t) = e_a(t) \tag{2.157}$$

where the inverse Laplace transform of $s\theta_m(s)$ is $d\theta_m(t)/dt$ or, alternately, $\omega_m(t)$.

If a dc voltage, $e_a$, is applied, the motor will turn at a constant angular velocity, $\omega_m$, with a constant torque, $T_m$. Hence, dropping the functional relationship based on time from Eq. (2.157), the following relationship exists when the motor is operating at steady state with a dc voltage input:

$$\frac{R_a}{K_t}T_m + K_b\omega_m = e_a \tag{2.158}$$

[13] The units for the electrical constants are $K_t$ = N-m-A (newton-meters/ampere), and $K_b$ = V-s/rad (volt-seconds/radian).

[14] If the values of the mechanical constants are not known, motor constants can be determined through laboratory testing using transient response or frequency response data. The concept of transient response is covered in Chapter 4; frequency response is covered in Chapter 10.

Solving for $T_m$ yields

$$T_m = -\frac{K_b K_t}{R_a}\omega_m + \frac{K_t}{R_a}e_a \tag{2.159}$$

Equation (2.159) is a straight line, $T_m$ vs. $\omega_m$, and is shown in Figure 2.38. This plot is called the *torque-speed curve*. The torque axis intercept occurs when the angular velocity reaches zero. That value of torque is called the *stall torque*, $T_{\text{stall}}$. Thus,

$$T_{\text{stall}} = \frac{K_t}{R_a}e_a \tag{2.160}$$

The angular velocity occurring when the torque is zero is called the *no-load speed*, $\omega_{\text{no-load}}$. Thus,

$$\omega_{\text{no-load}} = \frac{e_a}{K_b} \tag{2.161}$$

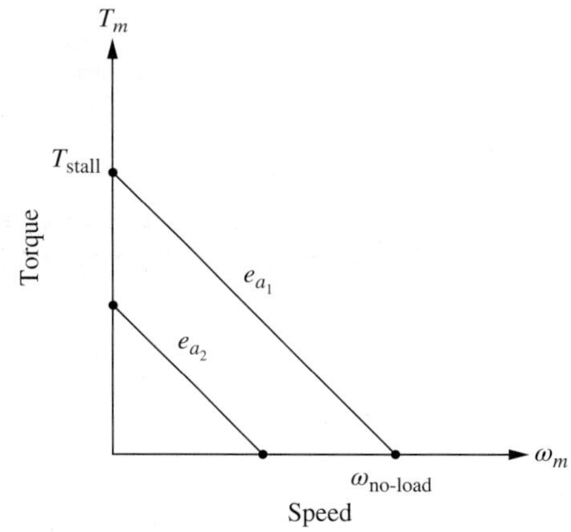


**FIGURE 2.38** Torque-speed curves with an armature voltage, $e_a$, as a parameter

The electrical constants of the motor's transfer function can now be found from Eqs. (2.160) and (2.161) as

$$\frac{K_t}{R_a} = \frac{T_{\text{stall}}}{e_a} \tag{2.162}$$

and

$$K_b = \frac{e_a}{\omega_{\text{no-load}}} \tag{2.163}$$

The electrical constants, $K_t/R_a$ and $K_b$, can be found from a dynamometer test of the motor, which would yield $T_{\text{stall}}$ and $\omega_{\text{no-load}}$ for a given $e_a$.

## Example 2.23 Transfer Function—DC Motor and Load

**PROBLEM:** Given the system and torque-speed curve of Figure 2.39(*a*) and (*b*), find the transfer function, $\theta_L(s)/E_a(s)$.

**SOLUTION:** Begin by finding the mechanical constants, $J_m$ and $D_m$, in Eq. (2.153). From Eq. (2.155), the total inertia at the armature of the motor is

$$J_m = J_a + J_L\left(\frac{N_1}{N_2}\right)^2 = 5 + 700\left(\frac{1}{10}\right)^2 = 12 \tag{2.164}$$

and the total damping at the armature of the motor is

$$D_m = D_a + D_L\left(\frac{N_1}{N_2}\right)^2 = 2 + 800\left(\frac{1}{10}\right)^2 = 10 \tag{2.165}$$

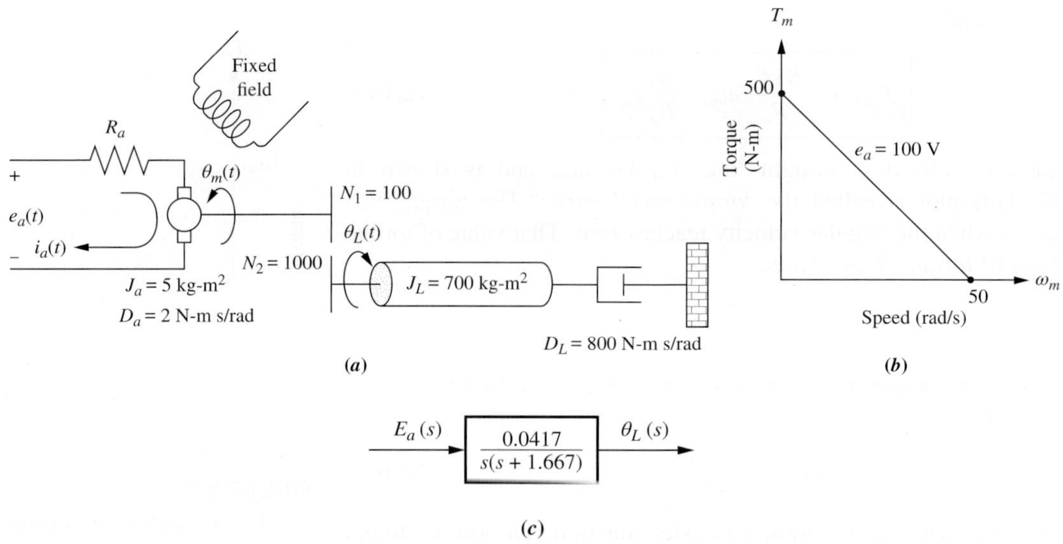


**FIGURE 2.39** **a.** DC motor and load; **b.** torque-speed curve; **c.** block diagram

Now we will find the electrical constants, $K_t/R_a$ and $K_b$. From the torque-speed curve of Figure 2.39(*b*),

$$T_{\text{stall}} = 500 \tag{2.166}$$

$$\omega_{\text{no-load}} = 50 \tag{2.167}$$

$$e_a = 100 \tag{2.168}$$

Hence the electrical constants are

$$\frac{K_t}{R_a} = \frac{T_{\text{stall}}}{e_a} = \frac{500}{100} = 5 \tag{2.169}$$

and

$$K_b = \frac{e_a}{\omega_{\text{no-load}}} = \frac{100}{50} = 2 \tag{2.170}$$

Substituting Eqs. (2.164), (2.165), (2.169), and (2.170) into Eq. (2.153) yield

$$\frac{\theta_m(s)}{E_a(s)} = \frac{5/12}{s\left\{s + \frac{1}{12}[10 + (5)(2)]\right\}} = \frac{0.417}{s(s+1.667)} \tag{2.171}$$

In order to find $\theta_L(s)/E_a(s)$, we use the gear ratio, $N_1/N_2 = 1/10$, and find

$$\frac{\theta_L(s)}{E_a(s)} = \frac{0.0417}{s(s+1.667)} \tag{2.172}$$

as shown in Figure 2.39(*c*).

## Skill-Assessment Exercise 2.11

**PROBLEM:** Find the transfer function, $G(s) = \theta_L(s)/E_a(s)$, for the motor and load shown in Figure 2.40. The torque-speed curve is given by $T_m = -8\omega_m + 200$ when the input voltage is 100 volts.

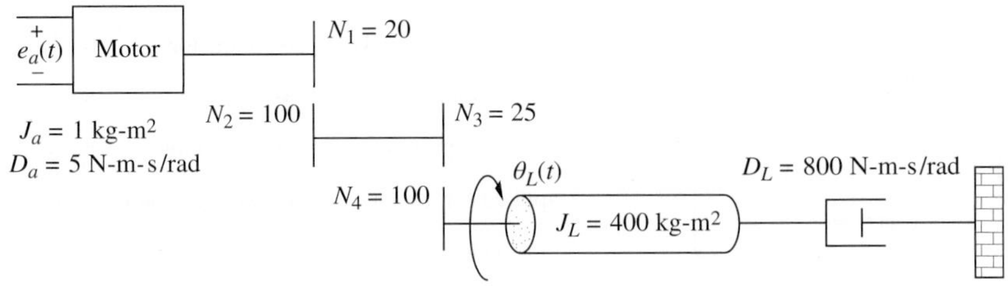


**FIGURE 2.40** Electromechanical system for Skill-Assessment Exercise 2.11

**ANSWER:** $G(s) = \dfrac{1/20}{s[s + (15/2)]}$

The complete solution is at www.wiley.com/college/nise.

# 2.9 Electric Circuit Analogs

In this section, we show the commonality of systems from the various disciplines by demonstrating that the mechanical systems with which we worked can be represented by equivalent electric circuits. We have pointed out the similarity between the equations resulting from Kirchhoff's laws for electrical systems and the equations of motion of mechanical systems. We now show this commonality even more convincingly by producing electric circuit equivalents for mechanical systems. The variables of the electric circuits behave exactly as the analogous variables of the mechanical systems. In fact, converting mechanical systems to electrical networks before writing the describing equations is a problem-solving approach that you may want to pursue.

An electric circuit that is analogous to a system from another discipline is called an electric circuit *analog*. Analogs can be obtained by comparing the describing equations, such as the equations of motion of a mechanical system, with either electrical mesh or nodal equations. When compared with mesh equations, the resulting electrical circuit is called a *series analog*. When compared with nodal equations, the resulting electrical circuit is called a *parallel analog*.

## Series Analog

Consider the translational mechanical system shown in Figure 2.41($a$), whose equation of motion is

$$(Ms^2 + f_v s + K)X(s) = F(s) \tag{2.173}$$

Kirchhoff's mesh equation for the simple series $RLC$ network shown in Figure 2.41($b$) is

$$\left(Ls + R + \frac{1}{Cs}\right)I(s) = E(s) \tag{2.174}$$

As we previously pointed out, Eq. (2.173) is not directly analogous to Eq. (2.174) because displacement and current are not analogous. We can create a direct analogy by

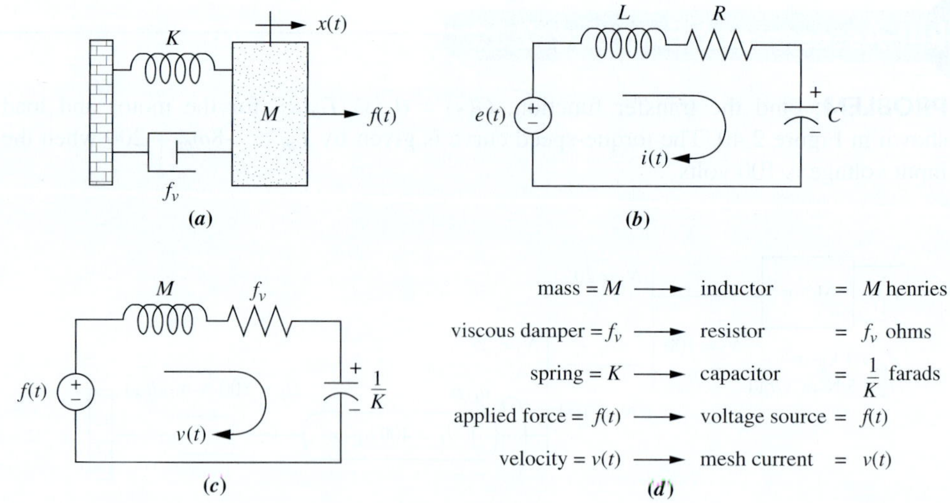


**FIGURE 2.41** Development of series analog: **a.** mechanical system; **b.** desired electrical representation; **c.** series analog; **d.** parameters for series analog

operating on Eq. (2.173) to convert displacement to velocity by dividing and multiplying the left-hand side by $s$, yielding

$$\frac{Ms^2+f_vs+K}{s}sX(s) = \left(Ms+f_v+\frac{K}{s}\right)V(s) = F(s) \tag{2.175}$$

Comparing Eqs. 2.174 and 2.175, we recognize the sum of impedances and draw the circuit shown in Figure 2.41($c$). The conversions are summarized in Figure 2.41($d$).

When we have more than one degree of freedom, the impedances associated with a motion appear as series electrical elements in a mesh, but the impedances between adjacent motions are drawn as series electrical impedances between the two corresponding meshes. We demonstrate with an example.

## Example 2.24 Converting a Mechanical System to a Series Analog

**PROBLEM:** Draw a series analog for the mechanical system of Figure 2.17($a$).

**SOLUTION:** Equations (2.118) are analogous to electrical mesh equations after conversion to velocity. Thus,

$$\left[M_1s+(f_{v_1}+f_{v_3})+\frac{(K_1+K_2)}{s}\right]V_1(s)-\left(f_{v_3}+\frac{K_2}{s}\right)V_2(s)=F(s) \tag{2.176a}$$

$$-\left(f_{v_3}+\frac{K_2}{s}\right)V_1(s)+\left[M_2s+(f_{v_2}+f_{v_3})+\frac{(K_2+K_3)}{s}\right]V_2(s)=0 \tag{2.176b}$$

Coefficients represent sums of electrical impedance. Mechanical impedances associated with $M_1$ form the first mesh, where impedances between the two masses are common to the two loops. Impedances associated with $M_2$ form the second mesh. The result is shown in Figure 2.42, where $v_1(t)$ and $v_2(t)$ are the velocities of $M_1$ and $M_2$, respectively.

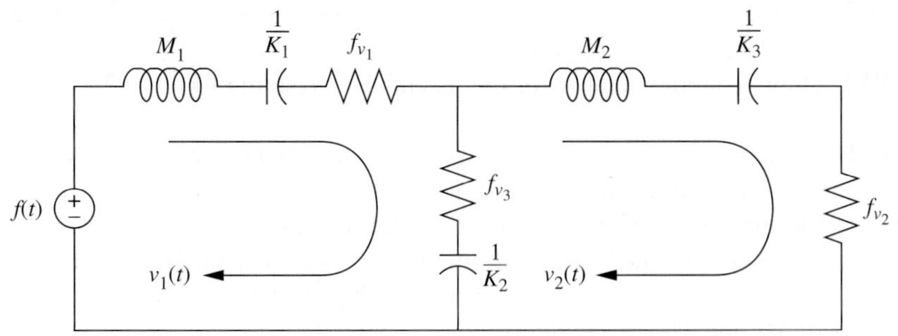


**FIGURE 2.42** Series analog of mechanical system of Figure 2.17($a$)

## Parallel Analog

A system can also be converted to an equivalent parallel analog. Consider the translational mechanical system shown in Figure 2.43($a$), whose equation of motion is given by Eq. (2.175). Kirchhoff's nodal equation for the simple parallel $RLC$ network shown in Figure 2.43($b$) is

$$\left(Cs + \frac{1}{R} + \frac{1}{Ls}\right)E(s) = I(s) \tag{2.177}$$

Comparing Eqs. (2.175) and (2.177), we identify the sum of admittances and draw the circuit shown in Figure 2.43($c$). The conversions are summarized in Figure 2.43($d$).

When we have more than one degree of freedom, the components associated with a motion appear as parallel electrical elements connected to a node. The components of adjacent motions are drawn as parallel electrical elements between two corresponding nodes. We demonstrate with an example.

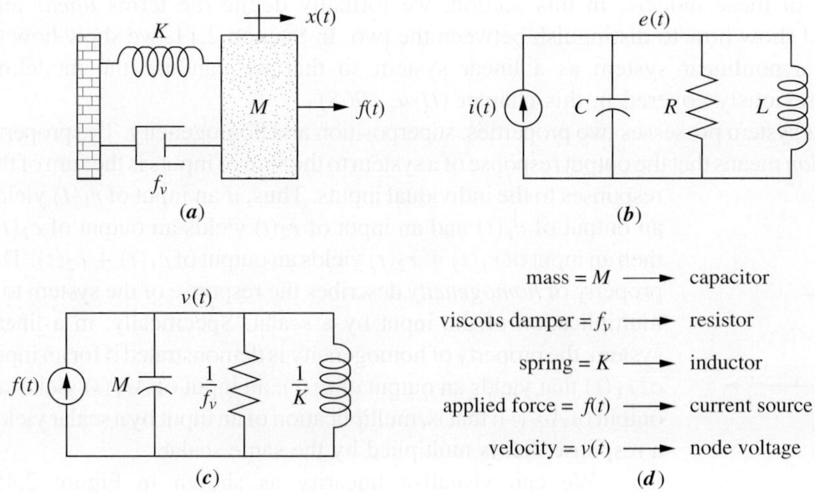


mass = $M$ ⟶ capacitor = $M$ farads

viscous damper = $f_v$ ⟶ resistor = $\frac{1}{f_v}$ ohms

spring = $K$ ⟶ inductor = $\frac{1}{K}$ henries

applied force = $f(t)$ ⟶ current source = $f(t)$

velocity = $v(t)$ ⟶ node voltage = $v(t)$

($d$)

**FIGURE 2.43** Development of parallel analog: **a.** mechanical system; **b.** desired electrical representation; **c.** parallel analog; **d.** parameters for parallel analog

## Example 2.25 Converting a Mechanical System to a Parallel Analog

**PROBLEM:** Draw a parallel analog for the mechanical system of Figure 2.17(*a*).

**SOLUTION:** Equation (2.176) is also analogous to electrical node equations. Coefficients represent sums of electrical admittances. Admittances associated with $M_1$ form the elements connected to the first node, where mechanical admittances between the two masses are common to the two nodes. Mechanical admittances associated with $M_2$ form the elements connected to the second node. The result is shown in Figure 2.44, where $v_1(t)$ and $v_2(t)$ are the velocities of $M_1$ and $M_2$, respectively.

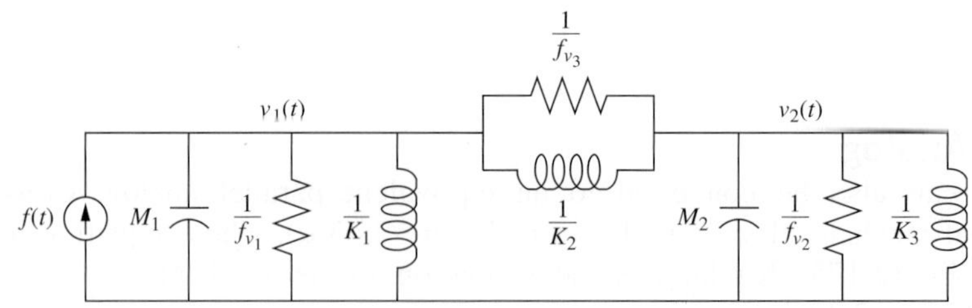


**FIGURE 2.44** Parallel analog of mechanical system of Figure 2.17(*a*)

## Skill-Assessment Exercise 2.12

**PROBLEM:** Draw a series and parallel analog for the rotational mechanical system of Figure 2.22.

**ANSWER:** The complete solution is at www.wiley.com/college/nise.

# 2.10 Nonlinearities

The models thus far are developed from systems that can be described approximately by linear, time-invariant differential equations. An assumption of *linearity* was implicit in the development of these models. In this section, we formally define the terms *linear* and *nonlinear* and show how to distinguish between the two. In Section 2.11, we show how to approximate a nonlinear system as a linear system so that we can use the modeling techniques previously covered in this chapter (*Hsu, 1968*).

A linear system possesses two properties: superposition and homogeneity. The property of *superposition* means that the output response of a system to the sum of inputs is the sum of the responses to the individual inputs. Thus, if an input of $r_1(t)$ yields an output of $c_1(t)$ and an input of $r_2(t)$ yields an output of $c_2(t)$, then an input of $r_1(t) + r_2(t)$ yields an output of $c_1(t) + c_2(t)$. The property of *homogeneity* describes the response of the system to a multiplication of the input by a scalar. Specifically, in a linear system, the property of homogeneity is demonstrated if for an input of $r_1(t)$ that yields an output of $c_1(t)$, an input of $Ar_1(t)$ yields an output of $Ac_1(t)$; that is, multiplication of an input by a scalar yields a response that is multiplied by the same scalar.

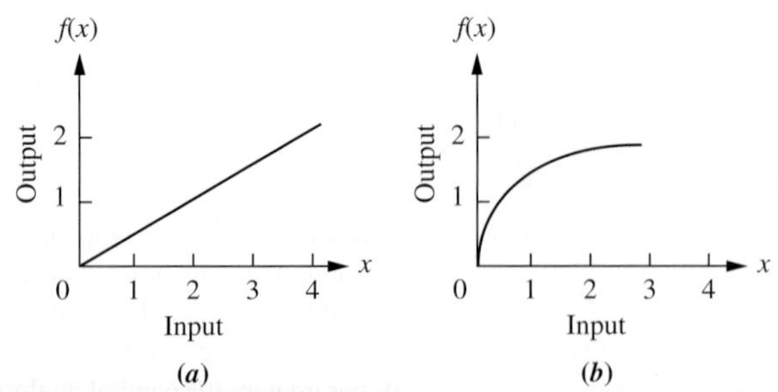


**FIGURE 2.45** **a.** Linear system; **b.** nonlinear system

We can visualize linearity as shown in Figure 2.45. Figure 2.45(*a*) is a linear system where the output is always

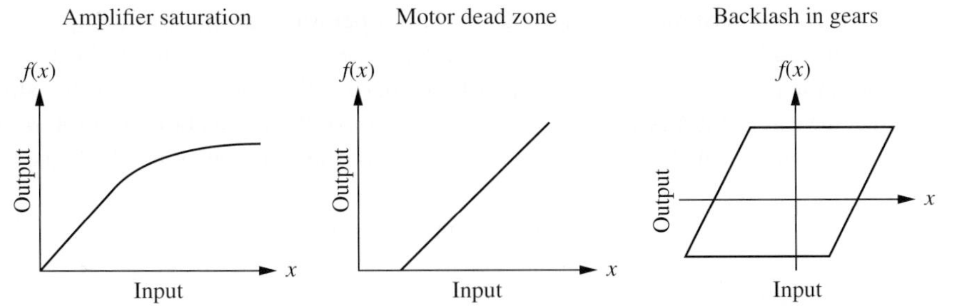


**FIGURE 2.46** Some physical nonlinearities

one half the input, or $f(x) = 0.5x$, regardless of the value of $x$. Thus each of the two properties of linear systems applies. For example, an input of 1 yields an output of $\frac{1}{2}$ and an input of 2 yields an output of 1. Using superposition, an input that is the sum of the original inputs, or 3, should yield an output that is the sum of the individual outputs, or 1.5. From Figure 2.45($a$), an input of 3 does indeed yield an output of 1.5.

To test the property of homogeneity, assume an input of 2, which yields an output of 1. Multiplying this input by 2 should yield an output of twice as much, or 2. From Figure 2.45($a$), an input of 4 does indeed yield an output of 2. The reader can verify that the properties of linearity certainly do not apply to the relationship shown in Figure 2.45($b$).

Figure 2.46 shows some examples of physical nonlinearities. An electronic amplifier is linear over a specific range but exhibits the nonlinearity called *saturation* at high input voltages. A motor that does not respond at very low input voltages due to frictional forces exhibits a nonlinearity called *dead zone*. Gears that do not fit tightly exhibit a nonlinearity called *backlash:* The input moves over a small range without the output responding. The reader should verify that the curves shown in Figure 2.46 do not fit the definitions of linearity over their entire range. Another example of a nonlinear subsystem is a phase detector, used in a phase-locked loop in an FM radio receiver, whose output response is the sine of the input.

A designer can often make a linear approximation to a nonlinear system. Linear approximations simplify the analysis and design of a system and are used as long as the results yield a good approximation to reality. For example, a linear relationship can be established at a point on the nonlinear curve if the range of input values about that point is small and the origin is translated to that point. Electronic amplifiers are an example of physical devices that perform linear amplification with small excursions about a point.

## 2.11 Linearization

The electrical and mechanical systems covered thus far were assumed to be linear. However, if any nonlinear components are present, we must linearize the system before we can find the transfer function. In the last section, we defined and discussed nonlinearities; in this section, we show how to obtain linear approximations to nonlinear systems in order to obtain transfer functions.

The first step is to recognize the nonlinear component and write the nonlinear differential equation. When we linearize a nonlinear differential equation, we linearize it for small-signal inputs about the steady-state solution when the small-signal input is equal to zero. This steady-state solution is called *equilibrium* and is selected as the second step in the linearization process. For example, when a pendulum is at rest, it is at equilibrium. The angular displacement is described by a nonlinear differential equation, but it can be expressed with a linear differential equation for small excursions about this equilibrium point.

Next we linearize the nonlinear differential equation, and then we take the Laplace transform of the linearized differential equation, assuming zero initial conditions. Finally, we separate input and output variables and form the transfer function. Let us first see how to linearize a function; later, we will apply the method to the linearization of a differential equation.

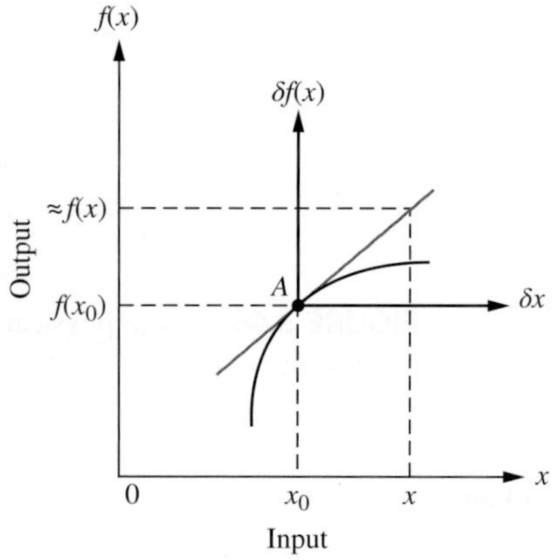


**FIGURE 2.47** Linearization about point $A$

If we assume a nonlinear system operating at point $A$, $[x_0, f(x_0)]$ in Figure 2.47, small changes in the input can be related to changes in the output about the point by way of the slope of the curve at the point $A$. Thus, if the slope of the curve at point $A$ is $m_a$, then small excursions of the input about point $A$, $\delta_x$, yield small changes in the output, $\delta f(x)$, related by the slope at point $A$. Thus,

$$[f(x) - f(x_0)] \approx m_a(x - x_0) \tag{2.178}$$

from which

$$\delta f(x) \approx m_a \delta x \tag{2.179}$$

and

$$f(x) \approx f(x_0) + m_a(x - x_0) \approx f(x_0) + m_a \delta x \tag{2.180}$$

This relationship is shown graphically in Figure 2.47, where a new set of axes, $\delta_x$ and $\delta f(x)$, is created at the point $A$, and $f(x)$ is approximately equal to $f(x_0)$, the ordinate of the new origin, plus small excursions, $m_a \delta x$, away from point $A$. Let us look at an example.

## Example 2.26 Linearizing a Function

**PROBLEM:** Linearize $f(x) = 5 \cos x$ about $x = \pi/2$.

**SOLUTION:** We first find that the derivative of $f(x)$ is $df/dx = (-5 \sin x)$. At $x = \pi/2$, the derivative is $-5$. Also $f(x_0) = f(\pi/2) = 5 \cos(\pi/2) = 0$. Thus, from Eq. (2.180), the system can be represented as $f(x) = -5\ \delta x$ for small excursions of $x$ about $\pi/2$. The process is shown graphically in Figure 2.48, where the cosine curve does indeed look like a straight line of slope $-5$ near $\pi/2$.

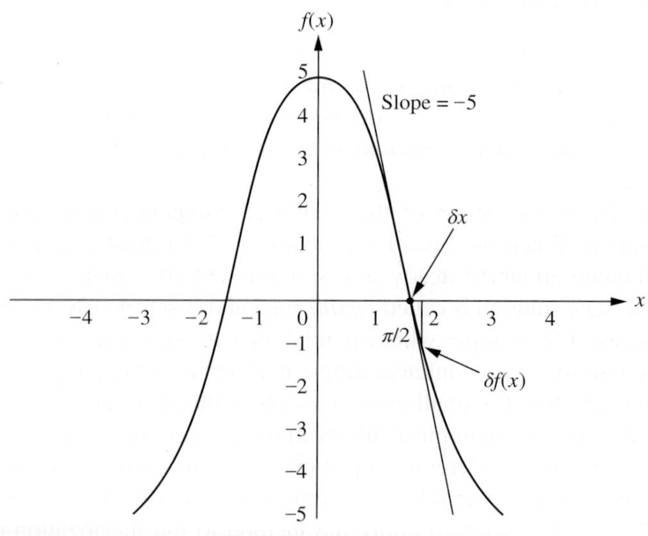


**FIGURE 2.48** Linearization of 5 cos $x$ about $x = \pi/2$

The previous discussion can be formalized using the Taylor series expansion, which expresses the value of a function in terms of the value of that function at a particular point, the excursion away from that point, and derivatives evaluated at that point. The Taylor series is shown in Eq. (2.181).

$$f(x) = f(x_0) + \left.\frac{df}{dx}\right|_{x=x_0} \frac{(x - x_0)}{1!} + \left.\frac{d^2f}{dx^2}\right|_{x=x_0} \frac{(x - x_0)^2}{2!} + \cdots \tag{2.181}$$

For small excursions of $x$ from $x_0$, we can neglect higher-order terms. The resulting approximation yields a straight-line relationship between the change in $f(x)$ and the excursions away from $x_0$. Neglecting the higher-order terms in Eq. (2.181), we get

$$f(x) - f(x_0) \approx \left.\frac{df}{dx}\right|_{x=x_0} (x - x_0) \tag{2.182}$$

or

$$\delta f(x) \approx m\big|_{x=x_0} \delta x \tag{2.183}$$

which is a linear relationship between $\delta f(x)$ and $\delta x$ for small excursions away from $x_0$. It is interesting to note that Eqs. (2.182) and (2.183) are identical to Eqs. (2.178) and (2.179), which we derived intuitively. The following examples illustrate linearization. The first example demonstrates linearization of a differential equation, and the second example applies linearization to finding a transfer function.

## Example 2.27 Linearizing a Differential Equation

**PROBLEM:** Linearize Eq. (2.184) for small excursions about $x = \pi/4$.

$$\frac{d^2x}{dt^2} + 2\frac{dx}{dt} + \cos x = 0 \tag{2.184}$$

**SOLUTION:** The presence of the term cos $x$ makes this equation nonlinear. Since we want to linearize the equation about $x = \pi/4$, we let $x = \delta x + \pi/4$, where $\delta x$ is the small excursion about $\pi/4$, and substitute $x$ into Eq. (2.184):

$$\frac{d^2\left(\delta x + \frac{\pi}{4}\right)}{dt^2} + 2\frac{d\left(\delta x + \frac{\pi}{4}\right)}{dt} + \cos\left(\delta x + \frac{\pi}{4}\right) = 0 \tag{2.185}$$

But

$$\frac{d^2\left(\delta x + \frac{\pi}{4}\right)}{dt^2} = \frac{d^2\delta x}{dt^2} \tag{2.186}$$

and

$$\frac{d\left(\delta x + \frac{\pi}{4}\right)}{dt} = \frac{d\delta x}{dt} \tag{2.187}$$

Finally, the term $\cos(\delta x + (\pi/4))$ can be linearized with the truncated Taylor series. Substituting $f(x) = \cos(\delta x + (\pi/4)), f(x_0) = f(\pi/4) = \cos(\pi/4)$, and $(x - x_0) = \delta x$ into Eq. (2.182) yields

$$\cos\left(\delta x + \frac{\pi}{4}\right) - \cos\left(\frac{\pi}{4}\right) = \left.\frac{d\cos x}{dx}\right|_{x=\frac{\pi}{4}} \delta x = -\sin\left(\frac{\pi}{4}\right)\delta x \tag{2.188}$$

Solving Eq. (2.188) for $\cos(\delta x + (\pi/4))$, we get

$$\cos\left(\delta x + \frac{\pi}{4}\right) = \cos\left(\frac{\pi}{4}\right) - \sin\left(\frac{\pi}{4}\right)\delta x = \frac{\sqrt{2}}{2} - \frac{\sqrt{2}}{2}\delta x \tag{2.189}$$

Substituting Eqs. (2.186), (2.187), and (2.189) into Eq. (2.185) yields the following linearized differential equation:

$$\frac{d^2\delta x}{dt^2} + 2\frac{d\delta x}{dt} - \frac{\sqrt{2}}{2}\delta x = -\frac{\sqrt{2}}{2} \tag{2.190}$$

This equation can now be solved for $\delta x$, from which we can obtain $x = \delta x + (\pi/4)$.

Even though the nonlinear Eq. (2.184) is homogeneous, the linearized Eq. (2.190) is not homogeneous. Eq. (2.190) has a forcing function on its right-hand side. This additional term can be thought of as an input to a system represented by Eq. (2.184).

Another observation about Eq. (2.190) is the negative sign on the left-hand side. The study of differential equations tells us that since the roots of the characteristic equation are positive, the homogeneous solution grows without bound instead of diminishing to zero. Thus, this system, linearized around $x = \pi/4$, is not stable.

## Example 2.28 Transfer Function—Nonlinear Electrical Network

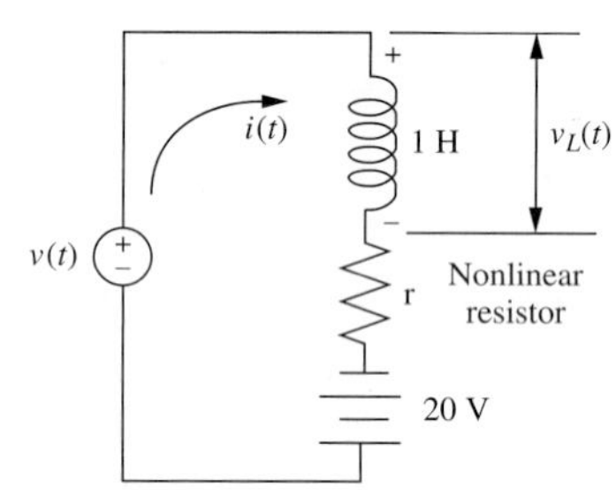


**FIGURE 2.49** Nonlinear electrical network

**PROBLEM:** Find the transfer function, $V_L(s)/V(s)$, for the electrical network shown in Figure 2.49, which contains a nonlinear resistor whose voltage-current relationship is defined by $i_r = 2e^{0.1v_r}$, where $i_r$ and $v_r$ are the resistor current and voltage, respectively. Also, $v(t)$ in Figure 2.49 is a small-signal source.

**SOLUTION:** We will use Kirchhoff's voltage law to sum the voltages in the loop to obtain the nonlinear differential equation, but first we must solve for the voltage across the nonlinear resistor. Taking the natural log of the resistor's current-voltage relationship, we get $v_r = 10\ln\frac{1}{2}i_r$. Applying Kirchhoff's voltage law around the loop, where $i_r = i$, yields

$$L\frac{di}{dt} + 10\ln\frac{1}{2}i - 20 = v(t) \tag{2.191}$$

Next, let us evaluate the equilibrium solution. First, set the small-signal source, $v(t)$, equal to zero. Now evaluate the steady-state current. With $v(t) = 0$, the circuit consists of a 20 V battery in series with the inductor and nonlinear resistor. In the steady state, the voltage across the inductor will be zero, since $v_L(t) = Ldi/dt$ and $di/dt$ is zero in the steady state, given a constant battery source. Hence, the resistor voltage, $v_r$, is 20 V. Using the characteristics of the resistor, $i_r = 2e^{0.1v_r}$, we find that $i_r = i = 14.78$ amps. This current, $i_0$, is the equilibrium value of the network current. Hence $i = i_0 + \delta i$. Substituting this current into Eq. (2.191) yields

$$L\frac{d(i_0 + \delta i)}{dt} + 10\ln\frac{1}{2}(i_0 + \delta i) - 20 = v(t) \tag{2.192}$$

Using Eq. (2.182) to linearize $\ln\frac{1}{2}(i_0 + \delta i)$, we get

$$\ln\frac{1}{2}(i_0 + \delta i) - \ln\frac{1}{2}i_0 = \frac{d\left(\ln\frac{1}{2}i\right)}{di}\Big|_{i=i_0}\delta i = \frac{1}{i}\Big|_{i=i_0}\delta i = \frac{1}{i_0}\delta i \tag{2.193}$$

or

$$\ln\frac{1}{2}(i_0 + \delta i) = \ln\frac{i_0}{2} + \frac{1}{i_0}\delta i \tag{2.194}$$

Substituting into Eq. (2.192), the linearized equation becomes

$$L\frac{d\delta i}{dt} + 10\left(\ln\frac{i_0}{2} + \frac{1}{i_0}\delta i\right) - 20 = v(t) \tag{2.195}$$

Letting $L = 1$ and $i_0 = 14.78$, the final linearized differential equation is

$$\frac{d\delta i}{dt} + 0.677\delta i = v(t) \tag{2.196}$$

Taking the Laplace transform with zero initial conditions and solving for $\delta i(s)$, we get

$$\delta i(s) = \frac{V(s)}{s + 0.677} \tag{2.197}$$

But the voltage across the inductor about the equilibrium point is

$$v_L(t) = L\frac{d}{dt}(i_0 + \delta i) = L\frac{d\delta i}{dt} \tag{2.198}$$

Taking the Laplace transform,

$$V_L(s) = Ls\delta i(s) = s\delta i(s) \tag{2.199}$$

Substituting Eq. (2.197) into Eq. (2.199) yields

$$V_L(s) = s\frac{V(s)}{s + 0.677} \tag{2.200}$$

from which the final transfer function is

$$\frac{V_L(s)}{V(s)} = \frac{s}{s + 0.677} \tag{2.201}$$

for small excursions about $i = 14.78$ or, equivalently, about $v(t) = 0$.

## Skill-Assessment Exercise 2.13

**PROBLEM:** Find the linearized transfer function, $G(s) = V(s)/I(s)$, for the electrical network shown in Figure 2.50. The network contains a nonlinear resistor whose voltage-current relationship is defined by $i_r = e^{v_r}$. The current source, $i(t)$, is a small-signal generator.

**ANSWER:** $G(s) = \frac{1}{s+2}$

The complete solution is at www.wiley.com/college/nise.

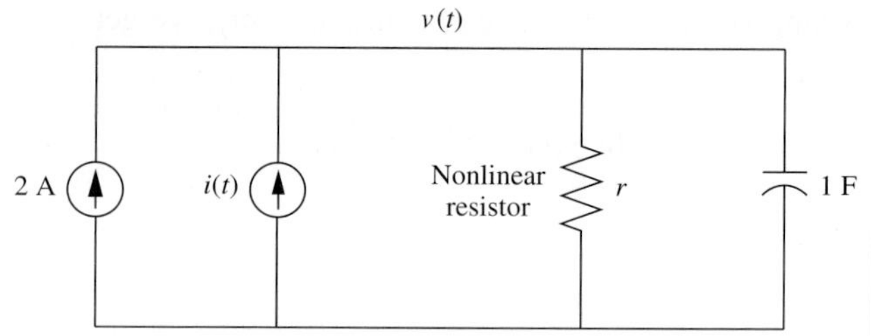


**FIGURE 2.50** Nonlinear electric circuit for Skill-Assessment Exercise 2.13

## Case Studies Antenna Control: Transfer Functions

This chapter showed that physical systems can be modeled mathematically with transfer functions. Typically, systems are composed of subsystems of different types, such as electrical, mechanical, and electromechanical.

The first case study uses our ongoing example of the antenna azimuth position control system to show how to represent each subsystem as a transfer function.

**PROBLEM:** Find the transfer function for each subsystem of the antenna azimuth position control system schematic shown on the front endpapers. Use Configuration 1.

**SOLUTION:** First, we identify the individual subsystems for which we must find transfer functions; they are summarized in Table 2.6. We proceed to find the transfer function for each subsystem.

**TABLE 2.6** Subsystems of the antenna azimuth position control system

| Subsystem | Input | Output |
|---|---|---|
| Input potentiometer | Angular rotation from user, $\theta_i(t)$ | Voltage to preamp, $v_i(t)$ |
| Preamp | Voltage from potentiometers, $v_e(t) = v_i(t) - v_0(t)$ | Voltage to power amp, $v_p(t)$ |
| Power amp | Voltage from preamp, $v_p(t)$ | Voltage to motor, $e_a(t)$ |
| Motor | Voltage from power amp, $e_a(t)$ | Angular rotation to load, $\theta_0(t)$ |
| Output potentiometer | Angular rotation from load, $\theta_0(t)$ | Voltage to preamp, $v_0(t)$ |

### Input Potentiometer; Output Potentiometer

Since the input and output potentiometers are configured in the same way, their transfer functions will be the same. We *neglect* the dynamics for the potentiometers and simply find the relationship between the output voltage and the input angular displacement. In the center position the output voltage is zero. Five turns toward either the positive 10 volts or the negative 10 volts yields a voltage change of 10 volts. Thus, the transfer function, $V_i(s)/\theta_i(s)$, for the potentiometers is found by dividing the voltage change by the angular displacement:

$$\frac{V_i(s)}{\theta_i(s)} = \frac{10}{10\pi} = \frac{1}{\pi} \tag{2.202}$$

### Preamplifier; Power Amplifier

The transfer functions of the amplifiers are given in the problem statement. Two phenomena are *neglected*. First, we *assume* that saturation is never reached. Second, the dynamics of the preamplifier are *neglected*, since its speed of response is typically much greater than that of the power amplifier. The transfer functions of both amplifiers are given in the problem statement and are the ratio of the Laplace transforms of the output voltage divided by the input voltage. Hence, for the preamplifier,

$$\frac{V_p(s)}{V_e(s)} = K \tag{2.203}$$

and for the power amplifier,

$$\frac{E_a(s)}{V_p(s)} = \frac{100}{s+100} \tag{2.204}$$

### Motor and Load

The motor and its load are next. The transfer function relating the armature displacement to the armature voltage is given in Eq. (2.153). The equivalent inertia, $J_m$, is

$$J_m = J_a + J_L\left(\frac{25}{250}\right)^2 = 0.02 + 1\frac{1}{100} = 0.03 \tag{2.205}$$

where $J_L = 1$ is the load inertia at $\theta_0$. The equivalent viscous damping, $D_m$, at the armature is

$$D_m = D_a + D_L\left(\frac{25}{250}\right)^2 = 0.01 + 1\frac{1}{100} = 0.02 \tag{2.206}$$

where $D_L$ is the load viscous damping at $\theta_0$. From the problem statement, $K_t = 0.5$ N-m/A, $K_b = 0.5$ V-s/rad, and the armature resistance $R_a = 8$ ohms. These quantities along with $J_m$ and $D_m$ are substituted into Eq. (2.153), yielding the transfer function of the motor from the armature voltage to the armature displacement, or

$$\frac{\theta_m(s)}{E_a(s)} = \frac{K_t/(R_aJ_m)}{s\left[s + \frac{1}{J_m}\left(D_m + \frac{K_tK_b}{R_a}\right)\right]} = \frac{2.083}{s(s+1.71)} \tag{2.207}$$

To complete the transfer function of the motor, we multiply by the gear ratio to arrive at the transfer function relating load displacement to armature voltage:

$$\frac{\theta_0(s)}{E_a(s)} = 0.1\frac{\theta_m(s)}{E_a(s)} = \frac{0.2083}{s(s+1.71)} \tag{2.208}$$

The results are summarized in the block diagram and table of block diagram parameters (Configuration 1) shown on the front endpapers. An animation PowerPoint presentation (PPT) demonstrating this system is available for instructors at www.wiley.com/college/nise. See *Antenna* (*Ch. 2*).

**CHALLENGE:** We now give you a problem to test your knowledge of this chapter's objectives: Referring to the antenna azimuth position control system schematic shown on the front endpapers, evaluate the transfer function of each subsystem. Use Configuration 2.

Record your results in the table of block diagram parameters shown on the front endpapers for use in subsequent chapters' case study challenges.

## Transfer Function of a Human Leg

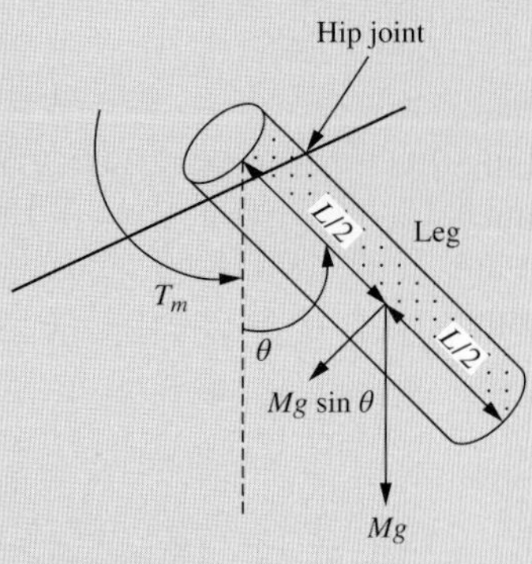


**FIGURE 2.51** Cylinder model of a human leg

In this case study we find the transfer function of a biological system. The system is a human leg, which pivots from the hip joint. In this problem, the component of weight is nonlinear, so the system requires linearization before the evaluation of the transfer function.

**PROBLEM:** The transfer function of a human leg relates the output angular rotation about the hip joint to the input torque supplied by the leg muscle. A simplified model for the leg is shown in Figure 2.51. The model *assumes* an applied muscular torque, $T_m(t)$, viscous damping, $D$, at the hip joint, and inertia, $J$, around the hip joint.[15] Also, a component of the weight of the leg, $Mg$, where $M$ is the mass of the leg and $g$ is the acceleration due to gravity, creates a nonlinear torque. If we *assume* that the leg is of uniform density, the weight can be applied at $L/2$, where $L$ is the length of the leg (*Milsum, 1966*). Do the following:

**a.** Evaluate the nonlinear torque.

**b.** Find the transfer function, $\theta(s)/T_m(s)$, for small angles of rotation, where $\theta(s)$ is the angular rotation of the leg about the hip joint.

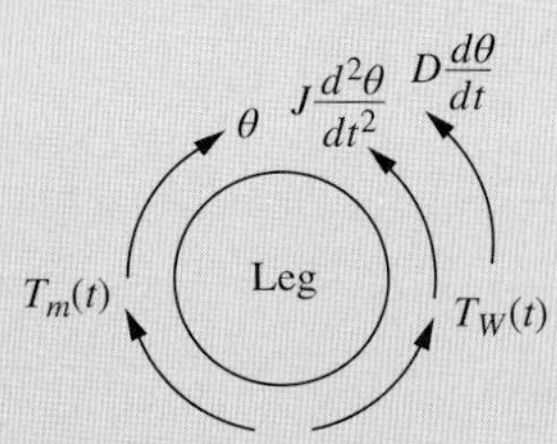


**FIGURE 2.52** Free-body diagram of leg model

**SOLUTION:** First, calculate the torque due to the weight. The total weight of the leg is $Mg$ acting vertically. The component of the weight in the direction of rotation is $Mg \sin \theta$. This force is applied at a distance $L/2$ from the hip joint. Hence the torque in the direction of rotation, $T_W(t)$, is $Mg(L/2) \sin \theta$. Next, draw a free-body diagram of the leg, showing the applied torque, $T_m(t)$, the torque due to the weight, $T_W(t)$, and the opposing torques due to inertia and viscous damping (see Figure 2.52).

Summing torques, we get

$$J\frac{d^2\theta}{dt^2} + D\frac{d\theta}{dt} + Mg\frac{L}{2}\sin\theta = T_m(t) \tag{2.209}$$

We linearize the system about the equilibrium point, $\theta = 0$, the vertical position of the leg. Using Eq. (2.182), we get

$$\sin\theta - \sin 0 = (\cos 0)\delta\theta \tag{2.210}$$

from which, $\sin\theta = \delta\theta$. Also, $Jd^2\theta/dt^2 = Jd^2\delta\theta/dt^2$ and $Dd\theta/dt = Dd\delta\theta/dt$. Hence Eq. (2.209) becomes

$$J\frac{d^2\delta\theta}{dt^2} + D\frac{d\delta\theta}{dt} + Mg\frac{L}{2}\delta\theta = T_m(t) \tag{2.211}$$

Notice that the torque due to the weight approximates a spring torque on the leg. Taking the Laplace transform with zero initial conditions yields

$$\left(Js^2 + Ds + Mg\frac{L}{2}\right)\delta\theta(s) = T_m(s) \tag{2.212}$$

[15] For emphasis, $J$ is not around the center of mass, as we previously assumed for inertia in mechanical rotation.

from which the transfer function is

$$\frac{\delta\theta(s)}{T_m(s)} = \frac{1/J}{s^2 + \frac{D}{J}s + \frac{MgL}{2J}} \tag{2.213}$$

for small excursions about the equilibrium point, $\theta = 0$.

**CHALLENGE:** We now introduce a case study challenge to test your knowledge of this chapter's objectives. Although the physical system is different from a human leg, the problem demonstrates the same principles: linearization followed by transfer function evaluation.

Given the nonlinear electrical network shown in Figure 2.53, find the transfer function relating the output nonlinear resistor voltage, $V_r(s)$, to the input source voltage, $V(s)$.

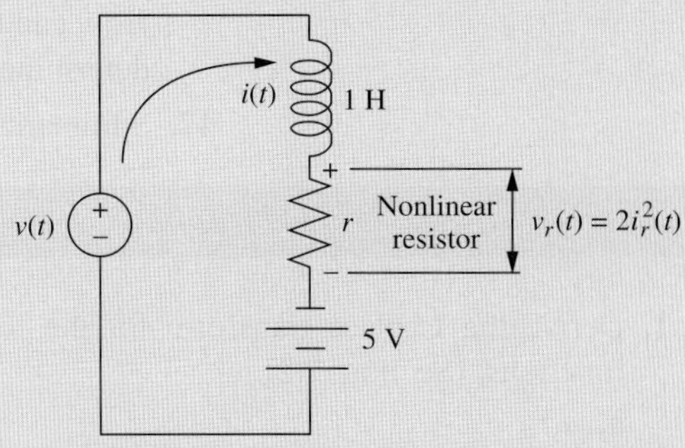


**FIGURE 2.53** Nonlinear electric circuit

## Summary

In this chapter, we discussed how to find a mathematical model, called a *transfer function*, for linear, time-invariant electrical, mechanical, and electromechanical systems. The transfer function is defined as $G(s) = C(s)/R(s)$, or the ratio of the Laplace transform of the output to the Laplace transform of the input. This relationship is algebraic and also adapts itself to modeling interconnected subsystems.

We realize that the physical world consists of more systems than we illustrated in this chapter. For example, we could apply transfer function modeling to hydraulic, pneumatic, heat, and even economic systems. Of course, we must assume these systems to be linear, or make linear approximations, in order to use this modeling technique.

Now that we have our transfer function, we can evaluate its response to a specified input. System response will be covered in Chapter 4. For those pursuing the state-space approach, we continue our discussion of modeling in Chapter 3, where we use the time domain rather than the frequency domain.

## Review Questions

1. What mathematical model permits easy interconnection of physical systems?
2. To what classification of systems can the transfer function be best applied?
3. What transformation turns the solution of differential equations into algebraic manipulations?
4. Define the transfer function.
5. What assumption is made concerning initial conditions when dealing with transfer functions?
6. What do we call the mechanical equations written in order to evaluate the transfer function?
7. If we understand the form the mechanical equations take, what step do we avoid in evaluating the transfer function?
8. Why do transfer functions for mechanical networks look identical to transfer functions for electrical networks?
9. What function do gears perform?

10. What are the component parts of the mechanical constants of a motor's transfer function?
11. The motor's transfer function relates armature displacement to armature voltage. How can the transfer function that relates load displacement and armature voltage be determined?
12. Summarize the steps taken to linearize a nonlinear system.

# Problems

1. Derive the Laplace transform for the following time functions: [Section: 2.2]
   a. $u(t)$
   b. $tu(t)$
   c. $\sin \omega t\, u(t)$
   d. $\cos \omega t\, u(t)$

2. Using the Laplace transform pairs of Table 2.1 and the Laplace transform theorems of Table 2.2, derive the Laplace transforms for the following time functions: [Section: 2.2]
   a. $e^{-at}\sin \omega t\, u(t)$
   b. $e^{-at}\cos \omega t\, u(t)$
   c. $t^3 u(t)$

3. Repeat Problem 20 in Chapter 1, using Laplace transforms. Use the following initial conditions for each part as follows: (a) $x(0) = 4$, $x'(0) = -4$; (b) $x(0)=4$, $x'(0)=1$; (c) $x(0)=2, x'(0)=3$, where $x'(0)=\frac{dx}{dt}(0)$. Assume that the forcing functions are zero prior to $t=0-$. [Section: 2.2]

4. Repeat Problem 21 in Chapter 1, using Laplace transforms. Assume that the forcing functions are zero prior to $t = 0-$. [Section: 2.2]

5. Use MATLAB and the Symbolic Math Toolbox to find the Laplace transform of the following time functions: [Section: 2.2] Symbolic Math **SM**
   a. $f(t) = 5t^2 \cos(3t + 45°)$
   b. $f(t) = 5te^{-2t} \sin(4t + 60°)$

6. Use MATLAB and the Symbolic Math Toolbox to find the inverse Laplace transform of the following frequency functions: [Section: 2.2] Symbolic Math **SM**
   a. $G(s) = \dfrac{(s^2 + 3s + 7)(s + 2)}{(s + 3)(s + 4)(s^2 + 2s + 100)}$
   b. $G(s) = \dfrac{s^3 + 4s^2 + 6s + 5}{(s + 8)(s^2 + 8s + 3)(s^2 + 5s + 7)}$

7. A system is described by the following differential equation:

$$\frac{d^3y}{dt^3} + 5\frac{d^2y}{dt^2} + 7\frac{dy}{dt} + y = \frac{d^3x}{dt^3} + 2\frac{d^2x}{dt^2} + 3\frac{dx}{dt} + 7x$$

Find the expression for the transfer function of the system, $Y(s)/X(s)$.

8. For each of the following transfer functions, write the corresponding differential equation. [Section: 2.3]
   a. $\dfrac{X(s)}{F(s)} = \dfrac{9}{s^2 + 7s + 11}$
   b. $\dfrac{X(s)}{F(s)} = \dfrac{7}{(s + 11)(s + 12)}$
   c. $\dfrac{X(s)}{F(s)} = \dfrac{s + 2}{s^3 + 10s^2 + 11s + 18}$

9. Write the differential equation for the system shown in Figure P2.1. [Section: 2.3]

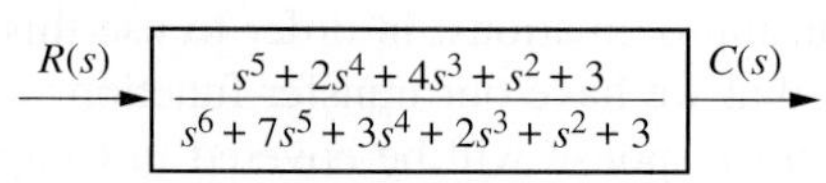


**FIGURE P2.1**

10. Write the differential equation that is mathematically equivalent to the block diagram shown in Figure P2.2. Assume that $r(t) = 3t^3$. [Section: 2.3]

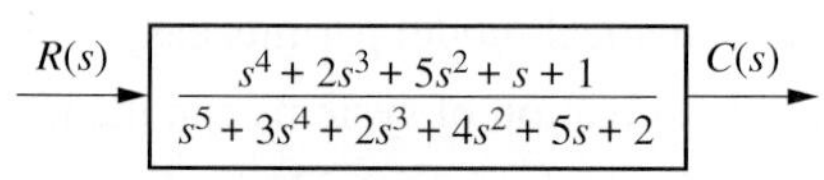


**FIGURE P2.2**

11. A system is described by the following differential equation: [Section 2.3]

$$\frac{d^2x}{dt^2} + 2\frac{dx}{dt} + 3x = 1$$

with the initial conditions $x(0) = 1$, $\dot{x}(0) = -1$. Show a block diagram of the system, giving its transfer function and all pertinent inputs and outputs. (Hint: the initial conditions will show up as added inputs to an effective system with zero initial conditions.)

12. Use MATLAB to generate the transfer function: [Section: 2.3] **MATLAB ML**

$$G(s)=\frac{5(s+15)(s+26)(s+72)}{s(s+55)(s^2+5s+30)(s+56)(s^2+27s+52)}$$

in the following ways:

**a.** the ratio of factors;

**b.** the ratio of polynomials.

13. Repeat Problem 12 for the following transfer function: [Section: 2.3] **MATLAB ML**

$$G(s)=\frac{s^4+25s^3+20s^2+15s+42}{s^5+13s^4+9s^3+37s^2+35s+50}$$

14. Use MATLAB to generate the partial-fraction expansion of the following function: [Section: 2.3] **MATLAB ML**

$$F(s)=\frac{10^4(s+10)(s+60)}{s(s+40)(s+50)(s^2+7s+100)(s^2+6s+90)}$$

15. Use MATLAB and the Symbolic Math Toolbox to input and form LTI objects in polynomial and factored form for the following frequency functions: [Section: 2.3] **Symbolic Math SM**

**a.** $G(s)=\dfrac{45(s^2+37s+74)(s^3+28s^2+32s+16)}{(s+39)(s+47)(s^2+2s+100)(s^3+27s^2+18s+15)}$

**b.** $G(s)=\dfrac{56(s+14)(s^3+49s^2+62s+53)}{(s^3+81s^2+76s+65)(s^2+88s+33)(s^2+56s+77)}$

16. Find the transfer function, $G(s)=V_o(s)/V_i(s)$, for each network shown in Figure P2.3. [Section: 2.4]

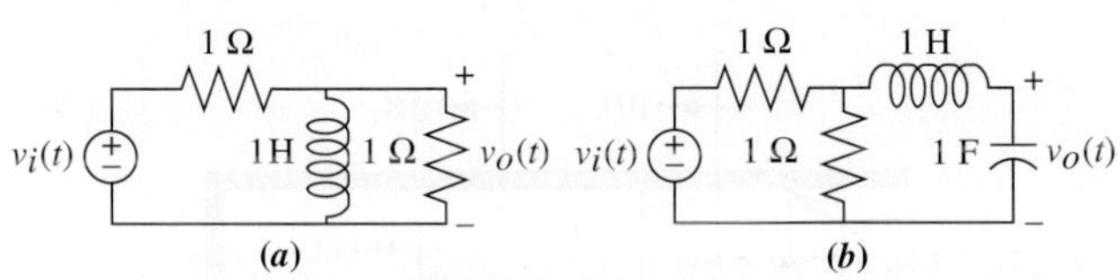

**FIGURE P2.3**

17. Find the transfer function, $G(s)=V_L(s)/V(s)$, for each network shown in Figure P2.4. [Section: 2.4]

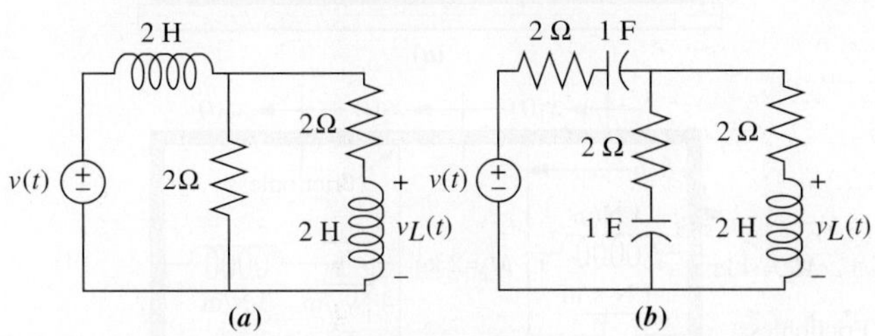

**FIGURE P2.4**

18. Find the transfer function, $G(s)=V_o(s)/V_i(s)$, for each network shown in Figure P2.5. Solve the problem using mesh analysis. [Section: 2.4]

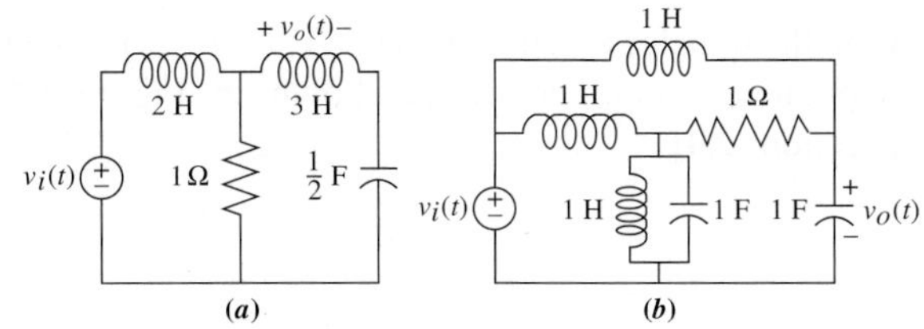

**FIGURE P2.5**

19. Repeat Problem 18 using nodal equations. [Section: 2.4]

20. **a.** Write, but do not solve, the mesh and nodal equations for the network of Figure P2.6. [Section: 2.4]

**b.** Use MATLAB, the Symbolic Math Toolbox, and the equations found in part a to solve for the transfer function, $G(s)=V_o(s)/V(s)$. Use both the mesh and nodal equations and show that either set yields the same transfer function. [Section: 2.4] **Symbolic Math SM**

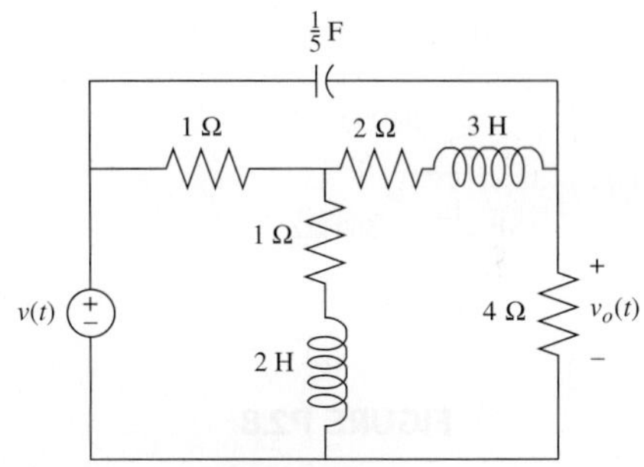

**FIGURE P2.6**

21. Find the transfer function, $G(s)=V_o(s)/V_i(s)$, for each operational amplifier circuit shown in Figure P2.7. [Section: 2.4]

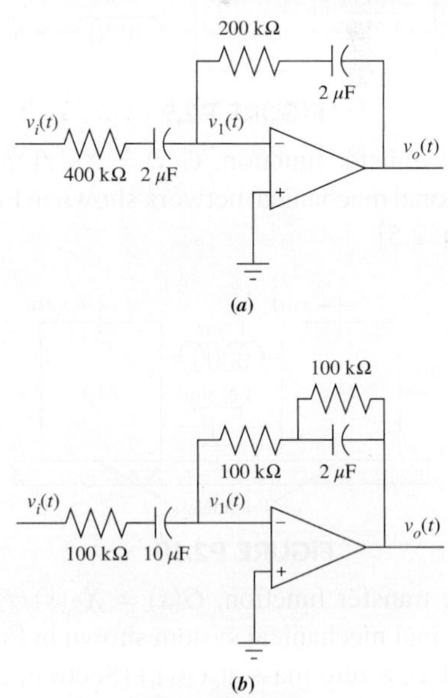

**FIGURE P2.7**

22. Find the transfer function, $G(s) = V_o(s)/V_i(s)$, for each operational amplifier circuit shown in Figure P2.8. [Section: 2.4]

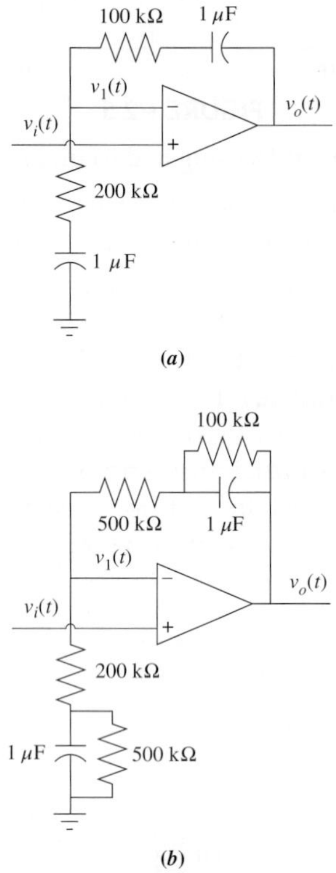


**FIGURE P2.8**

23. Find the transfer function, $G(s) = X_1(s)/F(s)$, for the translational mechanical system shown in Figure P2.9. [Section: 2.5]

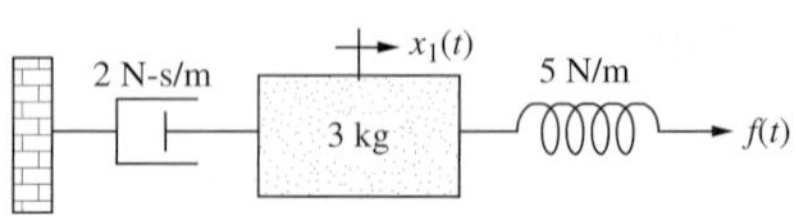


**FIGURE P2.9**

24. Find the transfer function, $G(s) = X_2(s)/F(s)$, for the translational mechanical network shown in Figure P2.10. [Section: 2.5]

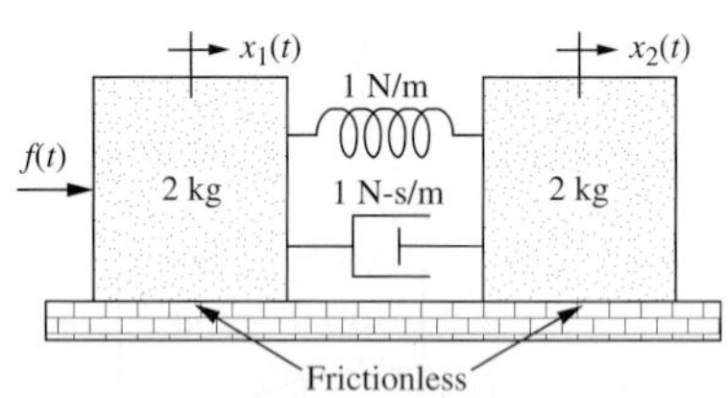


**FIGURE P2.10**

25. Find the transfer function, $G(s) = X_2(s)/F(s)$, for the translational mechanical system shown in Figure P2.11. (Hint: place a zero mass at $x_2(t)$.) [Section: 2.5]

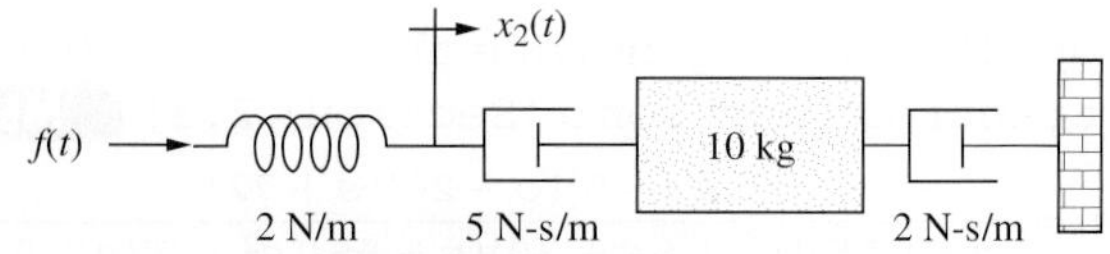


**FIGURE P2.11**

26. For the system of Figure P2.12 find the transfer function, $G(s) = X_1(s)/F(s)$. [Section: 2.5]

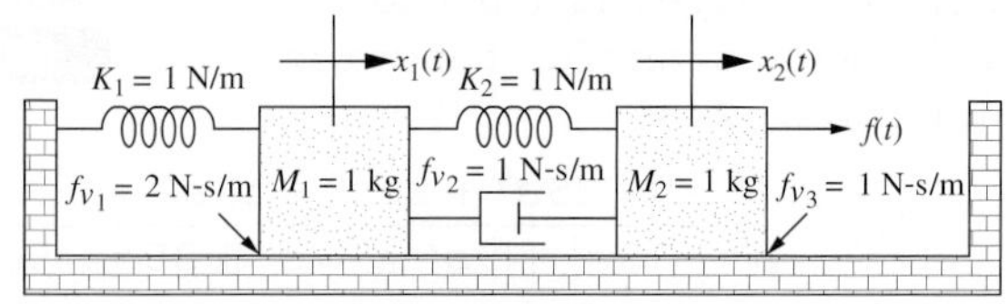


**FIGURE P2.12**

27. Find the transfer function, $G(s) = X_3(s)/F(s)$, for the translational mechanical system shown in Figure P2.13. [Section: 2.5]

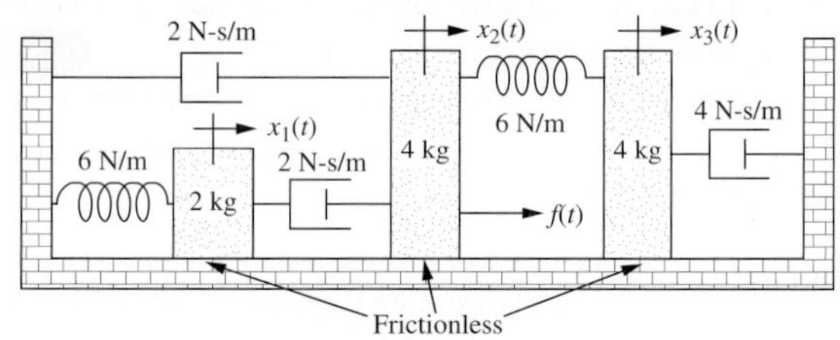


**FIGURE P2.13**

28. Find the transfer function, $X_3(s)/F(s)$, for each system shown in Figure P2.14. [Section: 2.5]

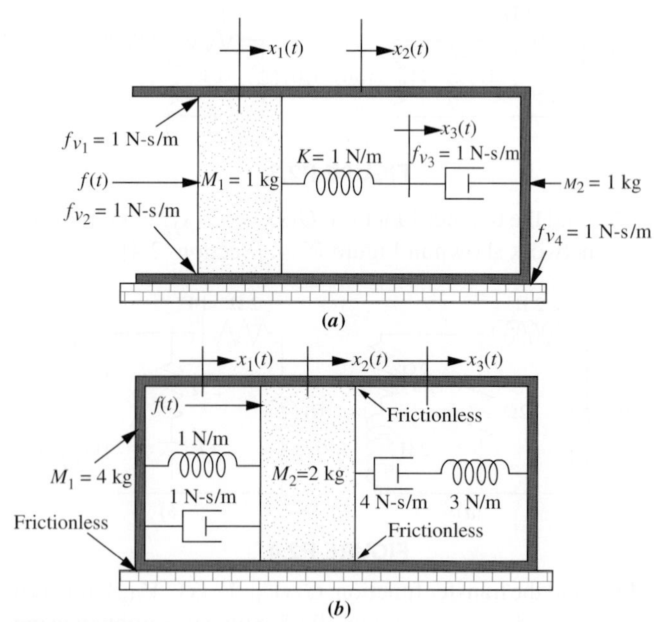


**FIGURE P2.14**

29. Write, but do not solve, the equations of motion for the translational mechanical system shown in Figure P2.15. [Section: 2.5]

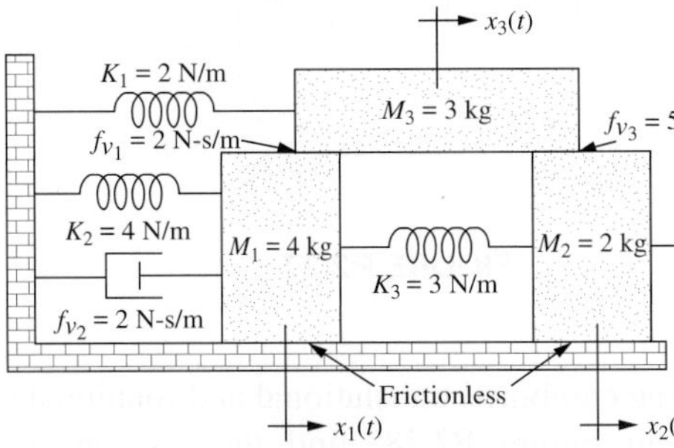


FIGURE P2.15

30. For the unexcited (no external force applied) system of Figure P2.16, do the following:

**a.** Write the differential equation that describes the system.

**b.** Assuming initial conditions $x(0) = x_0$ and $\dot{x}(0) = x_1$, write a Laplace transform expression for $X(s)$.

**c.** Find $x(t)$ by obtaining the inverse Laplace transform from the result in Part **b**.

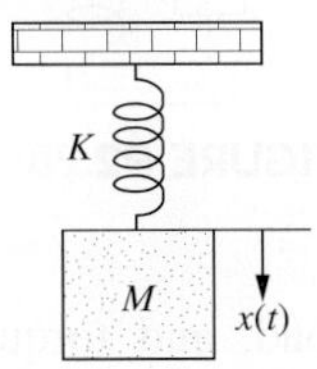


FIGURE P2.16

31. For each of the rotational mechanical systems shown in Figure P2.17, write, but do not solve, the equations of motion. [Section: 2.6]

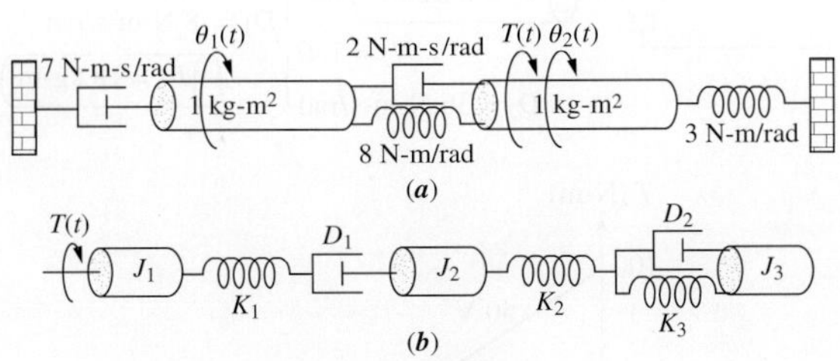


FIGURE P2.17

32. Find the transfer function, $\dfrac{\theta_1(s)}{T(s)}$, for the system shown in Figure P2.18.

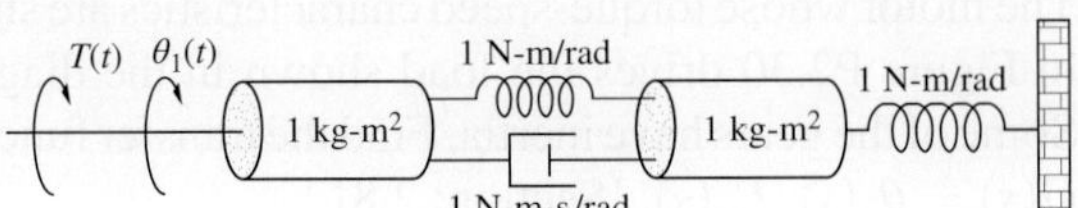


FIGURE P2.18

33. For the rotational mechanical system with gears shown in Figure P2.19, find the transfer function, $G(s) = \theta_3(s)/T(s)$. The gears have inertia and bearing friction as shown. [Section: 2.7]

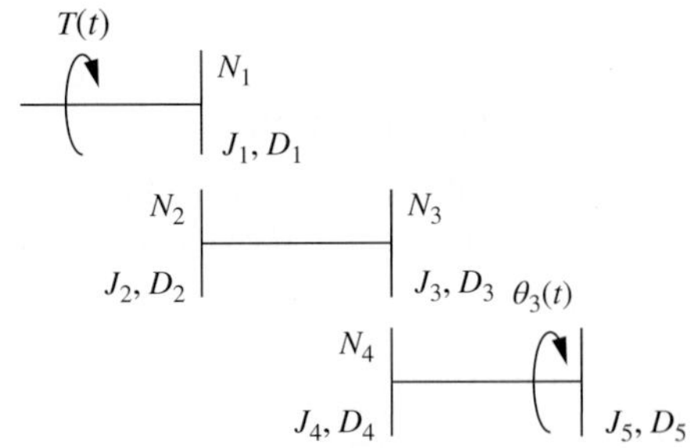


FIGURE P2.19

34. For the rotational system shown in Figure P2.20, find the transfer function, $G(s) = \theta_2(s)/T(s)$. [Section: 2.7]

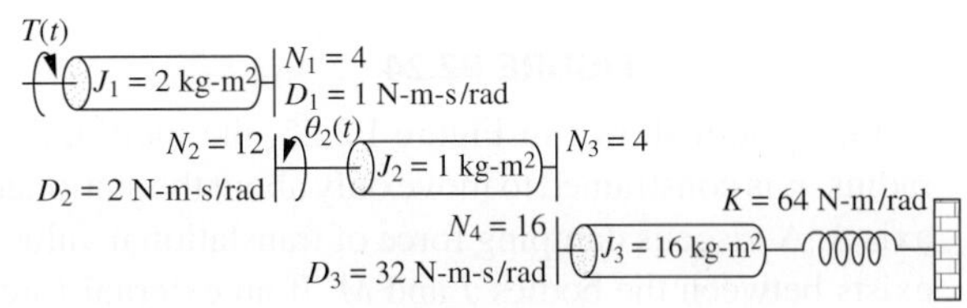


FIGURE P2.20

35. Find the transfer function, $G(s) = \theta_2(s)/T(s)$, for the rotational mechanical system shown in Figure P2.21. [Section: 2.7]

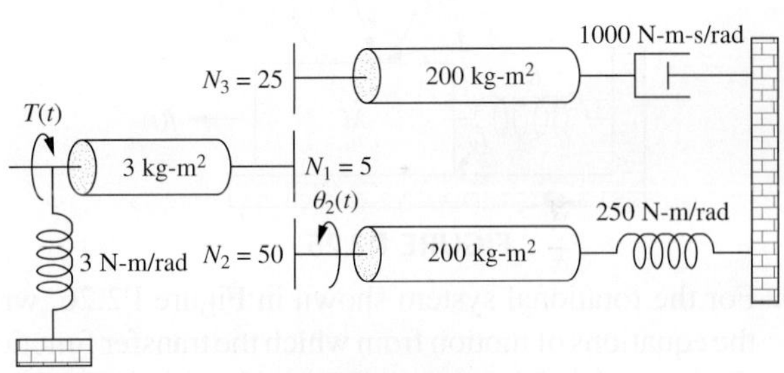


FIGURE P2.21

36. Find the transfer function, $G(s) = \theta_4(s)/T(s)$, for the rotational system shown in Figure P2.22. [Section: 2.7]

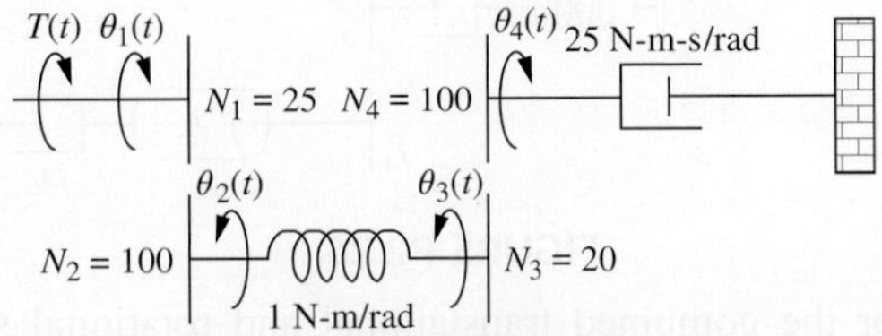


FIGURE P2.22

37. For the rotational system shown in Figure P2.23, find the transfer function, $G(s) = \theta_L(s)/T(s)$. [Section: 2.7]

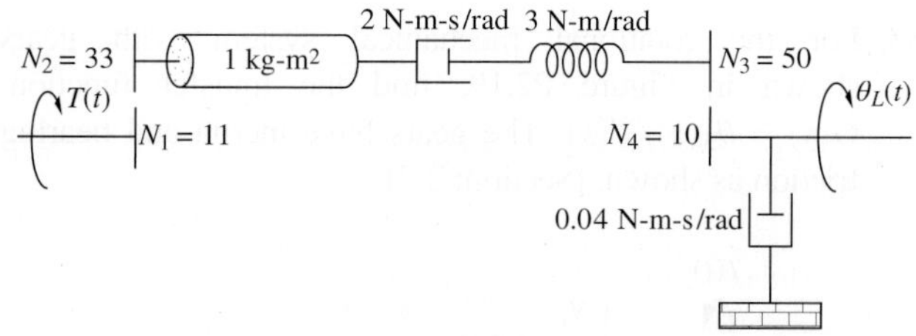


FIGURE P2.23

38. Given the rotational system shown in Figure P2.24, find the transfer function, $G(s) = \theta_6(s)/\theta_1(s)$. [Section: 2.7]

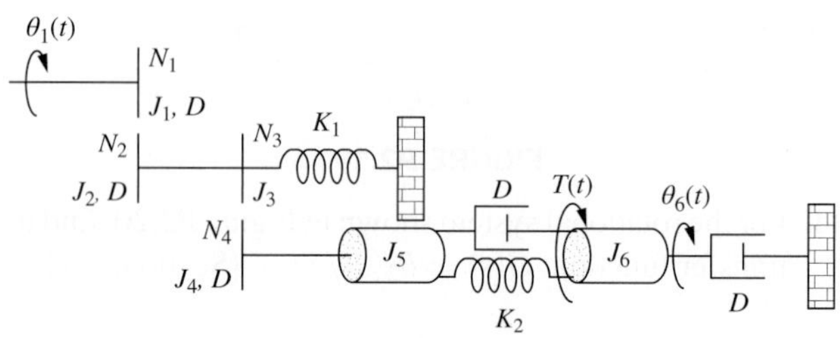


FIGURE P2.24

39. In the system shown in Figure P2.25, the inertia, $J$, of radius, $r$, is constrained to move only about the stationary axis $A$. A viscous damping force of translational value $f_v$ exists between the bodies $J$ and $M$. If an external force, $f(t)$, is applied to the mass, find the transfer function, $G(s) = \theta(s)/F(s)$. [Sections: 2.5; 2.6]

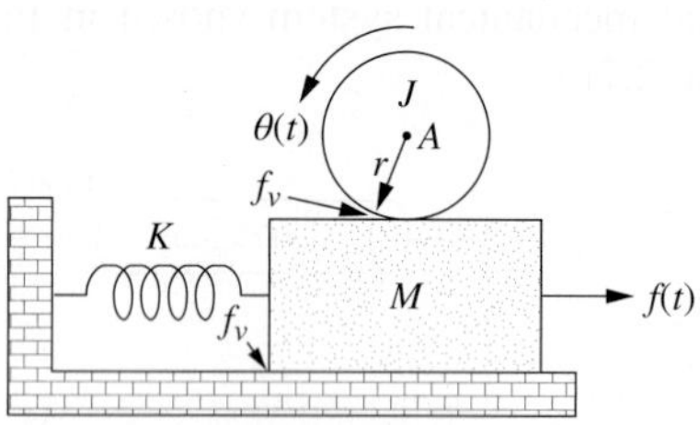


FIGURE P2.25

40. For the rotational system shown in Figure P2.26, write the equations of motion from which the transfer function, $G(s) = \theta_1(s)/T(s)$, can be found. [Section: 2.7]

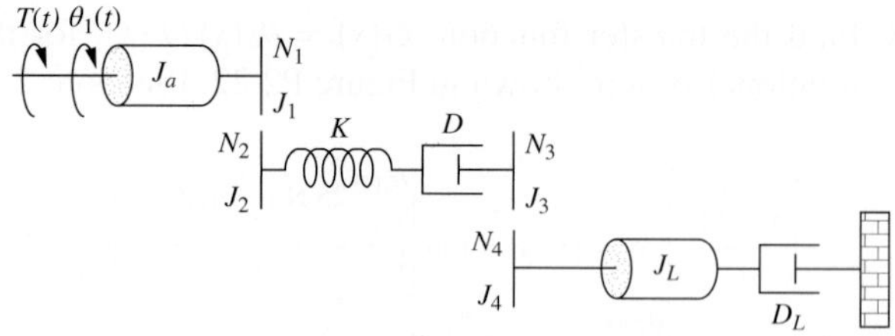


FIGURE P2.26

41. For the combined translational and rotational system shown in Figure P2.27, find the transfer function, $G(s) = X(s)/T(s)$. [Sections: 2.5; 2.6; 2.7]

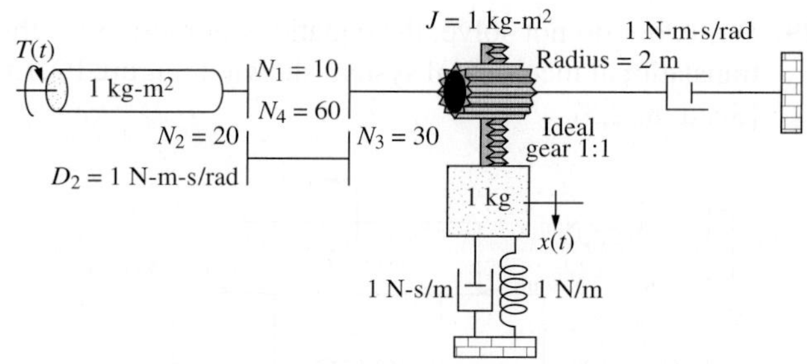


FIGURE P2.27

42. Given the combined translational and rotational system shown in Figure P2.28, find the transfer function, $G(s) = X(s)/T(s)$. [Sections: 2.5; 2.6]

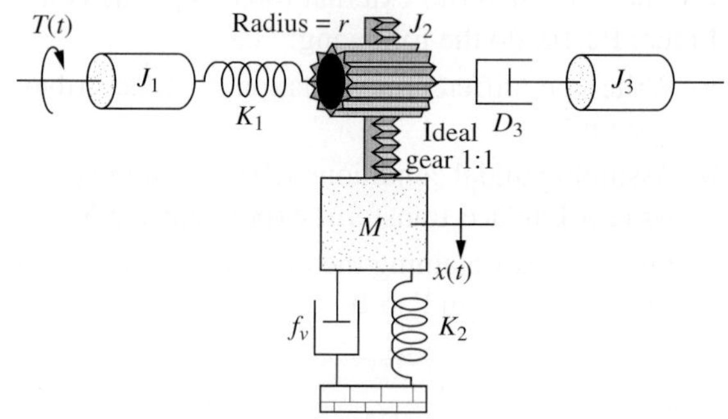


FIGURE P2.28

43. For the motor, load, and torque-speed curve shown in Figure P2.29, find the transfer function, $G(s) = \theta_L(s)/E_a(s)$. [Section: 2.8]

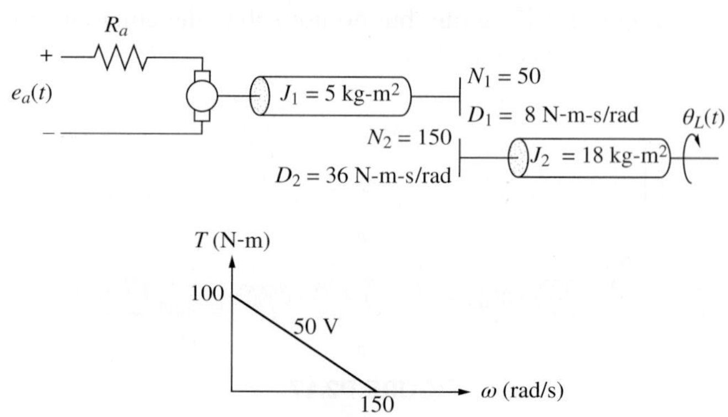


FIGURE P2.29

44. The motor whose torque-speed characteristics are shown in Figure P2.30 drives the load shown in the diagram. Some of the gears have inertia. Find the transfer function, $G(s) = \theta_2(s)/E_a(s)$. [Section: 2.8]

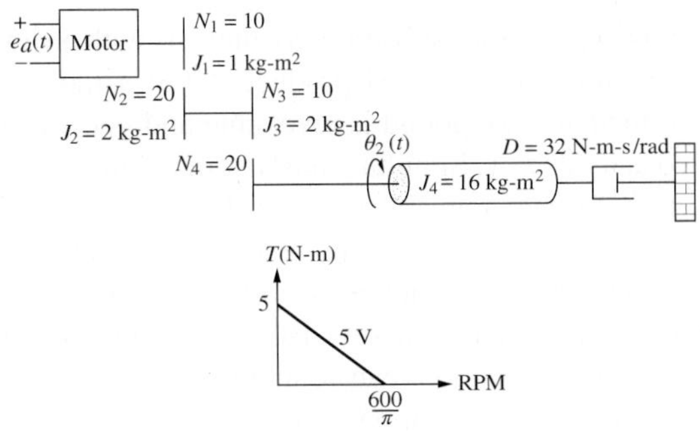


**FIGURE P2.30**

**45.** A dc motor develops 50 N-m of torque at a speed of 500 rad/s when 10 volts are applied. It stalls out at this voltage with 100 N-m of torque. If the inertia and damping of the armature are 5 kg-m$^2$ and 1 N-m-s/rad, respectively, find the transfer function, $G(s) = \theta_L(s)/E_a(s)$, of this motor if it drives an inertia load of 100 kg-m$^2$ through a gear train, as shown in Figure P2.31. [Section: 2.8]

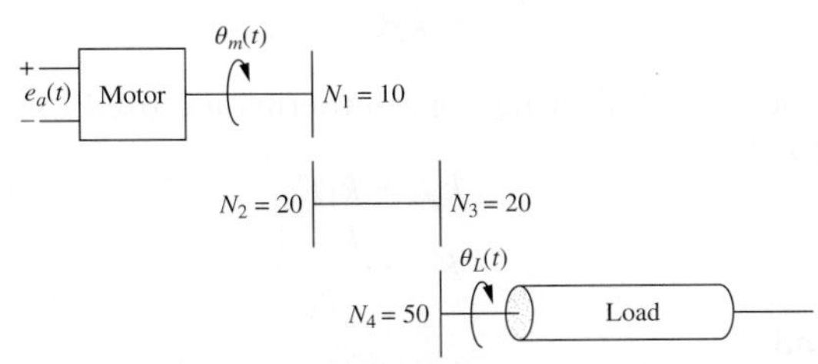


**FIGURE P2.31**

**46.** Find the transfer function, $G(s) = X(s)/E_a(s)$, for the system shown in Figure P2.32. [Sections: 2.5–2.8]

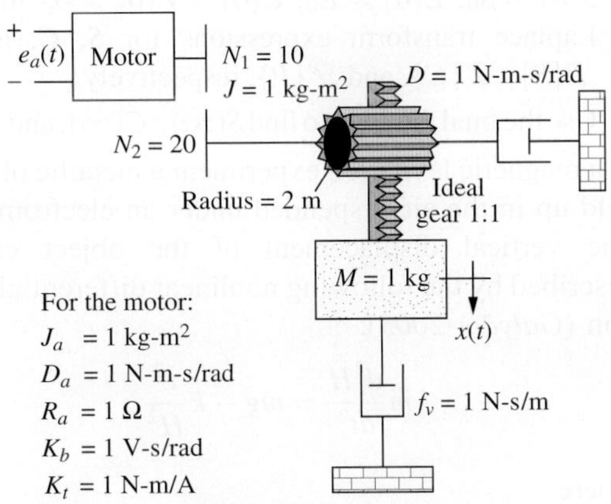


**FIGURE P2.32**

**47.** In this chapter, we derived the transfer function of a dc motor relating the angular displacement output to the armature voltage input. Often we want to control the output torque rather than the displacement. Derive the transfer function of the motor that relates output torque to input armature voltage. [Section: 2.8]

**48.** Find the series and parallel analogs for the rotational mechanical systems shown in Figure P2.17(*b*) in the problems. [Section: 2.9]

**49.** Find the series and parallel analogs for the translational mechanical system shown in Figure 2.20 in the text. [Section: 2.9]

**50.** A system's output, $c$, is related to the system's input, $r$, by the straight-line relationship, $c = 5r + 7$. Is the system linear? [Section: 2.10]

**51.** Consider the differential equation

$$\frac{d^2x}{dt^2} + 5\frac{dx}{dt} + 3x = f(x)$$

where $f(x)$ is the input and is a function of the output, $x$. If $f(x) = \sin x$, linearize the differential equation for small excursions. [Section: 2.10]

**a.** $x = 0$

**b.** $x = \pi$

**52.** Consider the differential equation

$$\frac{d^3x}{dt^3} + 10\frac{d^2x}{dt^2} + 31\frac{dx}{dt} + 30x = f(x)$$

where $f(x)$ is the input and is a function of the output, $x$. If $f(x) = e^{-x}$, linearize the differential equation for $x$ near 0. [Section: 2.10]

**53.** Many systems are *piecewise* linear. That is, over a *large* range of variable values, the system can be described linearly. A system with amplifier saturation is one such example. Given the differential equation

$$\frac{d^2x}{dt^2} + 15\frac{dx}{dt} + 50x = f(x)$$

assume that $f(x)$ is as shown in Figure P2.33. Write the differential equation for each of the following ranges of $x$: [Section: 2.10]

**a.** $-\infty < x < -2$

**b.** $-2 < x < 2$

**c.** $2 < x < \infty$

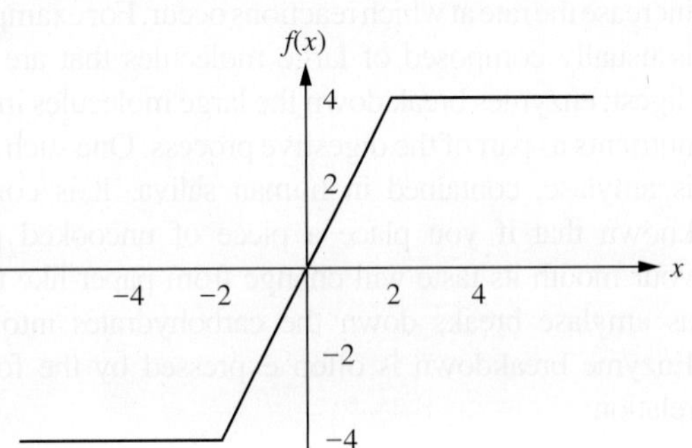


**FIGURE P2.33**

**54.** For the translational mechanical system with a nonlinear spring shown in Figure P2.34, find the transfer function, $G(s) = X(s)/F(s)$, for small excursions around $f(t) = 1$. The spring is defined by $x_s(t) = 1 - e^{-f_s(t)}$, where $x_s(t)$ is the spring displacement and $f_s(t)$ is the spring force. [Section: 2.10]

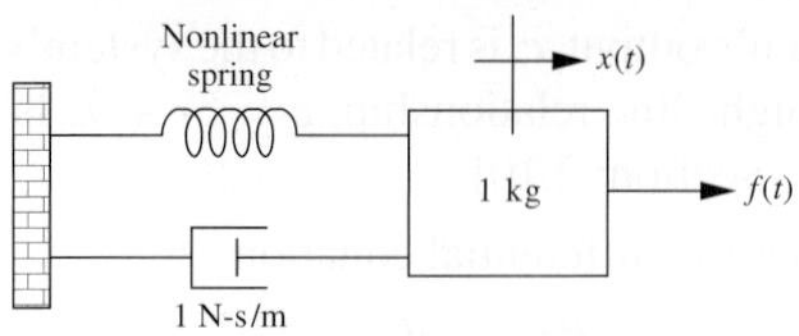


**FIGURE P2.34**

**55.** Consider the restaurant plate dispenser shown in Figure P2.35, which consists of a vertical stack of dishes supported by a compressed spring. As each plate is removed, the reduced weight on the dispenser causes the remaining plates to rise. Assume that the mass of the system minus the top plate is $M$, the viscous friction between the piston and the sides of the cylinder is $f_v$, the spring constant is $K$, and the weight of a single plate is $W_D$. Find the transfer function, $Y(s)/F(s)$, where $F(s)$ is the step reduction in force felt when the top plate is removed, and $Y(s)$ is the vertical displacement of the dispenser in an upward direction.

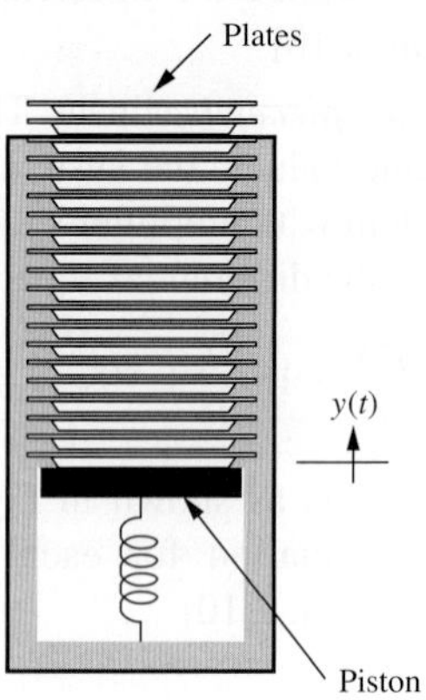


**FIGURE P2.35** Plate dispenser

**56.** Enzymes are large proteins that biological systems use to increase the rate at which reactions occur. For example, food is usually composed of large molecules that are hard to digest; enzymes break down the large molecules into small nutrients as part of the digestive process. One such enzyme is amylase, contained in human saliva. It is commonly known that if you place a piece of uncooked pasta in your mouth its taste will change from paper-like to sweet as amylase breaks down the carbohydrates into sugars. Enzyme breakdown is often expressed by the following relation:

$$S + E \underset{k_1}{\overset{k_s}{\rightleftarrows}} C \overset{k_1}{\rightarrow} P$$

In this expression a substrate ($S$) interacts with an enzyme ($E$) to form a combined product ($C$) at a rate $k_1$. The intermediate compound is reversible and gets disassociated at a rate $k_{-1}$. Simultaneously some of the compound is transformed into the final product ($P$) at a rate $k_2$. The kinetics describing this reaction are known as the Michaelis-Menten equations and consist of four nonlinear differential equations. However, under some conditions these equations can be simplified. Let $E_0$ and $S_0$ be the initial concentrations of enzyme and substrate, respectively. It is generally accepted that under some energetic conditions or when the enzyme concentration is very big ($E_0 \gg S_0$), the kinetics for this reaction are given by

$$\frac{dS}{dt} = k_\psi(\tilde{K}_s C - S)$$

$$\frac{dC}{dt} = k_\psi(S - \tilde{K}_M C)$$

$$\frac{dP}{dt} = k_2 C$$

where the following constant terms are used (*Schnell, 2004*):

$$k_\psi = k_1 E_0$$

$$\tilde{K}_s = \frac{k-1}{k_\psi}$$

and

$$\tilde{K}_M = \tilde{K}_s + \frac{k_2}{k_\psi}$$

**a.** Assuming the initial conditions for the reaction are $S(0) = S_0$, $E(0) = E_0$, $C(0) = P(0) = 0$, find the Laplace transform expressions for $S$, $C$, and $P$: $\mathscr{L}\{S\}$, $\mathscr{L}\{C\}$, and $\mathscr{L}\{P\}$, respectively.

**b.** Use the final theorem to find $S(\infty)$, $C(\infty)$, and $P(\infty)$.

**57.** In a magnetic levitation experiment a metallic object is held up in the air suspended under an electromagnet. The vertical displacement of the object can be described by the following nonlinear differential equation (*Galvão, 2003*):

$$m\frac{d^2H}{dt^2} = mg - k\frac{I^2}{H^2}$$

where

$m$ = mass of the metallic object
$g$ = gravity acceleration constant
$k$ = a positive constant
$H$ = distance between the electromagnet and the metallic object (output signal)
$I$ = electromagnet's current (input signal)

**a.** Show that a system's equilibrium will be achieved when $H_0 = I_0\sqrt{\frac{k}{mg}}$.

**b.** Linearize the equation about the equilibrium point found in Part **a** and show that the resulting transfer function obtained from the linearized differential equation can be expressed as

$$\frac{\delta H(s)}{\delta I(s)} = -\frac{a}{s^2 - b^2}$$

with $a > 0$. Hint: to perform the linearization, define $\delta H = H(t) - H_0$ and $\delta I = I(t) - I_0$; substitute into the original equation. This will give

$$m\frac{d^2(H_0 + \delta H)}{dt^2} = mg - k\frac{(I_0 + \delta I)^2}{(H_0 + \delta H)^2} = \gamma$$

Now get a first-order Taylor's series approximation on the right-hand side of the equation. Namely, calculate

$$m\frac{d^2\delta H}{dt^2} = \left.\frac{\partial \gamma}{\partial \delta H}\right|_{\delta H=0,\ \delta I=0} \delta H + \left.\frac{\partial \gamma}{\partial \delta I}\right|_{\delta H=0,\ \delta I=0} \delta I$$

**58.** Humans are able to stand on two legs through a complex feedback system that includes several sensory inputs—equilibrium and visual along with muscle actuation. In order to gain a better understanding of the workings of the postural feedback mechanism, an individual is asked to stand on a platform to which sensors are attached at the base. Vibration actuators are attached with straps to the individual's calves. As the vibration actuators are stimulated, the individual sways and movements are recorded. It was hypothesized that the human postural dynamics are analogous to those of a cart with a balancing standing pole attached (inverted pendulum). In that case, the dynamics can be described by the following two equations:

$$J\frac{d^2\theta}{dt^2} = mgl\sin\theta(t) + T_{\text{bal}} + T_d(t)$$

$$T_{\text{bal}}(t) = -mgl\sin\theta(t) + kJ\theta(t) - \eta J\dot{\theta}(t) - \rho J\int_0^t \theta(t)dt$$

where $m$ is the individual's mass; $l$ is the height of the individual's center of gravity; $g$ is the gravitational constant; $J$ is the individual's equivalent moment of inertia; $\eta$, $\rho$, and $k$ are constants given by the body's postural control system; $\theta(t)$ is the individual's angle with respect to a vertical line; $T_{\text{bal}}(t)$ is the torque generated by the body muscles to maintain balance; and $T_d(t)$ is the external torque input disturbance. Find the transfer function $\frac{\Theta(s)}{T_d(s)}$ (*Johansson, 1988*).

**59.** A muscle hanging from a beam is shown in Figure P2.36(*a*) (*Lessard, 2009*). The $\alpha$-motor neuron can be used to electrically stimulate the muscle to contract and pull the mass, $m$, which under static conditions causes the muscle to stretch. An equivalent mechanical system to this setup is shown in Figure P2.36(*b*). The force $F_{iso}$ will be exerted when the muscle contracts. Find an expression for the displacement $X_1(s)$ in terms of $F_1(s)$ and $F_{iso}(s)$.

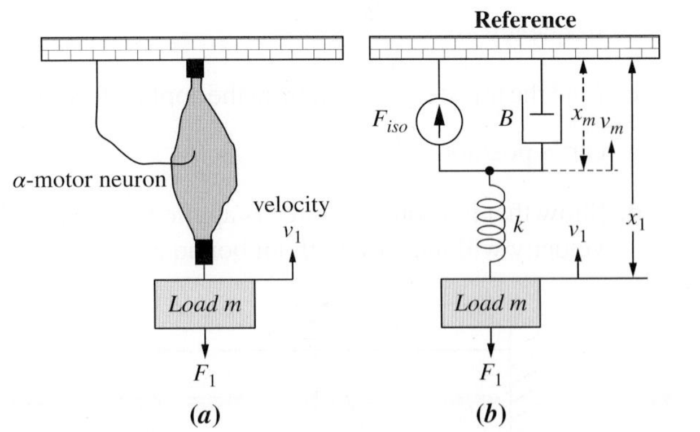


**FIGURE P2.36** **a.** Motor neuron stimulating a muscle;[16] **b.** equivalent circuit[17]

**60.** A photovoltaic system is used to capture solar energy to be converted to electrical energy. A control system is used to pivot the solar platform to track the sun's movements in order to maximize the captured energy. The system consists of a motor and load similar to that discussed in Section 2.8. A model has been proposed (*Agee, 2012*) that is different from the model developed in the chapter in the following ways: (1) the motor inductance was not neglected and (2) the load, in addition to having inertia and damping, has a spring. Find the transfer function, $\theta_m(s)/E_a(s)$, for this augmented system assuming all load impedances have already been reflected to the motor shaft.

**61.** Figure P2.37 shows a crane hoisting a load. Although the actual system's model is highly nonlinear, if the rope is considered to be stiff with a fixed length $L$, the system can be modeled using the following equations:

$$m_L\ddot{x}_{La} = m_L g\phi$$
$$m_T\ddot{x}_T = f_T - m_L g\phi$$
$$x_{La} = x_T - x_L$$
$$x_L = L\phi$$

where $m_L$ is the mass of the load, $m_T$ is the mass of the cart, $x_T$ and $x_L$ are displacements as defined in the figure, $\phi$ is the

[16] Lessard, C. D. *Basic Feedback Controls in Biomedicine*, Morgan & Claypool, San Rafael, CA, 2009. Figure 2.8, p. 12.

[17] Lessard, C. D. *Basic Feedback Controls in Biomedicine*, Morgan & Claypool, San Rafael, CA, 2009 Figure 2.9, p. 13.

rope angle with respect to the vertical, and $f_T$ is the force applied to the cart (*Marttinen, 1990*).

**a.** Obtain the transfer function from cart velocity to rope angle $\dfrac{\Phi(s)}{V_T(s)}$.

**b.** Assume that the cart is driven at a constant velocity $V_0$ and obtain an expression for the resulting $\phi(t)$. Show that under this condition, the load will sway with a frequency $\omega_0 = \sqrt{\dfrac{g}{L}}$.

**c.** Find the transfer function from the applied force to the cart's position, $\dfrac{X_T(s)}{F_T(s)}$.

**d.** Show that if a constant force is applied to the cart, its velocity will increase without bound as $t \to \infty$.

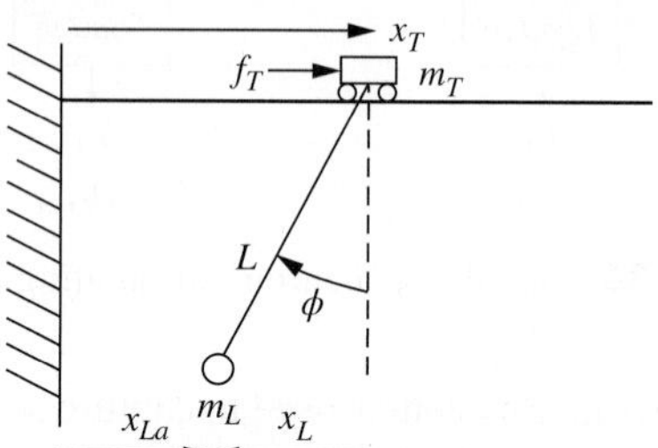


**FIGURE P2.37**[18]

**62.** In 1978, Malthus developed a model for human growth population that is also commonly used to model bacterial growth as follows. Let *N*(*t*) be the population density observed at time *t*. Let *K* be the rate of reproduction per unit time. Neglecting population deaths, the population density at a time $t + \Delta t$ (with small $\Delta t$) is given by

$$N(t + \Delta t) \approx N(t) + KN(t)\Delta t$$

which also can be written as

$$\frac{N(t + \Delta t) - N(t)}{\Delta t} = KN(t)$$

Since *N*(*t*) can be considered to be a very large number, letting $\Delta t \to 0$ gives the following differential equation (*Edelstein-Keshet, 2005*):

$$\frac{dN(t)}{dt} = KN(t)$$

**a.** Assuming an initial population $N(0) = N_0$, solve the differential equation by finding *N*(*t*).

**b.** Find the time at which the population is double the initial population.

**63.** In order to design an underwater vehicle that has the characteristics of both a long-range transit vehicle (torpedo-like) and a highly maneuverable low-speed vehicle (boxlike), researchers have developed a thruster that mimics that of squid jet locomotion (*Krieg, 2008*). It has been demonstrated there that the average normalized thrust due to a command step input, $U(s) = \dfrac{T_{ref}}{s}$ is given by:

$$T(t) = T_{ref}(1 - e^{-\lambda t}) + a\sin(2\pi f t)$$

where $T_{ref}$ is the reference or desired thrust, $\lambda$ is the system's damping constant, *a* is the amplitude of the oscillation caused by the pumping action of the actuator, *f* is the actuator frequency, and *T*(*t*) is the average resulting normalized thrust. Find the thruster's transfer function $\dfrac{T(s)}{U(s)}$. Show all steps.

**64.** In a paint mixing plant, two tanks supply fluids to a mixing cistern. The height, *h*, of the fluid in the cistern is dependent upon the difference between the input mass flow rate, *q*, and the output flow rate, $q_e$. A nonlinear differential equation describing this dependency is given by (*Schiop, 2010*)

$$\frac{dh}{dt} + \frac{A_e}{A}\sqrt{2gh} = \frac{q}{\rho A}$$

where $A$ = cross-sectional area of the cistern, $A_e$ = cross-sectional area of the exit pipe, $g$ = acceleration due to gravity, and $\rho$ = liquid density.

**a.** Linearize the nonlinear equation about the equilibrium point ($h_0$, $q_0$) and find the transfer function relating the output cistern fluid level, *H*(*s*), to the input mass flow rate, *Q*(*s*).

**b.** The color of the liquid in the cistern can be kept constant by adjusting the input flow rate, *q*, assuming the input flow's color is specifically controlled. Assuming an average height, $h_{av}$, of the liquid in the cistern, the following equation relates the net flow of color to the cistern to the color in the cistern.

$$e_1 q - eq_e = \frac{d}{dt}(\rho A e h_{av})$$

where $e_1$ = fractional part of flow representing color into the cistern, and $e$ = fractional part of the cistern

[18] Marttinen A., Virkkunen J., Salminen R.T. Control Study with Pilot Crane. *IEEE Transactions on Education*, Vol. 33, No.3, August 1990. Fig. 2. p. 300. IEEE Transactions on Education by Institute of Electrical and Electronics Engineers; IEEE Education Group; IEEE Education Society. Reproduced with permission of Institute of Electrical and Electronics Engineers, in the format Republish in a book via Copyright Clearance Center.

representing color in the cistern. Assume that the flow out of the cistern is constant and use the relationship, $q_e = \rho A_e \sqrt{2gh_{av}}$, along with the given equation above to find the transfer function, $E(s)/Q(s)$, that relates the color in the cistern to the input flow rate.

## PROGRESSIVE ANALYSIS AND DESIGN PROBLEMS

**65. Control of HIV/AIDS.** HIV inflicts its damage by infecting healthy CD4 + T cells (a type of white blood cell) that are necessary to fight infection. As the virus embeds in a T cell and the immune system produces more of these cells to fight the infection, the virus propagates in an opportunistic fashion. As we now develop a simple HIV model, refer to Figure P2.38. Normally T cells are produced at a rate $s$ and die at a rate $d$. The HIV virus is present in the bloodstream in the infected individual. These viruses in the bloodstream, called *free viruses*, infect healthy T cells at a rate $\beta$. Also, the viruses reproduce through the T cell multiplication process or otherwise at a rate $k$. Free viruses die at a rate $c$. Infected T cells die at a rate $\mu$.

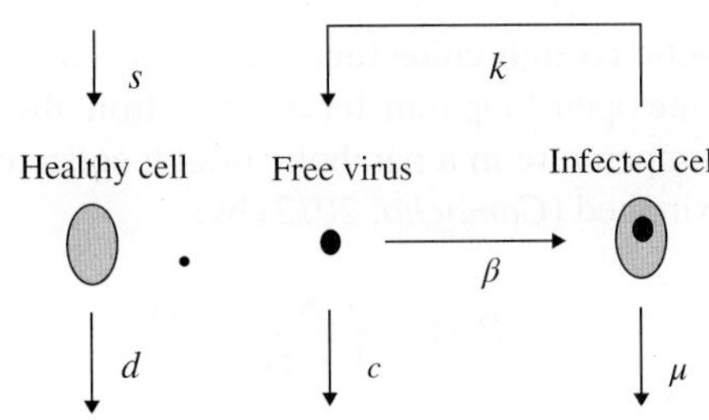


**FIGURE P2.38**[19]

A simple mathematical model that illustrates these interactions is given by the following equations (*Craig, 2004*):

$$\frac{dT}{dt} = s - dT - \beta T v$$

$$\frac{dT^*}{dt} = \beta T v - \mu T^*$$

$$\frac{dv}{dt} = kT^* - cv$$

[19] Craig, I. K., Xia, X., and Venter, J. W. Introducing HIV/AIDS Education Into the Electrical Engineering Curriculum at the University of Pretoria. *IEEE Transactions on Education*, vol. 47, no. 1, February 2004, pp. 65–73. Fig. 1, p. 66. IEEE transactions on education by Institute of Electrical and Electronics Engineers; IEEE Education Group; IEEE Education Society. Reproduced with permission of Institute of Electrical and Electronics Engineers, in the format Republish in a book via Copyright Clearance Center.

where

$T$ = number of healthy T cells
$T^*$ = number of infected T cells
$v$ = number of free viruses

**a.** The system is nonlinear; thus linearization is necessary to find transfer functions as you will do in subsequent chapters. The nonlinear nature of this model can be seen from the above equations. Determine which of these equations are linear, which are nonlinear, and explain why.

**b.** The system has two equilibrium points. Show that these are given by

$$(T_0, T_0^*, v_0) = \left(\frac{s}{d}, 0, 0\right)$$

and

$$(T_0, T_0^*, v_0) = \left(\frac{c\mu}{\beta k}, \frac{s}{\mu} - \frac{cd}{\beta k}, \frac{sk}{c\mu} - \frac{d}{\beta}\right)$$

**66. Hybrid vehicle.** Problem 23 in Chapter 1 discusses the cruise control of serial, parallel, and split-power hybrid electric vehicles (HEVs). The functional block diagrams developed for these HEVs indicated that the speed of a vehicle depends upon the balance between the motive forces (developed by the gasoline engine and/or the electric motor) and running resistive forces. The resistive forces include the aerodynamic drag, rolling resistance, and climbing resistance. Figure P2.39 illustrates the running resistances for a car moving uphill (*Bosch, 2007*).

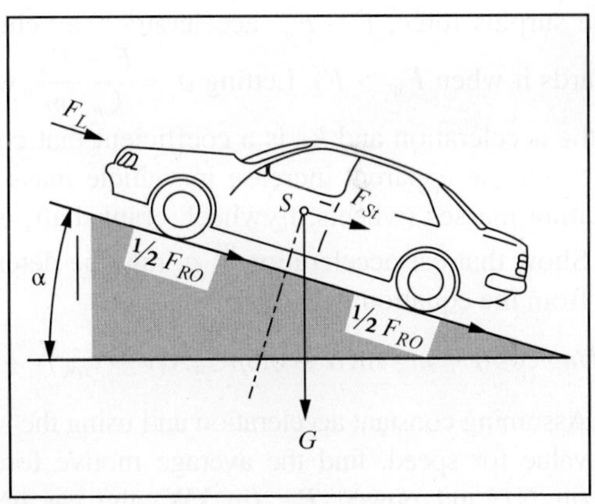


**FIGURE P2.39** Running resistances[20]

The total running resistance, $F_w$, is calculated as $F_w = F_{Ro} + F_L + F_{St}$, where $F_{Ro}$ is the rolling resistance, $F_L$ is the aerodynamic drag, and $F_{St}$ is the climbing resistance. The aerodynamic drag is

[20] Robert Bosch GmbH, *Bosch Automotive Handbook*, 7th ed. John Wiley & Sons Ltd. UK, 2007. P. 430. Figure at bottom left.

proportional to the square of the sum of car velocity, $v$, and the head-wind velocity, $v_{hw}$, or $v + v_{hw}$. The other two resistances are functions of car weight, $G$, and the gradient of the road (given by the gradient angle, $\alpha$), as seen from the following equations:

$$F_{Ro} = fG\cos\alpha = fmg\cos\alpha$$

where

$$\begin{aligned} f &= \text{coefficient of rolling resistance} \\ m &= \text{car mass, in kg} \\ g &= \text{gravitational acceleration, in m/s}^2 \\ F_L &= 0.5\rho C_w A(v + v_{\text{hw}})^2. \end{aligned}$$

and

$$\begin{aligned} \rho &= \text{air density, in kg/m}^3 \\ C_w &= \text{coefficient of aerodynamic drag} \\ A &= \text{largest cross-section of the car, in kg/m}^2 \\ F_{St} &= G\sin\alpha = mg\sin\alpha. \end{aligned}$$

The motive force, $F$, available at the drive wheels is:

$$F = \frac{Ti_{tot}}{r}\eta_{tot} = \frac{P\eta_{tot}}{v}$$

where

$$\begin{aligned} T &= \text{motive torque} \\ P &= \text{motive power} \\ i_{tot} &= \text{total transmission ratio} \\ r &= \text{tire radius} \\ \eta_{tot} &= \text{total drive-train efficiency.} \end{aligned}$$

The surplus force, $F - F_w$, accelerates the vehicle (or retards it when $F_w > F$). Letting $a = \dfrac{F - F_w}{k_m \cdot m}$, where $a$ is the acceleration and $k_m$ is a coefficient that compensates for the apparent increase in vehicle mass due to rotating masses (wheels, flywheel, crankshaft, etc.):

**a.** Show that car acceleration,[21] $a$, may be determined from the equation:

$$F = fmg\cos\alpha + mg\sin\alpha + 0.5\rho C_w A(v + v_{hw})^2 + k_m ma$$

**b.** Assuming constant acceleration and using the average value for speed, find the average motive force, $F_{av}$ (in N), and power, $P_{av}$ (in kW) the car needs to accelerate from 40 to 60 km/h in 4 seconds on a level road, ($\alpha = 0°$), under windless conditions, where $v_{hw} = 0$. You are given the following parameters: m = 1590 kg, A = 2 m$^2$, $f = 0.011$, $\rho = 1.2\text{ kg/m}^3$, $C_w = 0.3$, $\eta_{tot} = 0.9$, $k_m = 1.2$. Furthermore, calculate the additional power, $P_{add}$, the car needs after reaching 60 km/h to maintain its speed while climbing a hill with a gradient $\alpha = 5°$.

**c.** The equation derived in Part **a** describes the nonlinear car motion dynamics where $F(t)$ is the input to the system, and $v(t)$ the resulting output. Given that the aerodynamic drag is proportional to $v^2$ under windless conditions, linearize the resulting equation of motion around an average speed, $v_o = 50$ km/h, when the car travels on a level road,[22] where $\alpha = 0°$. (Hint: Expand $v^2 - v_0^2$ in a truncated Taylor series). Write that equation of motion and represent it with a block diagram in which the block $G_v$ represents the vehicle dynamics. The output of that block is the car speed, $v(t)$, and the input is the excess motive force, $F_e(t)$, defined as: $F_e = F - F_{St} - F_{Ro} + F_o$, where $F_o$ is the constant component of the linearized aerodynamic drag.

**d.** Use the equation in Part **c** to find the vehicle transfer function: $G_v(s) = V(s)/F_e(s)$.

**67. Parabolic trough collector.** In a significant number of cases, the open-loop transfer function from fluid flow to fluid temperature in a parabolic trough collector can be approximated (*Camacho, 2012*) by:

$$P(s) = \frac{K}{1 + \tau s)}e^{-sT}$$

**a.** Write an analytic expression for the unit step response of the open-loop system assuming that $h(t)$ represents the output temperature and $q(t)$ the input fluid flow.

**b.** Make a sketch of the unit step response of the open-loop system. Indicate on your figure the time delay, the settling time, the initial and final values of the response, and the value of the response when $t = \tau + T$.

**c.** Call the output temperature $h(t)$ and the input fluid flow $q(t)$. Find the differential equation that represents the open-loop system.

[21] Other quantities, such as top speed, climbing ability, etc., may also be calculated by manipulation from that equation.

[22] Note that on a level road the climbing resistance, $F_{st} = 0$, since $\sin\alpha = \sin 0° = 0$.

# Cyber Exploration Laboratory

## Experiment 2.1

**Objectives** To learn to use MATLAB to (1) generate polynomials, (2) manipulate polynomials, (3) generate transfer functions, (4) manipulate transfer functions, and (5) perform partial-fraction expansions.

**Minimum Required Software Packages** MATLAB and the Control System Toolbox

### Prelab

1. Calculate the following by hand or with a calculator:
   a. The roots of $P_1 = s^6 + 7s^5 + 2s^4 + 9s^3 + 10s^2 + 12s + 15$
   b. The roots of $P_2 = s^6 + 9s^5 + 8s^4 + 9s^3 + 12s^2 + 15s + 20$
   c. $P_3 = P_1 + P_2$; $P_4 = P_1 - P_2$; $P_5 = P_1P_2$
2. Calculate by hand or with a calculator the polynomial
$$P_6 = (s+7)(s+8)(s+3)(s+5)(s+9)(s+10)$$
3. Calculate by hand or with a calculator the following transfer functions:
   a. $G_1(s) = \dfrac{20(s+2)(s+3)(s+6)(s+8)}{s(s+7)(s+9)(s+10)(s+15)}$,
   represented as a numerator polynomial divided by a denominator polynomial.
   b. $G_2(s) = \dfrac{s^4 + 17s^3 + 99s^2 + 223s + 140}{s^5 + 32s^4 + 363s^3 + 2092s^2 + 5052s + 4320}$,
   expressed as factors in the numerator divided by factors in the denominator, similar to the form of $G_1(s)$ in Prelab 3**a**.
   c. $G_3(s) = G_1(s) + G_2(s)$; $G_4(s) = G_1(s) - G_2(s)$; $G_5(s) = G_1(s)G_2(s)$
   expressed as factors divided by factors and expressed as polynomials divided by polynomials.
4. Calculate by hand or with a calculator the partial-fraction expansion of the following transfer functions:
   a. $G_6 = \dfrac{5(s+2)}{s(s^2 + 8s + 15)}$
   b. $G_7 = \dfrac{5(s+2)}{s(s^2 + 6s + 9)}$
   c. $G_8 = \dfrac{5(s+2)}{s(s^2 + 6s + 34)}$

### Lab

1. Use MATLAB to find $P_3$, $P_4$, and $P_5$ in Prelab 1.
2. Use only one MATLAB command to find $P_6$ in Prelab 2.
3. Use only two MATLAB commands to find $G_1(s)$ in Prelab 3**a** represented as a polynomial divided by a polynomial.
4. Use only two MATLAB commands to find $G_2(s)$ expressed as factors in the numerator divided by factors in the denominator.

5. Using various combinations of $G_1(s)$ and $G_2(s)$, find $G_3(s)$, $G_4(s)$, and $G_5(s)$. Various combinations implies mixing and matching $G_1(s)$ and $G_2(s)$ expressed as factors and polynomials. For example, in finding $G_3(s)$, $G_1(s)$ can be expressed in factored form and $G_2(s)$ can be expressed in polynomial form. Another combination is $G_1(s)$ and $G_2(s)$ both expressed as polynomials. Still another combination is $G_1(s)$ and $G_2(s)$ both expressed in factored form.
6. Use MATLAB to evaluate the partial fraction expansions shown in Prelab 4.

### Postlab

1. Discuss your findings for Lab 5. What can you conclude?
2. Discuss the use of MATLAB to manipulate transfer functions and polynomials. Discuss any shortcomings in using MATLAB to evaluate partial fraction expansions.

## Experiment 2.2

**Objectives** To learn to use MATLAB and the Symbolic Math Toolbox to (1) find Laplace transforms for time functions, (2) find time functions from Laplace transforms, (3) create LTI transfer functions from symbolic transfer functions, and (4) perform solutions of symbolic simultaneous equations.

**Minimum Required Software Packages** MATLAB, the Symbolic Math Toolbox, and the Control System Toolbox

### Prelab

1. Using a hand calculation, find the Laplace transform of:

$$f(t) = 0.0075 - 0.00034e^{-2.5t}\cos(22t) + 0.087e^{-2.5t}\sin(22t) - 0.0072e^{-8t}$$

2. Using a hand calculation, find the inverse Laplace transform of

$$F(s) = \frac{2(s+3)(s+5)(s+7)}{s(s+8)(s^2+10s+100)}$$

3. Use a hand calculation to solve the circuit for the Laplace transforms of the loop currents shown in Figure P2.40.

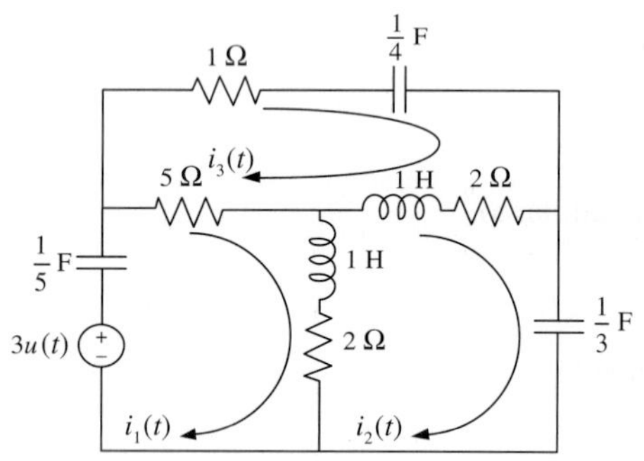


FIGURE P2.40

### Lab

1. Use MATLAB and the Symbolic Math Toolbox to
   a. Generate symbolically the time function $f(t)$ shown in Prelab 1.
   b. Generate symbolically $F(s)$ shown in Prelab 2. Obtain your result symbolically in both factored and polynomial forms.
   c. Find the Laplace transform of $f(t)$ shown in Prelab 1.
   d. Find the inverse Laplace transform of $F(s)$ shown in Prelab 2.
   e. Generate an LTI transfer function for your symbolic representation of $F(s)$ in Prelab 2 in both polynomial form and factored form. Start with the $F(s)$ you generated symbolically.
   f. Solve for the Laplace transforms of the loop currents in Prelab 3.

### Postlab

1. Discuss the advantages and disadvantages between the Symbolic Math Toolbox and MATLAB alone to convert a transfer function from factored form to polynomial form and vice versa.
2. Discuss the advantages and disadvantages of using the Symbolic Math Toolbox to generate LTI transfer functions.

3. Discuss the advantages of using the Symbolic Math Toolbox to solve simultaneous equations of the type generated by the electrical network in Prelab 3. Is it possible to solve the equations via MATLAB alone? Explain.
4. Discuss any other observations you had using the Symbolic Math Toolbox.

## Experiment 2.3

**Objectives** To learn to use LabVIEW to generate and manipulate polynomials and transfer functions.

**Minimum Required Software Packages** LabVIEW and the LabVIEW Control Design and Simulation Module.

### Prelab

1. Study Appendix D, Sections D.1 through Section D.4, Example D.1.
2. Perform by hand the calculations stated in Prelab 1 of Experiment 2.1.
3. Find by a hand calculation the polynomial whose roots are: $-7, -8, -3, -5, -9$, and $-10$.
4. Perform by hand a partial-fraction expansion of $G(s) = \dfrac{5s+10}{s^3+8s^2+15s}$.
5. Find by a hand calculation $G_1(s)+G_2(s)$, $G_1(s)-G_2(s)$, and $G_1(s)G_2(s)$, where $G_1(s) = \dfrac{1}{s^2+s+2}$ and $G_2(s) = \dfrac{s+1}{s^2+4s+3}$.

### Lab

1. Open the LabVIEW functions palette and select the **Mathematics/Polynomial** palette.
2. Generate the polynomials enumerated in Prelab 1a and lb of Experiment 2.1.
3. Generate the polynomial operations stated in Prelab 1c of Experiment 2.1.
4. Generate a polynomial whose roots are those stated in Prelab 3 of this experiment.
5. Generate the partial-fraction expansion of the transfer function given in Prelab 4 of this experiment.
6. Using the **Control Design and Simulation/Control Design/Model Construction** palette, construct the two transfer functions enumerated in Prelab 5.
7. Using the **Control Design and Simulation/Control Design/Model Interconnection** palette, display the results of the mathematical operations enumerated in Prelab 5 of this experiment.

### Postlab

1. Compare the polynomial operations obtained in Lab 3 to those obtained in Prelab 2.
2. Compare the polynomial displayed in Lab 4 with that calculated in Prelab 3.
3. Compare the partial-fraction expansion obtained in Lab 5 with that calculated in Prelab 4.
4. Compare the results of the mathematical operations found in Lab 7 to those calculated in Prelab 5.

# Hardware Interface Laboratory

***Note: Before performing experiments in this section, please study Appendix D—LabVIEW Tutorial, including the section discussing myDAQ.*** **When an experiment indicates a provided file, the file is obtained at www.wiley.com/college/nise.**

## Experiment 2.4 Programming with LabVIEW Part 1

### Objectives

1. To learn how to program LabVIEW, Part 1
2. To learn how to write basic LabVIEW programs and understand LabVIEW flow

**Material Required** Computer with LabVIEW Installed

**Prelab** Go to the website http://www.learnni.com/getting-started/. Complete modules 0-7.

### Lab

1. Write a LabVIEW program that executes an equivalent of the following C-like code, where x is an input and y is an output (Formula Nodes are not allowed):

```
if(abs(x)< 0.1)
        y = 1;
else
        if(x>=0)
             y = 0;
        else
             y = 2;
```

   Run your program for the following inputs: x= 0.05, −0.05, 1, −1.

2. Write a LabVIEW program that receives three colors representing a resistor's value and returns the numeric resistor value in ohms. Your interface should be similar to the one shown in Figure P2.41. The third band should include silver and gold colors.

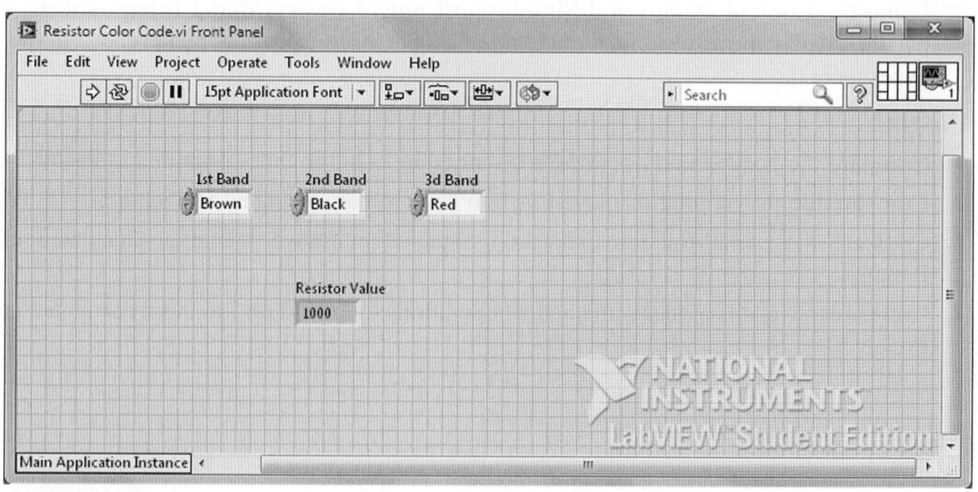


**FIGURE P2.41**

Run your program at least for the following inputs:

Red Red Black
Brown Black Orange
Orange White Gold

## Experiment 2.5 Programming with LabVIEW Part 2

### Objectives

1. To learn how to program LabVIEW, Part 2
2. To learn how to use loops, do basic math inside loops, and graph numerical information using LabVIEW

**Material Required** Computer with LabVIEW Installed

**Prelab** Go to the website http://www.learnni.com/getting-started/. Complete modules 8-10.

### Lab

1. It is well known that $1 + x + x^2 + x^3 + \cdots = \dfrac{1}{1-x}$ when $|x| < 1$.

   Write a LabVIEW program that takes as an input a value for $x$, and a number of iterations. The program will use a loop to calculate the sum of the geometric series for the specified number of operations. It will also calculate the closed-form expression for the series. The program will display the two results, and will also display the absolute error of the difference.

   Demonstrate your program with $x = 0.5$ and 3, 10, and 200 iterations.

2. Write a LabVIEW program that generates a 50% duty-cycle square-wave signal between 0 and $X_{\max}$ volts, where $X_{\max} < 10$ V, with a nonzero variable frequency. The amplitude and frequency will be inputs. Display the waveform on a waveform chart, which will be an output. In this part you are not allowed to use the LabVIEW-provided function generation blocks.

   Demonstrate this program for amplitudes of 1, 5, and 10 V and for 1 Hz and 5 Hz.

## Experiment 2.6 MyDAQ Programming

**Objectives** To become familiar with the data acquisition and signal generation capabilities of myDAQ

**Material Required** Computer with LabVIEW installed, and myDAQ

### Files Provided at www.wiley.com/college/nise

Battery Meter.ctl

### Prelab

Go to the website https://decibel.ni.com/content/docs/DOC-11624. Go over Unit 4—DAQ: Lesson 1. Then go over the measuring voltage tutorial in http://zone.ni.com/devzone/cda/epd/p/id/6436.

### Lab

1. Write a battery-tester program using LabVIEW and myDAQ as an acquisition device. The battery tester should work for three nominal values of batteries: 1.5 V, 6 V, and 9 V.

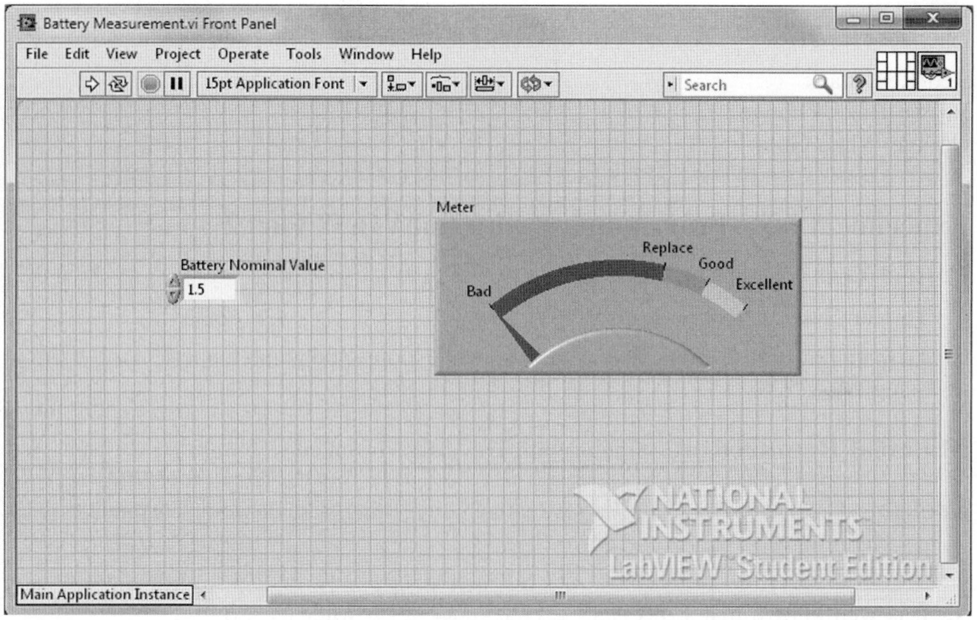


**FIGURE P2.42**

The batteries are considered dead for voltage values 20% or under the nominal. Between 20% and the nominal value the batteries are in a warning area, and for values above the nominal the batteries are OK. Your interface should be similar to the one shown Figure P2.42. A custom control accepting inputs from 0 to 120 has been created under the name **Battery Meter.ctl**.

2. Use the LabVIEW program that you wrote in Experiment 2.5 to generate a 50% duty-cycle square-wave signal. Output your signal through one of the myDAQ's analog channels and read the signal using the myDAQ oscilloscope function (available from myDAQ file: *NI ELVISmx Instrument Launcher*) to verify the generated signal. Print two examples using the scope's automatic measurements.

## Bibliography


Agee J. T., and Jimoh, A. A. Flat controller design for hardware-cost reduction in polar-axis photovoltaic systems. *Solar Energy*, vol. 86, pp. 452–462, 2012.

Aggarwal, J. K. *Notes on Nonlinear Systems.* Van Nostrand Reinhold, New York, 1972.

Bosch, R. GmbH, *Bosch Automotive Handbook*, 7th ed. John Wiley & Sons Ltd., UK, 2007.

Camacho, E. F., Berenguel, M., Rubio, F. R., and Martinez, D. *Control of Solar Energy Systems.* Springer-Verlag, London, 2012.

Cannon, R. H., Jr., *Dynamics of Physical Systems.* McGraw-Hill, New York, 1967.

Carlson, L. E., and Griggs, G. E. *Aluminum Catenary System Quarterly Report.* Technical Report Contract Number DOT-FR-9154, U.S. Department of Transportation, 1980.

Chignola, R., and Foroni, R. I. Estimating the Growth Kinetics of Experimental Tumors from as Few as Two Determinations of Tumor Size: Implications for Clinical Oncology. *IEEE Transactions on Biomedical Engineering*, vol. 52, no. 5, May 2005, pp. 808–815.

Cochin, I. *Analysis and Design of Dynamic Systems.* Harper and Row, New York, 1980.

Cook, P. A. *Nonlinear Dynamical Systems.* Prentice Hall, United Kingdom, 1986.

Craig, I. K., Xia, X., and Venter, J. W. Introducing HIV/AIDS Education into the Electrical Engineering Curriculum at the University of Pretoria. *IEEE Transactions on Education*, vol. 47, no. 1, February 2004, pp. 65–73.

Davis, S. A., and Ledgerwood, B. K. *Electromechanical Components for Servomechanisms*. McGraw-Hill, New York, 1961.

Doebelin, E. O. *Measurement Systems Application and Design*. McGraw-Hill, New York, 1983.

Dorf, R. *Introduction to Electric Circuits*, 2d ed. Wiley, New York, 1993.

D'Souza, A. *Design of Control Systems*. Prentice Hall, Englewood Cliffs, NJ, 1988.

Edelstein-Keshet, L. *Mathematical Models in Biology*. Society for Industrial and Applied Mathematics, Philadelphia, PA, 2005.

Elkins, J. A. *A Method for Predicting the Dynamic Response of a Pantograph Running at Constant Speed under a Finite Length of Overhead Equipment*. Technical Report TN DA36, British Railways, 1976.

Franklin, G. F., Powell, J. D., and Emami-Naeini, A. *Feedback Control of Dynamic Systems*. Addison-Wesley, Reading, MA, 1986.

Graovac D., and Katić V. Online Control of Current-Source-Type Active Rectifier Using Transfer Function Approach. *IEEE Transactions on Industrial Electronics*, vol. 48, no. 3, June 2001, pp. 526–535.

Hsu, J. C., and Meyer, A. U. *Modern Control Principles and Applications*. McGraw-Hill, New York, 1968.

Johansson, R., Magnusson, M., and Akesson, M. Identification of Human Postural Dynamics. *IEEE Transactions on Biomedical Engineering*, vol. 35, no. 10, October 1988, pp. 858–869.

Kailath, T. *Linear Systems*. Prentice Hall, Englewood Cliffs, NJ, 1980.

Kermurjian, A. From the Moon Rover to the Mars Rover. *The Planetary Report*, July/August 1990, pp. 4–11.

Krieg, M., and Mohseni, K. Developing a Transient Model for Squid Inspired Thrusters, and Incorporation into Underwater Robot Control Design. *2008 IEEE/RSJ Int. Conf. on Intelligent Robots and Systems, France*, September 2008.

Kuo, F. F. *Network Analysis and Synthesis*. Wiley, New York, 1966.

Lago, G., and Benningfield, L. M. *Control System Theory*. Ronald Press, New York, 1962.

Lessard, C. D. *Basic Feedback Controls in Biomedicine*. Morgan & Claypool, San Rafael, CA, 2009.

Mablekos, V E. *Electric Machine Theory for Power Engineers*. Harper & Row, Cambridge, MA, 1980.

Marttinen, A., Virkkunen, J., and Salminen, R. T. Control Study with Pilot Crane. *IEEE Transactions on Education*, vol. 33, no. 3, August 1990, pp. 298–305.

Minorsky, N. *Theory of Nonlinear Control Systems*. McGraw-Hill, New York, 1969.

Nilsson, J. W., and Riedel, S. A. *Electric Circuits*, 5th ed. Addison-Wesley, Reading, MA, 1996.

Ogata, K. *Modern Control Engineering*, 2d ed. Prentice Hall, Englewood Cliffs, NJ, 1990.

Raven, F. H. *Automatic Control Engineering*, 5th ed. McGraw-Hill, New York, 1995.

Schiop L., Gaiceanu M. Mathematical Modeling of Color Mixing Process and PLC Control Implementation by Using Human Machine Interface. *IEEE International Symposium on Electrical and Electronics Engineering*, 2010.

Van Valkenburg, M. E. *Network Analysis*. Prentice Hall, Englewood Cliffs, NJ, 1974.

Vidyasagar, M. *Nonlinear Systems Analysis*. Prentice Hall, Englewood Cliffs, NJ, 1978.

# 3

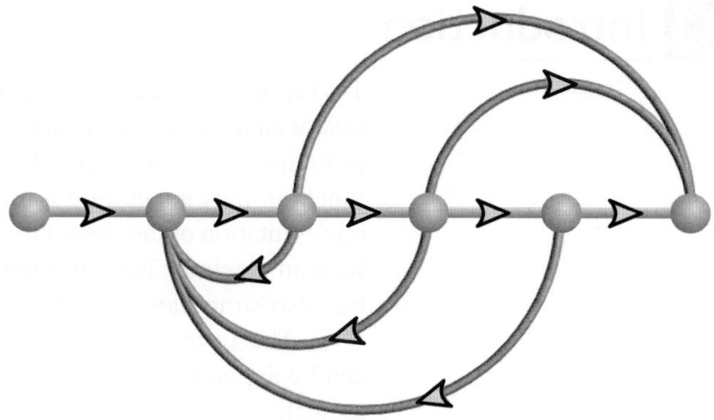

# Modeling in the Time Domain

This chapter covers only state-space methods.

State Space
**SS**

## Chapter Learning Outcomes

After completing this chapter, the student will be able to:

- Find a mathematical model, called a *state-space* representation, for a linear, time-invariant system (Sections 3.1–3.3)
- Model electrical and mechanical systems in state space (Section 3.4)
- Convert a transfer function to state space (Section 3.5)
- Convert a state-space representation to a transfer function (Section 3.6)
- Linearize a state-space representation (Section 3.7)

## ▶ Case Study Learning Outcomes

You will be able to demonstrate your knowledge of the chapter objectives with case studies as follows:

- Given the antenna azimuth position control system shown on the front endpapers, you will be able to find the state-space representation of each subsystem.
- Given a description of the way a pharmaceutical drug flows through a human being, you will be able to find the state-space representation to determine drug concentrations in specified compartmentalized blocks of the process and of the human body. You will also be able to apply the same concepts to an aquifer to find water level.

# 3.1 Introduction

Two approaches are available for the analysis and design of feedback control systems. The first, which we began to study in Chapter 2, is known as the *classical*, or *frequency-domain*, technique. This approach is based on converting a system's differential equation to a transfer function, thus generating a mathematical model of the system that *algebraically* relates a representation of the output to a representation of the input. Replacing a differential equation with an algebraic equation not only simplifies the representation of individual subsystems but also simplifies modeling interconnected subsystems.

The primary disadvantage of the classical approach is its limited applicability: It can be applied only to linear, time-invariant systems or systems that can be approximated as such.

A major advantage of frequency-domain techniques is that they rapidly provide stability and transient response information. Thus, we can immediately see the effects of varying system parameters until an acceptable design is met.

With the arrival of space exploration, requirements for control systems increased in scope. Modeling systems by using linear, time-invariant differential equations and subsequent transfer functions became inadequate. The *state-space* approach (also referred to as the *modern*, or *time-domain*, approach) is a unified method for modeling, analyzing, and designing a wide range of systems. For example, the state-space approach can be used to represent nonlinear systems that have backlash, saturation, and dead zone. Also, it can handle, conveniently, systems with nonzero initial conditions. Time-varying systems, (for example, missiles with varying fuel levels or lift in an aircraft flying through a wide range of altitudes) can be represented in state space. Many systems do not have just a single input and a single output. Multiple-input, multiple-output systems (such as a vehicle with input direction and input velocity yielding an output direction and an output velocity) can be compactly represented in state space with a model similar in form and complexity to that used for single-input, single-output systems. The time-domain approach can be used to represent systems with a digital computer in the loop or to model systems for digital simulation. With a simulated system, system response can be obtained for changes in system parameters—an important design tool. The state-space approach is also attractive because of the availability of numerous state-space software packages for the personal computer.

The time-domain approach can also be used for the same class of systems modeled by the classical approach. This alternate model gives the control systems designer another perspective from which to create a design. While the state-space approach can be applied to a wide range of systems, it is not as intuitive as the classical approach. The designer has to engage in several calculations before the physical interpretation of the model is apparent, whereas in classical control a few quick calculations or a graphic presentation of data rapidly yields the physical interpretation.

In this book, the coverage of state-space techniques is to be regarded as an introduction to the subject, a springboard to advanced studies, and an alternate approach to frequency-domain techniques. We will limit the state-space approach to linear, time-invariant systems or systems that can be linearized by the methods of Chapter 2. The study of other classes of systems is beyond the scope of this book. Since state-space analysis and design rely on matrices and matrix operations, you may want to review this topic in Appendix before continuing.

# 3.2 Some Observations

We proceed now to establish the state-space approach as an alternate method for representing physical systems. This section sets the stage for the formal definition of the state-space representation by making some observations about systems and their variables. In the discussion that follows, some of the development has been placed in footnotes to avoid clouding the main issues with an excess of equations and to ensure that the concept is clear. Although we use two electrical networks to illustrate the concepts, we could just as easily have used a mechanical or any other physical system.

We now demonstrate that for a system with many variables, such as inductor voltage, resistor voltage, and capacitor charge, we need to use differential equations only to solve for a selected subset of system variables because all other remaining system variables can be evaluated algebraically from the variables in the subset. Our examples take the following approach:

1. We select a particular *subset* of all possible system variables and call the variables in this subset *state variables*.
2. For an $n$th-order system, we write *n simultaneous, first-order differential equations* in terms of the state variables. We call this system of simultaneous differential equations *state equations*.
3. If we know the initial condition of all of the state variables at $t_0$ as well as the system input for $t \geq t_0$, we can solve the simultaneous differential equations for the state variables for $t \geq t_0$.
4. We *algebraically* combine the state variables with the system's input and find all of the other system variables for $t \geq t_0$. We call this algebraic equation the *output equation*.
5. We consider the state equations and the output equations a viable representation of the system. We call this representation of the system a *state-space representation*.

Let us now follow these steps through an example. Consider the *RL* network shown in Figure 3.1 with an initial current of $i(0)$.

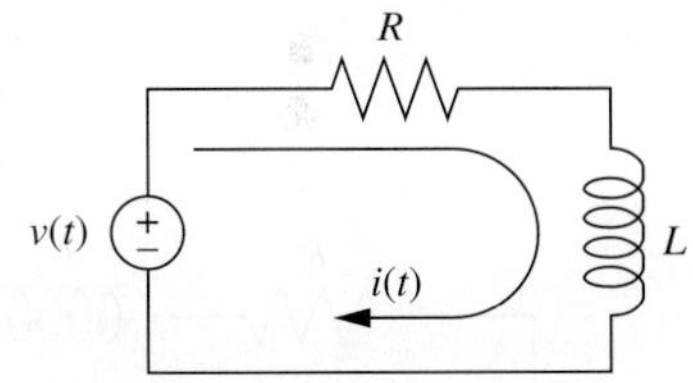

**FIGURE 3.1** *RL* network

1. We select the current, $i(t)$, for which we will write and solve a differential equation using Laplace transforms.
2. We write the loop equation,

$$L\frac{di}{dt} + Ri = v(t) \tag{3.1}$$

3. Taking the Laplace transform, using Table 2.2, Item 7, and including the initial conditions, yields

$$L[sI(s) - i(0)] + RI(s) = V(s) \tag{3.2}$$

Assuming the input, $v(t)$, to be a unit step, $u(t)$, whose Laplace transform is $V(s) = 1/s$, we solve for $I(s)$ and get

$$I(s) = \frac{1}{R}\left(\frac{1}{s} - \frac{1}{s + \frac{R}{L}}\right) + \frac{i(0)}{s + \frac{R}{L}} \tag{3.3}$$

from which

$$i(t) = \frac{1}{R}\left(1 - e^{-(R/L)t}\right) + i(0)e^{-(R/L)t} \tag{3.4}$$

The function $i(t)$ is a subset of all possible network variables that we are able to find from Eq. (3.4) if we know its initial condition, $i(0)$, and the input, $v(t)$. Thus, $i(t)$ is a state variable, and the differential equation (3.1) is a *state equation*.

4. We can now solve for all of the other network variables *algebraically* in terms of $i(t)$ and the applied voltage, $v(t)$. For example, the voltage across the resistor is

$$v_R(t) = Ri(t) \tag{3.5}$$

The voltage across the inductor is

$$v_L(t) = v(t) - Ri(t) \tag{3.6}$$ [1]

The derivative of the current is

$$\frac{di}{dt} = \frac{1}{L}[v(t) - Ri(t)] \tag{3.7}$$ [2]

Thus, knowing the state variable, $i(t)$, and the input, $v(t)$, we can find the value, or *state*, of any network variable at any time, $t \geq t_0$. Hence, the algebraic equations, Eqs. (3.5) through (3.7), are *output equations*.

5. Since the variables of interest are completely described by Eq. (3.1) and Eqs. (3.5) through (3.7), we say that the combined state equation (3.1) and the output equations (3.5 through 3.7) form a viable representation of the network, which we call a *state-space representation*.

Equation (3.1), which describes the dynamics of the network, is not unique. This equation could be written in terms of any other network variable. For example, substituting $i = v_R/R$ into Eq. (3.1) yields

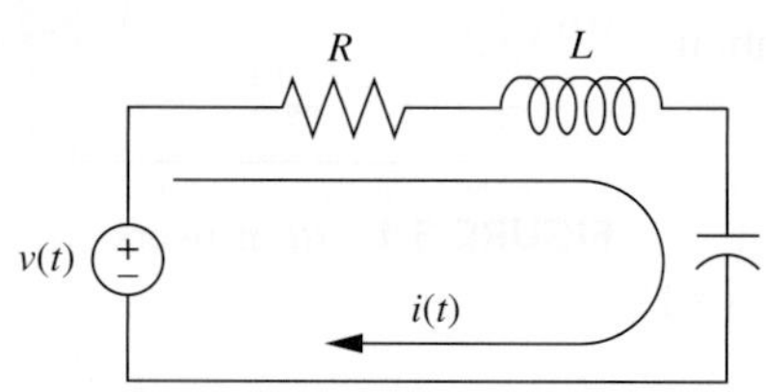


**FIGURE 3.2** *RLC* network

$$\frac{L}{R}\frac{dv_R}{dt} + v_R = v(t) \tag{3.8}$$

which can be solved knowing that the initial condition $v_R(0) = Ri(0)$ and knowing $v(t)$. In this case, the state variable is $v_R(t)$. Similarly, all other network variables can now be written in terms of the state variable, $v_R(t)$, and the input, $v(t)$. Let us now extend our observations to a second-order system, such as that shown in Figure 3.2.

1. Since the network is of second order, two simultaneous, first-order differential equations are needed to solve for two state variables. We select $i(t)$ and $q(t)$, the charge on the capacitor, as the two state variables.
2. Writing the loop equation yields

$$L\frac{di}{dt} + Ri + \frac{1}{C}\int i\,dt = v(t) \tag{3.9}$$

Converting to charge, using $i(t) = dq/dt$, we get

$$L\frac{d^2q}{dt^2} + R\frac{dq}{dt} + \frac{1}{C}q = v(t) \tag{3.10}$$

---

[1] Since $v_L(t) = v(t) - v_R(t) = v(t) - Ri(t)$.

[2] Since $\frac{di}{dt} = \frac{1}{L}v_L(t) = \frac{1}{L}[v(t) - Ri(t)]$.

But an $n$th-order differential equation can be converted to $n$ simultaneous first-order differential equations, with each equation of the form

$$\frac{dx_i}{dt} = a_{i1}x_1 + a_{i2}x_2 + \cdots + a_{in}x_n + b_i f(t) \tag{3.11}$$

where each $x_i$ is a state variable, and the $a_{ij}$'s and $b_i$ are constants for linear, time-invariant systems. We say that the right-hand side of Eq. (3.11) is a *linear combination* of the state variables and the input, $f(t)$.

We can convert Eq. (3.10) into two simultaneous, first-order differential equations in terms of $i(t)$ and $q(t)$. The first equation can be $dq/dt = i$. The second equation can be formed by substituting $\int i\,dt = q$ into Eq. (3.9) and solving for $di/dt$. Summarizing the two resulting equations, we get

$$\frac{dq}{dt} = i \tag{3.12a}$$

$$\frac{di}{dt} = -\frac{1}{LC}q - \frac{R}{L}i + \frac{1}{L}v(t) \tag{3.12b}$$

3. These equations are the state equations and can be solved simultaneously for the state variables, $q(t)$ and $i(t)$, using the Laplace transform and the methods of Chapter 2. In addition we must also know the input, $v(t)$, and the initial conditions for $q(t)$ and $i(t)$.
4. From these two state variables, we can solve for all other network variables. For example, the voltage across the inductor can be written in terms of the solved state variables and the input as

$$v_L(t) = -\frac{1}{C}q(t) - Ri(t) + v(t) \qquad (3.13)^3$$

Equation (3.13) is an *output equation*; we say that $v_L(t)$ is a *linear combination* of the state variables, $q(t)$ and $i(t)$, and the input, $v(t)$.
5. The combined state equations (3.12) and the output equation (3.13) form a viable representation of the network, which we call a *state-space representation*.

Another choice of two state variables can be made, for example, $v_R(t)$ and $v_C(t)$, the resistor and capacitor voltage, respectively. The resulting set of simultaneous, first-order differential equations follows:

$$\frac{dv_R}{dt} = -\frac{R}{L}v_R - \frac{R}{L}v_C + \frac{R}{L}v(t) \qquad (3.14a)^4$$

$$\frac{dv_C}{dt} = \frac{1}{RC}v_R \tag{3.14b}$$

Again, these differential equations can be solved for the state variables if we know the initial conditions along with $v(t)$. Further, all other network variables can be found as a linear combination of these state variables.

Is there a restriction on the choice of state variables? Yes! Typically, the minimum number of state variables required to describe a system equals the order of the differential equation. Thus, a second-order system requires a minimum of two state variables to describe it.

[3] Since $v_L(t) = L(di/dt) = -(1/C)q - Ri + v(t)$, where $di/dt$ can be found from Eq. (3.9), and $\int i\,dt = q$.
[4] Since $v_R(t) = i(t)R$, and $v_C(t) = (1/C)\int i\,dt$, differentiating $v_R(t)$ yields $dv_R/dt = R(di/dt) = (R/L)v_L = (R/L)[v(t) - v_R - v_C]$, and differentiating $v_C(t)$ yields $dv_C/dt = (1/C)i = (1/RC)v_R$.

We can define more state variables than the minimal set; however, within this minimal set the state variables must be linearly independent. For example, if $v_R(t)$ is chosen as a state variable, then $i(t)$ cannot be chosen, because $v_R(t)$ can be written as a linear combination of $i(t)$, namely $v_R(t) = Ri(t)$. Under these circumstances we say that the state variables are *linearly dependent*. State variables must be *linearly independent*; that is, no state variable can be written as a linear combination of the other state variables, or else we would not have enough information to solve for all other system variables, and we could even have trouble writing the simultaneous equations themselves.

The state and output equations can be written in vector-matrix form if the system is linear. Thus, Eq. (3.12), the state equations, can be written as

$$\dot{\mathbf{x}} = \mathbf{A}\mathbf{x} + \mathbf{B}u \tag{3.15}$$

where

$$\dot{\mathbf{x}} = \begin{bmatrix} dq/dt \\ di/dt \end{bmatrix}; \quad \mathbf{A} = \begin{bmatrix} 0 & 1 \\ -1/LC & -R/L \end{bmatrix}$$

$$\mathbf{x} = \begin{bmatrix} q \\ i \end{bmatrix}; \quad \mathbf{B} = \begin{bmatrix} 0 \\ 1/L \end{bmatrix}; \quad u = v(t)$$

Equation (3.13), the output equation, can be written as

$$y = \mathbf{C}\mathbf{x} + Du \tag{3.16}$$

where

$$y = v_L(t); \quad \mathbf{C} = [-1/C \quad -R]; \quad \mathbf{x} = \begin{bmatrix} q \\ i \end{bmatrix}; \quad D = 1; \quad u = v(t)$$

We call the combination of Eqs. (3.15) and (3.16) a *state-space representation* of the network of Figure 3.2. A state-space representation, therefore, consists of (1) the simultaneous, first-order differential equations from which the state variables can be solved and (2) the algebraic output equation from which all other system variables can be found. A state-space representation is not unique, since a different choice of state variables leads to a different representation of the same system.

In this section, we used two electrical networks to demonstrate some principles that are the foundation of the state-space representation. The representations developed in this section were for single-input, single-output systems, where $y, D$, and $u$ in Eqs. (3.15) and (3.16) are scalar quantities. In general, systems have multiple inputs and multiple outputs. For these cases, $y$ and $u$ become vector quantities, and $D$ becomes a matrix. In Section 3.3 we will generalize the representation for multiple-input, multiple-output systems and summarize the concept of the state-space representation.

## 3.3 The General State-Space Representation

Now that we have represented a physical network in state space and have a good idea of the terminology and the concept, let us summarize and generalize the representation for linear differential equations. First, we formalize some of the definitions that we came across in the last section.

*Linear combination*. A linear combination of $n$ variables, $x_i$, for $i = 1$ to $n$, is given by the following sum, $S$:

$$S = K_n x_n + K_{n-1} x_{n-1} + \cdots + K_1 x_1 \tag{3.17}$$

where each $K_i$ is a constant.

*Linear independence.* A set of variables is said to be linearly independent if none of the variables can be written as a linear combination of the others. For example, given $x_1, x_2$, and $x_3$, if $x_2 = 5x_1 + 6x_3$, then the variables are not linearly independent, since one of them can be written as a linear combination of the other two. Now, what must be true so that one variable cannot be written as a linear combination of the other variables? Consider the example $K_2x_2 = K_1x_1 + K_3x_3$. If no $x_i = 0$, then any $x_i$ can be written as a linear combination of other variables, unless all $K_i = 0$. Formally, then, variables $x_i$, for $i = 1$ to $n$, are said to be linearly independent if their linear combination, $S$, equals zero *only* if every $K_i = 0$ and *no* $x_i = 0$ for all $t \geq 0$.

*System variable.* Any variable that responds to an input or initial conditions in a system.

*State variables.* The smallest set of linearly independent system variables such that the values of the members of the set at time $t_0$ along with known forcing functions completely determine the value of all system variables for all $t \geq t_0$.

*State vector.* A vector whose elements are the state variables.

*State space.* The $n$-dimensional space whose axes are the state variables. This is a new term and is illustrated in Figure 3.3, where the state variables are assumed to be a resistor voltage, $v_R$, and a capacitor voltage, $v_C$. These variables form the axes of the *state space*. A trajectory can be thought of as being mapped out by the state vector, $\mathbf{x}(t)$, for a range of $t$. Also shown is the state vector at the particular time $t = 4$.

*State equations.* A set of $n$ simultaneous, first-order differential equations with $n$ variables, where the $n$ variables to be solved are the state variables.

*Output equation.* The algebraic equation that expresses the output variables of a system as linear combinations of the state variables and the inputs.

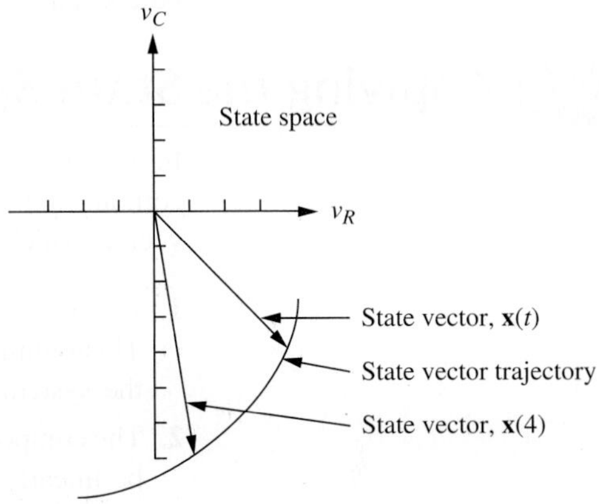


**FIGURE 3.3** Graphic representation of state space and a state vector

Now that the definitions have been formally stated, we define the state-space representation of a system. A system is represented in state space by the following equations:

$$\dot{\mathbf{x}} = \mathbf{Ax} + \mathbf{Bu} \tag{3.18}$$

$$\mathbf{y} = \mathbf{Cx} + \mathbf{Du} \tag{3.19}$$

for $t \geq t_0$ and initial conditions, $\mathbf{x}(t_0)$, where

$\mathbf{x}$ = state vector
$\dot{\mathbf{x}}$ = derivative of the state vector with respect to time
$\mathbf{y}$ = output vector
$\mathbf{u}$ = input or control vector
$\mathbf{A}$ = system matrix
$\mathbf{B}$ = input matrix
$\mathbf{C}$ = output matrix
$\mathbf{D}$ = feedforward matrix

Equation (3.18) is called the *state equation*, and the vector $\mathbf{x}$, the *state vector*, contains the state variables. Equation (3.18) can be solved for the state variables, which we demonstrate in Chapter 4. Equation (3.19) is called the *output equation*. This equation is used to calculate any other system variables. This representation of a system provides complete knowledge of all variables of the system at any $t \geq t_0$.

As an example, for a linear, time-invariant, second-order system with a single input $v(t)$, the state equations could take on the following form:

$$\frac{dx_1}{dt} = a_{11}x_1 + a_{12}x_2 + b_1 v(t) \tag{3.20a}$$

$$\frac{dx_2}{dt} = a_{21}x_1 + a_{22}x_2 + b_2 v(t) \tag{3.20b}$$

where $x_1$ and $x_2$ are the state variables. If there is a single output, the output equation could take on the following form:

$$y = c_1 x_1 + c_2 x_2 + d_1 v(t) \tag{3.21}$$

The choice of state variables for a given system is not unique. The requirement in choosing the state variables is that they be linearly independent and that a minimum number of them be chosen.

## 3.4 Applying the State-Space Representation

In this section, we apply the state-space formulation to the representation of more complicated physical systems. The first step in representing a system is to select the state vector, which must be chosen according to the following considerations:

1. A minimum number of state variables must be selected as components of the state vector. This minimum number of state variables is sufficient to describe completely the state of the system.
2. The components of the state vector (that is, this minimum number of state variables) must be linearly independent.

Let us review and clarify these statements.

### Linearly Independent State Variables

The components of the state vector must be linearly independent. For example, following the definition of linear independence in Section 3.3, if $x_1$, $x_2$, and $x_3$ are chosen as state variables, but $x_3 = 5x_1 + 4x_2$, then $x_3$ is not linearly independent of $x_1$ and $x_2$, since knowledge of the values of $x_1$ and $x_2$ will yield the value of $x_3$. Variables and their successive derivatives are linearly independent. For example, the voltage across an inductor, $v_L$, is linearly independent of the current through the inductor, $i_L$, since $v_L = L di_L/dt$. Thus, $v_L$ cannot be evaluated as a linear combination of the current, $i_L$.

### Minimum Number of State Variables

How do we know the minimum number of state variables to select? Typically, the minimum number required equals the order of the differential equation describing the system. For example, if a third-order differential equation describes the system, then three simultaneous, first-order differential equations are required along with three state variables. From the perspective of the transfer function, the order of the differential equation is the order of the denominator of the transfer function after canceling common factors in the numerator and denominator.

In most cases, another way to determine the number of state variables is to count the number of independent energy-storage elements in the system.[5] The number of

[5] Sometimes it is not apparent in a schematic how many independent energy-storage elements there are. It is possible that more than the minimum number of energy-storage elements could be selected, leading to a state vector whose components number more than the minimum required and are not linearly independent. Selecting additional dependent energy-storage elements results in a system matrix of higher order and more complexity than required for the solution of the state equations.

these energy-storage elements equals the order of the differential equation and the number of state variables. In Figure 3.2 there are two energy-storage elements, the capacitor and the inductor. Hence, two state variables and two state equations are required for the system.

If too few state variables are selected, it may be impossible to write particular output equations, since some system variables cannot be written as a linear combination of the reduced number of state variables. In many cases, it may be impossible even to complete the writing of the state equations, since the derivatives of the state variables cannot be expressed as linear combinations of the reduced number of state variables.

If you select the minimum number of state variables but they are not linearly independent, at best you may not be able to solve for all other system variables. At worst you may not be able to complete the writing of the state equations.

Often the state vector includes more than the minimum number of state variables required. Two possible cases exist. Often state variables are chosen to be physical variables of a system, such as position and velocity in a mechanical system. Cases arise where these variables, although linearly independent, are also *decoupled*. That is, some linearly independent variables are not required in order to solve for any of the other linearly independent variables or any other dependent system variable. Consider the case of a mass and viscous damper whose differential equation is $M\,dv/dt + Dv = f(t)$, where $v$ is the velocity of the mass. Since this is a first-order equation, one state equation is all that is required to define this system in state space with velocity as the state variable. Also, since there is only one energy-storage element, mass, only one state variable is required to represent this system in state space. However, the mass also has an associated position, which is linearly independent of velocity. If we want to include position in the state vector along with velocity, then we add position as a state variable that is linearly independent of the other state variable, velocity. Figure 3.4 illustrates what is happening. The first block is the transfer function equivalent to $Mdv(t)/dt + Dv(t) = f(t)$. The second block shows that we integrate the output velocity to yield output displacement (see Table 2.2, Item 10). Thus, if we want displacement as an output, the denominator, or characteristic equation, has increased in order to 2, the product of the two transfer functions. Many times, the writing of the state equations is simplified by including additional state variables.

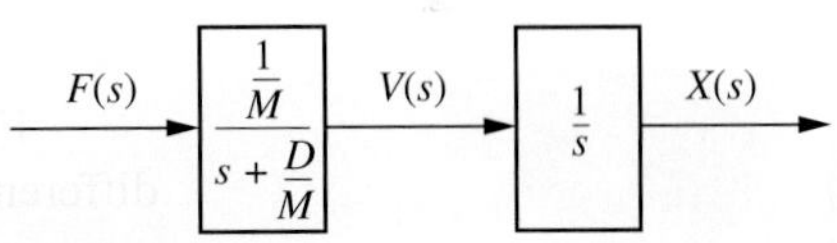


**FIGURE 3.4** Block diagram of a mass and damper

Another case that increases the size of the state vector arises when the added variable is not linearly independent of the other members of the state vector. This usually occurs when a variable is selected as a state variable but its dependence on the other state variables is not immediately apparent. For example, energy-storage elements may be used to select the state variables, and the dependence of the variable associated with one energy-storage element on the variables of other energy-storage elements may not be recognized. Thus, the dimension of the system matrix is increased unnecessarily, and the solution for the state vector, which we cover in Chapter 4, is more difficult. Also, adding dependent state variables affects the designer's ability to use state-space methods for design.[6]

We saw in Section 3.2 that the state-space representation is not unique. The following example demonstrates one technique for selecting state variables and representing a system in state space. Our approach is to write the simple derivative equation for each energy-storage element and solve for each derivative term as a linear combination of any of the system variables and the input that are present in the equation. Next we select each differentiated variable as a state variable. Then we express all other system variables in the equations in terms of the state variables and the input. Finally, we write the output variables as linear combinations of the state variables and the input.

[6] See Chapter 12 for state-space design techniques.

## Example 3.1 Representing an Electrical Network

**PROBLEM:** Given the electrical network of Figure 3.5, find a state-space representation if the output is the current through the resistor.

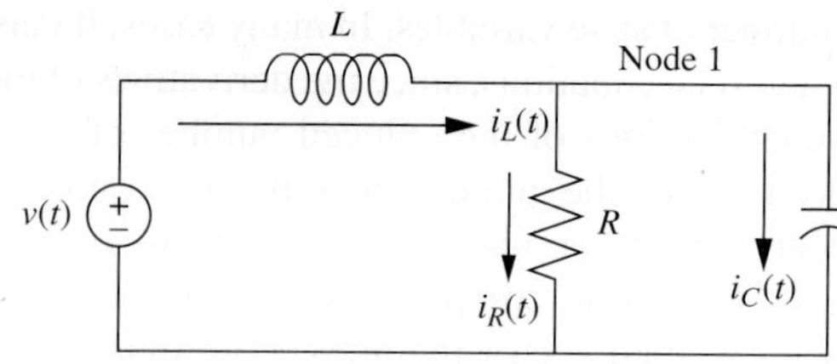


**FIGURE 3.5** Electrical network for representation in state space

**SOLUTION:** The following steps will yield a viable representation of the network in state space.

**Step 1** Label all of the branch currents in the network. These include $i_L$, $i_R$, and $i_C$, as shown in Figure 3.5.

**Step 2** Select the state variables by writing the derivative equation for all energy-storage elements, that is, the inductor and the capacitor. Thus,

$$C\frac{dv_C}{dt} = i_C \tag{3.22}$$

$$L\frac{di_L}{dt} = v_L \tag{3.23}$$

From Eqs. (3.22) and (3.23), choose the state variables as the quantities that are differentiated, namely $v_C$ and $i_L$. Using Eq. (3.20) as a guide, we see that the state-space representation is complete if the right-hand sides of Eqs. (3.22) and (3.23) can be written as linear combinations of the state variables and the input.

Since $i_C$ and $v_L$ are not state variables, our next step is to express $i_C$ and $v_L$ as linear combinations of the state variables, $v_C$ and $i_L$, and the input, $v(t)$.

**Step 3** Apply network theory, such as Kirchhoff's voltage and current laws, to obtain $i_C$ and $v_L$ in terms of the state variables, $v_C$ and $i_L$. At Node 1,

$$\begin{aligned} i_C &= -i_R + i_L \\ &= -\frac{1}{R}v_C + i_L \end{aligned} \tag{3.24}$$

which yields $i_C$ in terms of the state variables, $v_C$ and $i_L$.

Around the outer loop,

$$v_L = -v_C + v(t) \tag{3.25}$$

which yields $v_L$ in terms of the state variable, $v_C$, and the source, $v(t)$.

**Step 4** Substitute the results of Eqs. (3.24) and (3.25) into Eqs. (3.22) and (3.23) to obtain the following state equations:

$$C\frac{dv_C}{dt} = -\frac{1}{R}v_C + i_L \tag{3.26a}$$

$$L\frac{di_L}{dt} = -v_C + v(t) \tag{3.26b}$$

or

$$\frac{dv_C}{dt} = -\frac{1}{RC}v_C + \frac{1}{C}i_L \tag{3.27a}$$

$$\frac{di_L}{dt} = -\frac{1}{L}v_C + \frac{1}{L}v(t) \tag{3.27b}$$

**Step 5** Find the output equation. Since the output is $i_R(t)$,

$$i_R = \frac{1}{R}v_C \tag{3.28}$$

The final result for the state-space representation is found by representing Eqs. (3.27) and (3.28) in vector-matrix form as follows:

$$\begin{bmatrix} \dot{v}_C \\ \dot{i}_L \end{bmatrix} = \begin{bmatrix} -1/(RC) & 1/C \\ -1/L & 0 \end{bmatrix} \begin{bmatrix} v_C \\ i_L \end{bmatrix} + \begin{bmatrix} 0 \\ 1/L \end{bmatrix} v(t) \tag{3.29a}$$

$$i_R = \begin{bmatrix} 1/R & 0 \end{bmatrix} \begin{bmatrix} v_C \\ i_L \end{bmatrix} \tag{3.29b}$$

where the dot indicates differentiation with respect to time.

In order to clarify the representation of physical systems in state space, we will look at two more examples. The first is an electrical network with a dependent source. Although we will follow the same procedure as in the previous problem, this problem will yield increased complexity in applying network analysis to find the state equations. For the second example, we find the state-space representation of a mechanical system.

## Example 3.2 Representing an Electrical Network with a Dependent Source

**PROBLEM:** Find the state and output equations for the electrical network shown in Figure 3.6 if the output vector is $\mathbf{y} = [\,v_{R_2} \quad i_{R_2}\,]^T$, where $T$ means transpose.[7]

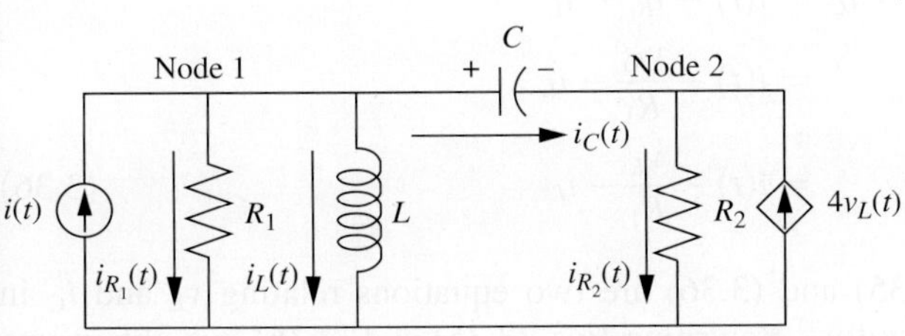


**FIGURE 3.6** Electrical network for Example 3.2

**SOLUTION:** Immediately notice that this network has a voltage-dependent current source.

[7] See Appendix G for a discussion of the transpose.

**Step 1** Label all of the branch currents on the network, as shown in Figure 3.6.

**Step 2** Select the state variables by listing the voltage-current relationships for all of the energy-storage elements:

$$L\frac{di_L}{dt} = v_L \tag{3.30a}$$

$$C\frac{dv_C}{dt} = i_C \tag{3.30b}$$

From Eqs. (3.30) select the state variables to be the differentiated variables. Thus, the state variables, $x_1$ and $x_2$, are

$$x_1 = i_L; \quad x_2 = v_C \tag{3.31}$$

**Step 3** Remembering that the form of the state equation is

$$\dot{\mathbf{x}} = \mathbf{Ax} + \mathbf{Bu} \tag{3.32}$$

we see that the remaining task is to transform the right-hand side of Eq. (3.30) into linear combinations of the state variables and input source current. Using Kirchhoff's voltage and current laws, we find $v_L$ and $i_C$ in terms of the state variables and the input current source.

Around the mesh containing $L$ and $C$,

$$v_L = v_C + v_{R_2} = v_C + i_{R_2}R_2 \tag{3.33}$$

But at Node 2, $i_{R_2} = i_C + 4v_L$. Substituting this relationship for $i_{R_2}$ into Eq. (3.33) yields

$$v_L = v_C + (i_C + 4v_L)R_2 \tag{3.34}$$

Solving for $v_L$, we get

$$v_L = \frac{1}{1 - 4R_2}(v_C + i_C R_2) \tag{3.35}$$

Notice that since $v_C$ is a state variable, we only need to find $i_C$ in terms of the state variables. We will then have obtained $v_L$ in terms of the state variables.

Thus, at Node 1 we can write the sum of the currents as

$$\begin{aligned} i_C &= i(t) - i_{R_1} - i_L \\ &= i(t) - \frac{v_{R_1}}{R_1} - i_L \\ &= i(t) - \frac{v_L}{R_1} - i_L \end{aligned} \tag{3.36}$$

where $v_{R_1} = v_L$. Equations (3.35) and (3.36) are two equations relating $v_L$ and $i_C$ in terms of the state variables $i_L$ and $v_C$. Rewriting Eqs. (3.35) and (3.36), we obtain two simultaneous equations yielding $v_L$ and $i_C$ as linear combinations of the state variables $i_L$ and $v_C$:

$$(1 - 4R_2)v_L - R_2 i_C = v_C \tag{3.37a}$$

$$-\frac{1}{R_1}v_L - i_C = i_L - i(t) \tag{3.37b}$$

Solving Eq. (3.37a) simultaneously for $v_L$ and $i_C$ yields

$$v_L = \frac{1}{\Delta}[R_2 i_L - v_C - R_2 i(t)] \tag{3.38}$$

and

$$i_C = \frac{1}{\Delta}\left[(1-4R_2)i_L + \frac{1}{R_1}v_C - (1-4R_2)i(t)\right] \tag{3.39}$$

where

$$\Delta = -\left[(1-4R_2) + \frac{R_2}{R_1}\right] \tag{3.40}$$

Substituting Eqs. (3.38) and (3.39) into (3.30), simplifying, and writing the result in vector-matrix form renders the following state equation:

$$\begin{bmatrix} \dot{i}_L \\ \dot{v}_C \end{bmatrix} = \begin{bmatrix} R_2/(L\Delta) & -1/(L\Delta) \\ (1-4R_2)/(C\Delta) & 1/(R_1 C\Delta) \end{bmatrix} \begin{bmatrix} i_L \\ v_C \end{bmatrix} + \begin{bmatrix} -R_2/(L\Delta) \\ -(1-4R_2)/(C\Delta) \end{bmatrix} i(t) \tag{3.41}$$

**Step 4** Derive the output equation. Since the specified output variables are $v_{R_2}$ and $i_{R_2}$, we note that around the mesh containing $C$, $L$, and $R_2$,

$$v_{R_2} = -v_C + v_L \tag{3.42a}$$

$$i_{R_2} = i_C + 4v_L \tag{3.42b}$$

Substituting Eqs. (3.38) and (3.39) into Eq. (3.42), $v_{R_2}$ and $i_{R_2}$ are obtained as linear combinations of the state variables, $i_L$ and $v_C$. In vector-matrix form, the output equation is

$$\begin{bmatrix} v_{R_2} \\ i_{R_2} \end{bmatrix} = \begin{bmatrix} R_2/\Delta & -(1+1/\Delta) \\ 1/\Delta & (1-4R_1)/(\Delta R_1) \end{bmatrix} \begin{bmatrix} i_L \\ v_C \end{bmatrix} + \begin{bmatrix} -R_2/\Delta \\ -1/\Delta \end{bmatrix} i(t) \tag{3.43}$$

In the next example, we find the state-space representation for a mechanical system. It is more convenient when working with mechanical systems to obtain the state equations directly from the equations of motion rather than from the energy-storage elements. For example, consider an energy-storage element such as a spring, where $F = Kx$. This relationship does not contain the derivative of a physical variable as in the case of electrical networks, where $i = C\,dv/dt$ for capacitors, and $v = L\,di/dt$ for inductors. Thus, in mechanical systems we change our selection of state variables to be the position and velocity of each point of linearly independent motion. In the example, we will see that although there are three energy-storage elements, there will be four state variables; an additional linearly independent state variable is included for the convenience of writing the state equations. It is left to the student to show that this system yields a fourth-order transfer function if we relate the displacement of either mass to the applied force, and a third-order transfer function if we relate the velocity of either mass to the applied force.

## Example 3.3 Representing a Translational Mechanical System

**PROBLEM:** Find the state equations for the translational mechanical system shown in Figure 3.7.

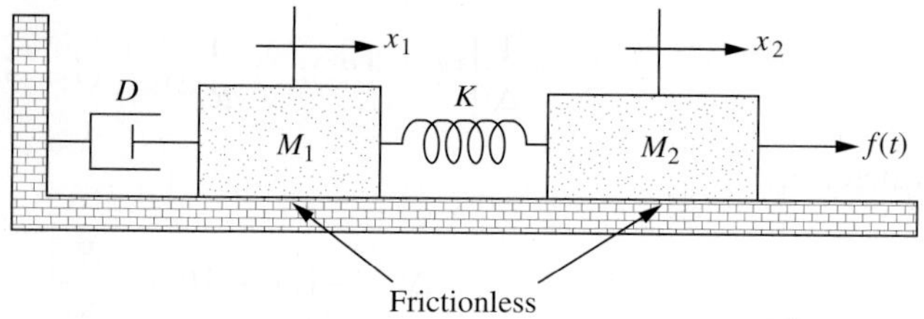


**FIGURE 3.7** Translational mechanical system

**SOLUTION:** First write the differential equations for the network in Figure 3.7, using the methods of Chapter 2 to find the Laplace-transformed equations of motion. Next take the inverse Laplace transform of these equations, assuming zero initial conditions, and obtain

$$M_1\frac{d^2x_1}{dt^2} + D\frac{dx_1}{dt} + Kx_1 - Kx_2 = 0 \tag{3.44}$$

$$-Kx_1 + M_2\frac{d^2x_2}{dt^2} + Kx_2 = f(t) \tag{3.45}$$

Now let $d^2x_1/dt^2 = dv_1/dt$, and $d^2x_2/dt^2 = dv_2/dt$, and then select $x_1$, $v_1$, $x_2$, and $v_2$ as state variables. Next form two of the state equations by solving Eq. (3.44) for $dv_1/dt$ and Eq. (3.45) for $dv_2/dt$. Finally, add $dx_1/dt = v_1$ and $dx_2/dt = v_2$ to complete the set of state equations. Hence,

$$\frac{dx_1}{dt} = \quad +v_1 \tag{3.46a}$$

$$\frac{dv_1}{dt} = -\frac{K}{M_1}x_1 - \frac{D}{M_1}v_1 + \frac{K}{M_1}x_2 \tag{3.46b}$$

$$\frac{dx_2}{dt} = \quad +v_2 \tag{3.46c}$$

$$\frac{dv_2}{dt} = +\frac{K}{M_2}x_1 - \frac{K}{M_2}x_2 + \frac{1}{M_2}f(t) \tag{3.46d}$$

In vector-matrix form,

$$\begin{bmatrix} \dot{x}_1 \\ \dot{v}_1 \\ \dot{x}_2 \\ \dot{v}_2 \end{bmatrix} = \begin{bmatrix} 0 & 1 & 0 & 0 \\ -K/M_1 & -D/M_1 & K/M_1 & 0 \\ 0 & 0 & 0 & 1 \\ K/M_2 & 0 & -K/M_2 & 0 \end{bmatrix} \begin{bmatrix} x_1 \\ v_1 \\ x_2 \\ v_2 \end{bmatrix} + \begin{bmatrix} 0 \\ 0 \\ 0 \\ 1/M_2 \end{bmatrix} f(t) \tag{3.47}$$

where the dot indicates differentiation with respect to time. What is the output equation if the output is $x(t)$?

## Skill-Assessment Exercise 3.1

**PROBLEM:** Find the state-space representation of the electrical network shown in Figure 3.8. The output is $v_o(t)$.

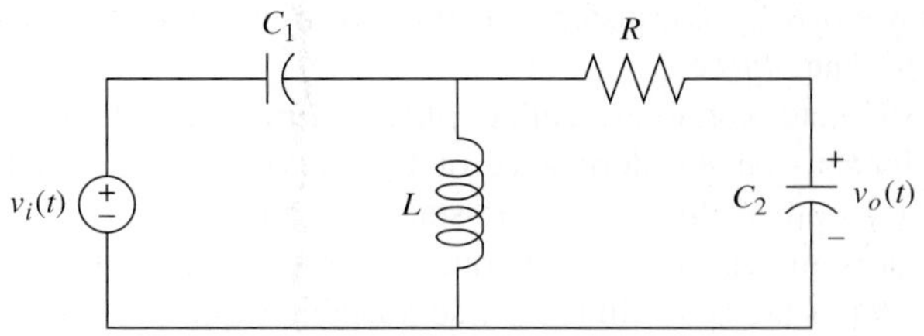


**FIGURE 3.8** Electric circuit for Skill-Assessment Exercise 3.1

**ANSWER:**

$$\dot{\mathbf{x}} = \begin{bmatrix} 1/C_1 & 1/C_1 & -1/C_1 \\ -1/L & 0 & 0 \\ 1/C_2 & 0 & -1/C_2 \end{bmatrix}\mathbf{x} + \begin{bmatrix} 0 \\ 1 \\ 0 \end{bmatrix} v_i(t)$$

$$y = [0 \quad 0 \quad 1]\mathbf{x}$$

The complete solution is at www.wiley.com/college/nise.

## Skill-Assessment Exercise 3.2

**PROBLEM:** Represent the translational mechanical system shown in Figure 3.9 in state space, where $x_3(t)$ is the output.

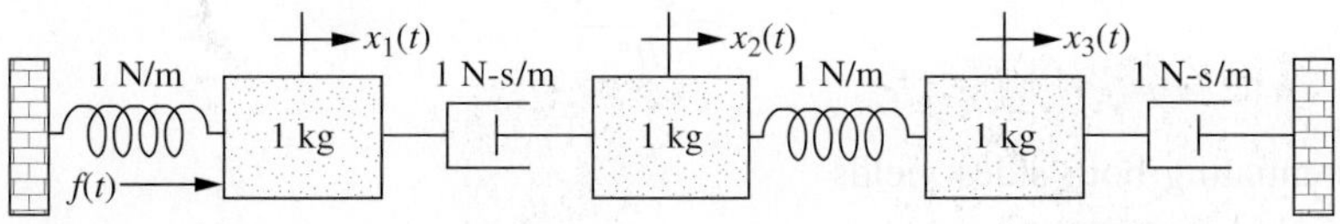


**FIGURE 3.9** Translational mechanical system for Skill-Assessment Exercise 3.2

**ANSWER:**

$$\dot{\mathbf{z}} = \begin{bmatrix} 0 & 1 & 0 & 0 & 0 & 0 \\ -1 & -1 & 0 & 1 & 0 & 0 \\ 0 & 0 & 0 & 1 & 0 & 0 \\ 0 & 1 & -1 & -1 & 1 & 0 \\ 0 & 0 & 0 & 0 & 0 & 1 \\ 0 & 0 & 1 & 0 & -1 & -1 \end{bmatrix}\mathbf{z} + \begin{bmatrix} 0 \\ 1 \\ 0 \\ 0 \\ 0 \\ 0 \end{bmatrix} f(t)$$

$$y = [\ 0 \quad 0 \quad 0 \quad 0 \quad 1 \quad 0\ ]\mathbf{z}$$

where

$$\mathbf{z} = [x_1 \quad \dot{x}_1 \quad x_2 \quad \dot{x}_2 \quad x_3 \quad \dot{x}_3]^T$$

The complete solution is at www.wiley.com/college/nise.

## 3.5 Converting a Transfer Function to State Space

In the last section, we applied the state-space representation to electrical and mechanical systems. We learn how to convert a transfer function representation to a state-space representation in this section. One advantage of the state-space representation is that it can be used for the simulation of physical systems on the digital computer. Thus, if we want to simulate a system that is represented by a transfer function, we must first convert the transfer function representation to state space.

At first we select a set of state variables, called *phase variables*, where each subsequent state variable is defined to be the derivative of the previous state variable. In Chapter 5 we show how to make other choices for the state variables.

Let us begin by showing how to represent a general, $n$th-order, linear differential equation with constant coefficients in state space in the phase-variable form. We will then show how to apply this representation to transfer functions.

Consider the differential equation

$$\frac{d^n y}{dt^n} + a_{n-1}\frac{d^{n-1}y}{dt^{n-1}} + \cdots + a_1\frac{dy}{dt} + a_0 y = b_0 u \tag{3.48}$$

A convenient way to choose state variables is to choose the output, $y(t)$, and its $(n-1)$ derivatives as the state variables. This choice is called the *phase-variable choice*. Choosing the state variables, $x_i$, we get

$$x_1 = y \tag{3.49a}$$

$$x_2 = \frac{dy}{dt} \tag{3.49b}$$

$$x_3 = \frac{d^2y}{dt^2} \tag{3.49c}$$

$$\vdots$$

$$x_n = \frac{d^{n-1}y}{dt^{n-1}} \tag{3.49d}$$

and differentiating both sides yields

$$\dot{x}_1 = \frac{dy}{dt} \tag{3.50a}$$

$$\dot{x}_2 = \frac{d^2y}{dt^2} \tag{3.50b}$$

$$\dot{x}_3 = \frac{d^3y}{dt^3} \tag{3.50c}$$

$$\vdots$$

$$\dot{x}_n = \frac{d^n y}{dt^n} \tag{3.50d}$$

where the dot above the $x$ signifies differentiation with respect to time.

Substituting the definitions of Eq. (3.49) into Eq. (3.50), the state equations are evaluated as

$$\dot{x}_1 = x_2 \quad (3.51a)$$

$$\dot{x}_2 = x_3 \quad (3.51b)$$

$$\vdots$$

$$\dot{x}_{n-1} = x_n \quad (3.51c)$$

$$\dot{x}_n = -a_0x_1 - a_1x_2 \cdots - a_{n-1}x_n + b_0u \quad (3.51d)$$

where Eq. (3.51d) was obtained from Eq. (3.48) by solving for $d^n y/dt^n$ and using Eq. (3.49). In vector-matrix form, Eq. (3.51) become

$$\begin{bmatrix} \dot{x}_1 \\ \dot{x}_2 \\ \dot{x}_3 \\ \vdots \\ \dot{x}_{n-1} \\ \dot{x}_n \end{bmatrix} = \begin{bmatrix} 0 & 1 & 0 & 0 & 0 & 0 & \cdots & 0 \\ 0 & 0 & 1 & 0 & 0 & 0 & \cdots & 0 \\ 0 & 0 & 0 & 1 & 0 & 0 & \cdots & 0 \\ \vdots & & & & & & & \\ 0 & 0 & 0 & 0 & 0 & 0 & \cdots & 1 \\ -a_0 & -a_1 & -a_2 & -a_3 & -a_4 & -a_5 & \cdots & -a_{n-1} \end{bmatrix} \begin{bmatrix} x_1 \\ x_2 \\ x_3 \\ \vdots \\ x_{n-1} \\ x_n \end{bmatrix} + \begin{bmatrix} 0 \\ 0 \\ 0 \\ \vdots \\ 0 \\ b_0 \end{bmatrix} u \quad (3.52)$$

Equation (3.52) is the phase-variable form of the state equations. This form is easily recognized by the unique pattern of 1's and 0's and the negative of the coefficients of the differential equation written in reverse order in the last row of the system matrix.

Finally, since the solution to the differential equation is $y(t)$, or $x_1$, the output equation is

$$y = [1 \quad 0 \quad 0 \quad \cdots \quad 0] \begin{bmatrix} x_1 \\ x_2 \\ x_3 \\ \vdots \\ x_{n-1} \\ x_n \end{bmatrix} \quad (3.53)$$

In summary, then, to convert a transfer function into state equations in phase-variable form, we first convert the transfer function to a differential equation by cross-multiplying and taking the inverse Laplace transform, assuming zero initial conditions. Then we represent the differential equation in state space in phase-variable form. An example illustrates the process.

## Example 3.4 Converting a Transfer Function with a Constant Term in the Numerator

**PROBLEM:** Find the state-space representation in phase-variable form for the transfer function shown in Figure 3.10(*a*).

**SOLUTION:**

**Step 1** Find the associated differential equation. Since

$$\frac{C(s)}{R(s)} = \frac{24}{(s^3 + 9s^2 + 26s + 24)} \quad (3.54)$$

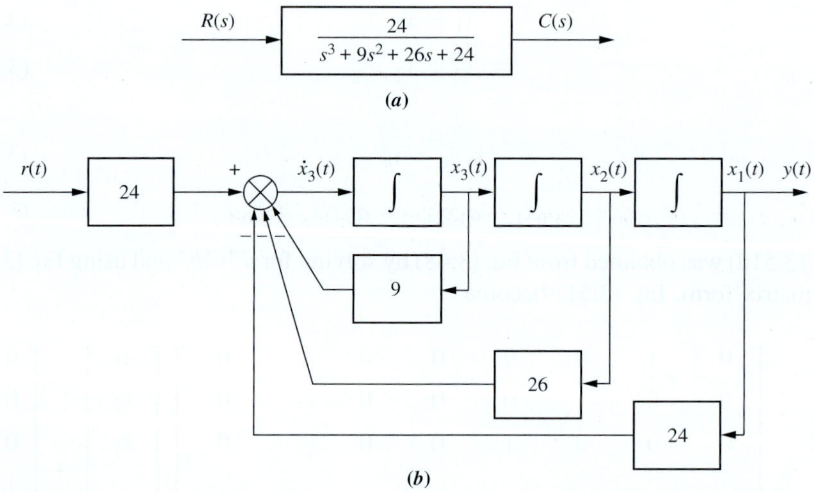


**FIGURE 3.10** **a.** Transfer function; **b.** equivalent block diagram showing phase variables. Note: $y(t) = c(t)$.

cross-multiplying yields

$$(s^3 + 9s^2 + 26s + 24)C(s) = 24R(s) \tag{3.55}$$

The corresponding differential equation is found by taking the inverse Laplace transform, assuming zero initial conditions:

$$\dddot{c} + 9\ddot{c} + 26\dot{c} + 24c = 24r \tag{3.56}$$

**Step 2** Select the state variables.
Choosing the state variables as successive derivatives, we get

$$x_1 = c \tag{3.57a}$$

$$x_2 = \dot{c} \tag{3.57b}$$

$$x_3 = \ddot{c} \tag{3.57c}$$

Differentiating both sides and making use of Eq. (3.57) to find $\dot{x}_1$ and $\dot{x}_2$, and Eq. (3.56) to find $\dddot{c} = \dot{x}_3$, we obtain the state equations. Since the output is $c = x_1$, the combined state and output equations are

$$\dot{x}_1 = x_2 \tag{3.58a}$$

$$\dot{x}_2 = x_3 \tag{3.58b}$$

$$\dot{x}_3 = -24x_1 - 26x_2 - 9x_3 + 24r \tag{3.58c}$$

$$y = c = x_1 \tag{3.58d}$$

In vector-matrix form,

$$\begin{bmatrix} \dot{x}_1 \\ \dot{x}_2 \\ \dot{x}_3 \end{bmatrix} = \begin{bmatrix} 0 & 1 & 0 \\ 0 & 0 & 1 \\ -24 & -26 & -9 \end{bmatrix} \begin{bmatrix} x_1 \\ x_2 \\ x_3 \end{bmatrix} + \begin{bmatrix} 0 \\ 0 \\ 24 \end{bmatrix} r \tag{3.59a}$$

$$y = [1 \quad 0 \quad 0] \begin{bmatrix} x_1 \\ x_2 \\ x_3 \end{bmatrix} \tag{3.59b}$$

Notice that the third row of the system matrix has the same coefficients as the denominator of the transfer function but negative and in reverse order.

At this point, we can create an equivalent block diagram of the system of Figure 3.10(*a*) to help visualize the state variables. We draw three integral blocks as shown in Figure 3.10(*b*) and label each output as one of the state variables, $x_i(t)$, as shown. Since the input to each integrator is $\dot{x}_i(t)$, use Eqs. (3.58a), (3.58b), and (3.58c) to determine the combination of input signals to each integrator. Form and label each input. Finally, use Eq. (3.58d) to form and label the output, $y(t) = c(t)$. The final result of Figure 3.10(*b*) is a system equivalent to Figure 3.10(*a*) that explicitly shows the state variables and gives a vivid picture of the state-space representation.

Students who are using MATLAB should now run ch3p1 through ch3p4 in Appendix B. You will learn how to represent the system matrix **A**, the input matrix **B**, and the output matrix **C** using MATLAB. You will learn how to convert a transfer function to the state-space representation in phase-variable form. Finally, Example 3.4 will be solved using MATLAB.

MATLAB
**ML**

The transfer function of Example 3.4 has a constant term in the numerator. If a transfer function has a polynomial in $s$ in the numerator that is of order less than the polynomial in the denominator, as shown in Figure 3.11(*a*), the numerator and denominator can be handled separately. First separate the transfer function into two cascaded transfer functions, as shown in Figure 3.11(*b*); the first is the denominator, and the second is just the numerator. The first transfer function with just the denominator is converted to the phase-variable representation in state space as demonstrated in the last example. Hence, phase variable $x_1$ is the output, and the rest of the phase variables are the internal variables of the first block, as shown in Figure 3.11(*b*). The second transfer function with just the numerator yields

$$Y(s) = C(s) = (b_2 s^2 + b_1 s + b_0)\, X_1(s) \tag{3.60}$$

where, after taking the inverse Laplace transform with zero initial conditions,

$$y(t) = b_2 \frac{d^2 x_1}{dt^2} + b_1 \frac{dx_1}{dt} + b_0 x_1 \tag{3.61}$$

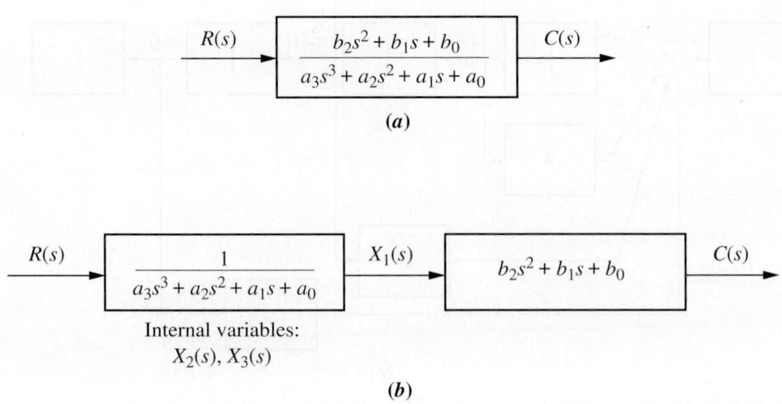


**FIGURE 3.11** Decomposing a transfer function

But the derivative terms are the definitions of the phase variables obtained in the first block. Thus, writing the terms in reverse order to conform to an output equation,

$$y(t) = b_0x_1 + b_1x_2 + b_2x_3 \tag{3.62}$$

Hence, the second block simply forms a specified linear combination of the state variables developed in the first block.

From another perspective, the denominator of the transfer function yields the state equations, while the numerator yields the output equation. The next example demonstrates the process.

## Example 3.5 Converting a Transfer Function with a Polynomial in the Numerator

**PROBLEM:** Find the state-space representation of the transfer function shown in Figure 3.12(*a*).

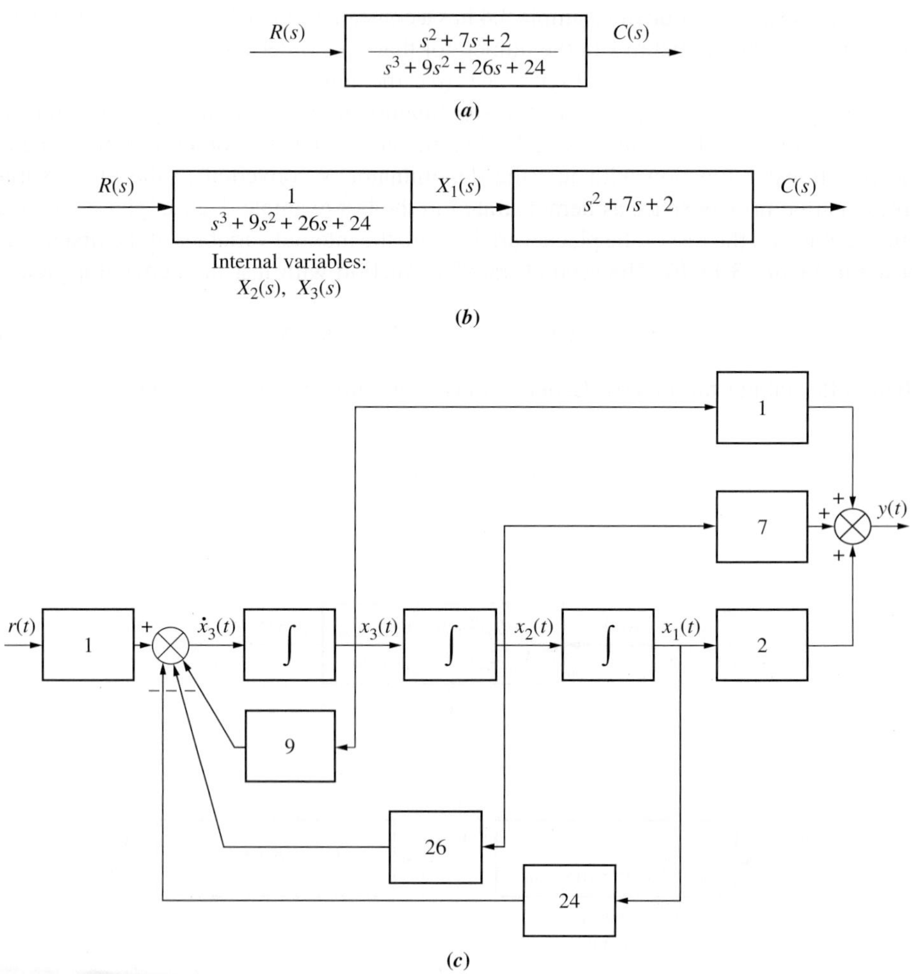


**FIGURE 3.12** **a.** Transfer function; **b.** decomposed transfer function; **c.** equivalent block diagram Note: $y(t) = c(t)$.

**SOLUTION:** This problem differs from Example 3.4, since the numerator has a polynomial in $s$ instead of just a constant term.

**Step 1** Separate the system into two cascaded blocks, as shown in Figure 3.12(*b*). The first block contains the denominator and the second block contains the numerator.

**Step 2** Find the state equations for the block containing the denominator. We notice that the first block's numerator is 1/24 that of Example 3.4. Thus, the state equations are the same except that this system's input matrix is 1/24 that of Example 3.4. Hence, the state equation is

$$\begin{bmatrix} \dot{x}_1 \\ \dot{x}_2 \\ \dot{x}_3 \end{bmatrix} = \begin{bmatrix} 0 & 1 & 0 \\ 0 & 0 & 1 \\ -24 & -26 & -9 \end{bmatrix} \begin{bmatrix} x_1 \\ x_2 \\ x_3 \end{bmatrix} + \begin{bmatrix} 0 \\ 0 \\ 1 \end{bmatrix} r \tag{3.63}$$

**Step 3** Introduce the effect of the block with the numerator. The second block of Figure 3.12(*b*), where $b_2 = 1$, $b_1 = 7$, and $b_0 = 2$, states that

$$C(s) = (b_2 s^2 + b_1 s + b_0)X_1(s) = (s^2 + 7s + 2)X_1(s) \tag{3.64}$$

Taking the inverse Laplace transform with zero initial conditions, we get

$$c = \ddot{x}_1 + 7\dot{x}_1 + 2x_1 \tag{3.65}$$

But,

$$\begin{aligned} x_1 &= x_1 \\ \dot{x}_1 &= x_2 \\ \ddot{x}_1 &= x_3 \end{aligned}$$

Hence,

$$y = c(t) = b_2 x_3 + b_1 x_2 + b_0 x_1 = x_3 + x_2 + 2x_1 \tag{3.66}$$

Thus, the last box of Figure 3.11(*b*) "collects" the states and generates the output equation. From Eq. (3.66),

$$y = \begin{bmatrix} b_0 & b_1 & b_2 \end{bmatrix} \begin{bmatrix} x_1 \\ x_2 \\ x_3 \end{bmatrix} = \begin{bmatrix} 2 & 7 & 1 \end{bmatrix} \begin{bmatrix} x_1 \\ x_2 \\ x_3 \end{bmatrix} \tag{3.67}$$

Although the second block of Figure 3.12(*b*) shows differentiation, this block was implemented without differentiation because of the partitioning that was applied to the transfer function. The last block simply collected derivatives that were already formed by the first block.

Once again we can produce an equivalent block diagram that vividly represents our state-space model. The first block of Figure 3.12(*b*) is the same as Figure 3.10(*a*) except for the different constant in the numerator. Thus, in Figure 3.12(*c*) we reproduce Figure 3.10(*b*) except for the change in the numerator constant, which appears as a change in the input multiplying factor. The second block of Figure 3.12(*b*) is represented using Eq. (3.66), which forms the output from a linear combination of the state variables, as shown in Figure 3.12(*c*).

**TryIt 3.1**

Use the following MATLAB statements to form an LTI state-space representation from the transfer function shown in Figure 3.12(a). The **A** matrix and **B** vector are shown in Eq. (3.63). The **C** vector is shown in Eq. (3.67).

```
num=[1 7 2];
den=[1 9 26 24];
[A,B,C,D]=tf2ss...
  (num, den);
P=[0 0 1;0 1 0;1 0 0];
A=inv(P)*A*P
B=inv(P)*B
C=C*P
```

**Skill-Assessment Exercise 3.3**

**PROBLEM:** Find the state equations and output equation for the phase-variable representation of the transfer function $G(s) = \dfrac{2s+1}{s^2+7s+9}$.

**ANSWER:**

$$\dot{\mathbf{x}} = \begin{bmatrix} 0 & 1 \\ -9 & -7 \end{bmatrix}\mathbf{x} + \begin{bmatrix} 0 \\ 1 \end{bmatrix} r(t)$$

$$y = [1 \quad 2]\mathbf{x}$$

The complete solution is at www.wiley.com/college.nise.

## 3.6 Converting from State Space to a Transfer Function

In Chapters 2 and 3, we have explored two methods of representing systems: the transfer function representation and the state-space representation. In the last section, we united the two representations by converting transfer functions into state-space representations. Now we move in the opposite direction and convert the state-space representation into a transfer function.

Given the state and output equations

$$\dot{\mathbf{x}} = \mathbf{Ax} + \mathbf{Bu} \tag{3.68a}$$

$$\mathbf{y} = \mathbf{Cx} + \mathbf{Du} \tag{3.68b}$$

take the Laplace transform assuming zero initial conditions:[8]

$$s\mathbf{X}(s) = \mathbf{AX}(s) + \mathbf{BU}(s) \tag{3.69a}$$

$$\mathbf{Y}(s) = \mathbf{CX}(s) + \mathbf{DU}(s) \tag{3.69b}$$

Solving for $\mathbf{X}(s)$ in Eq. (3.69a),

$$(s\mathbf{I} - \mathbf{A})\mathbf{X}(s) = \mathbf{BU}(s) \tag{3.70}$$

or

$$\mathbf{X}(s) = (s\mathbf{I} - \mathbf{A})^{-1}\mathbf{BU}(s) \tag{3.71}$$

where $\mathbf{I}$ is the identity matrix.

Substituting Eq. (3.71) into Eq. (3.69b) yields

$$\mathbf{Y}(s) = \mathbf{C}(s\mathbf{I} - \mathbf{A})^{-1}\mathbf{BU}(s) + \mathbf{DU}(s) = [\mathbf{C}(s\mathbf{I} - \mathbf{A})^{-1}\mathbf{B} + \mathbf{D}]\mathbf{U}(s) \tag{3.72}$$

We call the matrix $[\mathbf{C}(s\mathbf{I} - \mathbf{A})^{-1}\mathbf{B} + \mathbf{D}]$ the transfer function matrix, since it relates the output vector, $\mathbf{Y}(s)$, to the input vector, $\mathbf{U}(s)$. However, if $\mathbf{U}(s) = U(s)$ and $\mathbf{Y}(s) = Y(s)$ are scalars, we can find the transfer function. Thus,

[8] The Laplace transform of a vector is found by taking the Laplace transform of each component. Since $\dot{\mathbf{x}}$ consists of the derivatives of the state variables, the Laplace transform of $\dot{\mathbf{x}}$ with zero initial conditions yields each component with the form $sX_i(s)$, where $X_i(s)$ is the Laplace transform of the state variable. Factoring out the complex variable, $s$, in each component yields the Laplace transform of $\dot{\mathbf{x}}$ as $s\,\mathbf{X}(s)$, where $\mathbf{X}(s)$ is a column vector with components $X_i(s)$.

$$T(s) = \frac{Y(s)}{U(s)} = \mathbf{C}(s\mathbf{I} - \mathbf{A})^{-1}\mathbf{B} + \mathbf{D} \tag{3.73}$$

Let us look at an example.

## Example 3.6 State-Space Representation to Transfer Function

**PROBLEM:** Given the system defined by Eq. (3.74), find the transfer function, $T(s) = Y(s)/U(s)$, where $U(s)$ is the input and $Y(s)$ is the output.

$$\dot{\mathbf{x}} = \begin{bmatrix} 0 & 1 & 0 \\ 0 & 0 & 1 \\ -1 & -2 & -3 \end{bmatrix}\mathbf{x} + \begin{bmatrix} 10 \\ 0 \\ 0 \end{bmatrix}u \tag{3.74a}$$

$$y = [1 \quad 0 \quad 0]\mathbf{x} \tag{3.74b}$$

**SOLUTION:** The solution revolves around finding the term $(s\mathbf{I} - \mathbf{A})^{-1}$ in Eq. (3.73).[9] All other terms are already defined. Hence, first find $(s\mathbf{I} - \mathbf{A})$:

$$(s\mathbf{I} - \mathbf{A}) = \begin{bmatrix} s & 0 & 0 \\ 0 & s & 0 \\ 0 & 0 & s \end{bmatrix} - \begin{bmatrix} 0 & 1 & 0 \\ 0 & 0 & 1 \\ -1 & -2 & -3 \end{bmatrix} = \begin{bmatrix} s & -1 & 0 \\ 0 & s & -1 \\ 1 & 2 & s+3 \end{bmatrix} \tag{3.75}$$

Now form $(s\mathbf{I} - \mathbf{A})^{-1}$:

$$(s\mathbf{I} - \mathbf{A})^{-1} = \frac{\operatorname{adj}(s\mathbf{I} - \mathbf{A})}{\det(s\mathbf{I} - \mathbf{A})} = \frac{\begin{bmatrix} (s^2+3s+2) & s+3 & 1 \\ -1 & s(s+3) & s \\ -s & -(2s+1) & s^2 \end{bmatrix}}{s^3 + 3s^2 + 2s + 1} \tag{3.76}$$

Substituting $(s\mathbf{I} - \mathbf{A})^{-1}$, **B**, **C**, and **D** into Eq. (3.73), where

$$\mathbf{B} = \begin{bmatrix} 10 \\ 0 \\ 0 \end{bmatrix}$$

$$\mathbf{C} = [1 \quad 0 \quad 0]$$

$$\mathbf{D} = 0$$

we obtain the final result for the transfer function:

$$T(s) = \frac{10(s^2 + 3s + 2)}{s^3 + 3s^2 + 2s + 1} \tag{3.77}$$

[9] See Appendix G. It discusses the evaluation of the matrix inverse.

MATLAB
ML

Students who are using MATLAB should now run ch3p5 in Appendix B. You will learn how to convert a state-space representation to a transfer function using MATLAB. You can practice by writing a MATLAB program to solve Example 3.6.

Symbolic Math
SM

Students who are performing the MATLAB exercises and want to explore the added capability of MATLAB's Symbolic Math Toolbox should now run ch3sp1 in Appendix F located at www.wiley.com/college/nise. You will learn how to use the Symbolic Math Toolbox to write matrices and vectors. You will see that the Symbolic Math Toolbox yields an alternative way to use MATLAB to solve Example 3.6.

## Skill-Assessment Exercise 3.4

**TryIt 3.2**

Use the following MATLAB and the Control System Toolbox statements to obtain the transfer function shown in Skill-Assessment Exercise 3.4 from the state-space representation of Eq. (3.78).

```
A=[-4 -1.5;4 0];
B=[2 0]';
C=[1.5 0.625];
D=0;
T=ss(A,B,C,D);
T=tf(T)
```

**PROBLEM:** Convert the state and output equations shown in Eq. (3.78) to a transfer function.

$$\dot{\mathbf{x}} = \begin{bmatrix} -4 & -1.5 \\ 4 & 0 \end{bmatrix}\mathbf{x} + \begin{bmatrix} 2 \\ 0 \end{bmatrix}u(t) \tag{3.78a}$$

$$y = \begin{bmatrix} 1.5 & 0.625 \end{bmatrix}\mathbf{x} \tag{3.78b}$$

**ANSWER:**

$$G(s) = \frac{3s+5}{s^2+4s+6}$$

The complete solution is located at www.wiley.com/college/nise.

In Example 3.6, the state equations in phase-variable form were converted to transfer functions. In Chapter 5, we will see that other forms besides the phase-variable form can be used to represent a system in state space. The method of finding the transfer function representation for these other forms is the same as that presented in this section.

# 3.7 Linearization

A prime advantage of the state-space representation over the transfer function representation is the ability to represent systems with nonlinearities, such as the one shown in Figure 3.13. The ability to represent nonlinear systems does not imply the ability to solve their state equations for the state variables and the output. Techniques do exist for the solution of some nonlinear state equations, but this study is beyond the scope of this course. However, in Appendix H, you can see how to use the digital computer to solve state equations. This method also can be used for nonlinear state equations.

If we are interested in small perturbations about an equilibrium point, as we were when we studied linearization in Chapter 2, we can also linearize the state equations about the equilibrium point. The key to linearization about an equilibrium point is, once again, the Taylor series. In the following example, we write the state equations for a simple pendulum, showing that we can represent a nonlinear system in state space; then we linearize the pendulum about its equilibrium point, the vertical position with zero velocity.

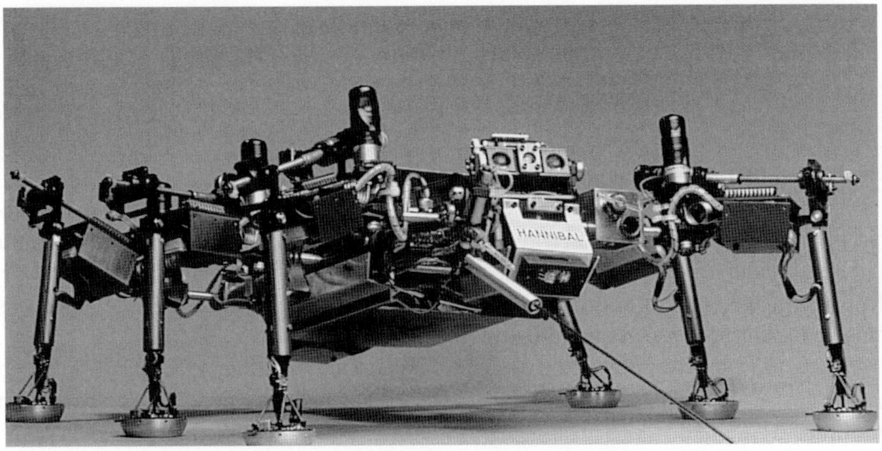




**FIGURE 3.13** Walking robots, such as *Hannibal* shown here, can be used to explore hostile environments and rough terrain, such as that found on other planets or inside volcanoes

## Example 3.7 Representing a Nonlinear System

**PROBLEM:** First represent the simple pendulum shown in Figure 3.14(*a*) (which could be a simple model for the leg of the robot shown in Figure 3.13) in state space: $Mg$ is the weight, $T$ is an applied torque in the $\theta$ direction, and $L$ is the length of the pendulum. Assume the mass is evenly distributed, with the center of mass at $L/2$. Then linearize the state equations about the pendulum's equilibrium point—the vertical position with zero angular velocity.

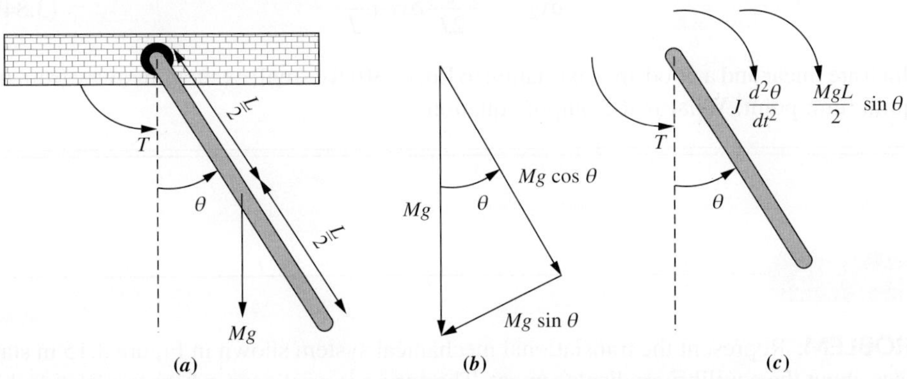


**FIGURE 3.14** **a.** Simple pendulum; **b.** force components of $Mg$; **c.** free-body diagram

**SOLUTION:** First draw a free-body diagram as shown in Figure 3.14(*c*). Summing the torques, we get

$$J\frac{d^2\theta}{dt^2}+\frac{MgL}{2}\sin\theta = T \tag{3.79}$$

where $J$ is the moment of inertia of the pendulum around the point of rotation. Select the state variables $x_1$ and $x_2$ as phase variables. Letting $x_1 = \theta$ and $x_2 = d\theta/dt$, we write the state equations as

$$\dot{x}_1 = x_2 \tag{3.80a}$$

$$\dot{x}_2 = -\frac{MgL}{2J}\sin x_1 + \frac{T}{J} \tag{3.80b}$$

where $\dot{x}_2 = d^2\theta/dt^2$ is evaluated from Eq. (3.79).

Thus, we have represented a nonlinear system in state space. It is interesting to note that the nonlinear Eqs. (3.80) represent a valid and complete model of the pendulum in state space even under nonzero initial conditions and even if parameters are time varying. However, if we want to apply classical techniques and convert these state equations to a transfer function, we must linearize them.

Let us proceed now to linearize the equation about the equilibrium point, $x_1 = 0, x_2 = 0$, that is, $\theta = 0$ and $d\theta/dt = 0$. Let $x_1$ and $x_2$ be perturbed about the equilibrium point, or

$$x_1 = 0 + \delta x_1 \tag{3.81a}$$

$$x_2 = 0 + \delta x_2 \tag{3.81b}$$

Using Eq. (2.182), we obtain

$$\sin x_1 - \sin 0 = \left.\frac{d(\sin x_1)}{dx_1}\right|_{x_1=0} \delta x_1 = \delta x_1 \tag{3.82}$$

from which

$$\sin x_1 = \delta x_1 \tag{3.83}$$

Substituting Eqs. (3.81) and (3.83) into Eq. (3.80) yields the following state equations:

$$\dot{\delta x}_1 = \delta x_2 \tag{3.84a}$$

$$\dot{\delta x}_2 = -\frac{MgL}{2J}\delta x_1 + \frac{T}{J} \tag{3.84b}$$

which are linear and a good approximation to Eq. (3.80) for small excursions away from the equilibrium point. What is the output equation?

## Skill-Assessment Exercise 3.5

**PROBLEM:** Represent the translational mechanical system shown in Figure 3.15 in state space about the equilibrium displacement. The spring is nonlinear, where the relationship between the spring force, $f_s(t)$, and the spring displacement, $x_s(t)$, is $f_s(t) = 2x_s^2(t)$. The applied force is $f(t) = 10 + \delta f(t)$, where $\delta f(t)$ is a small force about the 10 N constant value.

Assume the output to be the displacement of the mass, $x(t)$.

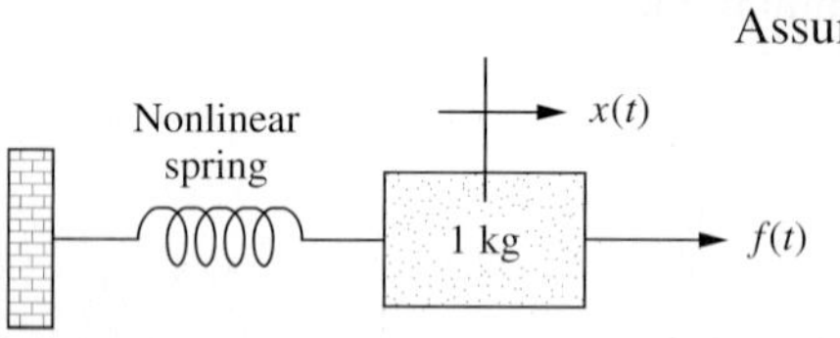


**FIGURE 3.15** Nonlinear translational mechanical system for Skill-Assessment Exercise 3.5

**ANSWER:**

$$\dot{\mathbf{x}} = \begin{bmatrix} 0 & 1 \\ -4\sqrt{5} & 0 \end{bmatrix}\mathbf{x} + \begin{bmatrix} 0 \\ 1 \end{bmatrix}\delta f(t)$$

$$y = \begin{bmatrix} 1 & 0 \end{bmatrix}\mathbf{x}$$

The complete solution is located at www.wiley.com/college/nise.

## Case Studies Antenna Control: State-Space Representation

We have covered the state-space representation of individual physical subsystems in this chapter. In Chapter 5, we will assemble individual subsystems into feedback control systems and represent the entire feedback system in state space. Chapter 5 also shows how the state-space representation, via signal-flow diagrams, can be used to interconnect these subsystems and permit the state-space representation of the whole closed-loop system. In the following case study, we look at the antenna azimuth position control system and demonstrate the concepts of this chapter by representing each subsystem in state space.

**PROBLEM:** Find the state-space representation in phase-variable form for each dynamic subsystem in the antenna azimuth position control system shown on the front endpapers, *Configuration 1*. By *dynamic*, we mean that the system does not reach the steady state instantaneously. For example, a system described by a differential equation of first order or higher is a dynamic system. A pure gain, on the other hand, is an example of a nondynamic system, since the steady state is reached instantaneously.

**SOLUTION:** In the case study problem of Chapter 2, each subsystem of the antenna azimuth position control system was identified. We found that the power amplifier and the motor and load were dynamic systems. The preamplifier and the potentiometers are pure gains and so respond instantaneously. Hence, we will find the state-space representations only of the power amplifier and of the motor and load.

**Power amplifier:**

The transfer function of the power amplifier is given on the front endpapers as $G(s) = 100/(s + 100)$. We will convert this transfer function to its state-space representation. Letting $v_p(t)$ represent the power amplifier input and $e_a(t)$ represent the power amplifier output,

$$G(s) = \frac{E_a(s)}{V_p(s)} = \frac{100}{(s + 100)} \tag{3.85}$$

Cross-multiplying, $(s + 100)E_a(s) = 100V_p(s)$, from which the differential equation can be written as

$$\frac{de_a}{dt} + 100e_a = 100v_p(t) \tag{3.86}$$

Rearranging Eq. (3.86) leads to the state equation with $e_a$ as the state variable:

$$\frac{de_a}{dt} = -100e_a + 100v_p(t) \tag{3.87}$$

Since the output of the power amplifier is $e_a(t)$, the output equation is

$$y = e_a \tag{3.88}$$

**Motor and load:**

We now find the state-space representation for the motor and load. We could of course use the motor and load block shown in the block diagram on the front endpapers to obtain the result. However, it is more informative to derive the state-space representation directly from the physics of the motor without first deriving the transfer function. The elements of the derivation were covered in Section 2.8 but are repeated here for continuity. Starting with Kirchhoff's voltage equation around the armature circuit, we find

$$e_a(t) = i_a(t)R_a + K_b\frac{d\theta_m}{dt} \tag{3.89}$$

where $e_a(t)$ is the armature input voltage, $i_a(t)$ is the armature current, $R_a$ is the armature resistance, $K_b$ is the armature constant, and $\theta_m$ is the angular displacement of the armature.

The torque, $T_m(t)$, delivered by the motor is related separately to the armature current and the load seen by the armature. From Section 2.8,

$$T_m(t) = K_t i_a(t) = J_m\frac{d^2\theta_m}{dt^2} + D_m\frac{d\theta_m}{dt} \tag{3.90}$$

where $J_m$ is the equivalent inertia as seen by the armature, and $D_m$ is the equivalent viscous damping as seen by the armature.

Solving Eq. (3.90) for $i_a(t)$ and substituting the result into Eq. (3.89) yields

$$e_a(t) = \left(\frac{R_a J_m}{K_t}\right)\frac{d^2\theta_m}{dt^2} + \left(\frac{D_m R_a}{K_t} + K_b\right)\frac{d\theta_m}{dt} \tag{3.91}$$

Defining the state variables $x_1$ and $x_2$ as

$$x_1 = \theta_m \tag{3.92a}$$

$$x_2 = \frac{d\theta_m}{dt} \tag{3.92b}$$

and substituting into Eq. (3.91), we get

$$e_a(t) = \left(\frac{R_a J_m}{K_t}\right)\frac{dx_2}{dt} + \left(\frac{D_m R_a}{K_t} + K_b\right)x_2 \tag{3.93}$$

Solving for $dx_2/dt$ yields

$$\frac{dx_2}{dt} = -\frac{1}{J_m}\left(D_m + \frac{K_t K_b}{R_a}\right)x_2 + \left(\frac{K_t}{R_a J_m}\right)e_a(t) \tag{3.94}$$

Using Eqs. (3.92) and (3.94), the state equations are written as

$$\frac{dx_1}{dt} = x_2 \tag{3.95a}$$

$$\frac{dx_2}{dt} = -\frac{1}{J_m}\left(D_m + \frac{K_t K_b}{R_a}\right)x_2 + \left(\frac{K_t}{R_a J_m}\right)e_a(t) \tag{3.95b}$$

The output, $\theta_o(t)$, is 1/10 the displacement of the armature, which is $x_1$. Hence, the output equation is

$$y = 0.1x_1 \tag{3.96}$$

In vector-matrix form,

$$\dot{\mathbf{x}} = \begin{bmatrix} 0 & 1 \\ 0 & -\frac{1}{J_m}\left(D_m + \frac{K_t K_b}{R_a}\right) \end{bmatrix}\mathbf{x} + \begin{bmatrix} 0 \\ \frac{K_t}{R_a J_m} \end{bmatrix} e_a(t) \tag{3.97a}$$

$$y = [0.1 \quad 0]\mathbf{x} \tag{3.97b}$$

But from the case study problem in Chapter 2, $J_m = 0.03$ and $D_m = 0.02$. Also, $K_t/R_a = 0.0625$ and $K_b = 0.5$. Substituting the values into Eq. (3.97a), we obtain the final state-space representation:

$$\dot{\mathbf{x}} = \begin{bmatrix} 0 & 1 \\ 0 & -1.71 \end{bmatrix}\mathbf{x} + \begin{bmatrix} 0 \\ 2.083 \end{bmatrix} e_a(t) \tag{3.98a}$$

$$y = \begin{bmatrix} 0.1 & 0 \end{bmatrix}\mathbf{x} \tag{3.98b}$$

**CHALLENGE:** You are now given a problem to test your knowledge of this chapter's objectives. Referring to the antenna azimuth position control system shown on the front endpapers, find the state-space representation of each dynamic subsystem. Use Configuration 2.

## Pharmaceutical Drug Absorption

An advantage of state-space representation over the transfer function representation is the ability to focus on component parts of a system and write $n$ simultaneous, first-order differential equations rather than attempt to represent the system as a single, $n$th-order differential equation, as we have done with the transfer function. Also, multiple-input, multiple-output systems can be conveniently represented in state space. This case study demonstrates both of these concepts.

**PROBLEM:** In the pharmaceutical industry we want to describe the distribution of a drug in the body. A simple model divides the process into compartments: the dosage, the absorption site, the blood, the peripheral compartment, and the urine. The rate of change of the amount of a drug in a compartment is equal to the input flow rate diminished by the output flow rate. Figure 3.16 summarizes the system. Here each $x_i$ is the amount of drug in that particular compartment (*Lordi, 1972*). Represent the system in state space, where the outputs are the amounts of drug in each compartment.

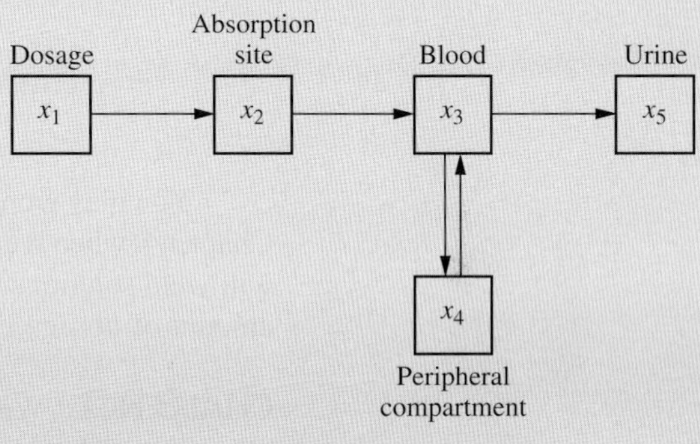


**FIGURE 3.16** Pharmaceutical drug-level concentrations in a human

**SOLUTION:** The flow rate of the drug into any given compartment is proportional to the concentration of the drug in the previous compartment, and the flow rate out of a given compartment is proportional to the concentration of the drug in its own compartment.

We now write the flow rate for each compartment. The dosage is released to the absorption site at a rate proportional to the dosage concentration, or

$$\frac{dx_1}{dt} = -K_1x_1 \tag{3.99}$$

The flow into the absorption site is proportional to the concentration of the drug at the dosage site. The flow from the absorption site into the blood is proportional to the concentration of the drug at the absorption site. Hence,

$$\frac{dx_2}{dt} = K_1x_1 - K_2x_2 \tag{3.100}$$

Similarly, the net flow rate into the blood and peripheral compartment is

$$\frac{dx_3}{dt} = K_2x_2 - K_3x_3 + K_4x_4 - K_5x_3 \tag{3.101}$$

$$\frac{dx_4}{dt} = K_5x_3 - K_4x_4 \tag{3.102}$$

where $(K_4x_4 - K_5x_3)$ is the net flow rate into the blood from the peripheral compartment. Finally, the amount of the drug in the urine is increased as the blood releases the drug to the urine at a rate proportional to the concentration of the drug in the blood. Thus,

$$\frac{dx_5}{dt} = K_3x_3 \tag{3.103}$$

Equations (3.99) through (3.103) are the state equations. The output equation is a vector that contains each of the amounts, $x_i$. Thus, in vector-matrix form,

$$\dot{\mathbf{x}} = \begin{bmatrix} -K_1 & 0 & 0 & 0 & 0 \\ K_1 & -K_2 & 0 & 0 & 0 \\ 0 & K_2 & -(K_3+K_5) & K_4 & 0 \\ 0 & 0 & K_5 & -K_4 & 0 \\ 0 & 0 & K_3 & 0 & 0 \end{bmatrix} \mathbf{x} \tag{3.104a}$$

$$\mathbf{y} = \begin{bmatrix} 1 & 0 & 0 & 0 & 0 \\ 0 & 1 & 0 & 0 & 0 \\ 0 & 0 & 1 & 0 & 0 \\ 0 & 0 & 0 & 1 & 0 \\ 0 & 0 & 0 & 0 & 1 \end{bmatrix} \mathbf{x} \tag{3.104b}$$

You may wonder how there can be a solution to these equations if there is no input. In Chapter 4, when we study how to solve the state equations, we will see that initial conditions will yield solutions without forcing functions. For this problem, an initial condition on the amount of dosage, $x_1$, will generate drug quantities in all other compartments.

**CHALLENGE:** We now give you a problem to test your knowledge of this chapter's objectives. The problem concerns the storage of water in aquifers. The principles are similar to those used to model pharmaceutical drug absorption.

Underground water supplies, called aquifers, are used in many areas for agricultural, industrial, and residential purposes. An aquifer system consists of a number of interconnected natural storage tanks. Natural water flows through the sand and sandstone of the aquifer system, changing the water levels in the tanks on its way to the sea. A water conservation policy can be established whereby water is pumped between tanks to prevent its loss to the sea.

A model for the aquifer system is shown in Figure 3.17. In this model, the aquifer is represented by three tanks, with water level $h_i$ called the *head*. Each $q_n$ is the natural water flow to the sea and is proportional to the difference in head between two adjoining tanks, or $q_n = G_n(h_n - h_{n-1})$, where $G_n$ is a constant of proportionality and the units of $q_n$ are m$^3$/yr.

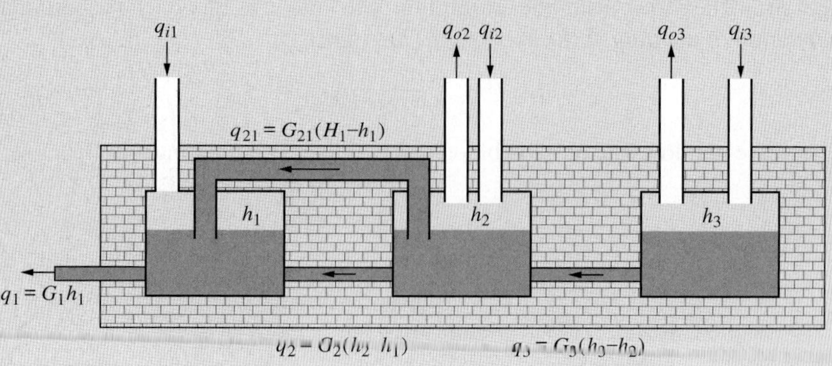


**FIGURE 3.17** Aquifer system model

The engineered flow consists of three components, also measured in $m^3/yr$: (1) flow from the tanks for irrigation, industry, and homes, $q_{on}$; (2) replenishing of the tanks from wells, $q_{in}$; and (3) flow, $q_{21}$, created by the water conservation policy to prevent loss to the sea. In this model, water for irrigation and industry will be taken only from Tank 2 and Tank 3. Water conservation will take place only between Tank 1 and Tank 2, as follows. Let $H_1$ be a reference head for Tank 1. If the water level in Tank 1 falls below $H_1$, water will be pumped from Tank 2 to Tank 1 to replenish the head. If $h_1$ is higher than $H_1$, water will be pumped back to Tank 2 to prevent loss to the sea. Calling this *flow for conservation* $q_{21}$, we can say this flow is proportional to the difference between the head of Tank 1, $h_1$, and the reference head, $H_1$, or $q_{21} = G_{21}(H_1 - h_1)$.

The net flow into a tank is proportional to the rate of change of head in each tank. Thus,

$$C_n dh_n/dt = q_{in} - q_{on} + q_{n+1} - q_n + q_{(n+1)n} - q_{n(n-1)}$$

(*Kandel, 1973*).

Represent the aquifer system in state space, where the state variables and the outputs are the heads of each tank.

## Summary

This chapter has dealt with the state-space representation of physical systems, which took the form of a state equation,

$$\dot{\mathbf{x}} = \mathbf{A}\mathbf{x} + \mathbf{B}\mathbf{u} \tag{3.105}$$

and an output equation,

$$\mathbf{y} = \mathbf{C}\mathbf{x} + \mathbf{D}\mathbf{u} \tag{3.106}$$

for $t \geq t_0$, and initial conditions $\mathbf{x}(t_0)$. Vector $\mathbf{x}$ is called the *state vector* and contains variables, called *state variables*. The state variables can be combined algebraically with the input to form the output equation, Eq. (3.106), from which any other system variables can be found. State variables, which can represent physical quantities such as current or voltage, are chosen to be linearly independent. The choice of state variables is not unique and affects how the matrices **A**, **B**, **C**, and **D** look. We will solve the state and output equations for **x** and **y** in Chapter 4.

In this chapter, transfer functions were represented in state space. The form selected was the phase-variable form, which consists of state variables that are successive derivatives of each other. In three-dimensional state space, the resulting system matrix, **A**, for the phase-variable representation is of the form

$$\begin{bmatrix} 0 & 1 & 0 \\ 0 & 0 & 1 \\ -a_0 & -a_1 & -a_2 \end{bmatrix} \tag{3.107}$$

where the $a_i$'s are the coefficients of the characteristic polynomial or denominator of the system transfer function. We also discussed how to convert from a state-space representation to a transfer function.

In conclusion, then, for linear, time-invariant systems, the state-space representation is simply another way of mathematically modeling them. One major advantage of applying the state-space representation to such linear systems is that it allows computer simulation. Programming the system on the digital computer and watching the system's response is an invaluable analysis and design tool. Simulation is covered in Appendix H located at www.wiley.com/college/nise.

## Review Questions

1. Give two reasons for modeling systems in state space.
2. State an advantage of the transfer function approach over the state-space approach.
3. Define *state variables*.
4. Define *state*.
5. Define *state vector*.
6. Define *state space*.
7. What is required to represent a system in state space?
8. An eighth-order system would be represented in state space with how many state equations?
9. If the state equations are a system of first-order differential equations whose solution yields the state variables, then the output equation performs what function?
10. What is meant by *linear independence*?
11. What factors influence the choice of state variables in any system?
12. What is a convenient choice of state variables for electrical networks?
13. If an electrical network has three energy-storage elements, is it possible to have a state-space representation with more than three state variables? Explain.
14. What is meant by the phase-variable form of the state equation?

## Problems

1. Represent the electrical network shown in Figure P3.1 in state space, where $v_o(t)$ is the output. [Section: 3.4]

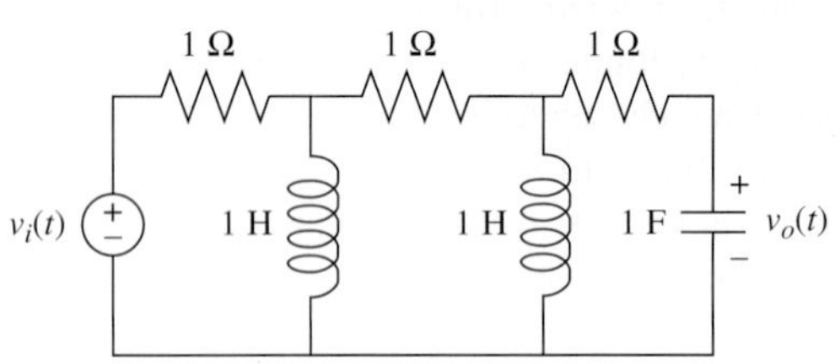


**FIGURE P3.1**

2. Represent the electrical network shown in Figure P3.2 in state space, where $i_R(t)$ is the output. [Section: 3.4]

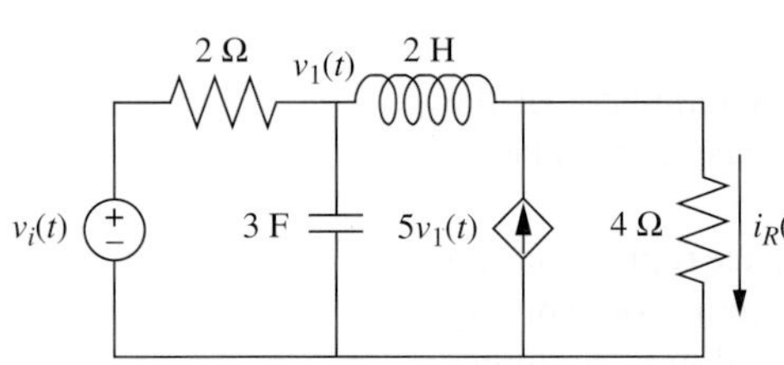


**FIGURE P3.2**

3. Find the state-space representation of the network shown in Figure P3.3 if the output is $v_o(t)$. [Section: 3.4]

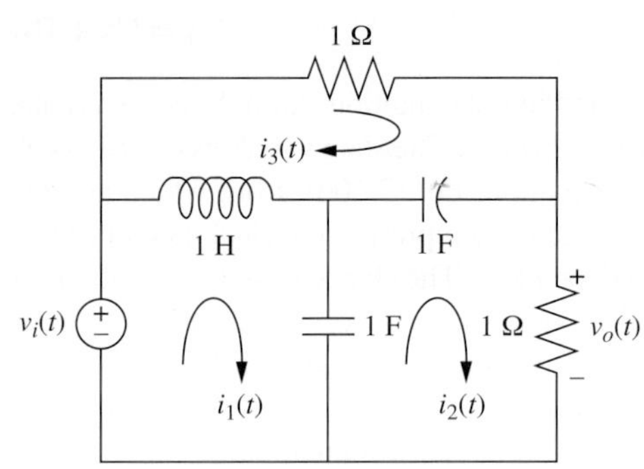


**FIGURE P3.3**

4. Represent the system shown in Figure P3.4 in state space where the output is $x_3(t)$. [Section: 3.4]

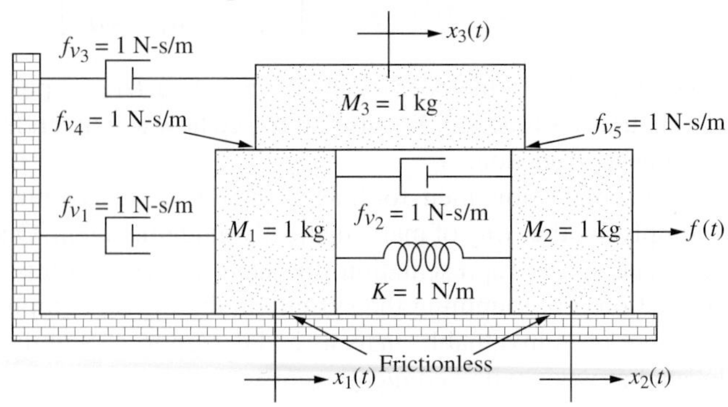


**FIGURE P3.4**

5. Represent the translational mechanical system shown in Figure P3.5 in state space, where $x_1(t)$ is the output. [Section: 3.4]

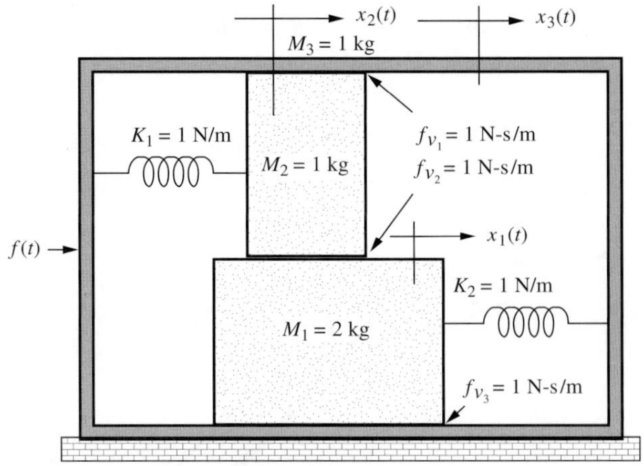


**FIGURE P3.5**

6. Represent the rotational mechanical system shown in Figure P3.6 in state space, where $\theta_1(t)$ is the output. [Section: 3.4]

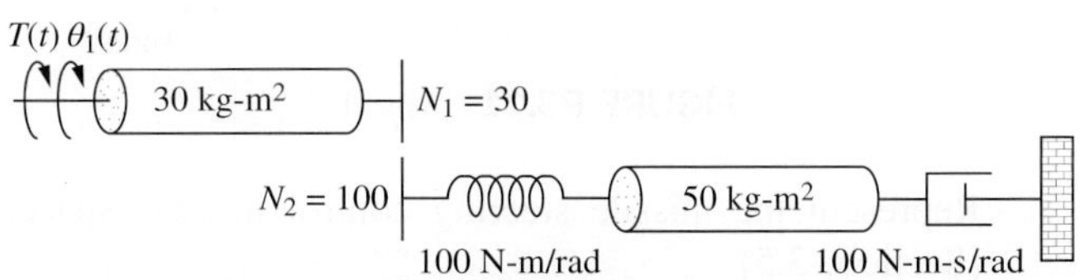


**FIGURE P3.6**

7. Represent the system shown in Figure P3.7 in state space where the output is $\theta_L(t)$. [Section: 3.4]

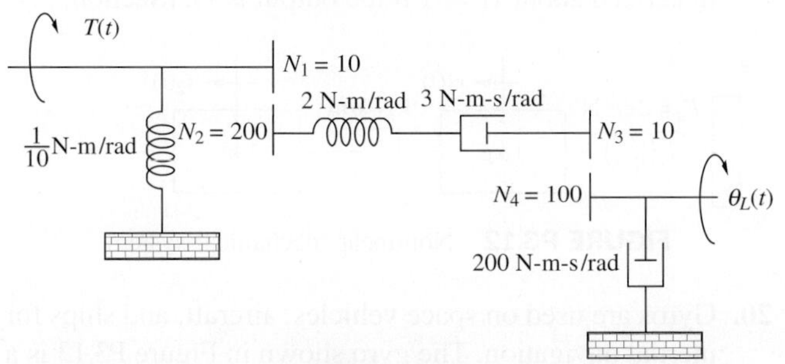


**FIGURE P3.7**

8. Show that the system of Figure 3.7 in the text yields a fourth-order transfer function if we relate the displacement of either mass to the applied force, and a third-order one if we relate the velocity of either mass to the applied force. [Section: 3.4]

9. Find the state-space representation in phase-variable form for each of the systems shown in Figure P3.8. [Section: 3.5]

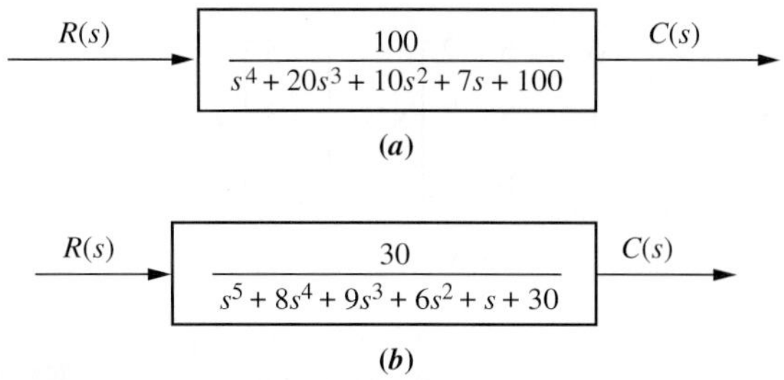


**FIGURE P3.8**

10. `Repeat Problem 9 using MATLAB. [Section: 3.5]` MATLAB ML

11. For each system shown in Figure P3.9, write the state equations and the output equation for the phase-variable representation. [Section: 3.5]

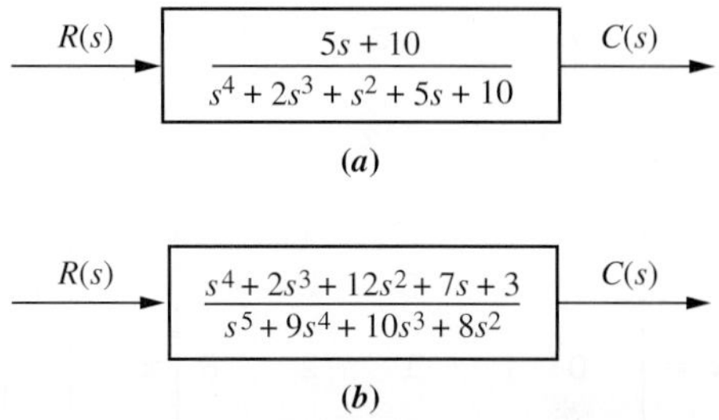


**FIGURE P3.9**

12. `Repeat Problem 11 using MATLAB. [Section: 3.5]` MATLAB ML

13. Represent the following transfer function in state space. Give your answer in vector-matrix form. [Section: 3.5]

$$T(s) = \frac{(s^2+3s+8)}{(s+1)(s^2+5s+5)}$$

14. Find the transfer function $G(s) = Y(s)/R(s)$ for each of the following systems represented in state space: [Section: 3.6]

**a.** $$\dot{\mathbf{x}} = \begin{bmatrix} 0 & 1 & 0 \\ 0 & 0 & 1 \\ -3 & -2 & -5 \end{bmatrix}\mathbf{x} + \begin{bmatrix} 0 \\ 0 \\ 10 \end{bmatrix} r$$

$$y = [1 \quad 0 \quad 0]\mathbf{x}$$

**b.** $$\dot{\mathbf{x}} = \begin{bmatrix} 2 & -3 & -8 \\ 0 & 5 & 3 \\ -3 & -5 & -4 \end{bmatrix}\mathbf{x} + \begin{bmatrix} 1 \\ 4 \\ 6 \end{bmatrix} r$$

$$y = [1 \quad 3 \quad 6]\mathbf{x}$$

**c.** $\dot{\mathbf{x}} = \begin{bmatrix} 3 & -5 & 2 \\ 1 & -8 & 7 \\ -3 & -6 & 2 \end{bmatrix}\mathbf{x} + \begin{bmatrix} 5 \\ -3 \\ 2 \end{bmatrix} r$

$y = [1 \quad -4 \quad 3]\mathbf{x}$

**15.** Use MATLAB to find the transfer function, $G(s) = Y(s)/R(s)$, for each of the following systems represented in state space: [Section: 3.6] MATLAB **ML**

**a.** $\dot{\mathbf{x}} = \begin{bmatrix} 0 & 1 & 3 & 0 \\ 0 & 0 & 1 & 0 \\ 0 & 0 & 0 & 1 \\ -7 & -9 & -2 & -3 \end{bmatrix}\mathbf{x} + \begin{bmatrix} 0 \\ 5 \\ 8 \\ 2 \end{bmatrix} r$

$y = [1 \quad 3 \quad 4 \quad 6]\mathbf{x}$

**b.** $\dot{\mathbf{x}} = \begin{bmatrix} 3 & 1 & 0 & 4 & -2 \\ -3 & 5 & -5 & 2 & -1 \\ 0 & 1 & -1 & 2 & 8 \\ -7 & 6 & -3 & -4 & 0 \\ -6 & 0 & 4 & -3 & 1 \end{bmatrix}\mathbf{x} + \begin{bmatrix} 2 \\ 7 \\ 6 \\ 5 \\ 4 \end{bmatrix} r$

$y = [1 \quad -2 \quad -9 \quad 7 \quad 6]\mathbf{x}$

**16.** Repeat Problem 15 using MATLAB, the Symbolic Math Toolbox, and Eq. (3.73). [Section: 3.6] Symbolic Math **SM**

**17.** Given the dc servomotor and load shown in Figure P3.10, represent the system in state space, where the state variables are the armature current, $i_a$, load displacement, $\theta_L$, and load angular velocity, $\omega_L$. Assume that the output is the angular displacement of the armature. Do not neglect armature inductance. [Section: 3.4]

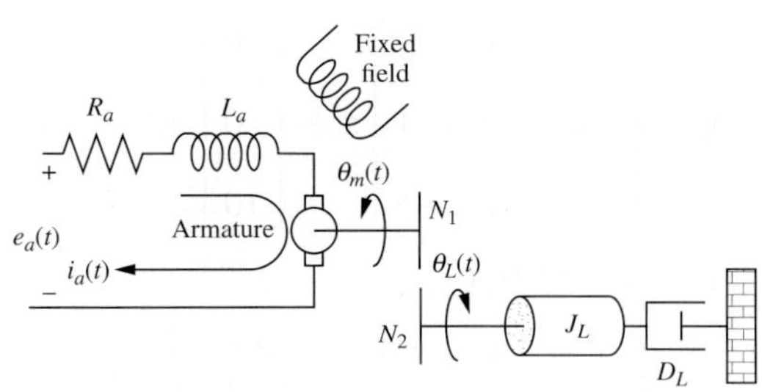


**FIGURE P3.10** Motor and load

**18.** A missile in flight, as shown in Figure P3.11, is subject to four forces: thrust, lift, drag, and gravity. The missile flies at an angle of attack, α, from its longitudinal axis, creating lift. For steering, the body angle from vertical, $\phi$, is controlled by rotating the engine at the tail. The transfer function relating the body angle, $\phi$, to the angular displacement, $\delta$, of the engine is of the form

$$\frac{\Phi(s)}{\delta(s)} = \frac{K_a s + K_b}{K_3 s^3 + K_2 s^2 + K_1 s + K_0}$$

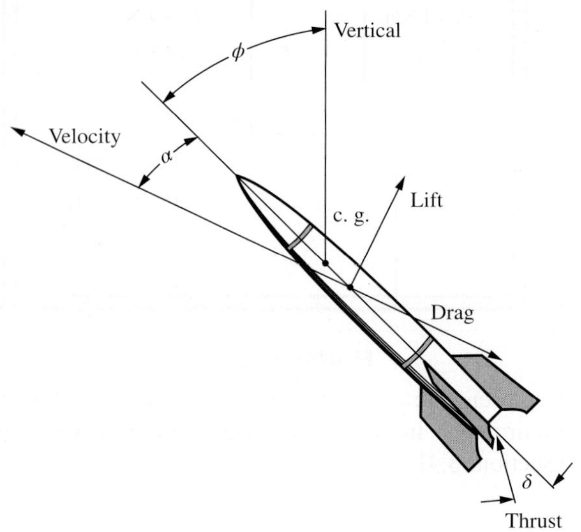


**FIGURE P3.11** Missile

Represent the missile steering control in state space. [Section: 3.5]

**19.** Consider the mechanical system of Figure P3.12. If the spring is nonlinear, and the force, $F_s$, required to stretch the spring is $F_s = 2x_1^2$, represent the system in state space linearized about $x_1 = 1$ if the output is $x_2$. [Section: 3.7]

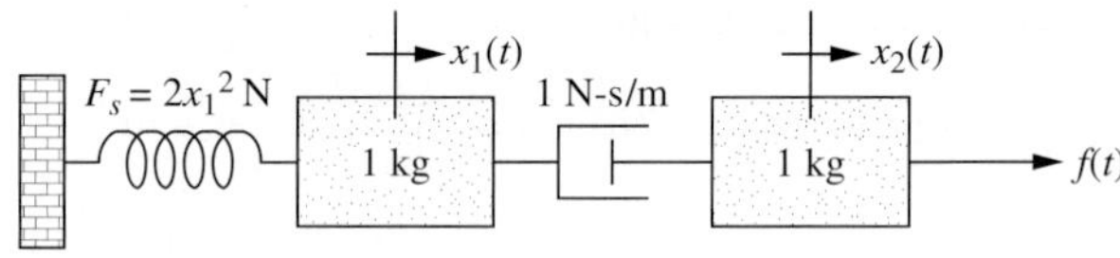


**FIGURE P3.12** Nonlinear mechanical system

**20.** Gyros are used on space vehicles, aircraft, and ships for inertial navigation. The gyro shown in Figure P3.13 is a rate gyro restrained by springs connected between the inner gimbal and the outer gimbal (frame) as shown. A rotational rate about the $z$-axis causes the rotating disk to precess about the $x$-axis. Hence, the input is a rotational rate about the $z$-axis, and the output is an angular displacement about the $x$-axis. Since the outer gimbal is secured to the vehicle, the displacement about the $x$-axis is a measure of the vehicle's angular rate about the $z$-axis. The equation of motion is

$$J_x \frac{d^2\theta_x}{dt^2} + D_x \frac{d\theta_x}{dt} + K_x\theta_x = J\omega \frac{d\theta_z}{dt}$$

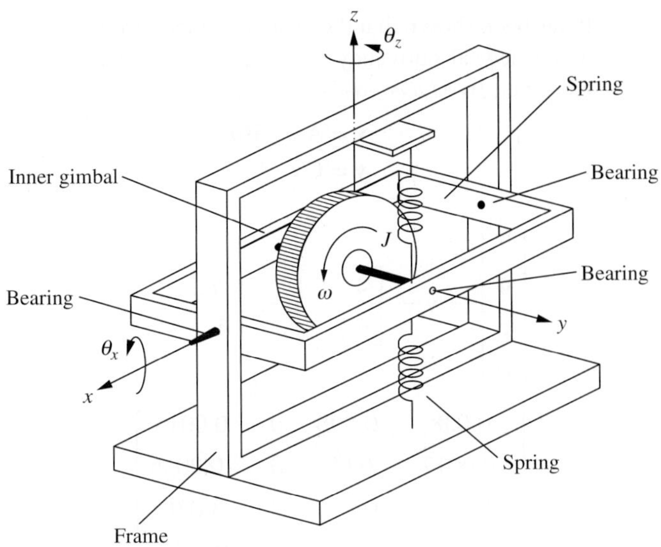


**FIGURE P3.13** Gyro system

21. Given the F4-E military aircraft shown in Figure P3.14($a$), where normal acceleration, $a_n$, and pitch rate, $q$, are controlled by elevator deflection, $\delta_e$, on the horizontal stabilizers and by canard deflection, $\delta_e$. A commanded deflection, $\delta_{com}$, as shown in Figure P3.14($b$), is used to effect a change in both $\delta_e$ and $\delta_c$. The relationships are

$$\frac{\delta_e(s)}{\delta_{com}(s)} = \frac{1/\tau}{s+1/\tau}$$

$$\frac{\delta_c(s)}{\delta_{com}(s)} = \frac{K_c/\tau}{s+1/\tau}$$

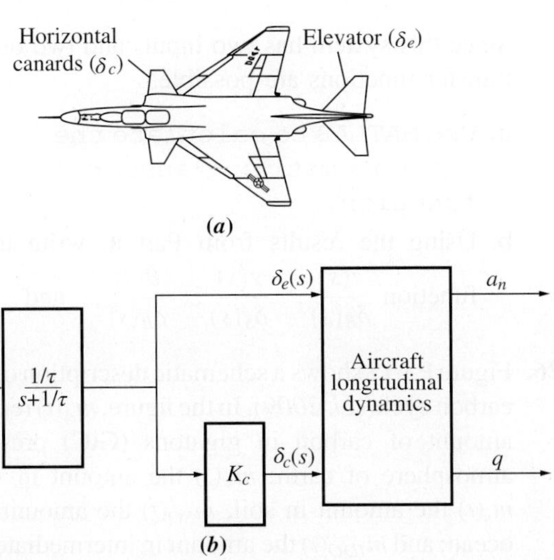


**FIGURE P3.14** **a.** F4-E with canards **b.** open-loop flight control system

These deflections yield, via the aircraft longitudinal dynamics, $a_n$ and $q$. The state equations describing the effect of $\delta_{com}$ on $a_n$ and $q$ is given by (*Cavallo, 1992*)

$$\begin{bmatrix} \dot{a}_n \\ \dot{q} \\ \dot{\delta}_e \end{bmatrix} = \begin{bmatrix} -1.702 & 50.72 & 263.38 \\ 0.22 & -1.418 & -31.99 \\ 0 & 0 & -14 \end{bmatrix} \begin{bmatrix} a_n \\ q \\ \delta_e \end{bmatrix} + \begin{bmatrix} -272.06 \\ 0 \\ 14 \end{bmatrix} \delta_{com}$$

Find the following transfer functions: [Section: 3.5]

$$G_1(s) = \frac{A_n(s)}{\delta_{com}(s)}$$

$$G_2(s) = \frac{Q(s)}{\delta_{com}(s)}$$

22. Modern robotic manipulators that act directly upon their target environments must be controlled so that impact forces as well as steady-state forces do not damage the targets. At the same time, the manipulator must provide sufficient force to perform the task. In order to develop a control system to regulate these forces, the robotic manipulator and target environment must be modeled. Assuming the model shown in Figure P3.15, represent in state space the manipulator and its environment under the following conditions (*Chiu, 1997*). [Section: 3.5]
   **a.** The manipulator is not in contact with its target environment.
   **b.** The manipulator is in constant contact with its target environment.

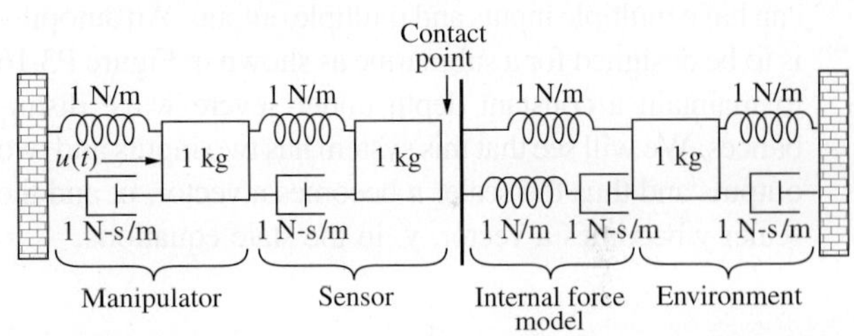


**FIGURE P3.15** Robotic manipulator and target environment[10]

23. Experiments to identify precision grip dynamics between the index finger and thumb have been performed using a ball-drop experiment. A subject holds a device with a small receptacle into which an object is dropped, and the response is measured (*Fagergren, 2000*). Assuming a step input, it has been found that the response of the motor subsystem together with the sensory system is of the form

$$G(s) = \frac{Y(s)}{R(s)} = \frac{s+c}{(s^2+as+b)(s+d)}$$

Convert this transfer function to a state-space representation.

[10] Based on Chiu, D. K., and Lee, S. Design and Experimentation of a Jump Impact Controller. *IEEE Control Systems*, June 1997, Figure 1, p. 99. 1997 IEEE.

**24.** A linear, time-invariant model of the hypothalamic-pituitary-adrenal axis of the endocrine system with five state variables has been proposed as follows (*Kyrylov, 2005*):

$$\frac{dx_0}{dt} = a_{00}x_0 + a_{02}x_2 + d_0$$

$$\frac{dx_1}{dt} = a_{10}x_0 + a_{11}x_1 + a_{12}x_2$$

$$\frac{dx_2}{dt} = a_{20}x_0 + a_{21}x_1 + a_{22}x_2 + a_{23}x_3 + a_{24}x_4$$

$$\frac{dx_3}{dt} = a_{32}x_2 + a_{33}x_3$$

$$\frac{dx_4}{dt} = a_{42}x_2 + a_{44}x_4$$

where each of the state variables represents circulatory concentrations as follows:

$x_0$ = corticotropin-releasing hormone
$x_1$ = corticotropin
$x_2$ = free cortisol
$x_3$ = albumin-bound cortisol
$x_4$ = corticosteroid-binding globulin
$d_0$ = an external generating factor

Express the system in the form $\dot{\mathbf{x}} = \mathbf{Ax} + \mathbf{Bu}$.

**25.** In this chapter, we described the state-space representation of single-input, single-output systems. In general, systems can have multiple inputs and multiple outputs. An autopilot is to be designed for a submarine as shown in Figure P3.16 to maintain a constant depth under severe wave disturbances. We will see that this system has two inputs and two outputs and thus the scaler $u$ becomes a vector, $\mathbf{u}$, and the scaler $y$ becomes a vector, $\mathbf{y}$, in the state equations.

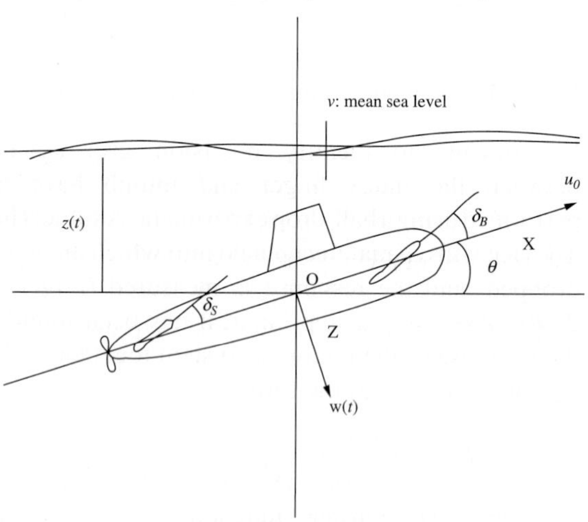


**FIGURE P3.16**[11]

[11] Liceaga-Castro E., van der Molen G.M. Submarine H∞ Depth Control Under Wave Disturbances IEEE Trans. on Control Systems Technology, Vol. 3, No. 3, 1995. Figure 1, p. 339.

It has been shown that the system's linearized dynamics under neutral buoyancy and at a given constant speed are given by (*Liceaga-Castro, 2009*):

$$\dot{\mathbf{x}} = \mathbf{Ax} + \mathbf{Bu}$$
$$\mathbf{y} = \mathbf{Cx}$$

where

$$\mathbf{x} = \begin{bmatrix} w \\ q \\ z \\ \theta \end{bmatrix}; \quad \mathbf{y} = \begin{bmatrix} z \\ \theta \end{bmatrix}; \quad \mathbf{u} = \begin{bmatrix} \delta_B \\ \delta_S \end{bmatrix}$$

$$\mathbf{A} = \begin{bmatrix} -0.038 & 0.896 & 0 & 0.0015 \\ 0.0017 & -0.092 & 0 & -0.0056 \\ 1 & 0 & 0 & -3.086 \\ 0 & 1 & 0 & 0 \end{bmatrix};$$

$$\mathbf{B} = \begin{bmatrix} -0.0075 & -0.023 \\ 0.0017 & -0.0022 \\ 0 & 0 \\ 0 & 0 \end{bmatrix}; \quad \mathbf{C} = \begin{bmatrix} 0 & 0 & 1 & 0 \\ 0 & 0 & 0 & 1 \end{bmatrix}$$

and where

$w$ = the heave velocity
$q$ = the pitch rate
$z$ = the submarine depth
$\theta$ = the pitch angle
$\delta_B$ = the bow hydroplane angle
$\delta_S$ = the stern hydroplane angle

Since this system has two inputs and two outputs, four transfer functions are possible.

**a.** Use MATLAB to calculate the system's matrix transfer function. MATLAB **ML**

**b.** Using the results from Part **a**, write the transfer function $\frac{z(s)}{\delta_B(s)}$, $\frac{z(s)}{\delta_S(s)}$, $\frac{\theta(s)}{\delta_B(s)}$, and $\frac{\theta(s)}{\delta_S(s)}$.

**26.** Figure P3.17 shows a schematic description of the global carbon cycle (*Li, 2009*). In the figure, $m_A(t)$ represents the amount of carbon in gigatons (GtC) present in the atmosphere of earth; $m_V(t)$ the amount in vegetation; $m_s(t)$ the amount in soil; $m_{SO}(t)$ the amount in surface ocean; and $m_{IDO}(t)$ the amount in intermediate and deep-ocean reservoirs. Let $u_E(t)$ stand for the human generated $CO_2$ emissions (GtC/yr). From the figure, the atmospheric mass balance in the atmosphere can be expressed as:

$$\frac{dm_A}{dt}(t) = u_E(t) - (k_{O1} + k_{L1})m_A(t) + k_{L2}m_V(t) + k_{O2}m_{SO}(t) + k_{L4}m_S(t)$$

where the $k$'s are exchange coefficients ($yr^{-1}$).

**a.** Write the remaining reservoir mass balances. Namely, write equations for $\frac{dm_{SO}(t)}{dt}$, $\frac{dm_{IDO}(t)}{dt}$, $\frac{dm_V(t)}{dt}$, and $\frac{dm_S(t)}{dt}$

**b.** Express the system in state-space form.

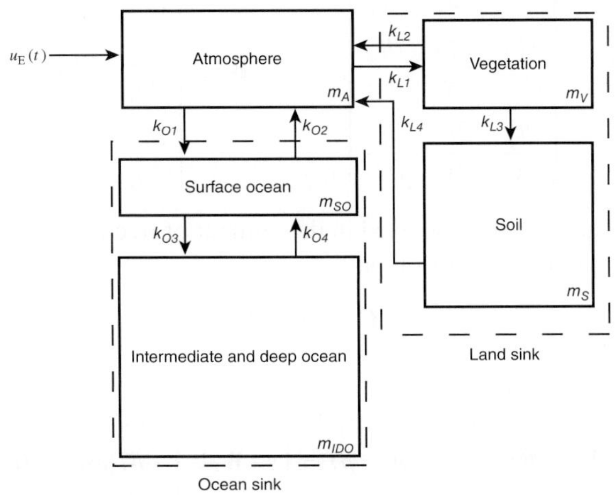


**FIGURE P3.17** Global carban cycle[12]

**27.** State-space representations are, in general, not unique. One system can be represented in several possible ways. For example, consider the following systems:

**a.** $\dot{x} = -5x + 3u$

$y = 7x$

**b.** $$\begin{bmatrix} \dot{x}_1 \\ \dot{x}_2 \end{bmatrix} = \begin{bmatrix} -5 & 0 \\ 0 & -1 \end{bmatrix}\begin{bmatrix} x_1 \\ x_2 \end{bmatrix} + \begin{bmatrix} 3 \\ 1 \end{bmatrix}u$$

$$y = [7 \quad 0]\begin{bmatrix} x_1 \\ x_2 \end{bmatrix}$$

**c.** $$\begin{bmatrix} \dot{x}_1 \\ \dot{x}_2 \end{bmatrix} = \begin{bmatrix} -5 & 0 \\ 0 & -1 \end{bmatrix}\begin{bmatrix} x_1 \\ x_2 \end{bmatrix} + \begin{bmatrix} 3 \\ 0 \end{bmatrix}u$$

$$y = [7 \quad 3]\begin{bmatrix} x_1 \\ x_2 \end{bmatrix}$$

Show that these systems will result in the same transfer function. We will explore this phenomenon in more detail in Chapter 5.

[12] Li, S., Jarvis, A.J., and Leedal, D.T. Are response function representations of the global carbon cycle ever interpretable? *Tellus*, vol. 61B, 2009, pp. 361–371. (Fig. 1 p. 363).

**28.** A single-pole oil cylinder valve contains a spool that regulates hydraulic pressure, which is then applied to a piston that drives a load. The transfer function relating piston displacement, $X_p(s)$ to spool displacement from equilibrium, $X_v(s)$, is given by (*Qu, 2010*):

$$G(s) = \frac{X_p(s)}{X_v(s)} = \frac{K_q\omega_h^2/A_1}{s(s^2 + 2\varsigma\omega_h s + \omega_h^2)}$$

where $A_1$ = effective area of a the valve's chamber, $K_q$ = rate of change of the load flow rate with a change in displacement, and $\omega_h$ = the natural frequency of the hydraulic system. Find the state-space representation of the system, where the state variables are the phase variables associated with the piston.

**29.** Figure P3.18 shows a free-body diagram of an inverted pendulum, mounted on a cart with a mass, $M$. The pendulum has a point mass, $m$, concentrated at the upper end of a rod with zero mass, a length, $l$, and a frictionless hinge. A motor drives the cart, applying a horizontal force, $u(t)$. A gravity force, $mg$, acts on $m$ at all times. The pendulum angle relative to the $y$-axis, $\theta$, its angular speed, $\dot{\theta}'$, the horizontal position of the cart, $x$, and its speed, $x'$, were selected to be the state variables. The state-space equations derived were heavily nonlinear.[13] They were then linearized around the stationary point, $\mathbf{x_0} = \mathbf{0}$ and $u_0 = 0$, and manipulated to yield the following open-loop model written in perturbation form:

$$\frac{d}{dt}\delta\mathbf{x} = \mathbf{A}\delta\mathbf{x} + \mathbf{B}\delta u$$

However, since $\mathbf{x_0} = \mathbf{0}$ and $u_0 = 0$, then let: $\mathbf{x} = \mathbf{x}_0 + \delta\mathbf{x} = \delta\mathbf{x}$ and $u = u_0 + \delta u = \delta u$. Thus the state equation may be rewritten as (*Prasad, 2012*):

$$\dot{\mathbf{x}} = \mathbf{A}\mathbf{x} + \mathbf{B}u$$

where

$$\mathbf{A} = \begin{bmatrix} 0 & 1 & 0 & 0 \\ \frac{(M+m)g}{Ml} & 0 & 0 & 0 \\ 0 & 0 & 0 & 1 \\ -\frac{mg}{M} & 0 & 0 & 0 \end{bmatrix} \text{ and } \mathbf{B} = \begin{bmatrix} 0 \\ -\frac{1}{Ml} \\ 0 \\ \frac{1}{M} \end{bmatrix}$$

Assuming the output to be the horizontal position of $m = x_m = x + l\sin\theta = x + l\theta$ for a small angle, $\theta$, the output equation becomes:

[13] As noted in the introduction to Section 3.7, the techniques for solving such nonlinear state equations are beyond the scope of this course.

$$y = l\theta + x = \mathbf{Cx} = [l \quad 0 \quad 1 \quad 0]\begin{bmatrix}\theta \\ \dot{\theta} \\ x \\ \dot{x}\end{bmatrix}$$

Given that: $M = 2.4$ kg, $m = 0.23$ kg, $l = 0.36$ m, $g = 9.81$ m/s$^2$, use MATLAB to find the transfer function, $G(s) = Y(s)/U(s) = X_m(s)/U(s)$.

MATLAB
ML

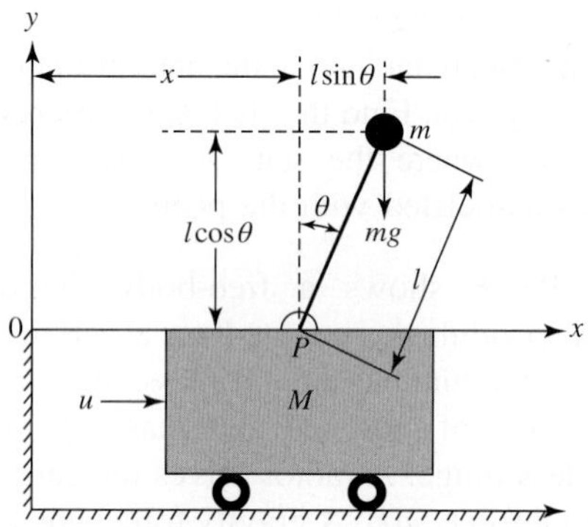


**FIGURE P3.18** Motor-driven inverted pendulum cart system[14]

## PROGRESSIVE ANALYSIS AND DESIGN PROBLEMS

**30. Control of HIV/AIDS.** Problem 65 in Chapter 2 introduced a model for HIV infection. If retroviral drugs, RTIs and PIs as discussed in Problem 22 in Chapter 1, are used, the model is modified as follows (*Craig, 2004*):

$$\frac{dT}{dt} = s - dT - (1 - u_1)\beta T v$$

$$\frac{dT^*}{dt} = (1 - u_1)\beta T v - \mu T^*$$

$$\frac{dv}{dt} = (1 - u_2)kT^* - cv$$

where $0 \le u_1 \le 1$, $0 \le u_2 \le 1$ represent the effectiveness of the RTI and PI medication, respectively.

**a.** Obtain a state-space representation of the HIV/AIDS model by linearizing the equations about the

$$(T_0,\ T_0^*,\ v_0) = \left(\frac{c\mu}{\beta k}, \frac{s}{\mu} - \frac{cd}{\beta k}, \frac{sk}{c\mu} - \frac{d}{\beta}\right)$$

equilibrium with $u_{10} = u_{20} = 0$. This equilibrium represents the asymptomatic HIV-infected patient. Note that each one of the above equations is of the form $\dot{x}_i = f_i(x_i, u_1, u_2)$, $i = 1, 2, 3$.

**b.** If Matrices **A** and **B** are given by

$$\mathbf{A} = \begin{bmatrix} \frac{\partial f_1}{\partial x_1} & \frac{\partial f_1}{\partial x_2} & \frac{\partial f_1}{\partial x_3} \\ \frac{\partial f_2}{\partial x_1} & \frac{\partial f_2}{\partial x_2} & \frac{\partial f_2}{\partial x_3} \\ \frac{\partial f_3}{\partial x_1} & \frac{\partial f_3}{\partial x_2} & \frac{\partial f_3}{\partial x_3} \end{bmatrix}_{T_0, T_0^*, v_0}; \quad \mathbf{B} = \begin{bmatrix} \frac{\partial f_1}{\partial u_1} & \frac{\partial f_1}{\partial u_2} \\ \frac{\partial f_2}{\partial u_1} & \frac{\partial f_2}{\partial u_2} \\ \frac{\partial f_3}{\partial u_1} & \frac{\partial f_3}{\partial u_2} \end{bmatrix}_{T_0, T_0^*, v_0}$$

and we are interested in the number of free HIV viruses as the system's output,

$$\mathbf{C} = [0 \quad 0 \quad 1]$$

show that

$$\mathbf{A} = \begin{bmatrix} -(d + \beta v_0) & 0 & -\beta T_0 \\ \beta v_0 & -\mu & \beta T_0 \\ 0 & k & -c \end{bmatrix}; \quad \mathbf{B} = \begin{bmatrix} \beta T_0 v_0 & 0 \\ -\beta T_0 v_0 & 0 \\ 0 & -kT_0^* \end{bmatrix}$$

**c.** Typical parameter values and descriptions for the HIV/AIDS model are shown in the following table.

Substitute the values from the table into your model and write as

$$\dot{\mathbf{x}} = \mathbf{Ax} + \mathbf{Bu}$$
$$\mathbf{y} = \mathbf{Cx}$$

**Table of HIV/AIDS Model Parameters**[15]

| | | |
|---|---|---|
| $t$ | Time | days |
| $d$ | Death of uninfected T cells | 0.02/day |
| $k$ | Rate of free viruses produced per infected T cell | 100 counts/cell |
| $s$ | Source term for uninfected T cells | 10/mm$^3$/day |
| $\beta$ | Infectivity rate of free virus particles | $2.4 \times 10^{-5}$/mm$^3$/day |
| $c$ | Death rate of viruses | 2.4/day |
| $\mu$ | Death rate of infected T cells | 0.24/day |

[14] Prasad, L., Tyagi, B.,and Gupta, H. Modeling & Simulation for Optimal Control of Nonlinear Inverted Pendulum Dynamical System using PID Controller & LQR. *IEEE Computer Society Sixth Asia Modeling Symposium*, 2012, pp. 138–143. Figure 1 p. 139. Reproduced with permission of IEEE in the format Republish in a book via Copyright Clearance Center.

[15] Craig, I. K., Xia, X., and Venter, J. W. Introducing HIV/AIDS Education Into the Electrical Engineering Curriculum at the University of Pretoria. *IEEE Transactions on Education*, vol. 47, no. 1, February 2004, pp. 65–73. Table II, p. 67. Modelling Symposium (AMS), 2012 Sixth Asia by IEEE. Reproduced with permission of IEEE in the format Republish in a book via Copyright Clearance Center.

**31. Hybrid vehicle.** For Problem 23 in Chapter 1 we developed the functional block diagrams for the cruise control of serial, parallel, and split-power hybrid electric vehicles (HEV). Those diagrams showed that the engine or electric motor or both may propel the vehicle. When electric motors are the sole providers of the motive force, the forward paths of all HEV topologies are similar. In general, such a forward path can be represented (*Preitl, 2007*) by a block diagram similar to the one of Figure P3.19.

Assume the motor to be an armature-controlled dc motor. In this diagram, $K_A$ is the power amplifier gain; $G_e(s)$ is the transfer function of the motor electric circuit and consists of a series inductor and resistor, $L_a$ and $R_a$, respectively; $K_t$ is the motor torque constant; $J_{tot}$, is the sum of the motor inertia, $J_m$, the inertias of the vehicle, $J_{veh}$, and the two driven wheels, $J_w$, both of which are reflected to the motor shaft; $k_f$ is the coefficient of viscous friction; and $k_b$ is the back emf constant.

The input variables are $u_c(t)$, the command voltage from the electronic control unit and $T_c(t)$, the load torque. The output variables in this block diagram are the motor angular speed, $\omega(t)$, and its armature current, $I_a(t)$.

**a.** Write the basic time-domain equations that characterize the relationships between the state, input, and output variables for the block diagram of Figure P3.19, given that the state variables are the motor armature current, $I_a(t)$, and angular speed, $\omega(t)$.

**b.** Write the resulting state-space equations and then represent them in matrix form. Regard the load torque $T_c(t)$ as an extra input to the system. Thus, in your resulting state-space representation, the system will have two inputs and two outputs.

**32. Parabolic trough collector.** A transfer function model from fluid flow to fluid temperature for a parabolic trough collector was introduced in Problem 67, Chapter 2. A more detailed model for the response of this system is given under specific operation conditions (*Camacho, 2012*) by:

$$\frac{H}{Q}(s) = \frac{137.2 \times 10^{-6}}{s^2 + 0.0224s + 196 \times 10^{-6}} e^{-39s}$$

Find an appropriate state-space representation for the system.

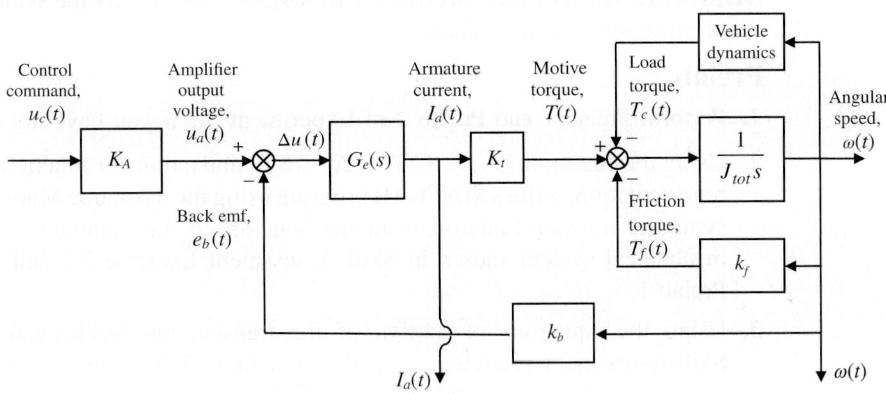


**FIGURE P3.19** Block diagram representation of an HEV forward path[16]

## Cyber Exploration Laboratory

### Experiment 3.1

**Objectives** To learn to use MATLAB to (1) generate an LTI state-space representation of a system and (2) convert an LTI state-space representation of a system to an LTI transfer function.

**Minimum Required Software Packages** MATLAB and the Control System Toolbox

[16] Preitl, Z., Bauer,P., and J. Bokor, J. A Simple Control Solution for Traction Motor Used in Hybrid Vehicles. *4th International Symposium on Applied Computational Intelligence and Informatics*. IEEE, 2007. Adapted from Figure 2, p. 2

### Prelab

1. Derive the state-space representation of the translational mechanical system shown in Skill-Assessment Exercise 3.2 if you have not already done so. Consider the output to be $x_3(t)$.
2. Derive the transfer function, $\dfrac{X_3(s)}{F(s)}$, from the equations of motion for the translational mechanical system shown in Skill-Assessment Exercise 3.2.

### Lab

1. Use MATLAB to generate the LTI state-space representation derived in Prelab 1.
2. Use MATLAB to convert the LTI state-space representation found in Lab 1 to the LTI transfer function found in Prelab 2.

### Postlab

1. Compare your transfer functions as found from Prelab 2 and Lab 2.
2. Discuss the use of MATLAB to create LTI state-space representations and the use of MATLAB to convert these representations to transfer functions.

## Experiment 3.2

**Objectives** To learn to use MATLAB and the Symbolic Math Toolbox to (1) find a symbolic transfer function from the state-space representation and (2) find a state-space representation from the equations of motion.

**Minimum Required Software Packages** MATLAB, the Symbolic Math Toolbox, and the Control System Toolbox

### Prelab

1. Perform Prelab 1 and Prelab 2 of Experiment 3.1 if you have not already done so.
2. Using the equation $T(s) = \mathbf{C}(s\mathbf{I} - \mathbf{A})^{-1}\mathbf{B}$ to find a transfer function from a state-space representation, write a MATLAB program using the Symbolic Math Toolbox to find the symbolic transfer function from the state-space representation of the translational mechanical system shown in Skill-Assessment Exercise 3.2 and found as a step in Prelab 1.
3. Using the equations of motion of the translational mechanical system shown in Skill-Assessment Exercise 3.2 and found in Prelab 1, write a symbolic MATLAB program to find the transfer function, $\frac{X_3(s)}{F(s)}$, for this system.

### Lab

1. Run the programs composed in Prelabs 2 and 3 and obtain the symbolic transfer functions by the two methods.

### Postlab

1. Compare the symbolic transfer function obtained from $T(s) = \mathbf{C}(s\mathbf{I} - \mathbf{A})^{-1}\mathbf{B}$ with the symbolic transfer function obtained from the equations of motion.
2. Discuss the advantages and disadvantages between the two methods.
3. Describe how you would obtain an LTI state-space representation and an LTI transfer function from your symbolic transfer function.

## Experiment 3.3

**Objectives** To learn to use LabVIEW to (1) generate state-space representations of transfer functions, (2) generate transfer functions from state-space representations, and (3) verify that there are multiple state-space representations for a transfer function.

**Minimum Required Software Packages** LabVIEW, the LabVIEW Control Design and Simulation Module, and the MathScript RT Module.

## Prelab

1. Study Appendix D, Sections D.1 through D.4, Example D.1.
2. Solve Skill-Assessment Exercise 3.3.
3. Use your solution to Prelab 2 and convert back to the transfer function.

## Lab

1. Use LabVIEW to convert the transfer function, $G(s) = \frac{2s+1}{s^2+7s+9}$, into a state-space representation using both the graphical and MathScript approaches. The front panel will contain controls for the entry of the transfer function and indicators of the transfer function and the two state-space results. Functions for this experiment can be found in the following palettes: (1) **Control Design and Simulation/Control Design/Model Construction,** (2) **Control Design and Simulation/Control Design/Model Conversion,** and (3) **Programming/Structures** Hint: Coefficients are entered in reverse order when using MathScript with MATLAB.
2. Use LabVIEW to convert all state-space representations found in Lab 1 to a transfer function. All state-space conversions should yield the transfer function given in Lab 1. The front panel will contain controls for entering state-space representations and indicators of the transfer function results as well as the state equations used.

## Postlab

1. Describe any correlation found between the results of Lab 1 and calculations made in the Prelab.
2. Describe and account for any differences between the results of Lab 1 and calculations made in the Prelab.
3. Explain the results of Lab 2 and draw conclusions from the results.

# Bibliography


Agee, J. T., and Jimoh, A. A. Flat controller design for hardware-cost reduction in polar-axis photovoltaic systems. *Solar Energy*, vol. 86, pp. 452–462, 2012.

Camacho, E. F., Berenguel, M., Rubio, F. R., and Martinez, D. *Control of Solar Energy Systems*. Springer-Verlag, London, 2012.

Carlson, L. E., and Griggs, G. E. *Aluminum Catenary System Quarterly Report*. Technical Report Contract Number DOT-FR-9154, U.S. Department of Transportation, 1980.

Cereijo, M. R. State Variable Formulations. *Instruments and Control Systems*, December 1969, pp. 87–88.

Chiu, D. K., and Lee, S. Design and Experimentation of a Jump Impact Controller. *IEEE Control Systems*, June 1997, pp. 99–106.

Cochin, I. *Analysis and Design of Dynamic Systems*. Harper & Row, New York, 1980.

Craig, I. K., Xia, X., and Venter, J. W. Introducing HIV/AIDS Education into the Electrical Engineering Curriculum at the University of Pretoria. *IEEE Transactions on Education*, vol. 47, no.1, February 2004, pp. 65–73.

Elkins, J. A. *A Method for Predicting the Dynamic Response of a Pantograph Running at Constant Speed under a Finite Length of Overhead Equipment*. Technical Report TN DA36, British Railways, 1976.

Fagergren, A., Ekeberg, Ö., and Forssberg, H. Precision Grip Force Dynamics: A System Identification Approach. *IEEE Transactions on Biomedical Engineering*, vol. 47, no.10, 2000, pp. 1366–1375.

Franklin, G. F., Powell, J. D., and Emami-Naeini, A. *Feedback Control of Dynamic Systems*. Addison-Wesley, Reading, MA, 1986.

Hong, J., Tan, X., Pinette, B., Weiss, R., and Riseman, E. M. Image-Based Homing. *IEEE Control Systems*, February 1992, pp. 38–45.

Inigo, R. M. Observer and Controller Design for D.C. Positional Control Systems Using State Variables. *Transactions, Analog/Hybrid Computer Educational Society*, December 1974, pp. 177–189.

Kailath, T. *Linear Systems*. Prentice Hall, Upper Saddle River, NJ, 1980.

Kandel, A. Analog Simulation of Groundwater Mining in Coastal Aquifers. *Transactions, Analog/Hybrid Computer Educational Society*, November 1973, pp. 175–183.

Kyrylov, V., Severyanova, L. A., and Vieira, A. Modeling Robust Oscillatory Behavior of the Hypothalamic-Pituitary-Adrenal Axis. *IEEE Transactions on Biomedical Engineering*, vol. 52, no.12, 2005, pp. 1977–1983.

Li, S., Jarvis, A.J., and Leedal, D.T., Are Response Function Representations of the Global Carbon Cycle Ever Interpretable? *Tellus*, vol. 61B, 2009, pp. 361–371.

Liceaga-Castro, E., van der Molen, G. M. Submarine $H^{\infty}$ Depth Control Under Wave Disturbances. *IEEE Transactions on Control Systems Technology*, vol. 3, no.3, 1995, pp. 338–346.

Lordi, N. G. Analog Computer Generated Lecture Demonstrations in Pharmacokinetics. *Transactions, Analog/Hybrid Computer Educational Society*, November 1972, pp. 217–222.

Philco Technological Center. *Servomechanism Fundamentals and Experiments*. Prentice Hall, Upper Saddle River, NJ, 1980.

Prasad, L., Tyagi, B., and Gupta, H. Modeling & Simulation for Optimal Control of Nonlinear Inverted Pendulum Dynamical System using PID Controller & LQR. *IEEE Computer Society Sixth Asia Modeling Symposium, 2012,* pp. 138–143.

Preitl, Z., Bauer, P., and Bokor, J. A Simple Control Solution for Traction Motor Used in Hybrid Vehicles. *Fourth International Symposium on Applied Computational Intelligence and Informatics. IEEE.* 2007.

Qu, S-G., Zhang, Y., Ye, Z., Shao, M., and Xia, W. Modeling and Simulation of the Proportional Valve Control System for the Turbocharger. *IEEE*, 2010.

Riegelman, S. et al. Shortcomings in Pharmacokinetic Analysis by Conceiving the Body to Exhibit Properties of a Single Compartment. *Journal of Pharmaceutical Sciences*, vol. 57, no.1, 1968, pp. 117–123.

Tarin, C., Teufel, E., Picó, J., Bondia, J., and Pfleiderer, H. J. Comprehensive Pharmacokinetic Model of Insulin Glargine and Other Insulin Formulations. *IEEE Transactions on Biomedical Engineering*, vol. 52, no.12, 2005, pp. 1994–2005.

Timothy, L. K., and Bona, B. E. *State Space Analysis: An Introduction*. McGraw-Hill, New York, 1968.

# 4

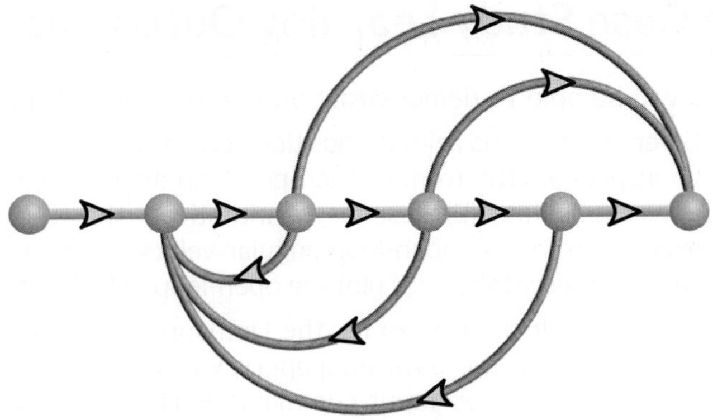

# Time Response

## Chapter Learning Outcomes

After completing this chapter the student will be able to:

- Use poles and zeros of transfer functions to determine the time response of a control system (Sections 4.1–4.2)
- Describe quantitatively the transient response of first-order systems (Section 4.3)
- Write the general response of second-order systems given the pole location (Section 4.4)
- Find the damping ratio and natural frequency of a second-order system (Section 4.5)
- Find the settling time, peak time, percent overshoot, and rise time for an underdamped second-order system (Section 4.6)
- Approximate higher-order systems and systems with zeros as first- or second-order systems (Sections 4.7–4.8)
- Describe the effects of nonlinearities on the system time response (Section 4.9)
- Find the time response from the state-space representation (Sections 4.10–4.11)

# ▶ Case Study Learning Outcomes

You will be able to demonstrate your knowledge of the chapter objectives with case studies as follows:

- Given the antenna azimuth position control system shown on the front endpapers, you will be able to (1) predict, by inspection, the form of the open-loop angular velocity response of the load to a step voltage input to the power amplifier; (2) describe quantitatively the transient response of the open-loop system; (3) derive the expression for the open-loop angular velocity output for a step voltage input; (4) obtain the open-loop state-space representation; (5) plot the open-loop velocity step response using a computer simulation.
- Given the block diagram for the Unmanned Free-Swimming Submersible (UFSS) vehicle's pitch control system shown on the back endpapers, you will be able to predict, find, and plot the response of the vehicle dynamics to a step input command. Further, you will be able to evaluate the effect of system zeros and higher-order poles on the response. You also will be able to evaluate the roll response of a ship at sea.

## 4.1 Introduction

In Chapter 2, we saw how transfer functions can represent linear, time-invariant systems. In Chapter 3, systems were represented directly in the time domain via the state and output equations. After the engineer obtains a mathematical representation of a subsystem, the subsystem is analyzed for its transient and steady-state responses to see if these characteristics yield the desired behavior. This chapter is devoted to the analysis of system transient response.

It may appear more logical to continue with Chapter 5, which covers the modeling of closed-loop systems, rather than to break the modeling sequence with the analysis presented here in Chapter 4. However, the student should not continue too far into system representation without knowing the application for the effort expended. Thus, this chapter demonstrates applications of the system representation by evaluating the transient response from the system model. Logically, this approach is not far from reality, since the engineer may indeed want to evaluate the response of a subsystem prior to inserting it into the closed-loop system.

After describing a valuable analysis and design tool, poles and zeros, we begin analyzing our models to find the step response of first- and second-order systems. The order refers to the order of the equivalent differential equation representing the system—the order of the denominator of the transfer function after cancellation of common factors in the numerator or the number of simultaneous first-order equations required for the state-space representation.

## 4.2 Poles, Zeros, and System Response

The output response of a system is the sum of two responses: the *forced response* and the *natural response*.[1] Although many techniques, such as solving a differential equation or taking the inverse Laplace transform, enable us to evaluate this output response, these techniques are laborious and time-consuming. Productivity is aided by analysis and design techniques that yield results in a minimum of time. If the technique is so rapid that we feel we derive the desired result by inspection, we sometimes use the attribute *qualitative* to describe the method. The use of poles and zeros and their relationship to the time response of a system is such a technique. Learning this relationship gives us a qualitative "handle" on problems. The concept of poles and zeros, fundamental to the analysis and design of control

[1] The forced response is also called the *steady-state response* or *particular solution*. The natural response is also called the *homogeneous solution*.

systems, simplifies the evaluation of a system's response. The reader is encouraged to master the concepts of poles and zeros and their application to problems throughout this book. Let us begin with two definitions.

## Poles of a Transfer Function

The *poles* of a transfer function are (1) the values of the Laplace transform variable, $s$, that cause the transfer function to become infinite or (2) any roots of the denominator of the transfer function that are common to roots of the numerator.

Strictly speaking, the poles of a transfer function satisfy part (1) of the definition. For example, the roots of the characteristic polynomial in the denominator are values of $s$ that make the transfer function infinite, so they are thus poles. However, if a factor of the denominator can be canceled by the same factor in the numerator, the root of this factor no longer causes the transfer function to become infinite. In control systems, we often refer to the root of the canceled factor in the denominator as a pole even though the transfer function will not be infinite at this value. Hence, we include part (2) of the definition.

## Zeros of a Transfer Function

The *zeros* of a transfer function are (1) the values of the Laplace transform variable, $s$, that cause the transfer function to become zero, or (2) any roots of the numerator of the transfer function that are common to roots of the denominator.

Strictly speaking, the zeros of a transfer function satisfy part (1) of this definition. For example, the roots of the numerator are values of $s$ that make the transfer function zero and are thus zeros. However, if a factor of the numerator can be canceled by the same factor in the denominator, the root of this factor no longer causes the transfer function to become zero. In control systems, we often refer to the root of the canceled factor in the numerator as a zero even though the transfer function will not be zero at this value. Hence, we include part (2) of the definition.

## Poles and Zeros of a First-Order System: An Example

Given the transfer function $G(s)$ in Figure 4.1($a$), a pole exists at $s = -5$, and a zero exists at $-2$. These values are plotted on the complex $s$-plane in Figure 4.1($b$), using an × for the pole and a ○ for the zero. To show the properties of the poles and zeros, let us find the unit step response of the system. Multiplying the transfer function of Figure 4.1($a$) by a step function yields

$$C(s) = \frac{(s+2)}{s(s+5)} = \frac{A}{s} + \frac{B}{s+5} = \frac{2/5}{s} + \frac{3/5}{s+5} \tag{4.1}$$

where

$$A = \left.\frac{(s+2)}{(s+5)}\right|_{s\to 0} = \frac{2}{5}$$

$$B = \left.\frac{(s+2)}{s}\right|_{s\to -5} = \frac{3}{5}$$

Thus,

$$c(t) = \frac{2}{5} + \frac{3}{5}e^{-5t} \tag{4.2}$$

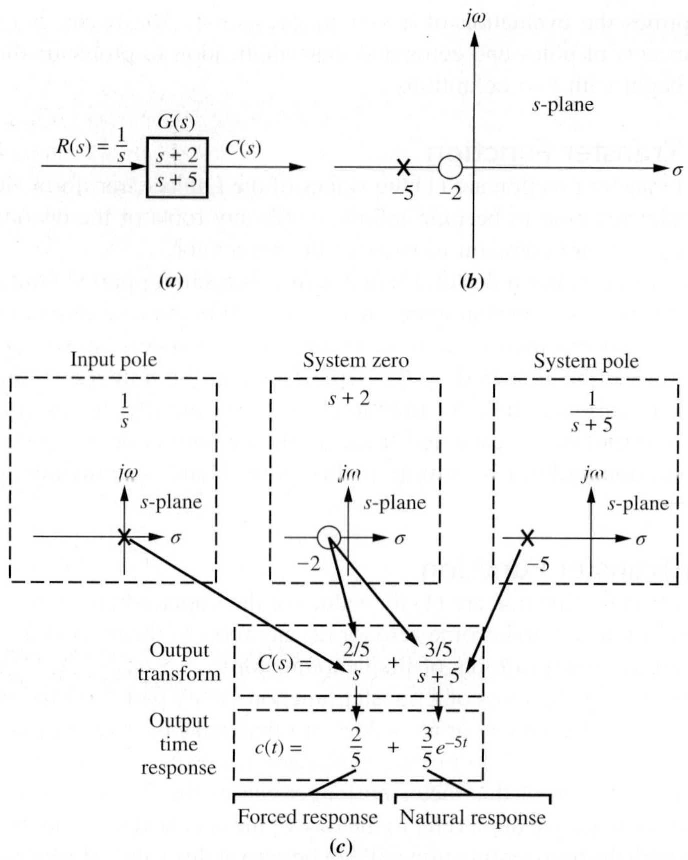


**FIGURE 4.1** **a.** System showing input and output; **b.** pole-zero plot of the system; **c.** evolution of a system response. Follow blue arrows to see the evolution of the response component generated by the pole or zero.

From the development summarized in Figure 4.1(*c*), we draw the following conclusions:

1. A pole of the input function generates the form of the *forced response* (that is, the pole at the origin generated a step function at the output).
2. A pole of the transfer function generates the form of the *natural response* (that is, the pole at $-5$ generated $e^{-5t}$).
3. A pole on the real axis generates an *exponential* response of the form $e^{-\alpha t}$, where $-\alpha$ is the pole location on the real axis. Thus, the farther to the left a pole is on the negative real axis, the faster the exponential transient response will decay to zero (again, the pole at $-5$ generated $e^{-5t}$; see Figure 4.2 for the general case).

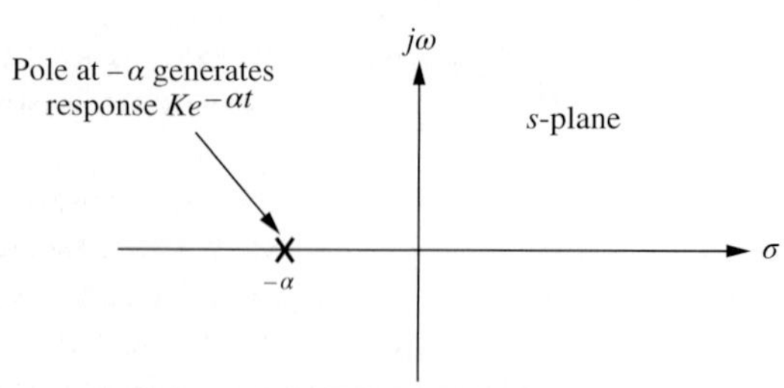


**FIGURE 4.2** Effect of a real-axis pole upon transient response.

4. The zeros and poles generate the *amplitudes* for both the forced and natural responses (this can be seen from the calculation of $A$ and $B$ in Eq. (4.1)).

Let us now look at an example that demonstrates the technique of using poles to obtain the form of the system response. We will learn to write the form of the response by inspection. Each pole of the system transfer function that is on the real axis generates an exponential response that is a component of the natural response. The input pole generates the forced response.

## Example 4.1 Evaluating Response Using Poles

**PROBLEM:** Given the system of Figure 4.3, write the output, $c(t)$, in general terms. Specify the forced and natural parts of the solution.

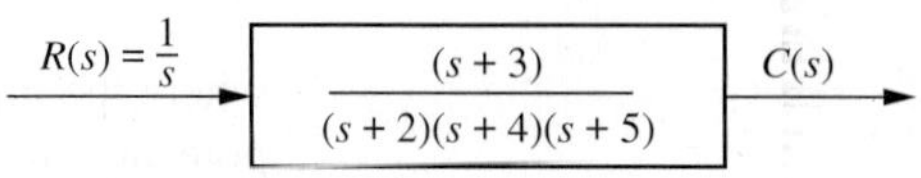


**FIGURE 4.3** System for Example 4.1

**SOLUTION:** By inspection, each system pole generates an exponential as part of the natural response. The input's pole generates the forced response. Thus,

$$C(s) \equiv \underbrace{\frac{K_1}{s}}_{\text{Forced response}} + \underbrace{\frac{K_2}{s+2} + \frac{K_3}{s+4} + \frac{K_4}{s+5}}_{\text{Natural response}} \tag{4.3}$$

Taking the inverse Laplace transform, we get

$$c(t) \equiv \underbrace{K_1}_{\text{Forced response}} \underbrace{+ K_2 e^{-2t} + K_3 e^{-4t} + K_4 e^{-5t}}_{\text{Natural response}} \tag{4.4}$$

## Skill-Assessment Exercise 4.1

**PROBLEM:** A system has a transfer function, $G(s) = \dfrac{10(s+4)(s+6)}{(s+1)(s+7)(s+8)(s+10)}$.

Write, by inspection, the output, $c(t)$, in general terms if the input is a unit step.

**ANSWER:** $c(t) \equiv A + Be^{-t} + Ce^{-7t} + De^{-8t} + Ee^{-10t}$

In this section, we learned that poles determine the nature of the time response: Poles of the input function determine the form of the forced response, and poles of the transfer function determine the form of the natural response. Zeros and poles of the input or transfer function contribute to the amplitudes of the component parts of the total response. Finally, poles on the real axis generate exponential responses.

## 4.3 First-Order Systems

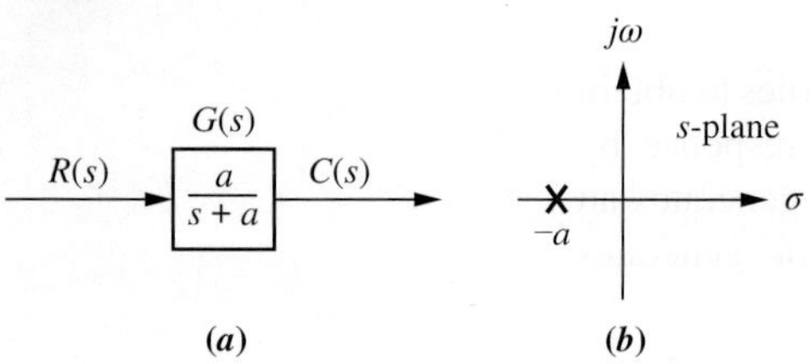


**FIGURE 4.4** **a.** First-order system; **b.** pole plot

We now discuss first-order systems without zeros to define a performance specification for such a system. A first-order system without zeros can be described by the transfer function shown in Figure 4.4(*a*). If the input is a unit step, where $R(s) = 1/s$, the Laplace transform of the step response is $C(s)$, where

$$C(s) = R(s)G(s) = \frac{a}{s(s+a)} \tag{4.5}$$

Taking the inverse transform, the step response is given by

$$c(t) = c_f(t) + c_n(t) = 1 - e^{-at} \tag{4.6}$$

where the input pole at the origin generated the forced response $c_f(t) = 1$, and the system pole at $-a$, as shown in Figure 4.4(*b*), generated the natural response $c_n(t) = -e^{-at}$. Equation (4.6) is plotted in Figure 4.5.

Let us examine the significance of parameter $a$, the only parameter needed to describe the transient response. When $t = 1/a$,

$$e^{-at}\big|_{t=1/a} = e^{-1} = 0.37 \tag{4.7}$$

or

$$c(t)\big|_{t=1/a} = 1 - e^{-at}\big|_{t=1/a} = 1 - 0.37 = 0.63 \tag{4.8}$$

We now use Eqs. (4.6), (4.7), and (4.8) to define three transient response performance specifications.

### Time Constant

We call $1/a$ the *time constant* of the response. From Eq. (4.7), the time constant can be described as the time for $e^{-at}$ to decay to 37% of its initial value. Alternately, from Eq. (4.8) the time constant is the time it takes for the step response to rise to 63% of its final value (see Figure 4.5).

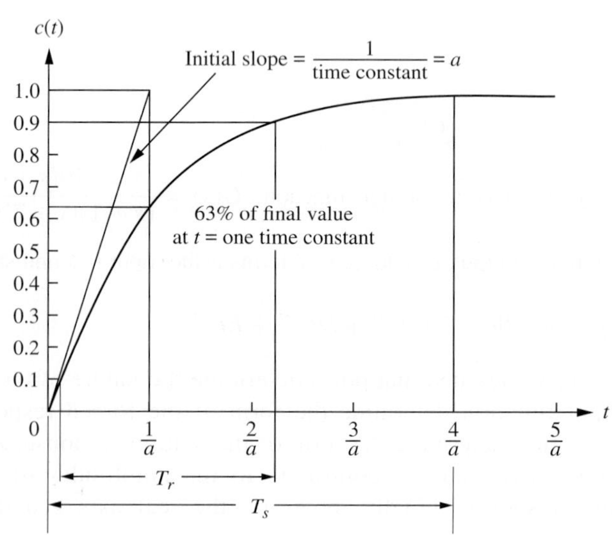


**FIGURE 4.5** First-order system response to a unit step

The reciprocal of the time constant has the units (1/seconds), or frequency. Thus, we can call the parameter $a$ the *exponential frequency*. Since the derivative of $e^{-at}$ is $-a$ when $t = 0$, $a$ is the initial rate of change of the exponential at $t = 0$. Thus, the time constant can be considered a transient response specification for a first-order system, since it is related to the speed at which the system responds to a step input.

The time constant can also be evaluated from the pole plot (see Figure 4.4($b$)). Since the pole of the transfer function is at $-a$, we can say the pole is located at the *reciprocal* of the time constant, and the farther the pole from the imaginary axis, the faster the transient response.

Let us look at other transient response specifications, such as rise time, $T_r$, and settling time, $T_s$, as shown in Figure 4.5.

## Rise Time, $T_r$

*Rise time* is defined as the time for the waveform to go from 0.1 to 0.9 of its final value. Rise time is found by solving Eq. (4.6) for the difference in time at $c(t) = 0.9$ and $c(t) = 0.1$. Hence,

$$T_r = \frac{2.31}{a} - \frac{0.11}{a} = \frac{2.2}{a} \tag{4.9}$$

## Settling Time, $T_s$

*Settling time* is defined as the time for the response to reach, and stay within, 2% of its final value.[2] Letting $c(t) = 0.98$ in Eq. (4.6) and solving for time, $t$, we find the settling time to be

$$T_s = \frac{4}{a} \tag{4.10}$$

## First-Order Transfer Functions via Testing

Often it is not possible or practical to obtain a system's transfer function analytically. Perhaps the system is closed, and the component parts are not easily identifiable. Since the transfer function is a representation of the system from input to output, the system's step response can lead to a representation even though the inner construction is not known. With a step input, we can measure the time constant and the steady-state value, from which the transfer function can be calculated.

Consider a simple first-order system, $G(s) = K/(s + a)$, whose step response is

$$C(s) = \frac{K}{s(s+a)} = \frac{K/a}{s} - \frac{K/a}{(s+a)} \tag{4.11}$$

If we can identify $K$ and $a$ from laboratory testing, we can obtain the transfer function of the system.

For example, assume the unit step response given in Figure 4.6. We determine that it has the first-order characteristics we have seen thus far, such as no overshoot and nonzero initial slope. From the response, we measure the time constant, that is, the time for the amplitude to reach 63% of its final value. Since the final value is about 0.72, the time constant is evaluated where the curve reaches $0.63 \times 0.72 = 0.45$, or about 0.13 second. Hence, $a = 1/0.13 = 7.7$.

[2] Strictly speaking, this is the definition of the 2% *setting time*. Other percentages, for example 5%, also can be used. We will use *settling time* throughout the book to mean 2% settling time.

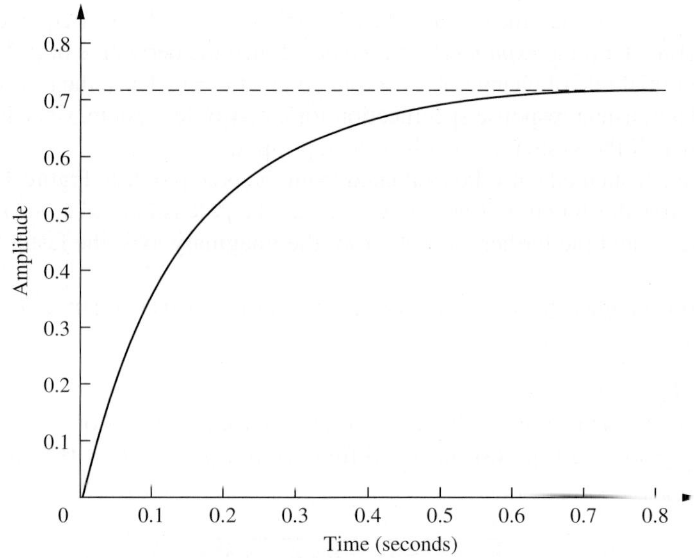


**FIGURE 4.6** Laboratory results of a system step response test

To find $K$, we realize from Eq. (4.11) that the forced response reaches a steady-state value of $K/a = 0.72$. Substituting the value of $a$, we find $K = 5.54$. Thus, the transfer function for the system is $G(s) = 5.54/(s + 7.7)$. It is interesting to note that the response of Figure 4.6 was generated using the transfer function $G(s) = 5/(s + 7)$.

## Skill-Assessment Exercise 4.2

**PROBLEM:** A system has a transfer function, $G(s) = \frac{50}{s + 50}$. Find the time constant, $T_c$, settling time, $T_s$, and rise time, $T_r$.

**ANSWER:** $T_c = 0.02$ s, $T_s = 0.08$ s, and $T_r = 0.044$ s.

The complete solution is located at www.wiley.com/college/nise.

# 4.4 Second-Order Systems: Introduction

Let us now extend the concepts of poles and zeros and transient response to second-order systems. Compared to the simplicity of a first-order system, a second-order system exhibits a wide range of responses that must be analyzed and described. Whereas varying a first-order system's parameter simply changes the speed of the response, changes in the parameters of a second-order system can change the *form* of the response. For example, a second-order system can display characteristics much like a first-order system, or, depending on component values, display damped or pure oscillations for its transient response.

To become familiar with the wide range of responses before formalizing our discussion in the next section, we take a look at numerical examples of the second-order system responses shown in Figure 4.7. All examples are derived from Figure 4.7($a$), the general case, which has two finite poles and no zeros. The term in the numerator is simply a scale or input multiplying factor that can take on any value without affecting the form of the derived results. By assigning appropriate values to parameters $a$ and $b$, we can show all possible second-order transient responses. The unit step response then can be found using

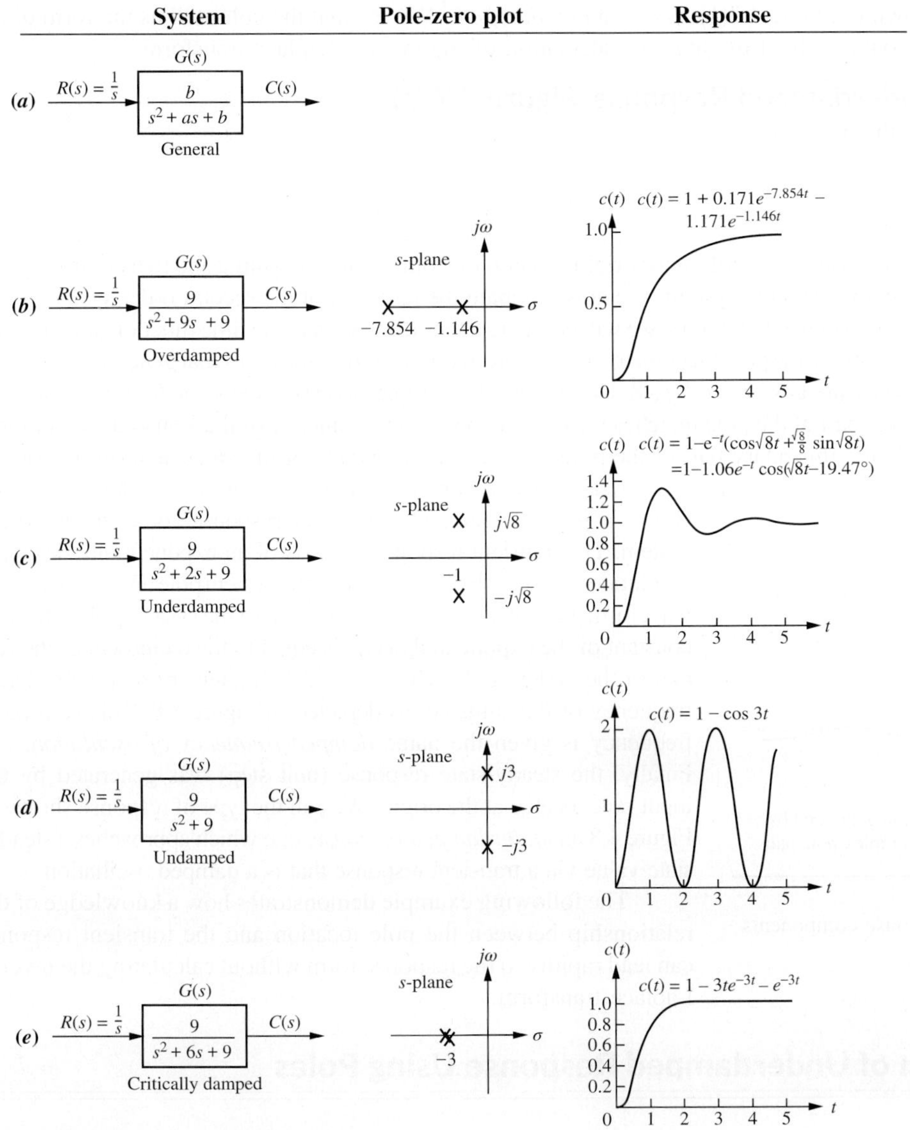


**FIGURE 4.7** Second-order systems, pole plots, and step responses

$C(s) = R(s)G(s)$, where $R(s) = 1/s$, followed by a partial-fraction expansion and the inverse Laplace transform. Details are left as an end-of-chapter problem, for which you may want to review Section 2.2.

We now explain each response and show how we can use the poles to determine the nature of the response without going through the procedure of a partial-fraction expansion followed by the inverse Laplace transform.

## Overdamped Response, Figure 4.7(*b*)

For this response,

$$C(s) = \frac{9}{s(s^2 + 9s + 9)} = \frac{9}{s(s + 7.854)(s + 1.146)} \tag{4.12}$$

This function has a pole at the origin that comes from the unit step input and two real poles that come from the system. The input pole at the origin generates the constant forced response; each of the two system poles on the real axis generates an exponential natural response whose exponential frequency is equal to the pole location. Hence, the output initially could have been written as $c(t) = K_1 + K_2e^{-7.854t} + K_3e^{-1.146t}$. This response, shown in Figure 4.7(*b*), is called *overdamped*.[3] We see that the poles tell us the form of the response without the tedious calculation of the inverse Laplace transform.

## Underdamped Response, Figure 4.7 (*c*)

For this response,

$$C(s) = \frac{9}{s(s^2 + 2s + 9)} \tag{4.13}$$

This function has a pole at the origin that comes from the unit step input and two complex poles that come from the system. We now compare the response of the second-order system to the poles that generated it. First we will compare the pole location to the time function, and then we will compare the pole location to the plot. From Figure 4.7(*c*), the poles that generate the natural response are at $s = -1 \pm j\sqrt{8}$. Comparing these values to $c(t)$ in the same figure, we see that the real part of the pole matches the exponential decay frequency of the sinusoid's amplitude, while the imaginary part of the pole matches the frequency of the sinusoidal oscillation.

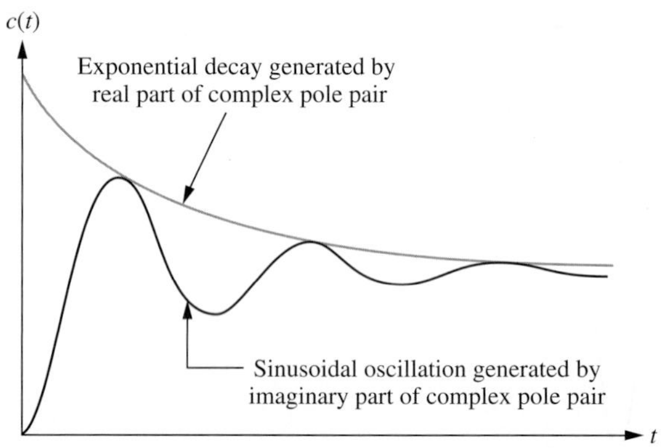


**FIGURE 4.8** Second-order step response components generated by complex poles

Let us now compare the pole location to the plot. Figure 4.8 shows a general, damped sinusoidal response for a second-order system. The transient response consists of an exponentially decaying amplitude generated by the real part of the system pole times a sinusoidal waveform generated by the imaginary part of the system pole. The time constant of the exponential decay is equal to the reciprocal of the real part of the system pole. The value of the imaginary part is the actual frequency of the sinusoid, as depicted in Figure 4.8. This sinusoidal frequency is given the name *damped frequency of oscillation*, $\omega_d$. Finally, the steady-state response (unit step) was generated by the input pole located at the origin. We call the type of response shown in Figure 4.8 an *underdamped response*, one which approaches a steady-state value via a transient response that is a damped oscillation.

The following example demonstrates how a knowledge of the relationship between the pole location and the transient response can lead rapidly to the response form without calculating the inverse Laplace transform.

## Example 4.2 Form of Underdamped Response Using Poles

**PROBLEM:** By inspection, write the form of the step response of the system in Figure 4.9.

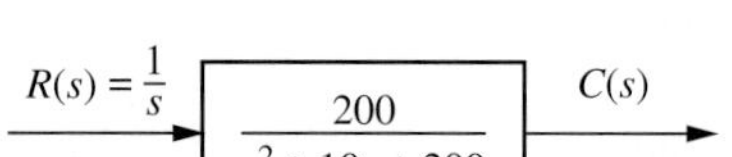


**FIGURE 4.9** System for Example 4.2

**SOLUTION:** First we determine that the form of the forced response is a step. Next we find the form of the natural response. Factoring the denominator of the transfer function in Figure 4.9, we find the poles to be $s = -5 \pm j13.23$. The real part, $-5$, is the exponential frequency for the damping. It is also the reciprocal of the time constant of the decay of the oscillations. The imaginary part, 13.23, is the radian frequency for the sinusoidal oscillations. Using our previous discussion and Figure 4.7(*c*) as a guide, we obtain $c(t) = K_1 + e^{-5t}\,(K_2 \cos 13.23t + K_3 \sin 13.23t) = K_1 + K_4 e^{-5t}(\cos 13.23t - \phi)$, where $\phi = \tan^{-1} K_3/K_2$, $K_4 = \sqrt{K_2^2 + K_3^2}$, and $c(t)$ is a constant plus an exponentially damped sinusoid.

[3] So named because *overdamped* refers to a large amount of energy absorption in the system, which inhibits the transient response from overshooting and oscillating about the steady-state value for a step input. As the energy absorption is reduced, an overdamped system will become underdamped and exhibit overshoot.

We will revisit the second-order underdamped response in Sections 4.5 and 4.6, where we generalize the discussion and derive some results that relate the pole position to other parameters of the response.

## Undamped Response, Figure 4.7(*d*)

For this response,

$$C(s) = \frac{9}{s(s^2+9)} \tag{4.14}$$

This function has a pole at the origin that comes from the unit step input and two imaginary poles that come from the system. The input pole at the origin generates the constant forced response, and the two system poles on the imaginary axis at $\pm j3$ generate a sinusoidal natural response whose frequency is equal to the location of the imaginary poles. Hence, the output can be estimated as $c(t) = K_1 + K_4\cos(3t - \phi)$. This type of response, shown in Figure 4.7(*d*), is called *undamped*. Note that the absence of a real part in the pole pair corresponds to an exponential that does not decay. Mathematically, the exponential is $e^{-0t} = 1$.

## Critically Damped Response, Figure 4.7 (*e*)

For this response,

$$C(s) = \frac{9}{s(s^2+6s+9)} = \frac{9}{s(s+3)^2} \tag{4.15}$$

This function has a pole at the origin that comes from the unit step input and two multiple real poles that come from the system. The input pole at the origin generates the constant forced response, and the two poles on the real axis at $-3$ generate a natural response consisting of an exponential and an exponential multiplied by time, where the exponential frequency is equal to the location of the real poles. Hence, the output can be estimated as $c(t) = K_1 + K_2e^{-3t} + K_3te^{-3t}$. This type of response, shown in Figure 4.7(*e*), is called *critically damped*. Critically damped responses are the fastest possible without the overshoot that is characteristic of the underdamped response.

We now summarize our observations. In this section we defined the following natural responses and found their characteristics:

1. *Overdamped responses*

   Poles: Two real at $-\sigma_1,\ -\sigma_2$

   Natural response: Two exponentials with time constants equal to the reciprocal of the pole locations, or

   $$c(t) = K_1e^{-\sigma_1 t} + K_2e^{-\sigma_2 t}$$

2. *Underdamped responses*

   Poles: Two complex at $-\sigma_d \pm j\omega_d$

   Natural response: Damped sinusoid with an exponential envelope whose time constant is equal to the reciprocal of the pole's real part. The radian frequency of the sinusoid, the damped frequency of oscillation, is equal to the imaginary part of the poles, or

   $$c(t) = Ae^{-\sigma_d t}\cos(\omega_d t - \phi)$$

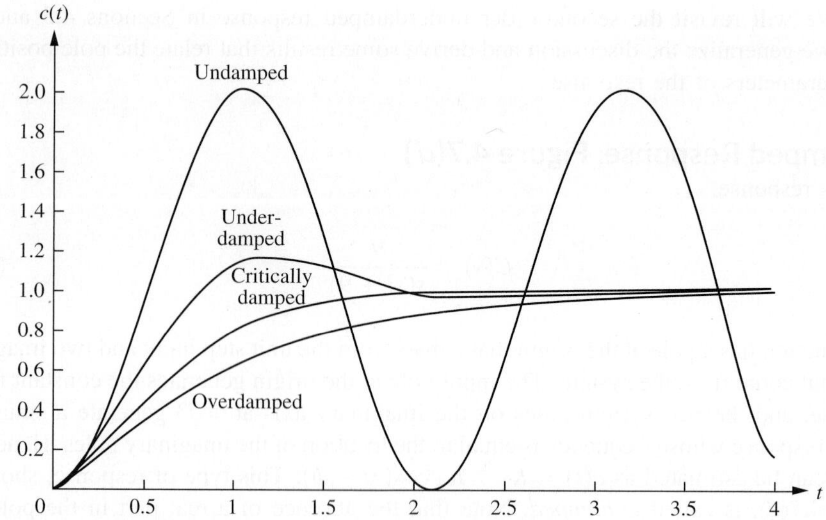


**FIGURE 4.10** Step responses for second-order system damping cases

**3.** *Undamped responses*

Poles: Two imaginary at $\pm j\omega_1$
Natural response: Undamped sinusoid with radian frequency equal to the imaginary part of the poles, or

$$c(t) = A\cos(\omega_1 t - \phi)$$

**4.** *Critically damped responses*

Poles: Two real at $-\sigma_1$

Natural response: One term is an exponential whose time constant is equal to the reciprocal of the pole location. Another term is the product of time, $t$, and an exponential with time constant equal to the reciprocal of the pole location, or

$$c(t) = K_1 e^{-\sigma_1 t} + K_2 t e^{-\sigma_1 t}$$

The step responses for the four cases of damping discussed in this section are superimposed in Figure 4.10. Notice that the critically damped case is the division between the overdamped cases and the underdamped cases and is the fastest response without overshoot.

## Skill-Assessment Exercise 4.3

**PROBLEM:** For each of the following transfer functions, write, by inspection, the general form of the step response:

**a.** $G(s) = \dfrac{400}{s^2 + 12s + 400}$

**b.** $G(s) = \dfrac{900}{s^2 + 90s + 900}$

**c.** $G(s) = \dfrac{225}{s^2 + 30s + 225}$

**d.** $G(s) = \dfrac{625}{s^2 + 625}$

**ANSWERS:**

**a.** $c(t) = A + Be^{-6t}\cos(19.08t + \phi)$

**b.** $c(t) = A + Be^{-78.54t} + Ce^{-11.46t}$

**c.** $c(t) = A + Be^{-15t} + Cte^{-15t}$

**d.** $c(t) = A + B\cos(25t + \phi)$

The complete solution is located at www.wiley.com/college/nise.

In the next section, we will formalize and generalize our discussion of second-order responses and define two specifications used for the analysis and design of second-order systems. In Section 4.6, we will focus on the *underdamped* case and derive some specifications unique to this response that we will use later for analysis and design.

## 4.5 The General Second-Order System

Now that we have become familiar with second-order systems and their responses, we generalize the discussion and establish quantitative specifications defined in such a way that the response of a second-order system can be described to a designer without the need for sketching the response. In this section, we define two physically meaningful specifications for second-order systems. These quantities can be used to describe the characteristics of the second-order transient response just as time constants describe the first-order system response. The two quantities are called natural frequency and damping ratio. Let us formally define them.

### Natural Frequency, $\omega_n$

The *natural frequency* of a second-order system is the frequency of oscillation of the system without damping. For example, the frequency of oscillation of a series *RLC* circuit with the resistance shorted would be the natural frequency.

### Damping Ratio, $\zeta$

Before we state our next definition, some explanation is in order. We have already seen that a second-order system's underdamped step response is characterized by damped oscillations. Our definition is derived from the need to quantitatively describe this damped oscillation regardless of the time scale. Thus, a system whose transient response goes through three cycles in a millisecond before reaching the steady state would have the same measure as a system that went through three cycles in a millennium before reaching the steady state. For example, the underdamped curve in Figure 4.10 has an associated measure that defines its shape. This measure remains the same even if we change the time base from seconds to microseconds or to millennia.

A viable definition for this quantity is one that compares the exponential decay frequency of the envelope to the natural frequency. This ratio is constant regardless of the time scale of the response. Also, the reciprocal, which is proportional to the ratio of the natural period to the exponential time constant, remains the same regardless of the time base.

We define the *damping ratio*, $\zeta$, to be

$$\zeta = \frac{\text{Exponential decay frequency}}{\text{Natural frequency (rad/second)}} = \frac{1}{2\pi}\frac{\text{Natural period (seconds)}}{\text{Exponential time constant}}$$

Let us now revise our description of the second-order system to reflect the new definitions. The general second-order system shown in Figure 4.7($a$) can be transformed to show the quantities $\zeta$ and $\omega_n$. Consider the general system

$$G(s) = \frac{b}{s^2 + as + b} \tag{4.16}$$

Without damping, the poles would be on the $j\omega$-axis, and the response would be an undamped sinusoid. For the poles to be purely imaginary, $a = 0$. Hence,

$$G(s) = \frac{b}{s^2 + b} \tag{4.17}$$

By definition, the natural frequency, $\omega_n$, is the frequency of oscillation of this system. Since the poles of this system are on the $j\omega$-axis at $\pm j\sqrt{b}$,

$$\omega_n = \sqrt{b} \tag{4.18}$$

Hence,

$$b = \omega_n^2 \tag{4.19}$$

Now what is the term $a$ in Eq. (4.16)? Assuming an underdamped system, the complex poles have a real part, $\sigma$, equal to $-a/2$. The magnitude of this value is then the exponential decay frequency described in Section 4.4. Hence,

$$\zeta = \frac{\text{Exponential decay frequency}}{\text{Natural frequency (rad/second)}} = \frac{|\sigma|}{\omega_n} = \frac{a/2}{\omega_n} \tag{4.20}$$

from which

$$a = 2\zeta\omega_n \tag{4.21}$$

Our general second-order transfer function finally looks like this:

$$G(s) = \frac{\omega_n^2}{s^2 + 2\zeta\omega_n s + \omega_n^2} \tag{4.22}$$

In the following example we find numerical values for $\zeta$ and $\omega_n$ by matching the transfer function to Eq. (4.22).

## Example 4.3 Finding $\zeta$ and $\omega_n$ For a Second-Order System

**PROBLEM:** Given the transfer function of Eq. (4.23), find $\zeta$ and $\omega_n$.

$$G(s) = \frac{36}{s^2 + 4.2s + 36} \tag{4.23}$$

**SOLUTION:** Comparing Eq. (4.23) to (4.22), $\omega_n^2 = 36$, from which $\omega_n = 6$. Also, $2\zeta\omega_n = 4.2$. Substituting the value of $\omega_n$, $\zeta = 0.35$.

Now that we have defined $\zeta$ and $\omega_n$, let us relate these quantities to the pole location. Solving for the poles of the transfer function in Eq. (4.22) yields

$$s_{1,2} = -\zeta\omega_n \pm \omega_n\sqrt{\zeta^2 - 1} \tag{4.24}$$

From Eq. (4.24) we see that the various cases of second-order response are a function of $\zeta$; they are summarized in Figure 4.11.[4]

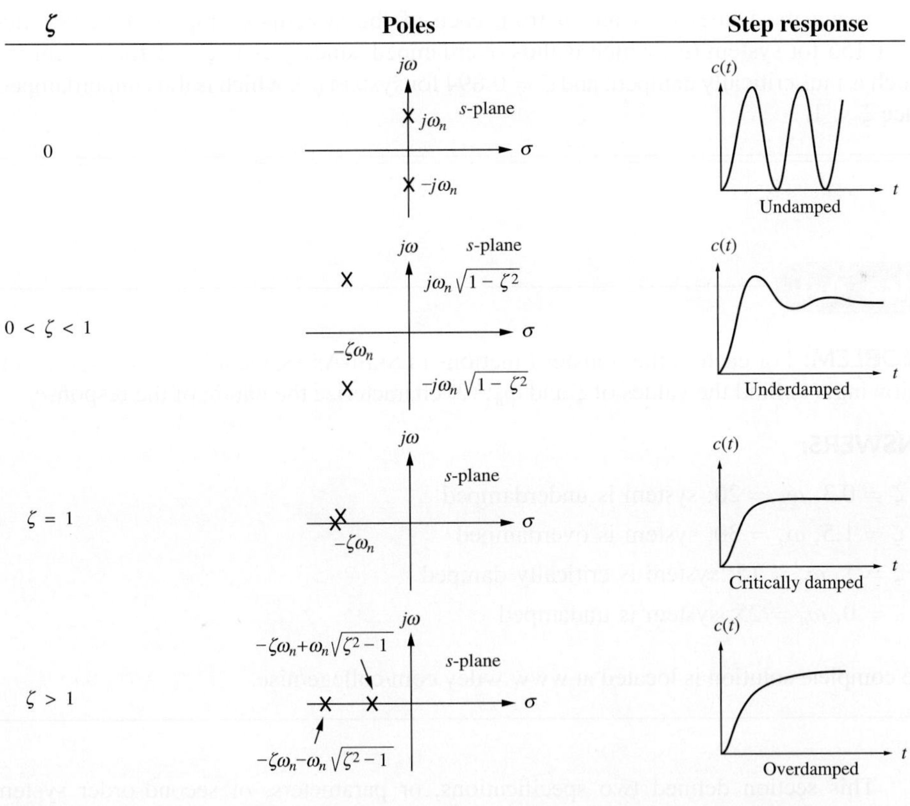


**FIGURE 4.11** Second-order response as a function of damping ratio

[4] The student should verify Figure 4.11 as an exercise.

In the following example we find the numerical value of $\zeta$ and determine the nature of the transient response.

## Example 4.4 Characterizing Response from the Value of $\zeta$

**PROBLEM:** For each of the systems shown in Figure 4.12, find the value of $\zeta$ and report the kind of response expected.

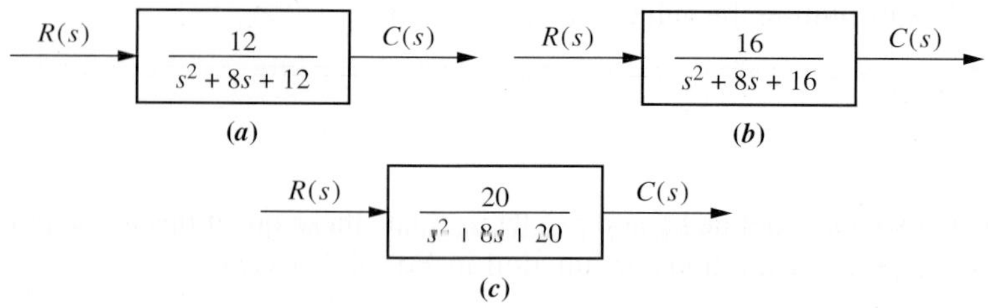


**FIGURE 4.12** Systems for Example 4.4

**SOLUTION:** First match the form of these systems to the forms shown in Eqs. (4.16) and (4.22). Since $a = 2\zeta\omega_n$ and $\omega_n = \sqrt{b}$,

$$\zeta = \frac{a}{2\sqrt{b}} \tag{4.25}$$

Using the values of $a$ and $b$ from each of the systems of Figure 4.12, we find $\zeta = 1.155$ for system ($a$), which is thus overdamped, since $\zeta > 1$; $\zeta = 1$ for system ($b$), which is thus critically damped; and $\zeta = 0.894$ for system ($c$), which is thus underdamped, since $\zeta < 1$.

## Skill-Assessment Exercise 4.4

**PROBLEM:** For each of the transfer functions in Skill-Assessment Exercise 4.3, do the following: (1) Find the values of $\zeta$ and $\omega_n$; (2) characterize the nature of the response.

**ANSWERS:**

**a.** $\zeta = 0.3$, $\omega_n = 20$; system is underdamped

**b.** $\zeta = 1.5$, $\omega_n = 30$; system is overdamped

**c.** $\zeta = 1$, $\omega_n = 15$; system is critically damped

**d.** $\zeta = 0$, $\omega_n = 25$; system is undamped

The complete solution is located at www.wiley.com/college/nise.

This section defined two specifications, or parameters, of second-order systems: natural frequency, $\omega_n$, and damping ratio, $\zeta$. We saw that the nature of the response obtained was related to the value of $\zeta$. Variations of damping ratio alone yield the complete range of overdamped, critically damped, underdamped, and undamped responses.

## 4.6 Underdamped Second-Order Systems

Now that we have generalized the second-order transfer function in terms of $\zeta$ and $\omega_n$, let us analyze the step response of an *underdamped* second-order system. Not only will this response be found in terms of $\zeta$ and $\omega_n$, but more specifications indigenous to the underdamped case will be defined. The underdamped second-order system, a common model for physical problems, displays unique behavior that must be itemized; a detailed description of the underdamped response is necessary for both analysis and design. Our first objective is to define transient specifications associated with underdamped responses. Next we relate these specifications to the pole location, drawing an association between pole location and the form of the underdamped second-order response. Finally, we tie the pole location to system parameters, thus closing the loop: Desired response generates required system components.

Let us begin by finding the step response for the general second-order system of Eq. (4.22). The transform of the response, $C(s)$, is the transform of the input times the transfer function, or

$$C(s) = \frac{\omega_n^2}{s(s^2 + 2\zeta\omega_n s + \omega_n^2)} = \frac{K_1}{s} + \frac{K_2 s + K_3}{s^2 + 2\zeta\omega_n s + \omega_n^2} \tag{4.26}$$

where it is assumed that $\zeta < 1$ (the underdamped case). Expanding by partial fractions, using the methods described in Section 2.2, Case 3, yields

$$C(s) = \frac{1}{s} - \frac{(s + \zeta\omega_n) + \dfrac{\zeta}{\sqrt{1-\zeta^2}}\omega_n\sqrt{1-\zeta^2}}{(s+\zeta\omega_n)^2 + \omega_n^2(1-\zeta^2)} \tag{4.27}$$

Taking the inverse Laplace transform, which is left as an exercise for the student, produces

$$\begin{aligned} c(t) &= 1 - e^{-\zeta\omega_n t}\left(\cos\omega_n\sqrt{1-\zeta^2}t + \frac{\zeta}{\sqrt{1-\zeta^2}}\sin\omega_n\sqrt{1-\zeta^2}t\right) \\ &= 1 - \frac{1}{\sqrt{1-\zeta^2}}e^{-\zeta\omega_n t}\cos(\omega_n\sqrt{1-\zeta^2}t - \phi) \end{aligned} \tag{4.28}$$

where $\phi = \tan^{-1}(\zeta/\sqrt{1-\zeta^2})$.

A plot of this response appears in Figure 4.13 for various values of $\zeta$, plotted along a time axis normalized to the natural frequency. We now see the relationship between the value of $\zeta$ and the type of response obtained: The lower the value of $\zeta$, the more oscillatory the response. The natural frequency is a time-axis scale factor and does not affect the nature of the response other than to scale it in time.

We have defined two parameters associated with second-order systems, $\zeta$ and $\omega_n$. Other parameters associated with the underdamped response are rise time, peak time, percent overshoot, and settling time. These specifications are defined as follows (see also Figure 4.14):

1. *Rise time*, $T_r$. The time required for the waveform to go from 0.1 of the final value to 0.9 of the final value.
2. *Peak time*, $T_P$. The time required to reach the first, or maximum, peak.

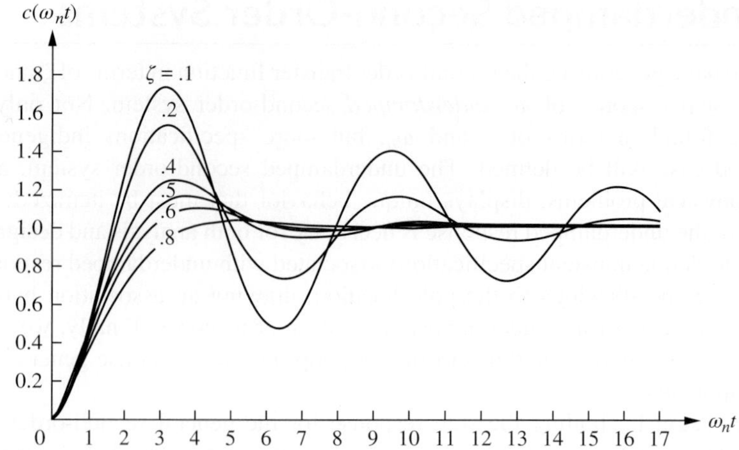


**FIGURE 4.13** Second-order underdamped responses for damping ratio values

3. *Percent overshoot, %OS*. The amount that the waveform overshoots the steady-state, or final, value at the peak time, expressed as a percentage of the steady-state value.
4. *Settling time*, $T_s$. The time required for the transient's damped oscillations to reach and stay within $\pm 2\%$ of the steady-state value.

Notice that the definitions for settling time and rise time are basically the same as the definitions for the first-order response. All definitions are also valid for systems of order higher than 2, although analytical expressions for these parameters cannot be found unless the response of the higher-order system can be approximated as a second-order system, which we do in Sections 4.7 and 4.8.

Rise time, peak time, and settling time yield information about the speed of the transient response. This information can help a designer determine if the speed and the nature of the response do or do not degrade the performance of the system. For example, the speed of an entire computer system depends on the time it takes for a hard drive head to reach steady state and read data; passenger comfort depends in part on the

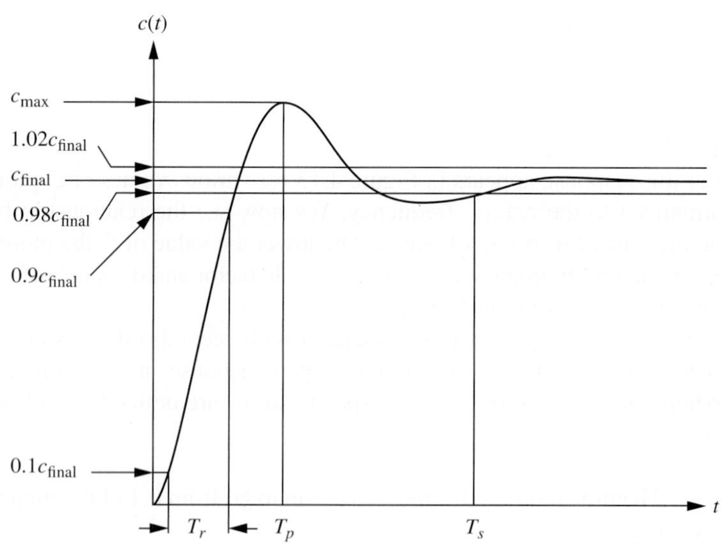


**FIGURE 4.14** Second-order underdamped response specifications

suspension system of a car and the number of oscillations it goes through after hitting a bump.

We now evaluate $T_p$, $\%OS$, and $T_s$ as functions of $\zeta$ and $\omega_n$. Later in this chapter we relate these specifications to the location of the system poles. A precise analytical expression for rise time cannot be obtained; thus, we present a plot and a table showing the relationship between $\zeta$ and rise time.

## Evaluation of $T_p$

$T_p$ is found by differentiating $c(t)$ in Eq. (4.28) and finding the first zero crossing after $t = 0$. This task is simplified by "differentiating" in the frequency domain by using Item 7 of Table 2.2. Assuming zero initial conditions and using Eq. (4.26), we get

$$\mathscr{L}[\dot{c}(t)] = sC(s) = \frac{\omega_n^2}{s^2 + 2\zeta\omega_n s + \omega_n^2} \tag{4.29}$$

Completing squares in the denominator, we have

$$\mathscr{L}[\dot{c}(t)] = \frac{\omega_n^2}{(s+\zeta\omega_n)^2 + \omega_n^2(1-\zeta^2)} = \frac{\dfrac{\omega_n}{\sqrt{1-\zeta^2}}\omega_n\sqrt{1-\zeta^2}}{(s+\zeta\omega_n)^2 + \omega_n^2(1-\zeta^2)} \tag{4.30}$$

Therefore,

$$\dot{c}(t) = \frac{\omega_n}{\sqrt{1-\zeta^2}} e^{-\zeta\omega_n t} \sin \omega_n\sqrt{1-\zeta^2}t \tag{4.31}$$

Setting the derivative equal to zero yields

$$\omega_n\sqrt{1-\zeta^2}t = n\pi \tag{4.32}$$

or

$$t = \frac{n\pi}{\omega_n\sqrt{1-\zeta^2}} \tag{4.33}$$

Each value of $n$ yields the time for local maxima or minima. Letting $n = 0$ yields $t = 0$, the first point on the curve in Figure 4.14 that has zero slope. The first peak, which occurs at the peak time, $T_p$, is found by letting $n = 1$ in Eq. (4.33):

$$\boxed{T_p = \frac{\pi}{\omega_n\sqrt{1-\zeta^2}}} \tag{4.34}$$

## Evaluation of $\%OS$

From Figure 4.14 the percent overshoot, $\%OS$, is given by

$$\%OS = \frac{c_{\max} - c_{\text{final}}}{c_{\text{final}}} \times 100 \tag{4.35}$$

The term $c_{\max}$ is found by evaluating $c(t)$ at the peak time, $c(T_p)$. Using Eq. (4.34) for $T_p$ and substituting into Eq. (4.28) yields

$$\begin{aligned} c_{\max} = c(T_p) &= 1 - e^{-(\zeta\pi/\sqrt{1-\zeta^2})}\left(\cos\pi + \frac{\zeta}{\sqrt{1-\zeta^2}}\sin\pi\right) \\ &= 1 + e^{-(\zeta\pi/\sqrt{1-\zeta^2})} \end{aligned} \tag{4.36}$$

For the unit step used for Eq. (4.28),

$$c_{\text{final}} = 1 \tag{4.37}$$

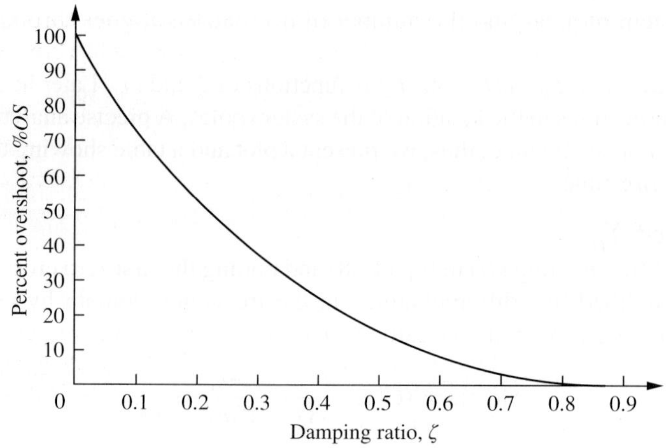


**FIGURE 4.15** Percent overshoot versus damping ratio

Substituting Eqs. (4.36) and (4.37) into Eq. (4.35), we finally obtain

$$\%OS = e^{-(\zeta\pi/\sqrt{1-\zeta^2})} \times 100 \tag{4.38}$$

Notice that the percent overshoot is a function only of the damping ratio, $\zeta$.

Whereas Eq. (4.38) allows one to find $\%OS$ given $\zeta$, the inverse of the equation allows one to solve for $\zeta$ given $\%OS$. The inverse is given by

$$\zeta = \frac{-\ln(\%OS/100)}{\sqrt{\pi^2 + \ln^2(\%OS/100)}} \tag{4.39}$$

The derivation of Eq. (4.39) is left as an exercise for the student. Equation (4.38) (or, equivalently, (4.39)) is plotted in Figure 4.15.

## Evaluation of $T_s$

In order to find the settling time, we must find the time for which $c(t)$ in Eq. (4.28) reaches and stays within $\pm 2\%$ of the steady-state value, $c_{\text{final}}$. Using our definition, the settling time is the time it takes for the amplitude of the decaying sinusoid in Eq. (4.28) to reach 0.02, or

$$e^{-\zeta\omega_n t}\frac{1}{\sqrt{1-\zeta^2}} = 0.02 \tag{4.40}$$

This equation is a conservative estimate, since we are assuming that $\cos(\omega_n\sqrt{1-\zeta^2}t - \phi) = 1$ at the settling time. Solving Eq. (4.40) for $t$, the settling time is

$$T_s = \frac{-\ln(0.02\sqrt{1-\zeta^2})}{\zeta\omega_n} \tag{4.41}$$

You can verify that the numerator of Eq. (4.41) varies from 3.91 to 4.74 as $\zeta$ varies from 0 to 0.9. Let us agree on an approximation for the settling time that will be used for all values of $\zeta$; let it be

$$T_s = \frac{4}{\zeta\omega_n} \tag{4.42}$$

## Evaluation of $T_r$

A precise analytical relationship between rise time and damping ratio, $\zeta$, cannot be found. However, using a computer and Eq. (4.28), the rise time can be found. We first designate $\omega_n t$

as the normalized time variable and select a value for $\zeta$. Using the computer, we solve for the values of $\omega_n t$ that yield $c(t) = 0.9$ and $c(t) = 0.1$. Subtracting the two values of $\omega_n t$ yields the normalized rise time, $\omega_n T_r$, for that value of $\zeta$. Continuing in like fashion with other values of $\zeta$, we obtain the results plotted in Figure 4.16.[5] Let us look at an example.

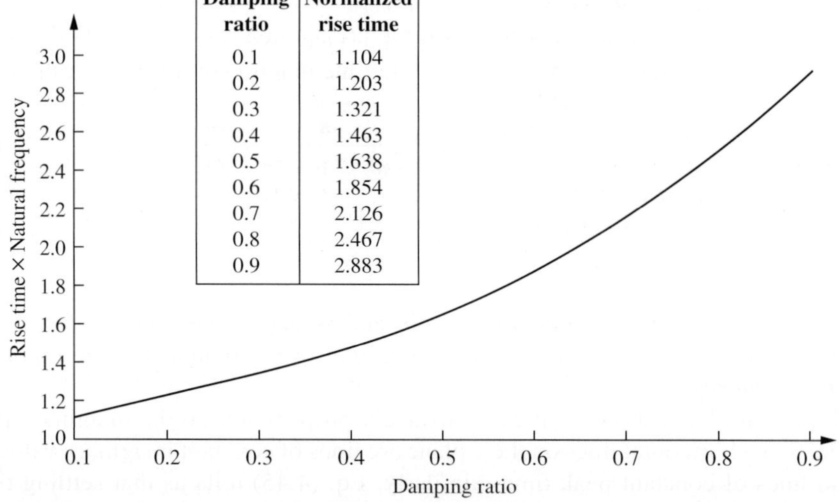


| Damping ratio | Normalized rise time |
|---|---|
| 0.1 | 1.104 |
| 0.2 | 1.203 |
| 0.3 | 1.321 |
| 0.4 | 1.463 |
| 0.5 | 1.638 |
| 0.6 | 1.854 |
| 0.7 | 2.126 |
| 0.8 | 2.467 |
| 0.9 | 2.883 |

**FIGURE 4.16** Normalized rise time versus damping ratio for a second-order underdamped response

## Example 4.5 Finding $T_p$, $\%OS$, $T_s$, and $T_r$ from a Transfer Function

**PROBLEM:** Given the transfer function

$$G(s) = \frac{100}{s^2 + 15s + 100} \tag{4.43}$$

find $T_p$, $\%OS$, $T_s$, and $T_r$.

**SOLUTION:** $\omega_n$ and $\zeta$ are calculated as 10 and 0.75, respectively. Now substitute $\zeta$ and $\omega_n$ into Eqs. (4.34), (4.38), and (4.42) and find, respectively, that $T_p = 0.475$ second, $\%OS = 2.838$, and $T_s = 0.533$ second. Using the table in Figure 4.16, the normalized rise time is approximately 2.3 seconds. Dividing by $\omega_n$ yields $T_r = 0.23$ second. This problem demonstrates that we can find $T_p$, $\%OS$, $T_s$, and $T_r$ without the tedious task of taking an inverse Laplace transform, plotting the output response, and taking measurements from the plot.

[5] Figure 4.16 can be approximated by the following polynomials: $\omega_n T_r = 1.76\zeta^3 - 0.417\zeta^2 + 1.039\zeta + 1$ (maximum error less than $\frac{1}{2}\%$ for $0 < \zeta < 0.9$), and $\zeta = 0.115(\omega_n T_r)^3 - 0.883(\omega_n T_r)^2 + 2.504(\omega_n T_r) - 1.738$ (maximum error less than 5% for $0.1 < \zeta < 0.9$). The polynomials were obtained using MATLAB's **polyfit** function.

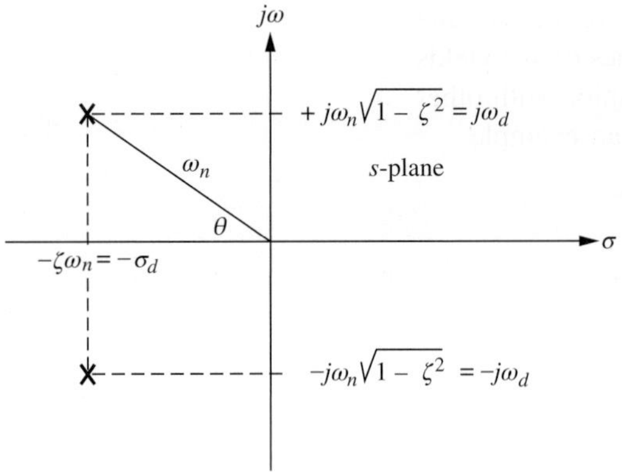


**FIGURE 4.17** Pole plot for an underdamped second-order system

We now have expressions that relate peak time, percent overshoot, and settling time to the natural frequency and the damping ratio. Now let us relate these quantities to the location of the poles that generate these characteristics.

The pole plot for a general, underdamped second-order system, previously shown in Figure 4.11, is reproduced and expanded in Figure 4.17 for focus. We see from the Pythagorean theorem that the radial distance from the origin to the pole is the natural frequency, $\omega_n$, and the $\cos\theta = \zeta$.

Now, comparing Eqs. (4.34) and (4.42) with the pole location, we evaluate peak time and settling time in terms of the pole location. Thus,

$$T_p = \frac{\pi}{\omega_n\sqrt{1-\zeta^2}} = \frac{\pi}{\omega_d} \tag{4.44}$$

$$T_s = \frac{4}{\zeta\omega_n} = \frac{\pi}{\sigma_d} \tag{4.45}$$

where $\omega_d$ is the imaginary part of the pole and is called the *damped frequency of oscillation*, and $\sigma_d$ is the magnitude of the real part of the pole and is the *exponential damping frequency*.

Equation (4.44) shows that $T_p$ is inversely proportional to the imaginary part of the pole. Since horizontal lines on the $s$-plane are lines of constant imaginary value, they are also lines of constant peak time. Similarly, Eq. (4.45) tells us that settling time is inversely proportional to the real part of the pole. Since vertical lines on the $s$-plane are lines of constant real value, they are also lines of constant settling time. Finally, since $\zeta = \cos\theta$, radial lines are lines of constant $\zeta$. Since percent overshoot is only a function of $\zeta$, radial lines are thus lines of constant percent overshoot, $\%OS$. These concepts are depicted in Figure 4.18, where lines of constant $T_p$, $T_s$, and $\%OS$ are labeled on the $s$-plane.

At this point, we can understand the significance of Figure 4.18 by examining the actual step response of comparative systems. Depicted in Figure 4.19($a$) are the step responses as the poles are moved in a vertical direction, keeping the real part the same.

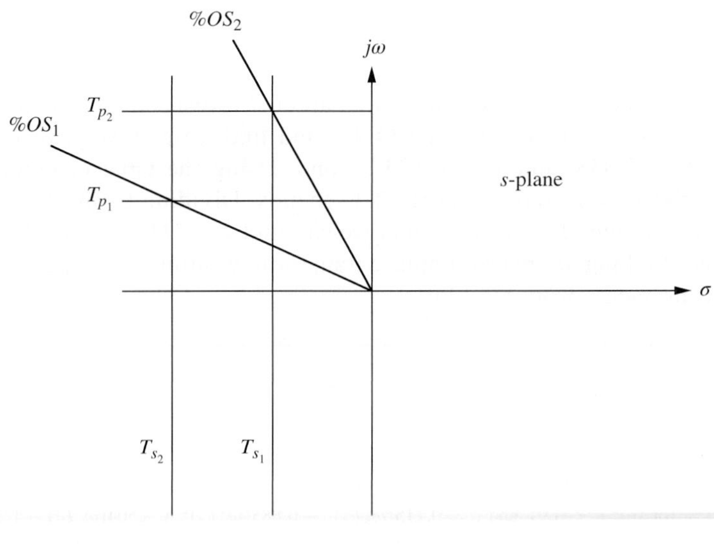


**FIGURE 4.18** Lines of constant peak time, $T_p$, settling time, $T_s$, and percent overshoot, $\%OS$. Note: $T_{s_2} < T_{s_1}$; $T_{p2} < T_{p1}$; $\%OS_1 < \%OS_2$.

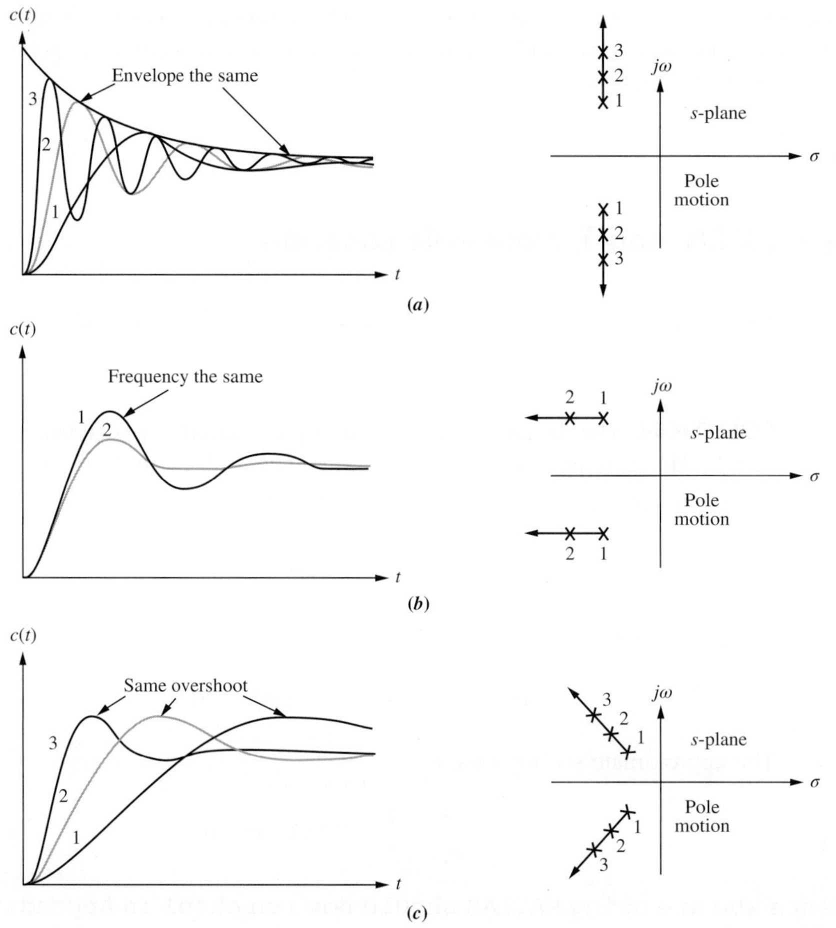


**FIGURE 4.19** Step responses of second-order underdamped systems as poles move: **a.** with constant real part; **b.** with constant imaginary part; **c.** with constant damping ratio

As the poles move in a vertical direction, the frequency increases, but the envelope remains the same since the real part of the pole is not changing. The figure shows a constant exponential envelope, even though the sinusoidal response is changing frequency. Since all curves fit under the same exponential decay curve, the settling time is virtually the same for all waveforms. Note that as overshoot increases, the rise time decreases.

Let us move the poles to the right or left. Since the imaginary part is now constant, movement of the poles yields the responses of Figure 4.19(*b*). Here the frequency is constant over the range of variation of the real part. As the poles move to the left, the response damps out more rapidly, while the frequency remains the same. Notice that the peak time is the same for all waveforms because the imaginary part remains the same.

Moving the poles along a constant radial line yields the responses shown in Figure 4.19(*c*). Here the percent overshoot remains the same. Notice also that the responses look exactly alike, except for their speed. The farther the poles are from the origin, the more rapid the response.

We conclude this section with some examples that demonstrate the relationship between the pole location and the specifications of the second-order underdamped response. The first example covers analysis. The second example is a simple design problem consisting of a physical system whose component values we want to design to meet a transient response specification. An animation PowerPoint presentation (PPT) demonstrating second-order principles is available for instructors at www.wiley.com/college/nise. See *Second-Order Step Response*.

## Example 4.6 Finding $T_p$, $\%OS$, and $T_s$ from Pole Location

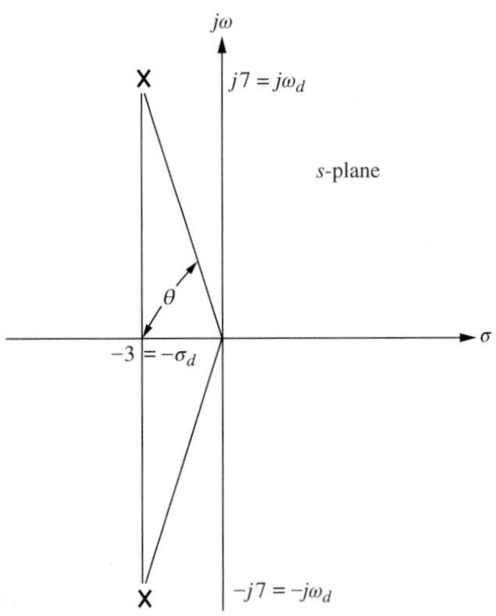


**FIGURE 4.20** Pole plot for Example 4.6

**PROBLEM:** Given the pole plot shown in Figure 4.20, find $\zeta$, $\omega_n$, $T_p$, $\%OS$, and $T_s$.

**SOLUTION:** The damping ratio is given by $\zeta = \cos\theta = \cos[\arctan(7/3)] = 0.394$. The natural frequency, $\omega_n$, is the radial distance from the origin to the pole, or $\omega_n = \sqrt{7^2 + 3^2} = 7.616$. The peak time is

$$T_p = \frac{\pi}{\omega_d} = \frac{\pi}{7} = 0.449 \text{ second} \tag{4.46}$$

The percent overshoot is

$$\%OS = e^{-(\zeta\pi/\sqrt{1-\zeta^2})} \times 100 = 26\% \tag{4.47}$$

The approximate settling time is

$$T_s = \frac{4}{\sigma_d} = \frac{4}{3} = 1.333 \text{ seconds} \tag{4.48}$$

MATLAB **ML**

Students who are using MATLAB should now run ch4p1 in Appendix B. You will learn how to generate a second-order polynomial from two complex poles as well as extract and use the coefficients of the polynomial to calculate $T_p$, $\%OS$, and $T_s$. This exercise uses MATLAB to solve the problem in Example 4.6.

## Example 4.7 Transient Response Through Component Design

Design **D**

**PROBLEM:** Given the system shown in Figure 4.21, find $J$ and $D$ to yield 20% overshoot and a settling time of 2 seconds for a step input of torque $T(t)$.

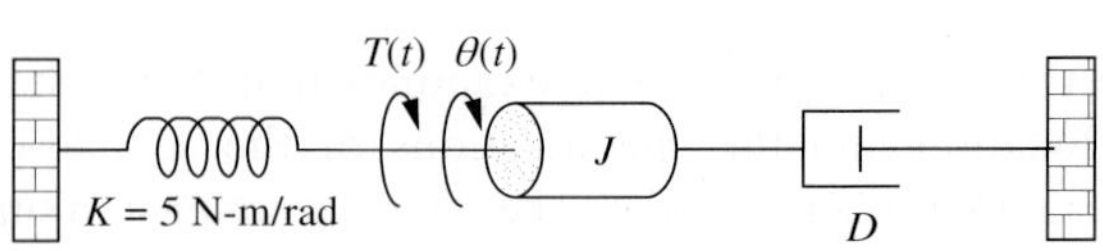


**FIGURE 4.21** Rotational mechanical system for Example 4.7

**SOLUTION:** First, the transfer function for the system is

$$G(s) = \frac{1/J}{s^2 + \frac{D}{J}s + \frac{K}{J}} \tag{4.49}$$

From the transfer function,

$$\omega_n = \sqrt{\frac{K}{J}} \tag{4.50}$$

and

$$2\zeta\omega_n = \frac{D}{J} \tag{4.51}$$

But, from the problem statement,

$$T_s = 2 = \frac{4}{\zeta\omega_n} \tag{4.52}$$

or $\zeta\omega_n = 2$. Hence,

$$2\zeta\omega_n = 4 = \frac{D}{J} \tag{4.53}$$

Also, from Eqs. (4.50) and (4.52),

$$\zeta = \frac{4}{2\omega_n} = 2\sqrt{\frac{J}{K}} \tag{4.54}$$

From Eq. (4.39), a 20% overshoot implies $\zeta = 0.456$. Therefore, from Eq. (4.54),

$$\zeta = 2\sqrt{\frac{J}{K}} = 0.456 \tag{4.55}$$

Hence,

$$\frac{J}{K} = 0.052 \tag{4.56}$$

From the problem statement, $K = 5$ N-m/rad. Combining this value with Eqs. (4.53) and (4.56), $D = 1.04$ N-m-s/rad, and $J = 0.26$ kg-m$^2$.

## Second-Order Transfer Functions via Testing

Just as we obtained the transfer function of a first-order system experimentally, we can do the same for a system that exhibits a typical underdamped second-order response. Again, we can measure the laboratory response curve for percent overshoot and settling time, from which we can find the poles and hence the denominator. The numerator can be found, as in the first-order system, from a knowledge of the measured and expected steady-state values. A problem at the end of the chapter illustrates the estimation of a second-order transfer function from the step response.

## Skill-Assessment Exercise 4.5

**TryIt 4.1**

Use the following MATLAB statements to calculate the answers to Skill-Assessment Exercise 4.5. Ellipses mean code continues on next line.

```
numg=361;
deng=[1 16 361];
omegan=sqrt(deng(3)...
  /deng(1))
zeta=(deng(2)/deng(1))...
  /(2*omegan)
Ts=4/(zeta*omegan)
Tp=pi/(omegan*sqrt...
  (1-zeta^2))
pos=100*exp(-zeta*...
  pi/sqrt(1-zeta^2))
Tr=(1.768*zeta^3 -...
  0.417*zeta^2+1.039*...
  zeta+1)/omegan
```

**PROBLEM:** Find $\zeta$, $\omega_n$, $T_s$, $T_p$, $T_r$, and $\%OS$ for a system whose transfer function is $G(s) = \dfrac{361}{s^2 + 16s + 361}$.

**ANSWERS:**

$\zeta = 0.421$, $\omega_n = 19$, $T_s = 0.5$ s, $T_p = 0.182$ s, $T_r = 0.079$ s, and $\%OS = 23.3\%$.

The complete solution is located at www.wiley.com/college/nise.

Now that we have analyzed systems with two poles, how does the addition of another pole affect the response? We answer this question in the next section.

# 4.7 System Response with Additional Poles

In the last section, we analyzed systems with one or two poles. It must be emphasized that the formulas describing percent overshoot, settling time, and peak time were derived only for a system with two complex poles and no zeros. If a system such as that shown in Figure 4.22 has more than two poles or has zeros, we cannot use the formulas to calculate the performance specifications that we derived. However, under certain conditions, a system with more than two poles or with zeros can be approximated as a second-order system that has just two complex *dominant poles*. Once we justify this approximation, the formulas for percent overshoot, settling time, and peak time can be applied to these higher-order systems by using the location of the dominant poles. In this section, we investigate the effect of an additional pole on the second-order response. In the next section, we analyze the effect of adding a zero to a two-pole system.

Let us now look at the conditions that would have to exist in order to approximate the behavior of a three-pole system as that of a two-pole system. Consider a three-pole system with complex poles and a third pole on the real axis. Assuming that the complex poles are at $-\zeta\omega_n \pm j\omega_n\sqrt{1-\zeta^2}$ and the real pole is at $-\alpha_r$, the step response of the system can be determined from a partial-fraction expansion. Thus, the output transform is

$$C(s) = \frac{A}{s} + \frac{B(s+\zeta\omega_n) + C\omega_d}{(s+\zeta\omega_n)^2 + \omega_d^2} + \frac{D}{s+\alpha_r} \tag{4.57}$$

or, in the time domain,

$$c(t) = Au(t) + e^{-\zeta\omega_n t}(B\cos\omega_d t + C\sin\omega_d t) + De^{-\alpha_r t} \tag{4.58}$$

The component parts of $c(t)$ are shown in Figure 4.23 for three cases of $\alpha_r$. For Case I, $\alpha_r = \alpha_{r_1}$ and is not much larger than $\zeta\omega_n$; for Case II, $\alpha_r = \alpha_{r_2}$ and is much larger than $\zeta\omega_n$; and for Case III, $\alpha_r = \infty$.

**FIGURE 4.22** Robot follows input commands from a human trainer

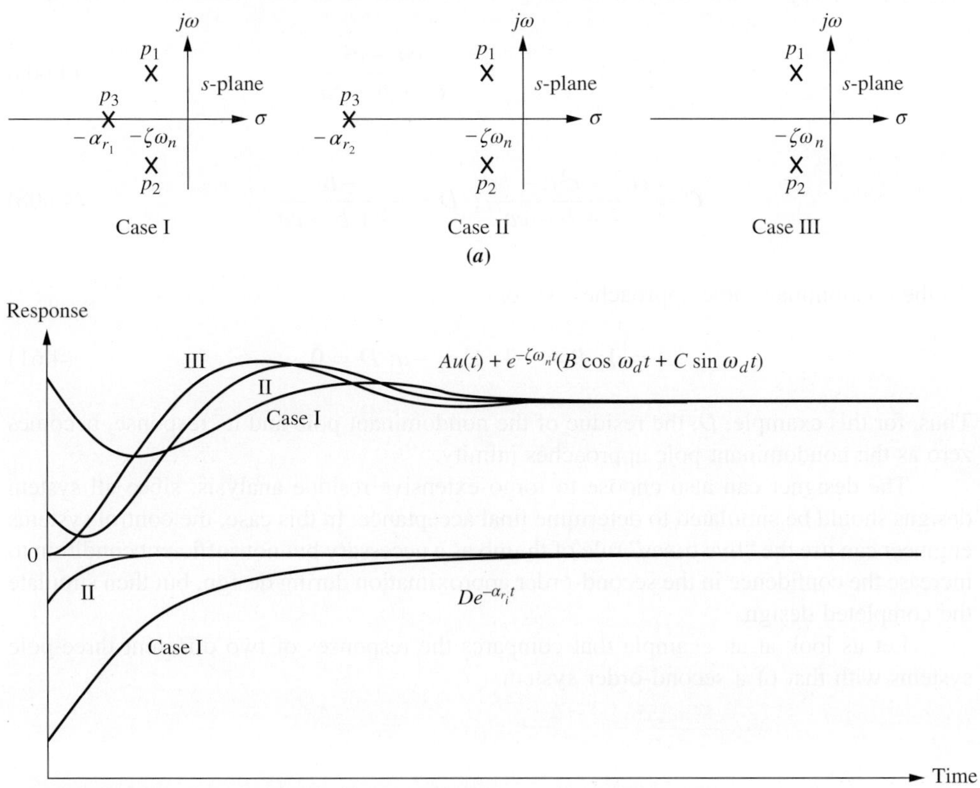


**FIGURE 4.23** Component responses of a three-pole system: **a.** pole plot; **b.** component responses: Nondominant pole is near dominant second-order pair (Case I), far from the pair (Case II), and at infinity (Case III)

Let us direct our attention to Eq. (4.58) and Figure 4.23. If $\alpha_r \gg \zeta\omega_n$ (Case II), the pure exponential will die out much more rapidly than the second-order underdamped step response. If the pure exponential term decays to an insignificant value at the time of the first overshoot, such parameters as percent overshoot, settling time, and peak time will be generated by the second-order underdamped step response component. Thus, the total response will approach that of a pure second-order system (Case III).

If $\alpha_r$ is not much greater than $\zeta\omega_n$ (Case I), the real pole's transient response will not decay to insignificance at the peak time or settling time generated by the second-order pair. In this case, the exponential decay is significant, and the system cannot be represented as a second-order system.

The next question is, How much farther from the dominant poles does the third pole have to be for its effect on the second-order response to be negligible? The answer of course depends on the accuracy for which you are looking. However, this book assumes that the exponential decay is negligible after five time constants. Thus, if the real pole is five times farther to the left than the dominant poles, we assume that the system is represented by its dominant second-order pair of poles.

What about the magnitude of the exponential decay? Can it be so large that its contribution at the peak time is not negligible? We can show, through a partial-fraction expansion, that the residue of the third pole, in a three-pole system with dominant second-order poles and no zeros, will actually decrease in magnitude as the third pole is moved farther into the left half-plane. Assume a step response, $C(s)$, of a three-pole system:

$$C(s) = \frac{bc}{s(s^2+as+b)(s+c)} = \frac{A}{s} + \frac{Bs+C}{s^2+as+b} + \frac{D}{s+c} \tag{4.59}$$

where we assume that the nondominant pole is located at $-c$ on the real axis and that the steady-state response approaches unity. Evaluating the constants in the numerator of each term,

$$A = 1; \qquad B = \frac{ca - c^2}{c^2 + b - ca} \tag{4.60a}$$

$$C = \frac{ca^2 - c^2a - bc}{c^2 + b - ca}; \quad D = \frac{-b}{c^2 + b - ca} \tag{4.60b}$$

As the nondominant pole approaches $\infty$, or $c \to \infty$,

$$A = 1; \; B = -1; \; C = -a; \; D = 0 \tag{4.61}$$

Thus, for this example, $D$, the residue of the nondominant pole and its response, becomes zero as the nondominant pole approaches infinity.

The designer can also choose to forgo extensive residue analysis, since all system designs should be simulated to determine final acceptance. In this case, the control systems engineer can use the "five times" rule of thumb as a necessary but not sufficient condition to increase the confidence in the second-order approximation during design, but then simulate the completed design.

Let us look at an example that compares the responses of two different three-pole systems with that of a second-order system.

## Example 4.8 Comparing Responses of Three-Pole Systems

**PROBLEM:** Find the step response of each of the transfer functions shown in Eqs. (4.62) through (4.64) and compare them.

$$T_1(s) = \frac{24.542}{s^2 + 4s + 24.542} \tag{4.62}$$

$$T_2(s) = \frac{245.42}{(s+10)(s^2 + 4s + 24.542)} \tag{4.63}$$

$$T_3(s) = \frac{73.626}{(s+3)(s^2 + 4s + 24.542)} \tag{4.64}$$

**SOLUTION:** The step response, $C_i(s)$, for the transfer function, $T_i(s)$, can be found by multiplying the transfer function by $1/s$, a step input, and using partial-fraction expansion followed by the inverse Laplace transform to find the response, $c_i(t)$. With the details left as an exercise for the student, the results are

$$c_1(t) = 1 - 1.09e^{-2t}\cos(4.532t - 23.8°) \tag{4.65}$$

$$c_2(t) = 1 - 0.29e^{-10t} - 1.189e^{-2t}\cos(4.532t - 53.34°) \tag{4.66}$$

$$c_3(t) = 1 - 1.14e^{-3t} + 0.707e^{-2t}\cos(4.532t + 78.63°) \tag{4.67}$$

The three responses are plotted in Figure 4.24. Notice that $c_2(t)$, with its third pole at $-10$ and farthest from the dominant poles, is the better approximation of $c_1(t)$, the pure second-order system response; $c_3(t)$, with a third pole close to the dominant poles, yields the most error.

MATLAB

ML

Students who are using MATLAB should now run ch4p2 in Appendix B. You will learn how to generate a step response for a transfer function and how to plot the response directly or collect the

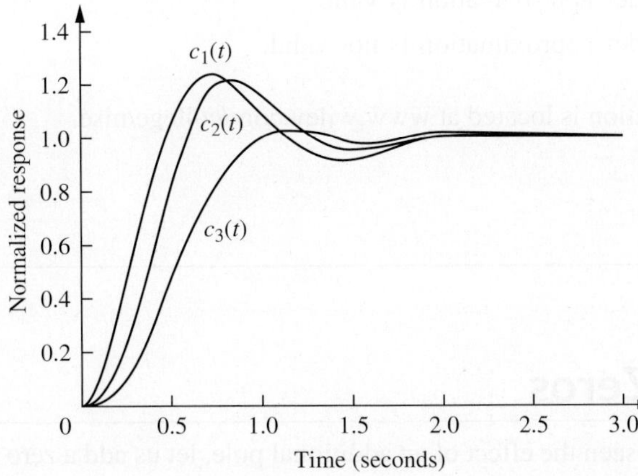


**FIGURE 4.24** Step responses of system $T_1(s)$, system $T_2(s)$, and system $T_3(s)$

points for future use. The example shows how to collect the points and then use them to create a multiple plot, title the graph, and label the axes and curves to produce the graph in Figure 4.24 to solve Example 4.8.

Simulink

SL

System responses can alternately be obtained using Simulink. Simulink is a software package that is integrated with MATLAB to provide a graphical user interface (GUI) for defining systems and generating responses. The reader is encouraged to study Appendix C, which contains a tutorial on Simulink as well as some examples. One of the illustrative examples, Example C.1, solves Example 4.8 using Simulink.

Gui Tool

GUIT

Another method to obtain systems responses is through the use of MATLAB's LTI Viewer. An advantage of the LTI Viewer is that it displays the values of settling time, peak time, rise time, maximum response, and the final value on the step response plot. The reader is encouraged to study Appendix E at www.wiley.com/college/nise, which contains a tutorial on the LTI Viewer as well as some examples. Example E.1 solves Example 4.8 using the LTI Viewer.

## Skill-Assessment Exercise 4.6

**PROBLEM:** Determine the validity of a second-order approximation for each of these two transfer functions:

**a.** $G(s) = \dfrac{700}{(s+15)(s^2+4s+100)}$

**b.** $G(s) = \dfrac{360}{(s+4)(s^2+2s+90)}$

**ANSWERS:**

**a.** The second-order approximation is valid.

**b.** The second-order approximation is not valid.

The complete solution is located at www.wiley.com/college/nise.

### TryIt 4.2

Use the following MATLAB and Control System Toolbox statements to investigate the effect of the additional pole in Skill-Assessment Exercise 4.6(*a*). Move the higher-order pole originally at −15 to other values by changing "a" in the code.

```
a=15
numga=100*a;
denga=conv([1 a],...
 [1 4 100]);
Ta=tf(numga,denga);
numg=100;
deng=[1 4 100];
T=tf(numg,deng);
step(Ta,'.',T,'-')
```

## 4.8 System Response with Zeros

Now that we have seen the effect of an additional pole, let us add a zero to the second-order system. In Section 4.2, we saw that the zeros of a response affect the residue, or amplitude, of a response component but do not affect the nature of the response—exponential, damped sinusoid, and so on. In this section, we add a real-axis zero to a two-pole system. The zero will be added first in the left half-plane and then in the right half-plane and its effects noted and analyzed. We conclude the section by talking about pole-zero cancellation.

Starting with a two-pole system with poles at $(-1 \pm j2.828)$, we consecutively add zeros at $-3$, $-5$, and $-10$. The results, normalized to the steady-state value, are plotted in Figure 4.25. We can see that the closer the zero is to the dominant poles, the greater its effect on the transient response. As the zero moves away from the dominant poles, the response approaches that of the two-pole system. This analysis can be reasoned via the partial-fraction expansion. If we assume a group of poles and a zero far from the poles, the residue of each pole will be affected the same by the zero. Hence, the relative amplitudes remain appreciably the same. For example, assume the partial-fraction expansion shown in Eq. (4.68):

$$T(s) = \frac{(s+a)}{(s+b)(s+c)} = \frac{A}{s+b} + \frac{B}{s+c} = \frac{(-b+a)/(-b+c)}{s+b} + \frac{(-c+a)/(-c+b)}{s+c} \tag{4.68}$$

If the zero is far from the poles, then $a$ is large compared to $b$ and $c$, and

$$T(s) \approx a\left[\frac{1/(-b+c)}{s+b} + \frac{1/(-c+b)}{s+c}\right] = \frac{a}{(s+b)(s+c)} \tag{4.69}$$

Hence, the zero looks like a simple gain factor and does not change the relative amplitudes of the components of the response.

Another way to look at the effect of a zero, which is more general, is as follows (*Franklin, 1991*): Let $C(s)$ be the response of a system, $T(s)$, with unity in the numerator. If we add a zero to the transfer function, yielding $(s+a)T(s)$, the Laplace transform of the response will be

$$(s+a)C(s) = sC(s) + aC(s) \tag{4.70}$$

Thus, the response of a system with a zero consists of two parts: the derivative of the original response and a scaled version of the original response. If $a$, the negative of the zero, is very large, the Laplace transform of the response is approximately $aC(s)$, or a scaled version of the original response. If $a$ is not very large, the response has an additional component consisting of the derivative of the original response. As $a$ becomes smaller, the derivative term contributes more to the response and has a greater effect. For step responses, the derivative is typically positive at the start of a step response. Thus, for small values of $a$, we can expect more overshoot in second-order systems because the derivative term will be additive around the first overshoot. This reasoning is borne out by Figure 4.25.

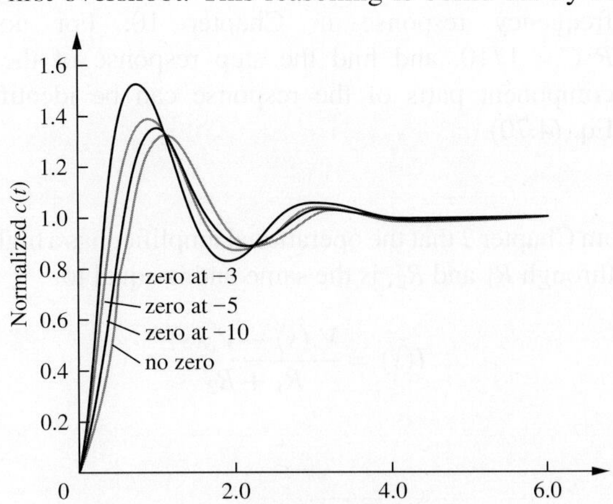


**FIGURE 4.25** Effect of adding a zero to a two-pole system

**TryIt 4.3**

Use the following MATLAB and Control System Toolbox statements to generate Figure 4.25.

```
deng=[1 2 9];
Ta=tf([1 3]*9/3,deng);
Tb=tf([1 5]*9/5,deng);
Tc=tf([1 10]*9/10,deng);
T=tf(9,deng);
step(T,Ta,Tb,Tc)
text(0.5,0.6,'no zero')
text(0.4,0.7,...
 'zero at -10')
text(0.35,0.8,...
 'zero at -5')
text(0.3,0.9,'zero at -3')
```

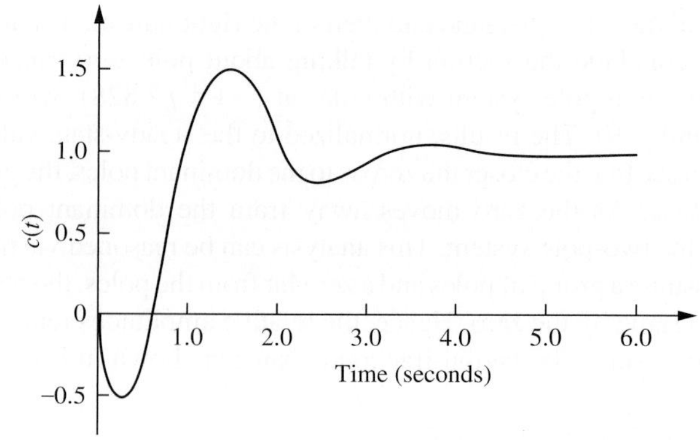


**FIGURE 4.26** Step response of a nonminimum-phase system

An interesting phenomenon occurs if $a$ is negative, placing the zero in the right half-plane. From Eq. (4.70) we see that the derivative term, which is typically positive initially, will be of opposite sign from the scaled response term. Thus, if the derivative term, $sC(s)$, is larger than the scaled response, $aC(s)$, the response will initially follow the derivative in the opposite direction from the scaled response. The result for a second-order system is shown in Figure 4.26, where the sign of the input was reversed to yield a positive steady-state value. Notice that the response begins to turn toward the negative direction even though the final value is positive. A system that exhibits this phenomenon is known as a *nonminimum-phase* system. If a motorcycle or airplane was a nonminimum-phase system, it would initially veer left when commanded to steer right.

Let us now look at an example of an electrical nonminimum-phase network.

## Example 4.9 Transfer Function of a Nonminimum-Phase System

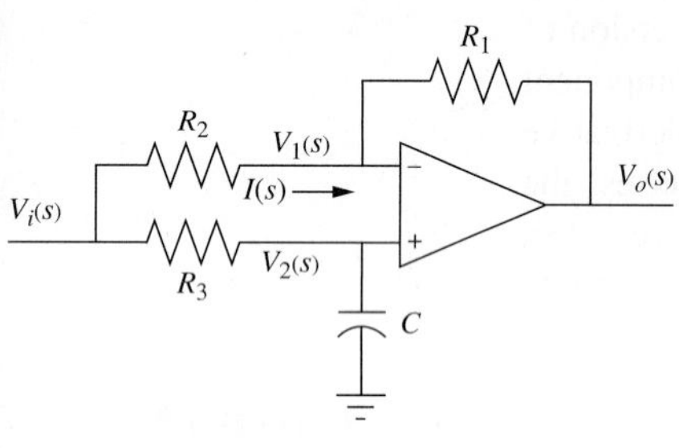


**FIGURE 4.27** Nonminimum-phase electric circuit[6]

**PROBLEM:**

**a.** Find the transfer function, $V_o(s)/V_i(s)$ for the operational amplifier circuit shown in Figure 4.27.

**b.** If $R_1 = R_2$, this circuit is known as an all-pass filter, since it passes sine waves of a wide range of frequencies without attenuating or amplifying their magnitude (*Dorf, 1993*). We will learn more about frequency response in Chapter 10. For now, let $R_1 = R_2$, $R_3C = 1/10$, and find the step response of the filter. Show that component parts of the response can be identified with those in Eq. (4.70).

**SOLUTION:**

**a.** Remembering from Chapter 2 that the operational amplifier has a high input impedance, the current, $I(s)$, through $R_1$ and $R_2$, is the same and is equal to

$$I(s) = \frac{V_i(s) - V_o(s)}{R_1 + R_2} \tag{4.71}$$

[6] Adapted from Dorf, R. C. *Introduction to Electric Circuits, 2nd ed.* (New York: John Wiley & Sons, 1989, 1993), p. 583. © 1989, 1993 John Wiley & Sons. Reprinted by permission of the publisher.

Also,

$$V_o(s) = A(V_2(s) - V_1(s)) \tag{4.72}$$

But,

$$V_1(s) = I(s)R_1 + V_o(s) \tag{4.73}$$

Substituting Eq. (4.71) into (4.73),

$$V_1(s) = \frac{1}{R_1 + R_2}(R_1 V_i(s) + R_2 V_0(s)) \tag{4.74}$$

Using voltage division,

$$V_2(s) = V_i(s)\frac{1/Cs}{R_3 + \dfrac{1}{Cs}} \tag{4.75}$$

Substituting Eqs. (4.74) and (4.75) into Eq. (4.72) and simplifying yields

$$\frac{V_o(s)}{V_i(s)} = \frac{A(R_2 - R_1 R_3 Cs)}{(R_3 Cs + 1)(R_1 + R_2(1 + A))} \tag{4.76}$$

Since the operational amplifier has a large gain, $A$, let $A$ approach infinity. Thus, after simplification

$$\frac{V_o(s)}{V_i(s)} = \frac{R_2 - R_1 R_3 Cs}{R_2 R_3 Cs + R_2} = -\frac{R_1}{R_2}\frac{\left(s - \dfrac{R_2}{R_1 R_3 C}\right)}{\left(s + \dfrac{1}{R_3 C}\right)} \tag{4.77}$$

**b.** Letting $R_1 = R_2$ and $R_3 C = 1/10$,

$$\frac{V_o(s)}{V_i(s)} = \frac{\left(s - \dfrac{1}{R_3 C}\right)}{\left(s + \dfrac{1}{R_3 C}\right)} = -\frac{(s - 10)}{(s + 10)} \tag{4.78}$$

For a step input, we evaluate the response as suggested by Eq. (4.70):

$$C(s) = -\frac{(s - 10)}{s(s + 10)} = -\frac{1}{s + 10} + 10\frac{1}{s(s + 10)} = sC_o(s) - 10C_o(s) \tag{4.79}$$

where

$$C_o(s) = -\frac{1}{s(s + 10)} \tag{4.80}$$

is the Laplace transform of the response without a zero. Expanding Eq. (4.79) into partial fractions,

$$C(s) = -\frac{1}{s + 10} + 10\frac{1}{s(s + 10)} = -\frac{1}{s + 10} + \frac{1}{s} - \frac{1}{s + 10} = \frac{1}{s} - \frac{2}{s + 10} \tag{4.81}$$

or the response with a zero is

$$c(t) = -e^{-10t} + 1 - e^{-10t} = 1 - 2e^{-10t} \tag{4.82}$$

Also, from Eq. (4.80),

$$C_o(s) = -\frac{1/10}{s} + \frac{1/10}{s+10} \tag{4.83}$$

or the response without a zero is

$$c_o(t) = -\frac{1}{10} + \frac{1}{10}e^{-10t} \tag{4.84}$$

The normalized responses are plotted in Figure 4.28. Notice the immediate reversal of the nonminimum-phase response, $c(t)$.

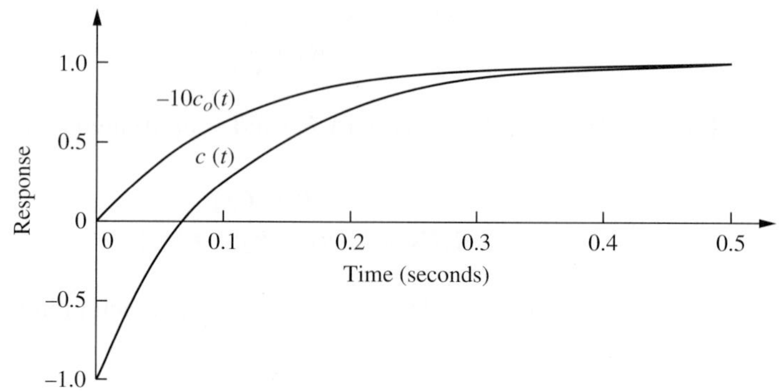


**FIGURE 4.28** Step response of the nonminimum-phase network of Figure 4.27 ($c(t)$) and normalized step response of an equivalent network without the zero ($-10\,c_o(t)$)

We conclude this section by talking about pole-zero cancellation and its effect on our ability to make second-order approximations to a system. Assume a three-pole system with a zero as shown in Eq. (4.85). If the pole term, $(s + p_3)$, and the zero term, $(s + z)$, cancel out, we are left with

$$T(s) = \frac{K\cancel{(s+z)}}{\cancel{(s+p_3)}(s^2 + as + b)} \tag{4.85}$$

as a second-order transfer function. From another perspective, if the zero at $-z$ is very close to the pole at $-p_3$, then a partial-fraction expansion of Eq. 4.85 will show that the residue of the exponential decay is much smaller than the amplitude of the second-order response. Let us look at an example.

## Example 4.10 Evaluating Pole-Zero Cancellation Using Residues

**PROBLEM:** For each of the response functions in Eqs. (4.86) and (4.87), determine whether there is cancellation between the zero and the pole closest to the zero. For any function for which pole-zero cancellation is valid, find the approximate response.

$$C_1(s) = \frac{26.25(s+4)}{s(s+3.5)(s+5)(s+6)} \tag{4.86}$$

$$C_2(s) = \frac{26.25(s+4)}{s(s+4.01)(s+5)(s+6)} \quad (4.87)$$

**SOLUTION:** The partial-fraction expansion of Eq. (4.86) is

$$C_1(s) = \frac{1}{s} - \frac{3.5}{s+5} + \frac{3.5}{s+6} - \frac{1}{s+3.5} \quad (4.88)$$

The residue of the pole at $-3.5$, which is closest to the zero at $-4$, is equal to 1 and is not negligible compared to the other residues. Thus, a second-order step response approximation cannot be made for $C_1(s)$. The partial-fraction expansion for $C_2(s)$ is

$$C_2(s) = \frac{0.87}{s} - \frac{5.3}{s+5} + \frac{4.4}{s+6} + \frac{0.033}{s+4.01} \quad (4.89)$$

The residue of the pole at $-4.01$, which is closest to the zero at $-4$, is equal to 0.033, about two orders of magnitude below any of the other residues. Hence, we make a second-order approximation by neglecting the response generated by the pole at $-4.01$:

$$C_2(s) \approx \frac{0.87}{s} - \frac{5.3}{s+5} + \frac{4.4}{s+6} \quad (4.90)$$

and the response $c_2(t)$ is approximately

$$c_2(t) \approx 0.87 - 5.3e^{-5t} + 4.4e^{-6t} \quad (4.91)$$

**TryIt 4.4**

Use the following MATLAB and Symbolic Math Toolbox statements to evaluate the effect of higher-order poles by finding the component parts of the time response of $c_1(t)$ and $c_2(t)$ in Example 4.10.

```
syms s
C1=26.25*(s+4)/...
  (s*(s+3.5)*...
  (s+5)*(s+6));
C2=26.25*(s+4)/...
  (s*(s+4.01)*...
  (s+5)*(s+6));
c1=ilaplace(C1);
'c1'
c1=vpa(c1,3)
c2=ilaplace(C2);
'c2'
c2=vpa(c2,3)
```

## Skill-Assessment Exercise 4.7

**PROBLEM:** Determine the validity of a second-order step-response approximation for each transfer function shown below.

**a.** $G(s) = \dfrac{185.71(s+7)}{(s+6.5)(s+10)(s+20)}$

**b.** $G(s) = \dfrac{197.14(s+7)}{(s+6.9)(s+10)(s+20)}$

**ANSWERS:**

**a.** A second-order approximation is not valid.

**b.** A second-order approximation is valid.

The complete solution is located at www.wiley.com/college/nise.

In this section, we have examined the effects of additional transfer function poles and zeros upon the response. In the next section we add nonlinearities of the type discussed in Section 2.10 and see what effects they have on system response.

# 4.9 Effects of Nonlinearities upon Time Response

In this section, we qualitatively examine the effects of nonlinearities upon the time response of physical systems. In the following examples, we insert nonlinearities, such as saturation, dead zone, and backlash, as shown in Figure 2.46, into a system to show the effects of these nonlinearities upon the linear responses.

The responses were obtained using Simulink, a simulation software package that is integrated with MATLAB to provide a graphical user interface (GUI). Readers who would like to learn how to use Simulink to generate nonlinear responses should consult the Simulink tutorial in Appendix C. Simulink block diagrams are included with all responses that follow.

Let us assume the motor and load from the Antenna Control Case Study of Chapter 2 and look at the load angular velocity, $\omega_o(s)$, where $\omega_o(s) = 0.1\, s\theta_m(s) = 0.2083\, \mathrm{E}_a(s)/(s + 1.71)$ from Eq. (2.208). If we drive the motor with a step input through an amplifier of unity gain that saturates at $\pm 5$ volts, Figure 4.29 shows that the effect of amplifier saturation is to limit the obtained velocity.

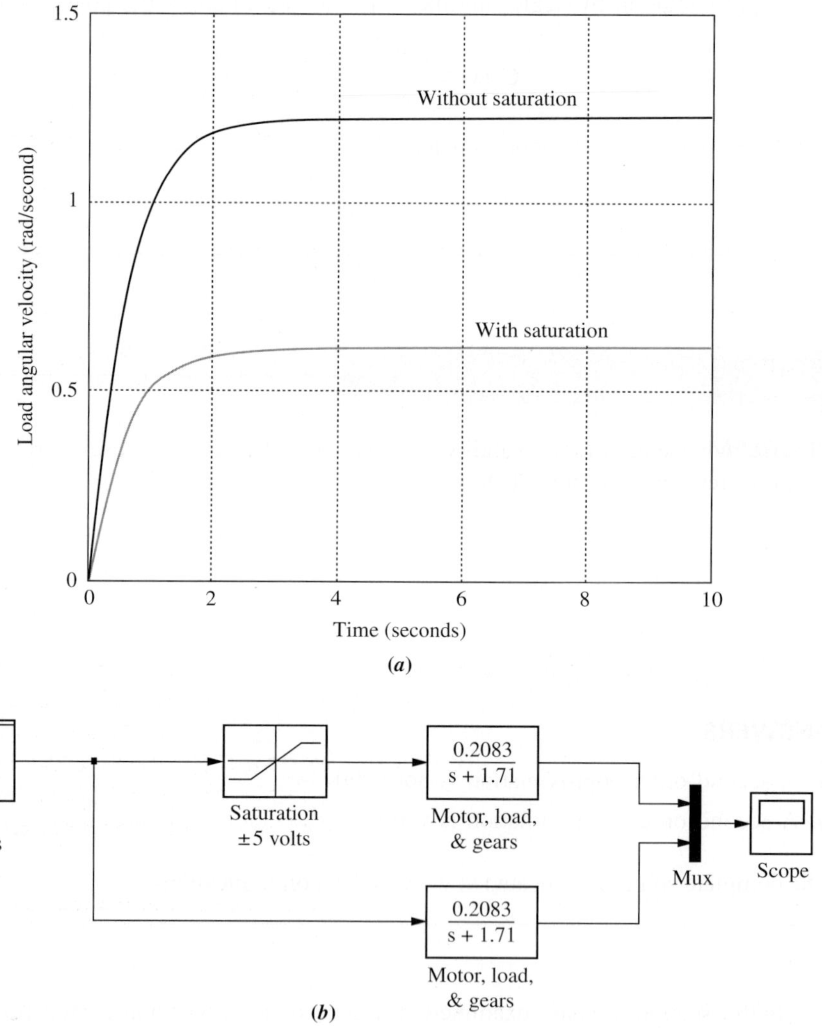


**FIGURE 4.29** **a.** Effect of amplifier saturation on load angular velocity response; **b.** Simulink block diagram

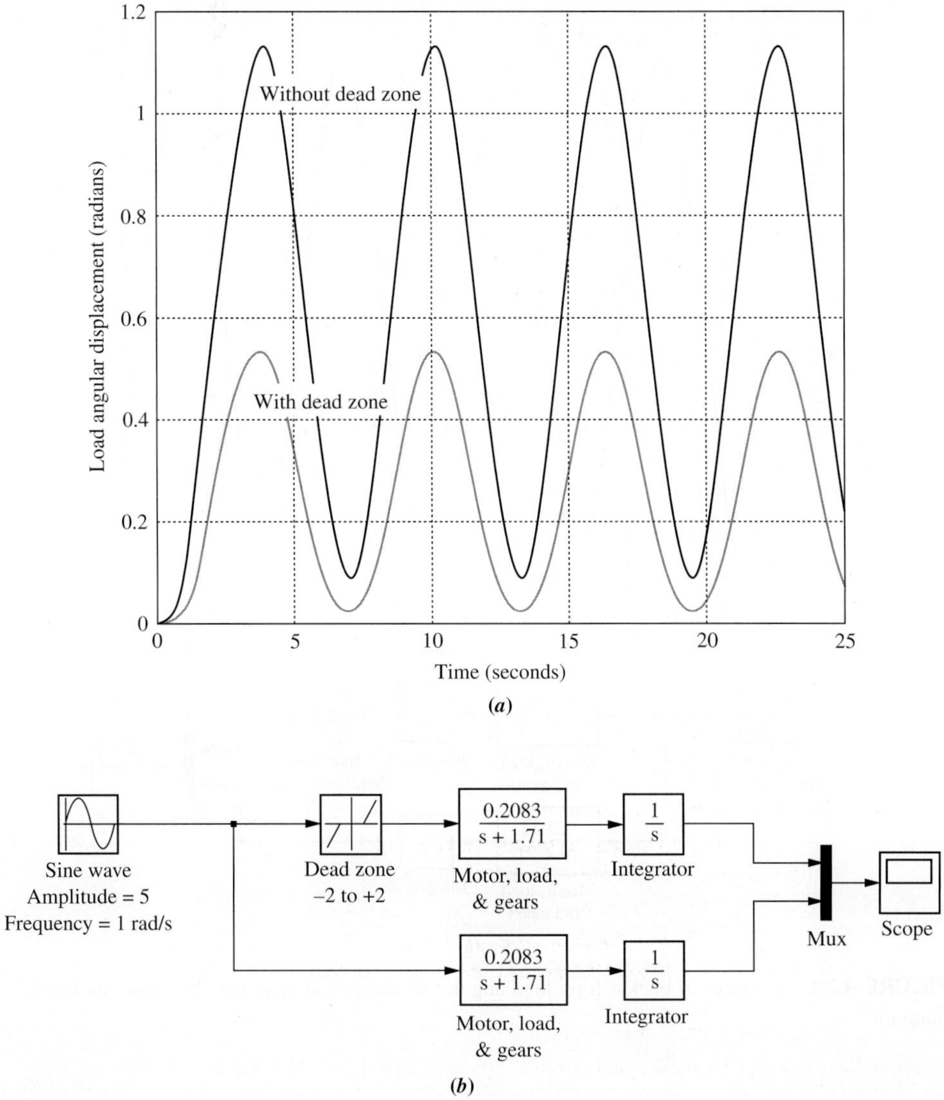


**FIGURE 4.30** **a.** Effect of dead zone on load angular displacement response; **b.** Simulink block diagram

The effect of dead zone on the output shaft driven by a motor and gears is shown in Figure 4.30. Here we once again assume the motor, load, and gears from Antenna Control Case Study of Chapter 2. Dead zone is present when the motor cannot respond to small voltages. The motor input is a sinusoidal waveform chosen to allow us to see the effects of dead zone vividly. The response begins when the input voltage to the motor exceeds a threshold. We notice a lower amplitude when dead zone is present.

The effect of backlash on the output shaft driven by a motor and gears is shown in Figure 4.31. Again we assume the motor, load, and gears from the Antenna Control Case Study of Chapter 2. The motor input is again a sinusoidal waveform, which is chosen to allow us to see vividly the effects of backlash in the gears driven by the motor. As the motor reverses direction, the output shaft remains stationary while the motor begins to reverse. When the gears finally connect, the output shaft itself begins to turn in the reverse direction. The resulting response is quite different from the linear response without backlash.

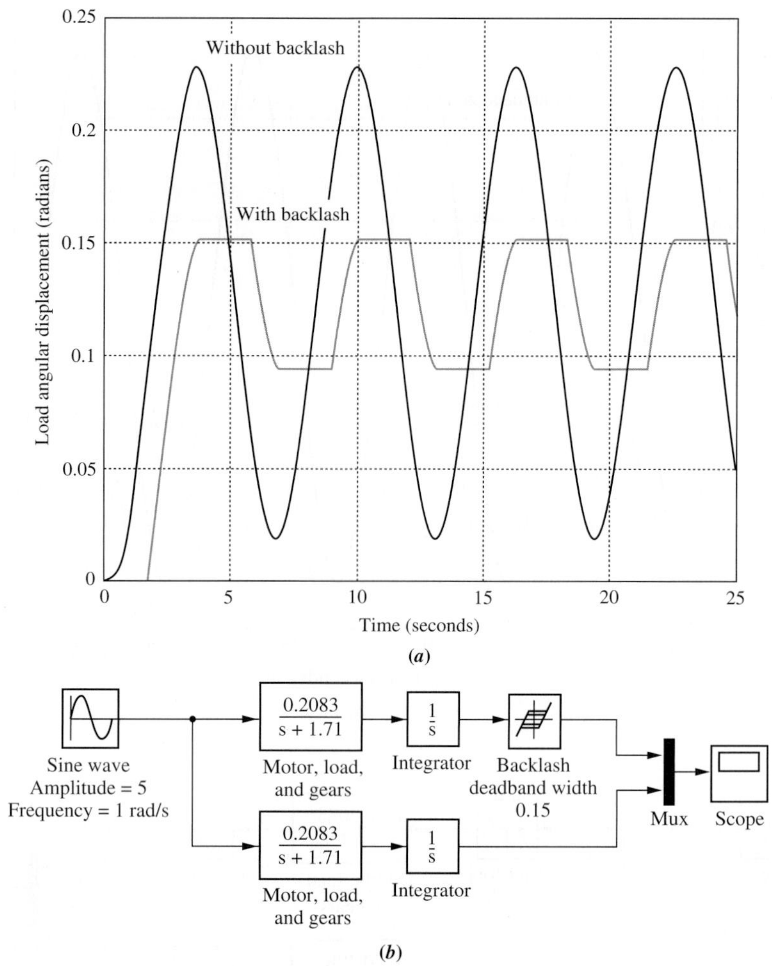


**FIGURE 4.31** **a.** Effect of backlash on load angular displacement response; **b.** Simulink block diagram

## Skill-Assessment Exercise 4.8

Simulink **SL**

**PROBLEM:** `Use MATLAB's Simulink to reproduce Figure 4.31.`

**ANSWER:** See Figure 4.31.

Now that we have seen the effects of nonlinearities on the time response, let us return to linear systems. Our coverage so far for linear systems has dealt with finding the time response by using the Laplace transform in the frequency domain. Another way to solve for the response is to use state-space techniques in the time domain. This topic is the subject of the next two sections.

# 4.10 Laplace Transform Solution of State Equations

State Space **SS**

In Chapter 3, systems were modeled in state space, where the state-space representation consisted of a state equation and an output equation. In this section, we use the Laplace transform to solve the state equations for the state and output vectors.

Consider the state equation

$$\dot{\mathbf{x}} = \mathbf{A}\mathbf{x} + \mathbf{B}\mathbf{u} \tag{4.92}$$

and the output equation

$$\mathbf{y} = \mathbf{C}\mathbf{x} + \mathbf{D}\mathbf{u} \tag{4.93}$$

Taking the Laplace transform of both sides of the state equation yields

$$s\mathbf{X}(s) - \mathbf{x}(0) = \mathbf{A}\mathbf{X}(s) + \mathbf{B}\mathbf{U}(s) \tag{4.94}$$

In order to separate $\mathbf{X}(s)$, replace $s\mathbf{X}(s)$ with $s\mathbf{I}\mathbf{X}(s)$, where $\mathbf{I}$ is an $n \times n$ identity matrix, and $n$ is the order of the system. Combining all of the $\mathbf{X}(s)$ terms, we get

$$(s\mathbf{I} - \mathbf{A})\mathbf{X}(s) = \mathbf{x}(0) + \mathbf{B}\mathbf{U}(s) \tag{4.95}$$

Solving for $\mathbf{X}(s)$ by premultiplying both sides of Eq. (4.95) by $(s\mathbf{I} - \mathbf{A})^{-1}$, the final solution for $\mathbf{X}(s)$ is

$$\begin{aligned}\mathbf{X}(s) &= (s\mathbf{I} - \mathbf{A})^{-1}\mathbf{x}(0) + (s\mathbf{I} - \mathbf{A})^{-1}\mathbf{B}\mathbf{U}(s) \\ &= \frac{\operatorname{adj}(s\mathbf{I} - \mathbf{A})}{\det(s\mathbf{I} - \mathbf{A})}[\mathbf{x}(0) + \mathbf{B}\mathbf{U}(s)]\end{aligned} \tag{4.96}$$

Taking the Laplace transform of the output equation yields

$$\mathbf{Y}(s) = \mathbf{C}\mathbf{X}(s) + \mathbf{D}\mathbf{U}(s) \tag{4.97}$$

## Eigenvalues and Transfer Function Poles

We saw that the poles of the transfer function determine the nature of the transient response of the system. Is there an equivalent quantity in the state-space representation that yields the same information? Section 5.8 formally defines the roots of $\det(s\mathbf{I} - \mathbf{A}) = 0$ (see the denominator of Eq. (4.96) to be *eigenvalues* of the system matrix, $\mathbf{A}$.[7] Let us show that the eigenvalues are equal to the poles of the system's transfer function. Let the output, $\mathbf{Y}(s)$, and the input, $\mathbf{U}(s)$, be scalar quantities $Y(s)$ and $U(s)$, respectively. Further, to conform to the definition of a transfer function, let $\mathbf{x}(0)$, the initial state vector, equal $\mathbf{0}$, the null vector. Substituting Eq. (4.96) into Eq. (4.97) and solving for the transfer function, $Y(s)/U(s)$, yields

$$\begin{aligned}\frac{Y(s)}{U(s)} &= \mathbf{C}\left[\frac{\operatorname{adj}(s\mathbf{I} - \mathbf{A})}{\det(s\mathbf{I} - \mathbf{A})}\right]\mathbf{B} + \mathbf{D} \\ &= \frac{\mathbf{C}\operatorname{adj}(s\mathbf{I} - \mathbf{A})\mathbf{B} + \mathbf{D}\det(s\mathbf{I} - \mathbf{A})}{\det(s\mathbf{I} - \mathbf{A})}\end{aligned} \tag{4.98}$$

The roots of the denominator of Eq. (4.98) are the poles of the system. Since the denominators of Eqs. (4.96) and (4.98) are identical, the system poles equal the eigenvalues. Hence, if a system is represented in state-space, we can find the poles from $\det(s\mathbf{I} - \mathbf{A}) = 0$. We will be more formal about these facts when we discuss stability in Chapter 6.

The following example demonstrates solving the state equations using the Laplace transform as well as finding the eigenvalues and system poles.

[7] Sometimes the symbol $\lambda$ is used in place of the complex variable $s$ when solving the state equations without using the Laplace transform. Thus, it is common to see the characteristic equation also written as $\det(\lambda\mathbf{I} - \mathbf{A}) = 0$.

## Example 4.11 Laplace Transform Solution; Eigenvalues and Poles

**PROBLEM:** Given the system represented in state space by Eqs. (4.99),

$$\dot{\mathbf{x}} = \begin{bmatrix} 0 & 1 & 0 \\ 0 & 0 & 1 \\ -24 & -26 & -9 \end{bmatrix}\mathbf{x} + \begin{bmatrix} 0 \\ 0 \\ 1 \end{bmatrix} e^{-t} \tag{4.99a}$$

$$y = \begin{bmatrix} 1 & 1 & 0 \end{bmatrix}\mathbf{x} \tag{4.99b}$$

$$\mathbf{x}(0) = \begin{bmatrix} 1 \\ 0 \\ 2 \end{bmatrix} \tag{4.99c}$$

do the following:

**a.** Solve the preceding state equation and obtain the output for the given exponential input.

**b.** Find the eigenvalues and the system poles.

**SOLUTION:**

**a.** We will solve the problem by finding the component parts of Eq. (4.96), followed by substitution into Eq. (4.97). First obtain **A** and **B** by comparing Eq. (4.99a) to Eq. (4.92). Since

$$s\mathbf{I} = \begin{bmatrix} s & 0 & 0 \\ 0 & s & 0 \\ 0 & 0 & s \end{bmatrix} \tag{4.100}$$

then

$$(s\mathbf{I} - \mathbf{A}) = \begin{bmatrix} s & -1 & 0 \\ 0 & s & -1 \\ 24 & 26 & s+9 \end{bmatrix} \tag{4.101}$$

and

$$(s\mathbf{I} - \mathbf{A})^{-1} = \frac{\begin{bmatrix} (s^2+9s+26) & (s+9) & 1 \\ -24 & s^2+9s & s \\ -24s & -(26s+24) & s^2 \end{bmatrix}}{s^3+9s^2+26s+24} \tag{4.102}$$

Since $\mathbf{U}(s)$ is $1/(s+1)$ (the Laplace transform for $e^{-t}$), $\mathbf{X}(s)$ can be calculated. Rewriting Eq. (4.96) as

$$\mathbf{X}(s) = (s\mathbf{I} - \mathbf{A})^{-1}[\mathbf{x}(0) + \mathbf{B}\mathbf{U}(s)] \tag{4.103}$$

and using **B** and $\mathbf{x}(0)$ from Eqs. (4.99a) and (4.99c), respectively, we get

$$X_1(s) = \frac{(s^3+10s^2+37s+29)}{(s+1)(s+2)(s+3)(s+4)} \tag{4.104a}$$

$$X_2(s) = \frac{(2s^2 - 21s - 24)}{(s+1)(s+2)(s+3)(s+4)} \tag{4.104b}$$

$$X_3(s) = \frac{s(2s^2 - 21s - 24)}{(s+1)(s+2)(s+3)(s+4)} \tag{4.104c}$$

The output equation is found from Eq. (4.99b). Performing the indicated addition yields

$$Y(s) = \begin{bmatrix} 1 & 1 & 0 \end{bmatrix} \begin{bmatrix} X_1(s) \\ X_2(s) \\ X_3(s) \end{bmatrix} = X_1(s) + X_2(s) \tag{4.105}$$

or

$$\begin{aligned} Y(s) &= \frac{(s^3 + 12s^2 + 16s + 5)}{(s+1)(s+2)(s+3)(s+4)} \\ &= \frac{-6.5}{s+2} + \frac{19}{s+3} - \frac{11.5}{s+4} \end{aligned} \tag{4.106}$$

where the pole at $-1$ canceled a zero at $-1$. Taking the inverse Laplace transform,

$$y(t) = -6.5e^{-2t} + 19e^{-3t} - 11.5e^{-4t} \tag{4.107}$$

**b.** The denominator of Eq. 4.102, which is $\det(s\mathbf{I} - \mathbf{A})$, is also the denominator of the system's transfer function. Thus, $\det(s\mathbf{I} - \mathbf{A}) = 0$ furnishes both the poles of the system and the eigenvalues $-2$, $-3$, and $-4$.

Symbolic Math
SM

Students who are performing the MATLAB exercises and want to explore the added capability of MATLAB's Symbolic Math Toolbox should now run ch4sp1 in Appendix F at www.wiley.com/college/nise. You will learn how to solve state equations for the output response using the Laplace transform. Example 4.11 will be solved using MATLAB and the Symbolic Math Toolbox.

## Skill-Assessment Exercise 4.9

**PROBLEM:** Given the system represented in state space by Eqs. (4.108),

$$\dot{\mathbf{x}} = \begin{bmatrix} 0 & 2 \\ -3 & -5 \end{bmatrix}\mathbf{x} + \begin{bmatrix} 0 \\ 1 \end{bmatrix} e^{-t} \tag{4.108a}$$

$$y = \begin{bmatrix} 1 & 3 \end{bmatrix}\mathbf{x} \tag{4.108b}$$

$$\mathbf{x}(0) = \begin{bmatrix} 2 \\ 1 \end{bmatrix} \tag{4.108c}$$

do the following:

**a.** Solve for $y(t)$ using state-space and Laplace transform techniques.

**b.** Find the eigenvalues and the system poles.

**TryIt 4.5**

Use the following MATLAB and Symbolic Math Toolbox statements to solve Skill-Assessment Exercise 4.9 .

```
syms s
A=[0 2;-3 -5]; B=[0;1];
C=[1 3];X0=[2;1];
U=1/(s+1);
I=[1 0;0 1];
X=((s*I-A)^-1)*...
  (X0+B*U);
Y=C*X; Y=simplify(Y);
y=ilaplace(Y);
pretty(y)
eig(A)
```

**ANSWERS:**

**a.** $y(t) = -0.5e^{-t} - 12e^{-2t} + 17.5e^{-3t}$

**b.** $-2,\ -3$

The complete solution is located at www.wiley.com/college/nise.

## 4.11 Time Domain Solution of State Equations

State Space

SS

We now look at another technique for solving the state equations. Rather than using the Laplace transform, we solve the equations directly in the time domain using a method closely allied to the classical solution of differential equations. We will find that the final solution consists of two parts that are different from the forced and natural responses.

The solution in the time domain is given directly by

$$\begin{aligned}\mathbf{x}(t) &= e^{\mathbf{A}t}\mathbf{x}(0) + \int_0^t e^{\mathbf{A}(t-\tau)}\mathbf{B}\mathbf{u}(\tau)d\tau \\ &= \mathbf{\Phi}(t)\mathbf{x}(0) + \int_0^t \Phi(t-\tau)\mathbf{B}\mathbf{u}(\tau)d\tau\end{aligned} \tag{4.109}$$

where $\mathbf{\Phi}(t) = e^{\mathbf{A}t}$ by definition, and which is called the *state-transition matrix*. Eq. (4.109) is derived in Appendix I located at www.wiley.com/college/nise. Readers who are not familiar with this equation or who may want to refresh their memory should consult Appendix I before proceeding.

Notice that the first term on the right-hand side of the equation is the response due to the initial state vector, $\mathbf{x}(0)$. Notice also that it is the only term dependent on the initial state vector and not the input. We call this part of the response the *zero-input response*, since it is the total response if the input is zero. The second term, called the *convolution integral*, is dependent only on the input, $\mathbf{u}$, and the input matrix, $\mathbf{B}$, not the initial state vector. We call this part of the response the *zero-state response*, since it is the total response if the initial state vector is zero. Thus, there is a partitioning of the response different from the forced/natural response we have seen when solving differential equations. In differential equations, the arbitrary constants of the natural response are evaluated based on the initial conditions and the initial values of the forced response and its derivatives. Thus, the natural response's amplitudes are a function of the initial conditions of the output and the input. In Eq. (4.109), the zero-input response is not dependent on the initial values of the input and its derivatives. It is dependent only on the initial conditions of the state vector. The next example vividly shows the difference in partitioning. Pay close attention to the fact that in the final result the zero-state response contains not only the forced solution but also pieces of what we previously called the natural response. We will see in the solution that the natural response is distributed through the zero-input response and the zero-state response.

Before proceeding with the example, let us examine the form the elements of $\mathbf{\Phi}(t)$ take for linear, time-invariant systems. The first term of Eq. (4.96), the Laplace transform of the response for unforced systems, is the transform of $\Phi(t)\mathbf{x}(0)$, the zero-input response from Eq. (4.109). Thus, for the unforced system

$$\mathscr{L}[\mathbf{x}(t)] = \mathscr{L}[\mathbf{\Phi}(t)\mathbf{x}(0)] = (s\mathbf{I} - \mathbf{A})^{-1}\mathbf{x}(0) \tag{4.110}$$

from which we can see that $(s\mathbf{I} - \mathbf{A})^{-1}$ *is the Laplace transform of the state-transition matrix*, $\mathbf{\Phi}(t)$. We have already seen that the denominator of $(s\mathbf{I} - \mathbf{A})^{-1}$ is a polynomial in $s$ whose roots are the system poles. This polynomial is found from the equation $\det(s\mathbf{I} - \mathbf{A}) = 0$. Since

$$\mathscr{L}^{-1}[(s\mathbf{I} - \mathbf{A})^{-1}] = \mathscr{L}^{-1}\left[\frac{\operatorname{adj}(s\mathbf{I} - \mathbf{A})}{\det(s\mathbf{I} - \mathbf{A})}\right] = \mathbf{\Phi}(t) \tag{4.111}$$

each term of $\Phi(t)$ would be the sum of exponentials generated by the system's poles.

Let us summarize the concepts with two numerical examples. The first example solves the state equations directly in the time domain. The second example uses the Laplace transform to solve for the state-transition matrix by finding the inverse Laplace transform of $(s\mathbf{I} - \mathbf{A})^{-1}$.

## Example 4.12 Time Domain Solution

**PROBLEM:** For the state equation and initial state vector shown in Eqs. (4.112), where $u(t)$ is a unit step, find the state-transition matrix and then solve for $\mathbf{x}(t)$.

$$\dot{\mathbf{x}}(t) = \begin{bmatrix} 0 & 1 \\ -8 & -6 \end{bmatrix}\mathbf{x}(t) + \begin{bmatrix} 0 \\ 1 \end{bmatrix}u(t) \tag{4.112a}$$

$$\mathbf{x}(0) = \begin{bmatrix} 1 \\ 0 \end{bmatrix} \tag{4.112b}$$

**SOLUTION:** Since the state equation is in the form

$$\dot{\mathbf{x}}(t) = \mathbf{A}\mathbf{x}(t) + \mathbf{B}u(t) \tag{4.113}$$

find the eigenvalues using $\det(s\mathbf{I} - \mathbf{A}) = 0$. Hence, $s^2 + 6s + 8 = 0$, from which $s_1 = -2$ and $s_2 = -4$. Since each term of the state-transition matrix is the sum of responses generated by the poles (eigenvalues), we assume a state-transition matrix of the form

$$\mathbf{\Phi}(t) = \begin{bmatrix} (K_1e^{-2t} + K_2e^{-4t}) & (K_3e^{-2t} + K_4e^{-4t}) \\ (K_5e^{-2t} + K_6e^{-4t}) & (K_7e^{-2t} + K_8e^{-4t}) \end{bmatrix} \tag{4.114}$$

In order to find the values of the constants, we make use of the properties of the state-transition matrix derived in Appendix J located at www.wiley.com/college/nise.

Since

$$\mathbf{\Phi}(0) = \mathbf{I} \tag{4.115}$$

then

$$K_1 + K_2 = 1 \tag{4.116a}$$

$$K_3 + K_4 = 0 \tag{4.116b}$$

$$K_5 + K_6 = 0 \tag{4.116c}$$

$$K_7 + K_8 = 1 \tag{4.116d}$$

And, since

$$\dot{\mathbf{\Phi}}(0) = \mathbf{A} \quad (4.117)$$

then

$$-2K_1 - 4K_2 = 0 \quad (4.118a)$$

$$-2K_3 - 4K_4 = 1 \quad (4.118b)$$

$$-2K_5 - 4K_6 = -8 \quad (4.118c)$$

$$-2K_7 - 4K_8 = -6 \quad (4.118d)$$

The constants are solved by taking two simultaneous equations four times. For example, Eq. (4.116a) can be solved simultaneously with Eq. (4.118*a*) to yield the values of $K_1$ and $K_2$. Proceeding similarly, all of the constants can be found. Therefore,

$$\mathbf{\Phi}(t) = \begin{bmatrix} (2e^{-2t} - e^{-4t}) & \left(\frac{1}{2}e^{-2t} - \frac{1}{2}e^{-4t}\right) \\ (-4e^{-2t} + 4e^{-4t}) & (-e^{-2t} + 2e^{-4t}) \end{bmatrix} \quad (4.119)$$

Also,

$$\mathbf{\Phi}(t-\tau)\mathbf{B} = \begin{bmatrix} \left(\frac{1}{2}e^{-2(t-\tau)} - \frac{1}{2}e^{-4(t-\tau)}\right) \\ \left(-e^{-2(t-\tau)} + 2e^{-4(t-\tau)}\right) \end{bmatrix} \quad (4.120)$$

Hence, the first term of Eq. (4.109) is

$$\mathbf{\Phi}(t)\mathbf{x}(0) = \begin{bmatrix} (2e^{-2t} - e^{-4t}) \\ (-4e^{-2t} + 4e^{-4t}) \end{bmatrix} \quad (4.121)$$

The last term of Eq. (4.109) is

$$\begin{aligned} \int_0^t \mathbf{\Phi}(t-\tau)\mathbf{B}\mathbf{u}(\tau)d\tau &= \begin{bmatrix} \frac{1}{2}e^{-2t}\int_0^t e^{2\tau}d\tau - \frac{1}{2}e^{-4t}\int_0^t e^{4\tau}d\tau \\ -e^{-2t}\int_0^t e^{2\tau}d\tau + 2e^{-4t}\int_0^t e^{4\tau}d\tau \end{bmatrix} \\ &= \begin{bmatrix} \frac{1}{8} - \frac{1}{4}e^{-2t} + \frac{1}{8}e^{-4t} \\ \frac{1}{2}e^{-2t} - \frac{1}{2}e^{-4t} \end{bmatrix} \end{aligned} \quad (4.122)$$

Notice, as promised, that Eq. (4.122), the zero-state response, contains not only the forced response, 1/8, but also terms of the form $Ae^{-2t}$ and $Be^{-4t}$ that are part of what we previously called the natural response. However, the coefficients, $A$ and $B$, are not dependent on the initial conditions.

The final result is found by adding Eqs. (4.121) and (4.122). Hence,

$$\mathbf{x}(t) = \mathbf{\Phi}(t)\mathbf{x}(0) + \int_0^t \Phi(t-\tau)\mathbf{B}\mathbf{u}(\tau)d\tau = \begin{bmatrix} \frac{1}{8} + \frac{7}{4}e^{-2t} - \frac{7}{8}e^{-4t} \\ -\frac{7}{2}e^{-2t} + \frac{7}{2}e^{-4t} \end{bmatrix} \quad (4.123)$$

## Example 4.13 State-Transition Matrix via Laplace Transform

**PROBLEM:** Find the state-transition matrix of Example 4.12, using $(s\mathbf{I} - \mathbf{A})^{-1}$.

**SOLUTION:** We use the fact that $\mathbf{\Phi}(t)$ is the inverse Laplace transform of $(s\mathbf{I} - \mathbf{A})^{-1}$. Thus, first find $(s\mathbf{I} - \mathbf{A})$ as

$$(s\mathbf{I} - \mathbf{A}) = \begin{bmatrix} s & -1 \\ 8 & (s+6) \end{bmatrix} \tag{4.124}$$

from which

$$(s\mathbf{I} - \mathbf{A})^{-1} = \frac{\begin{bmatrix} s+6 & 1 \\ -8 & s \end{bmatrix}}{s^2 + 6s + 8} = \begin{bmatrix} \dfrac{s+6}{s^2+6s+8} & \dfrac{1}{s^2+6s+8} \\ \dfrac{-8}{s^2+6s+8} & \dfrac{s}{s^2+6s+8} \end{bmatrix} \tag{4.125}$$

Expanding each term in the matrix on the right by partial fractions yields

$$(s\mathbf{I} - \mathbf{A})^{-1} = \begin{bmatrix} \left(\dfrac{2}{s+2} - \dfrac{1}{s+4}\right) & \left(\dfrac{1/2}{s+2} - \dfrac{1/2}{s+4}\right) \\ \left(\dfrac{-4}{s+2} + \dfrac{4}{s+4}\right) & \left(\dfrac{-1}{s+2} + \dfrac{2}{s+4}\right) \end{bmatrix} \tag{4.126}$$

Finally, taking the inverse Laplace transform of each term, we obtain

$$\mathbf{\Phi}(t) = \begin{bmatrix} (2e^{-2t} - e^{-4t}) & \left(\dfrac{1}{2}e^{-2t} - \dfrac{1}{2}e^{-4t}\right) \\ (-4e^{-2t} + 4e^{-4t}) & (-e^{-2t} + 2e^{-4t}) \end{bmatrix} \tag{4.127}$$

Symbolic Math SM

`Students who are performing the MATLAB exercises and want to explore the added capability of MATLAB's Symbolic Math Toolbox should now run ch4sp2 in Appendix F at www.wiley.com/college/nise. You will learn how to solve state equations for the output response using the convolution integral. Examples 4.12 and 4.13 will be solved using MATLAB and the Symbolic Math Toolbox.`

Systems represented in state space can be simulated on the digital computer. Programs such as MATLAB can be used for this purpose. Alternately, the user can write specialized programs, as discussed in Appendix H.1 at www.wiley.com/college/nise.

MATLAB ML

`Students who are using MATLAB should now run ch4p3 in Appendix B. This exercise uses MATLAB to simulate the step response of systems represented in state space. In addition to generating the step response, you will learn how to specify the range on the time axis for the plot.`

## Skill-Assessment Exercise 4.10

**PROBLEM:** Given the system represented in state space by Eqs. (4.128a):

$$\dot{\mathbf{x}} = \begin{bmatrix} 0 & 2 \\ -2 & -5 \end{bmatrix}\mathbf{x} + \begin{bmatrix} 0 \\ 1 \end{bmatrix}e^{-2t} \tag{4.128a}$$

$$y = \begin{bmatrix} 2 & 1 \end{bmatrix}\mathbf{x} \tag{4.128b}$$

$$\mathbf{x}(0) = \begin{bmatrix} 1 \\ 2 \end{bmatrix} \tag{4.128c}$$

do the following:

**a.** Solve for the state-transition matrix.

**b.** Solve for the state vector using the convolution integral.

**c.** Find the output, $y(t)$.

**ANSWERS:**

**a.** $\mathbf{\Phi}(t) = \begin{bmatrix} \left(\frac{4}{3}e^{-t} - \frac{1}{3}e^{-4t}\right) & \left(\frac{2}{3}e^{-t} - \frac{2}{3}e^{-4t}\right) \\ \left(-\frac{2}{3}e^{-t} + \frac{2}{3}e^{-4t}\right) & \left(-\frac{1}{3}e^{-t} + \frac{4}{3}e^{-4t}\right) \end{bmatrix}$

**b.** $\mathbf{x}(\mathrm{t}) = \begin{bmatrix} \left(\frac{10}{3}e^{-t} - e^{-2t} - \frac{4}{3}e^{-4t}\right) \\ \left(-\frac{5}{3}e^{-t} + e^{-2t} + \frac{8}{3}e^{-4t}\right) \end{bmatrix}$

**c.** $y(t) = 5e^{-t} - e^{-2t}$

The complete solution is located at www.wiley.com/college/nise.

## Case Studies

## Antenna Control: Open-Loop Response

In this chapter, we have made use of the transfer functions derived in Chapter 2 and the state equations derived in Chapter 3 to obtain the output response of an open-loop system. We also showed the importance of the poles of a system in determining the transient response. The following case study uses these concepts to analyze an open-loop portion of the antenna azimuth position control system. The open-loop function that we will deal with consists of a power amplifier and motor with load.

**PROBLEM:** For the schematic of the azimuth position control system shown on the front endpapers, Configuration 1, assume an open-loop system (feedback path disconnected).

**a.** Predict, by inspection, the form of the open-loop angular velocity response of the load to a step-voltage input to the power amplifier.

**b.** Find the damping ratio and natural frequency of the open-loop system.

**c.** Derive the complete analytical expression for the open-loop angular velocity response of the load to a step-voltage input to the power amplifier, using transfer functions.

**d.** Obtain the open-loop state and output equations.

State Space

**SS**

**e.** `Use MATLAB to obtain a plot of the open-loop angular velocity response to a step-voltage input.`

MATLAB

**ML**

**SOLUTION:** The transfer functions of the power amplifier, motor, and load as shown on the front endpapers, Configuration 1, were discussed in the Chapter 2 case study. The two subsystems are shown interconnected in Figure 4.32(*a*). Differentiating the angular position of the motor and load output by multiplying by $s$, we obtain the output angular velocity, $\omega_o$, as shown in Figure 4.32(*a*). The equivalent transfer function representing the three blocks in Figure 4.32(*a*) is the product of the individual transfer functions and is shown in Figure 4.32(b).[8]

**a.** Using the transfer function shown in Figure 4.32(*b*), we can predict the nature of the step response. The step response consists of the steady-state response generated by the step input and the transient response, which is the sum of two exponentials generated by each pole of the transfer function. Hence, the form of the response is

$$\omega_o(t) = A + Be^{-100t} + Ce^{-1.71t} \tag{4.129}$$

**b.** The damping ratio and natural frequency of the open-loop system can be found by expanding the denominator of the transfer function. Since the open-loop transfer function is

$$G(s) = \frac{20.83}{s^2 + 101.71s + 171} \tag{4.130}$$

$\omega_n = \sqrt{171} = 13.08$, and $\zeta = 3.89$ (overdamped).

**c.** In order to derive the angular velocity response to a step input, we multiply the transfer function of Eq. (4.130) by a step input, $1/s$, and obtain

$$\omega_o(s) = \frac{20.83}{s(s+100)(s+1.71)} \tag{4.131}$$

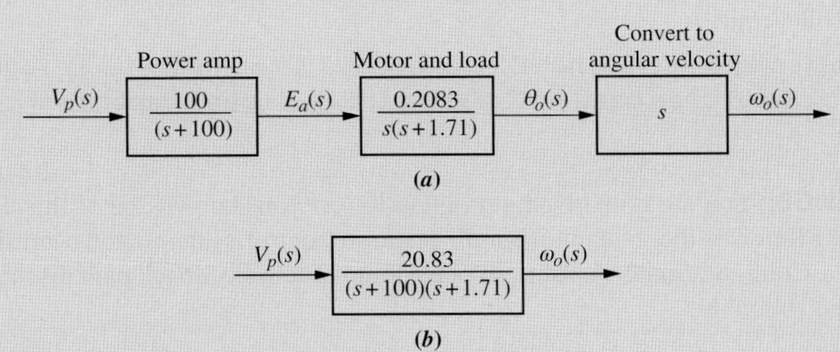


**FIGURE 4.32** Antenna azimuth position control system for angular velocity: **a.** forward path; **b.** equivalent forward path

[8] This product relationship will be derived in Chapter 5.

Expanding into partial fractions, we get

$$\omega_o(s) = \frac{0.122}{s} + \frac{2.12 \times 10^{-3}}{s+100} - \frac{0.124}{s+1.71} \quad (4.132)$$

Transforming to the time domain yields

$$\omega_o(t) = 0.122 + (2.12 \times 10^{-3})e^{-100t} - 0.124e^{-1.71t} \quad (4.133)$$

State Space
SS

**d.** First convert the transfer function into the state-space representation. Using Eq. (4.130), we have

$$\frac{\omega_o(s)}{V_p(s)} = \frac{20.83}{s^2 + 101.71s + 171} \quad (4.134)$$

Cross-multiplying and taking the inverse Laplace transform with zero initial conditions, we have

$$\ddot{\omega}_o + 101.71\dot{\omega}_o + 171\omega_o = 20.83v_p \quad (4.135)$$

Defining the phase variables as

$$x_1 = \omega_o \quad (4.136a)$$

$$x_2 = \dot{\omega}_o \quad (4.136b)$$

and using Eq. (4.135), the state equations are written as

$$\dot{x}_1 = x_2 \quad (4.137a)$$

$$\dot{x}_2 = -171x_1 - 101.71x_2 + 20.83v_p \quad (4.137b)$$

where $v_p = 1$, a unit step. Since $x_1 = \omega_o$ is the output, the output equation is

$$y = x_1 \quad (4.138)$$

Equations (4.137) and (4.138) can be programmed to obtain the step response using MATLAB or alternative methods described in Appendix H.1 at www.wiley.com/college/nise.

MATLAB
ML

**e.** `Students who are using MATLAB should now run ch4p4 in Appendix B. This exercise uses MATLAB to plot the step response.`

**CHALLENGE:** You are now given a problem to test your knowledge of this chapter's objectives. Refer to the antenna azimuth position control system shown on the front endpapers, Configuration 2. Assume an open-loop system (feedback path disconnected) and do the following:

**a.** Predict the open-loop angular velocity response of the power amplifier, motor, and load to a step voltage at the input to the power amplifier.

**b.** Find the damping ratio and natural frequency of the open-loop system.

**c.** Derive the open-loop angular velocity response of the power amplifier, motor, and load to a step-voltage input using transfer functions.

**d.** Obtain the open-loop state and output equations.

State Space
SS

**e.** `Use MATLAB to obtain a plot of the open-loop angular velocity response to a step-voltage input.`

MATLAB
ML

## Unmanned Free-Swimming Submersible Vehicle: Open-Loop Pitch Response

An Unmanned Free-Swimming Submersible (UFSS) vehicle is shown in Figure 4.33. The depth of the vehicle is controlled as follows. During forward motion, an elevator surface on the vehicle is deflected by a selected amount. This deflection causes the vehicle to rotate about the pitch axis. The pitch of the vehicle creates a vertical force that causes the vehicle to submerge or rise. The pitch control system for the vehicle is used here and in subsequent chapters as a case study to demonstrate the covered concepts. The block diagram for the pitch control system is shown in Figure 4.34 and on the back endpapers for future reference (*Johnson, 1980*). In this case study, we investigate the time response of the vehicle dynamics that relate the pitch angle output to the elevator deflection input.

Courtesy of the Naval Research Laboratory.

**FIGURE 4.33** Unmanned Free-Swimming Submersible (UFSS) vehicle

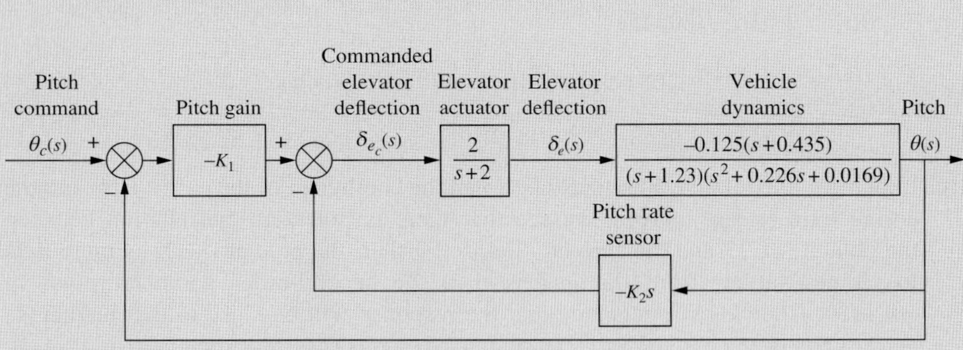


**FIGURE 4.34** Pitch control loop for the UFSS vehicle

**PROBLEM:** The transfer function relating pitch angle, $\theta(s)$, to elevator surface angle, $\delta_e(s)$, for the UFSS vehicle is

$$\frac{\theta(s)}{\delta_e(s)} = \frac{-0.125(s+0.435)}{(s+1.23)(s^2+0.226s+0.0169)} \tag{4.139}$$

**a.** Using only the second-order poles shown in the transfer function, predict percent overshoot, rise time, peak time, and settling time.

**b.** Using Laplace transforms, find the analytical expression for the response of the pitch angle to a step input in elevator surface deflection.

**c.** Evaluate the effect of the additional pole and zero on the validity of the second-order approximation.

**d.** Plot the step response of the vehicle dynamics and verify your conclusions found in (c). An animation PowerPoint presentation (PPT) demonstrating this system is available for instructors at www.wiley.com/college/nise. See *UFSS Vehicle*.

**SOLUTION:**

**a.** Using the polynomial $s^2+0.226s+0.0169$, we find that $\omega_n^2 = 0.0169$ and $2\zeta\omega_n = 0.226$. Thus, $\omega_n = 0.13$ rad/s and $\zeta = 0.869$. Hence, $\%OS = e^{-\zeta\pi/\sqrt{1-\zeta^2}}100 = 0.399\%$. From Figure 4.16, $\omega_n T_r = 2.75$, or $T_r = 21.2$ s. To find peak time, we use $T_p = \pi/\omega_n\sqrt{1-\zeta^2} = 48.9$ s. Finally, settling time is $T_s = 4/\zeta\omega_n = 35.4$ s.

**b.** In order to display a positive final value in Part **d.** we find the response of the system to a negative unit step, compensating for the negative sign in the transfer function. Using partial-fraction expansion, the Laplace transform of the response, $\theta(s)$, is

$$\begin{aligned}\theta(s) &= \frac{0.125(s+0.435)}{s(s+1.23)(s^2+0.226s+0.0169)} \\ &= 2.616\frac{1}{s} + 0.0645\frac{1}{s+1.23} \\ &\quad - \frac{2.68(s+0.113)+3.478\sqrt{0.00413}}{(s+0.113)^2+0.00413}\end{aligned} \tag{4.140}$$

Taking the inverse Laplace transform,

$$\begin{aligned}\theta(t) &= 2.616 + 0.0645e^{-1.23t} \\ &\quad - e^{-0.113t}(2.68\cos 0.0643t + 3.478\sin 0.0643t) \\ &= 2.616 + 0.0645e^{-1.23t} - 4.39e^{-0.113t}\cos(0.0643t + 52.38^\circ)\end{aligned} \tag{4.141}$$

**c.** Looking at the relative amplitudes between the coefficient of the $e^{-1.23t}$ term and the cosine term in Eq. (4.165), we see that there is pole-zero cancellation between the pole at $-1.23$ and the zero at $-0.435$. Further, the pole at $-1.23$ is more than five times farther from the $j\omega$ axis than the second-order dominant poles at $-0.113 \pm j0.0643$. We conclude that the response will be close to that predicted.

**d.** Plotting Eq. (4.141) or using a computer simulation, we obtain the step response shown in Figure 4.35. We indeed see a response close to that predicted.

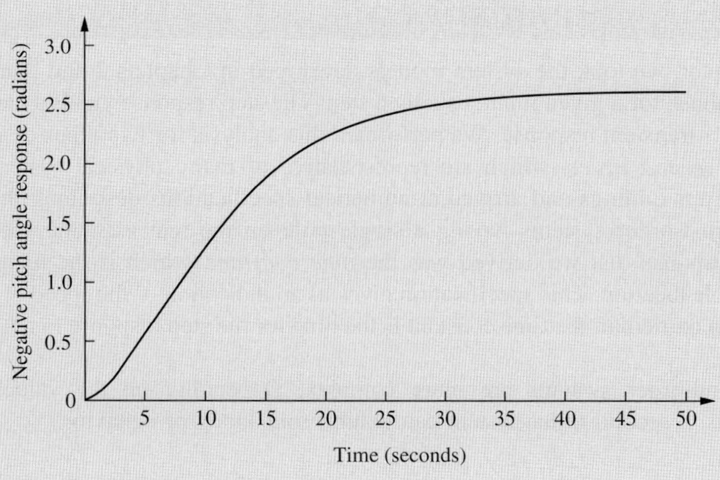


**FIGURE 4.35** Negative step response of pitch control for UFSS vehicle

MATLAB ML

`Students who are using MATLAB should now run ch4p5 in Appendix B. This exercise uses MATLAB to find` $\zeta$, $\omega$, $T_s$, $T_p$, `and` $T_r$ `and plot a step response. Table lookup is used to find` $T_r$. `The exercise applies the concepts to the problem above.`

**CHALLENGE:** You are now given a problem to test your knowledge of this chapter's objectives. This problem uses the same principles that were applied to the Unmanned Free-Swimming Submersible vehicle: Ships at sea undergo motion about their roll axis, as shown in Figure 4.36. Fins called *stabilizers* are used to reduce this rolling motion. The stabilizers can be positioned by a closed-loop roll control system that consists of components, such as fin actuators and sensors, as well as the ship's roll dynamics.

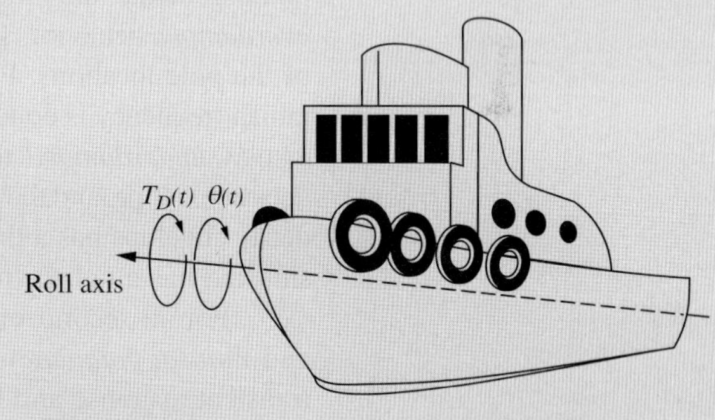


**FIGURE 4.36** A ship at sea, showing roll axis

Assume the roll dynamics, which relates the roll-angle output, $\theta(s)$, to a disturbance-torque input, $T_D(s)$, is

$$\frac{\theta(s)}{T_D(s)} = \frac{2.25}{(s^2 + 0.5s + 2.25)} \tag{4.142}$$

Do the following:

**a.** Find the natural frequency, damping ratio, peak time, settling time, rise time, and percent overshoot.

**b.** Find the analytical expression for the output response to a unit step input in voltage.

MATLAB ML

**c.** `Use MATLAB to solve` **`a`** `and` **`b`** `and to plot the response found in` **`b`**`.`

## Summary

In this chapter, we took the system models developed in Chapters 2 and 3 and found the output response for a given input, usually a step. The step response yields a clear picture of the system's transient response. We performed this analysis for two types of systems, *first order* and *second order*, which are representative of many physical systems. We then formalized our findings and arrived at numerical specifications describing the responses.

For first-order systems having a single pole on the real axis, the specification of transient response that we derived was the *time constant*, which is the reciprocal of the real-axis pole location. This specification gives us an indication of the speed of the transient response. In particular, the time constant is the time for the step response to reach 63% of its final value.

Second-order systems are more complex. Depending on the values of system components, a second-order system can exhibit four kinds of behavior:

1. Overdamped
2. Underdamped
3. Undamped
4. Critically damped

We found that the poles of the input generate the forced response, whereas the system poles generate the transient response. If the system poles are real, the system exhibits *overdamped* behavior. These exponential responses have time constants equal to the reciprocals of the pole locations. Purely imaginary poles yield *undamped* sinusoidal oscillations whose radian frequency is equal to the magnitude of the imaginary pole. Systems with complex poles display *underdamped* responses. The real part of the complex pole dictates the exponential decay envelope, and the imaginary part dictates the sinusoidal radian frequency. The exponential decay envelope has a time constant equal to the reciprocal of the real part of the pole, and the sinusoid has a radian frequency equal to the imaginary part of the pole.

For all second-order cases, we developed specifications called the *damping ratio*, $\zeta$, and *natural frequency*, $\omega_n$. The damping ratio gives us an idea about the nature of the transient response and how much overshoot and oscillation it undergoes, regardless of time scaling. The natural frequency gives an indication of the speed of the response.

We found that the value of $\zeta$ determines the form of the second-order natural response:

- If $\zeta = 0$, the response is undamped.
- If $\zeta < 1$, the response is underdamped.
- If $\zeta = 1$, the response is critically damped.
- If $\zeta > 1$, the response is overdamped.

The natural frequency is the frequency of oscillation if all damping is removed. It acts as a scaling factor for the response, as can be seen from Eq. (4.28), in which the independent variable can be considered to be $\omega_n t$.

For the underdamped case we defined several transient response specifications, including these:

- Percent overshoot, $\%OS$
- Peak time, $T_p$
- Settling time, $T_s$
- Rise time, $T_r$

The peak time is inversely proportional to the imaginary part of the complex pole. Thus, horizontal lines on the $s$-plane are lines of constant peak time. Percent overshoot is a function of only the damping ratio. Consequently, radial lines are lines of constant percent overshoot. Finally, settling time is inversely proportional to the real part of the complex pole. Hence, vertical lines on the $s$-plane are lines of constant settling time.

We found that peak time, percent overshoot, and settling time are related to pole location. Thus, we can design transient responses by relating a desired response to a pole location and then relating that pole location to a transfer function and the system's components.

The effects of nonlinearities, such as saturation, dead zone, and backlash, were explored using MATLAB's Simulink.

In this chapter, we also evaluated the time response using the state-space approach. The response found in this way was separated into the *zero-input response*, and the *zero-state response*, whereas the frequency response method yielded a total response divided into *natural response* and *forced response* components.

In the next chapter we will use the transient response specifications developed here to analyze and design systems that consist of the interconnection of multiple subsystems. We will see how to reduce these systems to a single transfer function in order to apply the concepts developed in Chapter 4.

## Review Questions

1. Name the performance specification for first-order systems.
2. What does the performance specification for a first-order system tell us?
3. In a system with an input and an output, what poles generate the steady-state response?
4. In a system with an input and an output, what poles generate the transient response?
5. The imaginary part of a pole generates what part of a response?
6. The real part of a pole generates what part of a response?
7. What is the difference between the natural frequency and the damped frequency of oscillation?
8. If a pole is moved with a constant imaginary part, what will the responses have in common?
9. If a pole is moved with a constant real part, what will the responses have in common?
10. If a pole is moved along a radial line extending from the origin, what will the responses have in common?
11. List five specifications for a second-order underdamped system.
12. For Question 11 how many specifications completely determine the response?
13. What pole locations characterize (1) the underdamped system, (2) the overdamped system, and (3) the critically damped system?
14. Name two conditions under which the response generated by a pole can be neglected.
15. How can you justify pole-zero cancellation?
16. Does the solution of the state equation yield the output response of the system? Explain. State Space SS
17. What is the relationship between $(s\mathbf{I} - \mathbf{A})$, which appeared during the Laplace transformation solution of the state equations, and the state-transition matrix, which appeared during the classical solution of the state equation? State Space SS
18. Name a major advantage of using time-domain techniques for the solution of the response. State Space SS
19. Name a major advantage of using frequency-domain techniques for the solution of the response.

State Space **SS**

20. What three pieces of information must be given in order to solve for the output response of a system using state-space techniques?

State Space **SS**

21. How can the poles of a system be found from the state equations?

## Problems

1. Derive the output responses for all parts of Figure 4.7. [Section: 4.4]

2. Find the output response, $c(t)$, for each of the systems shown in Figure P4.1. Also find the time constant, rise time, and settling time for each case. [Sections: 4.2, 4.3]

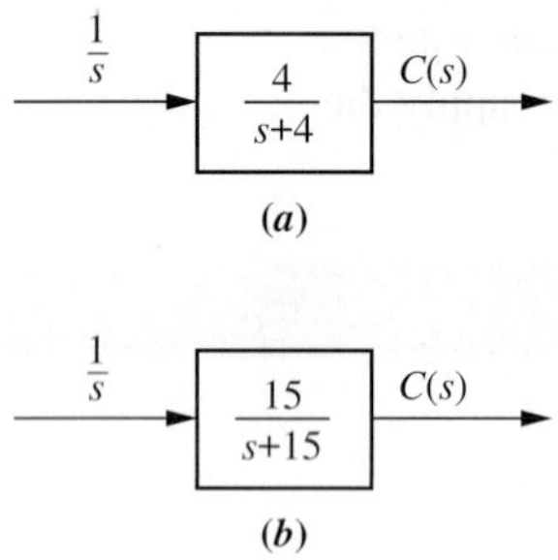


FIGURE P4.1

3. Plot the step responses for Problem 2 using MATLAB. (MATLAB **ML**)

4. Find the capacitor voltage in the network shown in Figure P4.2 if the switch closes at $t = 0$. Assume zero initial conditions. Also find the time constant, rise time, and settling time for the capacitor voltage. [Sections: 4.2, 4.3]

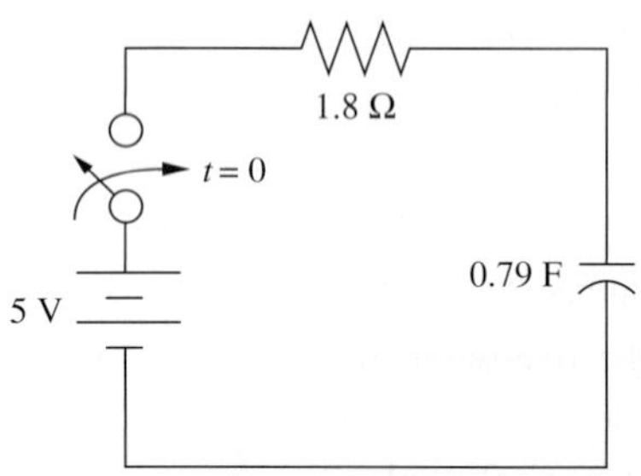


FIGURE P4.2

5. Plot the step response for Problem 4 using MATLAB. From your plots, find the time constant, rise time, and settling time. (MATLAB **ML**)

6. For the system shown in Figure P4.3, (a) find an equation that relates settling time of the velocity of the mass to $M$; (b) find an equation that relates rise time of the velocity of the mass to $M$. [Sections: 4.2, 4.3]

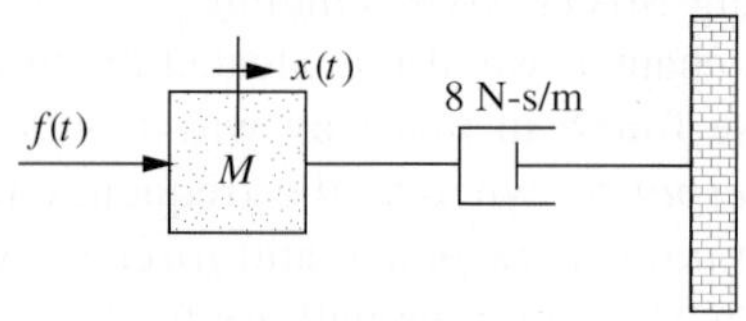


FIGURE P4.3

7. Plot the step response for Problem 6 using MATLAB. From your plots find the time constant, rise time, and settling time. Use $M = 1$ and $M = 2$. (MATLAB **ML**)

8. For each of the transfer functions shown below, find the locations of the poles and zeros, plot them on the $s$-plane, and then write an expression for the general form of the step response without solving for the inverse Laplace transform. State the nature of each response (overdamped, underdamped, and so on). [Sections: 4.3, 4.4]

   a. $T(s) = \dfrac{2}{s+2}$

   b. $T(s) = \dfrac{5}{(s+3)(s+6)}$

   c. $T(s) = \dfrac{10(s+7)}{(s+10)(s+20)}$

   d. $T(s) = \dfrac{20}{s^2+6s+144}$

   e. $T(s) = \dfrac{s+2}{s^2+9}$

   f. $T(s) = \dfrac{(s+5)}{(s+10)^2}$

9. Use MATLAB to find the poles of (MATLAB **ML**)

$$T(s) = \frac{s^2 + 2s + 2}{s^4 + 7s^3 + 3s^2 + 5s + 3}$$

10. Find the transfer function and poles of the system represented in state space here: [Section: 4.10]

$$\dot{\mathbf{x}} = \begin{bmatrix} 8 & -4 & 1 \\ -3 & 2 & 0 \\ 5 & 7 & -9 \end{bmatrix}\mathbf{x} + \begin{bmatrix} -4 \\ -3 \\ 4 \end{bmatrix}u(t)$$

$$y = [2 \quad 8 \quad -3]\mathbf{x}; \quad \mathbf{x}(0) = \begin{bmatrix} 0 \\ 0 \\ 0 \end{bmatrix}$$

**11.** `Repeat Problem 10 using MATLAB. [Section: 4.10].` MATLAB ML

**12.** Write the general form of the capacitor voltage for the electrical network shown in Figure P4.4. [Section: 4.4]

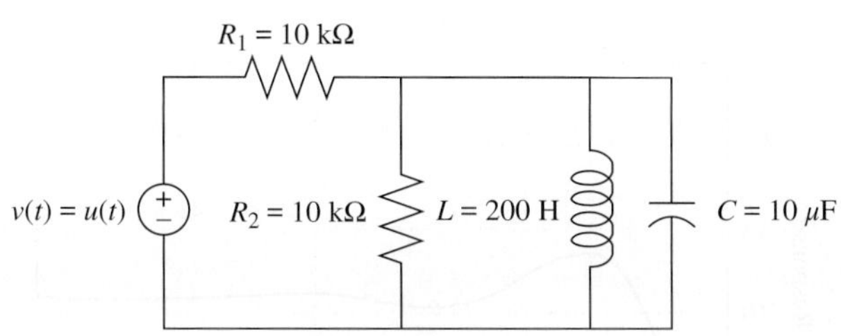


**FIGURE P4.4**

**13.** `Use MATLAB to plot the capacitor voltage in Problem 12. [Section: 4.4]` MATLAB ML

**14.** Solve for $x(t)$ in the system shown in Figure P4.5 if $f(t)$ is a unit step. [Section: 4.4].

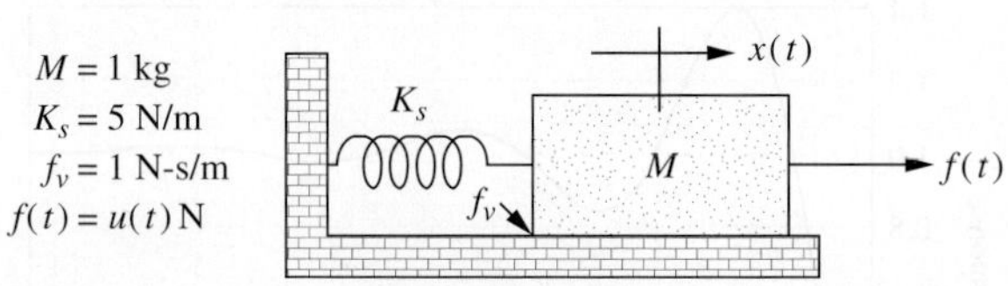


**FIGURE P4.5**

**15.** Calculate the exact response of each system of Problem 8 using Laplace transform techniques, and compare the results to those obtained in that problem. [Sections: 4.3, 4.4]

**16.** Find the damping ratio and natural frequency for each second-order system of Problem 8 and show that the value of the damping ratio conforms to the type of response (underdamped, overdamped, and so on) predicted in that problem. [Section: 4.5]

**17.** The system shown in Figure P4.6 has a unit step input. Find the output response as a function of time. Assume the system is underdamped. Notice that the result will be Eq. (4.28). [Section: 4.6]

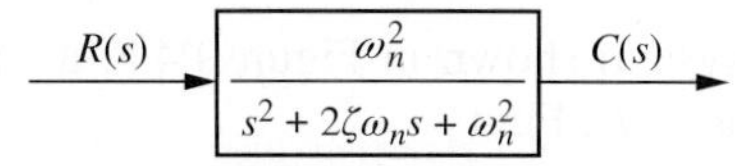


**FIGURE P4.6**

**18.** Derive the relationship for damping ratio as a function of percent overshoot, Eq. (4.39). [Section: 4.6]

**19.** A system has a damping ratio of 0.5, a natural frequency of 100 rad/s, and a dc gain of 1. Find the response of the system to a unit step input. [Section: 4.6].

**20.** For each of the second-order systems that follow, find $\zeta$, $\omega_n$, $T_s$, $T_p$, $T_r$, and $\%OS$. [Section: 4.6]

**a.** $T(s) = \dfrac{16}{s^2 + 3s + 16}$

**b.** $T(s) = \dfrac{0.04}{s^2 + 0.02s + 0.04}$

**c.** $T(s) = \dfrac{1.05 \times 10^7}{s^2 + 1.6 \times 10^3 s + 1.05 \times 10^7}$

**21.** `Repeat Problem 20 using MATLAB. Have the computer program estimate the given specifications and plot the step responses. Estimate the rise time from the plots. [Section: 4.6]` MATLAB ML

**22.** `Use MATLAB's LTI Viewer and obtain settling time, peak time, rise time, and percent overshoot for each of the systems in Problem 20. [Section: 4.6]` Gui Tool GUIT

**23.** For each pair of second-order system specifications that follow, find the location of the second-order pair of poles. [Section: 4.6].

**a.** $\%OS = 12\%$; $T_s = 0.2$ second

**b.** $\%OS = 15\%$; $T_p = 5$ seconds

**c.** $T_s = 7$ seconds; $T_p = 2$ seconds

**24.** Find the transfer function of a second-order system that yields a 12.3% overshoot and a settling time of 1 second. [Section: 4.6]

**25.** For the system shown in Figure P4.7, do the following: [Section: 4.6]

**a.** Find the transfer function $G(s) = X(s)/F(s)$.

**b.** Find $\zeta$, $\omega_n$, $\%OS$, $T_s$, $T_p$, and $T_r$.

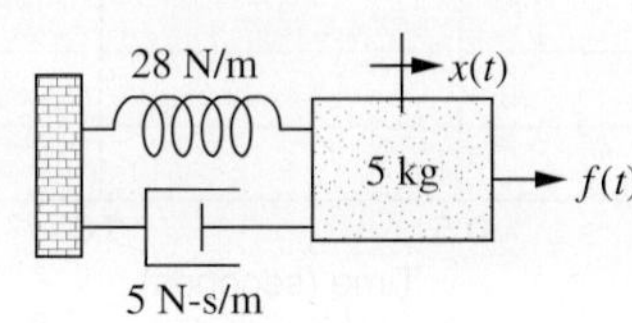


**FIGURE P4.7**

26. For the system shown in Figure P4.8, a step torque is applied at $\theta_1(t)$. Find
   a. The transfer function, $G(s) = \theta_2(s)/T(s)$.
   b. The percent overshoot, settling time, and peak time for $\theta_2(t)$. [Section: 4.6]

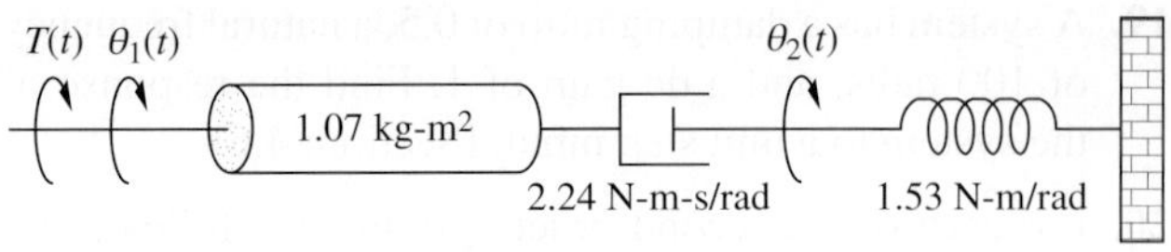


**FIGURE P4.8**

27. The derivation of Eq. (4.42) to calculate the settling time for a second-order system assumed an underdamped system ($\zeta < 1$). In this problem you will calculate a similar result for a critically damped system ($\zeta = 1$).
   a. Show that the unit-step response for a system with transfer function $\frac{C(s)}{R(s)} = \frac{a^2}{(s+a)^2}$ is $c(t) = 1 - e^{-at}(1+at)$. (Note: $\mathscr{L}\left\{\frac{1}{(s+a)^2}\right\} = te^{-at}$. Optional: You can derive this result similarly to Example 2.2.)
   b. Show that the settling time can be found by solving for $T_s$ in $e^{-aT_s}(1 + aT_s) = 0.02$.
   c. Use MATLAB to plot $e^{-x}(1 + x) = 0.02$ vs. $x$. Use the plot to show that $T_s = \frac{5.834}{a}$.

MATLAB ML

28. Figure P4.9 shows five step responses of an automatic voltage regulation system as one of the system parameters varies (*Gozde, 2011*). Assume for all five responses that they are those of a second-order system with an overshoot of 20%. Make a sketch of the positions of the poles in the complex plane for each one of the responses. Label the curves A through E from left to right.

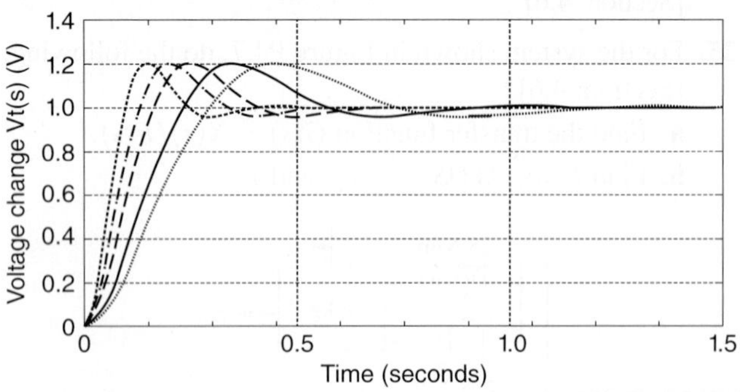


**FIGURE P4.9** Time responses for an automatic voltage regulation system

29. For each of the three unit step responses shown in Figure P4.10, find the transfer function of the system. [Sections: 4.3, 4.6]

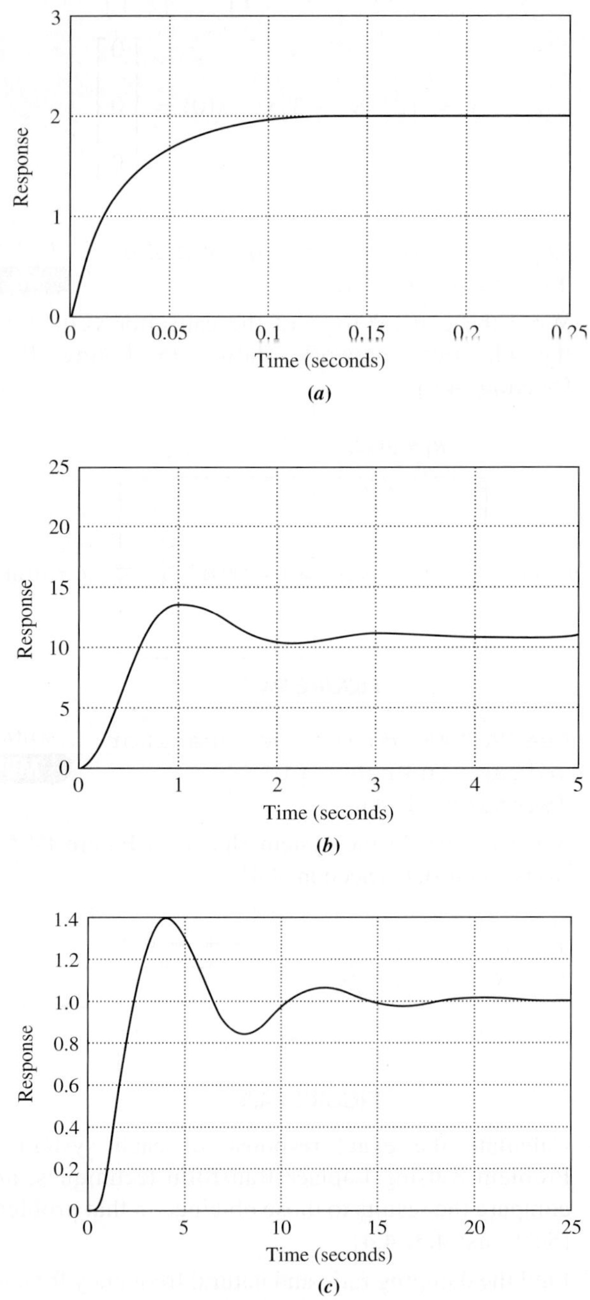


**FIGURE P4.10**

30. Find the percent overshoot, settling time, rise time, and peak time for

$$T(s) = \frac{14.145}{(s^2 + 0.842s + 2.829)(s + 5)}$$ [Section: 4.7]

31. Derive the unit step response for each transfer function in Example 4.8. [Section: 4.7]

**32.** For the following response functions, determine if pole-zero cancellation can be approximated. If it can, find percent overshoot, settling time, rise time, and peak time. [Section: 4.8].

**a.** $C(s) = \dfrac{(s+3)}{s(s+2)(s^2+3s+10)}$

**b.** $C(s) = \dfrac{(s+2.5)}{s(s+2)(s^2+4s+20)}$

**c.** $C(s) = \dfrac{(s+2.1)}{s(s+2)(s^2+s+5)}$

**d.** $C(s) = \dfrac{(s+2.01)}{s(s+2)(s^2+5s+20)}$

**33.** Using MATLAB, plot the time response of Problem 30a and from the plot determine percent overshoot, settling time, rise time, and peak time. [Section: 4.8] MATLAB **ML**

**34.** Find peak time, settling time, and percent overshoot for only those responses below that can be approximated as second-order responses. [Section: 4.8]

**a.** $c(t) = 0.003500 - 0.001524e^{-4t}$
$-0.001976e^{-3t}\cos(22.16t)$
$-0.0005427e^{-3t}\sin(22.16t)$

**b.** $c(t) = 0.05100 - 0.007353e^{-8t}$
$-0.007647e^{-6t}\cos(8t)$
$-0.01309e^{-6t}\sin(8t)$

**c.** $c(t) = 0.009804 - 0.0001857e^{-5.1t}$
$-0.009990e^{-2t}\cos(9.796t)$
$-0.001942e^{-2t}\sin(9.796t)$

**d.** $c(t) = 0.007000 - 0.001667e^{-10t}$
$-0.008667e^{-2t}\cos(9.951t)$
$-0.0008040e^{-2t}\sin(9.951t)$

**35.** For each of the following transfer functions with zeros, find the component parts of the unit step response: (1) the derivative of the response without a zero and (2) the response without a zero, scaled to the negative of the zero value. Also, find and plot the total response. Describe any nonminimum-phase behavior. [Section: 4.8]

**a.** $G(s) = \dfrac{s+2}{s^2+3s+36}$

**b.** $G(s) = \dfrac{s-2}{s^2+3s+36}$

**36.** Use MATLAB's Simulink to obtain the step response of a system, Simulink **SL**

$$G(s) = \frac{1}{s^2+3s+10}$$

under the following conditions: [Section: 4.9]

**a.** The system is linear and driven by an amplifier whose gain is 10.

**b.** An amplifier whose gain is 10 drives the system. The amplifier saturates at ±0.25 volts. Describe the effect of the saturation on the system's output.

**c.** An amplifier whose gain is 10 drives the system. The amplifier saturates at ±0.25 volts. The system drives a 1:1 gear train that has backlash. The deadband width of the backlash is 0.02 rad. Describe the effect of saturation and backlash on the system's output.

**37.** A system is represented by the state and output equations that follow. Without solving the state equation, find the poles of the system. [Section: 4.10] State Space **SS**

$$\dot{\mathbf{x}} = \begin{bmatrix} -2 & -1 \\ -3 & -5 \end{bmatrix}\mathbf{x} + \begin{bmatrix} 1 \\ 2 \end{bmatrix}u(t)$$

$$y = [3 \quad 2]\mathbf{x}$$

**38.** A system is represented by the state and output equations given below. Without solving the state equation, find

**a.** the characteristic equation

**b.** the poles of the system

$$\dot{\mathbf{x}} = \begin{bmatrix} 0 & 1 & 2 \\ 0 & 3 & 4 \\ 1 & 3 & 2 \end{bmatrix}\mathbf{x} + \begin{bmatrix} 0 \\ 0 \\ 1 \end{bmatrix}u(t)$$

$$y = [1 \quad 1 \quad 0]\mathbf{x}$$

**39.** Given the following state-space representation of a system, find $Y(s)$: [Section: 4.10] State Space **SS**

$$\dot{\mathbf{x}} = \begin{bmatrix} 1 & 3 \\ -3 & -1 \end{bmatrix}\mathbf{x} + \begin{bmatrix} 1 \\ 1 \end{bmatrix}\sin 5t$$

$$y = [1 \quad 2]\mathbf{x}; \ \mathbf{x}(0) = \begin{bmatrix} 3 \\ 0 \end{bmatrix}$$

**40.** Given the following system represented in state space, solve for $Y(s)$ using the Laplace transform method for solution of the state equation: [Section: 4.10] State Space SS

$$\dot{\mathbf{x}} = \begin{bmatrix} 0 & 1 & 0 \\ -2 & -4 & 1 \\ 0 & 0 & -6 \end{bmatrix}\mathbf{x} + \begin{bmatrix} 0 \\ 0 \\ 1 \end{bmatrix} e^{-t}$$

$$y = \begin{bmatrix} 0 & 0 & 1 \end{bmatrix}\mathbf{x}; \ \mathbf{x}(0) = \begin{bmatrix} 0 \\ 0 \\ 0 \end{bmatrix}$$

**41.** Solve the following state equation and output equation for $y(t)$, where $u(t)$ is the unit step. Use the Laplace transform method. [Section: 4.10] State Space SS

$$\dot{\mathbf{x}} = \begin{bmatrix} -2 & 0 \\ -1 & -1 \end{bmatrix}\mathbf{x} + \begin{bmatrix} 1 \\ 1 \end{bmatrix} u(t)$$

$$y = \begin{bmatrix} 0 & 2 \end{bmatrix}\mathbf{x}; \ \mathbf{x}(0) = \begin{bmatrix} 2 \\ 0 \end{bmatrix}$$

**42.** Solve for $y(t)$ for the following system represented in state space, where $u(t)$ is the unit step. Use the Laplace transform approach to solve the state equation. [Section: 4.10]

$$\dot{\mathbf{x}} = \begin{bmatrix} -3 & 1 & 0 \\ 0 & -6 & 1 \\ 0 & 0 & -5 \end{bmatrix}\mathbf{x} + \begin{bmatrix} 0 \\ 1 \\ 1 \end{bmatrix} u(t)$$

$$y = \begin{bmatrix} 0 & 1 & 1 \end{bmatrix}\mathbf{x}; \ \mathbf{x}(0) = \begin{bmatrix} 0 \\ 0 \\ 0 \end{bmatrix}$$

**43.** Repeat Problem 42 using MATLAB's Symbolic Math Toolbox and Eq. (4.96). In addition, run your program with an initial condition, Symbolic Math SM

$$\mathbf{x}(0) = \begin{bmatrix} 1 \\ 1 \\ 0 \end{bmatrix}. \text{[Section : 4.10]}$$

**44.** Use MATLAB to plot the step response of Problem 42. [Section: 4.10] MATLAB ML

**45.** Using classical (not Laplace) methods only, solve for the state-transition matrix, the state vector, and the output of the system represented here. [Section: 4.11] State Space SS

$$\dot{\mathbf{x}} = \begin{bmatrix} 0 & 1 \\ -1 & -5 \end{bmatrix}\mathbf{x}; \ y = \begin{bmatrix} 1 & 2 \end{bmatrix}\mathbf{x};$$

$$\mathbf{x}(0) = \begin{bmatrix} 1 \\ 0 \end{bmatrix}$$

**46.** Using classical (not Laplace) methods only, solve for the state-transition matrix, the state vector, and the output of the system represented here, where $u(t)$ is the unit step: [Section: 4.11]. State Space SS

$$\dot{\mathbf{x}} = \begin{bmatrix} 0 & 1 \\ -1 & 0 \end{bmatrix}\mathbf{x} + \begin{bmatrix} 0 \\ 1 \end{bmatrix} u(t)$$

$$y = \begin{bmatrix} 3 & 4 \end{bmatrix}\mathbf{x}; \ \mathbf{x}(0) = \begin{bmatrix} 0 \\ 0 \end{bmatrix}$$

**47.** Solve for $y(t)$ for the following system represented in state space, where $u(t)$ is the unit step. Use the classical approach to solve the state equation. [Section: 4.11] State Space SS

$$\dot{\mathbf{x}} = \begin{bmatrix} -2 & 1 & 0 \\ 0 & 0 & 1 \\ 0 & -6 & -1 \end{bmatrix}\mathbf{x} + \begin{bmatrix} 1 \\ 0 \\ 0 \end{bmatrix} u(t)$$

$$y = \begin{bmatrix} 1 & 0 & 0 \end{bmatrix}\mathbf{x}; \ \mathbf{x}(0) = \begin{bmatrix} 0 \\ 0 \\ 0 \end{bmatrix}$$

**48.** Repeat Problem 47 using MATLAB's Symbolic Math Toolbox and Eq. (4.109). In addition, run your program with an initial condition, Symbolic Math SM

$$\mathbf{x}(0) = \begin{bmatrix} 1 \\ 1 \\ 0 \end{bmatrix}. \text{[Section : 4.11]}$$

**49.** Using methods described in Appendix H.1 located at www.wiley.com/college/nise, simulate the following system and plot the step response. Verify the expected values of percent overshoot, peak time, and settling time. State Space SS

$$T(s) = \frac{1}{s^2 + 0.8s + 1}$$

**50.** Using methods described in Appendix H.1 located at www.wiley.com/college/nise, simulate the following system and plot the output, $y(t)$, for a step input:

State Space **SS**

$$\dot{\mathbf{x}} = \begin{bmatrix} 0 & 1 & 0 \\ -10 & -7 & 1 \\ 0 & 0 & -2 \end{bmatrix}\mathbf{x} + \begin{bmatrix} 0 \\ 0 \\ 1 \end{bmatrix}u(t)$$

$$y(t) = 4\begin{bmatrix} 1 & 1 & 0 \end{bmatrix}\mathbf{x}; \quad \mathbf{x}(0) = \begin{bmatrix} -1 \\ 0 \\ 0 \end{bmatrix}$$

**51.** Industrial robots are used for myriad applications. Figure P4.11 shows a robot used to move 55-pound bags of salt pellets; a vacuum head lifts the bags before positioning. The robot can move as many as 12 bags per minute (*Schneider, 1992*). Assume a model for the open-loop swivel controller and plant of

$$G_e(s) = \frac{\omega_o(s)}{V_i(s)} = \frac{K}{(s+10)(s^2+4s+10)}$$

where $\omega_o(s)$ is the Laplace transform of the robot's output swivel velocity and $V_i(s)$ is the voltage applied to the controller.

**a.** Evaluate percent overshoot, settling time, peak time, and rise time of the response of the open-loop swivel velocity to a step-voltage input. Justify all second-order assumptions.

**b.** Represent the open-loop system in state space.

State Space **SS**

**c.** `Use MATLAB or any other computer program to simulate the system and compare your results to a.`

MATLAB **ML**

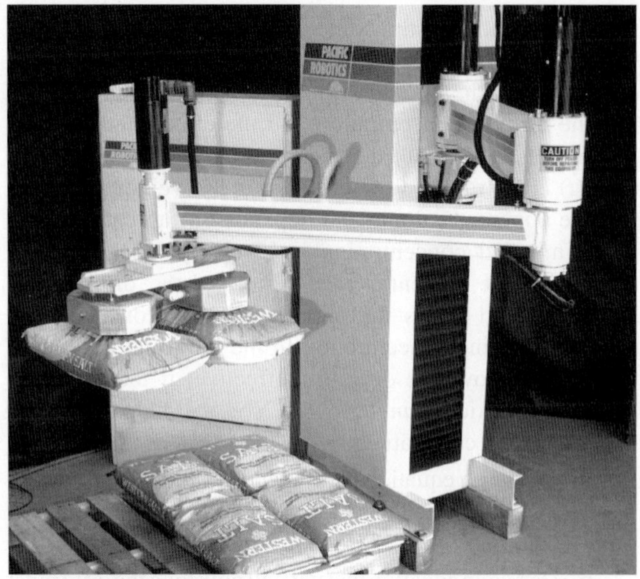

**FIGURE P4.11** Vacuum robot lifts two bags of salt.

**52.** A human responds to a visual cue with a physical response, as shown in Figure P4.12. The transfer function that relates the output physical response, $P(s)$, to the input visual command, $V(s)$, is (*Stefani, 1973*).

$$G(s) = \frac{P(s)}{V(s)} = \frac{(s+0.5)}{(s+2)(s+5)}$$

Do the following:

**a.** Evaluate the output response for a unit step input using the Laplace transform.

**b.** Represent the transfer function in state space.

State Space **SS**

**c.** `Use MATLAB to simulate the system and obtain a plot of the step response.`

MATLAB **ML**

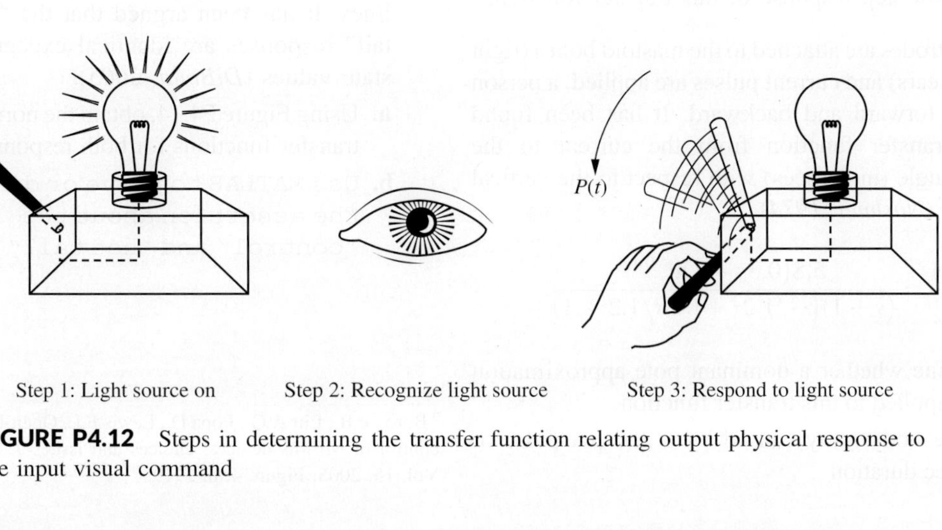


**FIGURE P4.12** Steps in determining the transfer function relating output physical response to the input visual command

53. To treat acute asthma, the drug theophylline is infused intravenously. The rate of change of the drug concentration in the blood is equal to the difference between the infused concentration and the eliminated concentration. The infused concentration is $i(t)/V_d$, where $i(t)$ is the rate of flow of the drug by weight and $V_d$ is the apparent volume and depends on the patient. The eliminated concentration is given by $k_{10}c(t)$, where $c(t)$ is the current concentration of the drug in the blood and $k_{10}$ is the elimination rate constant. The theophylline concentration in the blood is critical—if it is too low, the drug is ineffective; if too high, the drug is toxic (*Jannett, 1992*). You will help the doctor with your calculations.

   **a.** Derive an equation relating the desired blood concentration, $C_D$, to the required infusion rate by weight of the drug, $I_R$.

   **b.** Derive an equation that will tell how long the drug must be administered to reach the desired blood concentration. Use both rise time and settling time.

   **c.** Find the infusion rate of theophylline if $V_D = 600\ \text{ml}$, $k_{10} = 0.07\ \text{h}^{-1}$, and the required blood level of the drug is 12 mcg/ml ("mcg means micrograms). See (*Jannett, 1992*) for a description of parameter values.

   **d.** Find the rise and settling times for the constants in **c**.

54. Upper motor neuron disorder patients can benefit and regain useful function through the use of functional neuroprostheses. The design requires a good understanding of muscle dynamics. In an experiment to determine muscle responses, the identified transfer function was (*Zhou, 1995*)

$$M(s) = \frac{2.5e^{-0.008s}(1+0.172s)(1+0.008s)}{(1+0.07s)^2(1+0.05s)^2}$$

Find the unit step response of this transfer function.

55. When electrodes are attached to the mastoid bones (right behind the ears) and current pulses are applied, a person will sway forward and backward. It has been found that the transfer function from the current to the subject's angle (in degrees) with respect to the vertical is given by (*Nashner, 1974*)

$$\frac{\theta(s)}{I(s)} = \frac{5.8(0.3s+1)e^{-0.1s}}{(s+1)(s^2/1.2^2+0.6s/1.2+1)}$$

   **a.** Determine whether a dominant pole approximation can be applied to this transfer function.

   **b.** Find the body sway caused by a 250 μA pulse of 150 msec duration.

56. A MOEMS (optical MEMS) is a MEMS (Micro Electromechanical Systems) with an optical fiber channel that takes light generated from a laser diode. It also has a photodetector that measures light intensity variations and outputs voltage variations proportional to small mechanical device deflections. Additionally, a voltage input is capable of deflecting the device. The apparatus can be used as an optical switch or as a variable optical attenuator, and it does not exceed 2000 μm in any dimension. Figure P4.13 shows input-output signal pairs used to identify the parameters of the system. Assume a second-order transfer function and find the system's transfer function (*Borovic, 2005*).

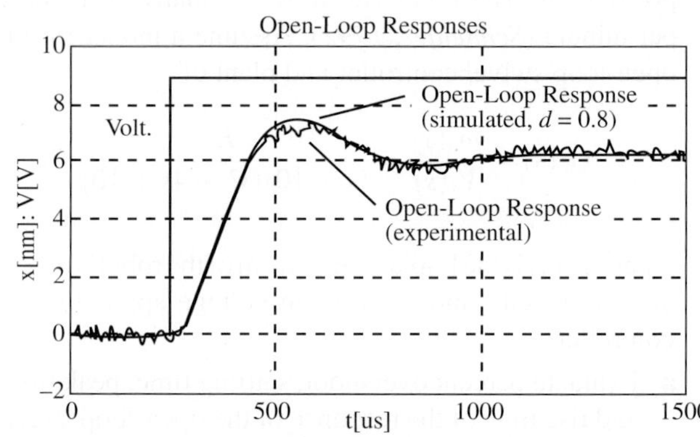


**FIGURE P4.13**[9]

57. Several factors affect the workings of the kidneys. For example, Figure P4.14 shows how a step change in arterial flow pressure affects renal blood flow in rats. In the "hot tail" part of the experiment, peripheral thermal receptor stimulation is achieved by inserting the rat's tail in heated water. Variations between different test subjects are indicated by the vertical lines. It has been argued that the "control" and "hot tail" responses are identical except for their steady-state values (*DiBona, 2005*).

   **a.** Using Figure P4.14, obtain the normalized ($C_{final} = 1$) transfer functions for both responses.

   **b.** Use MATLAB to prove or disprove the assertion about the "control" and "hot tail" responses. MATLAB ML

[9] Borovic B., Liu A.Q., Popa D., Lewis F.L. Open-loop versus closed-loop control of MEMS devices: choices and issues. *J. Micromech. Microeng.* Vol. 15, 2005. Figure 4, p. 1919.

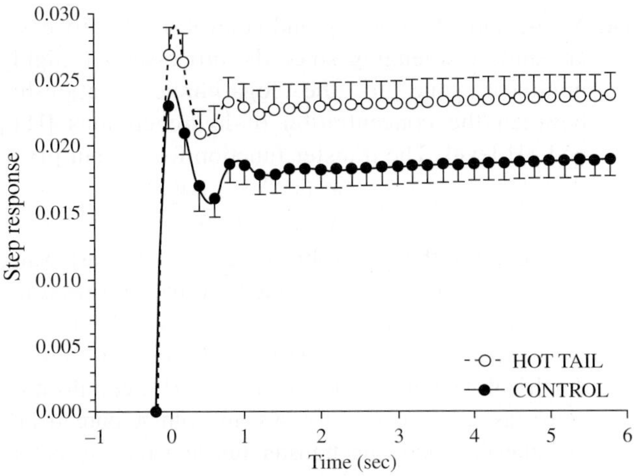


**FIGURE P4.14**[10]

58. The transfer function of a nano-positioning device capable of translating biological samples within a few μm uses a piezoelectric actuator and a linear variable differential transformer (LDVT) as a displacement sensor. The transfer function from input to displacement has been found to be (*Salapaka, 2002*)

$$G(s) = \frac{9.7 \times 10^4(s^2 - 14400s + 106.6 \times 10^6)}{(s^2 + 3800s + 23.86 \times 10^6)(s^2 + 240s + 2324.8 \times 10^3)}$$

Use a dominant-pole argument to find an equivalent transfer function with the same numerator but only three poles.

MATLAB **ML** `Use MATLAB to find the actual size and approximate system unit step responses, plotting them on the same graph.`

Explain the differences between both responses given that both pairs of poles are so far apart.

59. At some point in their lives most people will suffer from at least one onset of low back pain. This disorder can trigger excruciating pain and temporary disability, but its causes are hard to diagnose. It is well known that low back pain alters motor trunk patterns; thus it is of interest to study the causes for these alterations and their extent. Due to the different possible causes of this type of pain, a "control" group of people is hard to obtain for laboratory studies. However, pain can be stimulated in healthy people and muscle movement ranges can be compared. Controlled back pain can be induced by injecting saline solution directly into related muscles or ligaments. The transfer function from infusion rate to pain response was obtained experimentally by injecting a 5% saline solution at six different infusion rates over a period of 12 minutes. Subjects verbally rated their pain every 15 seconds on a scale from 0 to 10, with 0 indicating no pain and 10 unbearable pain. Several trials were averaged and the data was fitted to the following transfer function:

$$G(s) = \frac{9.72 \times 10^{-8}(s + 0.0001)}{(s + 0.009)^2(s^2 + 0.018s + 0.0001)}$$

For experimentation, it is desired to build an automatic dispensing system to make the pain level constant as shown in Figure P4.15. It follows that ideally the injection system transfer function has to be

$$M(s) = \frac{1}{G(s)}$$

to obtain an overall transfer function $M(s)G(s) \approx 1$. However, for implementation purposes $M(s)$ must have at least one more pole than zeros (*Zedka, 1999*). Find a suitable transfer function, $M(s)$ by inverting $G(s)$ and adding poles that are far from the imaginary axis.

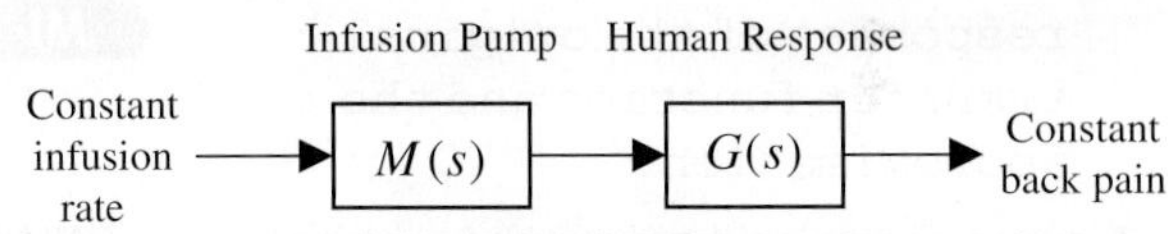


**FIGURE P4.15**

60. An artificial heart works in closed loop by varying its pumping rate according to changes in signals from the recipient's nervous system. For feedback compensation design it is important to know the heart's open-loop transfer function. To identify this transfer function, an artificial heart is implanted in a calf while the main parts of the original heart are left in place. Then the atrial pumping rate in the original heart is measured while step input changes are effected on the artificial heart. It has been found that, in general, the obtained response closely resembles that of a second-order system. In one such experiment it was found that the step response has a %OS = 30% and a time of first peak $T_p = 127$ sec (*Nakamura, 2002*). Find the corresponding transfer function.

61. An observed transfer function from voltage potential to force in skeletal muscles is given by (*Ionescu, 2005*)

$$T(s) = \frac{450}{(s + 5)(s + 20)}$$

**a.** Obtain the system's impulse response.

**b.** Integrate the impulse response to find the step response.

**c.** Verify the result in Part **b** by obtaining the step response using Laplace transform techniques.

[10] DiBona G.F. Physiology in perspective: The Wisdom of the Body. Neural control of the kidney. *Am. J. Physiol. Regul. Integr. Comp. Physiol* Vol. 289, 2005. Fig. 6, p. R639. Used with permission.

**62.** In typical conventional aircraft, longitudinal flight model linearization results in transfer functions with two pairs of complex conjugate poles. Consequently, the natural response for these airplanes has two modes in their natural response. The "short period" mode is relatively well-damped and has a high-frequency oscillation. The "plugoid mode" is lightly damped and its oscillation frequency is relatively low. For example, in a specific aircraft the transfer function from wing elevator deflection to nose angle (angle of attack) is (*McRuer, 1973*)

$$\frac{\theta(s)}{\delta_e(s)} = -\frac{26.12(s+0.0098)(s+1.371)}{(s^2+8.99\times10^{-3}s+3.97\times10^{-3})(s^2+4.21s+18.23)}$$

**a.** Find which of the poles correspond to the short period mode and which to the phugoid mode.

**b.** Peform a "phugoid approximation" (dominant-pole approximation), retaining the two poles and the zero closest to the; $\omega$-axis.

**c.** Use MATLAB to compare the step responses of the original transfer function and the approximation. MATLAB **ML**

**63.** A crosslapper is a machine that takes as an input a light fiber fabric and produces a heavier fabric by laying the original fabric in layers rotated by 90 degrees. A feedback system is required in order to maintain consistent product width and thickness by controlling its carriage velocity. The transfer function from servomotor torque, $T_m(s)$, to carriage velocity, $Y(s)$, was developed for such a machine (*Kuo, 2008*). Assume that the transfer function is: MATLAB **ML**

$$G(s) = \frac{Y(s)}{T_m(s)} = \frac{33s^4+202s^3+10061s^2+24332s+170704}{s^7+8s^6+464s^5+2411s^4+52899s^3+167829s^2+913599s+1076555}$$

**a.** Use MATLAB to find the partial fraction residues and poles of $G(s)$.

**b.** Find an approximation to $G(s)$ by neglecting the second-order terms found in **a**.

**c.** Use MATLAB to plot on one graph the step response of the transfer function given above and the approximation found in **b**. Explain the differences between the two plots.

**64.** Mathematical modeling and control of pH processes are quite challenging since the processes are highly nonlinear, due to the logarithmic relationship between the concentration of hydrogen ions [H+] and pH level. The transfer function from input pH to output pH is $G_a(s) = \frac{Y_a(s)}{X_a(s)} = \frac{14.49e^{-4s}}{1478.26s+1}$. $G_a(s)$ is a model for the anaerobic process in a wastewater treatment system in which methane bacteria need the pH to be maintained in its optimal range from 6.8 to 7.2 (*Jiayu, 2009*). Similarly, (*Elarafi, 2008*) used empirical techniques to model a pH neutralization plant as a second-order system with a pure delay, yielding the following transfer function relating output pH to input pH:

$$G_p(s) = \frac{Y_p(s)}{X_p(s)} = \frac{1.716\times10^{-5}e^{-30s}}{s^2+6.989\times10^{-3}s+1.185\times10^{-6}}$$

**a.** Find analytical expressions for the unit-step responses $y_a(t)$ and $y_p(t)$ for the two processes, $G_a(s)$ and $G_p(s)$. (Hint: Use the time shift theorem in Table 2.2).

**b.** Use Simulink to plot $y_a(t)$ and $y_p(t)$ on a single graph. Simulink **SL**

**65.** Figure P4.16 shows the step response of an electric vehicle's mechanical brakes when the input is the drive torque (N-m) and the output is the hydraulic brake pressure (bar) (*Ringdorfer, 2011*).

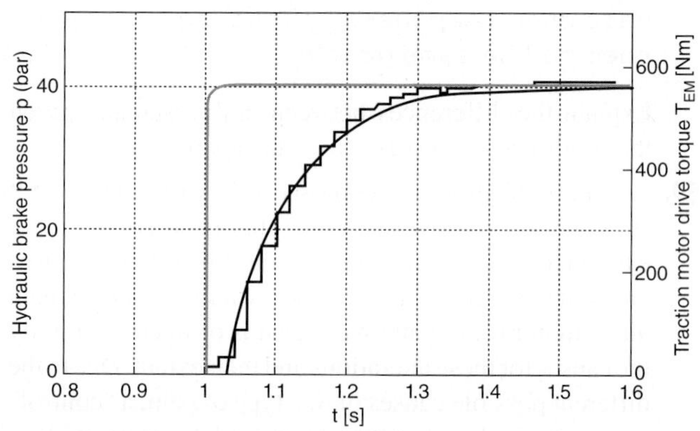


**FIGURE P4.16** Step response of an electric vehicle's mechanical brake[11]

[11] Ringdorfer M., and Horn M. Development of a Wheel Slip Actuator Controller for Electric Vehicles using Energy Recuperation and Hydraulic Brake Control, *2011 IEEE International Conference on Control Applications (CCA)*, Denver, CO, USA. September 28–30, 2011, pp. 313–318. Figure 4, p. 315. Modelling Symposium (AMS), 2012 Sixth Asia by IEEE. Reproduced with permission of IEEE in the format Republish in a book via Copyright Clearance Center.

**a.** Find the transfer function of the system.

**b.** Use the values of the parameters for the transfer function obtained in Part **a** to find an expression for the brake pressure as a function of time.

**c.** Find the output in bars of the system 0.2 sec after the input is applied. Check your result against Figure P4.16.

**66.** Figure P4.17 shows the free-body diagrams for planetary gear components used in the variable valve timing (VVT) system of an internal combustion engine (*Ren, 2011*). Here an electric motor is used to drive the carrier. Analysis showed that the electric motor with planetary gear load (Figure P4.17) may be represented by the following equation:

$$\Omega_c(s) = G_e(s)E_a(s) + G_m(s)T_{cam}(s)$$

where $\Omega_c(s)$ is the output carrier angular speed in rad/s, $E_a(s)$ is the input voltage applied to the armature, and $T_{cam}(s)$ is the input load torque. The voltage input transfer function, $G_e(s)$, is

$$G_e(s) \cong \frac{K_\tau}{R_m(Js+D)+K_\tau K_m} = \frac{45}{0.2s+1}$$

and the load torque input transfer function, $G_m(s)$, is

$$G_m(s) \cong \frac{-R_m k}{R_m(Js+D)+K_\tau K_m} = \frac{-5}{0.2s+1}$$

Find an analytical expression for the output carrier angular speed, $\omega_c(t)$, if a step voltage of 100 volts is applied at $t = 0$ followed by an equivalent load torque of 10 N-m, applied at $t = 0.4$ sec.

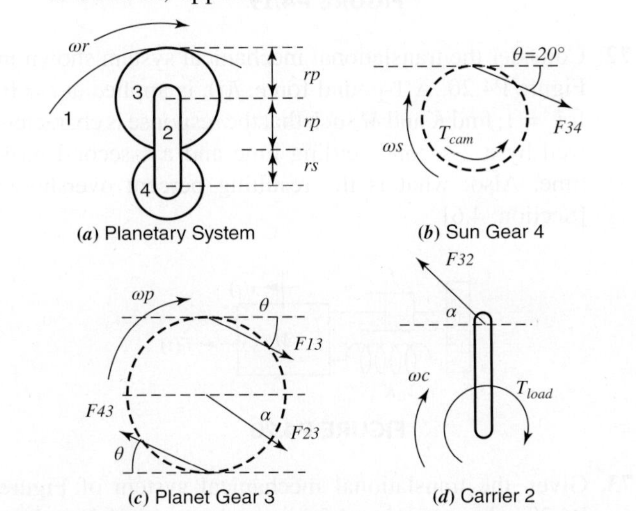


**FIGURE P4.17** Free-body diagrams of planetary gear system components[12]

[12] Ren Z., and Zhu G. G. Modeling and Control of an Electric Variable Valve Timing System for SI and HCCI Combustion Mode Transition. *American Control Conference*, San Francisco, CA, 2011, pp. 979–984. Figure 2, p. 980. Reproduced with permission of IEEE in the format Republish in a book via Copyright Clearance Center.

**67.** A drive system with elastically coupled load (Figure P4.18) has a motor that is connected to the load via a gearbox and a long shaft.

The system parameters are: $J_M$ = drive-side inertia = 0.0338 kg-m$^2$, $J_L$ = load-side inertia = 0.1287 kg-m$^2$, $K = C_T$ = torsional spring constant = 1700 N-m/rad, and $D$ = damping coefficient = 0.15 N-m-s/rad.

This system can be reduced to a simple two-inertia model that may be represented by the following transfer function, relating motor shaft speed output, $\Omega(s)$, to electromagnetic torque input (*Thomsen, 2011*):

$$G(s) = \frac{\Omega(s)}{T_{em}(s)} = \frac{1}{s(J_M+J_L)} \cdot \frac{\frac{J_L}{C_T}s^2 + \frac{D}{C_T}s + 1}{\frac{J_M J_L}{C_T(J_M+J_L)}s^2 + \frac{D}{C_T}s + 1}$$

MATLAB ML

Assume the system is at standstill at $t = 0$, when the electromagnetic torque, $T_{em}$, developed by the motor changes from zero to 50 N-m. Find and plot on one graph, using MATLAB or any other program, the motor shaft speed, $\omega(t)$, rad/sec, for the following two cases:

**a.** No load torque is applied and, thus, $\omega = \omega_{nl}$.

**b.** A load torque, $T_L = 0.2\,\omega(t)$ N-m is applied and $\omega = \omega_L$.

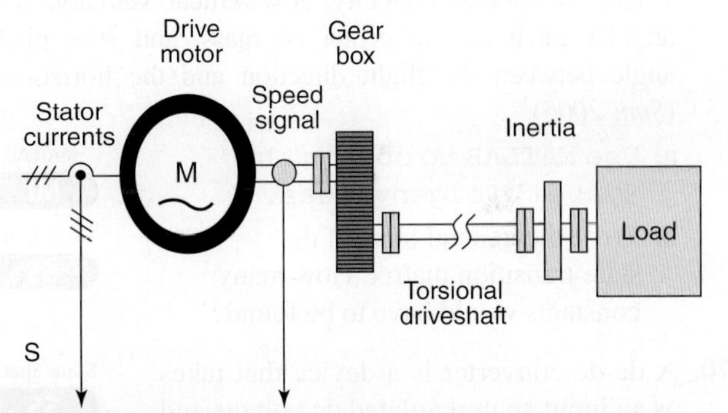


**FIGURE P4.18** Partial topology of a typical motor drive system[13]

**68.** An inverted pendulum mounted on a motor-driven cart was presented in Problem 29 of Chapter 3. The nonlinear state-space equations representing that system were linearized *(Prasad, 2012)* around a stationary point corresponding to the pendulum point-mass, $m$, being in

[13] Thomsen, S., Hoffmann, N., and Fuchs, F. W. PI Control, PI-Based State Space Control, and Model-Based Predictive Control for Drive Systems With Elastically Coupled Loads—A Comparative Study. *IEEE Transactions On Industrial Electronics*, Vol. 58, No. 8, August 2011, pp. 3647–3657. Portion of Figure 1, p. 3648. American Control Conference (ACC), 2011 by IEEE. Reproduced with permission of IEEE in the format Republish in a book via Copyright Clearance Center.

the upright position ($x_0 = 0$ at $t = 0$), when the force applied to the cart was zero ($u_0 = 0$). The state-space model developed in that problem is

$$\dot{\mathbf{x}} = \mathbf{A}\,\mathbf{x} + \mathbf{B}u$$

The state variables are the pendulum angle relative to the $y$-axis, $\theta$, its angular speed, $\theta'$, the horizontal position of the cart, $x$, and its speed, $x'$. The horizontal position of $m$ (for a small angle, $\theta$), $x_m = x + l\sin\theta = x + l\theta$, was selected to be the output variable.

MATLAB ML

Given the state-space model developed in that problem along with the output equation you developed in that problem, use MATLAB (or any other computer program) to find and plot the output, $x_m(t)$, in meters, for an input force, $u(t)$, equal to a unit impulse, $\delta(t)$, in Newtons.[14]

**69.** Using wind tunnel tests, insect flight dynamics can be studied in a very similar fashion to that of man-made aircraft. Linearized longitudinal flight equations for a bumblebee have been found in the unforced case to be

$$\begin{bmatrix} \dot{u} \\ \dot{w} \\ \dot{q} \\ \dot{\theta} \end{bmatrix} = \begin{bmatrix} -8.792 \times 10^{-3} & 0.56 \times 10^{-3} & -1.0 \times 10^{-3} & -13.79 \times 10^{-3} \\ -0.347 \times 10^{-3} & -11.7 \times 10^{-3} & -0.347 \times 10^{-3} & 0 \\ 0.261 & -20.8 \times 10^{-3} & -96.6 \times 10^{-3} & 0 \\ 0 & 0 & 1 & 0 \end{bmatrix} \begin{bmatrix} u \\ w \\ q \\ \theta \end{bmatrix}$$

where $u$ = forward velocity; $w$ = vertical velocity, $q$ = angular pitch rate at center of mass, and $\theta$ = pitch angle between the flight direction and the horizontal (*Sun, 2005*).

MATLAB ML

**a.** Use MATLAB to obtain the system's eigenvalues.

State Space SS

**b.** Write the general form of the state-transition matrix. How many constants would have to be found?

State Space SS

**70.** A dc-dc converter is a device that takes as an input an unregulated dc voltage and provides a regulated dc voltage as its output. The output voltage may be lower (buck converter), higher (boost converter), or the same as the input voltage. Switching dc-dc converters have a semiconductor active switch (BJT or FET) that is closed periodically with a duty cycle $d$ in a pulse width modulated (PWM) manner. For a boost converter, averaging techniques can be used to arrive at the following state equations (*Van Dijk, 1995*):

$$L\frac{di_L}{dt} = -(1-d)u_c + E_s$$

$$C\frac{du_C}{dt} = (1-d)i_L - \frac{u_C}{R}$$

where $L$ and $C$ are, respectively, the values of internal inductance and capacitance; $i_L$ is the current through the internal inductor; $R$ is the resistive load connected to the converter; $E_s$ is the dc input voltage; and the capacitor voltage, $u_C$, is the converter's output.

**a.** Write the converter's equations in the form

$$\dot{\mathbf{x}} = \mathbf{A}\mathbf{x} + \mathbf{B}\mathbf{u}$$
$$\mathbf{y} = \mathbf{C}\mathbf{x}$$

assuming $d$ is a constant.

**b.** Using the **A**, **B**, and **C** matrices of Part **a**, obtain the converter's transfer function $\dfrac{U_C(s)}{E_s(s)}$.

## DESIGN PROBLEMS

**71.** Find an equation that relates 2% settling time to the value of $f_v$ for the translational mechanical system shown in Figure P4.19. Neglect the mass of all components. [Section: 4.6]

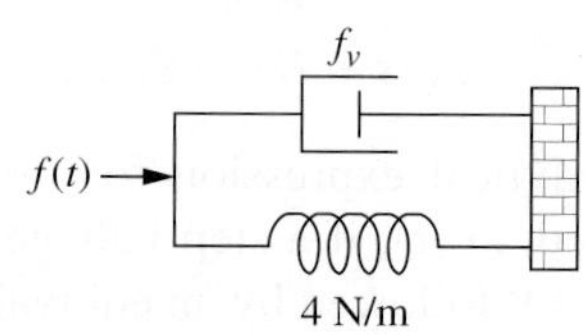


**FIGURE P4.19**

**72.** Consider the translational mechanical system shown in Figure P4.20. A 1-pound force, $f(t)$, is applied at $t = 0$. If $f_v = 1$, find $K$ and $M$ such that the response is characterized by a 2-second settling time and a 1-second peak time. Also, what is the resulting percent overshoot? [Section: 4.6]

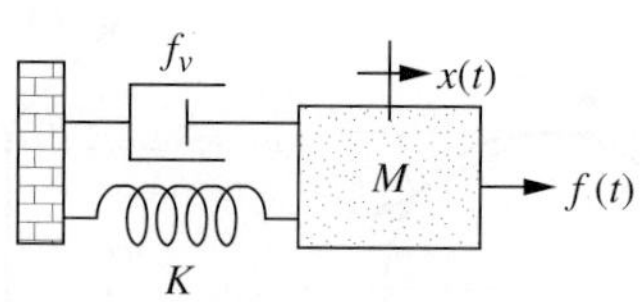


**FIGURE P4.20**

**73.** Given the translational mechanical system of Figure P4.20, where $K = 1$ and $f(t)$ is a unit step, find the values of $M$ and $f_v$ to yield a response with 30% overshoot and a settling time of 10 seconds. [Section: 4.6]

**74.** Find $J$ and $K$ in the rotational system shown in Figure P4.21 to yield a 20% overshoot and a settling time of 2 seconds for a step input in torque. [Section: 4.6]

[14] Hint: Use the command "impulseplot" over a time period from 0 to 11.0 seconds with a step of 0.1 seconds.

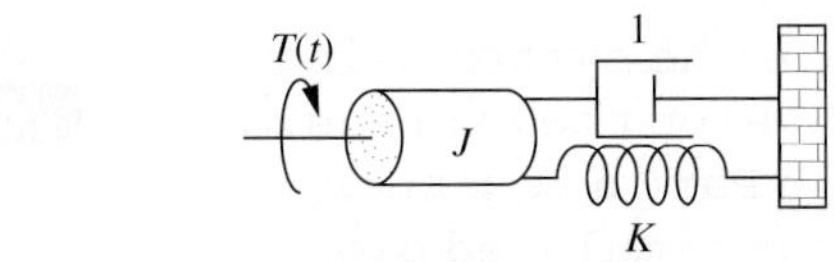


**FIGURE P4.21**

**75.** Given the system shown in Figure P4.22, find the damping, $D$, to yield a 20% overshoot in output angular displacement for a step input in torque. [Section: 4.6]

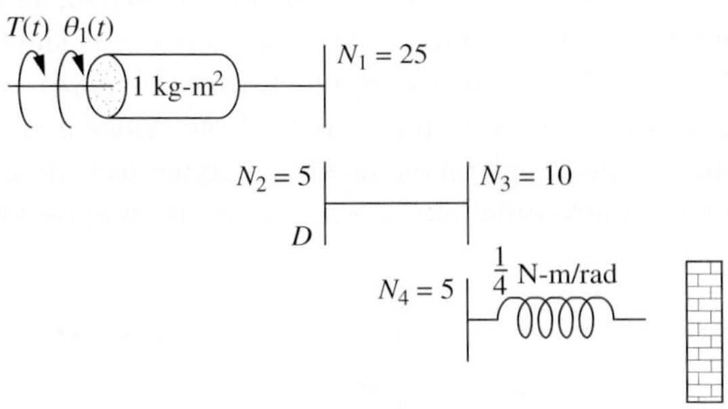


**FIGURE P4.22**

**76.** For the system shown in Figure P4.23, find $N_1/N_2$ so that the settling time for a step torque input is 16 seconds. [Section: 4.6]

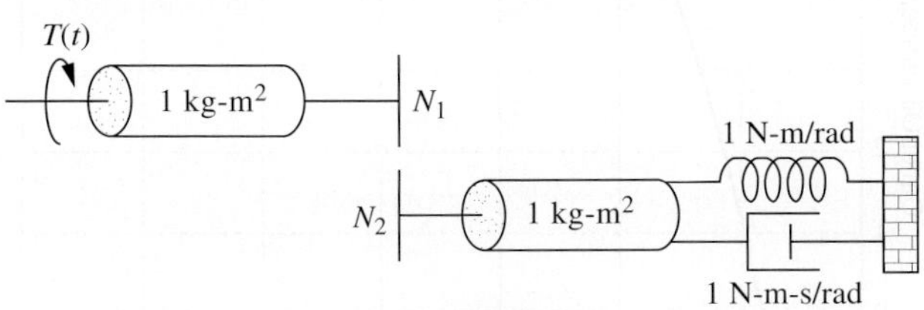


**FIGURE P4.23**

**77.** Find $M$ and $K$, shown in the system of Figure P4.24, to yield $x(t)$ with 10% overshoot and 15 seconds settling time for a step input in motor torque, $T_m(t)$. [Section: 4.6]

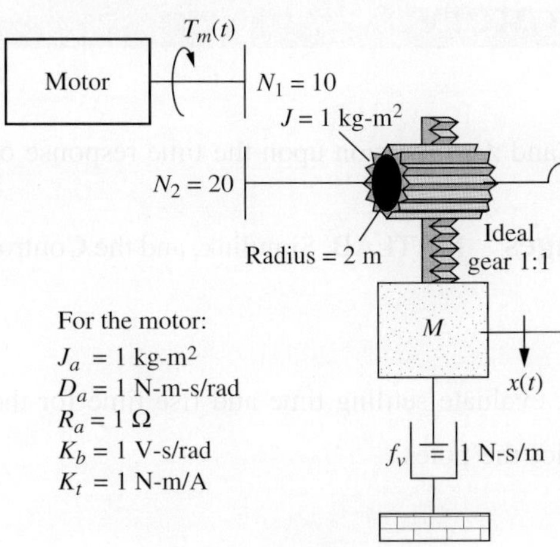


**FIGURE P4.24**

**78.** If $v_i(t)$ is a step voltage in the network shown in Figure P4.25, find the value of the resistor such that a 20% overshoot in voltage will be seen across the capacitor if $C = 10^{-6}$ F and $L = 0.5$ H. [Section: 4.6]

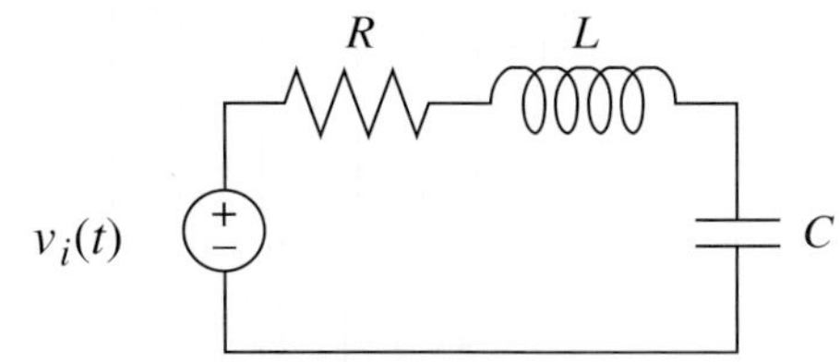


**FIGURE P4.25**

**79.** If $v_i(t)$ is a step voltage in the network shown in Figure P4.25, find the values of $R$ and $C$ to yield a 20% overshoot and a 1 ms settling time for $v_c(t)$ if $L = 1$ H. [Section: 4.6]

**80.** Given the circuit of Figure P4.25, where $C = 10\ \mu$F, find $R$ and $L$ to yield 10% overshoot with a settling time of 5 ms for the capacitor voltage. The input, $v(t)$, is a unit step. [Section: 4.6]

**81.** For the circuit shown in Figure P4.26, find the values of $R_2$ and $C$ to yield 8% overshoot with a settling time of 1 ms for the voltage across the capacitor, with $v_i(t)$ as a step input. [Section: 4.6]

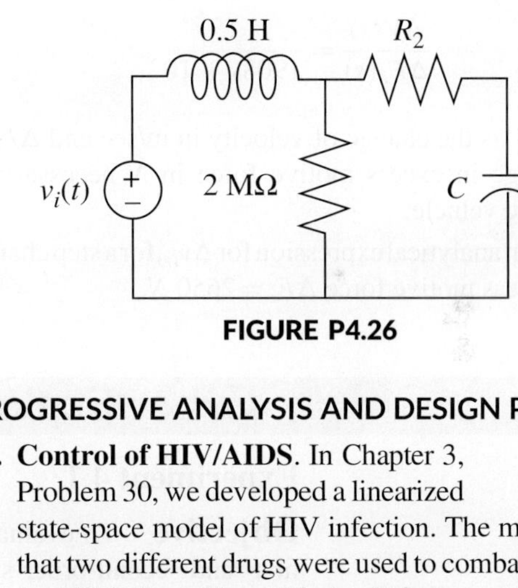


**FIGURE P4.26**

## PROGRESSIVE ANALYSIS AND DESIGN PROBLEMS

**82. Control of HIV/AIDS.** In Chapter 3, Problem 30, we developed a linearized state-space model of HIV infection. The model assumed that two different drugs were used to combat the spread of the HIV virus. Since this book focuses on single-input, single-output systems, only one of the two drugs will be considered. We will assume that only RTIs are used as an input. Thus, in the equations of Chapter 3, Problem 30, $u_2 = 0$ (*Craig, 2004*).

State Space **SS**

**a.** Show that when using only RTIs in the linearized system of Problem 30, Chapter 3, and substituting the typical parameter values given in the table of Problem 30**c**, Chapter 3, the resulting state-space representation for the system is given by

$$\begin{bmatrix} \dot{T} \\ \dot{T}^* \\ \dot{v} \end{bmatrix} = \begin{bmatrix} -0.04167 & 0 & -0.0058 \\ 0.0217 & -0.24 & 0.0058 \\ 0 & 100 & -2.4 \end{bmatrix} \times \begin{bmatrix} T \\ T^* \\ v \end{bmatrix} + \begin{bmatrix} 5.2 \\ -5.2 \\ 0 \end{bmatrix} u_1$$

$$y = [0 \quad 0 \quad 1] \begin{bmatrix} T \\ T^* \\ v \end{bmatrix}$$

**b.** Obtain the transfer function from RTI efficiency to virus count; namely, find $\frac{Y(s)}{U_1(s)}$.

**c.** Assuming RTIs are 100% effective, what will be the steady-state change of virus count in a given infected patient? Express your answer in virus copies per ml of plasma. Approximately how much time will the medicine take to reach its maximum possible effectiveness?

**83. Hybrid vehicle.** Assume that the car motive dynamics for a hybrid electric vehicle (HEV) can be described by the transfer function

$$\frac{\Delta V(s)}{\Delta F_e(s)} = \frac{1}{1908s + 10}$$

where $AV$ is the change of velocity in m/sec and $\Delta F_e$ is the change in excess motive force in $N$ necessary to propel the vehicle.

**a.** Find an analytical expression for $\Delta v_{(t)}$ for a step change in excess motive force $\Delta F_e = 2650\ N$.

**b.** `Simulate the system using MATLAB. Plot the expression found in Part a together with your simulated plot.`

MATLAB **ML**

**84. Parabolic trough collector.** Figure P4.27 illustrates the results of an open-loop step-response experiment performed on a parabolic trough collector setup (*Camacho, 2012*). In this experiment, the fluid flow on the system is suddenly decreased 1 liter/sec at $t = 0$ hours, resulting in a temperature increase as shown in Figure P4.27. Use the figure to find an *approximate* transfer function for the system. (Note: Since no further information is given on the system dynamics and due to visual approximations, several solutions are possible.)

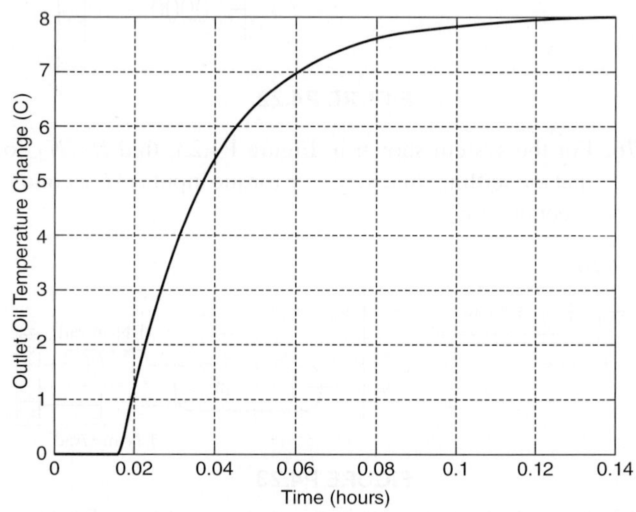


**FIGURE P4.27**

# Cyber Exploration Laboratory

## Experiment 4.1

**Objective** To evaluate the effect of pole and zero location upon the time response of first- and second-order systems.

**Minimum Required Software Packages** MATLAB, Simulink, and the Control System Toolbox

### Prelab

1. Given the transfer function $G(s) = \frac{a}{s + a}$, evaluate settling time and rise time for the following values of $a$: 1, 2, 3, 4. Also, plot the poles.

2. Given the transfer function $G(s) = \dfrac{b}{s^2 + as + b}$:

   a. Evaluate percent overshoot, settling time, peak time, and rise time for the following values: $a = 4$, $b = 25$. Also, plot the poles.

   b. Calculate the values of $a$ and $b$ so that the imaginary part of the poles remains the same but the real part is increased two times over that of Prelab 2**a**, and repeat Prelab 2**a**.

   c. Calculate the values of $a$ and $b$ so that the imaginary part of the poles remains the same but the real part is decreased by one half over that of Prelab 2**a**, and repeat Prelab 2**a**.

3. a. For the system of Prelab 2**a**, calculate the values of $a$ and $b$ so that the real part of the poles remains the same but the imaginary part is increased two times over that of Prelab 2**a**, and repeat Prelab 2**a**.

   b. For the system of Prelab 2**a**, calculate the values of $a$ and $b$ so that the real part of the poles remains the same but the imaginary part is increased four times over that of Prelab 2**a**, and repeat Prelab 2**a**.

4. a. For the system of Prelab 2**a**, calculate the values of $a$ and $b$ so that the damping ratio remains the same but the natural frequency is increased two times over that of Prelab 2**a**, and repeat Prelab 2**a**.

   b. For the system of Prelab 2**a**, calculate the values of $a$ and $b$ so that the damping ratio remains the same but the natural frequency is increased four times over that of Prelab 2**a**, and repeat Prelab 2**a**.

5. Briefly describe the effects on the time response as the poles are changed in each of Prelabs 2, 3, and 4.

## Lab

1. Using Simulink, set up the systems of Prelab 1 and plot the step response of each of the four transfer functions on a single graph by using the Simulink LTI Viewer (See Appendix E.6 online for tutorial). Also, record the values of settling time and rise time for each step response.
2. Using Simulink, set up the systems of Prelab 2 . Using the Simulink LTI Viewer, plot the step response of each of the three transfer functions on a single graph. Also, record the values of percent overshoot, settling time, peak time, and rise time for each step response.
3. Using Simulink, set up the systems of Prelab 2**a** and Prelab 3. Using the Simulink LTI Viewer, plot the step response of each of the three transfer functions on a single graph. Also, record the values of percent overshoot, settling time, peak time, and rise time for each step response.
4. Using Simulink, set up the systems of Prelab 2**a** and Prelab 4. Using the Simulink LTI Viewer, plot the step response of each of the three transfer functions on a single graph. Also, record the values of percent overshoot, settling time, peak time, and rise time for each step response.

## Postlab

1. For the first-order systems, make a table of calculated and experimental values of settling time, rise time, and pole location.
2. For the second-order systems of Prelab 2, make a table of calculated and experimental values of percent overshoot, settling time, peak time, rise time, and pole location.

3. For the second-order systems of Prelab 2**a** and Prelab 3, make a table of calculated and experimental values of percent overshoot, settling time, peak time, rise time, and pole location.
4. For the second-order systems of Prelab 2**a** and Prelab 4, make a table of calculated and experimental values of percent overshoot, settling time, peak time, rise time, and pole location.
5. Discuss the effects of pole location upon the time response for both first- and second-order systems. Discuss any discrepancies between your calculated and experimental values.

## Experiment 4.2

**Objective** To evaluate the effect of additional poles and zeros upon the time response of second-order systems.

**Minimum Required Software Packages** MATLAB, Simulink, and the Control System Toolbox

### Prelab

1. **a.** Given the transfer function $G(s) = \frac{25}{s^2 + 4s + 25}$, evaluate the percent overshoot, settling time, peak time, and rise time. Also, plot the poles.
   **b.** Add a pole at $-200$ to the system of Prelab 1**a**. Estimate whether the transient response in Prelab 1**a** will be appreciably affected.
   **c.** Repeat Prelab 1**b** with the pole successively placed at $-20$, $-10$, and $-2$.
2. A zero is added to the system of Prelab 1**a** at $-200$ and then moved to $-50$, $-20$, $-10$, $-5$, and $-2$. List the values of zero location in the order of the greatest to the least effect upon the pure second-order transient response.
3. Given the transfer function $G(s) = \frac{(25b/a)(s + a)}{(s + b)(s^2 + 4s + 25)}$, let $a = 3$ and $b = 3.01$, 3.1, 3.3, 3.5, and 4.0. Which values of $b$ will have minimal effect upon the pure second-order transient response?
4. Given the transfer function $G(s) = \frac{(2500b/a)(s + a)}{(s + b)(s^2 + 40s + 2500)}$, let $a = 30$ and $b = 30.01$, 30.1, 30.5, 31, 35, and 40. Which values of $b$ will have minimal effect upon the pure second-order transient response?

### Lab

1. Using Simulink, add a pole to the second-order system of Prelab 1**a** and plot the step responses of the system when the higher-order pole is nonexistent, at $-200$, $-20$, $-10$, and $-2$. Make your plots on a single graph, using the Simulink LTI Viewer. Normalize all plots to a steady-state value of unity. Record percent overshoot, settling time, peak time, and rise time for each response.
2. Using Simulink, add a zero to the second-order system of Prelab 1**a** and plot the step responses of the system when the zero is nonexistent, at $-200$, $-50$, $-20$, $-10$, $-5$, and $-2$. Make your plots on a single graph, using the Simulink LTI Viewer. Normalize all plots to a steady-state value of unity. Record percent overshoot, settling time, peak time, and rise time for each response.
3. Using Simulink and the transfer function of Prelab 3 with $a = 3$, plot the step responses of the system when the value of $b$ is 3, 3.01, 3.1, 3.3, 3.5, and 4.0. Make your plots on a single graph using the Simulink LTI Viewer. Record percent overshoot, settling time, peak time, and rise time for each response.

4. Using Simulink and the transfer function of Prelab 4 with $a = 30$, plot the step responses of the system when the value of $b$ is 30, 30.01, 30.1, 30.5, 31, 35, and 40. Make your plots on a single graph, using the Simulink LTI Viewer. Record percent overshoot, settling time, peak time, and rise time for each response.

### Postlab

1. Discuss the effect upon the transient response of the proximity of a higher-order pole to the dominant second-order pole pair.
2. Discuss the effect upon the transient response of the proximity of a zero to the dominant second-order pole pair. Explore the relationship between the length of the vector from the zero to the dominant pole and the zero's effect upon the pure second-order step response.
3. Discuss the effect of pole-zero cancellation upon the transient response of a dominant second-order pole pair. Allude to how close the canceling pole and zero should be and the relationships of (1) the distance between them and (2) the distance between the zero and the dominant second-order poles.

## Experiment 4.3

**Objective** To use LabVIEW Control Design and Simulation Module for time performance analysis of systems.

**Minimum Required Software Packages** LabVIEW with the Control Design and Simulation Module

**Prelab** One of the experimental direct drive robotic arms built at the MTT Artificial Intelligence Laboratory and the CMU Robotics Institute can be represented as a feedback control system with a desired angular position input for the robot's joint position and an angular position output representing the actual robot's joint position.

The forward path consists of three transfer functions in cascade; (1) a compensator, $G_c(s)$, to improve performance; (2) a power amplifier of gain, $K_a = 1$; and (3) the transfer function of the motor and load, $G(s) = 2292/s(s + 75.6)$. Assume a unity-feedback system. Initially the system will be controlled with $G_c(s) = 0.6234$, which is called a proportional controller (*McKerrow, 1991*).

1. Obtain the closed-loop system transfer function and use MATLAB to make a plot of the resulting unit step response.
2. Repeat with $G_c(s) = 3.05 + 0.04s$, which is called a PD controller.
3. Compare both responses and draw conclusions regarding their time domain specifications.

**Lab** Create a LabVIEW VI that uses a simulation loop to implement both controllers given in the Prelab. Plot the responses on the same graph for comparison purposes.

**Postlab** Compare the responses obtained using your LabVIEW VI with those obtained in the Prelab.

## Experiment 4.4

**Objective** To use the LabVIEW Control Design and Simulation Module to evaluate the effect of pole location upon the time response of second-order systems.

**Minimum Required Software Packages** LabVIEW with the Control Design and Simulation Module.

**Prelab** Solve the Cyber Exploration Laboratory Experiment 4.1 Prelab, Part 2.

**Lab** Build a LabVIEW VI to implement the functions studied in the Prelab of Cyber Exploration Laboratory 4.1, Part 2.

Specifically for Prelab Part 2**a**, your front panel will have the coefficients of the second-order transfer function as inputs. The front panel will also have the following indicators: (1) the transfer function; (2) the state-space representation; (3) the pole locations; (4) the step response graph; (5) the time response of the two states on the same graph; (6) the time response parametric data including rise time, peak time, settling time, percent overshoot, peak value, and final value.

For Prelab, Part 2**b**, your front panel will also have the following indicators: (1) the step response graph, and (2) the parametric data listed above for Prelab, Part 2**a**, but specific to Part 2**b**.

For Prelab, Part 2**c**, your front panel will also have the following indicators: (1) the step response graph, and (2) the parametric data listed above for Prelab, Part 2**a**, but specific to Part 2**c**.

Run the VI to obtain the data from the indicators.

**Postlab** Use your results to discuss the effect of pole location upon the step response.

## Hardware Interface Laboratory

### Experiment 4.5 Open-Loop Speed Control of a Motor

**Objectives** To control the speed of a motor in open-loop fashion and verify the functions of the motor control setup as preparation for future experiments.

**Material Required** Computer with LabVIEW Installed; myDAQ; dc brushed gear-motor with Hall Sensor quadrature encoder (–10 V to +10 V normal operation range); and motor control chip BA6886N or a transistor circuit substitute. (Note: For simplicity the input to the motor will be analog. PWM will be avoided as it adds an additional layer of complexity to these experiments. Plan accordingly if you decide to substitute the motor control chip.)

**File Provided at www.wiley.com/college/nise**
Open Loop Control.vi

**Prelab** Plan how you will wire your motor to the breadboard. A possibility is to solder a header to six matching color wires that will allow you to connect and disconnect the motor from the myDAQ in an efficient manner. You can also solder wires to the motor's cables.

**Lab**

**Software:** The front panel for the Open Loop Control VI is shown in Figure P4.28(*a*). The input to the system is the voltage applied to the motor. The output is the motor speed in revolutions per second (rps). The corresponding block diagram is shown in Figure P4.28(*b*).

Note the value indicated by the blue arrow in Figure P4.28(*b*). In order to get a meaningful reading for the speed of the motor, this value needs to be modified depending on the gear ratio of your motor and the counts of your encoder. To understand how this value is calculated, note that the DAQ Assistant block on top of the diagram reads the encoder input from the myDAQ. One would reasonably assume that the frequency of this signal is proportional to the speed of the motor, which is theoretically true. However, at very low speeds the DAQ assistant "times out" and fails to provide a reading if frequency is measured directly. To avoid this problem, a different method is used to calculate the signal frequency. The DAQ assistant measures the

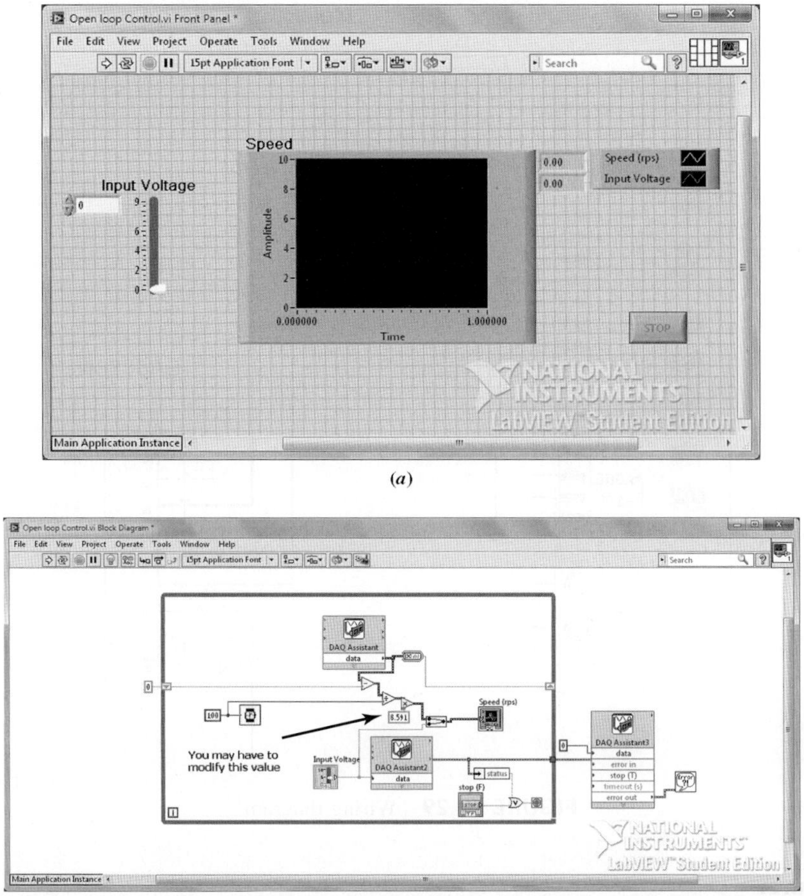


(*a*)

(*b*)

**FIGURE P4.28** Open loop control.vi: **a.** Front Panel; **b.** Block Diagram

rising edges of the encoder signal every 100 msec and subtracts that number from the ones accumulated during the previous 100 msec period. The frequency of the encoder signal (in edges/msec) is found by dividing the value of this subtraction by the period (100 msec). See the Block Diagram to understand how this algorithm was implemented.

We use an example to illustrate the calculation of the constant pointed to with the red arrow. If a 9.7:1 gear ratio is used with a 48 CPR encoder, with each revolution the encoder will generate a total of 48 edges in each of the encoder channels. Using one channel only and positive edges, there are 12 positive edges/rev of the motor shaft. The total number of counts (positive edges) generated by each revolution of the external shaft is $9.7 \times 12 = 116.4$ positive edges/rev.

In order to find the rotational speed, the frequency (edges/msec) of the signal is divided by the total number of counts generated by the external shaft, adjusting the units for time from msec to sec: Rotational Speed (rps) $= \text{Freq} \times 1000/(9.7 \times 12) = \text{Freq} \times 8.591$. This value has to be set as illustrated in the Block Diagram.

The DAQ Assistant2 block transmits the voltage from the control slider to the myDAQ and to the motor control chip. The DAQ Assistant3 block makes sure that the output to the chip is zeroed when the VI terminates.

**Hardware:** Connect the myDAQ, the motor, and the motor controller as shown in Figure P4.29.

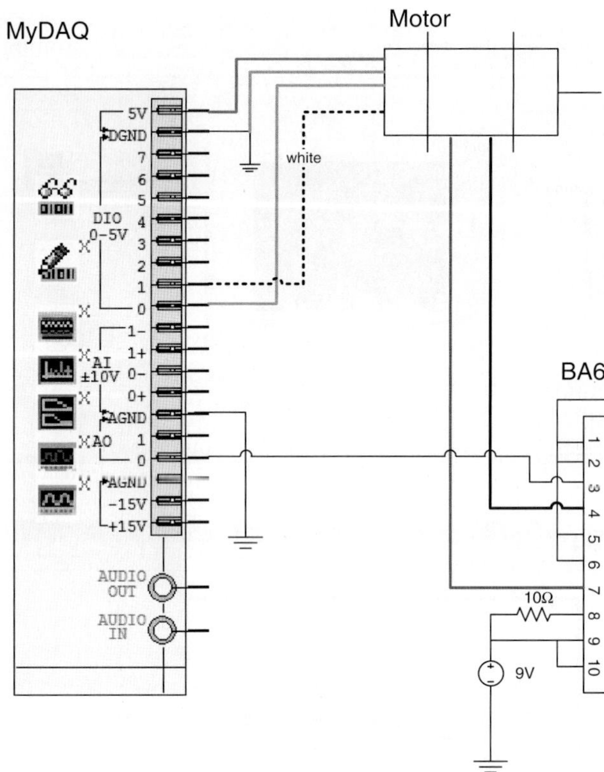


**FIGURE P4.29** Wiring diagram[15]

**Procedure:**

1. Verify the operation of your circuit by running the VI and changing the position of the slider. If everything is correct, the motor speed will vary as the slider's position changes.
2. Verify that you are using the correct scaling factor for your motor by setting your motor to rotate at 0.5 rps. Count the number of rotations in the shaft of the motor over 10 sec using a stopwatch. Repeat by setting the rotational speed at 1 rps. Your measurements must be consistent.
3. Perform the following measurements moving the slider:
   **a.** Increase the voltage starting at zero and record the minimum voltage for the motor to start rotating.
   **b.** Starting the slider at a rotating speed, reduce the voltage until the motor stops. Record this voltage.
   Are these values equal? These values are important and will be used in future labs. Keep them in a safe place so that you don't have to repeat these measurements again.
4. Make a graph where the $x$-axis is the input voltage, and the $y$-axis is the speed in rps. Include the results in Part 3.
5. Draw a functional block diagram of the system (similar to Chapter 1), labeling each of the components in the diagram.
6. The circuit and the VI above allow the motor to rotate in one direction only. Modify the VI and the circuit so that the motor direction and speed can be controlled from the VI.

[15] MyDAQ right slot shown on left is taken from Multisim program module NI myDAQ design and also reproduced in White-Paper 11423, Figure 2. Both Multisim and the White Paper are from National Instruments.

Note that the reference input to the chip can only accept positive voltage values. The motor control data sheet indicates that direction of rotation must be changed by flipping the logical values of Pin 2 and Pin 10 on the motor control chip. However, a careful reading of the data sheet indicates that there must be an instant of time (of unspecified duration) in which both inputs must be False before switching direction. You may want to use a LabVIEW ring to simulate a three-way switch. Use two of the selections of the ring to control the motor's rotation direction. The third selection in the ring should provide low inputs to the motor controller logical inputs to be able to stop the motor before switching direction.

## Experiment 4.6 Transfer Function Identification

**Objective** To identify the transfer function of a motor from voltage input to angular motor speed using myDAQ and LabView.

**Material Required** Computer with LabVIEW Installed; myDAQ; dc brushed gearmotor with Hall sensor quadrature encoder (–10 to +10 V normal operation range); and motor control chip B6886N (or a transistor circuit substitute).

### File Provided at www.wiley.com/college/nise

Plant Identification 2.vi

**Prelab** Answer the following questions:

1. What is the unit-step response of a system with transfer function $G(s) = \dfrac{K}{s\tau + 1}$, where $K$ and $\tau$ are constants $> 0$?
2. Make a hand-sketch of the response of the unit-step response of the system in Part 1.
3. What is the value of the step response of the system in Part 1 when $t = \tau$?
4. Find or derive the expression for the transfer function from voltage to angular speed of an unloaded permanent magnet dc motor. Compare this transfer function to the first-order system in Part 1.

**Lab** Connect the myDAQ, the motor, and the motor controller as shown in Figure P4.30. This setup is identical to the one that was used initially in Experiment 4.5, except that we have connected the two analog input channels to the two analog output channels. This will allow us to use the myDAQ oscilloscope for measurements. If you decide to use an external oscilloscope, these connections are not necessary.

1. Open the Oscilloscope and Plant Identification 2.vi shown in Figures P4.31 and P4.32, respectively. You can also choose to use an external oscilloscope. Use settings similar to the ones shown in Figure P4.31.
2. In your Plant Identification 2.vi choose the value of amplitude and offset shown in Figure P4.32. A LabView error will be generated if the square wave generates negative values as these are not allowed as inputs to the chip. The value of the frequency is irrelevant; you just have to make sure the input is slow enough so that the motor speed reaches steady state as shown in Figure P4.32.
3. Run the Plant Identification 2.vi and the Oscilloscope. Press the Stop button on the Oscilloscope as soon as it shows a full semi-cycle of positive speed, similar to Figure P4.31.
4. Click on the Log button in the Oscilloscope, give the file a name, and save it to disk. Open the file using a spreadsheet program.
5. Note that the response of the system in the Oscilloscope is in all likelihood that of a first-order system, consistent with theoretical expectations. Thus, the transfer function will be of the form: $\dfrac{\Omega(s)}{E_i(s)} = \dfrac{K}{s\tau + 1}$.

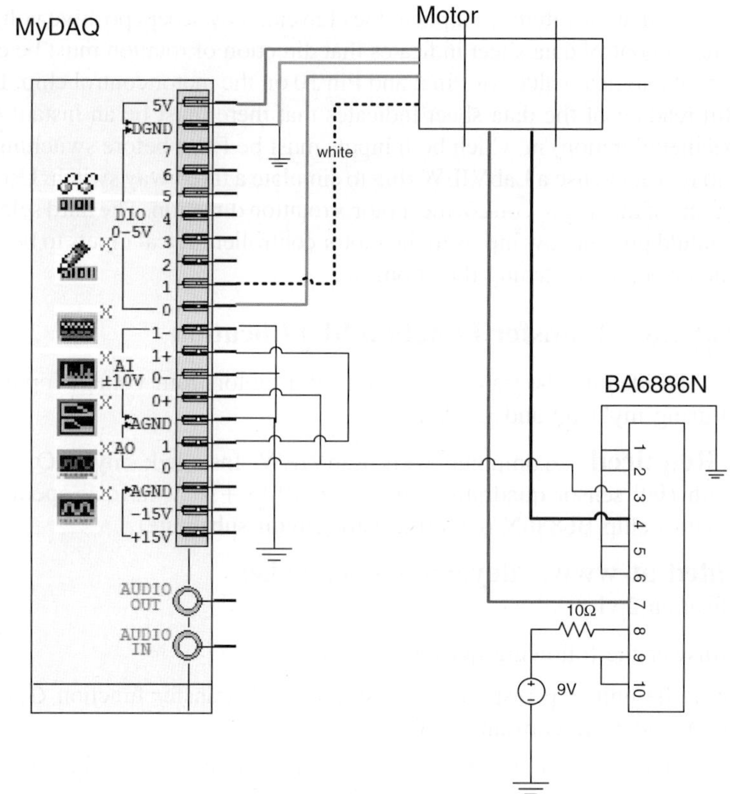


**FIGURE P4.30** Wiring diagram[16]

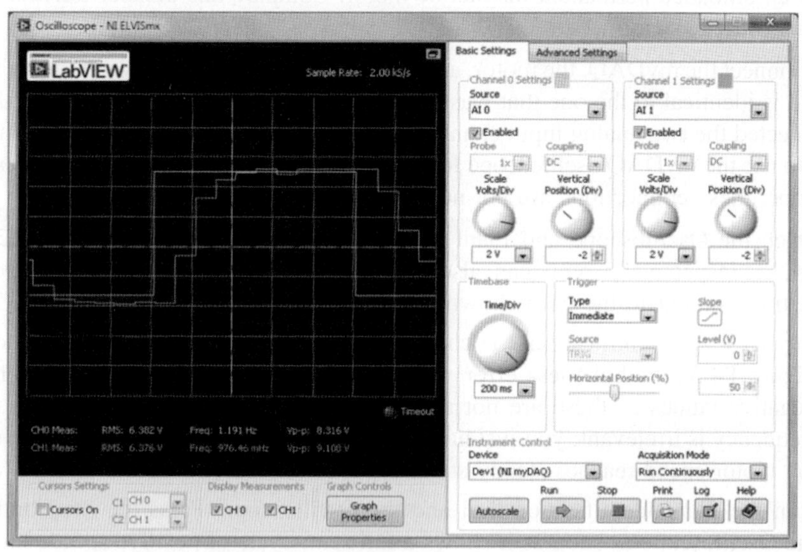




**FIGURE P4.31** LabVIEW Oscilloscope—NI ELVISmx

[16] MyDAQ right slot shown on left is taken from Multisim program module NI myDAQ design and also reproduced in White-Paper 11423, Figure 2. Both Multisim and the White Paper are from National Instruments.

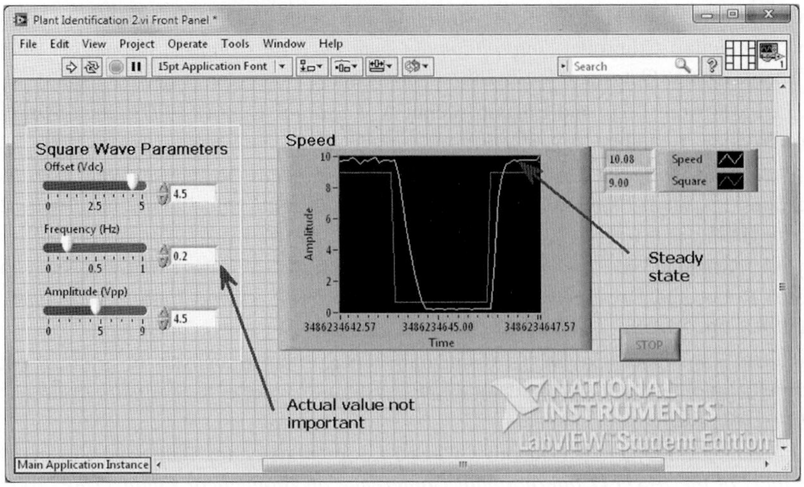


**FIGURE P4.32** Plant Identification 2.vi Front Panel

$K$ can readily be found from the Oscilloscope or the Plant Identification 2.vi. In the example shown, $K = \dfrac{9.71}{9} = 1.079$. We will use the spreadsheet data to find the time constant, $\tau$.

6. Use your spreadsheet data to find the time constant. For help on completing this task, go to www.wiley.com/college/nise.
7. Repeat the experiment for input voltages of 2 V, 5 V, and 9 V.

## Postlab

1. Is your system linear? How do you know?
2. If your system is linear for a range of inputs, find a judicious interpolation between the three transfer functions you found in Part 7 of the lab. Write down your final transfer function result and save it for use in subsequent experiments.

# Bibliography

Borovic, B., Liu, A. Q., Popa, D., and Lewis, F. L. Open-loop versus closed-loop control of MEMS devices: choices and issues. *Journal of Micromechanics, Microengineering*, vol. 15, 2005, pp. 1917–1924.

Camacho, E. F., Berenguel, M., Rubio, F. R., and Martinez, D. *Control of Solar Energy Systems*. Springer-Verlag, London, 2012.

Craig, I. K., Xia, X., and Venter, J. W.; Introducing HIV/AIDS Education into the Electrical Engineering Curriculum at the University of Pretoria. *IEEE Transactions on Education*, vol. 47, no. 1, February 2004, pp. 65–73.

DiBona, G. F. Physiology in perspective: The Wisdom of the Body. Neural control of the kidney. *American Journal of Physiology–Regulatory, Integrative and Comparative Physiology*, vol. 289, 2005, pp. R633–R641.

Dong, F., Heinemann, W., and Kasper, R. Development of a Row Guidance System for an Autonomous Robot for White Asparagus Harvesting. *Computers and Electronics in Agriculture*, vol. 79, pp. 216–225, 2011.

Dorf, R. C. *Introduction to Electric Circuits*, 2d ed. Wiley, New York, 1993.

Elarafi, M. G. M. K., and Hisham, S. B. Modeling and Control of pH Neutralization Using Neural Network Predictive Controller. *International Conference on Control, Automation and Systems 2008,* Seoul, Korea. Oct. pp. 14–17, 2008.

Franklin, G. F., Powell, J. D., and Emami-Naeini, A. *Feedback Control of Dynamic Systems*, 2d ed. Addison-Wesley, Reading, MA, 1991.

Glantz, A. S., and Tyberg, V. J. Determination of Frequency Response from Step Response: Application to Fluid-Filled Catheters. *American Journal of Physiology*, vol. 2, 1979, pp. H376–H378.

Good, M. C., Sweet, L. M., and Strobel, K. L. Dynamic Models for Control System Design of Integrated Robot and Drive Systems. *Journal of Dynamic Systems, Measurement, and Control*, March 1985, pp. 53–59.

Gozde, H., and Taplamacioglu, M. C. Comparative performance analysis of artificial bee colony algorithm for automatic voltage regulator (AVR) system. *Journal of the Franklin Institute*, vol. 348, pp. 1927–1946, 2011.

Ionescu, C., and De Keyser, R. Adaptive Closed-Loop Strategy for Paralyzed Skeletal Muscles. *Proceedings of the IASTED International Conference on Biomedical Engineering*, 2005.

Jiayu, K., Mengxiao, W., Linan, M., and Zhongjun, X. Cascade Control of pH in an Anaerobic Waste Water Treatment System, *3d International Conference on Bioinformatics and Biomedical Engineering*, 2009.

Johnson, H. et al. *Unmanned Free-Swimming Submersible (UFSS) System Description*. NRL Memorandum Report 4393. Naval Research Laboratory, Washington, D.C. 1980.

Kuo, B. C. *Automatic Control Systems*, 5th ed. Prentice Hall, Upper Saddle River, NJ, 1987.

Kuo, C-F. J., Tsai, C-C., and Tu, H-M. Carriage Speed Control of a Cross-lapper System for Nonwoven Web Quality. *Fibers and Polymers*, vol. 9, no. 4, 2008, pp. 495–502.

Mallavarapu, K., Newbury, K., and Leo, D. J. Feedback Control of the Bending Response of Ionic Polymer-Metal Composite Actuators. *Proceedings of the SPIE*, vol. 4329, 2001, pp. 301–310.

McKerrow, P. J. *Introduction to Robotics.* Addison-Wesley, Singapore, 1991.

McRuer, D., Ashkenas, I., and Graham, D. *Aircraft Dynamics and Automatic Control.* Princeton University Press, 1973.

Nakamura, M. et al. Transient Response of Remnant Atrial Heart Rate to Step Changes in Total Artificial Heart Output. *Journal of Artificial Organs*, vol. 5, 2002, pp. 6–12.

Nashner, L. M., and Wolfson, P. Influence of Head Position and Proprioceptive Cues on Short Latency Postural Reflexes Evoked by Galvanic Stimulation of the Human Labyrinth. *Brain Research*, vol. 67, 1974, pp. 255–268.

Ogata, K. *Modern Control Engineering*, 2d ed. Prentice Hall, Upper Saddle River, NJ, 1990.

Philips, C. L., and Nagle, H. T. *Digital Control Systems Analysis and Design*. Prentice Hall, Upper Saddle River, NJ, 1984.

Prasad, L., Tyagi, B., and Gupta, H. Modelling & Simulation for Optimal Control of Nonlinear Inverted Pendulum Dynamical System using PID Controller & LQR. *IEEE Computer Society Sixth Asia Modeling Symposium*, 2012, pp. 138–143.

Ren, Z., and Zhu, G. G. Modeling and Control of an Electric Variable Valve Timing System for SI and HCCI Combustion Mode Transition. *American Control Conference,* San Francisco, CA, 2011, pp. 979–984.

Ringdorfer M., and Horn M. Development of a Wheel Slip Actuator Controller for Electric Vehicles using Energy Recuperation and Hydraulic Brake Control. *2011 IEEE International Conference on Control Applications (CCA),* Denver, CO, September 28–30, 2011, pp. 313–318.

Salapaka, S., Sebastian, A., Cleveland, J. P., and Salapaka, M. V. High Bandwidth Nano-Positioner: A Robust Control Approach. *Review of Scientific Instruments*, vol. 73, No. 9, 2002, pp. 3232–3241.

Sawusch, M. R., and Summers, T. A. *1001 Things to Do with Your Macintosh*. TAB Books, Blue Ridge Summit, PA, 1984.

Stefani, R. T. *Modeling Human Response Characteristics*. COED Application Note No. 33. Computers in Education Division of ASEE, 1973.

Sun, M., and Xiong, Y. Dynamic Flight Stability of a Hovering Bumblebee. *Journal of Experimental Biology*, vol. 208, 2005, pp. 447–459.

Thomsen, S., Hoffmann, N., and Fuchs, F. W. PI Control, PI-Based State Space Control, and Model-Based Predictive Control for Drive Systems With Elastically Coupled Loads—A Comparative Study. *IEEE Transactions on Industrial Electronics*, vol. 58, no. 8, August 2011, pp. 3647–3657.

Timothy, L. K., and Bona, B. E. *State Space Analysis: An Introduction*. McGraw-Hill, New York, 1968.

Van Dijk, E., Spruijt, J. N., O'Sullivan, D. M., and Klaasens, J. B. PWM-Switch Modeling of DC-DC Converters. *IEEE Transactions on Power Electronics*, vol. 10, 1995, pp. 659–665.

Xue, D., and Chen, Y. D. Sub-Optimum $H_2$ Rational Approximations to Fractional Order Linear Systems. *Proceedings of IDET/CIE 2005*. ASME 2005 International Design Engineering Technical Conferences & Computers and Information in Engineering Conference, Long Beach. CA, 2005. pp. 1–10.

Zedka, M., Prochazka, A., Knight, B., Gillard, D., and Gauthier, M. Voluntary and Reflex Control of Human Back Muscles During Induced Pain. *Journal of Physiology*, vol. 520, 1999, pp. 591–604.

Zhou, B. H., Baratta, R. V., Solomonow, M., and D'Ambrosia, R. D. The Dynamic Response of the Cat Ankle Joint During Load-Moving Contractions. *IEEE Transactions on Biomedical Engineering*, vol. 42, no. 4, 1995, pp. 386–393.

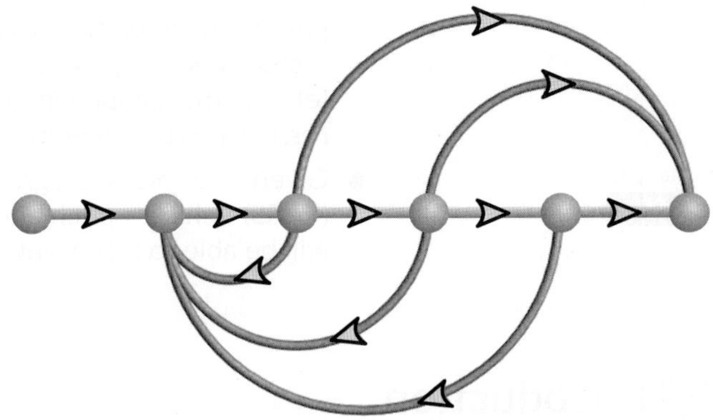

# 5

# Reduction of Multiple Subsystems

## Chapter Learning Outcomes

After completing this chapter the student will be able to:

- Reduce a block diagram of multiple subsystems to a single block representing the transfer function from input to output (Sections 5.1–5.2)
- Analyze and design transient response for a system consisting of multiple subsystems (Section 5.3)
- Convert block diagrams to signal-flow diagrams (Section 5.4)
- Find the transfer function of multiple subsystems using Mason's rule (Section 5.5)
- Represent state equations as signal-flow graphs (Section 5.6) — State Space **SS**
- Represent multiple subsystems in state space in cascade, parallel, controller canonical, and observer canonical forms (Section 5.7) — State Space **SS**
- Perform transformations between similar systems using transformation matrices; and diagonalize a system matrix (Section 5.8) — State Space **SS**

## ▶ Case Study Learning Outcomes

You will be able to demonstrate your knowledge of the chapter objectives with case studies as follows:

- Given the antenna azimuth position control system shown on the front endpapers, you will be able to (a) find the closed-loop transfer function that represents the system from input to output; (b) find a state-space representation for the closed-loop system; (c) predict, for a simplified system model, the percent overshoot, settling time, and peak time of the closed-loop system for a step input; (d) calculate the step response for the closed-loop system; and (e) for the simplified model, design the system gain to meet a transient response requirement.

State Space
SS

- Given the block diagrams for the Unmanned Free-Swimming Submersible (UFSS) vehicle's pitch and heading control systems on the back endpapers, you will be able to represent each control system in state space.

## 5.1 Introduction

We have been working with individual subsystems represented by a block with its input and output. More complicated systems, however, are represented by the interconnection of many subsystems. Since the response of a single transfer function can be calculated, we want to represent multiple subsystems as a single transfer function. We can then apply the analytical techniques of the previous chapters and obtain transient response information about the entire system.

In this chapter, multiple subsystems are represented in two ways: as block diagrams and as signal-flow graphs. Although neither representation is limited to a particular analysis and design technique, block diagrams are usually used for frequency-domain analysis and design, and signal-flow graphs for state-space analysis.

Signal-flow graphs represent transfer functions as lines, and signals as small-circular nodes. Summing is implicit. To show why it is convenient to use signal-flow graphs for state-space analysis and design, consider Figure 3.10. A graphical representation of a system's transfer function is as simple as Figure 3.10(*a*). However, a graphical representation of a system in state space requires representation of each state variable, as in Figure 3.10(*b*). In that example, a single-block transfer function requires seven blocks and a summing junction to show the state variables explicitly. Thus, signal-flow graphs have advantages over block diagrams, such as Figure 3.10(*b*): They can be drawn more quickly, they are more compact, and they emphasize the state variables.

We will develop techniques to reduce each representation to a single transfer function. Block diagram algebra will be used to reduce block diagrams and Mason's rule to reduce signal-flow graphs. Again, it must be emphasized that these methods are typically used as described. As we shall see, however, either method can be used for frequency-domain or state-space analysis and design.

## 5.2 Block Diagrams

As you already know, a subsystem is represented as a block with an input, an output, and a transfer function. Many systems are composed of multiple subsystems, as in Figure 5.1. When multiple subsystems are interconnected, a few more schematic elements must be added to the block diagram. These new elements are *summing junctions* and *pickoff points*. All component parts of a block diagram for a linear, time-invariant system are shown in Figure 5.2. The characteristic of the summing junction shown in Figure 5.2(*c*) is that the output signal, $C(s)$, is the algebraic sum of the input signals, $R_1(s)$, $R_2(s)$, and $R_3(s)$. The figure shows three inputs, but any number can be present. A pickoff point, as shown in Figure 5.2(*d*), distributes the input signal, $R(s)$, undiminished, to several output points.

We will now examine some common topologies for interconnecting subsystems and derive the single transfer function representation for each of them. These common topologies will form the basis for reducing more complicated systems to a single block.

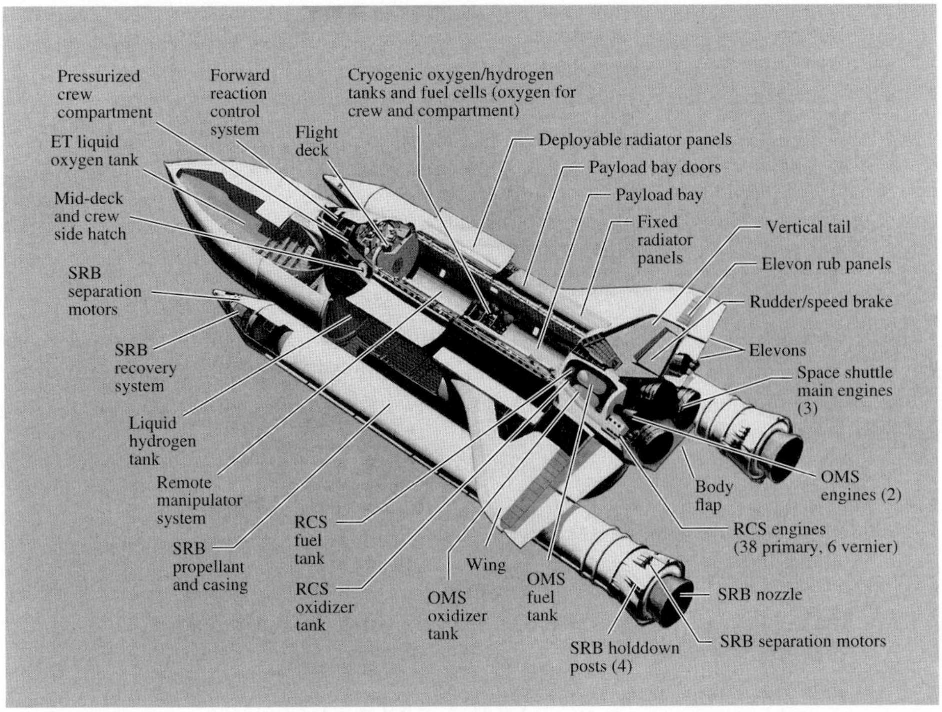


NASA-Houston.

**FIGURE 5.1** The recently retired space shuttle consisted of multiple subsystems. Can you identify those that are control systems or parts of control systems?

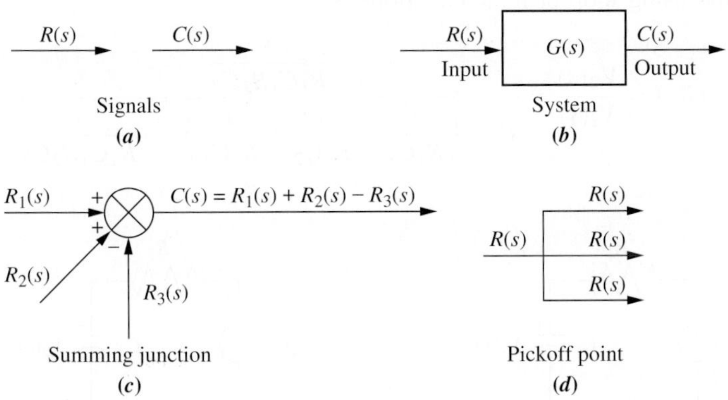


**FIGURE 5.2** Components of a block diagram for a linear, time-invariant system

## Cascade Form

Figure 5.3($a$) shows an example of cascaded subsystems. Intermediate signal values are shown at the output of each subsystem. Each signal is derived from the product of the input times the transfer function. The equivalent transfer function, $G_e(s)$, shown in Figure 5.3($b$), is the output Laplace transform divided by the input Laplace transform from Figure 5.3($a$), or

$$G_e(s) = G_3(s)G_2(s)G_1(s) \tag{5.1}$$

which is the product of the subsystems' transfer functions.

Equation (5.1) was derived under the assumption that interconnected subsystems do not load adjacent subsystems. That is, a subsystem's output remains the same whether or not

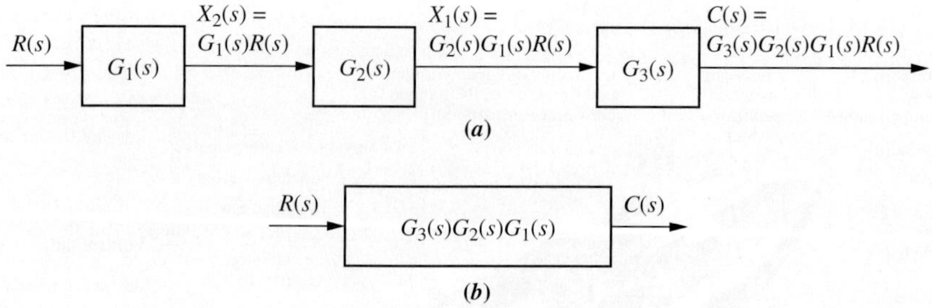


**FIGURE 5.3** **a.** Cascaded subsystems; **b.** equivalent transfer function

the subsequent subsystem is connected. If there is a change in the output, the subsequent subsystem loads the previous subsystem, and the equivalent transfer function is not the product of the individual transfer functions. The network of Figure 5.4(*a*) demonstrates this concept. Its transfer function is

$$G_1(s) = \frac{V_1(s)}{V_i(s)} = \frac{\dfrac{1}{R_1C_1}}{s + \dfrac{1}{R_1C_1}} \tag{5.2}$$

Similarly, the network of Figure 5.4(*b*) has the following transfer function:

$$G_2(s) = \frac{V_2(s)}{V_1(s)} = \frac{\dfrac{1}{R_2C_2}}{s + \dfrac{1}{R_2C_2}} \tag{5.3}$$

If the networks are placed in cascade, as in Figure 5.4(*c*), you can verify that the transfer function found using loop or node equations is

$$G(s) = \frac{V_2(s)}{V_i(s)} = \frac{\dfrac{1}{R_1C_1R_2C_2}}{s^2 + \left(\dfrac{1}{R_1C_1} + \dfrac{1}{R_2C_2} + \dfrac{1}{R_2C_1}\right)s + \dfrac{1}{R_1C_1R_2C_2}} \tag{5.4}$$

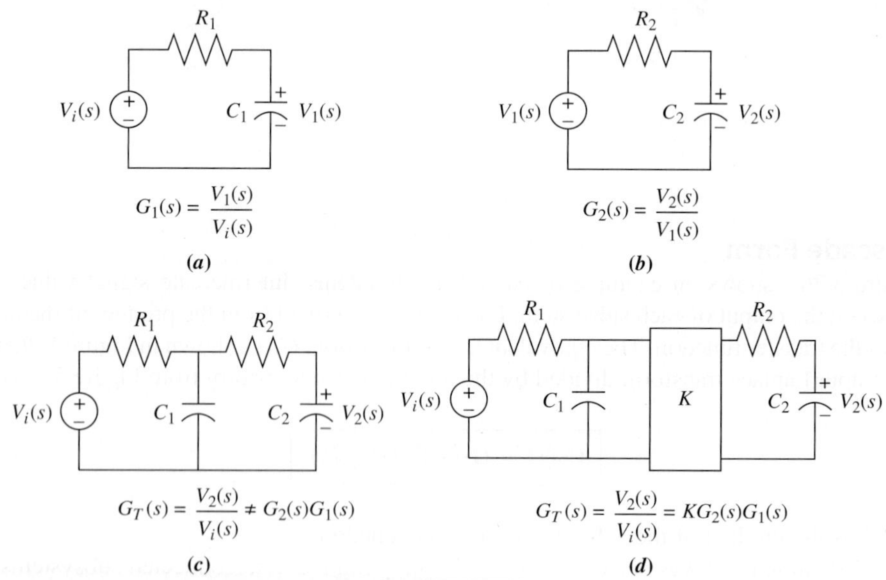


**FIGURE 5.4** Loading in cascaded systems

But, using Eq. (5.1),

$$G(s) = G_2(s)G_1(s) = \frac{\dfrac{1}{R_1C_1R_2C_2}}{s^2 + \left(\dfrac{1}{R_1C_1} + \dfrac{1}{R_2C_2}\right)s + \dfrac{1}{R_1C_1R_2C_2}} \tag{5.5}$$

Equations (5.4) and (5.5) are not the same: Eq. (5.4) has one more term for the coefficient of $s$ in the denominator and is correct.

One way to prevent loading is to use an amplifier between the two networks, as shown in Figure 5.4(*d*). The amplifier has a high-impedance input, so that it does not load the previous network. At the same time it has a low-impedance output, so that it looks like a pure voltage source to the subsequent network. With the amplifier included, the equivalent transfer function is the product of the transfer functions and the gain, $K$, of the amplifier.

## Parallel Form

Figure 5.5 shows an example of parallel subsystems. Again, by writing the output of each subsystem, we can find the equivalent transfer function. Parallel subsystems have a common input and an output formed by the algebraic sum of the outputs from all of the subsystems. The equivalent transfer function, $G_e(s)$, is the output transform divided by the input transform from Figure 5.5(*a*), or

$$G_e(s) = \pm G_1(s) \pm G_2(s) \pm G_3(s) \tag{5.6}$$

which is the algebraic sum of the subsystems' transfer functions; it appears in Figure 5.5(*b*).

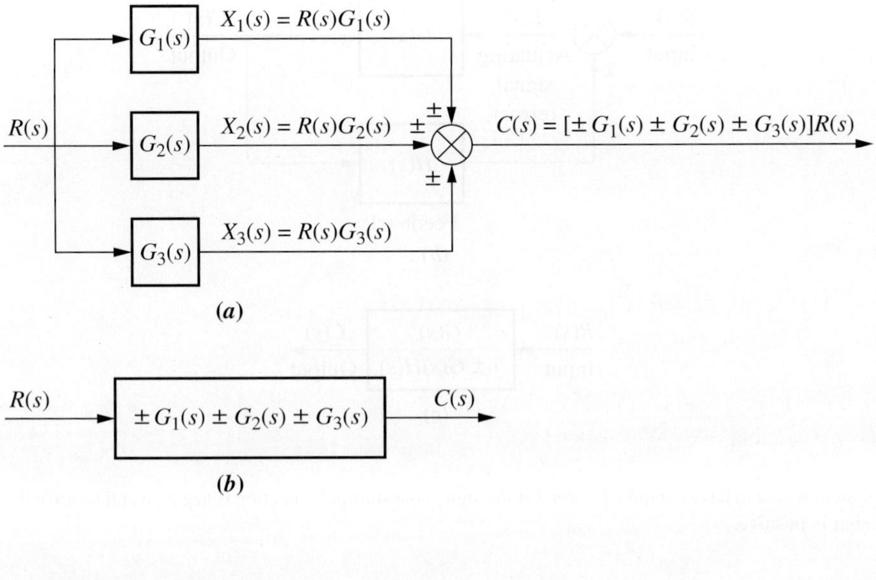


**FIGURE 5.5** **a.** Parallel subsystems; **b.** equivalent transfer function

## Feedback Form

The third topology is the feedback form, which will be seen repeatedly in subsequent chapters. The feedback system forms the basis for our study of control systems engineering. In Chapter 1, we defined open-loop and closed-loop systems and pointed out the advantage of closed-loop, or feedback control, systems over open-loop systems. As we move ahead, we will focus on the analysis and design of feedback systems.

Let us derive the transfer function that represents the system from its input to its output. The typical feedback system, described in detail in Chapter 1, is shown in Figure 5.6(*a*); a simplified model is shown in Figure 5.6(*b*).[1] Directing our attention to the simplified model,

$$E(s) = R(s) \mp C(s)H(s) \tag{5.7}$$

But since $C(s) = E(s)G(s)$,

$$E(s) = \frac{C(s)}{G(s)} \tag{5.8}$$

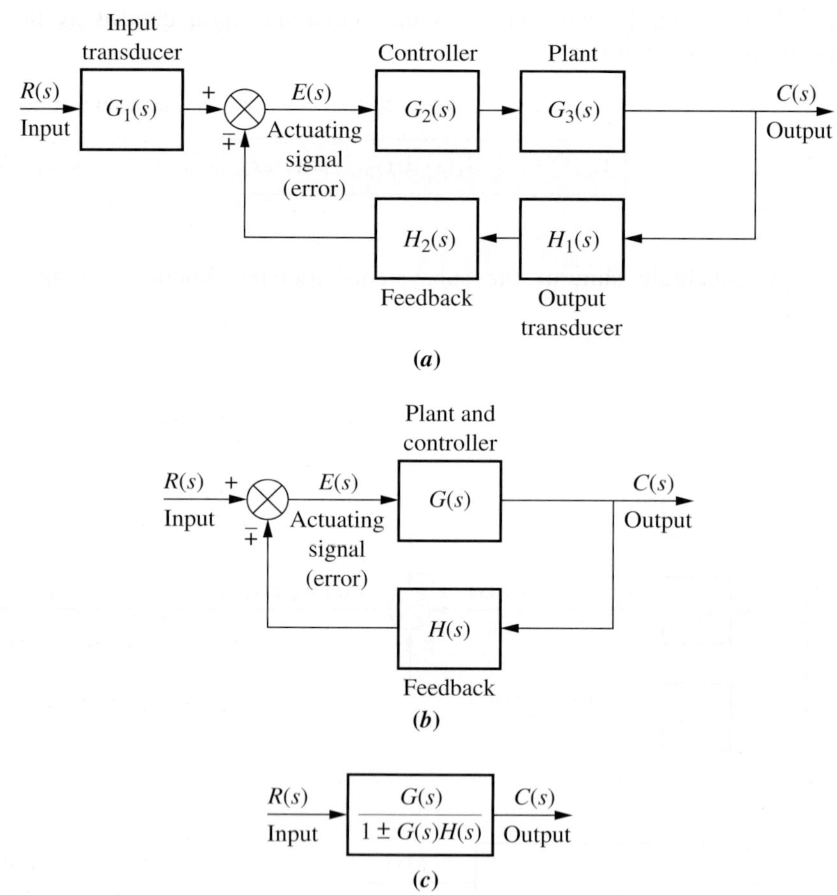


**FIGURE 5.6** **a.** Feedback control system; **b.** simplified model; **c.** equivalent transfer function

[1] The system is said to have *negative feedback* if the sign at the summing junction is negative and *positive feedback* if the sign is positive.

Substituting Eq. (5.8) into Eq. (5.7) and solving for the transfer function, $C(s)/R(s) = G_e(s)$, we obtain the equivalent, or *closed-loop*, transfer function shown in Figure 5.6(*c*),

$$G_e(s) = \frac{G(s)}{1 \pm G(s)H(s)} \tag{5.9}$$

The product, $G(s)H(s)$, in Eq. (5.9) is called the *open-loop transfer function*, or *loop gain*.

So far, we have explored three different configurations for multiple subsystems. For each, we found the equivalent transfer function. Since these three forms are combined into complex arrangements in physical systems, recognizing these topologies is a prerequisite to obtaining the equivalent transfer function of a complex system. In this section, we will reduce complex systems composed of multiple subsystems to single transfer functions.

## Moving Blocks to Create Familiar Forms

Before we begin to reduce block diagrams, it must be explained that the familiar forms (cascade, parallel, and feedback) are not always apparent in a block diagram. For example, in the feedback form, if there is a pickoff point after the summing junction, you cannot use the feedback formula to reduce the feedback system to a single block. That signal disappears, and there is no place to reestablish the pickoff point.

This subsection will discuss basic block moves that can be made in order to establish familiar forms when they almost exist. In particular, it will explain how to move blocks left and right past summing junctions and pickoff points.

Figure 5.7 shows equivalent block diagrams formed when transfer functions are moved left or right past a summing junction, and Figure 5.8 shows equivalent block diagrams formed when transfer functions are moved left or right past a pickoff point. In the diagrams the symbol ≡ means "equivalent to." These equivalences, along with the forms studied earlier in this section, can be used to reduce a block diagram to a single transfer function. In each case of Figures 5.7 and 5.8, the equivalence can be verified by tracing the signals at the input through to the output and recognizing that the output

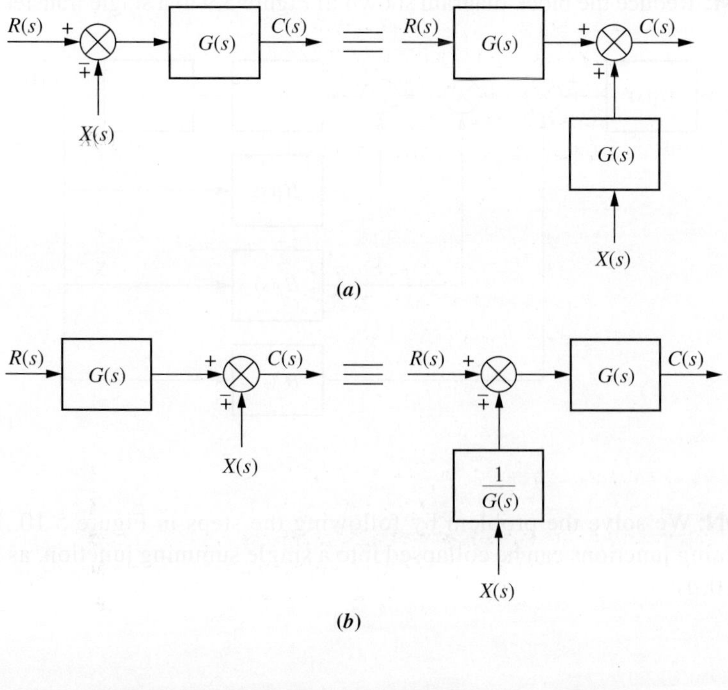


**FIGURE 5.7** Block diagram algebra for summing junctions—equivalent forms for moving a block **a.** to the left past a summing junction; **b.** to the right past a summing junction

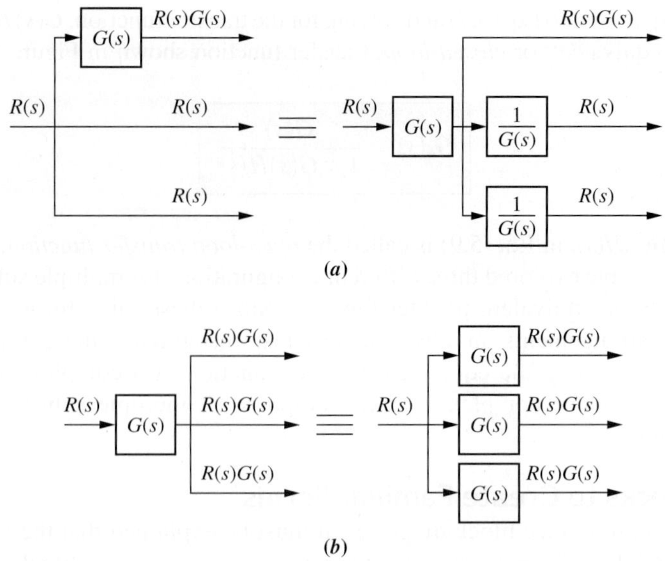


**FIGURE 5.8** Block diagram algebra for pickoff points—equivalent forms for moving a block **a.** to the left past a pickoff point; **b.** to the right past a pickoff point

signals are identical. For example, in Figure 5.7(*a*), signals $R(s)$ and $X(s)$ are multiplied by $G(s)$ before reaching the output. Hence, both block diagrams are equivalent, with $C(s) = R(s)G(s) \mp X(s)G(s)$. In Figure 5.7(*b*), $R(s)$ is multiplied by $G(s)$ before reaching the output, but $X(s)$ is not. Hence, both block diagrams in Figure 5.7(*b*) are equivalent, with $C(s) = R(s)G(s) \mp X(s)$. For pickoff points, similar reasoning yields similar results for the block diagrams of Figure 5.8(*a*) and (*b*).

Let us now put the whole story together with examples of block diagram reduction.

## Example 5.1 Block Diagram Reduction via Familiar Forms

**PROBLEM:** Reduce the block diagram shown in Figure 5.9 to a single transfer function.

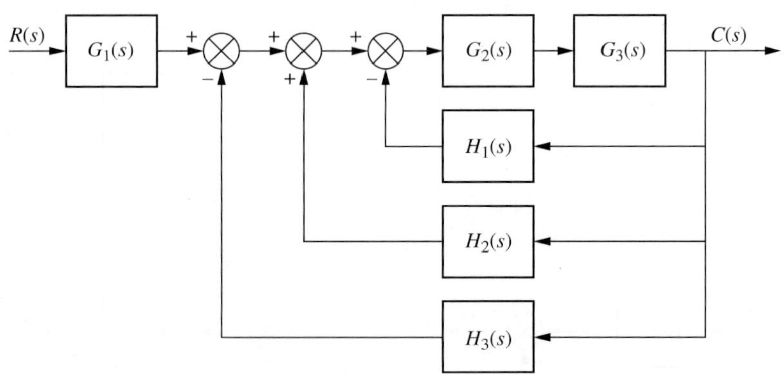


**FIGURE 5.9** Block diagram for Example 5.1

**SOLUTION:** We solve the problem by following the steps in Figure 5.10. First, the three summing junctions can be collapsed into a single summing junction, as shown in Figure 5.10(*a*).

Second, recognize that the three feedback functions, $H_1(s)$, $H_2(s)$, and $H_3(s)$, are connected in parallel. They are fed from a common signal source, and their outputs are summed. The equivalent function is $H_1(s) - H_2(s) + H_3(s)$. Also recognize that $G_2(s)$ and $G_3(s)$ are connected in cascade. Thus, the equivalent transfer function is the product, $G_3(s)G_2(s)$. The results of these steps are shown in Figure 5.10(*b*).

Finally, the feedback system is reduced and multiplied by $G_1(s)$ to yield the equivalent transfer function shown in Figure 5.10(*c*).

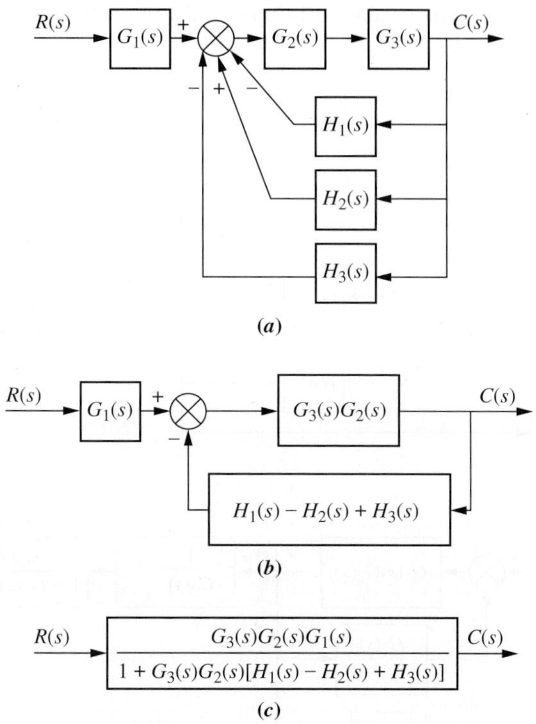


**FIGURE 5.10** Steps in solving Example 5.1: **a.** Collapse summing junctions; **b.** form equivalent cascaded system in the forward path and equivalent parallel system in the feedback path; **c.** form equivalent feedback system and multiply by cascaded $G_1(s)$

## Example 5.2 Block Diagram Reduction by Moving Blocks

**PROBLEM:** Reduce the system shown in Figure 5.11 to a single transfer function.

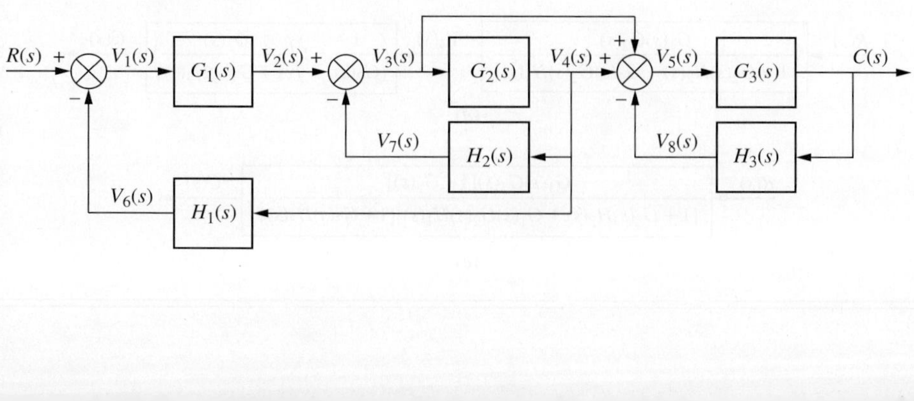


**FIGURE 5.11** Block diagram for Example 5.2

**SOLUTION:** In this example we make use of the equivalent forms shown in Figures 5.7 and 5.8. First, move $G_2(s)$ to the left past the pickoff point to create parallel subsystems, and reduce the feedback system consisting of $G_3(s)$ and $H_3(s)$. This result is shown in Figure 5.12(*a*).

Second, reduce the parallel pair consisting of $1/G_2(s)$ and unity, and push $G_1(s)$ to the right past the summing junction, creating parallel subsystems in the feedback. These results are shown in Figure 5.12(*b*).

Third, collapse the summing junctions, add the two feedback elements together, and combine the last two cascaded blocks. Figure 5.12(*c*) shows these results.

Fourth, use the feedback formula to obtain Figure 5.12(*d*).

Finally, multiply the two cascaded blocks and obtain the final result, shown in Figure 5.12(*e*).

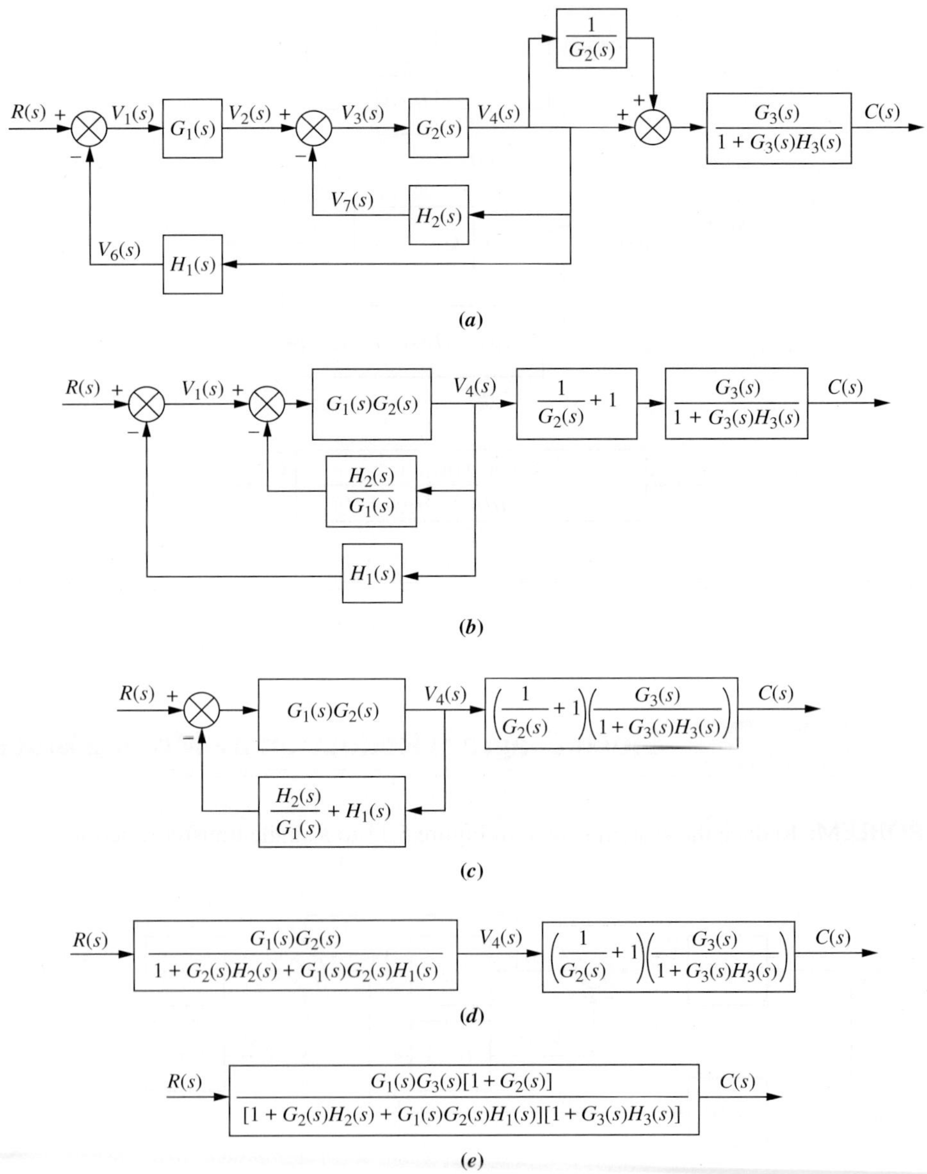


**FIGURE 5.12** Steps in the block diagram reduction for Example 5.2

Students who are using MATLAB should now run ch5p1 in Appendix B to perform block diagram reduction.

MATLAB

ML

## Skill-Assessment Exercise 5.1

**PROBLEM:** Find the equivalent transfer function, $T(s) = C(s)/R(s)$, for the system shown in Figure 5.13.

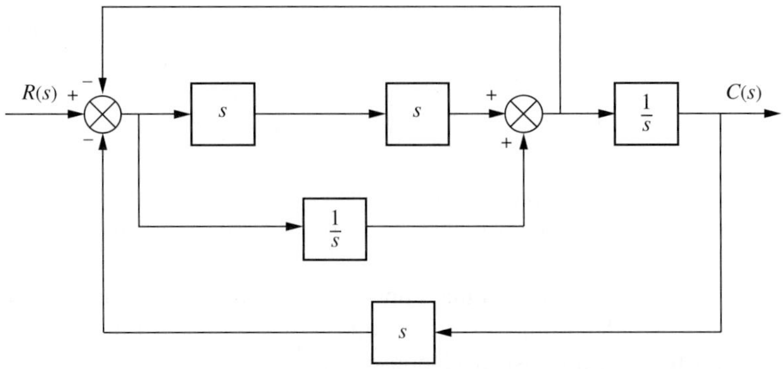


**FIGURE 5.13** Block diagram for Skill-Assessment Exercise 5.1

**ANSWER:**

$$T(s) = \frac{s^3 + 1}{2s^4 + s^2 + 2s}$$

The complete solution is at www.wiley.com/college/nise.

**TryIt 5.1**

Use the following MATLAB and Control System Toolbox statements to find the closed-loop transfer function of the system in Example 5.2 if all $G_i(s) = 1/(s+1)$ and all $H_i(s) = 1/s$.

```
G1=tf(1,[1 1]);
G2=G1;G3=G1;
H1=tf(1,[1 0]);
H2=H1;H3=H1;
System=append...
 (G1,G2,G3,H1,H2,H3);
input=1;output=3;
Q= [1  -4   0   0   0
    2   1  -5   0   0
    3   2   1  -5  -6
    4   2   0   0   0
    5   2   0   0   0
    6   3   0   0   0];
T=connect(System,...
 Q,input,output);
T=tf(T);T=minreal(T)
```

In this section, we examined the equivalence of several block diagram configurations containing signals, systems, summing junctions, and pickoff points. These configurations were the cascade, parallel, and feedback forms. During block diagram reduction, we attempt to produce these easily recognized forms and then reduce the block diagram to a single transfer function. In the next section, we will examine some applications of block diagram reduction.

# 5.3 Analysis and Design of Feedback Systems

An immediate application of the principles of Section 5.2 is the analysis and design of feedback systems that reduce to second-order systems. Percent overshoot, settling time, peak time, and rise time can then be found from the equivalent transfer function.

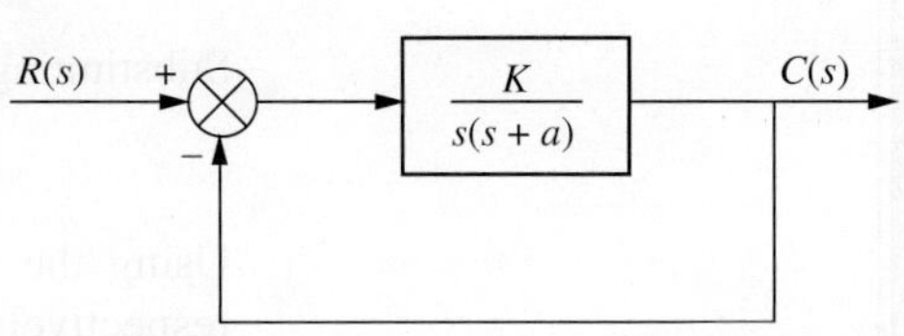


**FIGURE 5.14** Second-order feedback control system

Consider the system shown in Figure 5.14, which can model a control system such as the antenna azimuth position control system. For example, the transfer function, $K/s(s+a)$, can model the amplifiers, motor, load, and gears. From Eq. (5.9), the closed-loop transfer function, $T(s)$, for this system is

$$T(s) = \frac{K}{s^2 + as + K} \tag{5.10}$$

where $K$ models the amplifier gain, that is, the ratio of the output voltage to the input voltage. As $K$ varies, the poles move through the three ranges of operation of a second-order system:

overdamped, critically damped, and underdamped. For example, for $K$ between 0 and $a^2/4$, the poles of the system are real and are located at

$$s_{1,2} = -\frac{a}{2} \pm \frac{\sqrt{a^2 - 4K}}{2} \tag{5.11}$$

As $K$ increases, the poles move along the real axis, and the system remains overdamped until $K = a^2/4$. At that gain, or amplification, both poles are real and equal, and the system is critically damped.

For gains above $a^2/4$, the system is underdamped, with complex poles located at

$$s_{1,2} = -\frac{a}{2} \pm j\frac{\sqrt{4K - a^2}}{2} \tag{5.12}$$

Now as $K$ increases, the real part remains constant and the imaginary part increases. Thus, the peak time decreases and the percent overshoot increases, while the settling time remains constant.

Let us look at two examples that apply the concepts to feedback control systems. In the first example, we determine a system's transient response. In the second example, we design the gain to meet a transient response requirement.

## Example 5.3 Finding Transient Response

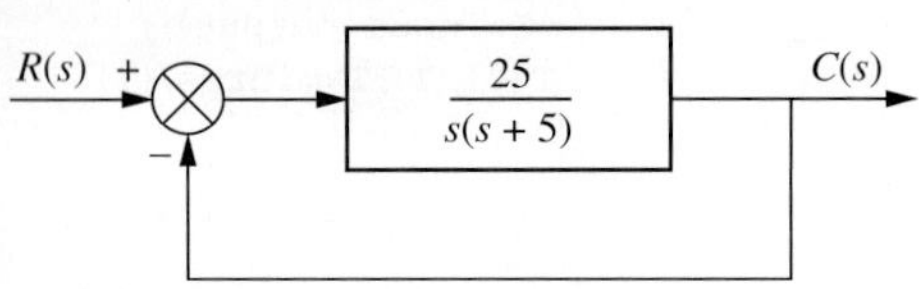


**FIGURE 5.15** Feedback system for Example 5.3

**PROBLEM:** For the system shown in Figure 5.15, find the peak time, percent overshoot, and settling time.

**SOLUTION:** The closed-loop transfer function found from Eq. (5.9) is

$$T(s) = \frac{25}{s^2 + 5s + 25} \tag{5.13}$$

From Eq. (4.18),

$$\omega_n = \sqrt{25} = 5 \tag{5.14}$$

From Eq. (4.21),

$$2\zeta\omega_n = 5 \tag{5.15}$$

Substituting Eq. (5.14) into (5.15) and solving for $\zeta$ yields

$$\zeta = 0.5 \tag{5.16}$$

Using the values for $\zeta$ and $\omega_n$ along with Eqs (4.34), (4.38), and (4.42), we find, respectively,

$$T_p = \frac{\pi}{\omega_n\sqrt{1-\zeta^2}} = 0.726 \text{ second} \tag{5.17}$$

$$\%OS = e^{-\zeta\pi/\sqrt{1-\zeta^2}} \times 100 = 16.303 \tag{5.18}$$

$$T_s = \frac{4}{\zeta\omega_n} = 1.6 \text{ seconds} \tag{5.19}$$

MATLAB
**ML**

Students who are using MATLAB should now run ch5p2 in Appendix B. You will learn how to perform block diagram reduction followed by an evaluation of the closed-loop system's transient response by finding, $T_p$, %$OS$, and $T_s$. Finally, you will learn how to use MATLAB to generate a closed-loop step response. This exercise uses MATLAB to do Example 5.3.

Simulink
**SL**

MATLAB's Simulink provides an alternative method of simulating feedback systems to obtain the time response. Students who are performing the MATLAB exercises and want to explore the added capability of MATLAB's Simulink should now consult Appendix C. Example C.3 includes a discussion about, and an example of, the use of Simulink to simulate feedback systems with nonlinearities.

## Example 5.4 Gain Design for Transient Response

**PROBLEM:** Design the value of gain. $K$, for the feedback control system of Figure 5.16 so that the system will respond with a 10% overshoot.

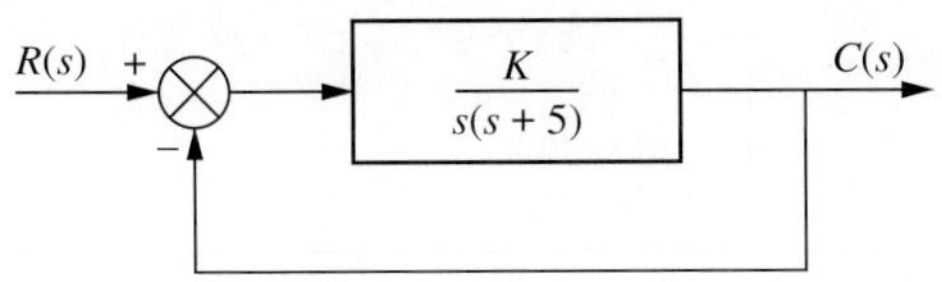


**FIGURE 5.16** Feedback system for Example 5.4

**SOLUTION:** The closed-loop transfer function of the system is

$$T(s) = \frac{K}{s^2 + 5s + K} \tag{5.20}$$

From Eq. (5.20),

$$2\zeta\omega_n = 5 \tag{5.21}$$

and

$$\omega_n = \sqrt{K} \tag{5.22}$$

Thus,

$$\zeta = \frac{5}{2\sqrt{K}} \tag{5.23}$$

Since percent overshoot is a function only of $\zeta$, Eq. (5.23) shows that the percent overshoot is a function of $K$.

A 10% overshoot implies that $\zeta = 0.591$. Substituting this value for the damping ratio into Eq. (5.23) and solving for $K$ yields

$$K = 17.9 \tag{5.24}$$

Although we are able to design for percent overshoot in this problem, we could not have selected settling time as a design criterion because, regardless of the value of $K$, the real parts, $-2.5$, of the poles of Eq. (5.20) remain the same.

## Skill-Assessment Exercise 5.2

**PROBLEM:** For a unity feedback control system with a forward-path transfer function $G(s) = \dfrac{16}{s(s+a)}$, design the value of $a$ to yield a closed-loop step response that has 5% overshoot.

**ANSWER:**

$$a = 5.52$$

The complete solution is at www.wiley.com/college/nise.

**TryIt 5.2**

Use the following MATLAB and Control System Toolbox statements to find $\zeta$, $\omega_n$, $\%OS$, $T_s$, $T_p$, and $T_r$ for the closed-loop unity feedback system described in Skill-Assessment Exercise 5.2. Start with $a = 2$ and try some other values. A step response for the closed-loop system will also be produced.

```
a=2;
numg=16;
deng=poly([0  -a]);
G=tf(numg, deng);
T=feedback(G, 1);
[numt, dent]=...
  tfdata(T,'v');
wn=sqrt(dent(3))
z=dent (2)/(2*wn)
Ts=4/(z*wn)
Tp=pi/(wn*...
  sqrt(1 -z^2))
pos=exp(-z*pi...
  /sqrt(1 -z^2))*100
Tr=(1.76*z^3-...
  0.417*z^2+1.039*...
  z+1)/wn
step(T)
```

## 5.4 Signal-Flow Graphs

Signal-flow graphs are an alternative to block diagrams. Unlike block diagrams, which consist of blocks, signals, summing junctions, and pickoff points, a signal-flow graph consists only of *branches*, which represent systems, and *nodes*, which represent signals. These elements are shown in Figures 5.17(*a*) and (*b*), respectively. A system is represented by a line with an arrow showing the direction of signal flow through the system. Adjacent to the line we write the transfer function. A signal is a node with the signal's name written adjacent to the node.

Figure 5.17(*c*) shows the interconnection of the systems and the signals. Each signal is the sum of signals flowing into it. For example, we see that the signal

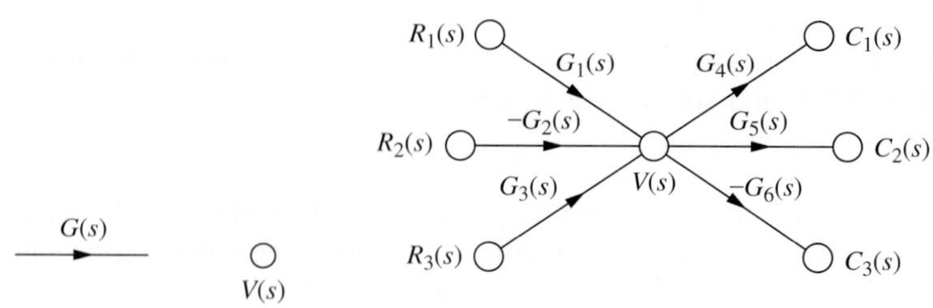


**FIGURE 5.17** Signal-flow graph components: **a.** system; **b.** signal; **c.** interconnection of systems and signals

$V(s) = R_1(s)G_1(s) - R_2(s)G_2(s) + R_3(s)G_3(s)$; the signal $C_2(s) = V(s)G_5(s) = R_1(s)G_1(s)G_5(s) - R_2(s)G_2(s)G_5(s) + R_3(s)G_3(s)G_5(s)$; and the signal $C_3(s) = -V(s)G_6(s) = -R_1(s)G_1(s)G_6(s) + R_2(s)G_2(s)G_6(s) - R_3(s)G_3(s)G_6(s)$. Notice that in summing negative signals we associate the negative sign with the system and not with a summing junction, as in the case of block diagrams.

To show the parallel between block diagrams and signal-flow graphs, we will take some of the block diagram forms from Section 5.2 and convert them to signal-flow graphs in Example 5.5. In each case, we will first convert the signals to nodes and then interconnect the nodes with system branches. In Example 5.6, we will convert an intricate block diagram to a signal-flow graph.

## Example 5.5 Converting Common Block Diagrams to Signal-Flow Graphs

**PROBLEM:** Convert the cascaded, parallel, and feedback forms of the block diagrams shown in Figures 5.3(*a*), 5.5(*a*), and 5.6(*b*), respectively, into signal-flow graphs.

**SOLUTION:** In each case, we start by drawing the signal nodes for that system. Next we interconnect the signal nodes with system branches. The signal nodes for the cascaded, parallel, and feedback forms are shown in Figure 5.18(*a*), (*c*), and (*e*), respectively. The interconnection of the nodes with branches that represent the subsystems is shown in Figure 5.18(*b*), (*d*), and (*f*) for the cascaded, parallel, and feedback forms, respectively.

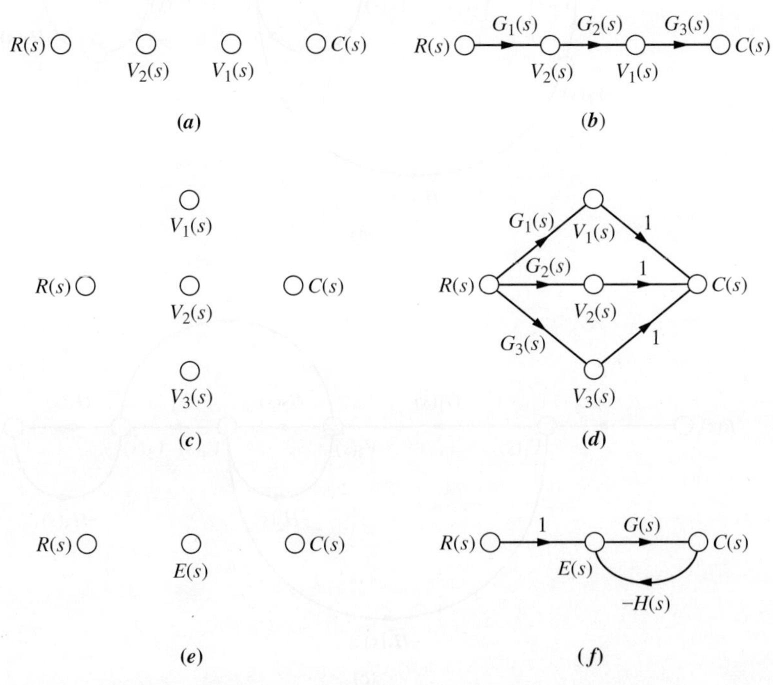


**FIGURE 5.18** Building signal-flow graphs: **a.** cascaded system nodes (from Figure 5.3(*a*)); **b.** cascaded system signal-flow graph; **c.** parallel system nodes (from Figure 5.5(*a*)); **d.** parallel system signal-flow graph; **e.** feedback system nodes (from Figure 5.6(*b*)); **f.** feedback system signal-flow graph

## Example 5.6 Converting a Block Diagram to a Signal-Flow Graph

**PROBLEM:** Convert the block diagram of Figure 5.11 to a signal-flow graph.

**SOLUTION:** Begin by drawing the signal nodes, as shown in Figure 5.19(*a*). Next, interconnect the nodes, showing the direction of signal flow and identifying each transfer function. The result is shown in Figure 5.19(*b*). Notice that the negative signs at the summing junctions of the block diagram are represented by the negative transfer functions of the signal-flow graph. Finally, if desired, simplify the signal-flow graph to the one shown in Figure 5.19(*c*) by eliminating signals that have a single flow in and a single flow out, such as $V_2(s)$, $V_6(s)$, $V_7(s)$, and $V_8(s)$.

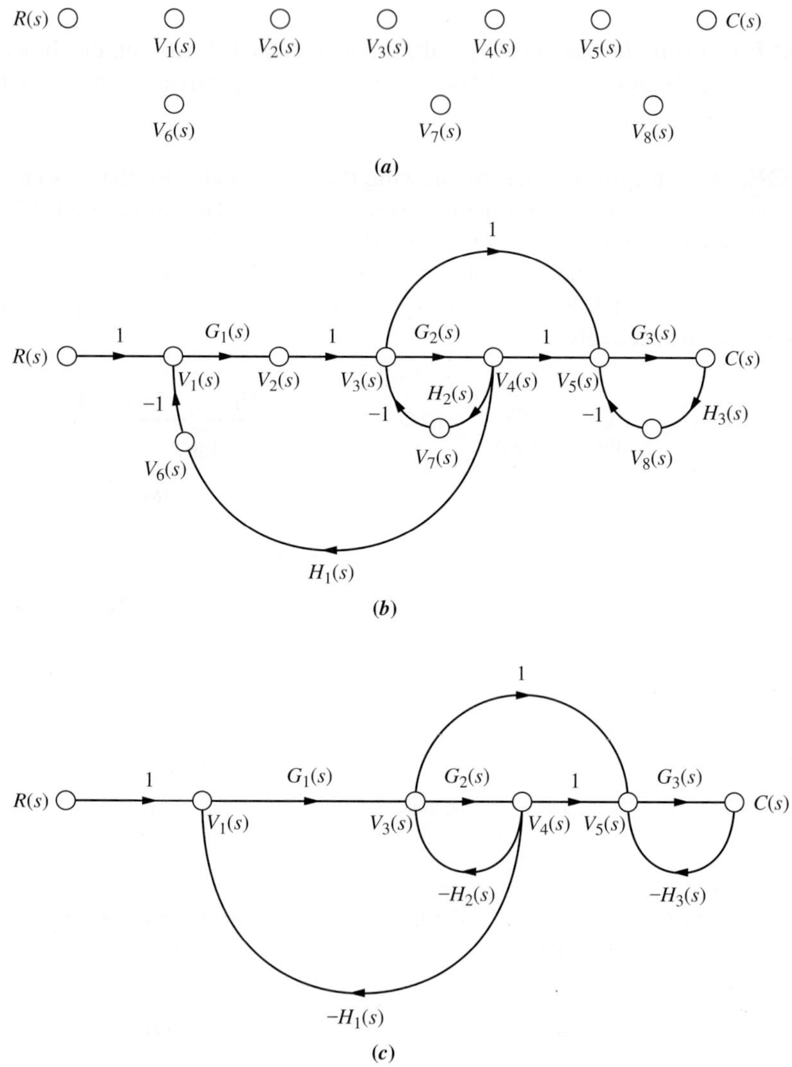


**FIGURE 5.19** Signal-flow graph development: **a.** signal nodes; **b.** signal-flow graph; **c.** simplified signal-flow graph

**Skill-Assessment Exercise 5.3**

**PROBLEM:** Convert the block diagram of Figure 5.13 to a signal-flow graph.

**ANSWER:** The complete solution is at www.wiley.com/college/nise.

# 5.5 Mason's Rule

Earlier in this chapter, we discussed how to reduce block diagrams to single transfer functions. Now we are ready to discuss a technique for reducing signal-flow graphs to single transfer functions that relate the output of a system to its input.

The block diagram reduction technique we studied in Section 5.2 requires successive application of fundamental relationships in order to arrive at the system transfer function. On the other hand, Mason's rule for reducing a signal-flow graph to a single transfer function requires the application of one formula. The formula was derived by S. J. Mason when he related the signal-flow graph to the simultaneous equations that can be written from the graph (*Mason, 1953*).

In general, it can be complicated to implement the formula without making mistakes. Specifically, the existence of what we will later call nontouching loops increases the complexity of the formula. However, many systems do not have non-touching loops. For these systems, you may find Mason's rule easier to use than block diagram reduction.

Mason's formula has several components that must be evaluated. First, we must be sure that the definitions of the components are well understood. Then we must exert care in evaluating the components. To that end, we discuss some basic definitions applicable to signal-flow graphs; then we state Mason's rule and do an example.

## Definitions

*Loop gain*. The product of branch gains found by traversing a path that starts at a node and ends at the same node, following the direction of the signal flow, without passing through any other node more than once. For examples of loop gains, see Figure 5.20. There are four loop gains:

1. $G_2(s)H_1(s)$ (5.25a)
2. $G_4(s)H_2(s)$ (5.25b)
3. $G_4(s)G_5(s)H_3(s)$ (5.25c)
4. $G_4(s)G_6(s)H_3(s)$ (5.25d)

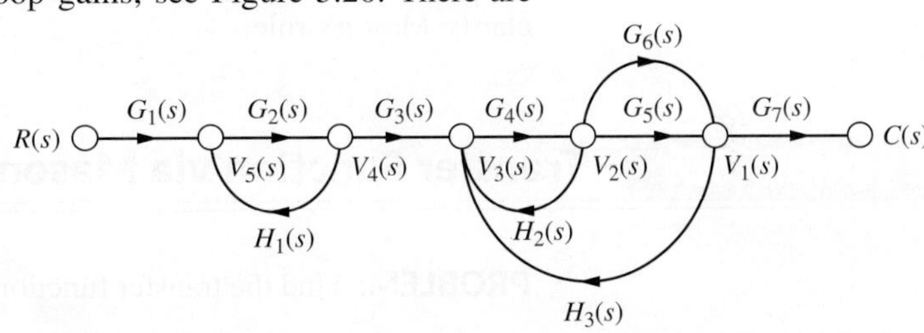


**FIGURE 5.20** Signal-flow graph for demonstrating Mason's rule

*Forward-path gain*. The product of gains found by traversing a path from the input node to the output node of the signal-flow graph in the direction of signal flow. Examples of forward-path gains are also shown in Figure 5.20. There are two forward-path gains:

1. $G_1(s)G_2(s)G_3(s)G_4(s)G_5(s)G_7(s)$ (5.26a)
2. $G_1(s)G_2(s)G_3(s)G_4(s)G_6(s)G_7(s)$ (5.26b)

*Nontouching loops*. Loops that do not have any nodes in common. In Figure 5.20, loop $G_2(s)H_1(s)$ does not touch loops $G_4(s)H_2(s)$, $G_4(s)G_5(s)H_3(s)$, and $G_4(s)G_6(s)H_3(s)$.

*Nontouching-loop gain*. The product of loop gains from nontouching loops taken two, three, four, or more at a time. In Figure 5.20 the product of loop gain $G_2(s)H_1(s)$ and loop gain

$G_4(s)H_2(s)$ is a nontouching-loop gain taken two at a time. In summary, all three of the nontouching-loop gains taken two at a time are

1. $[G_2(s)H_1(s)][G_4(s)H_2(s)]$ (5.27a)
2. $[G_2(s)H_1(s)][G_4(s)G_5(s)H_3(s)]$ (5.27b)
3. $[G_2(s)H_1(s)][G_4(s)G_6(s)H_3(s)]$ (5.27c)

The product of loop gains $[G_4(s)G_5(s)H_3(s)][G_4(s)G_6(s)H_3(s)]$ is not a nontouching-loop gain since these two loops have nodes in common. In our example there are no nontouching-loop gains taken three at a time since three nontouching loops do not exist in the example.

We are now ready to state Mason's rule.

## Mason's Rule

The transfer function, $C(s)/R(s)$, of a system represented by a signal-flow graph is

$$G(s) = \frac{C(s)}{R(s)} = \frac{\sum_k T_k \Delta_k}{\Delta} \tag{5.28}$$

where

$k$ = number of forward paths
$T_k$ = the $k$th forward-path gain
$\Delta$ = 1 − Σ loop gains + Σ nontouching-loop gains taken two at a time − Σ nontouching-loop gains taken three at a time + Σ nontouching-loop gains taken four at a time − ...
$\Delta_k$ = $\Delta$ − Σ loop gain terms in $\Delta$ that touch the $k$th forward path. In other words, $\Delta_k$ is formed by eliminating from $\Delta$ those loop gains that touch the $k$th forward path.

Notice the alternating signs for the components of $\Delta$. The following example will help clarify Mason's rule.

## Example 5.7 Transfer Function via Mason's Rule

**PROBLEM:** Find the transfer function, $C(s)/R(s)$, for the signal-flow graph in Figure 5.21.

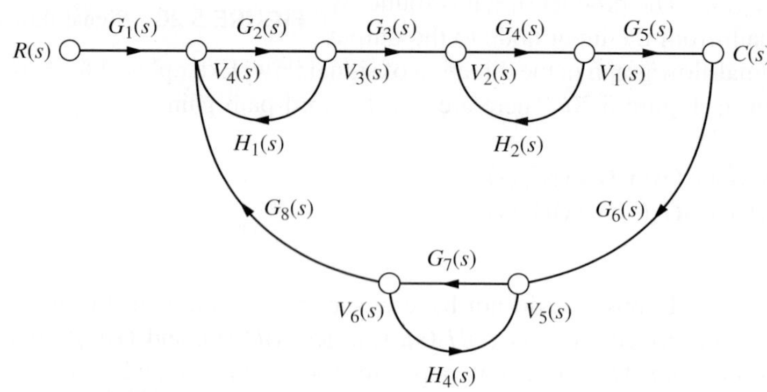


**FIGURE 5.21** Signal-flow graph for Example 5.7

**SOLUTION:** First, identify the *forward-path gains*. In this example there is only one:

$$G_1(s)G_2(s)G_3(s)G_4(s)G_5(s) \tag{5.29}$$

Second, identify the *loop gains*. There are four, as follows:

1. $G_2(s)H_1(s)$ (5.30a)
2. $G_4(s)H_2(s)$ (5.30b)
3. $G_7(s)H_4(s)$ (5.30c)
4. $G_2(s)G_3(s)G_4(s)G_5(s)G_6(s)G_7(s)G_8(s)$ (5.30d)

Third, identify the *nontouching loops taken two at a time*. From Eqs. (5.30) and Figure 5.21, we can see that loop 1 does not touch loop 2, loop 1 does not touch loop 3, and loop 2 does not touch loop 3. Notice that loops 1, 2, and 3 all touch loop 4. Thus, the combinations of nontouching loops taken two at a time are as follows:

$$\text{Loop 1 and loop 2: } G_2(s)H_1(s)G_4(s)H_2(s) \tag{5.31a}$$

$$\text{Loop 1 and loop 3: } G_2(s)H_1(s)G_7(s)H_4(s) \tag{5.31b}$$

$$\text{Loop 2 and loop 3: } G_4(s)H_2(s)G_7(s)H_4(s) \tag{5.31c}$$

Finally, the *nontouching loops taken three at a time* are as follows:

$$\text{Loops 1, 2, and 3: } G_2(s)H_1(s)G_4(s)H_2(s)G_7(s)H_4(s) \tag{5.32}$$

Now, from Eq. (5.28) and its definitions, we form $\Delta$ and $\Delta_k$. Hence,

$$\begin{aligned}\Delta = 1 &-[G_2(s)H_1(s) + G_4(s)H_2(s) + G_7(s)H_4(s) \\ &\qquad + G_2(s)G_3(s)G_4(s)G_5(s)G_6(s)G_7(s)G_8(s)] \\ &+[G_2(s)H_1(s)G_4(s)H_2(s) + G_2(s)H_1(s)G_7(s)H_4(s) \\ &\qquad + G_4(s)H_2(s)G_7(s)H_4(s)] \\ &-[G_2(s)H_1(s)G_4(s)H_2(s)G_7(s)H_4(s)]\end{aligned} \tag{5.33}$$

We form $\Delta_k$ by eliminating from $\Delta$ the loop gains that touch the $k$th forward path:

$$\Delta_1 = 1 - G_7(s)H_4(s) \tag{5.34}$$

Expressions (5.29), (5.33), and (5.34) are now substituted into Eq. (5.28), yielding the transfer function:

$$G(s) = \frac{T_1\Delta_1}{\Delta} = \frac{[G_1(s)G_2(s)G_3(s)G_4(s)G_5(s)][1 - G_7(s)H_4(s)]}{\Delta} \tag{5.35}$$

Since there is only one forward path, $G(s)$ consists of only one term, rather than a sum of terms, each coming from a forward path.

## Skill-Assessment Exercise 5.4

**PROBLEM:** Use Mason's rule to find the transfer function of the signal-flow diagram shown in Figure 5.19(*c*). Notice that this is the same system used in Example 5.2 to find the transfer function via block diagram reduction.

**ANSWER:**

$$T(s) = \frac{G_1(s)G_3(s)[1 + G_2(s)]}{[1 + G_2(s)H_2(s) + G_1(s)G_2(s)H_1(s)][1 + G_3(s)H_3(s)]}$$

The complete solution is at www.wiley.com/college/nise.

## 5.6 Signal-Flow Graphs of State Equations

State Space

SS

In this section, we draw signal-flow graphs from state equations. At first this process will help us visualize state variables. Later we will draw signal-flow graphs and then write alternate representations of a system in state space.

Consider the following state and output equations:

$$\dot{x}_1 = 2x_1 - 5x_2 + 3x_3 + 2r \tag{5.36a}$$

$$\dot{x}_2 = -6x_1 - 2x_2 + 2x_3 + 5r \tag{5.36b}$$

$$\dot{x}_3 = x_1 - 3x_2 - 4x_3 + 7r \tag{5.36c}$$

$$y = -4x_1 + 6x_2 + 9x_3 \tag{5.36d}$$

First, identify three nodes to be the three state variables, $x_1$, $x_2$, and $x_3$; also identify three nodes, placed to the left of each respective state variable, to be the derivatives of the state variables, as in Figure 5.22(*a*). Also identify a node as the input, $r$, and another node as the output, $y$.

Next interconnect the state variables and their derivatives with the defining integration, $1/s$, as shown in Figure 5.22(*b*). Then using Eqs. (5.36), feed to each node the indicated signals. For example, from Eq. (5.36*a*), $\dot{x}_1$ receives $2x_1 - 5x_2 + 3x_3 + 2r$, as shown in Figure 5.22(*c*). Similarly, $\dot{x}_2$ receives $-6x_1 - 2x_2 + 2x_3 + 5r$, as shown in Figure 5.22(*d*), and $\dot{x}_3$ receives $x_1 - 3x_2 - 4x_3 + 7r$, as shown in Figure 5.22(*e*). Finally, using Eq. (5.36*d*), the output, $y$, receives $-4x_1 + 6x_2 + 9x_3$, as shown in Figure 5.19(*f*), the final phase-variable representation, where the state variables are the outputs of the integrators.

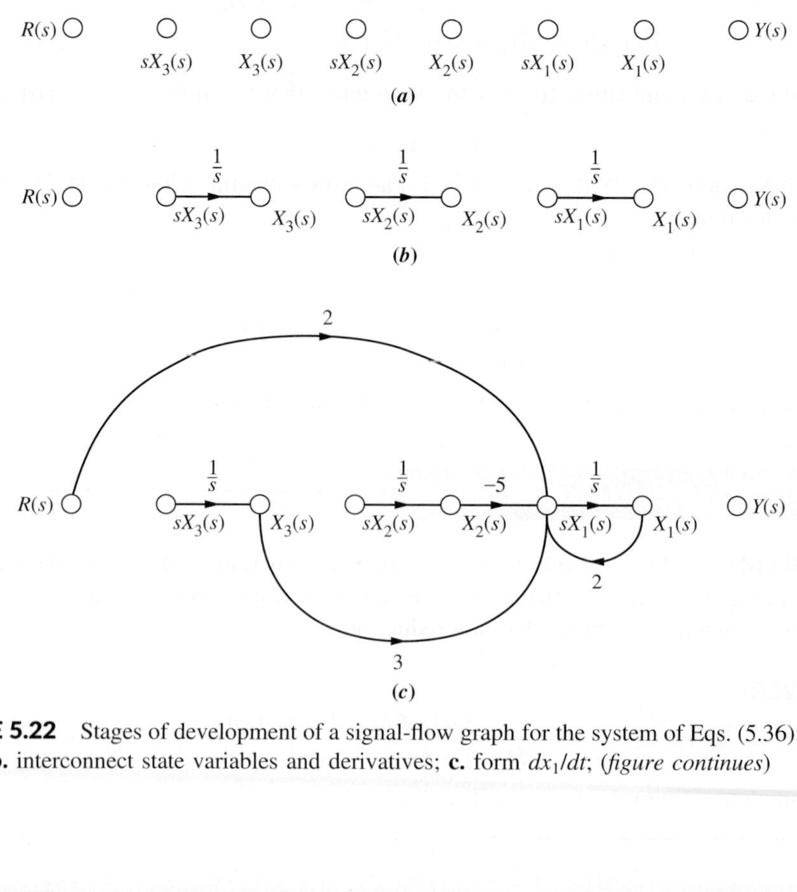


**FIGURE 5.22** Stages of development of a signal-flow graph for the system of Eqs. (5.36): **a.** Place nodes; **b.** interconnect state variables and derivatives; **c.** form $dx_1/dt$; (*figure continues*)

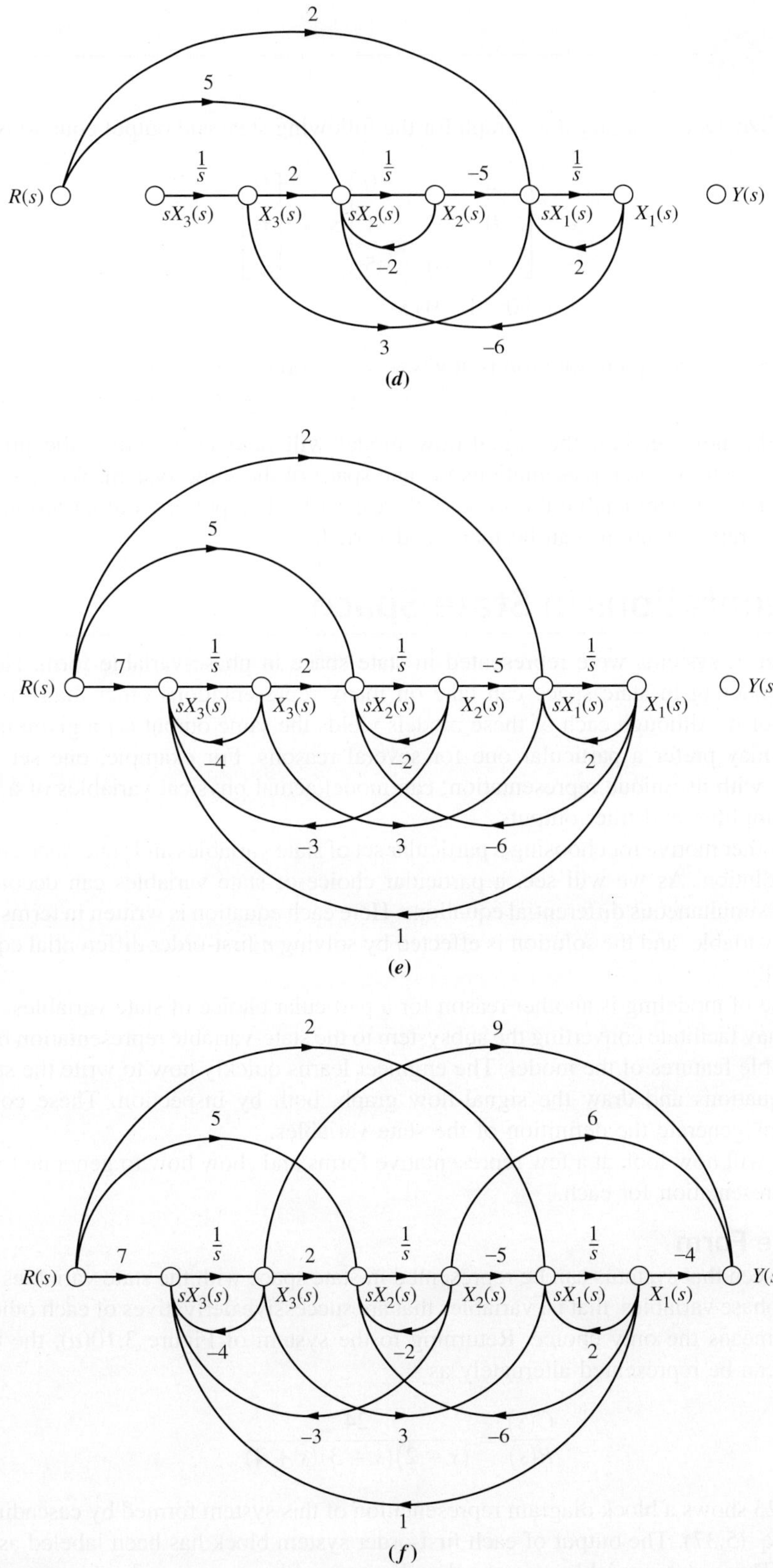


**FIGURE 5.22** (*Continued*) **d.** form $dx_2/dt$; **e.** form $dx_3/dt$; **f.** form output

**Skill-Assessment Exercise 5.5**

**PROBLEM:** Draw a signal-flow graph for the following state and output equations:

$$\dot{\mathbf{x}} = \begin{bmatrix} -2 & 1 & 0 \\ 0 & -3 & 1 \\ -3 & -4 & -5 \end{bmatrix} \mathbf{x} + \begin{bmatrix} 0 \\ 0 \\ 1 \end{bmatrix} r$$

$$y = \begin{bmatrix} 0 & 1 & 0 \end{bmatrix} \mathbf{x}$$

**ANSWER:** The complete solution is at www.wiley.com/college/nise.

In the next section, the signal-flow model will help us visualize the process of determining alternative representations in state space of the same system. We will see that even though a system can be the same with respect to its input and output terminals, the state-space representations can be many and varied.

## 5.7 Alternative Representations in State Space

State Space
SS

In Chapter 3, systems were represented in state space in phase-variable form. However, system modeling in state space can take on many representations other than the phase-variable form. Although each of these models yields the same output for a given input, an engineer may prefer a particular one for several reasons. For example, one set of state variables, with its unique representation, can model actual physical variables of a system, such as amplifier and filter outputs.

Another motive for choosing a particular set of state variables and state-space model is ease of solution. As we will see, a particular choice of state variables can decouple the system of simultaneous differential equations. Here each equation is written in terms of only one state variable, and the solution is effected by solving $n$ first-order differential equations individually.

Ease of modeling is another reason for a particular choice of state variables. Certain choices may facilitate converting the subsystem to the state-variable representation by using recognizable features of the model. The engineer learns quickly how to write the state and output equations and draw the signal-flow graph, both by inspection. These converted subsystems generate the definition of the state variables.

We will now look at a few representative forms and show how to generate the state-space representation for each.

### Cascade Form

We have seen that systems can be represented in state space with the state variables chosen to be the phase variables, that is, variables that are successive derivatives of each other. This is by no means the only choice. Returning to the system of Figure 3.10($a$), the transfer function can be represented alternately as

$$\frac{C(s)}{R(s)} = \frac{24}{(s+2)(s+3)(s+4)} \tag{5.37}$$

Figure 5.23 shows a block diagram representation of this system formed by cascading each term of Eq. (5.37). The output of each first-order system block has been labeled as a state variable. These state variables are not the phase variables.

We now show how the signal-flow graph can be used to obtain a state-space representation of this system. In order to write the state equations with our new set of

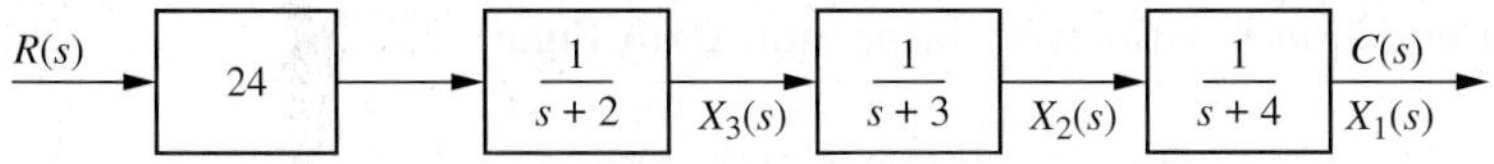


**FIGURE 5.23** Representation of Figure 3.10 system as cascaded first-order systems

state variables, it is helpful to draw a signal-flow graph first, using Figure 5.23 as a guide. The signal flow for each first-order system of Figure 5.23 can be found by transforming each block into an equivalent differential equation. Each first-order block is of the form

$$\frac{C_i(s)}{R_i(s)} = \frac{1}{(s + a_i)} \tag{5.38}$$

Cross-multiplying, we get

$$(s + a_i)C_i(s) = R_i(s) \tag{5.39}$$

After taking the inverse Laplace transform, we have

$$\frac{dc_i(t)}{dt} + a_i c_i(t) = r_i(t) \tag{5.40}$$

Solving for $dc_i(t)/dt$ yields

$$\frac{dc_i(t)}{dt} = -a_i c_i(t) + r_i(t) \tag{5.41}$$

Figure 5.24(*a*) shows the implementation of Eq. (5.41) as a signal-flow graph. Here again, a node was assumed for $c_i(t)$ at the output of an integrator, and its derivative was formed at the input.

Cascading the transfer functions shown in Figure 5.24(*a*), we arrive at the system representation shown in Figure 5.24(*b*).[2] Now write the state equations for the new representation of the system. Remember that the derivative of a state variable will be at the input to each integrator:

$$\dot{x}_1 = -4x_1 + x_2 \tag{5.42a}$$

$$\dot{x}_2 = -3x_2 + x_3 \tag{5.42b}$$

$$\dot{x}_3 = -2x_3 + 24r \tag{5.42c}$$

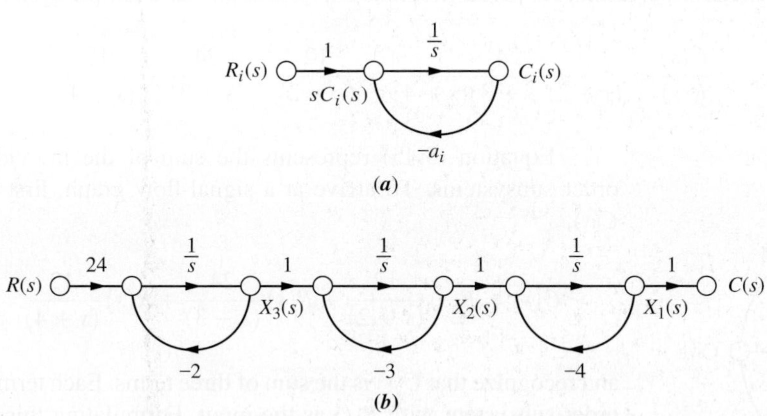


**FIGURE 5.24** **a.** First-order subsystem; **b.** Signal-flow graph for Figure 5.23 system

[2] Note that node $X_3(s)$ and the following node cannot be merged, or else the input to the first integrator would be changed by the feedback from $X_2(s)$, and the signal $X_3(s)$ would be lost. A similar argument can be made for $X_2(s)$ and the following node.

The output equation is written by inspection from Figure 5.24(*b*):

$$y = c(t) = x_1 \tag{5.43}$$

The state-space representation is completed by rewriting Eqs. (5.42) and (5.43) in vector-matrix form:

$$\dot{\mathbf{x}} = \begin{bmatrix} -4 & 1 & 0 \\ 0 & -3 & 1 \\ 0 & 0 & -2 \end{bmatrix} \mathbf{x} + \begin{bmatrix} 0 \\ 0 \\ 24 \end{bmatrix} r \tag{5.44a}$$

$$y = \begin{bmatrix} 1 & 0 & 0 \end{bmatrix} \mathbf{x} \tag{5.44b}$$

Comparing Eqs. (5.44) with Figure 5.24(*b*), you can form a vivid picture of the meaning of some of the components of the state equation. For the following discussion, please refer back to the general form of the state and output equations, Eqs. (3.18) and (3.19).

For example, the **B** matrix is the input matrix since it contains the terms that couple the input, $r(t)$, to the system. In particular, the constant 24 appears in both the signal-flow graph at the input, as shown in Figure 5.24(*b*), and the input matrix in Eqs. (5.44). The **C** matrix is the output matrix since it contains the constant that couples the state variable, $x_1$, to the output, $c(t)$. Finally, the **A** matrix is the system matrix since it contains the terms relative to the internal system itself. In the form of Eqs. (5.44), the system matrix actually contains the system poles along the diagonal.

Compare Eqs. (5.44) to the phase-variable representation in Eqs. (3.59). In that representation, the coefficients of the system's characteristic polynomial appeared along the last row, whereas in our current representation, the roots of the characteristic equation, the system poles, appear along the diagonal.

## Parallel Form

Another form that can be used to represent a system is the parallel form. This form leads to an **A** matrix that is purely diagonal, provided that no system pole is a repeated root of the characteristic equation.

Whereas the previous form was arrived at by cascading the individual first-order subsystems, the parallel form is derived from a partial-fraction expansion of the system transfer function. Performing a partial-fraction expansion on our example system, we get

$$\frac{C(s)}{R(s)} = \frac{24}{(s+2)(s+3)(s+4)} = \frac{12}{(s+2)} - \frac{24}{(s+3)} + \frac{12}{(s+4)} \tag{5.45}$$

Equation (5.45) represents the sum of the individual first-order subsystems. To arrive at a signal-flow graph, first solve for $C(s)$,

$$C(s) = R(s)\frac{12}{(s+2)} - R(s)\frac{24}{(s+3)} + R(s)\frac{12}{(s+4)} \tag{5.46}$$

and recognize that $C(s)$ is the sum of three terms. Each term is a first-order subsystem with $R(s)$ as the input. Formulating this idea as a signal-flow graph renders the representation shown in Figure 5.25.

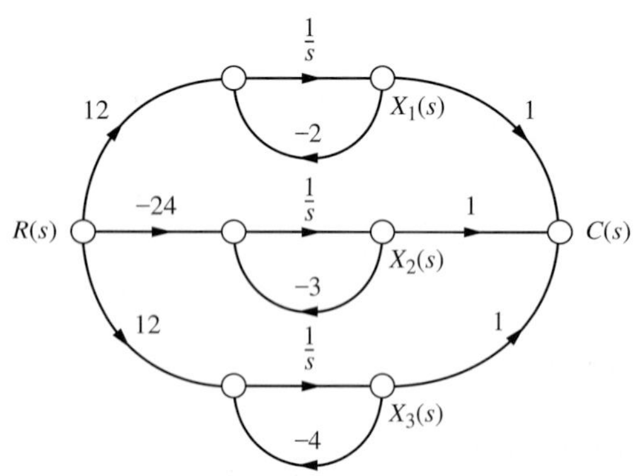

**FIGURE 5.25** Signal-flow representation of Eq. (5.45)

Once again, we use the signal-flow graph as an aid to obtaining the state equations. By inspection the state variables are the outputs of each integrator, where the derivatives of the state variables exist at the integrator inputs. We write the state equations by summing the signals at the integrator inputs:

$$\dot{x}_1 = -2x_1 \qquad\qquad\qquad +12r \tag{5.47a}$$

$$\dot{x}_2 = \qquad\quad -3x_2 \qquad -24r \tag{5.47b}$$

$$\dot{x}_3 = \qquad\qquad\qquad -4x_3 +12r \tag{5.47c}$$

The output equation is found by summing the signals that give $c(t)$:

$$y = c(t) = x_1 + x_2 + x_3 \tag{5.48}$$

In vector-matrix form, Eqs. (5.47) and (5.48) become

$$\dot{\mathbf{x}} = \begin{bmatrix} -2 & 0 & 0 \\ 0 & -3 & 0 \\ 0 & 0 & -4 \end{bmatrix}\mathbf{x} + \begin{bmatrix} 12 \\ -24 \\ 12 \end{bmatrix} r \tag{5.49}$$

and

$$y = [1 \quad 1 \quad 1]\mathbf{x} \tag{5.50}$$

Thus, our third representation of the system of Figure 3.10(*a*) yields a diagonal system matrix. What is the advantage of this representation? Each equation is a first-order differential equation in only one variable. Thus, we would solve these equations independently. The equations are said to be *decoupled*.

MATLAB
**ML**

`Students who are using MATLAB should now run ch5p3 in Appendix B. You will learn how to use MATLAB to convert a transfer function to state space in a specified form. The exercise solves the previous example by representing the transfer function in Eq. (5.45) by the state-space representation in parallel form of Eq. (5.49).`

If the denominator of the transfer function has repeated real roots, the parallel form can still be derived from a partial-fraction expansion. However, the system matrix will not be diagonal. For example, assume the system

$$\frac{C(s)}{R(s)} = \frac{(s+3)}{(s+1)^2(s+2)} \tag{5.51}$$

which can be expanded as partial fractions:

$$\frac{C(s)}{R(s)} = \frac{2}{(s+1)^2} - \frac{1}{(s+1)} + \frac{1}{(s+2)} \tag{5.52}$$

Proceeding as before, the signal-flow graph for Eq. (5.52) is shown in Figure 5.26. The term $-1/(s+1)$ was formed by creating the signal flow from $X_2(s)$ to $C(s)$. Now the state and output equations can be written by inspection from Figure 5.26 as follows:

$$\dot{x}_1 = -x_1 \quad + x_2 \tag{5.53a}$$

$$\dot{x}_2 = \qquad -x_2 \qquad + 2r \tag{5.53b}$$

$$\dot{x}_3 = \qquad\qquad -2x_3 + r \tag{5.53c}$$

$$y = c(t) = x_1 - \frac{1}{2}x_2 \qquad + x_3 \tag{5.53d}$$

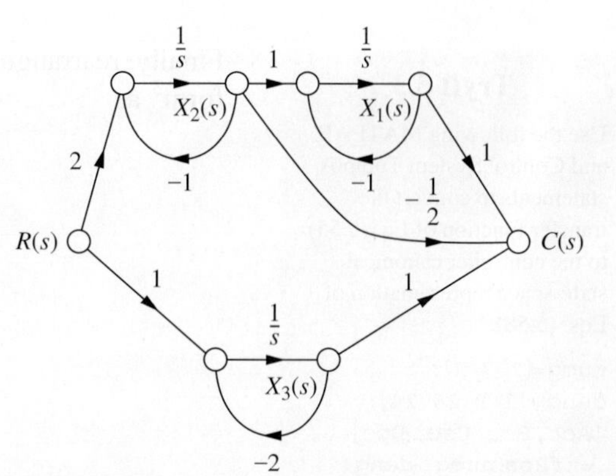


**FIGURE 5.26** Signal-flow representation of Eq. (5.52)

or, in vector-matrix form,

$$\dot{\mathbf{x}} = \begin{bmatrix} -1 & 1 & 0 \\ 0 & -1 & 0 \\ 0 & 0 & -2 \end{bmatrix} \mathbf{x} + \begin{bmatrix} 0 \\ 2 \\ 1 \end{bmatrix} r \tag{5.54a}$$

$$y = \begin{bmatrix} 1 & -\frac{1}{2} & 1 \end{bmatrix} \mathbf{x} \tag{5.54b}$$

This system matrix, although not diagonal, has the system poles along the diagonal. Notice the 1 off the diagonal for the case of the repeated root. The form of the system matrix is known as the *Jordan canonical form.*

## Controller Canonical Form

Another representation that uses phase variables is called the *controller canonical form*, so named for its use in the design of controllers, which is covered in Chapter 12. This form is obtained from the phase-variable form simply by ordering the phase variables in the reverse order. For example, consider the transfer function

$$G(s) = \frac{C(s)}{R(s)} = \frac{s^2 + 7s + 2}{s^3 + 9s^2 + 26s + 24} \tag{5.55}$$

The phase-variable form was derived in Example 3.5 as

$$\begin{bmatrix} \dot{x}_1 \\ \dot{x}_2 \\ \dot{x}_3 \end{bmatrix} = \begin{bmatrix} 0 & 1 & 0 \\ 0 & 0 & 1 \\ -24 & -26 & -9 \end{bmatrix} \begin{bmatrix} x_1 \\ x_2 \\ x_3 \end{bmatrix} + \begin{bmatrix} 0 \\ 0 \\ 1 \end{bmatrix} r \tag{5.56a}$$

$$y = \begin{bmatrix} 2 & 7 & 1 \end{bmatrix} \begin{bmatrix} x_1 \\ x_2 \\ x_3 \end{bmatrix} \tag{5.56b}$$

where $y = c(t)$. Renumbering the phase variables in reverse order yields

$$\begin{bmatrix} \dot{x}_3 \\ \dot{x}_2 \\ \dot{x}_1 \end{bmatrix} = \begin{bmatrix} 0 & 1 & 0 \\ 0 & 0 & 1 \\ -24 & -26 & -9 \end{bmatrix} \begin{bmatrix} x_3 \\ x_2 \\ x_1 \end{bmatrix} + \begin{bmatrix} 0 \\ 0 \\ 1 \end{bmatrix} r \tag{5.57a}$$

$$y = \begin{bmatrix} 2 & 7 & 1 \end{bmatrix} \begin{bmatrix} x_3 \\ x_2 \\ x_1 \end{bmatrix} \tag{5.57b}$$

**TryIt 5.3**

Use the following MATLAB and Control System Toolbox statements to convert the transfer function of Eq. (5.55) to the controller canonical state-space representation of Eqs. (5.58).

```
numg=[1 7 2];
deng=[1 9 26 24];
[Acc,Bcc,Ccc,Dcc]...
 =tf2ss(numg,deng)
```

Finally, rearranging Eqs. (5.57) in ascending numerical order yields the controller canonical form[3] as

$$\begin{bmatrix} \dot{x}_1 \\ \dot{x}_2 \\ \dot{x}_3 \end{bmatrix} = \begin{bmatrix} -9 & -26 & -24 \\ 1 & 0 & 0 \\ 0 & 1 & 0 \end{bmatrix} \begin{bmatrix} x_1 \\ x_2 \\ x_3 \end{bmatrix} + \begin{bmatrix} 1 \\ 0 \\ 0 \end{bmatrix} r \tag{5.58a}$$

$$y = \begin{bmatrix} 1 & 7 & 2 \end{bmatrix} \begin{bmatrix} x_1 \\ x_2 \\ x_3 \end{bmatrix} \tag{5.58b}$$

[3] Students who are using MATLAB to convert from transfer functions to state space using the command **tf2ss** will notice that MATLAB reports the results in controller canonical form.

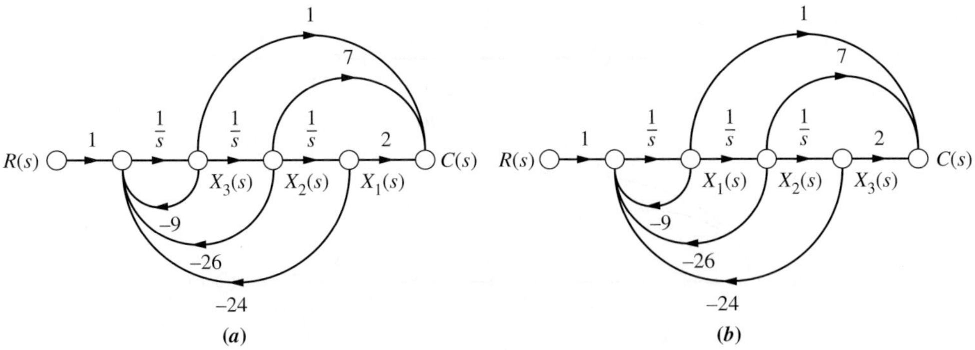


**FIGURE 5.27** Signal-flow graphs for obtaining forms for $G(s) = C(s)/R(s) = (s^2 + 7s + 2)/(s^3 + 9s^2 + 26s + 24)$: **a.** phase-variable form; **b.** controller canonical form

Figure 5.27 shows the steps we have taken on a signal-flow graph. Notice that the controller canonical form is obtained simply by renumbering the phase variables in the opposite order. Equations (5.56) can be obtained from Figure 5.27(*a*), and Eqs. (5.58) from Figure 5.27(*b*).

Notice that the phase-variable form and the controller canonical form contain the coefficients of the characteristic polynomial in the bottom row and in the top row, respectively. System matrices that contain the coefficients of the characteristic polynomial are called *companion matrices* to the characteristic polynomial. The phase-variable and controller canonical forms result in a lower and an upper companion system matrix, respectively. Companion matrices can also have the coefficients of the characteristic polynomial in the left or right column. In the next subsection, we discuss one of these representations.

## Observer Canonical Form

The *observer canonical form*, so named for its use in the design of observers (covered in Chapter 12), is a representation that yields a left companion system matrix. As an example, the system modeled by Eq. (5.55) will be represented in this form. Begin by dividing all terms in the numerator and denominator by the highest power of $s$, $s^3$, and obtain

$$\frac{C(s)}{R(s)} = \frac{\dfrac{1}{s} + \dfrac{7}{s^2} + \dfrac{2}{s^3}}{1 + \dfrac{9}{s} + \dfrac{26}{s^2} + \dfrac{24}{s^3}} \tag{5.59}$$

Cross-multiplying yields

$$\left[\frac{1}{s} + \frac{7}{s^2} + \frac{2}{s^3}\right] R(s) = \left[1 + \frac{9}{s} + \frac{26}{s^2} + \frac{24}{s^3}\right] C(s) \tag{5.60}$$

Combining terms of like powers of integration gives

$$C(s) = \frac{1}{s}[R(s) - 9C(s)] + \frac{1}{s^2}[7R(s) - 26C(s)] + \frac{1}{s^3}[2R(s) - 24C(s)] \tag{5.61}$$

or

$$C(s) = \frac{1}{s}\left[[R(s) - 9C(s)] + \frac{1}{s}\left([7R(s) - 26C(s)] + \frac{1}{s}[2R(s) - 24C(s)]\right)\right] \tag{5.62}$$

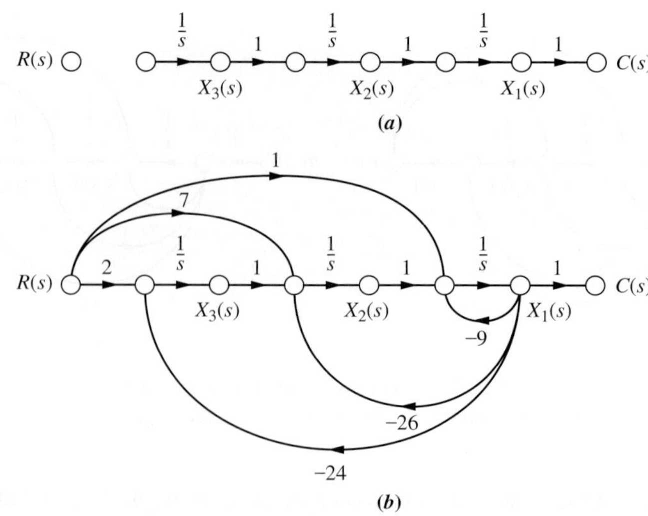


**FIGURE 5.28** Signal-flow graph for observer canonical form variables: **a.** planning; **b.** implementation

Equation (5.61) or (5.62) can be used to draw the signal-flow graph. Start with three integrations, as shown in Figure 5.28(*a*).

Using Eq. (5.61), the first term tells us that output $C(s)$ is formed, in part, by integrating $[R(s) - 9C(s)]$. We thus form $[R(s) - 9C(s)]$ at the input to the integrator closest to the output, $C(s)$, as shown in Figure 5.28(*b*). The second term tells us that the term $[7R(s) - 26C(s)]$ must be integrated twice. Now form $[7R(s) - 26C(s)]$ at the input to the second integrator. Finally, the last term of Eq. (5.61) says $[2R(s) - 24C(s)]$ must be integrated three times. Form $[2R(s) - 24C(s)]$ at the input to the first integrator.

Identifying the state variables as the outputs of the integrators, we write the following state equations:

$$\dot{x}_1 = -9x_1 + x_2 + r \tag{5.63a}$$

$$\dot{x}_2 = -26x_1 + x_3 + 7r \tag{5.63b}$$

$$\dot{x}_3 = -24x_1 + 2r \tag{5.63c}$$

The output equation from Figure 5.28(*b*) is

$$y = c(t) = x_1 \tag{5.64}$$

In vector-matrix form, Eqs. (5.63) and (5.64) become

$$\dot{\mathbf{x}} = \begin{bmatrix} -9 & 1 & 0 \\ -26 & 0 & 1 \\ -24 & 0 & 0 \end{bmatrix} \mathbf{x} + \begin{bmatrix} 1 \\ 7 \\ 2 \end{bmatrix} r \tag{5.65a}$$

$$y = \begin{bmatrix} 1 & 0 & 0 \end{bmatrix} \mathbf{x} \tag{5.65b}$$

**TryIt 5.4**

Use the following MATLAB and Control System Toolbox statements to convert the transfer function of Eq. (5.55) to the observer canonical state-space representation of Eqs. (5.65).

```
numg=[1 7 2];
deng=[1 9 26 24];
[Acc,Bcc,Ccc,Dcc]...
 =tf2ss(numg, deng);
Aoc=transpose(Acc)
Boc=transpose(Ccc)
Coc=transpose(Bcc)
```

Notice that the form of Eqs. (5.65) is similar to the phase-variable form, except that the coefficients of the denominator of the transfer function are in the first column, and the coefficients of the numerator form the input matrix, **B**. Also notice that the observer canonical form has an **A** matrix that is the transpose of the controller canonical form, a **B** vector that is the transpose of the controller canonical form's **C** vector, and a **C** vector that is the transpose of the controller canonical form's **B** vector. We therefore say that these two forms are *duals*. Thus, if a system is described by **A, B,** and **C**, its dual is described by

$\mathbf{A_D} = \mathbf{A}^T$, $\mathbf{B_D} = \mathbf{C}^T$, $\mathbf{C_D} = \mathbf{B}^T$. You can verify the significance of duality by comparing the signal-flow graphs of a system and its dual, Figures 5.27(*b*) and 5.28(*b*), respectively. The signal-flow graph of the dual can be obtained from that of the original by reversing all arrows, changing state variables to their derivatives and vice versa, and interchanging $C(s)$ and $R(s)$, thus reversing the roles of the input and the output.

We conclude this section with an example that demonstrates the application of the previously discussed forms to a feedback control system.

## Example 5.8 State-Space Representation of Feedback Systems

**PROBLEM:** Represent the feedback control system shown in Figure 5.29 in state space. Model the forward transfer function in cascade form.

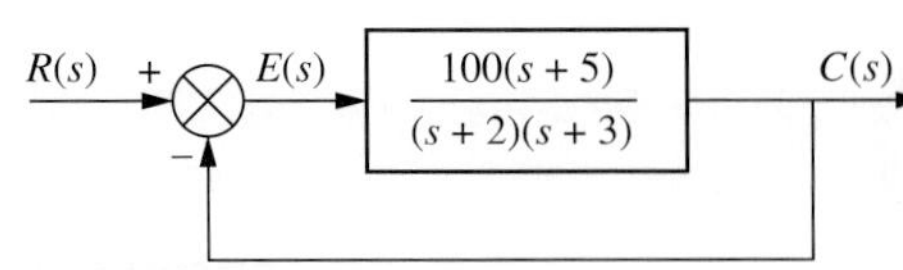


**FIGURE 5.29** Feedback control system for Example 5.8

**SOLUTION:** First we model the forward transfer function in cascade form. The gain of 100, the pole at $-2$, and the pole at $-3$ are shown cascaded in Figure 5.30(*a*). The zero at $-5$ was obtained using the method for implementing zeros for a system represented in phase-variable form, as discussed in Section 3.5.

Next add the feedback and input paths, as shown in Figure 5.30(*b*). Now, by inspection, write the state equations:

$$\dot{x}_1 = -3x_1 + x_2 \quad (5.66a)$$

$$\dot{x}_2 = -2x_2 + 100(r - c) \quad (5.66b)$$

But, from Figure 5.30(*b*),

$$c = 5x_1 + (x_2 - 3x_1) = 2x_1 + x_2 \quad (5.67)$$

Substituting Eq. (5.67) into Eq. (5.66b), we find the state equations for the system:

$$\dot{x}_1 = -3x_1 + x_2 \quad (5.68a)$$

$$\dot{x}_2 = -200x_1 - 102x_2 + 100r \quad (5.68b)$$

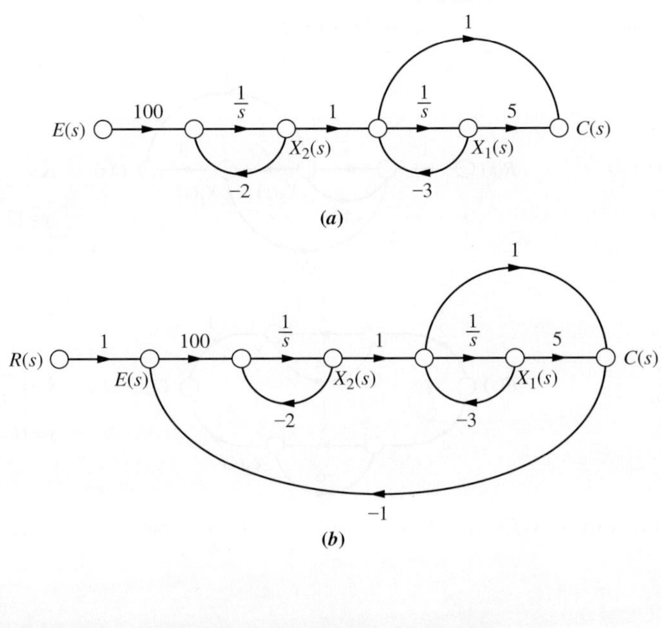


**FIGURE 5.30** Creating a signal-flow graph for the Figure 5.29 system: **a.** forward transfer function; **b.** complete system

The output equation is the same as Eq. (5.67), or

$$y = c(t) = 2x_1 + x_2 \tag{5.69}$$

In vector-matrix form

$$\dot{\mathbf{x}} = \begin{bmatrix} -3 & 1 \\ -200 & -102 \end{bmatrix}\mathbf{x} + \begin{bmatrix} 0 \\ 100 \end{bmatrix} r \tag{5.70a}$$

$$y = \begin{bmatrix} 2 & 1 \end{bmatrix}\mathbf{x} \tag{5.70b}$$

## Skill-Assessment Exercise 5.6

**PROBLEM:** Represent the feedback control system shown in Figure 5.29 in state space. Model the forward transfer function in controller canonical form.

**ANSWER:**

$$\dot{\mathbf{x}} = \begin{bmatrix} -105 & -506 \\ 1 & 0 \end{bmatrix}\mathbf{x} + \begin{bmatrix} 1 \\ 0 \end{bmatrix} r$$

$$y = \begin{bmatrix} 100 & 500 \end{bmatrix}\mathbf{x}$$

The complete solution is at www.wiley.com/college/nise.

In this section, we used transfer functions and signal-flow graphs to represent systems in parallel, cascade, controller canonical, and observer canonical forms, in addition to the phase-variable form. Using the transfer function $C(s)/R(s) = (s+3)/[(s+4)(s+6)]$ as an example, Figure 5.31 compares the aforementioned forms. Notice the duality of the controller and observer canonical forms, as demonstrated by their respective signal-flow graphs and state

| Form | Transfer function | Signal-flow diagram | State equations |
|---|---|---|---|
| Phase variable | $\frac{1}{(s^2+10s+24)} * (s+3)$ | 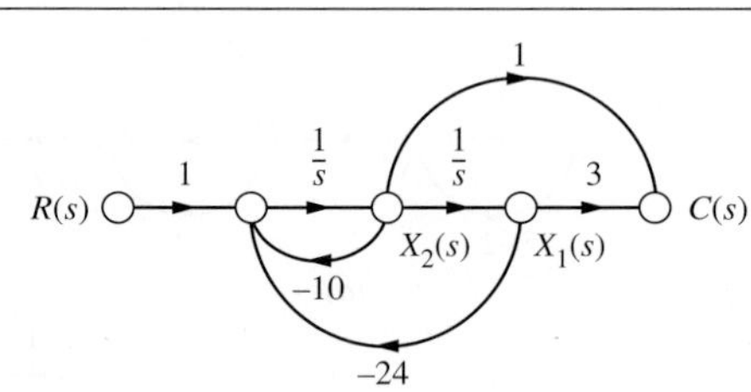 | $\dot{\mathbf{x}} = \begin{bmatrix} 0 & 1 \\ -24 & -10 \end{bmatrix}\mathbf{x} + \begin{bmatrix} 0 \\ 1 \end{bmatrix} r$<br>$y = [3 \;\; 1]\mathbf{x}$ |
| Parallel | $\frac{-1/2}{(s+4)} + \frac{3/2}{s+6}$ | 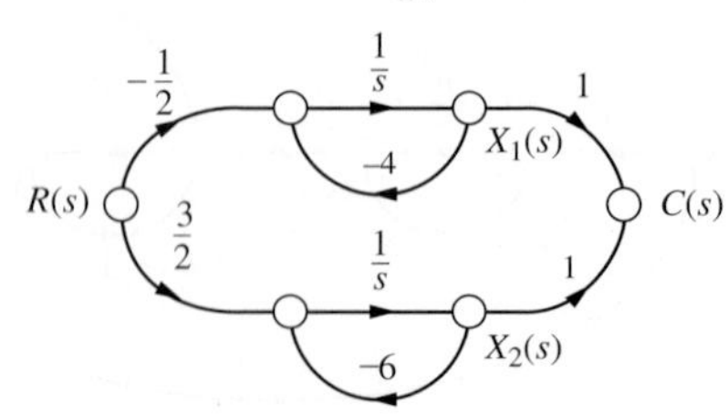 | $\dot{\mathbf{x}} = \begin{bmatrix} -4 & 0 \\ 0 & -6 \end{bmatrix}\mathbf{x} + \begin{bmatrix} -\frac{1}{2} \\ \frac{3}{2} \end{bmatrix} r$<br>$y = [1 \;\; 1]\mathbf{x}$ |

**FIGURE 5.31** State-space forms for $C(s)/R(s) = (s+3)/[(s+4)(s+6)]$. Note: $y = c(t)$ (*figure continues*)

| Form | Transfer function | Signal-flow diagram | State equations |
|---|---|---|---|
| Cascade | $\dfrac{1}{(s+4)} * \dfrac{(s+3)}{(s+6)}$ | 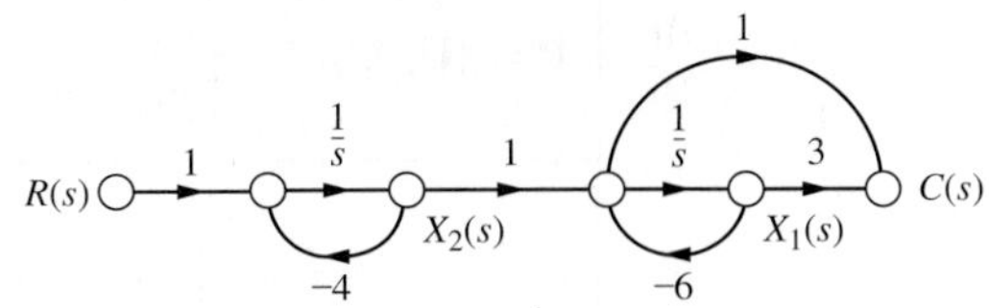 | $\dot{\mathbf{x}} = \begin{bmatrix} -6 & 1 \\ 0 & -4 \end{bmatrix}\mathbf{x} + \begin{bmatrix} 0 \\ 1 \end{bmatrix} r$<br>$y = [-3 \;\; 1]\,\mathbf{x}$ |
| Controller canonical | $\dfrac{1}{(s^2+10s+24)} * (s+3)$ | 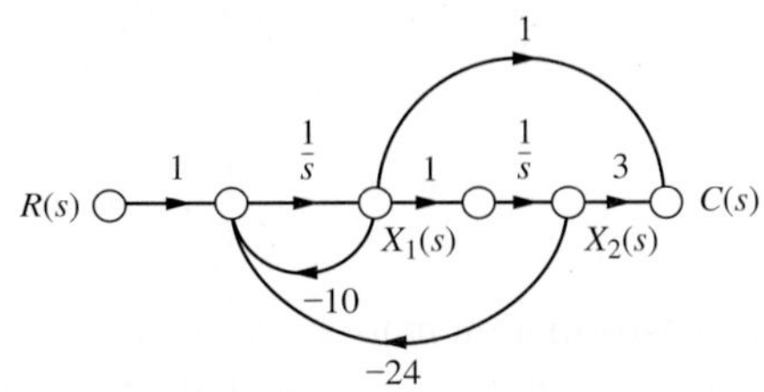 | $\dot{\mathbf{x}} = \begin{bmatrix} -10 & -24 \\ 1 & 0 \end{bmatrix}\mathbf{x} + \begin{bmatrix} 1 \\ 0 \end{bmatrix} r$<br>$y = [1 \;\; 3]\,\mathbf{x}$ |
| Observer canonical | $\dfrac{\frac{1}{s} + \frac{3}{s^2}}{1 + \frac{10}{s} + \frac{24}{s^2}}$ | 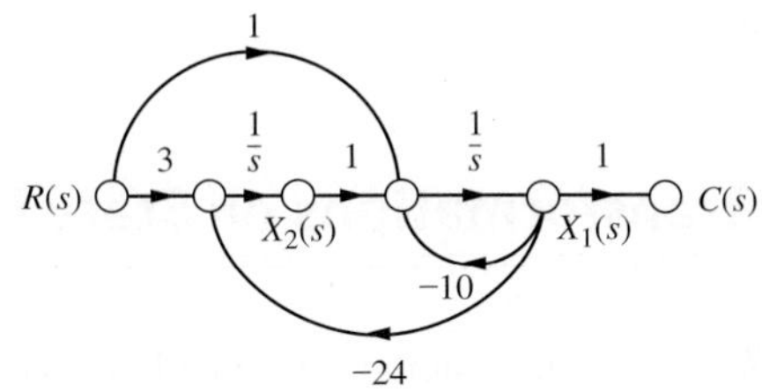 | $\dot{\mathbf{x}} = \begin{bmatrix} -10 & 1 \\ -24 & 0 \end{bmatrix}\mathbf{x} + \begin{bmatrix} 1 \\ 3 \end{bmatrix} r$<br>$y = [1 \;\; 0]\,\mathbf{x}$ |

**FIGURE 5.31** (*Continued*)

equations. In the next section, we will explore the possibility of transforming between representations without using transfer functions and signal-flow graphs.

## 5.8 Similarity Transformations

In Section 5.7, we saw that systems can be represented with different state variables even though the transfer function relating the output to the input remains the same. The various forms of the state equations were found by manipulating the transfer function, drawing a signal-flow graph, and then writing the state equations from the signal-flow graph. These systems are called *similar systems*. Although their state-space representations are different, similar systems have the same transfer function and hence the same poles and eigenvalues.

We can make transformations between similar systems from one set of state equations to another without using the transfer function and signal-flow graphs. The results are presented in this section along with examples. Students who have not broached this subject in the past or who wish to refresh their memories are encouraged to study Appendix L at www.wiley.com/college/nise for the derivation. The result of the derivation states: A system represented in state space as

$$\dot{\mathbf{x}} = \mathbf{A}\mathbf{x} + \mathbf{B}\mathbf{u} \tag{5.71a}$$

$$\mathbf{y} = \mathbf{C}\mathbf{x} + \mathbf{D}\mathbf{u} \tag{5.71b}$$

can be transformed to a similar system,

$$\dot{\mathbf{z}} = \mathbf{P}^{-1}\mathbf{A}\mathbf{P}\mathbf{z} + \mathbf{P}^{-1}\mathbf{B}\mathbf{u} \tag{5.72a}$$

$$\mathbf{y} = \mathbf{C}\mathbf{P}\mathbf{z} + \mathbf{D}\mathbf{u} \tag{5.72b}$$

where, for 2-space,

$$\mathbf{P} = [\mathbf{U}_{\mathbf{z}_1}\,\mathbf{U}_{\mathbf{z}_2}] = \begin{bmatrix} p_{11} & p_{12} \\ p_{21} & p_{22} \end{bmatrix} \tag{5.72c}$$

$$\mathbf{x} = \begin{bmatrix} p_{11} & p_{12} \\ p_{21} & p_{22} \end{bmatrix}\begin{bmatrix} z_1 \\ z_2 \end{bmatrix} = \mathbf{Pz} \tag{5.72d}$$

and

$$\mathbf{z} = \mathbf{P}^{-1}\mathbf{x} \tag{5.72e}$$

Thus, **P** is a transformation matrix whose columns are the coordinates of the basis vectors of the $z_1z_2$ space expressed as linear combinations of the $x_1x_2$ space. Let us look at an example.

## Example 5.9 Similarity Transformations on State Equations

**PROBLEM:** Given the system represented in state space by Eqs. (5.73),

$$\dot{\mathbf{x}} = \begin{bmatrix} 0 & 1 & 0 \\ 0 & 0 & 1 \\ -2 & -5 & -7 \end{bmatrix}\mathbf{x} + \begin{bmatrix} 0 \\ 0 \\ 1 \end{bmatrix}u \tag{5.73a}$$

$$y = [1 \quad 0 \quad 0]\mathbf{x} \tag{5.73b}$$

transform the system to a new set of state variables, **z**, where the new state variables are related to the original state variables, **x**, as follows:

$$z_1 = 2x_1 \tag{5.74a}$$

$$z_2 = 3x_1 + 2x_2 \tag{5.74b}$$

$$z_3 = x_1 + 4x_2 + 5x_3 \tag{5.74c}$$

**SOLUTION:** Expressing Eqs. (5.74) in vector-matrix form,

$$\mathbf{z} = \begin{bmatrix} 2 & 0 & 0 \\ 3 & 2 & 0 \\ 1 & 4 & 5 \end{bmatrix}\mathbf{x} = \mathbf{P}^{-1}\mathbf{x} \tag{5.75}$$

Using Eqs. (5.72) as a guide,

$$\mathbf{P}^{-1}\mathbf{AP} = \begin{bmatrix} 2 & 0 & 0 \\ 3 & 2 & 0 \\ 1 & 4 & 5 \end{bmatrix}\begin{bmatrix} 0 & 1 & 0 \\ 0 & 0 & 1 \\ -2 & -5 & -7 \end{bmatrix}\begin{bmatrix} 0.5 & 0 & 0 \\ -0.75 & 0.5 & 0 \\ 0.5 & -0.4 & 0.2 \end{bmatrix}$$

$$= \begin{bmatrix} -1.5 & 1 & 0 \\ -1.25 & 0.7 & 0.4 \\ -2.5 & 0.4 & -6.2 \end{bmatrix} \tag{5.76}$$

$$\mathbf{P}^{-1}\mathbf{B} = \begin{bmatrix} 2 & 0 & 0 \\ 3 & 2 & 0 \\ 1 & 4 & 5 \end{bmatrix} \begin{bmatrix} 0 \\ 0 \\ 1 \end{bmatrix} = \begin{bmatrix} 0 \\ 0 \\ 5 \end{bmatrix} \tag{5.77}$$

$$\mathbf{CP} = [1 \quad 0 \quad 0] \begin{bmatrix} 0.5 & 0 & 0 \\ -0.75 & 0.5 & 0 \\ 0.5 & -0.4 & 0.2 \end{bmatrix} = [0.5 \quad 0 \quad 0] \tag{5.78}$$

Therefore, the transformed system is

$$\dot{\mathbf{z}} = \begin{bmatrix} -1.5 & 1 & 0 \\ -1.25 & 0.7 & 0.4 \\ -2.55 & 0.4 & -6.2 \end{bmatrix} \mathbf{z} + \begin{bmatrix} 0 \\ 0 \\ 5 \end{bmatrix} u \tag{5.79a}$$

$$y = [0.5 \quad 0 \quad 0]\mathbf{z} \tag{5.79b}$$

Students who are using MATLAB should now run ch5p4 in Appendix B. You will learn how to perform similarity transformations. This exercise uses MATLAB to do Example 5.9.

MATLAB
**ML**

Thus far we have talked about transforming systems between basis vectors in a different state space. One major advantage of finding these similar systems is apparent in the transformation to a system that has a diagonal matrix.

## Diagonalizing a System Matrix

In Section 5.7, we saw that the parallel form of a signal-flow graph can yield a diagonal system matrix. A diagonal system matrix has the advantage that each state equation is a function of only one state variable. Hence, each differential equation can be solved independently of the other equations. We say that the equations are *decoupled*.

Rather than using partial fraction expansion and signal-flow graphs, we can decouple a system using matrix transformations. If we find the correct matrix, **P**, the transformed system matrix, $\mathbf{P}^{-1}\mathbf{AP}$, will be a diagonal matrix. Thus, we are looking for a transformation to another state space that yields a diagonal matrix in that space. This new state space also has basis vectors that lie along its state variables. We give a special name to any vectors that are collinear with the basis vectors of the new system that yields a diagonal system matrix: they are called *eigenvectors*. Thus, the coordinates of the eigenvectors form the columns of the transformation matrix, **P**, as we demonstrate in Eq. L.7 in Appendix L at www.wiley.com/college/nise.

First, let us formally define eigenvectors from another perspective and then show that they have the property just described. Then we will define eigenvalues. Finally, we will show how to diagonalize a matrix.

## Definitions

*Eigenvector*. The eigenvectors of the matrix **A** are all vectors, $\mathbf{x_i} \neq \mathbf{0}$, which under the transformation **A** become multiples of themselves; that is,

$$\mathbf{Ax_i} = \lambda_i \mathbf{x_i} \tag{5.80}$$

where $\lambda_i$'s are constants.

Figure 5.32 shows this definition of eigenvectors. If **Ax** is not collinear with **x** after the transformation, as in Figure 5.32(*a*), **x** is not an eigenvector. If **Ax** is collinear with **x** after the transformation, as in Figure 5.32(*b*), **x** is an eigenvector.

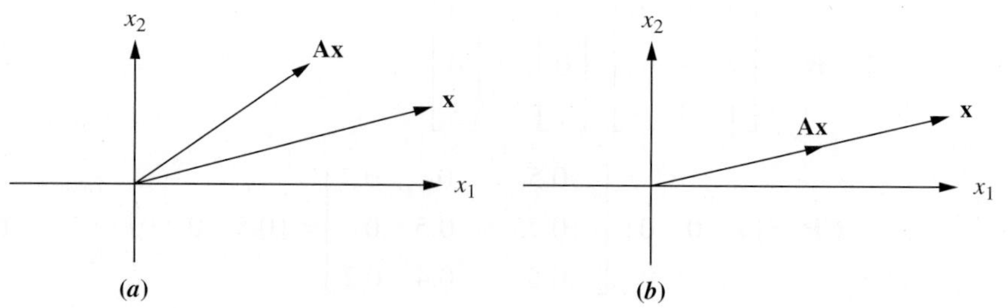


**FIGURE 5.32** To be an eigenvector, the transformation **Ax** must be collinear with **x**; thus, in (***a***), **x** is not an eigenvector; in (***b***), it is.

*Eigenvalue*. The eigenvalues of the matrix **A** are the values of $\lambda_i$ that satisfy Eq. (5.80) for $\mathbf{x_i} \neq \mathbf{0}$.

To find the eigenvectors, we rearrange Eq. (5.80). Eigenvectors, $\mathbf{x_i}$, satisfy

$$\mathbf{0} = (\lambda_i\mathbf{I} - \mathbf{A})\mathbf{x_i} \tag{5.81}$$

Solving for $\mathbf{x_i}$ by premultiplying both sides by $(\lambda_i\mathbf{I} - \mathbf{A})^{-1}$ yields

$$\mathbf{x_i} = (\lambda_i\mathbf{I} - \mathbf{A})^{-1}\mathbf{0} = \frac{\mathrm{adj}(\lambda_i\mathbf{I} - \mathbf{A})}{\det(\lambda_i\mathbf{I} - \mathbf{A})}\mathbf{0} \tag{5.82}$$

Since $\mathbf{x_i} \neq \mathbf{0}$, a nonzero solution exists if

$$\boxed{\det(\lambda_i\mathbf{I} - \mathbf{A}) = \mathbf{0}} \tag{5.83}$$

from which $\lambda_i$, the eigenvalues, can be found.

We are now ready to show how to find the eigenvectors, $\mathbf{x_i}$. First we find the eigenvalues, $\lambda_i$, using $\det(\lambda_i\mathbf{I} - \mathbf{A}) = \mathbf{0}$, and then we use Eq. (5.80) to find the eigenvectors.

## Example 5.10 Finding Eigenvectors

**PROBLEM:** Find the eigenvectors of the matrix

$$\mathbf{A} = \begin{bmatrix} -3 & 1 \\ 1 & -3 \end{bmatrix} \tag{5.84}$$

**SOLUTION:** The eigenvectors, $\mathbf{x_i}$, satisfy Eq. (5.81). First, use $\det(\lambda_i\mathbf{I} - \mathbf{A}) = 0$ to find the eigenvalues, $\lambda_i$, for Eq. (5.81):

$$\begin{aligned} \det(\lambda\mathbf{I} - \mathbf{A}) &= \left| \begin{bmatrix} \lambda & 0 \\ 0 & \lambda \end{bmatrix} - \begin{bmatrix} -3 & 1 \\ 1 & -3 \end{bmatrix} \right| \\ &= \begin{vmatrix} \lambda + 3 & -1 \\ -1 & \lambda + 3 \end{vmatrix} \\ &= \lambda^2 + 6\lambda + 8 \end{aligned} \tag{5.85}$$

from which the eigenvalues are $\lambda = -2$, and $-4$.

Using Eq. (5.80) successively with each eigenvalue, we have

$$\mathbf{A}\mathbf{x_i} = \lambda\mathbf{x_i}$$

$$\begin{bmatrix} -3 & 1 \\ 1 & -3 \end{bmatrix}\begin{bmatrix} x_1 \\ x_2 \end{bmatrix} = -2\begin{bmatrix} x_1 \\ x_2 \end{bmatrix} \tag{5.86}$$

or

$$-3x_1 + x_2 = -2x_1 \tag{5.87a}$$

$$x_1 - 3x_2 = -2x_2 \tag{5.87b}$$

from which $x_1 = x_2$. Thus,

$$\mathbf{x} = \begin{bmatrix} c \\ c \end{bmatrix} \tag{5.88}$$

Using the other eigenvalue, −4, we have

$$\mathbf{x} = \begin{bmatrix} c \\ -c \end{bmatrix} \tag{5.89}$$

Using Eqs. (5.88) and (5.89), one choice of eigenvectors is

$$\mathbf{x_1} = \begin{bmatrix} 1 \\ 1 \end{bmatrix} \quad \text{and} \quad \mathbf{x_2} = \begin{bmatrix} 1 \\ -1 \end{bmatrix} \tag{5.90}$$

We now show that if the eigenvectors of the matrix **A** are chosen as the basis vectors of a transformation, **P**, the resulting system matrix will be diagonal. Let the transformation matrix **P** consist of the eigenvectors of **A**, $\mathbf{x_i}$.

$$\mathbf{P} = [\mathbf{x}_1, \mathbf{x}_2, \mathbf{x}_3, \ldots, \mathbf{x_n}] \tag{5.91}$$

Since $\mathbf{x_i}$ are eigenvectors, $\mathbf{Ax_i} = \lambda_i \mathbf{x_i}$, which can be written equivalently as a set of equations expressed by

$$\mathbf{AP} = \mathbf{PD} \tag{5.92}$$

where **D** is a diagonal matrix consisting of $\lambda_i$'s, the eigenvalues, along the diagonal, and **P** is as defined in Eq. (5.91). Solving Eq. (5.92) for **D** by premultiplying by $\mathbf{P}^{-1}$, we get

$$\mathbf{D} = \mathbf{P}^{-1}\mathbf{AP} \tag{5.93}$$

which is the system matrix of Eq. (5.72).

In summary, under the transformation **P**, consisting of the eigenvectors of the system matrix, the transformed system is diagonal, with the eigenvalues of the system along the diagonal. The transformed system is identical to that obtained using partial-fraction expansion of the transfer function with distinct real roots.

In Example 5.10, we found eigenvectors of a second-order system. Let us continue with this problem and diagonalize the system matrix.

## Example 5.11 Diagonalizing a System in State Space

**PROBLEM:** Given the system of Eqs. (5.94), find the diagonal system that is similar.

$$\dot{\mathbf{x}} = \begin{bmatrix} -3 & 1 \\ 1 & -3 \end{bmatrix}\mathbf{x} + \begin{bmatrix} 1 \\ 2 \end{bmatrix}u \tag{5.94a}$$

$$y = [2 \quad 3]\,\mathbf{x} \tag{5.94b}$$

**SOLUTION:** First find the eigenvalues and the eigenvectors. This step was performed in Example 5.10. Next form the transformation matrix **P**, whose columns consist of the eigenvectors.

$$\mathbf{P} = \begin{bmatrix} 1 & 1 \\ 1 & -1 \end{bmatrix} \tag{5.95}$$

Finally, form the similar system's system matrix, input matrix, and output matrix, respectively.

$$\mathbf{P}^{-1}\mathbf{AP} = \begin{bmatrix} 1/2 & 1/2 \\ 1/2 & -1/2 \end{bmatrix}\begin{bmatrix} -3 & 1 \\ 1 & -3 \end{bmatrix}\begin{bmatrix} 1 & 1 \\ 1 & -1 \end{bmatrix} = \begin{bmatrix} -2 & 0 \\ 0 & -4 \end{bmatrix} \tag{5.96a}$$

$$\mathbf{P}^{-1}\mathbf{B} = \begin{bmatrix} 1/2 & 1/2 \\ 1/2 & -1/2 \end{bmatrix}\begin{bmatrix} 1 \\ 2 \end{bmatrix} = \begin{bmatrix} 3/2 \\ -1/2 \end{bmatrix} \tag{5.96b}$$

$$\mathbf{CP} = [2 \quad 3]\begin{bmatrix} 1 & 1 \\ 1 & -1 \end{bmatrix} = [5 \quad -1] \tag{5.96c}$$

Substituting Eqs. (5.96) into Eqs. (5.72), we get

$$\dot{\mathbf{z}} = \begin{bmatrix} -2 & 0 \\ 0 & -4 \end{bmatrix}\mathbf{z} + \begin{bmatrix} 3/2 \\ -1/2 \end{bmatrix}u \tag{5.97a}$$

$$y = [5 \quad -1]\mathbf{z} \tag{5.97b}$$

Notice that the system matrix is diagonal, with the eigenvalues along the diagonal.

MATLAB

ML

Students who are using MATLAB should now run ch5p5 in Appendix B. This problem, which uses MATLAB to diagonalize a system, is similar (but not identical) to Example 5.11.

## Skill-Assessment Exercise 5.7

**PROBLEM:** For the system represented in state space as follows:

$$\dot{\mathbf{x}} = \begin{bmatrix} 1 & 3 \\ -4 & -6 \end{bmatrix}\mathbf{x} + \begin{bmatrix} 1 \\ 3 \end{bmatrix}u$$
$$y = [1 \quad 4]\mathbf{x}$$

convert the system to one where the new state vector, **z**, is

$$\mathbf{z} = \begin{bmatrix} 3 & -2 \\ 1 & -4 \end{bmatrix}\mathbf{x}$$

**ANSWER:**

$$\dot{\mathbf{z}} = \begin{bmatrix} 6.5 & -8.5 \\ 9.5 & -11.5 \end{bmatrix}\mathbf{z} + \begin{bmatrix} -3 \\ -11 \end{bmatrix}u$$
$$y = [0.8 \quad -1.4]\mathbf{z}$$

The complete solution is at www.wiley.com/college/nise.

## Skill-Assessment Exercise 5.8

**PROBLEM:** For the original system of Skill-Assessment Exercise 5.7, find the diagonal system that is similar.

**ANSWER:**

$$\dot{\mathbf{z}} = \begin{bmatrix} -2 & 0 \\ 0 & -3 \end{bmatrix}\mathbf{z} + \begin{bmatrix} 18.39 \\ 20 \end{bmatrix}u$$
$$y = \begin{bmatrix} -2.121 & 2.6 \end{bmatrix}\mathbf{z}$$

The complete solution is at www.wiley.com/college/nise.

**TryIt 5.5**

Use the following MATLAB and Control System Toolbox statements to do Skill-Assessment Exercise 5.8.

```
A=[1 3; -4 -6];
B=[1; 3];
C=[1 4];
D=0;S=ss(A, B, C, D);
Sd=canon(S,'modal')
```

In this section, we learned how to move between different state-space representations of the same system via matrix transformations rather than transfer function manipulation and signal-flow graphs. These different representations are called *similar*. The characteristics of similar systems are that the transfer functions relating the output to the input are the same, as are the eigenvalues and poles. A particularly useful transformation was converting a system with distinct, real eigenvalues to a diagonal system matrix.

We now summarize the concepts of block diagram and signal-flow representations of systems, first through case study problems and then in a written summary. Our case studies include the antenna azimuth position control system and the Unmanned Free-Swimming Submersible vehicle (UFSS). Block diagram reduction is important for the analysis and design of these systems as well as the control systems on board *Alvin* (Figure 5.33), used to explore the wreckage of the *Titanic* 13,000 feet under the Atlantic in July 1986 (*Ballard, 1987*).

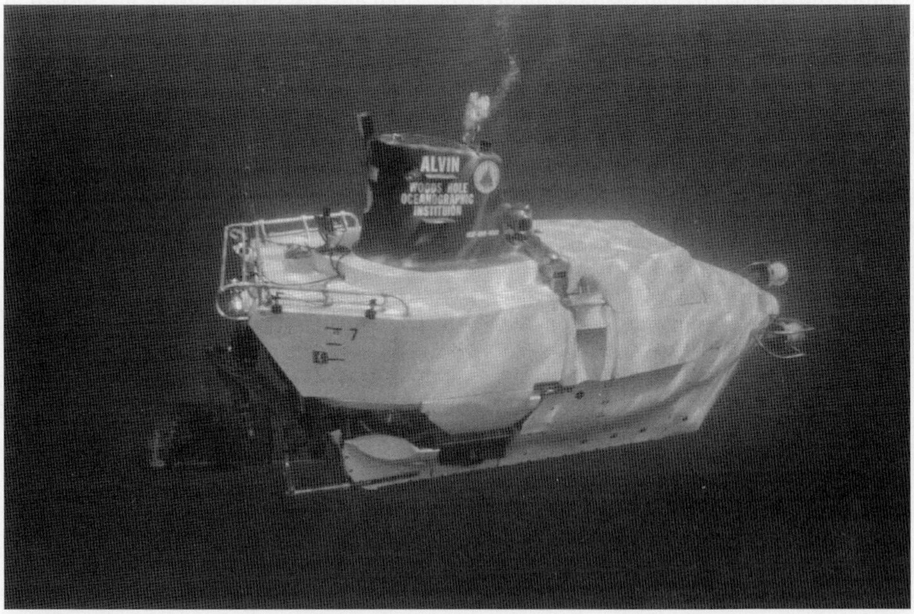


© Rob Catanach, Woods Hole Oceanographic Institution.

**FIGURE 5.33** *Alvin*, a manned submersible, explored the wreckage of the *Titanic* with a tethered robot, *Jason Junior*.

## Case Studies — Antenna Control: Designing a Closed-Loop Response

Design D

This chapter has shown that physical subsystems can be modeled mathematically with transfer functions and then interconnected to form a feedback system. The interconnected mathematical models can be reduced to a single transfer function representing the system from input to output. This transfer function, the closed-loop transfer function, is then used to determine the system response.

The following case study shows how to reduce the subsystems of the antenna azimuth position control system to a single, closed-loop transfer function in order to analyze and design the transient response characteristics.

**PROBLEM:** Given the antenna azimuth position control system shown on the front endpapers, Configuration 1, do the following:

**a.** Find the closed-loop transfer function using block diagram reduction.

State Space SS

**b.** Represent each subsystem with a signal-flow graph and find the state-space representation of the closed-loop system from the signal-flow graph.

**c.** Use the signal-flow graph found in **b** along with Mason's rule to find the closed-loop transfer function.

**d.** Replace the power amplifier with a transfer function of unity and evaluate the closed-loop peak time, percent overshoot, and settling time for $K = 1000$.

**e.** For the system of **d**, derive the expression for the closed-loop step response of the system.

**f.** For the simplified model of **d**, find the value of $K$ that yields a 10% overshoot.

**SOLUTION:** Each subsystem's transfer function was evaluated in the case study in Chapter 2. We first assemble them into the closed-loop, feedback control system block diagram shown in Figure 5.34(*a*).

**a.** The steps taken to reduce the block diagram to a single, closed-loop transfer function relating the output angular displacement to the input angular displacement are shown in Figure 5.34(*a–d*). In Figure 5.34(*b*), the input potentiometer was pushed to the right past the summing junction, creating a unity feedback system. In Figure 5.34(*c*), all the blocks of the forward transfer function are multiplied together, forming the equivalent forward transfer function. Finally, the feedback formula is applied, yielding the closed-loop transfer function in Figure 5.34(*d*).

State Space SS

**b.** In order to obtain the signal-flow graph of each subsystem, we use the state equations derived in the case study of Chapter 3. The signal-flow graph for the power amplifier is drawn from the state equations of Eqs. (3.87) and (3.88), and the signal-flow graph of the motor and load is drawn from the state equation of Eq. (3.98). Other subsystems are pure gains. The signal-flow graph for Figure 5.34(*a*) is shown in Figure 5.35 and consists of the interconnected subsystems.

The state equations are written from Figure 5.35. First define the state variables as the outputs of the integrators. Hence, the state vector is

$$\mathbf{x} = \begin{bmatrix} x_1 \\ x_2 \\ e_a \end{bmatrix} \tag{5.98}$$

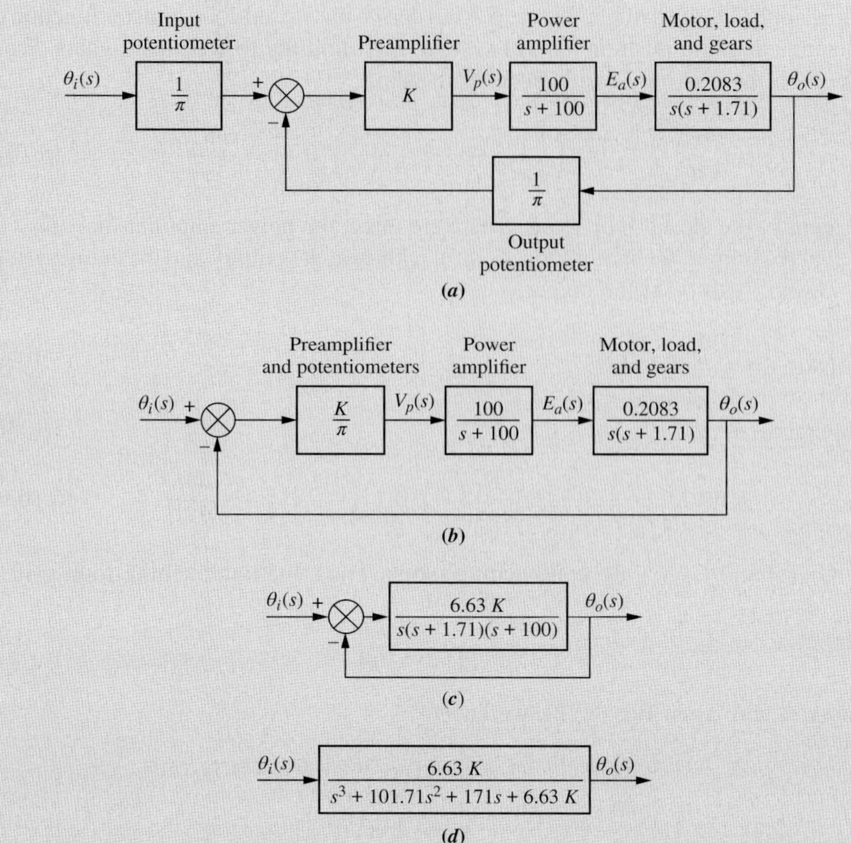


**FIGURE 5.34** Block diagram reduction for the antenna azimuth position control system: **a.** original; **b.** pushing input potentiometer to the right past the summing junction; **c.** showing equivalent forward transfer function; **d.** final closed-loop transfer function

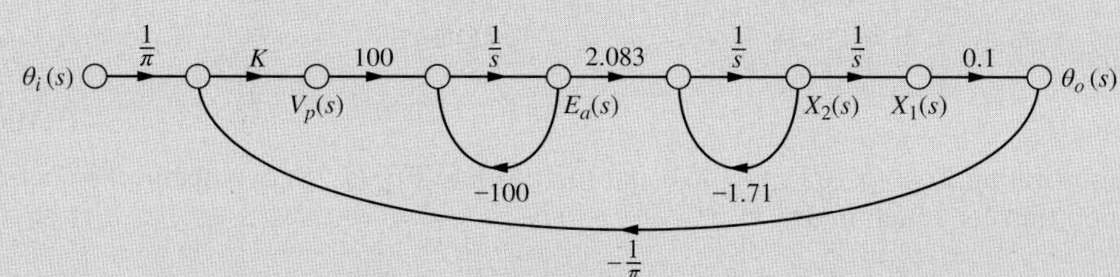


**FIGURE 5.35** Signal-flow graph for the antenna azimuth position control system

Using Figure 5.35, we write the state equations by inspection:

$$\dot{x}_1 = +x_2 \tag{5.99a}$$
$$\dot{x}_2 = -1.71x_2 + 2.083e_a \tag{5.99b}$$
$$\dot{e}_a = -3.18Kx_1 - 100e_a + 31.8K\theta_i \tag{5.99c}$$

along with the output equation,

$$y = \theta_o = 0.1x_1 \tag{5.100}$$

where $1/\pi = 0.318$.

In vector-matrix form,

$$\dot{\mathbf{x}} = \begin{bmatrix} 0 & 1 & 0 \\ 0 & -1.71 & 2.083 \\ -3.18K & 0 & -100 \end{bmatrix}\mathbf{x} + \begin{bmatrix} 0 \\ 0 \\ 31.8K \end{bmatrix}\theta_i \tag{5.101a}$$

$$y = [0.1 \quad 0 \quad 0]\,\mathbf{x} \tag{5.101b}$$

**c.** We now apply Mason's rule to Figure 5.35 to derive the closed-loop transfer function of the antenna azimuth position control system. First find the forward-path gains. From Figure 5.35 there is only one forward-path gain:

$$T_1 = \left(\frac{1}{\pi}\right)(K)(100)\left(\frac{1}{s}\right)(2.083)\left(\frac{1}{s}\right)\left(\frac{1}{s}\right)(0.1) = \frac{6.63K}{s^3} \tag{5.102}$$

Next identify the closed-loop gains. There are three: the power amplifier loop, $G_{L1}(s)$, with $e_a$ at the output; the motor loop, $G_{L2}(s)$, with $x_2$ at the output; and the entire system loop, $G_{L3}(s)$, with $\theta_0$ at the output.

$$G_{L1}(s) = \frac{-100}{s} \tag{5.103a}$$

$$G_{L2}(s) = \frac{-1.71}{s} \tag{5.103b}$$

$$G_{L3}(s) = (K)(100)\left(\frac{1}{s}\right)(2.083)\left(\frac{1}{s}\right)\left(\frac{1}{s}\right)(0.1)\left(\frac{-1}{\pi}\right) = \frac{-6.63K}{s^3} \tag{5.103c}$$

Only $G_{L1}(s)$ and $G_{L2}(s)$ are nontouching loops. Thus, the nontouching-loop gain is

$$G_{L1}(s)G_{L2}(s) = \frac{171}{s^2} \tag{5.104}$$

Forming $\Delta$ and $\Delta_k$ in Eq. (5.28), we have

$$\begin{aligned}\Delta &= 1 - [G_{L1}(s) + G_{L2}(s) + G_{L3}(s)] + [G_{L1}(s)G_{L2}(s)] \\ &= 1 + \frac{100}{s} + \frac{1.71}{s} + \frac{6.63K}{s^3} + \frac{171}{s^2}\end{aligned} \tag{5.105}$$

and

$$\Delta_1 = 1 \tag{5.106}$$

Substituting Eqs. (5.102), (5.105), and (5.106) into Eq. (5.28), we obtain the closed-loop transfer function as

$$T(s) = \frac{C(s)}{R(s)} = \frac{T_1\Delta_1}{\Delta} = \frac{6.63K}{s^3 + 101.71s^2 + 171s + 6.63K} \tag{5.107}$$

**d.** Replacing the power amplifier with unity gain and letting the preamplifier gain, $K$, in Figure 5.34(*b*) equal 1,000 yield a forward transfer function, $G(s)$, of

$$G(s) = \frac{66.3}{s(s + 1.71)} \tag{5.108}$$

Using the feedback formula to evaluate the closed-loop transfer function, we obtain

$$T(s) = \frac{66.3}{s^2 + 1.71s + 66.3} \tag{5.109}$$

From the denominator, $\omega_n = 8.14$, $\zeta = 0.105$. Using Eqs. (4.34), (4.38), and (4.42), the peak time = 0.388 second, the percent overshoot = 71.77%, and the settling time = 4.68 seconds.

**e.** The Laplace transform of the step response is found by first multiplying Eq. (5.109) by $1/s$, a unit-step input, and expanding into partial fractions:

$$C(s)=\frac{66.3}{s(s^2+1.71s+66.3)}=\frac{1}{s}-\frac{s+1.71}{s^2+1.71s+66.3}$$
$$=\frac{1}{s}-\frac{(s+0.855)+0.106(8.097)}{(s+0.855)^2+(8.097)^2} \quad (5.110)$$

Taking the inverse Laplace transform, we find

$$c(t)=1-e^{-0.855r}(\cos 8.097t+0.106\sin 8.097t) \quad (5.111)$$

**f.** For the simplified model we have

$$G(s)=\frac{0.0663K}{s(s+1.71)} \quad (5.112)$$

from which the closed-loop transfer function is calculated to be

$$T(s)=\frac{0.0663K}{s^2+1.71s+0.0663K} \quad (5.113)$$

From Eq. (4.39) a 10% overshoot yields $\zeta=0.591$. Using the denominator of Eq. (5.113), $\omega_n=\sqrt{0.0663K}$ and $2\zeta\omega_n=1.71$. Thus,

$$\zeta=\frac{1.71}{2\sqrt{0.0663K}}=0.591 \quad (5.114)$$

from which $K=31.6$.

**CHALLENGE:** You are now given a problem to test your knowledge of this chapter's objectives: Referring to the antenna azimuth position control system shown on the front endpapers, Configuration 2, do the following:

**a.** Find the closed-loop transfer function using block diagram reduction.

**b.** Represent each subsystem with a signal-flow graph and find the state-space representation of the closed-loop system from the signal-flow graph.

State Space
**SS**

**c.** Use the signal-flow graph found in (b) along with Mason's rule to find the closed-loop transfer function.

**d.** Replace the power amplifier with a transfer function of unity and evaluate the closed-loop percent overshoot, settling time, and peak time for $K=5$.

**e.** For the system used for (d), derive the expression for the closed-loop step response.

**f.** For the simplified model in (d), find the value of preamplifier gain, $\boldsymbol{K}$, to yield 15% overshoot.

## UFSS Vehicle: Pitch-Angle Control Representation

State Space
**SS**

We return to the Unmanned Free-Swimming Submersible (UFSS) vehicle introduced in the case studies in Chapter 5 (*Johnson, 1980*). We will represent in state space the pitch-angle control system that is used for depth control.

**PROBLEM:** Consider the block diagram of the pitch control loop of the UFSS vehicle shown on the back endpapers. The pitch angle, $\theta$, is controlled by a commanded pitch angle, $\theta_e$, which along with pitch-angle and pitch-rate feedback determines the elevator deflection, $\delta_e$, which acts through the vehicle dynamics to determine the pitch angle. Let $K_1 = K_2 = 1$ and do the following:

**a.** Draw the signal-flow graph for each subsystem, making sure that pitch angle, pitch rate, and elevator deflection are represented as state variables. Then interconnect the subsystems.

**b.** Use the signal-flow graph obtained in **a** to represent the pitch control loop in state space.

**SOLUTION:**

**a.** The vehicle dynamics are split into two transfer functions, from which the signal-flow graph is drawn. Figure 5.36 shows the division along with the elevator actuator. Each block is drawn in phase-variable form to meet the requirement that particular system variables be state variables. This result is shown in Figure 5.37(*a*). The feedback paths are then added to complete the signal-flow graph, which is shown in Figure 5.37(*b*).

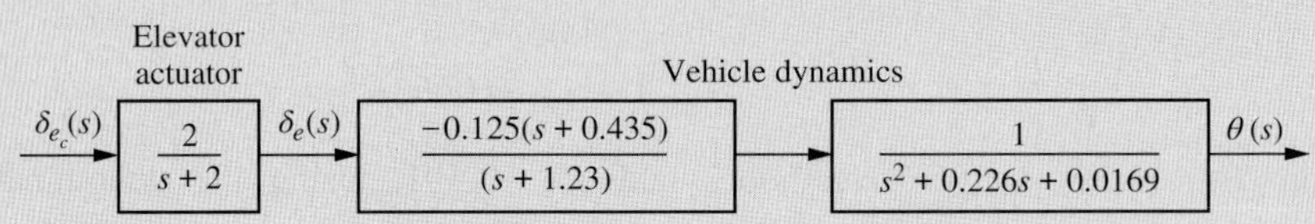


**FIGURE 5.36** Block diagram of the UFSS vehicle's elevator and vehicle dynamics, from which a signal-flow graph can be drawn

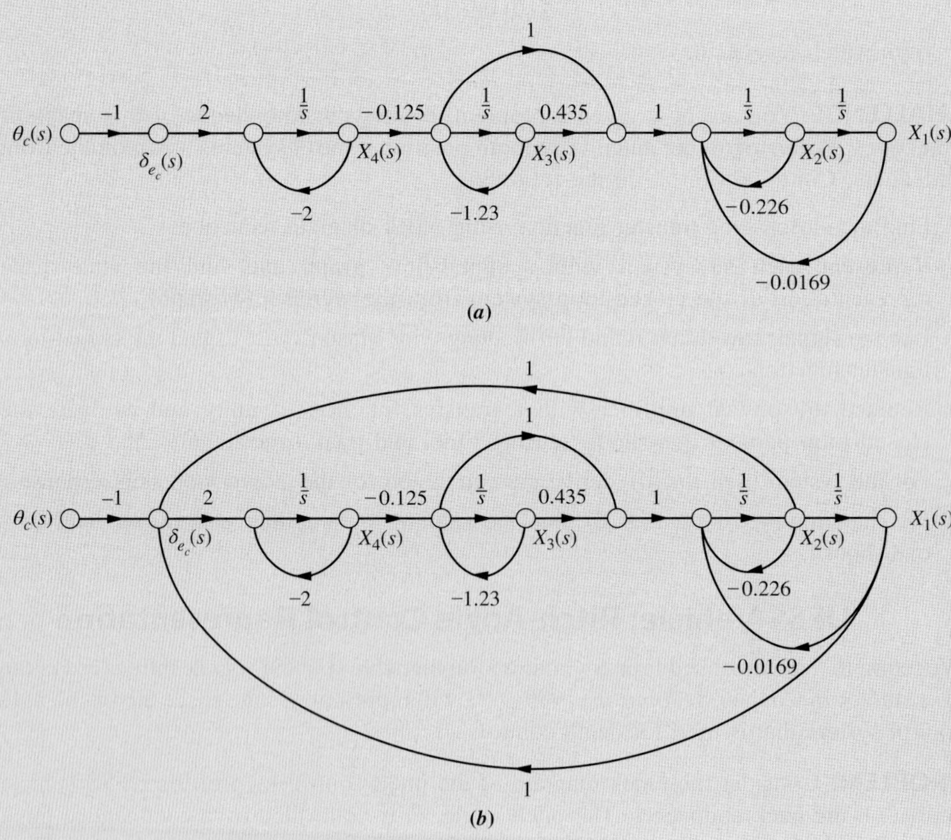


**FIGURE 5.37** Signal-flow graph representation of the UFSS vehicle's pitch control system: **a.** without position and rate feedback; **b.** with position and rate feedback. (Note: Explicitly required variables are $x_1 = \theta$, $x_2 = d\theta/dt$, and $x_4 = \delta_e$.)

**b.** By inspection, the derivatives of state variables $x_1$ through $x_4$ are written as

$$\dot{x}_1 = x_2 \quad (5.115a)$$

$$\dot{x}_2 = -0.0169x_1 - 0.226x_2 + 0.435x_3 - 1.23x_3 - 0.125x_4 \quad (5.115b)$$

$$\dot{x}_3 = -1.23x_3 - 0.125x_4 \quad (5.115c)$$

$$\dot{x}_4 = 2x_1 + 2x_2 - 2x_4 - 2\theta c \quad (5.115d)$$

Finally, the output $y = x_1$.

In vector-matrix form the state and output equations are

$$\dot{\mathbf{x}} = \begin{bmatrix} 0 & 1 & 0 & 0 \\ -0.0169 & -0.226 & -0.795 & -0.125 \\ 0 & 0 & -1.23 & -0.125 \\ 2 & 2 & 0 & -2 \end{bmatrix}\mathbf{x} + \begin{bmatrix} 0 \\ 0 \\ 0 \\ -2 \end{bmatrix}\theta_c \quad (5.116a)$$

$$y = [1 \quad 0 \quad 0 \quad 0]\,\mathbf{x} \quad (5.116b)$$

**CHALLENGE:** We now give you a problem to test your knowledge of this chapter's objectives. The UFSS vehicle steers via the heading control system shown in Figure 5.38 and repeated on the back endpapers. A heading command is the input. The input and feedback from the submersible's heading and yaw rate are used to generate a rudder command that steers the submersible (*Johnson, 1980*). Let $K_1 = K_2 = 1$ and do the following:

**a.** Draw the signal-flow graph for each subsystem, making sure that heading angle, yaw rate, and rudder deflection are represented as state variables. Then interconnect the subsystems.

**b.** Use the signal-flow graph obtained in **a** to represent the heading control loop in state space.

**c.** `Use MATLAB to represent the closed-loop UFSS heading control system in state space in controller canonical form.`

MATLAB
ML

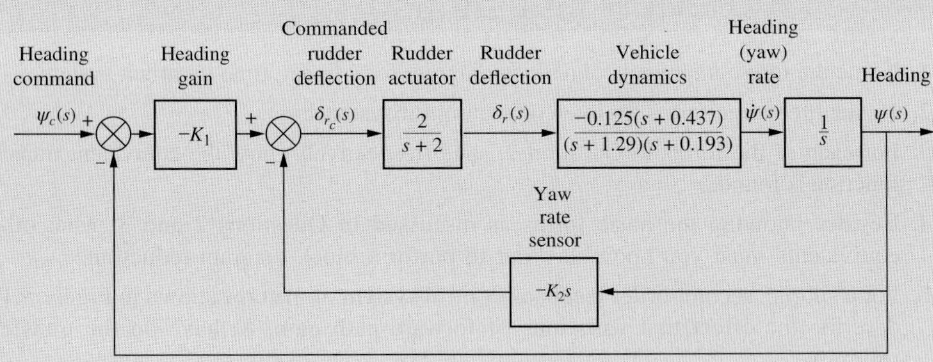


**FIGURE 5.38** Block diagram of the heading control system for the UFSS vehicle

# Summary

One objective of this chapter has been for you to learn how to represent multiple subsystems via block diagrams or signal-flow graphs. Another objective has been to be able to reduce either the block diagram representation or the signal-flow graph representation to a single transfer function.

We saw that the block diagram of a linear, time-invariant system consisted of four elements: *signals*, *systems*, *summing junctions*, and *pickoff points*. These elements were assembled into three basic forms: *cascade*, *parallel*, and *feedback*. Some basic operations were then derived: moving systems across summing junctions and across pickoff points.

Once we recognized the basic forms and operations, we could reduce a complicated block diagram to a single transfer function relating input to output. Then we applied the methods of Chapter 4 for analyzing and designing a second-order system for transient behavior. We saw that adjusting the gain of a feedback control system gave us partial control of the transient response.

The signal-flow representation of linear, time-invariant systems consists of two elements: nodes, which represent signals, and lines with arrows, which represent subsystems. Summing junctions and pickoff points are implicit in signal-flow graphs. These graphs are helpful in visualizing the meaning of the state variables. Also, they can be drawn first as an aid to obtaining the state equations for a system.

*Mason's rule* was used to derive the system's transfer function from the signal-flow graph. This formula replaced block diagram reduction techniques. Mason's rule seems complicated, but its use is simplified if there are no nontouching loops. In many of these cases, the transfer function can be written by inspection, with less labor than in the block diagram reduction technique.

Finally, we saw that systems in state space can be represented using different sets of variables. In the last three chapters, we have covered *phase-variable*, *cascade*, *parallel*, *controller canonical*, and *observer canonical* forms. A particular representation may be chosen because one set of state variables has a different physical meaning than another set, or because of the ease with which particular state equations can be solved.

In the next chapter, we discuss system stability. Without stability we cannot begin to design a system for the desired transient response. We will find out how to tell whether a system is stable and what effect parameter values have on a system's stability.

# Review Questions

1. Name the four components of a block diagram for a linear, time-invariant system.
2. Name three basic forms for interconnecting subsystems.
3. For each of the forms in Question 2, state (respectively) how the equivalent transfer function is found.
4. Besides knowing the basic forms as discussed in Questions 2 and 3, what other equivalents must you know in order to perform block diagram reduction?
5. For a simple, second-order feedback control system of the type shown in Figure 5.14, describe the effect that variations of forward-path gain, $K$, have on the transient response.
6. For a simple, second-order feedback control system of the type shown in Figure 5.14, describe the changes in damping ratio as the gain, $K$, is increased over the underdamped region.
7. Name the two components of a signal-flow graph.
8. How are summing junctions shown on a signal-flow graph?

9. If a forward path touched all closed loops, what would be the value of $\Delta_k$?

10. Name five representations of systems in state space. State Space SS

11. Which two forms of the state-space representation are found using the same method? State Space SS

12. Which form of the state-space representation leads to a diagonal matrix? State Space SS

13. When the system matrix is diagonal, what quantities lie along the diagonal? State Space SS

14. What terms lie along the diagonal for a system represented in Jordan canonical form? State Space SS

15. What is the advantage of having a system represented in a form that has a diagonal system matrix? State Space SS

16. Give two reasons for wanting to represent a system by alternative forms. State Space SS

17. For what kind of system would you use the observer canonical form? State Space SS

18. Describe state-vector transformations from the perspective of different bases. State Space SS

19. What is the definition of an eigenvector? State Space SS

20. Based upon your definition of an eigenvector, what is an eigenvalue? State Space SS

21. What is the significance of using eigenvectors as basis vectors for a system transformation? State Space SS

# Problems

1. Reduce the block diagram shown in Figure P5.1 to a single transfer function, $T(s) = C(s)/R(s)$. Use the following methods:
   a. Block diagram reduction [Section: 5.2]
   b. MATLAB

MATLAB ML

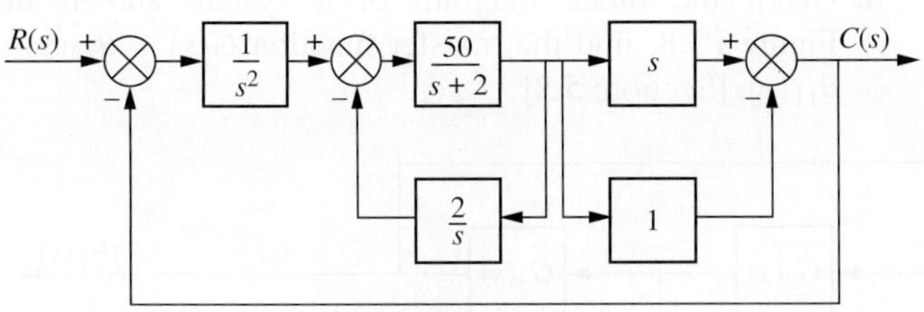


FIGURE P5.1

2. Find the equivalent transfer function, $T(s) = C(s)/R(s)$, for the system shown in Figure P5.2. [Section: 5.2]

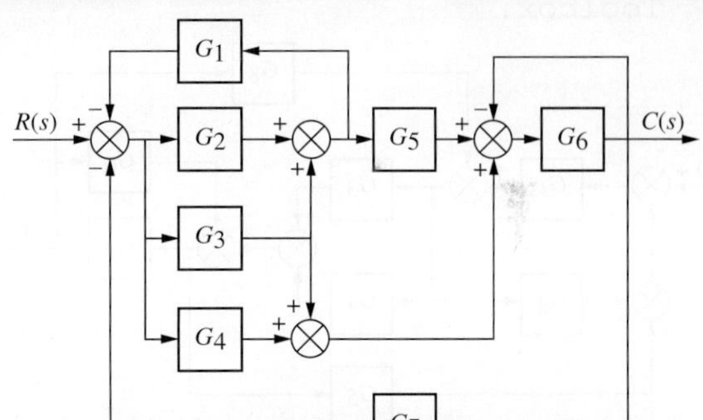


FIGURE P5.2

3. Find the closed-loop transfer function, $T(s) = C(s)/R(s)$ for the system shown in Figure P5.3, using block diagram reduction. [Section: 5.2]

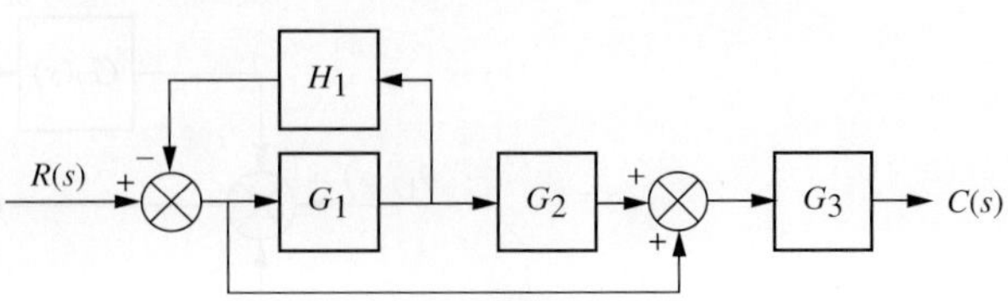


FIGURE P5.3

**4.** Reduce the system shown in Figure P5.4 to a single transfer function, $T(s) = C(s)/R(s)$. [Section: 5.2]

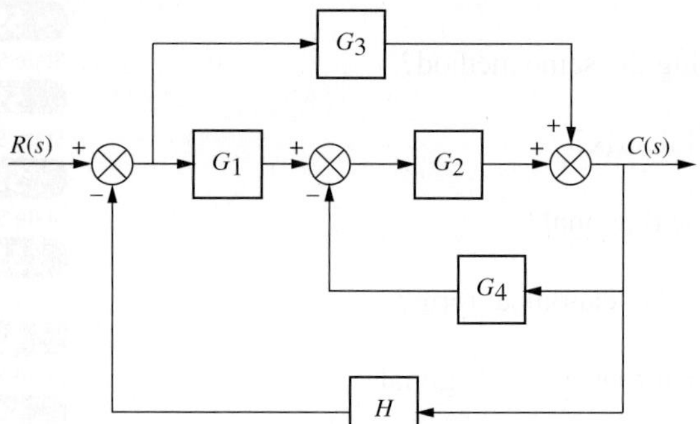


**FIGURE P5.4**

**5.** Find the transfer function, $T(s) = C(s)/R(s)$, for the system shown in Figure P5.5. Use the following methods:

**a.** Block diagram reduction [Section: 5.2]

**b.** MATLAB. Use the following transfer functions: MATLAB **ML**

$G_1(s) = 1/(s+7)$, $G_2(s) = 1/(s^2+2s+3)$, $G_3(s) = 1/(s+4)$, $G_4(s) = 1/s$, $G_5(s) = 5/(s+7)$, $G_6(s) = 1/(s^2+5s+10)$, $G_7(s) = 3/(s+2)$, $G_8(s) = 1/(s+6)$.

Hint: Use the **append** and **connect** commands in MATLAB's Control System Toolbox.

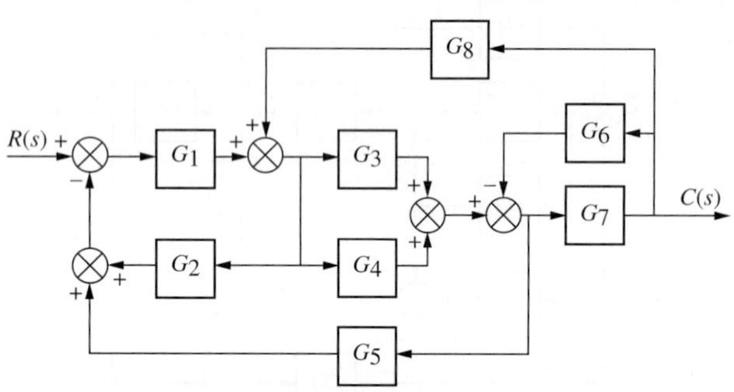


**FIGURE P5.5**

**6.** Reduce the block diagram shown in Figure P5.6 to a single block, $T(s) = C(s)/R(s)$. [Section: 5.2]

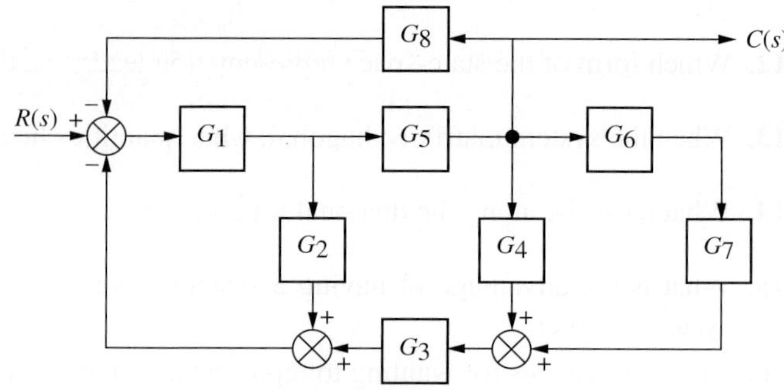


**FIGURE P5.6**

**7.** Find the unity feedback system that is equivalent to the system shown in Figure P5.7. [Section: 5.2]

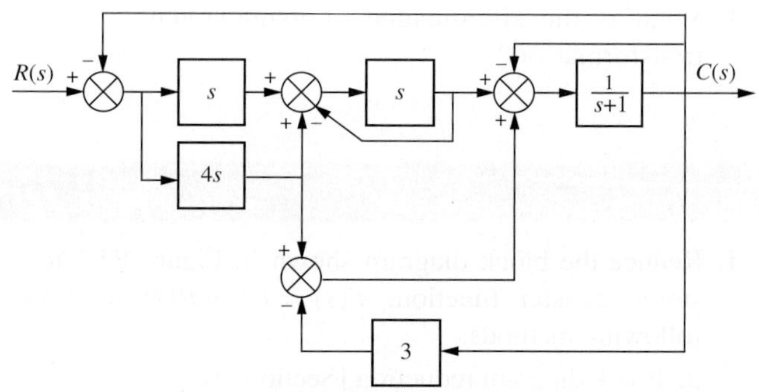


**FIGURE P5.7**

**8.** Given the block diagram of a system shown in Figure P5.8, find the transfer function $G(s) = \theta_{22}(s)/\theta_{11}(s)$. [Section: 5.2]

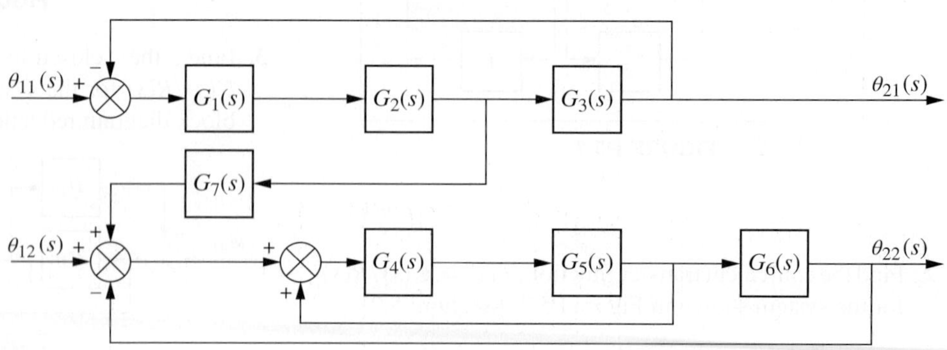


**FIGURE P5.8**

9. Reduce the block diagram shown in Figure P5.9 to a single block representing the transfer function, $T(s) = C(s)/R(s)$. [Section: 5.2]

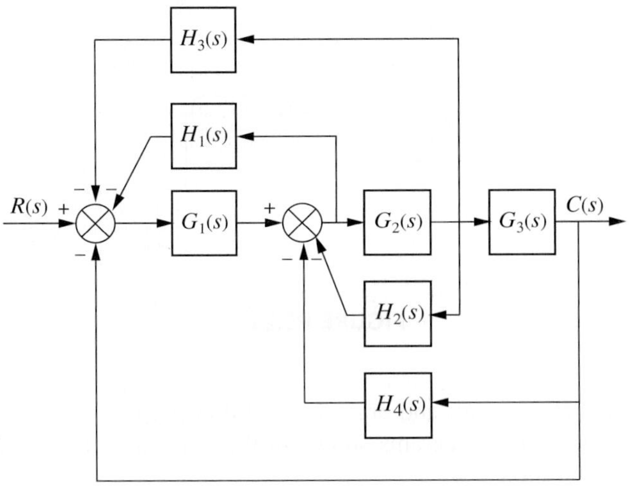


FIGURE P5.9

10. Reduce the block diagram shown in Figure P5.10 to a single transfer function, $T(s) = C(s)/R(s)$. [Section: 5.2]

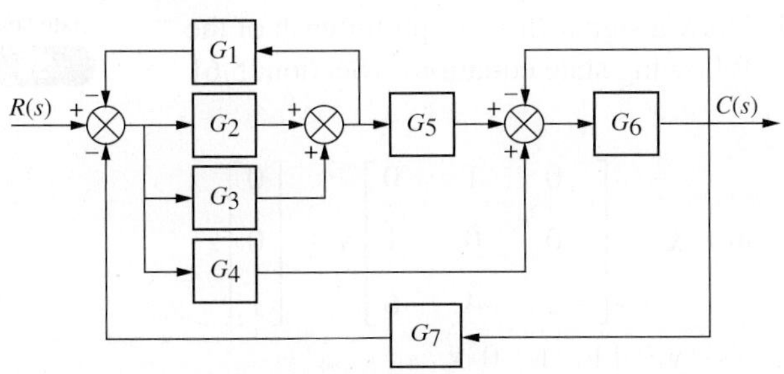


FIGURE P5.10

11. For the system shown in Figure P5.11, find the percent overshoot, settling time, and peak time for a step input if the system's response is underdamped. (Is it? Why?) [Section: 5.3]

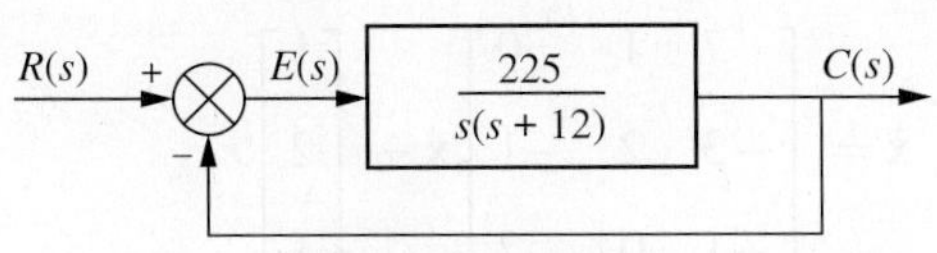


FIGURE P5.11

12. For the system shown in Figure P5.12, find the output, $c(t)$, if the input, $r(t)$, is a unit step. [Section: 5.3]

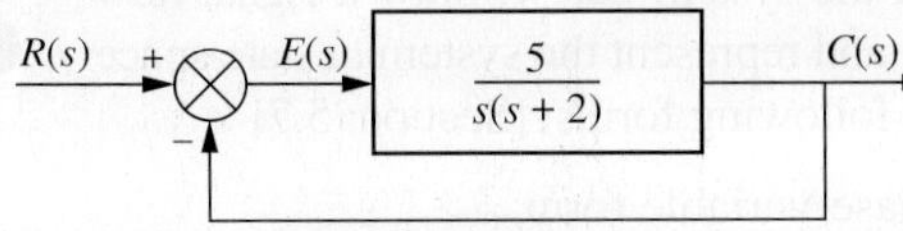


FIGURE P5.12

13. For the system shown in Figure P5.13, find the poles of the closed-loop transfer function, $T(s) = C(s)/R(s)$. [Section: 5.3]

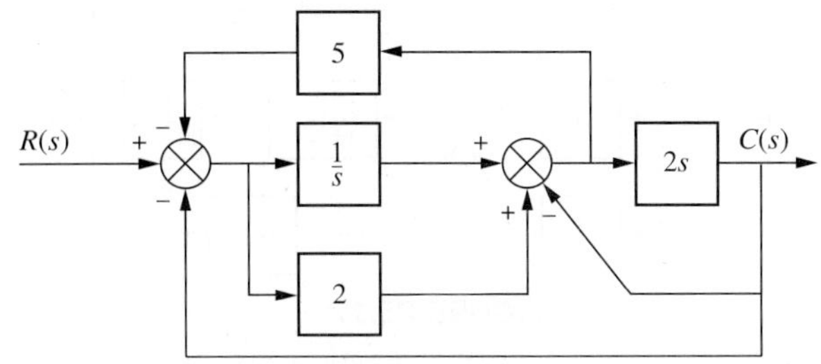


FIGURE P5.13

14. For the system of Figure P5.14, find the value of $K$ that yields 20% overshoot for a step input. [Section: 5.3]

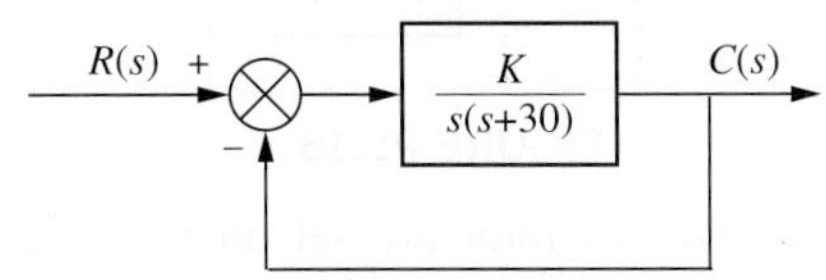


FIGURE P5.14

15. For the system shown in Figure P5.15, find $K$ and $\alpha$ to yield a settling time of 0.15 second and a 30% overshoot. [Section: 5.3]

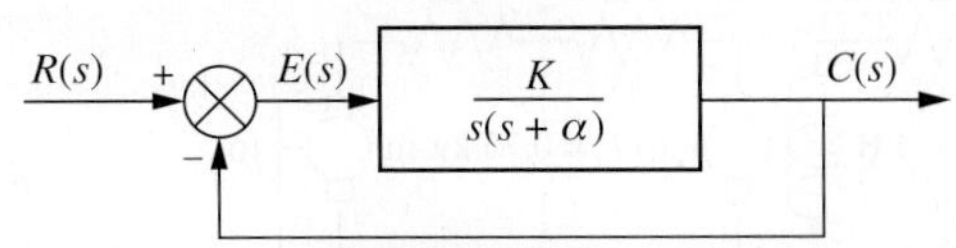


FIGURE P5.15

16. For the system of Figure P5.16, find the values of $K_1$ and $K_2$ to yield a peak time of 1.5 second and a settling time of 3.2 seconds for the closed-loop system's step response. [Section: 5.3]

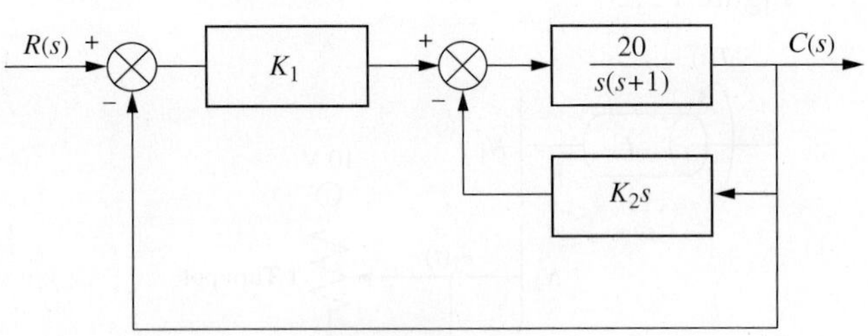


FIGURE P5.16

17. Find the following for the system shown in Figure P5.17: [Section: 5.3]

a. The equivalent single block that represents the transfer function, $T(s) = C(s)/R(s)$.

**b.** The damping ratio, natural frequency, percent overshoot, settling time, peak time, rise time, and damped frequency of oscillation.

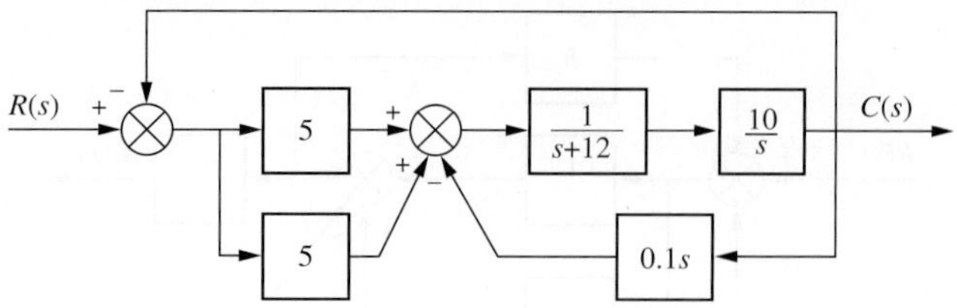


**FIGURE P5.17**

**18.** For the system shown in Figure P5.18, find $\zeta$, $\omega_n$, percent overshoot, peak time, rise time, and settling time. [Section: 5.3]

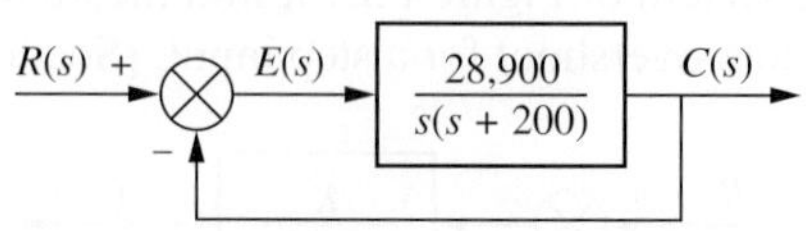


**FIGURE P5.18**

**19.** A motor and generator are set up to drive a load as shown in Figure P5.19. If the generator output voltage is $e_g(t) = K_f i_f(t)$, where $i_f(t)$ is the generator's field current, find the transfer function $G(s) = \theta_o(s)/E_i(s)$. For the generator, $K_f = 2\ \Omega$. For the motor, $K_t = 1$ N-m/A and $K_b = 1$ V-s/rad.

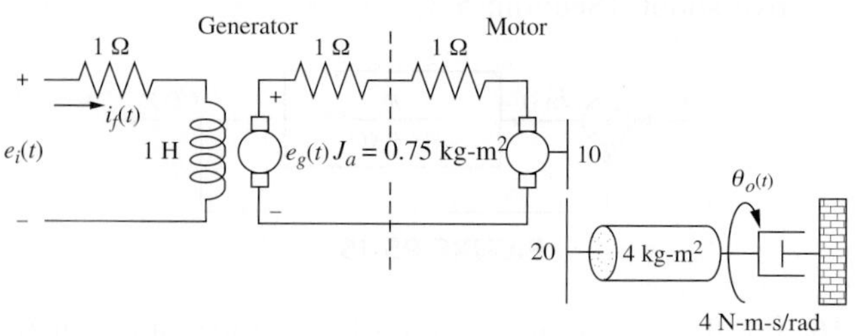


**FIGURE P5.19**

**20.** Find $G(s) = E_0(s)/T(s)$ for the system shown in Figure P5.20.

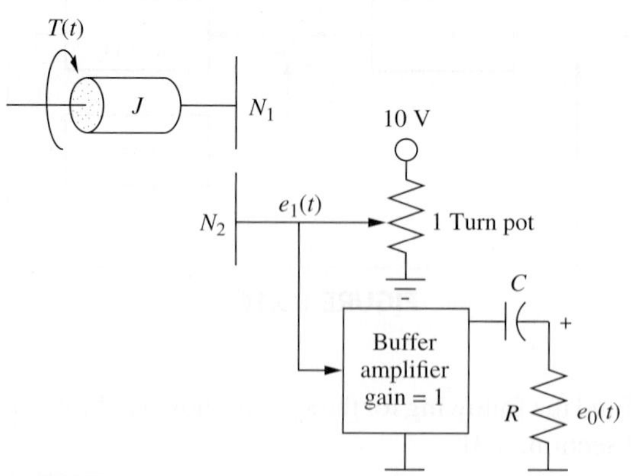


**FIGURE P5.20**

**21.** Find the transfer function $G(s) = E_o(s)/T(s)$ for the system shown in Figure P5.21.

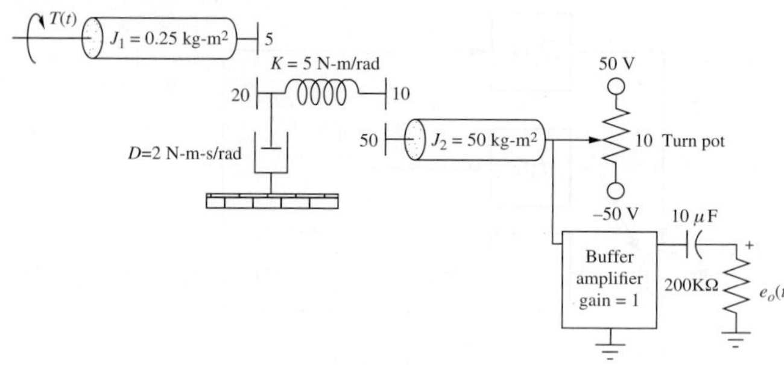


**FIGURE P5.21**

**22.** Label signals and draw a signal-flow graph for each of the block diagrams shown in the following problems: [Section: 5.4]

**a.** Problem 1

**b.** Problem 2

**c.** Problem 5

**23.** Draw a signal-flow graph for each of the following state equations: [Section: 5.6]

State Space **SS**

**a.** $$\dot{\mathbf{x}} = \begin{bmatrix} 0 & 1 & 0 \\ 0 & 0 & 1 \\ -2 & -4 & -6 \end{bmatrix}\mathbf{x} + \begin{bmatrix} 0 \\ 0 \\ 1 \end{bmatrix} r$$
$$y = [1 \quad 1 \quad 0]\mathbf{x}$$

**b.** $$\dot{\mathbf{x}} = \begin{bmatrix} 0 & 1 & 0 \\ 0 & -3 & 1 \\ -3 & -4 & -5 \end{bmatrix}\mathbf{x} + \begin{bmatrix} 0 \\ 1 \\ 1 \end{bmatrix} r$$
$$y = [1 \quad 2 \quad 0]\mathbf{x}$$

**c.** $$\dot{\mathbf{x}} = \begin{bmatrix} 7 & 1 & 0 \\ -3 & 2 & -1 \\ -1 & 0 & 2 \end{bmatrix}\mathbf{x} + \begin{bmatrix} 1 \\ 2 \\ 1 \end{bmatrix} r$$
$$y = [1 \quad 3 \quad 2]\mathbf{x}$$

**24.** Given the system below, draw a signal-flow graph and represent the system in state space in the following forms: [Section: 5.7]

State Space **SS**

**a.** Phase-variable form

**b.** Cascade form

$$G(s) = \frac{10}{(s+7)(s+8)(s+9)}$$

**25.** Repeat Problem 24 for

State Space **SS**

$$G(s) = \frac{20}{s(s-1)(s+3)(s+10)}$$

[Section: 5.7]

**26.** Use Mason's rule to find the transfer function of Figure 5.13 in the text. [Section: 5.5]

**27.** Use block diagram reduction to find the transfer function of Figure 5.21 in the text, and compare your answer with that obtained by Mason's rule. [Section: 5.5]

**28.** Using Mason's rule, find the transfer function, $T(s) = C(s)/R(s)$, for the system represented in Figure P5.22. [Section: 5.5]

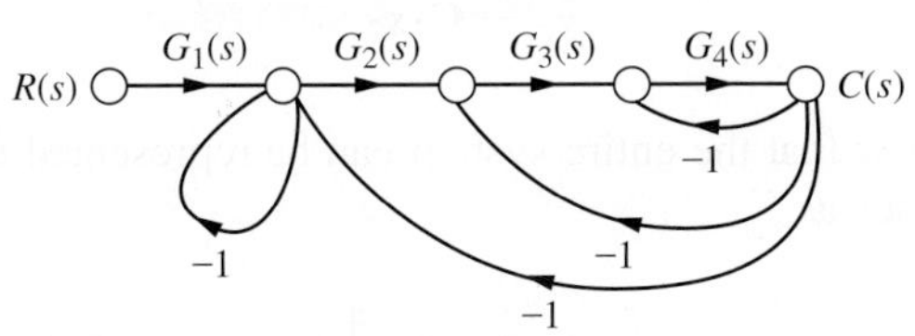


**FIGURE P5.22**

**29.** Using Mason's rule, find the transfer function, $T(s) = C(s)/R(s)$, for the system represented by Figure P5.23. [Section: 5.5]

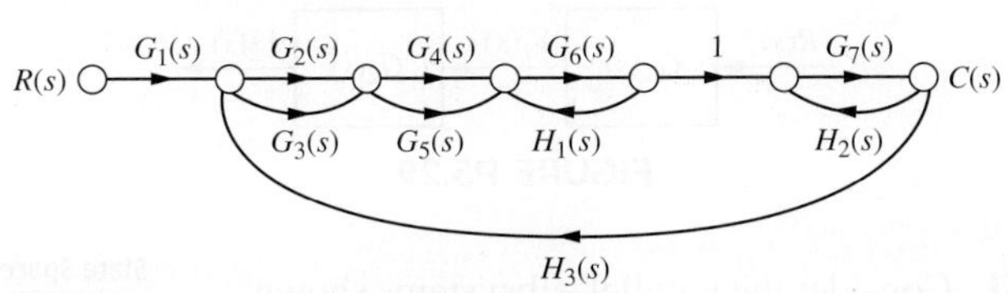


**FIGURE P5.23**

**30.** Represent the following systems in state space in Jordan canonical form. Draw the signal-flow graphs. [Section: 5.7]

State Space **SS**

**a.** $G(s) = \dfrac{(s+1)(s+2)}{(s+3)^2(s+4)}$

**b.** $G(s) = \dfrac{(s+2)}{(s+5)^2(s+7)^2}$

**c.** $G(s) = \dfrac{(s+3)}{(s+2)^2(s+4)(s+5)}$

**31.** Represent the systems below in state space in phase-variable form. Draw the signal-flow graphs. [Section: 5.7]

State Space **SS**

**a.** $G(s) = \dfrac{s+3}{s^2+2s+7}$

**b.** $G(s) = \dfrac{s^2+2s+6}{s^3+5s^2+2s+1}$

**c.** $G(s) = \dfrac{s^3+2s^2+7s+1}{s^4+3s^3+5s^2+6s+4}$

**32.** Repeat Problem 31 and represent each system in controller canonical and observer canonical forms. [Section: 5.7]

State Space **SS**

**33.** Represent the feedback control systems shown in Figure P5.24 in state space. When possible, represent the open-loop transfer functions separately in cascade and complete the feedback loop with the signal path from output to input. Draw your signal-flow graph to be in one-to-one correspondence to the block diagrams (as close as possible). [Section: 5.7]

State Space **SS**

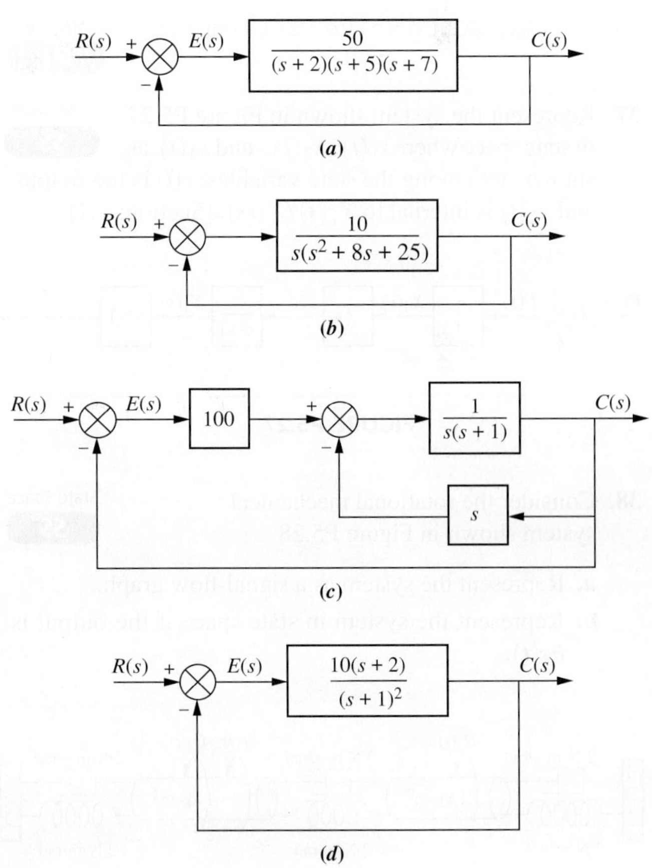


**FIGURE P5.24**

**34.** You are given the system shown in Figure P5.25. [Section: 5.7]

State Space SS

**a.** Represent the system in state space in phase-variable form.

**b.** Represent the system in state space in any other form besides phase-variable.

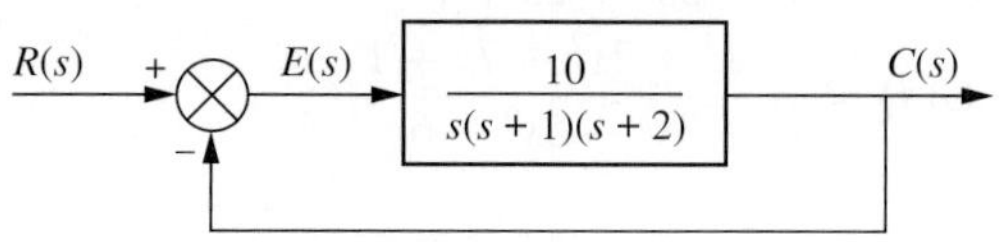


**FIGURE P5.25**

**35.** Repeat Problem 34 for the system shown in Figure P5.26. [Section: 5.7]

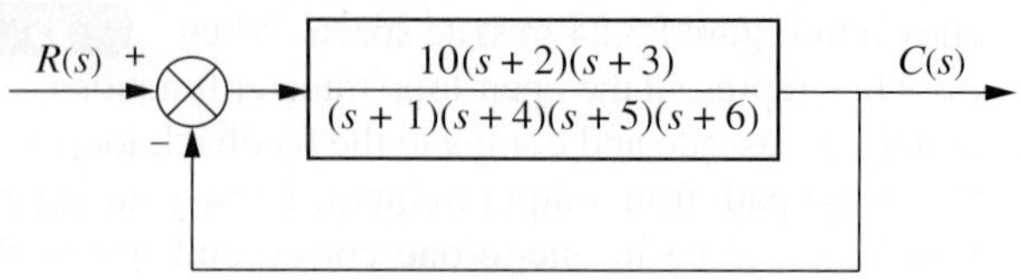


**FIGURE P5.26**

**36.** Use MATLAB to solve Problem 35.

MATLAB ML

**37.** Represent the system shown in Figure P5.27 in state space where $x_1(t)$, $x_3(t)$, and $x_4(t)$, as shown, are among the state variables, $c(t)$ is the output, and $x_2(t)$ is internal to $X_1(s)/X_3(s)$. [Section: 5.7]

State Space SS

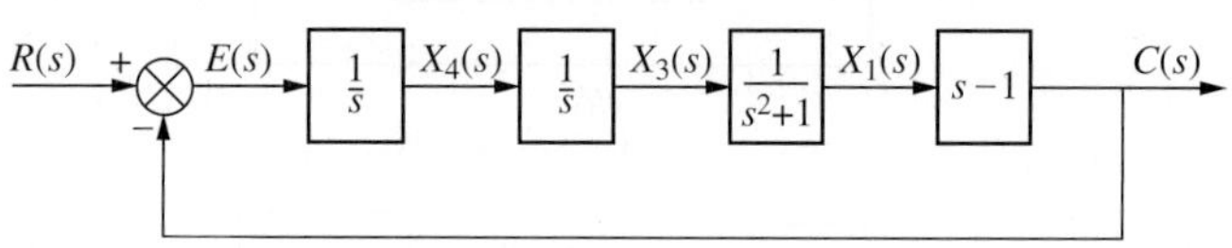


**FIGURE P5.27**

**38.** Consider the rotational mechanical system shown in Figure P5.28.

State Space SS

**a.** Represent the system as a signal-flow graph.

**b.** Represent the system in state space if the output is $\theta_2(t)$.

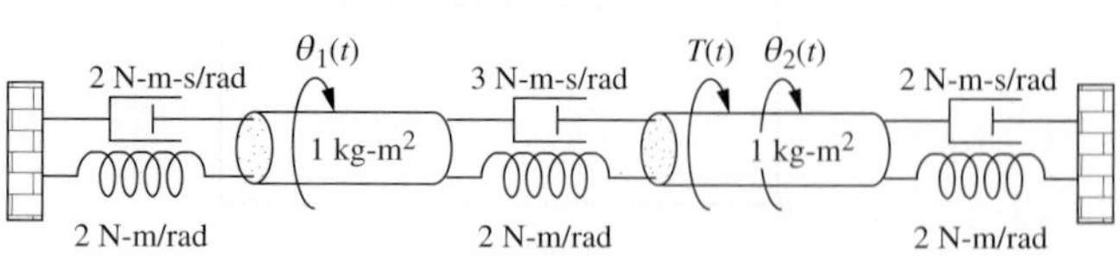


**FIGURE P5.28**

**39.** Given a unity feedback system with the forward-path transfer function

MATLAB ML

State Space SS

$$G(s) = \frac{7}{s(s+9)(s+12)}$$

use MATLAB to represent the closed loop system in state space in

**a.** phase-variable form;

**b.** parallel form.

**40.** Consider the cascaded subsystems shown in Figure P5.29. If $G_1(s)$ is represented in state space as

State Space SS

$$\dot{\mathbf{x}}_1 = \mathbf{A}_1\mathbf{x}_1 + \mathbf{B}_1 r$$
$$y_1 = \mathbf{C}_1\mathbf{x}_1$$

and $G_2(s)$ is represented in state space as

$$\dot{\mathbf{x}}_2 = \mathbf{A}_2\mathbf{x}_2 + \mathbf{B}_2 y_1$$
$$y_2 = \mathbf{C}_2\mathbf{x}_2$$

show that the entire system can be represented in state space as

$$\begin{bmatrix} \dot{\mathbf{x}}_1 \\ \dot{\mathbf{x}}_2 \end{bmatrix} = \begin{bmatrix} \mathbf{A}_1 & \mathbf{0} \\ \mathbf{B}_2\mathbf{C}_1 & \mathbf{A}_2 \end{bmatrix} \begin{bmatrix} \mathbf{x}_1 \\ \mathbf{x}_2 \end{bmatrix} + \begin{bmatrix} \mathbf{B}_1 \\ \mathbf{0} \end{bmatrix} r$$

$$y_2 = \begin{bmatrix} \mathbf{0} & \mathbf{C}_2 \end{bmatrix} \begin{bmatrix} \mathbf{x}_1 \\ \mathbf{x}_2 \end{bmatrix}$$

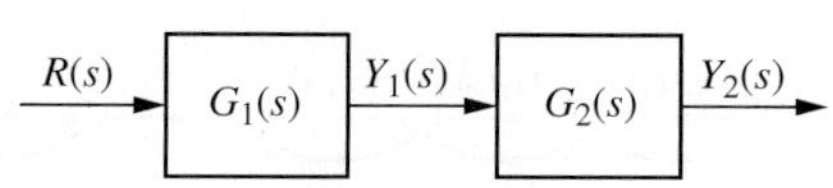


**FIGURE P5.29**

**41.** Consider the parallel subsystems shown in Figure P5.30. If $G_1(s)$ is represented in state space as

State Space SS

$$\dot{\mathbf{x}}_1 = \mathbf{A}_1\mathbf{x}_1 + \mathbf{B}_1 r$$
$$y_1 = \mathbf{C}_1\mathbf{x}_1$$

and $G_2(s)$ is represented in state space as

$$\dot{\mathbf{x}}_2 = \mathbf{A}_2\mathbf{x}_2 + \mathbf{B}_2 r$$
$$y_2 = \mathbf{C}_2\mathbf{x}_2$$

show that the entire system can be represented in state space as

$$\begin{bmatrix} \dot{\mathbf{x}}_1 \\ \cdots \\ \dot{\mathbf{x}}_2 \end{bmatrix} = \begin{bmatrix} \mathbf{A}_1 & \vdots & \mathbf{0} \\ \cdots & \vdots & \cdots \\ \mathbf{0} & \vdots & \mathbf{A}_2 \end{bmatrix} \begin{bmatrix} \mathbf{x}_1 \\ \cdots \\ \mathbf{x}_2 \end{bmatrix} + \begin{bmatrix} \mathbf{B}_1 \\ \cdots \\ \mathbf{B}_2 \end{bmatrix} r$$

$$y = \begin{bmatrix} \mathbf{C}_1 & \vdots & \mathbf{C}_2 \end{bmatrix} \begin{bmatrix} \mathbf{x}_1 \\ \cdots \\ \mathbf{x}_2 \end{bmatrix}$$

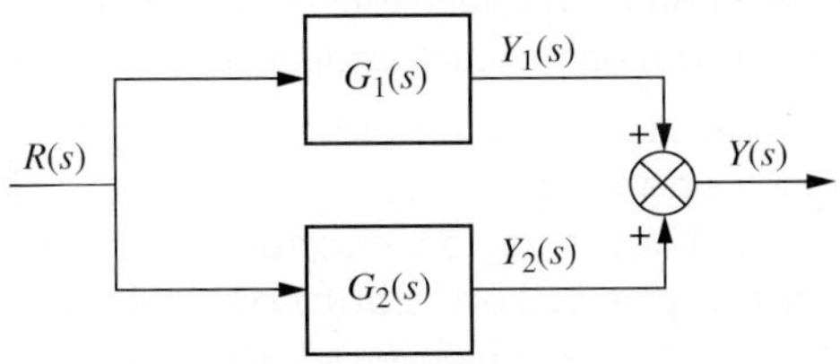


**FIGURE P5.30**

**42.** Consider the subsystems shown in Figure P5.31 and connected to form a feedback system. If $G(s)$ is represented in state space as

State Space
SS

$$\dot{\mathbf{x}}_1 = \mathbf{A}_1\mathbf{x}_1 + \mathbf{B}_1 e$$
$$y = \mathbf{C}_1\mathbf{x}_1$$

and $H_2(s)$ is represented in state space as

$$\dot{\mathbf{x}}_2 = \mathbf{A}_2\mathbf{x}_2 + \mathbf{B}_2 y$$
$$\rho = \mathbf{C}_2\mathbf{x}_2$$

show that the closed-loop system can be represented in state space as

$$\begin{bmatrix} \dot{\mathbf{x}}_1 \\ \cdots \\ \dot{\mathbf{x}}_2 \end{bmatrix} = \begin{bmatrix} \mathbf{A}_1 & \vdots & -\mathbf{B}_1\mathbf{C}_2 \\ \cdots & \vdots & \cdots \\ \mathbf{B}_2\mathbf{C}_1 & \vdots & \mathbf{A}_2 \end{bmatrix} \begin{bmatrix} \mathbf{x}_1 \\ \cdots \\ \mathbf{x}_2 \end{bmatrix} + \begin{bmatrix} \mathbf{B}_1 \\ \cdots \\ \mathbf{0} \end{bmatrix} r$$

$$y = \begin{bmatrix} \mathbf{C}_1 & \vdots & \mathbf{0} \end{bmatrix} \begin{bmatrix} \mathbf{x}_1 \\ \cdots \\ \mathbf{x}_2 \end{bmatrix}$$

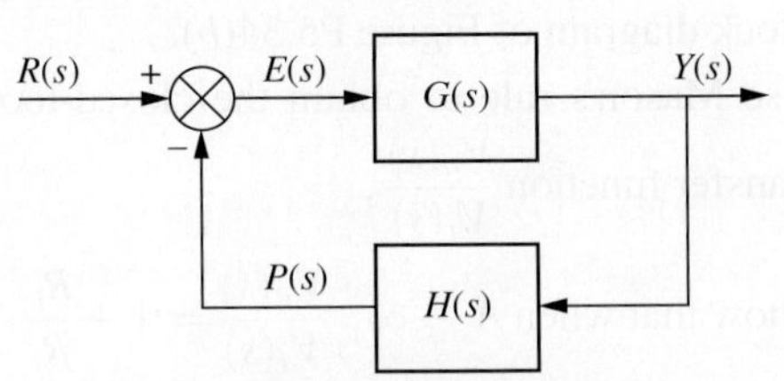


**FIGURE P5.31**

**43.** Given the system represented in state space as follows: [Section: 5.8]

State Space
SS

$$\dot{\mathbf{x}} = \begin{bmatrix} 2 & 0 & -1 \\ 0 & 4 & 6 \\ -6 & -5 & -8 \end{bmatrix}\mathbf{x} + \begin{bmatrix} 2 \\ 0 \\ 1 \end{bmatrix} r$$
$$y = \begin{bmatrix} 2 & 5 & 3 \end{bmatrix}\mathbf{x}$$

Convert the system to one where the new state vector, **z**, is

$$\mathbf{z} = \begin{bmatrix} 2 & 1 & -4 \\ 1 & -2 & 0 \\ 4 & 6 & 2 \end{bmatrix}\mathbf{x}$$

**44.** Repeat Problem 43 for the following system: [Section: 5.8]

State Space
SS

$$\dot{\mathbf{x}} = \begin{bmatrix} 1 & -1 & 1 \\ 2 & 1 & 3 \\ -2 & -1 & -3 \end{bmatrix}\mathbf{x} + \begin{bmatrix} 7 \\ 1 \\ -2 \end{bmatrix} r$$
$$y = \begin{bmatrix} 1 & -3 & 4 \end{bmatrix}\mathbf{x}$$

and the following state-vector transformation:

$$\mathbf{z} = \begin{bmatrix} 4 & -1 & 0 \\ 2 & 3 & -2 \\ 8 & 5 & 1 \end{bmatrix}\mathbf{x}$$

**45.** Diagonalize the following system: [Section: 5.8]

State Space
SS

$$\dot{\mathbf{x}} = \begin{bmatrix} -1 & 2.5 & 9 \\ -4 & -5 & -12 \\ 1.33 & -0.833 & 0 \end{bmatrix}\mathbf{x} + \begin{bmatrix} 1 \\ 3 \\ 2 \end{bmatrix} r$$
$$y = \begin{bmatrix} 1 & -2 & 4 \end{bmatrix}\mathbf{x}$$

**46.** `Diagonalize the system in Problem 45 using MATLAB.`

MATLAB
ML

**47.** Diagonalize the following system: [Section: 5.8]

State Space
SS

$$\dot{\mathbf{x}} = \begin{bmatrix} -5 & -5 & 4 \\ 2 & 0 & -2 \\ 0 & -2 & -1 \end{bmatrix}\mathbf{x} + \begin{bmatrix} -1 \\ 2 \\ -2 \end{bmatrix} r$$
$$y = \begin{bmatrix} -1 & 1 & 2 \end{bmatrix}\mathbf{x}$$

**48.** An AM radio modulator generates the product of a carrier waveform and a message waveform, as shown in Figure P5.32 (*Kurland, 1971*). Represent the system in state space if the carrier is a sinusoid of frequency $\omega = a$, and the message is a sinusoid of frequency $\omega = b$. Note that this system is nonlinear because of the multiplier.

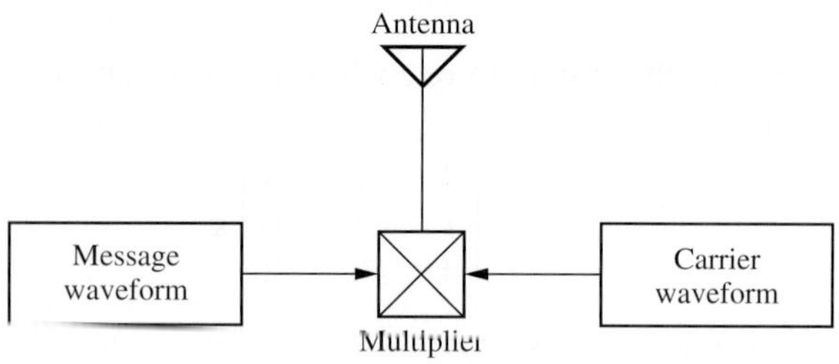


**FIGURE P5.32** AM modulator

**49.** A model for human eye movement consists of the closed-loop system shown in Figure P5.33, where an object's position is the input and the eye position is the output. The brain sends signals to the muscles that move the eye. These signals consist of the difference between the object's position and the position and rate information from the eye sent by the muscle spindles. The eye motion is modeled as an inertia and viscous damping and assumes no elasticity (spring) (*Milhorn, 1966*). Assuming that the delays in the brain and nervous system are negligible, find the closed-loop transfer function for the eye position control.

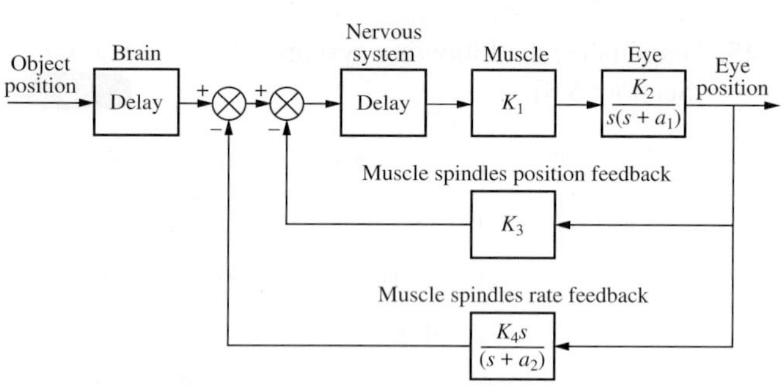


**FIGURE P5.33** Feedback control system representing human eye movement

**50.** Consider the F4-E aircraft of Problem 21, Chapter 3. If the open-loop transfer function relating normal acceleration, $A_n(s)$, to the input deflection command, $\delta_c(s)$, is approximated as

$$\frac{A_n(s)}{\delta_c(s)} = \frac{-272(s^2 + 1.9s + 84)}{(s+14)(s-1.8)(s+4.9)}$$

(*Cavallo, 1992*), find the state-space representation in

**a.** Phase-variable form
**b.** Controller canonical form
**c.** Observer canonical form
**d.** Cascade form
**e.** Parallel form

**51.** Find the closed-loop transfer function of the Unmanned Free-Swimming Submersible vehicle's pitch control system shown on the back endpapers (*Johnson, 1980*).

**52.** Use Simulink to plot the effects of nonlinearities upon the closed-loop step response of the antenna azimuth position control system shown on the front endpapers, Configuration 1. In particular, consider individually each of the following nonlinearities: saturation (±5 volts), backlash (deadband width 0.15), deadzone (−2 to +2), as well as the linear response. Assume the preamplifier gain is 100 and the step input is 2 radians.

**53.** Figure P5.34 shows a noninverting operational amplifier.

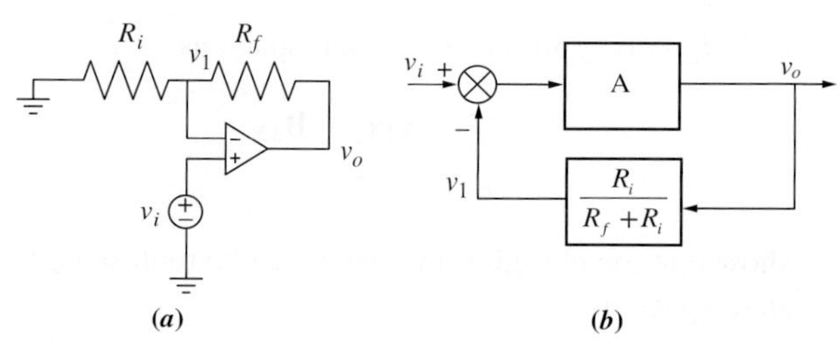


**FIGURE P5.34** **a.** Noninverting amplifier; **b.** block diagram

Assuming the operational amplifier is ideal,

**a.** Verify that the system can be described by the following two equations:

$$v_o = A(v_i - v_o)$$

$$v_1 = \frac{R_i}{R_i + R_f} v_o$$

**b.** Check that these equations can be described by the block diagram of Figure P5.34(*b*).

**c.** Use Mason's rule to obtain the closed-loop system transfer function $\frac{V_o(s)}{V_i(s)}$.

**d.** Show that when $A \to \infty$, $\frac{V_o(s)}{V_i(s)} = 1 + \frac{R_f}{R_i}$.

54. Figure P5.35 shows the diagram of an inverting operational amplifier.

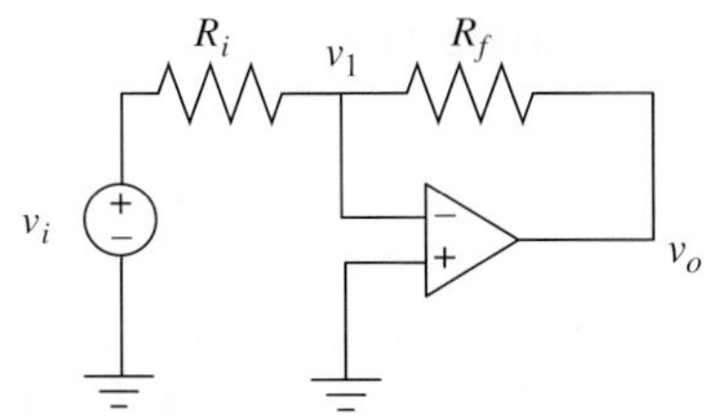


**FIGURE P5.35** Inverting operational amplifier

**a.** Assuming an ideal operational amplifier, use a similar procedure to the one outlined in Problem 53 to find the system equations.

**b.** Draw a corresponding block diagram and obtain the transfer function $\dfrac{V_o(s)}{V_i(s)}$.

**c.** Show that when $A \to \infty$, $\dfrac{V_o(s)}{V_i(s)} = -\dfrac{R_f}{R_i}$.

55. Figure P5.36(*a*) shows an $n$-channel enhancement-mode MOSFET source follower circuit. Figure P5.36(*b*) shows its small-signal equivalent (where $R_i = R_1 || R_2$) (*Neamen, 2001*).

**a.** Verify that the equations governing this circuit are

$$\frac{v_{in}}{v_i} = \frac{R_i}{R_i + R_s}; \quad v_{gs} = v_{in} - v_o; \; v_o = g_m(R_s || r_o)v_{gs}$$

**b.** Draw a block diagram showing the relations between the equations.

**c.** Use the block diagram in Part **b** to find $\dfrac{V_o(s)}{V_i(s)}$.

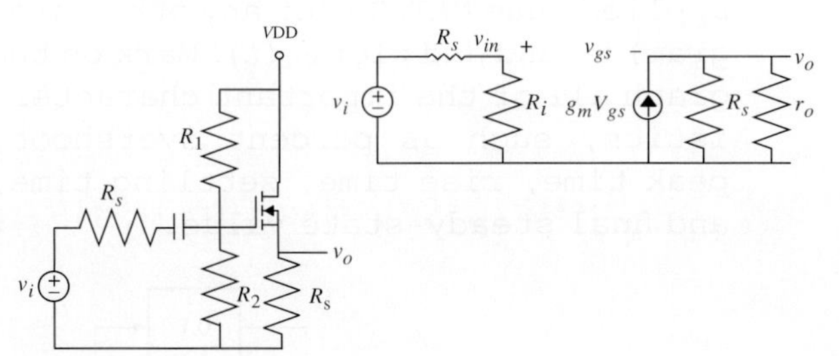


(*a*) (*b*)

**FIGURE P5.36** **a.** An $n$-channel enhancement-mode MOSFET source follower circuit; **b.** small-signal equivalent

56. A car active suspension system adds an active hydraulic actuator in parallel with the passive damper and spring to create a dynamic impedance that responds to road variations. The block diagram of Figure P5.37 depicts such an actuator with closed-loop control.

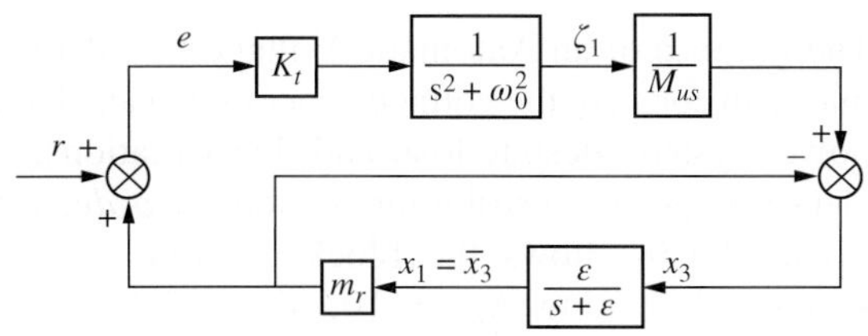


**FIGURE P5.37**[4]

In the figure, $K_t$ is the spring constant of the tire, $M_{US}$ is the wheel mass, $r$ is the road disturbance, $x_1$ is the vertical car displacement, $x_3$ is the wheel vertical displacement, $\omega_0^2 = \dfrac{K_t}{M_{US}}$ is the natural frequency of the unsprung system and $\varepsilon$ is a filtering parameter to be judiciously chosen (*Lin, 1997*). Find the two transfer functions of interest:

**a.** $\dfrac{X_3(s)}{R(s)}$

**b.** $\dfrac{X_1(s)}{R(s)}$

57. An electric ventricular assist device (EVAD) has been designed to help patients with diminished but still functional heart pumping action to work in parallel with the natural heart. The device consists of a brushless dc electric motor that actuates on a pusher plate. The plate movements help the ejection of blood in systole and sac filling in diastole. System dynamics during systolic mode have been found to be:

$$\begin{bmatrix} \dot{x} \\ \dot{v} \\ \dot{P}_{ao} \end{bmatrix} = \begin{bmatrix} 0 & 1 & 0 \\ 0 & -68.3 & -7.2 \\ 0 & 3.2 & -0.7 \end{bmatrix} \begin{bmatrix} x \\ v \\ P_{ao} \end{bmatrix} + \begin{bmatrix} 0 \\ 425.4 \\ 0 \end{bmatrix} e_m$$

The state variables in this model are $x$, the pusher plate position, $v$, the pusher plate velocity, and $P_{ao}$, the aortic blood pressure. The input to the system is $e_m$, the motor voltage (*Tasch, 1990*).

**a.** Use MATLAB to find a similarity transformation to diagonalize the system. MATLAB ML

**b.** Use MATLAB and the obtained similarity transformation of Part a to obtain a diagonalized expression for the system. MATLAB ML

[4] Lin, J.-S., Kanellakopoulos, I. "Nonlinear Design of Active Suspensions." *IEEE Control Systems Magazine*, Vol. 17, Issue 3, June 1997 pp. 45–59. Figure 3, p. 48. IEEE Control Systems by IEEE Control Systems Society. Reproduced with permission of Institute of Electrical and Electronics Engineers, in the format Republish in a book via Copyright Clearance Center.

**58.** The purpose of an Automatic Voltage Regulator is to maintain constant the voltage generated in an electrical power system, despite load and line variations, in an electrical power distribution system (*Gozde, 2011*). Figure P5.38 shows the block diagram of such a system. Assuming $K_a = 10$, $T_a = 0.1$, $K_e = 1$, $T_e = 0.4$, $K_g = 1$, $T_g = 1$, $K_s = 1$, $T_s = 0.001$, and the controller, $G_{PID}(s) = 1.6 + \frac{0.4}{s} + 0.3s$, find the closed-loop transfer function, $T(s) = \frac{\Delta V_t(s)}{\Delta V_{ref}(s)}$, of the system, expressing it as a rational function.

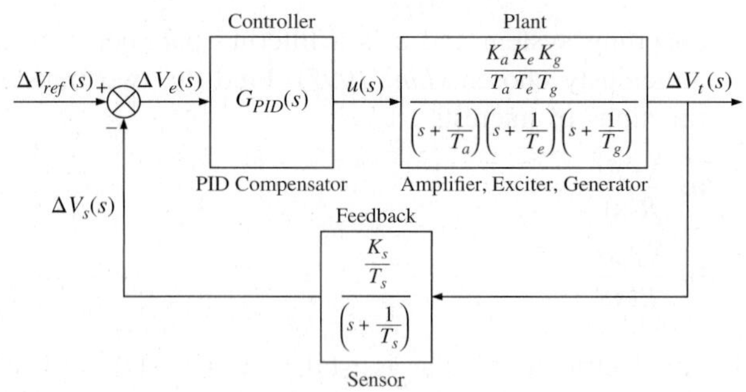


**FIGURE P5.38**

**59.** In an experiment to measure and identify postural arm reflexes, subjects hold in their hands a linear hydraulic manipulator. A load cell is attached to the actuator handle to measure resulting forces. At the application of a force, subjects try to maintain a fixed posture. Figure P5.39 shows a block diagram for the combined arm-environment system.

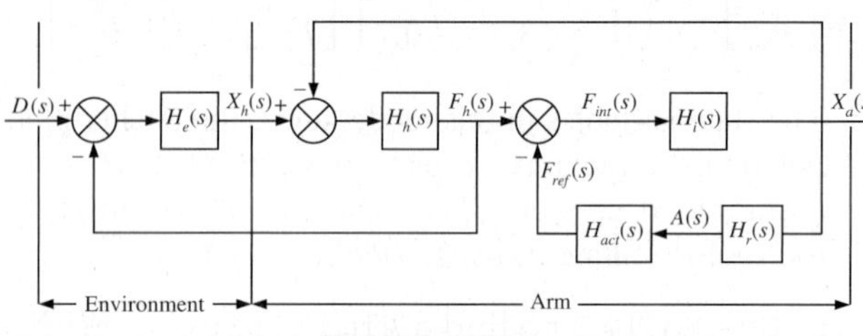


**FIGURE P5.39**

In the diagram, $H_r(s)$ represents the reflexive length and velocity feedback dynamics; $H_{act}(s)$ the activation dynamics, $H_i(s)$ the intrinsic act dynamics; $H_h(s)$ the hand dynamics; $H_e(s)$ the environmental dynamics; $X_a(s)$ the position of the arm; $X_h(s)$ the measured position of the hand; $F_h(s)$ the measured interaction force applied by the hand; $F_{int}(s)$ the intrinsic force; $F_{ref}(s)$ the reflexive force; $A(s)$ the reflexive activation; and $D(s)$ the external force perturbation (*de Vlugt, 2002*).

**a.** Obtain a signal-flow diagram from the block diagram.

**b.** Find $\frac{F_h(s)}{D(s)}$.

**60.** Use LabVIEW's Control Design and Simulation Module to obtain the controller and the observer canonical forms for:

State Space **SS**
LabVIEW **LV**

$$G(s) = \frac{s^2 + 7s + 2}{s^3 + 9s^2 + 26s + 24}$$

**61.** A drive system with an elastically coupled load was presented in Problem 67, Chapter 4. The mechanical part of this drive (*Thomsen, 2011*) was reduced to a two-inertia model. Using slightly different parameters, the following transfer function results:

$$G(s) = \frac{\Omega_L(s)}{T(s)} = \frac{25(s^2 + 1.2s + 12500)}{s(s^2 + 5.6s + 62000)}$$

Here, $T(s) = T_{em}(s) - T_L(s)$, where $T_{em}(s)$ = the electromagnetic torque developed by the motor, $T_L(s)$ = the load torque, and $\Omega_L(s)$ = the load speed.

The drive is shown in Figure P5.40 as the controlled unit in a feedback control loop, where $\Omega_r(s)$ = the desired (reference) speed. The controller transfer function is $G_C(s) = K_p + \frac{K_I}{s} = 4 + \frac{0.5}{s}$ and provides an output voltage = 0 – 5.0 volts. The motor and its power amplifier have a gain, $K_M$ = 10 N-m/volt.

**a.** Find the minor-loop transfer function, $D(s) = \frac{\Omega_L(s)}{T_{em}(s)}$ analytically or using MATLAB.

MATLAB **ML**

**b.** Given that at $t = 0$, the load speed $\omega_L(t) = 0$ rad/sec and a step reference input $\omega_r(t) = 260\ u(t)$, rad/sec, is applied, use MATLAB (or any other program) to find and plot $\omega_L(t)$. Mark on the graph all of the important characteristics, such as percent overshoot, peak time, rise time, settling time, and final steady-state value.

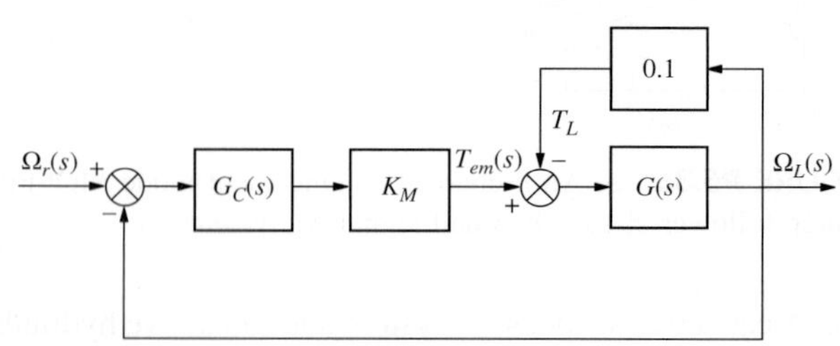


**FIGURE P5.40**

**62.** A virtual reality simulator with haptic (sense of touch) feedback was developed to simulate the control of a submarine driven through a joystick input. Operator haptic feedback is provided through joystick position constraints and simulator movement (*Karkoub, 2010*). Figure P5.41 shows the block diagram of the haptic feedback system in which the input $u_h$ is the force exerted by the muscle of the human arm; and the outputs are $y_s$, the position of the simulator, and $y_j$, the position of the joystick.

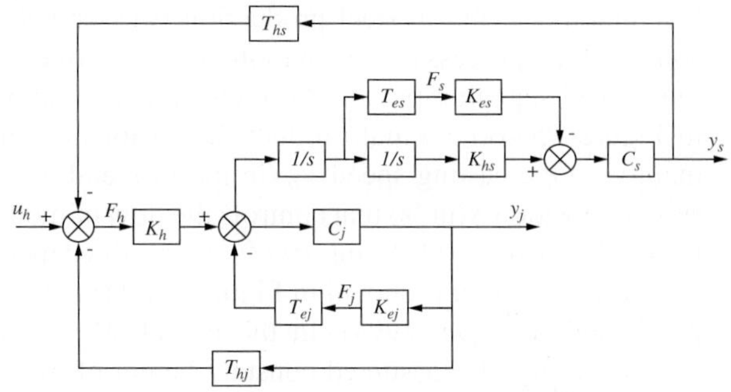


**FIGURE P5.41**[5]

**a.** Find the transfer function $\dfrac{Y_s(s)}{U_h(s)}$.

**b.** Find the transfer function $\dfrac{Y_j(s)}{U_h(s)}$.

**63.** Integrated circuits are manufactured through a lithographic process on a semiconductor wafer. In lithography, similarly to chemical photography, a semiconductor wafer is covered with a photosensitive emulsion and then selectively exposed to light to form the electronic components. Due to miniaturization, this process is to be performed with nanometer accuracy and at the highest possible speed. Sophisticated apparatus and methods have been developed for this purpose. Figure P5.42 shows the block diagram of a scanner dedicated to this purpose (*Butler, 2011*). Use Mason's Rule to find:

**a.** The transfer function $\dfrac{X_{ss}(s)}{R(s)}$.

**b.** The transfer function $\dfrac{X_{ls}(s)}{R(s)}$.

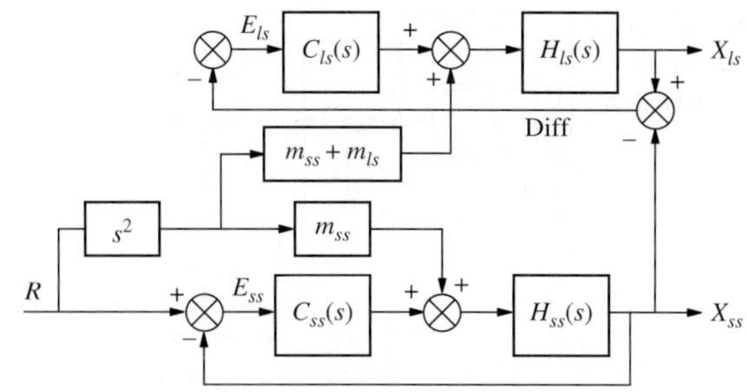


**FIGURE P5.42**[6]

**64.** Some medical procedures require the insertion of a needle under a patient's skin using CT scan monitoring guidance for precision. CT scans emit radiation, posing some cumulative risks for medical personnel. To avoid this problem, a remote control robot has been developed (*Piccin, 2009*). The robot controls the needle in position and angle in the constraint space of a CT scan machine and also provides the physician with force feedback commensurate with the insertion opposition encountered by the type of tissue in which the needle is inserted. The robot has other features that give the operator the similar sensations and maneuverability as if the needle was inserted directly. Figure P5.43 shows the block diagram of the force insertion mechanism, where $F_h$ is the input force and $X_h$ is the output displacement. Summing junction inputs are positive unless indicated with a negative sign. By way of explanation, $Z$ = impedance; $G$ = transfer function; $C_i$ = communication channel transfer functions; $F$ = force; $X$ = position. Subscripts $h$ and $m$ refer to the master manipulator. Subscripts $s$ and $e$ refer to the slave manipulator.

**a.** Assuming $Z_h = 0, C_1 = C_s, C_2 = 1 + C_6$ and $C_4 = -C_m$ use Mason's Rule to show that the transfer function from the operators force input $F_h$ to needle displacement $X_h$ is given by

$$Y(s) = \frac{X_h(s)}{F_h(s)} = \frac{Z_m^{-1}C_2(1 + G_sC_s)}{1 + G_sC_s + Z_m^{-1}(C_m + C_2Z_eG_sC_s)}$$

**b.** Now with $Z_h \neq 0$ show that $\dfrac{X_h(s)}{F_h(s)} = \dfrac{Y(s)}{1 + Y(s)Z_h}$

[5] Karkoub M., Her, M-G., and Chen, J.M. Design and control of a haptic interactive motion simulator for virtual entertainment systems, *Robotica*, vol. 28, 2010, Figure 8, p. 53. Reproduced by permission of Cambridge University Press.

[6] Butler, H. Position Control in Lithographic Equipment. *IEEE Control Systems Magazine*, October 2011, pp. 28–47. Figure 18, p. 37.

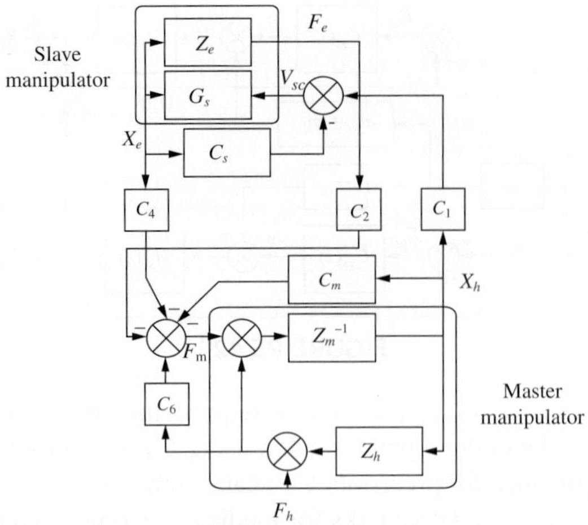


**FIGURE P5.43**[7]

**65.** A boost converter is a dc-to-dc switched power supply in which the voltage output is larger than the voltage input. A block diagram for a peak current mode controlled converter (*Chen, 2013*) is shown in Figure P5.44. Find the transfer function $\frac{\hat{v}_o(s)}{\hat{v}_i(s)}$.

**66.** A hybrid solar cell and diesel power distribution system has been proposed and tested (*Lee, 2007*). The system has been shown to have a very good uninterruptible power supply as well as line voltage regulation capabilities. Figure P5.45 shows a signal-flow diagram of the system. The output, $V_{Load}$, is the voltage across the load. The two inputs are $I_{Cf}$, the reference current, and $I_{Dist}$, the disturbance representing current changes in the supply.

**a.** Refer to Figure P5.45 and find the transfer function $\frac{V_{Load}(s)}{I_{Cf}(s)}$.

**b.** Find the transfer function $\frac{V_{Load}(s)}{I_{Dist}(s)}$.

**67.** Continuous casting in steel production is essentially a solidification process by which molten steel is solidified into a steel slab after passing through a mold, as shown in Figure P5.46(*a*). Final product dimensions depend mainly on the casting speed $V_p$ (in m/min), and on the stopper position $X$(in %) that controls the flow of molten material into the mold (*Kong, 1993*). A simplified model of a casting system is shown in Figure P5.46(*b*) (*Kong, 1993*) and (*Graebe, 1995*). In the model, $H_m$ = mold level (in mm); $H_t$ = assumed constant height of molten steel in the tundish; $D_z$ = mold thickness = depth of nozzle immerged into molten steel; and $W_t$ = weight of molten steel in the tundish.

For a specific setting let $A_m = 0.5$ and

$$G_x(s) = \frac{0.63}{s + 0.926}$$

Also assume that the valve positioning loop may be modeled by the following second-order transfer function:

$$G_V(s) = \frac{X(s)}{Y_C(s)} = \frac{100}{s^2 + 10s + 100}$$

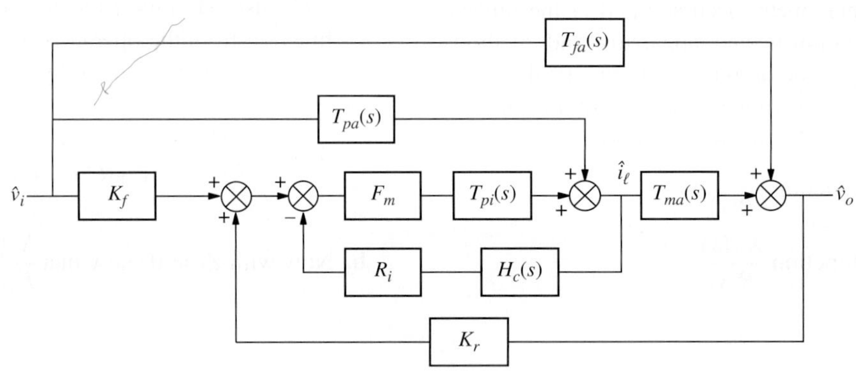


**FIGURE P5.44**[8]

[7] Piccin, O., Barbé L., Bayle B., and Mathelin M. A Force Feedback Teleoperated Needle Insertion Device for Percutaneous Procedures. *Int. J. of Robotics Research*, vol. 28, p. 1154. Figure 14. Copyright © 2009. Reprinted by Permission of SAGE.

[8] Chen, S-Y. Block diagrams and transfer functions of control to output and line-to-output for peak current-mode controlled boost converters, *IET Power Electron.*, Vol. 6, Iss. 1, pp. 60–66, 2013. Figure 4, p. 62. © The Institution of Engineering and Technology.

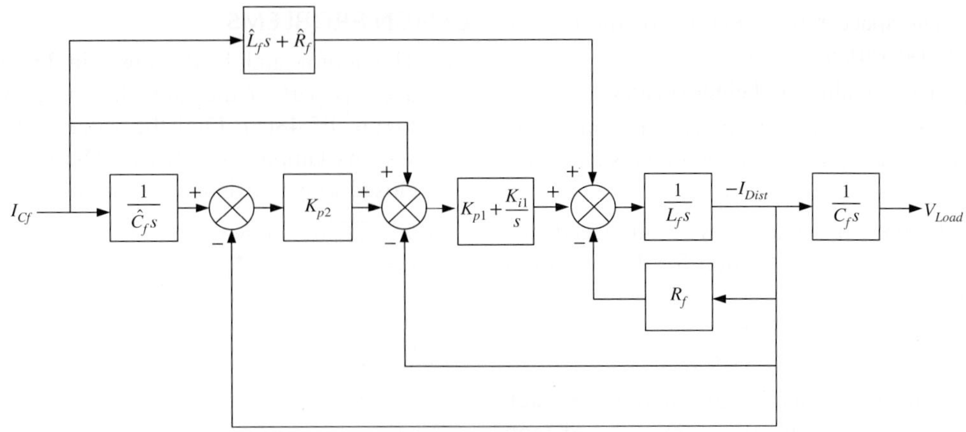


FIGURE P5.45

and the controller is modeled by the following transfer function:

$$G_C(s) = \frac{1.6(s^2 + 1.25s + 0.25)}{s}$$

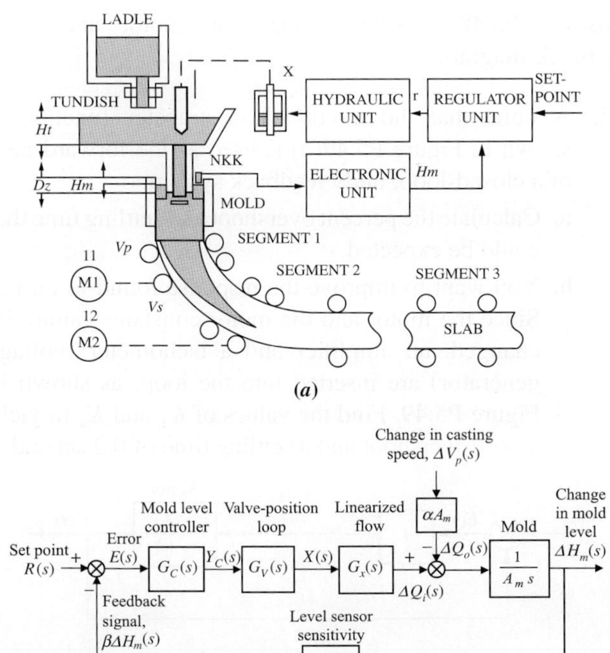


**FIGURE P5.46** Steel mold process: **a.** process;[9] **b.** block diagram

The sensitivity of the mold level sensor is $\beta = 0.5$ and the initial values of the system variables at $t = 0-$ are: $R(0-) = 0; Y_C(0-) = X(0-) = 41.2; \Delta H_m(0-) = 0; H_m(0-) = -75; \Delta V_p(0-) = 0$; and $V_p(0-) = 0$. Do the following:

**a.** Assuming $v_p(t)$ is constant $[\Delta v_p = 0]$, find the closed-loop transfer function $T(s) = \Delta H_m(s)/R(s)$.

**b.** `For` $r(t) = 5\,u(t)$, $v_p(t) = 0.97\,u(t)$, `and` $H_m(0-) = -75$ `mm, use Simulink to simulate the system. Record the time and mold level (in array format) by connecting them to` **`Workspace`** `sinks, each of which should carry the respective variable name. After the simulation ends, utilize MATLAB plot commands to obtain and edit the graph of` $h_m(t)$ `from` $t = 0$ `to 80 seconds.` Simulink **SL**

**68.** A simplified second-order transfer function model for bicycle dynamics is given by State Space **SS**

$$\frac{\varphi(s)}{\delta(s)} = \frac{aV}{bh}\frac{\left(s + \frac{V}{a}\right)}{\left(s^2 - \frac{g}{h}\right)}$$

The input is $\delta(s)$, the steering angle, and the output is $\varphi(s)$, the tilt angle (between the floor and the bicycle longitudinal plane). In the model parameter $a$ is the horizontal distance from the center of the back wheel to the bicycle center of mass; $b$ is the horizontal distance between the centers of both wheels; $h$ is the vertical distance from the center of mass to the floor; $V$ is the rear wheel velocity (assumed constant); and $g$ is the gravity constant. It is also assumed that the rider remains at a fixed position with respect to the bicycle so that the steer axis is vertical and that all angle deviations are small (*Åstrom, 2005*).

[9] Kong F., and de Keyser R. Identification and Control of the Mould Level in a Continuous Casting Machine. *Second IEEE Conference on Control Applications*, Vancouver, B.C., 1993. pp. 53–58. Figure 1. p. 53.

**a.** Obtain a state-space representation for the bicycle model in phase-variable form.

**b.** Find system eigenvalues and eigenvectors.

**c.** Find an appropriate similarity transformation matrix to diagonalize the system and obtain the state-space system's diagonal representation.

**69.** It is shown in Figure 5.6($c$) that when negative feedback is used, the overall transfer function for the system of Figure 5.6($b$) is

$$\frac{C(s)}{R(s)} = \frac{G(s)}{1+G(s)H(s)}$$

Develop the block diagram of an alternative feedback system that will result in the same closed-loop transfer function, $C(s)/R(s)$, with $G(s)$ unchanged and unmoved. In addition, your new block diagram must have unity gain in the feedback path. You can add input transducers and/or controllers in the main forward path as required.

**70.** A three-phase ac/dc converter supplies dc to a battery charging system or a dc motor (*Graovac, 2001*). Each phase had an ac filter represented by the equivalent circuit of Figure P5.47. You were asked to show that the following equation gives the $s$-domain relationship between the inductor current, $I_{acF}(s)$, and two active sources: a current source, $I_{acR}(s)$, representing a phase of the ac/dc converter, and the supply phase voltage, $V_a(s)$:

$$I_{acF}(s) = I_{acF1}(s) + I_{acF2}(s) = \frac{1+RCs}{LCs^2+RCs+1} I_{acR}(s) + \frac{Cs}{LCs^2+RCs+1} V_a(s)$$

**a.** Derive an $s$-domain equation for $V_c(s)$.

**b.** Given that $R = 1\,\Omega$, $L = 1$ mH, and $C = 20\ \mu$F, $i_{acR}(t) = 10\ u(t)$ amps, $v_a(t) = 20\ t\ u(t)$ volts,[10] and assuming zero initial conditions, use Simulink to model this system and plot the inductor current, $i_{acF}(t)$, and the capacitor voltage, $v_c(t)$, over a period from 0 to 15 ms.

Simulink SL

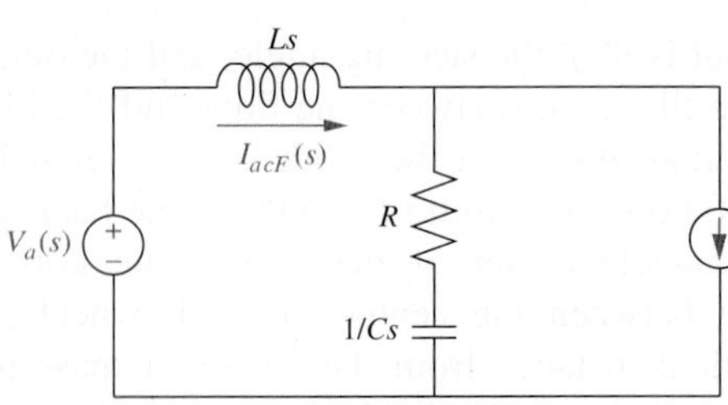


**FIGURE P5.47** AC filter equivalent circuit for a three-phase ac/dc converter

[10] Noting that a ramp is the integration of a step, we used an integrator with limits.

## DESIGN PROBLEMS

**71.** The motor and load shown in Figure P5.48($a$) are used as part of the unity feedback system shown in Figure P5.48($b$). Find the value of the coefficient of viscous damping, $D_L$, that must be used in order to yield a closed-loop transient response having a 10% overshoot.

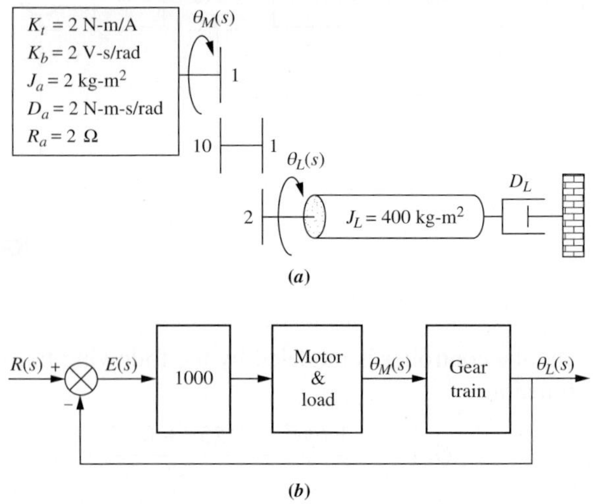


**FIGURE P5.48** Position control: **a.** motor and load; **b.** block diagram.

**72.** Assume that the motor whose transfer function is shown in Figure P5.49($a$) is used as the forward path of a closed-loop, unity feedback system.

**a.** Calculate the percent overshoot and settling time that could be expected.

**b.** You want to improve the response found in Part **a**. Since the motor and the motor constants cannot be changed, an amplifier and a tachometer (voltage generator) are inserted into the loop, as shown in Figure P5.49. Find the values of $K_1$ and $K_2$ to yield a 16% overshoot and a settling time of 0.2 second.

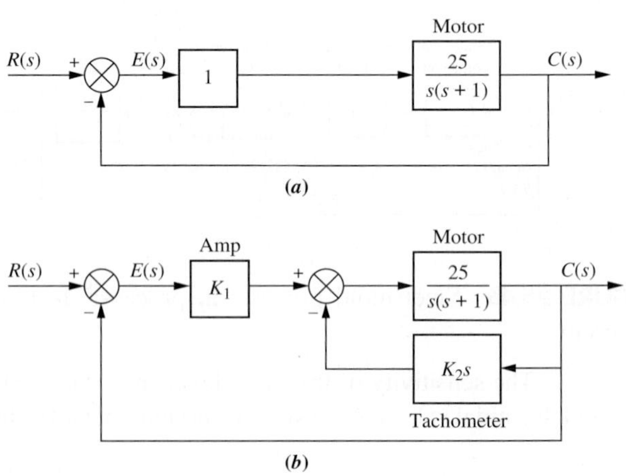


**FIGURE P5.49** **a.** Position control; **b.** position control with tachometer

73. The mechanical system shown in Figure P5.50($a$) is used as part of the unity feedback system shown in Figure P5.50($b$). Find the values of $M$ and $D$ to yield 20% overshoot and 2 seconds settling time.

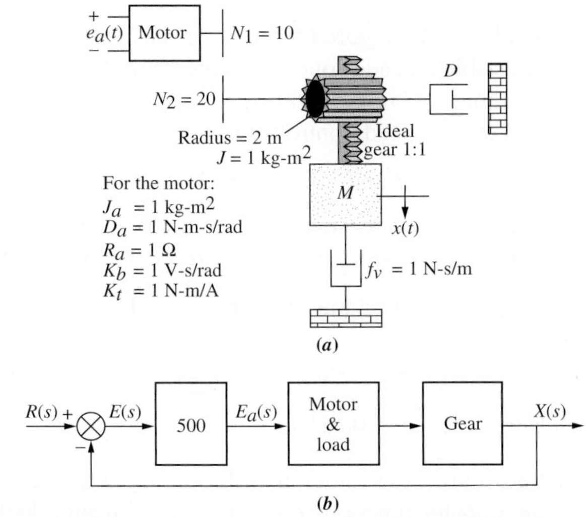


**FIGURE P5.50** **a.** Motor and load; **b.** motor and load in feedback system

74. The system shown in Figure P5.51, will have its transient response altered by adding a tachometer. Design $K$ and $K_2$ in the system to yield a damping ratio of 0.5. The natural frequency of the system before the addition of the tachometer is 10 rad/s.

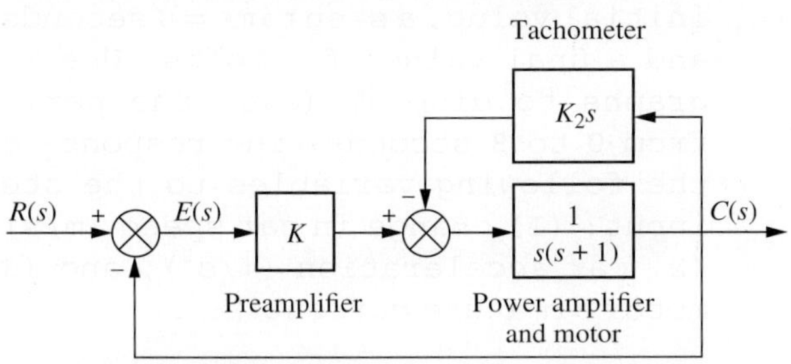


**FIGURE P5.51** Position control

75. A process is simulated by the second-order passive circuit, shown in Figure P5.52, where the feedback amplifier, controller, and final control element are represented by op-amp circuits.

**a.** Denoting the input and output as $R(s) = V_i(s)$ and $C(s) = V_o(s)$, with $R(s) - C(s) = E(s)$, and noting that the feedback amplifier has a unity gain, draw a block diagram for this feedback control system, where $G_C(s)$, $G_F(s)$, and $G_P(s)$ are the transfer functions of the controller, final control element, and the process, respectively.

**b.** Find the value of $R_P$ that makes the circuit representing the process critically damped.

**c.** Noting that the proportional controller is simply an amplifier, $G_C(s) = K_P$, find the value of its gain $K_P$ that results in dominant closed-loop poles with a damping ratio, $\zeta = 0.5$, and a settling time, $T_s = 4$ ms. Verify that the other pole is nondominant. What would be the appropriate value of the controller potentiometer, $R_F$, given that its tolerance is $\pm 10\%$?

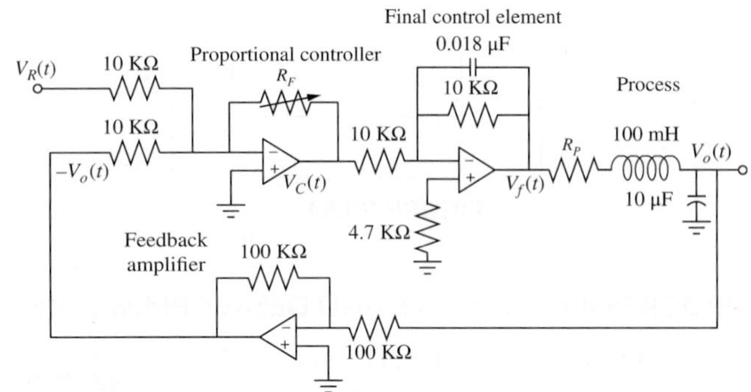


**FIGURE P5.52**

76. Assume ideal operational amplifiers in the circuit of Figure P5.53.

**a.** Show that the leftmost operational amplifier works as a subtracting amplifier. Namely, $v_1 = v_o - v_{in}$.

**b.** Draw a block diagram of the system, with the subtracting amplifier represented with a summing junction, and the circuit of the rightmost operational amplifier with a transfer function in the forward path. Keep $R$ as a variable.

**c.** Obtain the system's closed-loop transfer function.

**d.** For a unit step input, obtain the value of $R$ that will result in a settling time $T_s = 2$ msec.

**e.** Using the value of $R$ calculated in Part **d**, make a sketch of the resulting unit step response.

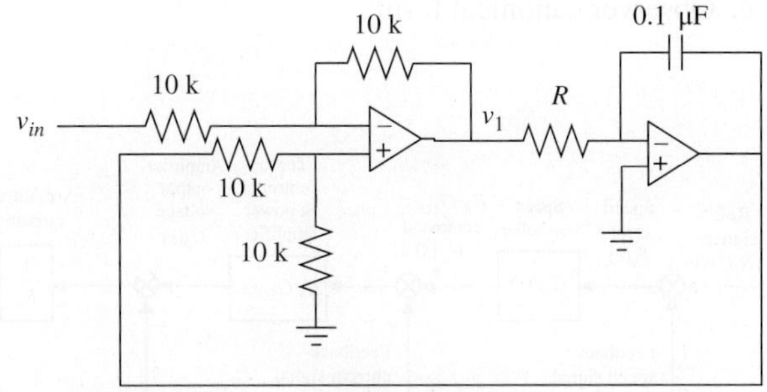


**FIGURE P5.53**

77. Given the rotational system shown in Figure P4.22 (Problem 76 of Chapter 4), do the following:

**a.** Using the transfer function you derived for that system, $G(s) = \Theta_1(s)/T(s)$, where $\Theta_1(s)$ is the angular displacement of the first shaft, find the value of $n = N_1/N_2$ that yields a settling time of 10 seconds for a step input in torque.

**b.** If this rotational system is the controlled unit, $G(s)$, in the feedback system of Figure P5.54, find the values of $\zeta$, $\omega_n$, $\%O.S.$, and $T_s$ for a controller gain $K = 4$ N-m/rad and $r(t) = u(t)$ radians.

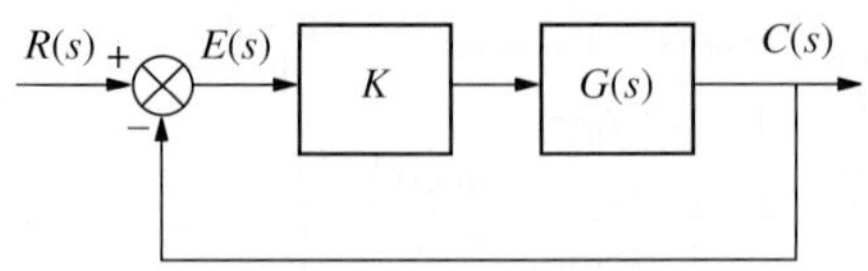


**FIGURE P5.54**

## PROGRESSIVE ANALYSIS AND DESIGN PROBLEMS

**78. Control of HIV/AIDS.** Given the HIV system of Problem 82 in Chapter 4 and repeated here for convenience (*Craig, 2004*):

State Space SS

$$\begin{bmatrix} \dot{T} \\ \dot{T}^* \\ \dot{v} \end{bmatrix} = \begin{bmatrix} -0.04167 & 0 & -0.0058 \\ 0.0217 & -0.24 & 0.0058 \\ 0 & 100 & -2.4 \end{bmatrix} \begin{bmatrix} T \\ T^* \\ v \end{bmatrix} + \begin{bmatrix} 5.2 \\ -5.2 \\ 0 \end{bmatrix} u_1$$

$$y = \begin{bmatrix} 0 & 0 & 1 \end{bmatrix} \begin{bmatrix} T \\ T^* \\ v \end{bmatrix}$$

Express the system in the following forms:

**a.** Phase-variable form

**b.** Controller canonical form

**c.** Observer canonical form

Finally,

**d.** `Use MATLAB to obtain the system's diagonalized representation.`

MATLAB ML

**79. Hybrid vehicle.** Figure P5.55 shows the block diagram of a possible cascade control scheme for an HEV driven by a dc motor (*Preitl, 2007*).

Let the speed controller $G_{SC}(s) = 100 + \dfrac{40}{s}$, the torque controller and power amp $K_A G_{TC}(s) = 10 + \dfrac{6}{s}$, the current sensor sensitivity $K_{CS} = 0.5$, the speed sensor sensitivity $K_{SS} = 0.0433$. Also, following the development in previous chapters $\dfrac{1}{R_a} = 1$; $\eta_{tot}K_t = 1.8$; $k_b = 2$; $D = k_f = 0.1$; $\dfrac{1}{J_{tot}} = \dfrac{1}{7.226}$; $\dfrac{r}{i_{tot}} = 0.0615$; and $\rho C_w A v_0 \dfrac{r}{i_{tot}} = 0.6154$.

**a.** Substitute these values in the block diagram, and find the transfer function, $T(s) = V(s)/R_v(s)$, using block-diagram reduction rules. [Hint: Start by moving the last $\dfrac{r}{i_{tot}}$ block to the right past the pickoff point.]

**b.** `Develop a Simulink model for the original system in Figure P5.54. Set the reference signal input,` $r_v(t) = 4\,u(t)$, `as a step input with a zero initial value, a step time = 0 seconds, and a final value of 4 volts. Use X-Y graphs to display (over the period from 0 to 8 seconds) the response of the following variables to the step input: (1) change in car speed (m/s), (2) car acceleration (m/s`$^2$`), and (3) motor armature current (A).`

State Space SS

MATLAB ML

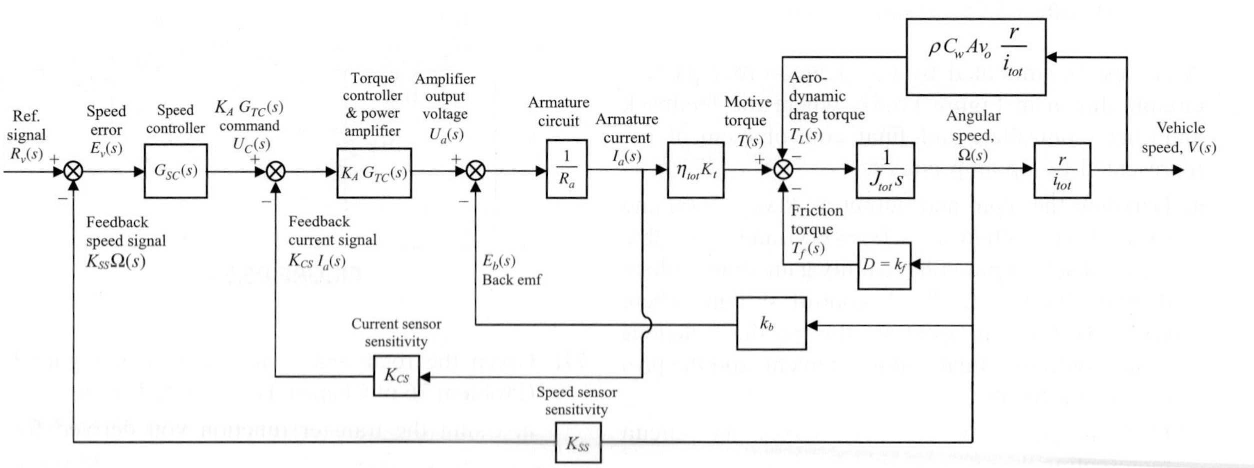


**FIGURE P5.55**

`To record the time and the above three variables (in array format), connect them to four` **`Workspace`** `sinks, each of which carries the respective variable name. After the simulation ends, utilize MATLAB plot commands to obtain and edit the three graphs of interest.`

80. **Parabolic trough collector.** Effective controller design for parabolic trough collector setups is an active area of research. One of the techniques used for controller design (*Camacho, 2012*) is Internal Model Control (IMC). Although complete details of IMC will not be presented here, Figure P5.56(*a*) shows a block diagram for the IMC setup. Use of IMC assumes a very good knowledge of the plant dynamics. In Figure P5.56(*a*), the actual plant is $P(s)$. $\tilde{P}(s)$ is a software model that mimics the plant functions. $G(s)$ is the controller to be designed. It is also assumed that all blocks represent linear time-invariant systems and thus the superposition theorem applies to the system.
 **a.** Use superposition (by assuming $D(s) = 0$) and Mason's gain formula to find the transfer function $\dfrac{C(s)}{R(s)}$ from command input to system output.
 **b.** Use superposition (by assuming $R(s) = 0$) and Mason's gain formula to find the transfer function $\dfrac{C(s)}{D(s)}$ from disturbance input to system output.
 **c.** Use the results of Parts **a** and **b** to find the combined output $C(s)$ due to both system inputs.
 **d.** Show that the system of Figure P5.56(*a*) has the same transfer function as the system in Figure P5.56(*b*) when $G_C(s) = \dfrac{G(s)}{1 - G(s)\tilde{P}(s)}$.

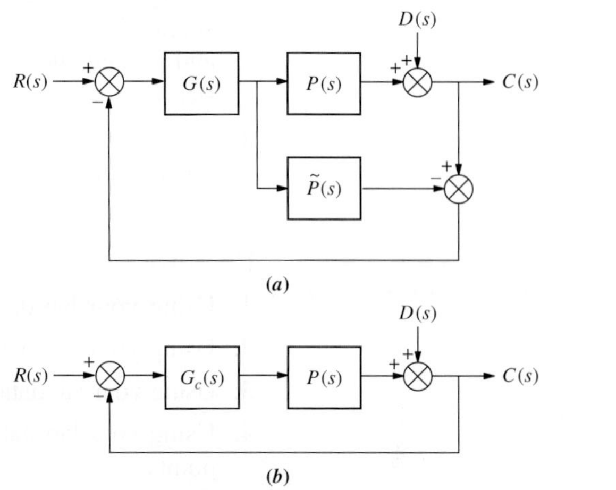


**FIGURE P5.56**

# Cyber Exploration Laboratory

## Experiment 5.1

**Objectives** To verify the equivalency of the basic forms, including cascade, parallel, and feedback forms. To verify the equivalency of the basic moves, including moving blocks past summing junctions, and moving blocks past pickoff points.

**Minimum Required Software Packages** MATLAB, Simulink, and the Control System Toolbox

### Prelab

1. Find the equivalent transfer function of three cascaded blocks, $G_1(s) = \dfrac{1}{s+1}$, $G_2(s) = \dfrac{1}{s+4}$, and $G_3(s) = \dfrac{s+3}{s+5}$.
2. Find the equivalent transfer function of three parallel blocks, $G_1(s) = \dfrac{1}{s+4}$, $G_2(s) = \dfrac{1}{s+4}$, and $G_3(s) = \dfrac{s+3}{s+5}$.
3. Find the equivalent transfer function of the negative feedback system of Figure P5.57 if $G(s) = \dfrac{s+1}{s(s+2)}$, and $H(s) = \dfrac{s+3}{s+4}$.
4. For the system of Prelab 3, push $H(s)$ to the left past the summing junction and draw the equivalent system.

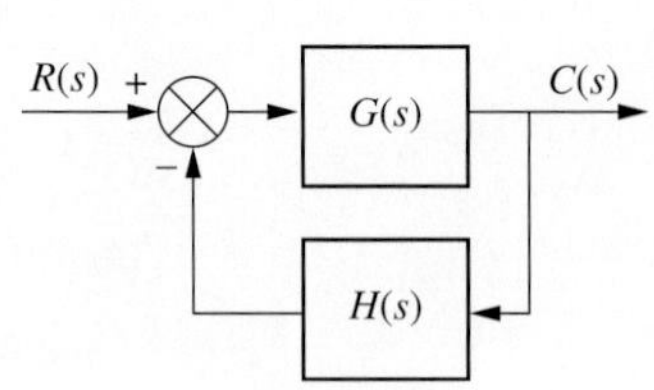


**FIGURE P5.57**

5. For the system of Prelab 3, push $H(s)$ to the right past the pickoff point and draw the equivalent system.

## Lab

1. Using Simulink, set up the cascade system of Prelab 1 and the equivalent single block. Make separate plots of the step response of the cascaded system and its equivalent single block. Record the values of settling time and rise time for each step response.
2. Using Simulink, set up the parallel system of Prelab 2 and the equivalent single block. Make separate plots of the step response of the parallel system and its equivalent single block. Record the values of settling time and rise time for each step response.
3. Using Simulink, set up the negative feedback system of Prelab 3 and the equivalent single block. Make separate plots of the step response of the negative feedback system and its equivalent single block. Record the values of settling time and rise time for each step response.
4. Using Simulink, set up the negative feedback systems of Prelabs 3, 4, and 5. Make separate plots of the step response of each of the systems. Record the values of settling time and rise time for each step response.

## Postlab

1. Using your lab data, verify the equivalent transfer function of blocks in cascade.
2. Using your lab data, verify the equivalent transfer function of blocks in parallel.
3. Using your lab data, verify the equivalent transfer function of negative feedback systems.
4. Using your lab data, verify the moving of blocks past summing junctions and pickoff points.
5. Discuss your results. Were the equivalencies verified?

## Experiment 5.2

**Objective** To use the various functions within LabVIEW's Control Design and Simulation Module to implement block diagram reduction.

**Minimum Required Software Packages** LabVIEW with the Control Design Simulation Module

**Prelab** Given the block diagram from Example 5.2, replace $G_1$, $G_2$, $G_3$, $H_1$, $H_2$, $H_3$ with the following transfer functions and obtain an equivalent transfer function.

$$G_1 = \frac{1}{s+10}; G_2 = \frac{1}{s+1}; G_3 = \frac{s+1}{s^2+4s+4}; H_1 = \frac{s+1}{s+2}; H_2 = 2; H_3 = 1$$

**Lab** Use LabVIEW to implement the block diagram from Example 5.2 using the transfer functions given in the Prelab.

**Postlab** Verify your calculations from the Prelab with that of the equivalent transfer function obtained with LabVIEW.

## Experiment 5.3

**Objective** To use the various functions within LabVIEW's Control Design and Simulation Module and the Mathematics/Polynomial palette to implement Mason's rule for block diagram reduction.

**Minimum Required Software Packages** LabVIEW with Control Design and Simulation Module, Math Script RT Module, and the Mathematics/Polynomial palette.

**Prelab** Given the block diagram created in the Prelab of Cyber Exploration Laboratory 5.2, use Mason's rule to obtain an equivalent transfer function.

**Lab** Use LabVIEW's Control Design and Simulation Module as well as the Mathematics/Polynomial functions to implement block diagram reduction using Mason's rule.

**Postlab** Verify your calculations from the Prelab with that of the equivalent transfer function obtained with LabVIEW.

# Bibliography


Åstrom, K., Klein, R. E., and Lennartsson, A. Bicycle Dynamics and Control. *IEEE Control Systems*, August 2005, pp. 26–47.

Ballard, R. D. *The Discovery of the Titanic*, Warner Books, New York, 1987. Also, figure caption source for Figure 5.33.

Butler, H. Position Control in Lithographic Equipment. *IEEE Control Systems*, October 2011, pp. 28–47.

Camacho, E. F., Berenguel, M., Rubio, F. R., and Martinez, D. *Control of Solar Energy Systems*. Springer-Verlag, London, 2012.

Chen, S.-Y. Block diagrams and transfer functions of control-to-output and line-to-output for peak current-mode controlled boost converters. *IET Power Electron.*, vol. 6, no. 1, pp. 60–66, 2013.

Craig, I. K., Xia, X., and Venter, J. W., Introducing HIV/AIDS Education into the Electrical Engineering Curriculum at the University of Pretoria. *IEEE Transactions on Education*, vol. 47, no. 1, February 2004, pp. 65–73.

de Vlugt, E., Schouten, A. C., and van derHelm, F.C.T. Adaptation of Reflexive Feedback during Arm Posture to Different Environments. *Biological Cybernetics*, vol. 87, 2002, pp. 10–26.

Ballard, R. D *The Discovery of the Titanic* (New York: Warner Books, Inc 1987).

Gozde, H., and Taplamacioglu, M. C. Comparative performance analysis of artificial bee colony algorithm for automatic voltage regulator (AVR) system. *Journal of the Franklin Institute*, vol. 348, pp. 1927–1946, 2011.

Graebe, S. F., Goodwin, G. C., and Elsley, G. Control Design and Implementation in Continuous Steel Casting. *IEEE Control Systems*, August 1995, pp. 64–71.

Hostetter, G. H., Savant, C. J., Jr. and Stefani, R. T. *Design of Feedback Control Systems*. 2d ed. Saunders College Publishing, New York, 1989.

Johnson, H. et al. *Unmanned Free-Swimming Submersible (UFSS) System Description*. NRL Memorandum Report 4393. Naval Research Laboratory, Washington, DC, 1980.

Karkoub, M., Her, M.-G., and Chen, J. M. Design and Control of a Haptic Interactive Motion Simulator for Virtual Entertainment Systems. *Robonica*, vol. 28, 2010, pp. 47–56.

Kong, F., and de Keyser, R. Identification and Control of the Mould Level in a Continuous Casting Machine. *Second IEEE Conference on Control Application*, Vancouver, B.C., 1993. pp. 53–58.

Lee, S.-R., Ko, S.-H., Dehbonei, H., Jeon, C.-H., and Kwon, O.-S., Operational Characteristics of PV/Diesel Hybrid Distributed Generation System Using Dual Voltage Source Inverter for Weak Grid, *ISIS 2007 Proceedings*, Eighth Symposium on Advanced Intelligent Systems. 2007.

Lin, J.-S., and Kanellakopoulos, I., Nonlinear Design of Active Suspensions. *IEEE Control Systems*, vol. 17, issue 3, June 1997, pp. 45–59.

Mason, S. J. Feedback Theory—Some Properties of Signal-Flow Graphs. *Proc. IRE*, September 1953, pp. 1144–1156.

Neamen, D. A. *Electronic Circuit Analysis and Design*. McGraw-Hill, 2d ed., 2001, p. 334.

Piccin, O., Barbé L., Bayle B., and de Mathelin, M. A Force Feedback Teleoperated Needle Insertion Device for Percutaneous Procedures. *Int. J. of Robotics Research*, vol. 28, 2009, p. 1154.

Preitl, Z., Bauer, P., and Bokor, J. A Simple Control Solution for Traction Motor Used in Hybrid Vehicles. *Fourth International Symposium on Applied Computational Intelligence and Informatics*. IEEE, 2007, pp. 157–162.

Tasch, U., Koontz, J. W., Ignatoski, M. A., and Geselowitz, D. B. An Adaptive Aortic Pressure Observer for the Penn State Electric Ventricular Assist Device. *IEEE Transactions on Biomedical Engineering*, vol. 37, 1990, pp. 374–383.

Timothy, L. K., and Bona, B. E. *State Space Analysis: An Introduction*. McGraw-Hill, New York, 1968.

Vaughan, N. D., and Gamble, J. B. The Modeling and Simulation of a Proportional Solenoid Valve. *Journal of Dynamic Systems, Measurements, and Control*, vol. 118, March 1996, pp. 120–125.

Yaniv, Y., Sivan, R., and Landesberg, A. Stability, Controllability and Observability of the "Four State" Model for the Sarcomeric Control of Contraction. *Annals of Biomedical Engineering*, vol. 34, 2006, pp. 778–789.

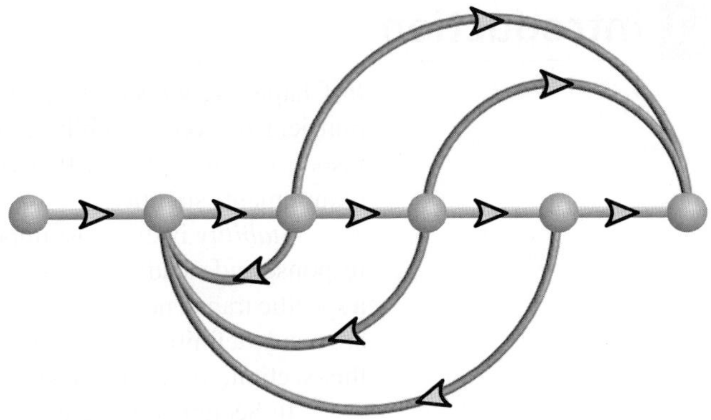

# 6 Stability

## Chapter Learning Outcomes

After completing this chapter the student will be able to:

- Make and interpret a basic Routh table to determine the stability of a system (Sections 6.1–6.2)
- Make and interpret a Routh table where either the first element of a row is zero or an entire row is zero (Sections 6.3–6.4)
- Use a Routh table to determine the stability of a system represented in state space (Section 6.5)

State Space SS

## ▶ Case Study Learning Outcomes

You will be able to demonstrate your knowledge of the chapter objectives with case studies as follows:

- Given the antenna azimuth position control system shown on the front endpapers, you will be able to find the range of preamplifier gain to keep the system stable.
- Given the block diagrams for the UFSS vehicle's pitch and heading control systems on the back endpapers, you will be able to determine the range of gain for stability of the pitch or heading control system.

## 6.1 Introduction

In Chapter 1, we saw that three requirements enter into the design of a control system: transient response, stability, and steady-state errors. Thus far we have covered transient response, which we will revisit in Chapter 8. We are now ready to discuss the next requirement, stability.

*Stability* is the most important system specification. If a system is unstable, transient response and steady-state errors are moot points. An unstable system cannot be designed for a specific transient response or steady-state error requirement. What, then, is stability? There are many definitions for stability, depending upon the kind of system or the point of view. In this section, we limit ourselves to linear, time-invariant systems.

In Section 1.5, we discussed that we can control the output of a system if the steady-state response consists of only the forced response. But the total response of a system is the sum of the forced and natural responses, or

$$c(t) = c_{\text{forced}}(t) + c_{\text{natural}}(t) \tag{6.1}$$

Using these concepts, we present the following definitions of stability, instability, and marginal stability:

A linear, time-invariant system is *stable* if the natural response approaches zero as time approaches infinity.

A linear, time-invariant system is *unstable* if the natural response grows without bound as time approaches infinity.

A linear, time-invariant system is *marginally stable* if the natural response neither decays nor grows but remains constant or oscillates as time approaches infinity.

Thus, the definition of stability implies that only the forced response remains as the natural response approaches zero.

These definitions rely on a description of the natural response. When one is looking at the total response, it may be difficult to separate the natural response from the forced response. However, we realize that if the input is bounded and the total response is not approaching infinity as time approaches infinity, then the natural response is obviously not approaching infinity. If the input is unbounded, we see an unbounded total response, and we cannot arrive at any conclusion about the stability of the system; we cannot tell whether the total response is unbounded because the forced response is unbounded or because the natural response is unbounded. Thus, our alternate definition of *stability*, one that regards the total response and implies the first definition based upon the natural response, is this:

A system is stable if *every* bounded input yields a bounded output.

We call this statement the bounded-input, bounded-output (BIBO) definition of stability.

Let us now produce an alternate definition for instability based on the total response rather than the natural response. We realize that if the input is bounded but the total response is unbounded, the system is unstable, since we can conclude that the natural response approaches infinity as time approaches infinity. If the input is unbounded, we will see an unbounded total response, and we cannot draw any conclusion about the stability of the system; we cannot tell whether the total response is unbounded because the forced response is unbounded or because the natural response is unbounded. Thus, our alternate definition of *instability*, one that regards the total response, is this:

A system is unstable if *any* bounded input yields an unbounded output.

These definitions help clarify our previous definition of *marginal stability*, which really means that the system is stable for some bounded inputs and unstable for others. For example, we will show that if the natural response is undamped, a bounded sinusoidal input of the same frequency yields a natural response of growing oscillations. Hence, the system appears stable for all bounded inputs except this one sinusoid. Thus, marginally stable systems by the natural response definitions are included as unstable systems under the BIBO definitions.

Let us summarize our definitions of stability for linear, time-invariant systems. Using the natural response:

1. A system is stable if the natural response approaches zero as time approaches infinity.
2. A system is unstable if the natural response approaches infinity as time approaches infinity.
3. A system is marginally stable if the natural response neither decays nor grows but remains constant or oscillates.

Using the total response (BIBO):

1. A system is stable if *every* bounded input yields a bounded output.
2. A system is unstable if *any* bounded input yields an unbounded output.

Physically, an unstable system whose natural response grows without bound can cause damage to the system, to adjacent property, or to human life. Many times systems are designed with limited stops to prevent total runaway. From the perspective of the time response plot of a physical system, instability is displayed by transients that grow without bound and, consequently, a total response that does not approach a steady-state value or other forced response.[1]

How do we determine if a system is stable? Let us focus on the natural response definitions of stability. Recall from our study of system poles that poles in the left half-plane (lhp) yield either pure exponential decay or damped sinusoidal natural responses. These natural responses decay to zero as time approaches infinity. Thus, if the closed-loop system poles are in the left half of the plane and hence have a negative real part, the system is stable. That is, *stable systems have closed-loop transfer functions with poles only in the left half-plane*.

Poles in the right half-plane (rhp) yield either pure exponentially increasing or exponentially increasing sinusoidal natural responses. These natural responses approach infinity as time approaches infinity. Thus, if the closed-loop system poles are in the right half of the $s$-plane and hence have a positive real part, the system is unstable. Also, poles of multiplicity greater than 1 on the imaginary axis lead to the sum of responses of the form $At^n \cos(\omega t + \phi)$, where $n = 1, 2, \ldots,$ where the amplitude approaches infinity as time approaches infinity. Thus, *unstable systems have closed-loop transfer functions with at least one pole in the right half-plane and/or poles of multiplicity greater than 1 on the imaginary axis*.

Finally, a system that has imaginary axis poles of multiplicity 1 yields pure sinusoidal oscillations as a natural response. These responses neither increase nor decrease in amplitude. Thus, *marginally stable systems have closed-loop transfer functions with only imaginary axis poles of multiplicity 1 and poles in the left half-plane*.

As an example, the unit step response of the stable system of Figure 6.1($a$) is compared to that of the unstable system of Figure 6.1($b$). The responses, also shown in

[1] Care must be taken here to distinguish between natural responses growing without bound and a forced response, such as a ramp or exponential increase, that also grows without bound. A system whose forced response approaches infinity is stable as long as the natural response approaches zero.

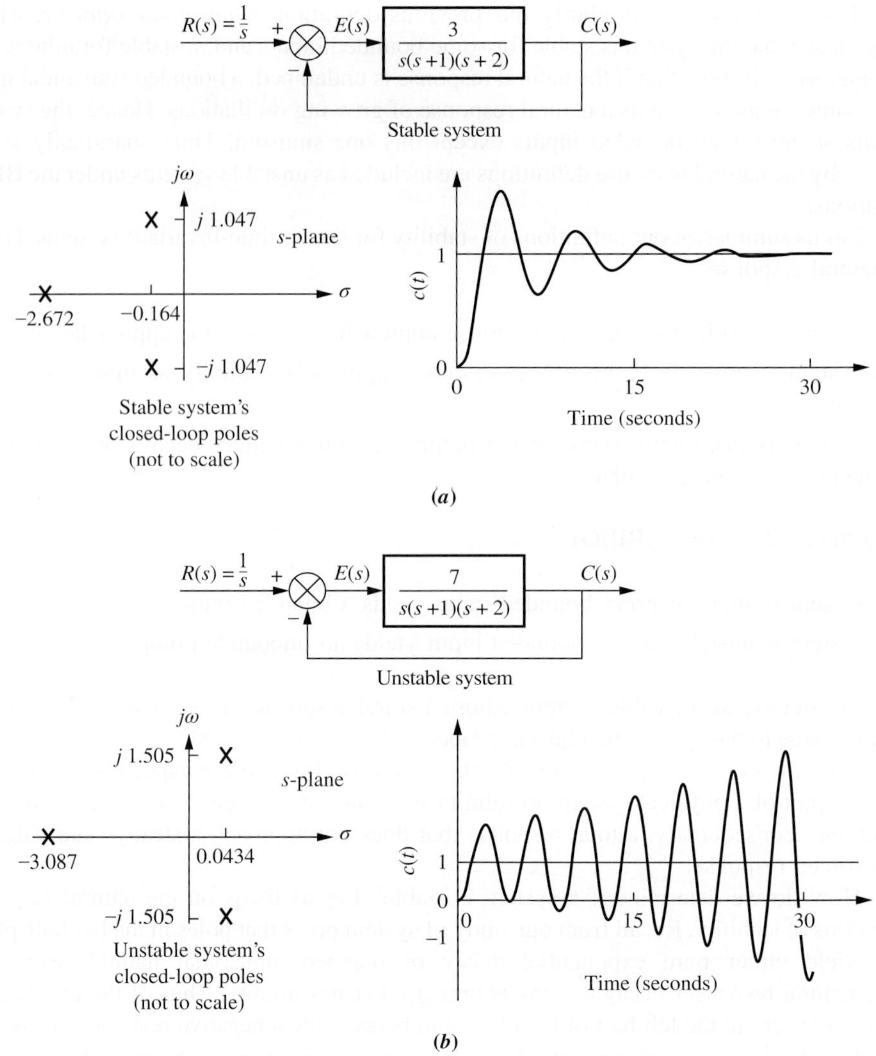


**FIGURE 6.1** Closed-loop poles and response: **a.** stable system; **b.** unstable system

Figure 6.1, shows that while the oscillations for the stable system diminish, those for the unstable system increase without bound. Also notice that the stable system's response in this case approaches a steady-state value of unity.

It is not always a simple matter to determine if a feedback control system is stable. Unfortunately, a typical problem that arises is shown in Figure 6.2. Although we know the poles of the forward transfer function in Figure 6.2(*a*), we do not know the location of the poles of the equivalent closed-loop system of Figure 6.2(*b*) without factoring or otherwise solving for the roots.

However, under certain conditions, we can draw some conclusions about the stability of the system. First, if the closed-loop transfer function has only left-half-plane poles, then the factors of the denominator of the closed-loop system transfer function consist of products of terms such as $(s + a_i)$, where $a_i$ is real and positive, or complex with a positive real part. The product of such terms is a polynomial with all positive coefficients.[2] No term of the polynomial can be missing, since that would imply cancellation between positive and negative coefficients or imaginary axis roots in the factors, which is not the case. Thus, a

[2] The coefficients can also be made all negative by multiplying the polynomial by −1. This operation does not change the root location.

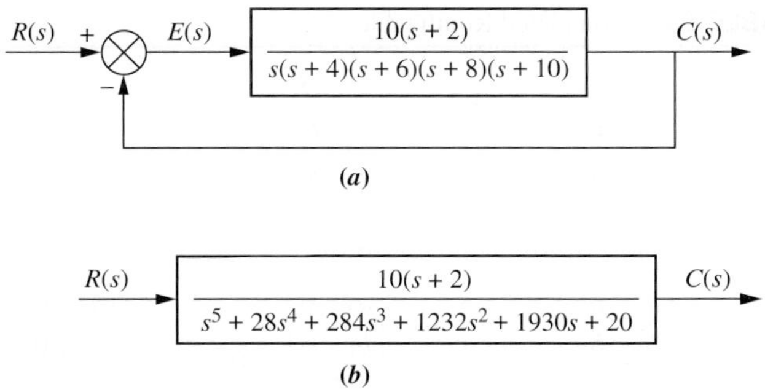


**FIGURE 6.2** Common cause of problems in finding closed-loop poles: **a.** original system; **b.** equivalent system

sufficient condition for a system to be unstable is that all signs of the coefficients of the denominator of the closed-loop transfer function are not the same. If powers of $s$ are missing, the system is either unstable or, at best, marginally stable. Unfortunately, if all coefficients of the denominator are positive and not missing, we do not have definitive information about the system's pole locations.

If the method described in the previous paragraph is not sufficient, then a computer can be used to determine the stability by calculating the root locations of the denominator of the closed-loop transfer function. Today some hand-held calculators can evaluate the roots of a polynomial. There is, however, another method to test for stability without having to solve for the roots of the denominator. We discuss this method in the next section.

## 6.2 Routh-Hurwitz Criterion

In this section, we learn a method that yields stability information without the need to solve for the closed-loop system poles. Using this method, we can tell how many closed-loop system poles are in the left half-plane, in the right half-plane, and on the $j\omega$-axis. (Notice that we say *how many*, not *where*.) We can find the number of poles in each section of the $s$-plane, but we cannot find their coordinates. The method is called the *Routh-Hurwitz criterion* for stability (*Routh, 1905*).

The method requires two steps: (1) Generate a data table called a *Routh table* and (2) interpret the Routh table to tell how many closed-loop system poles are in the left half-plane, the right half-plane, and on the $j\omega$-axis. You might wonder why we study the Routh-Hurwitz criterion when modern calculators and computers can tell us the exact location of system poles. The power of the method lies in design rather than analysis. For example, if you have an unknown parameter in the denominator of a transfer function, it is difficult to determine via a calculator the range of this parameter to yield stability. You would probably rely on trial and error to answer the stability question. We shall see later that the Routh-Hurwitz criterion can yield a closed-form expression for the range of the unknown parameter.

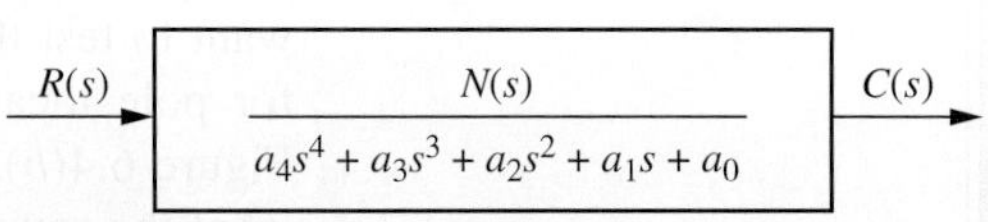


**FIGURE 6.3** Equivalent closed-loop transfer function

In this section, we make and interpret a basic Routh table. In the next section, we consider two special cases that can arise when generating this data table.

### Generating a Basic Routh Table

Look at the equivalent closed-loop transfer function shown in Figure 6.3. Since we are interested in the system poles, we focus our attention on the denominator. We first create the Routh table shown in Table 6.1. Begin by labeling the rows with powers of $s$ from the highest power of the denominator of the closed-loop transfer function to $s^0$. Next start with the coefficient of the highest power of $s$ in the denominator and list, horizontally in the first row, every other coefficient. In the second row, list horizontally, starting with the next highest power of $s$, every coefficient that was skipped in the first row.

**TABLE 6.1** Initial layout for Routh table

| | | | |
|---|---|---|---|
| $s^4$ | $a_4$ | $a_2$ | $a_0$ |
| $s^3$ | $a_3$ | $a_1$ | 0 |
| $s^2$ | | | |
| $s^1$ | | | |
| $s^0$ | | | |

**TABLE 6.2** Completed Routh table

| | | | |
|---|---|---|---|
| $s^4$ | $a_4$ | $a_2$ | $a_0$ |
| $s^3$ | $a_3$ | $a_1$ | $0$ |
| $s^2$ | $\dfrac{-\begin{vmatrix} a_4 & a_2 \\ a_3 & a_1 \end{vmatrix}}{a_3} = b_1$ | $\dfrac{-\begin{vmatrix} a_4 & a_0 \\ a_3 & 0 \end{vmatrix}}{a_3} = b_2$ | $\dfrac{-\begin{vmatrix} a_4 & 0 \\ a_3 & 0 \end{vmatrix}}{a_3} = 0$ |
| $s^1$ | $\dfrac{-\begin{vmatrix} a_3 & a_1 \\ b_1 & b_2 \end{vmatrix}}{b_1} = c_1$ | $\dfrac{-\begin{vmatrix} a_3 & 0 \\ b_1 & 0 \end{vmatrix}}{b_1} = 0$ | $\dfrac{-\begin{vmatrix} a_3 & 0 \\ b_1 & 0 \end{vmatrix}}{b_1} = 0$ |
| $s^0$ | $\dfrac{-\begin{vmatrix} b_1 & b_2 \\ c_1 & 0 \end{vmatrix}}{c_1} = d_1$ | $\dfrac{-\begin{vmatrix} b_1 & 0 \\ c_1 & 0 \end{vmatrix}}{c_1} = 0$ | $\dfrac{-\begin{vmatrix} b_1 & 0 \\ c_1 & 0 \end{vmatrix}}{c_1} = 0$ |

The remaining entries are filled in as follows. Each entry is a negative determinant of entries in the previous two rows divided by the entry in the first column directly above the calculated row. The left-hand column of the determinant is always the first column of the previous two rows, and the right-hand column is the elements of the column above and to the right. The table is complete when all of the rows are completed down to $s^0$. Table 6.2 is the completed Routh table. Let us look at an example.

## Example 6.1 Creating a Routh Table

**PROBLEM:** Make the Routh table for the system shown in Figure 6.4(*a*).

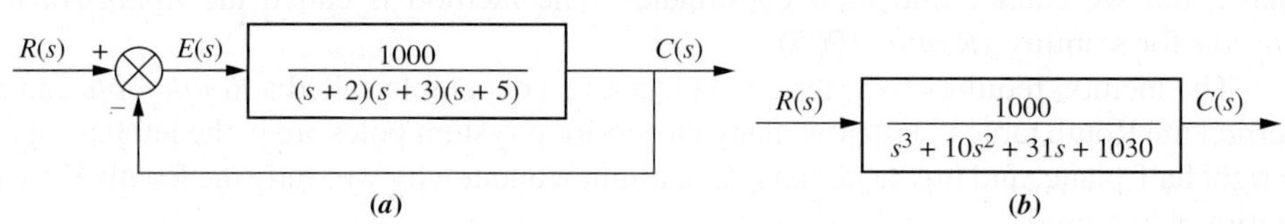


**FIGURE 6.4** **a.** Feedback system for Example 6.1; **b.** equivalent closed-loop system

**SOLUTION:** The first step is to find the equivalent closed-loop system because we want to test the denominator of this function, not the given forward transfer function, for pole location. Using the feedback formula, we obtain the equivalent system of Figure 6.4(*b*). The Routh-Hurwitz criterion will be applied to this denominator. First label the rows with powers of $s$ from $s^3$ down to $s^0$ in a vertical column, as shown in Table 6.3. Next form the first row of the table, using the coefficients of the denominator of the closed-loop transfer function. Start with the coefficient of the highest power and skip every other power of $s$. Now form the second row with the coefficients of the denominator skipped in the previous step. Subsequent rows are formed with determinants, as shown in Table 6.2.

For convenience, any row of the Routh table can be multiplied by a positive constant without changing the values of the rows below. This can be proved by examining the expressions for the entries and verifying that any multiplicative constant from a previous row cancels out. In the second row of Table 6.3, for example, the row was multiplied by 1/10. We see later that care must be taken not to multiply the row by a negative constant.

**TABLE 6.3** Completed Routh table for Example 6.1

| | | | |
|---|---|---|---|
| $s^3$ | 1 | 31 | 0 |
| $s^2$ | ~~10~~ 1 | ~~1030~~ 103 | 0 |
| $s^1$ | $\dfrac{-\begin{vmatrix}1 & 31\\1 & 103\end{vmatrix}}{1} = -72$ | $\dfrac{-\begin{vmatrix}1 & 0\\0 & 0\end{vmatrix}}{1} = 0$ | $\dfrac{-\begin{vmatrix}1 & 0\\1 & 0\end{vmatrix}}{1} = 0$ |
| $s^0$ | $\dfrac{-\begin{vmatrix}1 & 103\\-72 & 0\end{vmatrix}}{-72} = 103$ | $\dfrac{-\begin{vmatrix}1 & 0\\-72 & 0\end{vmatrix}}{-72} = 0$ | $\dfrac{-\begin{vmatrix}1 & 0\\-72 & 0\end{vmatrix}}{-72} = 0$ |

## Interpreting the Basic Routh Table

Now that we know how to generate the Routh table, let us see how to interpret it. The basic Routh table applies to systems with poles in the left and right half-planes. Systems with imaginary poles and the kind of Routh table that results will be discussed in the next section. Simply stated, the Routh-Hurwitz criterion declares that *the number of roots of the polynomial that are in the right half-plane is equal to the number of sign changes in the first column*.

If the closed-loop transfer function has all poles in the left half of the $s$-plane, the system is stable. Thus, a system is stable if there are no sign changes in the first column of the Routh table. For example, Table 6.3 has two sign changes in the first column. The first sign change occurs from 1 in the $s^2$ row to $-72$ in the $s^1$ row. The second occurs from $-72$ in the $s^1$ row to 103 in the $s^0$ row. Thus, the system of Figure 6.4 is unstable since two poles exist in the right half-plane.

### Skill-Assessment Exercise 6.1

**PROBLEM:** Make a Routh table and tell how many roots of the following polynomial are in the right half-plane and in the left half-plane.

$$P(s) = 3s^7 + 9s^6 + 6s^5 + 4s^4 + 7s^3 + 8s^2 + 2s + 6$$

**ANSWER:** Four in the right half-plane (rhp), three in the left half-plane (lhp).

The complete solution is at www.wiley.com/college/nise.

Now that we have described how to generate and interpret a basic Routh table, let us look at two special cases that can arise.

# 6.3 Routh-Hurwitz Criterion: Special Cases

Two special cases can occur: (1) The Routh table sometimes will have a zero *only in the first column* of a row, or (2) the Routh table sometimes will have an *entire row* that consists of zeros. Let us examine the first case.

## Zero Only in the First Column

If the first element of a row is zero, division by zero would be required to form the next row. To avoid this phenomenon, an epsilon, $\epsilon$, is assigned to replace the zero in the first column. The value $\epsilon$ is then allowed to approach zero from either the positive or the negative side, after which the signs of the entries in the first column can be determined. Let us look at an example.

### Example 6.2 Stability via Epsilon Method

**TryIt 6.1**

Use the following MATLAB statement to find the poles of the closed-loop transfer function in Eq. (6.1).

```
roots([1 2 3 6 5 3])
```

**PROBLEM:** Determine the stability of the closed-loop transfer function

$$T(s) = \frac{10}{s^5 + 2s^4 + 3s^3 + 6s^2 + 5s + 3} \tag{6.2}$$

**SOLUTION:** The solution is shown in Table 6.4. We form the Routh table by using the denominator of Eq. (6.2). Begin by assembling the Routh table down to the row where a zero appears *only* in the first column (the $s^3$ row). Next replace the zero by a small number, $\epsilon$, and complete the table. To begin the interpretation, we must first assume a sign, positive or negative, for the quantity $\epsilon$. Table 6.5 shows the first column of Table 6.4 along with the resulting signs for choices of $\epsilon$ positive and $\epsilon$ negative.

**TABLE 6.4** Completed Routh table for Example 6.2

| | | | |
|---|---|---|---|
| $s^5$ | 1 | 3 | 5 |
| $s^4$ | 2 | 6 | 3 |
| $s^3$ | $\cancel{0}\ \epsilon$ | $\frac{7}{2}$ | 0 |
| $s^2$ | $\frac{6\epsilon - 7}{\epsilon}$ | 3 | 0 |
| $s^1$ | $\frac{42\epsilon - 49 - 6\epsilon^2}{12\epsilon - 14}$ | 0 | 0 |
| $s^0$ | 3 | 0 | 0 |

**TABLE 6.5** Determining signs in first column of a Routh table with zero as first element in a row

| Label | First column | $\epsilon = +$ | $\epsilon = -$ |
|---|---|---|---|
| $s^5$ | 1 | + | + |
| $s^4$ | 2 | + | + |
| $s^3$ | $\cancel{0}\ \epsilon$ | + | − |
| $s^2$ | $\frac{6\epsilon - 7}{\epsilon}$ | − | + |
| $s^1$ | $\frac{42\epsilon - 49 - 6\epsilon^2}{12\epsilon - 14}$ | + | + |
| $s^0$ | 3 | + | + |

If $\epsilon$ is chosen positive, Table 6.5 will show a sign change from the $s^3$ row to the $s^2$ row, and there will be another sign change from the $s^2$ row to the $s^1$ row. Hence, the system is unstable and has two poles in the right half-plane.

Alternatively, we could choose $\epsilon$ negative. Table 6.5 would then show a sign change from the $s^4$ row to the $s^3$ row. Another sign change would occur from the $s^3$ row to the $s^2$ row. Our result would be exactly the same as that for a positive choice for $\epsilon$. Thus, the system is unstable, with two poles in the right half-plane.

Symbolic Math

**SM**

Students who are performing the MATLAB exercises and want to explore the added capability of MATLAB's Symbolic Math Toolbox should now run ch6sp1 in Appendix F at www.wiley.com/college/nise. You will learn how to use the Symbolic Math Toolbox to calculate the values of cells in a Routh table even if the table contains symbolic objects, such as $\epsilon$. You will see that the Symbolic Math Toolbox and MATLAB yield an alternate way to generate the Routh table for Example 6.2.

Another method that can be used when a zero appears only in the first column of a row is derived from the fact that a polynomial that has the reciprocal roots of the original polynomial has its roots distributed the same—right half-plane, left half-plane, or imaginary axis—because taking the reciprocal of the root value does not move it to another region. Thus, if we can find the polynomial that has the reciprocal roots of the original, it is possible that the Routh table for the new polynomial will not have a zero in the first column. This method is usually computationally easier than the epsilon method just described.

We now show that the polynomial we are looking for, the one with the reciprocal roots, is simply the original polynomial with its coefficients written in reverse order (*Phillips, 1991*). Assume the equation

$$s^n + a_{n-1}s^{n-1} + \cdots + a_1 s + a_0 = 0 \tag{6.3}$$

If $s$ is replaced by $1/d$, then $d$ will have roots which are the reciprocal of $s$. Making this substitution in Eq. (6.3),

$$\left(\frac{1}{d}\right)^n + a_{n-1}\left(\frac{1}{d}\right)^{n-1} + \cdots + a_1\left(\frac{1}{d}\right) + a_0 = 0 \tag{6.4}$$

Factoring out $(1/d)^n$,

$$\begin{aligned}\left(\frac{1}{d}\right)^n &\left[1 + a_{n-1}\left(\frac{1}{d}\right)^{-1} + \cdots + a_1\left(\frac{1}{d}\right)^{(1-n)} + a_0\left(\frac{1}{d}\right)^{-n}\right] \\ &= \left(\frac{1}{d}\right)^n \left[1 + a_{n-1}d + \cdots + a_1 d^{(n-1)} + a_0 d^n\right] = 0\end{aligned} \tag{6.5}$$

Thus, the polynomial with reciprocal roots is a polynomial with the coefficients written in reverse order. Let us redo the previous example to show the computational advantage of this method.

## Example 6.3 Stability via Reverse Coefficients

**PROBLEM:** Determine the stability of the closed-loop transfer function

$$T(s) = \frac{10}{s^5 + 2s^4 + 3s^3 + 6s^2 + 5s + 3} \tag{6.6}$$

**SOLUTION:** First write a polynomial that has the reciprocal roots of the denominator of Eq. (6.6). From our discussion, this polynomial is formed by writing the denominator of Eq. (6.6) in reverse order. Hence,

$$D(s) = 3s^5 + 5s^4 + 6s^3 + 3s^2 + 2s + 1 \tag{6.7}$$

We form the Routh table as shown in Table 6.6 using Eq. (6.7). Since there are two sign changes, the system is unstable and has two right-half-plane poles. This is the same as the result obtained in Example 6.2. Notice that Table 6.6 does not have a zero in the first column.

**TABLE 6.6** Routh table for Example 6.3

| | | | |
|---|---|---|---|
| $s^5$ | 3 | 6 | 2 |
| $s^4$ | 5 | 3 | 1 |
| $s^3$ | 4.2 | 1.4 | |
| $s^2$ | 1.33 | 1 | |
| $s^1$ | −1.75 | | |
| $s^0$ | 1 | | |

## Entire Row is Zero

We now look at the second special case. Sometimes while making a Routh table, we find that an entire row consists of zeros because there is an even polynomial that is a factor of the original polynomial. This case must be handled differently from the case of a zero in only the first column of a row. Let us look at an example that demonstrates how to construct and interpret the Routh table when an entire row of zeros is present.

## Example 6.4 Stability via Routh Table with Row of Zeros

**PROBLEM:** Determine the number of right-half-plane poles in the closed-loop transfer function

$$T(s) = \frac{10}{s^5 + 7s^4 + 6s^3 + 42s^2 + 8s + 56} \tag{6.8}$$

**SOLUTION:** Start by forming the Routh table for the denominator of Eq. (6.8) (see Table 6.7). At the second row we multiply through by 1/7 for convenience. We stop at the third row, since the entire row consists of zeros, and use the following procedure. First we return to the row immediately above the row of zeros and form an auxiliary polynomial, using the entries in that row as coefficients. The polynomial will start with the power of $s$ in the label column and continue by skipping every other power of $s$. Thus, the polynomial formed for this example is

$$P(s) = s^4 + 6s^2 + 8 \tag{6.9}$$

Next we differentiate the polynomial with respect to $s$ and obtain

$$\frac{dP(s)}{ds} = 4s^3 + 12s + 0 \tag{6.10}$$

Finally, we use the coefficients of Eq. (6.10) to replace the row of zeros. Again, for convenience, the third row is multiplied by 1/4 after replacing the zeros.

The remainder of the table is formed in a straightforward manner by following the standard form shown in Table 6.2. Table 6.7 shows that all entries in the first column are positive. Hence, there are no right-half-plane poles.

**TABLE 6.7** Routh table for Example 6.4

| | | | | | | | | | |
|---|---|---|---|---|---|---|---|---|---|
| $s^5$ | | | 1 | | | 6 | | | 8 |
| $s^4$ | | ~~7~~ | 1 | | ~~42~~ | 6 | | ~~56~~ | 8 |
| $s^3$ | ~~0~~ | ~~4~~ | 1 | ~~0~~ | ~~12~~ | 3 | ~~0~~ | ~~0~~ | 0 |
| $s^2$ | | | 3 | | | 8 | | | 0 |
| $s^1$ | | | $\frac{1}{3}$ | | | 0 | | | 0 |
| $s^0$ | | | 8 | | | 0 | | | 0 |

Let us look further into the case that yields an entire row of zeros. An entire row of zeros will appear in the Routh table when a purely even or purely odd polynomial is a factor of the original polynomial. For example, $s^4 + 5s^2 + 7$ is an even polynomial; it has only even powers of $s$. Even polynomials only have roots that are symmetrical about the origin.[3] This symmetry can occur under three conditions of root position: (1) The roots are symmetrical and real, (2) the roots are symmetrical and imaginary, or (3) the roots are quadrantal. Figure 6.5 shows examples of these cases. Each case or combination of these cases will generate an even polynomial.

It is this even polynomial that causes the row of zeros to appear. Thus, the row of zeros tells us of the existence of an even polynomial whose roots are symmetric about the origin. Some of these roots could be on the $j\omega$-axis. On the other hand, since $j\omega$ roots are symmetric about the origin, if we do not have a row of zeros, we cannot possibly have $j\omega$ roots.

Another characteristic of the Routh table for the case in question is that the row previous to the row of zeros contains the even polynomial that is a factor of the original polynomial. Finally, everything from the row containing the even polynomial down to the end of the Routh table is a test of only the even polynomial. Let us put these facts together in an example.

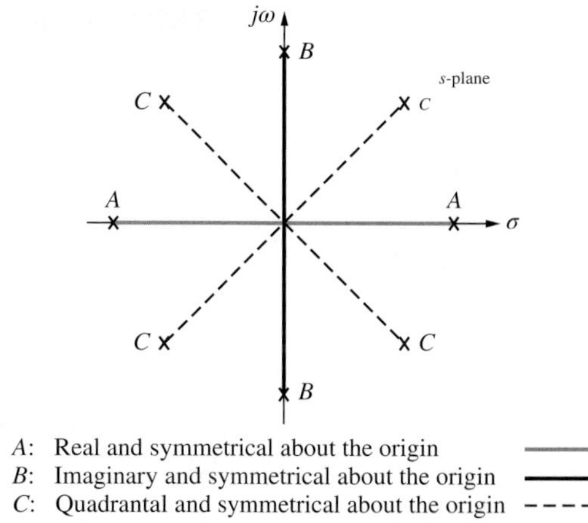


**FIGURE 6.5** Root positions to generate even polynomials: $A$, $B$, $C$, or any combination

## Example 6.5 Pole Distribution via Routh Table with Row of Zeros

**PROBLEM:** For the transfer function

$$T(s) = \frac{20}{s^8 + s^7 + 12s^6 + 22s^5 + 39s^4 + 59s^3 + 48s^2 + 38s + 20} \quad (6.11)$$

tell how many poles are in the right half-plane, in the left half-plane, and on the $j\omega$-axis.

**SOLUTION:** Use the denominator of Eq. (6.11) and form the Routh table in Table 6.8. For convenience the $s^6$ row is multiplied by 1/10, and the $s^5$ row is multiplied by 1/20. At the $s^3$ row we obtain a row of zeros. Moving back one row to $s^4$, we extract the even polynomial, $P(s)$, as

$$P(s) = s^4 + 3s^2 + 2 \quad (6.12)$$

This polynomial will divide evenly into the denominator of Eq. (6.11) and thus is a factor. Taking the derivative with respect to $s$ to obtain the coefficients that replace the row of zeros in the $s^3$ row, we find

$$\frac{dP(s)}{ds} = 4s^3 + 6s + 0 \quad (6.13)$$

Replace the row of zeros with 4, 6, and 0 and multiply the row by 1/2 for convenience. Finally, continue the table to the $s^0$ row, using the standard procedure.

[3] The polynomial $s^5 + 5s^3 + 7s$ is an example of an odd polynomial; it has only odd powers of $s$. Odd polynomials are the product of an even polynomial and an odd power of $s$. Thus, the constant term of an odd polynomial is always missing.

**TABLE 6.8** Routh table for Example 6.5

| | | | | | |
|---|---|---|---|---|---|
| $s^8$ | 1 | 12 | 39 | 48 | 20 |
| $s^7$ | 1 | 22 | 59 | 38 | 0 |
| $s^6$ | $-$~~10~~ $-1$ | $-$~~20~~ $-2$ | ~~10~~ 1 | ~~20~~ 2 | 0 |
| $s^5$ | ~~20~~ 1 | ~~60~~ 3 | ~~40~~ 2 | 0 | 0 |
| $s^4$ | 1 | 3 | 2 | 0 | 0 |
| $s^3$ | ~~0~~ ~~4~~ 2 | ~~0~~ ~~6~~ 3 | ~~0~~ ~~0~~ 0 | 0 | 0 |
| $s^2$ | ~~$\frac{3}{2}$~~ 3 | ~~2~~ 4 | 0 | 0 | 0 |
| $s^1$ | $\frac{1}{3}$ | 0 | 0 | 0 | 0 |
| $s^0$ | 4 | 0 | 0 | 0 | 0 |

How do we now interpret this Routh table? Since all entries from the even polynomial at the $s^4$ row down to the $s^0$ row are a test of the even polynomial, we begin to draw some conclusions about the roots of the even polynomial. No sign changes exist from the $s^4$ row down to the $s^0$ row. Thus, the even polynomial does not have right-half-plane poles. Since there are no right-half-plane poles, no left-half-plane poles are present because of the requirement for symmetry. Hence, the even polynomial, Eq. (6.12), must have all four of its poles on the $j\omega$-axis.[4] These results are summarized in the first column of Table 6.9.

The remaining roots of the total polynomial are evaluated from the $s^8$ row down to the $s^4$ row. We notice two sign changes: one from the $s^7$ row to the $s^6$ row and the other from the $s^6$ row to the $s^5$ row. Thus, the other polynomial must have two roots in the right half-plane. These results are included in Table 6.9 under **Other**. The final tally is the sum of roots from each component, the even polynomial and the other polynomial, as shown under **Total** in Table 6.9. Thus, the system has two poles in the right half-plane, two poles in the left half-plane, and four poles on the $j\omega$-axis; it is unstable because of the right-half-plane poles.

**TABLE 6.9** Summary of pole locations for Example 6.5

| | Polynomial | | |
|---|---|---|---|
| **Location** | **Even (fourth-order)** | **Other (fourth-order)** | **Total (eighth-order)** |
| Right half-plane | 0 | 2 | 2 |
| Left half-plane | 0 | 2 | 2 |
| $j\omega$ | 4 | 0 | 4 |

[4] A necessary condition for stability is that the $j\omega$ roots have unit multiplicity. The even polynomial must be checked for multiple $j\omega$ roots. For this case, the existence of multiple $j\omega$ roots would lead to a perfect, fourth-order square polynomial. Since Eq. (6.12) is not a perfect square, the four $j\omega$ roots are distinct.

We now summarize what we have learned about polynomials that generate entire rows of zeros in the Routh table. These polynomials have a purely even factor with roots that are symmetrical about the origin. The even polynomial appears in the Routh table in the row directly above the row of zeros. Every entry in the table from the even polynomial's row to the end of the chart applies only to the even polynomial. Therefore, the number of sign changes from the even polynomial to the end of the table equals the number of right-half-plane roots of the even polynomial. Because of the symmetry of roots about the origin, the even polynomial must have the same number of left-half-plane roots as it does right-half-plane roots. Having accounted for the roots in the right and left half-planes, we know the remaining roots must be on the $j\omega$-axis.

Every row in the Routh table from the beginning of the chart to the row containing the even polynomial applies only to the other factor of the original polynomial. For this factor, the number of sign changes, from the beginning of the table down to the even polynomial, equals the number of right-half-plane roots. The remaining roots are left-half-plane roots. There can be no $j\omega$ roots contained in the other polynomial.

## Skill-Assessment Exercise 6.2

**PROBLEM:** Use the Routh-Hurwitz criterion to find how many poles of the following closed-loop system, $T(s)$, are in the rhp, in the lhp, and on the $j\omega$-axis:

$$T(s) = \frac{s^3 + 7s^2 - 21s + 10}{s^6 + s^5 - 6s^4 + 0s^3 - s^2 - s + 6}$$

**ANSWER:** Two rhp, two lhp, and two $j\omega$

The complete solution is at www.wiley.com/college/nise.

Let us demonstrate the usefulness of the Routh-Hurwitz criterion with a few additional examples.

# 6.4 Routh-Hurwitz Criterion: Additional Examples

The previous two sections have introduced the Routh-Hurwitz criterion. Now we need to demonstrate the method's application to a number of analysis and design problems.

## Example 6.6 Standard Routh-Hurwitz

**PROBLEM:** Find the number of poles in the left half-plane, the right half-plane, and on the $j\omega$-axis for the system of Figure 6.6.

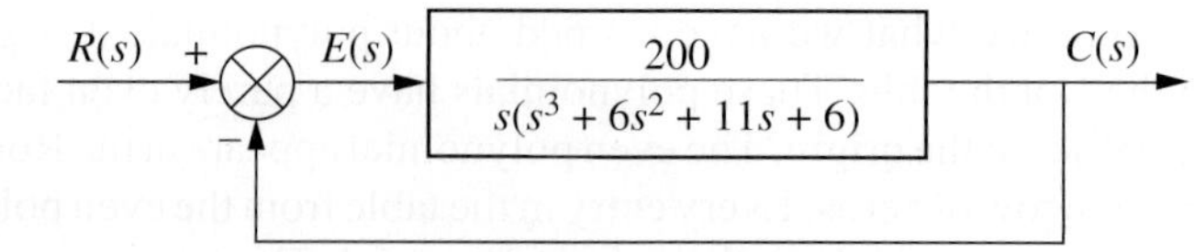


**FIGURE 6.6** Feedback control system for Example 6.6

**SOLUTION:** First, find the closed-loop transfer function as

$$T(s) = \frac{200}{s^4 + 6s^3 + 11s^2 + 6s + 200} \tag{6.14}$$

The Routh table for the denominator of Eq. (6.14) is shown as Table 6.10. For clarity, we leave most zero cells blank. At the $s^1$ row there is a negative coefficient; thus, there are two sign changes. The system is unstable, since it has two right-half-plane poles and two left-half-plane poles. The system cannot have $j\omega$ poles since a row of zeros did not appear in the Routh table.

**TABLE 6.10** Routh table for Example 6.6

| | | | |
|---|---|---|---|
| $s^4$ | 1 | 11 | 200 |
| $s^3$ | ~~6~~ 1 | ~~6~~ 1 | |
| $s^2$ | ~~10~~ 1 | ~~200~~ 20 | |
| $s^1$ | −19 | | |
| $s^0$ | 20 | | |

The next example demonstrates the occurrence of a zero in only the first column of a row.

## Example 6.7 Routh-Hurwitz with Zero in First Column

**PROBLEM:** Find the number of poles in the left half-plane, the right half-plane, and on the $j\omega$-axis for the system of Figure 6.7.

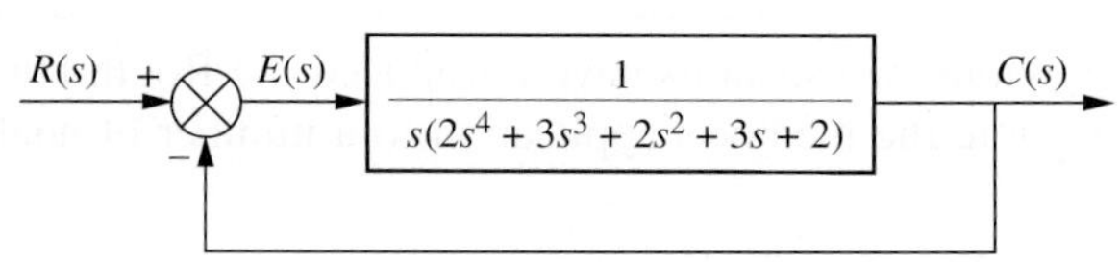


**FIGURE 6.7** Feedback control system for Example 6.7

**SOLUTION:** The closed-loop transfer function is

$$T(s) = \frac{1}{2s^5 + 3s^4 + 2s^3 + 3s^2 + 2s + 1} \tag{6.15}$$

Form the Routh table shown as Table 6.11, using the denominator of Eq. (6.15). A zero appears in the first column of the $s^3$ row. Since the entire row is not zero, simply replace the zero with a small quantity, $\epsilon$, and continue the table. Permitting $\epsilon$ to be a small, positive quantity, we find that the first term of the $s^2$ row is negative. Thus, there are two sign changes, and the system is unstable, with two poles in the right half-plane. The remaining poles are in the left half-plane.

**TABLE 6.11** Routh table for Example 6.7

| | | | |
|---|---|---|---|
| $s^5$ | 2 | 2 | 2 |
| $s^4$ | 3 | 3 | 1 |
| $s^3$ | ~~0~~ $\epsilon$ | $\frac{4}{3}$ | |
| $s^2$ | $\frac{3\epsilon - 4}{\epsilon}$ | 1 | |
| $s^1$ | $\frac{12\epsilon - 16 - 3\epsilon^2}{9\epsilon - 12}$ | | |
| $s^0$ | 1 | | |

We also can use the alternative approach, where we produce a polynomial whose roots are the reciprocal of the original. Using the denominator of Eq. (6.15), we form a polynomial by writing the coefficients in reverse order,

$$s^5 + 2s^4 + 3s^3 + 2s^2 + 3s + 2 \tag{6.16}$$

The Routh table for this polynomial is shown as Table 6.12. Unfortunately, in this case we also produce a zero only in the first column at the $s^2$ row. However, the table is easier to work with than Table 6.11. Table 6.12 yields the same results as Table 6.11: three poles in the left half-plane and two poles in the right-half-plane. The system is unstable.

**TABLE 6.12** Alternative Routh table for Example 6.7

| | | | |
|---|---|---|---|
| $s^5$ | 1 | 3 | 3 |
| $s^4$ | 2 | 2 | 2 |
| $s^3$ | 2 | 2 | |
| $s^2$ | ~~0~~ $\epsilon$ | 2 | |
| $s^1$ | $\frac{2\epsilon - 4}{\epsilon}$ | | |
| $s^0$ | 2 | | |

MATLAB

**ML**

`Students who are using MATLAB should now run ch6p1 in Appendix B. You will learn how to perform block diagram reduction to find` $T(s)$`, followed by an evaluation of the closed-loop system's poles to determine stability. This exercise uses MATLAB to do Example 6.7.`

In the next example, we see an entire row of zeros appear along with the possibility of imaginary roots.

## Example 6.8 Routh-Hurwitz with Row of Zeros

**PROBLEM:** Find the number of poles in the left half-plane, the right half-plane, and on the $j\omega$-axis for the system of Figure 6.8. Draw conclusions about the stability of the closed-loop system.

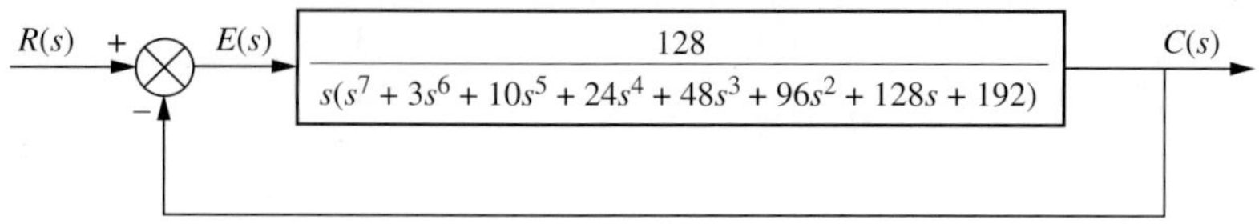


**FIGURE 6.8** Feedback control system for Example 6.8

**TryIt 6.2**

Use MATLAB, The Control System Toolbox, and the following statements to find the closed-loop transfer function, $T(s)$, for Figure 6.8 and the closed-loop poles.

```
numg=128;
deng=[1 3 10 24 ...
 48 96 128 192 0];
G=tf(numg, deng);
T=feedback(G, 1)
poles=pole(T)
```

**SOLUTION:** The closed-loop transfer function for the system of Figure 6.8 is

$$T(s) = \frac{128}{s^8 + 3s^7 + 10s^6 + 24s^5 + 48s^4 + 96s^3 + 128s^2 + 192s + 128} \tag{6.17}$$

Using the denominator, form the Routh table shown as Table 6.13. A row of zeros appears in the $s^5$ row. Thus, the closed-loop transfer function denominator must have an even polynomial as a factor. Return to the $s^6$ row and form the even polynomial:

$$P(s) = s^6 + 8s^4 + 32s^2 + 64 \tag{6.18}$$

**TABLE 6.13** Routh table for Example 6.8

| | | | | | |
|---|---|---|---|---|---|
| $s^8$ | 1 | 10 | 48 | 128 | 128 |
| $s^7$ | ~~3~~ 1 | ~~24~~ 8 | ~~96~~ 32 | ~~192~~ 64 | |
| $s^6$ | ~~2~~ 1 | ~~16~~ 8 | ~~64~~ 32 | ~~128~~ 64 | |
| $s^5$ | ~~0~~ ~~6~~ 3 | ~~0~~ ~~32~~ 16 | ~~0~~ ~~64~~ 32 | ~~0~~ ~~0~~ 0 | |
| $s^4$ | ~~$\frac{8}{3}$~~ 1 | ~~$\frac{64}{3}$~~ 8 | ~~64~~ 24 | | |
| $s^3$ | ~~8~~ −1 | ~~40~~ −5 | | | |
| $s^2$ | ~~3~~ 1 | ~~24~~ 8 | | | |
| $s^1$ | 3 | | | | |
| $s^0$ | 8 | | | | |

Differentiate this polynomial with respect to $s$ to form the coefficients that will replace the row of zeros:

$$\frac{dP(s)}{ds} = 6s^5 + 32s^3 + 64s + 0 \tag{6.19}$$

Replace the row of zeros at the $s^5$ row by the coefficients of Eq. (6.19) and multiply through by 1/2 for convenience. Then complete the table.

We note that there are two sign changes from the even polynomial at the $s^6$ row down to the end of the table. Hence, the even polynomial has two right-half-plane poles. Because of the symmetry about the origin, the even polynomial must have an equal number of left-half-plane poles. Therefore, the even polynomial has two left-half-plane poles. Since the even polynomial is of sixth order, the two remaining poles must be on the $j\omega$-axis.

There are no sign changes from the beginning of the table down to the even polynomial at the $s^6$ row. Therefore, the rest of the polynomial has no right-half-plane poles. The results are summarized in Table 6.14. The system has two poles in the right half-plane, four poles in the left half-plane, and two poles on the $j\omega$-axis, which are of unit multiplicity. The closed-loop system is unstable because of the right-half-plane poles.

**TABLE 6.14** Summary of pole locations for Example 6.8

| | Polynomial | | |
|---|---|---|---|
| **Location** | **Even (sixth-order)** | **Other (second-order)** | **Total (eighth-order)** |
| Right half-plane | 2 | 0 | 2 |
| Left half-plane | 2 | 2 | 4 |
| $j\omega$ | 2 | 0 | 2 |

The Routh-Hurwitz criterion gives vivid proof that changes in the gain of a feedback control system result in differences in transient response because of changes in closed-loop pole locations. The next example demonstrates this concept. We will see that for control systems, such as those shown in Figure 6.9, gain variations can move poles from stable regions of the $s$-plane onto the $j\omega$-axis and then into the right half-plane.

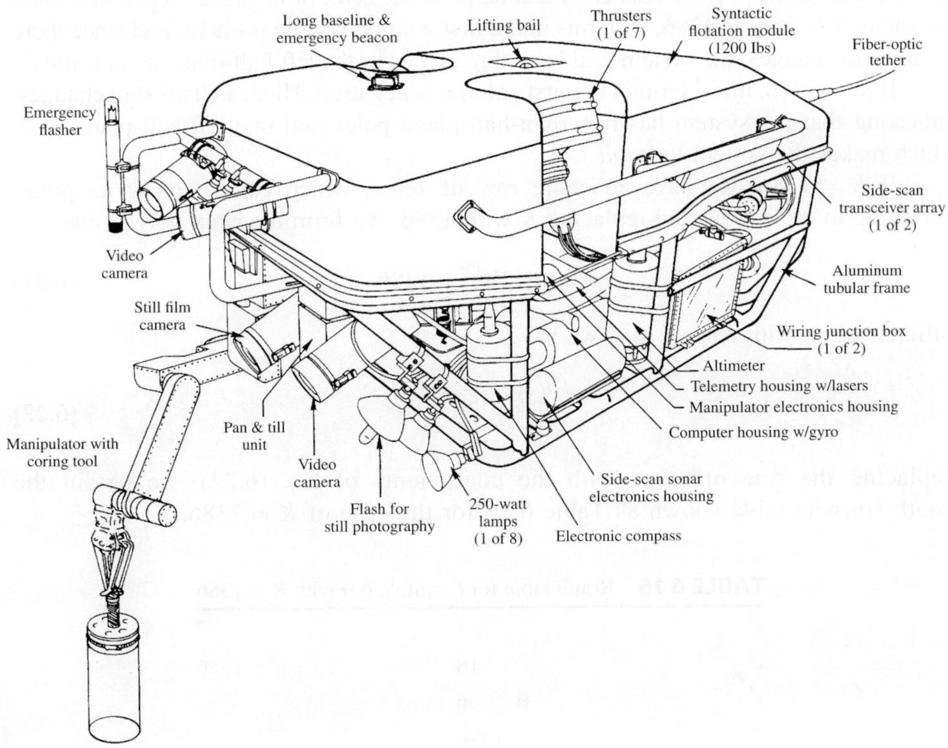


**FIGURE 6.9** *Jason* is an underwater, remote-controlled vehicle that has been used to explore the wreckage of the *Lusitania*. The manipulator and cameras comprise some of the vehicle's control systems (Courtesy of Woods Hole Oceanographic Institution)

## Example 6.9 Stability Design via Routh-Hurwitz

**PROBLEM:** Find the range of gain, $K$, for the system of Figure 6.10 that will cause the system to be stable, unstable, and marginally stable. Assume $K > 0$.

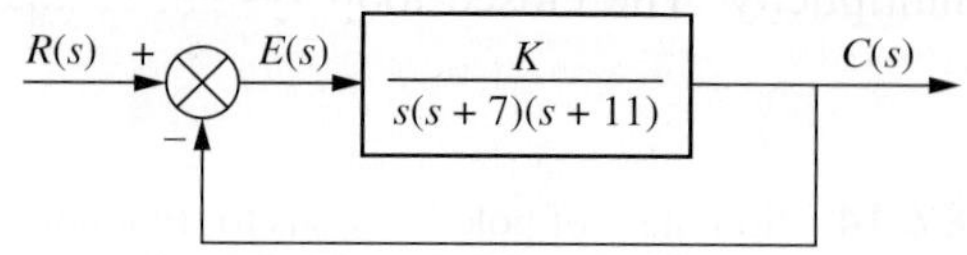


**FIGURE 6.10** Feedback control system for Example 6.9

**SOLUTION:** First find the closed-loop transfer function as

$$T(s) = \frac{K}{s^3 + 18s^2 + 77s + K} \tag{6.20}$$

Next form the Routh table shown as Table 6.15.

**TABLE 6.15** Routh table for Example 6.9

| | | |
|---|---|---|
| $s^3$ | 1 | 77 |
| $s^2$ | 18 | $K$ |
| $s^1$ | $\frac{1386 - K}{18}$ | |
| $s^0$ | $K$ | |

Since $K$ is assumed positive, we see that all elements in the first column are always positive except the $s^1$ row. This entry can be positive, zero, or negative, depending upon the value of $K$. If $K < 1386$, all terms in the first column will be positive, and since there are no sign changes, the system will have three poles in the left half-plane and be *stable*.

If $K > 1386$, the $s^1$ term in the first column is negative. There are two sign changes, indicating that the system has two right-half-plane poles and one left-half-plane pole, which makes the system *unstable*.

If $K = 1386$, we have an entire row of zeros, which could signify $j\omega$ poles. Returning to the $s^2$ row and replacing $K$ with 1386, we form the even polynomial

$$P(s) = 18s^2 + 1386 \tag{6.21}$$

Differentiating with respect to $s$, we have

$$\frac{dP(s)}{ds} = 36s + 0 \tag{6.22}$$

Replacing the row of zeros with the coefficients of Eq. (6.22), we obtain the Routh-Hurwitz table shown as Table 6.16 for the case of $K = 1386$.

**TABLE 6.16** Routh table for Example 6.9 with $K = 1386$

| | | |
|---|---|---|
| $s^3$ | 1 | 77 |
| $s^2$ | 18 | 1386 |
| $s^1$ | ~~0~~ 36 | |
| $s^0$ | 1386 | |

Since there are no sign changes from the even polynomial ($s^2$ row) down to the bottom of the table, the even polynomial has its two roots on the $j\omega$-axis of unit multiplicity. Since there are no sign changes above the even polynomial, the remaining root is in the left half-plane. Therefore the system is *marginally stable.*

MATLAB **ML**

Students who are using MATLAB should now run ch6p2 in Appendix B. You will learn how to set up a loop to search for the range of gain to yield stability. This exercise uses MATLAB to do Example 6.9.

Symbolic Math **SM**

Students who are performing the MATLAB exercises and want to explore the added capability of MATLAB's Symbolic Math Toolbox should now run ch6sp2 in Appendix F at www.wiley.com/college/nise. You will learn how to use the Symbolic Math Toolbox to calculate the values of cells in a Routh table even if the table contains symbolic objects, such as a variable gain, *K*. You will see that the Symbolic Math Toolbox and MATLAB yield an alternative way to solve Example 6.9.

The Routh-Hurwitz criterion is often used in limited applications to factor polynomials containing even factors. Let us look at an example.

## Example 6.10 Factoring via Routh-Hurwitz

**PROBLEM:** Factor the polynomial

$$s^4 + 3s^3 + 30s^2 + 30s + 200 \tag{6.23}$$

**SOLUTION:** Form the Routh table of Table 6.17. We find that the $s^1$ row is a row of zeros. Now form the even polynomial at the $s^2$ row:

$$P(s) = s^2 + 10 \tag{6.24}$$

**TABLE 6.17** Routh table for Example 6.10

| | | | |
|---|---|---|---|
| $s^4$ | 1 | 30 | 200 |
| $s^3$ | $\not{3}$ 1 | $\not{30}$ 10 | |
| $s^2$ | $\not{20}$ 1 | $\not{200}$ 10 | |
| $s^1$ | $\not{0}$ 2 | $\not{0}$ 0 | |
| $s^0$ | 10 | | |

This polynomial is differentiated with respect to *s* in order to complete the Routh table. However, since this polynomial is a factor of the original polynomial in Eq. (6.23), dividing Eq. (6.23) by (6.24) yields $(s^2 + 3s + 20)$ as the other factor. Hence,

$$\begin{aligned} s^4 + 3s^3 + 30s^2 + 30s + 200 &= (s^2 + 10)(s^2 + 3s + 20) \\ &= (s + j3.1623)(s - j3.1623) \\ &\quad \times (s + 1.5 + j4.213)(s + 1.5 - j4.213) \end{aligned} \tag{6.25}$$

## Skill-Assessment Exercise 6.3

**PROBLEM:** For a unity feedback system with the forward transfer function

$$G(s) = \frac{K(s+20)}{s(s+2)(s+3)}$$

find the range of $K$ to make the system stable.

**ANSWER:** $0 < K < 2$

The complete solution is at www.wiley.com/college/nise.

## 6.5 Stability in State Space

State Space

SS

Up to this point we have examined stability from the $s$-plane viewpoint. Now we look at stability from the perspective of state space. In Section 4.10, we mentioned that the values of the system's poles are equal to the eigenvalues of the system matrix, $\mathbf{A}$. We stated that the eigenvalues of the matrix $\mathbf{A}$ were solutions of the equation $\det(s\mathbf{I} - \mathbf{A}) = 0$, which also yielded the poles of the transfer function. Eigenvalues appeared again in Section 5.8, where they were formally defined and used to diagonalize a matrix. Let us now formally show that the eigenvalues and the system poles have the same values.

Reviewing Section 5.8, the eigenvalues of a matrix, $\mathbf{A}$, are values of $\lambda$ that permit a nontrivial solution (other than $\mathbf{0}$) for $\mathbf{x}$ in the equation

$$\mathbf{A}\mathbf{x} = \lambda\mathbf{x} \tag{6.26}$$

In order to solve for the values of $\lambda$ that do indeed permit a solution for $\mathbf{x}$, we rearrange Eq. (6.26) as follows:

$$\lambda\mathbf{x} - \mathbf{A}\mathbf{x} = \mathbf{0} \tag{6.27}$$

or

$$(\lambda\mathbf{I} - \mathbf{A})\mathbf{x} = \mathbf{0} \tag{6.28}$$

Solving for $\mathbf{x}$ yields

$$\mathbf{x} = (\lambda\mathbf{I} - \mathbf{A})^{-1}\mathbf{0} \tag{6.29}$$

or

$$\mathbf{x} = \frac{\mathrm{adj}(\lambda\mathbf{I} - \mathbf{A})}{\det(\lambda\mathbf{I} - \mathbf{A})}\mathbf{0} \tag{6.30}$$

We see that all solutions will be the null vector except for the occurrence of zero in the denominator. Since this is the only condition where elements of $\mathbf{x}$ will be $0/0$, or indeterminate, it is the only case where a nonzero solution is possible.

The values of $\lambda$ are calculated by forcing the denominator to zero:

$$\det(\lambda\mathbf{I} - \mathbf{A}) = 0 \tag{6.31}$$

This equation determines the values of $\lambda$ for which a nonzero solution for $\mathbf{x}$ in Eq. (6.26) exists. In Section 5.8, we defined $\mathbf{x}$ as *eigenvectors* and the values of $\lambda$ as the *eigenvalues* of the matrix $\mathbf{A}$.

Let us now relate the eigenvalues of the system matrix, $\mathbf{A}$, to the system's poles. In Chapter 3 we derived the equation of the system transfer function, Eq. (3.73), from the state equations. The system transfer function has $\det(s\mathbf{I}-\mathbf{A})$ in the denominator because of the presence of $(s\mathbf{I}-\mathbf{A})^{-1}$. Thus,

$$\det(s\mathbf{I}-\mathbf{A}) = 0 \tag{6.32}$$

is the characteristic equation for the system from which the system poles can be found.

Since Eqs. (6.31) and (6.32) are identical apart from a change in variable name, we conclude that the eigenvalues of the matrix $\mathbf{A}$ are identical to the system's poles before cancellation of common poles and zeroes in the transfer function. Thus, we can determine the stability of a system represented in state space by finding the eigenvalues of the system matrix, $\mathbf{A}$, and determining their locations on the $s$-plane.

## Example 6.11 Stability in State Space

**PROBLEM:** Given the system

$$\dot{\mathbf{x}} = \begin{bmatrix} 0 & 3 & 1 \\ 2 & 8 & 1 \\ -10 & -5 & -2 \end{bmatrix}\mathbf{x} + \begin{bmatrix} 10 \\ 0 \\ 0 \end{bmatrix}u \tag{6.33a}$$

$$y = [1 \quad 0 \quad 0]\mathbf{x} \tag{6.33b}$$

find out how many poles are in the left half-plane, in the right half-plane, and on the $j\omega$-axis.

**SOLUTION:** First form $(s\mathbf{I}-\mathbf{A})$:

$$(s\mathbf{I}-\mathbf{A}) = \begin{bmatrix} s & 0 & 0 \\ 0 & s & 0 \\ 0 & 0 & s \end{bmatrix} - \begin{bmatrix} 0 & 3 & 1 \\ 2 & 8 & 1 \\ -10 & -5 & -2 \end{bmatrix} = \begin{bmatrix} s & -3 & -1 \\ -2 & s-8 & -1 \\ 10 & 5 & s+2 \end{bmatrix} \tag{6.34}$$

Now find the $\det(s\mathbf{I}-\mathbf{A})$:

$$\det(s\mathbf{I}-\mathbf{A}) = s^3 - 6s^2 - 7s - 52 \tag{6.35}$$

Using this polynomial, form the Routh table of Table 6.18.

**TABLE 6.18** Routh table for Example 6.11

| | | |
|---|---|---|
| $s^3$ | 1 | −7 |
| $s^2$ | ~~−6~~ −3 | ~~−52~~ −26 |
| $s^1$ | ~~$\frac{47}{3}$~~ −1 | ~~0~~ 0 |
| $s^0$ | −26 | |

Since there is one sign change in the first column, the system has one right-half-plane pole and two left-half-plane poles. It is therefore unstable. Yet, you may question the possibility that if a nonminimum-phase zero cancels the unstable pole, the system will be stable. However, in practice, the nonminimum-phase zero or unstable pole will shift due to a slight change in the system's parameters. This change will cause the system to become unstable.

MATLAB

ML

Students who are using MATLAB should now run ch6p3 in Appendix B. You will learn how to determine the stability of a system represented in state space by finding the eigenvalues of the system matrix. This exercise uses MATLAB to do Example 6.11.

## Skill-Assessment Exercise 6.4

**TryIt 6.3**

Use the following MATLAB statements to find the eigenvalues of the system described in Skill-Assessment Exercise 6.4.

```
A=[2  1  1
   1  7  1
  -3  4  -5];
Eig=eig(A)
```

**PROBLEM:** For the following system represented in state space, find out how many poles are in the left half-plane, in the right half-plane, and on the $j\omega$-axis.

$$\dot{\mathbf{x}} = \begin{bmatrix} 2 & 1 & 1 \\ 1 & 7 & 1 \\ -3 & 4 & -5 \end{bmatrix} \mathbf{x} + \begin{bmatrix} 0 \\ 0 \\ 1 \end{bmatrix} r$$

$$y = [0 \quad 1 \quad 0]\mathbf{x}$$

**ANSWER:** Two rhp and one lhp.

The complete solution is at www.wiley.com/college/nise.

In this section, we have evaluated the stability of feedback control systems from the state-space perspective. Since the closed-loop poles and the eigenvalues of a system are the same, the stability requirement of a system represented in state space dictates that the eigenvalues cannot be in the right half of the $s$-plane or be multiple on the $j\omega$-axis.

We can obtain the eigenvalues from the state equations without first converting to a transfer function to find the poles: The equation $\det(s\mathbf{I} - \mathbf{A}) = 0$ yields the eigenvalues directly. If $\det(s\mathbf{I} - \mathbf{A})$, a polynomial in $s$, cannot be factored easily, we can apply the Routh-Hurwitz criterion to it to evaluate how many eigenvalues are in each region of the $s$-plane.

We now summarize this chapter, first with case studies and then with a written summary. Our case studies include the antenna azimuth position control system and the UFSS. Stability is as important to these systems as it is to the system shown in Figure 6.11.

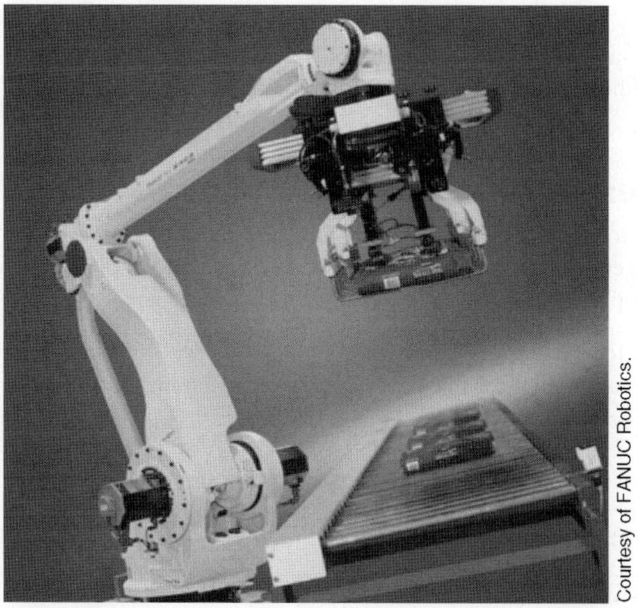


**FIGURE 6.11** A FANUC M-410*i*B$^{TM}$ has 4 axes of motion. It is seen here performing bag palletizing

## Case Studies Antenna Control: Stability Design via Gain

Design D

This chapter has covered the elements of stability. We saw that stable systems have their closed-loop poles in the left half of the $s$-plane. As the loop gain is changed, the locations of the poles are also changed, creating the possibility that the poles can move into the right half of the $s$-plane, which yields instability. Proper gain settings are essential for the stability of closed-loop systems. The following case study demonstrates the proper setting of the loop gain to ensure stability.

**PROBLEM:** You are given the antenna azimuth position control system shown on the front endpapers, Configuration 1. Find the range of preamplifier gain required to keep the closed-loop system stable.

**SOLUTION:** The closed-loop transfer function was derived in the case studies in Chapter 5 as

$$T(s) = \frac{6.63K}{s^3 + 101.71s^2 + 171s + 6.63K} \tag{6.36}$$

Using the denominator, create the Routh table shown as Table 6.19. The third row of the table shows that a row of zeros occurs if $K = 2623$. This value of $K$ makes the system marginally stable. Therefore, there will be no sign changes in the first column if $0 < K < 2623$. We conclude that, for stability, $0 < K < 2623$. An animation PowerPoint presentation (PPT) demonstrating this system is available for instructors at www.wiley.com/college/nise. See *Antenna (Ch. 6)*.

**TABLE 6.19** Routh table for antenna control case study

| | | |
|---|---|---|
| $s^3$ | 1 | 171 |
| $s^2$ | 101.71 | $6.63K$ |
| $s^1$ | $17392.41–6.63K$ | 0 |
| $s^0$ | $6.63K$ | |

**CHALLENGE:** We now give you a problem to test your knowledge of this chapter's objectives. Refer to the antenna azimuth position control system shown on the front endpapers, Configuration 2. Find the range of preamplifier gain required to keep the closed-loop system stable.

## UFSS Vehicle: Stability Design via Gain

Design
D

For this case study, we return to the UFSS vehicle and study the stability of the pitch control system, which is used to control depth. Specifically, we find the range of pitch gain that keeps the pitch control loop stable.

**PROBLEM:** The pitch control loop for the UFSS vehicle (*Johnson, 1980*) is shown on the back endpapers. Let $K_2 = 1$ and find the range of $K_1$ that ensures that the closed-loop pitch control system is stable.

**SOLUTION:** The first step is to reduce the pitch control system to a single, closed-loop transfer function. The equivalent forward transfer function, $G_e(s)$, is

$$G_e(s) = \frac{0.25K_1(s+0.435)}{s^4+3.456s^3+3.457s^2+0.719s+0.0416} \tag{6.37}$$

With unity feedback the closed-loop transfer function, $T(s)$, is

$$T(s) = \frac{0.25K_1(s+0.435)}{s^4+3.456s^3+3.457s^2+(0.719+0.25K_1)s+(0.0416+0.109K_1)} \tag{6.38}$$

The denominator of Eq. (6.38) is now used to form the Routh table shown as Table 6.20.

**TABLE 6.20** Routh table for UFSS case study

| | | | |
|---|---|---|---|
| $s^4$ | 1 | 3.457 | $0.0416+0.109K_1$ |
| $s^3$ | 3.456 | $0.719+0.25K_1$ | |
| $s^2$ | $11.228-0.25K_1$ | $0.144+0.377K_1$ | |
| $s^1$ | $\dfrac{-0.0625K_1^2+1.324K_1+7.575}{11.228-0.25K_1}$ | | |
| $s^0$ | $0.144+0.377K_1$ | | |

*Note:* Some rows have been multiplied by a positive constant for convenience.

Looking at the first column, the $s^4$ and $s^3$ rows are positive. Thus, all elements of the first column must be positive for stability. For the first column of the $s^2$ row to be positive, $-\infty < K_1 < 44.91$. For the first column of the $s^1$ row to be positive, the numerator must be positive, since the denominator is positive from the previous step. The solution to the quadratic term in the numerator yields roots of $K_1 = -4.685$ and $25.87$. Thus, for a positive numerator, $-4.685 < K_1 < 25.87$. Finally, for the first column of the $s^0$ row to be positive, $-0.382 < K_1 < \infty$. Using all three conditions, stability will be ensured if $-0.382 < K_1 < 25.87$.

**CHALLENGE:** You are now given a problem to test your knowledge of this chapter's objectives. For the UFSS vehicle (*Johnson, 1980*) heading control system shown on the back endpapers and introduced in the UFSS Case Study Challenge in Chapter 5, do the following:

MATLAB
ML

**a.** Find the range of heading gain that ensures the vehicle's stability. Let $K_2 = 1$

**b.** `Repeat Part a using MATLAB.`

In our case studies, we calculated the ranges of gain to ensure stability. The student should be aware that although these ranges yield stability, setting gain within these limits may not yield the desired transient response or steady-state error characteristics. In Chapters 9 and 11, we will explore design techniques, other than simple gain adjustment, that yield more flexibility in obtaining desired characteristics.

## Summary

In this chapter, we explored the concepts of system stability from both the classical and the state-space viewpoints. We found that for linear systems, *stability* is based on a natural response that decays to zero as time approaches infinity. On the other hand, if the natural response increases without bound, the forced response is overpowered by the natural response, and we lose control. This condition is known as *instability*. A third possibility exists: The natural response may neither decay nor grow without bound but oscillate. In this case, the system is said to be *marginally stable*.

We also used an alternative definition of stability when the natural response is not explicitly available. This definition is based on the total response and says that a system is stable if every bounded input yields a bounded output (BIBO) and unstable if any bounded input yields an unbounded output.

Mathematically, stability for linear, time-invariant systems can be determined from the location of the closed-loop poles:

- If the poles are only in the left half-plane, the system is stable.
- If any poles are in the right half-plane, the system is unstable.
- If the poles are on the $j\omega$-axis and in the left half-plane, the system is marginally stable as long as the poles on the $j\omega$-axis are of unit multiplicity; it is unstable if there are any multiple $j\omega$ poles.

Unfortunately, although the open-loop poles may be known, we found that in higher-order systems it is difficult to find the closed-loop poles without a computer program.

The *Routh-Hurwitz criterion* lets us find how many poles are in each section of the $s$-plane without giving us the coordinates of the poles. Just knowing that there are poles in the right half-plane is enough to determine that a system is unstable. Under certain limited conditions, when an even polynomial is present, the Routh table can be used to factor the system's characteristic equation.

Obtaining stability from the state-space representation of a system is based on the same concept—the location of the roots of the characteristic equation. These roots are equivalent to the eigenvalues of the system matrix and can be found by solving $\det(s\mathbf{I} - \mathbf{A}) = 0$. Again, the Routh-Hurwitz criterion can be applied to this polynomial. The point is that the state-space representation of a system need not be converted to a transfer function in order to investigate stability. In the next chapter, we will look at steady-state errors, the last of three important control system requirements we emphasize.

## Review Questions

1. What part of the output response is responsible for determining the stability of a linear system?
2. What happens to the response named in Question 1 that creates instability?
3. What would happen to a physical system that becomes unstable?

4. Why are marginally stable systems considered unstable under the BIBO definition of stability?
5. Where do system poles have to be to ensure that a system is not unstable?
6. What does the Routh-Hurwitz criterion tell us?
7. Under what conditions would the Routh-Hurwitz criterion easily tell us the actual location of the system's closed-loop poles?
8. What causes a zero to show up only in the first column of the Routh table?
9. What causes an entire row of zeros to show up in the Routh table?
10. Why do we sometimes multiply a row of a Routh table by a positive constant?
11. Why do we not multiply a row of a Routh table by a negative constant?
12. If a Routh table has two sign changes above the even polynomial and five sign changes below the even polynomial, how many right-half-plane poles does the system have?
13. Does the presence of an entire row of zeros always mean that the system has $j\omega$ poles?
14. If a seventh-order system has a row of zeros at the $s^3$ row and two sign changes below the $s^4$ row, how many $j\omega$ poles does the system have?

State Space SS

15. Is it true that the eigenvalues of the system matrix are the same as the closed-loop poles?

State Space SS

16. How do we find the eigenvalues?

# Problems

1. Tell how many roots of the following polynomial are in the right half-plane, in the left half-plane, and on the $j\omega$-axis: [Section: 6.2]

$$P(s) = s^5 + 2s^4 + 5s^3 + 4s^2 + s + 2$$

2. Tell how many roots of the following polynomial are in the right half-plane, in the left half-plane, and on the $j\omega$-axis: [Section: 6.3]

$$P(s) = s^5 + 6s^3 + 5s^2 + 8s + 20$$

3. Using the Routh table, tell how many poles of the following function are in the right half-plane, in the left half-plane, and on the $j\omega$-axis: [Section: 6.3]

$$T(s) = \frac{s+8}{s^5 - s^4 + 4s^3 - 4s^2 + 3s - 2}$$

4. The closed-loop transfer function of a system is [Section: 6.3]

$$T(s) = \frac{s^3 + 2s^2 + 7s + 21}{s^5 - 2s^4 + 3s^3 - 6s^2 + 2s - 4}$$

Determine how many closed-loop poles lie in the right half-plane, in the left half-plane, and on the $j\omega$-axis.

5. How many poles are in the right half-plane, the left half-plane, and on the $j\omega$-axis for the open-loop system of Figure P6.1?

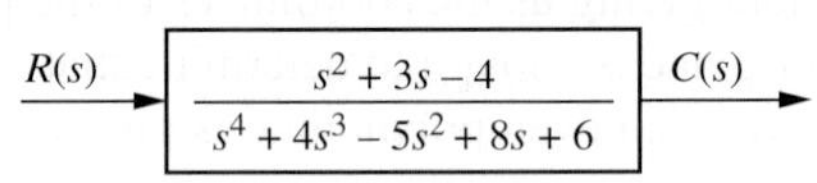


**FIGURE P6.1**

6. How many poles are in the right half-plane, the left half-plane, and on the $j\omega$-axis for the open-loop system of Figure P6.2? [Section: 6.3]

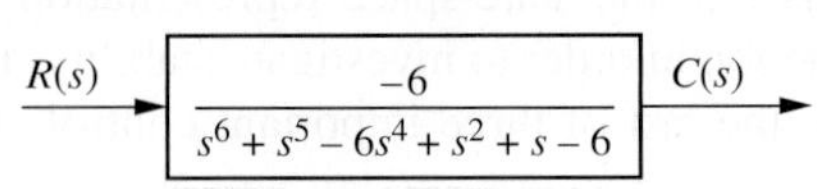


**FIGURE P6.2**

MATLAB ML

7. Use MATLAB to find the pole locations for the system of Problem 6.

Symbolic Math SM

8. Use MATLAB and the Symbolic Math Toolbox to generate a Routh table to solve Problem 3.

**9.** Determine whether the unity feedback system of Figure P6.3 is stable if [Section: 6.2]

$$G(s) = \frac{240}{(s+1)(s+2)(s+5)(s+6)}$$

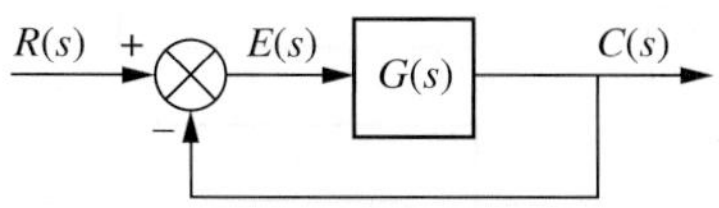


**FIGURE P6.3**

**10.** `Use MATLAB to find the pole locations for the system of Problem 9.` MATLAB ML

**11.** Consider the unity feedback system of Figure P6.3 with

$$G(s) = \frac{1}{4s^2(s^2+1)}$$

Using the Routh-Hurwitz criterion, find the region of the $s$-plane where the poles of the closed-loop system are located. [Section: 6.3].

**12.** In the system of Figure P6.3, let

$$G(s) = \frac{K(s+2)}{s(s-1)(s+3)}$$

Find the range of $K$ for closed-loop stability. [Section: 6.4]

**13.** Using the Routh-Hurwitz criterion and the unity feedback system of Figure P6.3 with

$$G(s) = \frac{2}{2s^4+s^3+s^2+2s}$$

tell whether or not the closed-loop system is stable. [Section: 6.2]

**14.** Given the unity feedback system of Figure P6.3 with

$$G(s) = \frac{84}{s(s^7+5s^6+12s^5+25s^4+45s^3+50s^2+82s+60)}$$

tell how many poles of the closed-loop transfer function lie in the right half-plane, in the left half-plane, and on the $j\omega$-axis. [Section: 6.3]

**15.** Given the unity feedback system of Figure P6.3 with

$$G(s) = \frac{8}{s(s^6-2s^5-s^4+2s^3+4s^2-8s-4)}$$

tell how many closed-loop poles are located in the right half-plane, in the left half-plane, and on the $j\omega$-axis. [Section: 6.3]

**16.** `Repeat Problem 15 using MATLAB.` MATLAB ML

**17.** Consider the following Routh table. Notice that the $s^5$ row was originally all zeros. Tell how many roots of the original polynomial were in the right half-plane, in the left half-plane, and on the $j\omega$-axis. [Section: 6.3]

| | | | | |
|---|---|---|---|---|
| $s^7$ | 1 | 2 | −1 | −2 |
| $s^6$ | 1 | 2 | −1 | −2 |
| $s^5$ | 3 | 4 | −1 | 0 |
| $s^4$ | 1 | −1 | −3 | 0 |
| $s^3$ | 7 | 8 | 0 | 0 |
| $s^2$ | −15 | −21 | 0 | 0 |
| $s^1$ | −9 | 0 | 0 | 0 |
| $s^0$ | −21 | 0 | 0 | 0 |

**18.** For the system of Figure P6.4, tell where the closed-loop poles are located (i.e., right half-plane, left half-plane, $j\omega$-axis). Notice that there is positive feedback. [Section: 6.3]

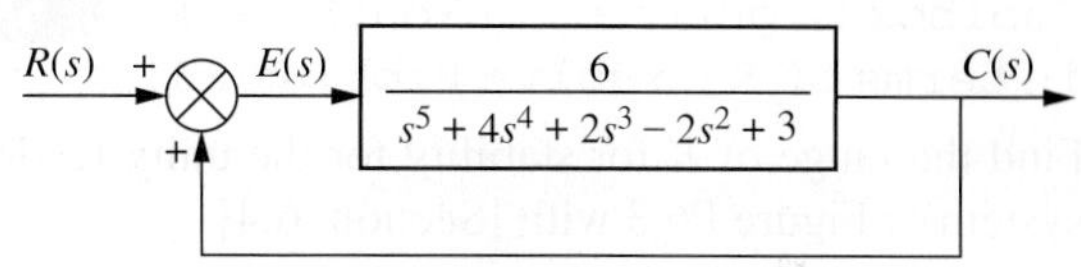


**FIGURE P6.4**

**19.** Using the Routh-Hurwitz criterion, tell how many closed-loop poles of the system shown in Figure P6.5 lie in the left half-plane, in the right half-plane, and on the $j\omega$-axis. [Section: 6.3]

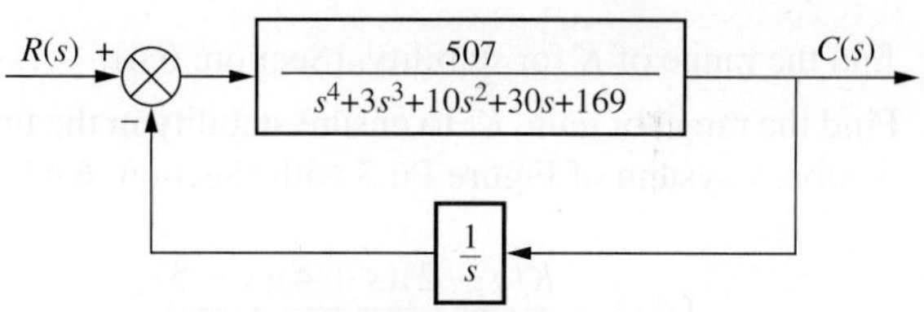


**FIGURE P6.5**

**20.** Determine if the unity feedback system of Figure P6.3 with

$$G(s) = \frac{K(s^2+1)}{(s+1)(s+2)}$$

can be unstable. [Section: 6.4]

**21.** For the unity feedback system of Figure P6.3 with

$$G(s) = \frac{K(s+6)}{s(s+1)(s+3)}$$

determine the range of $K$ to ensure stability. [Section: 6.4]

**22.** In the system of Figure P6.3, let

$$G(s) = \frac{K(s-a)}{s(s-b)}$$

Find the range of $K$ for closed-loop stability when: [Section: 6.4]

**a.** $a < 0, \quad b < 0$

**b.** $a < 0, \quad b > 0$

**c.** $a > 0, \quad b < 0$

**d.** $a > 0, \quad b > 0$

**23.** For the unity feedback system of Figure P6.3 with

$$G(s) = \frac{K(s+1)}{s(s+2)(s+3)(s+4)}$$

determine the range of $K$ for stability. [Section: 6.4]

**24.** `Repeat Problem 23 using MATLAB.` MATLAB ML

**25.** `Use MATLAB and the Symbolic Math Toolbox to generate a Routh table in terms of K to solve Problem 23.` Symbolic Math SM

**26.** Find the range of $K$ for stability for the unity feedback system of Figure P6.3 with [Section: 6.4]

$$G(s) = \frac{K(s+2)(s-2)}{(s^2+3)}$$

**27.** For the unity feedback system of Figure P6.3 with

$$G(s) = \frac{K(s+1)}{s^4(s+2)}$$

find the range of $K$ for stability. [Section: 6.4]

**28.** Find the range of gain, $K$, to ensure stability in the unity feedback system of Figure P6.3 with [Section: 6.4]

$$G(s) = \frac{K(s-2)(s+4)(s+5)}{(s^2+3)}$$

**29.** Find the range of gain, $K$, to ensure stability in the unity feedback system of Figure P6.3 with [Section: 6.4]

$$G(s) = \frac{K(s+3)}{(s^2+1)(s+5)(s-1)}$$

**30.** Using the Routh-Hurwitz criterion, find the value of $K$ that will yield oscillations for the unity feedback system of Figure P6.3 with [Section: 6.4]

$$G(s) = \frac{K}{(s+15)(s+27)(s+38)}$$

**31.** Use the Routh-Hurwitz criterion to find the range of $K$ for which the system of Figure P6.6 is stable. [Section: 6.4]

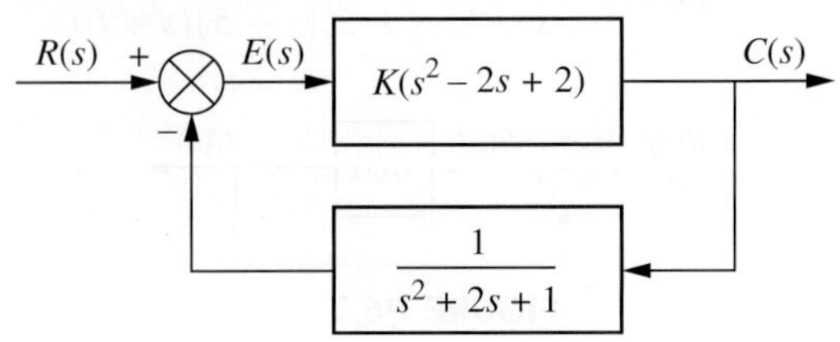


**FIGURE P6.6**

**32.** Repeat Problem 31 for the system of Figure P6.7. [Section: 6.4]

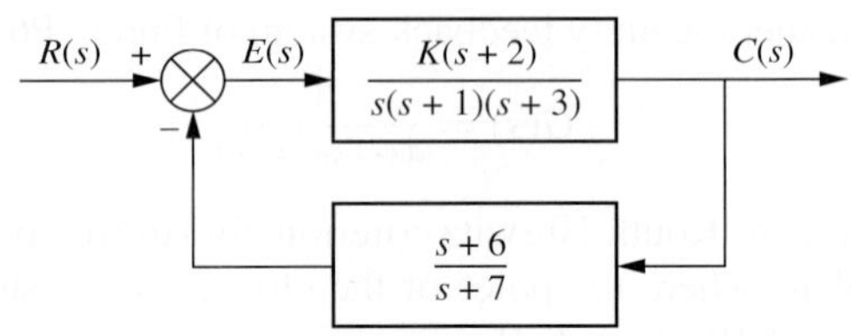


**FIGURE P6.7**

**33.** Given the unity feedback system of Figure P6.3 with

$$G(s) = \frac{K(s+5)}{s(s+1.2)(s+3)}$$

find the following: [Section: 6.4]

**a.** The range of $K$ that keeps the system stable

**b.** The value of $K$ that makes the system oscillate

**c.** The frequency of oscillation when $K$ is set to the value that makes the system oscillate

**34.** Given the unity feedback system of Figure P6.3 with

$$G(s) = \frac{K(s+4)}{s(s+1)(s+2)}$$

find the following: [Section: 6.4]

**a.** The range of $K$ that keeps the system stable

**b.** The value of $K$ that makes the system oscillate

**c.** The frequency of oscillation when $K$ is set to the value that makes the system oscillate

**35.** For the system shown in Figure P6.8, find the value of gain, $K$, that will make the system oscillate. Also, find the frequency of oscillation. [Section: 6.4]

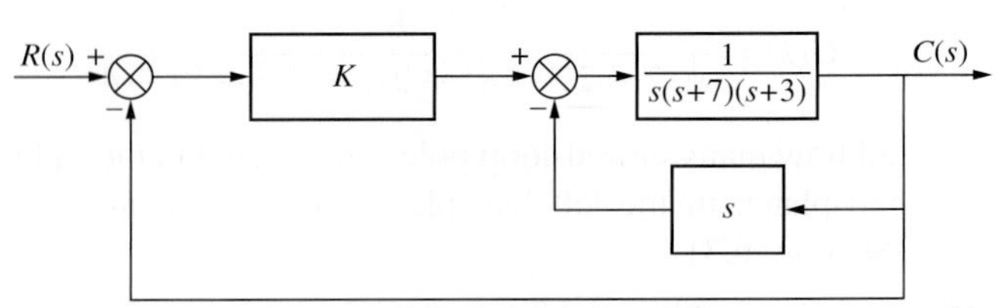


**FIGURE P6.8**

36. Given the unity feedback system of Figure P6.3 with [Section: 6.4]

$$G(s) = \frac{Ks(s+2)}{(s^2 - 4s + 8)(s+3)}$$

a. Find the range of $K$ for stability.

b. Find the frequency of oscillation when the system is marginally stable.

37. `Repeat Problem 36 using MATLAB.` MATLAB ML

38. For the unity feedback system of Figure P6.3 with

$$G(s) = \frac{K(s+2)}{(s^2+1)(s+4)(s-1)}$$

find the range of $K$ for which there will be only two closed-loop, right-half-plane poles. [Section: 6.4]

39. For the unity feedback system of Figure P6.3 with [Section: 6.4]

$$G(s) = \frac{K}{(s+1)^3(s+3)}$$

a. Find the range of $K$ for stability.

b. Find the frequency of oscillation when the system is marginally stable.

40. Given the unity feedback system of Figure P6.3 with [Section: 6.4]

$$G(s) = \frac{K}{(s+10)(s^2+4s+5)}$$

a. Find the range of $K$ for stability.

b. Find the frequency of oscillation when the system is marginally stable.

41. Using the Routh-Hurwitz criterion and the unity feedback system of Figure P6.3 with [Section: 6.4]

$$G(s) = \frac{K}{s(s+1)(s+2)(s+5)}$$

a. Find the range of $K$ for stability.

b. Find the value of $K$ for marginal stability.

c. Find the actual location of the closed-loop poles when the system is marginally stable.

42. Find the range of $K$ to keep the system shown in Figure P6.9 stable. [Section: 6.4]

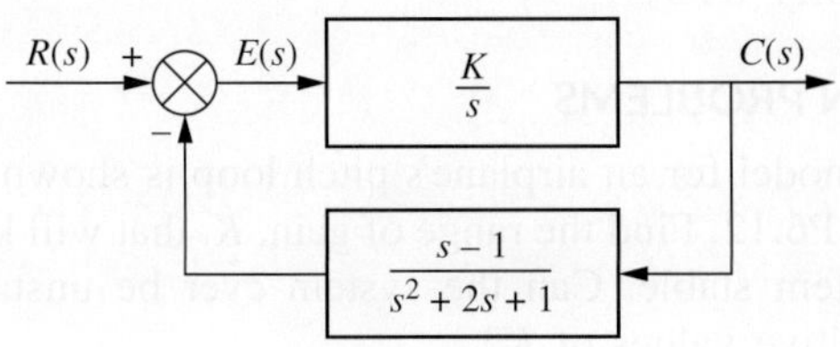


**FIGURE P6.9**

43. Find the value of $K$ in the system of Figure P6.10 that will place the closed-loop poles as shown. [Section: 6.4]

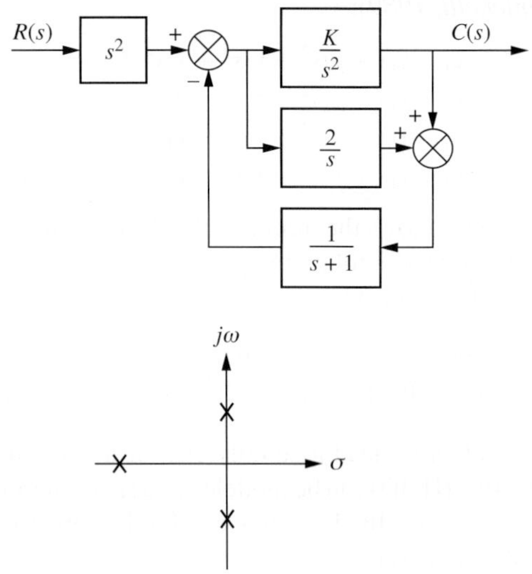


**FIGURE P6.10** Closed-loop system with pole plot

44. The closed-loop transfer function of a system is

$$T(s) = \frac{s^2 + K_1 s + K_2}{s^4 + K_1 s^3 + K_2 s^2 + 4s + 1}$$

Determine the range of $K_1$ in order for the system to be stable. What is the relationship between $K_1$ and $K_2$ for stability? [Section: 6.4]

45. For the transfer function below, find the constraints on $K_1$ and $K_2$ such that the function will have only two $j\omega$ poles. [Section: 6.4]

$$T(s) = \frac{K_1 s + K_2}{s^4 + K_1 s^3 + s^2 + K_2 s + 1}$$

46. The transfer function relating the output engine fan speed (rpm) to the input main burner fuel flow rate (lb/h) in a short takeoff and landing (STOL) fighter aircraft, ignoring the coupling between engine fan speed and the pitch control command, is (*Schierman, 1992*) [Section: 6.4]

$$G(s) = \frac{1.3s^7 + 90.5s^6 + 1970s^5 + 15{,}000s^4 + 3120s^3 - 41{,}300s^2 - 5000s - 1840}{s^8 + 103s^7 + 1180s^6 + 4040s^5 + 2150s^4 - 8960s^3 - 10{,}600s^2 - 1550s - 415}$$

a. Find how many poles are in the right half-plane, in the left half-plane, and on the $j\omega$-axis.

b. Is this open-loop system stable?

47. An interval polynomial is of the form

$$P(s) = a_0 + a_1 s + a_2 s^2 + a_3 s^3 + a_4 s^4 + a_5 s^5 + \cdots$$

with its coefficients belonging to intervals $x_i \le a_i \le y_i$, where $x_i, y_i$ are prescribed constants. Kharitonov's theorem says that an interval polynomial has all its roots in the left half-plane if each one of the following four polynomials has its roots in the left half-plane (*Minichelli, 1989*):

$$K_1(s) = x_0 + x_1 s + y_2 s^2 + y_3 s^3 + x_4 s^4 + x_5 s^5 + y_6 s^6 + \cdots$$
$$K_2(s) = x_0 + y_1 s + y_2 s^2 + x_3 s^3 + x_4 s^4 + y_5 s^5 + y_6 s^6 + \cdots$$
$$K_3(s) = y_0 + x_1 s + x_2 s^2 + y_3 s^3 + y_4 s^4 + x_5 s^5 + x_6 s^6 + \cdots$$
$$K_4(s) = y_0 + y_1 s + x_2 s^2 + x_3 s^3 + y_4 s^4 + y_5 s^5 + x_6 s^6 + \cdots$$

Use Kharitonov's theorem and the Routh-Hurwitz criterion to find if the following polynomial has any zeros in the right-half-plane.

$$P(s) = a_0 + a_1 s + a_2 s^2 + a_3 s^3$$
$$2 \le a_0 \le 3; \quad 1 \le a_1 \le 2; \quad 3 \le a_2 \le 5; \quad a_3 = 1$$

**48.** The read/write head assembly arm of a computer hard disk drive (HDD) can be modeled as a rigid rotating body with inertia $I_b$. Its dynamics can be described with the transfer function

$$P(s) = \frac{X(s)}{F(s)} = \frac{1}{I_b s^2}$$

where $X(s)$ is the displacement of the read/write head and $F(s)$ is the applied force (*Yan, 2003*). Show that if the HDD is controlled in the configuration shown in Figure P6.11, the arm will oscillate and cannot be positioned with any precision over a HDD track. Find the oscillation frequency.

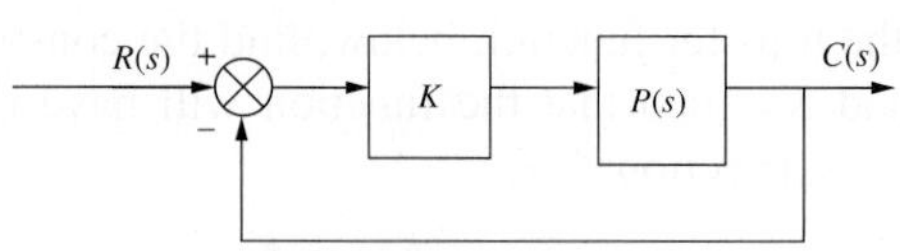


**FIGURE P6.11**

**49.** A linearized model of a torque-controlled crane hoisting a load with a fixed rope length is

$$P(s) = \frac{X_T(s)}{F_T(s)} = \frac{1}{m_T}\frac{s^2 + \omega_0^2}{s^2(s^2 + a\omega_0^2)}$$

where $\omega_0 = \sqrt{\frac{g}{L}}$, $L$ = the rope length, $m_T$ = the mass of the car, $a$ = the combined rope and car mass, $f_T$ = the force input applied to the car, and $x_T$ = the resulting rope displacement (*Marttinen, 1990*). If the system is controlled in a feedback configuration by placing it in a loop as shown in Figure P6.11, with $K > 0$, where will the closed-loop poles be located?

**50.** A system is represented in state space as

State Space SS

$$\dot{\mathbf{x}} = \begin{bmatrix} 0 & 1 & 2 \\ 3 & 1 & -4 \\ 1 & 1 & 3 \end{bmatrix}\mathbf{x} + \begin{bmatrix} 0 \\ 1 \\ 0 \end{bmatrix}u$$
$$y = [1 \quad 1 \quad 0]\mathbf{x}$$

Determine how many eigenvalues are in the right half-plane, in the left half-plane, and on the $j\omega$-axis. [Section: 6.5]

**51.** `Use MATLAB to find the eigenvalues of the following system:`

MATLAB ML

State Space SS

$$\dot{\mathbf{x}} = \begin{bmatrix} 0 & 1 & 0 \\ 0 & 1 & -4 \\ -1 & 1 & 3 \end{bmatrix}\mathbf{x} + \begin{bmatrix} 0 \\ 0 \\ 1 \end{bmatrix}u$$
$$y = [0 \quad 0 \quad 1]\mathbf{x}$$

**52.** The following system in state space represents the forward path of a unity feedback system. Use the Routh-Hurwitz criterion to determine if the closed-loop system is stable. [Section: 6.5]

State Space SS

$$\dot{\mathbf{x}} = \begin{bmatrix} 0 & 1 & 0 \\ 0 & 1 & 3 \\ -3 & -4 & -5 \end{bmatrix}\mathbf{x} + \begin{bmatrix} 0 \\ 0 \\ 1 \end{bmatrix}u$$
$$y = [0 \quad 1 \quad 1]\mathbf{x}$$

**53.** `Repeat Problem 52 using MATLAB.`

MATLAB ML

**54.** An inverted pendulum, mounted on a motor-driven cart was presented in Chapter 3, Problem 29. The system's state-space model was linearized around a stationary point, $\mathbf{x_0} = \mathbf{0}$, corresponding to the pendulum point-mass, $m$, being in the upright position at $t = 0$, when the force applied to the cart $u_0 = 0$ (*Prasad, 2012*). We'll modify that model here to have two output variables: the pendulum angle relative to the $y$-axis, $\theta$, and the horizontal position of the cart, $x$. The output equation becomes:

$$\mathbf{y} = \begin{bmatrix} \theta \\ x \end{bmatrix} = \mathbf{Cx} = \begin{bmatrix} 1 & 0 & 0 & 0 \\ 0 & 0 & 1 & 0 \end{bmatrix}\begin{bmatrix} \theta \\ \dot{\theta} \\ x \\ \dot{x} \end{bmatrix}$$

`Using MATLAB, find out how many eigenvalues are in the right half-plane, in the left half-plane, and on the` $j\omega$`-axis. What does that tell us about the stability of that unit? [Section: 6.5]`

MATLAB ML

## DESIGN PROBLEMS

**55.** A model for an airplane's pitch loop is shown in Figure P6.12. Find the range of gain, $K$, that will keep the system stable. Can the system ever be unstable for positive values of $K$?

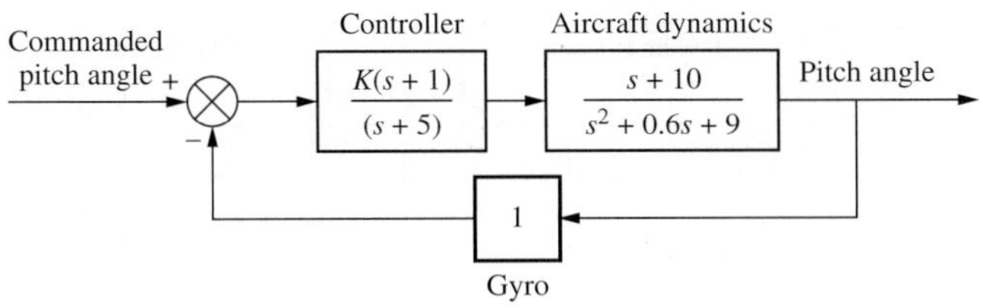


**FIGURE P6.12** Aircraft pitch loop model

**56.** A common application of control systems is in regulating the temperature of a chemical process (Figure P6.13). The flow of a chemical reactant to a process is controlled by an actuator and valve. The reactant causes the temperature in the vat to change. This temperature is sensed and compared to a desired set-point temperature in a closed loop, where the flow of reactant is adjusted to yield the desired temperature. In Chapter 9, we will learn how a PID controller is used to improve the performance of such process control systems. Figure P6.13 shows the control system prior to the addition of the PID controller. The PID controller is replaced by the shaded box with a gain of unity. For this system, prior to the design of the PID controller, find the range of amplifier gain, $K$, to keep the system stable.

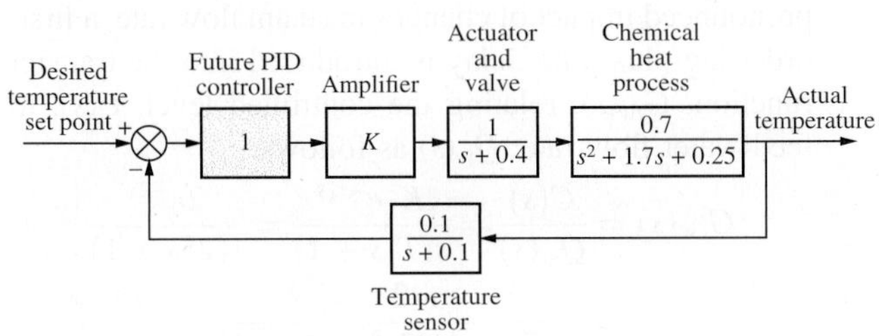


**FIGURE P6.13** Block diagram of a chemical process control system

**57.** Often an aircraft is required to tow another vehicle, such as a practice target or glider. To stabilize the towed vehicle and prevent it from rolling, pitching, and yawing, an autopilot is built into the towed vehicle. Assume the block diagram shown in Figure P6.14 represents the autopilot roll control system (*Cochran, 1992*). Find the range of $K$ to keep the roll angle stable.

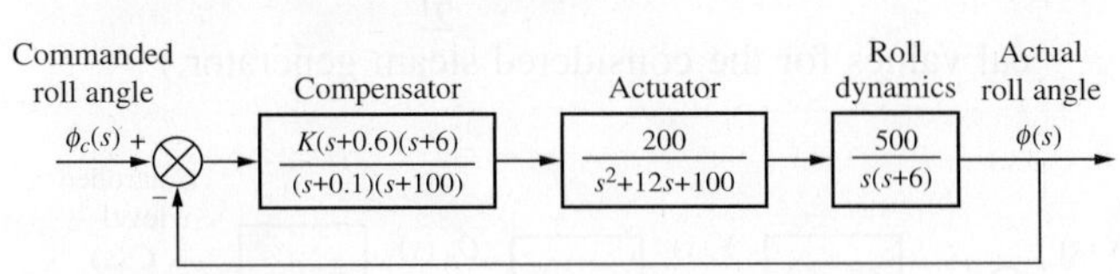


**FIGURE P6.14** Towed vehicle roll control

**58.** A transfer function from indoor radiator power, $\dot{Q}(s)$, to room temperature, $T(s)$, in an 11 m$^2$ room is

$$P(s)=\frac{T(s)}{\dot{Q}(s)}=\frac{1\times10^{-6}s^2+1.314\times10^{-9}s+2.66\times10^{-13}}{s^3+0.00163s^2+5.272\times10^{-7}s+3.538\times10^{-11}}$$

where $\dot{Q}$ is in watts and $T$ is in C (*Thomas, 2005*). The room's temperature will be controlled by embedding it in a closed loop, such as that of Figure P6.11. Find the range of $K$ for closed-loop stability.

**59.** Figure P5.38 shows the block diagram of an Automatic Voltage Regulator (*Gozde, 2011*). Assume in this diagram the following parameter values: $K_a = 10$, $T_a = 0.1$, $K_e = 1$, $T_e = 0.4$, $K_g = 1$, $T_g = 1$, $K_s = 1$, and $T_s = 0.001$. Also assume that the PID transfer function is substituted by a simple integrator, namely $G_{PID}(s) = \frac{K}{s}$. Find the range of $K$ for which the system is closed-loop stable.

**60.** During vertical spindle surface grinding, adjustments are made on a multi-axis computer numerical control (CNC) machine by measuring the applied force with a dynamometer and applying appropriate corrections. This feedback force control results in higher homogeneity and better tolerances in the resulting finished product. In a specific experiment with an extremely high feed rate, the transfer function from the desired depth of cut (DOC) to applied force was

$$\frac{F(s)}{DOC(s)}=\frac{K_C}{1+\frac{K_C}{ms^2+bs+k}-\frac{K_C}{K_f}\left(\frac{1}{Ts+1}\right)}$$

where $k = 2.1\times10^4$ N/m, $b = 0.78$ Ns/m, $m = 1.2\times10^{-4}$ kg, $K_C = 1.5\times10^4$ N/mm, and $T = 0.004$s. The parameter $K_f$ is varied to adjust the system. Find the range of $K_f$ under which the system is stable (*Hekman, 1999*).

**61.** It has been shown (*Pounds, 2011*) that an unloaded UAV helicopter is closed-loop stable and will have a characteristic equation given by

$$s^3+\left(\frac{mgh}{I}(q_2+kk_d)+q_1g\right)s^2$$
$$+k\frac{mgh}{I}s+\frac{mgh}{I}(kk_i+q_1)=0$$

where $m$ is the mass of the helicopter, $g$ is the gravitational constant, $I$ is the rotational inertia of the helicopter, $h$ is the height of the rotor plane above the center of gravity, $q_1$ and $q_2$ are stabilizer flapping parameters, $k$, $k_i$, and $k_d$

are controller parameters; all constants > 0. The UAV is supposed to pick up a payload; when this occurs, the mass, height, and inertia change to $m'$, $h'$, and $I'$, respectively, all still > 0. Show that the helicopter will remain stable as long as

$$\frac{m'gh'}{I'} > \frac{q_1 + kk_i - q_1 gk}{k(q_2 + kk_d)}$$

**62.** In order to obtain a low-cost lithium-ion battery charger, the feedback loop of Figure P6.3 is used, where $G(s) = G_c(s)P(s)$. The following transfer functions have been derived for $G(s)$ (*Tsang, 2009*):

$$P(s) = \frac{R_1R_2C_1C_2s^2 + (R_1C_1 + R_2C_1 + R_2C_2)s + 1}{C_1(1 + R_2C_2)s}$$

$$G_c(s) = K_p + \frac{K_I}{s}$$

If $R_1 = 0.15\Omega; R_2 = 0.44\Omega; C_1 = 7200\text{F}$; and $C_2 = 170\text{F}$, use the Routh-Hurwitz criteria to find the range of positive $K_P$ and $K_I$ for which the system is closed-loop stable.

**63.** Figure P6.15 shows the model of the dynamics of an economic system *(Wingrove, 2012)*. In this diagram $x$ represents the rate of growth in real Gross National Product (GNP), $x_0$ the long-term trend (dc value) of the GNP, $\Delta x$ the change over the long-term trend of the GNP, $r_x$ the real and psychological disturbance inputs that affect the economy, $r_m$ the random monetary inputs, and $\Delta u$ fluctuations in unemployment rate. The diagram has two feedback loops: one through Friedman's model in which the economy dynamics are approximated by

$$F(s) = \frac{K_x s}{\left(\frac{s}{\omega_n}\right)^2 + 2\xi\left(\frac{s}{\omega_n}\right) + 1}$$

and a second loop through Okun's law that relates the GNP to unemployment changes. Assuming the following parameter values: $K_x = 2$ years, $\omega_n = 1.5$ rad/year, $\zeta = 0.8$, $K_u = 0.4$ and $G_x = -0.4$. Find the range of $G_u$ for closed-loop stability.

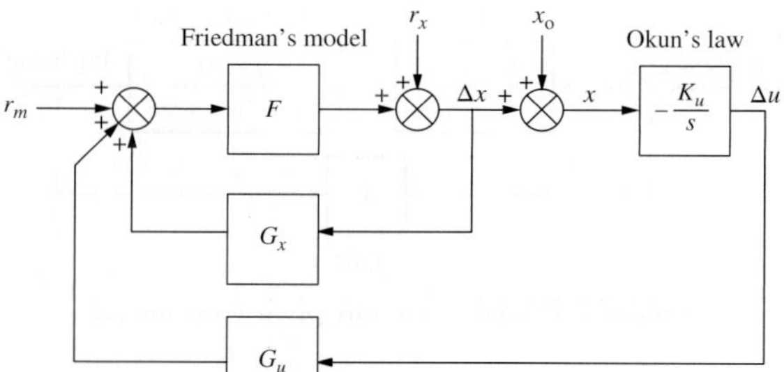


**FIGURE P6.15**

**64.** Figure P6.16 is a simplified and linearized block diagram of a cascade control system, which is used to control water level in a steam generator of a nuclear power plant (*Wang, 2009*).

In this system, the level controller, $G_{LC}(s)$, is the *master* controller and the feed-water flow controller, $G_{FC}(s)$, is the *slave* controller. Using mass balance equations, the water level would ordinarily be regarded as a simple integration process of water flow. In a steam generator, however, steam flow rate and the cooling effect of feed-water change the dynamics of that process. Taking the latter into account and ignoring the much-less pronounced impact of changes in steam flow rate, a first-order lag plus time delay is introduced into the transfer function, $G_{fw}(s)$, relating the controlled level, $C(s)$, to feed-water flow rate, $Q_w(s)$ as follows:

$$G_{fw}(s) = \frac{C(s)}{Q_w(s)} = \frac{K_1e^{-\tau_1 s}}{s(T_1s + 1)} = \frac{2e^{-2s}}{s(25s + 1)}$$

$$\approx \frac{2}{s(25s + 1)(2s^2 + 2s + 1)}$$

where $K_1 = 2$ is the process gain, $\tau_1 = 2$ is the pure time delay, and $T_1 = 25$ is the steam generator's time constant. (The expression $e^{-\tau_1 s}$ represents a time delay. This function can be represented by what is known as a *Pade approximation*. This approximation can take on many increasingly complicated forms, depending upon the degree of accuracy required. Here we use the Pade approximation, $e^{-x} \approx \frac{1}{1 + x + \frac{x^2}{2!}}$, and specific numerical values for the considered steam generator.)

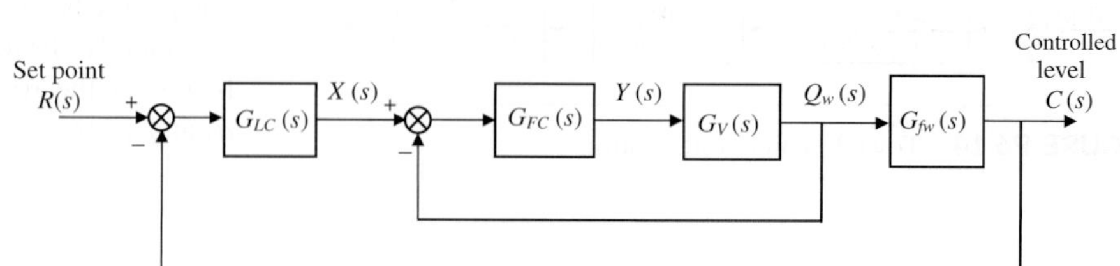


**FIGURE P6.16**

The dynamic characteristics of the control valve are approximated by the transfer function

$$G_v(s) = \frac{Q_w(s)}{Y(s)} = \frac{K_v}{T_v s + 1} = \frac{1}{3s + 1}$$

where $K_v$ is the valve gain and $T_v$ is its time constant.

Given that: $G_{FC}(s) = K_{P_{FC}} + K_{D_{FC}} s = 0.5 + 2s$ and $G_{LC}(s) = K_{P_{LC}} + K_{D_{LC}} s = 0.5 + Ks$, use the Routh-Hurwitz criterion to find the range of the level controller's derivative gain, $K_{D_{LC}} = K > 0$, that will keep the system stable.

**65.** The system shown in Figure P6.17 has $G_1(s) = 1/s(s+2)\ (s+4)$. Find the following:

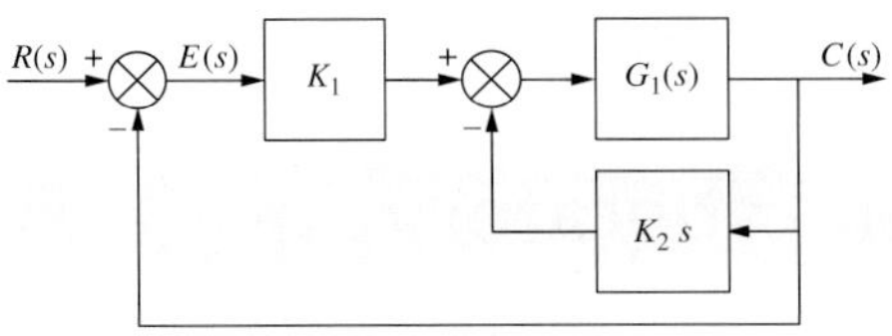


**FIGURE P6.17**

**a.** The value of $K_2$ for which the inner loop will have two equal negative real poles and the associated range of $K_1$ for system stability.

**b.** The value of $K_1$ at which the system oscillates and the associated frequency of oscillation.

**c.** The gain $K_1$ at which a real closed-loop pole is at $s = -5$. Can the step response, $c(t)$, be approximated by a second-order, underdamped response in this case? Why or why not?

**d.** If the response in Part **c** can be approximated as a second-order response, find the $\%OS$ and settling time, $T_s$, when the input is a unit step, $r(t) = u(t)$.

**66.** A drive system with an elastically coupled load was presented in Problems 67 and 61 in Chapters 4 and 5, respectively (*Thomsen, 2011*). That drive was shown in Figure P5.43 as the controlled unit in a feedback control system, where $\Omega_L(s)$ = the load speed and $\Omega_r(s)$ = the desired (reference) speed. If the controller transfer function is $G_C(s) = K_p + \frac{K_I}{s}$, while all of the other parameters and transfer functions are the same as in Problem 61 in Chapter 5, find the range of $K_P$ for stability of the system if $K_I = 0.1$.

## PROGRESSIVE ANALYSIS AND DESIGN PROBLEMS

**67. Control of HIV/AIDS.** The HIV infection linearized model developed in Problem 82, Chapter 4, can be shown to have the transfer function

$$P(s) = \frac{Y(s)}{U_1(s)} = \frac{-520s - 10.3844}{s^3 + 2.6817s^2 + 0.11s + 0.0126}$$

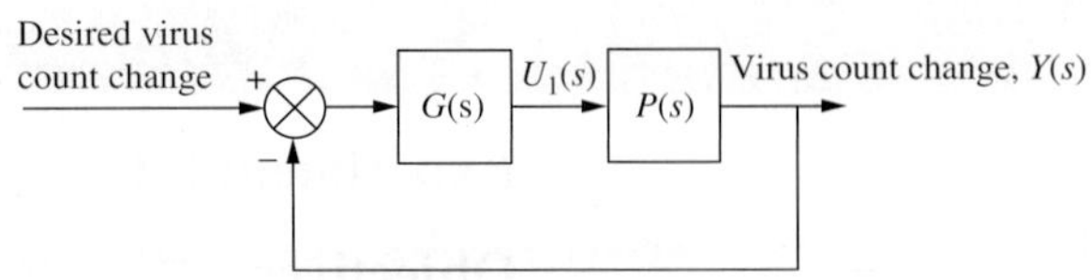


**FIGURE P6.18**

It is desired to develop a policy for drug delivery to maintain the virus count at prescribed levels. For the purpose of obtaining an appropriate $u_1(t)$, feedback will be used as shown in Figure P6.18 (*Craig, 2004*).

As a first approach, consider $G(s) = K$, a constant to be selected. Use the Routh-Hurwitz criteria to find the range of $K$ for which the system is closed-loop stable.

**68. Hybrid vehicle.** Figure P6.19 shows the HEV system presented in Chapter 5, where parameter values have been substituted. It is assumed here that the speed controller has a proportional gain, $K_p$, to be adjusted. Use the Routh-Hurwitz stability method to find the range of positive $K_p$ for which the system is closed-loop stable (*Graebe, 1995*).

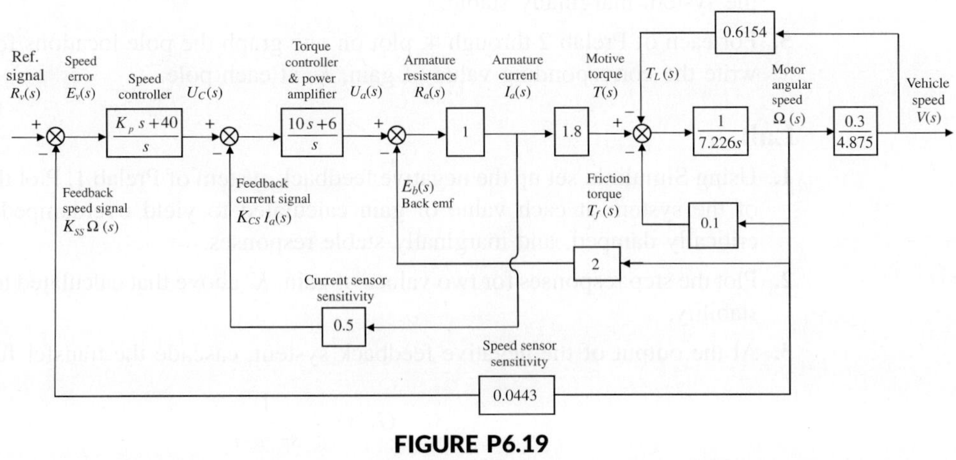


**FIGURE P6.19**

**69. Parabolic trough collector.** The fluid temperature of a parabolic trough collector (*Camacho, 2012*) will be controlled by using a unity feedback structure as shown in Figure P6.11. Assume the open-loop plant transfer function is given by

$$P(s) = \frac{137.2 \times 10^{-6}}{s^2 + 0.0224s + 196 \times 10^{-6}} e^{-39s}$$

Use the Routh-Hurwitz criteria to find the range of gain $K$ that will result in a closed-loop stable system. Note: Pure time-delay dynamics, such as the one in the transfer function of the parabolic trough collector, cannot be treated directly using the Routh-Hurwitz criterion because it is represented by a nonrational factor. However, a Padé approximation can be used for the nonrational component. The Padé approximation was introduced in Problem 64, but it can appear in different forms. Here, it is suggested you use a first-order approximation of the form

$$e^{-sT} \approx \frac{1 - \dfrac{T}{2}s}{1 + \dfrac{T}{2}s}$$

# Cyber Exploration Laboratory

## Experiment 6.1

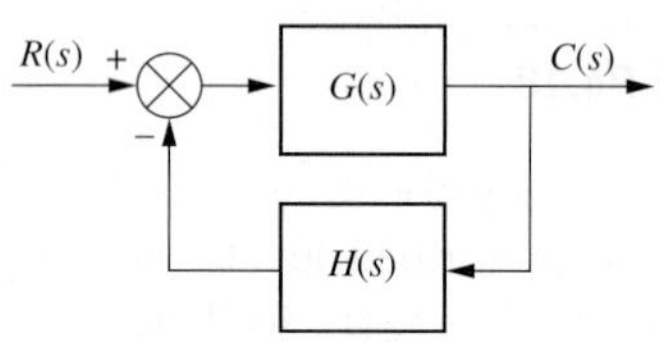


FIGURE P6.20

**Objectives** To verify the effect of pole location upon stability. To verify the effect upon stability of loop gain in a negative feedback system.

**Minimum Required Software Packages** MATLAB, Simulink, and the Control System Toolbox

### Prelab

1. Find the equivalent transfer function of the negative feedback system of Figure P6.20 if

$$G(s) = \frac{K}{s(s+2)^2} \quad \text{and} \quad H(s) = 1$$

2. For the system of Prelab 1, find two values of gain that will yield closed-loop, overdamped, second-order poles. Repeat for underdamped poles.
3. For the system of Prelab 1, find the value of gain, $K$, that will make the system critically damped.
4. For the system of Prelab 1, find the value of gain, $K$, that will make the system marginally stable. Also, find the frequency of oscillation at that value of $K$ that makes the system marginally stable.
5. For each of Prelab 2 through 4, plot on one graph the pole locations for each case and write the corresponding value of gain, $K$, at each pole.

### Lab

1. Using Simulink, set up the negative feedback system of Prelab 1. Plot the step response of the system at each value of gain calculated to yield overdamped, underdamped, critically damped, and marginally stable responses.
2. Plot the step responses for two values of gain, $K$, above that calculated to yield marginal stability.
3. At the output of the negative feedback system, cascade the transfer function

$$G_1(s) = \frac{1}{s^2 + 4}$$

Set the gain, $K$, at a value below that calculated for marginal stability and plot the step response. Repeat for $K$ calculated to yield marginal stability.

### Postlab

1. From your plots, discuss the conditions that lead to unstable responses.
2. Discuss the effect of gain upon the nature of the step response of a closed-loop system.

### Experiment 6.2

**Objective** To use the LabVIEW Control Design and Simulation Module for stability analysis.

**Minimum Required Software Package** LabVIEW with the Control Design and Simulation Module

### Prelab

1. Select six transfer functions of various orders and use Routh-Hurwitz to determine their stability.

### Lab

1. Create a LabVIEW VI that receives the order and the coefficients of the characteristic equation and outputs the location of the poles and information regarding stability.

### Postlab

1. Verify the stability of the systems from your Prelab.

## Bibliography

Ballard, R. D. The Riddle of the Lusitania. *National Geographic*, April 1994, National Geographic Society, Washington, D.C., 1994, pp. 68–85. Also, figure caption source for Figure 6.9.

Camacho, E. F., Berenguel, M., Rubio, F. R., and Martinez, D. *Control of Solar Energy Systems*. Springer-Verlag, London, 2012.

Craig, I. K., Xia, X., and Venter, J. W. Introducing HIV/AIDS Education into the Electrical Engineering Curriculum at the University of Pretoria, *IEEE Transactions on Education*, vol. 47, no. 1, February 2004, pp. 65–73.

D'Azzo, J., and Houpis, C. H. *Linear Control System Analysis and Design*, 3d ed. McGraw-Hill, New York, 1988.

Dorf, R. C. *Modern Control Systems*, 5th ed. Addison-Wesley, Reading, MA, 1989.

FANUC Robotics North America, Inc. Figure caption source for Figure 6.11.

Gozde, H., and Taplamacioglu, M. C. Comparative performance analysis of artificial bee colony algorithm for automatic voltage regulator (AVR) system. *Journal of the Franklin Institute*, vol. 348, pp. 1927–1946, 2011.

Graebe, S. F., Goodwin, G. C., and Elsley, G. Control Design and Implementation in Continuous Steel Casting. *IEEE Control Systems, August* 1995, pp. 64–71.

Hekman, K. A., and Liang, S. Y. Compliance Feedback Control for Part Parallelism in Grinding. *International Journal of Manufacturing Technology*, vol. 15, 1999, pp. 64–69.

Hostetter, G. H., Savant, C. J., Jr., and Stefani, R. T. *Design of Feedback Control Systems*, 2d ed. Saunders College Publishing, New York, 1989.

Johnson, H., et al. *Unmanned Free-Swimming Submersible (UFSS) System Description*. NRL Memorandum Report 4393. Naval Research Laboratory, Washington, D.C., 1980.

Martinnen, A., Virkkunen, J., and Salminen, R. T. Control Study with Pilot Crane. *IEEE Transactions on Education*, vol. 33, no. 3, August 1990, pp. 298–305.

Minnichelli, R. J., Anagnost, J. J., and Desoer, C. A. An Elementary Proof of Kharitonov's Stability Theorem with Extensions. *IEEE Transactions on Automatic Control*, vol. 34, 1989, pp. 995–998.

Özgüner, Ű., Űnyelioglu, K. A., and Haptipoĝlu, C. An Analytical Study of Vehicle Steering Control. Proceedings of the 4th IEEE Conference Control Applications, 1995, pp. 125–130.

Phillips, C. L., and Harbor, R. D. *Feedback Control Systems*, 2d ed. Prentice Hall, Upper Saddle River, NJ, 1991.

Pounds, P. E. I., Bersak, D. R., and Dollar, A . M. Grasping From the Air: Hovering Capture and Load Stability. *2011 IEEE International Conference on Robotics and Automation*, Shanghai International Conference Center, May 9–13, Shanghai, China, 2011.

Prasad, L., Tyagi, B., and Gupta, H. Modeling & Simulation for Optimal Control of Nonlinear Inverted Pendulum Dynamical System using PID Controller & LQR. *IEEE Computer Society Sixth Asia Modeling Symposium*, 2012, pp. 138–143.

Preitl, Z., Bauer, P., and Bokor, J. A Simple Control Solution for Traction Motor Used in Hybrid Vehicles. *4th International Symposium on Applied Computational Intelligence and Informatics*. IEEE. 2007, pp. 157–162.

Routh, E. J. *Dynamics of a System of Rigid Bodies*, 6th ed. Macmillan, London, 1905.

Schierman, J. D., and Schmidt, D. K. Analysis of Airframe and Engine Control Interactions and Integrated Flight/Propulsion Control. *Journal of Guidance, Control, and Dynamics*, vol. 15, no. 6, November–December 1992, pp. 1388–1396.

Thomas, B., Soleimani-Mosheni, M., and Fahlén, P. Feed-Forward in Temperature Control of Buildings. *Energy and Buildings*, vol. 37, 2005, pp. 755–761.

Thomsen, S., Hoffmann, N., and Fuchs, F. W. PI Control, PI-Based State Space Control, and Model-Based Predictive Control for Drive Systems With Elastically Coupled Loads — A Comparative Study. *IEEE Transactions on Industrial Electronics*, vol. 58, no. 8, August 2011, pp. 3647–3657.

Timothy, L. K., and Bona, B. E. *State Space Analysis: An Introduction*. McGraw-Hill, New York, 1968.

Tsang, K. M., and Chan, W. L. A Simple and Low-cost Charger for Lithium-Ion Batteries. *Journal of Power Sources*, vol. 191, 2009, pp. 633–635.

Wang, X.-K., Yang, X.-H., Liu, G., and Qian, H. Adaptive Neuro-Fuzzy Inference System PID Controller for Steam Generator Water Level of Nuclear Power Plant. *Procedings of the Eighth International Conference on Machine Learning and Cybernetics*, 2009, pp. 567–572.

Wingrove, R. C., and Da, R. E. Classical Linear-Control Analysis Applied to Business-Cycle Dynamics and Stability. *Comput. Econ.* vol. 39, Springer, 2012, pp. 77—98.

Yan, T., and Lin, R. Experimental Modeling and Compensation of Pivot Nonlinearly in Hard Disk Drives. *IEEE Transactions on Magnetics*, vol. 39, 2003, pp. 1064–1069.

# 7

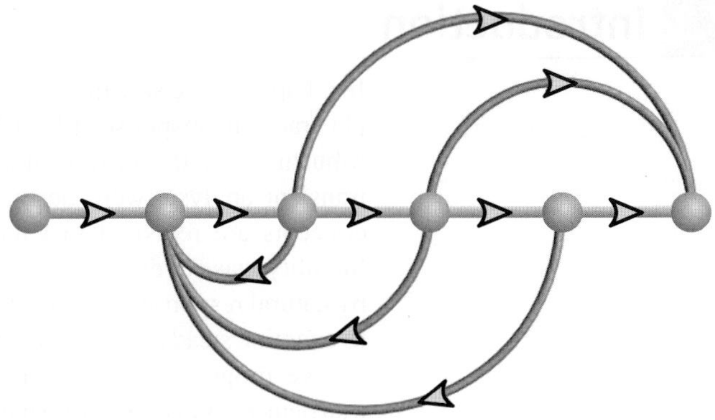

# Steady-State Errors

## Chapter Learning Outcomes

After completing this chapter the student will be able to:

- Find the steady-state error for a unity feedback system (Sections 7.1–7.2)
- Specify a system's steady-state error performance (Section 7.3)
- Design the gain of a closed-loop system to meet a steady-state error specification (Section 7.4)
- Find the steady-state error for disturbance inputs (Section 7.5)
- Find the steady-state error for nonunity feedback systems (Section 7.6)
- Find the steady-state error sensitivity to parameter changes (Section 7.7)
- Find the steady-state error for systems represented in state space (Section 7.8)

## ▶ Case Study Learning Outcomes

You will be able to demonstrate your knowledge of the chapter objectives with case studies as follows:

- Given the antenna azimuth position control system shown on the front endpapers, you will be able to find the preamplifier gain to meet steady-state error performance specifications.
- Given a video laser disc recorder, you will be able to find the gain required to permit the system to record on a warped disc.

# 7.1 Introduction

In Chapter 1, we saw that control systems analysis and design focus on three specifications: (1) transient response, (2) stability, and (3) steady-state errors, taking into account the robustness of the design along with economic and social considerations. Elements of transient analysis were derived in Chapter 4 for first- and second-order systems. These concepts are revisited in Chapter 8, where they are extended to higher-order systems. Stability was covered in Chapter 6, where we saw that forced responses were overpowered by natural responses that increase without bound if the system is unstable. Now we are ready to examine steady-state errors. We define the errors and derive methods of controlling them. As we progress, we find that control system design entails trade-offs between desired transient response, steady-state error, and the requirement that the system be stable.

## Definition and Test Inputs

*Steady-state error* is the difference between the input and the output for a prescribed test input as $t \to \infty$. Test inputs used for steady-state error analysis and design are summarized in Table 7.1.

In order to explain how these test signals are used, let us assume a position control system, where the output position follows the input commanded position. Step inputs represent constant position and thus are useful in determining the ability of the control system to position itself with respect to a stationary target, such as a satellite in geostationary orbit (see Figure 7.1). An antenna position control is an example of a system that can be tested for accuracy using step inputs.

Ramp inputs represent constant-velocity inputs to a position control system by their linearly increasing amplitude. These waveforms can be used to test a system's ability to follow a linearly increasing input or, equivalently, to track a constant-velocity target. For example, a position control system that tracks a satellite that moves across the sky at a constant angular velocity, as shown in Figure 7.1, would be tested with a ramp input to evaluate the steady-state error between the satellite's angular position and that of the control system.

**TABLE 7.1** Test waveforms for evaluating steady-state errors of position control systems

| Waveform | Name | Physical interpretation | Time function | Laplace transform |
|---|---|---|---|---|
| 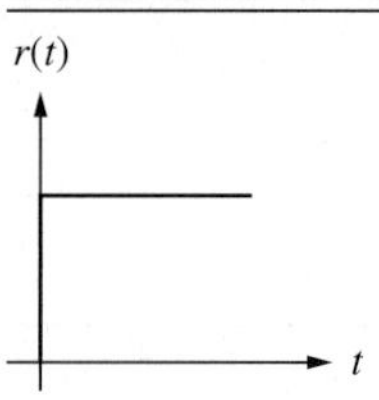 | Step | Constant position | 1 | $\frac{1}{s}$ |
| 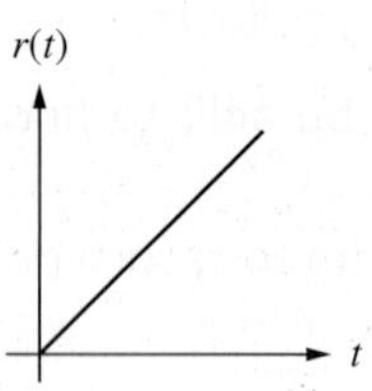 | Ramp | Constant velocity | $t$ | $\frac{1}{s^2}$ |
| 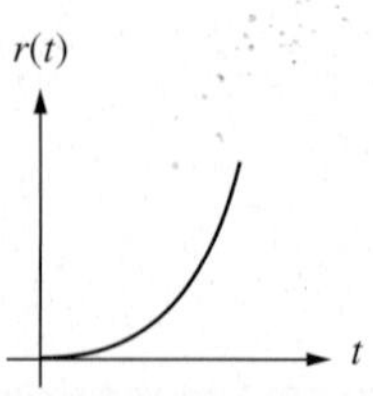 | Parabola | Constant acceleration | $\frac{1}{2}t^2$ | $\frac{1}{s^3}$ |

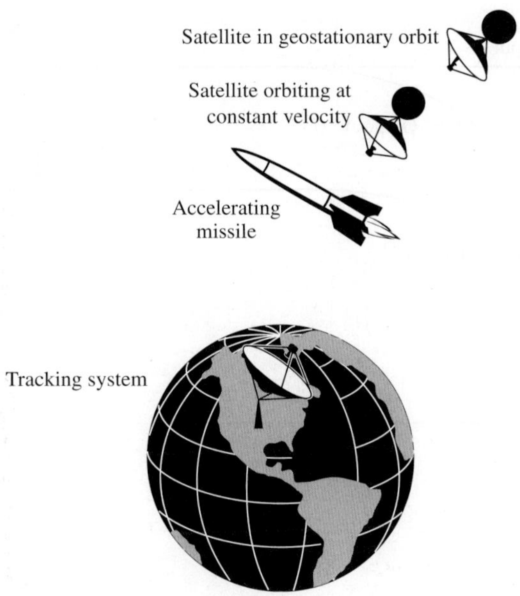


**FIGURE 7.1** Test inputs for steady-state error analysis and design vary with target type

Finally, parabolas, whose second derivatives are constant, represent constant-acceleration inputs to position control systems and can be used to represent accelerating targets, such as the missile in Figure 7.1, to determine the steady-state error performance.

## Application to Stable Systems

Since we are concerned with the difference between the input and the output of a feedback control system after the steady state has been reached, our discussion is limited to stable systems, where the natural response approaches zero as $t \rightarrow \infty$. Unstable systems represent loss of control in the steady state and are not acceptable for use at all. The expressions we derive to calculate the steady-state error can be applied erroneously to an unstable system. Thus, the engineer must check the system for stability while performing steady-state error analysis and design. However, in order to focus on the topic, we assume that all the systems in examples and problems in this chapter are stable. For practice, you may want to test some of the systems for stability.

## Evaluating Steady-State Errors

Let us examine the concept of steady-state errors. In Figure 7.2($a$) a step input and two possible outputs are shown. Output 1 has zero steady-state error, and Output 2 has a finite steady-state error, $e_2(\infty)$. A similar example is shown in Figure 7.2($b$), where a ramp input is compared with Output 1, which has zero steady-state error, and Output 2, which has a finite steady-state error, $e_2(\infty)$. Errors are measured vertically between the Input and Output 2 after the transients have died down. For the ramp input another possibility exists. If the output's slope is different from that of the input, then Output 3, shown in Figure 7.2($b$), results. Here the steady-state error is infinite as measured vertically between the Input and Output 3 after the transients have died down, and $t$ approaches infinity.

Let us now look at the error from the perspective of the most general block diagram. Since the error is the difference between the input and the output of a system, we assume a closed-loop transfer function, $T(s)$, and form the error, $E(s)$, by taking the difference between the input and the output, as shown in Figure 7.3($a$). Here we are interested in the steady-state, or final, value of $e(t)$. For unity feedback systems, $E(s)$ appears as shown in Figure 7.3($b$). In this chapter, we study and derive expressions for the steady-state error for unity feedback systems first and then expand to nonunity feedback systems.

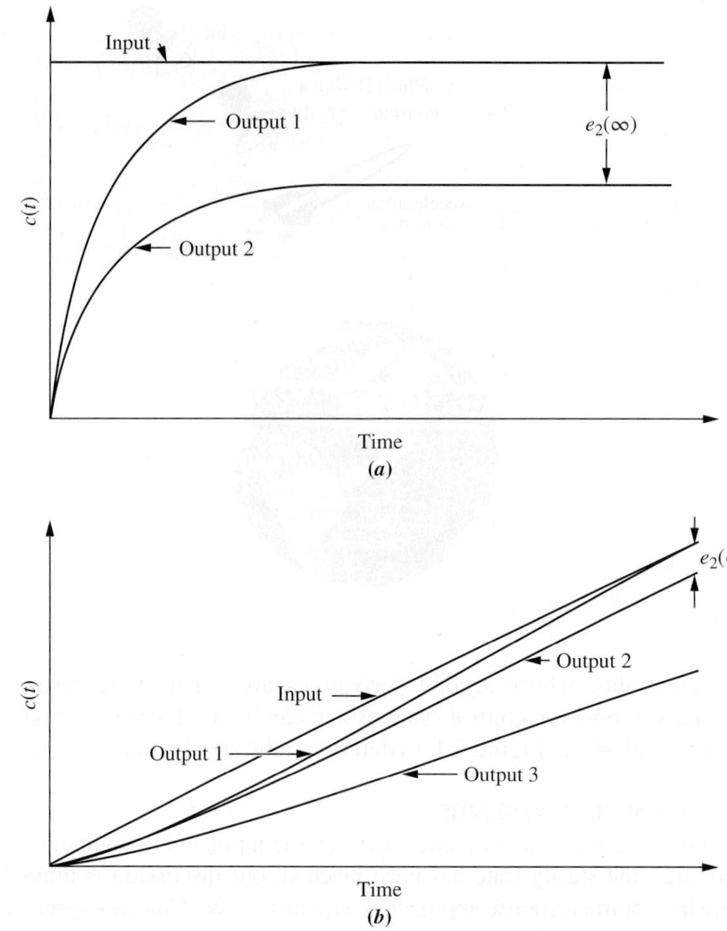


**FIGURE 7.2** Steady-state error: **a**. step input; **b**. ramp input

Before we begin our study of steady-state errors for unity feedback systems, let us look at the sources of the errors with which we deal.

## Sources of Steady-State Error

Many steady-state errors in control systems arise from nonlinear sources, such as backlash in gears or a motor that will not move unless the input voltage exceeds a threshold. Nonlinear behavior as a source of steady-state errors, although a viable topic for study is beyond the scope of a text on linear control systems. The steady-state errors we study here are errors that arise from the configuration of the system itself and the type of applied input.

For example, look at the system of Figure 7.4($a$), where $R(s)$ is the input, $C(s)$ is the output, and $E(s) = R(s) - C(s)$ is the error. Consider a step input. In the steady state, if $c(t)$ equals $r(t)$, $e(t)$ will be zero. But with a pure gain, $K$, the error, $e(t)$, cannot be zero if $c(t)$ is to be finite and nonzero. Thus, by virtue of the configuration of the system (a pure gain of $K$ in the forward path), an error must exist. If we call $c_{\text{steady-state}}$ the

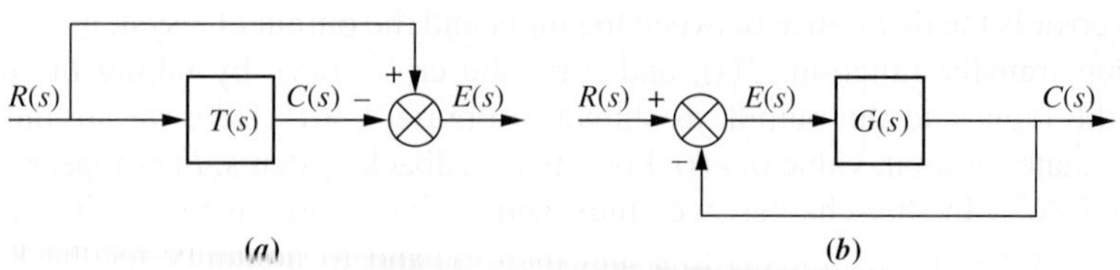


**FIGURE 7.3** Closed-loop control system error: **a.** general representation; **b.** representation for unity feedback systems

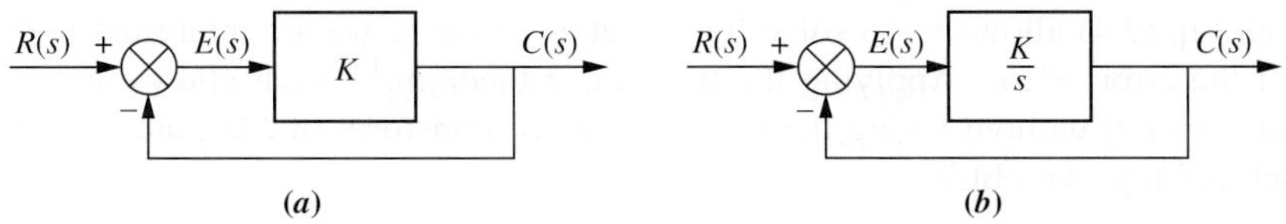


**FIGURE 7.4** System with **a.** finite steady-state error for a step input; **b.** zero steady-state error for step input

steady-state value of the output and $e_{\text{steady-state}}$ the steady-state value of the error, then $c_{\text{steady-state}} = Ke_{\text{steady-state}}$, or

$$e_{\text{steady-state}} = \frac{1}{K} c_{\text{steady-state}} \tag{7.1}$$

Thus, the larger the value of $K$, the smaller the value of $e_{\text{steady-state}}$ would have to be to yield a similar value of $c_{\text{steady-state}}$. The conclusion we can draw is that with a pure gain in the forward path, there will always be a steady-state error for a step input. This error diminishes as the value of $K$ increases.

If the forward-path gain is replaced by an integrator, as shown in Figure 7.4(*b*), there will be zero error in the steady state for a step input. The reasoning is as follows: As $c(t)$ increases, $e(t)$ will decrease, since $e(t) = r(t) - c(t)$. This decrease will continue until there is zero error, but there will still be a value for $c(t)$ since an integrator can have a constant output without any input. For example, a motor can be represented simply as an integrator. A voltage applied to the motor will cause rotation. When the applied voltage is removed, the motor will stop and remain at its present output position. Since it does not return to its initial position, we have an angular displacement output without an input to the motor. Therefore, a system similar to Figure 7.4(*b*), which uses a motor in the forward path, can have zero steady-state error for a step input.

We have examined two cases qualitatively to show how a system can be expected to exhibit various steady-state error characteristics, depending upon the system configuration. We now formalize the concepts and derive the relationships between the steady-state errors and the system configuration generating these errors.

## 7.2 Steady-State Error for Unity Feedback Systems

Steady-state error can be calculated from a system's closed-loop transfer function, $T(s)$, or the open-loop transfer function, $G(s)$, for unity feedback systems. We begin by deriving the system's steady-state error in terms of the closed-loop transfer function, $T(s)$, in order to introduce the subject and the definitions. Next we obtain insight into the factors affecting steady-state error by using the open-loop transfer function, $G(s)$, in unity feedback systems for our calculations. Later in the chapter we generalize this discussion to nonunity feedback systems.

### Steady-State Error in Terms of $T(s)$

Consider Figure 7.3(*a*). To find $E(s)$, the error between the input, $R(s)$, and the output, $C(s)$, we write

$$E(s) = R(s) - C(s) \tag{7.2}$$

But

$$C(s) = R(s)T(s) \tag{7.3}$$

Substituting Eq. (7.3) into Eq. (7.2), simplifying, and solving for $E(s)$ yields

$$E(s) = R(s)[1 - T(s)] \tag{7.4}$$

Although Eq. (7.4) allows us to solve for $e(t)$ at any time, $t$, we are interested in the final value of the error, $e(\infty)$. Applying the final value theorem,[1] which allows us to use the final value of $e(t)$ without taking the inverse Laplace transform of $E(s)$, and then letting $t$ approach infinity, we obtain

$$e(\infty) = \lim_{t\to\infty} e(t) = \lim_{s\to 0} sE(s) \tag{7.5}$$[2]

Substituting Eq. (7.4) into Eq. (7.5) yields

$$e(\infty) = \lim_{s\to 0} sR(s)[1 - T(s)] \tag{7.6}$$

Let us look at an example.

## Example 7.1 Steady-State Error in Terms of $T(s)$

**PROBLEM:** Find the steady-state error for the system of Figure 7.3($a$) if $T(s) = 5/(s^2 + 7s + 10)$ and the input is a unit step.

**SOLUTION:** From the problem statement, $R(s) = 1/s$ and $T(s) = 5/(s^2 + 7s + 10)$. Substituting into Eq. (7.4) yields

$$E(s) = \frac{s^2 + 7s + 5}{s(s^2 + 7s + 10)} \tag{7.7}$$

Since $T(s)$ is stable and, subsequently, $E(s)$ does not have right–half-plane poles or $j\omega$ poles other than at the origin, we can apply the final value theorem. Substituting Eq. (7.7) into Eq. (7.5) gives $e(\infty) = 1/2$.

### Steady-State Error in Terms of $G(s)$

Many times we have the system configured as a unity feedback system with a forward transfer function, $G(s)$. Although we can find the closed-loop transfer function, $T(s)$, and

[1] The final value theorem is derived from the Laplace transform of the derivative. Thus,

$$\mathscr{L}[\dot{f}(t)] = \int_{0-}^{\infty} \dot{f}(t)e^{-st}dt = sF(s) - f(0-)$$

As $s \to 0$,

$$\int_{0-}^{\infty} \dot{f}(t)dt = f(\infty) - f(0-) = \lim_{s\to 0} sF(s) - f(0-)$$

or

$$f(\infty) = \lim_{s\to 0} sF(s)$$

For finite steady-state errors, the final value theorem is valid only if $F(s)$ has poles only in the left half-plane and, at most, one pole at the origin. However, correct results that yield steady-state errors that are infinite can be obtained if $F(s)$ has more than one pole at the origin (see *D'Azzo and Houpis, 1988*). If $F(s)$ has poles in the right half-plane or poles on the imaginary axis other than at the origin, the final value theorem is invalid.

[2] Valid only if (1) $E(s)$ has poles only in the left half-plane and at the origin, and (2) the closed-loop transfer function, $T(s)$, is stable. Notice that by using Eq. (7.5), numerical results can be obtained for unstable systems. These results, however, are meaningless.

then proceed as in the previous subsection, we find more insight for analysis and design by expressing the steady-state error in terms of $G(s)$ rather than $T(s)$.

Consider the feedback control system shown in Figure 7.3(*b*). Since the feedback, $H(s)$, equals 1, the system has unity feedback. The implication is that $E(s)$ is actually the error between the input, $R(s)$, and the output, $C(s)$. Thus, if we solve for $E(s)$, we will have an expression for the error. We will then apply the final value theorem, Item 11 in Table 2.2, to evaluate the steady-state error.

Writing $E(s)$ from Figure 7.3(*b*), we obtain

$$E(s) = R(s) - C(s) \tag{7.8}$$

But

$$C(s) = E(s)G(s) \tag{7.9}$$

Finally, substituting Eq. (7.9) into Eq. (7.8) and solving for $E(s)$ yields

$$E(s) = \frac{R(s)}{1 + G(s)} \tag{7.10}$$

We now apply the final value theorem, Eq. (7.5). At this point in a numerical calculation, we must check to see whether the closed-loop system is stable, using, for example, the Routh-Hurwitz criterion. For now, though, assume that the closed-loop system is stable and substitute Eq. (7.10) into Eq. (7.5), obtaining

$$\boxed{e(\infty) = \lim_{s\to 0}\frac{sR(s)}{1 + G(s)}} \tag{7.11}$$

Equation (7.11) allows us to calculate the steady-state error, $e(\infty)$, given the input, $R(s)$, and the system, $G(s)$. We now substitute several inputs for $R(s)$ and then draw conclusions about the relationships that exist between the open-loop system, $G(s)$, and the nature of the steady-state error, $e(\infty)$.

The three test signals we use to establish specifications for a control system's steady-state error characteristics are shown in Table 7.1. Let us take each input and evaluate its effect on the steady-state error by using Eq. (7.11).

**Step Input.** Using Eq. (7.11) with $R(s) = 1/s$, we find

$$\boxed{e(\infty) = e_{\text{step}}(\infty) = \lim_{s\to 0}\frac{s(1/s)}{1 + G(s)} = \frac{1}{1 + \lim_{s\to 0} G(s)}} \tag{7.12}$$

The term

$$\lim_{s\to 0} G(s)$$

is the dc gain of the forward transfer function, since $s$, the frequency variable, is approaching zero. In order to have zero steady-state error,

$$\lim_{s\to 0} G(s) = \infty \tag{7.13}$$

Hence, to satisfy Eq. (7.13), $G(s)$ must take on the following form:

$$G(s) \equiv \frac{(s + z_1)(s + z_2)\cdots}{s^n(s + p_1)(s + p_2)\cdots} \tag{7.14}$$

For the limit to be infinite, the denominator must be equal to zero as $s$ goes to zero. Thus, $n \geq 1$; that is, at least one pole must be at the origin. Since division by $s$ in the frequency domain is integration in the time domain (see Table 2.2, Item 10), we are also saying that at least one pure integration must be present in the forward path. The steady-state response for this case of zero steady-state error is similar to that shown in Figure 7.2($a$), Output 1.

If there are no integrations, then $n = 0$. Using Eq. (7.14), we have

$$\lim_{s \to 0} G(s) = \frac{z_1 z_2 \cdots}{p_1 p_2 \cdots} \tag{7.15}$$

which is finite and yields a finite error from Eq. (7.12). Figure 7.2($a$), Output 2, is an example of this case of finite steady-state error.

In summary, for a step input to a unity feedback system, the steady-state error will be zero if there is at least one pure integration in the forward path. If there are no integrations, then there will be a nonzero finite error. This result is comparable to our qualitative discussion in Section 7.1, where we found that a pure gain yields a constant steady-state error for a step input, but an integrator yields zero error for the same type of input. We now repeat the development for a ramp input.

**Ramp Input.** Using Eq. (7.11), with $R(s) = 1/s^2$, we obtain

$$e(\infty) = e_{\text{ramp}}(\infty) = \lim_{s \to 0} \frac{s(1/s^2)}{1 + G(s)} = \lim_{s \to 0} \frac{1}{s + sG(s)} = \frac{1}{\lim\limits_{s \to 0} sG(s)} \tag{7.16}$$

To have zero steady-state error for a ramp input, we must have

$$\lim_{s \to 0} sG(s) = \infty \tag{7.17}$$

To satisfy Eq. (7.17), $G(s)$ must take the same form as Eq. (7.14), except that $n \geq 2$. In other words, there must be at least two integrations in the forward path. An example of zero steady-state error for a ramp input is shown in Figure 7.2($b$), Output 1.

If only one integration exists in the forward path, then, assuming Eq. (7.14),

$$\lim_{s \to 0} sG(s) = \frac{z_1 z_2 \cdots}{p_1 p_2 \cdots} \tag{7.18}$$

which is finite rather than infinite. Using Eq. (7.16), we find that this configuration leads to a constant error, as shown in Figure 7.2($b$), Output 2.

If there are no integrations in the forward path, then

$$\lim_{s \to 0} sG(s) = 0 \tag{7.19}$$

and the steady-state error would be infinite and lead to diverging ramps, as shown in Figure 7.2($b$), Output 3. Finally, we repeat the development for a parabolic input.

**Parabolic Input.** Using Eq. (7.11), with $R(s) = 1/s^3$, we obtain

$$e(\infty) = e_{\text{parabola}}(\infty) = \lim_{s \to 0} \frac{s(1/s^3)}{1 + G(s)} = \lim_{s \to 0} \frac{1}{s^2 + s^2 G(s)} = \frac{1}{\lim\limits_{s \to 0} s^2 G(s)} \tag{7.20}$$

In order to have zero steady-state error for a parabolic input, we must have

$$\lim_{s\to 0} s^2G(s) = \infty \tag{7.21}$$

To satisfy Eq. (7.21), $G(s)$ must take on the same form as Eq. (7.14), except that $n \geq 3$. In other words, there must be at least three integrations in the forward path.

If there are only two integrations in the forward path, then

$$\lim_{s\to 0} s^2G(s) = \frac{z_1 z_2 \cdots}{p_1 p_2 \cdots} \tag{7.22}$$

is finite rather than infinite. Using Eq. (7.20), we find that this configuration leads to a constant error.

If there is only one or less integrations in the forward path, then

$$\lim_{s\to 0} s^2G(s) = 0 \tag{7.23}$$

and the steady-state error is infinite. Two examples demonstrate these concepts.

## Example 7.2 Steady-State Errors for Systems with No Integrations

**PROBLEM:** Find the steady-state errors for inputs of $5u(t)$, $5tu(t)$, and $5t^2u(t)$ to the system shown in Figure 7.5. The function $u(t)$ is the unit step.

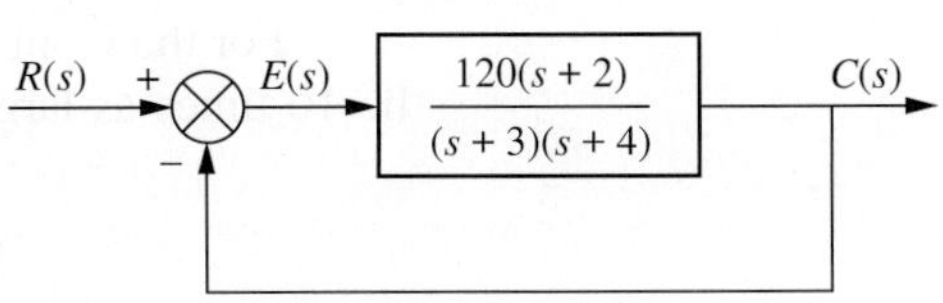


**FIGURE 7.5** Feedback control system for Example 7.2

**SOLUTION:** First we verify that the closed-loop system is indeed stable. For this example we leave out the details. Next, for the input $5u(t)$, whose Laplace transform is $5/s$, the steady-state error will be five times as large as that given by Eq. (7.12), or

$$e(\infty) = e_{\text{step}}(\infty) = \frac{5}{1 + \lim_{s\to 0} G(s)} = \frac{5}{1+20} = \frac{5}{21} \tag{7.24}$$

which implies a response similar to Output 2 of Figure 7.2(*a*).

For the input $5tu(t)$, whose Laplace transform is $5/s^2$, the steady-state error will be five times as large as that given by Eq. (7.16), or

$$e(\infty) = e_{\text{ramp}}(\infty) = \frac{5}{\lim_{s\to 0} sG(s)} = \frac{5}{0} = \infty \tag{7.25}$$

which implies a response similar to Output 3 of Figure 7.2(*b*).

For the input $5t^2u(t)$, whose Laplace transform is $10/s^3$, the steady-state error will be 10 times as large as that given by Eq. (7.20), or

$$e(\infty) = e_{\text{parabola}}(\infty) = \frac{10}{\lim_{s\to 0} s^2G(s)} = \frac{10}{0} = \infty \tag{7.26}$$

## Example 7.3 Steady-State Errors for Systems with One Integration

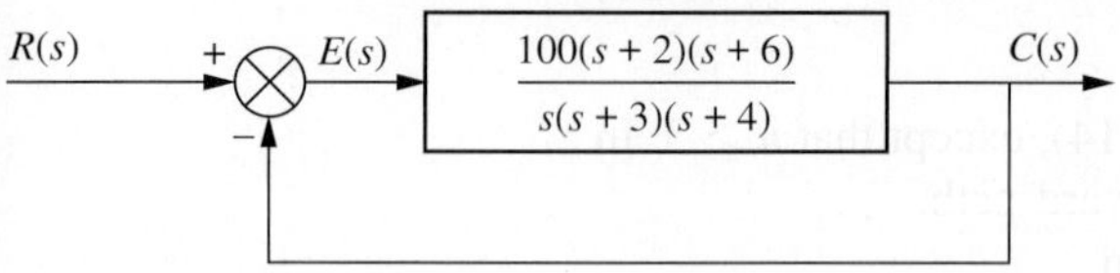


**FIGURE 7.6** Feedback control system for Example 7.3

**PROBLEM:** Find the steady-state errors for inputs of $5u(t)$, $5tu(t)$, and $5t^2u(t)$ to the system shown in Figure 7.6. The function $u(t)$ is the unit step.

**SOLUTION:** First verify that the closed-loop system is indeed stable. For this example we leave out the details. Next note that since there is an integration in the forward path, the steady-state errors for some of the input waveforms will be less than those found in Example 7.2. For the input $5u(t)$, whose Laplace transform is $5/s$, the steady-state error will be five times as large as that given by Eq. (7.12), or

$$e(\infty) = e_{\text{step}}(\infty) = \frac{5}{1 + \lim_{s\to 0} G(s)} = \frac{5}{\infty} = 0 \tag{7.27}$$

which implies a response similar to Output 1 of Figure 7.2($a$). Notice that the integration in the forward path yields zero error for a step input, rather than the finite error found in Example 7.2.

For the input $5tu(t)$, whose Laplace transform is $5/s^2$, the steady-state error will be five times as large as that given by Eq. (7.16), or

$$e(\infty) = e_{\text{ramp}}(\infty) = \frac{5}{\lim_{s\to 0} sG(s)} = \frac{5}{100} = \frac{1}{20} \tag{7.28}$$

which implies a response similar to Output 2 of Figure 7.2($b$). Notice that the integration in the forward path yields a finite error for a ramp input, rather than the infinite error found in Example 7.2.

For the input, $5t^2u(t)$, whose Laplace transform is $10/s^3$, the steady-state error will be 10 times as large as that given by Eq. (7.20), or

$$e(\infty) = e_{\text{parabola}}(\infty) = \frac{10}{\lim_{s\to 0} s^2G(s)} = \frac{10}{0} = \infty \tag{7.29}$$

Notice that the integration in the forward path does not yield any improvement in steady-state error over that found in Example 7.2 for a parabolic input.

## Skill-Assessment Exercise 7.1

**PROBLEM:** A unity feedback system has the following forward transfer function:

$$G(s) = \frac{10(s+20)(s+30)}{s(s+25)(s+35)}$$

**a.** Find the steady-state error for the following inputs: $15u(t)$, $15tu(t)$, and $15t^2u(t)$.

**b.** Repeat for

$$G(s) = \frac{10(s+20)(s+30)}{s^2(s+25)(s+35)(s+50)}$$

**ANSWERS:**

**a.** The closed-loop system is stable. For $15u(t)$, $e_{\text{step}}(\infty) = 0$; for $15tu(t)$, $e_{\text{ramp}}(\infty) = 2.1875$; for $15(t^2)u(t)$, $e_{\text{parabola}}(\infty) = \infty$.

**b.** The closed-loop system is unstable. Calculations cannot be made.

The complete solution is at www.wiley.com/college/nise.

# 7.3 Static Error Constants and System Type

We continue our focus on unity negative feedback systems and define parameters that we can use as steady-state error performance specifications. These definitions parallel our defining damping ratio, natural frequency, settling time, percent overshoot, and so on as performance specifications for the transient response. The steady-state error performance specifications are called *static error constants*. Let us see how they are defined, how to calculate them, and, in the next section, how to use them for design.

## Static Error Constants

In the previous section we derived the following relationships for steady-state error. For a step input, $u(t)$,

$$e(\infty) = e_{\text{step}}(\infty) = \frac{1}{1 + \lim_{s \to 0} G(s)} \tag{7.30}$$

For a ramp input, $tu(t)$,

$$e(\infty) = e_{\text{ramp}}(\infty) = \frac{1}{\lim_{s \to 0} sG(s)} \tag{7.31}$$

For a parabolic input, $\frac{1}{2}t^2u(t)$.

$$e(\infty) = e_{\text{parabola}}(\infty) = \frac{1}{\lim_{s \to 0} s^2G(s)} \tag{7.32}$$

The three terms in the denominator that are taken to the limit determine the steady-state error. We call these limits *static error constants*. Individually, their names are

*position constant*, $K_p$, where

$$K_p = \lim_{s \to 0} G(s) \tag{7.33}$$

*velocity constant*, $K_v$, where

$$K_v = \lim_{s \to 0} sG(s) \tag{7.34}$$

*acceleration constant*, $K_a$, where

$$K_a = \lim_{s \to 0} s^2G(s) \tag{7.35}$$

As we have seen, these quantities, depending upon the form of $G(s)$, can assume values of zero, finite constant, or infinity. Since the static error constant appears in the denominator of the steady-state error. Eqs. (7.30) through (7.32), the value of the steady-state error decreases as the static error constant increases.

In Section 7.2, we evaluated the steady-state error by using the final value theorem. An alternate method makes use of the static error constants. A few examples follow:

## Example 7.4 Steady-State Error via Static Error Constants

**PROBLEM:** For each system of Figure 7.7, evaluate the static error constants and find the expected error for the standard step, ramp, and parabolic inputs.

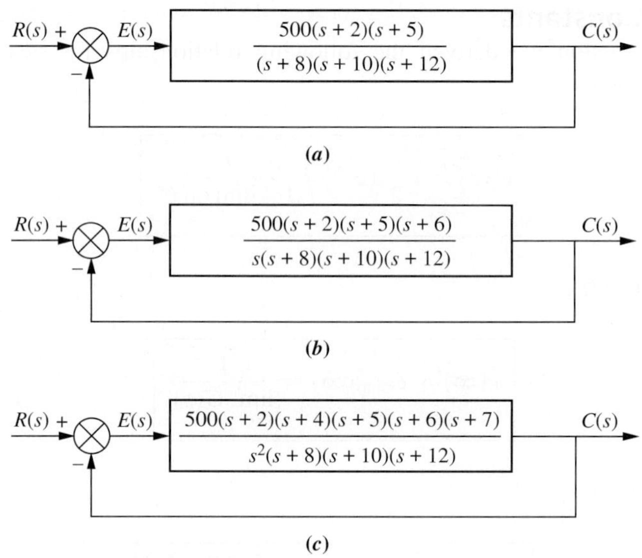


**FIGURE 7.7** Feedback control systems for Example 7.4

**SOLUTION:** First verify that all closed-loop systems shown are indeed stable. For this example we leave out the details. Next, for Figure 7.7(*a*),

$$K_p = \lim_{s\to 0} G(s) = \frac{500 \times 2 \times 5}{8 \times 10 \times 12} = 5.208 \tag{7.36}$$

$$K_v = \lim_{s\to 0} sG(s) = 0 \tag{7.37}$$

$$K_a = \lim_{s\to 0} s^2 G(s) = 0 \tag{7.38}$$

Thus, for a step input,

$$e(\infty) = \frac{1}{1+K_p} = 0.161 \tag{7.39}$$

For a ramp input,

$$e(\infty) = \frac{1}{K_v} = \infty \tag{7.40}$$

For a parabolic input,

$$e(\infty) = \frac{1}{K_a} = \infty \tag{7.41}$$

Now, for Figure 7.7(*b*),

$$K_p = \lim_{s\to 0} G(s) = \infty \tag{7.42}$$

$$K_v = \lim_{s\to 0} sG(s) = \frac{500 \times 2 \times 5 \times 6}{8 \times 10 \times 12} = 31.25 \tag{7.43}$$

and

$$K_a = \lim_{s\to 0} s^2G(s) = 0 \tag{7.44}$$

Thus, for a step input,

$$e(\infty) = \frac{1}{1+K_p} = 0 \tag{7.45}$$

For a ramp input,

$$e(\infty) = \frac{1}{K_v} = \frac{1}{31.25} = 0.032 \tag{7.46}$$

For a parabolic input,

$$e(\infty) = \frac{1}{K_a} = \infty \tag{7.47}$$

Finally, for Figure 7.7(*c*),

$$K_p = \lim_{s\to 0} G(s) = \infty \tag{7.48}$$

$$K_v = \lim_{s\to 0} sG(s) = \infty \tag{7.49}$$

and

$$K_a = \lim_{s\to 0} s^2G(s) = \frac{500 \times 2 \times 4 \times 5 \times 6 \times 7}{8 \times 10 \times 12} = 875 \tag{7.50}$$

Thus, for a step input,

$$e(\infty) = \frac{1}{1+K_p} = 0 \tag{7.51}$$

For a ramp input,

$$e(\infty) = \frac{1}{K_v} = 0 \tag{7.52}$$

For a parabolic input,

$$e(\infty) = \frac{1}{K_a} = \frac{1}{875} = 1.14 \times 10^{-3} \tag{7.53}$$

Students who are using MATLAB should now run ch7p1 in Appendix B. You will learn how to test the system for stability, evaluate static error constants, and calculate steady-state error using MATLAB. This exercise applies MATLAB to solve Example 7.4 with System (b).

MATLAB

**ML**

## System Type

Let us continue to focus on a unity negative feedback system. The values of the static error constants, again, depend upon the form of $G(s)$, especially the number of pure integrations in the forward path. Since steady-state errors are dependent upon the number of integrations in the forward path, we give a name to this system attribute. Given the system in

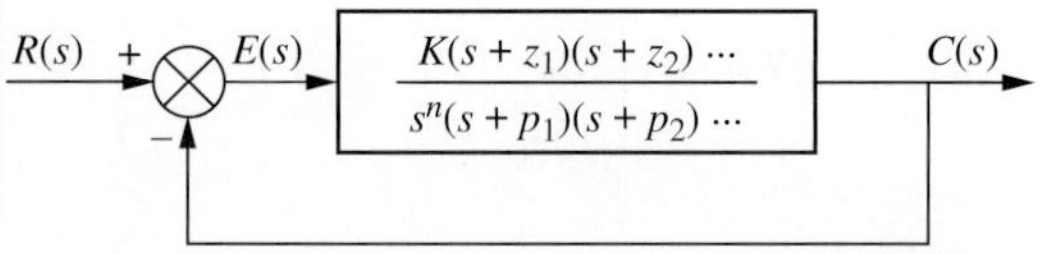


**FIGURE 7.8** Feedback control system for defining system type

Figure 7.8, we define *system type* to be the value of $n$ in the denominator or, equivalently, the number of pure integrations in the forward path. Therefore, a system with $n = 0$ is a Type 0 system. If $n = 1$ or $n = 2$, the corresponding system is a Type 1 or Type 2 system, respectively.

Table 7.2 ties together the concepts of steady-state error, static error constants, and system type. The table shows the static error constants and the steady-state errors as functions of input waveform and system type.

**TABLE 7.2** Relationships between input, system type, static error constants, and steady-state errors

| | | Type 0 | | Type 1 | | Type 2 | |
|---|---|---|---|---|---|---|---|
| **Input** | **Steady-state error formula** | **Static error constant** | **Error** | **Static error constant** | **Error** | **Static error constant** | **Error** |
| Step, $u(t)$ | $\frac{1}{1+K_p}$ | $K_p =$ Constant | $\frac{1}{1+K_p}$ | $K_p = \infty$ | 0 | $K_p = \infty$ | 0 |
| Ramp, $tu(t)$ | $\frac{1}{K_v}$ | $K_v = 0$ | $\infty$ | $K_v =$ Constant | $\frac{1}{K_v}$ | $K_v = \infty$ | 0 |
| Parabola, $\frac{1}{2}t^2u(t)$ | $\frac{1}{K_a}$ | $K_a = 0$ | $\infty$ | $K_a = 0$ | $\infty$ | $K_a =$ Constant | $\frac{1}{K_a}$ |

## Skill-Assessment Exercise 7.2

**TryIt 7.1**

Use MATLAB, the Control System Toolbox, and the following statements to find $K_p$, $e_{step}(\infty)$, and the closed-loop poles to check for stability for the system of Skill-Assessment Exercise 7.2.

```
numg=1000*[1 8];
deng=poly([-7 -9]);
G=tf(numg,deng);
Kp=dcgain(G)
estep=1/(1+Kp)
T=feedback(G,1);
poles=pole(T)
```

**PROBLEM:** A unity feedback system has the following forward transfer function:

$$G(s) = \frac{1000(s+8)}{(s+7)(s+9)}$$

**a.** Evaluate system type, $K_p$, $K_v$, and $K_a$.

**b.** Use your answers to **a.** to find the steady-state errors for the standard step, ramp, and parabolic inputs.

**ANSWERS:**

**a.** The closed-loop system is stable. System type = Type 0. $K_p = 127$, $K_v = 0$, and $K_a = 0$.

**b.** $e_{step}(\infty) = 7.8 \times 10^{-3}$, $e_{ramp}(\infty) = \infty$, and $e_{parabola}(\infty) = \infty$

The complete solution is at www.wiley.com/college/nise.

In this section, we defined steady-state errors, static error constants, and system type. Now the specifications for a control system's steady-state errors will be formulated, followed by some examples.

# 7.4 Steady-State Error Specifications

Static error constants can be used to specify the steady-state error characteristics of control systems, such as that shown in Figure 7.9. Just as damping ratio, $\zeta$, settling time, $T_s$, peak time, $T_p$, and percent overshoot, $\%OS$, are used as specifications for a control system's transient response, so the position constant, $K_p$, velocity constant, $K_v$, and acceleration constant, $K_a$, can

Chuck O'Rear/Westlight/Corbis Images.

**FIGURE 7.9** A robot used in the manufacturing of semiconductor random-access memories (RAMs) similar to those in personal computers. Steady-state error is an important design consideration for assembly-line robots

be used as specifications for a control system's steady-state errors. We will soon see that a wealth of information is contained within the specification of a static error constant.

For example, if a control system has the specification $K_v = 1000$, we can draw several conclusions:

1. The system is stable.
2. The system is of Type 1, since only Type 1 systems have $K_v$'s that are finite constants. Recall that $K_v = 0$ for Type 0 systems, whereas $K_v = \infty$ for Type 2 systems.
3. A ramp input is the test signal. Since $K_v$ is specified as a finite constant, and the steady-state error for a ramp input is inversely proportional to $K_v$, we know the test input is a ramp.
4. The steady-state error between the input ramp and the output ramp is $1/K_v$ per unit of input slope.

Let us look at two examples that demonstrate analysis and design using static error constants.

## Example 7.5 Interpreting the Steady-State Error Specification

**PROBLEM:** What information is contained in the specification $K_p = 1000$?

**SOLUTION:** The system is stable. The system is Type 0, since only a Type 0 system has a finite $K_p$. Type 1 and Type 2 systems have $K_p = \infty$. The input test signal is a step, since $K_p$ is specified. Finally, the error per unit step is

$$e(\infty) = \frac{1}{1 + K_p} = \frac{1}{1 + 1000} = \frac{1}{1001} \tag{7.54}$$

## Example 7.6 Gain Design to Meet a Steady-State Error Specification

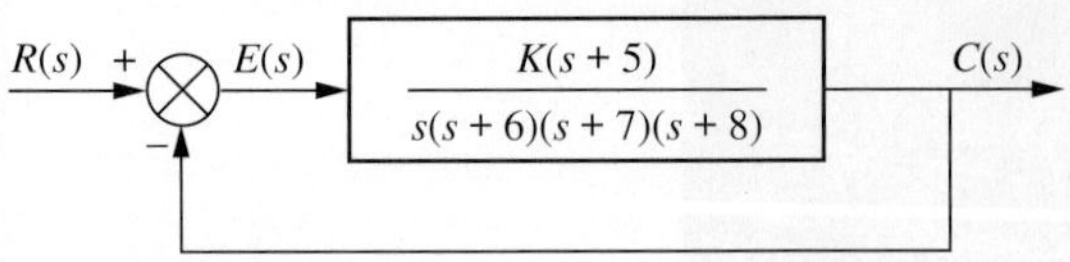


**FIGURE 7.10** Feedback control system for Example 7.6

**PROBLEM:** Given the control system in Figure 7.10, find the value of $K$ so that there is 10% error in the steady state.

**SOLUTION:** Since the system is Type 1, the error stated in the problem must apply to a ramp input; only a ramp yields a finite error in a Type 1 system. Thus,

$$e(\infty) = \frac{1}{K_v} = 0.1 \tag{7.55}$$

Therefore,

$$K_v = 10 = \lim_{s\to 0} sG(s) = \frac{K \times 5}{6 \times 7 \times 8} \tag{7.56}$$

which yields

$$K = 672 \tag{7.57}$$

Applying the Routh-Hurwitz criterion, we see that the system is stable at this gain.

Although this gain meets the criteria for steady-state error and stability, it may not yield a desirable transient response. In Chapter 9 we will design feedback control systems to meet all three specifications.

MATLAB

ML

Students who are using MATLAB should now run ch7p2 in Appendix B. You will learn how to find the gain to meet a steady-state error specification using MATLAB. This exercise solves Example 7.6 using MATLAB.

## Skill-Assessment Exercise 7.3

**TryIt 7.2**

Use MATLAB, the Control System Toolbox, and the following statements to solve Skill-Assessment Exercise 7.3 and check the resulting system for stability.

```
numg=[1 12];
deng=poly([-14 -18]);
G=tf(numg,deng);
Kpdk=dcgain(G);
estep=0.1;
K=(1/estep-1)/Kpdk
T=feedback(G,1);
poles=pole(T)
```

**PROBLEM:** A unity feedback system has the following forward transfer function:

$$G(s) = \frac{K(s+12)}{(s+14)(s+18)}$$

Find the value of $K$ to yield a 10% error in the steady state.

**ANSWER:** $K = 189$

The complete solution is at www.wiley.com/college/nise.

This example and exercise complete our discussion of unity feedback systems. In the remaining sections, we will deal with the steady-state errors for disturbances and the steady-state errors for feedback control systems in which the feedback is not unity.

## 7.5 Steady-State Error for Disturbances

Feedback control systems are used to compensate for disturbances or unwanted inputs that enter a system. The advantage of using feedback is that regardless of these disturbances, the system can be designed to follow the input with small or zero error, as we now demonstrate. Figure 7.11 shows a feedback control system with a disturbance, $D(s)$, injected between the controller and the plant. We now re-derive the expression for steady-state error with the disturbance included.

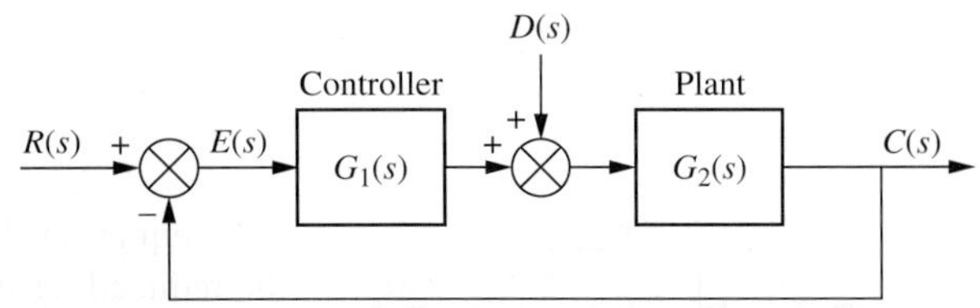


**FIGURE 7.11** Feedback control system showing disturbance

The transform of the output is given by

$$C(s) = E(s)G_1(s)G_2(s) + D(s)G_2(s) \tag{7.58}$$

But

$$C(s) = R(s) - E(s) \tag{7.59}$$

Substituting Eq. (7.59) into Eq. (7.58) and solving for $E(s)$, we obtain

$$E(s) = \frac{1}{1 + G_1(s)G_2(s)}R(s) - \frac{G_2(s)}{1 + G_1(s)G_2(s)}D(s) \tag{7.60}$$

where we can think of $1/[1 + G_1(s)G_2(s)]$ as a transfer function relating $E(s)$ to $R(s)$ and $-G_2(s)/[1 + G_1(s)G_2(s)]$ as a transfer function relating $E(s)$ to $D(s)$.

To find the steady-state value of the error, we apply the final value theorem[3] to Eq. (7.60) and obtain

$$\begin{aligned} e(\infty) = \lim_{s\to 0} sE(s) &= \lim_{s\to 0}\frac{s}{1 + G_1(s)G_2(s)}R(s) - \lim_{s\to 0}\frac{sG_2(s)}{1 + G_1(s)G_2(s)}D(s) \\ &= e_R(\infty) + e_D(\infty) \end{aligned} \tag{7.61}$$

where

$$e_R(\infty) = \lim_{s\to 0}\frac{s}{1 + G_1(s)G_2(s)}R(s)$$

and

$$e_D(\infty) = -\lim_{s\to 0}\frac{sG_2(s)}{1 + G_1(s)G_2(s)}D(s)$$

The first term, $e_R(\infty)$, is the steady-state error due to $R(s)$, which we have already obtained. The second term, $e_D(\infty)$, is the steady-state error due to the disturbance. Let us explore the conditions on $e_D(\infty)$ that must exist to reduce the error due to the disturbance.

[3] Remember that the final value theorem can be applied only if the system is stable, with the roots of $[1 + G_1(s)G_2(s)]$ in the left–half–plane.

At this point, we must make some assumptions about $D(s)$, the controller, and the plant. First we assume a step disturbance, $D(s) = 1/s$. Substituting this value into the second term of Eq. (7.61), $e_D(\infty)$, the steady-state error component due to a step disturbance is found to be

$$e_D(\infty) = -\frac{1}{\lim_{s\to 0}\frac{1}{G_2(s)} + \lim_{s\to 0} G_1(s)} \tag{7.62}$$

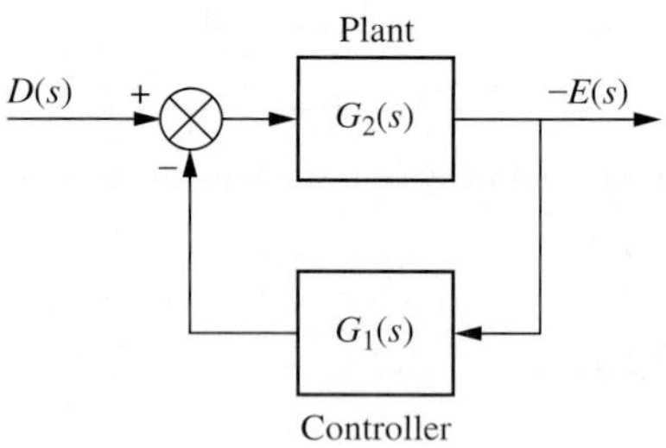


**FIGURE 7.12** Figure 7.11 system rearranged to show disturbance as input and error as output, with $R(s) = 0$

This equation shows that the steady-state error produced by a step disturbance can be reduced by increasing the dc gain of $G_1(s)$ or decreasing the dc gain of $G_2(s)$.

This concept is shown in Figure 7.12, where the system of Figure 7.11 has been rearranged so that the disturbance, $D(s)$, is depicted as the input and the error, $E(s)$, as the output, with $R(s)$ set equal to zero. If we want to minimize the steady-state value of $E(s)$, shown as the output in Figure 7.12, we must either increase the dc gain of $G_1(s)$ so that a lower value of $E(s)$ will be fed back to match the steady-state value of $D(s)$, or decrease the dc value of $G_2(s)$, which then yields a smaller value of $e(\infty)$ as predicted by the feedback formula.

Let us look at an example and calculate the numerical value of the steady-state error that results from a disturbance.

## Example 7.7 Steady-State Error Due to Step Disturbance

**PROBLEM:** Find the steady-state error component due to a step disturbance for the system of Figure 7.13.

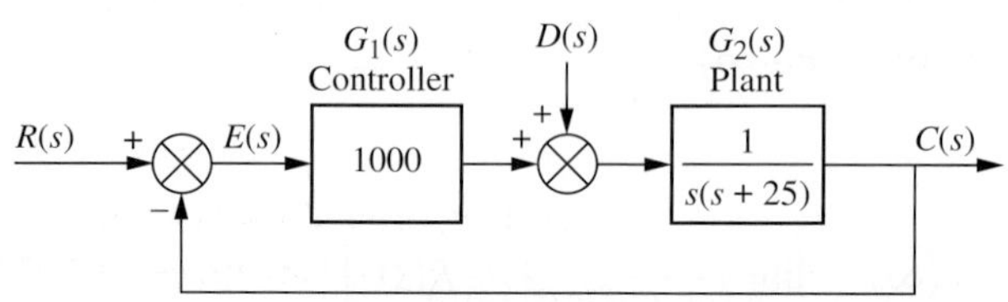


**FIGURE 7.13** Feedback control system for Example 7.7

**SOLUTION:** The system is stable. Using Figure 7.12 and Eq. (7.62), we find

$$e_D(\infty) = -\frac{1}{\lim_{s\to 0}\frac{1}{G_2(s)} + \lim_{s\to 0} G_1(s)} = -\frac{1}{0 + 1000} = -\frac{1}{1000} \tag{7.63}$$

The result shows that the steady-state error produced by the step disturbance is inversely proportional to the dc gain of $G_1(s)$. The dc gain of $G_2(s)$ is infinite in this example.

**Skill-Assessment Exercise 7.4**

**PROBLEM:** Evaluate the steady-state error component due to a step disturbance for the system of Figure 7.14.

**ANSWER:** $e_D(\infty) = -9.98 \times 10^{-4}$

The complete solution is at www.wiley.com/college/nise.

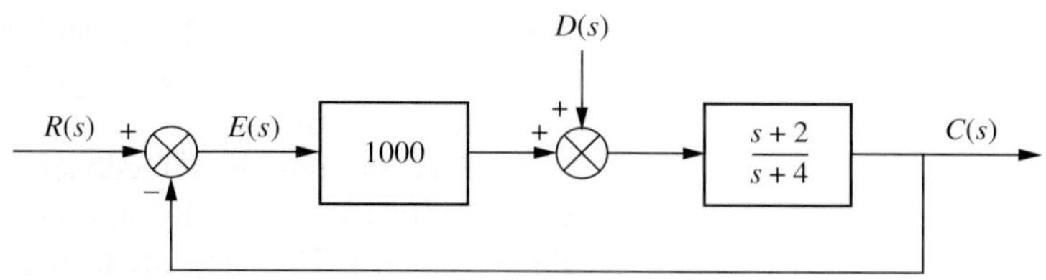


**FIGURE 7.14** System for Skill-Assessment Exercise 7.4

## 7.6 Steady-State Error for Nonunity Feedback Systems

Control systems often do not have unity feedback because of the compensation used to improve performance or because of the physical model for the system. The feedback path can be a pure gain other than unity or have some dynamic representation.

A general feedback system, showing the input transducer, $G_1(s)$, controller and plant, $G_2(s)$, and feedback, $H_1(s)$, is shown in Figure 7.15($a$). Pushing the input transducer to the right past the summing junction yields the general nonunity feedback system shown in

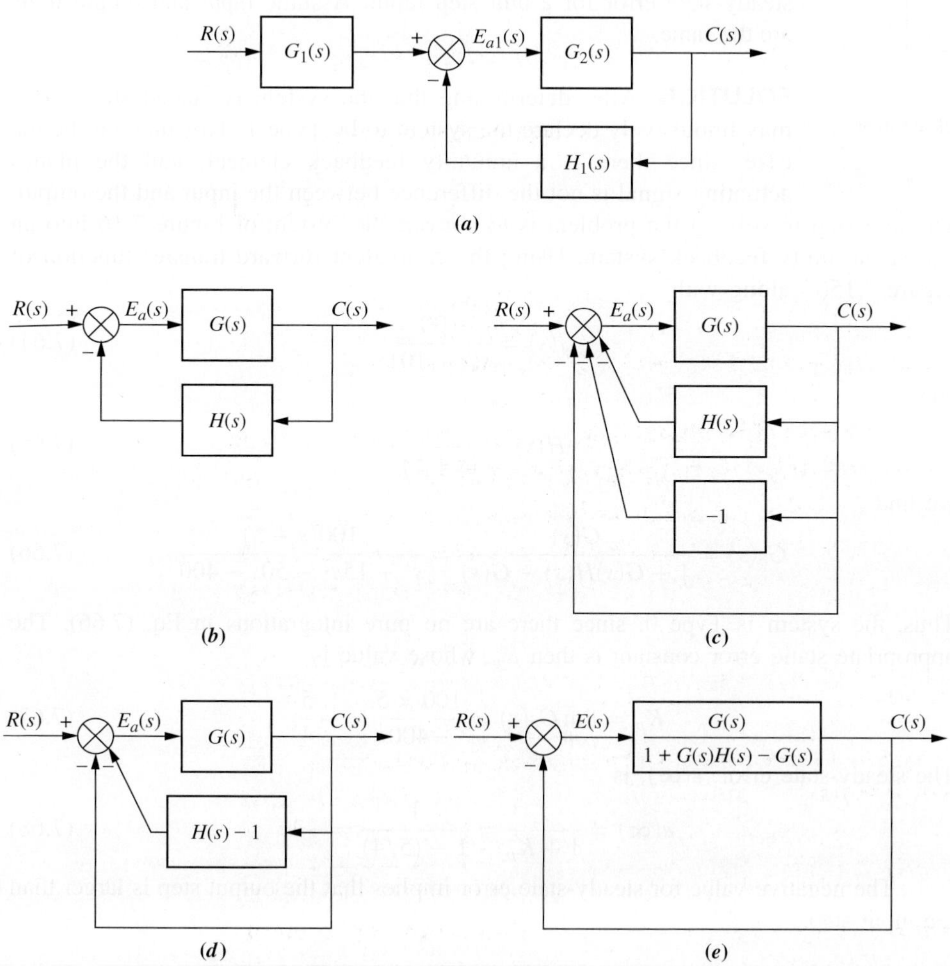


**FIGURE 7.15** Forming an equivalent unity feedback system from a general nonunity feedback system

Figure 7.15(*b*), where $G(s) = G_1(s)G_2(s)$ and $H(s) = H_1(s)/G_1(s)$. Notice that unlike a unity feedback system, where $H(s) = 1$, the error is not the difference between the input and the output. For this case we call the signal at the output of the summing junction the *actuating signal,* $E_a(s)$. If $r(t)$ and $c(t)$ have the same units, we can find the steady-state error, $e(\infty) = r(\infty) - c(\infty)$. The first step is to show explicitly $E(s) = R(s) - C(s)$ on the block diagram.

Take the nonunity feedback control system shown in Figure 7.15(*b*) and form a unity feedback system by adding and subtracting unity feedback paths, as shown in Figure 7.15(*c*). This step requires that input and output units be the same. Next combine $H(s)$ with the negative unity feedback, as shown in Figure 7.15(*d*). Finally, combine the feedback system consisting of $G(s)$ and $[H(s) - 1]$, leaving an equivalent forward path and a unity feedback, as shown in Figure 7.15(*e*). Notice that the final figure shows $E(s) = R(s) - C(s)$ explicitly.

The following example summarizes the concepts of steady-state error, system type, and static error constants for nonunity feedback systems.

## Example 7.8 Steady-State Error for Nonunity Feedback Systems

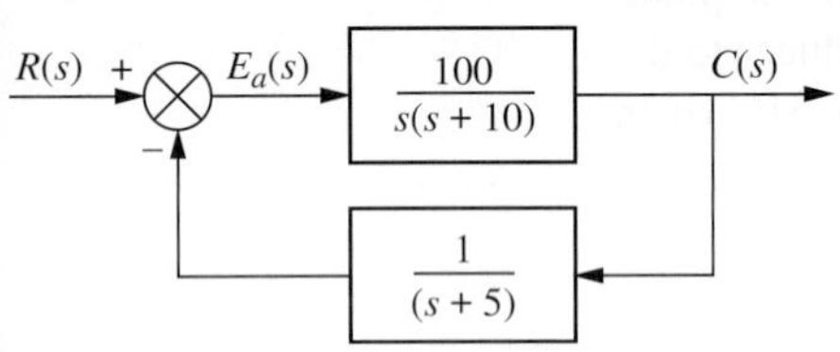


**FIGURE 7.16** Nonunity feedback control system for Example 7.8

**PROBLEM:** For the system shown in Figure 7.16, find the system type, the appropriate error constant associated with the system type, and the steady-state error for a unit step input. Assume input and output units are the same.

**SOLUTION:** After determining that the system is indeed stable, one may impulsively declare the system to be Type 1. This may not be the case, since there is a nonunity feedback element, and the plant's actuating signal is not the difference between the input and the output. The first step in solving the problem is to convert the system of Figure 7.16 into an equivalent unity feedback system. Using the equivalent forward transfer function of Figure 7.15(*e*) along with

$$G(s) = \frac{100}{s(s+10)} \tag{7.64}$$

and

$$H(s) = \frac{1}{(s+5)} \tag{7.65}$$

we find

$$G_e(s) = \frac{G(s)}{1 + G(s)H(s) - G(s)} = \frac{100(s+5)}{s^3 + 15s^2 - 50s - 400} \tag{7.66}$$

Thus, the system is Type 0, since there are no pure integrations in Eq. (7.66). The appropriate static error constant is then $K_p$, whose value is

$$K_p = \lim_{s \to 0} G_e(s) = \frac{100 \times 5}{-400} = -\frac{5}{4} \tag{7.67}$$

The steady-state error, $e(\infty)$, is

$$e(\infty) = \frac{1}{1 + K_p} = \frac{1}{1 - (5/4)} = -4 \tag{7.68}$$

The negative value for steady-state error implies that the output step is larger than the input step

### TryIt 7.3

Use MATLAB, the Control System Toolbox, and the following statements to find $G_e(s)$ in Example 7.8.

```
G=zpk([],[0 -10],100);
H=zpk([], -5,1);
Ge=feedback...
  (G, (H-1));
'Ge(s)'
Ge=tf(Ge)
T=feedback(Ge,1);
'Poles of T(s)'
pole(T)
```

To continue our discussion of steady-state error for systems with nonunity feedback, let us look at the general system of Figure 7.17, which has both a disturbance and nonunity feedback. We will derive a general equation for the steady-state error and then determine the parameters of the system in order to drive the error to zero for step inputs and step disturbances.[4]

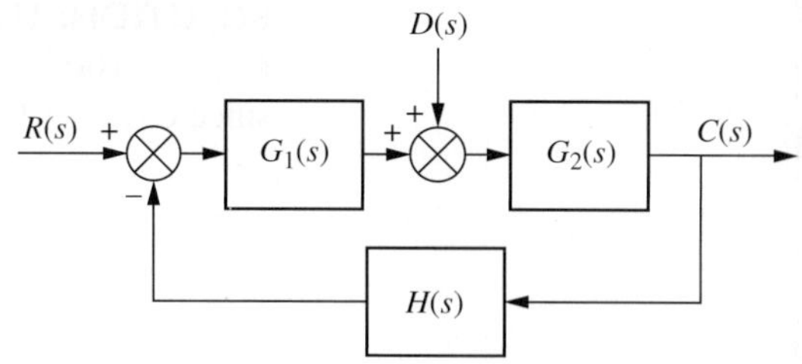


**FIGURE 7.17** Nonunity feedback control system with disturbance

The steady-state error for this system, $e(\infty) = r(\infty) - c(\infty)$, is

$$e(\infty) = \lim_{s\to 0} sE(s) = \lim_{s\to 0} s\left\{\left[1 - \frac{G_1(s)G_2(s)}{1+G_1(s)G_2(s)H(s)}\right]R(s) - \left[\frac{G_2(s)}{1+G_1(s)G_2(s)H(s)}D(s)\right]\right\} \tag{7.69}$$

Now limiting the discussion to step inputs and step disturbances, where $R(s) = D(s) = 1/s$, Eq. (7.69) becomes

$$e(\infty) = \lim_{s\to 0} sE(s) = \left\{\left[1 - \frac{\lim_{s\to 0}[G_1(s)G_2(s)]}{\lim_{s\to 0}[1+G_1(s)G_2(s)H(s)]}\right] - \left[\frac{\lim_{s\to 0} G_2(s)}{\lim_{s\to 0}[1+G_1(s)G_2(s)H(s)]}\right]\right\} \tag{7.70}$$

For zero error,

$$\frac{\lim_{s\to 0}[G_1(s)G_2(s)]}{\lim_{s\to 0}[1+G_1(s)G_2(s)H(s)]} = 1 \quad \text{and} \quad \frac{\lim_{s\to 0} G_2(s)}{\lim_{s\to 0}[1+G_1(s)G_2(s)H(s)]} = 0 \tag{7.71}$$

The two equations in Eq. (7.71) can always be satisfied if (1) the system is stable, (2) $G_1(s)$ is a Type 1 system, (3) $G_2(s)$ is a Type 0 system, and (4) $H(s)$ is a Type 0 system with a dc gain of unity.

To conclude this section, we discuss finding the steady-state value of the actuating signal, $E_{a1}(s)$, in Figure 7.15(*a*). For this task there is no restriction that the input and output units be the same, since we are finding the steady-state difference between signals at the summing junction, which do have the same units.[5] The steady-state actuating signal for Figure 7.15(*a*) is

$$e_{a1}(\infty) = \lim_{s\to 0} \frac{sR(s)G_1(s)}{1+G_2(s)H_1(s)} \tag{7.72}$$

The derivation is left to the student in the problem set at the end of this chapter.

## Example 7.9 Steady-State Actuating Signal for Nonunity Feedback Systems

**PROBLEM:** Find the steady-state actuating signal for the system of Figure 7.16 for a unit step input. Repeat for a unit ramp input.

[4] Details of the derivation are included as a problem at the end of this chapter.

[5] For clarity, steady-state error is the steady-state difference between the input and the output. Steady-state actuating signal is the steady-state difference at the output of the summing junction. In questions asking for steady-state error in problems, examples, and skill-assessment exercises, it will be assumed that input and output units are the same.

**SOLUTION:** Use Eq. (7.72) with $R(s) = 1/s$, a unit step input, $G_1(s) = 1$, $G_2(s) = 100/[s(s+10)]$, and $H_1(s) = 1/(s+5)$. Also, realize that $e_{a1}(\infty) = e_a(\infty)$, since $G_1(s) = 1$. Thus,

$$e_a(\infty) = \lim_{s \to 0} \frac{s\left(\frac{1}{s}\right)}{1 + \left(\frac{100}{s(s+10)}\right)\left(\frac{1}{(s+5)}\right)} = 0 \quad (7.73)$$

Now use Eq. (7.72) with $R(s) = 1/s^2$, a unit ramp input, and obtain

$$e_a(\infty) = \lim_{s \to 0} \frac{s\left(\frac{1}{s^2}\right)}{1 + \left(\frac{100}{s(s+10)}\right)\left(\frac{1}{(s+5)}\right)} = \frac{1}{2} \quad (7.74)$$

## Skill-Assessment Exercise 7.5

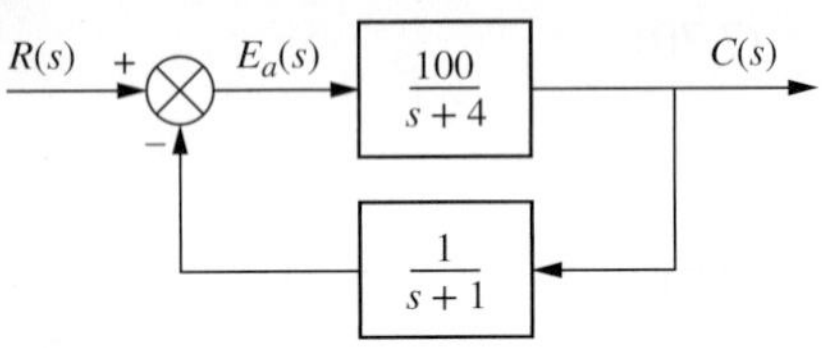


**FIGURE 7.18** Nonunity feedback system for Skill-Assessment Exercise 7.5

**PROBLEM:**

**a.** Find the steady-state error, $e(\infty) = r(\infty) - c(\infty)$, for a unit step input given the nonunity feedback system of Figure 7.18. Repeat for a unit ramp input. Assume input and output units are the same.

**b.** Find the steady-state actuating signal, $e_a(\infty)$, for a unit step input given the nonunity feedback system of Figure 7.18. Repeat for a unit ramp input.

**ANSWERS:**

**a.** $e_{\text{step}}(\infty) = 3.846 \times 10^{-2}$; $e_{\text{ramp}}(\infty) = \infty$

**b.** For a unit step input, $e_a(\infty) = 3.846 \times 10^{-2}$; for a unit ramp input, $e_a(\infty) = \infty$

The complete solution is at www.wiley.com/college/nise.

In this section, we have applied steady-state error analysis to nonunity feedback systems. When nonunity feedback is present, the plant's actuating signal is not the actual error or difference between the input and the output. With nonunity feedback we may choose to (1) find the steady-state error for systems where the input and output units are the same or (2) find the steady-state actuating signal.

We also derived a general expression for the steady-state error of a nonunity feedback system with a disturbance. We used this equation to determine the attributes of the subsystems so that there was zero error for step inputs and step disturbances.

Before concluding this chapter, we will discuss a topic that is not only significant for steady-state errors but generally useful throughout the control systems design process.

# 7.7 Sensitivity

During the design process, the engineer may want to consider the extent to which changes in system parameters affect the behavior of a system. Ideally, parameter changes due to heat or other causes should not appreciably affect a system's performance. The degree to which changes in system parameters affect system transfer functions, and hence performance, is called *sensitivity*. A system with zero sensitivity (that is, changes in the system parameters have no effect on the transfer function) is ideal. The greater the sensitivity, the less desirable the effect of a parameter change.

For example, assume the function $F = K/(K + a)$. If $K = 10$ and $a = 100$, then $F = 0.091$. If parameter $a$ triples to 300, then $F = 0.032$. We see that a fractional change in parameter $a$ of $(300 - 100)/100 = 2$ (a 200% change) yields a change in the function $F$ of $(0.032 - 0.091)/0.091 = -0.65$ ($-65$% change). Thus, the function $F$ has reduced sensitivity to changes in parameter $a$. As we proceed, we will see that another advantage of feedback is that in general it affords reduced sensitivity to parameter changes.

Based upon the previous discussion, let us formalize a definition of sensitivity: *Sensitivity* is the ratio of the fractional change in the function to the fractional change in the parameter as the fractional change of the parameter approaches zero. That is,

$$
\begin{aligned}
S_{F:P} &= \lim_{\Delta P \to 0} \frac{\text{Fractional change in the function, } F}{\text{Fractional change in the parameter, } P} \\
&= \lim_{\Delta P \to 0} \frac{\Delta F/F}{\Delta P/P} \\
&= \lim_{\Delta P \to 0} \frac{P \Delta F}{F \Delta P}
\end{aligned}
$$

which reduces to

$$
\boxed{S_{F:P} = \frac{P}{F}\frac{\delta F}{\delta P}} \tag{7.75}
$$

Let us now apply the definition, first to a closed-loop transfer function and then to the steady-state error.

## Example 7.10 Sensitivity of a Closed-Loop Transfer Function

**PROBLEM:** Given the system of Figure 7.19, calculate the sensitivity of the closed-loop transfer function to changes in the parameter $a$. How would you reduce the sensitivity?

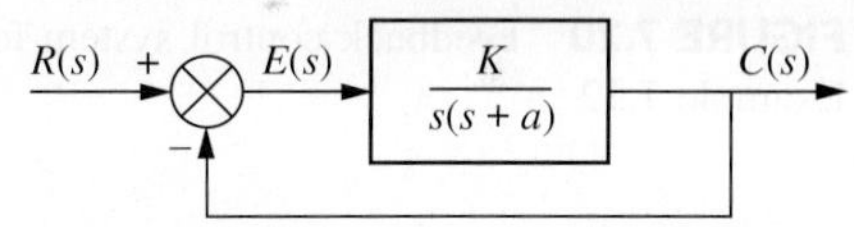


**FIGURE 7.19** Feedback control system for Examples 7.10 and 7.11

**SOLUTION:** The closed-loop transfer function is

$$
T(s) = \frac{K}{s^2 + as + K} \tag{7.76}
$$

Using Eq. (7.75), the sensitivity is given by

$$
S_{T:a} = \frac{a}{T}\frac{\delta T}{\delta a} = \frac{a}{\left(\dfrac{K}{s^2 + as + K}\right)} \left(\frac{-Ks}{(s^2 + as + K)^2}\right) = \frac{-as}{s^2 + as + K} \tag{7.77}
$$

which is, in part, a function of the value of $s$. For any value of $s$, however, an increase in $K$ reduces the sensitivity of the closed-loop transfer function to changes in the parameter $a$.

## Example 7.11 Sensitivity of Steady-State Error with Ramp Input

**PROBLEM:** For the system of Figure 7.19, find the sensitivity of the steady-state error to changes in parameter $K$ and parameter $a$ with ramp inputs.

**SOLUTION:** The steady-state error for the system is

$$e(\infty) = \frac{1}{K_v} = \frac{a}{K} \tag{7.78}$$

The sensitivity of $e(\infty)$ to changes in parameter $a$ is

$$S_{e:a} = \frac{a}{e}\frac{\delta e}{\delta a} = \frac{a}{a/K}\left[\frac{1}{K}\right] = 1 \tag{7.79}$$

The sensitivity of $e(\infty)$ to changes in parameter $K$ is

$$S_{e:K} = \frac{K}{e}\frac{\delta e}{\delta K} = \frac{K}{a/K}\left[\frac{-a}{K^2}\right] = -1 \tag{7.80}$$

Thus, changes in either parameter $a$ or parameter $K$ are directly reflected in $e(\infty)$, and there is no reduction or increase in sensitivity. The negative sign in Eq. (7.80) indicates a decrease in $e(\infty)$ for an increase in $K$. Both of these results could have been obtained directly from Eq. (7.78) since $e(\infty)$ is directly proportional to parameter $a$ and inversely proportional to parameter $K$.

## Example 7.12 Sensitivity of Steady-State Error with Step Input

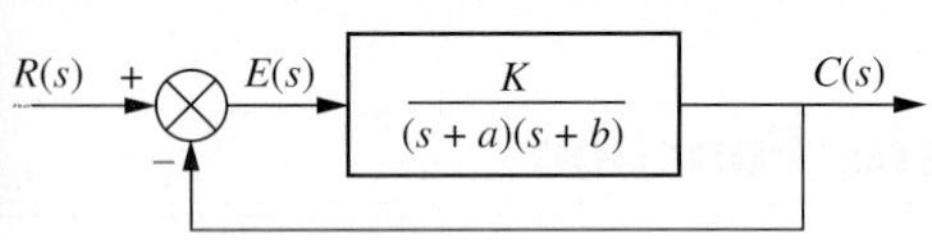


**FIGURE 7.20** Feedback control system for Example 7.12

**PROBLEM:** Find the sensitivity of the steady-state error to changes in parameter $K$ and parameter $a$ for the system shown in Figure 7.20 with a step input.

**SOLUTION:** The steady-state error for this Type 0 system is

$$e(\infty) = \frac{1}{1+K_p} = \frac{1}{1+\dfrac{K}{ab}} = \frac{ab}{ab+K} \tag{7.81}$$

The sensitivity of $e(\infty)$ to changes in parameter $a$ is

$$S_{e:a} = \frac{a}{e}\frac{\delta e}{\delta a} = \frac{a}{\left(\dfrac{ab}{ab+K}\right)}\frac{(ab+K)b - ab^2}{(ab+K)^2} = \frac{K}{ab+K} \tag{7.82}$$

The sensitivity of $e(\infty)$ to changes in parameter $K$ is

$$S_{e:K} = \frac{K}{e}\frac{\delta e}{\delta K} = \frac{K}{\left(\dfrac{ab}{ab+K}\right)}\frac{-ab}{(ab+K)^2} = \frac{-K}{ab+K} \tag{7.83}$$

Equations (7.82) and (7.83) show that the sensitivity to changes in parameter $K$ and parameter $a$ is less than unity for positive $a$ and $b$. Thus, feedback in this case yields reduced sensitivity to variations in both parameters.

### TryIt 7.4

Use MATLAB, the Symbolic Math Toolbox, and the following statements to find $S_{e:a}$ in Example 7.12.

```
syms K a b s
G=K/((s+a)*(s+b));
Kp=subs(G,s,0);
e=1/(1+Kp);
Sea=(a/e)*diff(e,a);
Sea=simplify(Sea);
'Sea'
pretty(Sea)
```

**Skill-Assessment Exercise 7.6**

**PROBLEM:** Find the sensitivity of the steady-state error to changes in $K$ for the system of Figure 7.21.

**ANSWER:** $S_{e:k} = \dfrac{-7K}{10 + 7K}$

The complete solution is at www.wiley.com/college/nise.

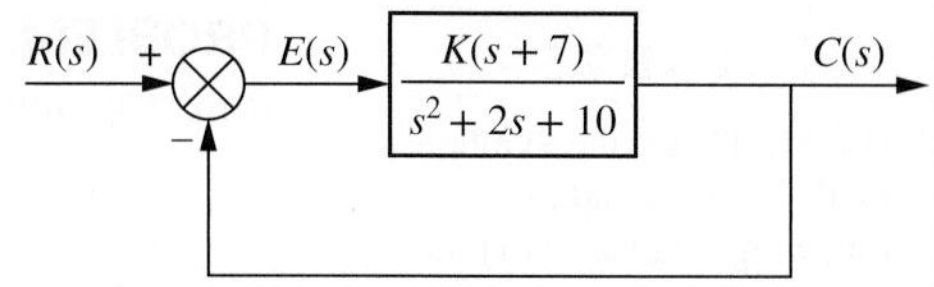


**FIGURE 7.21** System for Skill-Assessment Exercise 7.6

In this section, we defined sensitivity and showed that in some cases feedback reduces the sensitivity of a system's steady-state error to changes in system parameters. The concept of sensitivity can be applied to other measures of control system performance, as well; it is not limited to the sensitivity of the steady-state error performance.

# 7.8 Steady-State Error for Systems in State Space

Up to this point, we have evaluated the steady-state error for systems modeled as transfer functions. In this section, we will discuss how to evaluate the steady-state error for systems represented in state space. Two methods for calculating the steady-state error will be covered: (1) analysis via final value theorem and (2) analysis via input substitution. We will consider these methods individually.

## Analysis via Final Value Theorem

A single-input, single-output system represented in state space can be analyzed for steady-state error using the final value theorem and the closed-loop transfer function, Eq. (3.73), derived in terms of the state-space representation. Consider the closed-loop system represented in state space:

$$\dot{\mathbf{x}} = \mathbf{A}\mathbf{x} + \mathbf{B}r \tag{7.84a}$$

$$y = \mathbf{C}\mathbf{x} \tag{7.84b}$$

The Laplace transform of the error is

$$E(s) = R(s) - Y(s) \tag{7.85}$$

But

$$Y(s) = R(s)T(s) \tag{7.86}$$

where $T(s)$ is the closed-loop transfer function. Substituting Eq. (7.86) into (7.85), we obtain

$$E(s) = R(s)[1 - T(s)] \tag{7.87}$$

Using Eq. (3.73) for $T(s)$, we find

$$E(s) = R(s)[1 - \mathbf{C}(s\mathbf{I} - \mathbf{A})^{-1}\mathbf{B}] \tag{7.88}$$

Applying the final value theorem, we have

$$\lim_{s\to 0} sE(s) = \lim_{s\to 0} sR(s)[1 - \mathbf{C}(s\mathbf{I} - \mathbf{A})^{-1}\mathbf{B}] \tag{7.89}$$

Let us apply the result to an example.

## Example 7.13 Steady-State Error Using the Final Value Theorem

**TryIt 7.5**

Use MATLAB, the Symbolic Math Toolbox, and the following statements to find the steady-state error for a step input to the system of Example 7.13.

```
syms s
A=[-5 1 0
   0 -2 1
   20 -10 1];
B=[0; 0; 1];
C=[-1 1 0];
I=[1 0 0
   0 1 0
   0 0 1];
E=(1/s)*[1-C*...
  [(s*I-A)^-1]*B];
%New command:
%subs(X,old,new):
%Replace old in...
%X(old) with new.
error=subs(s*E,s,0)
```

**PROBLEM:** Evaluate the steady-state error for the system described by Eqs. (7.90) for unit step and unit ramp inputs. Use the final value theorem.

$$\mathbf{A}=\begin{bmatrix}-5 & 1 & 0\\ 0 & -2 & 1\\ 20 & -10 & 1\end{bmatrix};\quad \mathbf{B}=\begin{bmatrix}0\\0\\1\end{bmatrix};\quad \mathbf{C}=\begin{bmatrix}-1 & 1 & 0\end{bmatrix} \tag{7.90}$$

**SOLUTION:** Substituting Eqs. (7.90) into (7.89), we obtain

$$\begin{aligned} e(\infty) &= \lim_{s\to 0} sR(s)\left(1-\frac{s+4}{s^3+6s^2+13s+20}\right)\\ &= \lim_{s\to 0} sR(s)\left(\frac{s^3+6s^2+12s+16}{s^3+6s^2+13s+20}\right)\end{aligned} \tag{7.91}$$

For a unit step, $R(s)=1/s$, and $e(\infty)=4/5$. For a unit ramp, $R(s)=1/s^2$, and $e(\infty)=\infty$. Notice that the system behaves like a Type 0 system.

### Analysis via Input Substitution

Another method for steady-state analysis avoids taking the inverse of $(s\mathbf{I}-\mathbf{A})$ and can be expanded to multiple-input, multiple-output systems; it substitutes the input along with an assumed solution into the state equations (*Hostetter, 1989*). We will derive the results for unit step and unit ramp inputs.

**Step Inputs.** Given the state Eqs. (7.84), if the input is a unit step where $r=1$, a steady-state solution, $\mathbf{x}_{ss}$, for $\mathbf{x}$, is

$$\mathbf{x}_{ss}=\begin{bmatrix}V_1\\V_2\\ \vdots\\ V_n\end{bmatrix}=\mathbf{V} \tag{7.92}$$

where $V_i$ is constant. Also,

$$\dot{\mathbf{x}}_{ss}=\mathbf{0} \tag{7.93}$$

Substituting $r=1$, a unit step, along with Eqs. (7.92) and (7.93), into Eqs. (7.84) yields

$$\mathbf{0}=\mathbf{AV}+\mathbf{B} \tag{7.94a}$$

$$y_{ss}=\mathbf{CV} \tag{7.94b}$$

where $y_{ss}$ is the steady-state output. Solving for $\mathbf{V}$ yields

$$\mathbf{V}=-\mathbf{A}^{-1}\mathbf{B} \tag{7.95}$$

But the steady-state error is the difference between the steady-state input and the steady-state output. The final result for the steady-state error for a unit step input into a system represented in state space is

$$\boxed{e(\infty) = 1 - y_{ss} = 1 - \mathbf{CV} = 1 + \mathbf{CA}^{-1}\mathbf{B}} \tag{7.96}$$

**Ramp Inputs.** For unit ramp inputs, $r = t$, a steady-state solution for $\mathbf{x}$ is

$$\mathbf{x_{ss}} = \begin{bmatrix} V_1 t + W_1 \\ V_2 t + W_2 \\ \vdots \\ V_n t + W_n \end{bmatrix} = \mathbf{V}t + \mathbf{W} \tag{7.97}$$

where $V_i$ and $W_i$ are constants. Hence,

$$\dot{\mathbf{x}}_{\mathbf{ss}} = \begin{bmatrix} V_1 \\ V_2 \\ \vdots \\ V_n \end{bmatrix} = \mathbf{V} \tag{7.98}$$

Substituting $r = t$ along with Eqs. (7.97) and (7.98) into Eqs. (7.84) yields

$$\mathbf{V} = \mathbf{A}(\mathbf{V}t + \mathbf{W}) + \mathbf{B}t \tag{7.99a}$$

$$y_{ss} = \mathbf{C}(\mathbf{V}t + \mathbf{W}) \tag{7.99b}$$

In order to balance Eq. (7.99a), we equate the matrix coefficients of $t$, $\mathbf{AV} = -\mathbf{B}$, or

$$\mathbf{V} = -\mathbf{A}^{-1}\mathbf{B} \tag{7.100}$$

Equating constant terms in Eq. (7.99a), we have $\mathbf{AW} = \mathbf{V}$, or

$$\mathbf{W} = \mathbf{A}^{-1}\mathbf{V} \tag{7.101}$$

Substituting Eqs. (7.100) and (7.101) into (7.99b) yields

$$y_{ss} = \mathbf{C}[-\mathbf{A}^{-1}\mathbf{B}t + \mathbf{A}^{-1}(-\mathbf{A}^{-1}\mathbf{B})] = -\mathbf{C}[\mathbf{A}^{-1}\mathbf{B}t + (\mathbf{A}^{-1})^2\mathbf{B}] \tag{7.102}$$

The steady-state error is therefore

$$\boxed{e(\infty) = \lim_{t\to\infty}(t - y_{ss}) = \lim_{t\to\infty}[(1 + \mathbf{CA}^{-1}\mathbf{B})t + \mathbf{C}(\mathbf{A}^{-1})^2\mathbf{B}]} \tag{7.103}$$

Notice that in order to use this method, $\mathbf{A}^{-1}$ must exist. That is, $\det\mathbf{A} \neq 0$.

We now demonstrate the use of Eqs. (7.96) and (7.103) to find the steady-state error for step and ramp inputs.

## Example 7.14 Steady-State Error Using Input Substitution

**PROBLEM:** Evaluate the steady-state error for the system described by the three equations in Eq. (7.90) for unit step and unit ramp inputs. Use input substitution.

**SOLUTION:** For a unit step input, the steady-state error given by Eq. (7.96) is

$$e(\infty) = 1 + \mathbf{CA}^{-1}\mathbf{B} = 1 - 0.2 = 0.8 \tag{7.104}$$

where **C**, **A**, and **B** are as follows:

$$\mathbf{A} = \begin{bmatrix} -5 & 1 & 0 \\ 0 & -2 & 1 \\ 20 & -10 & 1 \end{bmatrix}; \quad \mathbf{B} = \begin{bmatrix} 0 \\ 0 \\ 1 \end{bmatrix}; \quad \mathbf{C} = \begin{bmatrix} -1 & 1 & 0 \end{bmatrix} \tag{7.105}$$

For a ramp input, using Eq. (7.103), we have

$$e(\infty) = \left[\lim_{t\to\infty}[(1 + \mathbf{CA}^{-1}\mathbf{B})]t + \mathbf{C}(\mathbf{A}^{-1})^2\mathbf{B}\right] = \lim_{t\to\infty}(0.8t + 0.08) = \infty \tag{7.106}$$

## Skill-Assessment Exercise 7.7

**PROBLEM:** Find the steady-state error for a step input given the system represented in state space below. Calculate the steady-state error using both the final value theorem and input substitution methods.

$$\mathbf{A} = \begin{bmatrix} 0 & 1 \\ -3 & -6 \end{bmatrix}; \quad \mathbf{B} = \begin{bmatrix} 0 \\ 1 \end{bmatrix}; \quad \mathbf{C} = \begin{bmatrix} 1 & 1 \end{bmatrix}$$

**ANSWER:**

$$e_{\text{step}}(\infty) = \frac{2}{3}$$

The complete solution is at www.wiley.com/college/nise.

In this chapter, we covered the evaluation of steady-state error for systems represented by transfer functions as well as systems represented in state space. For systems represented in state space, two methods were presented: (1) final value theorem and (2) input substitution.

## Case Studies Antenna Control: Steady-State Error Design via Gain

Design D

This chapter showed how to find steady-state errors for step, ramp, and parabolic inputs to a closed-loop feedback control system. We also learned how to evaluate the gain to meet a steady-state error requirement. This ongoing case study uses our antenna azimuth position control system to summarize the concepts.

**PROBLEM:** For the antenna azimuth position control system shown on the front endpapers, Configuration 1,

**a.** Find the steady-state error in terms of gain, $K$, for step, ramp, and parabolic inputs.

**b.** Find the value of gain, $K$, to yield a 10% error in the steady state.

**SOLUTION:**

**a.** The simplified block diagram for the system is shown on the front endpapers. The steady-state error is given by

$$e(\infty) = \lim_{s\to 0} sE(s) = \lim_{s\to 0} \frac{sR(s)}{1+G(s)} \tag{7.107}$$

From the block diagram, after pushing the potentiometer to the right past the summing junction, the equivalent forward transfer function is

$$G(s) = \frac{6.63K}{s(s+1.71)(s+100)} \tag{7.108}$$

To find the steady-state error for a step input, use $R(s) = 1/s$ along with Eq. (7.108), and substitute these in Eq. (7.107). The result is $e(\infty) = 0$.

To find the steady-state error for a ramp input, use $R(s) - 1/s^2$ along with Eq. (7.108), and substitute these in Eq. (7.107). The result is $e(\infty) = 25.79/K$.

To find the steady-state error for a parabolic input, use $R(s) = 1/s^3$ along with Eq. (7.108), and substitute these in Eq. (7.107). The result is $e(\infty) = \infty$.

**b.** Since the system is Type 1, a 10% error in the steady state must refer to a ramp input. This is the only input that yields a finite, nonzero error. Hence, for a unit ramp input,

$$e(\infty) = 0.1 = \frac{1}{K_v} = \frac{(1.71)(100)}{6.63K} = \frac{25.79}{K} \tag{7.109}$$

from which $K = 257.9$. You should verify that the value of $K$ is within the range of gains that ensures system stability. In the antenna control case study in the last chapter, the range of gain for stability was found to be $0 < K < 2623.29$. Hence, the system is stable for a gain of 257.9.

**CHALLENGE:** You are now given a problem to test your knowledge of this chapter's objectives: Referring to the antenna azimuth position control system shown on the front endpapers, Configuration 2, do the following:

**a.** Find the steady-state errors in terms of gain, $K$, for step, ramp, and parabolic inputs.

**b.** Find the value of gain, $K$, to yield a 20% error in the steady state.

## Video Laser Disc Recorder: Steady-State Error Design via Gain

As a second case study, let us look at a video laser disc focusing system for recording.

Design
D

**PROBLEM:** In order to record on a video laser disc, a 0.5 $\mu$m laser spot must be focused on the recording medium to burn pits that represent the program material. The small laser spot requires that the focusing lens be positioned to an accuracy of $\pm 0.1$ $\mu$m. A model of the feedback control system for the focusing lens is shown in Figure 7.22.

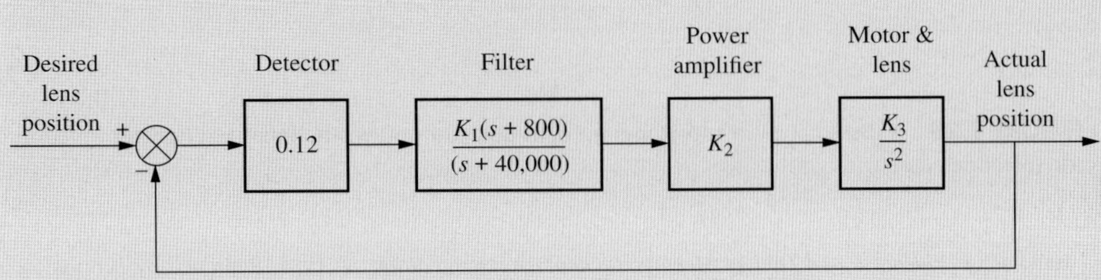


**FIGURE 7.22** Video laser disc recording: control system for focusing write beam

The detector detects the distance between the focusing lens and the video disc by measuring the degree of focus as shown in Figure 7.23($a$). Laser light reflected from the disc, $D$, is split by beam splitters $B_1$ and $B_2$ and focused behind aperture $A$. The remainder is reflected by the mirror and focuses in front of aperture $A$. The amount of light of each beam that passes through the aperture depends on how far the beam's focal point is from the aperture. Each side of the split photodiode, $P$, measures the intensity of each beam. Thus, as the disc's distance from the recording objective lens changes, so does the focal point of each beam. As a result, the relative voltage detected by each part of the split photodiode changes. When the beam is out of focus, one side of the photodiode outputs a larger voltage. When the beam is in focus, the voltage outputs from both sides of the photodiode are equal.

A simplified model for the detector is a straight line relating the differential voltage output from the two elements to the distance of the laser disc from nominal focus. A linearized plot of the detector input-output relationship is shown in Figure 7.23($b$)

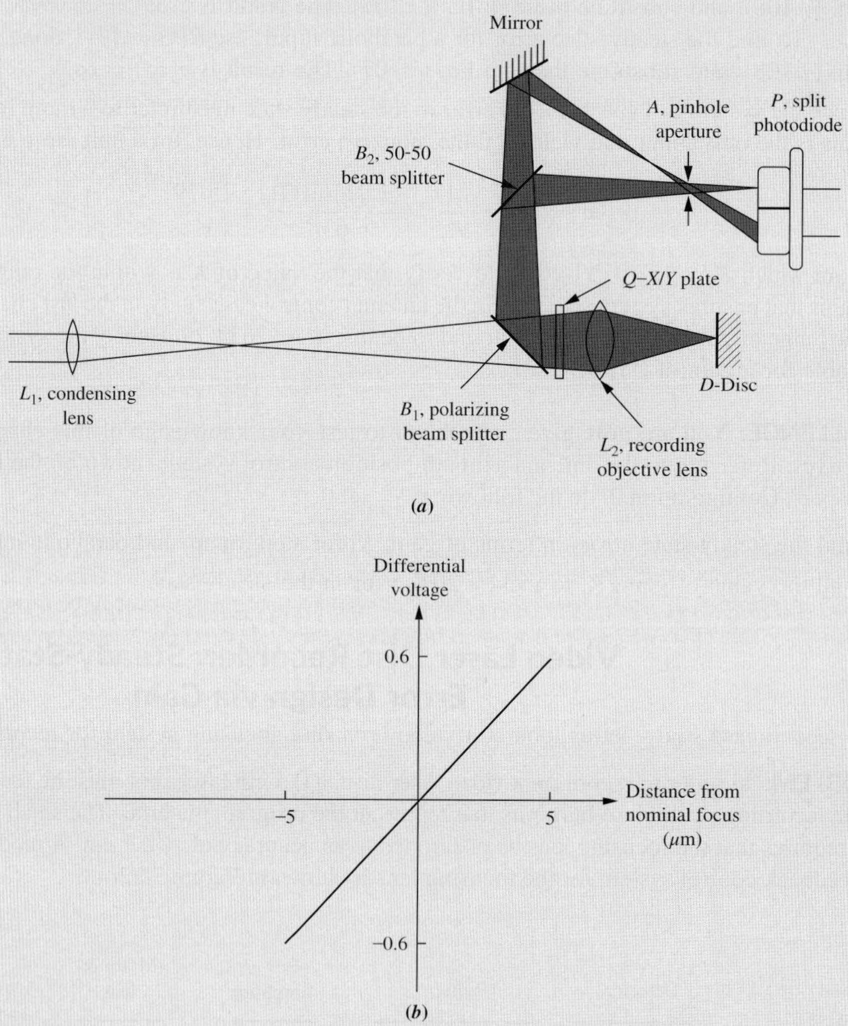


**FIGURE 7.23** Video disc laser recording: **a.** focus detector optics;[6] **b.** linearized transfer function for focus detector[6]

[6] Isailović, J. *Videodisc and Optical Memory Technologies, 1st Edition*, © 1985. Reprinted by permission of Pearson Education, Inc., Upper Saddle River, NJ.

(*Isailović, 1985*). Assume that a warp on the disc yields a worst-case disturbance in the focus of $10t^2$ $\mu$m. Find the value of $K_1K_2K_3$ in order to meet the focusing accuracy required by the system.

**SOLUTION:** Since the system is Type 2, it can respond to parabolic inputs with finite error. We can assume that the disturbance has the same effect as an input of $10t^2$ $\mu$m. The Laplace transform of $10t^2$ is $20/s^3$, or 20 units greater than the unit acceleration used to derive the general equation of the error for a parabolic input. Thus, $e(\infty) = 20/K_a$. But $K_a = \lim_{s\to 0} s^2G(s)$.

From Figure 7.22, $K_a = 0.0024K_1K_2K_3$. Also, from the problem statement, the error must be no greater than 0.1 $\mu$m. Hence, $e(\infty) = 8333.33/K_1K_2K_3 = 0.1$. Thus, $K_1K_2K_3 \geq 83333.3$, and the system is stable.

**CHALLENGE:** You are now given a problem to test your knowledge of this chapter's objectives: Given the video laser disc recording system whose block diagram is shown in Figure 7.24, do the following:

**a.** If the focusing lens needs to be positioned to an accuracy of $\pm 0.005$ $\mu$m, find the value of $K_1K_2K_3$ if the warp on the disc yields a worst-case disturbance in the focus of $15t^2$ $\mu$m.

**b.** Use the Routh-Hurwitz criterion to show that the system is stable when the conditions of **a.** are met.

**c.** Use MATLAB to show that the system is stable when the conditions of **a.** are met. MATLAB ML

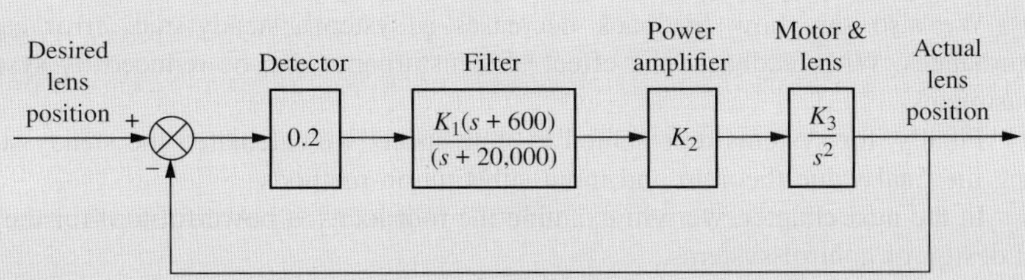


**FIGURE 7.24** Video laser disc recording focusing system

## Summary

This chapter covered the analysis and design of feedback control systems for steady-state errors. The steady-state errors studied resulted strictly from the system configuration. On the basis of a system configuration and a group of selected test signals, namely steps, ramps, and parabolas, we can analyze or design for the system's steady-state error performance. The greater the number of pure integrations a system has in the forward path, the higher the degree of accuracy, assuming the system is stable.

The steady-state errors depend upon the type of test input. Applying the final value theorem to stable systems, the steady-state error for unit step inputs is

$$e(\infty) = \frac{1}{1 + \lim_{s\to 0} G(s)} \tag{7.110}$$

The steady-state error for ramp inputs of unit velocity is

$$e(\infty) = \frac{1}{\lim_{s \to 0} sG(s)} \tag{7.111}$$

and for parabolic inputs of unit acceleration, it is

$$e(\infty) = \frac{1}{\lim_{s \to 0} s^2G(s)} \tag{7.112}$$

The terms taken to the limit in Eqs. (7.110) through (7.112) are called *static error constants*. Beginning with Eq. (7.110), the terms in the denominator taken to the limit are called the *position constant, velocity constant*, and *acceleration constant*, respectively. The static error constants are the steady-state error specifications for control systems. By specifying a static error constant, one is stating the number of pure integrations in the forward path, the test signal used, and the expected steady-state error.

Another definition covered in this chapter was that of *system type*. The system type is the number of pure integrations in the forward path, assuming a unity feedback system. Increasing the system type decreases the steady-state error as long as the system remains stable.

Since the steady-state error is, for the most part, inversely proportional to the static error constant, the larger the static error constant, the smaller the steady-state error. Increasing system gain increases the static error constant. Thus, in general, increasing system gain decreases the steady-state error as long as the system remains stable.

Nonunity feedback systems were handled by deriving an equivalent unity feedback system whose steady-state error characteristics followed all previous development. The method was restricted to systems where input and output units are the same.

We also saw how feedback decreases a system's steady-state error caused by disturbances. With feedback, the effect of a disturbance can be reduced by system gain adjustments.

Finally, for systems represented in state space, we calculated the steady-state error using the final value theorem and input substitution methods.

In the next chapter, we will examine the root locus, a powerful tool for the analysis and design of control systems.

## Review Questions

1. Name two sources of steady-state errors.
2. A position control, tracking with a constant difference in velocity, would yield how much position error in the steady state?
3. Name the test inputs used to evaluate steady-state error.
4. How many integrations in the forward path are required in order for there to be zero steady-state error for each of the test inputs listed in Question 3?
5. Increasing system gain has what effect upon the steady-state error?
6. For a step input, the steady-state error is approximately the reciprocal of the static error constant if what condition holds true?
7. What is the exact relationship between the static error constants and the steady-state errors for ramp and parabolic inputs?
8. What information is contained in the specification $K_p = 10{,}000$?
9. Define *system type*.

10. The forward transfer function of a control system has three poles at $-1, -2$, and $-3$. What is the system type?
11. What effect does feedback have upon disturbances?
12. For a step input disturbance at the input to the plant, describe the effect of controller and plant gain upon minimizing the effect of the disturbance.
13. Is the forward-path actuating signal the system error if the system has nonunity feedback?
14. How are nonunity feedback systems analyzed and designed for steady-state errors?
15. Define, in words, *sensitivity* and describe the goal of feedback-control-system engineering as it applies to sensitivity.
16. Name two methods for calculating the steady-state error for systems represented in state space.

# Problems

1. For the unity feedback system shown in Figure P7.1, where

$$G(s) = \frac{500(s+8)(s+10)(s+15)}{s(s+38)(s^2+2s+25)}$$

find the steady-state errors for the following test inputs: $25u(t)$, $37tu(t)$, $47t^2u(t)$. [Section: 7.2]

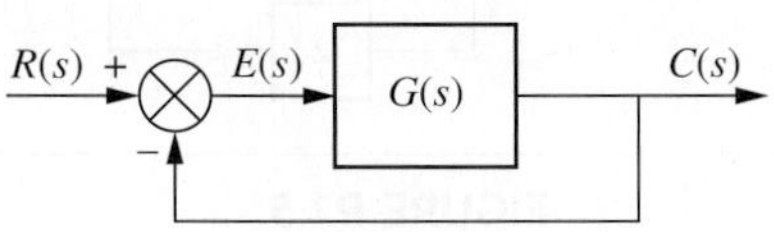


**FIGURE P7.1**

2. Figure P7.2 shows the ramp input $r(t)$ and the output $c(t)$ of a system. Assuming the output's steady state can be approximated by a ramp, find [Section: 7.1]
   **a.** the steady-state error;
   **b.** the steady-state error if the input becomes $r(t) = tu(t)$.

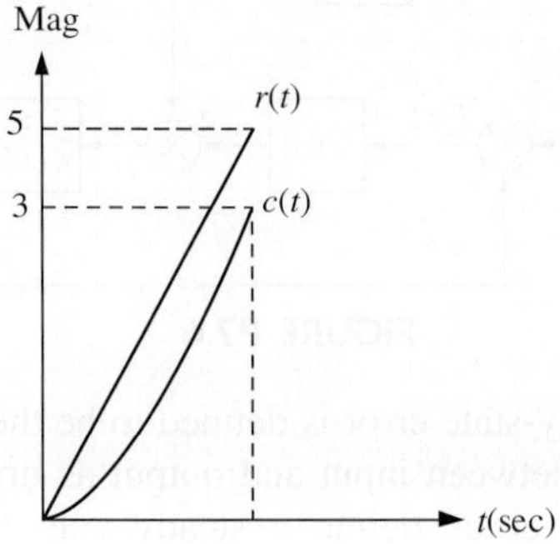


**FIGURE P7.2**

3. For the unity feedback system shown in Figure P7.1, where

$$G(s) = \frac{20(s+3)(s+4)(s+8)}{s^2(s+2)(s+15)}$$

find the steady-state error if the input is $30t^2u(t)$. [Section: 7.2]

4. For the system shown in Figure P7.3, what steady-state error can be expected for the following test inputs: $15u(t)$, $15tu(t)$, $15t^2u(t)$. [Section: 7.2]

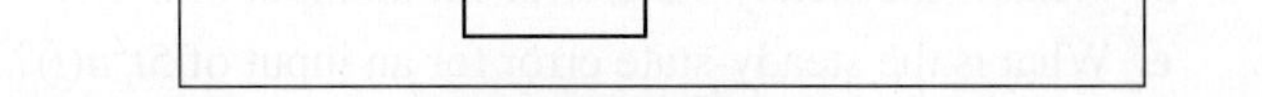


**FIGURE P7.3**

5. For the unity feedback system shown in Figure P7.1, where

$$G(s) = \frac{500}{(s+20)(s^2+4s+10)}$$

find the steady-state error for inputs of $40u(t)$, $70tu(t)$, and $80t^2u(t)$. [Section: 7.3]

6. An input of $12t^3u(t)$ is applied to the input of a Type 3 unity feedback system, as shown in Figure P7.1,

where

$$G(s) = \frac{200(s+2)(s+5)(s+7)(s+9)}{s^3(s+3)(s+10)(s+15)}$$

Find the steady-state error in position. [Section: 7.3]

**7.** The steady-state error in velocity of a system is defined to be

$$\left(\frac{dr}{dt} - \frac{dc}{dt}\right)\Bigg|_{t\to\infty}$$

where $r$ is the system input, and $c$ is the system output. Find the steady-state error in velocity for an input of $t^3u(t)$ to a unity feedback system with a forward transfer function of [Section: 7.2]

$$G(s) = \frac{100(s+1)(s+2)}{s^2(s+3)(s+10)}$$

**8.** What is the steady-state error for a step input of 15 units applied to the unity feedback system of Figure P7.1, where [Section: 7.3]

$$G(s) = \frac{1000(s+12)(s+25)(s+32)}{(s+61)(s+73)(s+87)}$$

**9.** A system has $K_p = 3$. What steady-state error can be expected for inputs of $8u(t)$ and $8tu(t)$? [Section 7.3]

**10.** For the unity feedback system shown in Figure P7.1, where

$$G(s) = \frac{1250}{s(s+50)}$$

**a.** What is the expected percent overshoot for a unit step input?

**b.** What is the settling time for a unit step input?

**c.** What is the steady-state error for an input of $5u(t)$?

**d.** What is the steady-state error for an input of $5tu(t)$?

**e.** What is the steady-state error for an input of $5t^2u(t)$?

**11.** Given the unity feedback system shown in Figure P7.1, where

$$G(s) = \frac{10^5(s+3)(s+10)(s+20)}{s(s+25)(s+\alpha)(s+30)}$$

find the value of $\alpha$ to yield a $K_v = 10^4$. [Section: 7.4]

**12.** For the unity feedback system of Figure P7.1, where

$$G(s) = \frac{K(s+2)(s+4)(s+6)}{s^2(s+5)(s+7)}$$

find the value of $K$ to yield a static error constant of 10,000. [Section: 7.4]

**13.** For the system shown in Figure P7.4, [Section: 7.3]

**a.** Find $K_p$, $K_v$, and $K_a$.

**b.** Find the steady-state error for an input of $50u(t)$, $50tu(t)$, and $50t^2u(t)$.

**c.** State the system type.

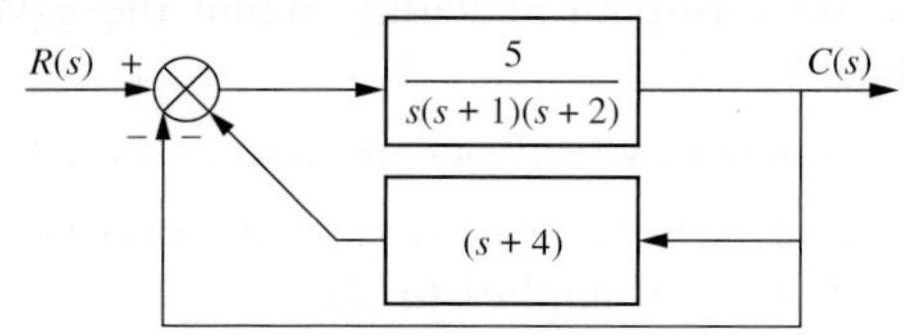


**FIGURE P7.4**

**14.** A Type 3 unity feedback system has $r(t) = t^3$ applied to its input. Find the steady-state position error for this input if the forward transfer function is [Section: 7.3]

$$G(s) = \frac{1000(s^2+4s+20)(s^2+20s+15)}{s^3(s+2)(s+10)}$$

**15.** Find the system type for the system of Figure P7.5.

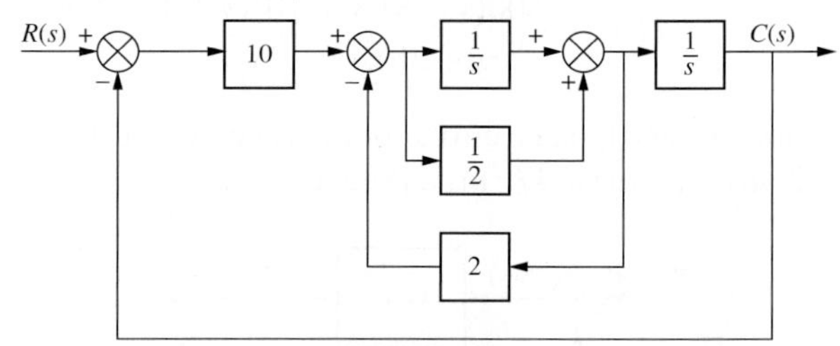


**FIGURE P7.5**

**16.** What are the restrictions on the feedforward transfer function $G_2(s)$ in the system of Figure P7.6 to obtain zero steady-state error for step inputs if: [Section: 7.3]

**a.** $G_1(s)$ is a Type 0 transfer function;

**b.** $G_1(s)$ is a Type 1 transfer function;

**c.** $G_1(s)$ is a Type 2 transfer function?

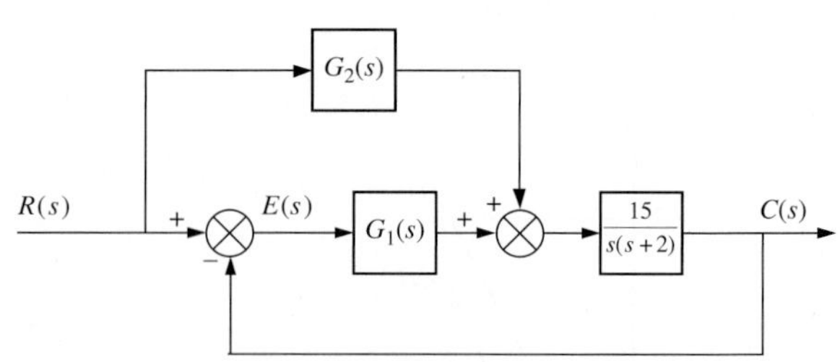


**FIGURE P7.6**

**17.** The steady-state error is defined to be the difference in position between input and output as time approaches infinity. Let us define a steady-state velocity error, which is the difference in velocity between input and output. Derive an expression for the error in velocity,

$\dot{e}(\infty) = \dot{r}(\infty) - \dot{c}(\infty)$, and complete Table P7.1 for the error in velocity. [Sections: 7.2, 7.3]

**TABLE P7.1**

| Input | Type 0 | Type 1 | Type 2 |
|---|---|---|---|
| Step | | | |
| Ramp | | | |
| Parabola | | | |

**18.** For the system shown in Figure P7.7, [Section: 7.4]

**a.** What value of $K$ will yield a steady-state error in position of 0.01 for an input of $(1/10)t$?

**b.** What is the $K_v$ for the value of $K$ found in Part **a**?

**c.** What is the minimum possible steady-state position error for the input given in Part **a**?

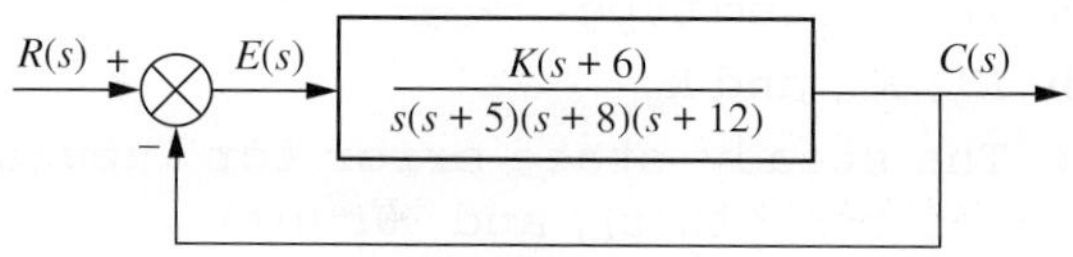


**FIGURE P7.7**

**19.** Given the unity feedback system of Figure P7.1, where

$$G(s) = \frac{K(s+a)}{s(s+2)(s+13)}$$

find the value of $Ka$ so that a ramp input of slope 40 will yield an error of 0.006 in the steady state when compared to the output. [Section: 7.4]

**20.** Given the system of Figure P7.8, design the value of $K$ so that for an input of $100tu(t)$, there will be a 0.01 error in the steady state. [Section: 7.4]

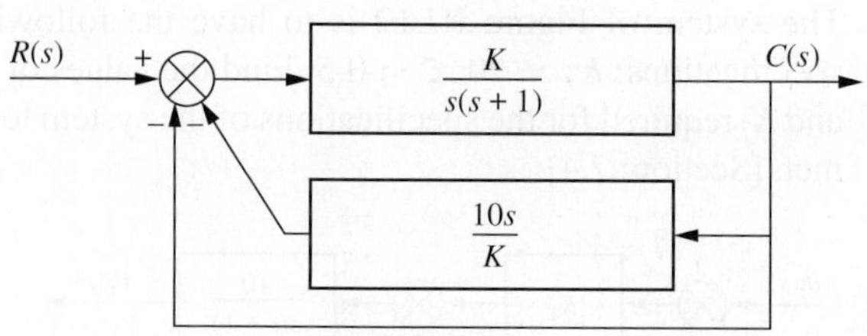


**FIGURE P7.8**

**21.** Find the value of $K$ for the unity feedback system shown in Figure P7.1, where

$$G(s) = \frac{K(s+2)}{s^2(s+4)}$$

if the input is $10t^2u(t)$, and the desired steady-state error is 0.01 for this input. [Section: 7.4]

**22.** The unity feedback system of Figure P7.1, where

$$G(s) = \frac{K(s^2+3s+30)}{s^n(s+5)}$$

is to have 1/6000 error between an input of $10tu(t)$ and the output in the steady state. [Section: 7.4]

**a.** Find $K$ and $n$ to meet the specification.

**b.** What are $K_p$, $K_v$, and $K_a$?

**23.** For the unity feedback system of Figure P7.1, where [Section: 7.3]

$$G(s) = \frac{K(s^2+2s+5)}{(s+2)^2(s+3)}$$

**a.** Find the system type.

**b.** What error can be expected for an input of $10u(t)$?

**c.** What error can be expected for an input of $10tu(t)$?

**24.** For the unity feedback system of Figure P7.1, where

$$G(s) = \frac{K(s+10)(s+15)}{s(s+3)(s+7)(s+20)}$$

find the value of $K$ to yield a steady-state error of 0.1 for a ramp input of $25tu(t)$. [Section: 7.4]

**25.** Given the unity feedback system of Figure P7.1, where

$$G(s) = \frac{K(s+4)}{(s+1)(s^2+10s+26)}$$

find the value of $K$ to yield a steady-state error of 5%. [Section: 7.4]

**26.** For the unity feedback system of Figure P7.1, where

$$G(s) = \frac{K}{s(s+4)(s+8)}$$

find the minimum possible steady-state position error if a unit ramp is applied. What places the constraint upon the error?

**27.** The unity feedback system of Figure P7.1, where

$$G(s) = \frac{K(s+\alpha)}{(s+\beta)^2}$$

is to be designed to meet the following specifications: steady-state error for a unit step input = 0.1; damping ratio = 0.5; natural frequency = $\sqrt{10}$. Find $K$, $\alpha$, and $\beta$. [Section: 7.4]

**28.** A second-order, unity feedback system is to follow a ramp input with the following specifications: the steady-state output position shall differ from the input position by 0.01 of the input velocity; the natural frequency of the closed-loop system shall be 5 rad/s. Find the following:

a. The system type

b. The exact expression for the forward-path transfer function

c. The closed-loop system's damping ratio

**29.** The unity feedback system of Figure P7.1, where

$$G(s) = \frac{K(s+\alpha)}{s(s+\beta)}$$

is to be designed to meet the following requirements: The steady-state position error for a unit ramp input equals 1/10; the closed-loop poles will be located at $-1 \pm j1$. Find $K$, $\alpha$, and $\beta$ in order to meet the specifications. [Section: 7.4]

**30.** The unity feedback system of Figure P7.1 has a transfer function $G(s) = \dfrac{C(s)}{E(s)} = \dfrac{K}{s(s+\alpha)}$ and is to follow a ramp input, $r(t) = tu(t)$, so that the steady-state output position differs from the input position by 0.01 of the input velocity (e.g., $e(\infty) = \dfrac{1}{K_v} = 0.01$). The natural frequency of the closed-loop system will be $\omega_n = 5$ rad/s. [Section: 7.4] Find the following:

a. The system type

b. The values of $K$ and $\alpha$

c. The closed-loop system's damping ratio, $\zeta$

d. If $K$ is reduced to 4 and $\alpha = 0.4$, find the corresponding new values of $e(\infty)$, $\omega_n$, and $\zeta$.

**31.** Given the unity feedback control system of Figure P7.1, where

$$G(s) = \frac{K}{s^n(s+a)}$$

find the values of $n$, $K$, and $a$ in order to meet specifications of 10% overshoot and $K_v = 100$. [Section: 7.4]

**32.** Given the unity feedback control system of Figure P7.1, where

$$G(s) = \frac{K}{s(s+a)}$$

find the following: [Section: 7.4]

a. $K$ and $a$ to yield $K_v = 500$ and a 20% overshoot

b. $K$ and $a$ to yield a 2% error in the steady state and a 10% overshoot

**33.** Given the system in Figure P7.9, find the following: [Section: 7.3]

a. The closed-loop transfer function

b. The system type

c. The steady-state error for an input of $5u(t)$

d. The steady-state error for an input of $5tu(t)$

e. Discuss the validity of your answers to Parts **c** and **d**.

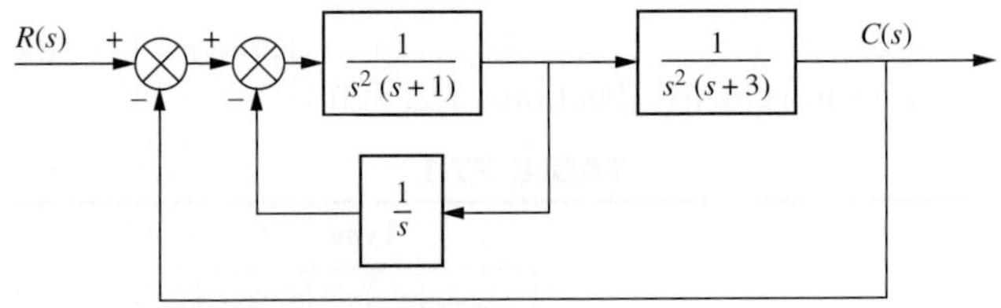


**FIGURE P7.9**

**34.** Repeat Problem 33 for the system shown in Figure P7.10. [Section: 7.3]

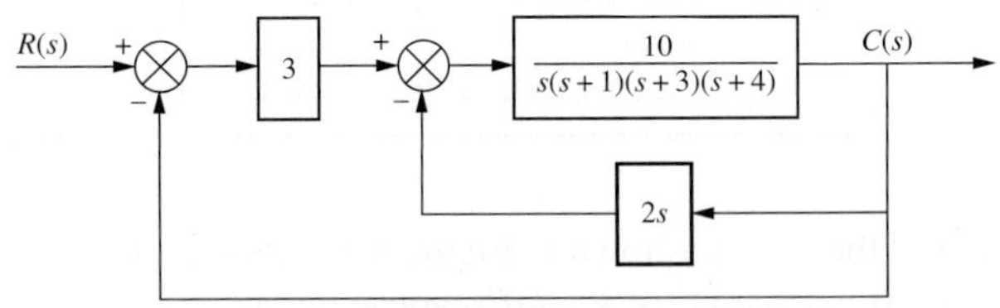


**FIGURE P7.10**

**35.** For the system shown in Figure P7.11, use MATLAB to find the following: [Section: 7.3]

MATLAB ML

a. The system type

b. $K_p$, $K_v$, and $K_a$

c. The steady-state error for inputs of $30u(t)$, $30tu(t)$, and $30t^2u(t)$

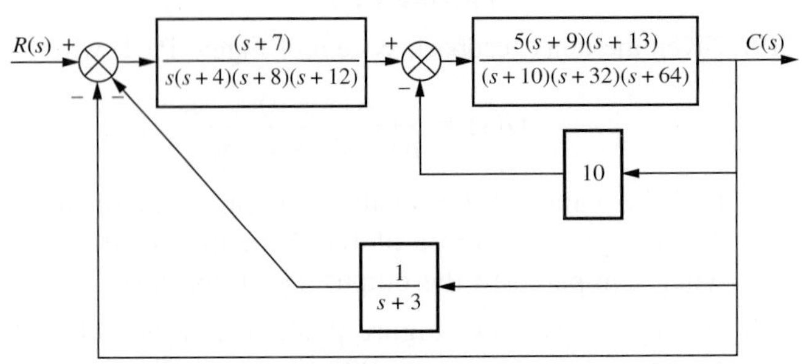


**FIGURE P7.11**

**36.** The system of Figure P7.12 is to have the following specifications: $K_v = 10$; $\zeta = 0.5$. Find the values of $K_1$ and $K_f$ required for the specifications of the system to be met. [Section: 7.4]

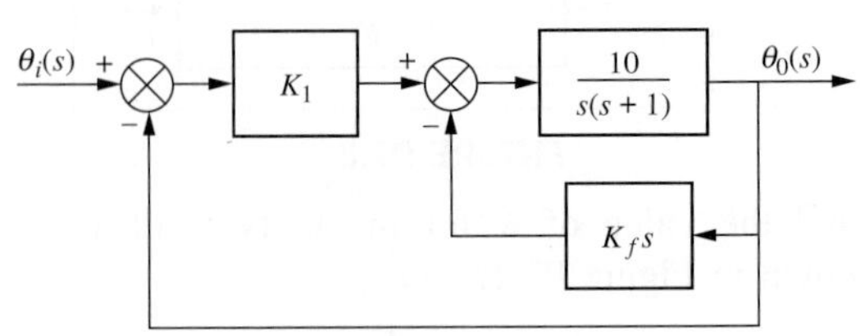


**FIGURE P7.12**

**37.** The transfer function from elevator deflection to altitude change in a Tower Trainer 60 Unmanned Aerial Vehicle is

$$P(s) = \frac{h(s)}{\delta(s)_e}$$
$$= \frac{-34.16s^3 - 144.4s^2 + 7047s + 557.2}{s^5 + 13.18s^4 + 95.93s^3 + 14.61s^2 + 31.94s}$$

An autopilot is built around the aircraft as shown in Figure P7.13, with $F(s) = H(s) = 1$ and

$$G(s) = \frac{0.00842(s + 7.895)(s^2 + 0.108s + 0.3393)}{(s + 0.07895)(s^2 + 4s + 8)}$$

(*Barkana, 2005*). The steady-state error for a ramp input in this system is $e_{ss} = 25$. Find the slope of the ramp input.

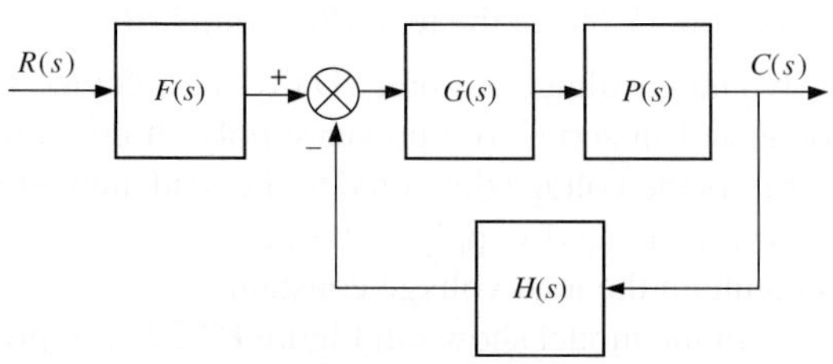


**FIGURE P7.13**

**38.** Find the total steady-state error due to a unit step input and a unit step disturbance in the system of Figure P7.14. [Section: 7.5]

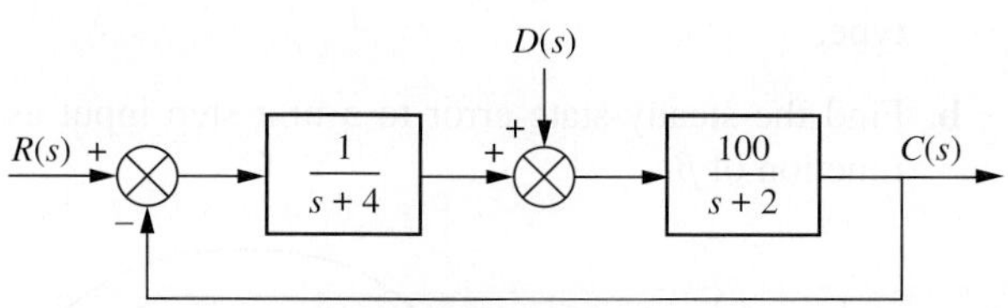


**FIGURE P7.14**

**39.** Design the values of $K_1$ and $K_2$ in the system of Figure P7.15 to meet the following specifications: Steady-state error component due to a unit step disturbance is $-0.000012$; steady-state error component due to a unit ramp input is 0.003. [Section: 7.5]

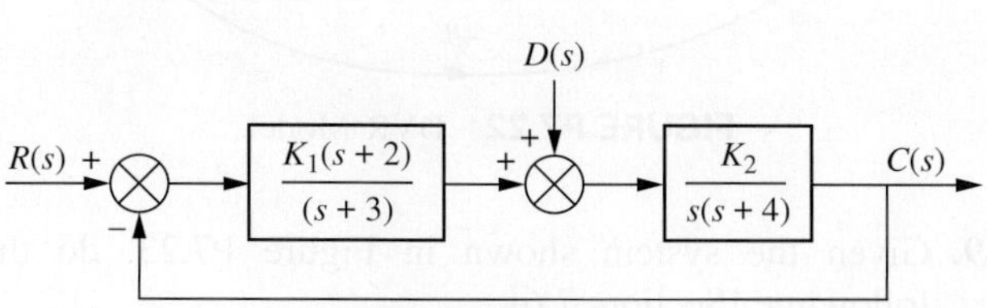


**FIGURE P7.15**

**40.** In Figure P7.16, let $G(s) = 5$ and $P(s) = \dfrac{7}{s+2}$.

**a.** Calculate the steady-state error due to a command input $R(s) = \dfrac{3}{s}$ with $D(s) = 0$.

**b.** `Verify the result of Part a using Simulink.` Simulink SL

**c.** Calculate the steady-state error due to a disturbance input $D(s) = -\dfrac{1}{s}$ with $R(s) = 0$.

**d.** `Verify the result of Part c using Simulink.` Simulink SL

**e.** Calculate the total steady-state error due to a command input $R(s) = \dfrac{3}{s}$ and a disturbance $D(s) = -\dfrac{1}{s}$ applied simultaneously.

**f.** `Verify the result of Part e using Simulink.` Simulink SL

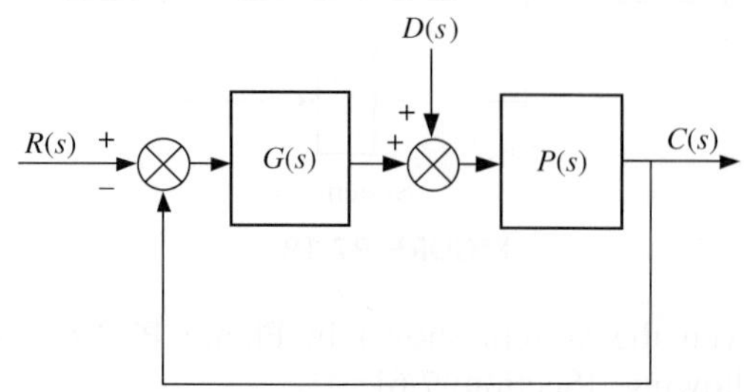


**FIGURE P7.16**

**41.** Derive Eq. (7.72) in the text, which is the final value of the actuating signal for nonunity feedback systems. [Section: 7.6]

**42.** For each of the systems shown in Figure P7.17, find the following: [Section: 7.6]

**a.** The system type

**b.** The appropriate static error constant

**c.** The input waveform to yield a constant error

**d.** The steady-state error for a unit input of the waveform found in Part **c**

**e.** The steady-state value of the actuating signal

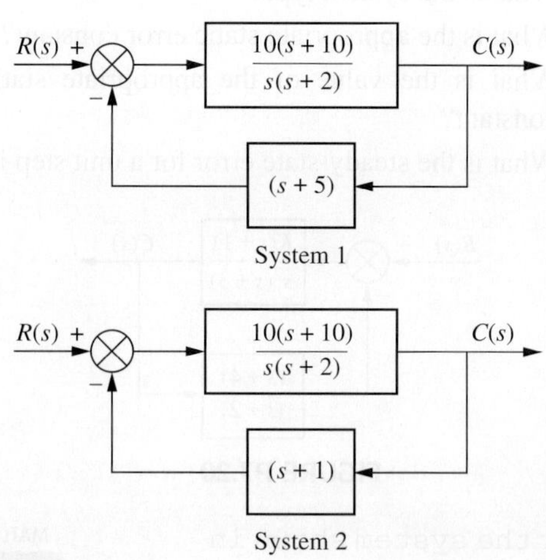


**FIGURE P7.17** Closed-loop systems with nonunity feedback

**43.** For each of the systems shown in Figure P7.18, find the appropriate static error constant as well as the steady-state error, $r(\infty) - c(\infty)$, for unit step, ramp, and parabolic inputs. [Section: 7.6]

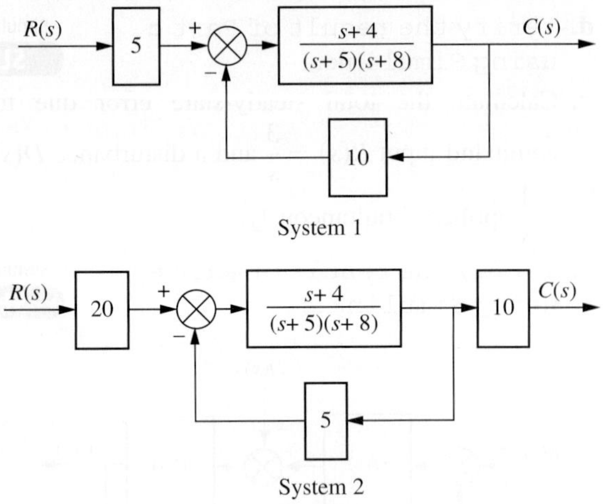


**FIGURE P7.18**

**44.** Given the system shown in Figure P7.19, find the following: [Section: 7.6]

**a.** The system type

**b.** The value of $K$ to yield 0.1% error in the steady state

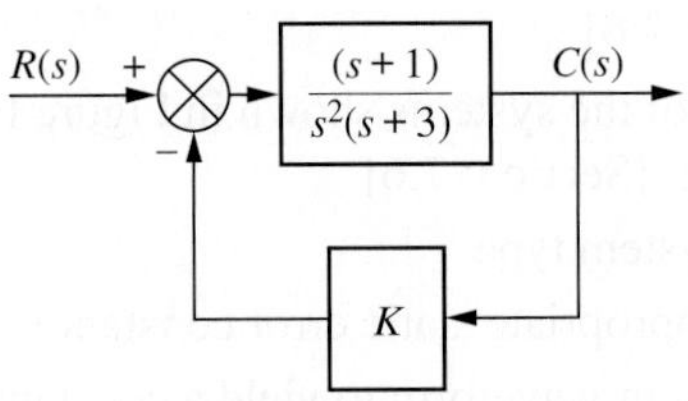


**FIGURE P7.19**

**45.** For the system shown in Figure P7.20, [Section: 7.6]

**a.** What is the system type?

**b.** What is the appropriate static error constant?

**c.** What is the value of the appropriate static error constant?

**d.** What is the steady-state error for a unit step input?

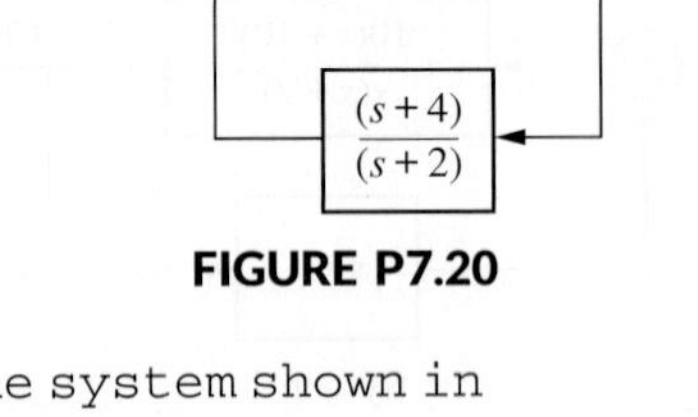


**FIGURE P7.20**

**46.** For the system shown in Figure P7.21, use MATLAB to find the following for $K = 10$, and $K = 10^6$: [Section: 7.6]

MATLAB ML

**a.** The system type

**b.** $K_p$, $K_v$, and $K_a$

**c.** The steady-state error for inputs of $30u(t)$, $30tu(t)$, and $30t^2u(t)$

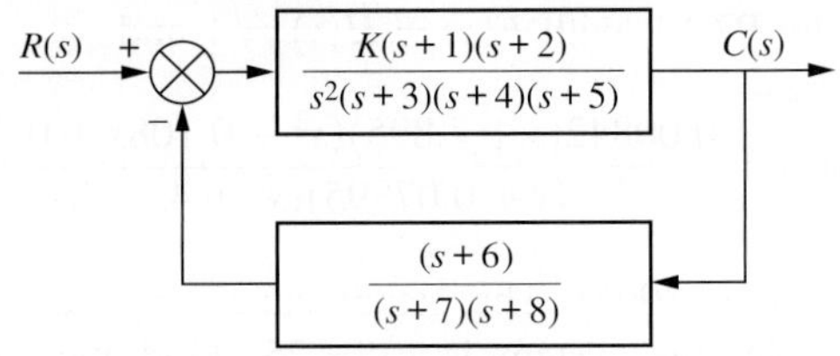


**FIGURE P7.21**

**47.** Derive Eq. (7.69) in the text. [Section: 7.6]

**48.** A dynamic voltage restorer (DVR) is a device that is connected in series to a power supply. It continuously monitors the voltage delivered to the load, and compensates voltage sags by applying the necessary extra voltage to maintain the load voltage constant.

In the model shown in Figure P7.22, $u_r$ represents the desired reference voltage, $u_o$ is the output voltage, and $Z_L$ is the load impedance. All other parameters are internal to the DVR (*Lam, 2004*).

**a.** Assuming $Z_L = \frac{1}{sC_L}$, and $\beta \neq 1$, find the system's type.

**b.** Find the steady-state error to a unit step input as a function of $\beta$.

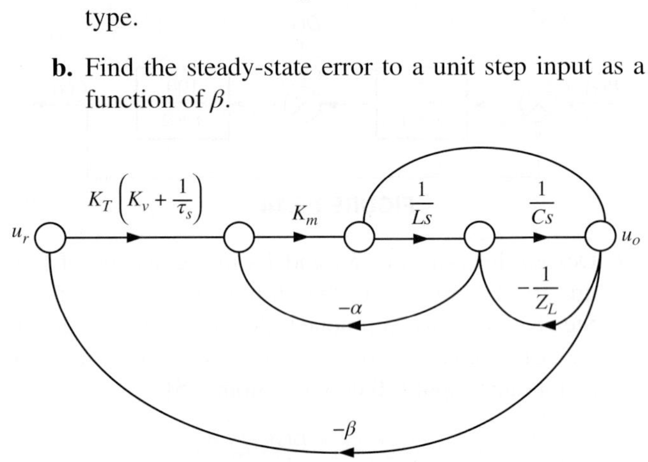


**FIGURE P7.22** DVR Model

**49.** Given the system shown in Figure P7.23, do the following: [Section: 7.6]

**a.** Derive the expression for the error, $E(s) = R(s) - C(s)$, in terms of $R(s)$ and $D(s)$.

**b.** Derive the steady-state error, $e(\infty)$, if $R(s)$ and $D(s)$ are unit step functions.

**c.** Determine the attributes of $G_1(s)$, $G_2(s)$, and $H(s)$ necessary for the steady-state error to become zero.

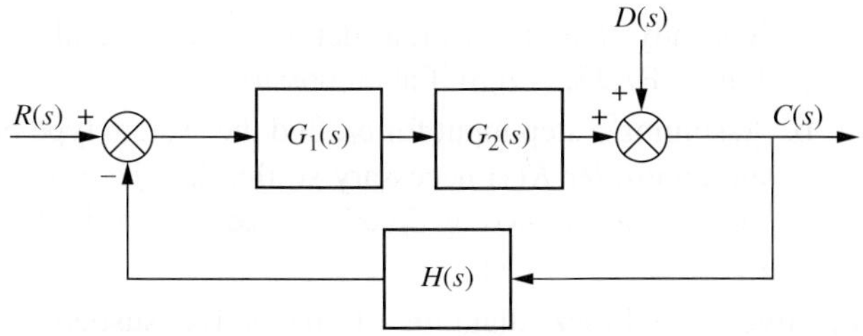


**FIGURE P7.23** System with input and disturbance

**50.** Given the system shown in Figure P7.24, find the sensitivity of the steady-state error to parameter $a$. Assume a step input. Plot the sensitivity as a function of parameter $a$. [Section: 7.7]

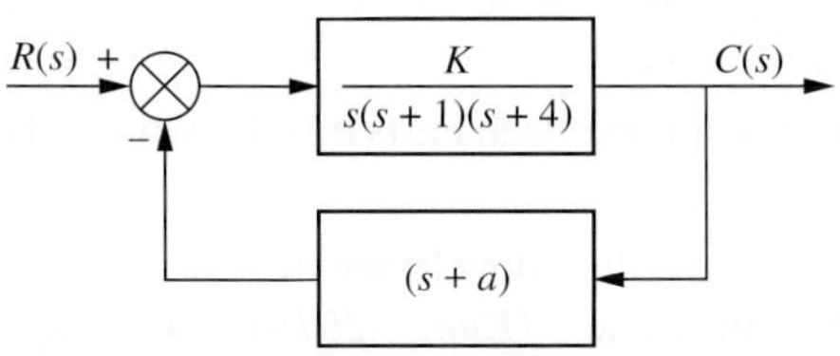


**FIGURE P7.24**

**51. a.** Show that the sensitivity to plant changes in the system of Figure P7.13 is

$$S_{T:P} = \frac{P}{T}\frac{\delta T}{\delta P} = \frac{1}{1+L(s)}$$

where $L(s) = G(s)P(s)H(s)$ and

$$T(s) = \frac{C(s)}{R(s)} = \frac{F(s)}{H(s)} \cdot \frac{L(s)}{1+L(s)}.$$

**b.** Show that $S_{T:P}(s) + \dfrac{T(s)H(s)}{F(s)} = 1$ for all values of $s$.

**52.** In Figure P7.13, $P(s) = \dfrac{2}{s}$, $H(s) = 1$,

$$T(s) = \frac{C(s)}{R(s)} = \frac{14K}{(s+1)(s+2)(s^2+5s+14)},$$

and

$$S_{T:P} = \frac{P}{T}\frac{\delta T}{\delta P} = \frac{s^2+5s}{s^2+5s+14}.$$

**a.** Find $F(s)$ and $G(s)$.

**b.** Find the value of $K$ that will result in zero steady-state error for a unit step input.

**53.** For the system shown in Figure P7.25, find the sensitivity of the steady-state error for changes in $K_1$ and in $K_2$, when $K_1 = 100$ and $K_2 = 0.1$. Assume step inputs for both the input and the disturbance. [Section: 7.7]

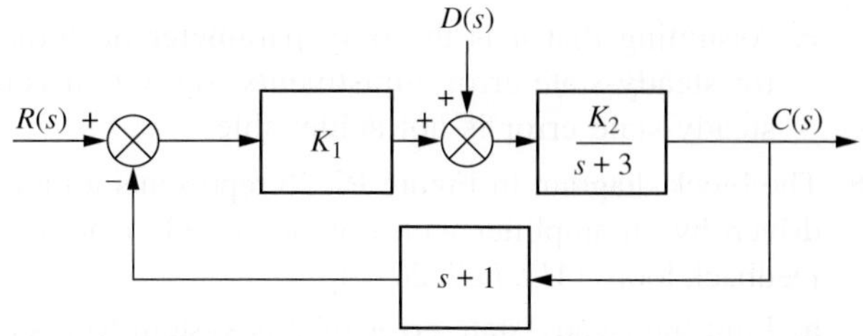


**FIGURE P7.25** System with input and disturbance

**54.** For each of the following closed-loop systems, find the steady-state error for unit step and unit ramp inputs. Use both the final value theorem and input substitution methods. [Section: 7.8]

**a.** $\dot{\mathbf{x}} = \begin{bmatrix} -5 & -4 & -2 \\ -3 & -10 & 0 \\ -1 & 1 & -5 \end{bmatrix}\mathbf{x} + \begin{bmatrix} 1 \\ 1 \\ 0 \end{bmatrix} r;\ y = [-1 \ \ 2 \ \ 1]\mathbf{x}$

**b.** $\dot{\mathbf{x}} = \begin{bmatrix} 0 & 1 & 0 \\ -5 & -9 & 7 \\ -1 & 0 & 0 \end{bmatrix}\mathbf{x} + \begin{bmatrix} 0 \\ 0 \\ 1 \end{bmatrix} r;\ y = [1 \ \ 0 \ \ 0]\mathbf{x}$

**c.** $\dot{\mathbf{x}} = \begin{bmatrix} -9 & -5 & -1 \\ 1 & 0 & -2 \\ -3 & -2 & -5 \end{bmatrix}\mathbf{x} + \begin{bmatrix} 2 \\ 3 \\ 5 \end{bmatrix} r;\ y = [1 \ \ -2 \ \ 4]\mathbf{x}$

**55.** Glycolysis is a feedback process through which living cells use glucose to generate adenosine triphosphate (ATP), necessary for cell operations. A linearized glycolysis model (*Chandra, 2011*) is given by

$$\begin{bmatrix} \Delta\dot{x} \\ \Delta\dot{y} \end{bmatrix} = \begin{bmatrix} -k & a+g+h \\ (q+1)k & -qa - g(q+1) + qh \end{bmatrix}\begin{bmatrix} \Delta x \\ \Delta y \end{bmatrix} + \begin{bmatrix} 0 \\ 1 \end{bmatrix}\delta$$

where $\delta$ is the perturbation (disturbance input) on ATP production, $\Delta y$ is the change in ATP level (output). $\alpha > 0$ is the cooperativity of ATP binding to PFK, $g > 0$ is the feedback strength of ATP on PK (PFK and PK are two different types of glycolytic enzymes), $k > 0$ is the intermediate reaction rate, $q > 0$ is the autocatalytic stoichiometry, and $h > 0$ is the feedback strength of ATP on the PFK enzyme.

**a.** Since in this system $\delta$ is a disturbance input, zero steady-state error is achieved when $\dfrac{\Delta y}{\delta} = 0$. Show that in steady state ($\Delta\dot{x} = 0$, $\Delta\dot{y} = 0$), $\dfrac{\Delta y}{\delta} = \dfrac{1}{a-h}$.

**b.** Use the Routh-Hurwitz stability criterion to show that the system will be closed-loop stable as long as

$$0 < h - a < \frac{k + g(q+1)}{q}$$

**c.** Assuming that $h$ is the only parameter of choice for steady-state error adjustments, show that zero steady-state error is not achievable.

**56.** The block diagram in Figure P7.26 represents a motor driven by an amplifier with double-nested tachometer feedback loops (*Mitchell, 2010*).

**a.** Find the steady-state error of this system to a step input.

**b.** What is the system type?

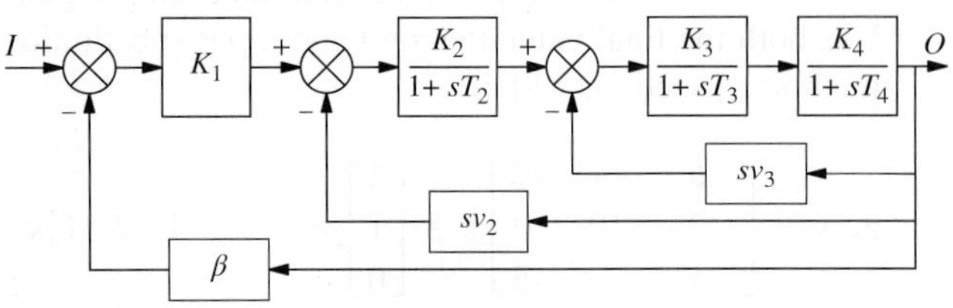


**FIGURE P7.26**[7]

**57.** A simplified model of the steering of a four-wheel drive vehicle is shown in Figure P7.27.

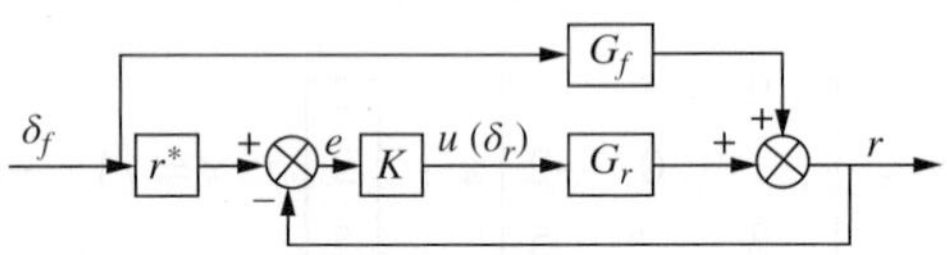


**FIGURE P7.27** Steering model for a four-wheel drive vehicle[8]

In this block diagram, the output $r$ is the vehicle's yaw rate, while $\delta_f$ and $\delta_r$ are the steering angles of the front and rear tires, respectively. In this model,

$$r^*(s) = \frac{\frac{s}{300} + 0.8}{\frac{s}{10} + 1}, \quad G_f(s) = \frac{h_1 s + h_2}{s^2 + a_1 s + a_2}$$

$$G_r(s) = \frac{h_3 s + b_1}{s^2 + a_1 s + a_2}$$

and $K(s)$ is a controller to be designed. (*Yin, 2007*).

**a.** Assuming a step input for $\delta_f$, find the minimum system type of the controller $K(s)$ necessary so that in steady-state the error, as defined by the signal $e$ in Figure P7.27, is zero if at all possible.

**b.** Assuming a step input for $\delta_f$, find the system type of the controller $K(s)$ necessary so that in steady state the error as defined by $\delta_f(\infty) - r(\infty)$ is zero if at all possible.

**58.** Given the block diagram of the active suspension system shown in Figure P5.37, (*Lin, 1997*)

**a.** Find the transfer function from a road disturbance $r$ to the error signal $e$.

**b.** Use the transfer function in Part **a** to find the steady-state value of $e$ for a unit step road disturbance.

**c.** Use the transfer function in Part **a** to find the steady-state value of $e$ for a unit ramp road disturbance.

**d.** From your results in Parts **b** and **c**, what is the system's type for $e$?

**59.** As part of the development of a textile cross-lapper machine (*Kuo, 2010*), a torque input, $u(t) = \left\{ \begin{array}{ll} 1 & 0 \le t < 50 \\ -1 & 50 \le t < 100 \end{array} \right\}$, is applied to the motor of one of the movable racks embedded in a feedback loop. The corresponding velocity output response is shown in Figure P7.28.

**a.** What is the open-loop system's type?

**b.** What is the steady-state error?

**c.** What would be the steady-state error for a ramp input?

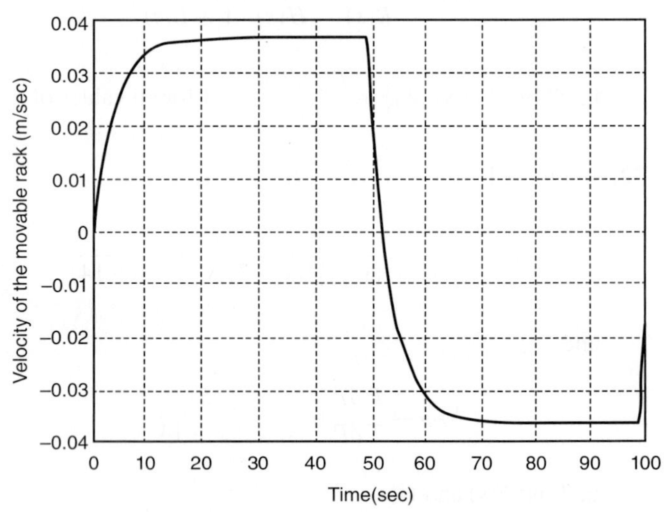


**FIGURE P7.28** Velocity output response[9]

[7] Mitchell, R. J. More Nested Velocity Feedback Control. *IEEE 9th International Conference on Cybernetic Intelligent Systems (CIS)*, 2010. Figure 5, page 3 of the paper.

[8] Yin, G., Chen, N., and Li, P. Improving Handling Stability Performance of Four-Wheel Steering Vehicle via $\mu$-Synthesis Robust Control. *IEEE Transactions on Vehicular Technology*, vol. 56, no. 5, 2007, pp. 2432–2439. Fig. 2, p. 2434. IEEE Transactions on Vehicular Technology by Vehicular Technology Society; Institute of Electrical and Electronics Engineers; IEEE Vehicular Technology Group. Reproduced with permission of Institute of Electrical and Electronics Engineers, in the format Republish in a book via Copyright Clearance Center.

[9] Kuo, C-F. J., Tu, H-M., and Liu, C-H. "Dynamic Modeling and Control of a Current New Horizontal Type Cross-Lapper Machine," *Textile Research Journal*, Vol. 80 (19), pp. 2016–2027, Figure 5. Copyright © 2010. Reprinted by Permission of SAGE.

**60.** PID control, which is discussed in Chapter 9, may be recommended for Type 3 systems when the output in a feedback system is required to perfectly track a parabolic as well as step and ramp reference signals (*Papadopoulos, 2013*). In the system of Figure P7.29, the transfer functions of the plant, $G_P(s)$, and the recommended controller, $G_C(s)$, are given by:

$$G_P(s) = \frac{127e^{-0.2s}}{s(s+1)(s+2)(s+5)^2(s+10)}$$

$$G_C(s) = \frac{(92.9s^2 + 13.63s + 1)}{97.6s^2(0.1s+1)}$$

Simulink SL

Use Simulink to model this system and plot its response (from 0 to 300 seconds) to a unit-step reference input, *r(t)*, applied at *t* = 0, and (on the same graph) to a disturbance, *d(t)* = 0.25 *r(t)*, applied at *t* = 150 seconds. What are the values of the steady-state error due to the reference input and due to the disturbance? What about the relative stability of this Type 3 system as evidenced by the percent overshoot in response to the unit-step reference input?

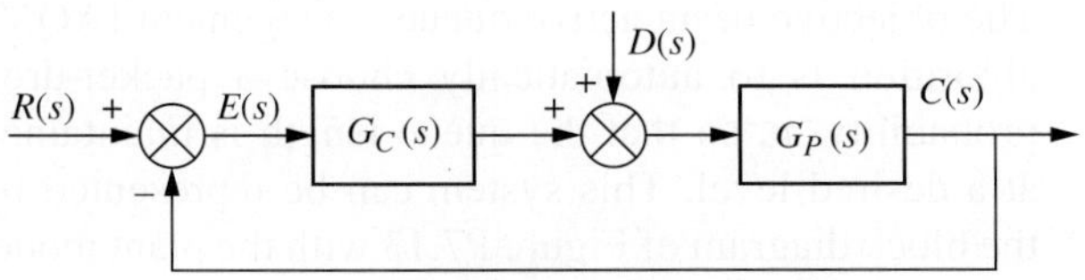


**FIGURE P7.29**

Simulink SL

**61.** A Type 3 feedback control system (*Papadopoulos, 2013*) is presented in Problem 60. Modify the Simulink model you develope in that problem to plot its response (from 0 to 100 seconds) to a unit-ramp reference input, *r(t)* = *tu(t)*, applied at *t* = 0, and (on the same graph) to a disturbance, *d(t)* = 0.25*tu(t)*, applied at *t* = 50 seconds. What are the values of the steady-state position error due to the reference-input and disturbance ramps?

Copy this model and paste it in the same file. Then, in that copy, change the reference input to a unit parabola, *r(t)* = 0.5 $t^2$ *u(t)*, applied at *t* = 0, and the disturbance to *d(t)* = 0.125$t^2$*u(t)*, applied at *t* = 50 seconds, and plot, on a new graph (Scope 1), the system's response to these parabolic signals.

## DESIGN PROBLEMS

**62.** The following specification applies to a position control: $K_v = 10$. On hand is an amplifier with a variable gain, $K_2$, with which to drive a motor. Two one-turn pots to convert shaft position into voltage are also available, where $\pm 3\pi$ volts are placed across the pots. A motor is available whose transfer function is

$$\frac{\theta_m(s)}{E_a(s)} = \frac{K}{s(s+\alpha)}$$

where $\theta_m(s)$ is the motor armature position and $E_a(s)$ is the armature voltage. The components are interconnected as shown in Figure P7.30.

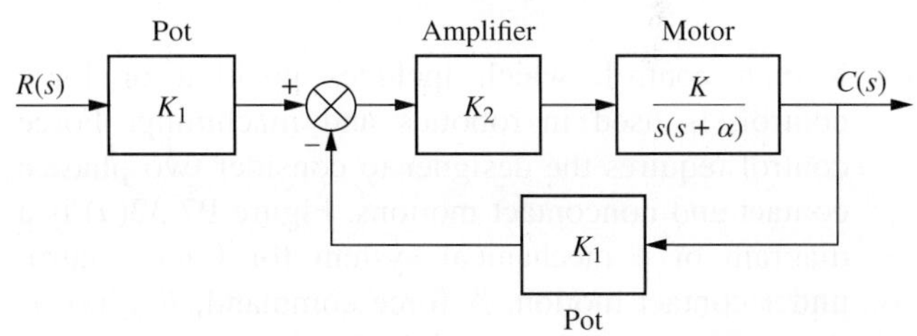


**FIGURE P7.30** Position control system

The transfer function of the motor is found experimentally as follows. The motor and load are driven separately by applying a large, short square wave (a unit impulse) to the armature. An oscillograph of the response shows that the motor reached 63% of its final output value 1 second after application of the impulse. Furthermore, with 10 volts dc applied to the armature, the constant output speed was 100 rad/s. Draw the completed block diagram of the system, specifying the transfer function of each component of the block diagram.

**63.** A boat is circling a ship that is using a tracking radar. The speed of the boat is 20 knots, and it is circling the ship at a distance of 1 nautical mile, as shown in Figure P7.31(*a*). A simplified model of the tracking system is shown in Figure P7.31(*b*). Find the value of *K* so that the boat is kept in the center of the radar beam with no more than 0.1 degree error.

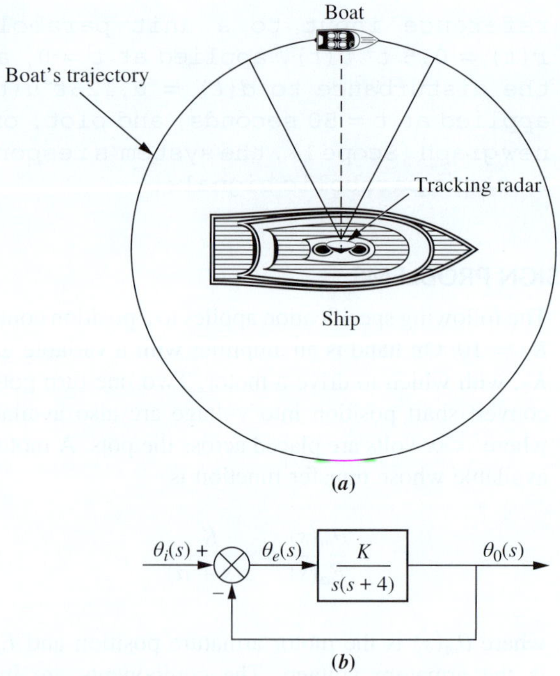


**FIGURE P7.31** Boat tracked by ship's radar: **a.** physical arrangement; **b.** block diagram of tracking system

**64.** Motion control, which includes position or force control, is used in robotics and machining. Force control requires the designer to consider two phases: contact and noncontact motions. Figure P7.32(*a*) is a diagram of a mechanical system for force control under contact motion. A force command, $F_{\text{cmd}}(s)$, is the input to the system, while the output, $F(s)$, is the controlled contact force.

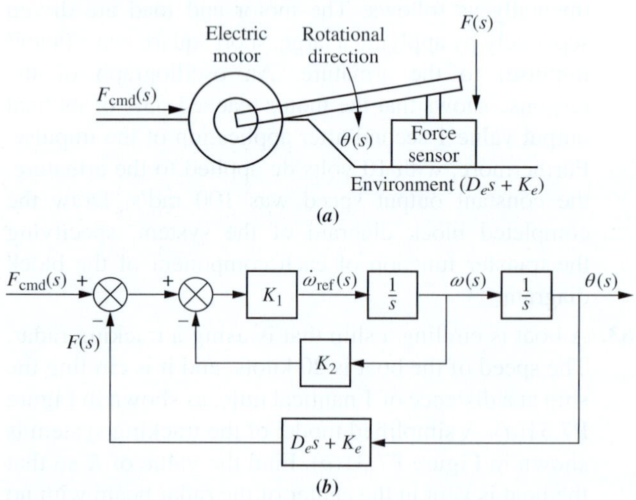


**FIGURE P7.32** **a.** Force control mechanical loop under contact motion;[10] **b.** block diagram[10]

In the figure a motor is used as the force actuator. The force output from the actuator is applied to the object through a force sensor. A block diagram representation of the system is shown in Figure P7.32(*b*). $K_2$ is velocity feedback used to improve the transient response. The loop is actually implemented by an electrical loop (not shown) that controls the armature current of the motor to yield the desired torque at the output. Recall that $T_m = K_t i_a$ (*Ohnishi, 1996*). Find an expression for the range of $K_2$ to keep the steady-state force error below 10% for ramp inputs of commanded force.

**65.** Packet information flow in a router working under TCP/IP can be modeled using the linearized transfer function

$$P(s) = \frac{Q(s)}{f(s)} = \frac{\dfrac{C^2}{2N} e^{-sR}}{\left(s + \dfrac{2N}{R^2 C}\right)\left(s + \dfrac{1}{R}\right)}$$

where

$C$ = link capacity (packets/second)

$N$ = load factor (number of TCP sessions)

$Q$ = expected queue length

$R$ = round trip time (second)

$p$ = probability of a packet drop

The objective of an active queue management (AQM) algorithm is to automatically choose a packet-drop probability, $p$, so that the queue length is maintained at a desired level. This system can be represented by the block diagram of Figure P7.13 with the plant model in the $P(s)$ block, the AQM algorithm in the $G(s)$ block, and $F(s) = H(s) = 1$. Several AQM algorithms are available, but one that has received special attention in the literature is the random early detection (RED) algorithm. This algorithm can be approximated with $G(s) = \dfrac{LK}{s+K}$, where $L$ and $K$ are constants (*Hollot, 2001*). Find the value of $L$ required to obtain a 10%

[10] Ohnishi, K.; Shibata, M.; and Murakami, T. Motion Controlfor Advanced Mechatronics, *IEEE/ASME Transactions on Mechatronics*, vol. 1, no. 1, March 1996, Figures 14 & 16, p. 62. IEEE/ASME transactions on mechatronics: a joint publication of the IEEE Industrial Electronics Society and the ASME Dynamic Systems and Control Division by Institute of Electrical and Electronics Engineers; IEEE Industrial Electronics Society; American Society of Mechanical Engineers. Dynamic Systems and Control Division Reproduced with permission of IEEE in the format Republish in a book via Copyright Clearance Center.

steady-state error for a unit step input when $C = 3750$ packets/s, $N = 50$ TCP sessions, $R = 0.1$ s, and $K = 0.005$.

## PROGRESSIVE ANALYSIS AND DESIGN PROBLEMS

**66. Control of HIV/AIDS.** Consider the HIV infection model of Problem 67 in Chapter 6 and its block diagram in Figure P6.17 (*Craig, 2004*).

**a.** Find the system's type if $G(s)$ is a constant.

**b.** It was shown in Problem 67, Chapter 6, that when $G(s) = K$ the system will be stable when $K < 2.04 \times 10^{-4}$. What value of $K$ will result in a unit step input steady-state error of 10%?

**c.** It is suggested that to reduce the steady-state error the system's type should be augmented by making $G(s) = \frac{K}{s}$. Is this a wise choice? What is the resulting stability range for $K$?

**67. Hybrid vehicle.** Figure P7.33 shows the block diagram of the speed control of an HEV taken from Figure P5.54, and rearranged as a unity feedback system (*Preitl, 2007*). Here the system output is $C(s) = K_{SS}V(s)$, the output voltage of the speed sensor/transducer.

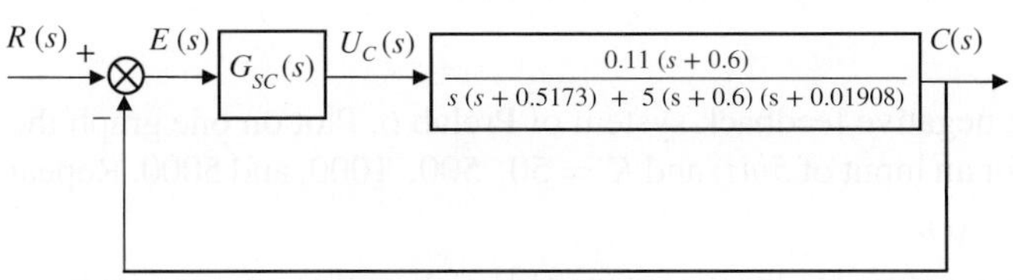


FIGURE P7.33

**a.** Assume the speed controller is given as $G_{SC}(s) = K_{P_{SC}}$. Find the gain, $K_{P_{SC}}$, that yields a steady-state error, $e_{\text{step}}(\infty) = 1\%$.

**b.** Now assume that in order to reduce the steady-state error for step inputs, integration is added to the controller yielding $G_{SC}(s) = K_{P_{SC}} + (K_{I_{SC}}/s) = 100 + (K_{I_{SC}}/s)$. Find the value of the integral gain, $K_{I_{SC}}$, that results in a steady-state error, $e_{\text{ramp}}(\infty) = 2.5\%$.

**c.** In Parts **a** and **b**, the HEV was assumed to be driven on level ground. Consider the case when, after reaching a steady-state speed with a controller given by $G_{SC}(s) = 100 + \frac{40}{s}$, the car starts climbing up a hill with a gradient angle, $\alpha = 5°$. For small angles $\sin\alpha = \alpha$ (in radians) and, hence, when reflected to the motor shaft the climbing torque is

$$T_{st} = \frac{F_{st}r}{i_{tot}} = \frac{mgr}{i_{tot}}\sin\alpha = \frac{mgr\alpha}{i_{tot}}$$

$$= \frac{1590 \times 9.8 \times 0.3 \times 5}{4.875 \times 57.3} = 83.7 \text{ Nm}.$$

The block diagram in Figure P7.34 represents the control system of the HEV rearranged for Part **c**.

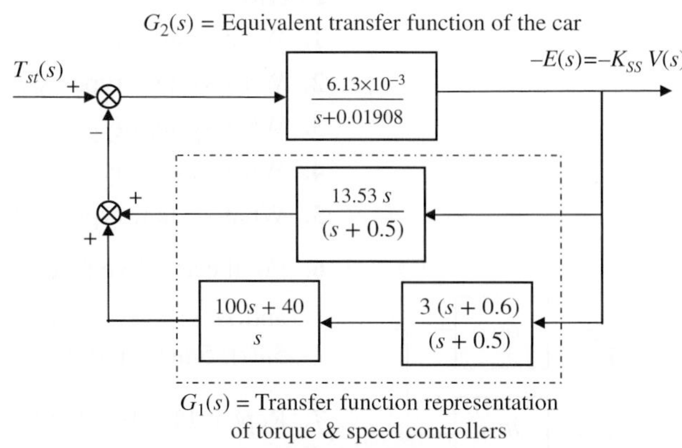


FIGURE P7.34

In this diagram, the input is $T_{st}(t) = 83.7\,u(t)$, corresponding to $\alpha = 5°$, and the output is the negative error, $-e(t) = -c(t) = -K_{SS}v(t)$, proportional to the change in car speed, $v(t)$. Find the steady-state error $e(\infty)$ due to a step change in the disturbance; e.g., the climbing torque, $T_{st}(t) = 83.7\,u(t)$.

**68. Parabolic trough collector.** The parabolic trough collector (*Camacho, 2012*) is embedded in a unit feedback configuration as shown in Figure P7.1, where $G(s) = G_C(s)P(s)$ and

$$P(s) = \frac{137.2 \times 10^{-6}}{S^2 + 0.0224s + 196 \times 10^{-6}}e^{-39s}$$

**a.** Assuming $G_C(s) = K$, find the value of $K$ required for a unit-step input steady-state error of 3%. Use the result you obtained in Problem 69, Chapter 6, to verify that the system is closed-loop stable when that value of $K$ is used.

**b.** What is the minimum unit-step input steady-state error achievable with $G_C(s) = K$?

**c.** What is the simplest compensator, $G_C(s)$, that can be used to achieve a steady-state error of 0%?

# Cyber Exploration Laboratory

## Experiment 7.1

**Objective** To verify the effect of input waveform, loop gain, and system type upon steady-state errors.

**Minimum Required Software Packages** MATLAB, Simulink, and the Control System Toolbox

### Prelab

1. What system types will yield zero steady-state error for step inputs?
2. What system types will yield zero steady-state error for ramp inputs?
3. What system types will yield infinite steady-state error for ramp inputs?
4. What system types will yield zero steady-state error for parabolic inputs?
5. What system types will yield infinite steady-state error for parabolic inputs?
6. For the negative feedback system of Figure P7.35, where $G(s) = \frac{K(s+6)}{(s+4)(s+7)(s+9)(s+12)}$ and $H(s) = 1$, calculate the steady-state error in terms of $K$ for the following inputs: $5u(t)$, $5tu(t)$, and $5t^2u(t)$.

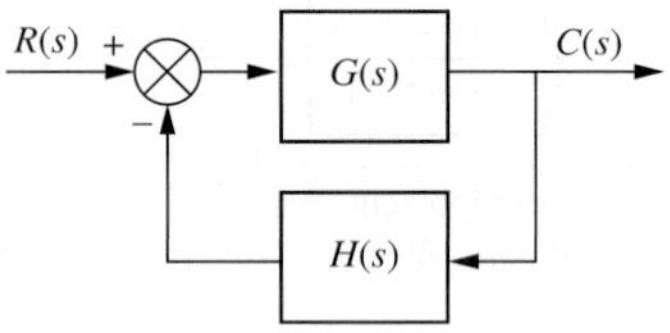


FIGURE P7.35

7. Repeat Prelab 6 for $G(s) = \dfrac{K(s+6)(s+8)}{s(s+4)(s+7)(s+9)(s+12)}$ and $H(s) = 1$.
8. Repeat Prelab 6 for $G(s) = \dfrac{K(s+1)(s+6)(s+8)}{s^2(s+4)(s+7)(s+9)(s+12)}$ and $H(s) = 1$.

### Lab

1. Using Simulink, set up the negative feedback system of Prelab 6. Plot on one graph the error signal of the system for an input of $5u(t)$ and $K = 50, 500, 1000,$ and $5000$. Repeat for inputs of $5tu(t)$ and $5t^2u(t)$.
2. Using Simulink, set up the negative feedback system of Prelab 7. Plot on one graph the error signal of the system for an input of $5u(t)$ and $K = 50, 500, 1000,$ and $5000$. Repeat for inputs of $5tu(t)$ and $5t^2u(t)$.
3. Using Simulink, set up the negative feedback system of Prelab 8. Plot on one graph the error signal of the system for an input of $5u(t)$ and $K = 200, 400, 800,$ and $1000$. Repeat for inputs of $5tu(t)$ and $5t^2u(t)$.

### Postlab

1. Use your plots from Lab 1 and compare the expected steady-state errors to those calculated in the Prelab. Explain the reasons for any discrepancies.
2. Use your plots from Lab 2 and compare the expected steady-state errors to those calculated in the Prelab. Explain the reasons for any discrepancies.
3. Use your plots from Lab 3 and compare the expected steady-state errors to those calculated in the Prelab. Explain the reasons for any discrepancies.

## Experiment 7.2

**Objective** To use the LabVIEW Control Design and Simulation Module for analysis of steady-state performance for step and ramp inputs.

**Minimum Required Software Package** LabVIEW with the Control Design and Simulation Module

**Prelab** You are given the model of a single joint of a robotic manipulator shown in Figure P7.36 (*Spong, 2005*), where $B$ is the coefficient of viscous friction, $\theta_d(s)$ is the desired angle, $\theta(s)$ is the output angle, and $D(s)$ is the disturbance. We want to track the joint angle using a PD controller, which we will study in Chapter 9. Assume $J = B = 1$. Find the step and ramp responses of this system for the following combinations of PD gains $(K_P, K_D)$: (16, 7), (64, 15), and (144, 23).

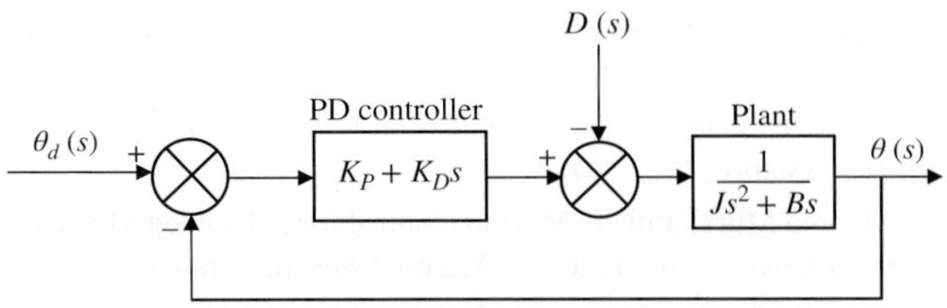


**FIGURE P7.36**

## Lab

1. Create a LabVIEW VI to simulate the response of this system to a step and a ramp inputs, under no-disturbance conditions. Use the functions available in the **Control Design and Simulation/Control Design** palette.
2. Create a LabVIEW VI using the functions available in the **Control Design and Simulation/Simulation** palette, to track an input set-point of 10 under a disturbance of D = 40.

**Postlab** Compare your results with those of the Prelab. What conclusions can you draw from the various responses of this system to different inputs and different PD parameters? What is the system type? Does the steady-state behavior corroborate the theory you learned regarding system type and the steady-state error for various inputs? Explain your answer.

# Bibliography


Barkana, I. Classical and Simple Adaptive Control of Nonminimum Phase Autopilot Design. *Journal of Guidance, Control, and Dynamics*, vol. 28, 2005, pp. 631–638.

Bylinkski, G. *Silicon Valley High Tech: Window to the Future*. Intercontinental Publishing Corp, Ltd., Hong Kong, 1985. Figure caption source for Figure 7.9.

Camacho, E. F., Berenguel, M., Rubio, F. R., and Martinez, D. *Control of Solar Energy Systems*. Springer-Verlag, London, 2012.

Chandra, F. A., Buzi, G., and Doyle, J. C. Glycolytic Oscillations and Limits on Robust Efficiency. *Science,* vol. 333, American Association for the Advancement of Science, July 8, 2011, pp. 187–192.

Craig, I. K., Xia, X., and Venter, J. W. Introducing HIV/AIDS Education into the Electrical Engineering Curriculum at the University of Pretoria. *IEEE Transactions on Education*, vol. 47, no. 1, February 2004, pp. 65–73.

D'Azzo, J. J., and Houpis, C. H. *Feedback Control System Analysis and Design Conventional and Modern*, 3d ed. McGraw-Hill, New York, 1988.

Hollot, C. V., Misra, V., Towsley, D., and Gong, W. A Control Theoretic Analysis of RED. *Proceedings of IEEE INFOCOM*, 2001, pp. 1510–1519.

Hostetter, G. H., Savant, C. J., Jr., and Stefani, R. T. *Design of Feedback Control Systems*, 2d ed. Saunders College Publishing, New York, 1989.

Isailović, J. *Videodisc and Optical Memory Systems*. Prentice Hall, Upper Saddle River, NJ, 1985.

Kuo, C.-F. J., Tu, H.-M., and Liu, C.-H. Dynamic Modeling and Control of a Current New Horizontal Type Cross-Lapper Machine. *Textile Research Journal*, vol. 80, no. 19, Sage, 2010, pp. 2016–2027.

Lam, C. S., Wong, M. C., and Han, Y. D. Stability Study on Dynamic Voltage Restorer (DVR). *Power Electronics Systems and Applications, 2004; Proceedings First International Conference on Power Electronics*, 2004, pp. 66–71.

Lin, J.-S., and Kanellakopoulos, I. Nonlinear Design of Active Suspensions. *IEEE Control Systems*, vol. 17, issue 3, June 1997, pp. 45–59.

Low, K. H., Wang, H., Liew, K. M., and Cai, Y. Modeling and Motion Control of Robotic Hand for Telemanipulation Application. *International Journal of Software Engineering and Knowledge Engineering*, vol. 15, 2005, pp. 147–152.

Mitchell, R. J. More Nested Velocity Feedback Control. *IEEE 9th International Conference on Cybernetic Intelligent Systems (CIS)*, 2010.

Ohnishi, K., Shibata, M., and Murakami, T. Motion Control for Advanced Mechatronics. *IEEE/ASME Transactions on Mechatronics*, vol. 1, no. 1, March 1996, pp. 56–67.

Papadopoulos, K . G., Papastefanaki, E. N., and Margaris, N. I. Explicit Analytical PID Tuning Rules for the Design of Type-III Control Loops. *IEEE Transactions on Industrial Electronics*, vol. 60, no. 10, October 2013, pp. 4650–4664.

Preitl, Z., Bauer, P., and Bokor, J. A Simple Control Solution for Traction Motor Used in Hybrid Vehicles. *Fourth International Symposium on Applied Computational Intelligence and Informatics*, IEEE, 2007, pp. 157–162.

Schneider, R. T. Pneumatic Robots Continue to Improve. *Hydraulics & Pneumatics*, October 1992, pp. 38–39.

Spong, M., Hutchinson, S., and Vidyasagar, M. *Robot Modeling and Control*. John Wiley & Sons. Hoboken, NJ, 2006.

Yin, G., Chen, N., and Li, P. Improving Handling Stability Performance of Four-Wheel Steering Vehicle via $\mu$-Synthesis Robust Control. *IEEE Transactions on Vehicular Technology*, vol. 56, no. 5, 2007, pp. 2432–2439.

# 8

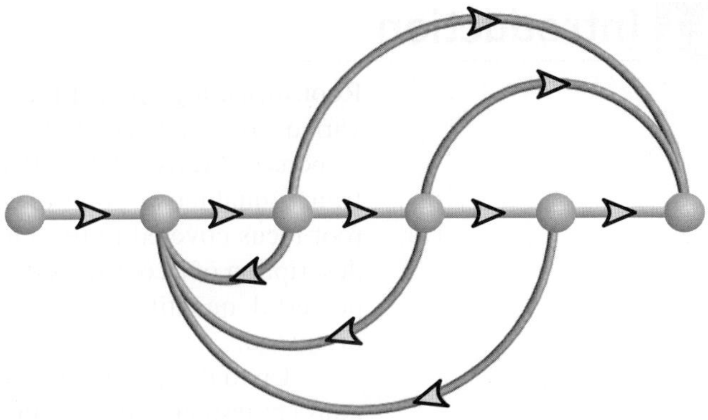

# Root Locus Techniques

## Chapter Learning Outcomes

After completing this chapter the student will be able to:

- Define a root locus (Sections 8.1–8.2)
- State the properties of a root locus (Section 8.3)
- Sketch a root locus (Section 8.4)
- Find the coordinates of points on the root locus and their associated gains (Sections 8.5–8.6)
- Use the root locus to design a parameter value to meet a transient response specification for systems of order 2 and higher (Sections 8.7–8.8)
- Sketch the root locus for positive-feedback systems (Section 8.9)
- Find the root sensitivity for points along the root locus (Section 8.10)

## ▶ Case Study Learning Outcomes

You will be able to demonstrate your knowledge of the chapter objectives with case studies as follows:

- Given the antenna azimuth position control system shown on the front endpapers, you will be able to find the preamplifier gain to meet a transient response specification.
- Given the pitch or heading control system for the Unmanned Free-Swimming Submersible vehicle shown on the back endpapers, you will be able to plot the root locus and design the gain to meet a transient response specification. You will then be able to evaluate other performance characteristics.

# 8.1 Introduction

Root locus, a graphical presentation of the closed-loop poles as a system parameter is varied, is a powerful method of analysis and design for stability and transient response (*Evans, 1948, 1950*). Feedback control systems are difficult to comprehend from a qualitative point of view, and hence they rely heavily upon mathematics. The root locus covered in this chapter is a graphical technique that gives us the qualitative description of a control system's performance that we are looking for and also serves as a powerful quantitative tool that yields more information than the methods already discussed.

Up to this point, gains and other system parameters were designed to yield a desired transient response for only first- and second-order systems. Even though the root locus can be used to solve the same kind of problem, its real power lies in its ability to provide solutions for systems of order higher than 2. For example, under the right conditions, a fourth-order system's parameters can be designed to yield a given percent overshoot and settling time using the concepts learned in Chapter 4.

The root locus can be used to describe qualitatively the performance of a system as various parameters are changed. For example, the effect of varying gain upon percent overshoot, settling time, and peak time can be vividly displayed. The qualitative description can then be verified with quantitative analysis.

Besides transient response, the root locus also gives a graphical representation of a system's stability. We can clearly see ranges of stability, ranges of instability, and the conditions that cause a system to break into oscillation.

Before presenting root locus, let us review two concepts that we need for the ensuing discussion: (1) the control system problem and (2) complex numbers and their representation as vectors.

## The Control System Problem

We have previously encountered the control system problem in Chapter 6: Whereas the poles of the open-loop transfer function are easily found (typically, they are known by inspection and do not change with changes in system gain), the poles of the closed-loop transfer function are more difficult to find (typically, they cannot be found without factoring the closed-loop system's characteristic polynomial, the denominator of the closed-loop transfer function), and further, the closed-loop poles change with changes in system gain.

A typical closed-loop feedback control system is shown in Figure 8.1($a$). The open-loop transfer function was defined in Chapter 5 as $KG(s)H(s)$. Ordinarily, we can determine the poles of $KG(s)H(s)$, since these poles arise from simple cascaded first- or second-order subsystems. Further, variations in $K$ do not affect the location of any pole of this function. On the other hand, we cannot determine the poles of $T(s) = KG(s)/[1 + KG(s)H(s)]$ unless we factor the denominator. Also, the poles of $T(s)$ change with $K$.

Let us demonstrate. Letting

$$G(s) = \frac{N_G(s)}{D_G(s)} \tag{8.1}$$

and

$$H(s) = \frac{N_H(s)}{D_H(s)} \tag{8.2}$$

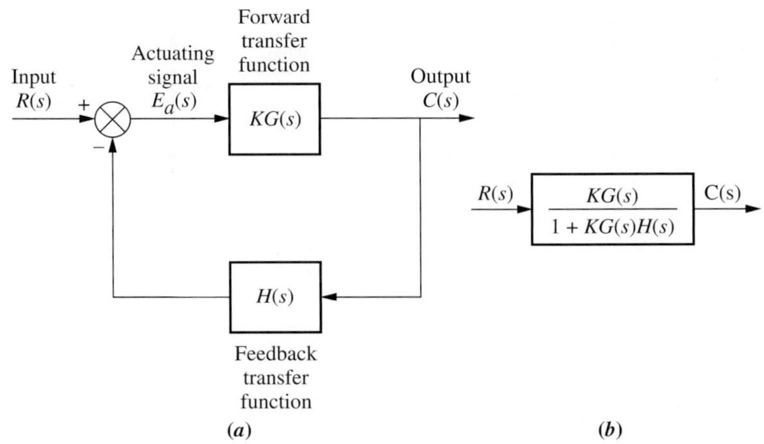


**FIGURE 8.1** **a.** Closed-loop system; **b.** equivalent transfer function

then

$$T(s) = \frac{KN_G(s)D_H(s)}{D_G(s)D_H(s) + KN_G(s)N_H(s)} \tag{8.3}$$

where $N$ and $D$ are factored polynomials and signify numerator and denominator terms, respectively. We observe the following: Typically, we know the factors of the numerators and denominators of $G(s)$ and $H(s)$. Also, the zeros of $T(s)$ consist of the zeros of $G(s)$ and the poles of $H(s)$. The poles of $T(s)$ are not immediately known and in fact can change with $K$. For example, if $G(s) = (s+1)/[s(s+2)]$ and $H(s) = (s+3)/(s+4)$, the poles of $KG(s)H(s)$ are $0, -2$, and $-4$. The zeros of $KG(s)H(s)$ are $-1$ and $-3$. Now, $T(s) = K(s+1)(s+4)/[s^3 + (6+K)s^2 + (8+4K)s + 3K]$. Thus, the zeros of $T(s)$ consist of the zeros of $G(s)$ and the poles of $H(s)$. The poles of $T(s)$ are not immediately known without factoring the denominator, and they are a function of $K$. Since the system's transient response and stability are dependent upon the poles of $T(s)$, we have no knowledge of the system's performance unless we factor the denominator for specific values of $K$. The root locus will be used to give us a vivid picture of the poles of $T(s)$ as $K$ varies.

## Vector Representation of Complex Numbers

Any *complex number*, $\sigma + j\omega$, described in Cartesian coordinates can be graphically represented by a vector, as shown in Figure 8.2($a$). The complex number also can be described in polar form with magnitude $M$ and angle $\theta$, as $M\angle\theta$. If the complex number is substituted into a complex function, $F(s)$, another complex number will result. For example, if $F(s) = (s+a)$, then substituting the complex number $s = \sigma + j\omega$ yields $F(s) = (\sigma + a) + j\omega$, another complex number. This number is shown in Figure 8.2($b$). Notice that $F(s)$ has a zero at $-a$. If we translate the vector $a$ units to the left, as in Figure 8.2($c$), we have an alternate representation of the complex number that originates at the zero of $F(s)$ and terminates on the point $s = \sigma + j\omega$.

We conclude that $(s+a)$ *is a complex number and can be represented by a vector drawn from the zero of the function to the point s*. For example, $(s+7)|_{s\to 5+j2}$ is a complex number drawn from the zero of the function, $-7$, to the point $s$, which is $5 + j2$, as shown in Figure 8.2($d$).

Now let us apply the concepts to a complicated function. Assume a function

$$F(s) = \frac{\prod_{i=1}^{m}(s+z_i)}{\prod_{j=1}^{n}(s+p_j)} = \frac{\prod \text{numerator's complex factors}}{\prod \text{denominator's complex factors}} \tag{8.4}$$

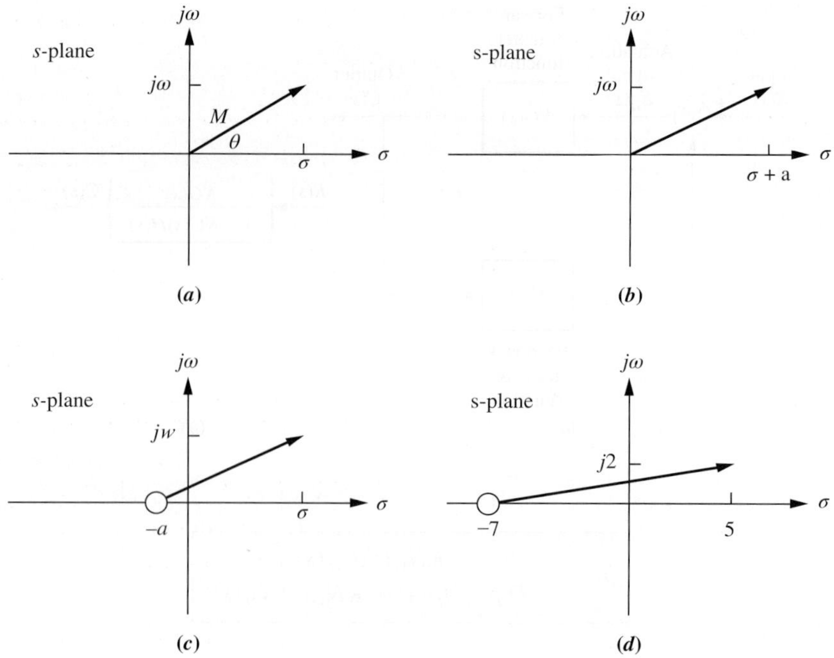


**FIGURE 8.2** Vector representation of complex numbers: **a.** $s = \sigma + j\omega$; **b.** $(s + a)$; **c.** alternate representation of $(s + a)$; **d.** $(s + 7)|_{s \rightarrow 5+j2}$

where the symbol Π means "product," $m$ = number of zeros, and $n$ = number of poles. Each factor in the numerator and each factor in the denominator is a complex number that can be represented as a vector. The function defines the complex arithmetic to be performed in order to evaluate $F(s)$ at any point, $s$. Since each complex factor can be thought of as a vector, the magnitude, $M$, of $F(s)$ at any point, $s$, is

$$M = \frac{\Pi \text{ zero lengths}}{\Pi \text{ pole lengths}} = \frac{\prod_{i=1}^{m} |(s + z_i)|}{\prod_{j=1}^{n} |(s + p_j)|} \tag{8.5}$$

where a zero length, $|(s + z_i)|$, is the magnitude of the vector drawn from the zero of $F(s)$ at $-z_i$ to the point $s$, and a pole length, $|(s + p_j)|$, is the magnitude of the vector drawn from the pole of $F(s)$ at $-p_j$ to the point $s$. The angle, $\theta$, of $F(s)$ at any point, $s$, is

$$\begin{aligned} \theta &= \sum \text{zero angles} - \sum \text{pole angles} \\ &= \sum_{i=1}^{m} \angle(s + z_i) - \sum_{j=1}^{n} \angle(s + p_j) \end{aligned} \tag{8.6}$$

where a zero angle is the angle, measured from the positive extension of the real axis, of a vector drawn from the zero of $F(s)$ at $-z_i$ to the point $s$, and a pole angle is the angle, measured from the positive extension of the real axis, of the vector drawn from the pole of $F(s)$ at $-p_j$ to the point $s$.

As a demonstration of Eqs. (8.5) and (8.6), consider the following example.

## Example 8.1 Evaluation of a Complex Function via Vectors

**PROBLEM:** Given

$$F(s) = \frac{(s+1)}{s(s+2)} \tag{8.7}$$

find $F(s)$ at the point $s = -3 + j4$.

**SOLUTION:** The problem is graphically depicted in Figure 8.3, where each vector, $(s+\alpha)$, of the function is shown terminating on the selected point $s = -3 + j4$. The vector originating at the zero at $-1$ is

$$\sqrt{20}\angle 116.6^\circ \tag{8.8}$$

The vector originating at the pole at the origin is

$$5\angle 126.9^\circ \tag{8.9}$$

The vector originating at the pole at $-2$ is

$$\sqrt{17}\angle 104.0^\circ \tag{8.10}$$

Substituting Eqs. (8.8) through (8.10) into Eqs. (8.5) and (8.6) yields

$$M\angle\theta = \frac{\sqrt{20}}{5\sqrt{17}}\angle 116.6^\circ - 126.9^\circ - 104.0^\circ = 0.217\angle - 114.3^\circ \tag{8.11}$$

as the result for evaluating $F(s)$ at the point $-3 + j4$.

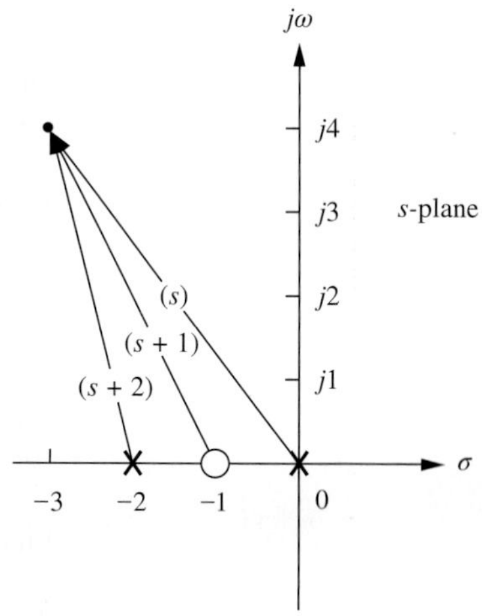


**FIGURE 8.3** Vector representation of Eq. 8.7

## Skill-Assessment Exercise 8.1

**PROBLEM:** Given

$$F(s) = \frac{(s+2)(s+4)}{s(s+3)(s+6)}$$

find $F(s)$ at the point $s = -7 + j9$ the following ways:

**a.** Directly substituting the point into $F(s)$

**b.** Calculating the result using vectors

**ANSWER:**

$$-0.0339 - j0.0899 = 0.096\angle - 110.7^\circ$$

The complete solution is at www.wiley.com/college/nise.

**TryIt 8.1**

Use the following MATLAB statements to solve the problem given in Skill-Assessment Exercise 8.1.

```
s=-7+9j;
G=(s+2)*(s+4)/...
 (s*(s+3)*(s+6));
Theta=(180/pi)*...
 angle(G)
M=abs(G)
```

We are now ready to begin our discussion of the root locus.

## 8.2 Defining the Root Locus

A security camera system similar to that shown in Figure 8.4(*a*) can automatically follow a subject. The tracking system monitors pixel changes and positions the camera to center the changes.

The root locus technique can be used to analyze and design the effect of loop gain upon the system's transient response and stability. Assume the block diagram representation of a tracking system as shown in Figure 8.4(*b*), where the closed-loop poles of the system change location as the gain, $K$, is varied. Table 8.1, which was formed by applying the quadratic formula to the denominator of the transfer function in Figure 8.4(*c*), shows the variation of pole location for different values of gain, $K$. The data of Table 8.1 is graphically displayed in Figure 8.5(*a*), which shows each pole and its gain.

As the gain, $K$, increases in Table 8.1 and Figure 8.5(*a*), the closed-loop pole, which is at $-10$ for $K = 0$, moves toward the right, and the closed-loop pole, which is at 0 for $K = 0$, moves toward the left. They meet at $-5$, break away from the real axis, and move into the complex plane. One closed-loop pole moves upward while the other moves downward. We cannot tell which pole moves up or which moves down. In Figure 8.5(*b*), the individual closed-loop pole locations are removed and their paths are represented with solid lines. It is this *representation of the paths of the closed-loop poles as the gain is varied* that we call a *root locus*. For most of our work, the discussion will be limited to positive gain, or $K \geq 0$.

The root locus shows the changes in the transient response as the gain, $K$, varies. First of all, the poles are real for gains less than 25. Thus, the system is overdamped. At a gain of 25, the poles are real and multiple and hence critically damped. For gains above 25, the system is underdamped. Even though these preceding conclusions were available

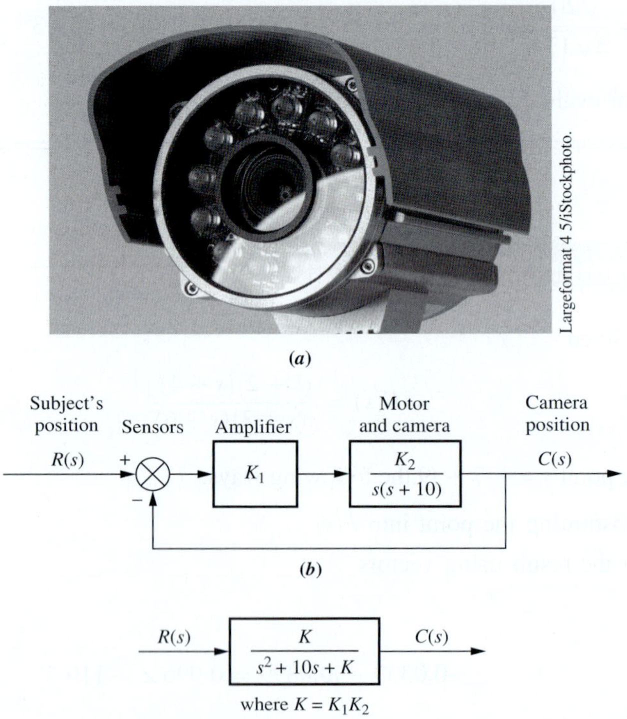


**FIGURE 8.4** **a.** Security cameras with auto tracking can be used to follow moving objects automatically; **b.** block diagram; **c.** closed-loop transfer function

**TABLE 8.1** Pole location as function of gain for the system of Figure 8.4

| $K$ | Pole 1 | Pole 2 |
|---|---|---|
| 0 | $-10$ | 0 |
| 5 | $-9.47$ | $-0.53$ |
| 10 | $-8.87$ | $-1.13$ |
| 15 | $-8.16$ | $-1.84$ |
| 20 | $-7.24$ | $-2.76$ |
| 25 | $-5$ | $-5$ |
| 30 | $-5+j2.24$ | $-5-j2.24$ |
| 35 | $-5+j3.16$ | $-5-j3.16$ |
| 40 | $-5+j3.87$ | $-5-j3.87$ |
| 45 | $-5+j4.47$ | $-5-j4.47$ |
| 50 | $-5+j5$ | $-5-j5$ |

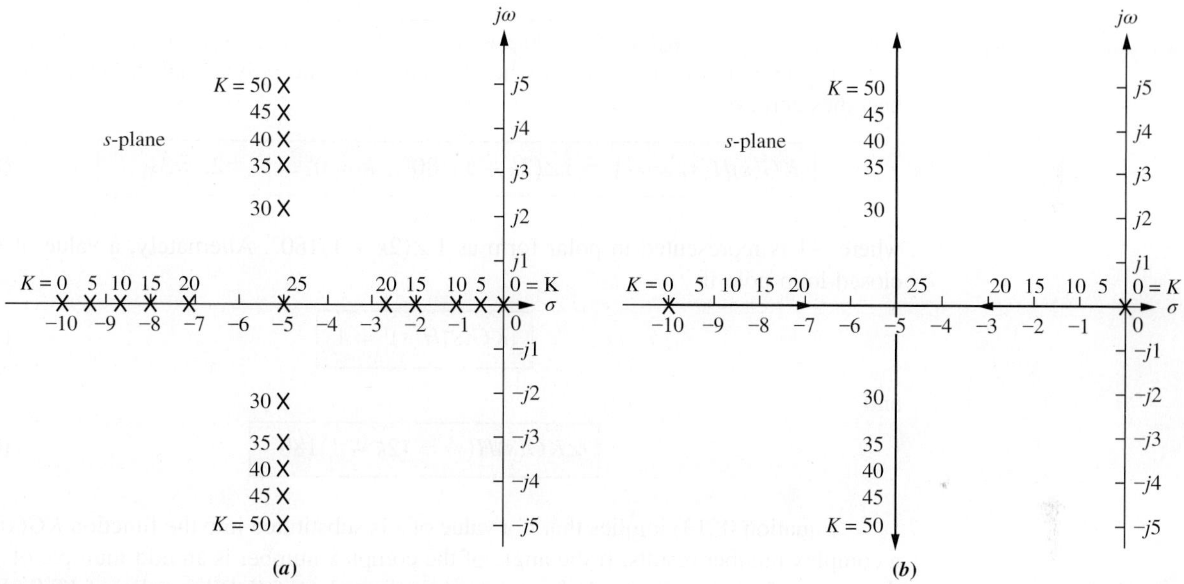


**FIGURE 8.5** **a.** Pole plot from Table 8.1; **b.** root locus

through the analytical techniques covered in Chapter 4, the following conclusions are graphically demonstrated by the root locus.

Directing our attention to the underdamped portion of the root locus, we see that regardless of the value of gain, the real parts of the complex poles are always the same. Since the settling time is inversely proportional to the real part of the complex poles for this second-order system, the conclusion is that regardless of the value of gain, the settling time for the system remains the same under all conditions of underdamped responses.

Also, as we increase the gain, the damping ratio diminishes, and the percent overshoot increases. The damped frequency of oscillation, which is equal to the imaginary part of the pole, also increases with an increase in gain, resulting in a reduction of the peak time. Finally, since the root locus never crosses over into the right half-plane, the system is always stable, regardless of the value of gain, and can never break into a sinusoidal oscillation.

These conclusions for such a simple system may appear to be trivial. What we are about to see is that the analysis is applicable to systems of order higher than two. For these systems, it is difficult to tie transient response characteristics to the pole location. The root locus will allow us to make that association and will become an important technique in the analysis and design of higher-order systems.

## 8.3 Properties of the Root Locus

In Section 8.2, we arrived at the root locus by factoring the second-order polynomial in the denominator of the transfer function. Consider what would happen if that polynomial were of fifth or tenth order. Without a computer, factoring the polynomial would be quite a problem for numerous values of gain.

We are about to examine the properties of the root locus. From these properties we will be able to make a rapid *sketch* of the root locus for higher-order systems without having to factor the denominator of the closed-loop transfer function.

The properties of the root locus can be derived from the general control system of Figure 8.1(*a*). The closed-loop transfer function for the system is

$$T(s) = \frac{KG(s)}{1 + KG(s)H(s)} \tag{8.12}$$

From Eq. (8.12), a pole, $s$, exists when the characteristic polynomial in the denominator becomes zero, or

$$KG(s)H(s) = -1 = 1\angle(2k+1)180^\circ \quad k = 0, \pm 1, \pm 2, \pm 3, \ldots \tag{8.13}$$

where $-1$ is represented in polar form as $1\angle(2k+1)180^\circ$. Alternately, a value of $s$ is a closed-loop pole if

$$|KG(s)H(s)| = 1 \tag{8.14}$$

and

$$\angle KG(s)H(s) = (2k+1)180^\circ \tag{8.15}$$

Equation (8.13) implies that if a value of $s$ is substituted into the function $KG(s)H(s)$, a complex number results. If the angle of the complex number is an odd multiple of 180°, that value of $s$ is a system pole for some particular value of $K$. What value of $K$? Since the angle criterion of Eq. (8.15) is satisfied, all that remains is to satisfy the magnitude criterion, Eq. (8.14). Thus,

$$K = \frac{1}{|G(s)||H(s)|} \tag{8.16}$$

We have just found that a pole of the closed-loop system causes the angle of $KG(s)H(s)$, or simply $G(s)H(s)$ since $K$ is a scalar, to be an odd multiple of 180°. Furthermore, the magnitude of $KG(s)H(s)$ must be unity, implying that the value of $K$ is the reciprocal of the magnitude of $G(s)H(s)$ when the pole value is substituted for $s$.

Let us demonstrate this relationship for the second-order system of Figure 8.4. The fact that closed-loop poles exist at $-9.47$ and $-0.53$ when the gain is 5 has already been established in Table 8.1. For this system,

$$KG(s)H(s) = \frac{K}{s(s+10)} \tag{8.17}$$

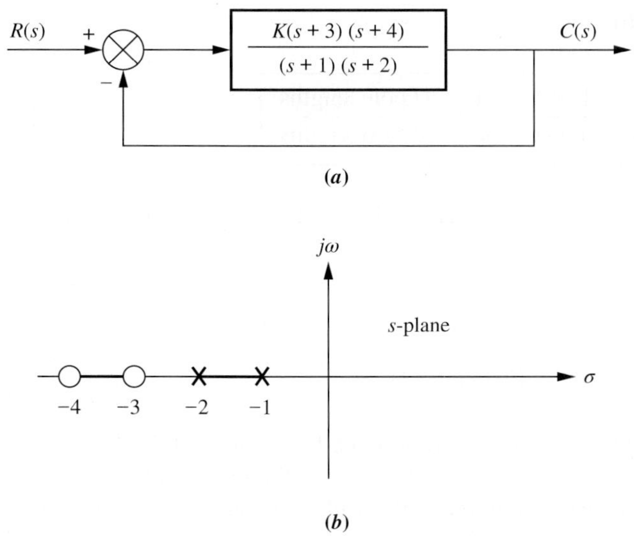


**FIGURE 8.6** **a.** Example system; **b.** pole-zero plot of $G(s)$

Substituting the pole at $-9.47$ for $s$ and 5 for $K$ yields $KG(s)H(s) = -1$. The student can repeat the exercise for other points in Table 8.1 and show that each case yields $KG(s)H(s) = -1$.

It is helpful to visualize graphically the meaning of Eq. (8.15). Let us apply the complex number concepts reviewed in Section 8.1 to the root locus of the system shown in Figure 8.6. For this system the open-loop transfer function is

$$KG(s)H(s) = \frac{K(s+3)(s+4)}{(s+1)(s+2)} \tag{8.18}$$

The closed-loop transfer function, $T(s)$, is

$$T(s) = \frac{K(s+3)(s+4)}{(1+K)s^2 + (3+7K)s + (2+12K)} \tag{8.19}$$

If point $s$ is a closed-loop system pole for some value of gain, $K$, then $s$ must satisfy Eqs. (8.14) and (8.15).

Consider the point $-2+j3$. If this point is a closed-loop pole for some value of gain, then the angles of the zeros minus the angles of the poles must equal an odd multiple of 180°. From Figure 8.7,

$$\begin{aligned}\theta_1 + \theta_2 - \theta_3 - \theta_4 &= 56.31° + 71.57° - 90° - 108.43° \\ &= -70.55° \end{aligned} \tag{8.20}$$

Therefore, $-2+j3$ is not a point on the root locus, or alternatively, $-2+j3$ is not a closed-loop pole for any gain.

If these calculations are repeated for the point $-2+j(\sqrt{2}/2)$, the angles do add up to 180°. That is, $-2+j(\sqrt{2}/2)$ is a point on the root locus for some value of gain. We now proceed to evaluate that value of gain.

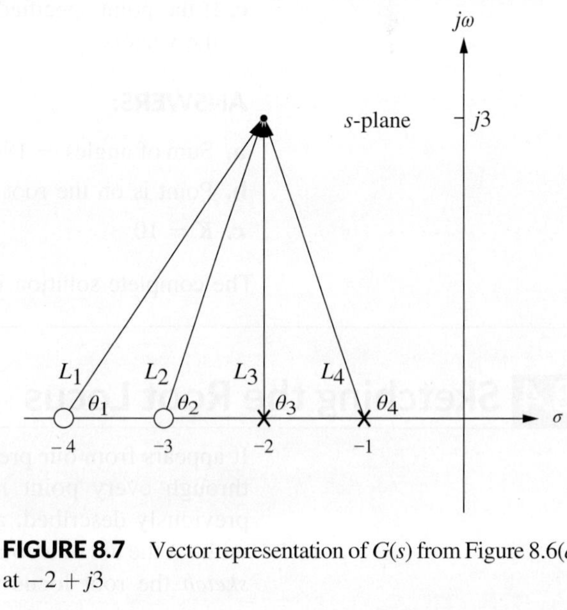


**FIGURE 8.7** Vector representation of $G(s)$ from Figure 8.6($a$) at $-2+j3$

From Eqs. (8.5) and (8.16),

$$K = \frac{1}{|G(s)H(s)|} = \frac{1}{M} = \frac{\Pi \text{ pole lengths}}{\Pi \text{ zero lengths}} \tag{8.21}$$

Looking at Figure 8.7 with the point $-2 + j3$ replaced by $-2 + j(\sqrt{2}/2)$, the gain, $K$, is calculated as

$$K = \frac{L_3 L_4}{L_1 L_2} = \frac{\frac{\sqrt{2}}{2}(1.22)}{(2.12)(1.22)} = 0.33 \tag{8.22}$$

Thus, the point $-2 + j(\sqrt{2}/2)$ is a point on the root locus for a gain of 0.33.

We summarize what we have found as follows: Given the poles and zeros of the open-loop transfer function, $KG(s)H(s)$, a point in the $s$-plane is on the root locus for a particular value of gain, $K$, if the angles of the zeros minus the angles of the poles, all drawn to the selected point on the $s$-plane, add up to $(2k + 1)180°$. Furthermore, gain $K$ at that point for which the angles add up to $(2k + 1)180°$ is found by dividing the product of the pole lengths by the product of the zero lengths.

## Skill-Assessment Exercise 8.2

**TryIt 8.2**

Use MATLAB and the following statements to solve Skill-Assessment Exercise 8.2.

```
s=-3+0j;
G=(s+2)/(s^2+4*s+13);
Theta=(180/pi)*...
 angle(G)
M=abs(G);
K=1/M
```

**PROBLEM:** Given a unity feedback system that has the forward transfer function

$$G(s) = \frac{K(s+2)}{(s^2 + 4s + 13)}$$

do the following:

**a.** Calculate the angle of $G(s)$ at the point $(-3 + j0)$ by finding the algebraic sum of angles of the vectors drawn from the zeros and poles of $G(s)$ to the given point.

**b.** Determine if the point specified in **a** is on the root locus.

**c.** If the point specified in **a** is on the root locus, find the gain, $K$, using the lengths of the vectors.

**ANSWERS:**

**a.** Sum of angles $= 180°$

**b.** Point is on the root locus

**c.** $K = 10$

The complete solution is at www.wiley.com/college/nise.

## 8.4 Sketching the Root Locus

It appears from our previous discussion that the root locus can be obtained by sweeping through every point in the $s$-plane to locate those points for which the angles, as previously described, add up to an odd multiple of 180°. Although this task is tedious without the aid of a computer, the concept can be used to develop rules that can be used to *sketch* the root locus without the effort required to *plot* the locus. Once a sketch is obtained, it is possible to accurately plot just those points that are of interest to us for a particular problem.

The following five rules allow us to sketch the root locus using minimal calculations. The rules yield a sketch that gives intuitive insight into the behavior of a control system. In the next section, we refine the sketch by finding actual points or angles on the root locus. These refinements, however, require some calculations or the use of computer programs, such as MATLAB.

1. **Number of branches.** Each closed-loop pole moves as the gain is varied. If we define a *branch* as the path that one pole traverses, then there will be one branch for each closed-loop pole. Our first rule, then, defines the number of branches of the root locus:

   *The number of branches of the root locus equals the number of closed-loop poles.*

   As an example, look at Figure 8.5(*b*), where the two branches are shown. One originates at the origin, the other at $-10$.

2. **Symmetry.** If complex closed-loop poles do not exist in conjugate pairs, the resulting polynomial, formed by multiplying the factors containing the closed-loop poles, would have complex coefficients. Physically realizable systems cannot have complex coefficients in their transfer functions. Thus, we conclude:

   *The root locus is symmetrical about the real axis.*

   An example of symmetry about the real axis is shown in Figure 8.5(*b*).

3. **Real-axis segments.** Let us make use of the angle property, Eq. (8.15), of the points on the root locus to determine where the real-axis segments of the root locus exist. Figure 8.8 shows the poles and zeros of a general open-loop system. If an attempt is made to calculate the angular contribution of the poles and zeros at each point, $P_1$, $P_2$, $P_3$, and $P_4$, along the real axis, we observe the following: (1) At each point the angular contribution of a pair of open-loop complex poles or zeros is zero, and (2) the contribution of the open-loop poles and open-loop zeros to the left of the respective point is zero. The conclusion is that the only contribution to the angle at any of the points comes from the open-loop, real-axis poles and zeros that exist to the right of the respective point. If we calculate the angle at each point using only the open-loop, real-axis poles and zeros to the right of each point, we note the following: (1) The angles on the real axis alternate between 0° and 180°, and (2) the angle is 180° for regions of the real axis that exist to the left of an odd number of poles and/or zeros. The following rule summarizes the findings:

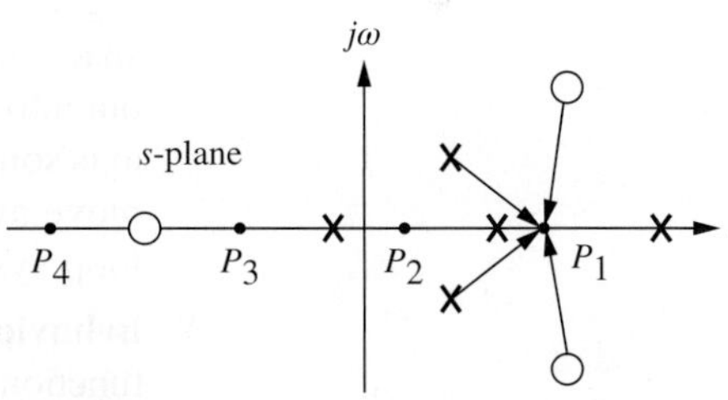

**FIGURE 8.8** Poles and zeros of a general open-loop system with test points, $P_i$, on the real axis

   *On the real axis, for $K > 0$ the root locus exists to the left of an odd number of real-axis, finite open-loop poles and/or finite open-loop zeros.*

   Examine Figure 8.6(*b*). According to the rule just developed, the real-axis segments of the root locus are between $-1$ and $-2$ and between $-3$ and $-4$ as shown in Figure 8.9.

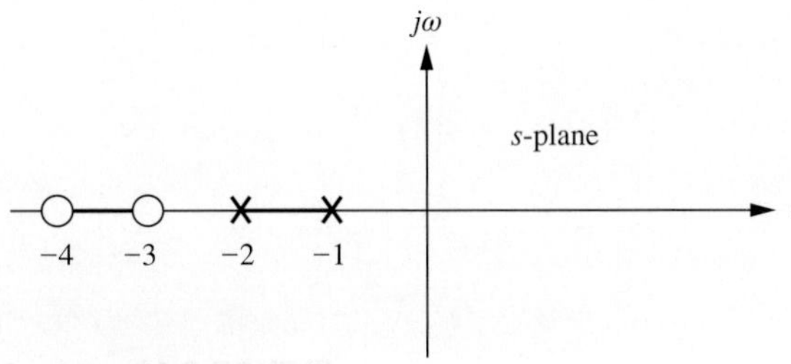

**FIGURE 8.9** Real-axis segments of the root locus for the system of Figure 8.6

4. **Starting and ending points.** Where does the root locus begin (zero gain) and end (infinite gain)? The answer to this question will enable us to expand the sketch of the root locus beyond the real-axis segments. Consider the closed-loop transfer function,

$T(s)$, described by Eq. (8.3). $T(s)$ can now be evaluated for both large and small gains, $K$. As $K$ approaches zero (small gain),

$$T(s) \approx \frac{KN_G(s)D_H(s)}{D_G(s)D_H(s) + \epsilon} \tag{8.23}$$

From Eq. (8.23) we see that the closed-loop system poles at small gains approach the combined poles of $G(s)$ and $H(s)$. We conclude that the root locus begins at the poles of $G(s)H(s)$, the open-loop transfer function.

At high gains, where $K$ is approaching infinity,

$$T(s) \approx \frac{KN_G(s)D_H(s)}{\epsilon + KN_G(s)N_H(s)} \tag{8.24}$$

From Eq. (8.24) we see that the closed-loop system poles at large gains approach the combined zeros of $G(s)$ and $H(s)$. Now we conclude that the root locus ends at the zeros of $G(s)H(s)$, the open-loop transfer function.

Summarizing what we have found:

*The root locus begins at the finite and infinite poles of G(s)H(s) and ends at the finite and infinite zeros of G(s)H(s).*

Remember that these poles and zeros are the open-loop poles and zeros.

In order to demonstrate this rule, look at the system in Figure 8.6(*a*), whose real-axis segments have been sketched in Figure 8.9. Using the rule just derived, we find that the root locus begins at the poles at −1 and −2 and ends at the zeros at −3 and −4 (see Figure 8.10). Thus, the poles start out at −1 and −2 and move through the real-axis space between the two poles. They meet somewhere between the two poles and break out into the complex plane, moving as complex conjugates. The poles return to the real axis somewhere between the zeros at −3 and −4, where their path is completed as they move away from each other, and end up, respectively, at the two zeros of the open-loop system at −3 and −4.

5. **Behavior at infinity.** Consider applying Rule 4 to the following open-loop transfer function:

$$KG(s)H(s) = \frac{K}{s(s+1)(s+2)} \tag{8.25}$$

There are three finite poles, at $s = 0, -1$, and $-2$, and no finite zeros.

A function can also have *infinite* poles and zeros. If the function approaches infinity as $s$ approaches infinity, then the function has a pole at infinity. If the function approaches zero

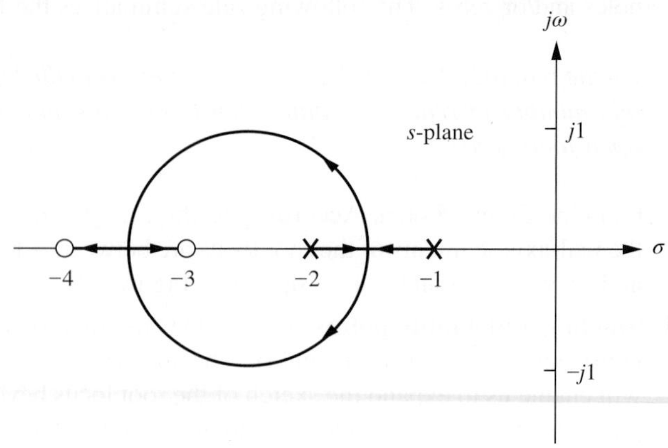


**FIGURE 8.10** Complete root locus for the system of Figure 8.6

as $s$ approaches infinity, then the function has a zero at infinity. For example, the function $G(s) = s$ has a pole at infinity, since $G(s)$ approaches infinity as $s$ approaches infinity. On the other hand, $G(s) = 1/s$ has a zero at infinity, since $G(s)$ approaches zero as $s$ approaches infinity.

Every function of $s$ has an equal number of poles and zeros if we include the infinite poles and zeros as well as the finite poles and zeros. In this example, Eq. (8.25) contains three finite poles and three infinite zeros. To illustrate, let $s$ approach infinity. The open-loop transfer function becomes

$$KG(s)H(s) \approx \frac{K}{s^3} = \frac{K}{s \cdot s \cdot s} \tag{8.26}$$

Each $s$ in the denominator causes the open-loop function, $KG(s)H(s)$, to become zero as that $s$ approaches infinity. Hence, Eq. (8.26) has three zeros at infinity.

Thus, for Eq. (8.25), the root locus begins at the finite poles of $KG(s)H(s)$ and ends at the infinite zeros. The question remains: Where are the infinite zeros? We must know where these zeros are in order to show the locus moving from the three finite poles to the three infinite zeros. Rule 5 helps us locate these zeros at infinity. Rule 5 also helps us locate poles at infinity for functions containing more finite zeros than finite poles.[1]

We now state Rule 5, which will tell us what the root locus looks like as it approaches the zeros at infinity or as it moves from the poles at infinity. The derivation can be found in Appendix M.1 at www.wiley.com/college/nise.

*The root locus approaches straight lines as asymptotes as the locus approaches infinity. Further, the equation of the asymptotes is given by the real-axis intercept, $\sigma_a$ and angle, $\theta_a$ as follows:*

$$\sigma_a = \frac{\sum \text{finite poles} - \sum \text{finite zeros}}{\#\,\text{finite poles} - \#\,\text{finite zeros}} \tag{8.27}$$

$$\theta_a = \frac{(2k+1)\pi}{\#\,\text{finite poles} - \#\,\text{finite zeros}} \tag{8.28}$$

*where $k = 0, \pm 1, \pm 2, \pm 3$ and the angle is given in radians with respect to the positive extension of the real axis.*

Notice that the running index, $k$, in Eq. (8.28) yields a multiplicity of lines that account for the many branches of a root locus that approach infinity. Let us demonstrate the concepts with an example.

## Example 8.2 Sketching a Root Locus with Asymptotes

**PROBLEM:** Sketch the root locus for the system shown in Figure 8.11.

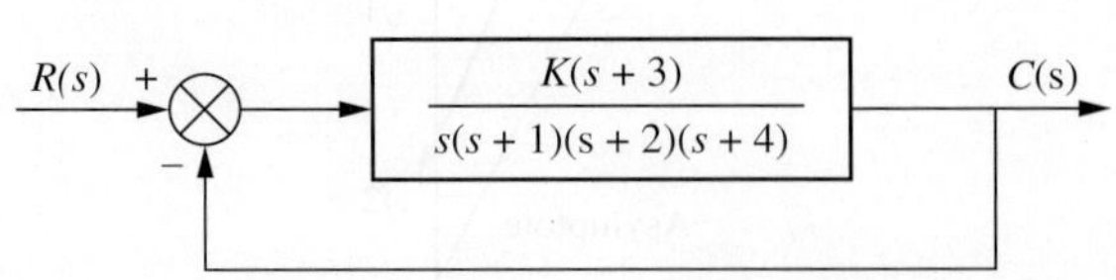


**FIGURE 8.11** System for Example 8.2

[1] Physical systems, however, have more finite poles than finite zeros, since the implied differentiation yields infinite output for discontinuous input functions, such as step inputs.

**SOLUTION:** Let us begin by calculating the asymptotes. Using Eq. (8.27), the real-axis intercept is evaluated as

$$\sigma_a = \frac{(-1-2-4)-(-3)}{4-1} = -\frac{4}{3} \tag{8.29}$$

The angles of the lines that intersect at $-4/3$, given by Eq. (8.28), are

$$\theta_a = \frac{(2k+1)\pi}{\#\text{ finite poles} - \#\text{ finite zeros}} \tag{8.30a}$$

$$= \pi/3 \quad \text{for } k = 0 \tag{8.30b}$$

$$= \pi \quad \text{for } k = 1 \tag{8.30c}$$

$$= 5\pi/3 \quad \text{for } k = 2 \tag{8.30d}$$

If the value for $k$ continued to increase, the angles would begin to repeat. The number of lines obtained equals the difference between the number of finite poles and the number of finite zeros.

Rule 4 states that the locus begins at the open-loop poles and ends at the open-loop zeros. For the example there are more open-loop poles than open-loop zeros. Thus, there must be zeros at infinity. The asymptotes tell us how we get to these zeros at infinity.

Figure 8.12 shows the complete root locus as well as the asymptotes that were just calculated. Notice that we have made use of all the rules learned so far. The real-axis segments lie to the left of an odd number of poles and/or zeros. The locus starts at the open-loop poles and ends at the open-loop zeros. For the example there is only one open-loop finite zero and three infinite zeros. Rule 5, then, tells us that the three zeros at infinity are at the ends of the asymptotes.

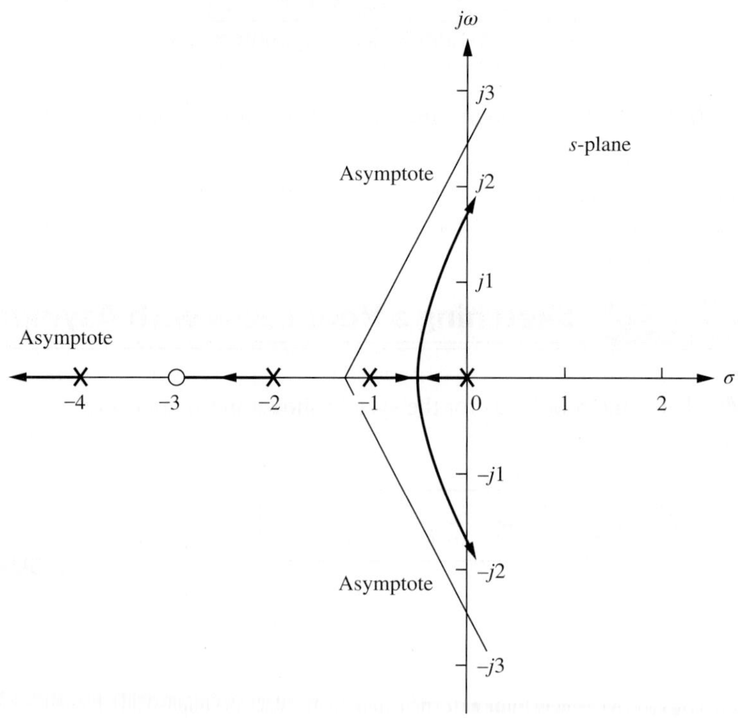


**FIGURE 8.12** Root locus and asymptotes for the system of Figure 8.11

**Skill-Assessment Exercise 8.3**

**PROBLEM:** Sketch the root locus and its asymptotes for a unity feedback system that has the forward transfer function

$$G(s) = \frac{K}{(s+2)(s+4)(s+6)}$$

**ANSWER:** The complete solution is at www.wiley.com/college/nise.

## 8.5 Refining the Sketch

The rules covered in the previous section permit us to sketch a root locus rapidly. If we want more detail, we must be able to accurately find important points on the root locus along with their associated gain. Points on the real axis where the root locus enters or leaves the complex plane—real-axis breakaway and break-in points—and the $j\omega$-axis crossings are candidates. We can also derive a better picture of the root locus by finding the angles of departure and arrival from complex poles and zeros, respectively.

In this section, we discuss the calculations required to obtain specific points on the root locus. Some of these calculations can be made using the basic root locus relationship that the sum of the zero angles minus the sum of the pole angles equals an odd multiple of 180°, and the gain at a point on the root locus is found as the ratio of (1) the product of pole lengths drawn to that point to (2) the product of zero lengths drawn to that point. We have yet to address how to implement this task. In the past, an inexpensive tool called a Spirule™ added the angles together rapidly and then quickly multiplied and divided the lengths to obtain the gain. Today we can rely on hand-held or programmable calculators as well as personal computers.

Students pursuing MATLAB will learn how to apply it to the root locus at the end of Section 8.6. Other alternatives are discussed in Appendix H.2 at www.wiley.com/college/nise. The discussion can be adapted to programmable hand-held calculators. All readers are encouraged to select a computational aid at this point. Root locus calculations can be labor intensive if hand calculations are used.

We now discuss how to refine our root locus sketch by calculating real-axis breakaway and break-in points, $j\omega$-axis crossings, angles of departure from complex poles, and angles of arrival to complex zeros. We conclude by showing how to find accurately any point on the root locus and calculate the gain.

### Real-Axis Breakaway and Break-In Points

Numerous root loci appear to break away from the real axis as the system poles move from the real axis to the complex plane. At other times the loci appear to return to the real axis as a pair of complex poles becomes real. We illustrate this in Figure 8.13. This locus is sketched using the first four rules: (1) number of branches, (2) symmetry, (3) real-axis segments, and (4) starting and ending points. The figure shows a root locus leaving the real axis between $-1$ and $-2$ and returning to the real axis between $+3$ and $+5$. The point where the locus leaves the real axis, $-\sigma_1$, is called the *breakaway point*, and the point where the locus returns to the real axis, $\sigma_2$, is called the *break-in point*.

At the breakaway or break-in point, the branches of the root locus form an angle of $180°/n$ with the real axis, where $n$ is the number of closed-loop poles arriving at or departing from the single breakaway or break-in point on the real axis (*Kuo, 1991*). Thus, for the two poles shown in Figure 8.13, the branches at the breakaway point form 90° angles with the real axis.

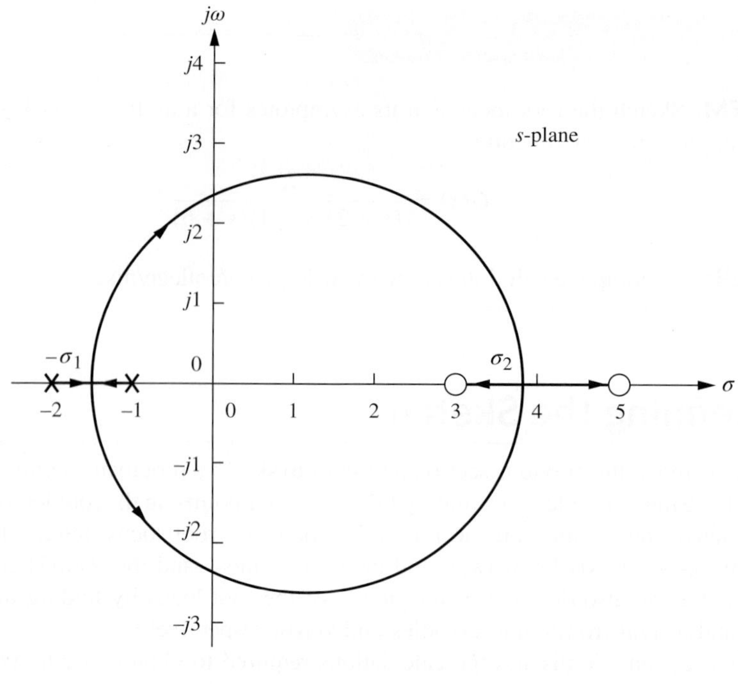


**FIGURE 8.13** Root locus example showing real-axis breakaway ($-\sigma_1$) and break-in points ($\sigma_2$)

We now show how to find the breakaway and break-in points. As the two closed-loop poles, which are at $-1$ and $-2$ when $K = 0$, move toward each other, the gain increases from a value of zero. We conclude that the gain must be maximum along the real axis at the point where the breakaway occurs, somewhere between $-1$ and $-2$. Naturally, the gain increases above this value as the poles move into the complex plane. We conclude that the breakaway point occurs at a point of maximum gain on the real axis between the open-loop poles.

Now let us turn our attention to the break-in point somewhere between $+3$ and $+5$ on the real axis. When the closed-loop complex pair returns to the real axis, the gain will continue to increase to infinity as the closed-loop poles move toward the open-loop zeros. It must be true, then, that the gain at the break-in point is the minimum gain found along the real axis between the two zeros.

The sketch in Figure 8.14 shows the variation of real-axis gain. The breakaway point is found at the maximum gain between $-1$ and $-2$, and the break-in point is found at the minimum gain between $+3$ and $+5$.

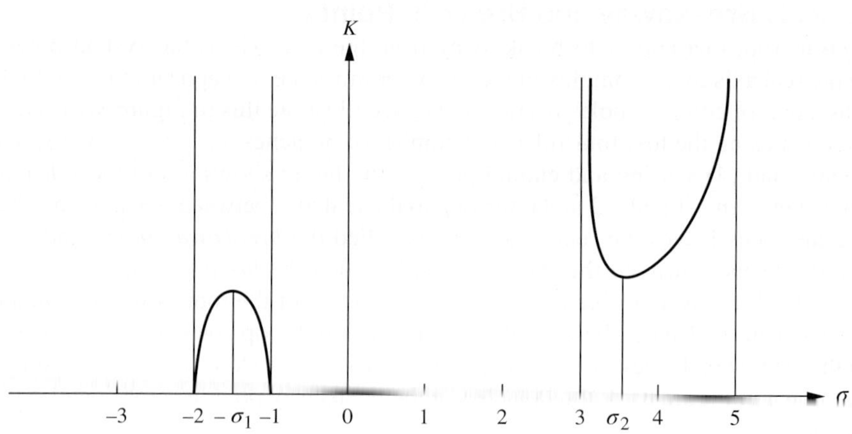


**FIGURE 8.14** Variation of gain along the real axis for the root locus of Figure 8.13

There are three methods for finding the points at which the root locus breaks away from and breaks into the real axis. The first method is to maximize and minimize the gain, $K$, using differential calculus. For all points on the root locus, Eq. (8.13) yields

$$K = -\frac{1}{G(s)H(s)} \tag{8.31}$$

For points along the real-axis segment of the root locus where breakaway and break-in points could exist, $s = \sigma$. Hence, along the real axis Eq. (8.31) becomes

$$K = -\frac{1}{G(\sigma)H(\sigma)} \tag{8.32}$$

This equation then represents a curve of $K$ versus $\sigma$ similar to that shown in Figure 8.14. Hence, if we differentiate Eq. (8.32) with respect to $\sigma$ and set the derivative equal to zero, we can find the points of maximum and minimum gain and hence the breakaway and break-in points. Let us demonstrate.

## Example 8.3 Breakaway and Break-in Points via Differentiation

**PROBLEM:** Find the breakaway and break-in points for the root locus of Figure 8.13, using differential calculus.

**SOLUTION:** Using the open-loop poles and zeros, we represent the open-loop system whose root locus is shown in Figure 8.13 as follows:

$$KG(s)H(s) = \frac{K(s-3)(s-5)}{(s+1)(s+2)} = \frac{K(s^2-8s+15)}{(s^2+3s+2)} \tag{8.33}$$

But for all points along the root locus, $KG(s)H(s) = -1$, and along the real axis, $s = \sigma$. Hence,

$$\frac{K(\sigma^2-8\sigma+15)}{(\sigma^2+3\sigma+2)} = -1 \tag{8.34}$$

Solving for $K$, we find

$$K = \frac{-(\sigma^2+3\sigma+2)}{(\sigma^2-8\sigma+15)} \tag{8.35}$$

Differentiating $K$ with respect to $\sigma$ and setting the derivative equal to zero yields

$$\frac{dK}{d\sigma} = \frac{(11\sigma^2-26\sigma-61)}{(\sigma^2-8\sigma+15)^2} = 0 \tag{8.36}$$

Solving for $\sigma$, we find $\sigma = -1.45$ and $3.82$, which are the breakaway and break-in points.

The second method is a variation on the differential calculus method. Called the *transition method*, it eliminates the step of differentiation (*Franklin, 1991*). This method, derived in Appendix M.2 at www.wiley.com/college/nise, is now stated:

*Breakaway and break-in points satisfy the relationship*

$$\sum_{1}^{m} \frac{1}{\sigma + z_i} = \sum_{1}^{n} \frac{1}{\sigma + p_i} \tag{8.37}$$

*where $z_i$ and $p_i$ are the negative of the zero and pole values, respectively, of $G(s)H(s)$.*

Solving Eq. (8.37) for $\sigma$, the real-axis values that minimize or maximize $K$, yields the breakaway and break-in points without differentiating. Let us look at an example.

## Example 8.4 Breakaway and Break-in Points Without Differentiation

**PROBLEM:** Repeat Example 8.3 without differentiating.

**SOLUTION:** Using Eq. (8.37),

$$\frac{1}{\sigma - 3} + \frac{1}{\sigma - 5} = \frac{1}{\sigma + 1} + \frac{1}{\sigma + 2} \tag{8.38}$$

Simplifying,

$$11\sigma^2 - 26\sigma - 61 = 0 \tag{8.39}$$

Hence, $\sigma = -1.45$ and $3.82$, which agrees with Example 8.3.

For the third method, the root locus program discussed in Appendix H.2 at www.wiley.com/college/nise can be used to find the breakaway and break-in points. Simply use the program to search for the point of maximum gain between $-1$ and $-2$ and to search for the point of minimum gain between $+3$ and $+5$. Table 8.2 shows the results of the search. The locus leaves the axis at $-1.45$, the point of maximum gain

**TABLE 8.2** Data for breakaway and break-in points for the root locus of Figure 8.13

| Real-axis value | Gain | | Comment |
|---|---|---|---|
| −1.41 | 0.008557 | | |
| −1.42 | 0.008585 | | |
| −1.43 | 0.008605 | | |
| −1.44 | 0.008617 | | |
| −1.45 | 0.008623 | ← | Max. gain: breakaway |
| −1.46 | 0.008622 | | |
| 3.3 | 44.686 | | |
| 3.4 | 37.125 | | |
| 3.5 | 33.000 | | |
| 3.6 | 30.667 | | |
| 3.7 | 29.440 | | |
| 3.8 | 29.000 | ← | Min. gain: break-in |
| 3.9 | 29.202 | | |

between $-1$ and $-2$, and reenters the real axis at $+3.8$, the point of minimum gain between $+3$ and $+5$. These results are the same as those obtained using the first two methods. MATLAB also has the capability of finding breakaway and break-in points.

## The $j\omega$-Axis Crossings

We now further refine the root locus by finding the imaginary-axis crossings. The importance of the $j\omega$-axis crossings should be readily apparent. Looking at Figure 8.12, we see that the system's poles are in the left half-plane up to a particular value of gain. Above this value of gain, two of the closed-loop system's poles move into the right half-plane, signifying that the system is unstable. The $j\omega$-axis crossing is a point on the root locus that separates the stable operation of the system from the unstable operation. The value of $\omega$ at the axis crossing yields the frequency of oscillation, while the gain at the $j\omega$-axis crossing yields, for this example, the maximum positive gain for system stability. We should note here that other examples illustrate instability at small values of gain and stability at large values of gain. These systems have a root locus starting in the right–half-plane (unstable at small values of gain) and ending in the left–half-plane (stable for high values of gain).

To find the $j\omega$-axis crossing, we can use the Routh-Hurwitz criterion, covered in Chapter 6, as follows: Forcing a row of zeros in the Routh table will yield the gain; going back one row to the even polynomial equation and solving for the roots yields the frequency at the imaginary-axis crossing.

## Example 8.5 Frequency and Gain at Imaginary-Axis Crossing

**PROBLEM:** For the system of Figure 8.11, find the frequency and gain, $K$, for which the root locus crosses the imaginary axis. For what range of $K$ is the system stable?

**SOLUTION:** The closed-loop transfer function for the system of Figure 8.11 is

$$T(s) = \frac{K(s+3)}{s^4 + 7s^3 + 14s^2 + (8+K)s + 3K} \tag{8.40}$$

Using the denominator and simplifying some of the entries by multiplying any row by a constant, we obtain the Routh array shown in Table 8.3.

A complete row of zeros yields the possibility for imaginary axis roots. For positive values of gain, those for which the root locus is plotted, only the $s^1$ row can yield a row of zeros. Thus,

$$-K^2 - 65K + 720 = 0 \tag{8.41}$$

**TABLE 8.3** Routh table for Eq (8.40)

| | | | |
|---|---|---|---|
| $s^4$ | 1 | 14 | $3K$ |
| $s^3$ | 7 | $8+K$ | |
| $s^2$ | $90-K$ | $21K$ | |
| $s^1$ | $\dfrac{-K^2-65K+720}{90-K}$ | | |
| $s^0$ | $21K$ | | |

From this equation $K$ is evaluated as

$$K = 9.65 \tag{8.42}$$

Forming the even polynomial by using the $s^2$ row with $K = 9.65$, we obtain

$$(90 - K)s^2 + 21K = 80.35s^2 + 202.7 = 0 \tag{8.43}$$

and $s$ is found to be equal to $\pm j1.59$. Thus the root locus crosses the $j\omega$-axis at $\pm j1.59$ at a gain of 9.65. We conclude that the system is stable for $0 \le K < 9.65$.

Another method for finding the $j\omega$-axis crossing (or any point on the root locus, for that matter) uses the fact that at the $j\omega$-axis crossing, the sum of angles from the finite open-loop poles and zeros must add to $(2k + 1)180°$. Thus, we can search $j\omega$-axis until we find the point that meets this angle condition. A computer program, such as the root locus program discussed in Appendix H.2 at www.wiley.com/college/nise or MATLAB, can be used for this purpose. Subsequent examples in this chapter use this method to determine the $j\omega$-axis crossing.

## Angles of Departure and Arrival

In this subsection, we further refine our sketch of the root locus by finding angles of departure and arrival from complex poles and zeros. Consider Figure 8.15, which shows the open-loop poles and zeros, some of which are complex. The root locus starts at the open-loop poles and ends at the open-loop zeros. In order to sketch the root locus more accurately, we want to calculate the root locus departure angle from the complex poles and the arrival angle to the complex zeros.

If we assume a point on the root locus $\epsilon$ close to a complex pole, the sum of angles drawn from all finite poles and zeros to this point is an odd multiple of 180°. Except for the pole that is $\epsilon$ close to the point, we assume all angles drawn from all other poles and zeros are drawn directly to the pole that is near the point. Thus, the only unknown angle in the sum is the angle drawn from the pole that is $\epsilon$ close. We can solve for this unknown angle, which is also the angle of departure from this complex pole. Hence, from Figure 8.15($a$),

$$-\theta_1 + \theta_2 + \theta_3 - \theta_4 - \theta_5 + \theta_6 = (2k + 1)180° \tag{8.44a}$$

or

$$\theta_1 = \theta_2 + \theta_3 - \theta_4 - \theta_5 + \theta_6 - (2k + 1)180° \tag{8.44b}$$

If we assume a point on the root locus $\epsilon$ close to a complex zero, the sum of angles drawn from all finite poles and zeros to this point is an odd multiple of 180°. Except for the zero that is $\epsilon$ close to the point, we can assume all angles drawn from all other poles and zeros are drawn directly to the zero that is near the point. Thus, the only unknown angle in the sum is the angle drawn from the zero that is $\epsilon$ close. We can solve for this unknown angle, which is also the angle of arrival to this complex zero. Hence, from Figure 8.15($b$),

$$-\theta_1 + \theta_2 + \theta_3 - \theta_4 - \theta_5 + \theta_6 = (2k + 1)180° \tag{8.45a}$$

or

$$\theta_2 = \theta_1 - \theta_3 + \theta_4 + \theta_5 - \theta_6 + (2k + 1)180° \tag{8.45b}$$

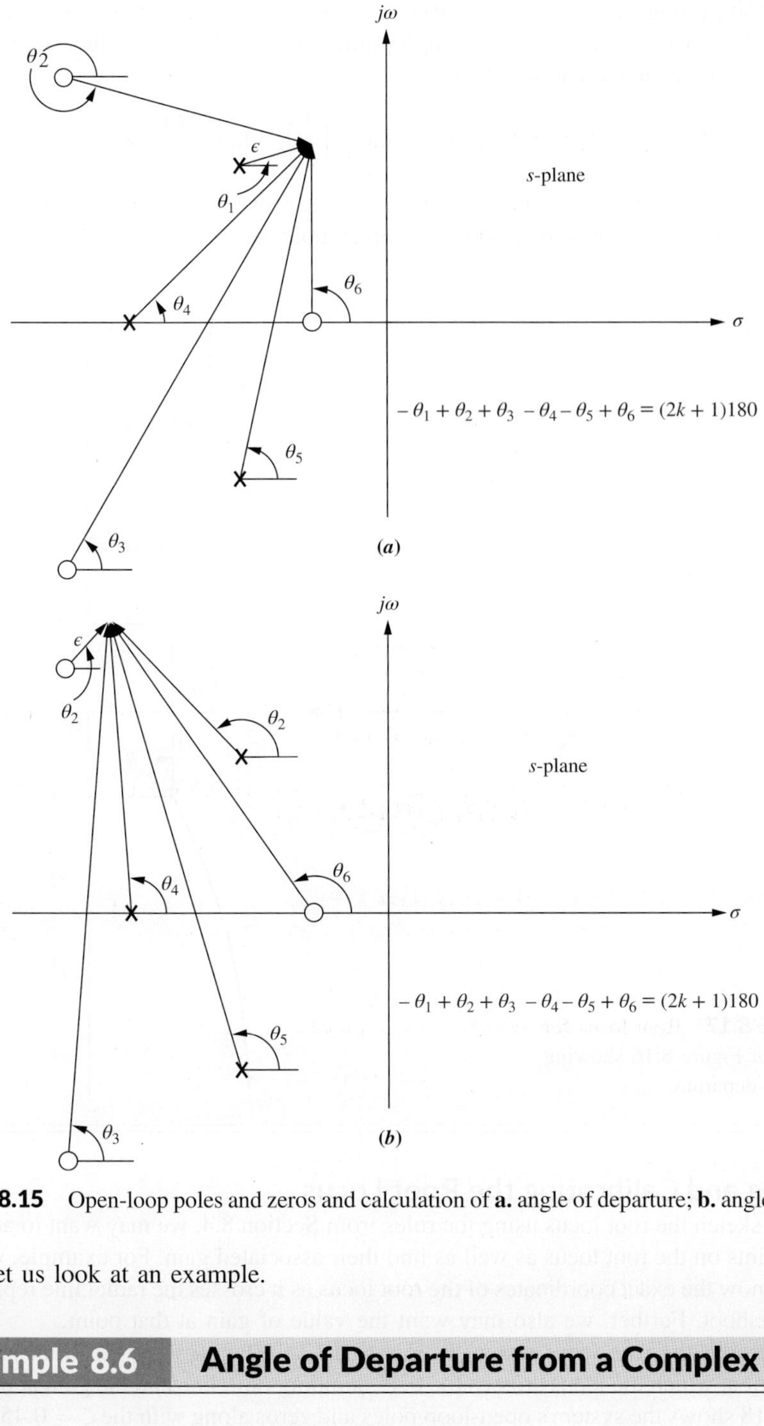


**FIGURE 8.15** Open-loop poles and zeros and calculation of **a.** angle of departure; **b.** angle of arrival

Let us look at an example.

## Example 8.6 Angle of Departure from a Complex Pole

**PROBLEM:** Given the unity feedback system of Figure 8.16, find the angle of departure from the complex poles and sketch the root locus.

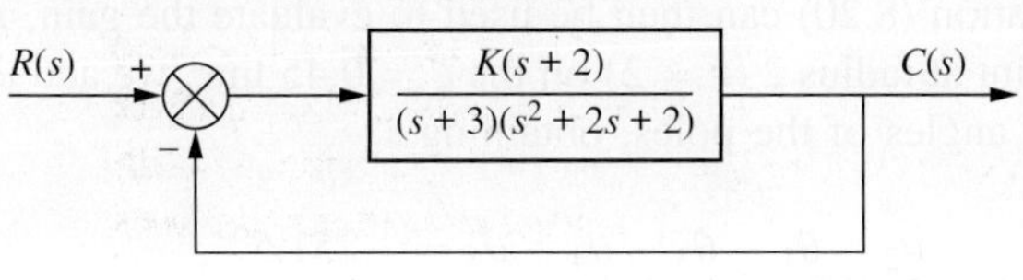


**FIGURE 8.16** Unity feedback system with complex poles

**SOLUTION:** Using the poles and zeros of $G(s) = (s+2)/[(s+3)(s^2+2s+2)]$ as plotted in Figure 8.17, we calculate the sum of angles drawn to a point $\epsilon$ close to the complex pole, $-1+j1$, in the second quadrant. Thus,

$$-\theta_1 - \theta_2 + \theta_3 - \theta_4 = -\theta_1 - 90^\circ + \tan^{-1}\left(\frac{1}{1}\right) - \tan^{-1}\left(\frac{1}{2}\right) = 180^\circ \qquad (8.46)$$

from which $\theta = -251.6^\circ = 108.4^\circ$. A sketch of the root locus is shown in Figure 8.17. Notice how the departure angle from the complex poles helps us to refine the shape.

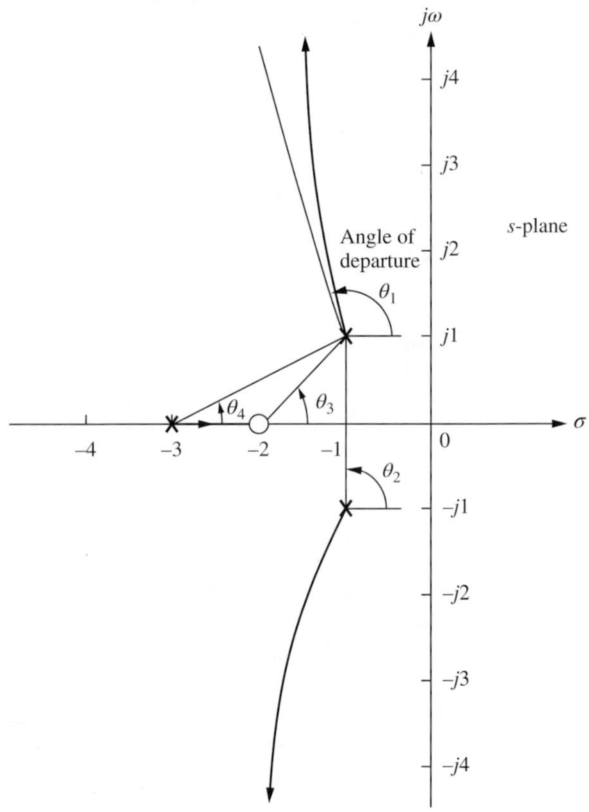


**FIGURE 8.17** Root locus for system of Figure 8.16 showing angle of departure

## Plotting and Calibrating the Root Locus

Once we sketch the root locus using the rules from Section 8.4, we may want to accurately locate points on the root locus as well as find their associated gain. For example, we might want to know the exact coordinates of the root locus as it crosses the radial line representing 20% overshoot. Further, we also may want the value of gain at that point.

Consider the root locus shown in Figure 8.12. Let us assume we want to find the exact point at which the locus crosses the 0.45 damping ratio line and the gain at that point. Figure 8.18 shows the system's open-loop poles and zeros along with the $\zeta = 0.45$ line. If a few test points along the $\zeta = 0.45$ line are selected, we can evaluate their angular sum and locate that point where the angles add up to an odd multiple of 180°. It is at this point that the root locus exists. Equation (8.20) can then be used to evaluate the gain, $K$, at that point.

Selecting the point at radius 2 ($r = 2$) on the $\zeta = 0.45$ line, we add the angles of the zeros and subtract the angles of the poles, obtaining

$$\theta_2 - \theta_1 - \theta_3 - \theta_4 - \theta_5 = -251.5^\circ \qquad (8.47)$$

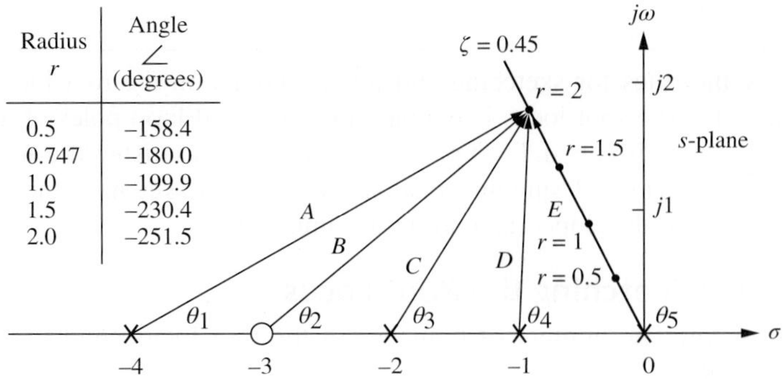


| Radius $r$ | Angle $\angle$ (degrees) |
|---|---|
| 0.5 | −158.4 |
| 0.747 | −180.0 |
| 1.0 | −199.9 |
| 1.5 | −230.4 |
| 2.0 | −251.5 |

**FIGURE 8.18** Finding and calibrating exact points on the root locus of Figure 8.12

Since the sum is not equal to an odd multiple of 180°, the point at radius = 2 is not on the root locus. Proceeding similarly for the points at radius = 1.5, 1, 0.747, and 0.5, we obtain the table shown in Figure 8.18. This table lists the points, giving their radius, $r$, and the sum of angles indicated by the symbol $\angle$. From the table we see that the point at radius 0.747 is on the root locus, since the angles add up to −180°. Using Eq. (8.21), the gain, $K$, at this point is

$$K = \frac{|A||C||D||E|}{|B|} = 1.71 \tag{8.48}$$

In summary, *we search a given line for the point yielding a summation of angles (zero angles–pole angles) equal to an odd multiple of 180°*. We conclude that the point is on the root locus. The gain at that point is then found by *multiplying the pole lengths drawn to that point and dividing by the product of the zero lengths drawn to that point*. A computer program, such as that discussed in Appendix H.2 at www.wiley.com/college/nise or MATLAB, can be used.

## Skill-Assessment Exercise 8.4

**PROBLEM:** Given a unity feedback system that has the forward transfer function

$$G(s) = \frac{K(s+2)}{(s^2 - 4s + 13)}$$

do the following:

**a.** Sketch the root locus.

**b.** Find the imaginary-axis crossing.

**c.** Find the gain, $K$, at the $j\omega$-axis crossing.

**d.** Find the break-in point.

**e.** Find the angle of departure from the complex poles.

**ANSWERS:**

**a.** See solution at www.wiley.com/college/nise.

**b.** $s = \pm j\sqrt{21}$

**c.** $K = 4$

**d.** Break-in point = −7

**e.** Angle of departure = −233.1°

The complete solution is at www.wiley.com/college/nise.

# 8.6 An Example

We now review the rules for sketching and finding points on the root locus, as well as present an example. The root locus is the path of the closed-loop poles of a system as a parameter of the system is varied. Each point on the root locus satisfies the angle condition, $\angle G(s)H(s) = (2k+1)180°$. Using this relationship, rules for sketching and finding points on the root locus were developed and are now summarized.

## Basic Rules for Sketching the Root Locus

***Number of branches*** The number of branches of the root locus equals the number of closed-loop poles.

***Symmetry*** The root locus is symmetrical about the real axis.

***Real-axis segments*** On the real axis, for $K > 0$ the root locus exists to the left of an odd number of real-axis, finite open-loop poles and/or finite open-loop zeros.

***Starting and ending points*** The root locus begins at the finite and infinite poles of $G(s)H(s)$ and ends at the finite and infinite zeros of $G(s)H(s)$.

***Behavior at infinity*** The root locus approaches straight lines as asymptotes as the locus approaches infinity. Further, the equations of the asymptotes are given by the real-axis intercept and angle in radians as follows:

$$\sigma_a = \frac{\sum \text{finite poles} - \sum \text{finite zeros}}{\#\,\text{finite poles} - \#\,\text{finite zeros}} \tag{8.49}$$

$$\theta_a = \frac{(2k+1)\pi}{\#\,\text{finite poles} - \#\,\text{finite zeros}} \tag{8.50}$$

where $k = 0, \pm 1, \pm 2, \pm 3, \ldots$

## Additional Rules for Refining the Sketch

***Real-axis breakaway and break-in points*** The root locus breaks away from the real axis at a point where the gain is maximum and breaks into the real axis at a point where the gain is minimum.

***Calculation of jω-axis crossings*** The root locus crosses the $j\omega$-axis at the point where $\angle G(s)H(s) = (2k+1)180°$. Routh-Hurwitz or a search of the $j\omega$-axis for $(2k+1)180°$ can be used to find the $j\omega$-axis crossing.

***Angles of departure and arrival*** The root locus departs from complex, open-loop poles and arrives at complex, open-loop zeros at angles that can be calculated as follows. Assume a point $\epsilon$ close to the complex pole or zero. Add all angles drawn from all open-loop poles and zeros to this point. The sum equals $(2k+1)180°$. The only unknown angle is that drawn from the $\epsilon$ close pole or zero, since the vectors drawn from all other poles and zeros can be considered drawn to the complex pole or zero that is $\epsilon$ close to the point. Solving for the unknown angle yields the angle of departure or arrival.

***Plotting and calibrating the root locus*** All points on the root locus satisfy the relationship $\angle G(s)H(s) = (2k+1)180°$. The gain, $K$, at any point on the root locus is given by

$$K = \frac{1}{|G(s)H(s)|} = \frac{1}{M} = \frac{\Pi \text{ finite pole lengths}}{\Pi \text{ finite zero lengths}} \tag{8.51}$$

Two animation PowerPoint presentations (PPTs) demonstrating root locus plotting are available for instructors at www.wiley.com/college/nise. See *Root-Locus Plotter* and *Dynamic Root-Locus*.

Let us now look at a summary example.

## Example 8.7 Sketching a Root Locus and Finding Critical Points

**PROBLEM:** Sketch the root locus for the system shown in Figure 8.19($a$) and find the following:

**a.** The exact point and gain where the locus crosses the 0.45 damping ratio line

**b.** The exact point and gain where the locus crosses the $j\omega$-axis

**c.** The breakaway point on the real axis

**d.** The range of $K$ within which the system is stable

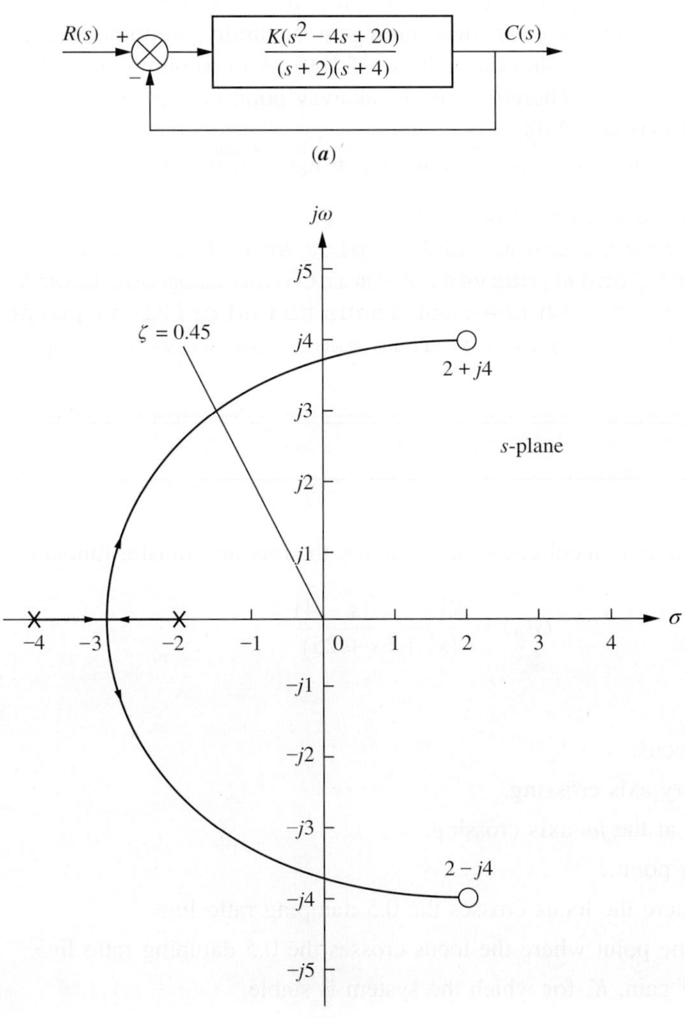


**FIGURE 8.19** **a.** System for Example 8.7; **b.** root locus sketch

**SOLUTION:** The problem solution is shown, in part, in Figure 8.19(*b*). First sketch the root locus. Using Rule 3, the real-axis segment is found to be between −2 and −4. Rule 4 tells us that the root locus starts at the open-loop poles and ends at the open-loop zeros. These two rules alone give us the general shape of the root locus.

**a.** To find the exact point where the locus crosses the $\zeta = 0.45$ line, we can use the root locus program discussed in Appendix H.2 at www.wiley.com/college/nise to search along the line

$$\theta = 180^\circ - \cos^{-1} 0.45 = 116.7^\circ \tag{8.52}$$

for the point where the angles add up to an odd multiple of 180°. Searching in polar coordinates, we find that the root locus crosses the $\zeta = 0.45$ line at $3.4 \angle 116.7^\circ$ with a gain, $K$, of 0.417.

**b.** To find the exact point where the locus crosses the $j\omega$-axis, use the root locus program to search along the line

$$\theta = 90^\circ \tag{8.53}$$

for the point where the angles add up to an odd multiple of 180°. Searching in polar coordinates, we find that the root locus crosses the $j\omega$-axis at $\pm j3.9$ with a gain of $K = 1.5$.

**c.** To find the breakaway point, use the root locus program to search the real axis between −2 and −4 for the point that yields maximum gain. Naturally, all points will have the sum of their angles equal to an odd multiple of 180°. A maximum gain of 0.0248 is found at the point −2.88. Therefore, the breakaway point is between the open-loop poles on the real axis at −2.88.

**d.** From the answer to **b**, the system is stable for $K$ between 0 and 1.5.

MATLAB

**ML**

Students who are using MATLAB should now run ch8p1 in Appendix B. You will learn how to use MATLAB to plot and title a root locus, overlay constant $\zeta$ and $\omega_n$ curves, zoom into and zoom out from a root locus, and interact with the root locus to find critical points as well as gains at those points. This exercise solves Example 8.7 using MATLAB.

## Skill-Assessment Exercise 8.5

**TryIt 8.3**

Use MATLAB, the Control System Toolbox, and the following statements to plot the root locus for Skill-Assessment Exercise 8.5. Solve the remaining parts of the problem by clicking on the appropriate points on the plotted root locus.

```
numg=poly([2 4]);
deng=[1 6 25];
G=tf(numg,deng)
rlocus(G)
z=0.5
sgrid(z,0)
```

**PROBLEM:** Given a unity feedback system that has the forward transfer function

$$G(s) = \frac{K(s-2)(s-4)}{(s^2+6s+25)}$$

do the following:

**a.** Sketch the root locus.

**b.** Find the imaginary-axis crossing.

**c.** Find the gain, $K$, at the $j\omega$-axis crossing.

**d.** Find the break-in point.

**e.** Find the point where the locus crosses the 0.5 damping ratio line.

**f.** Find the gain at the point where the locus crosses the 0.5 damping ratio line.

**g.** Find the range of gain, $K$, for which the system is stable.

**ANSWERS:**

**a.** See solution at www.wiley.com/college/nise.

**b.** $s = \pm j4.06$

**c.** $K = 1$

**d.** Break-in point $= +2.89$

**e.** $s = -2.42 + j4.18$

**f.** $K = 0.108$

**g.** $K < 1$

The complete solution is at www.wiley.com/college/nise.

## 8.7 Transient Response Design via Gain Adjustment

Now that we know how to sketch a root locus, we show how to use it for the design of transient response. In the last section we found that the root locus crossed the 0.45 damping ratio line with a gain of 0.417. Does this mean that the system will respond with 20.5% overshoot, the equivalent to a damping ratio of 0.45? It must be emphasized that the formulas describing percent overshoot, settling time, and peak time were derived only for a system with two closed-loop complex poles and no closed-loop zeros. The effect of additional poles and zeros and the conditions for justifying an approximation of a two-pole system were discussed in Sections 4.7 and 4.8 and apply here to closed-loop systems and their root loci. The conditions justifying a second-order approximation are restated here:

1. Higher-order poles are much farther into the left half of the $s$-plane than the dominant second-order pair of poles. The response that results from a higher-order pole does not appreciably change the transient response expected from the dominant second-order poles.
2. Closed-loop zeros near the closed-loop second-order pole pair are nearly canceled by the close proximity of higher-order closed-loop poles.
3. Closed-loop zeros not canceled by the close proximity of higher-order closed-loop poles are far removed from the closed-loop second-order pole pair.

The first condition as it applies to the root locus is shown graphically in Figure 8.20($a$) and ($b$). Figure 8.20($b$) would yield a much better second-order approximation than Figure 8.20($a$), since closed-loop pole $p_3$ is farther from the dominant, closed-loop second-order pair, $p_1$ and $p_2$.

The second condition is shown graphically in Figure 8.20($c$) and ($d$). Figure 8.20($d$) would yield a much better second-order approximation than Figure 8.20($c$), since closed-loop pole $p_3$ is closer to canceling the closed-loop zero.

Summarizing the design procedure for higher-order systems, we arrive at the following:

1. Sketch the root locus for the given system.
2. Assume the system is a second-order system without any zeros and then find the gain to meet the transient response specification.
3. Justify your second-order assumption by finding the location of all higher-order poles and evaluating the fact that they are much farther from the $j\omega$-axis than the dominant

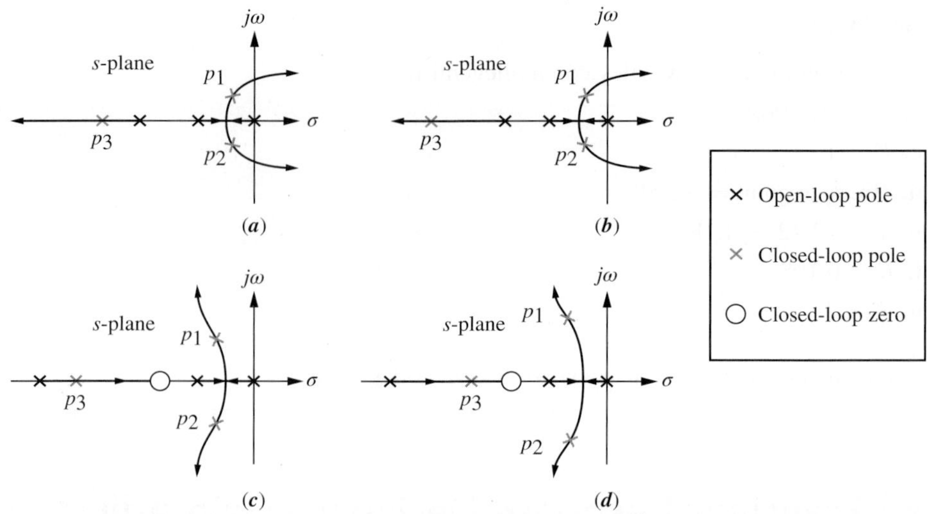


**FIGURE 8.20** Making second-order approximations

second-order pair. As a rule of thumb, this textbook assumes a factor of five times farther. Also, verify that closed-loop zeros are approximately canceled by higher-order poles. If closed-loop zeros are not canceled by higher-order closed-loop poles, be sure that the zero is far removed from the dominant second-order pole pair to yield approximately the same response obtained without the finite zero.

4. If the assumptions cannot be justified, your solution will have to be simulated in order to be sure it meets the transient response specification. It is a good idea to simulate all solutions, anyway.

We now look at a design example to show how to make a second-order approximation and then verify whether or not the approximation is valid.

## Example 8.8 Third-Order System Gain Design

**PROBLEM:** Consider the system shown in Figure 8.21. Design the value of gain, $K$, to yield 1.52% overshoot. Also estimate the settling time, peak time, and steady-state error.

**SOLUTION:** The root locus is shown in Figure 8.22. Notice that this is a third-order system with one zero. Breakaway points on the real axis can occur between 0 and $-1$ and between $-1.5$ and $-10$, where the gain reaches a peak. Using the root locus program and searching in these regions for the peaks in gain, breakaway points are found at $-0.62$ with a gain of 2.511 and at $-4.4$ with a gain of 28.89. A break-in point on the real axis can occur between $-1.5$ and $-10$, where the gain reaches a local minimum. Using the root locus program and searching in these regions for the local minimum gain, a break-in point is found at $-2.8$ with a gain of 27.91.

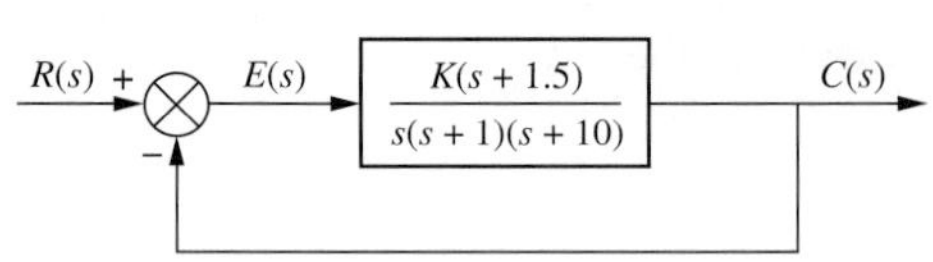


**FIGURE 8.21** System for Example 8.8

Next assume that the system can be approximated by a second-order, under-damped system without any zeros. A 1.52% overshoot corresponds to a damping ratio of 0.8. Sketch this damping ratio line on the root locus, as shown in Figure 8.22.

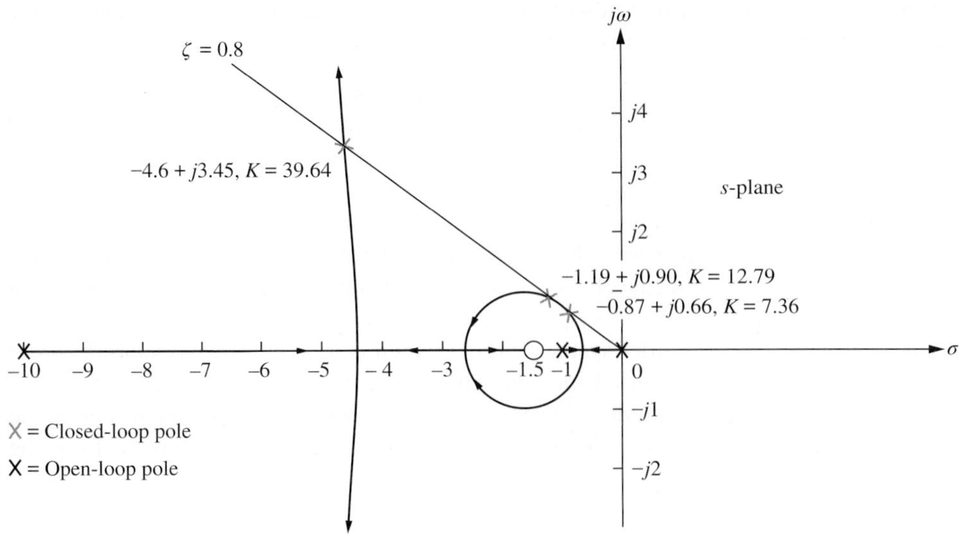


**FIGURE 8.22** Root locus for Example 8.8

Use the root locus program to search along the 0.8 damping ratio line for the point where the angles from the open-loop poles and zeros add up to an odd multiple of 180°. This is the point where the root locus crosses the 0.8 damping ratio or 1.52 percent overshoot line. Three points satisfy this criterion: $-0.87 \pm j0.66$, $-1.19 \pm j0.90$, and $-4.6 \pm j3.45$ with respective gains of 7.36, 12.79, and 39.64. For each point the settling time and peak time are evaluated using

$$T_s = \frac{4}{\zeta\omega_n} \tag{8.54}$$

where $\zeta\omega_n$ is the real part of the closed-loop pole, and also using

$$T_p = \frac{\pi}{\omega_n\sqrt{1-\zeta^2}} \tag{8.55}$$

where $\omega_n\sqrt{1-\zeta^2}$ is the imaginary part of the closed-loop pole.

To test our assumption of a second-order system, we must calculate the location of the third pole. Using the root locus program, search along the negative extension of the real axis between the zero at $-1.5$ and the pole at $-10$ for points that match the value of gain found at the second-order dominant poles. For each of the three crossings of the 0.8 damping ratio line, the third closed-loop pole is at $-9.25$, $-8.6$, and $-1.8$, respectively. The results are summarized in Table 8.4.

Finally, let us examine the steady-state error produced in each case. Note that we have little control over the steady-state error at this point. When the gain is set to meet

**TABLE 8.4** Characteristics of the system of Example 8.8

| Case | Closed-loop poles | Closed-loop zero | Gain | Third closed-loop pole | Settling time | Peak time | $K_v$ |
|---|---|---|---|---|---|---|---|
| 1 | $-0.87 \pm j0.66$ | $-1.5 + j0$ | 7.36 | $-9.25$ | 4.60 | 4.76 | 1.1 |
| 2 | $-1.19 \pm j0.90$ | $-1.5 + j0$ | 12.79 | $-8.61$ | 3.36 | 3.49 | 1.9 |
| 3 | $-4.60 \pm j3.45$ | $-1.5 + j0$ | 39.64 | $-1.80$ | 0.87 | 0.91 | 5.9 |

the transient response, we have also designed the steady-state error. For the example, the steady-state error specification is given by $K_v$ and is calculated as

$$K_v = \lim_{s \to 0} sG(s) = \frac{K(1.5)}{(1)(10)} \tag{8.56}$$

The results for each case are shown in Table 8.4.

How valid are the second-order assumptions? From Table 8.4, Cases 1 and 2 yield third closed-loop poles that are relatively far from the closed-loop zero. For these two cases there is no pole-zero cancellation, and a second-order system approximation is not valid. In Case 3, the third closed-loop pole and the closed-loop zero are relatively close to each other, and a second-order system approximation can be considered valid. In order to show this, let us make a partial-fraction expansion of the closed-loop step response of Case 3 and see that the amplitude of the exponential decay is much less than the amplitude of the underdamped sinusoid. The closed-loop step response, $C_3(s)$, formed from the closed-loop poles and zeros of Case 3 is

$$\begin{aligned} C_3(s) &= \frac{39.64(s+1.5)}{s(s+1.8)(s+4.6+j3.45)(s+4.6-j3.45)} \\ &= \frac{39.64(s+1.5)}{s(s+1.8)(s^2+9.2s+33.06)} \\ &= \frac{1}{s} + \frac{0.3}{s(s+18)} - \frac{1.3(s+4.6)+1.6(3.45)}{(s+4.6)^2+3.45^2} \end{aligned} \tag{8.57}$$

Thus, the amplitude of the exponential decay from the third pole is 0.3, and the amplitude of the underdamped response from the dominant poles is $\sqrt{1.3^2 + 1.6^2} = 2.06$. Hence, the dominant pole response is 6.9 times as large as the nondominant exponential response, and we assume that a second-order approximation is valid.

Using a simulation program, we obtain Figure 8.23, which shows comparisons of step responses for the problem we have just solved. Cases 2 and 3 are plotted for both the third-order response and a second-order response, assuming just the dominant pair of poles calculated in the design problem. Again, the second-order approximation was justified for Case 3, where there is a small difference in percent overshoot. The second-order approximation is not valid for Case 2. Other than the excess overshoot, Case 3 responses are similar.

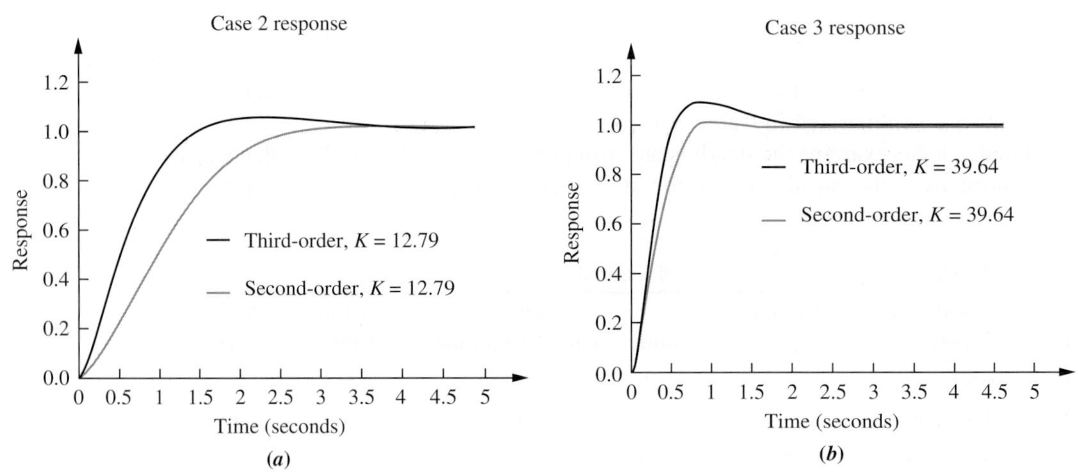


**FIGURE 8.23** Second- and third-order responses for Example 8.8. **a.** Case 2, **b.** Case 3

MATLAB ML

Students who are using MATLAB should now run ch8p2 in Appendix B. You will learn how to use MATLAB to enter a value of percent overshoot from the keyboard. MATLAB will then draw the root locus and overlay the percent overshoot line requested. You will then interact with MATLAB and select the point of intersection of the root locus with the requested percent overshoot line. MATLAB will respond with the value of gain, all closed-loop poles at that gain, and a closed-loop step response plot corresponding to the selected point. This exercise solves Example 8.8 using MATLAB.

MATLAB ML

Students who are using MATLAB may want to explore the SISO Design Tool described in Appendix E at www.wiley.com/college/nise. The SISO Design Tool is a convenient and intuitive way to obtain, view, and interact with a system's root locus. Section D.7 describes the advantages of using the tool, while Section D.8 describes how to use it. For practice, you may want to apply the SISO Design Tool to some of the problems at the end of this chapter.

## Skill-Assessment Exercise 8.6

**PROBLEM:** Given a unity feedback system that has the forward-path transfer function

$$G(s) = \frac{K}{(s+2)(s+4)(s+6)}$$

do the following:

**a.** Sketch the root locus.

**b.** Using a second-order approximation, design the value of $K$ to yield 10% overshoot for a unit-step input.

**c.** Estimate the settling time, peak time, rise time, and steady-state error for the value of $K$ designed in (b).

**d.** Determine the validity of your second-order approximation.

**ANSWERS:**

**a.** See solution located at www.wiley.com/college/nise.

**b.** $K = 45.55$

**c.** $T_s = 1.97$ s, $T_p = 1.13$ s, $T_r = 0.53$ s, and $e_{\text{step}}(\infty) = 0.51$

**d.** Second-order approximation is not valid.

The complete solution is located at www.wiley.com/college/nise.

# 8.8 Generalized Root Locus

Up to this point we have always drawn the root locus as a function of the forward-path gain, $K$. The control system designer must often know how the closed-loop poles change as a function of another parameter. For example, in Figure 8.24, the parameter of interest is the open-loop pole at $-p_1$. How can we obtain a root locus for variations of the value of $p_1$?

If the function $KG(s)H(s)$ is formed as

$$KG(s)H(s) = \frac{10}{(s+2)(s+p_1)} \tag{8.58}$$

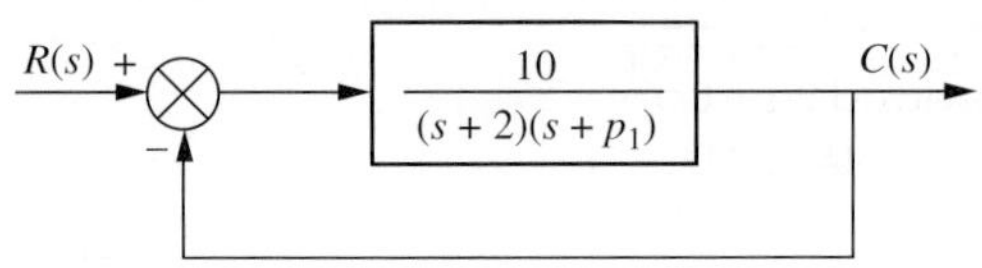


**FIGURE 8.24** System requiring a root locus calibrated with $p_1$ as a parameter

the problem is that $p_1$ is not a multiplying factor of the function, as the gain, $K$, was in all of the previous problems. The solution to this dilemma is to create an equivalent system where $p_1$ appears as the forward-path gain. Since the closed-loop transfer function's denominator is $1 + KG(s)H(s)$, we effectively want to create an equivalent system whose denominator is $1 + p_1G(s)H(s)$.

For the system of Figure 8.24, the closed-loop transfer function is

$$T(s) = \frac{KG(s)}{1+KG(s)H(s)} = \frac{10}{s^2+(p_1+2)s+2p_1+10} \tag{8.59}$$

Isolating $p_1$, we have

$$T(s) = \frac{10}{s^2+2s+10+p_1(s+2)} \tag{8.60}$$

Converting the denominator to the form $[1 + p_1G(s)H(s)]$ by dividing numerator and denominator by the term not included with $p_1$, $s^2 + 2s + 10$, we obtain

$$T(s) = \frac{\dfrac{10}{s^2+2s+10}}{1+\dfrac{p_1(s+2)}{s^2+2s+10}} \tag{8.61}$$

Conceptually, Eq. (8.61) implies that we have a system for which

$$KG(s)H(s) = \frac{p_1(s+2)}{s^2+2s+10} \tag{8.62}$$

The root locus can now be sketched as a function of $p_1$, assuming the open-loop system of Eq. (8.62). The final result is shown in Figure 8.25.

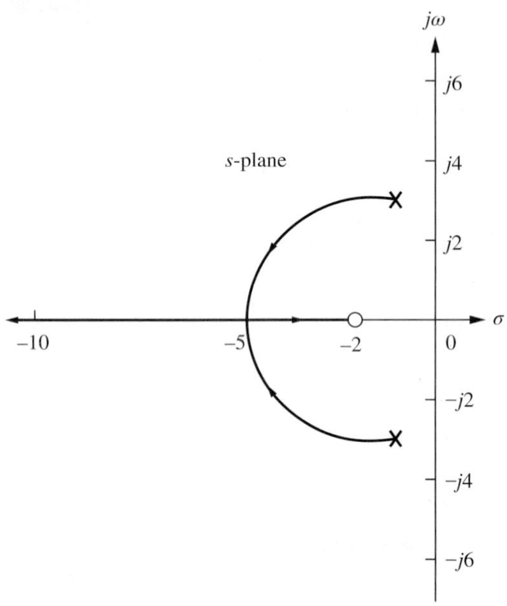


**FIGURE 8.25** Root locus for the system of Figure 8.24, with $p_1$ as a parameter

## Skill-Assessment Exercise 8.7

**PROBLEM:** Sketch the root locus for variations in the value of $p_1$, for a unity feedback system that has the following forward transfer function:

$$G(s) = \frac{100}{s(s+p_1)}$$

**ANSWER:** The complete solution is at www.wiley.com/college/nise.

In this section, we learned to plot the root locus as a function of any system parameter. In the next section we will learn how to plot root loci for positive-feedback systems.

# 8.9 Root Locus for Positive-Feedback Systems

The properties of the root locus were derived from the system of Figure 8.1. This is a negative-feedback system because of the negative summing of the feedback signal to the input signal. The properties of the root locus change dramatically if the feedback signal is added to the input rather than subtracted. A positive-feedback system can be thought of as a negative-feedback system with a negative value of $H(s)$. Using this concept, we find that the transfer function for the positive-feedback system shown in Figure 8.26 is

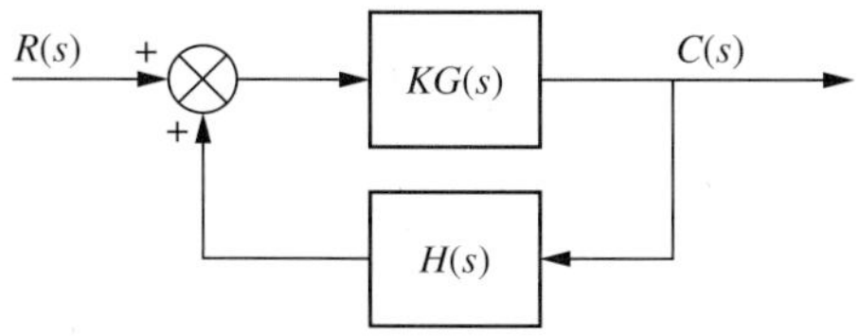


**FIGURE 8.26** Positive-feedback system

$$T(s) = \frac{KG(s)}{1 - KG(s)H(s)} \tag{8.63}$$

We now retrace the development of the root locus for the denominator of Eq. (8.63). Obviously, a pole, $s$, exists when

$$KG(s)H(s) = 1 = 1\angle k360^\circ \quad k = 0, \pm 1, \pm 2, \pm 3, \ldots \tag{8.64}$$

Therefore, the root locus for positive-feedback systems consists of all points on the $s$-plane where the angle of $KG(s)H(s) = k360^\circ$. How does this relationship change the rules for sketching the root locus presented in Section 8.4?

1. **Number of branches.** The same arguments as for negative feedback apply to this rule. There is no change.
2. **Symmetry.** The same arguments as for negative feedback apply to this rule. There is no change.
3. **Real-axis segments.** The development in Section 8.4 for the real-axis segments led to the fact that the angles of $G(s)H(s)$ along the real axis added up to either an odd multiple of 180° or a multiple of 360°. Thus, for positive-feedback systems the root locus exists on the real axis along sections where the locus for negative-feedback systems does not exist. The rule follows:

   *Real-axis segments: On the real axis, the root locus for positive-feedback systems exists to the left of an even number of real-axis, finite open-loop poles and/or finite open-loop zeros.*

   The change in the rule is the word *even;* for negative-feedback systems the locus existed to the left of an *odd* number of real-axis, finite open-loop poles and/or zeros.
4. **Starting and ending points.** You will find no change in the development in Section 8.4 if Eq. (8.63) is used instead of Eq. (8.12). Therefore, we have the following rule.

   *Starting and ending points: The root locus for positive-feedback systems begins at the finite and infinite poles of $G(s)H(s)$ and ends at the finite and infinite zeros of $G(s)H(s)$.*
5. **Behavior at infinity.** The changes in the development of the asymptotes begin at Eq. (M.4) in Appendix M at www.wiley.com/college/nise since positive-feedback systems follow the relationship in Eq. (8.64). That change yields a different slope for the asymptotes. The value of the real-axis intercept for the asymptotes remains unchanged. The student is encouraged to go through the development in detail and show that the behavior at infinity for positive-feedback systems is given by the following rule:

   *The root locus approaches straight lines as asymptotes as the locus approaches infinity. Further, the equations of the asymptotes for positive-feedback systems are given by the real-axis intercept, $\sigma_a$, and angle, $\theta_a$, as follows:*

$$\sigma_a = \frac{\sum \text{finite poles} - \sum \text{finite zeros}}{\#\ \text{finite poles} - \#\ \text{finite zeros}} \tag{8.65}$$

$$\theta_a = \frac{k2\pi}{\#\ \text{finite poles} - \#\ \text{finite zeros}} \tag{8.66}$$

*where $k = 0, \pm 1, \pm 2, \pm 3, \ldots$, and the angle is given in radians with respect to the positive extension of the real axis.*

The change we see is that the numerator of Eq. (8.66) is $k2\pi$ instead of $(2k+1)\pi$.

What about other calculations? The imaginary-axis crossing can be found using the root locus program. In a search of the $j\omega$-axis, you are looking for the point where the angles add up to a multiple of 360° instead of an odd multiple of 180°. The breakaway points are found by looking for the maximum value of $K$. The break-in points are found by looking for the minimum value of $K$.

When we were discussing *negative*-feedback systems, we always made the root locus plot for positive values of gain. Since *positive*-feedback systems can also be thought of as *negative*-feedback systems with negative gain, the rules developed in this section apply equally to *negative*-feedback systems with negative gain. Let us look at an example.

## Example 8.9 Root Locus for a Positive-Feedback System

**PROBLEM:** Sketch the root locus as a function of negative gain, $K$, for the system shown in Figure 8.11.

**SOLUTION:** The equivalent positive-feedback system found by pushing $-1$, associated with $K$, to the right past the pickoff point is shown in Figure 8.27(*a*). Therefore, as the gain of the equivalent system goes through positive values of $K$, the root locus will be equivalent to that generated by the gain, $K$, of the original system in Figure 8.11 as it goes through negative values.

The root locus exists on the real axis to the left of an even number of real, finite open-loop poles and/or zeros. Therefore, the locus exists on the entire positive extension of the real axis, between $-1$ and $-2$ and between $-3$ and $-4$. Using Eq. (8.27), the $\sigma_a$ intercept is found to be

$$\sigma_a = \frac{(-1-2-4)-(-3)}{4-1} = -\frac{4}{3} \tag{8.67}$$

The angles of the lines that intersect at $-4/3$ are given by

$$\theta_a = \frac{k2\pi}{\#\ \text{finite poles} - \#\ \text{finite zeros}} \tag{8.68a}$$

$$= 0 \qquad \text{for } k = 0 \tag{8.68b}$$

$$= 2\pi/3 \qquad \text{for } k = 1 \tag{8.68c}$$

$$= 4\pi/3 \qquad \text{for } k = 2 \tag{8.68d}$$

The final root locus sketch is shown in Figure 8.27(*b*).

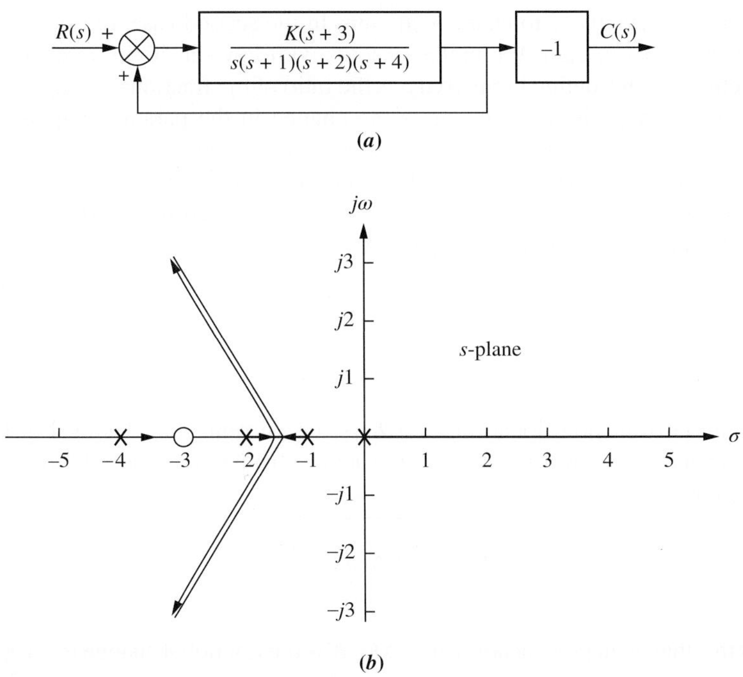


**FIGURE 8.27** **a.** Equivalent positive-feedback system for Example 8.9; **b.** root locus

## Skill-Assessment Exercise 8.8

**PROBLEM:** Sketch the root locus for the positive-feedback system whose forward transfer function is

$$G(s) = \frac{K(s+4)}{(s+1)(s+2)(s+3)}$$

The system has unity feedback.

**ANSWER:** The complete solution is at www.wiley.com/college/nise.

# 8.10 Pole Sensitivity

The root locus is a plot of the closed-loop poles as a system parameter is varied. Typically, that system parameter is gain. Any change in the parameter changes the closed-loop poles and, subsequently, the performance of the system. Many times the parameter changes against our wishes, due to heat or other environmental conditions. We would like to find out the extent to which changes in parameter values affect the performance of our system.

The root locus exhibits a nonlinear relationship between gain and pole location. Along some sections of the root locus, (1) very small changes in gain yield very large changes in pole location and hence performance; along other sections of the root locus, (2) very large changes in gain yield very small changes in pole location. In the first case we say that the system has a high sensitivity to changes in gain. In the second case, the system has a low sensitivity to changes in gain. We prefer systems with low sensitivity to changes in gain.

In Section 7.7, we defined sensitivity as the ratio of the fractional change in a function to the fractional change in a parameter as the change in the parameter approaches zero. Applying the same definition to the closed-loop poles of a system that vary with a parameter, we define *root sensitivity* as the ratio of the fractional change in a closed-loop pole to the fractional change in a system parameter, such as gain. Using Eq. (7.75), we calculate the sensitivity of a closed-loop pole, $s$, to gain, $K$:

$$S_{s:K} = \frac{K}{s}\frac{\delta s}{\delta K} \tag{8.69}$$

where $s$ is the current pole location, and $K$ is the current gain. Using Eq. (8.69) and converting the partials to finite increments, the actual change in the closed-loop poles can be approximated as

$$\Delta s = s(S_{s:K})\frac{\Delta K}{K} \tag{8.70}$$

where $\Delta s$ is the change in pole location, and $\Delta K/K$ is the fractional change in the gain, $K$. Let us demonstrate with an example. We begin with the characteristic equation from which $\delta s/\delta K$ can be found. Then, using Eq. (8.69) with the current closed-loop pole, $s$, and its associated gain, $K$, we can find the sensitivity.

## Example 8.10 Root Sensitivity of a Closed-Loop System to Gain Variations

**PROBLEM:** Find the root sensitivity of the system in Figure 8.4 at $s = -9.47$ and $-5 + j5$. Also calculate the change in the pole location for a 10% change in $K$.

**SOLUTION:** The system's characteristic equation, found from the closed-loop transfer function denominator, is $s^2 + 10s + K = 0$. Differentiating with respect to $K$, we have

$$2s\frac{\delta s}{\delta K} + 10\frac{\delta s}{\delta K} + 1 = 0 \tag{8.71}$$

from which

$$\frac{\delta s}{\delta K} = \frac{-1}{2s + 10} \tag{8.72}$$

Substituting Eq. (8.72) into Eq. (8.69), the sensitivity is found to be

$$S_{s:K} = \frac{K}{s} * \frac{-1}{2s + 10} \tag{8.73}$$

For $s = -9.47$, Table 8.1 shows $K = 5$. Substituting these values into Eq. (8.73) yields $S_{s:K} = -0.059$. The change in the pole location for a 10% change in $K$ can be found using Eq. (8.70), with $s = -9.47$, $\Delta K/K = 0.1$, and $S_{s:K} = -0.059$. Hence, $\Delta s = 0.056$, or the pole will move to the right by 0.056 units for a 10% change in $K$.

For $s = -5 + j5$, Table 8.1 shows $K = 50$. Substituting these values into Eq. (8.73) yields $S_{s:K} = 1/(1 + j1) = (1/\sqrt{2})\angle -45°$. The change in the pole location for a 10% change in $K$ can be found using Eq. (8.70), with $s = -5 + j5$, $\Delta K/K = 0.1$, and

$S_{s:K} = (1/\sqrt{2})\angle - 45°$. Hence, $\Delta s = -j5$, or the pole will move vertically by 0.5 unit for a 10% change in $K$.

In summary, then, at $K = 5, S_{s:K} = -0.059$. At $K = 50$, $S_{s:K} = (1/\sqrt{2})\angle - 45°$. Comparing magnitudes, we conclude that the root locus is less sensitive to changes in gain at the lower value of $K$. Notice that root sensitivity is a complex quantity possessing both the magnitude and direction information from which the change in poles can be calculated.

## Skill-Assessment Exercise 8.9

**PROBLEM:** A negative unity feedback system has the forward transfer function

$$G(s) = \frac{K(s+1)}{s(s+2)}$$

If $K$ is set to 20, find the changes in closed-loop pole location for a 5% change in $K$.

**ANSWER:** For the closed-loop pole at $-21.05$, $\Delta s = -0.9975$; for the closed-loop pole at $-0.95$, $\Delta s = -0.0025$.

The complete solution is at www.wiley.com/college/nise.

## Case Studies Antenna Control: Transient Design via Gain

Design D

The main thrust of this chapter is to demonstrate design of higher-order systems (higher than two) through gain adjustment. Specifically, we are interested in determining the value of gain required to meet transient response requirements, such as percent overshoot, settling time, and peak time. The following case study emphasizes this design procedure, using the root locus.

**PROBLEM:** Given the antenna azimuth position control system shown on the front endpapers, Configuration 1, find the preamplifier gain required for 25% overshoot.

**SOLUTION:** The block diagram for the system was derived in the Case Studies section in Chapter 5 and is shown in Figure 5.34(*c*), where $G(s) = 6.63K/[s(s+1.71)(s+100)]$.

First a sketch of the root locus is made to orient the designer. The real-axis segments are between the origin and $-1.71$ and from $-100$ to infinity. The locus begins at the open-loop poles, which are all on the real axis at the origin, $-1.71$, and $-100$. The locus then moves toward the zeros at infinity by following asymptotes that, from Eqs. (8.27) and (8.28), intersect the real axis at $-33.9$ at angles of 60°, 180°, and $-60°$. A portion of the root locus is shown in Figure 8.28.

From Eq. (4.39), 25% overshoot corresponds to a damping ratio of 0.404. Now draw a radial line from the origin at an angle of $\cos^{-1}\zeta = 113.8$. The intersection of this line with the root locus locates the system's dominant, second-order closed-loop poles. Using the root locus program discussed in Appendix H.2 at www.wiley.com/college/nise

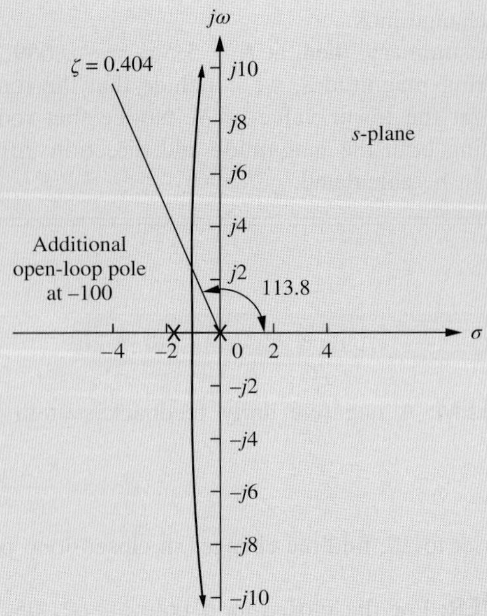


**FIGURE 8.28** Portion of the root locus for the antenna control system

to search the radial line for 180° yields the closed-loop dominant poles as $2.063 \angle 113.8^\circ = -0.833 \pm j1.888$. The gain value yields $6.63K = 425.7$, from which $K = 64.21$.

Checking our second-order assumption, the third pole must be to the left of the open-loop pole at $-100$ and is thus greater than five times the real part of the dominant pole pair, which is $-0.833$. The second-order approximation is thus valid.

The computer simulation of the closed-loop system's step response in Figure 8.29 shows that the design requirement of 25% overshoot is met.

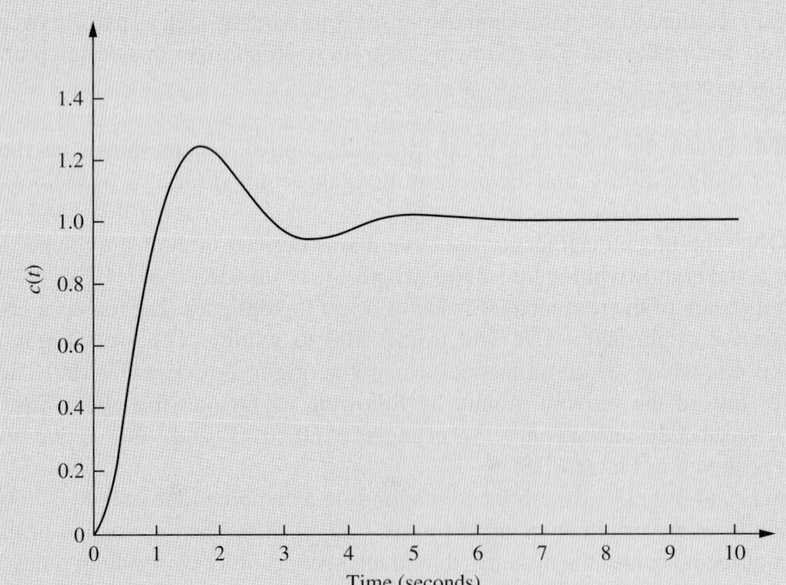


**FIGURE 8.29** Step response of the gain-adjusted antenna control system

**CHALLENGE:** You are now given a problem to test your knowledge of this chapter's objectives. Referring to the antenna azimuth position control system shown on the front endpapers, Configuration 2, do the following:

**a.** Find the preamplifier gain, $K$, required for an 8-second settling time.

**b.** `Repeat, using MATLAB.`

MATLAB
ML

## UFSS Vehicle: Transient Design via Gain

Design
D

In this case study, we apply the root locus to the UFSS vehicle pitch control loop. The pitch control loop is shown with both rate and position feedback on the back endpapers. In the example that follows, we plot the root locus without the rate feedback and then with the rate feedback. We will see the stabilizing effect that rate feedback has upon the system.

**PROBLEM:** Consider the block diagram of the pitch control loop for the UFSS vehicle shown on the back endpapers (*Johnson, 1980*).

**a.** If $K_2 = 0$ (no rate feedback), plot the root locus for the system as a function of pitch gain, $K_1$, and estimate the settling time and peak time of the closed-loop response with 20% overshoot.

**b.** Let $K_2 = K_1$ (add rate feedback) and repeat **a**.

**SOLUTION:**

**a.** Letting $K_2 = 0$, the open-loop transfer function is

$$G(s)H(s) = \frac{0.25K_1(s+0.435)}{(s+1.23)(s+2)(s^2+0.226s+0.0169)} \tag{8.74}$$

from which the root locus is plotted in Figure 8.30. Searching along the 20% overshoot line evaluated from Eq. (4.39), we find the dominant second-order poles to be $-0.202 \pm j0.394$ with a gain of $K = 0.25K_1 = 0.706$, or $K_1 = 2.824$.

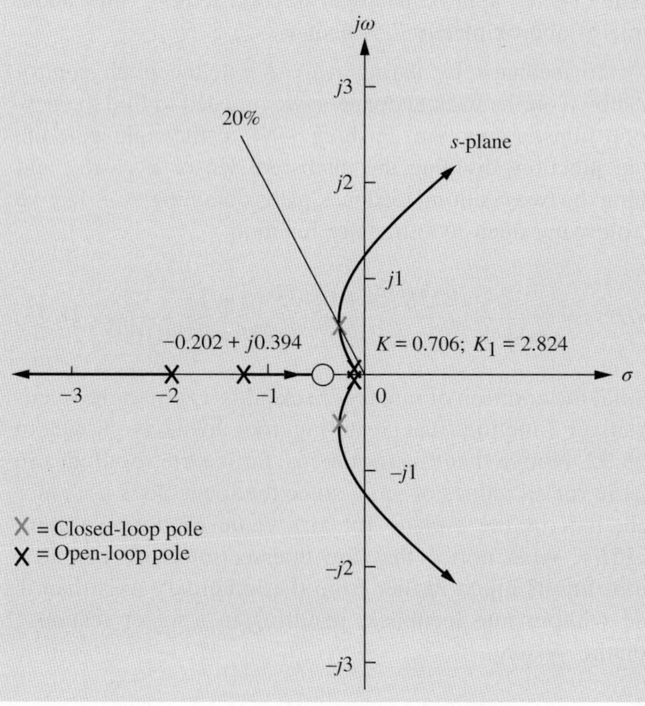


**FIGURE 8.30** Root locus of pitch control loop without rate feedback, UFSS vehicle

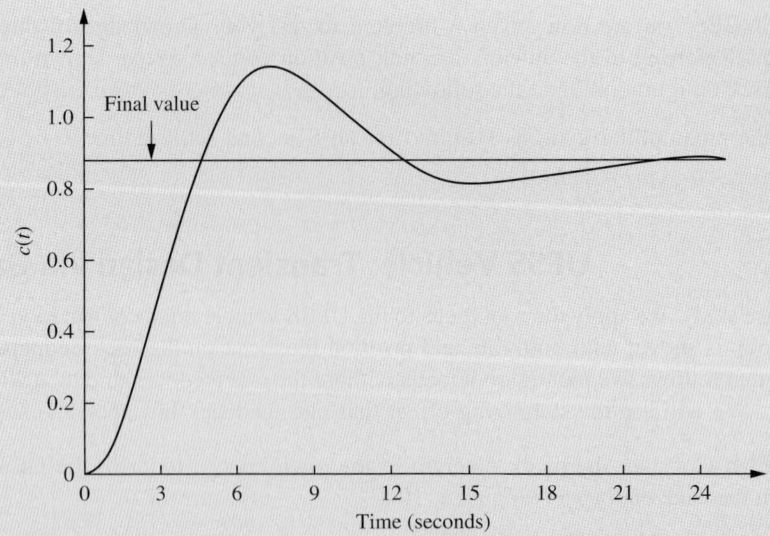


**FIGURE 8.31** Computer simulation of step response of pitch control loop without rate feedback, UFSS vehicle

From the real part of the dominant pole, the settling time is estimated to be $T_s = 4/0.202 = 19.8$ seconds. From the imaginary part of the dominant pole, the peak time is estimated to be $T_p = \pi/0.394 = 7.97$ seconds. Since our estimates are based upon a second-order assumption, we now test our assumption by finding the third closed-loop pole location between $-0.435$ and $-1.23$ and the fourth closed-loop pole location between $-2$ and infinity. Searching each of these regions for a gain of $K = 0.706$, we find the third and fourth poles at $-0.784$ and $-2.27$, respectively. The third pole, at $-0.784$, may not be close enough to the zero at $-0.435$, and thus the system should be simulated. The fourth pole, at $-2.27$, is 11 times as far from the imaginary axis as the dominant poles and thus meets the requirement of at least five times the real part of the dominant poles.

A computer simulation of the step response for the system, which is shown in Figure 8.31, shows a 29% overshoot above a final value of 0.88, approximately 20-second settling time, and a peak time of approximately 7.5 seconds.

**b.** Adding rate feedback by letting $K_2 = K_1$ in the pitch control system shown on the back endpapers, we proceed to find the new open-loop transfer function. Pushing $-K_1$ to the right past the summing junction, dividing the pitch rate sensor by $-K_1$, and combining the two resulting feedback paths obtaining $(s+1)$ give us the following open-loop transfer function:

$$G(s)H(s) = \frac{0.25K_1(s+0.435)(s+1)}{(s+1.23)(s+2)(s^2+0.226s+0.0169)} \tag{8.75}$$

Notice that the addition of rate feedback adds a zero to the open-loop transfer function. The resulting root locus is shown in Figure 8.32. Notice that this root locus, unlike the root locus in **a**, is stable for all values of gain, since the locus does not enter the right half of the $s$-plane for any value of positive gain, $K = 0.25K_1$. Also notice that the intersection with the 20% overshoot line is much farther from the imaginary axis than is the case without rate feedback, resulting in a faster response time for the system.

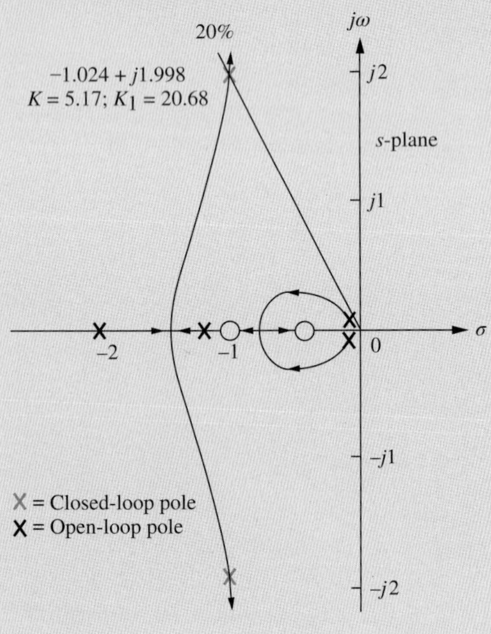


**FIGURE 8.32** Root locus of pitch control loop with rate feedback, UFSS vehicle

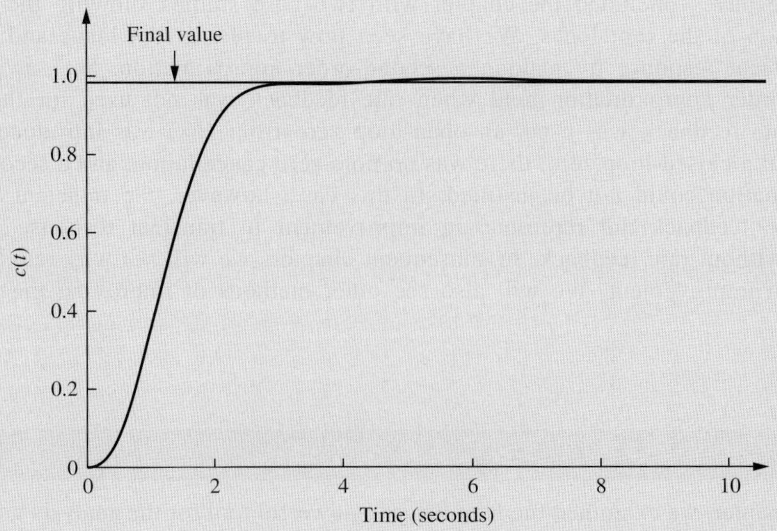


**FIGURE 8.33** Computer simulation of step response of pitch control loop with rate feedback, UFSS vehicle

The root locus intersects the 20% overshoot line at $-1.024 \pm j1.998$ with a gain of $K = 0.25K_1 = 5.17$, or $K_1 = 20.68$. Using the real and imaginary parts of the dominant pole location, the settling time is predicted to be $T_s = 4/1.024 = 3.9$ seconds, and the peak time is estimated to be $T_p = \pi/1.998 = 1.57$ seconds. The new estimates show considerable improvement in the transient response as compared to the system without the rate feedback.

Now we test our second-order approximation by finding the location of the third and fourth poles between $-0.435$ and $-1$. Searching this region for a gain of $K = 5.17$, we locate the third and fourth poles at approximately $-0.5$ and $-0.91$. Since the zero at $-1$ is a zero of $H(s)$, the student can verify that this zero is not a zero of the closed-loop transfer function. Thus, although there may be pole-zero cancellation between the closed-loop pole at $-0.5$ and the closed-loop zero at $-0.435$, there is no *closed-loop* zero to cancel the closed-loop pole at $-0.91$.[2] Our second-order approximation is not valid.

A computer simulation of the system with rate feedback is shown in Figure 8.33. Although the response shows that our second-order approximation is invalid, it still represents a considerable improvement in performance over the system without rate feedback; the percent overshoot is small, and the settling time is about 6 seconds instead of about 20 seconds.

**CHALLENGE:** You are now given a problem to test your knowledge of this chapter's objectives. For the UFSS vehicle (*Johnson, 1980*) heading control system shown on the back endpapers, and introduced in the case study challenge in Chapter 5, do the following:

**a.** Let $K_2 = K_1$ and find the value of $K_1$ that yields 10% overshoot.

**b.** `Repeat, using MATLAB.`

MATLAB
**ML**

[2] The zero at $-1$ shown on the root locus plot of Figure 8.32 is an open-loop zero since it comes from the numerator of $H(s)$.

We have concluded the chapter with two case studies showing the use and application of the root locus. We have seen how to plot a root locus and estimate the transient response by making a second-order approximation. We saw that the second-order approximation held when rate feedback was not used for the UFSS. When rate feedback was used, an open-loop zero from $H(s)$ was introduced. Since it was not a closed-loop zero, there was no pole-zero cancellation, and a second-order approximation could not be justified. In this case, however, the transient response with rate feedback did represent an improvement in transient response over the system without rate feedback. In subsequent chapters we will see why rate feedback yields an improvement. We will also see other methods of improving the transient response.

## Summary

In this chapter, we examined the *root locus*, a powerful tool for the analysis and design of control systems. The root locus empowers us with qualitative and quantitative information about the stability and transient response of feedback control systems. The root locus allows us to find the poles of the closed-loop system by starting from the open-loop system's poles and zeros. It is basically a graphical root-solving technique.

We looked at ways to sketch the root locus rapidly, even for higher-order systems. The sketch gave us qualitative information about changes in the transient response as parameters were varied. From the locus we were able to determine whether a system was unstable for any range of gain.

Next we developed the criterion for determining whether a point in the $s$-plane was on the root locus: The angles from the open-loop zeros, minus the angles from the open-loop poles drawn to the point in the $s$-plane, add up to an odd multiple of 180°.

The computer program discussed in Appendix G.2 at www.wiley.com/college/nise helps us to search rapidly for points on the root locus. This program allows us to find points and gains to meet certain transient response specifications as long as we are able to justify a second-order assumption for higher-order systems. Other computer programs, such as MATLAB, plot the root locus and allow the user to interact with the display to determine transient response specifications and system parameters.

Our method of design in this chapter is gain adjustment. We are limited to transient responses governed by the poles on the root locus. Transient responses represented by pole locations outside of the root locus cannot be obtained by a simple gain adjustment. Further, once the transient response has been established, the gain is set, and so is the steady-state error performance. In other words, by a simple gain adjustment, we have to trade off between a specified transient response and a specified steady-state error. Transient response and steady-state error cannot be designed independently with a simple gain adjustment.

We also learned how to plot the root locus against system parameters other than gain. In order to make this root locus plot, we must first convert the closed-loop transfer function into an equivalent transfer function that has the desired system parameter in the same position as the gain. The chapter discussion concluded with positive-feedback systems and how to plot the root loci for these systems.

The next chapter extends the concept of the root locus to the design of compensation networks. These networks have as an advantage the separate design of transient performance and steady-state error performance.

## Review Questions

1. What is a root locus?
2. Describe two ways of obtaining the root locus.
3. If $KG(s)H(s) = 5\angle 180^\circ$, for what value of gain is $s$ a point on the root locus?
4. Do the zeros of a system change with a change in gain?
5. Where are the zeros of the closed-loop transfer function?
6. What are two ways to find where the root locus crosses the imaginary axis?
7. How can you tell from the root locus if a system is unstable?
8. How can you tell from the root locus if the settling time does not change over a region of gain?
9. How can you tell from the root locus that the natural frequency does not change over a region of gain?
10. How would you determine whether or not a root locus plot crossed the real axis?
11. Describe the conditions that must exist for all closed-loop poles and zeros in order to make a second-order approximation.
12. What rules for plotting the root locus are the same whether the system is a positive- or a negative-feedback system?
13. Briefly describe how the zeros of the open-loop system affect the root locus and the transient response.

## Problems

1. For each of the root loci shown in Figure P8.1, tell whether or not the sketch can be a root locus. If the sketch cannot be a root locus, explain why. Give *all* reasons. [Section: 8.4]

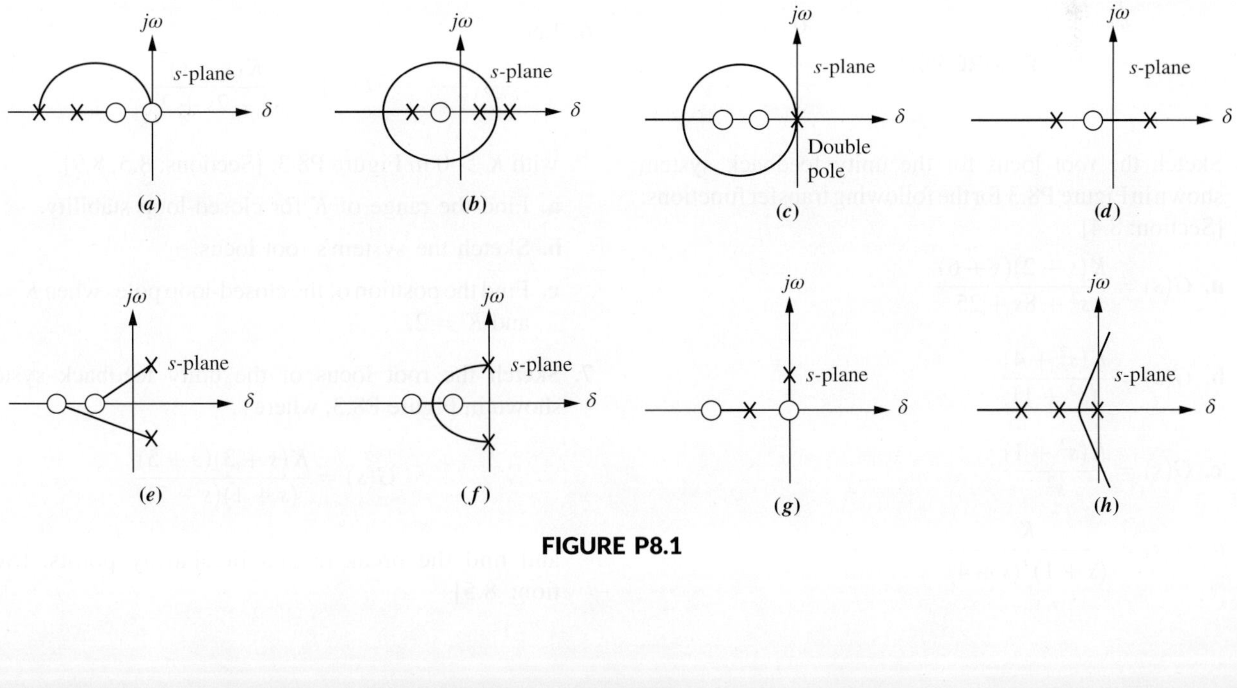


**FIGURE P8.1**

**2.** Sketch the general shape of the root locus for each of the open-loop pole-zero plots shown in Figure P8.2. [Section: 8.4]

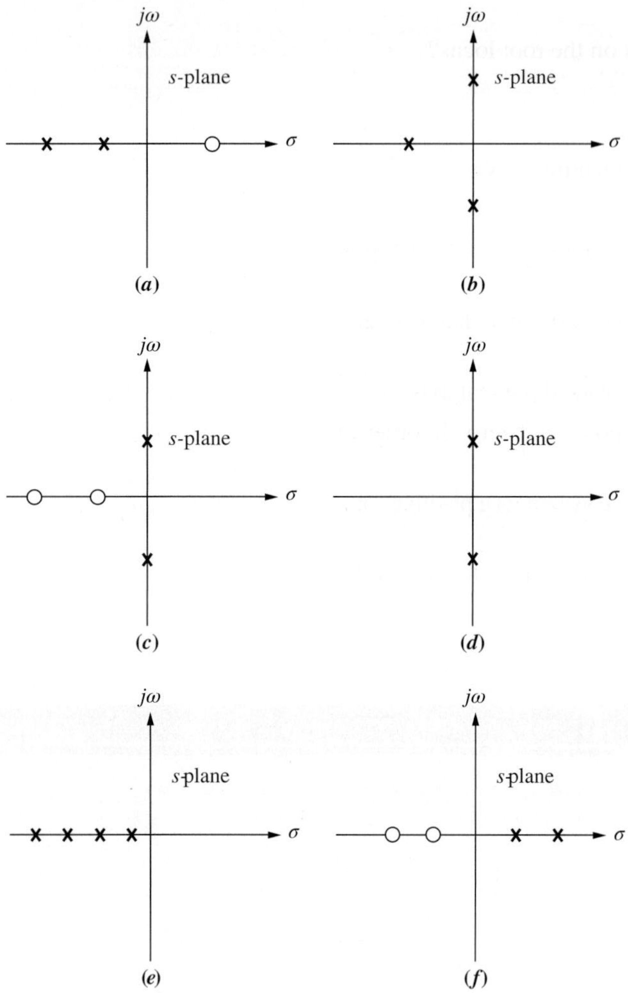


**FIGURE P8.2**

**3.** Sketch the root locus for the unity feedback system shown in Figure P8.3 for the following transfer functions: [Section: 8.4]

**a.** $G(s) = \dfrac{K(s+2)(s+6)}{s^2+8s+25}$

**b.** $G(s) = \dfrac{K(s^2+4)}{(s^2+1)}$

**c.** $G(s) = \dfrac{K(s^2+1)}{s^2}$

**d.** $G(s) = \dfrac{K}{(s+1)^3(s+4)}$

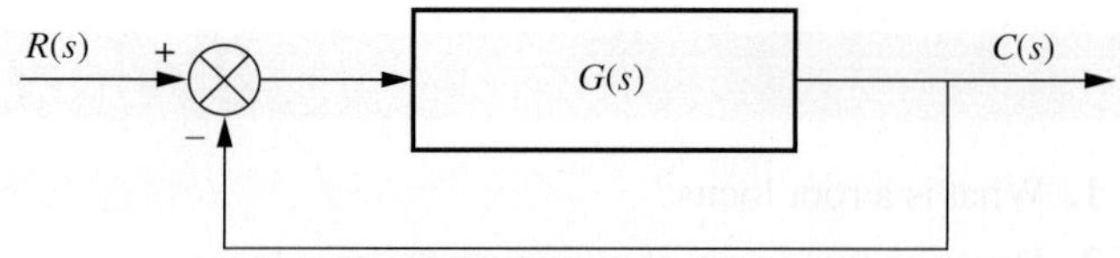


**FIGURE P8.3**

**4.** Let

$$G(s) = \frac{K(s+1)}{s^2(s+9)}$$

in Figure P8.3. [Section: 8.5]

**a.** Plot the root locus.

**b.** Write an expression for the closed-loop transfer function at the point where the three closed-loop poles meet.

**5.** For the open-loop pole-zero plot shown in Figure P8.4, sketch the root locus and find the break-in point. [Section: 8.5]

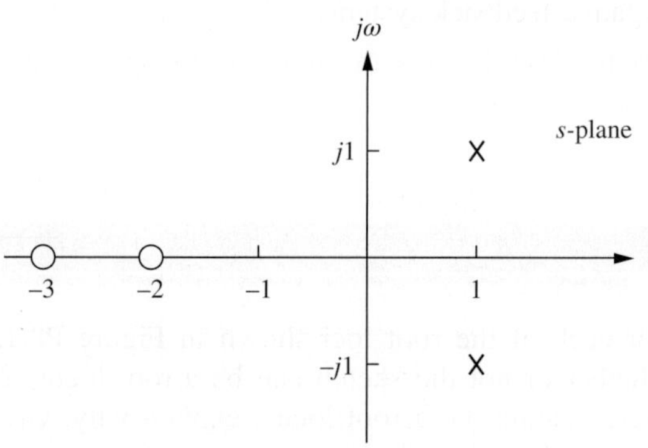


**FIGURE P8.4**

**6.** Let

$$G(s) = \frac{-K(s+1)^2}{s^2+2s+2}$$

with $K > 0$ in Figure P8.3. [Sections: 8.5, 8.9]

**a.** Find the range of $K$ for closed-loop stability.

**b.** Sketch the system's root locus.

**c.** Find the position of the closed-loop poles when $K = 1$ and $K = 2$.

**7.** Sketch the root locus of the unity feedback system shown in Figure P8.3, where

$$G(s) = \frac{K(s+3)(s+5)}{(s+1)(s-7)}$$

and find the break-in and breakaway points. [Section: 8.5]

**8.** The characteristic polynomial of a feedback control system, which is the denominator of the closed-loop transfer function, is given by $s^3 + 3s^2 + (K+2)s + 10K$. Sketch the root locus for this system.

**9.** Figure P8.5 shows open-loop poles and zeros. There are two possibilities for the sketch of the root locus. Sketch each of the two possibilities. Be aware that only one can be the *real* locus for specific open-loop pole and zero values. [Section: 8.4]

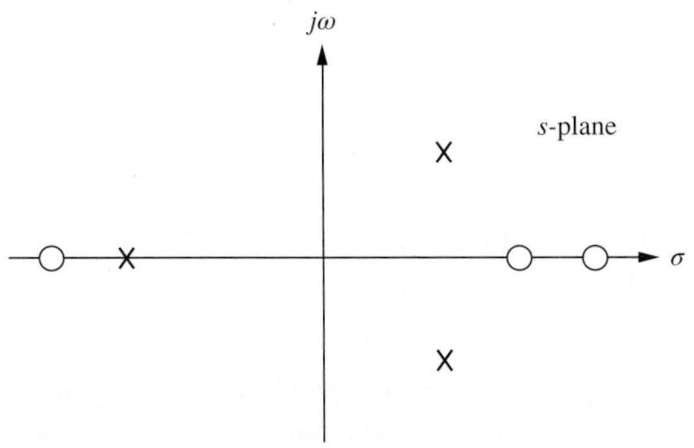


**FIGURE P8.5**

**10.** Plot the root locus for the unity feedback system shown in Figure P8.3, where

$$G(s) = \frac{K(s+1)(s^2+2)}{(s+3)(s-3)}$$

For what range of $K$ will the poles be in the right half-plane? [Section: 8.5]

**11.** For the unity feedback system shown in Figure P8.3, where

$$G(s) = \frac{K(s^2-9)}{(s^2+4)}$$

sketch the root locus and tell for what values of $K$ the system is stable and unstable. [Section: 8.5]

**12.** Sketch the root locus for the unity feedback system shown in Figure P8.3, where

$$G(s) = \frac{K(s^2+4)}{(s+3)(s+4)}$$

Give the values for all critical points of interest. Is the system ever unstable? If so, for what range of $K$? [Section: 8.5]

**13.** Sketch the root locus and find the range of $K$ for stability for the unity feedback system shown in Figure P8.3 for the following conditions: [Section: 8.5]

**a.** $G(s) = \dfrac{K(s^2+1)}{(s-1)(s+2)(s+3)}$

**b.** $G(s) = \dfrac{K(s^2-2s+2)}{s(s+1)(s+2)}$

**14.** For each system shown in Figure P8.6, make an accurate plot of the root locus and find the following: [Section: 8.5]

**a.** The breakaway and break-in points

**b.** The range of $K$ to keep the system stable

**c.** The value of $K$ that yields a stable system with critically damped second-order poles

**d.** The value of $K$ that yields a stable system with a pair of second-order poles that have a damping ratio of 0.707

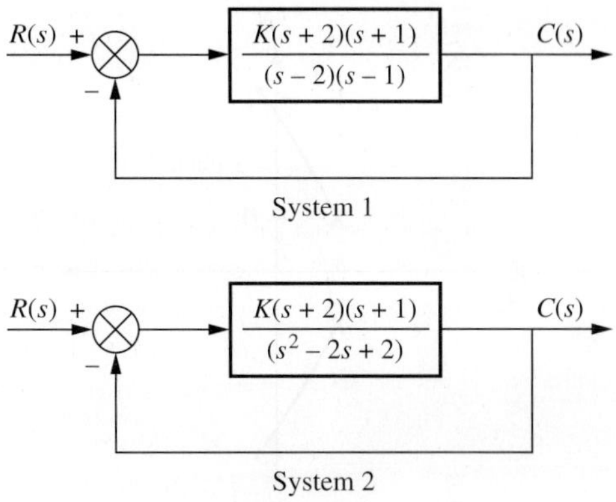


**FIGURE P8.6**

**15.** For the unity feedback system of Figure P8.3, where

$$G(s) = \frac{K(s+3)}{(s^2+2)(s-2)(s+5)}$$

sketch the root locus and find the range of $K$ such that there will be only two right–half-plane poles for the closed-loop system. [Section: 8.5]

**16.** For the unity feedback system of Figure P8.3, where

$$G(s) = \frac{K}{s(s+6)(s+9)}$$

plot the root locus and calibrate your plot for gain. Find all the critical points, such as breakaways, asymptotes, $j\omega$-axis crossing, and so forth. [Section: 8.5]

**17.** Given the unity feedback system of Figure P8.3, make an accurate plot of the root locus for the following:

**a.** $G(s) = \dfrac{K(s^2 - 2s + 2)}{(s+1)(s+2)}$

**b.** $G(s) = \dfrac{K(s-1)(s-2)}{(s+1)(s+2)}$

Also, do the following:

Calibrate the gain for at least four points for each case. Also find the breakaway points, the $j\omega$-axis crossing, and the range of gain for stability for each case. Find the angles of arrival for Part **a**. [Section: 8.5]

**18.** Given the root locus shown in Figure P8.7, [Section: 8.5]

**a.** Find the value of gain that will make the system marginally stable.

**b.** Find the value of gain for which the closed-loop transfer function will have a pole on the real axis at −5.

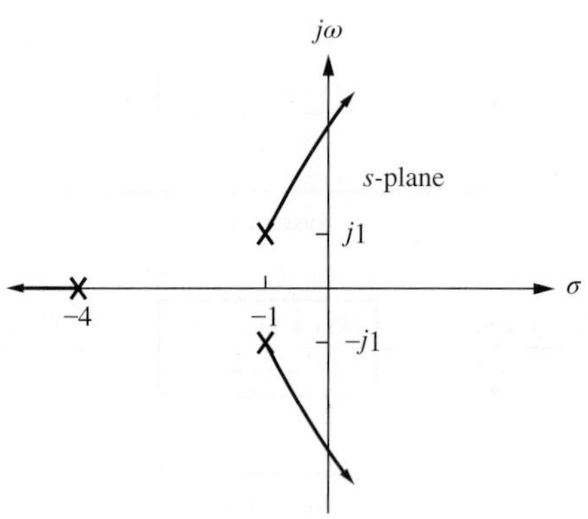


**FIGURE P8.7**

**19.** Given the unity feedback system of Figure P8.3, where

$$G(s) = \frac{K(s+1)}{s(s+2)(s+3)(s+4)}$$

do the following: [Section: 8.5]

**a.** Sketch the root locus.

**b.** Find the asymptotes.

**c.** Find the value of gain that will make the system marginally stable.

**d.** Find the value of gain for which the closed-loop transfer function will have a pole on the real axis at −0.5.

**20.** For the unity feedback system of Figure P8.3, where

$$G(s) = \frac{K(s+\alpha)}{s(s+3)(s+6)}$$

find the values of $\alpha$ and $K$ that will yield a second-order closed-loop pair of poles at $-1 \pm j100$. [Section: 8.5]

**21.** For the unity feedback system of Figure P8.3, where

$$G(s) = \frac{K(s-1)(s-2)}{s(s+1)}$$

sketch the root locus and find the following: [Section: 8.5]

**a.** The breakaway and break-in points

**b.** The $j\omega$-axis crossing

**c.** The range of gain to keep the system stable

**d.** The value of $K$ to yield a stable system with second-order complex poles, with a damping ratio of 0.5

**22.** For the unity feedback system shown in Figure P8.3, where

$$G(s) = \frac{K(s+10)(s+20)}{(s+30)(s^2 - 20s + 200)}$$

do the following: [Section: 8.7]

**a.** Sketch the root locus.

**b.** Find the range of gain, $K$, that makes the system stable.

**c.** Find the value of $K$ that yields a damping ratio of 0.707 for the system's closed-loop dominant poles.

**d.** Find the value of $K$ that yields closed-loop critically damped dominant poles.

**23.** For the system of Figure P8.8(*a*), sketch the root locus and find the following: [Section: 8.7]

**a.** Asymptotes

**b.** Breakaway points

**c.** The range of $K$ for stability

**d.** The value of $K$ to yield a 0.7 damping ratio for the dominant second-order pair

To improve stability, we desire the root locus to cross the $j\omega$-axis at $j5.5$. To accomplish this, the open-loop function is cascaded with a zero, as shown in Figure P8.8(*b*).

**e.** Find the value of $\alpha$ and sketch the new root locus.

**f.** Repeat Part **c** for the new locus.

**g.** Compare the results of Part **c** and Part **f**. What improvement in transient response do you notice?

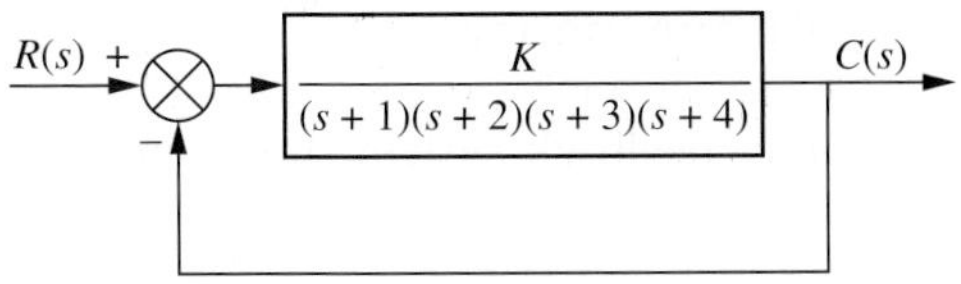


(*a*)

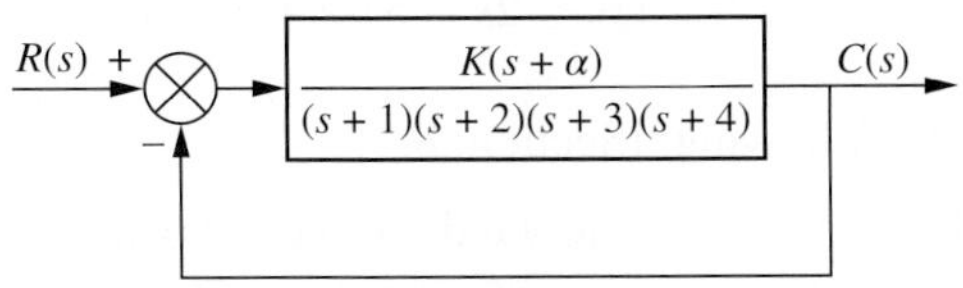


(*b*)

**FIGURE P8.8**

**24.** Root loci are usually plotted for variations in the gain. Sometimes we are interested in the variation of the closed-loop poles as other parameters are changed. For the system shown in Figure P8.9, sketch the root locus as $\alpha$ is varied. [Section: 8.8]

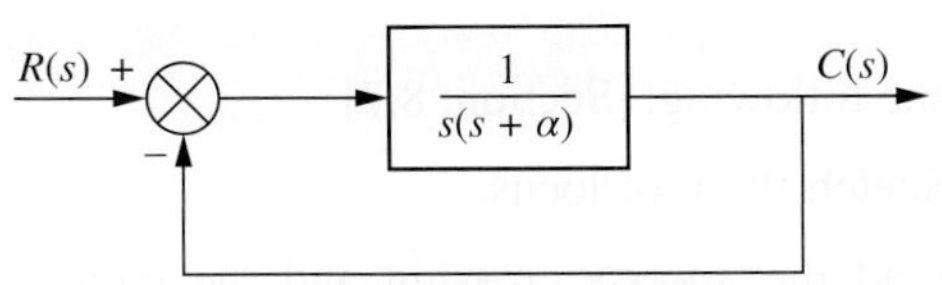


**FIGURE P8.9**

**25.** Sketch the root locus for the positive-feedback system shown in Figure P8.10. [Section: 8.9]

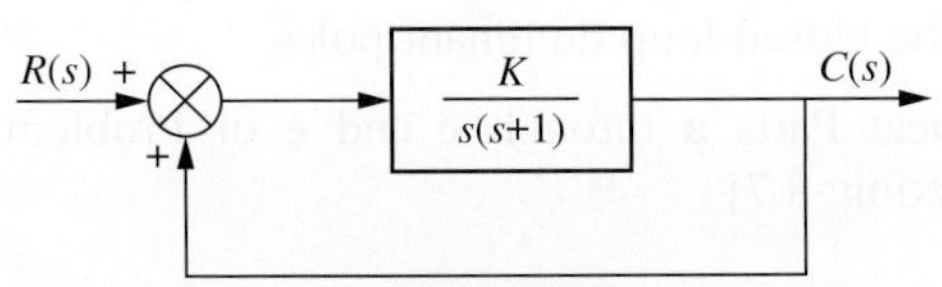


**FIGURE P8.10**

**26.** Given the unity feedback system shown in Figure P8.3, where

$$G(s) = \frac{K}{(s+1)(s+2)(s+4)}$$

do the following problem parts by first making a second-order approximation. After you are finished with all of the parts, justify your second-order approximation. [Section: 8.7]

**a.** Sketch the root locus.

**b.** Find $K$ for 20% overshoot.

**c.** For $K$ found in Part **b**, what is the settling time, and what is the peak time?

**d.** Find the locations of higher-order poles for $K$ found in Part **b**.

**e.** Find the range of $K$ for stability.

**27.** For the unity feedback system shown in Figure P8.3, where

$$G(s) = \frac{K(s^2 - 2s + 2)}{(s+2)(s+4)(s+5)(s+6)}$$

do the following: [Section: 8.7]

**a.** Sketch the root locus.

**b.** Find the asymptotes.

**c.** Find the range of gain, $K$, that makes the system stable.

**d.** Find the breakaway points.

**e.** Find the value of $K$ that yields a closed-loop step response with 25% overshoot.

**f.** Find the location of higher-order closed-loop poles when the system is operating with 25% overshoot.

**g.** Discuss the validity of your second-order approximation.

**h.** `Use MATLAB to obtain the closed-loop step response to validate or refute your second-order approximation.` MATLAB **ML**

**28.** The unity feedback system shown in Figure P8.3, where

$$G(s) = \frac{K(s+2)(s+4)}{s(s+1)}$$

is to be designed for minimum damping ratio. Find the following: [Section: 8.7]

**a.** The value of $K$ that will yield minimum damping ratio

**b.** The estimated percent overshoot for that case

**c.** The estimated settling time and peak time for that case

**d.** The justification of a second-order approximation (discuss)

**e.** The expected steady-state error for a unit ramp input for the case of minimum damping ratio

**29.** For the unity feedback system shown in Figure P8.3, where

$$G(s) = \frac{K}{s(s+1)(s+2)}$$

find $K$ to yield closed-loop complex poles with a damping ratio of 0.7. Does your solution require a justification of a second-order approximation? Explain.

**30.** For the unity feedback system shown in Figure P8.3, where

$$G(s) = \frac{K(s+\alpha)}{s(s+1)(s+10)}$$

find the value of $\alpha$ so that the system will have a settling time of 4 seconds for large values of $K$. Sketch the resulting root locus. [Section: 8.8]

**31.** For the unity feedback system shown in Figure P8.3, where

$$G(s) = \frac{K(s+6)}{(s^2+10s+26)(s+1)^2(s+\alpha)}$$

design $K$ and $\alpha$ so that the dominant complex poles of the closed-loop function have a damping ratio of 0.45 and a natural frequency of 9/8 rad/s.

**32.** For the unity feedback system shown in Figure P8.3, where

$$G(s) = \frac{K}{s(s+1)(s+5)(s+6)}$$

do the following: [Section: 8.7]

**a.** Sketch the root locus.

**b.** Find the value of $K$ that will yield a 10% overshoot.

**c.** Locate all nondominant poles. What can you say about the second-order approximation that led to your answer in Part **b.**

**d.** Find the range of $K$ that yields a stable system.

**33.** Repeat Problem 32 using MATLAB. Use one program to do the following: MATLAB ML

**a.** Display a root locus and pause.

**b.** Draw a close-up of the root locus where the axes go from −1 to 0 on the real axis and −2 to 2 on the imaginary axis.

**c.** Overlay the 10% overshoot line on the close-up root locus.

**d.** Allow you to select interactively the point where the root locus crosses the 10% overshoot line, and respond with the gain at that point as well as all of the closed-loop poles at that gain.

**e.** Generate the step response at the gain for 10% overshoot.

**34.** For the unity feedback system shown in Figure P8.3, where

$$G(s) = \frac{K(s^2+4s+6)}{(s^2+2s+5)(s+3)(s+5)}$$

do the following: [Section: 8.7]

**a.** Find the gain, $K$, to yield a 1-second peak time if one assumes a second-order approximation.

**b.** Check the accuracy of the second-order approximation using MATLAB to simulate the system. MATLAB ML

**35.** For the unity feedback system shown in Figure P8.3, where

$$G(s) = \frac{K(s+2)(s+3)}{(s^2+2s+2)(s+4)(s+5)(s+6)}$$

do the following: [Section: 8.7]

**a.** Sketch the root locus.

**b.** Find the $j\omega$-axis crossing and the gain, $K$, at the crossing.

**c.** Find all breakaway and break-in points.

**d.** Find angles of departure from the complex poles.

**e.** Find the gain, $K$, to yield a damping ratio of 0.3 for the closed-loop dominant poles.

**36.** Repeat Parts **a** through **c** and **e** of Problem 35 for [Section: 8.7]

$$G(s) = \frac{K(s+8)}{s(s+2)(s+4)(s+6)}$$

**37.** For the unity feedback system shown in Figure P8.3, where

$$G(s) = \frac{K}{(s+4)(s^2+4s+6)}$$

do the following: [Section: 8.7]

**a.** Find the location of the closed-loop dominant poles if the system is operating with 15% overshoot.

**b.** Find the gain for Part **a.**

**c.** Find all other closed-loop poles.

**d.** Evaluate the accuracy of your second-order approximation.

**38.** For the system shown in Figure P8.11, do the following: [Section: 8.7]

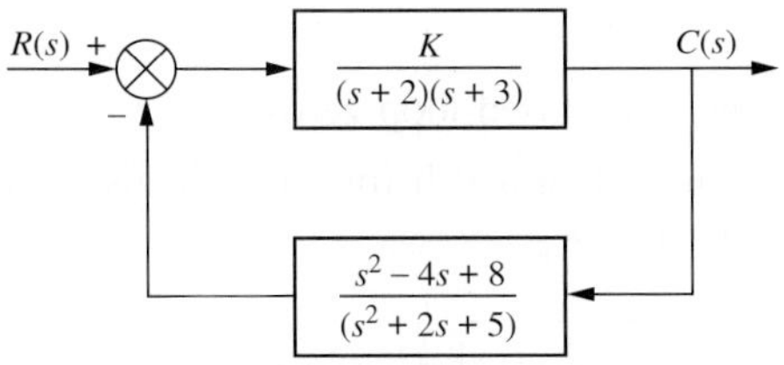


**FIGURE P8.11**

**a.** Sketch the root locus.

**b.** Find the $j\omega$-axis crossing and the gain, $K$, at the crossing.

**c.** Find the real-axis breakaway to two-decimal-place accuracy.

**d.** Find angles of arrival to the complex zeros.

**e.** Find the closed-loop zeros.

**f.** Find the gain, $K$, for a closed-loop step response with 30% overshoot.

**g.** Discuss the validity of your second-order approximation.

**39.** Sketch the root locus for the system of Figure P8.12 and find the following: [Section: 8.7]

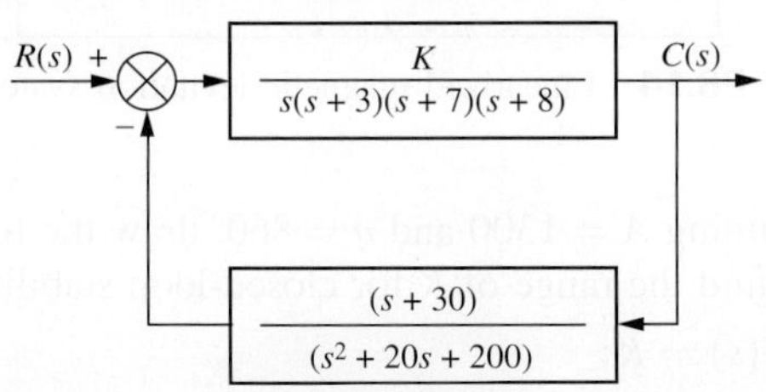


**FIGURE P8.12**

**a.** The range of gain to yield stability

**b.** The value of gain that will yield a damping ratio of 0.707 for the system's dominant poles

**c.** The value of gain that will yield closed-loop poles that are critically damped

**40.** `Repeat Problem 39 using MATLAB. The program will do the following in one program:` MATLAB ML

**a.** `Display a root locus and pause.`

**b.** `Display a close-up of the root locus where the axes go from –2 to 2 on the real axis and –2 to 2 on the imaginary axis.`

**c.** `Overlay the 0.707 damping ratio line on the close-up root locus.`

**d.** `Allow you to select interactively the point where the root locus crosses the 0.707 damping ratio line, and respond by displaying the gain at that point as well as all of the closed-loop poles at that gain. The program will then allow you to select interactively the imaginary-axis crossing and respond with a display of the gain at that point as well as all of the closed-loop poles at that gain. Finally, the program will repeat the evaluation for critically damped dominant closed-loop poles.`

**e.** `Generate the step response at the gain for 0.707 damping ratio.`

**41.** Given the unity feedback system shown in Figure P8.3, where

$$G(s) = \frac{K(s+z)}{s^2(s+20)}$$

do the following: [Section: 8.7]

**a.** If $z = 6$, find $K$ so that the damped frequency of oscillation of the transient response is 10 rad/s.

**b.** For the system of Part **a**, what static error constant (finite) can be specified? What is its value?

**c.** The system is to be redesigned by changing the values of $z$ and $K$. If the new specifications are $\%OS = 4.32\%$ and $T_s = 0.4$ s, find the new values of $z$ and $K$.

**42.** Given the unity feedback system shown in Figure P8.3, where

$$G(s) = \frac{K}{(s+1)(s+3)(s+6)^2}$$

find the following: [Section: 8.7]

**a.** The value of gain, $K$, that will yield a settling time of 4 seconds

**b.** The value of gain, $K$, that will yield a critically damped system

**43.** You are given the unity-feedback system of Figure P8.3, where (MATLAB ML)

$$G(s) = \frac{K(s+0.02)}{s^2(s+4)(s+10)(s+25)}$$

Use MATLAB to plot the root locus. Use a closeup of the locus (from −5 to 0 and −1 to 6) to find the gain, $K$, that yields a closed-loop unit-step response, $c(t)$, with 20.5% overshoot and a settling time of $T_s = 3$ seconds. Mark on the time response graph all other relevant characteristics, such as the peak time, rise time, and final steady-state value.

**44.** Let

$$G(s) = \frac{K(s-1)}{(s+2)(s+3)}$$

in Figure P8.3. [Section: 8.7].

**a.** Find the range of $K$ for closed-loop stability.

**b.** Plot the root locus for $K > 0$.

**c.** Plot the root locus for $K < 0$.

**d.** Assuming a step input, what value of $K$ will result in the smallest attainable settling time?

**e.** Calculate the system's $e_{ss}$ for a unit step input assuming the value of $K$ obtained in Part **d**.

**f.** Make an approximate hand sketch of the unit step response of the system if $K$ has the value obtained in Part **d**.

**45.** Given the unity feedback system shown in Figure P8.3, where

$$G(s) = \frac{K}{s(s+1)(s+5)}$$

evaluate the pole sensitivity of the closed-loop system if the second-order, underdamped closed-loop poles are set for [Section: 8.10]

**a.** $\zeta = 0.591$

**b.** $\zeta = 0.456$

**c.** Which of the two previous cases has more desirable sensitivity?

**46.** Repeat Problem 3 but sketch your root loci for negative values of $K$. [Section: 8.9]

**47.** During ascent, the automatic steering program aboard the space shuttle provides the interface between the low-rate processing of guidance (commands) and the high-rate processing of flight control (steering in response to the commands). The function performed is basically that of smoothing. A simplified representation of a maneuver smoother linearized for coplanar maneuvers is shown in Figure P8.13. Here $\theta_{CB}(s)$ is the commanded body angle as calculated by guidance, and $\theta_{CB}(s)$ is the desired body angle sent to flight control after smoothing.[3] Using the methods of Section 8.8, do the following:

**a.** Sketch a root locus where the roots vary as a function of $K_3$.

**b.** Locate the closed-loop zeros.

**c.** Repeat Parts **a** and **b** for a root locus sketched as a function of $K_2$.

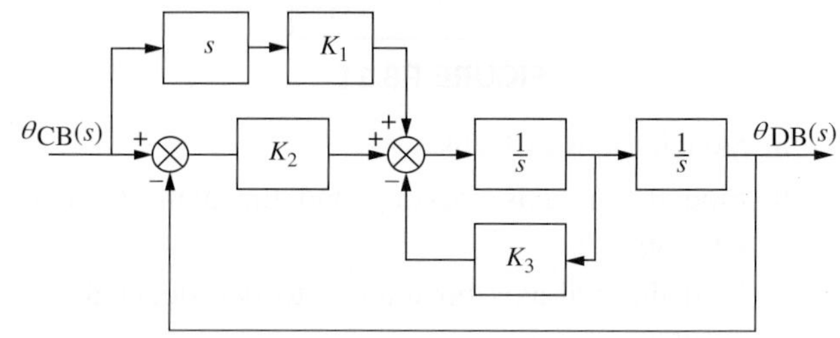


**FIGURE P8.13** Block diagram of smoother

**48.** Figure P8.14 shows the block diagram of a closed-loop control of a linearized magnetic levitation system (*Galvão, 2003*).

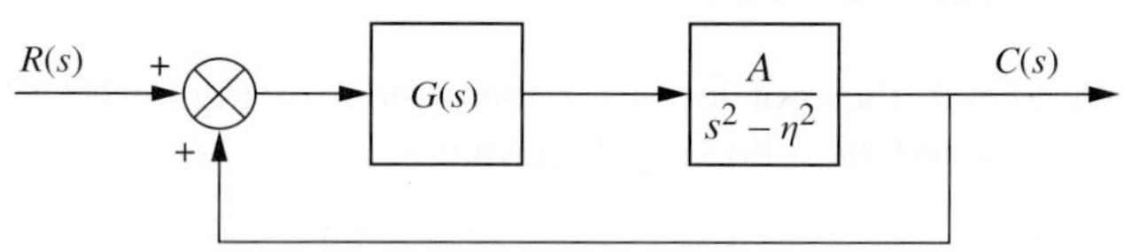


**FIGURE P8.14** Linearized magnetic levitation system block diagram

Assuming $A = 1300$ and $\eta = 860$, draw the root locus and find the range of $K$ for closed-loop stability when:

**a.** $G(s) = K$;

**b.** $G(s) = \dfrac{K(s+200)}{s+1000}$

**49.** The simplified transfer function model from steering angle $\delta(s)$ to tilt angle $\varphi(s)$ in a bicycle is given by

$$G(s) = \frac{\varphi(s)}{\delta(s)} = \frac{aV}{bh} * \frac{s + \frac{V}{a}}{s^2 - \frac{g}{h}}$$

[3] Source: Rockwell International.

In this model, $h$ represents the vertical distance from the center of mass to the floor, so it can be readily verified that the model is open-loop unstable. (*Åström, 2005*). Assume that for a specific bicycle, $a = 0.6$ m, $b = 1.5$ m, $h = 0.8$ m, and $g = 9.8$ m/sec. In order to stabilize the bicycle, it is assumed that the bicycle is placed in the closed-loop configuration shown in Figure P8.3 and that the only available control variable is $V$, the rear wheel velocity.

**a.** Find the range of $V$ for closed-loop stability.

**b.** Explain why the methods presented in this chapter cannot be used to obtain the root locus.

**c.** `Use MATLAB to obtain the system's root locus.` MATLAB ML

**50.** A technique to control the steering of a vehicle that follows a line located in the middle of a lane is to define a look-ahead point and measure vehicle deviations with respect to the point. A linearized model for such a vehicle is

$$\begin{bmatrix} \dot{V} \\ \dot{r} \\ \dot{\psi} \\ \dot{Y}_g \end{bmatrix} = \begin{bmatrix} a_{11} & a_{12} & -b_1 K & \dfrac{b_1 K}{d} \\ a_{21} & a_{22} & -b_2 K & \dfrac{b_2 K}{d} \\ 0 & 1 & 0 & 0 \\ 1 & 0 & U & 0 \end{bmatrix} \begin{bmatrix} V \\ r \\ \psi \\ Y_g \end{bmatrix}$$

where $V$ = vehicle's lateral velocity, $r$ = vehicle's yaw velocity, $\psi$ = vehicle's yaw position, and $Y_g$ = the $y$-axis coordinate of the vehicle's center of gravity. $K$ is a parameter to be varied depending upon trajectory changes. In a specific vehicle traveling at a speed of $U = -10$ m/sec, the parameters are $a_{11} = -11.6842$, $a_{12} = 6.7632$, $b_1 = -61.5789$, $a_{21} = -3.5143$, $a_{22} = 24.0257$, and $b_2 = 66.8571$. $d = 5$ m is the look-ahead distance (*Ünyelioğlu, 1997*). Assuming the vehicle will be controlled in closed loop:

**a.** Find the system's characteristic equation as a function of $K$.

**b.** Find the system's root locus as $K$ is varied.

**c.** Using the root locus found in Part **b**, show that the system will be unstable for all values $K$.

**51.** It is known that mammals have hormonal regulation mechanisms that help maintain almost constant calcium plasma levels (0.08–0.1 g/L in dairy cows). This control is necessary to maintain healthy functions, as calcium is responsible for diverse physiological functions, such as bone formation, intracellular communications, and blood clotting. It has been postulated that the mechanism of calcium control resembles that of a PI (proportional-plus-integral) controller. PI controllers (discussed in detail in Chapter 9) are placed in cascade with the plant and used to improve steady-state error. Assume that the PI controller has the form $G_c(s) = \left[K_P + \dfrac{K_I}{S}\right]$ where $K_P$ and $K_I$ are constants. Also assume that the mammal's system accumulates calcium in an integrator-like fashion, namely $P(s) = \dfrac{1}{V_s}$, where $V$ is the plasma volume. The closed-loop model is similar to that of Figure P8.3, where $G(s) = G_c(s)P(s)$ (*Khammash, 2004*).

**a.** Sketch the system's root locus as a function of $K_P$, assuming $K_I > 0$ is constant.

**b.** Sketch the system's root locus as a function of $K_I$, assuming $K_P > 0$ is constant.

**52.** The closed-loop vehicle response in stopping a train depends on the train's dynamics and the driver, who is an integral part of the feedback loop. In Figure P8.3, let the input be $R(s) = v_r$ the reference velocity, and the output $C(s) = v$, the actual vehicle velocity. (*Yamazaki, 2008*) shows that such dynamics can be modeled by $G(s) = G_d(s)G_t(s)$ where

$$G_d(s) = h\left(1 + \frac{K}{s}\right)\frac{s - \dfrac{L}{2}}{s + \dfrac{L}{2}}$$

represents the driver dynamics with $h$, $K$, and $L$ parameters particular to each individual driver. We assume here that $h = 0.003$ and $L = 1$. The train dynamics are given by

$$G_t(s) = \frac{k_b f K_p}{M(1 + k_e)s(\tau s + 1)}$$

where $M = 8000$ kg, the vehicle mass; $k_e = 0.1$ the inertial coefficient; $k_b = 142.5$, the brake gain; $K_p = 47.5$, the pressure gain; $\tau = 1.2$ sec, a time constant; and $f = 0.24$, the normal friction coefficient.

**a.** Make a root locus plot of the system as a function of the driver parameter $K$.

**b.** Discuss why this model may not be an accurate description of a real driver-train situation.

**53.** For the dynamic voltage restorer (DVR) discussed in Problem 48, Chapter 7, do the following:

**a.** When $Z_L = \dfrac{1}{sC_L}$, a pure capacitance, the system is more inclined toward instability. Find the system's characteristic equation for this case.

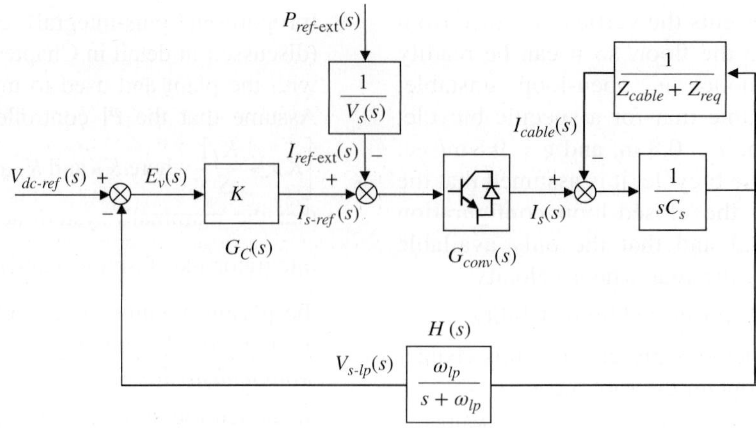


**FIGURE P8.15**[4]

**b.** Using the characteristic equation found in Part **a**, sketch the root locus of the system as a function of $C_L$. Let $L = 7.6\,\text{mH}$, $C = 11\,\mu\text{F}, \alpha = 26.4$, $\beta = 1$, $K_m = 25$, $K_v = 15$, $K_T = 0.09565$, and $\tau = 2$ ms (*Lam, 2004*).

**54.** It has been suggested that the use of closed-loop feedback in ventilators can highly reduce mortality and health problems in patients in need of respiratory treatments (*Hahn, 2012*). A good knowledge of the transfer functions involved is necessary for the design of an appropriate controller. In a study with 18 patients it was found that the open-loop transfer function from minute ventilation (MV) to end-tidal carbon dioxide partial pressure ($P_{ET}CO_2$) can be nominally modeled as:

$$G(s) = \frac{0.415k_c(s+0.092)(s+0.25)}{s(s+0.007)(s+0.207)}$$

**a.** Make a sketch of the root locus of the system indicating the breakaway points and the value $k_c$ takes in each of them.

**b.** In the design of ventilators it is very important to have negligible overshoot with the fastest possible settling time. It has been suggested that a value of $k_c = 5.35$ will achieve these specifications. Mark the position of the closed-loop poles for this value of $k_c$ and explain why this is a reasonable gain choice.

**55.** Voltage droop control is a technique in which loads are driven at lower voltages than those provided by the source. In general, the voltage is decreased as current demand increases in the load. The advantage of voltage droop is that it results in lower sensitivity to load current variations. Voltage droop can be applied to the power distribution of several generators and loads linked through a dc bus. In (*Karlsson, 2003*) generators and loads are driven by 3-phase ac power, so they are interfaced to the bus through ac/dc converters. Since each one of the loads works independently, a feedback system shown in Figure P8.15 is used in each to respond equally to bus voltage variations. Given that $C_s = C_r = 8{,}000\ \mu\text{F}$, $L_{cable} = 50\ \mu\text{H}$, $R_{cable} = 0.06\ \Omega$, $Z_r = R_r = 5\ \Omega$, $\omega_{lp} = 200$ rad/s, $G_{conv}(s) = 1$, $V_{dc\text{-}ref} = 750$ V, and $P_{ref\text{-}ext} = 0$, do the following:

**a.** If $Z_{r_{eq}}$ is the parallel combination of $R_r$ and $C_r$, and $G_{conv}(s) = 1$, find

$$G(s) = \frac{V_s(s)}{I_s(s)} = \frac{V_s(s)}{I_{s-ref}(s)}$$

**b.** `Write a MATLAB M-file to plot and copy the full root locus for that system, then zoom-in the locus by setting the x-axis (real-axis) limits to −150 to 0 and the y-axis (imaginary-axis) limits to −150 to 150. Copy that plot, too, and find and record the following:` MATLAB **ML**

**(1)** `The gain, K, at which the system would have complex-conjugate`

[4] Karlsson, P., and Svensson, J. DC Bus Voltage Control for a Distributed Power System, *IEEE Trans. Power Electronics*, vol. 18, no. 6, 2003, pp. 1405–1412. Fig. 4, p. 1406. *IEEE Transactions on Power Electronics* by Institute of Electrical and Electronics Engineers; Power Electronics Council (Institute of Electrical and Electronic Engineers); IEEE Power Electronics Society. Reproduced with permission of Institute of Electrical and Electronics Engineers, in the format Republish in a book via Copyright Clearance Center.

closed-loop dominant poles with a damping ratio $\zeta = 0.707$

**(2)** The coordinates of the corresponding point selected on the root-locus

**(3)** The values of all closed-loop poles at that gain

**(4)** The output voltage $v_s(t)$ for a step input voltage $v_{dc\text{-}ref}(t) = 750\ u(t)$ volts

**c.** Plot that step response and use the MATLAB **Characteristics** tool (in the graph window) to note on the curve the following parameters:

MATLAB ML

**(1)** The actual percent overshoot and the corresponding peak time, $T_p$

**(2)** The rise time, $T_r$, and the settling time, $T_s$

**(3)** The final steady-state value in volts

**56.** Figure P8.16 shows a simplified drawing of a feedback system that includes the drive system

$$G(s) = \frac{25\,(s^2 + 1.2s + 12500)}{s\,(s^2 + 5.6s + 62000)}$$

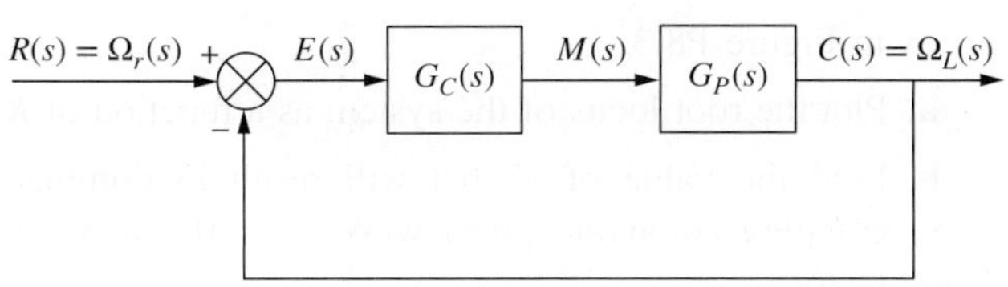


**FIGURE P8.16**

presented in Problem 61, Chapter 5 (*Thomsen, 2011*). Referring to Figures P5.40 and P8.16, $G_p(s)$ in Figure P8.16 is given by:

$$G_p(s) = K_M \frac{G(s)}{1 + 0.1G(s)}$$

Given that the controller is proportional, that is, $G_C(s) = K_P$, use MATLAB and a procedure similar to that developed in Problem 40 in this chapter to plot the root locus[5] and obtain the output response, $c(t) = \omega_L(t)$, when a step input, $r(t) = \omega_r(t) = 260\ u(t)$ rad/sec, is applied at $t = 0$. Mark on the time response graph, $c(t)$, all relevant characteristics, such as the percent overshoot (which should not exceed 16%), peak time, rise time, settling time, and final steady-state value.

MATLAB ML

## DESIGN PROBLEMS

**57.** A disk drive is a position control system in which a read/write head is positioned over a magnetic disk. The system responds to a command from a computer to position itself at a particular track on the disk. A physical representation of the system and a block diagram are shown in Figure P8.17.

**a.** Find $K$ to yield a settling time of 0.1 second.

**b.** What is the resulting percent overshoot?

**c.** What is the range of $K$ that keeps the system stable?

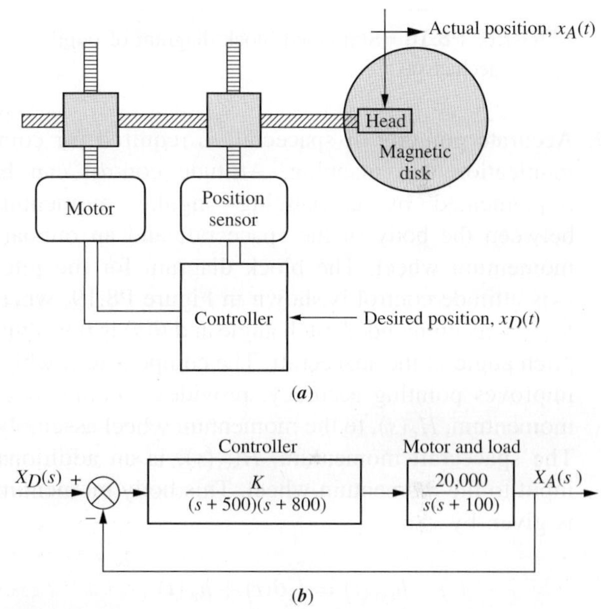


**FIGURE P8.17** Disk drive: **a.** physical representation; **b.** block diagram

**58.** A simplified block diagram of a human pupil servomechanism is shown in Figure P8.18. The term $e^{-0.18s}$ represents a time delay. This function can be approximated by what is known as a Padé approximation. This approximation can take on many increasingly complicated forms, depending upon the degree of accuracy required. If we use the Padé approximation

$$e^{-x} = \frac{1}{1 + x + \dfrac{x^2}{2!}}$$

[5] Select a point on the close-up of the root locus that corresponds to a gain between 1 and 5.

then

$$e^{-0.18s} = \frac{61.73}{s^2 + 11.11s + 61.73}$$

Since the retinal light flux is a function of the opening of the iris, oscillations in the amount of retinal light flux imply oscillations of the iris (*Guy, 1976*). Find the following:

**a.** The value of $K$ that will yield oscillations

**b.** The frequency of these oscillations

**c.** The settling time for the iris if $K$ is such that the eye is operating with 20% overshoot

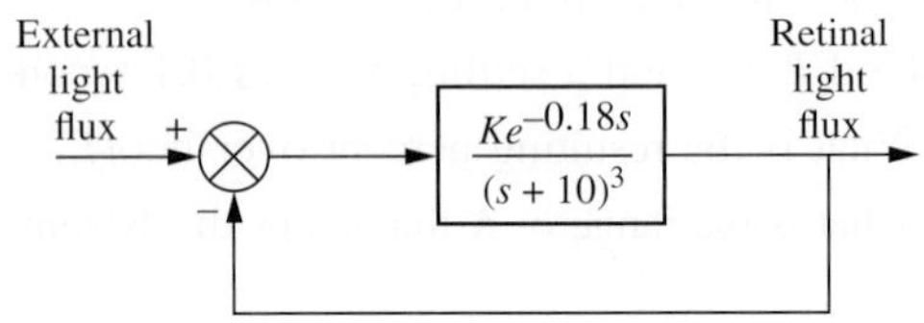


**FIGURE P8.18** Simplified block diagram of pupil servomechanism

**59.** Accurate pointing of spacecraft is required for communication and mapping. Attitude control can be implemented by exchanging angular momentum between the body of the spacecraft and an onboard momentum wheel. The block diagram for the pitch axis attitude control is shown in Figure P8.19, where $\theta_c(s)$ is a commanded pitch angle and $\theta(s)$ is the actual pitch angle of the spacecraft. The compensator, which improves pointing accuracy, provides a commanded momentum, $H_c(s)$, to the momentum wheel assembly. The spacecraft momentum, $H_{sys}(s)$, is an additional input to the momentum wheel. This body momentum is given by

$$h_{sys}(t) = I_2\dot{\theta}(t) + h_w(t)$$

where $I_2$ is the spacecraft moment of inertia about the pitch axis and $h_w(t)$ is the momentum of the wheel. The total torque output from the momentum wheel, $T_w$, as shown in Figure P8.19, is

$$T_w(t) = \frac{h_{sys}(t) - h_w(t) + h_c(t)}{\tau}$$

If $\tau = 23$ seconds and $I_2 = 9631$ in-lb-s$^2$, do the following (*Piper, 1992*):

**a.** Sketch the root locus for the pitch axis control system.

**b.** Find the value of $K$ to yield a closed-loop step response with 25% overshoot.

**c.** Evaluate the accuracy of any second-order approximations that were made.

**60.** A hard disk drive (HDD) arm has an open-loop unstable transfer function,

$$P(s) = \frac{X(s)}{F(s)} = \frac{1}{I_b s^2}$$

where $X(s)$ is arm displacement and $F(s)$ is the applied force (*Yan, 2003*). Assume the arm has an inertia of $I_b = 3 \times 10^{-5}$ kg-m$^2$ and that a lead controller, $G_c(s)$ (used to improve transient response and discussed in Chapter 9), is placed in cascade to yield

$$P(s)G_c(s) = G(s) = \frac{K}{I_b}\frac{(s+1)}{s^2(s+10)}$$

as in Figure P8.3.

**a.** Plot the root locus of the system as a function of $K$.

**b.** Find the value of $K$ that will result in dominant complex conjugate poles with a $\zeta = 0.7$ damping factor.

**61.** Wind turbines, such as the one shown in Figure P8.20(*a*), are becoming popular as a way of generating electricity. Feedback control loops are designed to control the output power of the turbine, given an input power demand. Blade-pitch control may be used as part of the control loop for a constant-speed, pitch-controlled wind turbine, as shown in Figure P8.20(*b*). The drivetrain, consisting of the windmill rotor, gearbox, and electric generator (see Figure P8.20(c)), is part of the control loop. The torque

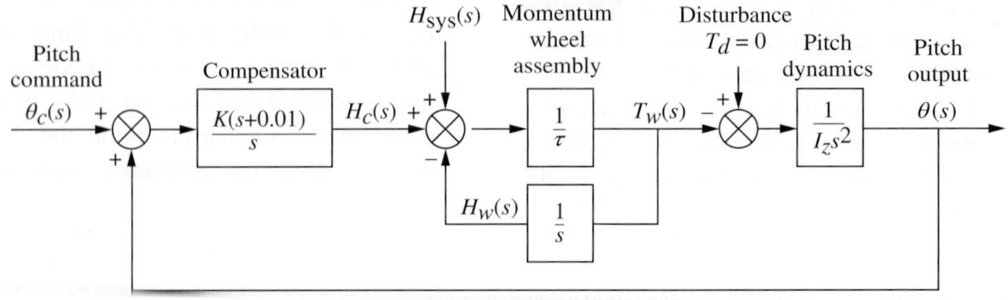


**FIGURE P8.19** Pitch axis attitude control system utilizing momentum wheel

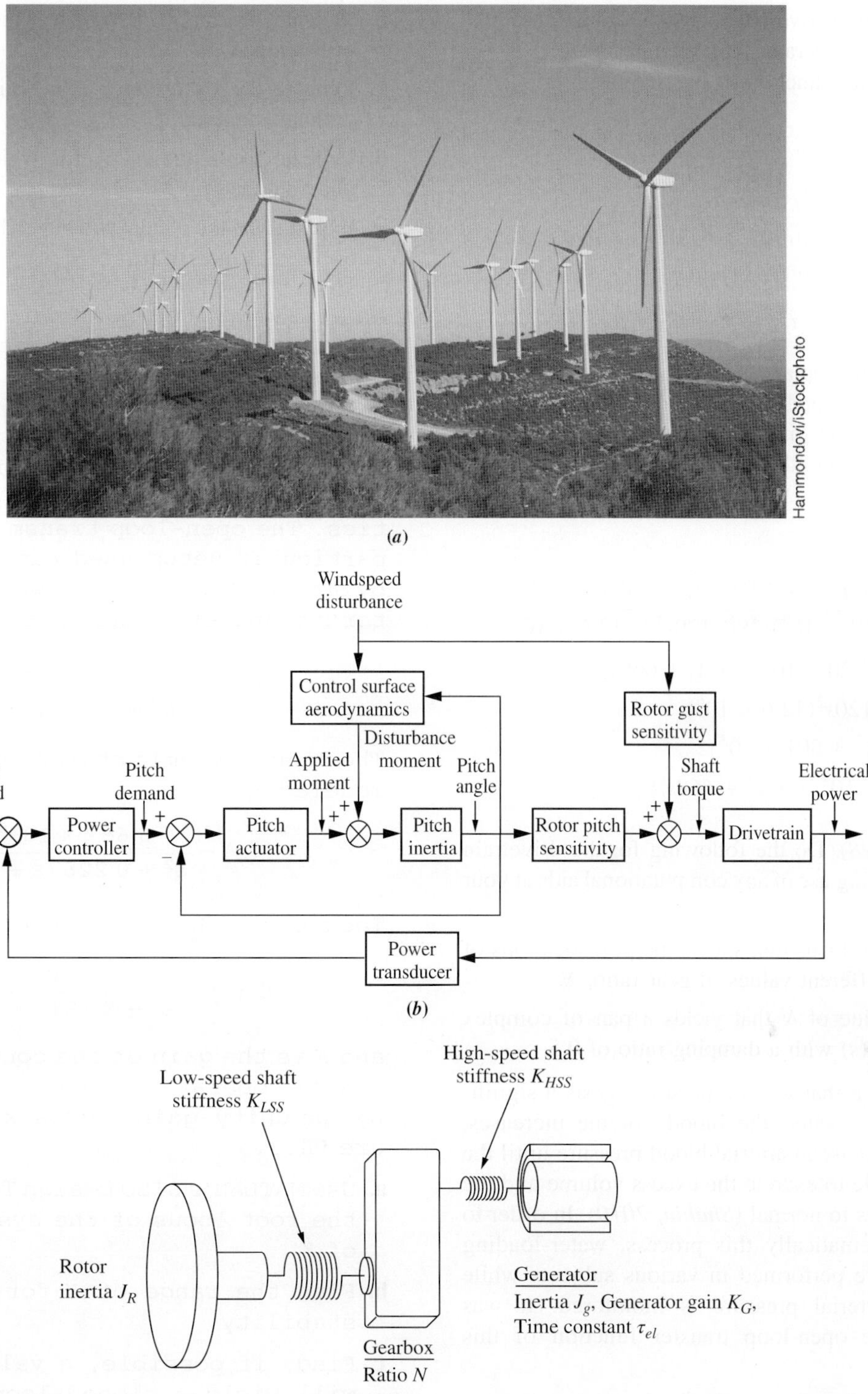


**FIGURE P8.20** **a.** Wind turbines generating electricity near Palm Springs, California; **b.** control loop for a constant-speed pitch-controlled wind turbine;[6] **c.** drivetrain[6]

[6] Adapted from Anderson, C. G., Richon, J-B., and Campbell, T. J. An Aerodynamic Moment-Controlled Surface for Gust Load Alleviation on Wind Turbine Rotors, *IEEE Transactions on Control System Technology*, vol. 6, no. 5, September 1998, pp. 577–595. © 1998 IEEE.

created by the wind drives the rotor. The windmill rotor is connected to the generator through a gearbox.

The transfer function of the drivetrain is

$$\frac{P_o(s)}{T_R(s)} = G_{dt}(s)$$
$$= \frac{3.92K_{LSS}K_{HSS}K_G N^2 s}{\{N^2 K_{HSS}(J_R s^2 + K_{LSS})(J_G s^2[\tau_{el}s + 1] + K_G s) + J_R s^2 K_{LSS}[(J_G s^2 + K_{HSS})(\tau_{el}s + 1) + K_G s]\}}$$

where $P_o(s)$ is the Laplace transform of the output power from the generator and $T_R(s)$ is the Laplace transform of the input torque on the rotor. Substituting typical numerical values into the transfer function yields

$$\frac{P_o(s)}{T_R(s)} = G_{dt}(s)$$
$$= \frac{(3.92)(12.6\times10^6)(301\times10^3)(688)N^2 s}{\{N^2(301\times10^3)(190,120s^2 + 12.6\times10^6) \times (3.8s^2[20\times10^{-3}s+1] + 668s) + 190,120s^2(12.6\times10^6) \times [(3.8s^2 + 301\times10^3) \times (20\times10^{-3}s+1) + 668s]\}}$$

(*Anderson, 1998*). Do the following for the drivetrain dynamics, making use of any computational aids at your disposal:

**a.** Sketch a root locus that shows the pole locations of $G_{dt}(s)$ for different values of gear ratio, $N$.

**b.** Find the value of $N$ that yields a pair of complex poles of $G_{dt}(s)$ with a damping ratio of 0.5.

**62.** It is well known that when a person ingests a significant amount of water, the blood volume increases, causing an increase in arterial blood pressure until the kidneys are able to excrete the excess volume and the pressure returns to normal (*Shahin, 2010*). In order to describe mathematically this process, water-loading experiments are performed in various subjects while their mean arterial pressure is monitored. It was found that the open-loop transfer function of this process is

$$G(s) = \frac{b_p(1.759s^3 + 2.318s^2 + 2.173\times10^{-4})}{3.362s^3 + 11.34s^2 + 7.803s + 0.00293}$$

where $b_p$ is an autonomous nervous activity parameter.

**a.** Make a sketch of the root locus of the system, indicating the breakaway points and the value of $b_p$ for each point.

**b.** Indicate the range of $b_p$ for which the system is overdamped.

**c.** Indicate the values of $b_p$ for which the system is critically damped.

**d.** Indicate the range of $b_p$ for which the system is underdamped.

**e.** Explain why the system will have a larger settling time for larger values of $b_p$.

**63.** An active system for the elimination of floor vibrations due to human presence is presented in (*Nyawako, 2009*). The system consists of a sensor that measures the floor's vertical acceleration and an actuator that changes the floor characteristics. The open-loop transmission of the particular setup used can be described by $G(s) = KG_a(s)F(s)G_m(s)$, where the actuator's transfer function is

GUI Tool **GUIT**

$$G_a(s) = \frac{10.26}{s^2 + 11.31s + 127.9}$$

The floor's dynamic charactristics can be modeled by

$$F(s) = \frac{6.667\times10^{-5}s^2}{s^2 + 0.2287s + 817.3}$$

The sensor's transfer function is

$$G_m(s) = \frac{s}{s^2 + 5.181s + 22.18}$$

and $K$ is the gain of the controller. The system operations can be described by the unity-gain feedback loop of Figure P8.3.

**a.** Use MATLAB's SISO Design Tool to obtain the root locus of the system in terms of $K$.

**b.** Find the range of $K$ for closed-loop stability.

**c.** Find, if possible, a value of $K$ that will yield a closed-loop overdamped response.

**64.** One of the treatments for Parkinson's disease in some patients is Deep Brain Stimulation (DBS) (*Davidson, 2012*). In DBS a set of electrodes is surgically implanted and a vibrating current is applied to the subthalamic nucleus, also known as a brain pacemaker. Root locus has been used on a linearized model of the system to help explain the dynamics of DBS. The DBS model can be obtained by substituting $G(s) = \frac{ks}{(s-b)^2}$ $(b > 0)$ in the unity-feedback diagram of Figure P8.3.

**a.** Make a sketch of the resulting root locus as a function of $k$ and find the break-in point and its corresponding value of gain.

**b.** Find the range of $k$ for closed-loop stability in terms of $b$.

**c.** Find the frequency of oscillation when the system has closed-loop poles on the $j\omega$ axis.

**65.** An automatic regulator is used to control the field current of a three-phase synchronous machine with identical symmetrical armature windings (*Stapleton, 1964*). The purpose of the regulator is to maintain the system voltage constant within certain limits. The transfer function of the synchronous machine is MATLAB ML

$$G_{sm}(s) = \frac{\Delta\delta(s)}{\Delta P_m(s)} = \frac{M(s-z_1)(s-z_2)}{(s-p_1)(s-p_2)(s-p_3)}$$

which relates the variation of rotor angle, $\Delta\delta(s)$, to the change in the synchronous machine's shaft power, $\Delta P_m(s)$. The closed-loop system is shown in Figure P8.3, where $G(s) = KG_c(s)G_{sm}(s)$ and $K$ is a gain to be adjusted. The regulator's transfer function, $G_c(s)$, is given by

$$G_c(s) = \frac{\mu/T_e}{s + \frac{1}{T_e}}$$

Assume the following parameter values:

$\mu = 4, M = 0.117, T_e = 0.5, z_{1,2} = -0.071 \pm j6.25,$ $p_1 = -0.047$, and $p_{2,3} = -0.262 \pm j5.1,$

and do the following:

Write a MATLAB M-file to plot the root locus for the system and to find the following:

**a.** The gain $K$ at which the system becomes marginally stable

**b.** The closed-loop poles, $p$, and transfer function, $T(s)$, corresponding to a 16% overshoot

**c.** The coordinates of the point selected on the root-locus corresponding to 16% overshoot

**d.** A simulation of the unit-step response of the closed-loop system corresponding to your 16% overshoot design. Note in your simulation the following values: (1) actual percent overshoot, (2) corresponding peak time, $T_p$, (3) rise time, $T_r$, (4) settling time, $T_s$, and (5) final steady-state value.

**66.** A linear dynamic model of the $\alpha$-subsystem of a grid-connected voltage-source converter (VSC) using a Y-Y transformer is shown in Figure P8.21(*a*) (*Mahmood,*

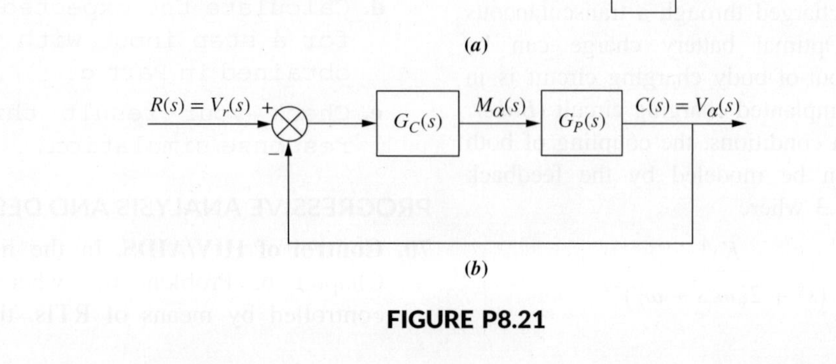


FIGURE P8.21

*2012*). Here, $C = 135\ \mu\text{F}$, $R_1 = 0.016\ \Omega$, $L_1 = 0.14\ \text{mH}$, $R_2 = 0.014\ \Omega$, $L_2 = 10\ \mu\text{H}$, $R_g = 1.1\Omega$, and $L_g = 0.5\ \text{mH}$.

**a.** Find the transfer function $G_P(s) = \dfrac{V_a(s)}{M_a(s)}$.

**b.** If $G_P(s)$ is the plant in Figure P8.21(*b*) and $G_C(s) = K$, use MATLAB to plot the root locus. On a closeup of the locus (from −300 to 0 on the real axis and from −50 to 5000 on the imaginary axis), find *K* and the coordinates of the dominant poles, which correspond to $\zeta = 0.012$. Plot the output response, $c(t) = v_\alpha(t)$, at that value of the gain when a step input, $r(t) = v_r(t) = 208\,u(t)$ volts, is applied at $t = 0$. Mark on the time response graph, $c(t)$, all relevant characteristics, such as the percent overshoot, peak time, rise time, settling time, and final steady-state value.

MATLAB **ML**

**67.** A robotic manipulator together with a cascade PI controller (used to improve steady-state response and discussed in Chapter 9) has a transfer function (*Low, 2005*)

$$G(s) = \left(K_p + \frac{K_1}{s}\right)\frac{48,500}{s^2 + 2.89s}$$

Assume the robot's joint will be controlled in the configuration shown in Figure P8.3.

**a.** Find the value of $K_I$ that will result in $e_{ss} = 2\%$ for a parabolic input.

**b.** Using the value of $K_I$ found in Part **a**, plot the root locus of the system as a function of $K_P$.

**c.** Find the value of $K_P$ that will result in a real pole at −1. Find the location of the other two poles.

**68.** Many implantable medical devices such as pacemakers, retinal implants, deep brain stimulators, and spinal cord stimulators are powered by an in-body battery that can be charged through a transcutaneous inductive device. Optimal battery charge can be obtained when the out-of-body charging circuit is in resonance with the implanted charging circuit (*Baker, 2007*). Under certain conditions, the coupling of both resonant circuits can be modeled by the feedback system in Figure P8.3 where

$$G(s) = \frac{Ks^4}{(s^2 + 2\zeta\omega_n s + \omega_n^2)^2}$$

The gain *K* is related to the magnetic coupling between the external and in-body circuits. *K* may vary due to positioning, skin conditions, and other variations. For this problem let $\zeta = 0.5$ and $\omega_n = 1$.

**a.** Find the range of *K* for closed-loop stability.

**b.** Draw the corresponding root locus.

**69.** It is important to precisely control the amount of organic fertilizer applied to a specific crop area in order to provide specific nutrient quantities and to avoid unnecessary environmental pollution. A precise delivery liquid manure machine has been developed for this purpose (*Saeys, 2008*). The system consists of a pressurized tank, a valve, and a rheological flow sensor. After simplification, the system can be modeled as a closed-loop negative-feedback system with a forward-path transfer function

MATLAB **ML** GUI Tool **GUIT**

$$G(s) = \frac{2057.38K(s^2 - 120s + 4800)}{s(s + 13.17)(s^2 + 120s + 4800)}$$

consisting of an electrohydraulic system in cascade with the gain of the manure flow valve and a variable gain, *K*. The feedback path is comprised of

$$H(s) = \frac{10(s^2 - 4s + 5.333)}{(s + 10)(s^2 + 4s + 5.333)}$$

**a.** Use the SISO Design Tool in MATLAB to obtain the root locus of the system.

**b.** Use the SISO Design Tool to find the range of *K* for closed-loop stability.

**c.** Find the value of *K* that will result in the smallest settling time for this system.

**d.** Calculate the expected settling time for a step input with the value of *K* obtained in Part **c**.

**e.** Check your result through a step-response simulation.

## PROGRESSIVE ANALYSIS AND DESIGN PROBLEMS

**70. Control of HIV/AIDS.** In the linearized model of Chapter 6, Problem 67, where virus levels are controlled by means of RTIs, the open-loop plant transfer function was shown to be

$$P(s) = \frac{Y(s)}{U_1(s)} = \frac{-520s - 10.3844}{s^3 + 2.6817s^2 + 0.11s + 0.0126}$$

The amount of RTIs delivered to the patient will automatically be calculated by embedding the patient in the control loop as $G(s)$ shown in Figure P6.19 (*Craig, 2004*).

**a.** In the simplest case, $G(s) = K$, with $K > 0$. Note that this effectively creates a positive-feedback loop because the negative sign in the numerator of $P(s)$ cancels out with the negative-feedback sign in the summing junction. Use positive-feedback rules to plot the root locus of the system.

**b.** Now assume $G(s) = -K$ with $K > 0$. The system is now a negative-feedback system. Use negative-feedback rules to draw the root locus. Show that in this case the system will be closed-loop stable for all $K > 0$.

**71. Hybrid vehicle.** In Chapter 7, Figure P7.32 shows the block diagram of the speed control of an HEV rearranged as a unity feedback system (*Preitl*, 2007). MATLAB ML

Let the transfer function of the speed controller be

$$G_{SC}(s) = K_{P_{SC}} + \frac{K_{I_{SC}}}{s} = \frac{K_{P_{SC}}\left(s + \frac{K_{I_{SC}}}{K_{P_{SC}}}\right)}{s}$$

**a.** Assume first that the speed controller is configured as a proportional controller ($K_{I_{SC}} = 0$ and $G_{SC}(s) = K_{P_{SC}}$. Calculate the forward-path open-loop poles. Now use MATLAB to plot the system's root locus and find the gain, $K_{P_{SC}}$ that yields a critically damped closed-loop response. Finally, plot the time-domain response, $c(t)$, for a unit-step input using MATLAB. Note on the curve the rise time, $T_r$, and settling time, $T_s$. MATLAB ML

**b.** Now add an integral gain, $K_{I_{SC}}$, to the controller, such that $K_{I_{SC}}/K_{P_{SC}} = 0.4$. Use MATLAB to plot the root locus and find the proportional gain, $K_{P_{SC}}$, that could lead to a closed-loop unit-step response with 10% overshoot. Plot $c(t)$ using MATLAB and note on the curve the peak time, $T_p$, and settling time, $T_s$. Does the response obtained resemble a second-order under-damped response? MATLAB ML

**72. Parabolic trough collector.** Consider the fluid temperature control of a parabolic trough collector (*Camocho, 2012*) embedded in the unity feedback structure as shown in Figure P8.3, where the open-loop plant transfer function is given by

$$G(s) = \frac{137.2 \times 10^{-6}K}{s^2 + 0.0224s + 196 \times 10^{-6}}e^{-39s}$$

Approximating the time-delay term with $e^{-sT} \approx \dfrac{1 - \frac{T}{2}s}{1 + \frac{T}{2}s}$, make a sketch of the resulting root locus (Note: After substituting the approximation, $G(\infty) < 0$, the positive feedback rules of Section 8.9 must be used). Mark where appropriate in the plot and find:

**a.** The asymptotes and their intersection with the real axis;

**b.** The break-in and breakaway points. (The procedures presented in Section 8.5 are also valid for positive feedback systems);

**c.** The range of $K$ for closed-loop stability;

**d.** The value of $K$ that will make the system oscillate and the oscillation frequency.

# Cyber Exploration Laboratory

## Experiment 8.1

**Objectives** To verify the effect of open-loop poles and zeros upon the shape of the root locus. To verify the root locus as a tool for estimating the effect of open-loop gain upon the transient response of closed-loop systems.

**Minimum Required Software Packages** MATLAB and the Control System Toolbox

### Prelab

1. Sketch two possibilities for the root locus of a unity negative-feedback system with the open-loop pole-zero configuration shown in Figure P8.22.

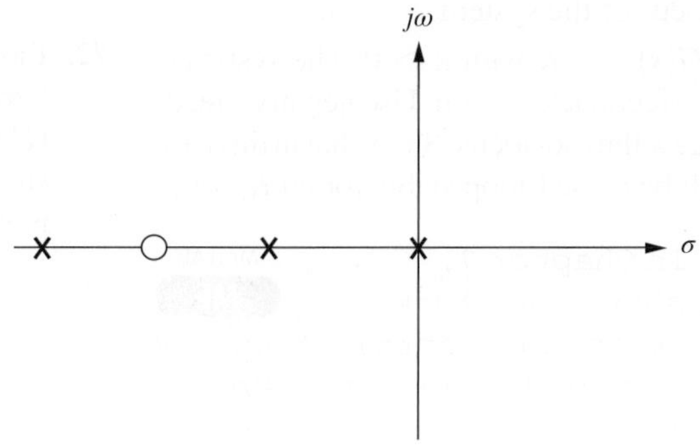


**FIGURE P8.22**

2. If the open-loop system of Prelab 1 is $G(s) = \dfrac{K(s+1.5)}{s(s+0.5)(s+10)}$, estimate the percent overshoot at the following values of gain, $K$: 20, 50, 85, 200, and 700.

### Lab

1. Using Matlab's SISO Design Tool, set up a negative unity feedback system with

$$G(s) = \frac{K(s+6)}{s(s+0.5)(s+10)}$$

to produce a root locus. For convenience, set up the zero at $-6$ using SISO Design Tool's compensator function by simply dragging a zero to $-6$ on the resulting root locus. Print the root locus for the zero at $-6$. Move the zero to the following locations and print out a root locus at each location: $-2$, $-1.5$, $-1.37$, and $-1.2$.

2. Using Matlab's SISO Design Tool, set up a negative unity feedback system with

$$G(s) = \frac{K(s+1.5)}{s(s+0.5)(s+10)}$$

to produce a root locus. Open the LTI Viewer for SISO Design Tool to show step responses. Using the values of K specified in Prelab 2, record the percent overshoot and settling time and print the root loci and step response for each value of K.

### Postlab

1. Discuss your findings from Prelab 1 and Lab 1. What conclusions can you draw?
2. Make a table comparing percent overshoot and settling time from your calculations in Prelab 2 and your experimental values found in Lab 2. Discuss the reasons for any discrepancies. What conclusions can you draw?

## Experiment 8.2

**Objective** To use MATLAB to design the gain of a controller via root locus.

**Minimum Required Software Package** MATLAB with the Control Systems Toolbox.

**Prelab** The open-loop system dynamics model for the NASA eight-axis Advanced Research Manipulator II (ARM II) electromechanical shoulder joint/link, actuated by an armature-controlled dc servomotor is shown in Figure P8.23.

The ARM II shoulder joint constant parameters are $K_a = 12$, $L = 0.006$ H, $R = 1.4\ \Omega$, $K_b = 0.00867$, $n = 200$, $K_m = 4.375$, $J = J_m + J_L/n^2$, $D = D_m + D_L/n^2$, $J_L = 1$, $D_L = 0.5$, $J_m = 0.00844$, and $D_m = 0.00013$ (*Craig, 2005*), (*Nyzen, 1999*), and (*Williams, 1994*).

**a.** Obtain the equivalent open-loop transfer function, $G(s) = \dfrac{\theta_L(s)}{V_{ref}(s)}$.

**b.** The loop is to be closed by cascading a controller, $G_c(s) = K_D s + K_P$, with $G(s)$ in the forward path forming an equivalent forward-transfer function, $G_e(s) = G_c(s)G(s)$. Parameters of $G_c(s)$ will be used to design a desired transient performance. The input to the closed-loop system is a voltage, $V_I(s)$, representing the desired angular displacement of the robotic joint with a ratio of 1 volt equals 1 radian. The output of the closed-loop system is the actual angular displacement of the joint, $\theta_L(s)$. An encoder in the feedback path, $K_e$, converts the actual joint displacement to a voltage with a ratio of 1 radian equals 1 volt. Draw the closed-loop system showing all transfer functions.

**c.** Find the closed-loop transfer function.

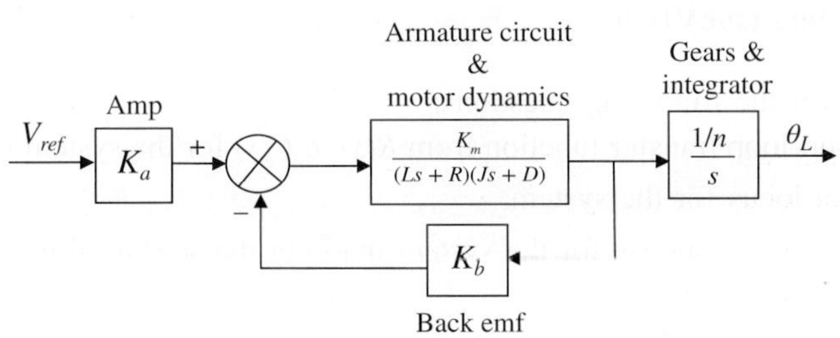


**FIGURE P8.23** Open-loop model for ARM II

**Lab** Let $\dfrac{K_P}{K_D} = 4$ and use MATLAB to design the value of $K_D$ to yield a step response with a maximum percent overshoot of 0.2%.

**Postlab**

1. Discuss the success of your design.
2. Is the steady-state error what you would expect? Give reasons for your answer.

## Experiment 8.3

**Objective** To use LabVIEW to design the gain of a controller via root locus.

**Minimum Required Software Package** LabVIEW with the Control Design and Simulation Module, and the MathScript RT Module.

**Prelab** Complete the Prelab to Experiment 8.2 if you have not already done so.

**Lab** Let $\dfrac{K_P}{K_D} = 4$. Use LabVIEW to open and customize the Interactive Root Locus VI from the Examples in order to implement a design of $K_D$ to yield a step response with a maximum percent overshoot of 0.2%. Use a hybrid graphical/MathScript approach.

### Postlab

1. Discuss the success of your design.
2. Is the steady-state error what you would expect? Give reasons for your answer.

## Hardware Interface Laboratory

### Experiment 8.4 Speed Control Using Gain Adjustment

**Objective** To control the speed of a motor in closed-loop using gain compensation. To make observations about tradeoffs between the compensated transient response and the steady-state error.

**Material Required** Computer with LabVIEW Installed; myDAQ; dc brushed gearmotor with Hall Sensor quadrature encoder (−10 V to +10 V normal operation range); and motor control chip BA6886N or a transistor circuit substitute

**Files Provided at www.wiley.com/college/nise**
Speed P Control Incomplete.vi
Signal Conditioning (subVI).vi

**Prelab** Answer the following questions:

1. Find the closed-loop transfer function from *R(s)* to *C(s)* for the system in Figure P8.24.
2. Draw the root locus for the system.
3. Draw the unit-step response for the system marking the settling time, peak time, and maximum output.
4. Find an expression for the steady-state error to a unit-step input for the system.

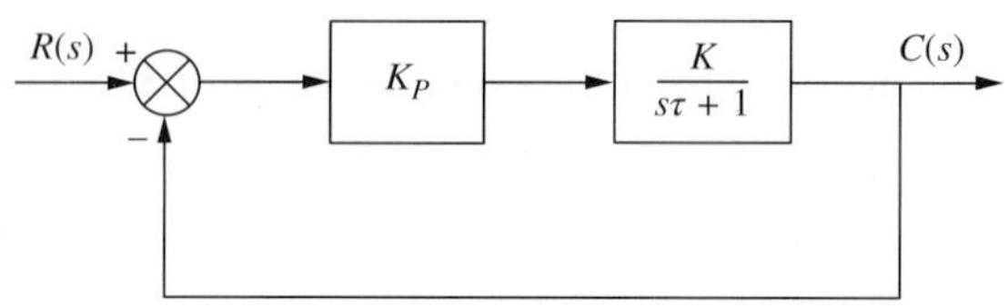


FIGURE P8.24

### Lab

**Software:** The **Speed P Control Incomplete.vi** is provided and illustrated in Figure P8.25. You need to modify it as follows before it becomes operational.

1. You need to change the constant on the left to fit your motor's gear ratio and encoder CPR (counts per revolution) as shown in Figure P8.25(*b*).
2. You need to write a SubVI for a P controller and place it where the arrow indicates it in Figure P8.25(*b*). The function of a P controller is $u = K_P e$. Your SubVI has two inputs, the system's error $e$ and the proportional constant $K_P$. It will have one output $u$.

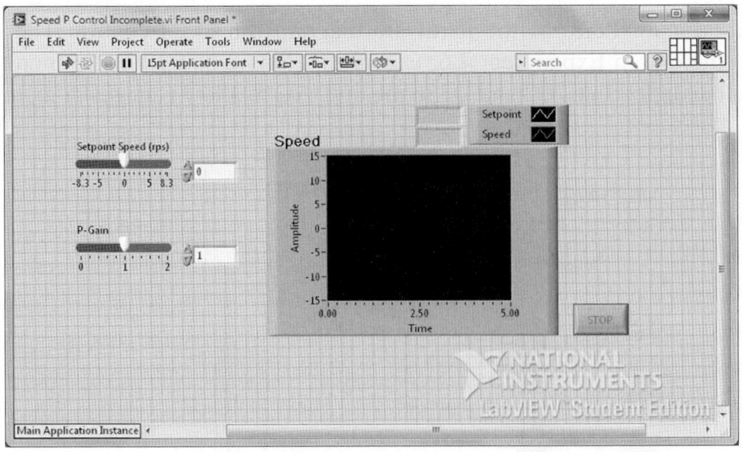


(*a*)

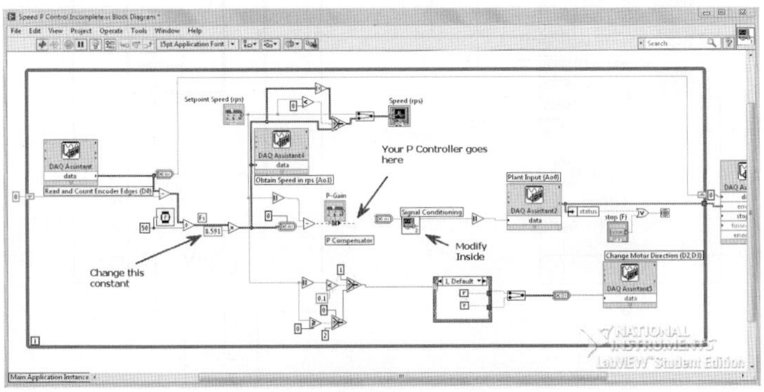


(*b*)

**FIGURE P8.25 Speed P Control Incomplete.vi: a. Front Panel; b. Block Diagram**

3. Double-clicking on the **Signal Conditioning (SubVI)**, Figure P8.25(*b*), you get Figure P8.26. Modify the indicated constant to reflect the dead zone parameter of your motor. This SubVI limits the input voltage to the motor controller and eliminates the dead zone by offsetting the input to the motor controller.

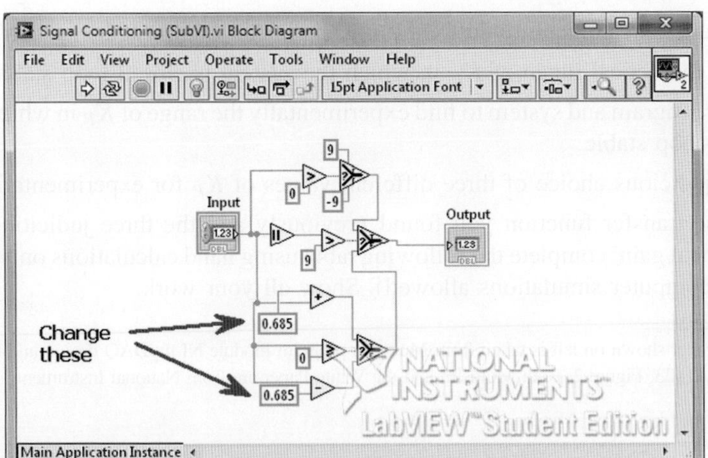


**FIGURE P8.26 Signal Conditioning (SubVI).vi Block Diagram**

**Hardware:** Figure P8.27 is the hardware diagram for speed control. The diagram is identical to the one in Experiment 4.6, except that Pins 2 and 10 in the motor controller chip connect to digital lines D2 and D3 in the myDAQ to allow changes in motor direction.

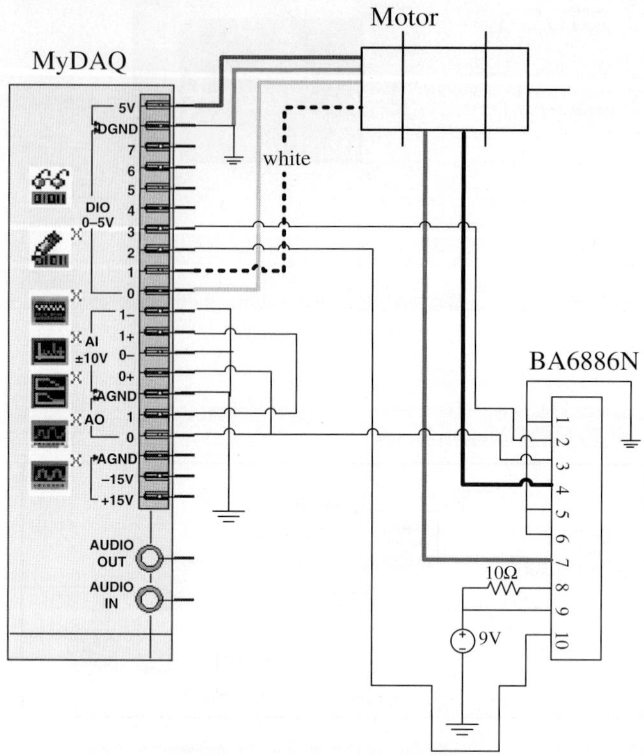


**FIGURE P8.27** Wiring diagram[7]

**Procedure:**

1. Verify the operation of your closed-loop system.
2. Draw a functional block diagram (similar to the ones shown in control systems textbooks) of the system. Do not include the signal conditioning functions, nor the change-of-direction signals.
3. Using the transfer function you found in Experiment 4.6, draw the system's root locus.
4. Find the theoretical range of $K_P$ in which the system is closed-loop stable.
5. Run your program and system to find experimentally the range of $K_P$ in which the system is closed-loop stable.
6. Make a judicious choice of three different values of $K_P$ for experimentation.
7. Using the transfer function you found previously and the three judicious choices of proportional gain, complete the following table using hand calculations only (calculators OK, no computer simulations allowed). Show all your work.

[7] MyDAQ right slot shown on left is taken from Multisim program module NI myDAQ design and also reproduced in White-Paper 11423, Figure 2. Both Multisim and the White Paper are from National Instruments.

| | | | |
|---|---|---|---|
| $K_P$ | | | |
| $T_P$—Peak time | | | |
| %OS—Percent overshoot | | | |
| $T_s$—Settling time | | | |
| $e_{ss}$—Steady state error (step input) | | | |

**Theoretical**

8. For each one of the three values of $K_P$, perform step-input experiments; use one single value of step-input amplitude for the three values. Make sure that your oscilloscope captures contain the system's transient response in its entirety. Show measurements of all the parameters in the table above and fill in the following table. Please note that $T_s$, the settling time, is hard to measure in the current setting because of the limited number of analog channels present. Instead of measuring $T_s$, mark in your oscilloscope its theoretical value using the scope cursors.

| | | | |
|---|---|---|---|
| $K_P$ | | | |
| $T_P$—Peak time | | | |
| %OS—Percent overshoot | | | |
| $e_{ss}$—Steady-state error (step input) | | | |

**Experimental**

**Postlab** Make a detailed comparison of your theoretical and experimental results. Discuss similarities and discrepancies between experimental and theoretical values and give possible reasons.

## Experiment 8.5 Position Control Using Gain Adjustment

**Objectives** To control the angular position of the shaft of a permanent-magnet dc motor in closed-loop using gain compensation. To make observations about tradeoffs between the compensated transient response and the steady-state error.

**Material Required** Computer with LabVIEW Installed; myDAQ; dc brushed gearmotor with Hall Sensor quadrature encoder ($-10$ V to $+10$ V normal operation range); and motor control chip BA6886N or a transistor circuit substitute.

### Files Provided at www.wiley.com/college/nise

Position control.vi
Signal Conditioning (SubVI).vi
P Controller (SubVI).vi

**Prelab** Answer the following questions:

1. For a given permanent-magnet dc motor it has been found that the transfer function from armature voltage $E_a(s)$ to angular velocity $\Omega(s)$ is $\dfrac{\Omega(s)}{E_a(s)} = \dfrac{K}{\tau s + 1}$. Find the transfer function of the motor from armature voltage to angular position $\dfrac{\Theta(s)}{E_a(s)}$.
2. Draw the root locus for the system in Figure P8.28.
3. Draw the unit-step response for the system marking the settling time, peak time, and maximum output. Find all the possibilities: overdamped, critically damped, and underdamped.

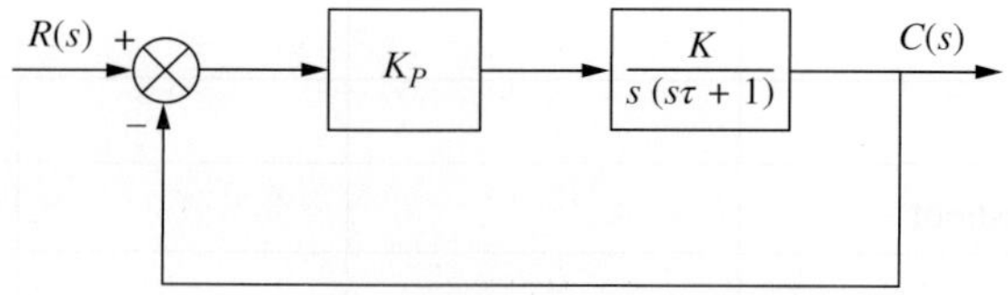


**FIGURE P8.28**

4. Find an expression for the steady-state error to a unit-step input for the system.

## Lab

**Software:** The front panel and block diagram of the **Position control.vi** are shown in Figure P8.29. Change the constants inside the **Signal Conditioning SubVI** to match your

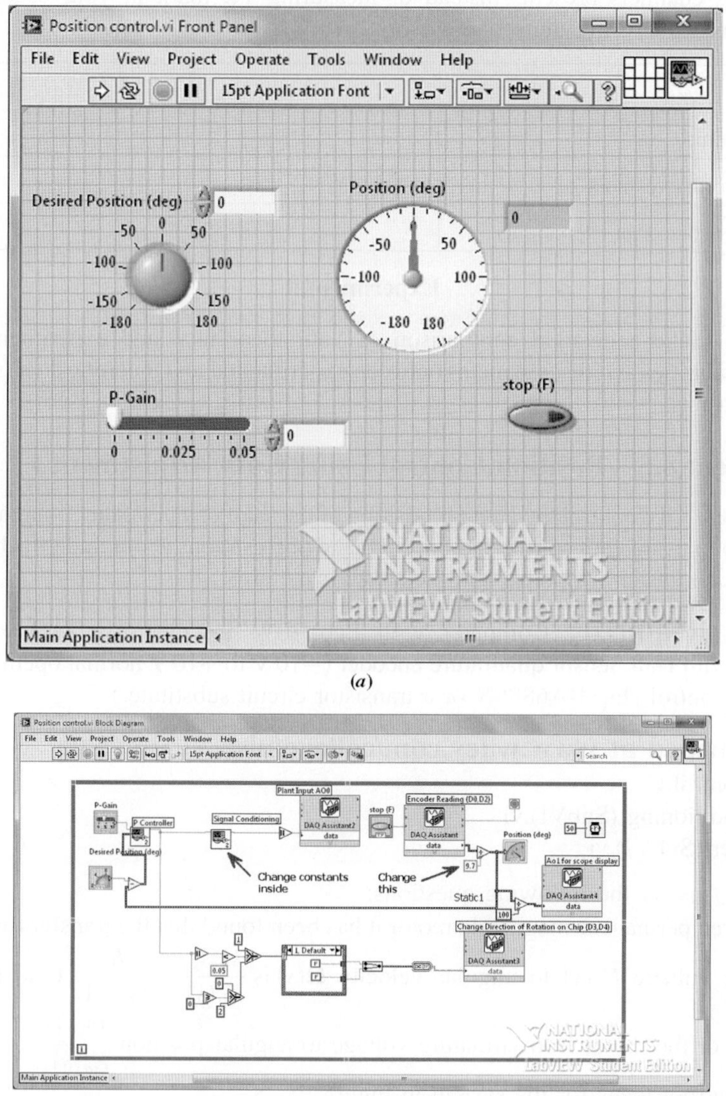


(*a*)

(*b*)

**FIGURE P8.29** **Position control.vi: a. Front Panel; b. Block Diagram**

dead zone parameters. The constant on the right of the diagram must be modified to match your motor's gear ratio.

**Hardware:** Make the following changes to the wiring diagram shown in Figure P8.27: Move the connections from D1, D2, and D3 to D2, D3, and D4, respectively. All the other connections remain the same.

**Procedure:**

1. Choose a small $P$ gain. Verify the operation of your closed-loop system. The motor should be able to move in both directions and through the full-scale range.
2. Using the transfer function found in Experiment 4.6, calculate the motor's transfer function from armature voltage to angular position $\frac{\Theta(s)}{E_a(s)}$.
3. Draw a functional block diagram of the system. Do not include the signal conditioning functions, nor the change-of-direction signals. Label all the pertinent signals.
4. Using the transfer function $\frac{\Theta(s)}{E_a(s)}$ you just calculated, draw the system's root locus.
5. Find the theoretical range of $K_P$ for which the system is closed-loop stable.
6. Run your program and system to find experimentally the range of $K_P$ for which the system is closed-loop stable.
7. Make a judicious choice of three different values of $K_P$ for experimentation.
8. Using the transfer function you calculated above and the three judicious choices of proportional gain, complete the following table using hand calculations only (calculators OK, no computer simulations allowed). Show all your work.

| $K_P$ | | | |
|---|---|---|---|
| $T_P$—Peak time | | | |
| %OS—Percent overshoot | | | |
| $T_s$—Settling time | | | |
| $e_{ss}$—Steady-state error (step input) | | | |

**Theoretical**

9. For each one of the three values of $K_P$, perform step-input experiments using one value of step input for the three values. Make sure that your oscilloscope captures contain the system's transient response in its entirety. Show measurements of all the parameters in the table above and fill in the following table. Please note that $T_s$, the settling time, is hard to measure in the current setting because of the limited number of analog channels available. Instead of measuring $T_s$, mark in your oscilloscope its theoretical value using the scope cursors.

| $K_P$ | | | |
|---|---|---|---|
| $T_P$—Peak time | | | |
| %OS—Percent overshoot | | | |
| $e_{ss}$—Steady-state error (step input) | | | |

**Experimental**

**Postlab** Make a detailed comparison of your theoretical and experimental tables. Discuss similarities and discrepancies between experimental and theoretical results and give possible reasons.

# Bibliography

Anderson, C. G., Richon, J.-B., and Campbell, T. J. An Aerodynamic Moment-Controlled Surface for Gust Load Alleviation on Wind Turbine Rotors. *IEEE Transactions on Control System Technology*, vol. 6, no. 5, September 1998, pp. 577–595.

Åström, K., Klein, R. E., and Lennartsson, A. Bicycle Dynamics and Control. *IEEE Control Systems*, August 2005, pp. 26–47.

Baker, M. W., and Sarpeshkar, R. Feedback Analysis and Design of RF Power Links for Low-Power Bionic Systems. *IEEE Transactions on Biomedical, Circuits and Systems*. vol. 1, 2007, pp. 28–38.

Camacho, E. F., Berenguel, M., Rubio, F. R., and Martinez, D. *Control of Solar Energy Systems*. Springer-Verlag, London, 2012.

Craig, I. K., Xia, X., and Venter, J. W. Introducing HIV/AIDS Education into the Electrical Engineering Curriculum at the University of Pretoria. *IEEE Transactions on Education*, vol. 47, no. 1, February 2004, pp. 65–73.

Craig, J. J. *Introduction to Robotics: Mechanics and Control*, 3d ed. Prentice Hall, Upper Saddle River, NJ, 2005.

Davidson, C. M., de Paor, A. M., and Lowery, M. M. Insights from Control Theory into Deep Brain Stimulation for Relief from Parkinson's Disease. *IEEE, Proc. of the 9th Int. Conf. Elektro*, 2012, pp. 2–7.

Dorf, R. C. *Modern Control Systems*, 5th ed. Addison-Wesley, Reading, MA., 1989.

Evans, W. R. Control System Synthesis by Root Locus Method. *AIEE Transactions*, vol. 69, 1950, pp. 66–69.

Evans, W. R. Graphical Analysis of Control Systems. *AIEE Transactions*, vol. 67, 1948, pp. 547–551.

Franklin, G. F., Powell, J. D., and Emami-Naeini, A. *Feedback Control of Dynamic Systems*, 2d ed. Addison-Wesley, Reading, MA., 1991.

Galvão, K. H. R., Yoneyama, T., and de Araújo, F. M. U. A Simple Technique for Identifying a Linearized Model for a Didactic Magnetic Levitation System. *IEEE Transactions on Education*, vol. 46, no. 1, February 2003, pp. 22–25.

Guy, W., *The Human Pupil Servomechanism*. Computers in Education Division of ASEE, Application Note No. 45, 1976.

Hahn, J. O., Dumont, G. A., and Ansermino, J. M. System Identification and Closed-Loop Control of End-Tidal $CO_2$ in Mechanically Ventilated Patients. *IEEE Transactions on Information Technology in Biomedicine*, vol. 16, no. 6, November 2012, pp. 1176–2012.

Hollot, C. V., Misra, V., Towsley, D., and Gong, W. A Control Theoretic Analysis of RED. *Proceedings of IEEE INFOCOM*, 2001, pp. 1510–1519.

Johnson, H., et al. *Unmanned Free-Swimming Submersible (UFSS) System Description*. NRL Memorandum Report 4393. Naval Research Laboratory, Washington, D.C., 1980.

Karlsson, P., and Svesson, J. DC Bus Voltage Control for a Distributed Power System. *IEEE Trans. on Power Electronics*, vol. 18, no. 6, 2003, pp. 1405–1412.

Khammash, M., and El-Samad, H. Systems Biology: From Physiology to Gene Regulation. *IEEE Control Systems*, August 2004, pp. 62–76.

Kuo, B. C. *Automatic Control Systems*, 6th ed. Prentice Hall, Upper Saddle River, NJ, 1991.

Lam, C. S., Wong, M. C., and Han, Y. D. Stability Study on Dynamic Voltage Restorer (DVR). *Power Electronics Systems and Applications 2004; Proceedings of the First International Conference on Power Electronics*, 2004, pp. 66–71.

Low, K. H., Wang, H., Liew, K. M., and Cai, Y. Modeling and Motion Control of Robotic Hand for Telemanipulation Application. *International Journal of Software Engineering and Knowledge Engineering*, vol. 15, 2005, pp. 147–152.

Mahmood, H., and Jiang, J. Modeling and Control System Design of a Grid Connected VSC Considering the Effect of the Interface Transformer Type. *IEEE Transactions on Smart Grid*, vol. 3, no. 1, March 2012, pp. 122–134.

Nyawako, D., and Reynolds, P. Response-Dependent Velocity Feedback Control for Mitigation of Human-Induced Floor Vibrations. *Smart Materials and Structures*, vol. 18, IOP Publishing Ltd., UK, 2009, pp. 1–13.

Nyzen, R. J. *Analysis and Control of an Eight-Degree-of-Freedom Manipulator.* Ohio University Masters Thesis, Mechanical Engineering, Dr. Robert L. Williams II, advisor, August 1999.

Preitl, Z., Bauer, P., and Bokor, J. A Simple Control Solution for Traction Motor Used in Hybrid Vehicles. *Fourth International Symposium on Applied Computational Intelligence and Informatics,* IEEE, 2007, pp. 157–162.

Saeys, W., Deblander, J., Ramon, H., and Anthonis, J. High-Performance Flow Control for Site-Specific Application of Liquid Manure. *Biosystems Engineering*, vol. 99, 2008, pp. 22–34.

Shahin, M., and Maka, S. A Transfer Function Method for the Assessment of Nervous System Modulation of Long-Term Dynamics of Blood Pressure. *IEEE International Conf. on Communication, Control, and Computing*, IEEE 2010, pp. 560–564.

Spong, M., Hutchinson, S., and Vidyasagar, M. *Robot Modeling and Control.* John Wiley & Sons, Hoboken, NJ, 2006.

Stapleton, C.A. Root-Locus Study of Synchronous-Machine Regulation. *IEE Proceedings*, vol. 111, issue 4, 1964, pp. 761–768.

Thomsen, S., Hoffmann, N., and Fuchs, F. W. PI Control, PI-Based State Space Control, and Model-Based Predictive Control for Drive Systems With Elastically Coupled Loads—A Comparative Study. *IEEE Transactions on Industrial Electronics*, vol. 58, no. 8, August 2011, pp. 3647–3657.

Ünyelioğlu, K. A., Hatopoğlu, C., and Özgüner, Ü. Design and Stability Analysis of a Lane Following Controller. *IEEE Transactions on Control Systems Technology*, vol. 5, 1997, pp. 127–134.

Williams, R. L. II. Local Performance Optimization for a Class of Redundant Eight-Degree-of-Freedom Manipulators. *NASA Technical Paper 3417*, NASA Langley Research Center, Hampton VA, March 1994.

Yamazaki, H., Marumo, Y., Iizuka, Y., and Tsunashima, H. Driver Model Simulation for Railway Brake Systems. *Fourth IET Int. Conf. on Railway Condition Monitoring*, 2008.

Yan, T., and Lin, R. Experimental Modeling and Compensation of Pivot Nonlinearity in Hard Disk Drives. *IEEE Transactions on Magnetics*, vol. 39, 2003, pp. 1064–1069.

# 9

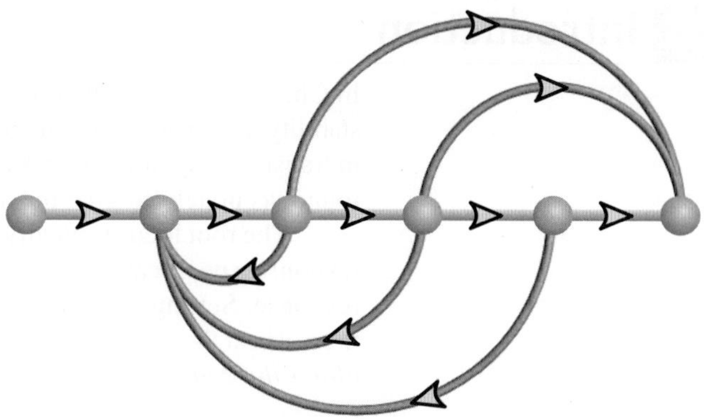

# Design via Root Locus

## Chapter Learning Outcomes

After completing this chapter the student will be able to:

- Use the root locus to design cascade compensators to improve the steady-state error (Sections 9.1–9.2)
- Use the root locus to design cascade compensators to improve the transient response (Section 9.3)
- Use the root locus to design cascade compensators to improve both the steady-state error and the transient response (Section 9.4)
- Use the root locus to design feedback compensators to improve the transient response (Section 9.5)
- Realize the designed compensators physically (Section 9.6)

## ▶ Case Study Learning Outcomes

You will be able to demonstrate your knowledge of the chapter objectives with case studies as follows:

- Given the antenna azimuth position control system shown on the front endpapers, you will be able to design a cascade compensator to meet transient response and steady-state error specifications.
- Given the pitch or heading control system for the UFSS vehicle shown on the back endpapers, you will be able to design a cascade or feedback compensator to meet transient response specifications.

# 9.1 Introduction

In Chapter 8, we saw that the root locus graphically displayed both transient response and stability information. The locus can be sketched quickly to get a general idea of the changes in transient response generated by changes in gain. Specific points on the locus also can be found accurately to give quantitative design information.

The root locus typically allows us to choose the proper loop gain to meet a transient response specification. As the gain is varied, we move through different regions of response. Setting the gain at a particular value yields the transient response dictated by the poles at that point on the root locus. Thus, *we are limited to those responses that exist along the root locus.*

## Improving Transient Response

Flexibility in the design of a desired transient response can be increased if we can design for transient responses that are not on the root locus. Figure 9.1(*a*) illustrates the concept. Assume that the desired transient response, defined by percent overshoot and settling time, is represented by point *B*. Unfortunately, on the current root locus at the specified percent overshoot, we only can obtain the settling time represented by point *A* after a simple gain adjustment. Thus, our goal is to speed up the response at *A* to that of *B*, without affecting the percent overshoot. This increase in speed cannot be accomplished by a simple gain adjustment, since point *B* does not lie on the root locus. Figure 9.1(*b*) illustrates the improvement in the transient response we seek: The faster response has the same percent overshoot as the slower response.

One way to solve our problem is to replace the existing system with a system whose root locus intersects the desired design point, *B*. Unfortunately, this replacement is expensive and counterproductive. Most systems are chosen for characteristics other than transient

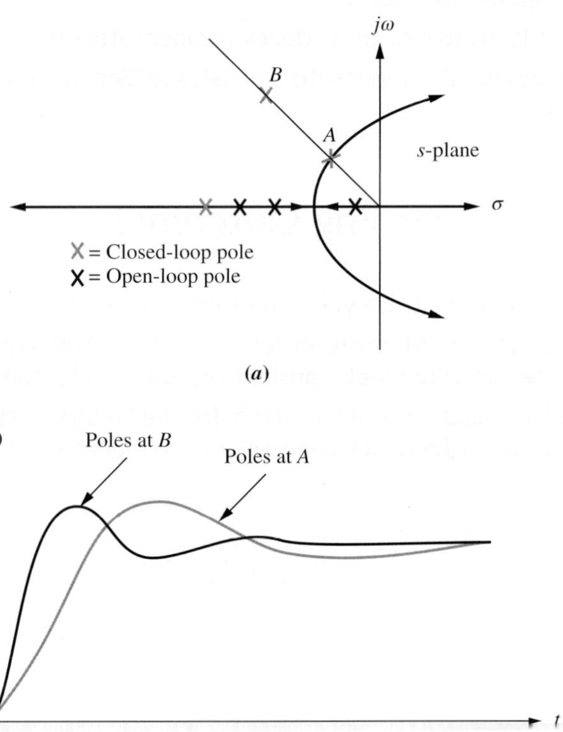


**FIGURE 9.1** **a.** Sample root locus, showing possible design point via gain adjustment (*A*) and desired design point that cannot be met via simple gain adjustment (*B*); **b.** responses from poles at *A* and *B*

response. For example, an elevator cage and motor are chosen for speed and power. Components chosen for their transient response may not necessarily meet, for example, power requirements.

Rather than change the existing system, we augment, or *compensate*, the system with *additional* poles and zeros, so that the compensated system has a root locus that goes through the desired pole location for some value of gain. One of the advantages of compensating a system in this way is that additional poles and zeros can be added at the low-power end of the system before the plant. Addition of compensating poles and zeros need not interfere with the power output requirements of the system or present additional load or design problems. The compensating poles and zeros can be generated with a passive or an active network.

A possible disadvantage of compensating a system with additional open-loop poles and zeros is that the system order can increase, with a subsequent effect on the desired response. In Chapters 4 and 8, we discussed the effect of additional closed-loop poles and zeros on the transient response. At the beginning of the design process discussed in this chapter, we determine the proper location of additional *open-loop* poles and zeros to yield the desired second-order *closed-loop* poles. However, we do not know the location of the higher-order *closed-loop* poles until the end of the design. Thus, we should evaluate the transient response through simulation after the design is complete to be sure the requirements have been met.

In Chapter 12, when we discuss state-space design, the disadvantage of finding the location of higher-order closed-loop poles after the design will be eliminated. Techniques that allow the designer to specify and design the location of all the closed-loop poles at the beginning of the design process.

One method of compensating for transient response, that will be discussed later, is to insert a differentiator in the forward path in parallel with the gain. We can visualize the operation of the differentiator with the following example. Assuming a position control with a step input, we note that the error undergoes an initial large change. Differentiating this rapid change yields a large signal that drives the plant. The output from the differentiator is much larger than the output from the pure gain. This large, initial input to the plant produces a faster response. As the error approaches its final value, its derivative approaches zero, and the output from the differentiator becomes negligible compared to the output from the gain.

## Improving Steady-State Error

Compensators are not only used to improve the transient response of a system; they are also used *independently* to improve the steady-state error characteristics. Previously, when the system gain was adjusted to meet the transient response specification, steady-state error performance deteriorated, since both the transient response and the static error constant were related to the gain. The higher the gain, the smaller the steady-state error, but the larger the percent overshoot. On the other hand, reducing gain to reduce overshoot increased the steady-state error. If we use dynamic compensators, compensating networks can be designed that will allow us to meet transient and steady-state error specifications *simultaneously*.[1] We no longer need to compromise between transient response and steady-state error, as long as the system operates in its linear range.

In Chapter 7, we learned that steady-state error can be improved by adding an open-loop pole at the origin in the forward path, thus increasing the system type and driving the associated steady-state error to zero. This additional pole at the origin requires an integrator for its realization.

[1] The word *dynamic* describes compensators with noninstantaneous transient response. The transfer functions of such compensators are functions of the Laplace variable, $s$, rather than pure gain.

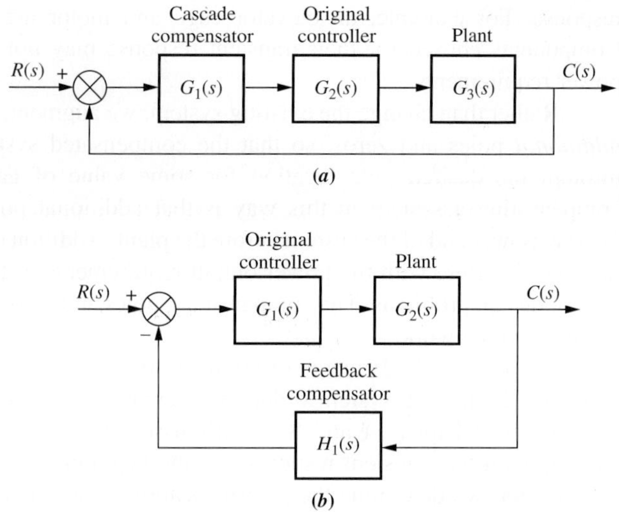


**FIGURE 9.2** Compensation techniques: **a.** cascade; **b.** feedback

In summary, then, transient response is improved with the addition of differentiation, and steady-state error is improved with the addition of integration in the forward path.

## Configurations

Two configurations of compensation are covered in this chapter: cascade compensation and feedback compensation. These methods are modeled in Figure 9.2. With cascade compensation, the compensating network, $G_1(s)$, is placed at the low-power end of the forward path in cascade with the plant. If feedback compensation is used, the compensator, $H_1(s)$, is placed in the feedback path. Both methods change the open-loop poles and zeros, thereby creating a new root locus that goes through the desired closed-loop pole location.

## Compensators

Compensators that use pure integration for improving steady-state error or pure differentiation for improving transient response are defined as *ideal compensators*. Ideal compensators must be implemented with active networks, which, in the case of electric networks, require the use of active amplifiers and possible additional power sources. An advantage of ideal integral compensators is that steady-state error is reduced to zero. Electromechanical ideal compensators, such as tachometers, are often used to improve transient response, since they can be conveniently interfaced with the plant.

Other design techniques that preclude the use of active devices for compensation can be adopted. These compensators, which can be implemented with passive elements such as resistors and capacitors, do not use pure integration and differentiation and are not ideal compensators. Advantages of passive networks are that they are less expensive and do not require additional power sources for their operation. Their disadvantage is that the steady-state error is not driven to zero in cases where ideal compensators yield zero error.

Thus, the choice between an active or a passive compensator revolves around cost, weight, desired performance, transfer function, and the interface between the compensator and other hardware. In Sections 9.2, 9.3, and 9.4, we first discuss cascade compensator design using ideal compensation and follow with cascade compensation using compensators that are not implemented with pure integration and differentiation.

## 9.2 Improving Steady-State Error via Cascade Compensation

In this section, we discuss two ways to improve the steady-state error of a feedback control system using cascade compensation. One objective of this design is to improve the steady-state error without appreciably affecting the transient response.

The first technique is *ideal integral compensation*, which uses a pure integrator to place an open-loop, forward-path pole at the origin, thus increasing the system type and reducing the error to zero. The second technique does not use pure integration. This compensation technique places the pole near the origin, and although it does not drive the steady-state error to zero, it does yield a measurable reduction in steady-state error.

While the first technique reduces the steady-state error to zero, the compensator must be implemented with active networks, such as amplifiers. The second technique, although it does not reduce the error to zero, does have the advantage that it can be implemented with a less expensive passive network that does not require additional power sources.

The names associated with the compensators come either from the method of implementing the compensator or from the compensator's characteristics. Systems that feed the error forward to the plant are called *proportional control systems*. Systems that feed the integral of the error to the plant are called *integral control systems*. Finally, systems that feed the derivative of the error to the plant are called *derivative control systems*. Thus, in this section we call the ideal integral compensator a *proportional-plus-integral* (*PI*) *controller*, since the implementation, as we will see, consists of feeding the error (proportional) plus the integral of the error forward to the plant. The second technique uses what we call a *lag compensator*. The name of this compensator comes from its frequency response characteristics, which will be discussed in Chapter 11. Thus, we use the name *PI controller* interchangeably with *ideal integral compensator*, and we use the name *lag compensator* when the cascade compensator does not employ pure integration.

### Ideal Integral Compensation (PI)

Steady-state error can be improved by placing an open-loop pole at the origin, because this increases the system type by one. For example, a Type 0 system responding to a step input with a finite error responds with zero error if the system type is increased by one. Active circuits can be used to place poles at the origin. Later in this chapter, we show how to build an integrator with active electronic circuits.

To see how to improve the steady-state error without affecting the transient response, look at Figure 9.3(*a*). Here we have a system operating with a desirable transient response generated by the closed-loop poles at *A*. If we add a pole at the origin to increase the system type, the angular contribution of the open-loop poles at point *A* is no longer 180°, and the root locus no longer goes through point *A*, as shown in Figure 9.3(*b*).

To solve the problem, we also add a zero close to the pole at the origin, as shown in Figure 9.3(*c*). Now the angular contribution of the compensator zero and compensator pole cancel out, point *A* is still on the root locus, and the system type has been increased. Furthermore, the required gain at the dominant pole is about the same as before compensation, since the ratio of lengths from the compensator pole and the compensator zero is approximately unity. Thus, we have improved the steady-state error without appreciably affecting the transient response.

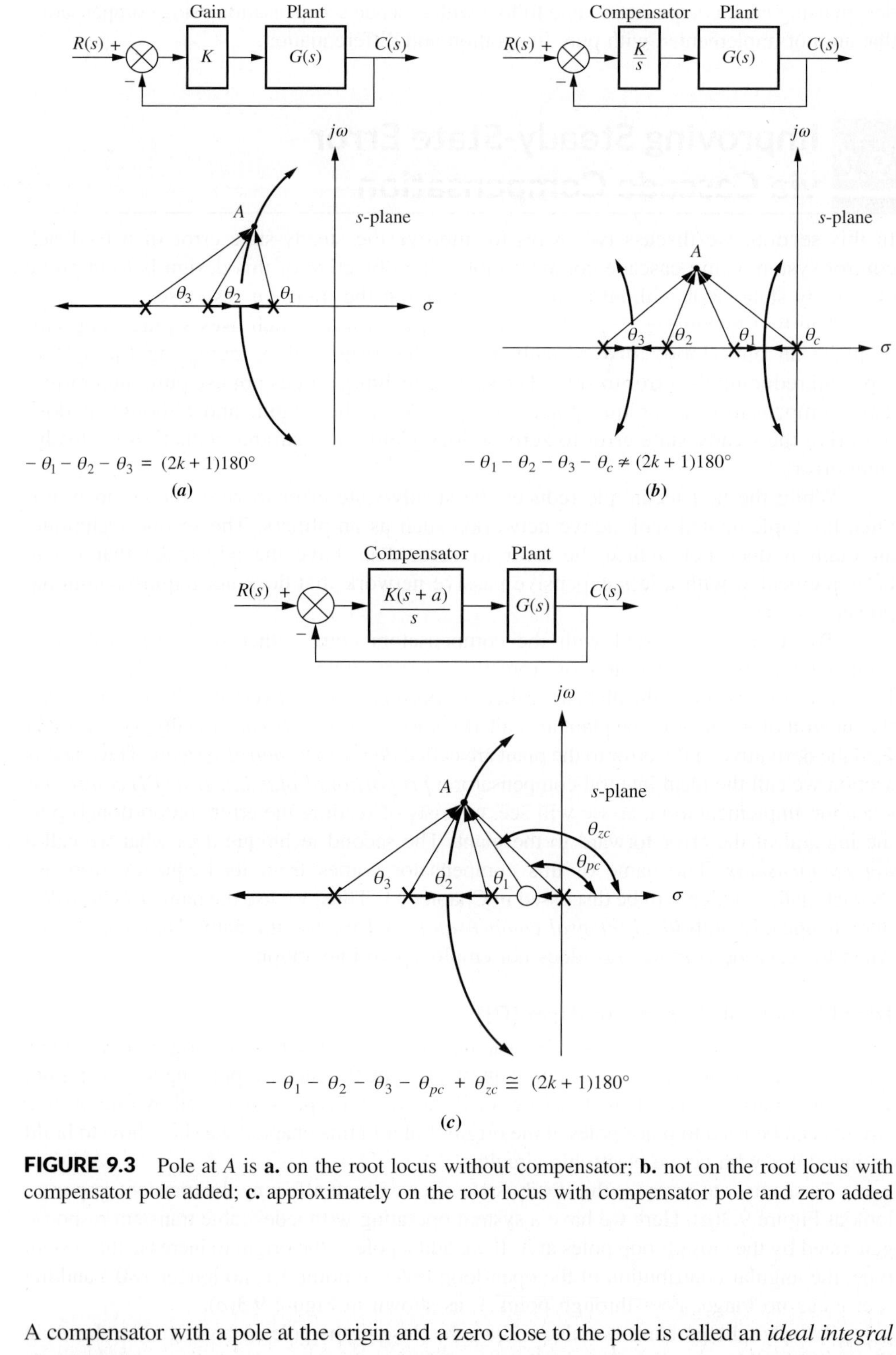


**FIGURE 9.3** Pole at $A$ is **a.** on the root locus without compensator; **b.** not on the root locus with compensator pole added; **c.** approximately on the root locus with compensator pole and zero added

A compensator with a pole at the origin and a zero close to the pole is called an *ideal integral compensator*.

In the example that follows, we demonstrate the effect of ideal integral compensation. An open-loop pole will be placed at the origin to increase the system type and drive the steady-state error to zero. An open-loop zero will be placed very close to the open-loop pole at the origin so that the original closed-loop poles on the original root locus still remain at approximately the same points on the compensated root locus.

## Example 9.1 Effect of an Ideal Integral Compensator

**PROBLEM:** Given the system of Figure 9.4(*a*), operating with a damping ratio of 0.174, show that the addition of the ideal integral compensator shown in Figure 9.4(*b*) reduces the steady-state error to zero for a step input without appreciably affecting transient response. The compensating network is chosen with a pole at the origin to increase the system type and a zero at −0.1, close to the compensator pole, so that the angular contribution of the compensator evaluated at the original, dominant, second-order poles is approximately zero. Thus, the original, dominant, second-order closed-loop poles are still approximately on the new root locus.

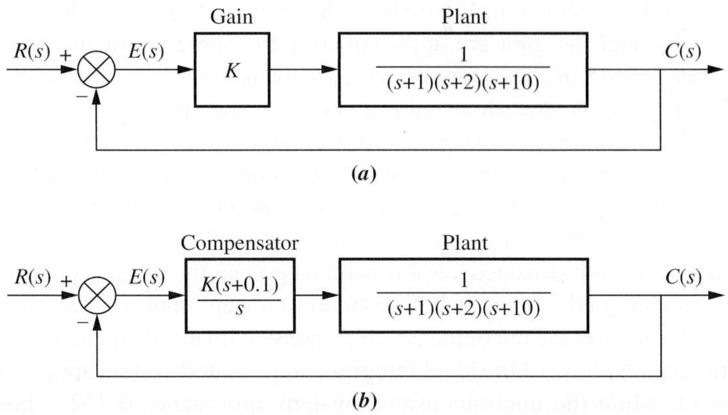


**FIGURE 9.4** Closed-loop system for Example 9.1: **a.** before compensation; **b.** after ideal integral compensation

**SOLUTION:** We first analyze the uncompensated system and determine the location of the dominant, second-order poles. Next we evaluate the uncompensated steady-state error for a unit step input. The root locus for the uncompensated system is shown in Figure 9.5.

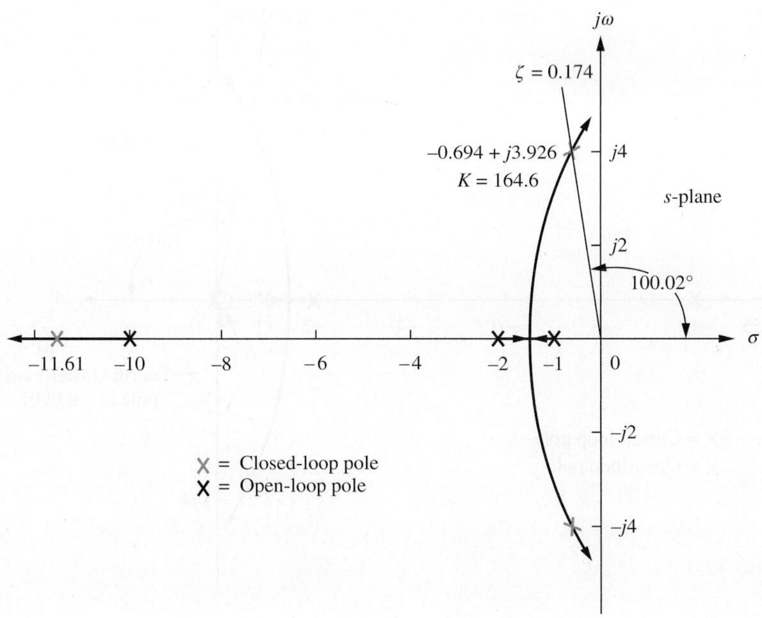


**FIGURE 9.5** Root locus for uncompensated system of Figure 9.4(*a*)

A damping ratio of 0.174 is represented by a radial line drawn on the $s$-plane at 100.02°. Searching along this line with the root locus program discussed in Appendix H at www.wiley.com/college/nise, we find that the dominant poles are $0.694 \pm j3.926$ for a gain, $K$, of 164.6. Now look for the third pole on the root locus beyond $-10$ on the real axis. Using the root locus program and searching for the same gain as that of the dominant pair, $K = 164.6$, we find that the third pole is approximately at $-11.61$. This gain yields $K_p = 8.23$. Hence, the steady-state error is

$$e(\infty) = \frac{1}{1+K_p} = \frac{1}{1+8.23} = 0.108 \tag{9.1}$$

Adding an ideal integral compensator with a zero at $-0.1$, as shown in Figure 9.4($b$), we obtain the root locus shown in Figure 9.6. The dominant second-order poles, the third pole beyond $-10$, and the gain are approximately the same as for the uncompensated system. Another section of the compensated root locus is between the origin and $-0.1$. Searching this region for the same gain at the dominant pair, $K = 158.2$, the fourth closed-loop pole is found at $-0.0902$, close enough to the zero to cause pole-zero cancellation. Thus, the compensated system's closed-loop poles and gain are approximately the same as the uncompensated system's closed-loop poles and gain, which indicates that the transient response of the compensated system is about the same as the uncompensated system. However, the compensated system, with its pole at the origin, is a Type 1 system; unlike the uncompensated system, it will respond to a step input with zero error.

Figure 9.7 compares the uncompensated response with the ideal integral compensated response. The step response of the ideal integral compensated system approaches unity in the steady state, while the uncompensated system approaches 0.892. Thus, the ideal integral compensated system responds with zero steady-state error. The transient response of both the uncompensated and the ideal integral compensated systems is the same up to approximately 3 seconds. After that time, the integrator in the compensator,

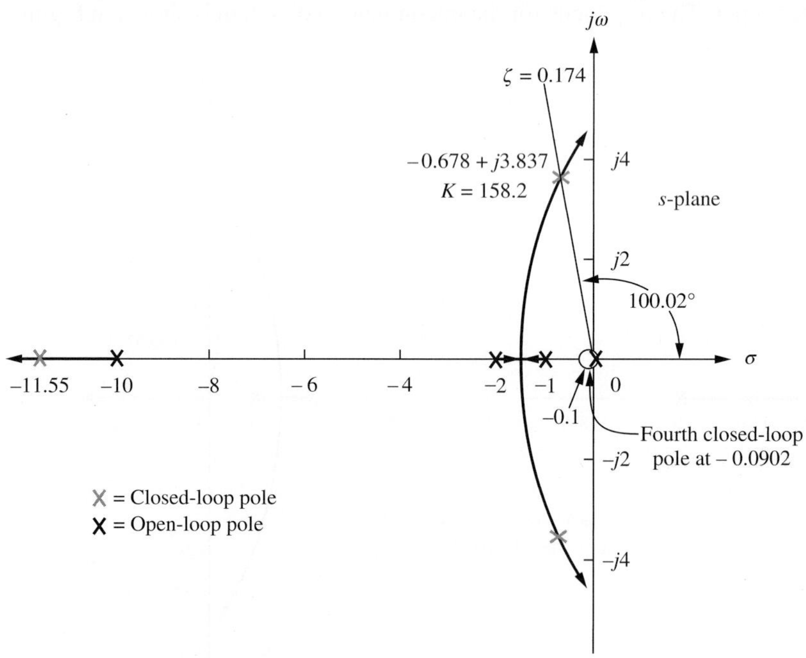


**FIGURE 9.6** Root locus for compensated system of Figure 9.4($b$)

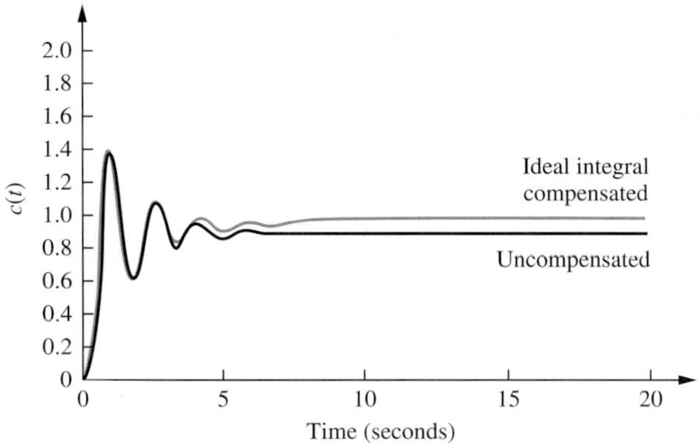


**FIGURE 9.7** Ideal integral compensated system response and the uncompensated system response of Example 9.1

shown in Figure 9.4(*b*), slowly compensates for the error until zero error is finally reached. The simulation shows that it takes 18 seconds for the compensated system to reach to within $\pm 2\%$ of the final value of unity, while the uncompensated system takes about 6 seconds to settle to within $\pm 2\%$ of its final value of 0.892. The compensation at first may appear to yield deterioration in the settling time. However, notice that the compensated system reaches the uncompensated system's final value in about the same time. The remaining time is used to improve the steady-state error over that of the uncompensated system.

A method of implementing an ideal integral compensator is shown in Figure 9.8. The compensating network precedes $G(s)$ and is an ideal integral compensator since

$$G_c(s) = K_1 + \frac{K_2}{s} = \frac{K_1\left(s + \frac{K_2}{K_1}\right)}{s} \tag{9.2}$$

The value of the zero can be adjusted by varying $K_2/K_1$. In this implementation, the error and the integral of the error are fed forward to the plant, $G(s)$. Since Figure 9.8 has both proportional and integral control, the ideal integral controller, or compensator, is given the alternate name *PI controller*. Later in the chapter we will see how to implement each block, $K_1$ and $K_2/s$.

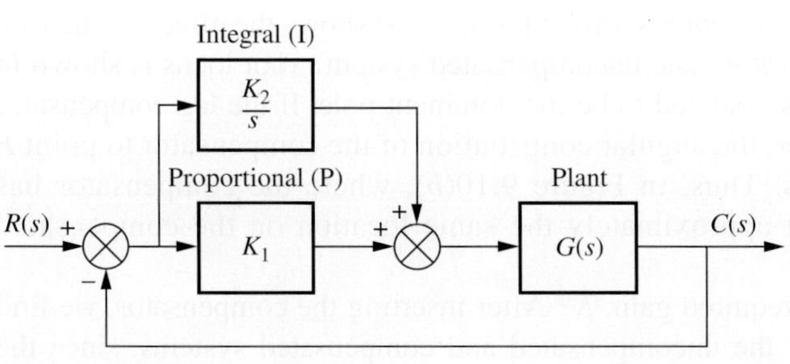


**FIGURE 9.8** PI controller

## Lag Compensation

Ideal integral compensation, with its pole on the origin, requires an active integrator. If we use passive networks, the pole and zero are moved to the left, close to the origin, as shown in Figure 9.9(*c*). One may guess that this placement of the pole, although it does not increase the system type, does yield an improvement in the static error constant over an uncompensated system. Without loss of generality, we demonstrate that this improvement is indeed realized for a Type 1 system.

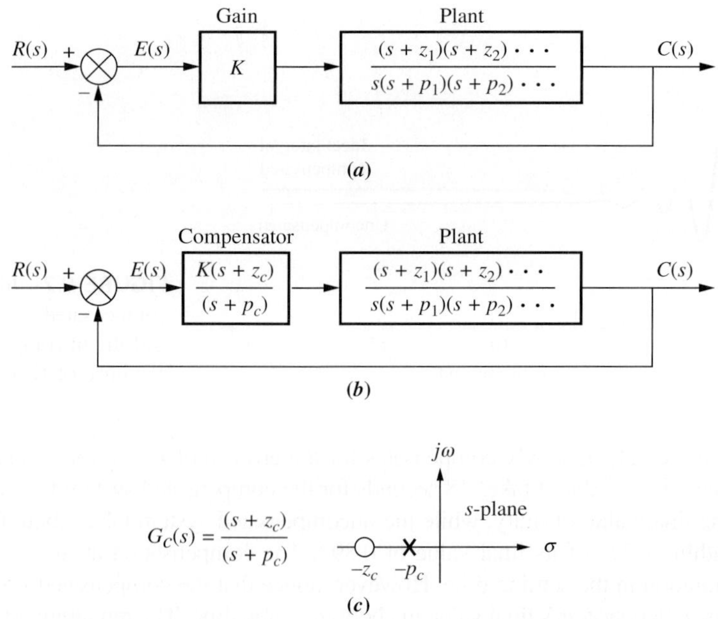


**FIGURE 9.9** **a.** Type 1 uncompensated system; **b.** Type 1 compensated system; **c.** compensator pole-zero plot

Assume the uncompensated system shown in Figure 9.9(*a*). The static error constant, $K_{v_O}$, for the system is

$$K_{v_O} = \frac{K z_1 z_2 \cdots}{p_1 p_2 \cdots} \tag{9.3}$$

Assuming the lag compensator shown in Figure 9.9(*b*) and (*c*), the new static error constant is

$$K_{v_N} = \frac{(K z_1 z_2 \cdots)(z_c)}{(p_1 p_2 \cdots)(p_c)} \tag{9.4}$$

What is the effect on the transient response? Figure 9.10 shows the effect on the root locus of adding the lag compensator. The uncompensated system's root locus is shown in Figure 9.10(*a*), where point *P* is assumed to be the dominant pole. If the lag compensator pole and zero are close together, the angular contribution of the compensator to point *P* is approximately zero degrees. Thus, in Figure 9.10(*b*), where the compensator has been added, point *P* is still at approximately the same location on the compensated root locus.

What is the effect on the required gain, *K*? After inserting the compensator, we find that *K* is virtually the same for the uncompensated and compensated systems, since the lengths of the vectors drawn from the lag compensator are approximately equal and all other vectors have not changed appreciably.

Now, what improvement can we expect in the steady-state error? Since we established that the gain, *K*, is about the same for the uncompensated and compensated systems, we can substitute Eq. (9.3) into (9.4) and obtain

$$K_{v_N} = K_{v_O} \frac{z_c}{p_c} > K_{v_O} \tag{9.5}$$

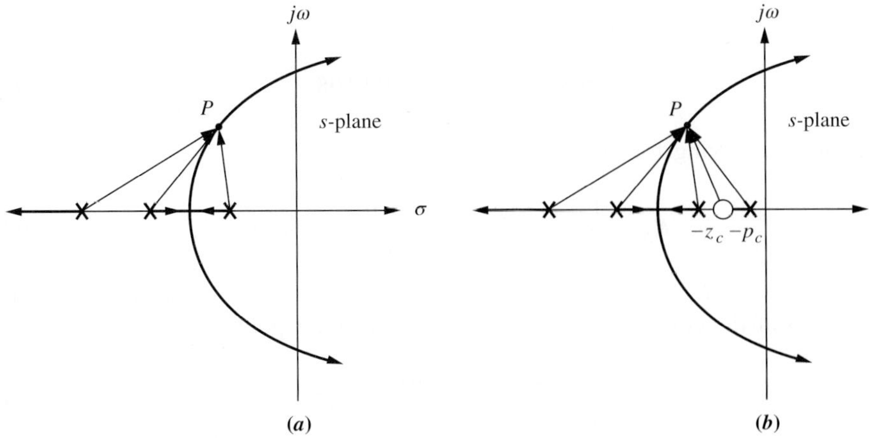


**FIGURE 9.10** Root locus: **a.** before lag compensation; **b.** after lag compensation

Equation (9.5) shows that the improvement in the compensated system's $K_v$ over the uncompensated system's $K_v$ is equal to the ratio of the magnitude of the compensator zero to the compensator pole. In order to keep the transient response unchanged, we know the compensator pole and zero must be close to each other. The only way the ratio of $z_c$ to $p_c$ can be large in order to yield an appreciable improvement in steady-state error and, simultaneously, have the compensator's pole and zero close to each other to minimize the angular contribution, is to place the compensator's pole-zero pair close to the origin. For example, the ratio of $z_c$ to $p_c$ can be equal to 10 if the pole is at $-0.001$ and the zero is at $-0.01$. Thus, the ratio is 10, yet the pole and zero are very close, and the angular contribution of the compensator is small.

In conclusion, although the ideal compensator drives the steady-state error to zero, a lag compensator with a pole that is not at the origin will improve the static error constant by a factor equal to $z_c/p_c$. There also will be a minimal effect upon the transient response if the pole-zero pair of the compensator is placed close to the origin. Later in the chapter we show circuit configurations for the lag compensator. These circuit configurations can be obtained with passive networks and thus do not require the active amplifiers and possible additional power supplies that are required by the ideal integral (PI) compensator. In the following example we design a lag compensator to yield a specified improvement in steady-state error.

## Example 9.2 Lag Compensator Design

**PROBLEM:** Compensate the system of Figure 9.4(*a*), whose root locus is shown in Figure 9.5, to improve the steady-state error by a factor of 10 if the system is operating with a damping ratio of 0.174.

**SOLUTION:** The uncompensated system error from Example 9.1 was 0.108 with $K_p = 8.23$. A tenfold improvement means a steady-state error of

$$e(\infty) = \frac{0.108}{10} = 0.0108 \tag{9.6}$$

Since

$$e(\infty) = \frac{1}{1+K_p} = 0.0108 \tag{9.7}$$

rearranging and solving for the required $K_p$ yields

$$K_p = \frac{1-e(\infty)}{e(\infty)} = \frac{1-0.0108}{0.0108} = 91.59 \tag{9.8}$$

The improvement in $K_p$ from the uncompensated system to the compensated system is the required ratio of the compensator zero to the compensator pole, or

$$\frac{z_c}{p_c} = \frac{K_{p_N}}{K_{p_O}} = \frac{91.59}{8.23} = 11.13 \tag{9.9}$$

Arbitrarily selecting

$$p_c = 0.01 \tag{9.10}$$

we use Eq. (9.9) and find

$$z_c = 11.13p_c \approx 0.111 \tag{9.11}$$

Let us now compare the compensated system, shown in Figure 9.11, with the uncompensated system. First sketch the root locus of the compensated system, as shown in Figure 9.12. Next search along the $\zeta = 0.174$ line for a multiple of 180° and find that the

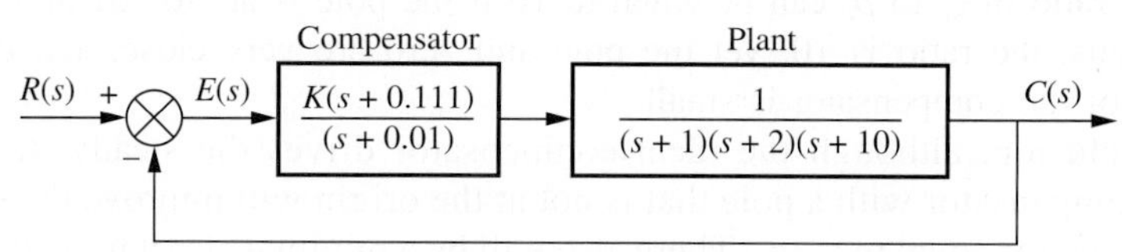


**FIGURE 9.11** Compensated system for Example 9.2

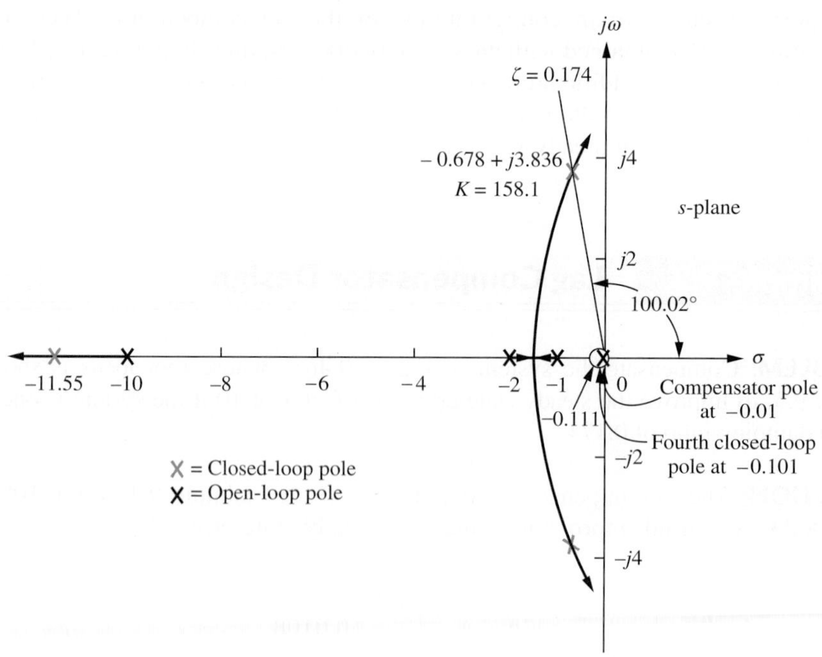


**FIGURE 9.12** Root locus for compensated system of Figure 9.11

**TABLE 9.1** Predicted characteristics of uncompensated and lag-compensated systems for Example 9.2

| Parameter | Uncompensated | Lag-compensated |
|---|---|---|
| Plant and compensator | $\frac{K}{(s+1)(s+2)(s+10)}$ | $\frac{K(s+0.111)}{(s+1)(s+2)(s+10)(s+0.01)}$ |
| $K$ | 164.6 | 158.1 |
| $K_p$ | 8.23 | 87.75 |
| $e(\infty)$ | 0.108 | 0.011 |
| Dominant second-order poles | $-0.694 \pm j3.926$ | $-0.678 \pm j3.836$ |
| Third pole | $-11.61$ | $-11.55$ |
| Fourth pole | None | $-0.101$ |
| Zero | None | $-0.111$ |

second-order dominant poles are at $-0.678 \pm j3.836$ with a gain, $K$, of 158.1. The third and fourth closed-loop poles are at $-11.55$ and $-0.101$, respectively, and are found by searching the real axis for a gain equal to that of the dominant poles. All transient and steady-state results for both the uncompensated and the compensated systems are shown in Table 9.1.

The fourth pole of the compensated system cancels its zero. This leaves the remaining three closed-loop poles of the compensated system very close in value to the three closed-loop poles of the uncompensated system. Hence, the transient response of both systems is approximately the same, as is the system gain, but notice that the steady-state error of the compensated system is 1/9.818 that of the uncompensated system and is close to the design specification of a tenfold improvement.

Figure 9.13 shows the effect of the lag compensator in the time domain. Even though the transient responses of the uncompensated and lag-compensated systems are the same, the lag-compensated system exhibits less steady-state error by approaching unity more closely than the uncompensated system.

We now examine another design possibility for the lag compensator and compare the response to Figure 9.13. Let us assume a lag compensator whose pole and zero are 10 times as close to the origin as in the previous design. The results are compared in Figure 9.14. Even though both responses will eventually reach approximately the same steady-state value, the lag compensator previously designed, $G_c(s) = (s+0.111)/(s+0.01)$, approaches the final value faster than the proposed lag compensator, $G_c(s) = (s+0.0111)/(s+0.001)$. We can explain this phenomenon as follows. From Table 9.1, the previously designed lag compensator has a fourth closed-loop pole at $-0.101$.

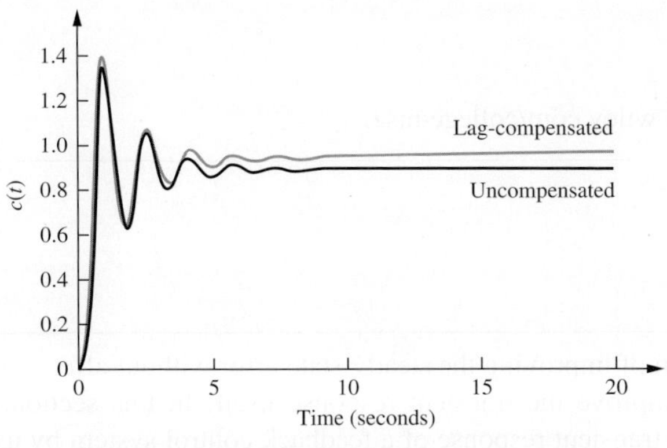


**FIGURE 9.13** Step responses of uncompensated and lag-compensated systems for Example 9.2

**TryIt 9.1**

Use the following MATLAB and Control System Toolbox statements to reproduce Figure 9.13.

```
Gu=zpk([],...
 [-1 -2 -10],164.6);
Gc=zpk([-0.111],...
 [-0.01],1);
Gce=Gu*Gc;
Tu=feedback(Gu,1);
Tc=feedback(Gce,1);
step(Tu)
hold
step(Tc)
```

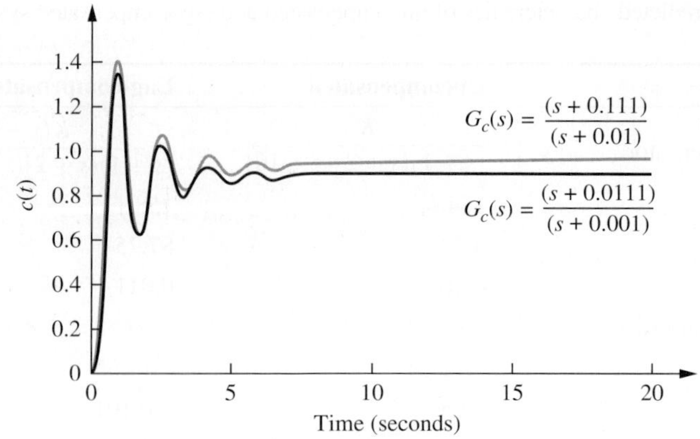


**FIGURE 9.14** Step responses of the system for Example 9.2 using different lag compensators

Using the same analysis for the new lag compensator with its open-loop pole 10 times as close to the imaginary axis, we find its fourth closed-loop pole at $-0.01$. Thus, the new lag compensator has a closed-loop pole closer to the imaginary axis than the original lag compensator. This pole at $-0.01$ will produce a longer transient response than the original pole at $-0.101$, and the steady-state value will not be reached as quickly.

## Skill-Assessment Exercise 9.1

**PROBLEM:** A unity feedback system with the forward transfer function

$$G(s) = \frac{K}{s(s+7)}$$

is operating with a closed-loop step response that has 15% overshoot. Do the following:

**a.** Evaluate the steady-state error for a unit ramp input.

**b.** Design a lag compensator to improve the steady-state error by a factor of 20.

**c.** Evaluate the steady-state error for a unit ramp input to your compensated system.

**d.** Evaluate how much improvement in steady-state error was realized.

**ANSWERS:**

**a.** $e_{\text{ramp}}(\infty) = 0.1527$

**b.** $G_{\text{lag}}(s) = \frac{s+0.2}{s+0.01}$

**c.** $e_{\text{ramp}}(\infty) = 0.0078$

**d.** 19.58 times improvement

The complete solution is at www.wiley.com/college/nise.

# 9.3 Improving Transient Response via Cascade Compensation

Since we have solved the problem of improving the steady-state error without affecting the transient response, let us now improve the transient response itself. In this section, we discuss two ways to improve the transient response of a feedback control system by using cascade compensation. Typically, the objective is to design a response that has a desirable percent overshoot and a shorter settling time than the uncompensated system.

The first technique we will discuss is *ideal derivative compensation*. With ideal derivative compensation, a pure differentiator is added to the forward path of the feedback control system. We will see that the result of adding differentiation is the addition of a zero to the forward-path transfer function. This type of compensation requires an active network for its realization. Further, differentiation is a noisy process; although the level of the noise is low, the frequency of the noise is high compared to the signal. Thus, differentiating high-frequency noise yields a large, unwanted signal.

The second technique does not use pure differentiation. Instead, it approximates differentiation with a passive network by adding a zero and a more distant pole to the forward-path transfer function. The zero approximates pure differentiation as described previously.

As with compensation to improve steady-state error, we introduce names associated with the implementation of the compensators. We call an ideal derivative compensator a *proportional-plus-derivative* (*PD*) *controller*, since the implementation, as we will see, consists of feeding the error (proportional) plus the derivative of the error forward to the plant. The second technique uses a passive network called a *lead compensator*. As with the lag compensator, the name comes from its frequency response, which is discussed in Chapter 11. Thus, we use the name *PD controller* interchangeably with *ideal derivative compensator*, and we use the name *lead compensator* when the cascade compensator does not employ pure differentiation.

## Ideal Derivative Compensation (PD)

The transient response of a system can be selected by choosing an appropriate closed-loop pole location on the $s$-plane. If this point is on the root locus, then a simple gain adjustment is all that is required in order to meet the transient response specification. If the closed-loop pole location is not on the root locus, then the root locus must be reshaped so that the compensated (new) root locus goes through the selected closed-loop pole location. In order to accomplish the latter task, poles and zeros can be added in the forward path to produce a new open-loop function whose root locus goes through the design point on the $s$-plane. One way to speed up the original system that generally works is to add a single zero to the forward path.

This zero can be represented by a compensator whose transfer function is

$$G_c(s) = s + z_c \tag{9.12}$$

This function, the sum of a differentiator and a pure gain, is called an *ideal derivative*, or *PD controller*. Judicious choice of the position of the compensator zero can quicken the response over the uncompensated system. In summary, transient responses unattainable by a simple gain adjustment can be obtained by augmenting the system's poles and zeros with an ideal derivative compensator.

We now show that ideal derivative compensation speeds up the response of a system. Several simple examples are shown in Figure 9.15, where the uncompensated system of Figure 9.15($a$), operating with a damping ratio of 0.4, becomes a compensated system by the addition of a compensating zero at −2, −3, and −4 in Figures 9.15($b$), ($c$), and ($d$), respectively. In each design, the zero is moved to a different position, and the root locus is shown. For each compensated case, the dominant, second-order poles are farther out along the 0.4 damping ratio line than the uncompensated system.

Each of the compensated cases has dominant poles with the same damping ratio as the uncompensated case. Thus, we predict that the percent overshoot will be the same for each case.

Also, the compensated, dominant, closed-loop poles have more negative real parts than the uncompensated, dominant, closed-loop poles. Hence, we predict that the settling times for the compensated cases will be shorter than for the uncompensated case. The compensated,

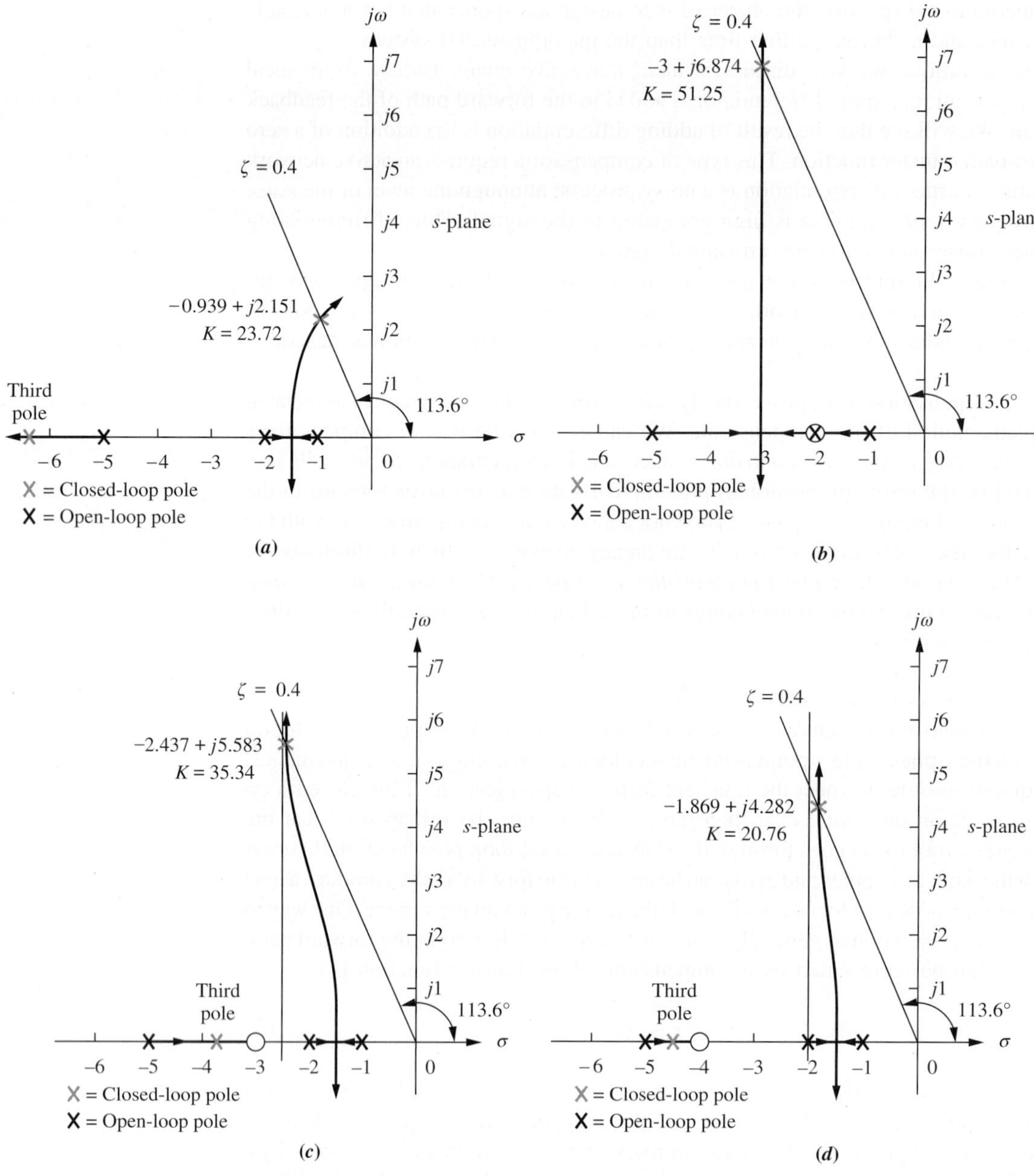


**FIGURE 9.15** Using ideal derivative compensation: **a.** uncompensated; **b.** compensator zero at −2; **c.** compensator zero at −3; **d.** compensator zero at −4

dominant, closed-loop poles with the more negative real parts will have the shorter settling times. The system in Figure 9.15(*b*) will have the shortest settling time.

All of the compensated systems will have smaller peak times than the uncompensated system, since the imaginary parts of the compensated systems are larger. The system of Figure 9.15(*b*) will have the smallest peak time.

Also notice that as the zero is placed farther from the dominant poles, the closed-loop, compensated dominant poles move closer to the origin and to the uncompensated, dominant closed-loop poles. Table 9.2 summarizes the results obtained from the root locus of each of the design cases shown in Figure 9.15.

**TABLE 9.2** Predicted characteristics for the systems of Figure 9.15

| | Uncompensated | Compensation b | Compensation c | Compensation d |
|---|---|---|---|---|
| Plant and compensator | $\frac{K}{(s+1)(s+2)(s+5)}$ | $\frac{K(s+2)}{(s+1)(s+2)(s+5)}$ | $\frac{K(s+3)}{(s+1)(s+2)(s+5)}$ | $\frac{K(s+4)}{(s+1)(s+2)(s+5)}$ |
| Dom, poles | $-0.939 \pm j2.151$ | $-3 \pm j6.874$ | $-2.437 \pm j5.583$ | $-1.869 \pm j4.282$ |
| $K$ | 23.72 | 51.25 | 35.34 | 20.76 |
| $\zeta$ | 0.4 | 0.4 | 0.4 | 0.4 |
| $\omega_n$ | 2.347 | 7.5 | 6.091 | 4.673 |
| $\%OS$ | 25.38 | 25.38 | 25.38 | 25.38 |
| $T_s$ | 4.26 | 1.33 | 1.64 | 2.14 |
| $T_p$ | 1.46 | 0.46 | 0.56 | 0.733 |
| $K_p$ | 2.372 | 10.25 | 10.6 | 8.304 |
| $e(\infty)$ | 0.297 | 0.089 | 0.086 | 0.107 |
| Third pole | −6.123 | None | −3.127 | −4.262 |
| Zero | None | None | −3 | −4 |
| Comments | Second-order approx. OK | Pure second-order | Second-order approx. OK | Second-order approx. OK |

In summary, although compensation methods *c* and *d* yield slower responses than method *b*, the addition of ideal derivative compensation shortened the response time in each case while keeping the percent overshoot the same. This change can best be seen in the settling time and peak time, where there is at least a doubling of speed across all of the cases of compensation. An added benefit is the improvement in the steady-state error, even though lag compensation was not used. Here the steady-state error of the compensated system is at least one-third that of the uncompensated system, as seen by $e(\infty)$ and $K_p$. All systems in Table 9.2 are Type 0, and some steady-state error is expected. The reader must not assume that, in general, improvement in transient response always yields an improvement in steady-state error.

The time response of each case in Table 9.2 is shown in Figure 9.16. We see that the compensated responses are faster and exhibit less error than the uncompensated response.

Now that we have seen what ideal derivative compensation can do, we are ready to design our own ideal derivative compensator to meet a transient response specification. Basically, we will evaluate the sum of angles from the open-loop poles and zeros to a design point that is the closed-loop pole that yields the desired transient response. The difference between 180° and the calculated angle must be the angular contribution of the compensator zero. Trigonometry is then used to locate the position of the zero to yield the required difference in angle.

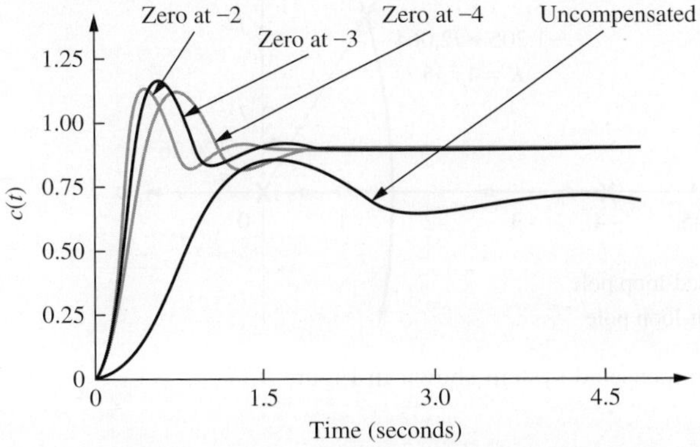


**FIGURE 9.16** Uncompensated system and ideal derivative compensation solutions from Table 9.2

## Example 9.3 Ideal Derivative Compensator Design

**PROBLEM:** Given the system of Figure 9.17, design an ideal derivative compensator to yield a 16% overshoot, with a threefold reduction in settling time.

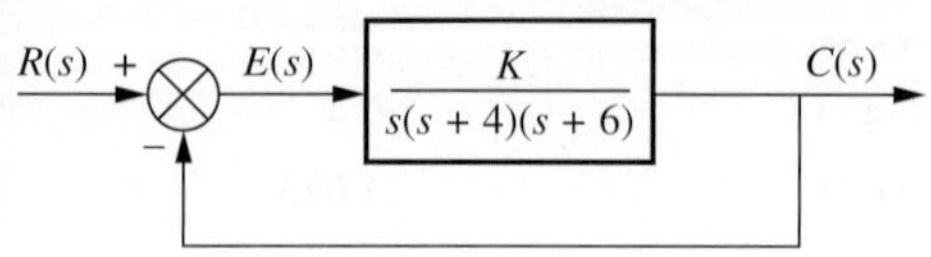


**FIGURE 9.17** Feedback control system for Example 9.3

**SOLUTION:** Let us first evaluate the performance of the uncompensated system operating with 16% overshoot. The root locus for the uncompensated system is shown in Figure 9.18. Since 16% overshoot is equivalent to $\zeta = 0.504$, we search along that damping ratio line for an odd multiple of 180° and find that the dominant, second-order pair of poles is at $-1.205 \pm j2.064$. Thus, the settling time of the uncompensated system is

$$T_s = \frac{4}{\zeta\omega_n} = \frac{4}{1.205} = 3.320 \tag{9.13}$$

Since our evaluation of percent overshoot and settling time is based upon a second-order approximation, we must check the assumption by finding the third pole and justifying the second-order approximation. Searching beyond −6 on the real axis for a gain equal to the gain of the dominant, second-order pair, 43.35, we find a third pole at −7.59, which is over six times as far from the $j\omega$-axis as the dominant, second-order pair. We conclude that our approximation is valid. The transient and steady-state error characteristics of the uncompensated system are summarized in Table 9.3.

Now we proceed to compensate the system. First we find the location of the compensated system's dominant poles. In order to have a threefold reduction in the settling time, the compensated system's settling time will be one-third of Eq. (9.13). The new settling time will be 1.107. Therefore, the real part of the compensated system's dominant, second-order pole is

$$\sigma = \frac{4}{T_s} = \frac{4}{1.107} = 3.613 \tag{9.14}$$

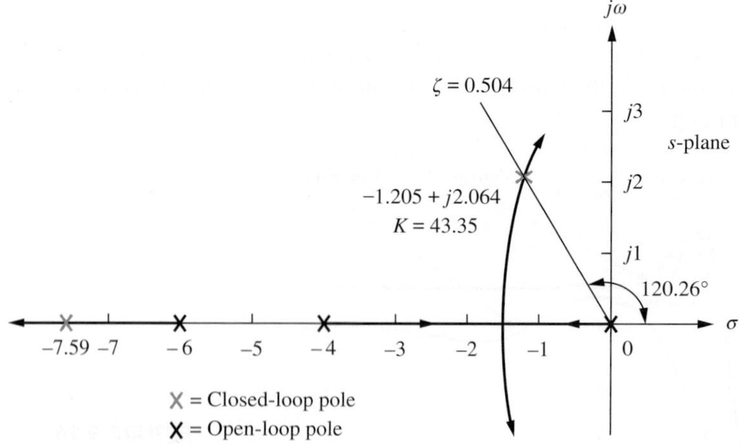


**FIGURE 9.18** Root locus for uncompensated system shown in Figure 9.17

**TABLE 9.3** Uncompensated and compensated system characteristic of Example 9.3

| | Uncompensated | Simulation | Compensated | Simulation |
|---|---|---|---|---|
| Plant and compensator | $\frac{K}{s(s+4)(s+6)}$ | | $\frac{K(s+3.006)}{s(s+4)(s+6)}$ | |
| Dominant poles | $-1.205 \pm j2.064$ | | $-3.613 \pm j6.193$ | |
| $K$ | 43.35 | | 47.45 | |
| $\zeta$ | 0.504 | | 0.504 | |
| $\omega_n$ | 2.39 | | 7.17 | |
| $\%OS$ | 16 | 14.8 | 16 | 11.8 |
| $T_s$ | 3.320 | 3.6 | 1.107 | 1.2 |
| $T_p$ | 1.522 | 1.7 | 0.507 | 0.5 |
| $K_v$ | 1.806 | | 5.94 | |
| $e(\infty)$ | 0.554 | | 0.168 | |
| Third pole | −7.591 | | −2.775 | |
| Zero | None | | −3.006 | |
| Comments | Second-order approx. OK | | Pole-zero not canceling | |

Figure 9.19 shows the designed dominant, second-order pole, with a real part equal to −3.613 and an imaginary part of

$$\omega_d = 3.613 \tan(180^\circ - 120.26^\circ) = 6.193 \tag{9.15}$$

Next we design the location of the compensator zero. Input the uncompensated system's poles and zeros in the root locus program as well as the design point $-3.613 \pm j6.193$ as a test point. The result is the sum of the angles to the design point of all the poles and zeros of the compensated system except for those of the compensator zero itself. The difference between the result obtained and 180° is the angular contribution required of the compensator zero. Using the open-loop poles shown in Figure 9.19 and the test point, $-3.613 + j6.193$, which is the desired dominant second-order pole, we obtain the sum of the angles as −275.6°. Hence, the angular contribution required from the compensator zero for the test point to be on the root locus is $+275.6^\circ - 180^\circ = 95.6^\circ$. The

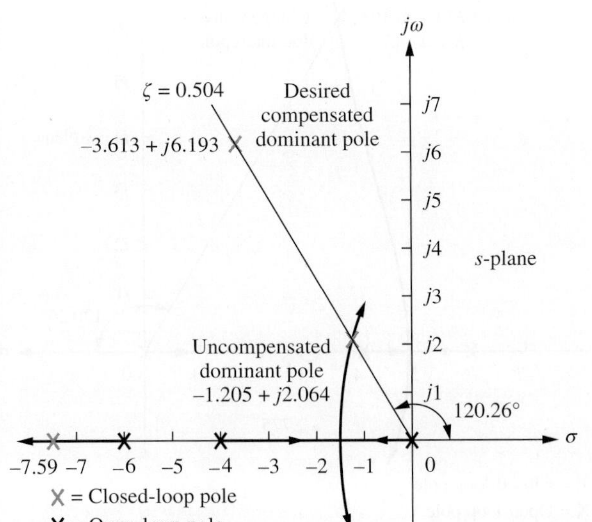


**FIGURE 9.19** Compensated dominant pole superimposed over the uncompensated root locus for Example 9.3

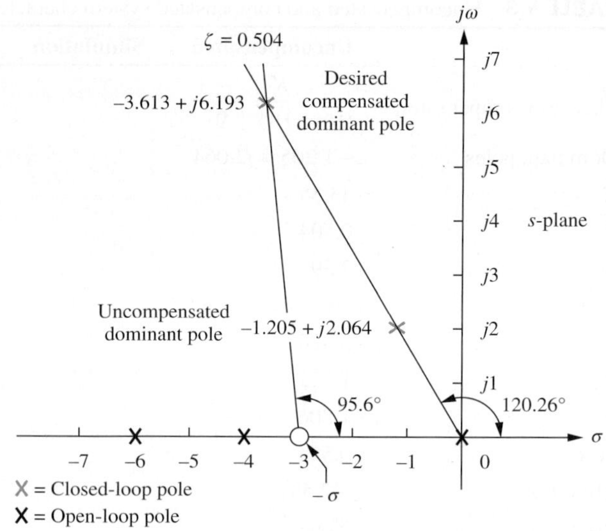


**FIGURE 9.20** Evaluating the location of the compensating zero for Example 9.3

geometry is shown in Figure 9.20, where we now must solve for $-\sigma$, the location of the compensator zero.

From the figure,

$$\frac{6.193}{3.613 - \sigma} = \tan(180^\circ - 95.6^\circ) \tag{9.16}$$

Thus, $\sigma = 3.006$. The complete root locus for the compensated system is shown in Figure 9.21.

Table 9.3 summarizes the results for both the uncompensated system and the compensated system. For the uncompensated system, the estimate of the transient response is accurate since the third pole is at least five times the real part of the dominant, second-order pair. The second-order approximation for the compensated system, however, may be invalid because there is no approximate closed-loop third-pole and zero cancellation

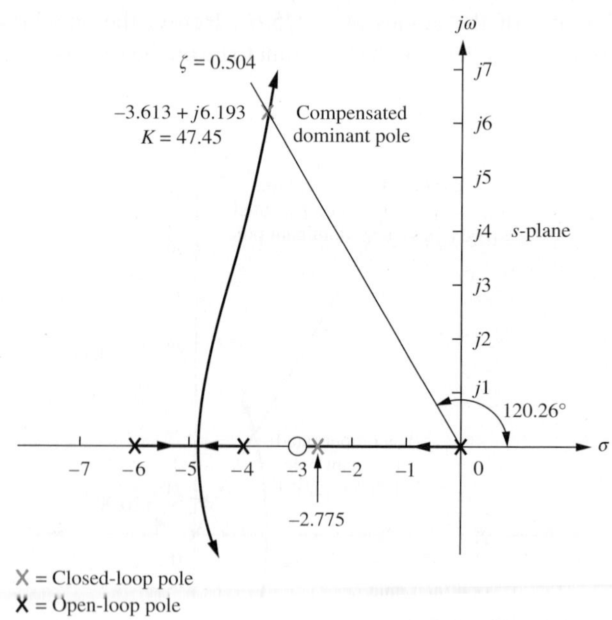


**FIGURE 9.21** Root locus for the compensated system of Example 9.3

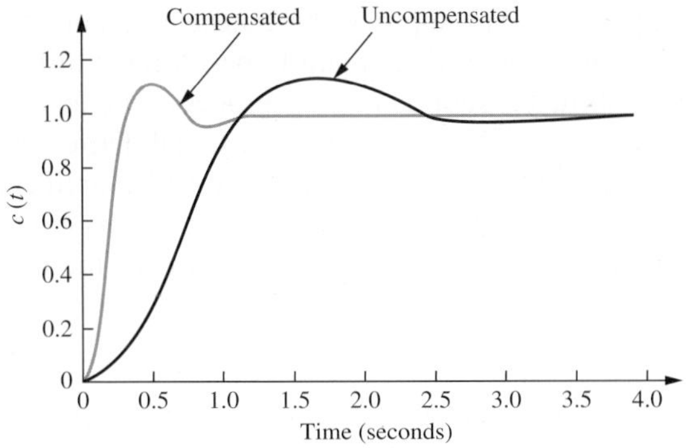


**FIGURE 9.22** Uncompensated and compensated system step responses of Example 9.3

between the closed-loop pole at $-2.775$ and the closed-loop zero at $-3.006$. A simulation or a partial-fraction expansion of the closed-loop response to compare the residue of the pole at $-2.775$ to the residues of the dominant poles at $-3.613 \pm j6.193$ is required. The results of a simulation are shown in the table's second column for the uncompensated system and the fourth column for the compensated system. The simulation results can be obtained using MATLAB (discussed at the end of this example) or a program like the state-space step-response program described in Appendix H.1 at www.wiley.com/college/nise. The percent overshoot differs by 3% between the uncompensated and compensated systems, while there is approximately a threefold improvement in speed as evaluated from the settling time.

The final results are displayed in Figure 9.22, which compares the uncompensated system and the faster compensated system.

MATLAB **ML**

`Students who are using MATLAB should now run ch9p1 in Appendix B. MATLAB will be used to design a PD controller. You will input the desired percent overshoot from the keyboard. MATLAB will plot the root locus of the uncompensated system and the percent overshoot line. You will interactively select the gain, after which MATLAB will display the performance characteristics of the uncompensated system and plot its step response. Using these characteristics, you will input the desired settling time. MATLAB will design the PD controller, enumerate its performance characteristics, and plot a step response. This exercise solves Example 9.3 using MATLAB.`

Once we decide on the location of the compensating zero, how do we implement the ideal derivative, or PD controller? The ideal integral compensator that improved steady-state error was implemented with a proportional-plus-integral (PI) controller. The ideal derivative compensator used to improve the transient response is implemented with a proportional-plus-derivative (PD) controller. For example, in Figure 9.23 the transfer function of the controller is

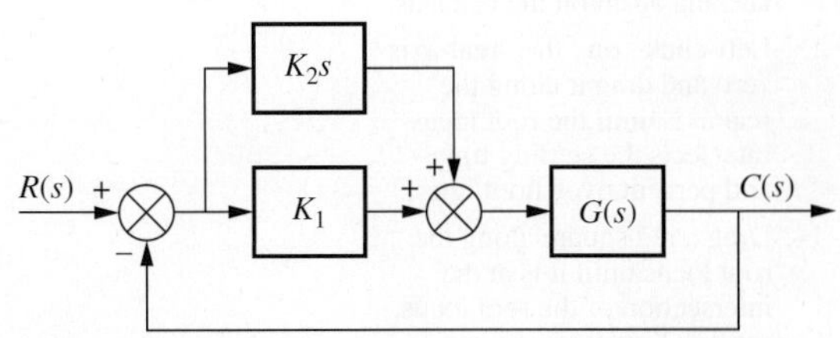


**FIGURE 9.23** PD controller

$$G_c(s) = K_2 s + K_1 = K_2\left(s + \frac{K_1}{K_2}\right) \tag{9.17}$$

Hence, $K_1/K_2$ is chosen to equal the negative of the compensator zero, and $K_2$ is chosen to contribute to the required loop-gain value. Later in the chapter, we will study circuits that can be used to approximate differentiation and produce gain.

**TryIt 9.2**

Use MATLAB, the Control System Toolbox, and the following steps to use SISOTOOL to perform the design of Example 9.3.

1. Type **SISOTOOL** in the MATLAB **Command Window**.
2. Select **Import** in the File menu of the **SISO Design for SISO Design Task Window**.
3. In the **Data** field for G, type `zpk([ ],[0,-4,-6],1)` and hit **ENTER** on the keyboard. Click **OK**.
4. On the **Edit** menu choose **SISO Tool Preferences . . .** and select **Zero/pole/gain:** under the **Options** tab. Click **OK**.
5. Right-click on the root locus white space and choose **Design Requirements/ New . . .**
6. Choose **Percent overshoot** and type in 16. Click **OK**.
7. Right-click on the root locus white space and choose **Design Requirements/ New . . .**
8. Choose **Settling time** and click **OK**.
9. Drag the settling time vertical line to the intersection of the root locus and 16% overshoot radial line.
10. Read the settling time at the bottom of the window.
11. Drag the settling time vertical line to a settling time that is 1/3 of the value found in Step 9.
12. Click on a red zero icon in the menu bar. Place the zero on the root locus real axis by clicking again on the real axis.
13. Left-click on the real-axis zero and drag it along the real axis until the root locus intersects the settling time and percent overshoot lines.
14. Drag a red square along the root locus until it is at the intersection of the root locus, settling time line, and the percent overshoot line.
15. Click the **Compensator Editor** tab of the **Control and Estimation Tools Manager** window to see the resulting compensator, including the gain.

While the ideal derivative compensator can improve the transient response of the system, it has two drawbacks. First, it requires an active circuit to perform the differentiation. Second, as previously mentioned, differentiation is a noisy process: The level of the noise is low, but the frequency of the noise is high compared to the signal. Differentiation of high frequencies can lead to large unwanted signals or saturation of amplifiers and other components. The lead compensator is a passive network used to overcome the disadvantages of ideal differentiation and still retain the ability to improve the transient response.

## Lead Compensation

Just as the active ideal integral compensator can be approximated with a passive lag network, an active ideal derivative compensator can be approximated with a passive lead compensator. When passive networks are used, a single zero cannot be produced; rather, a compensator zero and a pole result. However, if the pole is farther from the imaginary axis than the zero, the angular contribution of the compensator is still positive and thus approximates an equivalent single zero. In other words, the angular contribution of the compensator pole subtracts from the angular contribution of the zero. This deduction does not preclude the use of the compensator to improve transient response, since the net angular contribution is positive, just as for a single PD controller zero.

The advantages of a passive lead network over an active PD controller are that (1) no additional power supplies are required and (2) noise due to differentiation is reduced. The disadvantage is that the additional pole does not reduce the number of branches of the root locus that cross the imaginary axis into the right half-plane. On the other hand, the addition of the single zero of the PD controller tends to reduce the number of branches of the root locus that cross into the right half-plane.

Let us first look at the concept behind lead compensation. If we select a desired dominant, second-order pole on the $s$-plane, the sum of the angles from the uncompensated system's poles and zeros to the design point can be found. The difference between 180° and the sum of the angles must be the angular contribution required of the compensator.

For example, looking at Figure 9.24, we see that

$$\theta_2 - \theta_1 - \theta_3 - \theta_4 + \theta_5 = (2k+1)180^\circ \tag{9.18}$$

where $(\theta_2 - \theta_1) = \theta_c$ is the angular contribution of the lead compensator. From Figure 9.24 we see that $\theta_c$ is the angle of a ray extending from the design point and intersecting the real axis at the pole value and zero value of the compensator. Now visualize this ray rotating about the desired closed-loop pole location and intersecting the real axis at the compensator pole and zero, as illustrated in Figure 9.25. We realize that an infinite number of lead compensators could be used to meet the transient response requirement.

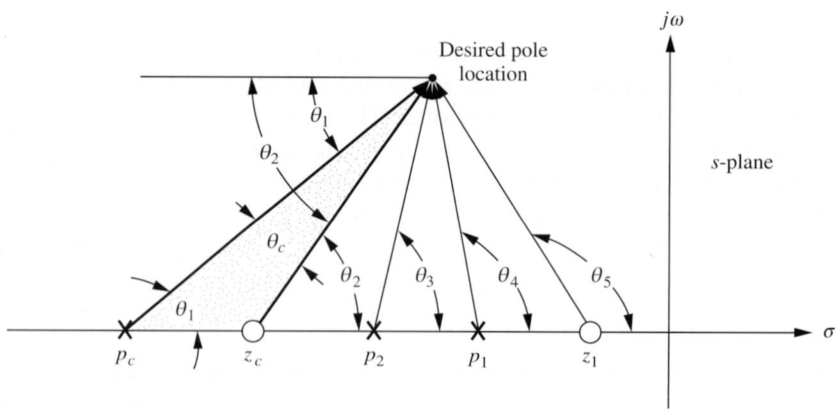


**FIGURE 9.24** Geometry of lead compensation

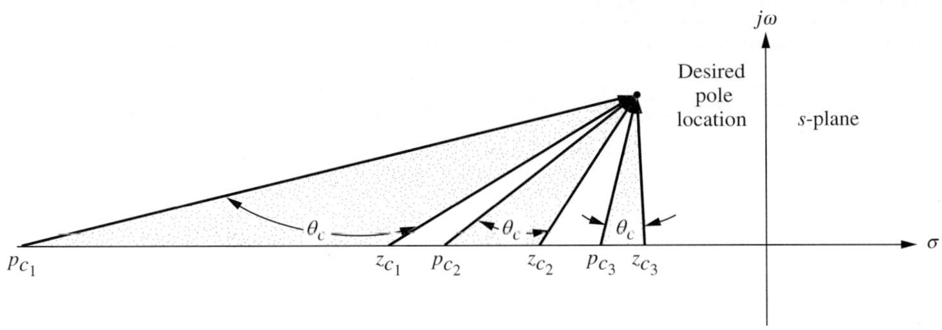


**FIGURE 9.25** Three of the infinite possible lead compensator solutions

How do the possible lead compensators differ? The differences are in the values of static error constants, the gain required to reach the design point on the compensated root locus, the difficulty in justifying a second-order approximation when the design is complete, and the ensuing transient response.

For design, we arbitrarily select either a lead compensator pole or zero and find the angular contribution at the design point of this pole or zero along with the system's open-loop poles and zeros. The difference between this angle and 180° is the required contribution of the remaining compensator pole or zero. Let us look at an example.

## Example 9.4 Lead Compensator Design

**PROBLEM:** Design three lead compensators for the system of Figure 9.17 that will reduce the settling time by a factor of 2 while maintaining 30% overshoot. Compare the system characteristics between the three designs.

**SOLUTION:** First determine the characteristics of the uncompensated system operating at 30% overshoot to see what the uncompensated settling time is. Since 30% overshoot is equivalent to a damping ratio of 0.358, we search along the $\zeta = 0.358$ line for the uncompensated dominant poles on the root locus, as shown in Figure 9.26. From the pole's real part, we calculate the uncompensated settling time as $T_s = 4/1.007 = 3.972$ seconds. The remaining characteristics of the uncompensated system are summarized in Table 9.4.

Next we find the design point. A twofold reduction in settling time yields $T_s = 3.972/2 = 1.986$ seconds, from which the real part of the desired pole location is $-\zeta\omega_n = -4/T_s = -2.014$. The imaginary part is $\omega_d = -2.014\tan(110.98°) = 5.252$.

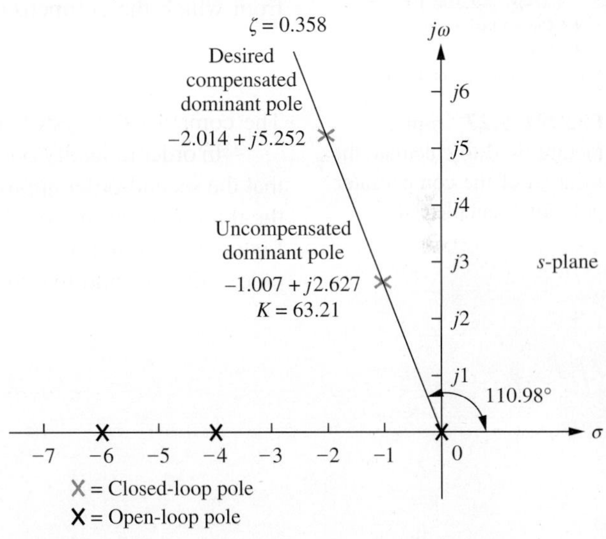


**FIGURE 9.26** Lead compensator design, showing evaluation of uncompensated and compensated dominant poles for Example 9.4

We continue by designing the lead compensator. Arbitrarily assume a compensator zero at −5 on the real axis as a possible solution. Using the root locus program, sum the angles from both this zero and the uncompensated system's poles and zeros, using the design point as a test point. The resulting angle is −172.69°. The difference between this angle and 180° is the angular contribution required from the compensator pole in order to place the design point on the root locus. Hence, an angular contribution of −7.31° is required from the compensator pole.

**TABLE 9.4** Comparison of lead compensation designs for Example 9.4

| | Uncompensated | Compensation a | Compensation b | Compensation c |
|---|---|---|---|---|
| Plant and compensator | $\frac{K}{s(s+4)(s+6)}$ | $\frac{K(s+5)}{s(s+4)(s+6)(s+42.96)}$ | $\frac{K(s+4)}{s(s+4)(s+6)(s+20.09)}$ | $\frac{K(s+2)}{s(s+4)(s+6)(s+8.971)}$ |
| Dominant poles | $-1.007 \pm j2.627$ | $-2.014 \pm j5.252$ | $-2.014 \pm j5.252$ | $-2.014 \pm j5.252$ |
| $K$ | 63.21 | 1423 | 698.1 | 345.6 |
| $\zeta$ | 0.358 | 0.358 | 0.358 | 0.358 |
| $\omega_n$ | 2.813 | 5.625 | 5.625 | 5.625 |
| $\%OS^*$ | 30 (28) | 30 (30.7) | 30 (28.2) | 30 (14.5) |
| $T_s^*$ | 3.972 (4) | 1.986 (2) | 1.986 (2) | 1.986 (1.7) |
| $T_p^*$ | 1.196 (1.3) | 0.598 (0.6) | 0.598 (0.6) | 0.598 (0.7) |
| $K_v$ | 2.634 | 6.9 | 5.791 | 3.21 |
| $e(\infty)$ | 0.380 | 0.145 | 0.173 | 0.312 |
| Other poles | −7.986 | −43.8, −5.134 | −22.06 | −13.3, −1.642 |
| Zero | None | −5 | None | −2 |
| Comments | Second-order approx. OK | Second-order approx. OK | Second-order approx. OK | No pole-zero cancellation |

*Simulation results are shown in parentheses.

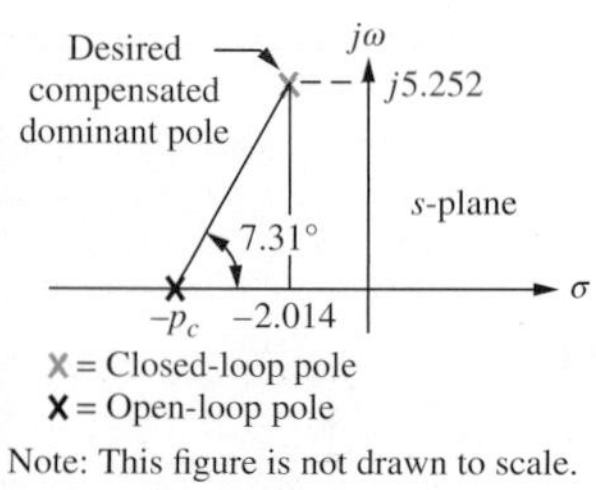


**FIGURE 9.27** $s$-plane picture used to calculate the location of the compensator pole for Example 9.4

The geometry shown in Figure 9.27 is used to calculate the location of the compensator pole. From the figure,

$$\frac{5.252}{p_c - 2.014} = \tan 7.31^\circ \tag{9.19}$$

from which the compensator pole is found to be

$$p_c = 42.96 \tag{9.20}$$

The compensated system root locus is sketched in Figure 9.28.

In order to justify our estimates of percent overshoot and settling time, we must show that the second-order approximation is valid. To perform this validity check, we search for the third and fourth closed-loop poles found beyond −42.96 and between −5 and −6 in Figure 9.28. Searching these regions for the gain equal to that of the compensated dominant pole, 1423, we find that the third and fourth poles are at −43.8 and −5.134, respectively.

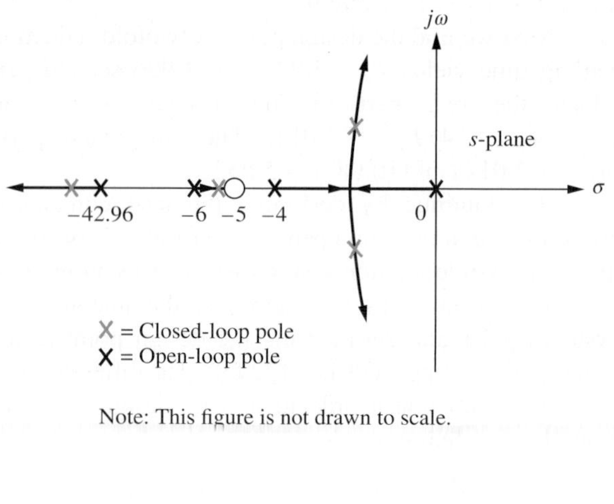


**FIGURE 9.28** Compensated system root locus

Since $-43.8$ is more than 20 times the real part of the dominant pole, the effect of the third closed-loop pole is negligible. Since the closed-loop pole at $-5.134$ is close to the zero at $-5$, we have pole-zero cancellation, and the second-order approximation is valid.

All results for this design and two other designs, which place the compensator zero arbitrarily at $-2$ and $-4$ and follow similar design techniques, are summarized in Table 9.4. Each design should be verified by a simulation, which could consist of using MATLAB (discussed at the end of this example) or the state-space model and the step-response program discussed in Appendix II.1 at www.wiley.com/college/nise. We have performed a simulation for this design problem, and the results are shown by parenthetical entries next to the estimated values in the table. The only design that disagrees with the simulation is the case where the compensator zero is at $-2$. For this case the closed-loop pole and zero do not cancel.

A sketch of the root locus, which you should generate, shows why the effect of the zero is pronounced, causing the response to be different from that predicted. Placing the zero to the right of the pole at $-4$ creates a portion of the root locus that is between the origin and the zero. In other words, there is a closed-loop pole closer to the origin than the dominant poles, with little chance of pole-zero cancellation except at high gain. Thus, a quick sketch of the root locus gives us information from which we can make better design decisions. For this example, we want to place the zero on, or to the left of, the pole at $-4$, which gives a better chance for pole-zero cancellation and for a higher-order pole that is to the left of the dominant poles and subsequently faster. This is verified by the fact that our results show good second-order approximations for the cases where the zero was placed at $-4$ and $-5$. Again, decisions about where to place the zero are based on simple rules of thumb and must be verified by simulations at the end of the design.

Let us now summarize the results shown in Table 9.4. First we notice differences in the following:

1. The position of the arbitrarily selected zero
2. The amount of improvement in the steady-state error
3. The amount of required gain, $K$
4. The position of the third and fourth poles and their relative effect upon the second-order approximation. This effect is measured by their distance from the dominant poles or the degree of cancellation with the closed-loop zero.

Once a simulation verifies desired performance, the choice of compensation can be based upon the amount of gain required or the improvement in steady-state error that can be obtained without a lag compensator.

The results of Table 9.4 are supported by simulations of the step response, shown in Figure 9.29 for the uncompensated system and the three lead compensation solutions.

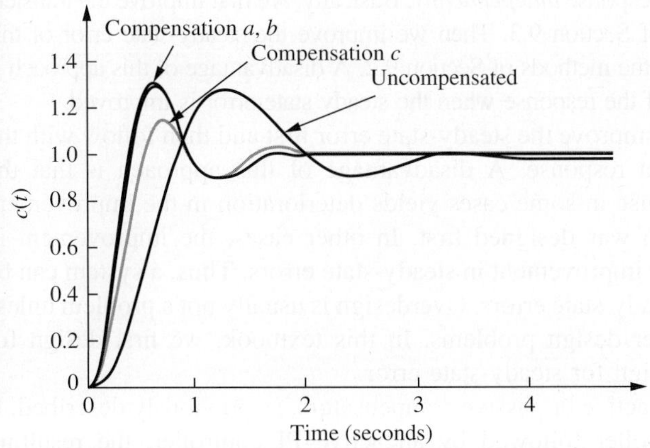


**FIGURE 9.29** Uncompensated system and lead compensation responses for Example 9.4

MATLAB

ML

Students who are using MATLAB should now run ch9p2 in Appendix B. MATLAB will be used to design a lead compensator. You will input the desired percent overshoot from the keyboard. MATLAB will plot the root locus of the uncompensated system and the percent overshoot line. You will interactively select the gain, after which MATLAB will display the performance characteristics of the uncompensated system and plot its step response. Using these characteristics, you will input the desired settling time and a zero value for the lead compensator. You will then interactively select a value for the compensator pole. MATLAB will respond with a root locus. You can then continue selecting pole values until the root locus goes through the desired point. MATLAB will display the lead compensator, enumerate its performance characteristics, and plot a step response. This exercise solves Example 9.4 using MATLAB.

### Skill-Assessment Exercise 9.2

**PROBLEM:** A unity feedback system with the forward transfer function

$$G(s) = \frac{K}{s(s+7)}$$

is operating with a closed-loop step response that has 15% overshoot. Do the following:

**a.** Evaluate the settling time.

**b.** Design a lead compensator to decrease the settling time by three times. Choose the compensator's zero to be at $-10$.

**ANSWER:**

**a.** $T_s = 1.143$ s

**b.** $G_{\text{lead}}(s) = \dfrac{s+10}{s+25.52}, \quad K = 476.3$

The complete solution is at www.wiley.com/college/nise.

## 9.4 Improving Steady-State Error and Transient Response

We now combine the design techniques covered in Sections 9.2 and 9.3 to obtain improvement in steady-state error and transient response *independently*. Basically, we first improve the transient response by using the methods of Section 9.3. Then we improve the steady-state error of this compensated system by applying the methods of Section 9.2. A disadvantage of this approach is the slight decrease in the speed of the response when the steady-state error is improved.

As an alternative, we can improve the steady-state error first and then follow with the design to improve the transient response. A disadvantage of this approach is that the improvement in transient response in some cases yields deterioration in the improvement of the steady-state error, which was designed first. In other cases, the improvement in transient response yields further improvement in steady-state errors. Thus, a system can be overdesigned with respect to steady-state errors. Overdesign is usually not a problem unless it affects cost or produces other design problems. In this textbook, we first design for transient response and then design for steady-state error.

The design can use either active or passive compensators, as previously described. If we design an active PD controller followed by an active PI controller, the resulting

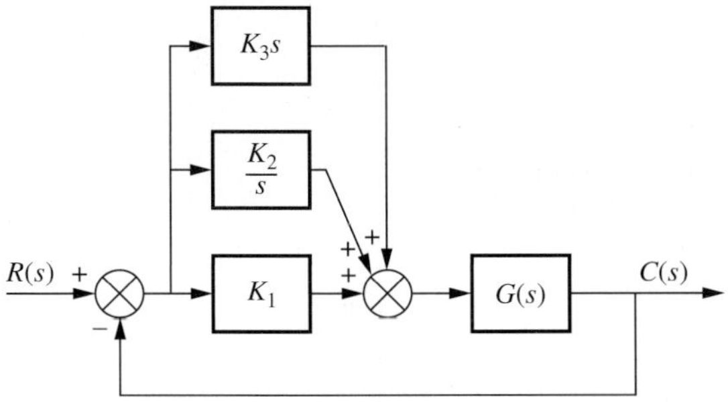


**FIGURE 9.30** PID controller

compensator is called a *proportional-plus-integral-plus-derivative* (PID) *controller*. If we first design a passive lead compensator and then design a passive lag compensator, the resulting compensator is called a *lag-lead compensator*.

## PID Controller Design

A PID controller is shown in Figure 9.30. Its transfer function is

$$G_c(s) = K_1 + \frac{K_2}{s} + K_3 s = \frac{K_1 s + K_2 + K_3 s^2}{s} = \frac{K_3\left(s^2 + \frac{K_1}{K_3}s + \frac{K_2}{K_3}\right)}{s} \tag{9.21}$$

which has two zeros plus a pole at the origin. One zero and the pole at the origin can be designed as the ideal integral compensator; the other zero can be designed as the ideal derivative compensator.

The design technique, which is demonstrated in Example 9.5, consists of the following steps:

1. Evaluate the performance of the uncompensated system to determine how much improvement in transient response is required.
2. Design the PD controller to meet the transient response specifications. The design includes the zero location and the loop gain.
3. Simulate the system to be sure all requirements have been met.
4. Redesign if the simulation shows that requirements have not been met.
5. Design the PI controller to yield the required steady-state error.
6. Determine the gains, $K_1$, $K_2$, and $K_3$, in Figure 9.30.
7. Simulate the system to be sure all requirements have been met.
8. Redesign if simulation shows that requirements have not been met.

## Example 9.5 PID Controller Design

**PROBLEM:** Given the system of Figure 9.31, design a PID controller so that the system can operate with a peak time that is two-thirds that of the uncompensated system at 20% overshoot and with zero steady-state error for a step input.

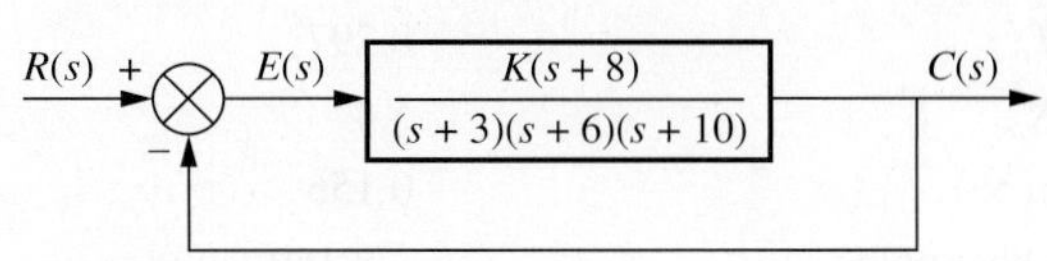


**FIGURE 9.31** Uncompensated feedback control system for Example 9.5

**SOLUTION:** Note that our solution follows the eight-step procedure described earlier.

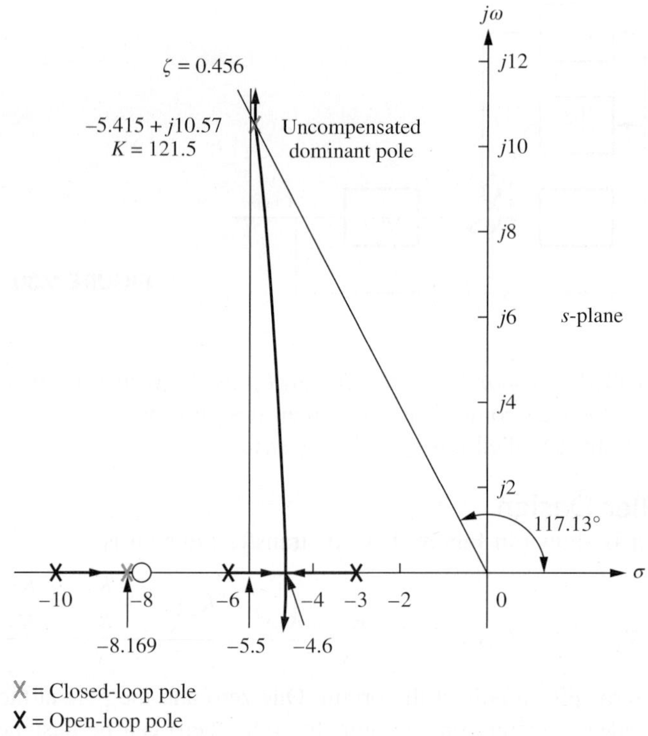


**FIGURE 9.32** Root locus for the uncompensated system of Example 9.5

**Step 1** Let us first evaluate the uncompensated system operating at 20% overshoot. Searching along the 20% overshoot line ($\zeta = 0.456$) in Figure 9.32, we find the dominant poles to be $-5.415 \pm j10.57$ with a gain of 121.5. A third pole, which exists at $-8.169$, is found by searching the region between $-8$ and $-10$ for a gain equivalent to that at the dominant poles. The complete performance of the uncompensated system is shown in the first column of Table 9.5, where we

**TABLE 9.5** Predicted characteristics of uncompensated, PD-, and PID-compensated systems of Example 9.5

| | Uncompensated | PD-compensated | PID-compensated |
|---|---|---|---|
| Plant and compensator | $\frac{K(s+8)}{(s+3)(s+6)(s+10)}$ | $\frac{K(s+8)(s+55.92)}{(s+3)(s+6)(s+10)}$ | $\frac{K(s+8)(s+55.92)(s+0.5)}{(s+3)(s+6)(s+10)s}$ |
| Dominant poles | $-5.415 \pm j10.57$ | $-8.13 \pm j15.87$ | $-7.516 \pm j14.67$ |
| $K$ | 121.5 | 5.34 | 4.6 |
| $\zeta$ | 0.456 | 0.456 | 0.456 |
| $\omega_n$ | 11.88 | 17.83 | 16.49 |
| $\%OS$ | 20 | 20 | 20 |
| $T_s$ | 0.739 | 0.492 | 0.532 |
| $T_p$ | 0.297 | 0.198 | 0.214 |
| $K_p$ | 5.4 | 13.27 | $\infty$ |
| $e(\infty)$ | 0.156 | 0.070 | 0 |
| Other poles | $-8.169$ | $-8.079$ | $-8.099$, $-0.468$ |
| Zeros | $-8$ | $-8$, $-55.92$ | $-8$, $-55.92$, $-0.5$ |
| Comments | Second-order approx. OK | Second-order approx. OK | Zeros at $-55.92$ and $-0.5$ not canceled |

compare the calculated values to those obtained through simulation (Figure 9.35). We estimate that the uncompensated system has a peak time of 0.297 second at 20% overshoot.

**Step 2** To compensate the system to reduce the peak time to two-thirds of that of the uncompensated system, we must first find the compensated system's dominant pole location. The imaginary part of the compensated dominant pole is

$$\omega_d = \frac{\pi}{T_p} = \frac{\pi}{(2/3)(0.297)} = 15.87 \tag{9.22}$$

Thus, the real part of the compensated dominant pole is

$$\sigma = \frac{\omega_d}{\tan 117.13^\circ} = -8.13 \tag{9.23}$$

Next we design the compensator. Using the geometry shown in Figure 9.33, we calculate the compensating zero's location. Using the root locus program, we find the sum of angles from the uncompensated system's poles and zeros to the desired compensated dominant pole to be $-198.37^\circ$. Thus, the contribution required from the compensator zero is $198.37^\circ - 180^\circ = 18.37^\circ$. Assume that the compensator zero is located at $-z_c$, as shown in Figure 9.33. Since

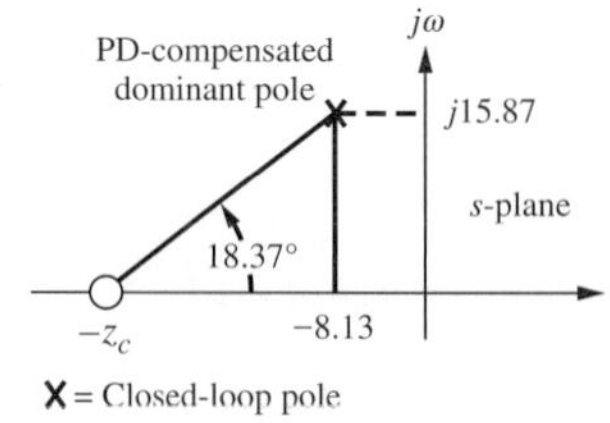

Note: This figure is not drawn to scale.

**FIGURE 9.33** Calculating the PD compensator zero for Example 9.5

$$\frac{15.87}{z_c - 8.13} = \tan 18.37^\circ \tag{9.24}$$

then

$$z_c = 55.92 \tag{9.25}$$

Thus, the PD controller is

$$G_{\text{PD}}(s) = (s + 55.92) \tag{9.26}$$

The complete root locus for the PD-compensated system is sketched in Figure 9.34. Using a root locus program, the gain at the design point is 5.34. Complete specifications for ideal derivative compensation are shown in the third column of Table 9.5.

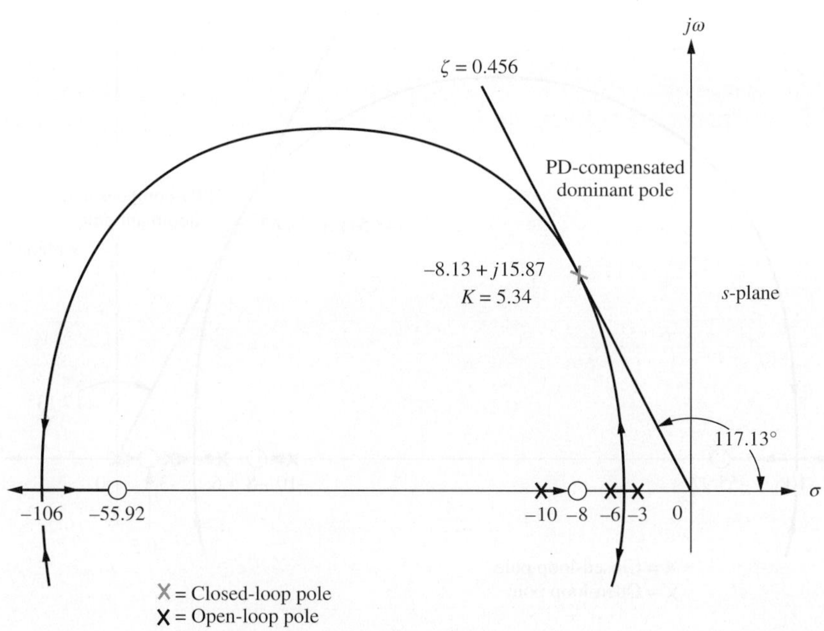

Note: This figure is not drawn to scale.

**FIGURE 9.34** Root locus for PD-compensated system of Example 9.5

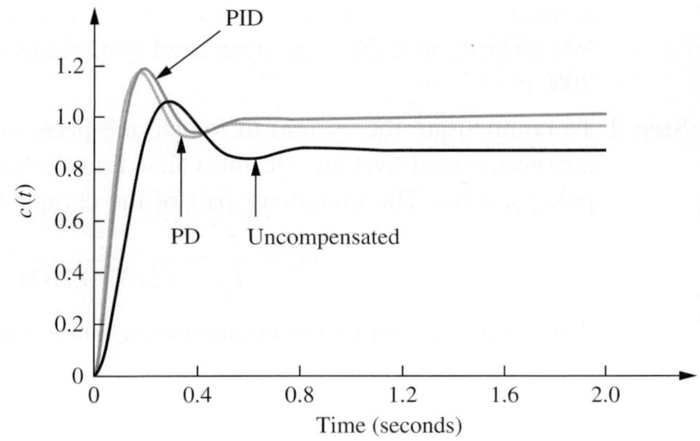


**FIGURE 9.35** Step responses for uncompensated, PD-compensated, and PID-compensated systems of Example 9.5

**Steps 3 & 4** We simulate the PD-compensated system, as shown in Figure 9.35. We see the reduction in peak time and the improvement in steady-state error over the uncompensated system.

**Step 5** After we design the PD controller, we design the ideal integral compensator to reduce the steady-state error to zero for a step input. Any ideal integral compensator zero will work, as long as the zero is placed close to the origin. Choosing the ideal integral compensator to be

$$G_{\mathrm{PI}}(s) = \frac{s+0.5}{s} \tag{9.27}$$

we sketch the root locus for the PID-compensated system, as shown in Figure 9.36. Searching the 0.456 damping ratio line, we find the dominant, second-order poles to be

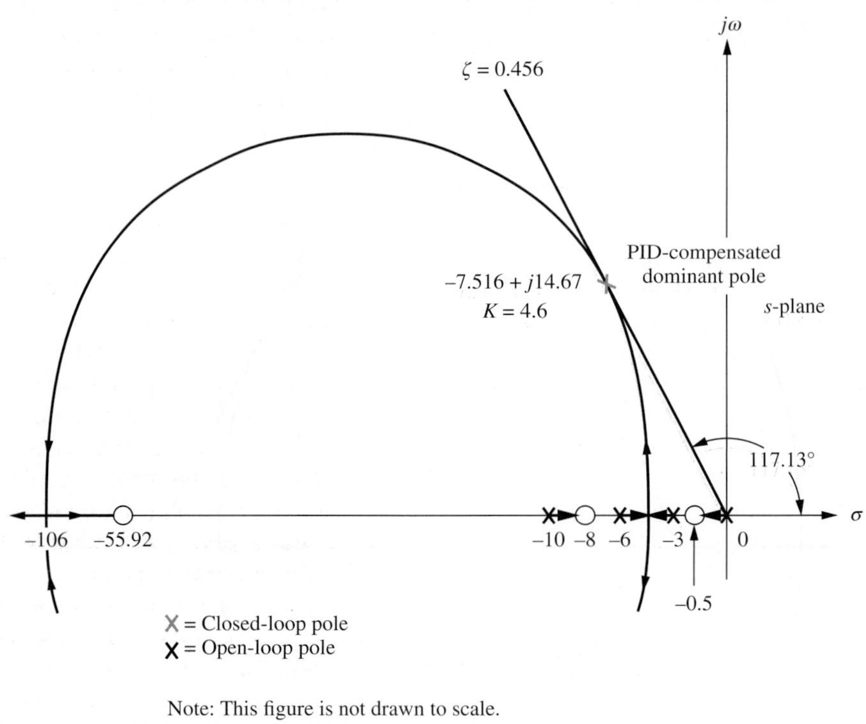


**FIGURE 9.36** Root locus for PID-compensated system of Example 9.5

$-7.516 \pm j14.67$, with an associated gain of 4.6. The remaining characteristics for the PID-compensated system are summarized in the fourth column of Table 9.5.

**Step 6** Now we determine the gains, $K_1$, $K_2$, and $K_3$, in Figure 9.30. From Eqs. (9.26) and (9.27), the product of the gain and the PID controller is

$$G_{\text{PID}}(s) = \frac{K(s+55.92)(s+0.5)}{s} = \frac{4.6(s+55.92)(s+0.5)}{s}$$

$$= \frac{4.6(s^2+56.42s+27.96)}{s} \tag{9.28}$$

Matching Eqs. (9.21) and (9.28), $K_1 = 259.5$, $K_2 = 128.6$, and $K_3 = 4.6$.

**Steps 7 & 8** Returning to Figure 9.35, we summarize the results of our design. PD compensation improved the transient response by decreasing the time required to reach the first peak as well as yielding some improvement in the steady-state error. The complete PID controller further improved the steady-state error without appreciably changing the transient response designed with the PD controller. As we have mentioned before, the PID controller exhibits a slower response, reaching the final value of unity at approximately 3 seconds. If this is undesirable, the speed of the system must be increased by redesigning the ideal derivative compensator or moving the PI controller zero farther from the origin. Simulation plays an important role in this type of design since our derived equation for settling time is not applicable for this part of the response, where there is a slow correction of the steady-state error.

## Lag-Lead Compensator Design

In the previous example, we serially combined the concepts of ideal derivative and ideal integral compensation to arrive at the design of a PID controller that improved both the transient response and the steady-state error performance. In the next example, we improve both transient response and the steady-state error by using a lead compensator and a lag compensator rather than the ideal PID. Our compensator is called a *lag-lead compensator*.

We first design the lead compensator to improve the transient response. Next we evaluate the improvement in steady-state error still required. Finally, we design the lag compensator to meet the steady-state error requirement. Later in the chapter we show circuit designs for the passive network. The following steps summarize the design procedure:

1. Evaluate the performance of the uncompensated system to determine how much improvement in transient response is required.
2. Design the lead compensator to meet the transient response specifications. The design includes the zero location, pole location, and the loop gain.
3. Simulate the system to be sure all requirements have been met.
4. Redesign if the simulation shows that requirements have not been met.
5. Evaluate the steady-state error performance for the lead-compensated system to determine how much more improvement in steady-state error is required.
6. Design the lag compensator to yield the required steady-state error.
7. Simulate the system to be sure all requirements have been met.
8. Redesign if the simulation shows that requirements have not been met.

## Example 9.6 Lag-Lead Compensator Design

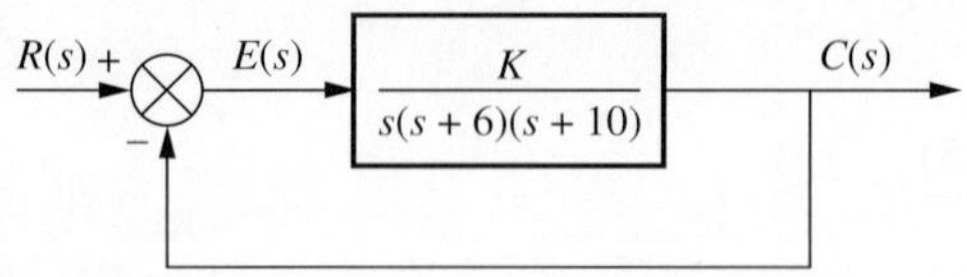


**FIGURE 9.37** Uncompensated system for Example 9.6

**PROBLEM:** Design a lag-lead compensator for the system of Figure 9.37 so that the system will operate with 20% overshoot and a twofold reduction in settling time. Further, the compensated system will exhibit a tenfold improvement in steady-state error for a ramp input.

**SOLUTION:** Again, our solution follows the steps just described.

**Step 1** First we evaluate the performance of the uncompensated system. Searching along the 20% overshoot line ($\zeta = 0.456$) in Figure 9.38, we find the dominant poles at $-1.794 \pm j3.501$, with a gain of 192.1. The performance of the uncompensated system is summarized in Table 9.6.

**Step 2** Next we begin the lead compensator design by selecting the location of the compensated system's dominant poles. In order to realize a twofold reduction in settling time, the real part of the dominant pole must be increased by a factor of 2,

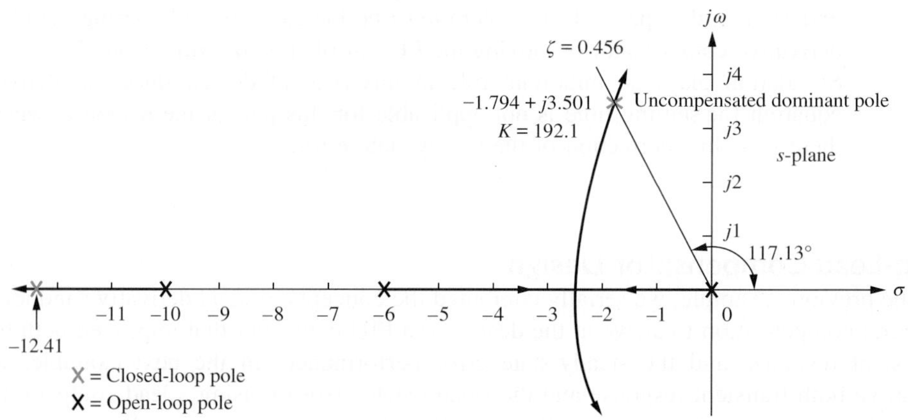


**FIGURE 9.38** Root locus for uncompensated system of Example 9.6

**TABLE 9.6** Predicted characteristics of uncompensated, lead-compensated, and lag-lead-compensated systems of Example 9.6

| | Uncompensated | Lead-compensated | Lag-lead-compensated |
|---|---|---|---|
| Plant and compensator | $\frac{K}{s(s+6)(s+10)}$ | $\frac{K}{s(s+10)(s+29.1)}$ | $\frac{K(s+0.04713)}{s(s+10)(s+29.1)(s+0.01)}$ |
| Dominant poles | $-1.794 \pm j3.501$ | $-3.588 \pm j7.003$ | $-3.574 \pm j6.976$ |
| $K$ | 192.1 | 1977 | 1971 |
| $\zeta$ | 0.456 | 0.456 | 0.456 |
| $\omega_n$ | 3.934 | 7.869 | 7.838 |
| $\%OS$ | 20 | 20 | 20 |
| $T_s$ | 2.230 | 1.115 | 1.119 |
| $T_p$ | 0.897 | 0.449 | 0.450 |
| $K_v$ | 3.202 | 6.794 | 31.92 |
| $e(\infty)$ | 0.312 | 0.147 | 0.0313 |
| Third pole | $-12.41$ | $-31.92$ | $-31.91, -0.0474$ |
| Zero | None | None | $-0.04713$ |
| Comments | Second-order approx. OK | Second-order approx. OK | Second-order approx. OK |

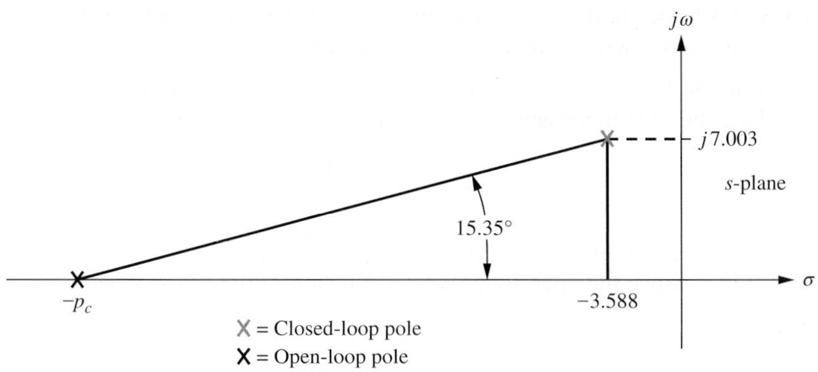


**FIGURE 9.39** Evaluating the compensator pole for Example 9.6

since the settling time is inversely proportional to the real part. Thus,

$$-\zeta\omega_n = -2(1.794) = -3.588 \tag{9.29}$$

The imaginary part of the design point is

$$\omega_d = \zeta\omega_n \tan 117.13^\circ = 3.588 \tan 117.13^\circ = 7.003 \tag{9.30}$$

Now we design the lead compensator. Arbitrarily select a location for the lead compensator zero. For this example, we select the location of the compensator zero coincident with the open-loop pole at −6. This choice will eliminate a zero and leave the lead-compensated system with three poles, the same number as the uncompensated system.

We complete the design by finding the location of the compensator pole. Using the root locus program, sum the angles to the design point from the uncompensated system's poles and zeros and the compensator zero and get −164.65°. The difference between 180° and this quantity is the angular contribution required from the compensator pole, or −15.35°. Using the geometry shown in Figure 9.39,

$$\frac{7.003}{p_c - 3.588} = \tan 15.35^\circ \tag{9.31}$$

from which the location of the compensator pole, $p_c$, is found to be −29.1.

The complete root locus for the lead-compensated system is sketched in Figure 9.40. The gain setting at the design point is found to be 1977.

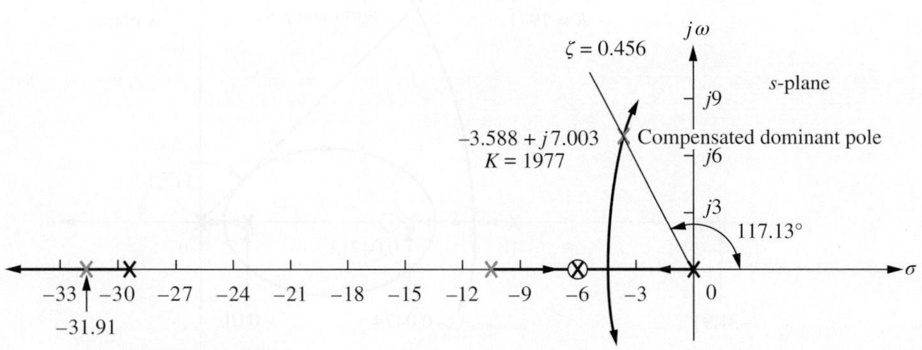


**FIGURE 9.40** Root locus for lead-compensated system of Example 9.6

**Steps 3 & 4** Check the design with a simulation. (The result for the lead-compensated system is shown in Figure 9.42 and is satisfactory.)

**Step 5** Continue by designing the lag compensator to improve the steady-state error. Since the uncompensated system's open-loop transfer function is

$$G(s) = \frac{192.1}{s(s+6)(s+10)} \tag{9.32}$$

the static error constant, $K_v$, which is inversely proportional to the steady-state error, is 3.201. Since the open-loop transfer function of the lead-compensated system is

$$G_{LC}(s) = \frac{1977}{s(s+10)(s+29.1)} \tag{9.33}$$

the static error constant, $K_v$, which is inversely proportional to the steady-state error, is 6.794. Thus, the addition of lead compensation has improved the steady-state error by a factor of 2.122. Since the requirements of the problem specified a tenfold improvement, the lag compensator must be designed to improve the steady-state error by a factor of 4.713 $(10/2.122 = 4.713)$ over the lead-compensated system.

**Step 6** We arbitrarily choose the lag compensator pole at 0.01, which then places the lag compensator zero at 0.04713, yielding

$$G_{lag}(s) = \frac{(s+0.04713)}{(s+0.01)} \tag{9.34}$$

as the lag compensator. The lag-lead-compensated system's open-loop transfer function is

$$G_{LLC}(s) = \frac{K(s+0.04713)}{s(s+10)(s+29.1)(s+0.01)} \tag{9.35}$$

where the uncompensated system pole at $-6$ canceled the lead compensator zero at $-6$. By drawing the complete root locus for the lag-lead-compensated system and by searching along the 0.456 damping ratio line, we find the dominant, closed-loop poles to be at $-3.574 \pm j6.976$, with a gain of 1971. The lag-lead-compensated root locus is shown in Figure 9.41.

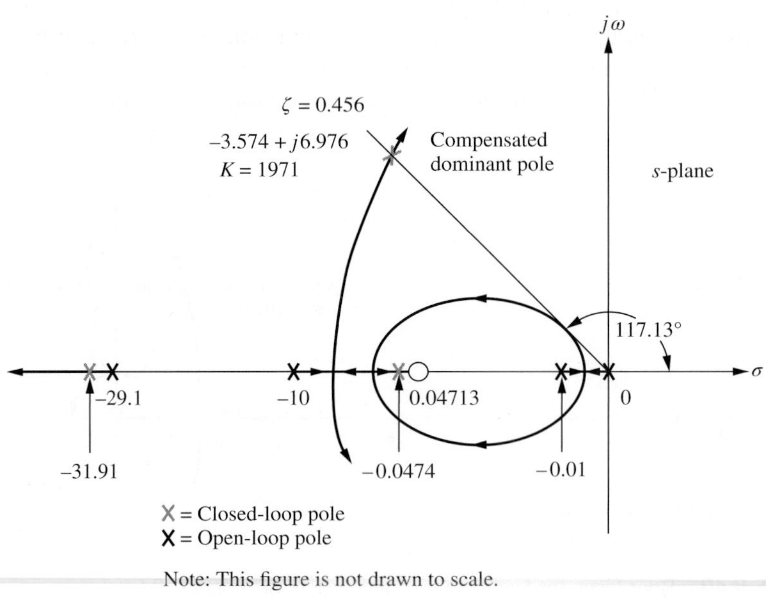


**FIGURE 9.41** Root locus for lag-lead-compensated system of Example 9.6

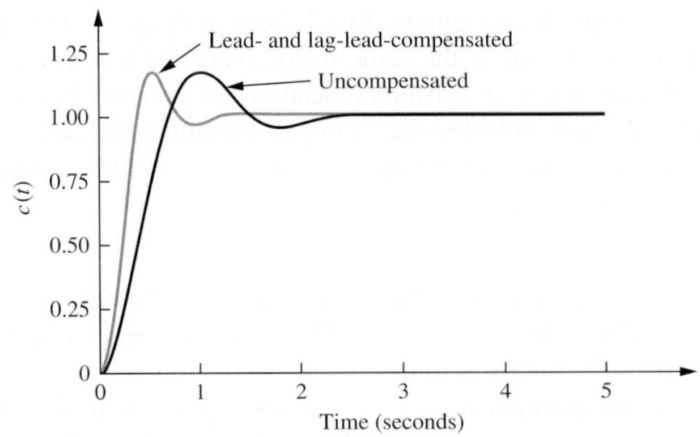


**FIGURE 9.42** Improvement in step response for lag-lead-compensated system of Example 9.6

A summary of our design is shown in Table 9.6. Notice that the lag-lead compensation has indeed increased the speed of the system, as witnessed by the settling time or the peak time. The steady-state error for a ramp input has also decreased by about 10 times, as seen from $e(\infty)$.

**Step 7** The final proof of our designs is shown by the simulations of Figures 9.42 and 9.43. The improvement in the transient response is shown in Figure 9.42, where we see the peak time occurring sooner in the lag-lead-compensated system. Improvement in the steady-state error for a ramp input is seen in Figure 9.43, where each step of our design yields more improvement. The improvement for the lead-compensated system is shown in Figure 9.43(*a*), and the final improvement due to the addition of the lag is shown in Figure 9.43(*b*).

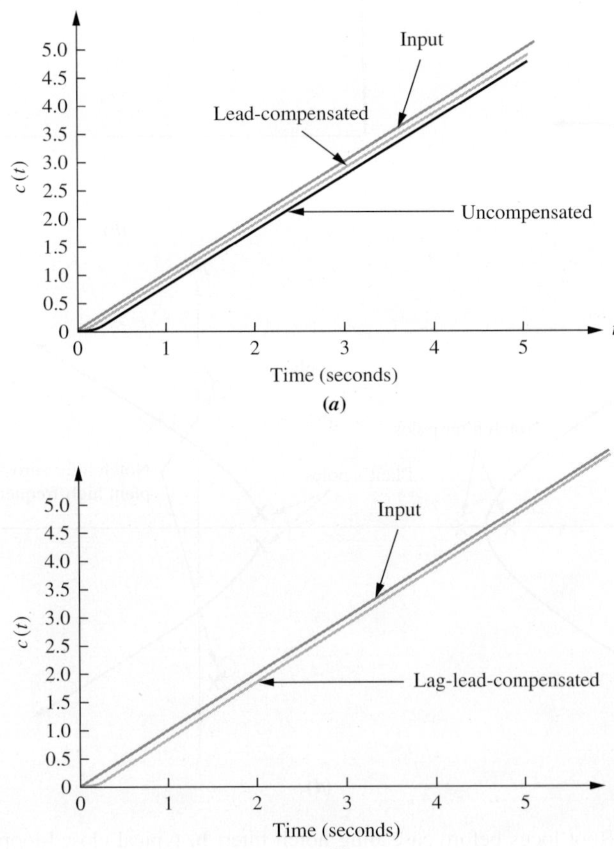


**FIGURE 9.43** Improvement in ramp response error for the system of Example 9.6:
**a.** lead-compensated;
**b.** lag-lead-compensated

In the previous example, we canceled the system pole at −6 with the lead compensator zero. The design technique is the same if you place the lead compensator zero at a different location. Placing a zero at a different location and not canceling the open-loop pole yields a system with one more pole than the example. This increased complexity could make it more difficult to justify a second-order approximation. In any case, simulations should be used at each step to verify performance.

## Notch Filter

If a plant, such as a mechanical system, has high-frequency vibration modes, then a desired closed-loop response may be difficult to obtain. These high-frequency vibration modes can be modeled as part of the plant's transfer function by pairs of complex poles near the imaginary axis. In a closed-loop configuration, these poles can move closer to the imaginary axis or even cross into the right half-plane, as shown in Figure 9.44(*a*). Instability or high-frequency oscillations superimposed over the desired response can result (see Figure 9.44(*b*)).

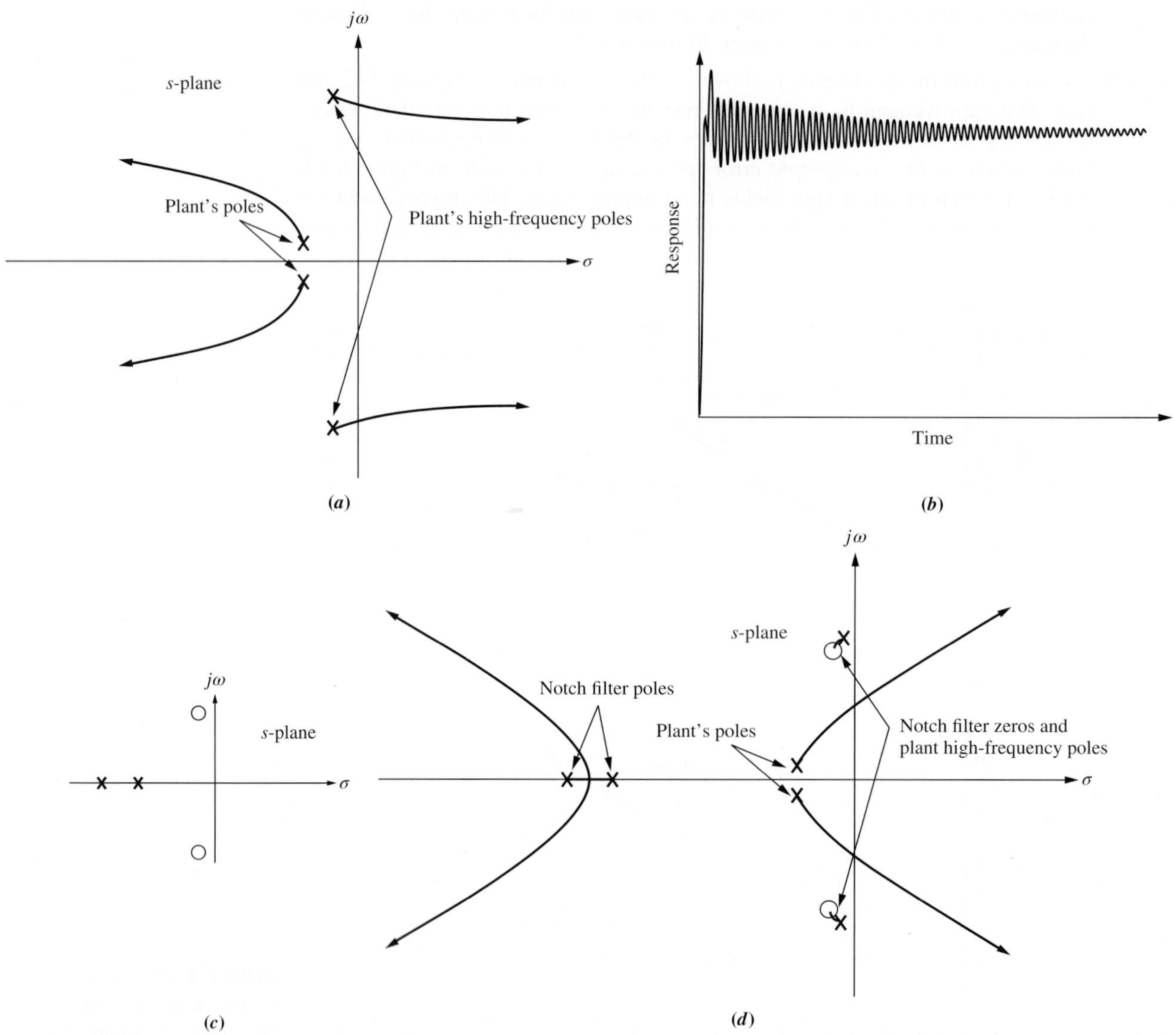


**FIGURE 9.44** **a.** Root locus before cascading notch filter; **b.** typical closed-loop step response before cascading notch filter; **c.** pole-zero plot of a notch filter; **d.** root locus after cascading notch filter; (*figure continues*)

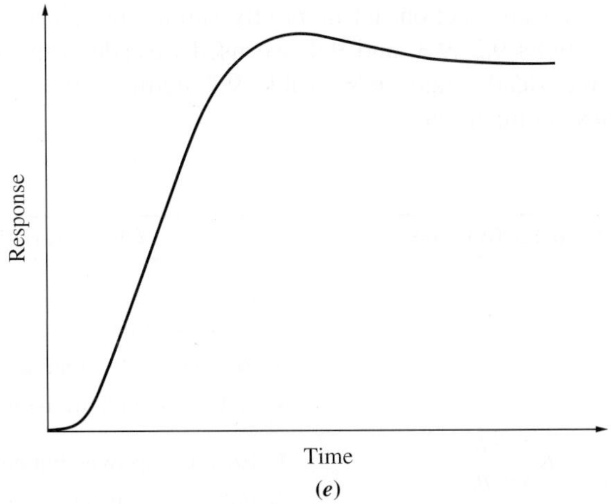


**FIGURE 9.44** (*Continued*) **e.** closed-loop step response after cascading notch filter

One way of eliminating the high-frequency oscillations is to cascade a *notch filter*[2] with the plant (*Kuo, 1995*), as shown in Figure 9.44(*c*). The notch filter has zeros close to the low-damping-ratio poles of the plant as well as two real poles. Figure 9.44(*d*) shows that the root locus branch from the high-frequency poles now goes a short distance from the high-frequency pole to the notch filter's zero. The high-frequency response will now be negligible because of the pole-zero cancellation (see Figure 9.44(*e*)). Other cascade compensators can now be designed to yield a desired response.

## Skill-Assessment Exercise 9.3

**PROBLEM:** A unity feedback system with forward transfer function

$$G(s) = \frac{K}{s(s+7)}$$

is operating with a closed-loop step response that has 20% overshoot. Do the following:

**a.** Evaluate the settling time.

**b.** Evaluate the steady-state error for a unit ramp input.

**c.** Design a lag-lead compensator to decrease the settling time by 2 times and decrease the steady-state error for a unit ramp input by 10 times. Place the lead zero at $-3$.

**ANSWERS:**

**a.** $T_s = 1.143$ s

**b.** $e_{\text{ramp}}(\infty) = 0.1189$

**c.** $G_c(s) = \dfrac{(s+3)(s+0.092)}{(s+9.61)(s+0.01)}, \quad K = 205.4$

The complete solution is at www.wiley.com/college/nise.

[2] The name of this filter comes from the shape of its magnitude frequency response characteristics, which shows a dip near the damped frequency of the high-frequency poles. Magnitude frequency response is discussed in Chapter 10.

Before concluding this section, let us briefly summarize our discussion of cascade compensation. In Sections 9.2, 9.3, and 9.4, we used cascade compensators to improve transient response and steady-state error. Table 9.7 itemizes the types, functions, and characteristics of these compensators.

**TABLE 9.7** Types of cascade compensators

| Function | Compensator | Transfer function | Characteristics |
|---|---|---|---|
| Improve steady-state error | PI | $K\frac{s+z_c}{s}$ | **1.** Increases system type.<br>**2.** Error becomes zero.<br>**3.** Zero at $-z_c$ is small and negative.<br>**4.** Active circuits are required to implement. |
| Improve steady-state error | Lag | $K\frac{s+z_c}{s+p_c}$ | **1.** Error is improved but not driven to zero.<br>**2.** Pole at $-p_c$ is small and negative.<br>**3.** Zero at $-z_c$ is close to, and to the left of, the pole at $-p_c$.<br>**4.** Active circuits are not required to implement. |
| Improve transient response | PD | $K(s+z_c)$ | **1.** Zero at $-z_c$ is selected to put design point on root locus.<br>**2.** Active circuits are required to implement.<br>**3.** Can cause noise and saturation; implement with rate feedback or with a pole (lead). |
| Improve transient response | Lead | $K\frac{s+z_c}{s+p_c}$ | **1.** Zero at $-z_c$ and pole at $-p_c$ are selected to put design point on root locus.<br>**2.** Pole at $-p_c$ is more negative than zero at $-z_c$.<br>**3.** Active circuits are not required to implement. |
| Improve steady-state error and transient response | PID | $K\frac{(s+z_{\text{lag}})(s+z_{\text{lead}})}{s}$ | **1.** Lag zero at $-z_{\text{lag}}$ and pole at origin improve steady-state error.<br>**2.** Lead zero at $-z_{\text{lead}}$ improves transient response.<br>**3.** Lag zero at $-z_{\text{lag}}$ is close to, and to the left of, the origin.<br>**4.** Lead zero at $-z_{\text{lead}}$ is selected to put design point on root locus.<br>**5.** Active circuits required to implement.<br>**6.** Can cause noise and saturation; implement with rate feedback or with an additional pole. |
| Improve steady-state error and transient response | Lag-lead | $K\frac{(s+z_{\text{lag}})(s+z_{\text{lead}})}{(s+p_{\text{lag}})(s+p_{\text{lead}})}$ | **1.** Lag pole at $-p_{\text{lag}}$ and lag zero at $-z_{\text{lag}}$ are used to improve steady-state error.<br>**2.** Lead pole at $-p_{\text{lead}}$ and lead zero at $-z_{\text{lead}}$ are used to improve transient response.<br>**3.** Lag pole at $-p_{\text{lag}}$ is small and negative.<br>**4.** Lag zero at $-z_{\text{lag}}$ is close to, and to the left of, lag pole at $-p_{\text{lag}}$.<br>**5.** Lead zero at $-z_{\text{lead}}$ and lead pole at $-p_{\text{lead}}$ are selected to put design point on root locus.<br>**6.** Lead pole at $-p_{\text{lead}}$ is more negative than lead zero at $-z_{\text{lead}}$.<br>**7.** Active circuits are not required to implement. |

# 9.5 Feedback Compensation

In Section 9.4, we used cascade compensation as a way to improve transient response and steady-state response independently. Cascading a compensator with the plant is not the only way to reshape the root locus to intersect the closed-loop $s$-plane poles that yield a desired transient response. Transfer functions designed to be placed in a feedback path can also reshape the root locus. Figure 9.45 is a generic configuration showing a compensator, $H_c(s)$, placed in the *minor loop* of a feedback control system. Other configurations arise if we consider $K$ unity, $G_2(s)$ unity, or both unity.

The design procedures for feedback compensation can be more complicated than for cascade compensation. On the other hand, feedback compensation can yield faster responses. Thus, the engineer has the luxury of designing faster responses into portions of a control loop in order to provide isolation. For example, the transient response of the ailerons and rudder control systems of an aircraft can be designed separately to be fast in order to reduce the effect of their dynamic response on the steering control loop. Feedback compensation can be used in cases where noise problems preclude the use of cascade compensation. Also, feedback compensation may not require additional amplification, since the signal passing through the compensator originates at the high-level output of the forward path and is delivered to a low-level input in the forward path. For example, let $K$ and $G_2(s)$ in Figure 9.45 be unity. The input to the feedback compensator, $K_fH_c(s)$, is from the high-level output of $G_1(s)$, while the output of $K_fH_c(s)$ is one of the low-level inputs into $K_1$. Thus, there is a reduction in level through $K_fH_c(s)$, and amplification is usually not required.

A popular feedback compensator is a rate sensor that acts as a differentiator. In aircraft and ship applications, the rate sensor can be a rate gyro that responds with an output voltage proportional to the input angular velocity. In many other systems this rate sensor is implemented with a tachometer. A tachometer is a voltage generator that yields a voltage output proportional to input rotational speed. This compensator can easily be geared to the position output of a system. Figure 9.46 is a position control

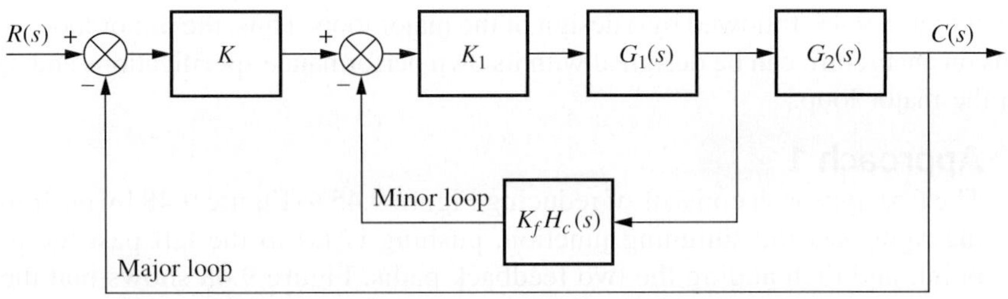


**FIGURE 9.45** Generic control system with feedback compensation.

Photo by Mark E. Van Dusen.

**FIGURE 9.46** A position control system that uses a tachometer as a differentiator in the feedback path. Can you see the similarity between this system and the schematic on the front endpapers?

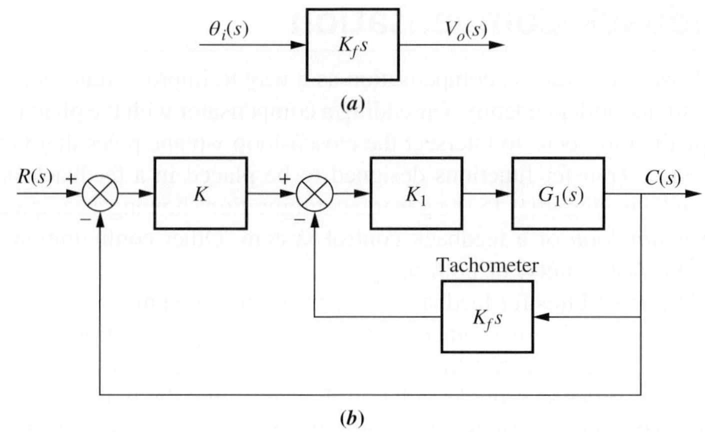


**FIGURE 9.47** **a.** Transfer function of a tachometer; **b.** tachometer feedback compensation

system showing the gearing of the tachometer to the motor. You can see the input and output potentiometers as well as the motor and inertial load. The block diagram representation of a tachometer is shown in Figure 9.47($a$), and its typical position within a control loop is shown in Figure 9.47($b$).

While this section shows methods for designing systems using rate feedback, it also sets the stage for compensation techniques in Chapter 12, where not only rate but all states including position will be fed back for proper control system performance.

We now discuss design procedures. Typically, the design of feedback compensation consists of finding the gains, such as $K$, $K_1$, and $K_f$ in Figure 9.45, after establishing a dynamic form for $H_c(s)$. There are two approaches. The first is similar to cascade compensation. Assume a typical feedback system, where $G(s)$ is the forward path and $H(s)$ is the feedback. Now consider that a root locus is plotted from $G(s)H(s)$. With cascade compensation we added poles and zeros to $G(s)$. With feedback compensation, poles and zeros are added via $H(s)$.

With the second approach, we design a specified performance for the minor loop, shown in Figure 9.45, followed by a design of the major loop. Thus, the minor loop, such as ailerons on an aircraft, can be designed with its own performance specifications and operate within the major loop.

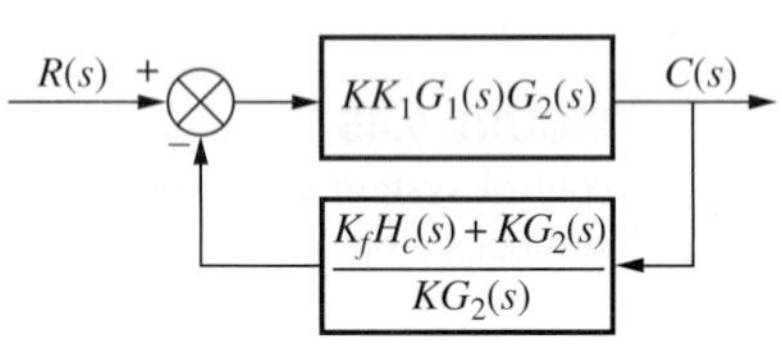


**FIGURE 9.48** Equivalent block diagram of Figure 9.45

## Approach 1

The first approach consists of reducing Figure 9.45 to Figure 9.48 by pushing $K$ to the right past the summing junction, pushing $G_2(s)$ to the left past the pickoff point, and then adding the two feedback paths. Figure 9.48 shows that the loop gain, $G(s)H(s)$, is

$$G(s)H(s) = K_1G_1(s)[K_fH_c(s) + KG_2(s)] \tag{9.36}$$

Without feedback, $K_fH_c(s)$, the loop gain is

$$G(s)H(s) = KK_1G_1(s)G_2(s) \tag{9.37}$$

Thus, the effect of adding feedback is to replace the poles and zeros of $G_2(s)$ with the poles and zeros of $[K_fH_c(s) + KG_2(s)]$. Hence, this method is similar to cascade compensation in that we add new poles and zeros via $H(s)$ to reshape the root locus to go through the design point. However, one must remember that zeros of the equivalent feedback shown in Figure 9.48, $H(s) = [K_fH_c(s) + KG_2(s)]/KG_2(s)$, are not closed-loop zeros.

For example, if $G_2(s) = 1$ and the minor-loop feedback, $K_f H_c(s)$, is a rate sensor, $K_f H_c(s) = K_f s$, then from Eq. (9.36) the loop gain is

$$G(s)H(s) = K_f K_1 G_1(s)\left(s + \frac{K}{K_f}\right) \tag{9.38}$$

Thus, a zero at $-K/K_f$ is added to the existing open-loop poles and zeros. This zero reshapes the root locus to go through the desired design point. A final adjustment of the gain, $K_1$, yields the desired response. Again, you should verify that this zero is not a closed-loop zero. Let us look at a numerical example.

## Example 9.7 Compensating Zero via Rate Feedback

**PROBLEM:** Given the system of Figure 9.49($a$), design rate feedback compensation, as shown in Figure 9.49($b$), to reduce the settling time by a factor of 4 while continuing to operate the system with 20% overshoot.

**SOLUTION:** First design a PD compensator. For the uncompensated system, search along the 20% overshoot line ($\zeta = 0.456$) and find that the dominant poles are at $-1.809 \pm j3.531$, as shown in Figure 9.50. The estimated specifications for the uncompensated system

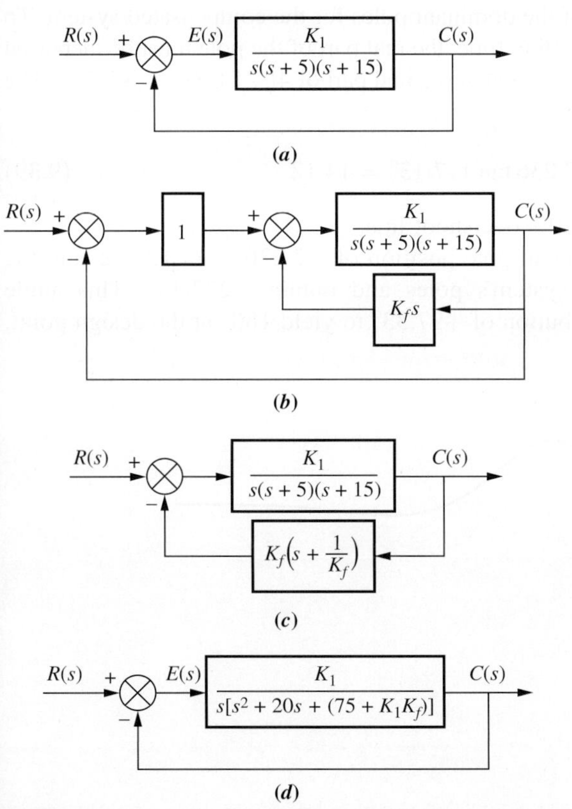


**FIGURE 9.49** **a.** System for Example 9.7; **b.** system with rate feedback compensation; **c.** equivalent compensated system; **d.** equivalent compensated system showing unity feedback

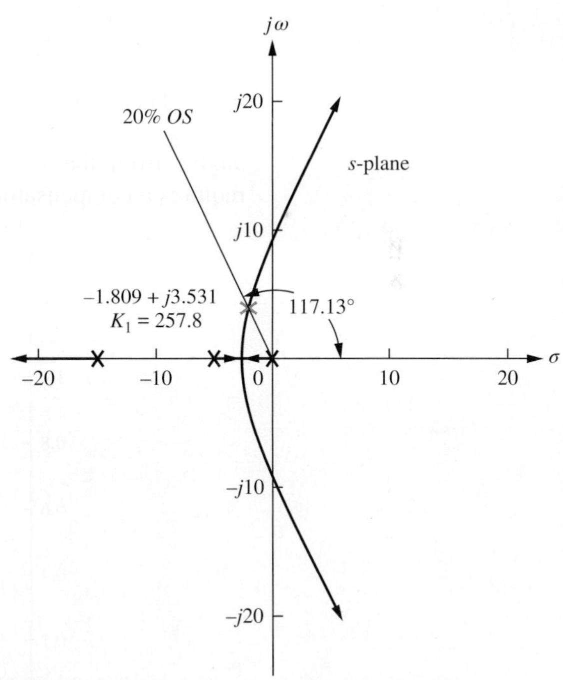


**FIGURE 9.50** Root locus for uncompensated system of Example 9.7

**TABLE 9.8** Predicted characteristics of uncompensated and compensated systems of Example 9.7

| | Uncompensated | Compensated |
|---|---|---|
| Plant and compensator | $\frac{K_1}{s(s+5)(s+15)}$ | $\frac{K_1}{s(s+5)(s+15)}$ |
| Feedback | 1 | $0.185(s+5.42)$ |
| Dominant poles | $-1.809 \pm j3.531$ | $-7.236 \pm j14.12$ |
| $K_1$ | 257.8 | 1388 |
| $\zeta$ | 0.456 | 0.456 |
| $\omega_n$ | 3.97 | 15.87 |
| %OS | 20 | 20 |
| $T_s$ | 2.21 | 0.55 |
| $T_p$ | 0.89 | 0.22 |
| $K_v$ | 3.44 | 4.18 |
| $e(\infty)$ (ramp) | 0.29 | 0.24 |
| Other poles | −16.4 | −5.53 |
| Zero | None | None |
| Comments | Second-order approx. OK | Simulate |

are shown in Table 9.8, and the step response is shown in Figure 9.51. The settling time is 2.21 seconds and must be reduced by a factor of 4 to 0.55 second.

Next determine the location of the dominant poles for the compensated system. To achieve a fourfold decrease in the settling time, the real part of the pole must be increased by a factor of 4. Thus, the compensated pole has a real part of $4(-1.809) = -7.236$. The imaginary part is then

$$\omega_d = -7.236 \tan 117.13° = 14.12 \tag{9.39}$$

where 117.13° is the angle of the 20% overshoot line.

Using the compensated dominant pole position of $-7.236 \pm j14.12$, we sum the angles from the uncompensated system's poles and obtain $-277.33°$. This angle requires a compensator zero contribution of $+97.33°$ to yield 180° at the design point.

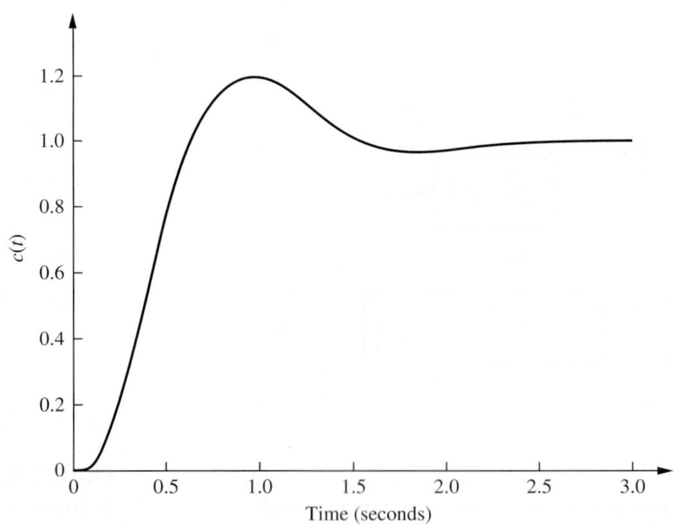


**FIGURE 9.51** Step response for uncompensated system of Example 9.7

The geometry shown in Figure 9.52 leads to the calculation of the compensator's zero location. Hence,

$$\frac{14.12}{7.236 - z_c} = \tan(180° - 97.33°) \tag{9.40}$$

from which $z_c = 5.42$.

The root locus for the equivalent compensated system of Figure 9.49(*c*) is shown in Figure 9.53. The gain at the design point, which is $K_1K_f$ from Figure 9.49(*c*), is found to be 256.7. Since $K_f$ is the reciprocal of the compensator zero, $K_f = 0.185$. Thus, $K_1 = 1388$.

In order to evaluate the steady-state error characteristic, $K_v$ is found from Figure 9.49(*d*) to be

$$K_v = \frac{K_1}{75 + K_1K_f} = 4.18 \tag{9.41}$$

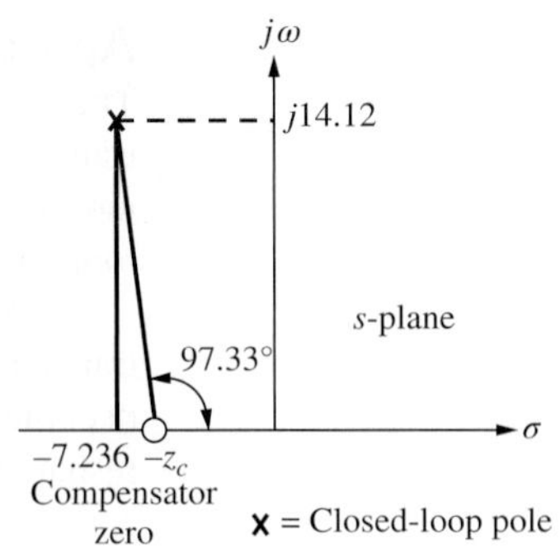


**FIGURE 9.52** Finding the compensator zero in Example 9.7

Predicted performance for the compensated system is shown in Table 9.8. Notice that the higher-order pole is not far enough away from the dominant poles and thus cannot be neglected. Further, from Figure 9.49(*d*), we see that the closed-loop transfer function is

$$T(s) = \frac{G(s)}{1 + G(s)H(s)} = \frac{K_1}{s^3 + 20s^2 + (75 + K_1K_f)s + K_1} \tag{9.42}$$

Thus, as predicted, the open-loop zero is not a closed-loop zero, and there is no pole-zero cancellation. Hence, the design must be checked by simulation.

The results of the simulation are shown in Figure 9.54 and show an over-damped response with a settling time of 0.75 second, compared to the uncompensated system's settling time of approximately 2.2 seconds. Although not meeting the design requirements, the response still represents an improvement over the uncompensated system of Figure 9.51. Typically, less overshoot is acceptable. The system should be redesigned for further reduction in settling time.

You may want to do Problem 8 at the end of this chapter, where you can repeat this example using PD cascade compensation. You will see that the compensator zero for cascade compensation is a closed-loop zero, yielding the possibility of pole-zero cancellation. However, PD compensation is usually noisy and not always practical.

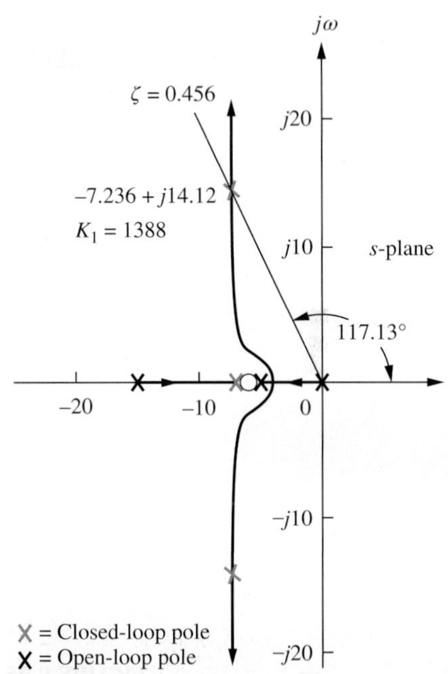


**FIGURE 9.53** Root locus for the compensated system of Example 9.7

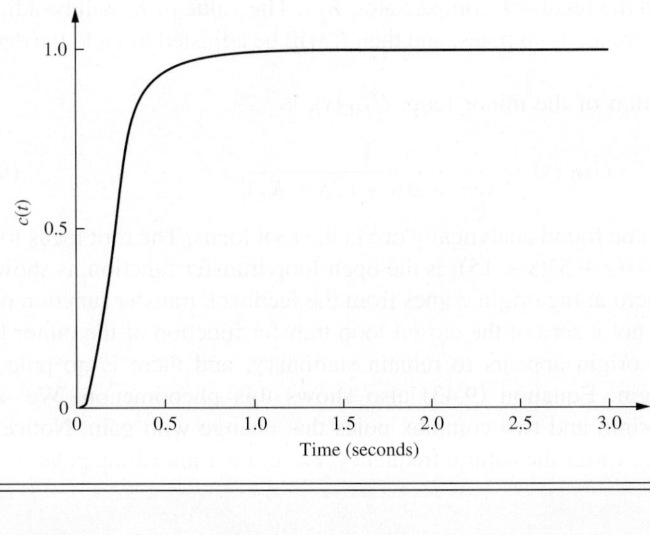


**FIGURE 9.54** Step response for the compensated system of Example 9.7

### Approach 2

The second approach allows us to use feedback compensation to design a minor loop's transient response separately from the closed-loop system response. In the case of an aircraft, the minor loop may control the position of the aerosurfaces, while the entire closed-loop system may control the entire aircraft's pitch angle.

We will see that the minor loop of Figure 9.45 basically represents a forward-path transfer function whose poles can be adjusted with the minor-loop gain. These poles then become the open-loop poles for the entire control system. In other words, rather than reshaping the root locus with additional poles and zeros, as in cascade compensation, we can actually change the plant's poles through a gain adjustment. Finally, the closed-loop poles are set by the loop gain, as in cascade compensation.

## Example 9.8 Minor-Loop Feedback Compensation

**PROBLEM:** For the system of Figure 9.55(*a*), design minor-loop feedback compensation, as shown in Figure 9.55(*b*), to yield a damping ratio of 0.8 for the minor loop and a damping ratio of 0.6 for the closed-loop system.

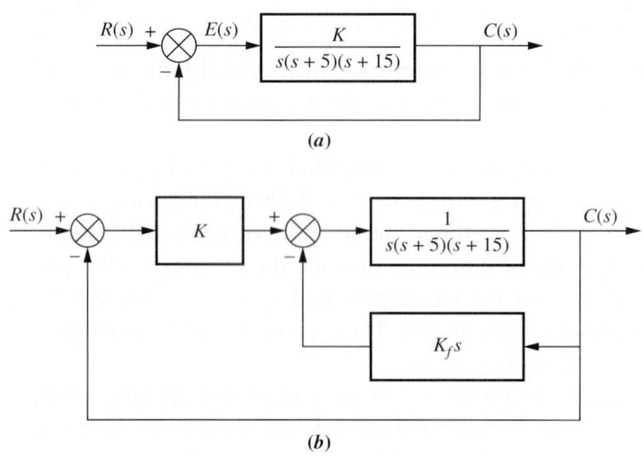


**FIGURE 9.55** **a.** Uncompensated system and **b.** feedback-compensated system for Example 9.8

**SOLUTION:** The minor loop is defined as the loop containing the plant, $1/[s(s+5)(s+15)]$, and the feedback compensator, $K_f s$. The value of $K_f$ will be adjusted to set the location of the minor-loop poles, and then $K$ will be adjusted to yield the desired closed-loop response.

The transfer function of the minor loop, $G_{ML}(s)$, is

$$G_{ML}(s) = \frac{1}{s[s^2 + 20s + (75 + K_f)]} \tag{9.43}$$

The poles of $G_{ML}(s)$ can be found analytically or via the root locus. The root locus for the minor loop, where $K_f s/[s(s+5)(s+15)]$ is the open-loop transfer function, is shown in Figure 9.56. Since the zero at the origin comes from the feedback transfer function of the minor loop, this zero is not a zero of the closed-loop transfer function of the minor loop. Hence, the pole at the origin appears to remain stationary, and there is no pole-zero cancellation at the origin. Equation (9.43) also shows this phenomenon. We see a stationary pole at the origin and two complex poles that change with gain. Notice that the compensator gain, $K_f$, varies the natural frequency, $\omega_n$, of the minor-loop poles as seen

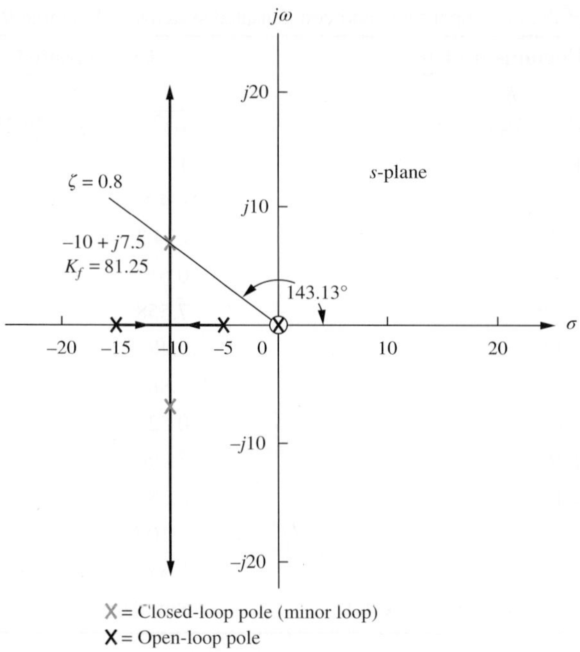


**FIGURE 9.56** Root locus for minor loop of Example 9.8

from Eq. (9.43). Since the real parts of the complex poles are constant at $-\zeta\omega_n = -10$, the damping ratio must also be varying to keep $2\zeta\omega_n = 20$, a constant. Drawing the $\zeta = 0.8$ line in Figure 9.56 yields the complex poles at $-10 \pm j7.5$. The gain, $K_f$, which equals 81.25, places the minor-loop poles in a position to meet the specifications. The poles just found, $-10 \pm j7.5$, as well as the pole at the origin (Eq. (9.43)), act as open-loop poles that generate a root locus for variations of the gain, $K$.

The final root locus for the system is shown in Figure 9.57. The $\zeta = 0.6$ damping ratio line is drawn and searched. The closed-loop complex poles are found to be $-4.535 \pm j6.046$, with a required gain of 624.3. A third pole is at $-10.93$.

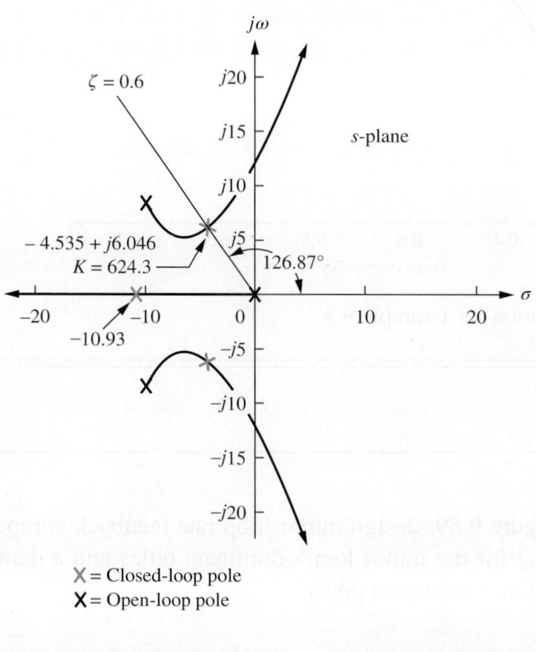


**FIGURE 9.57** Root locus for closed-loop system of Example 9.8

**TABLE 9.9** Predicted characteristics of the uncompensated and compensated systems of Example 9.8

| | Uncompensated | Compensated |
|---|---|---|
| Plant and compensator | $\dfrac{K_1}{s(s+5)(s+15)}$ | $\dfrac{K}{s(s^2+20s+156.25)}$ |
| Feedback | 1 | 1 |
| Dominant poles | $-1.997 \pm j2.662$ | $-4.535 \pm j6.046$ |
| $K$ | 177.3 | 624.3 |
| $\zeta$ | 0.6 | 0.6 |
| $\omega_n$ | 3.328 | 7.558 |
| $\%OS$ | 9.48 | 9.48 |
| $T_s$ | 2 | 0.882 |
| $T_p$ | 1.18 | 0.52 |
| $K_v$ | 2.364 | 3.996 |
| $e(\infty)$(ramp) | 0.423 | 0.25 |
| Other poles | −16 | −10.93 |
| Zero | None | None |
| Comments | Second-order approx. OK | Simulate |

The results are summarized in Table 9.9. We see that the compensated system, although having the same damping ratio as the uncompensated system, is much faster and also has a smaller steady-state error. The results, however, are predicted results and must be simulated to verify percent overshoot, settling time, and peak time, since the third pole is not far enough from the dominant poles. The step response is shown in Figure 9.58 and closely matches the predicted performance.

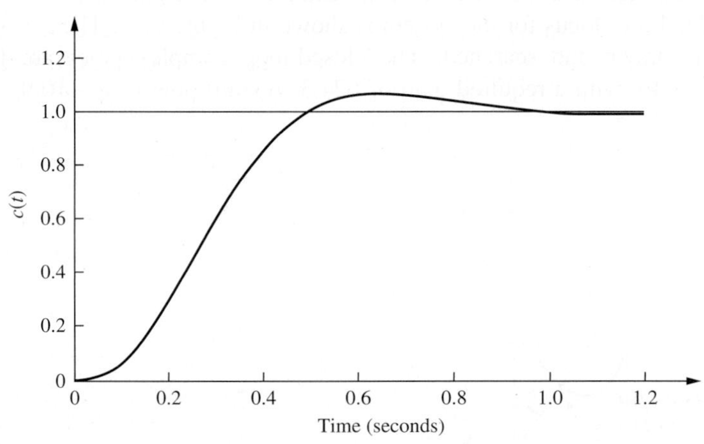


**FIGURE 9.58** Step response simulation for Example 9.8

## Skill-Assessment Exercise 9.4

**PROBLEM:** For the system of Figure 9.59, design minor-loop rate feedback compensation to yield a damping ratio of 0.7 for the minor loop's dominant poles and a damping ratio of 0.5 for the closed-loop system's dominant poles.

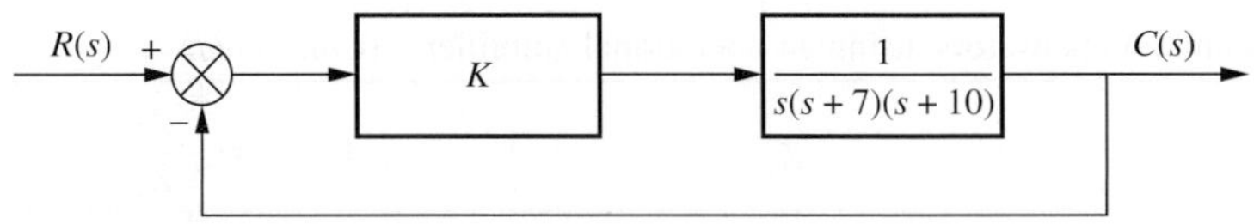


**FIGURE 9.59** System for Skill-Assessment Exercise 9.4

**ANSWER:** The system is configured similar to Figure 9.55(*b*) with $K_f = 77.42$ and $K = 626.3$.

The complete solution is at www.wiley.com/college/nise.

Our discussion of compensation methods is now complete. We studied both cascade and feedback compensation and compared and contrasted them. We are now ready to show how to physically realize the controllers and compensators we designed.

# 9.6 Physical Realization of Compensation

In this chapter, we derived compensation to improve transient response and steady-state error in feedback control systems. Transfer functions of compensators used in cascade with the plant or in the feedback path were derived. These compensators were defined by their pole-zero configurations. They were either active PI, PD, or PID controllers or passive lag, lead, or lag-lead compensators. In this section, we show how to implement the active controllers and the passive compensators.

## Active-Circuit Realization

In Chapter 2, we derived

$$\frac{V_o(s)}{V_i(s)} = -\frac{Z_2(s)}{Z_1(s)} \tag{9.44}$$

as the transfer function of an inverting operational amplifier whose configuration is repeated here in Figure 9.60. By judicious choice of $Z_1(s)$ and $Z_2(s)$, this circuit can be used as a building block to implement the compensators and controllers, such as PID controllers, discussed in this chapter. Table 9.10 summarizes the realization of PI, PD, and PID controllers as well as lag, lead,

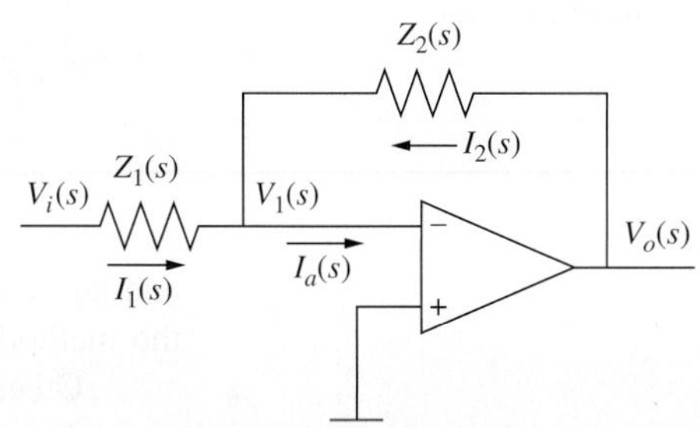


**FIGURE 9.60** Operational amplifier configured for transfer function realization

**TABLE 9.10** Active realization of controllers and compensators, using an operational amplifier

| Function | $Z_1(s)$ | $Z_2(s)$ | $G_c(s) = -\frac{Z_2(s)}{Z_1(s)}$ |
|---|---|---|---|
| Gain | $R_1$ | $R_2$ | $-\frac{R_2}{R_1}$ |
| Integration | $R$ | $C$ | $-\frac{\frac{1}{RC}}{s}$ |
| Differentiation | $C$ | $R$ | $-RCs$ |

*(table continues)*

**TABLE 9.10** Active realization of controllers and compensators, using an operational amplifier *(Continued)*

| Function | $Z_1(s)$ | $Z_2(s)$ | $G_c(s) = -\frac{Z_2(s)}{Z_1(s)}$ |
|---|---|---|---|
| PI controller | 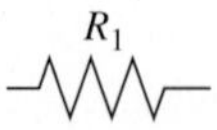 |  | $-\frac{R_2}{R_1}\frac{\left(s+\frac{1}{R_2C}\right)}{s}$ |
| PD controller | 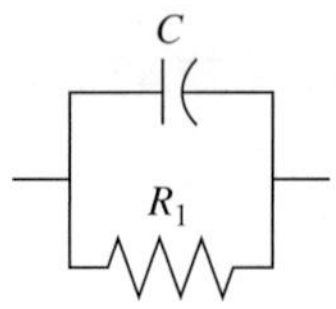 | 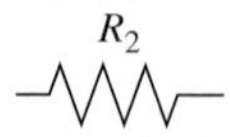 | $-R_2C\left(s+\frac{1}{R_1C}\right)$ |
| PID controller | 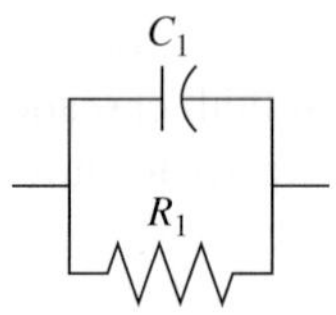 | 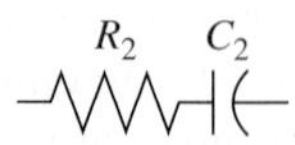 | $-\left[\left(\frac{R_2}{R_1}+\frac{C_1}{C_2}\right)+R_2C_1s+\frac{\frac{1}{R_1C_2}}{s}\right]$ |
| Lag compensation | 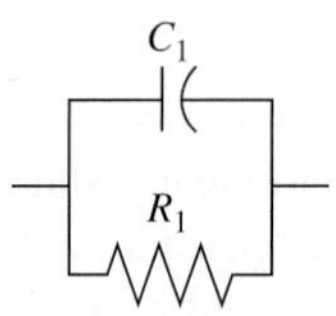 | 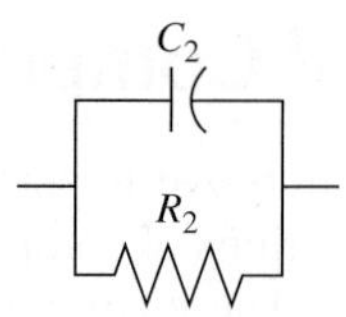 | $-\frac{C_1}{C_2}\frac{\left(s+\frac{1}{R_1C_1}\right)}{\left(s+\frac{1}{R_2C_2}\right)}$ where $R_2C_2 > R_1C_1$ |
| Lead compensation | 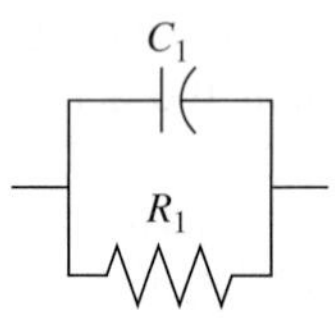 | 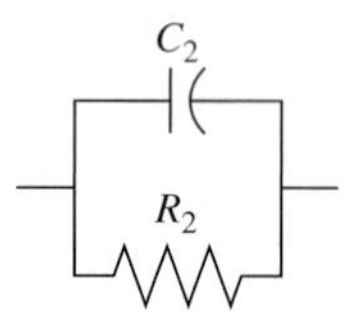 | $-\frac{C_1}{C_2}\frac{\left(s+\frac{1}{R_1C_1}\right)}{\left(s+\frac{1}{R_2C_2}\right)}$ where $R_1C_1 > R_2C_2$ |

and lag-lead compensators using operational amplifiers. You can verify the table by using the methods of Chapter 2 to find the impedances.

Other compensators can be realized by cascading compensators shown in the table. For example, a lag-lead compensator can be formed by cascading the lag compensator with the lead compensator, as shown in Figure 9.61. As an example, let us implement one of the controllers we designed earlier in the chapter.

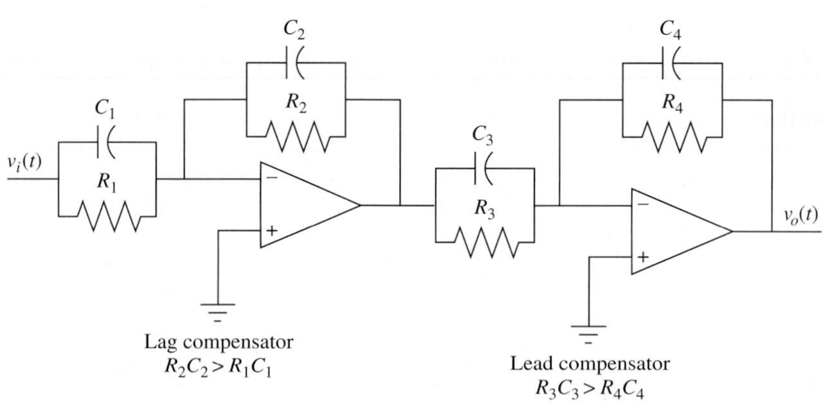


**FIGURE 9.61** Lag-lead compensator implemented with operational amplifiers

## Example 9.9 Implementing a PID Controller

**PROBLEM:** Implement the PID controller of Example 9.5.

**SOLUTION:** The transfer function of the PID controller is

$$G_c(s) = \frac{(s+55.92)(s+0.5)}{s} \tag{9.45}$$

which can be put in the form

$$G_c(s) = s + 56.42 + \frac{27.96}{s} \tag{9.46}$$

Comparing the PID controller in Table 9.10 with Eq. (9.46), we obtain the following three relationships:

$$\frac{R_2}{R_1} + \frac{C_1}{C_2} = 56.42 \tag{9.47}$$

$$R_2C_1 = 1 \tag{9.48}$$

and

$$\frac{1}{R_1C_2} = 27.96 \tag{9.49}$$

Since there are four unknowns and three equations, we arbitrarily select a practical value for one of the elements. Selecting $C_2 = 0.1\ \mu\text{F}$, the remaining values are found to be $R_1 = 357.65\ \text{k}\Omega$, $R_2 = 178,891\ \text{k}\Omega$, and $C_1 = 5.59\ \mu\text{F}$.

The complete circuit is shown in Figure 9.62, where the circuit element values have been rounded off.

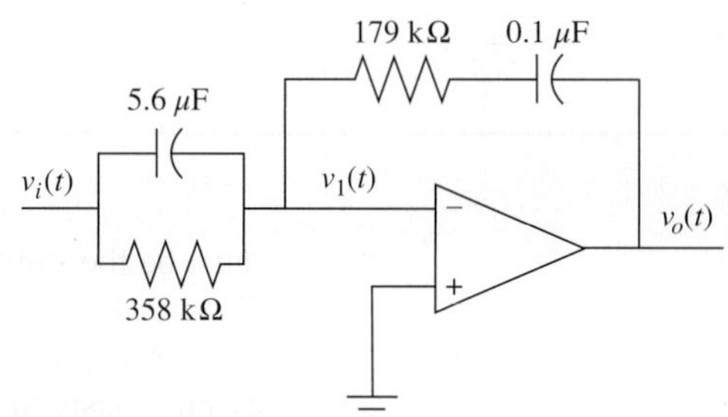


**FIGURE 9.62** PID controller

## Passive-Circuit Realization

Lag, lead, and lag-lead compensators can also be implemented with passive networks. Table 9.11 summarizes the networks and their transfer functions. The transfer functions can be derived with the methods of Chapter 2.

The lag-lead transfer function can be put in the following form:

$$G_c(s) = \frac{\left(s+\frac{1}{T_1}\right)\left(s+\frac{1}{T_2}\right)}{\left(s+\frac{1}{\alpha T_1}\right)\left(s+\frac{\alpha}{T_2}\right)} \tag{9.50}$$

where $\alpha < 1$. Thus, the terms with $T_1$ form the lead compensator, and the terms with $T_2$ form the lag compensator. Equation (9.50) shows a restriction inherent in using this passive realization. We see that the ratio of the lead compensator zero to the lead compensator pole

**TABLE 9.11** Passive realization of compensators

| Function | Network | Transfer function, $\dfrac{V_o(s)}{V_i(s)}$ |
|---|---|---|
| Lag compensation | 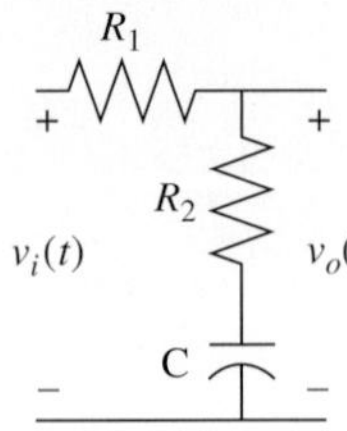 | $\dfrac{R_2}{R_1+R_2}\dfrac{s+\dfrac{1}{R_2C}}{s+\dfrac{1}{(R_1+R_2)C}}$ |
| Lead compensation | 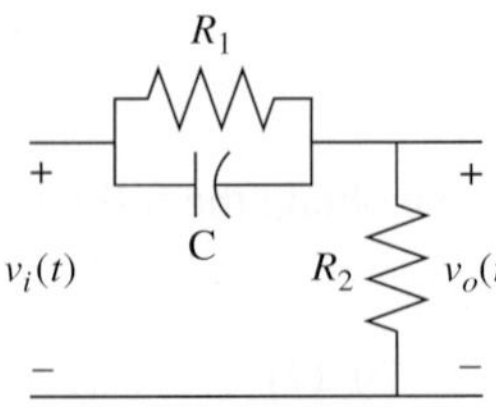 | $\dfrac{s+\dfrac{1}{R_1C}}{s+\dfrac{1}{R_1C}+\dfrac{1}{R_2C}}$ |
| Lag-lead compensation | 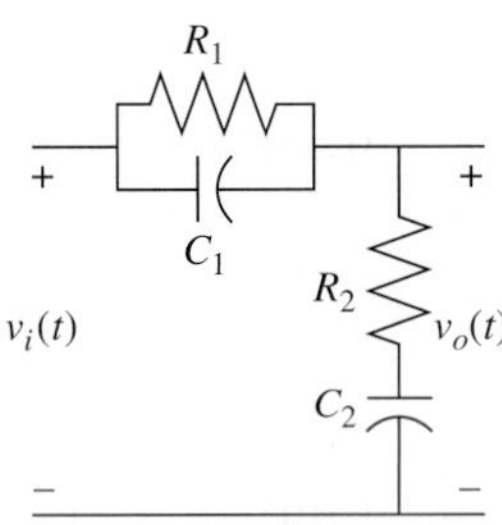 | $\dfrac{\left(s+\dfrac{1}{R_1C_1}\right)\left(s+\dfrac{1}{R_2C_2}\right)}{s^2+\left(\dfrac{1}{R_1C_1}+\dfrac{1}{R_2C_2}+\dfrac{1}{R_2C_1}\right)s+\dfrac{1}{R_1R_2C_1C_2}}$ |

must be the same as the ratio of the lag compensator pole to the lag compensator zero. In Chapter 11 we design a lag-lead compensator with this restriction.

A lag-lead compensator without this restriction can be realized with an active network as previously shown or with passive networks by cascading the lead and lag networks shown in Table 9.11. Remember, though, that the two networks must be isolated to ensure that one network does not load the other. If the networks load each other, the transfer function will not be the product of the individual transfer functions. A possible realization using the passive networks uses an operational amplifier to provide isolation. The circuit is shown in Figure 9.63. Example 9.10 demonstrates the design of a passive compensator.

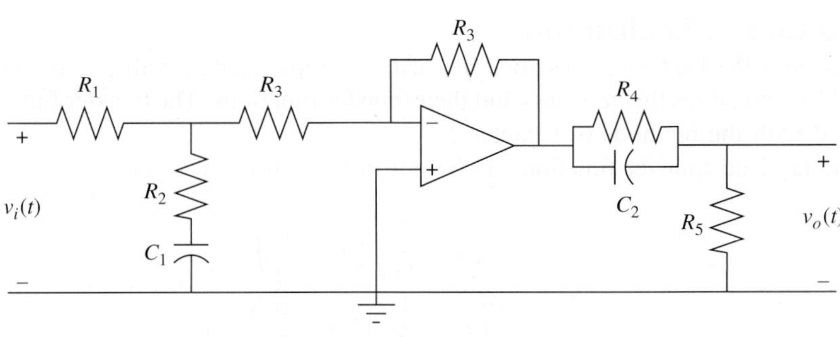


**FIGURE 9.63** Lag-lead compensator implemented with cascaded lag and lead networks with isolation

## Example 9.10 Realizing a Lead Compensator

**PROBLEM:** Realize the lead compensator designed in Example 9.4 (Compensator *b*).

**SOLUTION:** The transfer function of the lead compensator is

$$G_c(s) = \frac{s+4}{s+20.09} \tag{9.51}$$

Comparing the transfer function of a lead network shown in Table 9.11 with Eq. (9.51), we obtain the following two relationships:

$$\frac{1}{R_1C} = 4 \tag{9.52}$$

and

$$\frac{1}{R_1C} + \frac{1}{R_2C} = 20.09 \tag{9.53}$$

Hence, $R_1C = 0.25$, and $R_2C = 0.0622$. Since there are three network elements and two equations, we may select one of the element values arbitrarily. Letting $C = 1\ \mu\text{F}$, then $R_1 = 250\ \text{k}\Omega$ and $R_2 = 62.2\ \text{k}\Omega$.

## Skill-Assessment Exercise 9.5

**PROBLEM:** Implement the compensators shown in **a.** and **b.** below. Choose a passive realization if possible.

**a.** $G_c(s) = \dfrac{(s+0.1)(s+5)}{s}$

**b.** $G_c(s) = \dfrac{(s+0.1)(s+2)}{(s+0.01)(s+20)}$

**ANSWERS:**

**a.** $G_c(s)$ is a PID controller and thus requires active realization. Use Figure 9.60 with the PID controller circuits shown in Table 9.10. One possible set of approximate component values is

$$C_1 = 10\ \mu\text{F}, \quad C_2 = 100\ \mu\text{F}, \quad R_1 = 20\ \text{k}\Omega, \quad R_2 = 100\ \text{k}\Omega$$

**b.** $G_c(s)$ is a lag-lead compensator that can be implemented with a passive network because the ratio of the lead pole to zero is the inverse of the ratio of the lag pole to zero. Use the lag-lead compensator circuit shown in Table 9.11. One possible set of approximate component values is

$$C_1 = 100\ \mu\text{F}, \quad C_2 = 900\ \mu\text{F}, \quad R_1 = 100\ \text{k}\Omega, \quad R_2 = 560\ \Omega$$

The complete solution is at www.wiley.com.college/nise.

## Case Studies Antenna Control: Lag-Lead Compensation

Design D

For the antenna azimuth position control system case study in Chapter 8, we obtained a 25% overshoot using a simple gain adjustment. Once this percent overshoot was obtained, the settling time was determined. If we try to improve the settling time by increasing the gain, the percent overshoot also increases. In this section, we continue with the antenna azimuth position control by designing a cascade compensator that yields 25% overshoot at a reduced settling time. Further, we effect an improvement in the steady-state error performance of the system.

**PROBLEM:** Given the antenna azimuth position control system shown on the front endpapers, Configuration 1, design cascade compensation to meet the following requirements: (1) 25% overshoot, (2) 2-second settling time, and (3) $K_v = 20$.

**SOLUTION:** For the case study in Chapter 8, a preamplifier gain of 64.21 yielded 25% overshoot, with the dominant, second-order poles at $-0.833 \pm j1.888$. The settling time is thus $4/\zeta\omega_n = 4/.833 = 4.8$ seconds. The open-loop function for the system as derived in the case study in Chapter 5 is $G(s) = 6.63K/[s(s+1.71)(s+100)]$. Hence $K_v = 6.63K/(1.71 \times 100) = 2.49$. Comparing these values to this example's problem statement, we want to improve the settling time by a factor of 2.4, and we want approximately an eightfold improvement in $K_v$.

**Lead compensator design to improve transient response:** First locate the dominant second-order pole. To obtain a settling time, $T_s$, of 2 seconds and a percent overshoot of 25%, the real part of the dominant second-order pole should be at $-4/T_s = -2$. Locating the pole on the 113.83° line ($\zeta = 0.404$, corresponding to 25% overshoot) yields an imaginary part of 4.529 (see Figure 9.64).

Second, assume a lead compensator zero and find the compensator pole. Assuming a compensator zero at $-2$, along with the uncompensated system's open-loop poles and zeros, use the root locus program in Appendix H.2 at www.wiley.com/college/nise to find that there is an angular contribution of $-120.14°$ at the design point of $-2 \pm j4.529$. Therefore, the compensator's pole must contribute $120.14° - 180° = -59.86°$ for the design point to be on the compensated system's root locus. The geometry is shown in Figure 9.64. To calculate the compensator pole, we use $4.529/(p_c - 2) = \tan 59.86°$ or $p_c = 4.63$.

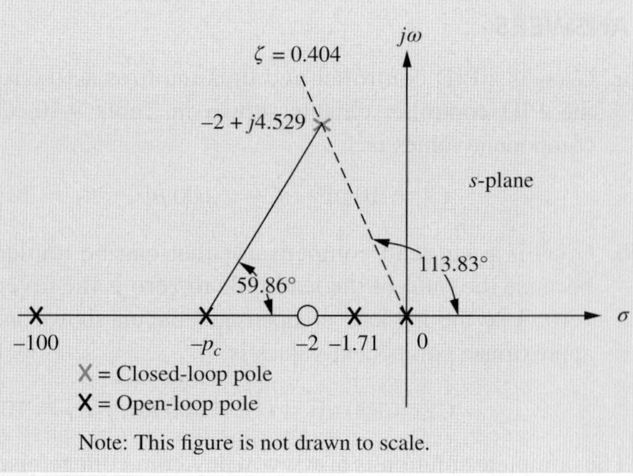


**FIGURE 9.64** Locating compensator pole

Now determine the gain. Using the lead-compensated system's open-loop function,

$$G(s) = \frac{6.63K(s+2)}{s(s+1.71)(s+100)(s+4.63)} \tag{9.54}$$

and the design point $-2 + j4.529$ as the test point in the root locus program, the gain, $6.63K$, is found to be 2549.

**Lag compensator design to improve the steady-state error:** $K_v$ for the lead-compensated system is found using Eq. (9.54). Hence,

$$K_v = \frac{2549(2)}{(1.71)(100)(4.63)} = 6.44 \tag{9.55}$$

Since we want $K_v = 20$, the amount of improvement required over the lead-compensated system is $20/6.44 = 3.1$. Choose $p_c = -0.01$ and calculate $z_c = 0.031$, which is 3.1 times larger.

**Determine gain:** The complete lag-lead-compensated open-loop function, $G_{LLC}(s)$, is

$$G_{LLC}(s) = \frac{6.63K(s+2)(s+0.031)}{s(s+.01)(s+1.71)(s+4.63)(s+100)} \tag{9.56}$$

Using the root locus program in Appendix H.2 at www.wiley.com/college/nise and the poles and zeros of Eq. (9.56), search along the 25% overshoot line (113.83°) for the design point. This point has moved slightly with the addition of the lag compensator to $-1.99 \pm j4.51$. The gain at this point equals 2533, which is $6.63K$. Solving for $K$ yields $K = 382.1$.

**Realization of the compensator:** A realization of the lag-lead compensator is shown in Figure 9.63. From Table 9.11 the lag portion has the following transfer function:

$$G_{lag}(s) = \frac{R_2}{R_1 + R_2} \frac{s + \frac{1}{R_2 C}}{s + \frac{1}{(R_1 + R_2)C}} = \frac{R_2}{R_1 + R_2} \frac{(s+0.031)}{(s+0.01)} \tag{9.57}$$

Selecting $C = 10\,\mu\text{F}$, we find $R_2 = 3.2\ \text{M}\Omega$ and $R_1 = 6.8\ \text{M}\Omega$.

From Table 9.11 the lead compensator portion has the following transfer function:

$$G_{lead}(s) = \frac{s + \frac{1}{R_1 C}}{s + \frac{1}{R_1 C} + \frac{1}{R_2 C}} = \frac{(s+2)}{(s+4.63)} \tag{9.58}$$

Selecting $C = 10\,\mu\text{F}$, we find $R_1 = 50\ \text{k}\Omega$ and $R_2 = 38\ \text{k}\Omega$.

The total loop gain required by the system is 2533. Hence,

$$6.63K \frac{R_2}{R_1 + R_2} = 2533 \tag{9.59}$$

where $K$ is the gain of the preamplifier, and $R_2/(R_1 + R_2)$ is the gain of the lag portion. Using the values of $R_1$ and $R_2$ found during the realization of the lag portion, we find $K = 1194$.

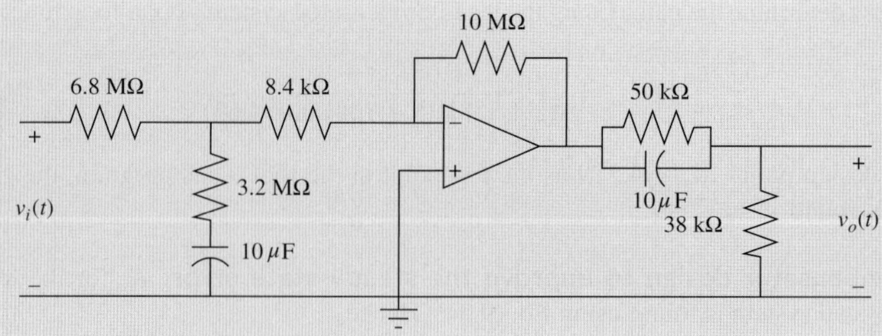


**FIGURE 9.65** Realization of lag-lead compensator

The final circuit is shown in Figure 9.65, where the preamplifier is implemented with an operational amplifier whose feedback and input resistor ratio approximately equals 1194, the required preamplifier gain. The preamplifier isolates the lag and lead portions of the compensator.

**Summary of the design results:** Using Eq. (9.56) along with $K = 382.1$ yields the compensated value of $K_v$. Thus,

$$K_v = \lim_{s \to 0} sG_{\text{LLC}}(s) = \frac{2533(2)(0.031)}{(0.01)(1.71)(4.63)(100)} = 19.84 \tag{9.60}$$

which is an improvement over the gain-compensated system in the case study of Chapter 8, where $K_v = 2.49$. This value is calculated from the uncompensated $G(s)$ by letting $K = 64.21$, as found in the Case Study of Chapter 8.

Finally, checking the second-order approximation via simulation, we see in Figure 9.66 the actual transient response. Compare this to the gain-compensated system response of Figure 8.29 to see the improvement effected by cascade compensation over simple gain adjustment. The gain-compensated system yielded 25%, with a settling time of about 4 seconds. The lag-lead-compensated system yields 28% overshoot, with a settling time of about 2 seconds. If the results are not adequate for the application, the system should be redesigned to reduce the percent overshoot.

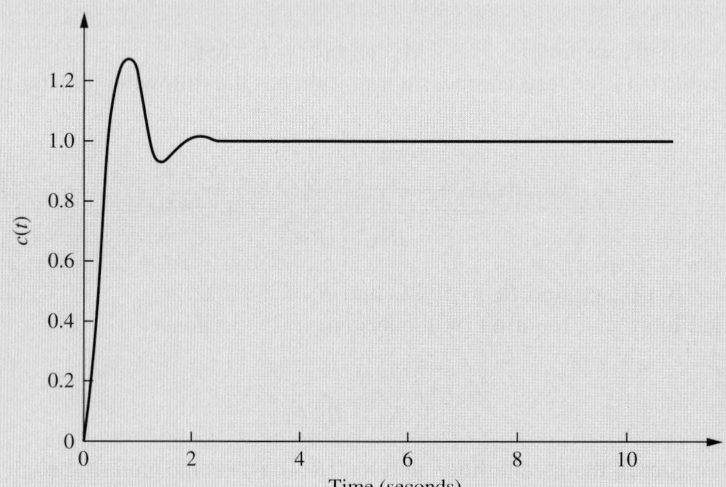


**FIGURE 9.66** Step response of lag-lead-compensated antenna control

**CHALLENGE:** You are now given a problem to test your knowledge of this chapter's objectives. You are given the antenna azimuth position control system shown on the front endpapers, Configuration 2. In the challenge in Chapter 8, you were asked to design, via gain adjustment, an 8-second settling time.

**a.** For your solution to the challenge in Chapter 8, evaluate the percent overshoot and the value of the appropriate static error constant.

**b.** Design a cascade compensator to reduce the percent overshoot by a factor of 4 and the settling time by a factor of 2. Also, improve the appropriate static error constant by a factor of 2.

**c.** Repeat Part **b** using MATLAB.

MATLAB
ML

## UFSS Vehicle: Lead and Feedback Compensation

As a final look at this case study, we redesign the pitch control loop for the UFSS vehicle. For the case study in Chapter 8, we saw that rate feedback improved the transient response. In this chapter's case study, we replace the rate feedback with a cascade compensator.

**PROBLEM:** Given the pitch control loop without rate feedback ($K_2 = 0$) for the UFSS vehicle shown on the back endpapers, design a compensator to yield 20% overshoot and a settling time of 4 seconds (*Johnson, 1980*).

**SOLUTION:** First determine the location of the dominant closed-loop poles. Using the required 20% overshoot and a 4-second settling time, a second-order approximation shows the dominant closed-loop poles are located at $-1 \pm j1.951$. From the uncompensated system analyzed in the Chapter 8 case study, the estimated settling time was 19.8 seconds for dominant closed-loop poles of $-0.202 \pm j0.394$. Hence, a lead compensator is required to speed up the system.

Arbitrarily assume a lead compensator zero at $-1$. Using the root locus program in Appendix H.2 at www.wiley.com/college/nise, we find that this compensator zero, along with the open-loop poles and zeros of the system, yields an angular contribution at the design point, $-1 + j1.951$, of $-178.92°$. The difference between this angle and 180°, or $-1.08°$, is the angular contribution required from the compensator pole.

Using the geometry shown in Figure 9.67, where $-p_c$ is the compensator pole location, we find that

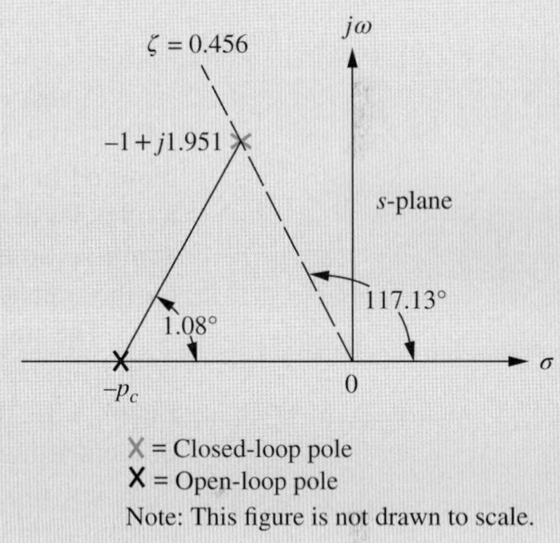


**FIGURE 9.67** Locating compensator pole

$$\frac{1.951}{p_c - 1} = \tan 1.08° \tag{9.61}$$

from which $p_c = 104.5$. The compensated open-loop transfer function is thus

$$G(s) = \frac{0.25K_1(s + 0.435)(s + 1)}{(s + 1.23)(s + 2)(s^2 + 0.226s + 0.0169)(s + 104.5)} \tag{9.62}$$

where the compensator is

$$G_c(s) = \frac{(s + 1)}{(s + 104.5)} \tag{9.63}$$

Using all poles and zeros shown in Eq. (9.62), the root locus program shows that a gain of 516.5 is required at the design point, $-1 \pm j1.951$. The root locus of the compensated system is shown in Figure 9.68.

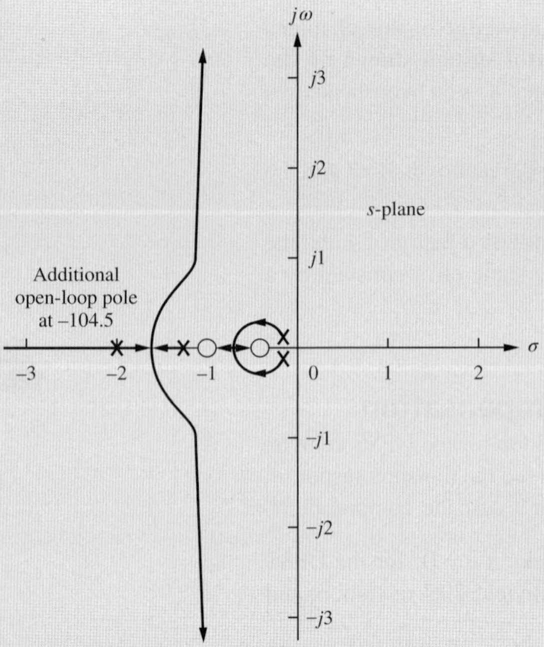


**FIGURE 9.68** Root locus for lead-compensated system

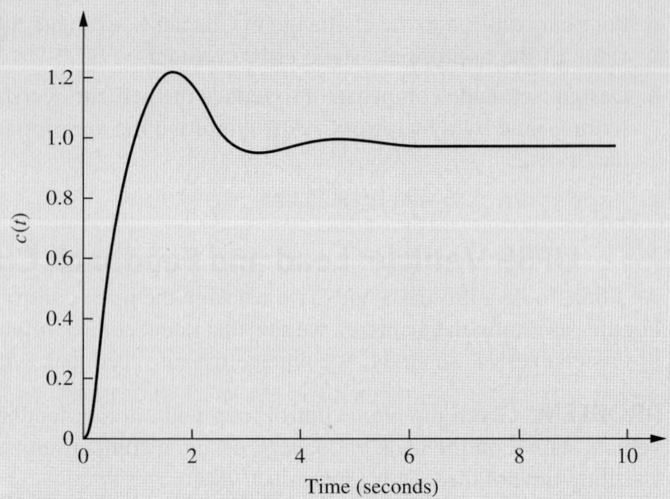


**FIGURE 9.69** Step response of lead-compensated UFSS vehicle

A test of the second-order approximation shows three more closed-loop poles at $-0.5$, $-0.9$, and $-104.5$. Since the open-loop zeros are at $-0.435$ and $-1$, simulation is required to see if there is effectively closed-loop pole-zero cancellation with the closed-loop poles at $-0.5$ and $-0.9$, respectively. Further, the closed-loop pole at $-104.5$ is more than five times the real part of the dominant closed-loop pole, $-1 \pm j1.951$, and its effect on the transient response is therefore negligible.

The step response of the closed-loop system is shown in Figure 9.69, where we see a 26% overshoot and a settling time of about 4.5 seconds. Comparing this response with Figure 8.31, the response of the uncompensated system, we see considerable improvement in the settling time and steady-state error. However, the transient response performance does not meet the design requirements. Thus, a redesign of the system to reduce the percent overshoot is suggested if required by the application.

**CHALLENGE:** You are now given a problem to test your knowledge of this chapter's objectives. The heading control system for the UFSS vehicle is shown on the back endpapers. The minor loop contains the rudder and vehicle dynamics, and the major loop relates output and input headings (*Johnson, 1980*).

**a.** Find the values of $K_1$ and $K_2$ so that the minor-loop dominant poles have a damping ratio of 0.6 and the major-loop dominant poles have a damping ratio of 0.5.

MATLAB
ML

**b.** `Repeat, using MATLAB.`

## Summary

In this chapter, we learned how to design a system to meet transient and steady-state specifications. These design techniques overcame limitations in the design methodology covered in Chapter 8, whereby a transient response could be created only if the poles generating that response were on the root locus. Subsequent gain adjustment yielded the desired response. Since this value of gain dictated the amount of steady-state error in the response, a trade-off was required between the desired transient response and the desired steady-state error.

*Cascade* or *feedback compensation* is used to overcome the disadvantages of gain adjustment as a compensating technique. In this chapter, we saw that the transient response and the steady-state error can be designed separately from each other. No longer was a trade-off between these two specifications required. Further, we were able to design for a transient response that was not represented on the original root locus.

The transient response design technique covered in this chapter is based upon reshaping the root locus to go through a desired transient response point, followed by a gain adjustment. Typically, the resulting gain is much higher than the original if the compensated system response is faster than the uncompensated response.

The root locus is reshaped by adding additional poles and zeros via a cascade or feedback compensator. The additional poles and zeros must be checked to see that any second-order approximations used in the design are valid. All poles, besides the dominant second-order pair, must yield a response that is much faster than the designed response. Thus, nondominant poles must be at least five times as far from the imaginary axis as the dominant pair. Further, any zeros of the system must be close to a nondominant pole for pole-zero cancellation, or far from the dominant pole pair. The resulting system can then be approximated by two dominant poles.

The steady-state response design technique is based upon placing a pole on or near the origin in order to increase or nearly increase the system type. Then a zero is placed near this pole so that the effect upon the transient response is negligible. However, final reduction of steady-state error occurs with a long-time constant. The same arguments about other poles yielding fast responses and about zeros being cancelled in order to validate a second-order approximation also hold true for this technique. If the second-order approximations cannot be justified, then a simulation is required to make sure the design is within tolerance.

Steady-state design compensators are implemented via *PI controllers* or *lag compensators*. PI controllers add a pole at the origin, thereby increasing the system type. Lag compensators, usually implemented with passive networks, place the pole off the origin but near it. Both methods add a zero very close to the pole in order not to affect the transient response.

The transient response design compensators are implemented through *PD controllers* or *lead compensators*. PD controllers add a zero to compensate the transient response; they are considered *ideal*. Lead compensators, on the other hand, are not ideal since they add a pole along with the zero. Lead compensators are usually passive networks.

We can correct both transient response and steady-state error with a *PID* or *lag-lead compensator*. Both of these are simply combinations of the previously described compensators. Table 9.7 summarized the types of cascade compensators.

Feedback compensation can also be used to improve the transient response. Here the compensator is placed in the feedback path. The feedback gain is used to change the compensator zero or the system's open-loop poles, giving the designer a wide choice of various root loci. The system gain is then varied to move along the selected root locus to the design point. An advantage of feedback compensation is the ability to design a fast response into a subsystem independently of the system's total response.

In the next chapter, we look at another method of design, frequency response, which is an alternate method to the root locus.

## Review Questions

1. Briefly distinguish between the design techniques in Chapter 8 and Chapter 9.
2. Name two major advantages of the design techniques of Chapter 9 over the design techniques of Chapter 8.

3. What kind of compensation improves the steady-state error?
4. What kind of compensation improves transient response?
5. What kind of compensation improves both steady-state error and transient response?
6. Cascade compensation to improve the steady-state error is based upon what pole-zero placement of the compensator? Also, state the reasons for this placement.
7. Cascade compensation to improve the transient response is based upon what pole-zero placement of the compensator? Also, state the reasons for this placement.
8. What difference on the $s$-plane is noted between using a PD controller or using a lead network to improve the transient response?
9. In order to speed up a system without changing the percent overshoot, where must the compensated system's poles on the $s$-plane be located in comparison to the uncompensated system's poles?
10. Why is there more improvement in steady-state error if a PI controller is used instead of a lag network?
11. When compensating for steady-state error, what effect is sometimes noted in the transient response?
12. A lag compensator with the zero 25 times as far from the imaginary axis as the compensator pole will yield approximately how much improvement in steady-state error?
13. If the zero of a feedback compensator is at $-3$ and a closed-loop system pole is at $-3.001$, can you say there will be pole-zero cancellation? Why?
14. Name two advantages of feedback compensation.

## Problems

1. Design a PI controller to drive the step response error to zero for the unity feedback system shown in Figure P9.1, where

$$G(s) = \frac{K}{(s+1)(s+3)(s+10)}$$

The system operates with a damping ratio of 0.5. Compare the specifications of the uncompensated and compensated systems. [Section: 9.2]

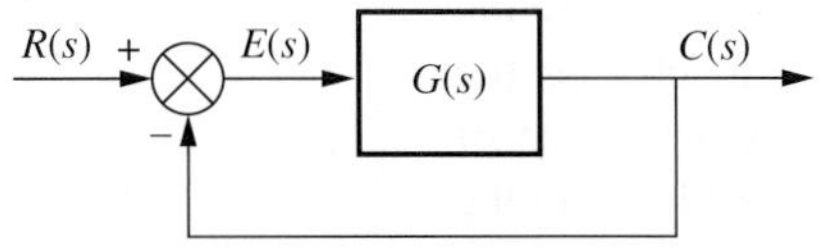


FIGURE P9.1

2. Consider the unity feedback system shown in Figure P9.1, where

$$G(s) = \frac{K}{s(s+2)(s+5)}$$

a. Design a PI controller to drive the ramp response error to zero for any $K$ that yields stability. [Section: 9.2]

b. Use MATLAB to simulate your design for $K=1$. Show both the input ramp and the output response on the same plot. MATLAB ML

3. The unity feedback system shown in Figure P9.1 with

$$G(s) = \frac{K}{(s+1)(s+3)(s+5)}$$

is operating with 10% overshoot. [Section: 9.2]

a. What is the value of the appropriate static error constant?

b. Find the transfer function of a lag network so that the appropriate static error constant equals 4 without appreciably changing the dominant poles of the uncompensated system.

c. Use MATLAB or any other computer program to simulate the system to see the effect of your compensator. MATLAB ML

4. Repeat Problem 3 for $G(s) = \dfrac{K}{s(s+3)(s+5)}$. [Section: 9.2]

5. Consider the unity feedback system shown in Figure P9.1 with

$$G(s)=\frac{K}{(s+2)(s+4)(s+6)}$$

a. Design a compensator that will yield $K_p = 20$ without appreciably changing the dominant pole location that yields a 10% overshoot for the uncompensated system. [Section: 9.2]

b. Use MATLAB or any other computer program to simulate the uncompensated and compensated systems. MATLAB ML

c. Use MATLAB or any other computer program to determine how much time it takes the slow response of the lag compensator to bring the output to within 2% of its final compensated value. MATLAB ML

6. The unity feedback system shown in Figure P9.1 with

$$G(s)=\frac{K(s+6)}{(s+2)(s+3)(s+5)}$$

is operating with a dominant-pole damping ratio of 0.707. Design a PD controller so that the settling time is reduced by a factor of 2. Compare the transient and steady-state performance of the uncompensated and compensated systems. Describe any problems with your design. [Section: 9.3]

7. Redo Problem 6 using MATLAB in the following way: MATLAB ML

a. MATLAB will generate the root locus for the uncompensated system along with the 0.707 damping ratio line. You will interactively select the operating point. MATLAB will then inform you of the coordinates of the operating point, the gain at the operating point, as well as the estimated %OS, $T_s$, $T_p$, $\zeta$, $\omega_n$, and $K_p$ represented by a second-order approximation at the operating point.

b. MATLAB will display the step response of the uncompensated system.

c. Without further input, MATLAB will calculate the compensated design point and will then ask you to input a value for the PD compensator zero from the keyboard. MATLAB will respond with a plot of the root locus showing the compensated design point. MATLAB will then allow you to keep changing the PD compensator value from the keyboard until a root locus is plotted that goes through the design point.

d. For the compensated system, MATLAB will inform you of the coordinates of the operating point, the gain at the operating point, as well as the estimated %OS, $T_s$, $T_p$, $\zeta$, $\omega_n$, and $K_p$ represented by a second-order approximation at the operating point.

e. MATLAB will then display the step response of the compensated system.

8. Design a PD controller for the system shown in Figure P9.2 to reduce the settling time by a factor of 4 while continuing to operate the system with 20% overshoot. Compare the performance of the compensated system to that of the uncompensated system. Summarize the results in a table similar to that in Example 9.7.

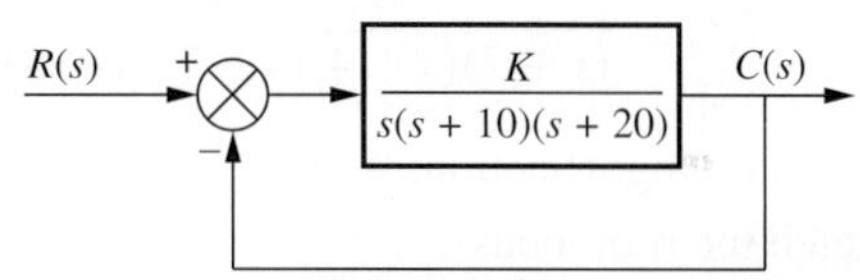


FIGURE P9.2

9. Consider the unity feedback system shown in Figure P9.1 with [Section: 9.3]

$$G(s)=\frac{K}{(s+4)^3}$$

a. Find the location of the dominant poles to yield a 1.6 second settling time and an overshoot of 25%.

b. If a compensator with a zero at −1 is used to achieve the conditions of Part **a**, what must the angular contribution of the compensator pole be?

c. Find the location of the compensator pole.

d. Find the gain required to meet the requirements stated in Part **a**.

e. Find the location of other closed-loop poles for the compensated system.

f. Discuss the validity of your second-order approximation.

g. Use MATLAB or any other computer program to simulate the compensated system to check your design. MATLAB ML

**10.** The unity feedback system shown in Figure P9.1 with

$$G(s) = \frac{K}{s^2}$$

is to be designed for a settling time of 1.818 seconds and a 16.3% overshoot. If the compensator zero is placed at −1, do the following: [Section: 9.3]

**a.** Find the coordinates of the dominant poles.

**b.** Find the compensator pole.

**c.** Find the system gain.

**d.** Find the location of all nondominant poles.

**e.** Estimate the accuracy of your second-order approximation.

**f.** Evaluate the steady-state error characteristics.

**g.** Use MATLAB or any other computer program to simulate the system and evaluate the actual transient response characteristics for a step input. MATLAB ML

**11.** Given the unity feedback system of Figure P9.1, with

$$G(s) = \frac{K(s+6)}{(s+2)(s+4)(s+7)(s+8)}$$

do the following: [Section: 9.3]

**a.** Sketch the root locus.

**b.** Find the coordinates of the dominant poles for which $\zeta = 0.8$.

**c.** Find the gain for which $\zeta = 0.8$.

**d.** If the system is to be cascade-compensated so that $T_s = 1$ second and $\zeta = 0.8$, find the compensator pole if the compensator zero is at −4.5.

**e.** Discuss the validity of your second-order approximation.

**f.** Use MATLAB or any other computer program to simulate the compensated and uncompensated systems and compare the results to those expected. MATLAB ML

**12.** Redo Problem 11 using MATLAB in the following way: MATLAB ML

**a.** MATLAB will generate the root locus for the uncompensated system along with the 0.8 damping-ratio line. You will interactively select the operating point. MATLAB will then inform you of the coordinates of the operating point, the gain at the operating point, as well as the estimated $\%OS$, $T_s$, $T_p$, $\zeta$, $\omega_n$, and $K_p$ represented by a second-order approximation at the operating point.

**b.** MATLAB will display the step response of the uncompensated system.

**c.** Without further input, MATLAB will calculate the compensated design point and will then ask you to input a value for the lead compensator pole from the keyboard. MATLAB will respond with a plot of the root locus showing the compensated design point. MATLAB will then allow you to keep changing the lead compensator pole value from the keyboard until a root locus is plotted that goes through the design point.

**d.** For the compensated system, MATLAB will inform you of the coordinates of the operating point, the gain at the operating point, as well as the estimated $\%OS$, $T_s$, $T_p$, $\zeta$, $\omega_n$, and $K_p$ represented by a second-order approximation at the operating point.

**e.** MATLAB will then display the step response of the compensated system.

**f.** Change the compensator's zero location a few times and collect data on the compensated system to see if any other choices of compensator zero yield advantages over the original design.

**13.** Consider the unity feedback system of Figure P9.1 with

$$G(s) = \frac{K}{s(s+10)(s+30)}$$

The system is operating at 15% overshoot. Design a compensator to decrease the settling time by a factor of 3 without affecting the percent overshoot and do the following: [Section: 9.3]

**a.** Evaluate the uncompensated system's dominant poles, gain, and settling time.

**b.** Evaluate the compensated system's dominant poles and settling time.

**c.** Evaluate the compensator's pole and zero. Find the required gain.

**d.** Use MATLAB or any other computer program to simulate the compensated and uncompensated systems' step response. MATLAB ML

**14.** For the unity feedback system of Figure P9.1 with

$$G(s) = \frac{K}{s(s+1)(s^2+10s+26)}$$

do the following: [Section: 9.3]

**a.** Find the settling time for the system if it is operating with 15% overshoot.

**b.** Find the zero of a compensator and the gain, $K$, so that the settling time is 7 seconds. Assume that the pole of the compensator is located at $-15$.

**c.** Use MATLAB or any other computer program to simulate the system's step response to test the compensator. MATLAB ML

**15.** The unity feedback system shown in Figure P9.1 with

$$G(s) = \frac{K}{(s+15)(s^2+6s+13)}$$

is operating with 30% overshoot. [Section: 9.3]

**a.** Find the transfer function of a cascade compensator, the system gain, and the dominant pole location that will cut the settling time in half if the compensator zero is at $-7$.

**b.** Find other poles and zeros and discuss your second-order approximation.

**c.** Use MATLAB or any other computer program to simulate both the uncompensated and compensated systems to see the effect of your compensator. MATLAB ML

**16.** A unity feedback control system has the following forward transfer function: [Section: 9.3]

$$G(s) = \frac{K}{s^2(s+4)(s+15)}$$

**a.** Design a lead compensator to yield a closed-loop step response with 20.5% overshoot and a settling time of 3 seconds. Be sure to specify the value of $K$.

**b.** Is your second-order approximation valid?

**c.** Use MATLAB or any other computer program to simulate and compare the transient response of the compensated system to the predicted transient response. MATLAB ML

**17.** For the unity feedback system of Figure P9.1, with

$$G(s) = \frac{K}{(s^2+20s+101)(s+20)}$$

the damping ratio for the dominant poles is to be 0.4, and the settling time is to be 0.5 second. [Section: 9.3]

**a.** Find the coordinates of the dominant poles.

**b.** Find the location of the compensator zero if the compensator pole is at –15.

**c.** Find the required system gain.

**d.** Compare the performance of the uncompensated and compensated systems.

**e.** Use MATLAB or any other computer program to simulate the system to check your design. Redesign if necessary. MATLAB ML

**18.** Consider the unity feedback system of Figure P9.1, with

$$G(s) = \frac{K}{(s+3)(s+6)}$$

**a.** Show that the system cannot operate with a settling time of 2/3 second and a percent overshoot of 1.5% with a simple gain adjustment.

**b.** Design a lead compensator so that the system meets the transient response characteristics of Part **a.** Specify the compensator's pole, zero, and the required gain.

**19.** Given the unity feedback system of Figure P9.1 with

$$G(s) = \frac{K}{(s+2)(s+4)(s+6)(s+8)}$$

Find the transfer function of a lag-lead compensator that will yield a settling time 0.5 second shorter than that of the uncompensated system. The compensated system also will have a damping ratio of 0.5, and improve the steady-state error by a factor of 30. The compensator zero is at $-5$. Also, find the compensated system's gain. Justify any second-order approximations or verify the design through simulation. [Section: 9.4]

**20.** Redo Problem 19 using a lag-lead compensator and MATLAB in the following way: MATLAB ML

**a.** MATLAB will generate the root locus for the uncompensated system along with the 0.5 damping-ratio line. You will interactively select the operating point. MATLAB will then proceed to inform you of the coordinates of the operating point, the gain at the operating point, as well as the estimated %OS, $T_s$, $T_p$, $\zeta$, $\omega_n$, and $K_p$ represented by a second-order approximation at the operating point.

**b.** MATLAB will display the step response of the uncompensated system.

**c.** Without further input, MATLAB will calculate the compensated design point and will then ask you to input a value for the lead compensator pole from the keyboard. MATLAB will respond with a plot of the root locus showing the compensated design point. MATLAB will then allow you to keep changing the lead compensator pole value from the keyboard until a root locus is plotted that goes through the design point.

**d.** For the compensated system, MATLAB will inform you of the coordinates of the operating point, the gain at the operating point, as well as the estimated %OS, $T_s$, $T_p$, $\zeta$, $\omega_n$, and $K_p$ represented by a second-order approximation at the operating point.

**e.** MATLAB will then display the step response of the compensated system.

**f.** Change the compensator's zero location a few times and collect data on the compensated system to see if any other choices of the compensator zero yield advantages over the original design.

**g.** Using the steady-state error of the uncompensated system, add a lag compensator to yield an improvement of 30 times over the uncompensated system's steady-state error, with minimal effect on the designed transient response. Have MATLAB plot the step response. Try several values for the lag compensator's pole and see the effect on the step response.

**21.** Given the uncompensated unity feedback system of Figure P9.1, with

$$G(s) = \frac{K}{s(s+1)(s+3)}$$

do the following: [Section: 9.4]

**a.** Design a compensator to yield the following specifications: settling time = 2.86 seconds; percent overshoot = 4.32%; the steady-state error is to be improved by a factor of 2 over the uncompensated system.

**b.** Compare the transient and steady-state error specifications of the uncompensated and compensated systems.

**c.** Compare the gains of the uncompensated and compensated systems.

**d.** Discuss the validity of your second-order approximation.

**e.** Use MATLAB or any other computer program to simulate the uncompensated and compensated systems and verify the specifications. MATLAB **ML**

**22.** For the unity feedback system given in Figure P9.1 with

$$G(s) = \frac{K}{s(s+5)(s+12)}$$

do the following: [Section: 9.4]

**a.** Find the gain, $K$, for the uncompensated system to operate with 30% overshoot.

**b.** Find the peak time and $K_v$ for the uncompensated system.

**c.** Design a lag-lead compensator to decrease the peak time by a factor of 2, decrease the percent overshoot by a factor of 2, and improve the steady-state error by a factor of 30. Specify all poles, zeros, and gains.

**23.** The unity feedback system shown in Figure P9.1 with

$$G(s) = \frac{K}{(s^2+4s+8)(s+10)}$$

is to be designed to meet the following specifications:

Overshoot: Less than 25%

Settling time: Less than 1 second

$K_p = 10$

Do the following: [Section: 9.4]

**a.** Evaluate the performance of the uncompensated system operating at 10% overshoot.

**b.** Design a passive compensator to meet the desired specifications.

**c.** Use MATLAB to simulate the compensated system. Compare the response with the desired specifications. MATLAB **ML**

**24.** Consider the unity feedback system in Figure P9.1, with

$$G(s) = \frac{K}{(s+2)(s+4)}$$

The system is operated with 4.32% overshoot. In order to improve the steady-state error, $K_p$ is to be increased by at least a factor of 5. A lag compensator of the form

$$G_c(s) = \frac{(s+0.5)}{(s+0.1)}$$

is to be used. [Section: 9.4]

**a.** Find the gain required for both the compensated and the uncompensated systems.

**b.** Find the value of $K_p$ for both the compensated and the uncompensated systems.

**c.** Estimate the percent overshoot and settling time for both the compensated and the uncompensated systems.

**d.** Discuss the validity of the second-order approximation used for your results in Part **c**.

**e.** Use MATLAB or any other computer program to simulate the step response for the uncompensated and compensated systems. What do you notice about the compensated system's response?

MATLAB ML

**f.** Design a lead compensator that will correct the objection you notice in Part **e**.

**25.** For the unity feedback system in Figure P9.1, with

$$G(s) = \frac{K}{(s+1)(s+4)}$$

design a PID controller that will yield a peak time of 1.047 seconds and a damping ratio of 0.8, with zero error for a step input. [Section: 9.4]

**26.** For the unity feedback system in Figure P9.1, with

$$G(s) = \frac{K}{(s+4)(s+6)(s+10)}$$

do the following:

**a.** Design a controller that will yield no more than 25% overshoot and no more than a 2-second settling time for a step input and zero steady-state error for step and ramp inputs.

**b.** Use MATLAB and verify your design.

MATLAB ML

**27.** Redo Problem 26 using MATLAB in the following way:

MATLAB ML

**a.** MATLAB will ask for the desired percent overshoot, settling time, and PI compensator zero.

**b.** MATLAB will design the PD controller's zero.

**c.** MATLAB will display the root locus of the PID-compensated system with the desired percent overshoot line.

**d.** The user will interactively select the intersection of the root locus and the desired percent overshoot line.

**e.** MATLAB will display the gain and transient response characteristics of the PID-compensated system.

**f.** MATLAB will display the step response of the PID-compensated system.

**g.** MATLAB will display the ramp response of the PID-compensated system.

**28.** If the system of Figure P9.3 operates with a damping ratio of 0.517 for the dominant second-order poles, find the location of all closed-loop poles and zeros.

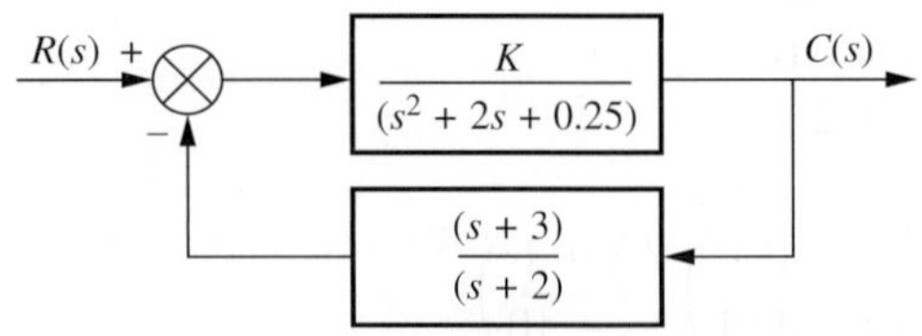


**FIGURE P9.3**

**29.** Given the system of Figure P9.4: [Section: 9.5]

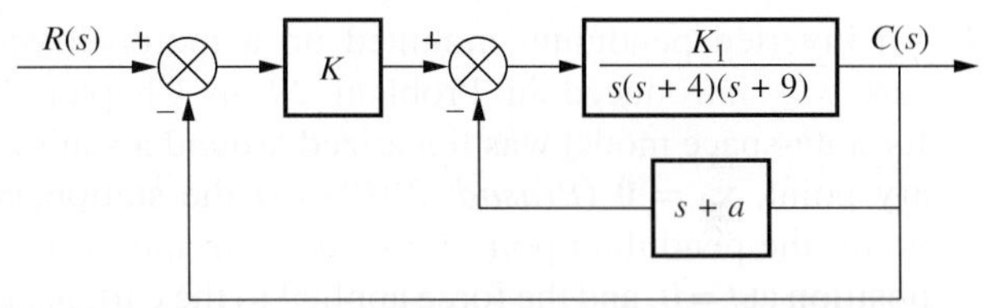


**FIGURE P9.4**

**a.** Design the value of $K_1$, as well as $a$ in the feedback path of the minor loop, to yield a settling time of 1 second with 5% overshoot for the step response.

**b.** Design the value of $K$ to yield a major-loop response with 10% overshoot for a step input.

**c.** Use MATLAB or any other computer program to simulate the step response to the entire closed-loop system.

MATLAB ML

**d.** Add a PI compensator to reduce the major-loop steady-state error to zero and simulate the step response using MATLAB or any other computer program.

MATLAB ML

**30.** For the unity feedback system in Figure P9.1, with

$$G(s) = \frac{K}{s(s+2)(s+4)(s+6)}$$

do the following: [Section: 9.5]

**a.** Design rate feedback to yield a step response with no more than 15% overshoot and no more than 3 seconds settling time. Use Approach 1.

**b.** Use MATLAB and simulate your compensated system. MATLAB **ML**

**31.** Identify and realize the following controllers with operational amplifiers. [Section: 9.6]

**a.** $\dfrac{s+0.02}{s}$

**b.** $s+4$

**32.** Identify and realize the following compensators with passive networks. [Section: 9.6]

**a.** $\dfrac{s+0.1}{s+0.01}$

**b.** $\dfrac{s+2}{s+5}$

**c.** $\left(\dfrac{s+0.1}{s+0.01}\right)\left(\dfrac{s+1}{s+10}\right)$

**33.** Repeat Problem 32 using operational amplifiers. [Section: 9.6]

**34.** An inverted pendulum mounted on a motor-driven cart was introduced in Problem 29 in Chapter 3. Its state-space model was linearized around a stationary point, $\mathbf{x}_0 = \mathbf{0}$ (*Prasad, 2012*). At the stationary point, the pendulum point-mass, $m$, is in the upright position at $t = 0$, and the force applied to the cart, $u_0$, is 0. Its model was then modified in Problem 54 in Chapter 6 to have two output variables: the pendulum angle relative to the $y$-axis, $\theta(t)$, and the horizontal position of the cart, $x(t)$. MATLAB was then used to find its eigenvalues. Noting that only one pole (out of four) is located in the left half of the $s$-plane, we concluded that this unit requires stabilization.

To accomplish stability and design an appropriate control system, do the following:

**a.** Draw a signal-flow diagram for that unit and use it to develop Simulink models for two feedback systems: one to control the cart position, $x(t)$, and the other to control the pendulum angle, $\theta(t)$. Set the upper and lower saturation limits of the second integrator in the angle loop to 100 and −100, respectively, and those limits in the position loop to 10 and −10. Simulink **SL**

**b.** Use rate feedback with a gain of 12.5 to stabilize the pendulum angle control system. The forward path of each of these systems should include a PD (proportional plus derivative) controller that adjusts the force applied to the cart, $u(t)$. These controllers[3] may be configured with the following settings:[4]

$P = 5$, $I = 0$, and $D = 5$ for the cart position controller

$P = 2$, $I = 0$, and $D = 10$ for the pendulum angle controller

**c.** Utilizing scopes to capture the two output responses, use a unit-impulse[5] as the reference input in the angle control loop and a unit-step source (configured to start at $t = 0$ with a final value of 1) as the input for the cart position control loop.

**d.** If either of the responses has a steady-state error, $e(\infty) > 2\%$, a peak time, $T_P > 1.2$ seconds, or a percent overshoot, $\%OS > 20.5\%$, suggest appropriate changes to its controller settings.

**35.** Design a PI controller to drive the step-response error to zero for the unity feedback system shown in Figure P9.1, where

$$G(s) = \frac{K}{(s+1)(s+3)(s+10)}$$

The system operates with a damping factor of 0.4. Design for each of the following two cases: (1) compensator zero at −0.1, and (2) compensator zero at −0.7.

Compare the specifications of the uncompensated and each one of the compensated systems. Simulate each one of the systems using any software program.

## DESIGN PROBLEMS

**36.** Figure P9.5 shows a two-tank system. The liquid inflow to the upper tank can be controlled using a valve and is represented by $F_0$. The upper tank's outflow equals the lower tank's inflow and is represented by $F_1$. The outflow of the lower tank is $F_2$. The objective of the design is to control the liquid level, $y(t)$, in the lower tank. The open-loop transmission for this system is $\dfrac{Y(s)}{F_o(s)} = \dfrac{a_2 a_3}{s^2 + (a_1 + a_4)s + a_1 a_4}$

[3] These are PID controllers, in which the integral actions are set to zero to avoid any negative effect on stability.

[4] Note that a high value for the derivative action of the angle controller and a low value for its proportional gain have been selected to further stabilize the pendulum.

[5] To create a unit-impulse, use a unit-step source followed by a derivative block.

(*Romagnoli, 2006*). The system will be controlled in a loop analogous to that of Figure P9.1, where the lower liquid level will be measured and compared to a set point. The resulting error will be fed to a controller, which in turn will open or close the valve feeding the upper tank.

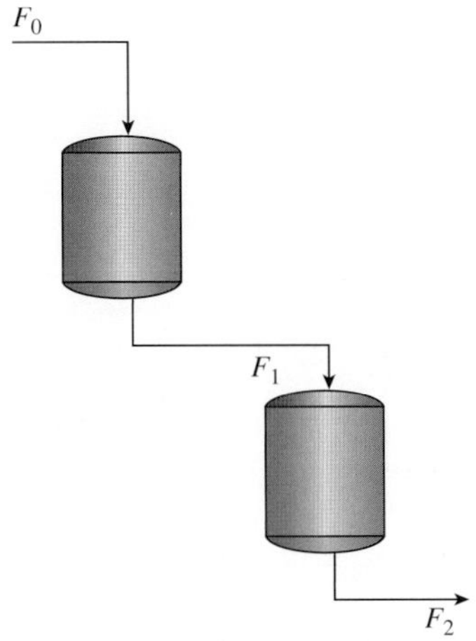


**FIGURE P9.5**[6]

**a.** Assuming $a_1 = 0.04$, $a_2 = 0.0187$, $a_3 = 1$, and $a_4 = 0.227$, design a lag compensator to obtain a step-response steady-state error of 10% without affecting the system's transient response appreciably.

**b.** `Verify your design through MATLAB simulations.` MATLAB **ML**

**37.** The room temperature of an 11 $m^2$ room is to be controlled by varying the power of an indoor radiator. For this specific room, the open-loop transfer function from radiator power, $\dot{Q}(s)$, to temperature, $T(s)$, is (*Thomas, 2005*)

$$G(s) = \frac{T(s)}{\dot{Q}(s)} = \frac{(1 \times 10^{-6})s^2 + (1.314 \times 10^{-9})s + (2.66 \times 10^{-13})}{s^3 + 0.00163s^2 + (5.272 \times 10^{-7})s + (3.538 \times 10^{-11})}$$

The system is assumed to be in the closed-loop configuration shown in Figure P9.1.

**a.** For a unit step input, calculate the steady-state error of the system.

**b.** Try using the procedure of Section 9.2 to design a PI controller to obtain zero steady-state error for step inputs without appreciably changing the transient response. Then explain why it is not possible to do so.

**c.** Design a PI controller of the form $G_c(s) = \dfrac{K(s+z)}{s}$ that will reduce the step-response error to zero while not changing significantly the transient response. (Hint: Place the zero of the compensator in a position where the closed-loop poles of the uncompensated root locus will not be affected significantly.)

**d.** `Use Simulink to simulate the systems of Parts` **b** `and` **c** `and to verify the correctness of your design in Part` **c**. Simulink **SL**

**38.** Figure P9.6(*a*) shows a heat-exchanger process whose purpose is to maintain the temperature of a liquid at a prescribed temperature.

The temperature is measured using a sensor and a transmitter, TT 22, that sends the measurement to a corresponding controller, TC 22, that compares the actual temperature with a desired temperature set point, SP. The controller automatically opens or closes a valve to allow or prevent the flow of steam to change the temperature in the tank. The corresponding block diagram for this system is shown in Figure P9.6(*b*) (*Smith, 2002*). Assume the following transfer functions:

$$G_v(s) = \frac{0.02}{4s+1}; \quad G_1(s) = \frac{70}{50s+1}; \quad H(s) = \frac{1}{12s+1}$$

**a.** Assuming $G_c(s) = K$, find the value of $K$ that will result in a dominant pole with $\zeta = 0.7$. Obtain the corresponding $T_s$.

**b.** Design a PD controller to obtain the same damping factor as Part **a** but with a settling time 20% smaller.

**c.** `Verify your results through MATLAB simulation.` MATLAB **ML**

**39.** Repeat Problem 38, Parts **b** and **c**, using a lead compensator.

**40. a.** Find the transfer function of a motor whose torque-speed curve and load are given in Figure P9.7.

**b.** Design a tachometer compensator to yield a damping ratio of 0.5 for a position control employing a power amplifier of gain 1 and a preamplifier of gain 5000.

**c.** Compare the transient and steady-state characteristics of the uncompensated system and the compensated system.

[6] Romagnoli, J.A., and Palazoglu. A. *Introduction to Process Control.* CRC Press, Boca Raton, 2006. P. 44, Figure 3.4.

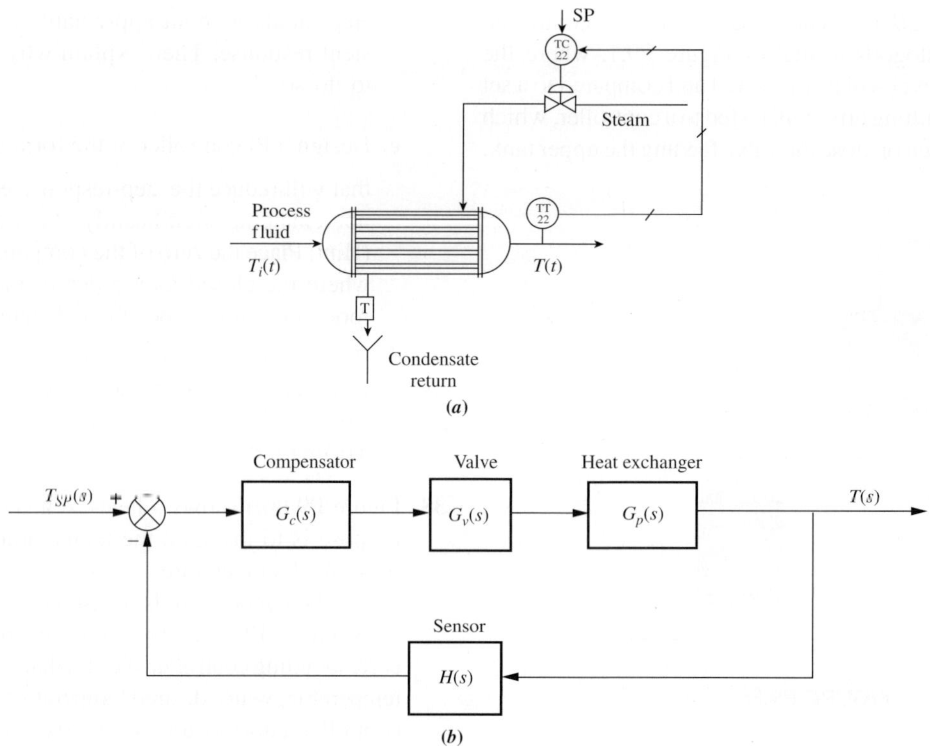


**FIGURE P9.6** **a.** Heat-exchanger process;[7] **b.** block diagram

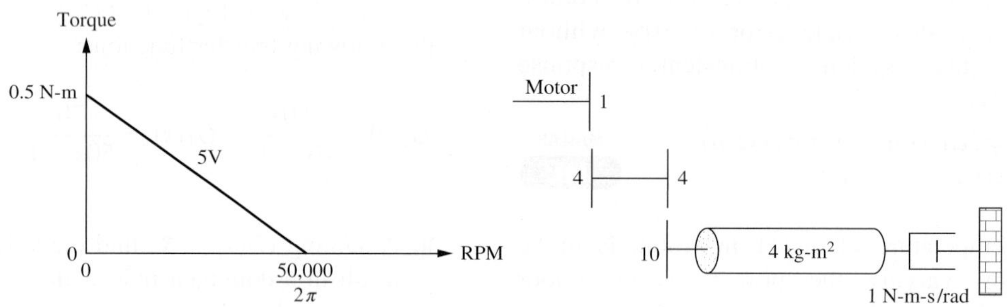


**FIGURE P9.7**

41. You are given the motor whose transfer function is shown in Figure P9.8(*a*).

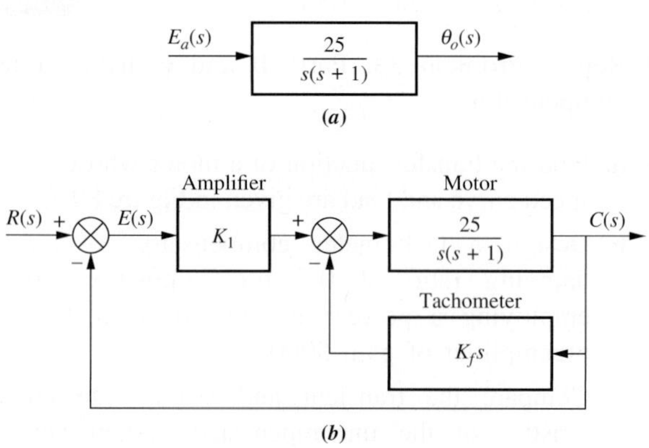


**FIGURE P9.8**

**a.** If this motor were the forward transfer function of a unity feedback system, calculate the percent overshoot and settling time that could be expected.

**b.** You want to improve the closed-loop response. Since the motor constants cannot be changed and you cannot use a different motor, an amplifier and tachometer are inserted into the loop as shown in Figure P9.8(*b*). Find the values of $K_1$ and $K_f$ to yield a percent overshoot of 25% and a settling time of 0.2 second.

**c.** Evaluate the steady-state error specifications for both the uncompensated and the compensated systems.

[7] Smith, C.A. *Automated Continuous Process Control.* John Wiley & Sons, New York, NY, 2002, p. 128, Figure 6-1.1.

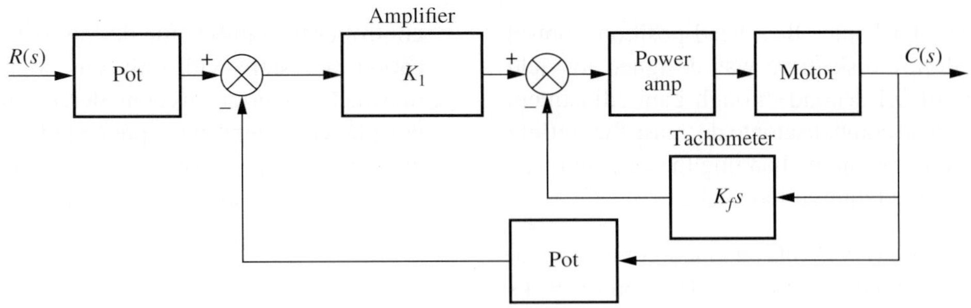


**FIGURE P9.9**

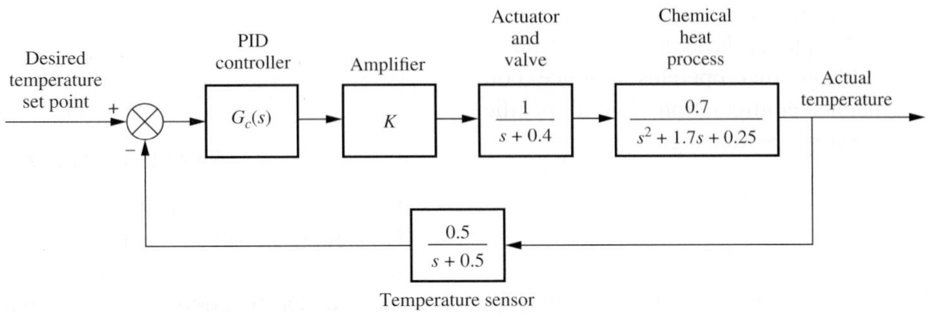


**FIGURE P9.10** Chemical process temperature control system

**42.** A position control is to be designed with a 20% overshoot and a settling time of 2 seconds. You have on hand an amplifier and a power amplifier whose cascaded transfer function is $K_1/(s+25)$ with which to drive the motor. Two 10-turn pots are available to convert shaft position into voltage. A voltage of $\pm 5\pi$ volts is placed across the pots. A dc motor whose transfer function is of the form

$$\frac{\theta_o(s)}{E_a(s)} = \frac{K}{s(s+a)}$$

is also available. The transfer function of the motor is found experimentally as follows. The motor and geared load are driven open-loop by applying a large, short, rectangular pulse to the armature. An oscillo-gram of the response shows that the motor reached 63% of its final output value at 0.4 second after the application of the pulse. Further, with a constant 10 volts dc applied to the armature, the constant output speed was 100 rad/s.

**a.** Draw a complete block diagram of the system, specifying the transfer function of each component when the system is operating with 20% overshoot.

**b.** What will the steady-state error be for a unit ramp input?

**c.** Determine the transient response characteristics.

**d.** If tachometer feedback is used around the motor, as shown in Figure P9.9, find the tachometer and the amplifier gain to meet the original specifications. Summarize the transient and steady-state characteristics.

**43.** Consider the temperature control system for a chemical process shown in Figure P9.10. The uncompensated system is operating with a rise time approximately the same as a second-order system with a peak time of 16 seconds and 5% overshoot. There is also considerable steady-state error. Design a PID controller so that the compensated system will have a rise time approximately equivalent to a second-order system with a peak time of 8 seconds and 5% overshoot, and zero steady-state error for a step input.

**44.** Given the system shown in Figure P9.11, find the values of $K$ and $K_f$ so that the closed-loop dominant poles will have a damping ratio of 0.5 and the under-damped poles of the minor loop will have a damping ratio of 0.8.

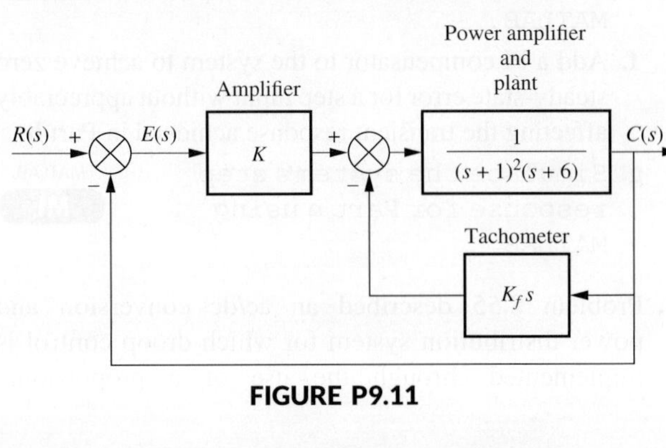


**FIGURE P9.11**

**45.** In Problem 58 of Chapter 8, a head-position control system for a floppy disk drive was designed to yield a settling time of 0.1 second through gain adjustment alone. Design a lead compensator to decrease the settling time to 0.05 second without changing the percent overshoot. Also, find the required loop gain.

**46.** Digital versatile disc (DVD) players incorporate several control systems for their operations. The control tasks include (1) keeping the laser beam focused on the disc surface, (2) fast track selection, (3) disc rotation speed control, and (4) following a track accurately. In order to follow a track, the pickup-head radial position is controlled via a voltage that operates a voice coil embedded in a magnet configuration. For a specific DVD player, the transfer function is given by

$$P(s) = \frac{X(s)}{V(s)} = \frac{0.63}{\left(1 + \frac{0.36}{305.4}s + \frac{s^2}{305.4^2}\right)\left(1 + \frac{0.04}{248.2}s + \frac{s^2}{248.2^2}\right)}$$

where $x(t)$ = radial pickup position and $v(t)$ = the coil input voltage (*Bittanti, 2002*).

**a.** Assume that the system will be controlled in a closed-loop configuration, such as the one shown in Figure P9.1. Assuming that the plant, $P(s)$, is cascaded with a proportional compensator, $G_c(s) = K$, plot the root locus of the system.

**b.** Repeat Part **a** using MATLAB if your root locus plot was created by any other tool. MATLAB ML

**c.** Find the range of $K$ for closed-loop stability, the resulting damping factor range, and the smallest settling time.

**d.** Design a notch filter compensator so that the system's dominant poles have a damping factor of $\zeta = 0.7$ with a closed-loop settling time of 0.1 second.

**e.** Simulate the system's step response for Part **c** using MATLAB. MATLAB ML

**f.** Add a PI compensator to the system to achieve zero steady-state error for a step input without appreciably affecting the transient response achieved in Part **b**.

**g.** Simulate the system's step response for Part **e** using MATLAB. MATLAB ML

**47.** Problem 8.55 described an ac/dc conversion and power distribution system for which droop control is implemented through the use of a proportional controller to stabilize the dc-bus voltage. For simplification, a system with only one source converter and one load converter was considered. The parameters and design considerations presented in that problem, along with some solution results, allow us to represent the block-diagram of that system as shown in the Figure P9.12.

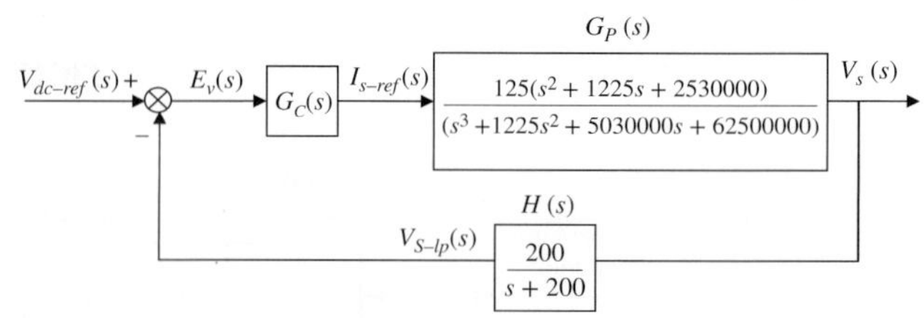


**FIGURE P9.12**

Here $G_c(s)$ is the transfer function of the controller, $G_p(s)$ represents the forward path of the controlled plant (a conversion and power distribution unit), and $H(s)$ is the transfer function of the feedback low-pass filter (*Karlsson, 2003*).

Prepare a table, such as Table 9.5, where the first column, headed *Uncompensated*, is filled in with your results from the proportional design of Problem 8.55, assuming an input step, $v_{dc-ref}(t) = 750\,u(t)$.

Follow Steps 2–8 as described in Section 9.4 (Example 9.5), to design a proportional-plus-integral-plus-derivative (PID) controller so that the system can operate with a percent overshoot $\leq$4.4%, a peak time 20% smaller than that of the uncompensated system, and zero steady-state error, $e_{V\text{step}}(\infty) = 0$. Fill in the remaining two columns of your table, *PD-compensated* and *PID-compensated*.

**48.** Testing of hypersonic flight is performed in wind tunnels where maintaining a constant air pressure is important. Air pressure control is accomplished in several stages. For a specific setup, a simplified transfer function has been found to be (*Varghese, 2009*)

$$\frac{P(s)}{M(s)} = \frac{-2.369 \times 10^6 s^2 + 7.897 \times 10^7 s + 4.21 \times 10^5}{0.015s^5 + 0.7802s^4 + 9.89s^3 + 18.46s^2 + 3.377s + 0.01937}$$

where $M(s)$ is the stem movement of a valve feeding compressed air into the storage tank, and $P(s)$ is the settling chamber pressure.

In order to achieve steady-state error, design a PI controller that operates with a damping factor of 0.4. Compare the characteristics of the uncompensated and compensated systems, and use a computer program to simulate the step response to the compensated system.

**49.** The transfer function for an AFTI/F-16 aircraft relating angle of attack, $\alpha(t)$, to elevator deflection, $\delta_e(t)$, is given by

$$G(s) = \frac{\alpha(s)}{\delta_e(s)} = 0.072\frac{(s+23)(s^2+0.05s+0.04)}{(s-0.7)(s+1.7)(s^2+0.08s+0.04)}$$

Assume the block diagram shown in Figure P9.13 for controlling the angle of attack, $\alpha$, and do the following (*Monahemi, 1992*):

**a.** Find the range of $K$ for stability.

**b.** Plot or sketch a root locus.

**c.** Design a cascade compensator to yield zero steady-state error, a settling time of about 0.05 second, and a percent overshoot not greater than 20%.

**d.** `Use MATLAB or any other computer program to simulate the system to check your design.` MATLAB **ML**

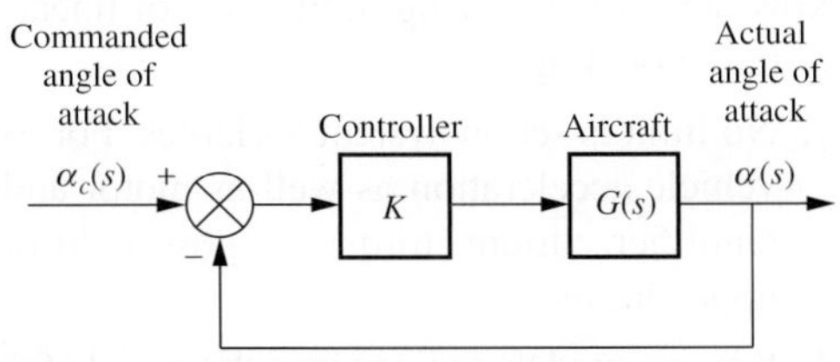


**FIGURE P9.13** Simplified block diagram for angle of attack control

**50.** Figure P9.14 is a simplified block diagram of a self-guiding vehicle's bearing angle control. Design a lead compensator to yield a closed-loop step response with 10% overshoot and a settling time of 1.5 seconds.

**51.** A *metering pump* is a pump capable of delivering a precise flow rate of fluid. Most metering pumps consist of an electric motor that varies the strike length of a shaft, allowing more or less fluid to pass through its body. The control of such a valve has been considered and the open-loop transfer function of the pump was found to be (*Yu, 2011*)

$$\frac{U(s)}{Y(s)} = \frac{1.869}{s^2+12.32s+0.4582}$$

where the output of the system, $Y(s)$, represents the liquid flow, and the input, $U(s)$, is the command signal to the motor that varies the pump's plunger strike length.

**a.** Design a PID controller to reduce the steady-state error to zero with a maximum 5% overshoot and a 20% reduction in uncompensated settling time.

**b.** Find the characteristics of the uncompensated system.

**c.** Design a PID controller to reduce the steady-state error to zero; achieve a maximum 5% overshoot and a 20% reduction in uncompensated settling time.

**d.** Simulate the step response for the compensated system to verify the validity of your design.

**52.** Steam-driven power generators rotate at a constant speed via a governor that maintains constant steam pressure in the turbine. In addition, automatic generation control (AGC) or load frequency control (LFC) is added to ensure reliability and consistency despite load variations or other disturbances that can affect the distribution line frequency output. A specific turbine-governor system can be described only using the block diagram of Figure P9.1 in which $G(s) = G_c(s)G_g(s)G_t(s)G_m(s)$, where (*Khodabakhshian, 2005*)

$G_g(s) = \dfrac{1}{0.2s+1}$ is the governor's transfer function

$G_t(s) = \dfrac{1}{0.5s+1}$ is the turbine transfer function

$G_m(s) = \dfrac{1}{10s+0.8}$ represents the machine and load transfer functions

$G_c(s)$ is the LFC compensation to be designed

**a.** Assuming $G_c(s) = K$, find the value of $K$ that will result in a dominant pole with $\zeta = 0.7$. Obtain the corresponding $T_s$.

**b.** Design a PID controller to obtain the same damping factor as in Part **a**, but with a settling time of 2 seconds and zero steady-state error to step input commands.

**c.** `Verify your results using a MATLAB simuation.` MATLAB **ML**

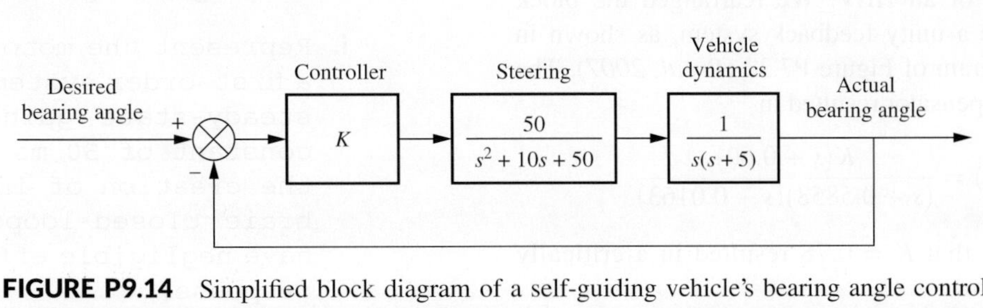


**FIGURE P9.14** Simplified block diagram of a self-guiding vehicle's bearing angle control

**53.** Repeat Problem 52 using a lag-lead compensator instead of a PID controller. Design for a steady-state error of 1% for a step input command.

**54.** Design a PID controller for the drive system of Problem 56, Chapter 8, and shown in Figure P8.16 (*Thomsen, 2011*). Obtain an output response, $\omega_L(t)$, with an overshoot $\leq 15\%$, a settling time of preferably 0.2 second, but not more than 5 seconds, a zero steady-state position error, and a velocity error of $< 2\%$, for a step input, $\omega_r(t) = 260\,u(t)$ rad/sec, applied at $t = 0$.

## PROGRESSIVE ANALYSIS AND DESIGN PROBLEMS

**55. Control of HIV/AIDS.** It was shown in Chapter 6, Problem 67, that when the virus levels in an HIV/AIDS patient are controlled using RTIs the linearized plant model is

$$P(s) = \frac{Y(s)}{U_1(s)} = \frac{-520s - 10.3844}{s^3 + 2.6817s^2 + 0.11s + 0.0126}$$

Assume that the system is embedded in a configuration, such as the one shown in Figure P9.1, where $G(s) = G_c(s)\,P(s)$. Here, $G_c(s)$ is a cascade compensator. For simplicity in this problem, choose the dc gain of $G_c(s)$ less than zero to obtain a negative-feedback system (the negative signs of $G_c(s)$ and $P(s)$ cancel out) (*Craig, I. K., 2004*).

**a.** Consider the uncompensated system with $G_c(s) = -K$. Find the value of $K$ that will place all closed-loop poles on the real axis.

**b.** Use MATLAB to simulate the unit-step response of the gain-compensated system. Note the %*OS* and the $T_s$ from the simulation. MATLAB ML

**c.** Design a PI compensator so that the steady-state error for step inputs is zero. Choose a gain value to make all poles real.

**d.** Use MATLAB to simulate the design in Part c for a unit step input. Compare the simulation to Part **b**. MATLAB ML

**56. Hybrid vehicle.** In the previous chapter, we used the root locus to design a proportional controller for the speed control of an HEV. We rearranged the block diagram to be a unity feedback system, as shown in the block diagram of Figure P7.32 (*Preitl, 2007*). The plant and compensator resulted in

$$G(s) = \frac{K(s + 0.60)}{(s + 0.5858)(s + 0.0163)}$$

and we found that $K = 0.78$ resulted in a critically damped system.

**a.** Use this design to itemize the performance specifications by filling in a table, similar to Table 9.5, under the column *Uncompensated*. Take advantage of the results from Chapter 8 or use MATLAB to find the entries. Plot $c(t)$ for $r(t) = 4\,u(t)$ volts.

**b.** Now assume that the system specifications require zero steady-state error for step inputs, a steady-state error for ramp inputs $\leq 2\%$, a %OS $\leq 4.32\%$, and a settling time $\leq 4$ sec. It should be evident that this is not accomplished with a proportional controller. Thus, start by designing a PI controller to meet the requirements. If necessary, add a PD mode to get a PID controller. Simulate your final design using MATLAB. Fill in the results of this design in the second column of your table with the heading *Compensated*. MATLAB ML

**c.** Now note the following limitations of linear control system modeling:

1. No limit is set on system variables. For example, vehicle acceleration as well as motor and power amplifier current, torque or power do not have upper limits.
2. It is assumed that to improve the speed of response in Part **b**, we could place the PI controller's zero on top of the pole closest to the origin. Realistically, such pole-zero cancellation is not always possible to maintain.

If you do not expand your model beyond the described limitations if required for accuracy, unrealistic response characteristics, such as rise and settling times could result. Look at your design results including response curves. Are they realistic? If not, revise your Simulink model, which you developed for Problem 5.79, according to the following 4 steps: Simulink SL

**i.** Represent the motor armature as a first-order system with a unity steady-state gain and a time constant of 50 ms, which avoids the creation of internal algebraic closed-loops and should have negligible effect on system response;

ii. `Add a saturation element at the output of the motor armature and set it to an upper limit of 250 A;`

iii. `Use the following PI settings. The PI settings of the speed controller are P = 61 and I = 0.795. The PI settings of the torque controller are P = 10 and I = 6;`

iv. `Run the modified model and comment on the graphs obtained for motor current, car acceleration, and speed.`

57. **Parabolic trough collector.** The parabolic trough collector (*Camacho, 2012*) is a Type 0 system as can be seen from its transfer function,

$$G(s) = \frac{137.2 \times 10^{-6} K}{s^2 + 0.0224s + 196 \times 10^{-6}} e^{-39s}$$

We want the system operating in the critically damped mode, but with reduced steady-state error. Using the root locus and a Padé approximation, $e^{-sT} \approx \frac{1 - \frac{T}{2}s}{1 + \frac{T}{2}s}$, do the following:

a. Substitute the Padé approximation for the delay and find the gain necessary to have the system operating with a damping factor, $\zeta = 0.5$. Also, find the corresponding steady-state error.

b. Design a PI compensator to obtain a zero steady-state error while maintaining $\zeta = 0.5$.

c. `Simulate the resulting design using MATLAB to verify your design.` MATLAB ML

# Cyber Exploration Laboratory

## Experiment 9.1

**Objectives** To perform a trade-off study for lead compensation. To design a PI controller and see its effect upon steady-state error.

**Minimum Required Software Packages** MATLAB, and the Control System Toolbox

### Prelab

1. How many lead compensator designs will meet the transient response specifications of a system?
2. What differences do the lead compensators of Prelab 1 make?
3. Design a lead compensator for a unity negative feedback system with a forward transfer function of $G(s) = \frac{K}{s(s+3)(s+6)}$ to meet the following specifications: percent overshoot = 20%; settling time = 2 seconds. Specify the required gain, $K$. Estimate the validity of the second-order approximation.
4. What is the total angular contribution of the lead compensator of Prelab 3?
5. Determine the pole and zero of two more lead compensators that will meet the requirements of Prelab 3.
6. What is the expected steady-state error for a step input for each of the lead-compensated systems?
7. What is the expected steady-state error for a ramp input for each of the lead-compensated systems?
8. Select one of the lead compensator designs and specify a PI controller that can be cascaded with the lead compensator that will produce a system with zero steady-state error for both step and ramp inputs.

### Lab

1. Using the SISO Design Tool, create the design in Prelab 3 and plot the root locus, step response, and ramp response. Take data to determine the percent overshoot, settling time, and step and ramp steady-state errors. Record the gain, $K$.
2. Repeat Lab 1 for each of the designs in Prelab 5.
3. For the design selected in Prelab 8, use the SISO Design Tool and insert the PI controller. Plot the step response and measure the percent overshoot, settling time, and steady-state error. Also, plot the ramp response for the design and measure the steady-state error.
4. Plot the step and ramp responses for two more values of the PI controller zero.

### Postlab

1. Make a table showing calculated and actual values for percent overshoot, settling time, gain, $K$, steady-state error for step inputs, and steady-state error for ramp inputs. Use the three systems without the PI controller and the single system with the PI controller from Lab 3.
2. Itemize the benefits of each system without the PI controller.
3. Choose a final design and discuss the reasons for your choice.

### Experiment 9.2

**Objective** To design a PID controller via LabVIEW

**Minimum Required Software Packages** LabVIEW with the Control Design and Simulation Module

### Prelab

1. Perform Cyber Exploration Laboratory Experiment 8.3.
2. Use the system described in Cyber Exploration Laboratory Experiment 8.3 and replace the controller described there, $G_c(s) = K_D s + K_P$, with a PID controller.
3. Design the controller to meet the following requirements: (1) shorten the settling time found in the design of Cyber Exploration Laboratory Experiment 8.3 to less than 1 sec., and (2) limit the percent overshoot to no more than 5%.
4. Design a LabVIEW VI to test your design. The front panel inputs will be the PID gains and the numerator and denominator of the plant. The indicators will be the transfer functions of the plant, PID controller, and closed-loop system. Finally, provide an indicator for the step-response graph.

**Lab** Run your LabVIEW VI and obtain the step response of the closed-loop system.

**Postlab** Compare the transient and steady-state error performance between the closed-loop step responses of Cyber Exploration Laboratory Experiment 8.3 and this experiment.

## Hardware Interface Laboratory

### Experiment 9.3 Speed Control Using PI Control

**Objectives** To control the speed of the motor in closed loop using integral control and to investigate the tradeoffs of this approach

**Material Required** Computer with LabView installed; myDAQ; dc brushed gearmotor with Hall Sensor quadrature encoder (−10V to +10V normal operation range); and motor control chip BA6886N (or a transistor circuit substitute)

## Files Provided at www.wiley.com/college/nise

Speed PI Control.vi
Signal Conditioning (SubVI).vi
PI Controller (SubVI).vi

**Prelab** Answer the following questions:

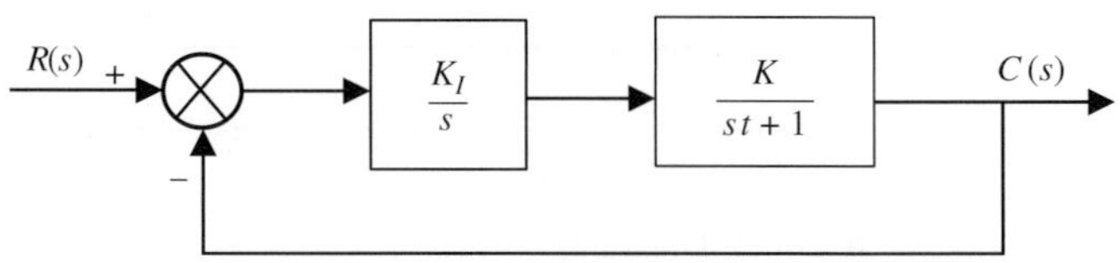


**FIGURE P9.15**

For the system shown in Figure P9.15, do the following:

1. Find the closed-loop transfer function from $R(s)$ to $C(s)$.
2. Draw the root locus as a function of $K_I$.
3. Draw the unit-step response marking the settling time, peak time, and maximum output. Find all the possibilities: overdamped, critically damped, and underdamped.
4. Find an expression for the steady-state error for a unit-step input.

## Lab

**Software:** Use the Speed PI Control.vi and change the constant on the left to fit your motor's gear ratio and encoder CPR as shown in Figure P9.16(*b*). Also change the constants wired to the PI controller. These values should be the maximum motor voltage-dead-zone constant. Change the dead-zone constants inside the signal conditioning block just as you did in Experiment 8.4.

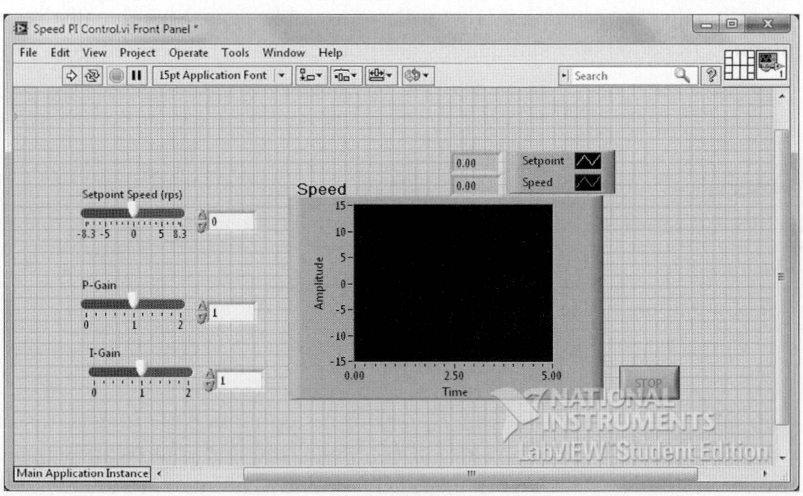


(*a*)

**FIGURE P9.16** Speed PI Control.vi: **a.** Front Panel; (*figure continues*)

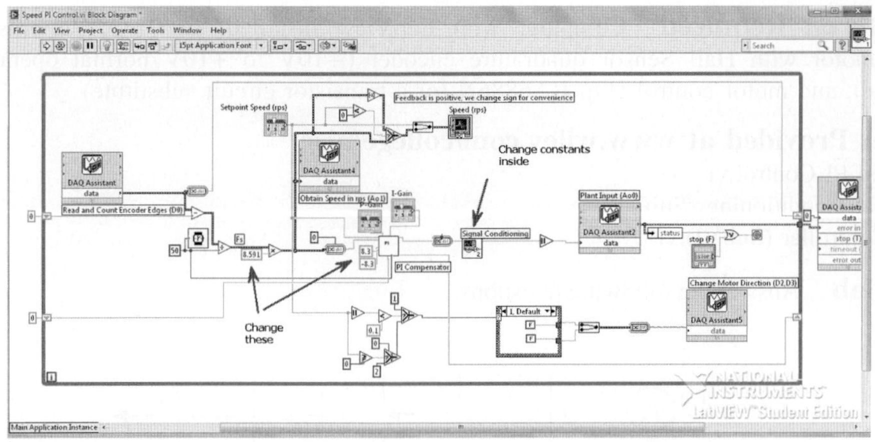


(*b*)

**FIGURE P9.16** (*Continued*) **b.** Block Diagram

**Hardware:** Same as Experiment 8.4

**Procedure:**

1. Make the P gain = 0, and choose a small I gain. Verify the operation of your closed-loop system. In this experiment we will keep the P gain at 0.
2. Draw a functional block diagram (similar to that presented in Chapter 1) of the system. Do not include the signal conditioning functions, nor the change-of-direction signals.
3. Using the transfer function you found in Experiment 4.6, draw the system's root locus.
4. Find the theoretical range of $K_I$ in which the system is closed-loop stable.
5. Run your program and system to find experimentally the range of $K_I$ in which the system is closed-loop stable.
6. Make a judicious choice of three different values of $K_I$ for experimentation.
7. Using the transfer function you found in Experiment 8.4 and the three judicious choices of proportional gain, complete the following table using hand calculations only (calculators OK, computer simulations are not acceptable). Show all your work.

| $K_I$ | | | |
|---|---|---|---|
| $T_P$—Peak time | | | |
| %OS—Percent overshoot | | | |
| $T_s$—Settling time | | | |
| $e_{ss}$—Steady-state error (step input) | | | |

**Theoretical**

8. For each one of the three values of $K_I$, perform step-input experiments. Use a single value of step input for the three values of $K_I$. Make sure that your oscilloscope captures contain the system's transient response in its entirety. Show measurements of all the parameters in the Theoretical table above and fill in the following Experimental table. Please note that $T_s$, the settling time, is hard to measure in the current setting because of

the limited number of analog channels available. Instead of measuring $T_s$, mark on your oscilloscope the theoretical value using the scope cursors. (**Important:** In this experiment stop the VI before restarting it every time you apply a step input. This action will reset the integrator).

| | | | |
|---|---|---|---|
| $K_I$ | | | |
| $T_p$—Peak time | | | |
| %OS—Percent overshoot | | | |
| $e_{ss}$—Steady-state error (step input) | | | |

**Experimental**

**Postlab** Make a detailed comparison of your theoretical and experimental tables. Discuss similarities and discrepancies between experimental and theoretical and give possible reasons.

# Bibliography


Bittanti, S., Dell'Orto, F., Di Carlo, A., and Savaresi, S. M. Notch Filtering and Multirate Control for Radial Tracking in High Speed DVD-Players. *IEEE Transactions on Consumer Electronics*, vol. 48, 2002, pp. 56–62.

Budak, A. *Passive and Active Network Analysis and Synthesis*. Houghton Mifflin, Boston, MA, 1974.

Camacho, E. F., Berenguel, M., Rubio, F. R., and Martinez, D. *Control of Solar Energy Systems*. Springer-Verlag, London, 2012.

Craig, I. K., Xia, X., and Venter, J. W. Introducing HIV/AIDS Education into the Electrical Engineering Curriculum at the University of Pretoria. *IEEE Transactions on Education*, vol. 47, no. 1, February 2004, pp. 65–73.

Craig, J. J. *Introduction to Robotics. Mechanics and Control*, 3d ed. Prentice Hall, Upper Saddle River, NJ, 2005.

D'Azzo, J. J., and Houpis, C. H. *Feedback Control System Analysis and Synthesis*, 2d ed. McGraw-Hill, New York, 1966.

Dorf, R. C. *Modern Control Systems*, 5th ed. Addison-Wesley, Reading, MA, 1989.

Hostetter, G. H., Savant, C. J., Jr., and Stefani, R. T. *Design of Feedback Control Systems*, 2d ed. Saunders College Publishing, New York, 1989.

Johnson, H. et al. *Unmanned Free-Swimming Submersible (UFSS) System Description*. NRL Memorandum Report 4393. Naval Research Laboratory, Washington, D.C., 1980.

Karlsson, P., and Svensson, J. DC Bus Voltage Control for a Distributed Power System, *IEEE Trans. on Power Electronics*, vol. 18, no. 6, 2003, pp. 1405–1412.

Khodabakhshian, A., and Golbon, N. Design of a New Load Frequency PID Controller Using QFT. *Proceedings of the 13th Mediterranean Conference on Control and Automation*, 2005, pp. 970–975.

Kuo, B. C. *Automatic Control Systems*, 7th ed. Prentice Hall, Upper Saddle River, NJ, 1995.

Mahmood, H., and Jiang, J. Modeling and Control System Design of a Grid Connected VSC Considering the Effect of the Interface Transformer Type. *IEEE Transactions on Smart Grid*, vol. 3, no. 1, March 2012, pp. 122–134.

Ogata, K. *Modern Control Engineering*, 2d ed. Prentice Hall, Upper Saddle River, NJ, 1990.

Özel, T. Precision Tracking Control of a Horizontal Arm Coordinate Measuring Machine. *Proceedings of the IEEE Conference on Control Applications*, 2003, pp. 103–108.

Pounds, P. E. I., Mahony, R. E., and Corke, P. I. Design of a Static Thruster for Microair Vehicle Rotorcraft. *Journal of Aerospace Engineering*, vol. 22, no. 1, 2009, pp. 85–94.

Prasad, L., Tyagi, B., and Gupta, H. Modeling & Simulation for Optimal Control of Nonlinear Inverted Pendulum Dynamical System using PID Controller & LQR. *IEEE Computer Society Sixth Asia Modeling Symposium*, 2012, pp. 138–143.

Preitl, Z., Bauer, P., and Bokor, J. A Simple Control Solution for Traction Motor Used in Hybrid Vehicles. *Fourth International Symposium on Applied Computational Intelligence and Informatics*. IEEE. 2007, pp. 157–162.

Romagnoli, J. A., and Palazoglu, A. *Introduction to Process Control*. CRC Press, Boca Raton, FL, 2006.

Smith, C. A. *Automated Continuous Process Control*. Wiley, New York, 2002.

Thomas, B., Soleimani-Mosheni, M., and Fahlén, P. Feed-forward in Temperature Control of Buildings. *Energy and Buildings*, vol. 37, 2005, pp. 755–761.

Thomsen, S., Hoffmann, N., and Fuchs, F. W. PI Control, PI-Based State Space Control, and Model-Based Predictive Control for Drive Systems With Elastically Coupled Loads—A Comparative Study. *IEEE Transactions On Industrial Electronics*, vol. 58, no. 8, August 2011, pp. 3647–3657.

Van de Vegte, J. *Feedback Control Systems*, 2d ed. Prentice Hall, Upper Saddle River, NJ, 1990.

Varghese, J., and Binu, L. S. Adaptive Fuzzy PI Controller for Hypersonic Wind Tunnel Pressure Regulation. *10th National Conference on Technological Trends (NCTT09)*, Nov. 6–7, 2009, pp. 184–187.

Yu, S., Wu, J., and Zhang, D. Robust Non-fragile PID Controller Design for the Stroke Regulation of Metering Pumps. *Chinese Journal of Chemical Engineering*, vol. 19, no. 1, 2011, pp. 83–88.

# 10

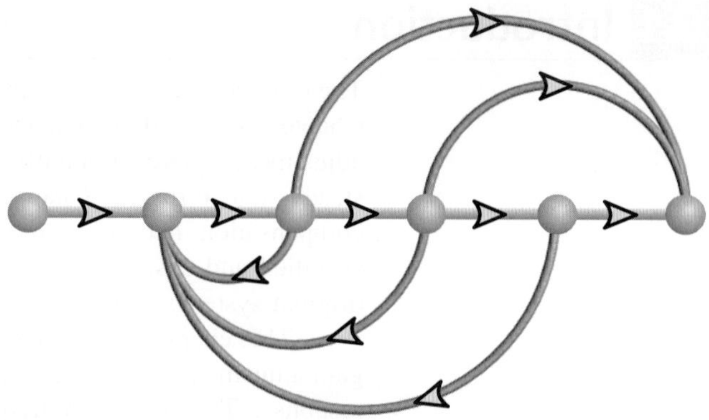

# Frequency Response Techniques

## Chapter Learning Outcomes

After completing this chapter the student will be able to:

- Define and plot the frequency response of a system (Section 10.1)
- Plot asymptotic approximations to the frequency response of a system (Section 10.2)
- Sketch a Nyquist diagram (Sections 10.3–10.4)
- Use the Nyquist criterion to determine the stability of a system (Section 10.5)
- Find stability and gain and phase margins using Nyquist diagrams and Bode plots (Sections 10.6–10.7)
- Find the bandwidth, peak magnitude, and peak frequency of a closed-loop frequency response given the closed-loop time response parameters of peak time, settling time, and percent overshoot (Section 10.8)
- Find the closed-loop frequency response given the open-loop frequency response (Section 10.9)
- Find the closed-loop time response parameters of peak time, settling time, and percent overshoot given the open-loop frequency response (Section 10.10)

## ▶ Case Study Learning Outcomes

You will be able to demonstrate your knowledge of the chapter objectives with a case study as follows:

- Given the antenna azimuth position control system shown on the front endpapers and using frequency response methods, you will be able to find the range of gain, $K$, for stability. You will also be able to find percent overshoot, settling time, peak time, and rise time, given $K$.

## 10.1 Introduction

The root locus method for transient design, steady-state design, and stability was covered in Chapters 8 and 9. In Chapter 8, we covered the simple case of design through gain adjustment, where a trade-off was made between a desired transient response and a desired steady-state error. In Chapter 9, the need for this trade-off was eliminated by using compensation networks so that transient and steady-state errors could be separately specified and designed. Further, a desired transient response no longer had to be on the original system's root locus.

This chapter and Chapter 11 present the design of feedback control systems through gain adjustment and compensation networks from another perspective—that of frequency response. The results of frequency response compensation techniques are not new or different from the results of root locus techniques.

Frequency response methods, developed by Nyquist and Bode in the 1930s, are older than the root locus method, which was discovered by Evans in 1948 (*Nyquist, 1932*; *Bode, 1945*). The older method, which is covered in this chapter, is not as intuitive as the root locus. However, frequency response yields a new vantage point from which to view feedback control systems. This technique has distinct advantages in the following situations:

1. When modeling transfer functions from physical data, as shown in Figure 10.1
2. When designing lead compensators to meet a steady-state error requirement and a transient response requirement
3. When finding the stability of nonlinear systems
4. In settling ambiguities when sketching a root locus

We first discuss the concept of frequency response, define frequency response, derive analytical expressions for the frequency response, plot the frequency response, develop ways of sketching the frequency response, and then apply the concept to control system analysis and design.

**FIGURE 10.1** National Instruments PXI, Compact RIO, Compact DAQ, and USB hardware plantforms (shown from left to right) coupled with NI LabVIEW software to provide stimulus and acquire signals from physical systems. NI LabVIEW can then be used to analyze data, determine the mathematical model, and prototype and deploy a controller for the physical system

## The Concept of Frequency Response

In the steady state, sinusoidal inputs to a linear system generate sinusoidal responses of the same frequency. Even though these responses are of the same frequency as the input, they differ in amplitude and phase angle from the input. These differences are functions of frequency.

Before defining frequency response, let us look at a convenient representation of sinusoids. Sinusoids can be represented as complex numbers called *phasors*. The magnitude of the complex number is the amplitude of the sinusoid, and the angle of the complex number is the phase angle of the sinusoid. Thus, $M_1 \cos(\omega t + \phi_1)$ can be represented as $M_1 \angle \phi_1$ where the frequency, $\omega$, is implicit.

Since a system causes both the amplitude and phase angle of the input to be changed, we can think of the system itself as represented by a complex number, defined so that the product of the input phasor and the system function yields the phasor representation of the output.

Consider the mechanical system of Figure 10.2(*a*). If the input force, $f(t)$, is sinusoidal, the steady-state output response, $x(t)$, of the system is also sinusoidal and at the same frequency as the input. In Figure 10.2(*b*) the input and output sinusoids are represented by complex numbers, or phasors, $M_i(\omega) \angle \phi_i(\omega)$ and $M_o(\omega) \angle \phi_o(\omega)$, respectively. Here, the $M$'s are the amplitudes of the sinusoids and the $\phi$'s are the phase angles of the sinusoids as shown in Figure 10.2(*c*). Assume that the system is represented by the complex number, $M(\omega) \angle \phi(\omega)$. The output steady-state sinusoid is found by multiplying the complex number representation of the input by the complex number representation of the system. Thus, the steady-state output sinusoid is

$$M_o(\omega)\angle\phi_o(\omega) = M_i(\omega)M(\omega)\angle[\phi_i(\omega) + \phi(\omega)] \tag{10.1}$$

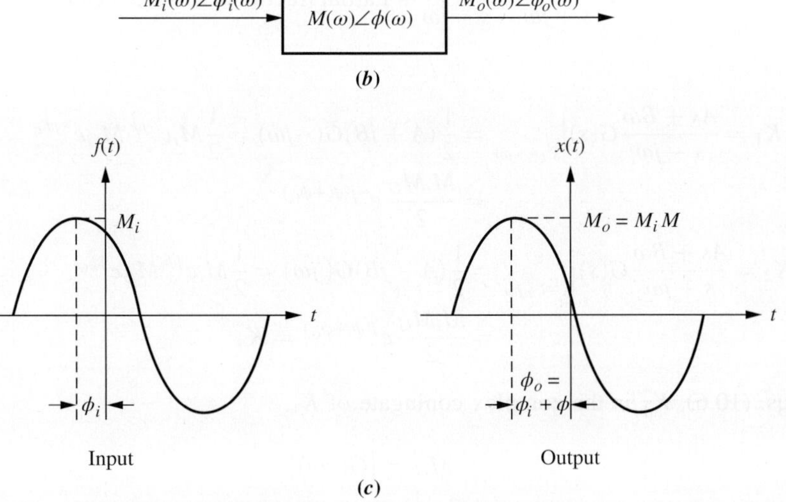


**FIGURE 10.2** Sinusoidal frequency response: **a.** system; **b.** transfer function; **c.** input and output waveforms

From Eq. (10.1) we see that the system function is given by

$$M(\omega) = \frac{M_o(\omega)}{M_i(\omega)} \tag{10.2}$$

and

$$\phi(\omega) = \phi_o(\omega) - \phi_i(\omega) \tag{10.3}$$

Equations (10.2) and (10.3) form our definition of frequency response. We call $M(\omega)$ the *magnitude frequency response* and $\phi(\omega)$ the *phase frequency response*. The combination of the magnitude and phase frequency responses is called the *frequency response* and is $M(\omega)\angle\phi(\omega)$.

In other words, we define the magnitude frequency response to be the ratio of the output sinusoid's magnitude to the input sinusoid's magnitude. We define the phase response to be the difference in phase angle between the output and the input sinusoids. Both responses are a function of frequency and apply only to the steady-state sinusoidal response of the system.

## Analytical Expressions for Frequency Response

Now that we have defined frequency response, let us obtain the analytical expression for it (*Nilsson, 1990*). Later in the chapter, we will use this analytical expression to determine stability, transient response, and steady-state error. Figure 10.3 shows a system, $G(s)$, with the Laplace transform of a general sinusoid, $r(t) = A\cos\omega t + B\sin\omega t = \sqrt{A^2+B^2}\cos[\omega t - \tan^{-1}(B/A)]$ as the input. We can represent the input as a phasor in three ways: (1) in polar form, $M_i\angle\phi_i$, where $M_i = \sqrt{A^2+B^2}$ and $\phi_i = -\tan^{-1}(B/A)$; (2) in rectangular form, $A - jB$; and (3) using Euler's formula, $M_i e^{j\phi_i}$.

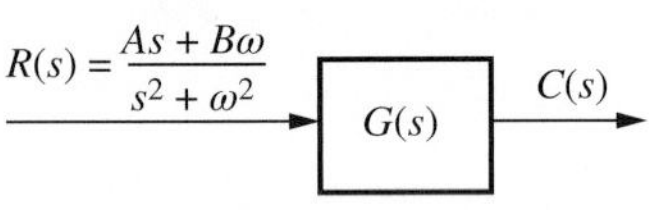


**FIGURE 10.3** System with sinusoidal input

We now solve for the forced response portion of $C(s)$, from which we evaluate the frequency response. From Figure 10.3,

$$C(s) = \frac{As + B\omega}{(s^2+\omega^2)}G(s) \tag{10.4}$$

We separate the forced solution from the transient solution by performing a partial-fraction expansion on Eq. (10.4). Thus,

$$\begin{aligned} C(s) &= \frac{As + B\omega}{(s + j\omega)(s - j\omega)}G(s) \\ &= \frac{K_1}{s + j\omega} + \frac{K_2}{s - j\omega} + \text{Partial fraction terms from } G(s) \end{aligned} \tag{10.5}$$

where

$$\begin{aligned} K_1 = \left.\frac{As + B\omega}{s - j\omega}G(s)\right|_{s\to -j\omega} &= \frac{1}{2}(A + jB)G(-j\omega) = \frac{1}{2}M_i e^{-j\phi_i} M_G e^{-j\phi_G} \\ &= \frac{M_i M_G}{2} e^{-j(\phi_i + \phi_G)} \end{aligned} \tag{10.6a}$$

$$\begin{aligned} K_2 = \left.\frac{As + B\omega}{s + j\omega}G(s)\right|_{s\to +j\omega} &= \frac{1}{2}(A - jB)G(j\omega) = \frac{1}{2}M_i e^{j\phi_i} M_G e^{j\phi_G} \\ &= \frac{M_i M_G}{2} e^{j(\phi_i + \phi_G)} = K_1^* \end{aligned} \tag{10.6b}$$

For Eqs. (10.6), $K_1^*$ is the complex conjugate of $K_1$,

$$M_G = |G(j\omega)| \tag{10.7}$$

and

$$\phi_G = \text{angle of } G(j\omega) \tag{10.8}$$

The steady-state response is that portion of the partial-fraction expansion that comes from the input waveform's poles, or just the first two terms of Eq. (10.5). Hence, the sinusoidal steady-state output, $C_{ss}(s)$, is

$$C_{ss}(s) = \frac{K_1}{s + j\omega} + \frac{K_2}{s - j\omega} \tag{10.9}$$

Substituting Eqs. (10.6) into Eq. (10.9), we obtain

$$C_{ss}(s) = \frac{\frac{M_i M_G}{2} e^{-j(\phi_i + \phi_G)}}{s + j\omega} + \frac{\frac{M_i M_G}{2} e^{j(\phi_i + \phi_G)}}{s - j\omega} \tag{10.10}$$

Taking the inverse Laplace transformation, we obtain

$$\begin{aligned} c(t) &= M_i M_G \left( \frac{e^{-j(\omega t + \phi_i + \phi_G)} + e^{j(\omega t + \phi_i + \phi_G)}}{2} \right) \\ &= M_i M_G \cos(\omega t + \phi_i + \phi_G) \end{aligned} \tag{10.11}$$

which can be represented in phasor form as $M_o\angle\phi_o = (M_1\angle\phi_1)(M_G\angle\phi_G)$, where $M_G\angle\phi_G$ is the frequency response function. But from Eqs. 10.7 and 10.8, $M_G\angle\phi_G = G(j\omega)$. In other words, the frequency response of a system whose transfer function is $G(s)$ is

$$\boxed{G(j\omega) = G(s)|_{s \to j\omega}} \tag{10.12}$$

## Plotting Frequency Response

$G(j\omega) = M_G(\omega) < \phi_G(\omega)$ can be plotted in several ways; two of them are (1) as a function of frequency, with separate magnitude and phase plots; and (2) as a polar plot, where the phasor length is the magnitude and the phasor angle is the phase. When plotting separate magnitude and phase plots, the magnitude curve can be plotted in decibels (dB) vs. log $\omega$, where dB $= 20 \log M$.[1] The phase curve is plotted as phase angle vs. log $\omega$. The motivation for these plots is shown in Section 10.2.

Using the concepts covered in Section 8.1, data for the plots also can be obtained using vectors on the $s$-plane drawn from the poles and zeros of $G(s)$ to the imaginary axis. Here the magnitude response at a particular frequency is the product of the vector lengths from the zeros of $G(s)$ divided by the product of the vector lengths from the poles of $G(s)$, drawn to points on the imaginary axis. The phase response is the sum of the angles from the zeros of $G(s)$ minus the sum of the angles from the poles of $G(s)$ drawn to points on the imaginary axis. Performing this operation for successive points along the imaginary axis yields the data for the frequency response. Remember, each point is equivalent to substituting that point, $s = j\omega_1$, into $G(s)$ and evaluating its value.

The plots also can be made from a computer program that calculates the frequency response. For example, the root locus program discussed in Appendix H at www.wiley.com/college/nise can be used with test points that are on the imaginary axis. The calculated $K$ value at each frequency is the reciprocal of the scaled magnitude response, and the calculated angle is, directly, the phase angle response at that frequency.

The following example demonstrates how to obtain an analytical expression for frequency response and make a plot of the result.

[1] Throughout this book, "log" is used to mean $\log_{10}$, or logarithm to the base 10.

## Example 10.1 Frequency Response from the Transfer Function

**PROBLEM:** Find the analytical expression for the magnitude frequency response and the phase frequency response for a system $G(s) = 1/(s+2)$. Also, plot both the separate magnitude and phase diagrams and the polar plot.

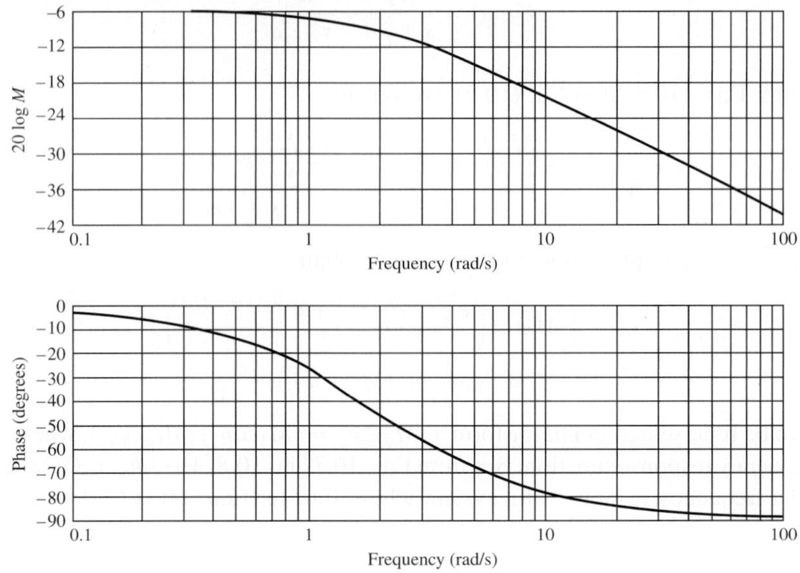


**FIGURE 10.4** Frequency response plots for $G(s) = 1/(s+2)$: separate magnitude and phase diagrams

**SOLUTION:** First substitute $s = j\omega$ in the system function and obtain $G(j\omega) = 1/(j\omega + 2) = (2 - j\omega)/(\omega^2 + 4)$. The magnitude of this complex number, $|G(j\omega)| = M(\omega) = 1/\sqrt{(\omega^2 + 4)}$, is the magnitude frequency response. The phase angle of $G(j\omega)$, $\phi(\omega) = -\tan^{-1}(\omega/2)$, is the phase frequency response.

$G(j\omega)$ can be plotted in two ways: (1) in separate magnitude and phase plots and (2) in a polar plot. Figure 10.4 shows separate magnitude and phase diagrams, where the magnitude diagram is $20 \log M(\omega) = 20 \log\left(1/\sqrt{\omega^2 + 4}\right)$ vs. $\log \omega$, and the phase diagram is $\phi(\omega) = -\tan^{-1}(\omega/2)$ vs. $\log \omega$. The polar plot, shown in Figure 10.5, is a plot of $M(\omega) < \phi(\omega) = 1/\sqrt{\omega^2 + 4} < -\tan^{-1}(\omega/2)$ for different $\omega$.

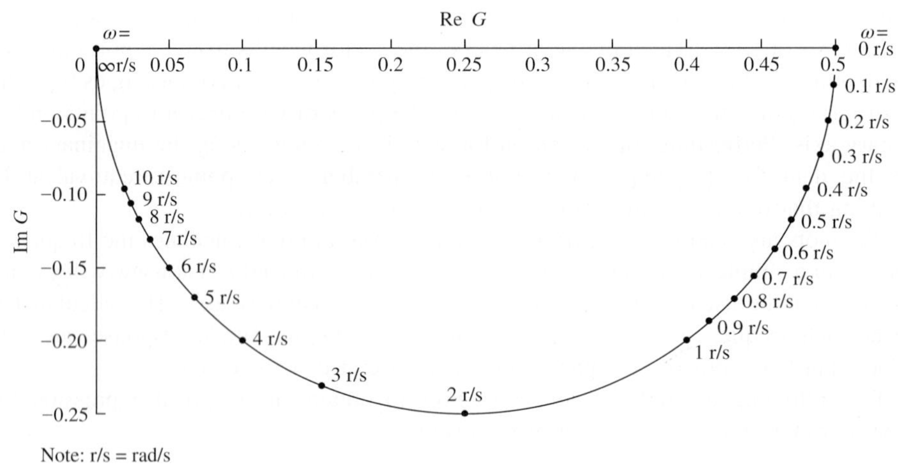


**FIGURE 10.5** Frequency response plot for $G(s) = 1/(s+2)$: polar plot

In the previous example, we plotted the separate magnitude and phase responses, as well as the polar plot, using the mathematical expression for the frequency response. Either of these frequency response presentations can also be obtained from the other. You should practice this conversion by looking at Figure 10.4 and obtaining Figure 10.5 using successive points. For example, at a frequency of 1 rad/s in Figure 10.4, the magnitude is approximately $-7$ dB, or $10^{-7/20} = 0.447$. The phase plot at 1 rad/s tells us that the phase is about $-26°$. Thus on the polar plot, a point of radius 0.447 at an angle of $-26°$ is plotted and identified as 1 rad/s. Continuing in like manner for other frequencies in Figure 10.4, you can obtain Figure 10.5.

Similarly, Figure 10.4 can be obtained from Figure 10.5 by selecting a sequence of points in Figure 10.5 and translating them to separate magnitude and phase values. For example, drawing a vector from the origin to the point at 2 rad/s in Figure 10.5, we see that the magnitude is $20 \log 0.35 = -9.12$ dB and the phase angle is about $-45°$. The magnitude and phase angle are then plotted at 2 rad/s in Figure 10.4 on the separate magnitude and phase curves.

## Skill-Assessment Exercise 10.1

**PROBLEM:**

**a.** Find analytical expressions for the magnitude and phase responses of

$$G(s) = \frac{1}{(s+2)(s+4)}$$

**b.** Make plots of the log-magnitude and the phase, using log-frequency in rad/s as the ordinate.

**c.** Make a polar plot of the frequency response.

**ANSWERS:**

**a.** $M(\omega) = \dfrac{1}{\sqrt{(8-\omega^2)^2 + (6\omega)^2}}$; for $\omega \le \sqrt{8}$: $\phi(\omega) = -\arctan\left(\dfrac{6\omega}{8-\omega^2}\right)$, for $\omega > \sqrt{8}$: $\phi(\omega) = -\left[\pi + \arctan\left(\dfrac{6\omega}{8-\omega^2}\right)\right]$

**b.** See the answer at www.wiley.com/college/nise.

**c.** See the answer at www.wiley.com/college/nise.

The complete solution is at www.wiley.com/college/nise.

In this section, we defined frequency response and saw how to obtain an analytical expression for the frequency response of a system simply by substituting $s = j\omega$ into $G(s)$. We also saw how to make a plot of $G(j\omega)$. The next section shows how to approximate the magnitude and phase plots in order to sketch them rapidly.

# 10.2 Asymptotic Approximations: Bode Plots

The log-magnitude and phase frequency response curves as functions of $\log \omega$ are called Bode plots or Bode diagrams. Sketching Bode plots can be simplified because they can be approximated as a sequence of straight lines. Straight-line approximations simplify the evaluation of the magnitude and phase frequency response.

Consider the following transfer function:

$$G(s)=\frac{K(s+z_1)(s+z_2)\cdots(s+z_k)}{s^m(s+p_1)(s+p_2)\cdots(s+p_n)} \tag{10.13}$$

The magnitude frequency response is the product of the magnitude frequency responses of each term, or

$$|G(j\omega)|=\left.\frac{K|(s+z_1)||(s+z_2)|\cdots|(s+z_k)|}{|s^m||(s+p_1)||(s+p_2)|\cdots|(s+p_n)|}\right|_{s\to j\omega} \tag{10.14}$$

Thus, if we know the magnitude response of each pole and zero term, we can find the total magnitude response. The process can be simplified by working with the logarithm of the magnitude, since the zero terms' magnitude responses would be added and the pole terms' magnitude responses subtracted, rather than, respectively, multiplied or divided, to yield the logarithm of the total magnitude response. Converting the magnitude response into dB, we obtain

$$\begin{aligned}20\log|G(j\omega)| &= 20\log K+20\log|(s+z_1)|+20\log|(s+z_2)|\\ &\quad+\cdots-20\log|s^m|-20\log|(s+p_1)|-\cdots|_{s\to j\omega}\end{aligned} \tag{10.15}$$

Thus, if we knew the response of each term, the algebraic sum would yield the total response in dB. Further, if we could make an approximation of each term that would consist only of straight lines, graphical addition of terms would be greatly simplified.

Before proceeding, let us look at the phase response. From Eq. (10.13), the phase frequency response is the *sum* of the phase frequency response curves of the zero terms minus the *sum* of the phase frequency response curves of the pole terms. Again, since the phase response is the sum of individual terms, straight-line approximations to these individual responses simplify graphical addition.

Let us now show how to approximate the frequency response of simple pole and zero terms by straight-line approximations. Later we show how to combine these responses to sketch the frequency response of more complicated functions. In subsequent sections, after a discussion of the Nyquist stability criterion, we learn how to use the Bode plots for the analysis and design of stability and transient response.

## Bode Plots for $G(s) = (s + a)$

Consider a function, $G(s)=(s+a)$, for which we want to sketch separate logarithmic magnitude and phase response plots. Letting $s=j\omega$, we have

$$G(j\omega)=(j\omega+a)=a\left(j\frac{\omega}{a}+1\right) \tag{10.16}$$

At low frequencies when $\omega$ approaches zero,

$$G(j\omega)\approx a \tag{10.17}$$

The magnitude response in dB is

$$20\log M=20\log a \tag{10.18}$$

where $M=|G(j\omega)|$ and is a constant. Eq. (10.18) is shown plotted in Figure 10.6(*a*) from $\omega=0.01a$ to $a$.

At high frequencies where $\omega \gg a$, Eq. (10.16) becomes

$$G(j\omega)\approx a\left(\frac{j\omega}{a}\right)=a\left(\frac{\omega}{a}\right)\angle 90^\circ=\omega\angle 90^\circ \tag{10.19}$$

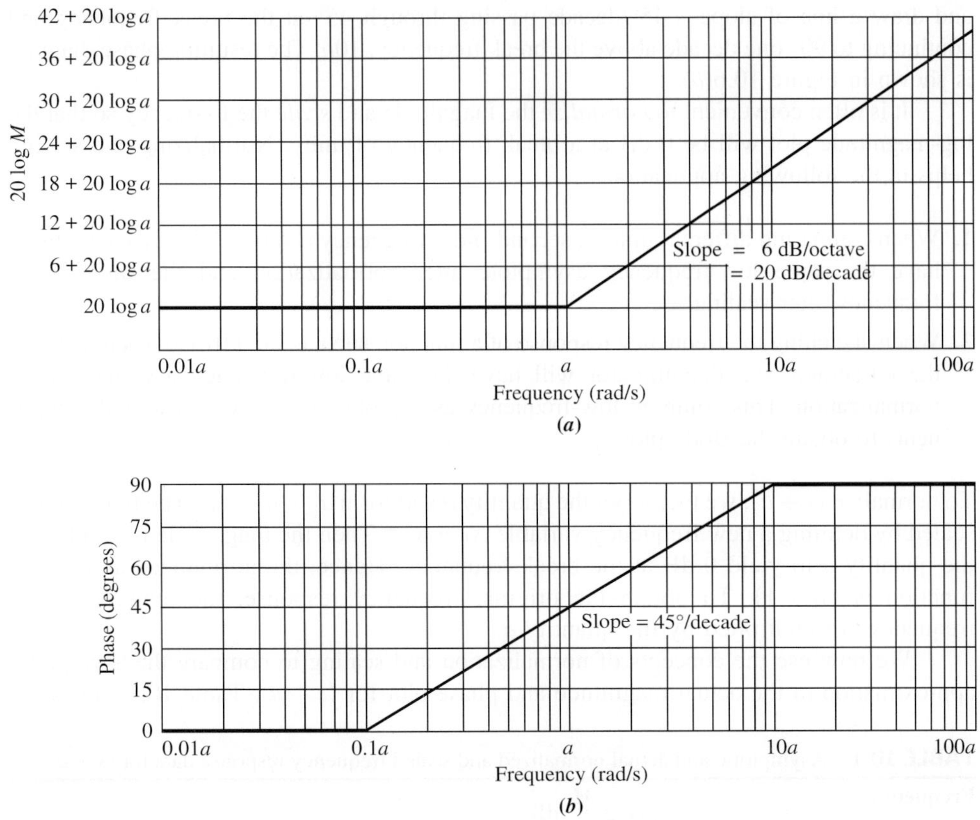


**FIGURE 10.6** Bode plots of $(s + a)$: **a.** magnitude plot; **b.** phase plot

The magnitude response in dB is

$$20 \log M = 20 \log a + 20 \log \frac{\omega}{a} = 20 \log \omega \tag{10.20}$$

where $a < \omega < \infty$. Notice from the middle term that the high-frequency approximation is equal to the low-frequency approximation when $\omega = a$, and increases for $\omega > a$.

If we plot dB, 20 log $M$, against log $\omega$, Eq. (10.20) becomes a straight line:

$$y = 20x \tag{10.21}$$

where $y = 20 \log M$, and $x = \log \omega$. The line has a slope of 20 when plotted as dB vs. log $\omega$.

Since each doubling of frequency causes 20 log $\omega$ to increase by 6 dB, the line rises at an equivalent slope of 6 dB/octave, where an *octave* is a doubling of frequency. This rise begins at $\omega = a$, where the low-frequency approximation equals the high-frequency approximation.

We call the straight-line approximations *asymptotes*. The low-frequency approximation is called the *low-frequency asymptote*, and the high-frequency approximation is called the *high-frequency asymptote*. The frequency, $a$, is called the *break frequency* because it is the break between the low- and the high-frequency asymptotes.

Many times it is convenient to draw the line over a decade rather than an octave, where a *decade* is 10 times the initial frequency. Over one decade, 20 log $\omega$ increases by 20 dB. Thus, a slope of 6 dB/octave is equivalent to a slope of 20 dB/decade. The plot is shown in Figure 10.6(*a*) from $\omega = 0.01a$ to $100a$.

Let us now turn to the phase response, which can be drawn as follows. At the break frequency, $a$, Eq. (10.16) shows the phase to be 45°. At low frequencies, Eq. (10.17) shows that the phase is 0°. At high frequencies, Eq. (10.19) shows that the phase is 90°. To draw the curve, start one decade (1/10) below the break frequency, $0.1a$, with 0° phase, and draw a line of slope $+45^\circ$/decade passing through $45^\circ$ at the break frequency and continuing to $90^\circ$ one decade above the break frequency, $10a$. The resulting phase diagram is shown in Figure 10.6(*b*).

It is often convenient to *normalize* the magnitude and *scale* the frequency so that the log-magnitude plot will be 0 dB at a break frequency of unity. Normalizing and scaling helps in the following applications:

1. When comparing different first- or second-order frequency response plots, each plot will have the same low-frequency asymptote after normalization and the same break frequency after scaling.
2. When sketching the frequency response of a function such as Eq. (10.13), each factor in the numerator and denominator will have the same low-frequency asymptote after normalization. This common low-frequency asymptote makes it easier to add components to obtain the Bode plot.

To normalize $(s + a)$, we factor out the quantity $a$ and form $a[(s/a) + 1]$. The frequency is scaled by defining a new frequency variable, $s_1 = s/a$. Then the magnitude is divided by the quantity $a$ to yield 0 dB at the break frequency. Hence, the normalized and scaled function is $(s_1 + 1)$. To obtain the original frequency response, the magnitude and frequency are multiplied by the quantity $a$.

We now use the concepts of normalization and scaling to compare the asymptotic approximation to the actual magnitude and phase plot for $(s + a)$. Table 10.1 shows the

**TABLE 10.1** Asymptotic and actual normalized and scaled frequency response data for $(s + a)$

| Frequency $\frac{\text{}}{a}$ (rad/s) | $20 \log \frac{M}{a}$ (dB) | | Phase (degrees) | |
|---|---|---|---|---|
| | **Asymptotic** | **Actual** | **Asymptotic** | **Actual** |
| 0.01 | 0 | 0.00 | 0.00 | 0.57 |
| 0.02 | 0 | 0.00 | 0.00 | 1.15 |
| 0.04 | 0 | 0.01 | 0.00 | 2.29 |
| 0.06 | 0 | 0.02 | 0.00 | 3.43 |
| 0.08 | 0 | 0.03 | 0.00 | 4.57 |
| 0.1 | 0 | 0.04 | 0.00 | 5.71 |
| 0.2 | 0 | 0.17 | 13.55 | 11.31 |
| 0.4 | 0 | 0.64 | 27.09 | 21.80 |
| 0.6 | 0 | 1.34 | 35.02 | 30.96 |
| 0.8 | 0 | 2.15 | 40.64 | 38.66 |
| 1 | 0 | 3.01 | 45.00 | 45.00 |
| 2 | 6 | 6.99 | 58.55 | 63.43 |
| 4 | 12 | 12.30 | 72.09 | 75.96 |
| 6 | 15.56 | 15.68 | 80.02 | 80.54 |
| 8 | 18 | 18.13 | 85.64 | 82.87 |
| 10 | 20 | 20.04 | 90.00 | 84.29 |
| 20 | 26.02 | 26.03 | 90.00 | 87.14 |
| 40 | 32.04 | 32.04 | 90.00 | 88.57 |
| 60 | 35.56 | 35.56 | 90.00 | 89.05 |
| 80 | 38.06 | 38.06 | 90.00 | 89.28 |
| 100 | 40 | 40.00 | 90.00 | 89.43 |

comparison for the normalized and scaled frequency response of $(s+a)$. Notice that the actual magnitude curve is never greater than 3.01 dB from the asymptotes. This maximum difference occurs at the break frequency. The maximum difference for the phase curve is 5.71°, which occurs at the decades above and below the break frequency. For convenience, the data in Table 10.1 is plotted in Figures 10.7 and 10.8.

We now find the Bode plots for other common transfer functions.

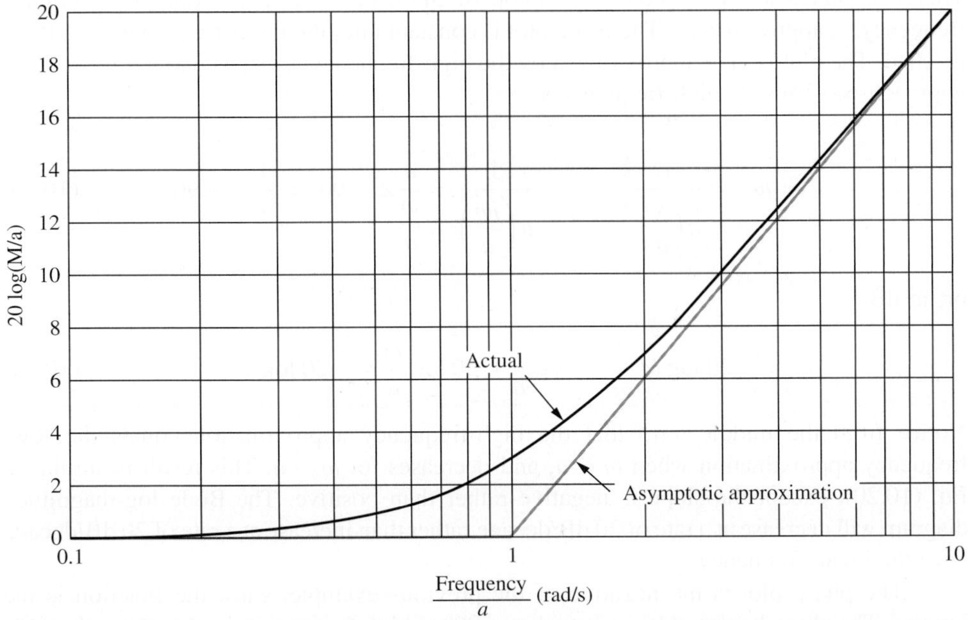


**FIGURE 10.7** Asymptotic and actual normalized and scaled magnitude response of $(s+a)$

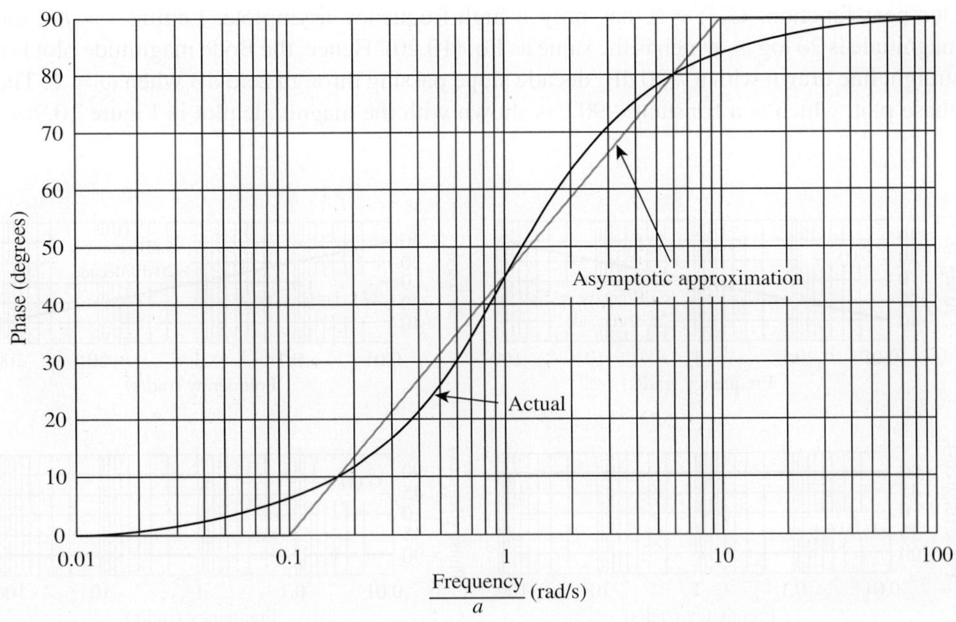


**FIGURE 10.8** Asymptotic and actual normalized and scaled phase response of $(s+a)$

## Bode Plots for $G(s) = 1/(s + a)$

Let us find the Bode plots for the transfer function

$$G(s) = \frac{1}{(s+a)} = \frac{1}{a\left(\frac{s}{a}+1\right)} \tag{10.22}$$

This function has a low-frequency asymptote of 20 log $(1/a)$, which is found by letting the frequency, $s$, approach zero. The Bode plot is constant until the break frequency, $a$ rad/s, is reached. The plot is then approximated by the high-frequency asymptote found by letting $s$ approach $\infty$. Thus, at high frequencies

$$G(j\omega) = \left.\frac{1}{a\left(\frac{s}{a}\right)}\right|_{s\to j\omega} = \frac{1}{a\left(\frac{j\omega}{a}\right)} = \frac{\frac{1}{a}}{\frac{\omega}{a}}\angle -90^\circ = \frac{1}{\omega}\angle -90^\circ \tag{10.23}$$

or, in dB,

$$20 \log M = 20 \log \frac{1}{a} - 20 \log \frac{\omega}{a} = -20 \log \omega \tag{10.24}$$

Notice from the middle term that the high-frequency approximation equals the low-frequency approximation when $\omega = a$, and decreases for $\omega > a$. This result is similar to Eq. (10.20), except the slope is negative rather than positive. The Bode log-magnitude diagram will decrease at a rate of 20 dB/decade rather than increase at a rate of 20 dB/decade after the break frequency.

The phase plot is the negative of the previous example, since the function is the inverse. The phase begins at 0° and reaches −90° at high frequencies, going through −45° at the break frequency. Both the Bode normalized and scaled log-magnitude and phase plot are shown in Figure 10.9(*d*).

## Bode Plots for $G(s) = s$

Our next function, $G(s) = s$, has only a high-frequency asymptote. Letting $s = j\omega$, the magnitude is 20 log $\omega$, which is the same as Eq. (10.20). Hence, the Bode magnitude plot is a straight line drawn with a +20 dB/decade slope passing through zero dB when $\omega = 1$. The phase plot, which is a constant +90°, is shown with the magnitude plot in Figure 10.9(*a*).

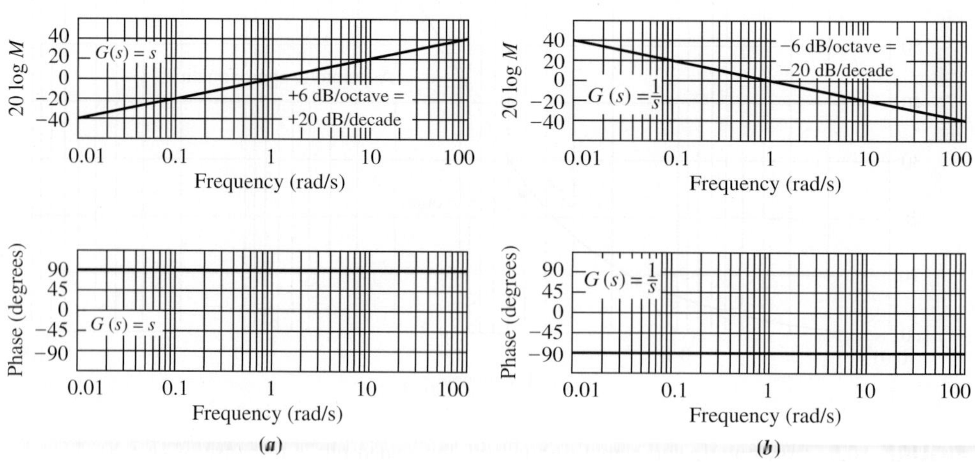


**FIGURE 10.9** Normalized and scaled Bode plots for
**a.** $G(s) = s$;
**b.** $G(s) = 1/s$;
(*figure continues*)

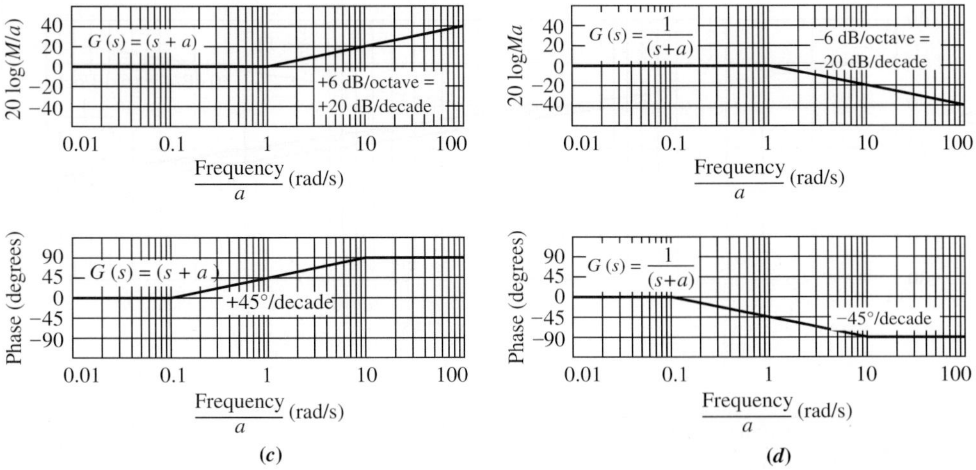


**FIGURE 10.9** (*Continued*)
**c.** $G(s) = (s + a)$;
**d.** $G(s) = 1/(s + a)$

## Bode Plots for $G(s) = 1/s$

The frequency response of the inverse of the preceding function, $G(s) = 1/s$, is shown in Figure 10.9(*b*) and is a straight line with a $-20$ dB/decade slope passing through zero dB at $\omega = 1$. The Bode phase plot is equal to a constant $-90°$.

We have covered four functions that have first-order polynomials in $s$ in the numerator or denominator. Before proceeding to second-order polynomials, let us look at an example of drawing the Bode plots for a function that consists of the product of first-order polynomials in the numerator and denominator. The plots will be made by adding together the individual frequency response curves.

## Example 10.2 Bode Plots for Ratio of First-Order Factors

**PROBLEM:** Draw the Bode plots for the system shown in Figure 10.10, where $G(s) = K(s + 3)/[s(s + 1)(s + 2)]$.

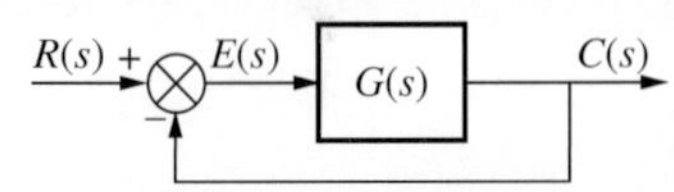


**FIGURE 10.10** Closed-loop unity feedback system

**SOLUTION:** We will make a Bode plot for the open-loop function $G(s) = K(s + 3)/[s(s + 1)(s + 2)]$. The Bode plot is the sum of the Bode plots for each first-order term. Thus, it is convenient to use the normalized plot for each of these terms so that the low-frequency asymptote of each term, except the pole at the origin, is at 0 dB, making it easier to add the components of the Bode plot. We rewrite $G(s)$ showing each term normalized to a low-frequency gain of unity. Hence,

$$G(s) = \frac{\frac{3}{2}K\left(\frac{s}{3} + 1\right)}{s(s + 1)\left(\frac{s}{2} + 1\right)} \tag{10.25}$$

Now determine that the break frequencies are at 1, 2, and 3. The magnitude plot should begin a decade below the lowest break frequency and extend a decade above the highest

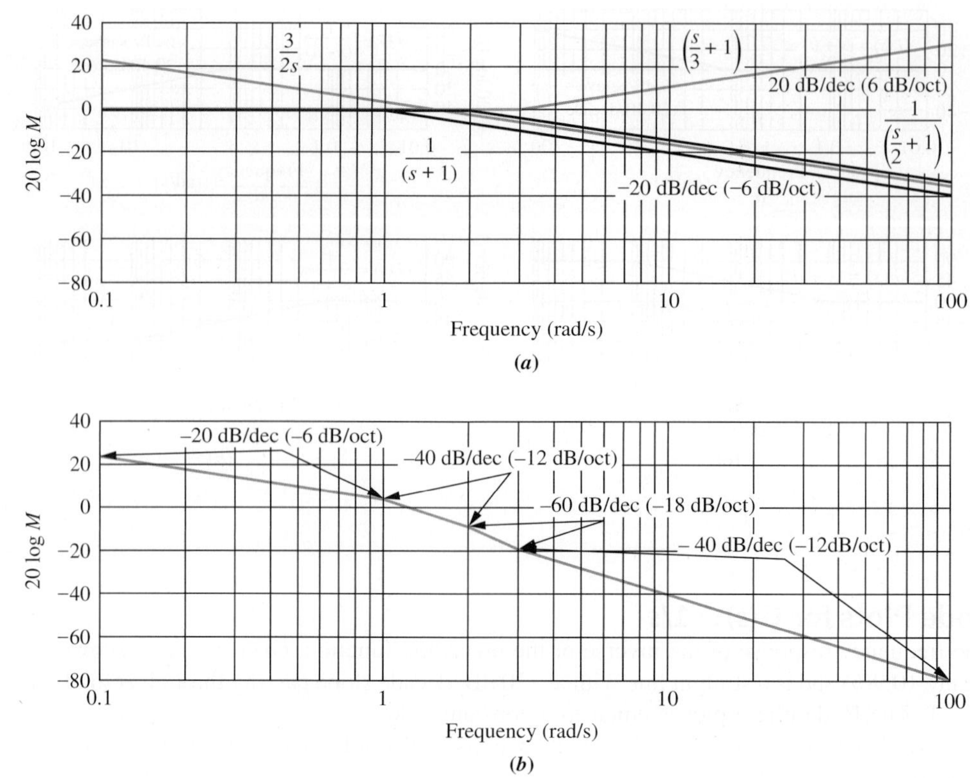


**FIGURE 10.11** Bode log-magnitude plot for Example 10.2:
**a.** components;
**b.** composite

break frequency. Hence, we choose 0.1 radian to 100 radians, or three decades, as the extent of our plot.

At $\omega = 0.1$ the low-frequency value of the function is found from Eq. (10.25) using the low-frequency values for all of the $[(s/a)+1]$ terms (that is, $s = 0$) and the actual value for the $s$ term in the denominator. Thus, $G(j0.1) \approx \frac{3}{2}K/0.1 = 15\,K$. The effect of $K$ is to move the magnitude curve up (increasing $K$) or down (decreasing $K$) by the amount of 20 log $K$. $K$ has no effect upon the phase curve. If we choose $K = 1$, the magnitude plot can be denormalized later for any value of $K$ that is calculated or known.

Figure 10.11($a$) shows each component of the Bode log-magnitude frequency response. Summing the components yields the composite plot shown in Figure 10.11($b$). The results are summarized in Table 10.2, which can be used to obtain the slopes. Each

**TABLE 10.2** Bode magnitude plot: slope contribution from each pole and zero in Example 10.2

| | Frequency (rad/s) | | | |
|---|---|---|---|---|
| **Description** | **0.1 (Start: Pole at 0)** | **1 (Start: Pole at −1)** | **2 (Start: Pole at −2)** | **3 (Start: Zero at −3)** |
| Pole at 0 | −20 | −20 | −20 | −20 |
| Pole at −1 | 0 | −20 | −20 | −20 |
| Pole at −2 | 0 | 0 | −20 | −20 |
| Zero at −3 | 0 | 0 | 0 | 20 |
| Total slope (dB/dec) | −20 | −40 | −60 | −40 |

pole and zero is itemized in the first column. Reading across the table shows its contribution at each frequency. The last row is the sum of the slopes and correlates with Figure 10.11(*b*). The Bode magnitude plot for $K = 1$ starts at $\omega = 0.1$ with a value of $20 \log 15 = 23.52$ dB, and decreases immediately at a rate of $-20$ dB/decade, due to the $s$ term in the denominator. At $\omega = 1$, the $(s + 1)$ term in the denominator begins its 20 dB/decade downward slope and causes an additional 20 dB/decade negative slope, or a total of $-40$ dB/decade. At $\omega = 2$, the term $[(s/2) + 1]$ begins its $-20$ dB/decade slope, adding yet another $-20$ dB/decade to the resultant plot, or a total of $-60$ dB/decade slope that continues until $\omega = 3$. At this frequency, the $[(s/3) + 1]$ term in the numerator begins its positive 20 dB/decade slope. The resultant magnitude plot, therefore, changes from a slope of $-60$ dB/decade to $-40$ dB/decade at $\omega = 3$, and continues at that slope, since there are no other break frequencies.

The slopes are easily drawn by sketching straight-line segments decreasing by 20 dB over a decade. For example, the initial $-20$ dB/decade slope is drawn from 23.52 dB at $\omega = 0.1$, to 3.52 dB (a 20 dB decrease) at $\omega = 1$. The $-40$ dB/decade slope starting at $\omega = 1$ is drawn by sketching a line segment from 3.52 dB at $\omega = 1$, to $-36.48$ dB (a 40 dB decrease) at $\omega = 10$, and using only the portion from $\omega = 1$ to $\omega = 2$. The next slope of $-60$ dB/decade is drawn by first sketching a line segment from $\omega = 2$ to $\omega = 20$ (1 decade) that drops down by 60 dB, and using only that portion of the line from $\omega = 2$ to $\omega = 3$. The final slope is drawn by sketching a line segment from $\omega = 3$ to $\omega = 30$ (1 decade) that drops by 40 dB. This slope continues to the end of the plot.

Phase is handled similarly. However, the existence of breaks, a decade below and a decade above the break frequency, requires a little more bookkeeping. Table 10.3 shows the starting and stopping frequencies of the 45°/decade slope for each of the poles and zeros. For example, reading across for the pole at $-2$, we see that the $-45°$ slope starts at a frequency of 0.2 and ends at 20. Filling in the rows for each pole and then summing the columns yields the slope portrait of the resulting phase plot. Looking at the row marked *Total slope*, we see that the phase plot will have a slope of $-45°$/decade from a frequency of 0.1 to 0.2. The slope will then increase to $-90°$/decade from 0.2 to 0.3. The slope will return to $-45°$/decade from 0.3 to 10 rad/s. A slope of 0 ensues from 10 to 20 rad/s, followed by a slope of $+45°$/decade from 20 to 30 rad/s. Finally, from 30 rad/s to infinity, the slope is 0°/decade.

The resulting component and composite phase plots are shown in Figure 10.12. Since the pole at the origin yields a constant $-90°$ phase shift, the plot begins at $-90°$ and follows the slope portrait just described.

**TABLE 10.3** Bode phase plot: slope contribution from each pole and zero in Example 10.2

| | Frequency (rad/s) | | | | | |
|---|---|---|---|---|---|---|
| **Description** | **0.1 (Start: Pole at −1)** | **0.2 (Start: Pole at −2)** | **0.3 (Start: Pole at −3)** | **0 (End: Pole at −1)** | **20 (End: Pole at −2)** | **30 (End: Zero at −3)** |
| Pole at −1 | −45 | −45 | −45 | 0 | | |
| Pole at −2 | | −45 | −45 | −45 | 0 | |
| Zero at −3 | | | 45 | 45 | 45 | 0 |
| Total slope (deg/dec) | −45 | −90 | −45 | 0 | 45 | 0 |

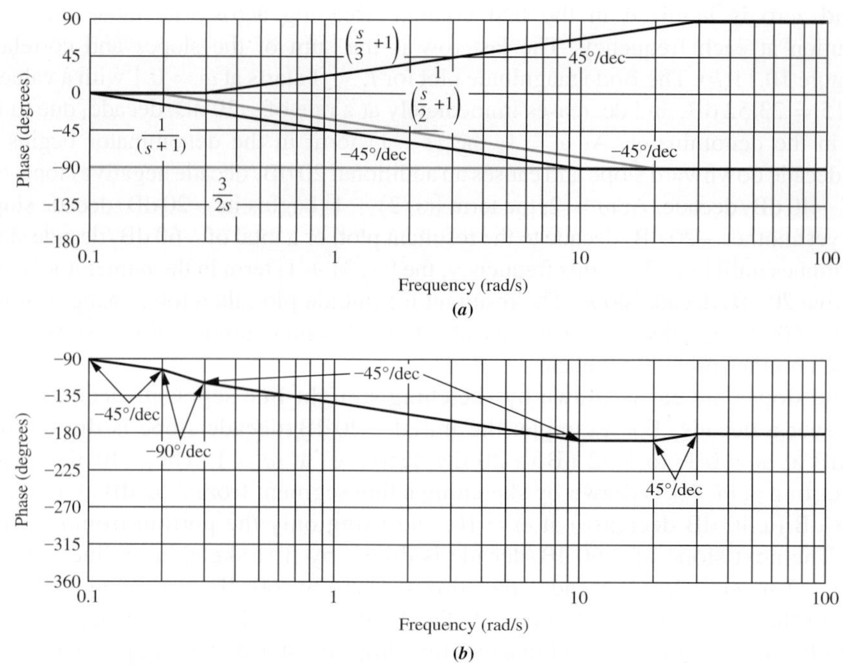


**FIGURE 10.12** Bode phase plot for Example 10.2: **a.** components; **b.** composite

## Bode Plots for $G(s) = s^2 + 2\zeta\omega_n s + \omega_n^2$

Now that we have covered Bode plots for first-order systems, we turn to the Bode log-magnitude and phase plots for second-order polynomials in $s$. The second-order polynomial is of the form

$$G(s) = s^2 + 2\zeta\omega_n s + \omega_n^2 = \omega_n^2\left(\frac{s^2}{\omega_n^2} + 2\zeta\frac{s}{\omega_n} + 1\right) \tag{10.26}$$

Unlike the first-order frequency response approximation, the difference between the asymptotic approximation and the actual frequency response can be great for some values of $\zeta$. A correction to the Bode diagrams can be made to improve the accuracy. We first derive the asymptotic approximation and then show the difference between the asymptotic approximation and the actual frequency response curves.

At low frequencies, Eq. (10.26) becomes

$$G(s) \approx \omega_n^2 = \omega_n^2\angle 0^\circ \tag{10.27}$$

The magnitude, $M$, in dB at low frequencies therefore is

$$20\log M = 20\log|G(j\omega)| = 20\log\omega_n^2 \tag{10.28}$$

At high frequencies,

$$G(s) \approx s^2 \tag{10.29}$$

or

$$G(j\omega) \approx -\omega^2 = \omega^2\angle 180^\circ \tag{10.30}$$

The log-magnitude is

$$20\log M = 20\log|G(j\omega)| = 20\log\omega^2 = 40\log\omega \tag{10.31}$$

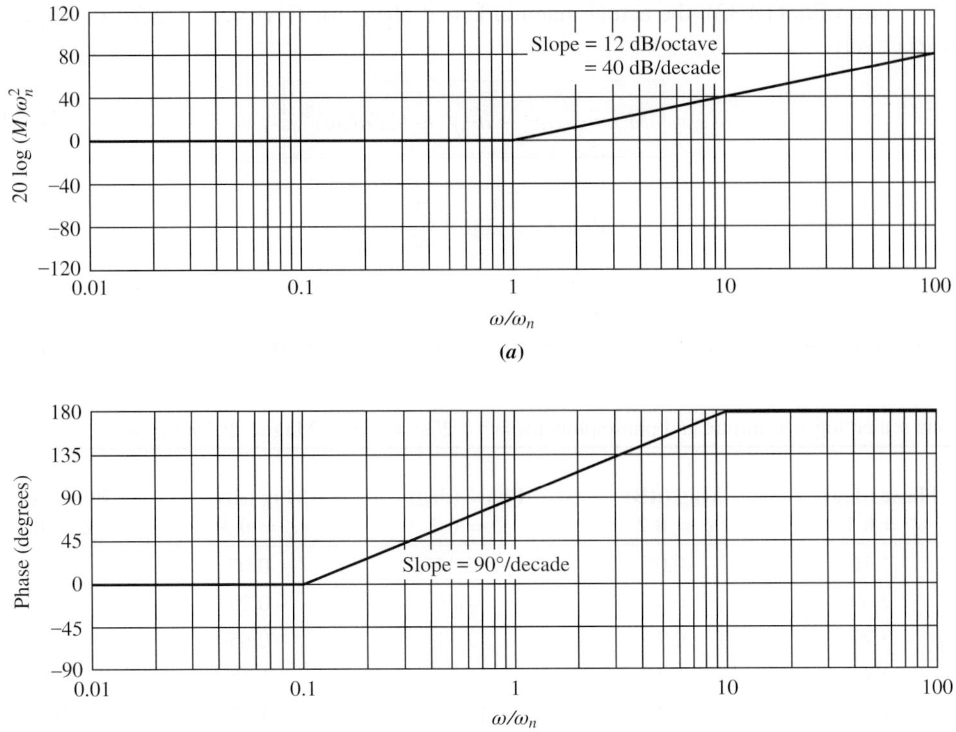


**FIGURE 10.13** Bode asymptotes for normalized and scaled $G(s) = s^2 + 2\zeta\omega_n s + \omega_n^2$: **a.** magnitude; **b.** phase

Equation (10.31) is a straight line with twice the slope of a first-order term (Eq. (10.20)). Its slope is 12 dB/octave, or 40 dB/decade.

The low-frequency asymptote (Eq. (10.27)) and the high-frequency asymptote (Eq. (10.31)) are equal when $\omega = \omega_n$. Thus, $\omega_n$ is the break frequency for the second-order polynomial.

For convenience in representing systems with different $\omega_n$, we normalize and scale our findings before drawing the asymptotes. Using the normalized and scaled term of Eq. (10.26), we normalize the magnitude, dividing by $\omega_n^2$, and scale the frequency, dividing by $\omega_n$. Thus, we plot $G(s_1)/\omega_n^2 = s_1^2 + 2\zeta s_1 + 1$, where $s_1 = s/\omega_n$. $G(s_1)$ has a low-frequency asymptote at 0 dB and a break frequency of 1 rad/s. Figure 10.13($a$) shows the asymptotes for the normalized and scaled magnitude plot.

We now draw the phase plot. It is 0° at low frequencies (Eq. (10.27)) and 180° at high frequencies (Eq. (10.30)). To find the phase at the natural frequency, first evaluate $G(j\omega)$:

$$G(j\omega) = s^2 + 2\zeta\omega_n s + \omega_n^2|_{s\rightarrow j\omega} = (\omega_n^2 - \omega^2) + j2\zeta\omega_n\omega \tag{10.32}$$

Then find the function value at the natural frequency by substituting $\omega = \omega_n$. Since the result is $j2\zeta\omega_n^2$, the phase at the natural frequency is $+90°$. Figure 10.13($b$) shows the phase plotted with frequency scaled by $\omega_n$. The phase plot increases at a rate of 90°/decade from 0.1 to 10 and passes through 90° at 1.

## Corrections to Second-Order Bode Plots

Let us now examine the error between the actual response and the asymptotic approximation of the second-order polynomial. Whereas the first-order polynomial has a disparity of no more than 3.01 dB magnitude and 5.71° phase, the second-order function may have a greater disparity, which depends upon the value of $\zeta$.

From Eq. (10.32), the actual magnitude and phase for $G(s) = s^2 + 2\zeta\omega_n s + \omega_n^2$ are, respectively,

$$M = \sqrt{(\omega_n^2 - \omega^2)^2 + (2\zeta\omega_n\omega)^2} \tag{10.33}$$

$$\text{Phase} = \tan^{-1}\frac{2\zeta\omega_n\omega}{\omega_n^2 - \omega^2} \tag{10.34}$$

These relationships are tabulated in Table 10.4 for a range of values of $\zeta$ and plotted in Figures 10.14 and 10.15 along with the asymptotic approximations for normalized

**TABLE 10.4** Data for normalized and scaled log-magnitude and phase plots for $(s^2 + 2\zeta\omega_n s + \omega_n^2)$. Mag $= 20\log(M/\omega_n^2)$

| Freq. $\frac{\omega}{\omega_n}$ | Mag (dB) $\zeta = 0.1$ | Phase (deg) $\zeta = 0.1$ | Mag (dB) $\zeta = 0.2$ | Phase (deg) $\zeta = 0.2$ | Mag (dB) $\zeta = 0.3$ | Phase (deg) $\zeta = 0.3$ |
|---|---|---|---|---|---|---|
| 0.10 | −0.09 | 1.16 | −0.08 | 2.31 | −0.07 | 3.47 |
| 0.20 | −0.35 | 2.39 | −0.32 | 4.76 | −0.29 | 7.13 |
| 0.30 | −0.80 | 3.77 | −0.74 | 7.51 | −0.65 | 11.19 |
| 0.40 | −1.48 | 5.44 | −1.36 | 10.78 | −1.17 | 15.95 |
| 0.50 | −2.42 | 7.59 | −2.20 | 14.93 | −1.85 | 21.80 |
| 0.60 | −3.73 | 10.62 | −3.30 | 20.56 | −2.68 | 29.36 |
| 0.70 | −5.53 | 15.35 | −4.70 | 28.77 | −3.60 | 39.47 |
| 0.80 | −8.09 | 23.96 | −6.35 | 41.63 | −4.44 | 53.13 |
| 0.90 | −11.64 | 43.45 | −7.81 | 62.18 | −4.85 | 70.62 |
| 1.00 | −13.98 | 90.00 | −7.96 | 90.00 | −4.44 | 90.00 |
| 1.10 | −10.34 | 133.67 | −6.24 | 115.51 | −3.19 | 107.65 |
| 1.20 | −6.00 | 151.39 | −3.73 | 132.51 | −1.48 | 121.43 |
| 1.30 | −2.65 | 159.35 | −1.27 | 143.00 | 0.35 | 131.50 |
| 1.40 | 0.00 | 163.74 | 0.92 | 149.74 | 2.11 | 138.81 |
| 1.50 | 2.18 | 166.50 | 2.84 | 154.36 | 3.75 | 144.25 |
| 1.60 | 4.04 | 168.41 | 4.54 | 157.69 | 5.26 | 148.39 |
| 1.70 | 5.67 | 169.80 | 6.06 | 160.21 | 6.64 | 151.65 |
| 1.80 | 7.12 | 170.87 | 7.43 | 162.18 | 7.91 | 154.26 |
| 1.90 | 8.42 | 171.72 | 8.69 | 163.77 | 9.09 | 156.41 |
| 2.00 | 9.62 | 172.41 | 9.84 | 165.07 | 10.19 | 158.20 |
| 3.00 | 18.09 | 175.71 | 18.16 | 171.47 | 18.28 | 167.32 |
| 4.00 | 23.53 | 176.95 | 23.57 | 173.91 | 23.63 | 170.91 |
| 5.00 | 27.61 | 177.61 | 27.63 | 175.24 | 27.67 | 172.87 |
| 6.00 | 30.89 | 178.04 | 30.90 | 176.08 | 30.93 | 174.13 |
| 7.00 | 33.63 | 178.33 | 33.64 | 176.66 | 33.66 | 175.00 |
| 8.00 | 35.99 | 178.55 | 36.00 | 177.09 | 36.01 | 175.64 |
| 9.00 | 38.06 | 178.71 | 38.07 | 177.42 | 38.08 | 176.14 |
| 10.00 | 39.91 | 178.84 | 39.92 | 177.69 | 39.93 | 176.53 |

(*table continues*)

**TABLE 10.4** Data for normalized and scaled log-magnitude and phase plots for $(s^2 + 2\zeta\omega_n s + \omega_n^2)$. Mag $= 20\log(M/\omega_n^2)$ *(Continued)*

| Freq. $\frac{\omega}{\omega_n}$ | Mag (dB) $\zeta = 0.5$ | Phase (deg) $\zeta = 0.5$ | Mag (dB) $\zeta = 0.7$ | Phase (deg) $\zeta = 0.7$ | Mag (dB) $\zeta = 1.0$ | Phase (deg) $\zeta = 1.0$ |
|---|---|---|---|---|---|---|
| 0.10 | −0.04 | 5.77 | 0.00 | 8.05 | 0.09 | 11.42 |
| 0.20 | −0.17 | 11.77 | 0.00 | 16.26 | 0.34 | 22.62 |
| 0.30 | −0.37 | 18.25 | 0.02 | 24.78 | 0.75 | 33.40 |
| 0.40 | −0.63 | 25.46 | 0.08 | 33.69 | 1.29 | 43.60 |
| 0.50 | −0.90 | 33.69 | 0.22 | 43.03 | 1.94 | 53.13 |
| 0.60 | −1.14 | 43.15 | 0.47 | 52.70 | 2.67 | 61.93 |
| 0.70 | −1.25 | 53.92 | 0.87 | 62.51 | 3.46 | 69.98 |
| 0.80 | −1.14 | 65.77 | 1.41 | 72.18 | 4.30 | 77.32 |
| 0.90 | −0.73 | 78.08 | 2.11 | 81.42 | 5.15 | 83.97 |
| 1.00 | 0.00 | 90.00 | 2.92 | 90.00 | 6.02 | 90.00 |
| 1.10 | 0.98 | 100.81 | 3.83 | 97.77 | 6.89 | 95.45 |
| 1.20 | 2.13 | 110.14 | 4.79 | 104.68 | 7.75 | 100.39 |
| 1.30 | 3.36 | 117.96 | 5.78 | 110.76 | 8.60 | 104.86 |
| 1.40 | 4.60 | 124.44 | 6.78 | 116.10 | 9.43 | 108.92 |
| 1.50 | 5.81 | 129.81 | 7.76 | 120.76 | 10.24 | 112.62 |
| 1.60 | 6.98 | 134.27 | 8.72 | 124.85 | 11.03 | 115.99 |
| 1.70 | 8.10 | 138.03 | 9.66 | 128.45 | 11.80 | 119.07 |
| 1.80 | 9.17 | 141.22 | 10.56 | 131.63 | 12.55 | 121.89 |
| 1.90 | 10.18 | 143.95 | 11.43 | 134.46 | 13.27 | 124.48 |
| 2.00 | 11.14 | 146.31 | 12.26 | 136.97 | 13.98 | 126.87 |
| 3.00 | 18.63 | 159.44 | 19.12 | 152.30 | 20.00 | 143.13 |
| 4.00 | 23.82 | 165.07 | 24.09 | 159.53 | 24.61 | 151.93 |
| 5.00 | 27.79 | 168.23 | 27.96 | 163.74 | 28.30 | 157.38 |
| 6.00 | 31.01 | 170.27 | 31.12 | 166.50 | 31.36 | 161.08 |
| 7.00 | 33.72 | 171.70 | 33.80 | 168.46 | 33.98 | 163.74 |
| 8.00 | 36.06 | 172.76 | 36.12 | 169.92 | 36.26 | 165.75 |
| 9.00 | 38.12 | 173.58 | 38.17 | 171.05 | 38.28 | 167.32 |
| 10.00 | 39.96 | 174.23 | 40.00 | 171.95 | 40.09 | 168.58 |

magnitude and scaled frequency. In Figure 10.14, which is normalized to the square of the natural frequency, the normalized log-magnitude at the scaled natural frequency is $+20\log 2\zeta$. The student should verify that the actual magnitude at the unscaled natural frequency is $+20\log 2\zeta\omega_n^2$. Table 10.4 and Figures 10.14 and 10.15 can be used to improve accuracy when drawing Bode plots. For example, a magnitude correction of $+20\log 2\zeta$ can be made at the natural, or break, frequency on the Bode asymptotic plot.

## Bode Plots for $G(s) = 1/(s^2 + 2\zeta\omega_n s + \omega_n^2)$

Bode plots for $G(s) = 1/(s^2 + 2\zeta\omega_n s + \omega_n^2)$ can be derived similarly to those for $G(s) = s^2 + 2\zeta\omega_n s + \omega_n^2$. We find that the magnitude curve breaks at the natural frequency

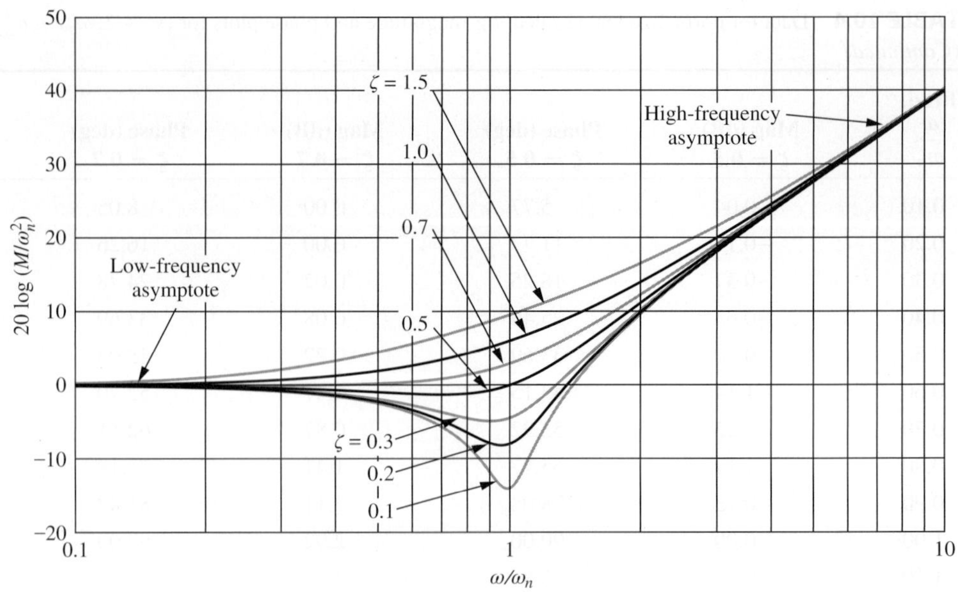


**FIGURE 10.14** Normalized and scaled log-magnitude response for $(s^2 + 2\zeta\omega_n s + \omega_n^2)$

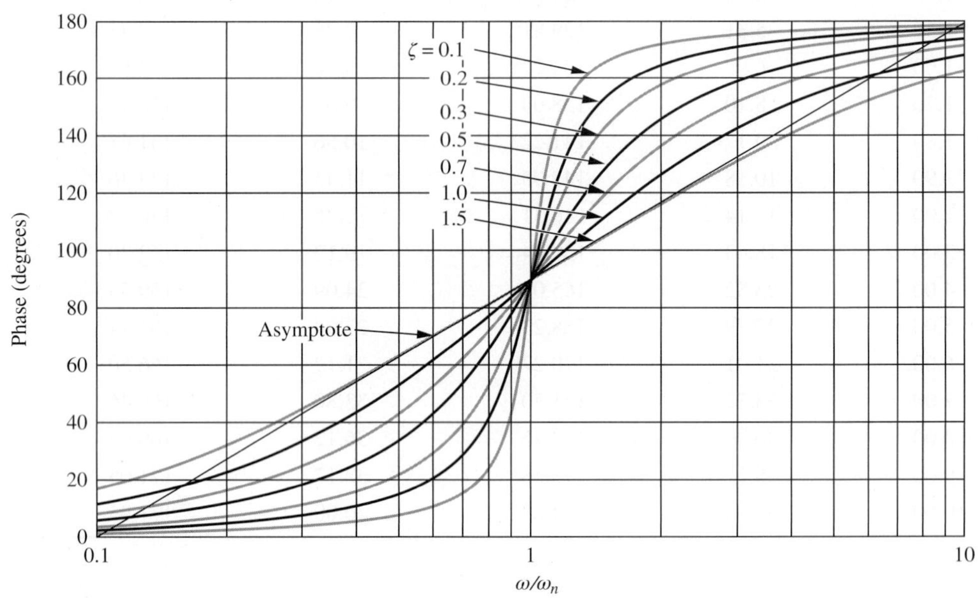


**FIGURE 10.15** Scaled phase response for $(s^2 + 2\zeta\omega_n s + \omega_n^2)$

and decreases at a rate of $-40$ dB/decade. The phase plot is 0° at low frequencies. At $0.1\omega_n$ it begins a decrease of $-90°$/decade and continues until $\omega = 10\omega_n$, where it levels off at $-180°$.

The exact frequency response also follows the same derivation as that of $G(s) = s^2 + 2\zeta\omega_n s + \omega_n^2$. The results are summarized in Table 10.5, as well as Figures 10.16 and 10.17. The exact magnitude is the reciprocal of Eq. (10.33), and the exact phase is the negative of Eq. (10.34). The normalized magnitude at the scaled natural frequency is $-20 \log 2\zeta$, which can be used as a correction at the break frequency on the Bode asymptotic plot.

**TABLE 10.5** Data for normalized and scaled log-magnitude and phase plots for $1/(s^2 + 2\zeta\omega_n s + \omega_n^2)$. Mag $= 20\log(M/\omega_n^2)$

| Freq. $\frac{\omega}{\omega_n}$ | Mag (dB) $\zeta = 0.1$ | Phase (deg) $\zeta = 0.1$ | Mag (dB) $\zeta = 0.2$ | Phase (deg) $\zeta = 0.2$ | Mag (dB) $\zeta = 0.3$ | Phase (deg) $\zeta = 0.3$ |
|---|---|---|---|---|---|---|
| 0.10 | 0.09 | −1.16 | 0.08 | −2.31 | 0.07 | −3.47 |
| 0.20 | 0.35 | −2.39 | 0.32 | −4.76 | 0.29 | −7.13 |
| 0.30 | 0.80 | −3.77 | 0.74 | −7.51 | 0.65 | −11.19 |
| 0.40 | 1.48 | −5.44 | 1.36 | −10.78 | 1.17 | −15.95 |
| 0.50 | 2.42 | −7.59 | 2.20 | −14.93 | 1.85 | −21.80 |
| 0.60 | 3.73 | −10.62 | 3.30 | −20.56 | 2.68 | −29.36 |
| 0.70 | 5.53 | −15.35 | 4.70 | −28.77 | 3.60 | −39.47 |
| 0.80 | 8.09 | −23.96 | 6.35 | −41.63 | 4.44 | −53.13 |
| 0.90 | 11.64 | −43.45 | 7.81 | −62.18 | 4.85 | −70.62 |
| 1.00 | 13.98 | −90.00 | 7.96 | −90.00 | 4.44 | −90.00 |
| 1.10 | 10.34 | −133.67 | 6.24 | −115.51 | 3.19 | −107.65 |
| 1.20 | 6.00 | −151.39 | 3.73 | −132.51 | 1.48 | −121.43 |
| 1.30 | 2.65 | −159.35 | 1.27 | −143.00 | −0.35 | −131.50 |
| 1.40 | 0.00 | −163.74 | −0.92 | −149.74 | −2.11 | −138.81 |
| 1.50 | −2.18 | −166.50 | −2.84 | −154.36 | −3.75 | −144.25 |
| 1.60 | −4.04 | −168.41 | −4.54 | −157.69 | −5.26 | −148.39 |
| 1.70 | −5.67 | −169.80 | −6.06 | −160.21 | −6.64 | −151.65 |
| 1.80 | −7.12 | −170.87 | −7.43 | −162.18 | −7.91 | −154.26 |
| 1.90 | −8.42 | −171.72 | −8.69 | −163.77 | −9.09 | −156.41 |
| 2.00 | −9.62 | −172.41 | −9.84 | −165.07 | −10.19 | −158.20 |
| 3.00 | −18.09 | −175.71 | −18.16 | −171.47 | −18.28 | −167.32 |
| 4.00 | −23.53 | −176.95 | −23.57 | −173.91 | −23.63 | −170.91 |
| 5.00 | −27.61 | −177.61 | −27.63 | −175.24 | −27.67 | −172.87 |
| 6.00 | −30.89 | −178.04 | −30.90 | −176.08 | −30.93 | −174.13 |
| 7.00 | −33.63 | −178.33 | −33.64 | −176.66 | −33.66 | −175.00 |
| 8.00 | −35.99 | −178.55 | −36.00 | −177.09 | −36.01 | −175.64 |
| 9.00 | −38.06 | −178.71 | −38.07 | −177.42 | −38.08 | −176.14 |
| 10.00 | −39.91 | −178.84 | −39.92 | −177.69 | −39.93 | −176.53 |

*(table continues)*

**TABLE 10.5** Data for normalized and scaled log-magnitude and phase plots for $1/(s^2 + 2\zeta\omega_n s + \omega_n^2)$. Mag $= 20\log(M/\omega_n^2)$ (*Continued*)

| Freq. $\frac{\omega}{\omega_n}$ | Mag (dB) $\zeta = 0.5$ | Phase (deg) $\zeta = 0.5$ | Mag (dB) $\zeta = 0.7$ | Phase (deg) $\zeta = 0.7$ | Mag (dB) $\zeta = 1.0$ | Phase (deg) $\zeta = 1.0$ |
|---|---|---|---|---|---|---|
| 0.10 | 0.04 | −5.77 | 0.00 | −8.05 | −0.09 | −11.42 |
| 0.20 | 0.17 | −11.77 | 0.00 | −16.26 | −0.34 | −22.62 |
| 0.30 | 0.37 | −18.25 | −0.02 | −24.78 | −0.75 | −33.40 |
| 0.40 | 0.63 | −25.46 | −0.08 | −33.69 | −1.29 | −43.60 |
| 0.50 | 0.90 | −33.69 | −0.22 | −43.03 | −1.94 | −53.13 |
| 0.60 | 1.14 | −43.15 | −0.47 | −52.70 | −2.67 | −61.93 |
| 0.70 | 1.25 | −53.92 | −0.87 | −62.51 | −3.46 | −69.98 |
| 0.80 | 1.14 | −65.77 | −1.41 | −72.18 | −4.30 | −77.32 |
| 0.90 | 0.73 | −78.08 | −2.11 | −81.42 | −5.15 | −83.97 |
| 1.00 | 0.00 | −90.00 | −2.92 | −90.00 | −6.02 | −90.00 |
| 1.10 | −0.98 | −100.81 | −3.93 | −97.77 | −6.89 | −95.45 |
| 1.20 | −2.13 | −110.14 | −4.79 | −104.68 | −7.75 | −100.39 |
| 1.30 | −3.36 | −117.96 | −5.78 | −110.76 | −8.60 | −104.86 |
| 1.40 | −4.60 | −124.44 | −6.78 | −116.10 | −9.43 | −108.92 |
| 1.50 | −5.81 | −129.81 | −7.76 | −120.76 | −10.24 | −112.62 |
| 1.60 | −6.98 | −134.27 | −8.72 | −124.85 | −11.03 | −115.99 |
| 1.70 | −8.10 | −138.03 | −9.66 | −128.45 | −11.80 | −119.07 |
| 1.80 | −9.17 | −141.22 | −10.56 | −131.63 | −12.55 | −121.89 |
| 1.90 | −10.18 | −143.95 | −11.43 | −134.46 | −13.27 | −124.48 |
| 2.00 | −11.14 | −146.31 | −12.26 | −136.97 | −13.98 | −126.87 |
| 3.00 | −18.63 | −159.44 | −19.12 | −152.30 | −20.00 | −143.13 |
| 4.00 | −23.82 | −165.07 | −24.09 | −159.53 | −24.61 | −151.93 |
| 5.00 | −27.79 | −168.23 | −27.96 | −163.74 | −28.30 | −157.38 |
| 6.00 | −31.01 | −170.27 | −31.12 | −166.50 | −31.36 | −161.08 |
| 7.00 | −33.72 | −171.70 | −33.80 | −168.46 | −33.98 | −163.74 |
| 8.00 | −36.06 | −172.76 | −36.12 | −169.92 | −36.26 | −165.75 |
| 9.00 | −38.12 | −173.58 | −38.17 | −171.05 | −38.28 | −167.32 |
| 10.00 | −39.96 | −174.23 | −40.00 | −171.95 | −40.09 | −168.58 |

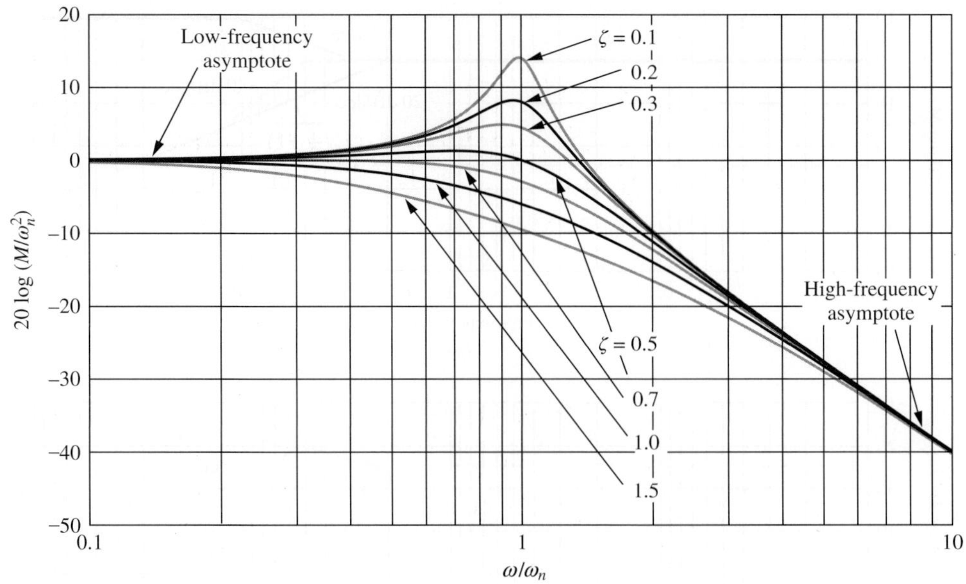


**FIGURE 10.16** Normalized and scaled log-magnitude response for $1/(s^2 + 2\zeta\omega_n s + \omega_n^2)$

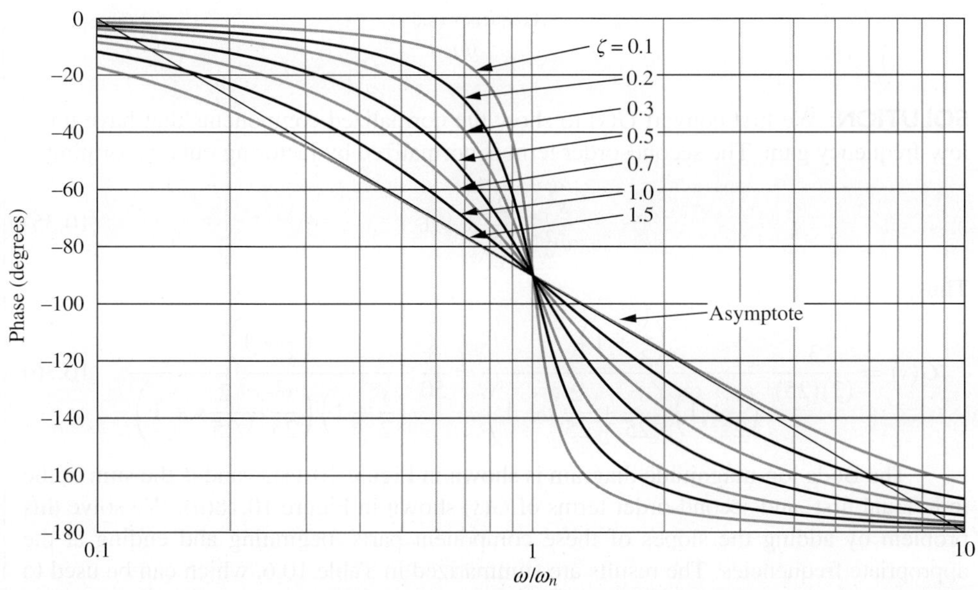


**FIGURE 10.17** Scaled phase response for $1/(s^2 + 2\zeta\omega_n s + \omega_n^2)$

Let us now look at an example of drawing Bode plots for transfer functions that contain second-order factors.

## Example 10.3 Bode Plots for Ratio of First- and Second-Order Factors

**PROBLEM:** Draw the Bode log-magnitude and phase plots of $G(s)$ for the unity feedback system shown in Figure 10.10, where $G(s) = (s + 3)/[(s + 2)(s^2 + 2s + 25)]$.

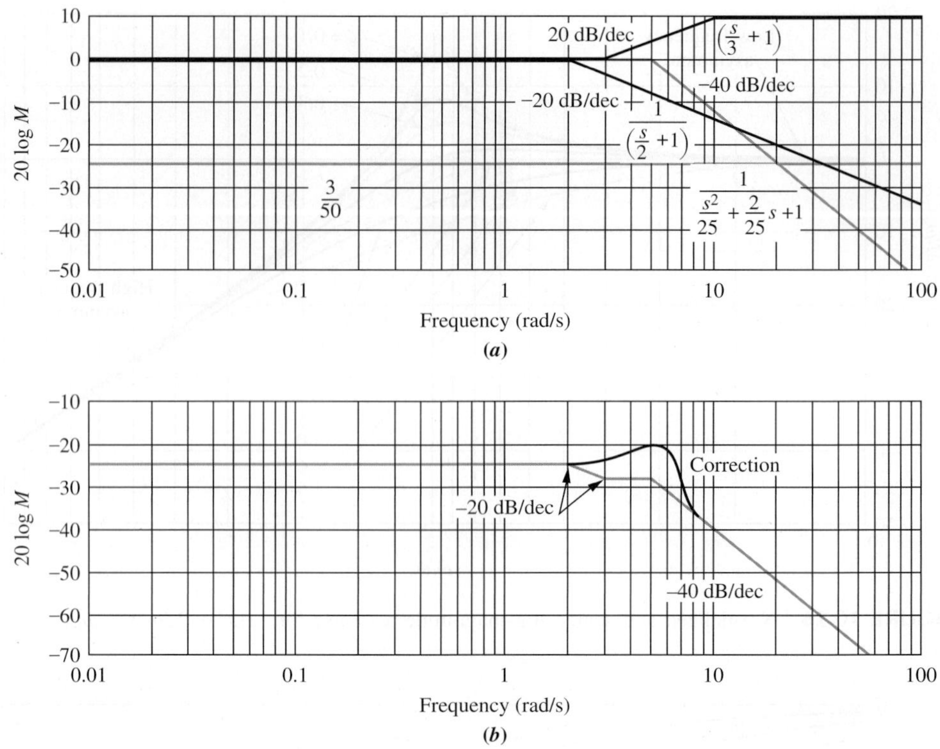


**FIGURE 10.18** Bode magnitude plot for $G(s) = (s+3)/[(s+2)(s^2+2s+25)]$: **a.** components; **b.** composite

**SOLUTION:** We first convert $G(s)$ to show the normalized components that have unity low-frequency gain. The second-order term is normalized by factoring out $\omega_n^2$, forming

$$\frac{s^2}{\omega_n^2} + \frac{2\zeta}{\omega_n}s + 1 \tag{10.35}$$

Thus,

$$G(s) = \frac{3}{(2)(25)} \frac{\left(\frac{s}{3}+1\right)}{\left(\frac{s}{2}+1\right)\left(\frac{s^2}{25}+\frac{2}{25}s+1\right)} = \frac{3}{50} \frac{\left(\frac{s}{3}+1\right)}{\left(\frac{s}{2}+1\right)\left(\frac{s^2}{25}+\frac{2}{25}s+1\right)} \tag{10.36}$$

The Bode log-magnitude diagram is shown in Figure 10.18(*b*) and is the sum of the individual first- and second-order terms of $G(s)$ shown in Figure 10.18(*a*). We solve this problem by adding the slopes of these component parts, beginning and ending at the appropriate frequencies. The results are summarized in Table 10.6, which can be used to

**TABLE 10.6** Magnitude diagram slopes for Example 10.3

| | Frequency (rad/s) | | | |
|---|---|---|---|---|
| **Description** | **0.01 (Start: Plot)** | **2 (Start: Pole at −2)** | **3 (Start: Zero at −3)** | **5 (Start: $\omega_n = 5$)** |
| Pole at −2 | 0 | −20 | −20 | −20 |
| Zero at −3 | 0 | 0 | 20 | 20 |
| $\omega_n = 5$ | 0 | 0 | 0 | −40 |
| Total slope (dB/dec) | 0 | −20 | 0 | −40 |

obtain the slopes. The low-frequency value for $G(s)$, found by letting $s = 0$, is 3/50, or $-24.44$ dB. The Bode magnitude plot starts out at this value and continues until the first break frequency at 2 rad/s. Here the pole at $-2$ yields a $-20$ dB/decade slope downward until the next break at 3 rad/s. The zero at $-3$ causes an upward slope of $+20$ dB/decade, which, when added to the previous $-20$ dB/decade curve, gives a net slope of 0. At a frequency of 5 rad/s, the second-order term initiates a $-40$ dB/decade downward slope, which continues to infinity.

The correction to the log-magnitude curve due to the underdamped second-order term can be found by plotting a point $-20 \log 2\zeta$ above the asymptotes at the natural frequency. Since $\zeta = 0.2$ for the second-order term in the denominator of $G(s)$, the correction is 7.96 dB. Points close to the natural frequency can be corrected by taking the values from the curves of Figure 10.16.

**TABLE 10.7** Phase diagram slopes for Example 10.3

| | **Frequency (rad/s)** | | | | | |
|---|---|---|---|---|---|---|
| **Description** | **0.2 (Start: Pole at $-2$)** | **0.3 (Start: Zero at $-3$)** | **0.5 (Start: $\omega_n$ at $-5$)** | **20 (End: Pole at $-2$)** | **30 (End: Zero at $-3$)** | **50 (End: $\omega_n = 5$)** |
| Pole at $-2$ | $-45$ | $-45$ | $-45$ | 0 | | |
| Zero at $-3$ | | 45 | 45 | 45 | 0 | |
| $\omega_n = 5$ | | | $-90$ | $-90$ | $-90$ | 0 |
| Total slope (dB/dec) | $-45$ | 0 | $-90$ | $-45$ | $-90$ | 0 |

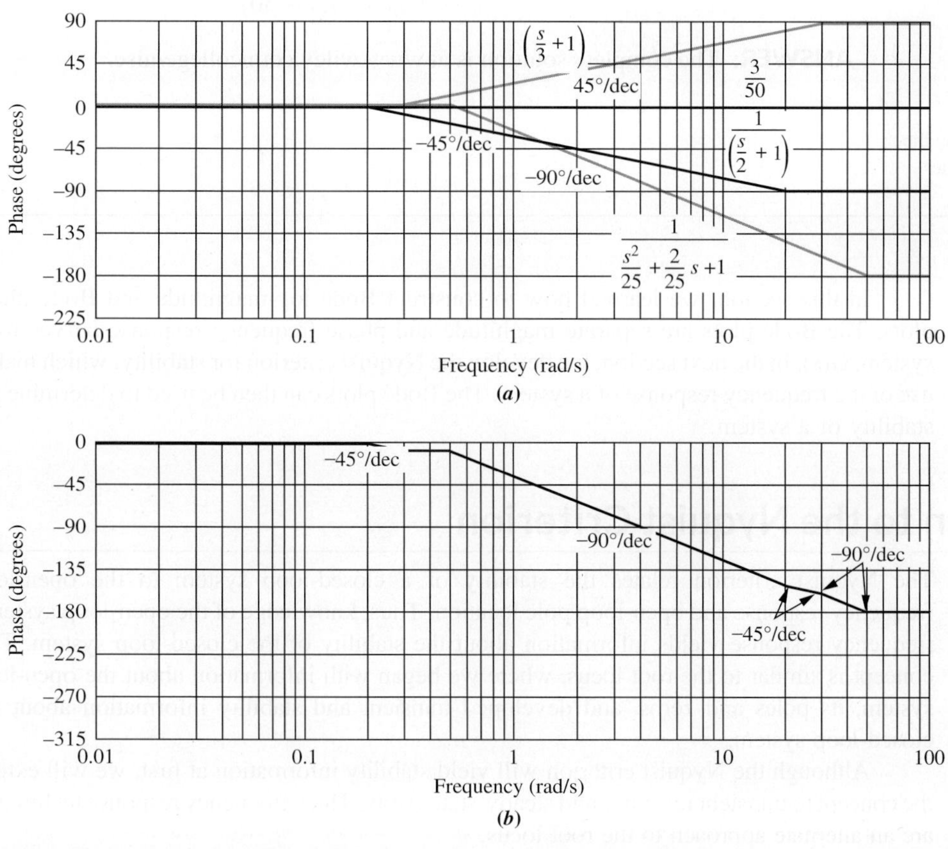


**FIGURE 10.19** Bode phase plot for $G(s) = (s+3)/[(s+2)(s^2+2s+25)]$:
**a.** components;
**b.** composite

We now turn to the phase plot. Table 10.7 is formed to determine the progression of slopes on the phase diagram. The first-order pole at $-2$ yields a phase angle that starts at $0^\circ$ and ends at $-90^\circ$ via a $-45^\circ$/decade slope starting a decade below its break frequency and ending a decade above its break frequency. The first-order zero yields a phase angle that starts at $0^\circ$ and ends at $+90^\circ$ via a $+45^\circ$/decade slope starting a decade below its break frequency and ending a decade above its break frequency. The second-order poles yield a phase angle that starts at $0^\circ$ and ends at $-180^\circ$ via a $-90^\circ$/decade slope starting a decade below their natural frequency ($\omega_n = 5$) and ending a decade above their natural frequency. The slopes, shown in Figure 10.19(*a*), are summed over each frequency range, and the final Bode phase plot is shown in Figure 10.19(*b*).

MATLAB
**ML**

Students who are using MATLAB should now run ch10p1 in Appendix B. You will learn how to use MATLAB to make Bode plots and list the points on the plots. This exercise solves Example 10.3 using MATLAB.

## Skill-Assessment Exercise 10.2

**TryIt 10.1**

Use MATLAB, the Control System Toolbox, and the following statements to obtain the Bode plots for the system of Skill-Assessment Exercise 10.2

```
G=zpk([-20],[-1,-7,...
  -50], 1)
bode (G); grid on
```

After the Bode plots appear, click on the curve and drag to read the coordinates.

**PROBLEM:** Draw the Bode log-magnitude and phase plots for the system shown in Figure 10.10, where

$$G(s) = \frac{(s+20)}{(s+1)(s+7)(s+50)}$$

**ANSWER:** The complete solution is at www.wiley.com/college/nise.

In this section, we learned how to construct Bode log-magnitude and Bode phase plots. The Bode plots are separate magnitude and phase frequency response curves for a system, $G(s)$. In the next section, we develop the Nyquist criterion for stability, which makes use of the frequency response of a system. The Bode plots can then be used to determine the stability of a system.

# 10.3 Introduction to the Nyquist Criterion

The Nyquist criterion relates the stability of a closed-loop system to the open-loop frequency response and open-loop pole location. Thus, knowledge of the open-loop system's frequency response yields information about the stability of the closed-loop system. This concept is similar to the root locus, where we began with information about the open-loop system, its poles and zeros, and developed transient and stability information about the closed-loop system.

Although the Nyquist criterion will yield stability information at first, we will extend the concept to transient response and steady-state errors. Thus, frequency response techniques are an alternate approach to the root locus.

## Derivation of the Nyquist Criterion

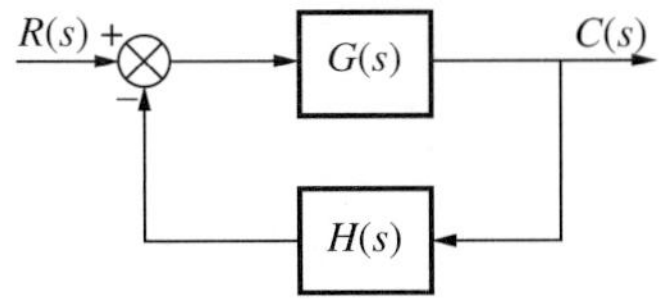


**FIGURE 10.20** Closed-loop control system

Consider the system of Figure 10.20. The Nyquist criterion can tell us how many closed-loop poles are in the right half-plane. Before deriving the criterion, let us establish four important concepts that will be used during the derivation: (1) the relationship between the poles of $1 + G(s)H(s)$ and the poles of $G(s)H(s)$; (2) the relationship between the zeros of $1 + G(s)H(s)$ and the poles of the closed-loop transfer function, $T(s)$; (3) the concept of *mapping* points; and (4) the concept of mapping *contours*.

Letting

$$G(s) = \frac{N_G}{D_G} \tag{10.37a}$$

$$H(s) = \frac{N_H}{D_H} \tag{10.37b}$$

we find

$$G(s)H(s) = \frac{N_G N_H}{D_G D_H} \tag{10.38a}$$

$$1 + G(s)H(s) = 1 + \frac{N_G N_H}{D_G D_H} = \frac{D_G D_H + N_G N_H}{D_G D_H} \tag{10.38b}$$

$$T(s) = \frac{G(s)}{1 + G(s)H(s)} = \frac{N_G D_H}{D_G D_H + N_G N_H} \tag{10.38c}$$

From Eqs. (10.38), we conclude that (1) *the poles of* $1 + G(s)H(s)$ *are the same as the poles of* $G(s)H(s)$*, the open-loop system, and* (2) *the zeros of* $1 + G(s)H(s)$ *are the same as the poles of* $T(s)$*, the closed-loop system.*

Next, let us define the term *mapping*. If we take a complex number on the $s$-plane and substitute it into a function, $F(s)$, another complex number results. This process is called *mapping*. For example, substituting $s = 4 + j3$ into the function $(s^2 + 2s + 1)$ yields $16 + j30$. We say that $4 + j3$ maps into $16 + j30$ through the function $(s^2 + 2s + 1)$.

Finally, we discuss the concept of mapping *contours*. Consider the collection of points, called a *contour*, shown in Figure 10.21 as contour A. Also, assume that

$$F(s) = \frac{(s - z_1)(s - z_2)\ldots}{(s - p_1)(s - p_2)\ldots} \tag{10.39}$$

Contour $A$ can be mapped through $F(s)$ into contour $B$ by substituting each point of contour $A$ into the function $F(s)$ and plotting the resulting complex numbers. For example, point $Q$ in Figure 10.21 maps into point $Q'$ through the function $F(s)$.

The vector approach to performing the calculation, covered in Section 8.1, can be used as an alternative. Some examples of contour mapping are shown in Figure 10.22 for some simple $F(s)$. The mapping of each point is defined by complex arithmetic, where the resulting complex number, $R$, is evaluated from the complex numbers

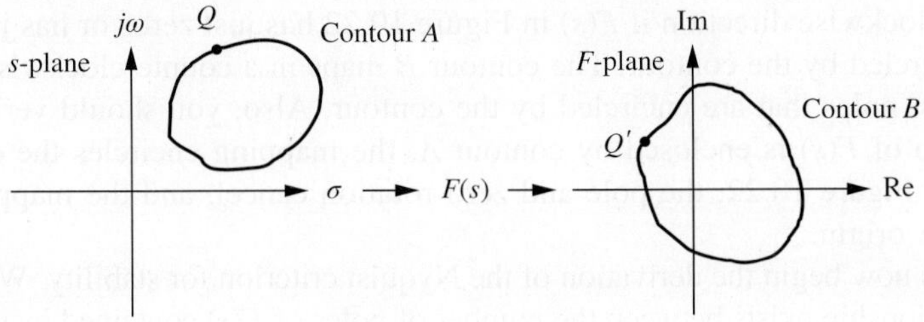


**FIGURE 10.21** Mapping contour $A$ through function $F(s)$ to contour $B$

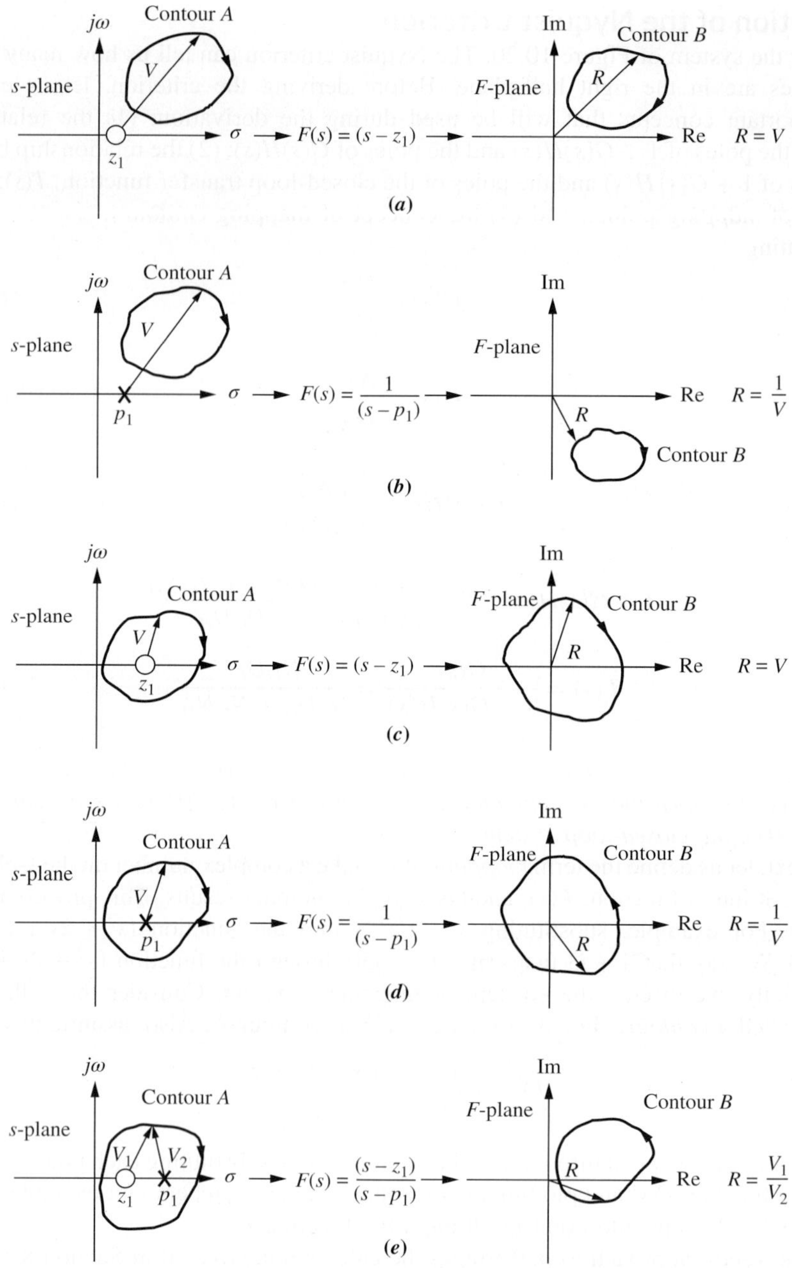


**FIGURE 10.22** Examples of contour mapping

represented by $V$, as shown in the last column of Figure 10.22. You should verify that if we assume a clockwise direction for mapping the points on contour $A$, then contour $B$ maps in a clockwise direction if $F(s)$ in Figure 10.22 has just zeros or has just poles that are not encircled by the contour. The contour $B$ maps in a counterclockwise direction if $F(s)$ has just poles that are encircled by the contour. Also, you should verify that if the pole or zero of $F(s)$ is enclosed by contour $A$, the mapping encircles the origin. In the last case of Figure 10.22, the pole and zero rotation cancel, and the mapping does not encircle the origin.

Let us now begin the derivation of the Nyquist criterion for stability. We show that a unique relationship exists between the number of poles of $F(s)$ contained inside contour $A$, the number of zeros of $F(s)$ contained inside contour $A$, and the number of counterclockwise

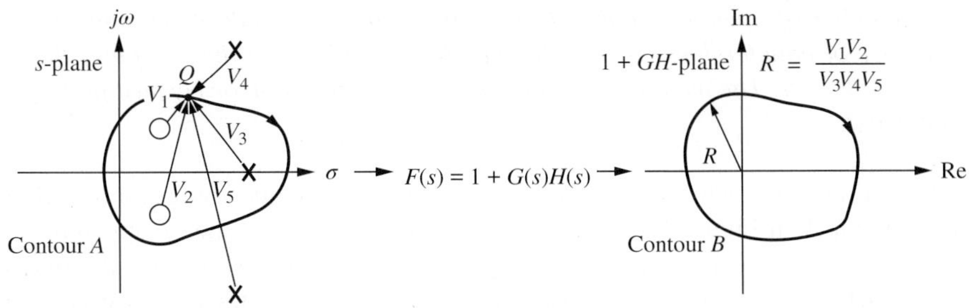


**FIGURE 10.23** Vector representation of mapping

encirclements of the origin for the mapping of contour $B$. We then show how this interrelationship can be used to determine the stability of closed-loop systems. This method of determining stability is called the *Nyquist criterion*.

Let us first assume that $F(s) = 1 + G(s)H(s)$, with the picture of the poles and zeros of $1 + G(s)H(s)$ as shown in Figure 10.23 near contour $A$. Hence, $R = (V_1V_2)/(V_3V_4V_5)$. As each point $Q$ of the contour $A$ is substituted into $1 + G(s)H(s)$, a mapped point results on contour $B$. Assuming that $F(s) = 1 + G(s)H(s)$ has two zeros and three poles, each parenthetical term of Eq. (10.39) is a vector in Figure 10.23. As we move around contour $A$ in a clockwise direction, each vector of Eq. (10.39) that lies inside contour $A$ will appear to undergo a complete rotation, or a change in angle of 360°. On the other hand, each vector drawn from the poles and zeros of $1 + G(s)H(s)$ that exists outside contour $A$ will appear to oscillate and return to its previous position, undergoing a net angular change of 0°.

Each pole or zero factor of $1 + G(s)H(s)$ whose vector undergoes a complete rotation around contour $A$ must yield a change of 360° in the resultant, $R$, or a complete rotation of the mapping of contour $B$. If we move in a clockwise direction along contour $A$, each zero inside contour $A$ yields a rotation in the clockwise direction, while each pole inside contour $A$ yields a rotation in the counterclockwise direction, since poles are in the denominator of Eq. (10.39).

Thus, $N = P - Z$, where $N$ equals the number of counterclockwise rotations of contour $B$ about the origin; $P$ equals the number of poles of $1 + G(s)H(s)$ inside contour $A$, and $Z$ equals the number of zeros of $1 + G(s)H(s)$ inside contour $A$.

Since the poles shown in Figure 10.23 are poles of $1 + G(s)H(s)$, we know from Eqs. (10.38) that they are also the poles of $G(s)H(s)$ and are known. But, since *the zeros shown in Figure 10.23 are the zeros of* $1 + G(s)H(s)$, we know from Eqs. (10.38) that *they are also the poles of the closed-loop system and are not known*. Thus, $P$ equals the number of enclosed open-loop poles, and $Z$ equals the number of enclosed closed-loop poles. Hence, $N = P - Z$, or alternately, $Z = P - N$, tells us that the number of closed-loop poles inside the contour (which is the same as the zeros inside the contour) equals the number of open-loop poles of $G(s)H(s)$ inside the contour minus the number of counterclockwise rotations of the mapping about the origin.

If we extend the contour to include the entire right half-plane, as shown in Figure 10.24, we can count the number of right-half-plane closed-loop poles inside contour $A$ and determine a system's stability. Since we can count the number of open-loop poles, $P$, inside the contour, which are the same as the right-half-plane poles of $G(s)H(s)$, the only problem remaining is how to obtain the mapping and find $N$.

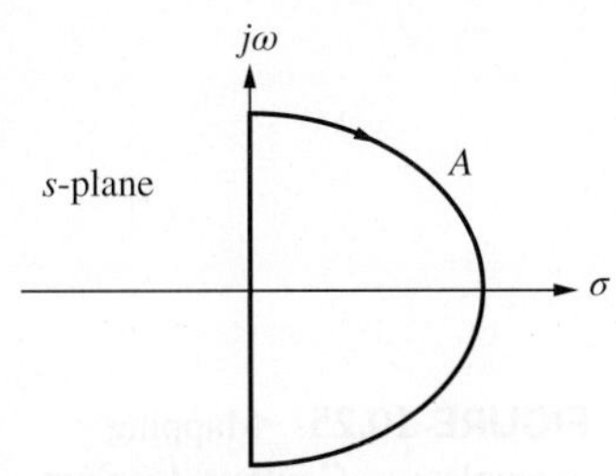


**FIGURE 10.24** Contour enclosing right half-plane to determine stability

Since all of the poles and zeros of $G(s)H(s)$ are known, what if we map through $G(s)H(s)$ instead of $1 + G(s)H(s)$? The resulting contour is the same as a mapping through $1 + G(s)H(s)$, except that it is translated one unit to the left. Thus, we count rotations about $-1$ instead of rotations about the origin. Hence, the final statement of the Nyquist stability criterion is as follows:

*If a contour, A, that encircles the entire right half-plane is mapped through G(s)H(s), then the number of closed-loop poles, Z, in the right half-plane equals*

*the number of open-loop poles, P, that are in the right half-plane minus the number of counterclockwise revolutions, N, around −1 of the mapping; that is, Z = P − N. The mapping is called the Nyquist diagram, or Nyquist plot, of G(s)H(s).*

We can now see why this method is classified as a frequency response technique. Around contour $A$ in Figure 10.24, the mapping of the points on the $j\omega$-axis through the function $G(s)H(s)$ is the same as substituting $s = j\omega$ into $G(s)H(s)$ to form the frequency response function $G(j\omega)H(j\omega)$. We are thus finding the frequency response of $G(s)H(s)$ over that part of contour $A$ on the positive $j\omega$-axis. In other words, part of the Nyquist diagram is the polar plot of the frequency response of $G(s)H(s)$.

## Applying the Nyquist Criterion to Determine Stability

Before describing how to sketch a Nyquist diagram, let us look at some typical examples that use the Nyquist criterion to determine the stability of a system. These examples give us a perspective prior to engaging in the details of mapping. Figure 10.25($a$) shows a contour $A$ that does not enclose closed-loop poles, that is, the zeros of $1 + G(s)H(s)$. The contour thus maps through $G(s)H(s)$ into a Nyquist diagram that does not encircle $-1$. Hence, $P = 0$, $N = 0$, and $Z = P - N = 0$. Since $Z$ is the number of closed-loop poles inside contour $A$, which encircles the right half-plane, this system has no right-half-plane poles and is stable.

On the other hand, Figure 10.25($b$) shows a contour $A$ that, while it does not enclose open-loop poles, does generate two clockwise encirclements of $-1$. Thus, $P = 0$, $N = -2$, and the system is unstable; it has two closed-loop poles in the right half-plane, since $Z = P - N = 2$. The two closed-loop poles are shown inside contour $A$ in Figure 10.25($b$) as zeros of $1 + G(s)H(s)$. You should keep in mind that the existence of these poles is not known a priori.

In this example, notice that clockwise encirclements imply a negative value for $N$. The number of encirclements can be determined by drawing a test radius from $-1$ in any convenient direction and counting the number of times the Nyquist diagram crosses the test radius. Counterclockwise crossings are positive, and clockwise crossings are negative. For example, in Figure 10.25($b$), contour $B$ crosses the test radius twice in a clockwise direction. Hence, there are $-2$ encirclements of the point $-1$.

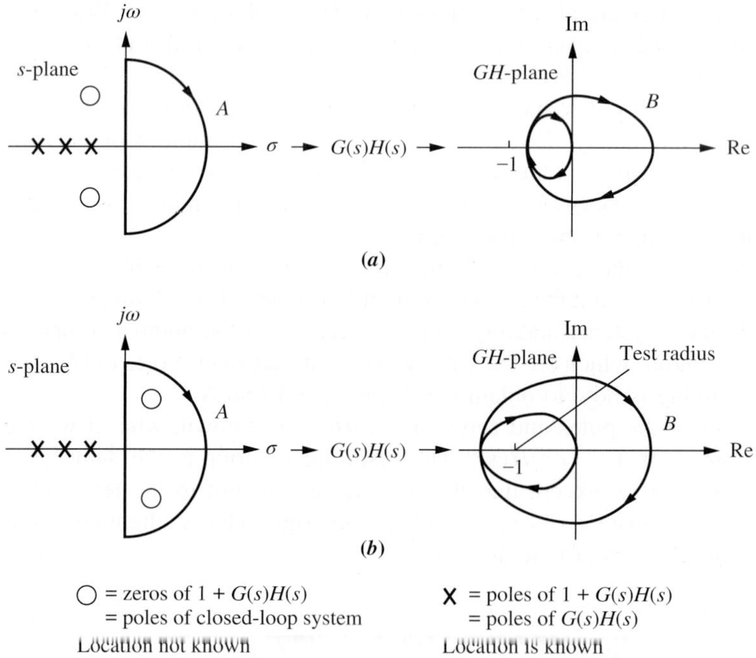


**FIGURE 10.25** Mapping examples: **a.** Contour does not enclose closed-loop poles; **b.** contour does enclose closed-loop poles

Before applying the Nyquist criterion to other examples in order to determine a system's stability, we must first gain experience in sketching Nyquist diagrams. The next section covers the development of this skill.

## 10.4 Sketching the Nyquist Diagram

The contour that encloses the right half-plane can be mapped through the function $G(s)H(s)$ by substituting points along the contour into $G(s)H(s)$. The points along the positive extension of the imaginary axis yield the polar frequency response of $G(s)H(s)$. Approximations can be made to $G(s)H(s)$ for points around the infinite semicircle by assuming that the vectors originate at the origin. Thus, their length is infinite, and their angles are easily evaluated.

However, most of the time a simple sketch of the Nyquist diagram is all that is needed. A sketch can be obtained rapidly by looking at the vectors of $G(s)H(s)$ and their motion along the contour. In the examples that follow, we stress this rapid method for sketching the Nyquist diagram. However, the examples also include analytical expressions for $G(s)H(s)$ for each section of the contour to aid you in determining the shape of the Nyquist diagram.

### Example 10.4 Sketching a Nyquist Diagram

**PROBLEM:** Speed controls find wide application throughout industry and the home. Figure 10.26(*a*) shows one application: output frequency control of electrical power from a turbine and generator pair. By regulating the speed, the control system ensures that the generated frequency remains within tolerance. Deviations from the desired speed are sensed, and a steam valve is changed to compensate for the speed error. The system block diagram is shown in Figure 10.26(*b*). Sketch the Nyquist diagram for the system of Figure 10.26.

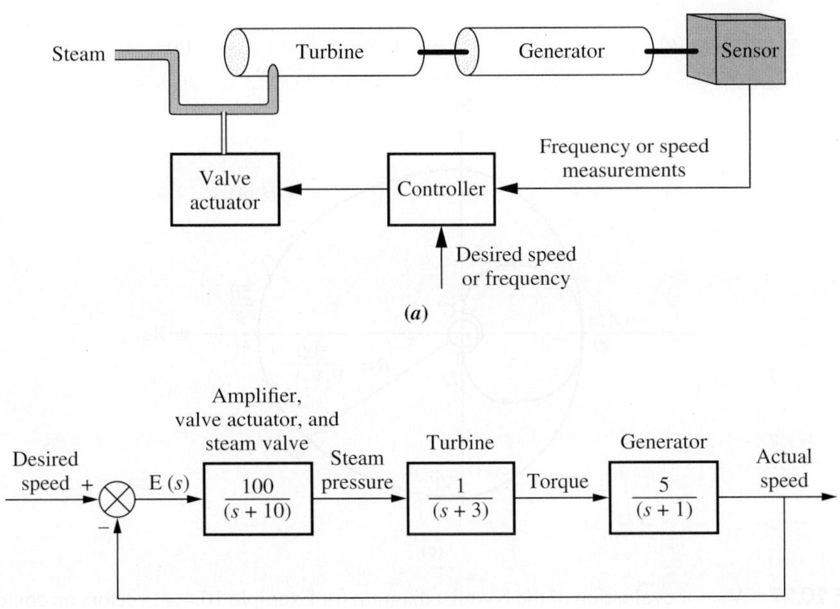


**FIGURE 10.26** **a.** Turbine and generator; **b.** block diagram of speed control system for Example 10.4

**SOLUTION:** Conceptually, the Nyquist diagram is plotted by substituting the points of the contour shown in Figure 10.27(*a*) into $G(s) = 500/[(s+1)(s+3)(s+10)]$. This process is equivalent to performing complex arithmetic using the vectors of $G(s)$ drawn to the points of the contour as shown in Figure 10.27(*a*) and (*b*). Each pole and zero term of $G(s)$ shown in Figure 10.26(*b*) is a vector in Figure 10.27(*a*) and (*b*). The resultant vector, $R$, found at any point along the contour is in general the product of the zero vectors divided by the product of the pole vectors (see Figure 10.27(*c*)). Thus, the magnitude of the resultant is the product of the zero lengths divided by the product of the pole lengths, and the angle of the resultant is the sum of the zero angles minus the sum of the pole angles.

As we move in a clockwise direction around the contour from point $A$ to point $C$ in Figure 10.27(*a*), the resultant angle goes from 0° to $-3 \times 90° = -270°$, or from $A'$ to $C'$ in Figure 10.27(*c*). Since the angles emanate from poles in the denominator of $G(s)$, the rotation or increase in angle is really a decrease in angle of the function $G(s)$; the poles gain 270° in a counterclockwise direction, which explains why the function loses 270°.

While the resultant moves from $A'$ to $C'$ in Figure 10.27(*c*), its magnitude changes as the product of the zero lengths divided by the product of the pole lengths. Thus, the

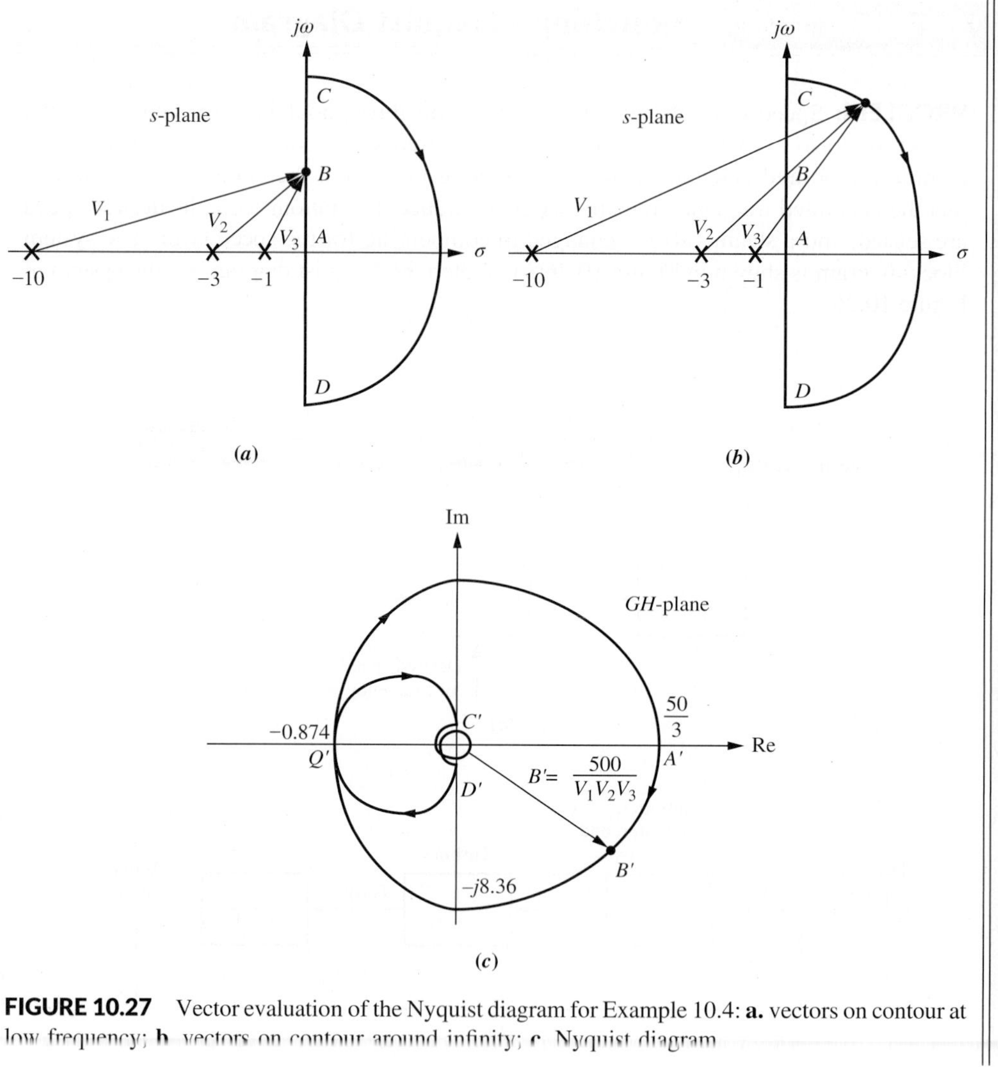


**FIGURE 10.27** Vector evaluation of the Nyquist diagram for Example 10.4: **a.** vectors on contour at low frequency; **b.** vectors on contour around infinity; **c.** Nyquist diagram

resultant goes from a finite value at zero frequency (at point $A$ of Figure 10.27($a$), there are three finite pole lengths) to zero magnitude at infinite frequency at point $C$ (at point $C$ of Figure 10.27($a$), there are three infinite pole lengths).

The mapping from point $A$ to point $C$ can also be explained analytically. From $A$ to $C$ the collection of points along the contour is imaginary. Hence, from $A$ to $C$, $G(s) = G(j\omega)$, or from Figure 10.26($b$),

$$G(j\omega) = \left.\frac{500}{(s+1)(s+3)(s+10)}\right|_{s\to j\omega} = \frac{500}{(-14\omega^2+30)+j(43\omega-\omega^3)} \tag{10.40}$$

Multiplying the numerator and denominator by the complex conjugate of the denominator, we obtain

$$G(j\omega) = 500\frac{(-14\omega^2+30)-j(43\omega-\omega^3)}{(-14\omega^2+30)^2+(43\omega-\omega^3)^2} \tag{10.41}$$

At zero frequency, $G(j\omega) = 500/30 = 50/3$. Thus, the Nyquist diagram starts at $50/3$ at an angle of 0°. As $\omega$ increases, the real part remains positive and the imaginary part remains negative. At $\omega = \sqrt{30/14}$, the real part becomes negative. At $\omega = \sqrt{43}$, the Nyquist diagram crosses the negative real axis, since the imaginary term goes to zero. The real value at the axis crossing, point $Q'$ in Figure 10.27($c$), found by substituting into Eq. (10.41), is $-0.874$. Continuing toward $\omega = \infty$, the real part is negative, and the imaginary part is positive. At infinite frequency $G(j\omega) \approx j500/\omega^3$, or approximately zero at 90°.

Around the infinite semicircle from point $C$ to point $D$ shown in Figure 10.27($b$), the vectors rotate clockwise, each by 180°. Hence, the resultant undergoes a counterclockwise rotation of $3 \times 180°$, starting at point $C'$ and ending at point $D'$ of Figure 10.27($c$). Analytically, we can see this by assuming that around the infinite semicircle, the vectors originate approximately at the origin and have infinite length. For any point on the $s$-plane, the value of $G(s)$ can be found by representing each complex number in polar form, as follows:

$$G(s) = \frac{500}{(R_{-1}e^{j\theta_{-1}})(R_{-3}e^{j\theta_{-3}})(R_{-10}e^{j\theta_{-10}})} \tag{10.42}$$

where $R_{-i}$ is the magnitude of the complex number $(s+i)$, and $\theta_{-i}$ is the angle of the complex number $(s+i)$. Around the infinite semicircle, all $R_{-i}$ are infinite, and we can use our assumption to approximate the angles as if the vectors originated at the origin. Thus, around the infinite semicircle,

$$G(s) = \frac{500}{\infty\angle(\theta_{-1}+\theta_{-3}+\theta_{-10})} = 0\angle-(\theta_{-1}+\theta_{-3}+\theta_{-10}) \tag{10.43}$$

At point $C$ in Figure 10.27($b$), the angles are all 90°. Hence, the resultant is $0\angle-270°$, shown as point $C'$ in Figure 10.27($c$). Similarly, at point $D$, $G(s) = 0\angle+270°$ and maps into point $D'$. You can select intermediate points to verify the spiral whose radius vector approaches zero at the origin, as shown in Figure 10.27($c$).

The negative imaginary axis can be mapped by realizing that the real part of $G(j\omega)H(j\omega)$ is always an even function, whereas the imaginary part of $G(j\omega)H(j\omega)$ is an odd function. That is, the real part will not change sign when negative values of $\omega$ are used, whereas the imaginary part will change sign. Thus, the mapping of the negative imaginary axis is a mirror image of the mapping of the positive imaginary axis. The mapping of the section of the contour from points $D$ to $A$ is drawn as a mirror image about the real axis of the mapping of points $A$ to $C$.

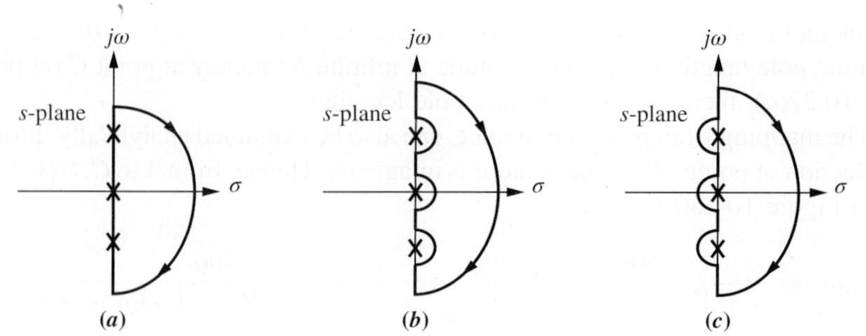


**FIGURE 10.28** Detouring around open-loop poles: **a.** poles on contour; **b.** detour right; **c.** detour left

In the previous example, there were no open-loop poles situated along the contour enclosing the right half-plane. If such poles exist, then a detour around the poles on the contour is required; otherwise, the mapping would go to infinity in an undetermined way, without angular information. Subsequently, a complete sketch of the Nyquist diagram could not be made, and the number of encirclements of $-1$ could not be found.

Let us assume a $G(s)H(s) = N(s)/sD(s)$ where $D(s)$ has imaginary roots. The $s$ term in the denominator and the imaginary roots of $D(s)$ are poles of $G(s)H(s)$ that lie on the contour, as shown in Figure 10.28($a$). To sketch the Nyquist diagram, the contour must detour around each open-loop pole lying on its path. The detour can be to the right of the pole, as shown in Figure 10.28($b$), which makes it clear that each pole's vector rotates through $+180°$ as we move around the contour near that pole. This knowledge of the angular rotation of the poles on the contour permits us to complete the Nyquist diagram. Of course, our detour must carry us only an infinitesimal distance into the right half-plane, or else some closed-loop, right-half-plane poles will be excluded in the count.

We can also detour to the left of the open-loop poles. In this case, each pole rotates through an angle of $-180°$ as we detour around it. Again, the detour must be infinitesimally small, or else we might include some left-half-plane poles in the count. Let us look at an example.

## Example 10.5 Nyquist Diagram for Open-Loop Function with Poles on Contour

**PROBLEM:** Sketch the Nyquist diagram of the unity feedback system of Figure 10.10, where $G(s) = (s+2)/s^2$.

**SOLUTION:** The system's two poles at the origin are on the contour and must be bypassed, as shown in Figure 10.29($a$). The mapping starts at point $A$ and continues in a clockwise direction. Points $A$, $B$, $C$, $D$, $E$, and $F$ of Figure 10.29($a$) map, respectively, into points $A'$, $B'$, $C'$, $D'$, $E'$, and $F'$ of Figure 10.29($b$).

At point $A$, the two open-loop poles at the origin contribute $2 \times 90° = 180°$, and the zero contributes 0°. The total angle at point $A$ is thus $-180°$. Close to the origin, the function is infinite in magnitude because of the close proximity to the two open-loop poles. Thus, point $A$ maps into point $A'$, located at infinity at an angle of $-180°$.

Moving from point $A$ to point $B$ along the contour yields a net change in angle of $+90°$ from the zero alone. The angles of the poles remain the same. Thus, the mapping changes by $+90°$ in the counterclockwise direction. The mapped vector goes from $-180°$ at $A'$ to $-90°$ at $B'$. At the same time, the magnitude changes from infinity to zero, since at point $B$ there is one infinite length from the zero divided by two infinite lengths from the poles.

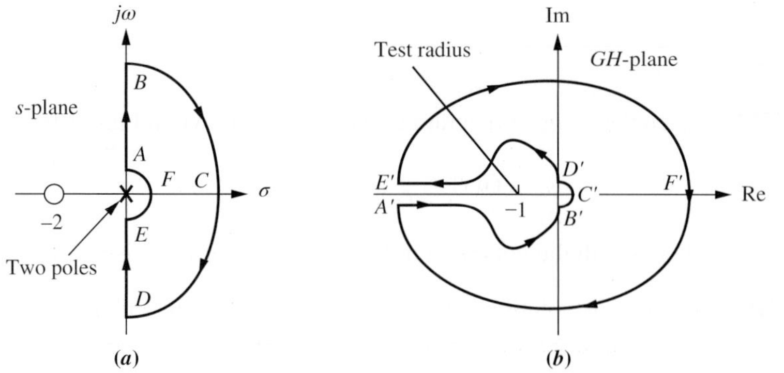


**FIGURE 10.29** **a.** Contour for Example 10.5; **b.** Nyquist diagram for Example 10.5

Alternately, the frequency response can be determined analytically from $G(j\omega) = (2 + j\omega)/(-\omega^2)$, considering $\omega$ going from 0 to $\infty$. At low frequencies, $G(j\omega) \approx 2/(-\omega^2)$, or $\infty \angle 180°$. At high frequencies, $G(j\omega) \approx j/(-\omega)$, or $0\angle - 90°$. Also, the real and imaginary parts are always negative.

As we travel along the contour *BCD*, the function magnitude stays at zero (one infinite zero length divided by two infinite pole lengths). As the vectors move through *BCD*, the zero's vector and the two poles' vectors undergo changes of $-180°$ each. Thus, the mapped vector undergoes a net change of $+180°$, which is the angular change of the zero minus the sum of the angular changes of the poles $\{-180 - [2(-180)] = +180\}$. The mapping is shown as $B'\,C'\,D'$, where the resultant vector changes by $+180°$ with a magnitude of $\varepsilon$ that approaches zero.

From the analytical point of view,

$$G(s) = \frac{R_{-2}\angle\theta_{-2}}{(R_0\angle\theta_0)(R_0\angle\theta_0)} \tag{10.44}$$

anywhere on the *s*-plane where $R_{-2}\angle\theta_{-2}$ is the vector from the zero at $-2$ to any point on the *s*-plane, and $R_0\angle\theta_0$ is the vector from a pole at the origin to any point on the *s*-plane. Around the infinite semicircle, all $R_{-i} = \infty$, and all angles can be approximated as if the vectors originated at the origin. Thus at point *B*, $G(s) = 0\angle - 90°$, since all $\theta_{-i} = 90°$ in Eq. (10.44). At point *C*, all $R_{-i} = \infty$, and all $\theta_{-i} = 0°$ in Eq. (10.44). Thus, $G(s) = 0\angle 0°$. At point *D*, all $R_{-i} = \infty$, and all $\theta_{-i} = -90°$ in Eq. (10.44). Thus, $G(s) = 0\angle 90°$.

The mapping of the section of the contour from *D* to *E* is a mirror image of the mapping of *A* to *B*. The result is $D'$ to $E'$.

Finally, over the section *EFA*, the resultant magnitude approaches infinity. The angle of the zero does not change, but each pole changes by $+180°$. This change yields a change in the function of $-2 \times 180° = -360°$. Thus, the mapping from $E'$ to $A'$ is shown as infinite in length and rotating $-360°$. Analytically, we can use Eq. (10.44) for the points along the contour *EFA*. At *E*, $G(s) = (2\angle 0°)/[(\epsilon\angle - 90°)(\epsilon\angle - 90°)] = \infty\angle 180°$. At *F*, $G(s) = (2\angle 0°)/[(\epsilon\angle 0°)(\epsilon\angle 0°)] = \infty\angle 0°$. At *A*, $G(s) = (2\angle 0°)/[(\epsilon\angle 90°)(\epsilon\angle 90°)] = \infty\angle - 180°$.

The Nyquist diagram is now complete, and a test radius drawn from $-1$ in Figure 10.29(*b*) shows one counterclockwise revolution, and one clockwise revolution, yielding zero encirclements.

Students who are using MATLAB should now run ch10p2 in Appendix B. You will learn how to use MATLAB to make a Nyquist plot and list the points on the plot. You will also learn how to specify a range for frequency. This exercise solves Example 10.5 using MATLAB.

MATLAB
**ML**

## Skill-Assessment Exercise 10.3

**PROBLEM:** Sketch the Nyquist diagram for the system shown in Figure 10.10 where

$$G(s) = \frac{1}{(s+2)(s+4)}$$

Compare your sketch with the polar plot in Skill-Assessment Exercise 10.1(c).

**ANSWER:** The complete solution is located at www.wiley.com/college/nise.

In this section, we learned how to sketch a Nyquist diagram. We saw how to calculate the value of the intersection of the Nyquist diagram with the negative real axis. This intersection is important in determining the number of encirclements of $-1$. Also, we showed how to sketch the Nyquist diagram when open-loop poles exist on the contour; this case required detours around the poles. In the next section, we apply the Nyquist criterion to determine the stability of feedback control systems.

# 10.5 Stability via the Nyquist Diagram

**TryIt 10.2**

Use MATLAB, the Control System Toolbox, and the following statements to plot the Nyquist diagram of the system shown in Figure 10.30(*a*).

```
G=zpk([-3,-5],...
 [2,4],1)
nyquist(G)
```

After the Nyquist diagram appears, click on the curve and drag to read the coordinates.

We now use the Nyquist diagram to determine a system's stability, using the simple equation $Z = P - N$. The values of $P$, the number of open-loop poles of $G(s)H(s)$ enclosed by the contour, and $N$, the number of encirclements the Nyquist diagram makes about $-1$, are used to determine $Z$, the number of right-half-plane poles of the closed-loop system.

If the closed-loop system has a variable gain in the loop, one question we would like to ask is, "For what range of gain is the system stable?" This question, previously answered by the root locus method and the Routh-Hurwitz criterion, is now answered via the Nyquist criterion. The general approach is to set the loop gain equal to unity and draw the Nyquist diagram. Since gain is simply a multiplying factor, the effect of the gain is to multiply the resultant by a constant anywhere along the Nyquist diagram.

For example, consider Figure 10.30, which summarizes the Nyquist approach for a system with variable gain, $K$. As the gain is varied, we can visualize the Nyquist diagram in Figure 10.30($c$) expanding (increased gain) or shrinking (decreased gain) like a balloon.

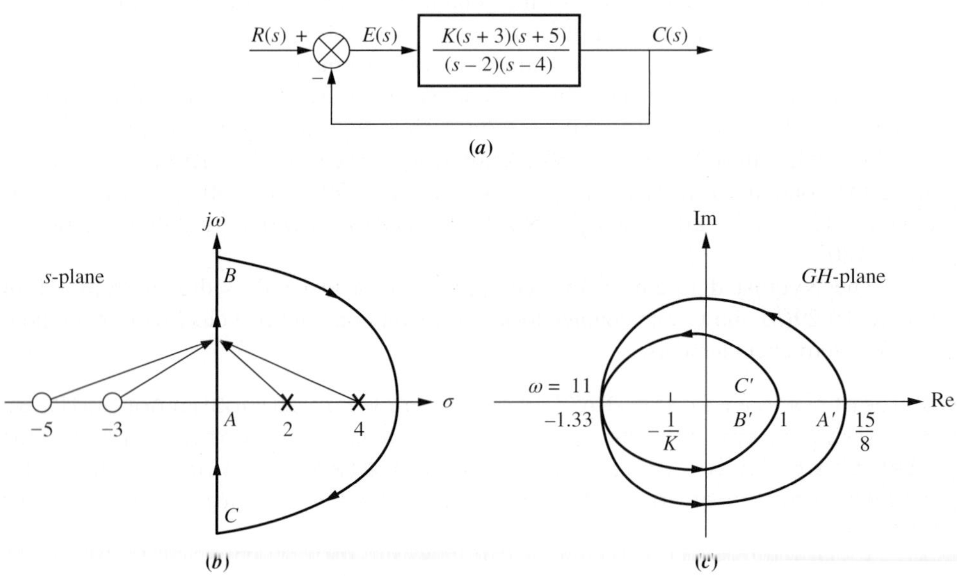


**FIGURE 10.30** Demonstrating Nyquist stability: **a.** system; **b.** contour; **c.** Nyquist diagram

This motion could move the Nyquist diagram past the $-1$ point, changing the stability picture. For this system, since $P = 2$, the critical point must be encircled by the Nyquist diagram to yield $N = 2$ and a stable system. A reduction in gain would place the critical point outside the Nyquist diagram where $N = 0$, yielding $Z = 2$, an unstable system.

From another perspective we can think of the Nyquist diagram as remaining stationary and the $-1$ point moving along the real axis. In order to do this, we set the gain to unity and position the critical point at $-1/K$ rather than $-1$. Thus, the critical point appears to move closer to the origin as $K$ increases.

Finally, if the Nyquist diagram intersects the real axis at $-1$, then $G(j\omega)H(j\omega) = -1$. From root locus concepts, when $G(s)H(s) = -1$, the variable $s$ is a closed-loop pole of the system. Thus, the frequency at which the Nyquist diagram intersects $-1$ is the same frequency at which the root locus crosses the $j\omega$-axis. Hence, the system is marginally stable if the Nyquist diagram intersects the real axis at $-1$.

In summary, then, if the open-loop system contains a variable gain, $K$, set $K = 1$ and sketch the Nyquist diagram. Consider the critical point to be at $-1/K$ rather than at $-1$. Adjust the value of $K$ to yield stability, based upon the Nyquist criterion.

## Example 10.6 Range of Gain for Stability via the Nyquist Criterion

**PROBLEM:** For the unity feedback system of Figure 10.10, where $G(s) = K/[s(s+3)\,(s+5)]$, find the range of gain, $K$, for stability, instability, and the value of gain for marginal stability. For marginal stability also find the frequency of oscillation. Use the Nyquist criterion.

**SOLUTION:** First set $K = 1$ and sketch the Nyquist diagram for the system, using the contour shown in Figure 10.31($a$). For all points on the imaginary axis,

$$G(j\omega)H(j\omega) = \left.\frac{K}{s(s+3)(s+5)}\right|_{\substack{K=1\\ s=j\omega}} = \frac{-8\omega^2 - j(15\omega - \omega^3)}{64\omega^4 + \omega^2(15-\omega^2)^2} \tag{10.45}$$

At $\omega = 0$, $G(j\omega)H(j\omega) = -0.0356 - j\infty$.

Next find the point where the Nyquist diagram intersects the negative real axis. Setting the imaginary part of Eq. 10.45 equal to zero, we find $\omega = \sqrt{15}$. Substituting this

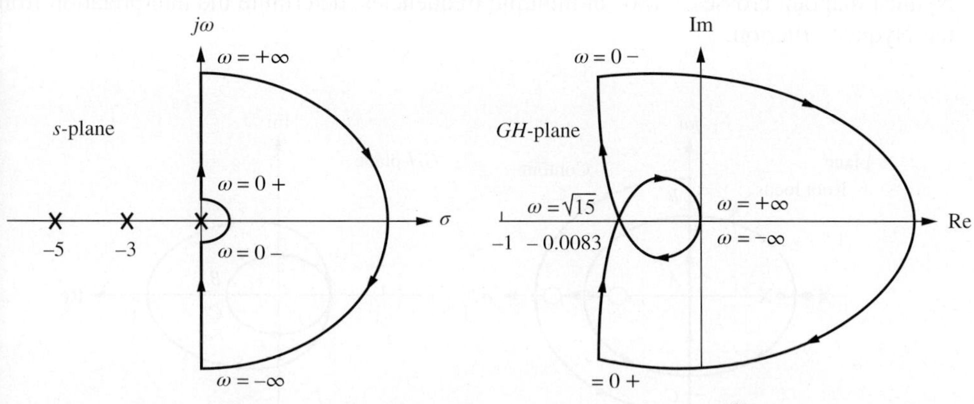


**FIGURE 10.31** **a.** Contour for Example 10.6; **b.** Nyquist diagram

value of $\omega$ back into Eq. (10.45) yields the real part of $-0.0083$. Finally, at $\omega = \infty$, $G(j\omega)H(j\omega) = G(s)H(s)|_{s\to j\infty} = 1/(j\infty)^3 = 0\angle -270°$.

From the contour of Figure 10.31($a$), $P = 0$; for stability $N$ must then be equal to zero. From Figure 10.31($b$), the system is stable if the critical point lies outside the contour ($N = 0$), so that $Z = P - N = 0$. Thus, $K$ can be increased by $1/0.0083 = 120.5$ before the Nyquist diagram encircles $-1$. Hence, for stability, $K < 120.5$. For marginal stability $K = 120.5$. At this gain the Nyquist diagram intersects $-1$, and the frequency of oscillation is $\sqrt{15}$ rad/s.

Now that we have used the Nyquist diagram to determine stability, we can develop a simplified approach that uses only the mapping of the positive $j\omega$-axis.

## Stability via Mapping Only the Positive $j\omega$-Axis

Once the stability of a system is determined by the Nyquist criterion, continued evaluation of the system can be simplified by using just the mapping of the positive $j\omega$-axis. This concept plays a major role in the next two sections, where we discuss stability margin and the implementation of the Nyquist criterion with Bode plots.

Consider the system shown in Figure 10.32, which is stable at low values of gain and unstable at high values of gain. Since the contour does not encircle open-loop poles, the Nyquist criterion tells us that we must have no encirclements of $-1$ for the system to be stable. We can see from the Nyquist diagram that the encirclements of the critical point can be determined from the mapping of the positive $j\omega$-axis alone. If the gain is small, the mapping will pass to the right of $-1$, and the system will be stable. If the gain is high, the mapping will pass to the left of $-1$, and the system will be unstable. Thus, this system is stable for the range of loop gain, $K$, that ensures that the *open-loop magnitude is less than unity at that frequency where the phase angle is* 180° (*or, equivalently*, $-180°$). This statement is thus an alternative to the Nyquist criterion for this system.

Now consider the system shown in Figure 10.33, which is unstable at low values of gain and stable at high values of gain. Since the contour encloses two open-loop poles, two counterclockwise encirclements of the critical point are required for stability. Thus, for this case the system is stable if the *open-loop magnitude is greater than unity at that frequency where the phase angle is* 180° (*or, equivalently*, $-180°$).

In summary, first determine stability from the Nyquist criterion and the Nyquist diagram. Next, interpret the Nyquist criterion and determine whether the mapping of just the positive imaginary axis should have a gain of less than or greater than unity at 180°. If the Nyquist diagram crosses $\pm 180°$ at multiple frequencies, determine the interpretation from the Nyquist criterion.

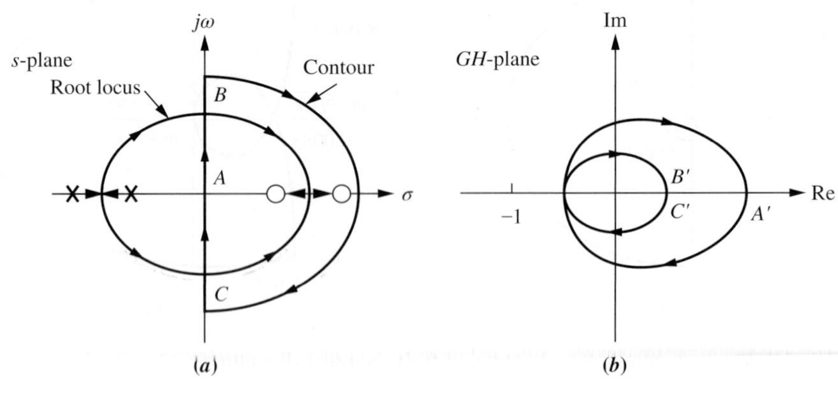


**FIGURE 10.32** **a.** Contour and root locus of system that is stable for small gain and unstable for large gain; **b.** Nyquist diagram

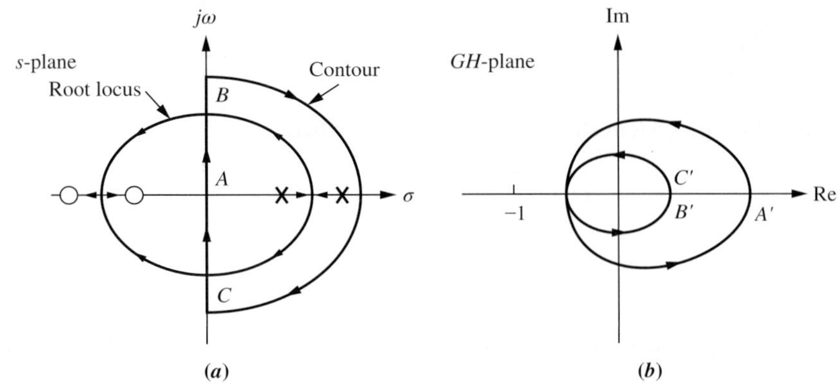


**FIGURE 10.33** **a.** Contour and root locus of system that is unstable for small gain and stable for large gain; **b.** Nyquist diagram

## Example 10.7 Stability Design via Mapping Positive $j\omega$-Axis

**PROBLEM:** Find the range of gain for stability and instability, and the gain for marginal stability, for the unity feedback system shown in Figure 10.10, where $G(s) = K/[(s^2 + 2s + 2)(s + 2)]$. For marginal stability, find the radian frequency of oscillation. Use the Nyquist criterion and the mapping of only the positive imaginary axis.

**SOLUTION:** Since the open-loop poles are only in the left-half-plane, the Nyquist criterion tells us that we want no encirclements of $-1$ for stability. Hence, a gain less than unity at $\pm 180°$ is required. Begin by letting $K = 1$ and draw the portion of the contour along the positive imaginary axis as shown in Figure 10.34(*a*). In Figure 10.34(*b*), the intersection with the negative real axis is found by letting $s = j\omega$ in $G(s)H(s)$, setting the imaginary part equal to zero to find the frequency, and then substituting the frequency into the real part of $G(j\omega)H(j\omega)$. Thus, for any point on the positive imaginary axis,

$$\begin{aligned} G(j\omega)H(j\omega) &= \left.\frac{1}{(s^2+2s+2)(s+2)}\right|_{s\to j\omega} \\ &= \frac{4(1-\omega^2) - j\omega(6-\omega^2)}{16(1-\omega^2)^2 + \omega^2(6-\omega^2)^2} \end{aligned} \tag{10.46}$$

Setting the imaginary part equal to zero, we find $\omega = \sqrt{6}$. Substituting this value back into Eq. (10.46) yields the real part, $-(1/20) = (1/20)\angle 180°$.

This closed-loop system is stable if the magnitude of the frequency response is less than unity at 180°. Hence, the system is stable for $K < 20$, unstable for $K > 20$, and marginally stable for $K = 20$. When the system is marginally stable, the radian frequency of oscillation is $\sqrt{6}$.

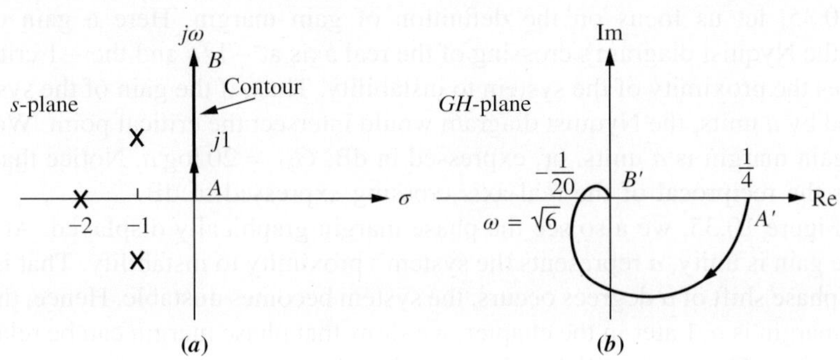


**FIGURE 10.34** **a.** Portion of contour to be mapped for Example 10.7; **b.** Nyquist diagram of mapping of positive imaginary axis

## Skill-Assessment Exercise 10.4

**PROBLEM:** For the system shown in Figure 10.10, where

$$G(s) = \frac{K}{(s+2)(s+4)(s+6)}$$

do the following:

**a.** Plot the Nyquist diagram.

**b.** Use your Nyquist diagram to find the range of gain, $K$, for stability.

**ANSWERS:**

**a.** See the answer at www.wiley.com/college/nise.
**b.** Stable for $K < 480$

The complete solution is at www.wiley.com/college/nise.

# 10.6 Gain Margin and Phase Margin via the Nyquist Diagram

Now that we know how to sketch and interpret a Nyquist diagram to determine a closed-loop system's stability, let us extend our discussion to concepts that will eventually lead us to the design of transient response characteristics via frequency response techniques.

Using the Nyquist diagram, we define two quantitative measures of how stable a system is. These quantities are called *gain margin* and *phase margin*. Systems with greater gain and phase margins can withstand greater changes in system parameters before becoming unstable. In a sense, gain and phase margins can be qualitatively related to the root locus, in that systems whose poles are farther from the imaginary axis have a greater degree of stability.

In the last section, we discussed stability from the point of view of gain at 180° phase shift. This concept leads to the following definitions of gain margin and phase margin:

*Gain margin,* $G_M$. The gain margin is the change in open-loop gain, expressed in decibels (dB), required at 180° of phase shift to make the closed-loop system unstable.

*Phase margin,* $\Phi_M$. The phase margin is the change in open-loop phase shift required at unity gain to make the closed-loop system unstable.

These two definitions are shown graphically on the Nyquist diagram in Figure 10.35.

Assume a system that is stable if there are no encirclements of $-1$. Using Figure 10.35, let us focus on the definition of gain margin. Here a gain difference between the Nyquist diagram's crossing of the real axis at $-1/a$ and the $-1$ critical point determines the proximity of the system to instability. Thus, if the gain of the system were multiplied by $a$ units, the Nyquist diagram would intersect the critical point. We then say that the gain margin is $a$ units, or, expressed in dB, $G_M = 20 \log a$. Notice that the gain margin is the reciprocal of the real-axis crossing expressed in dB.

In Figure 10.35, we also see the phase margin graphically displayed. At point $Q'$, where the gain is unity, $\alpha$ represents the system's proximity to instability. That is, at unity gain, if a phase shift of $\alpha$ degrees occurs, the system becomes unstable. Hence, the amount of phase margin is $\alpha$. Later in the chapter, we show that phase margin can be related to the damping ratio. Thus, we will be able to relate frequency response characteristics to

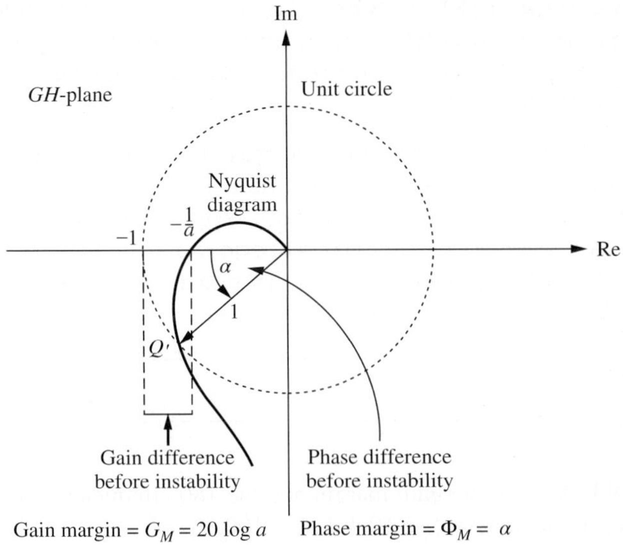


**FIGURE 10.35** Nyquist diagram showing gain and phase margins

transient response characteristics as well as stability. We will also show that the calculations of gain and phase margins are more convenient if Bode plots are used rather than a Nyquist diagram, such as that shown in Figure 10.35.

For now, let us look at an example that shows the calculation of the gain and phase margins.

## Example 10.8 Finding Gain and Phase Margins

**PROBLEM:** Find the gain and phase margin for the system of Example 10.7 if $K = 6$.

**SOLUTION:** To find the gain margin, first find the frequency where the Nyquist diagram crosses the negative real axis. Finding $G(j\omega)H(j\omega)$, we have

$$G(j\omega)H(j\omega) = \left.\frac{6}{(s^2+2s+2)(s+2)}\right|_{s\to j\omega} = \frac{6[4(1-\omega^2) - j\omega(6-\omega^2)]}{16(1-\omega^2)^2 + \omega^2(6-\omega^2)^2} \tag{10.47}$$

The Nyquist diagram crosses the real axis at a frequency of $\sqrt{6}$ rad/s. The real part is calculated to be $-0.3$. Thus, the gain can be increased by $(1/0.3) = 3.33$ before the real part becomes $-1$. Hence, the gain margin is

$$G_M = 20 \log 3.33 = 10.45 \text{ dB} \tag{10.48}$$

To find the phase margin, find the frequency in Eq. (10.47) for which the magnitude is unity. As the problem stands, this calculation requires computational tools, such as a function solver or the program described in Appendix H.2. Later in the chapter we will simplify the process by using Bode plots. Equation (10.47) has unity gain at a frequency of 1.253 rad/s. At this frequency, the phase angle is $-112.3°$. The difference between this angle and $-180°$ is $67.7°$, which is the phase margin.

MATLAB
**ML**

Students who are using MATLAB should now run ch10p3 in Appendix B. You will learn how to use MATLAB to find gain margin, phase margin, zero dB frequency, and 180° frequency. This exercise solves Example 10.8 using MATLAB.

GUI Tool
**GUIT**

MATLAB's LTI Viewer, with the Nyquist diagram selected, is another method that may be used to find gain margin, phase margin, zero dB frequency, and 180° frequency. You are encouraged to study Appendix E, at www.wiley.com/college/nise, which contains a tutorial on the LTI Viewer as well as some examples. Example E.2 solves Example 10.8 using the LTI Viewer.

## Skill-Assessment Exercise 10.5

**TryIt 10.3**

Use MATLAB, the Control System Toolbox, and the following statements to find the gain and phase margins of $G(s)H(s) = 100/[(s+2)(s+4)(s+6)]$ using the Nyquist diagram.

```
G=zpk([],[-2,-4,-6],100)
nyquist(G)
```

After the Nyquist diagram appears:

1. Right-click in the graph area.
2. Select **Characteristics**.
3. Select **All Stability Margins**.
4. Let the mouse rest on the margin points to read the gain and phase margins.

**PROBLEM:** Find the gain margin and the 180° frequency for the problem in Skill-Assessment Exercise 10.4 if $K = 100$.

**ANSWERS:** Gain margin = 13.62 dB; 180° frequency = 6.63 rad/s

The complete solution is at www.wiley.com/college/nise.

In this section, we defined gain margin and phase margin and calculated them via the Nyquist diagram. In the next section, we show how to use Bode diagrams to implement the stability calculations performed in Sections 10.5 and 10.6 using the Nyquist diagram. We will see that the Bode plots reduce the time and simplify the calculations required to obtain results.

# 10.7 Stability, Gain Margin, and Phase Margin via Bode Plots

In this section, we determine stability, gain and phase margins, and the range of gain required for stability. All of these topics were covered previously in this chapter, using Nyquist diagrams as the tool. Now we use Bode plots to determine these characteristics. Bode plots are subsets of the complete Nyquist diagram but in another form. They are a viable alternative to Nyquist plots, since they are easily drawn without the aid of the computational devices or long calculations required for the Nyquist diagram and root locus. You should remember that all calculations applied to stability were derived from and based upon the Nyquist stability criterion. The Bode plots are an alternate way of visualizing and implementing the theoretical concepts.

## Determining Stability

Let us look at an example and determine the stability of a system, implementing the Nyquist stability criterion using Bode plots. We will draw a Bode log-magnitude plot and then determine the value of gain that ensures that the magnitude is less than 0 dB (unity gain) at that frequency where the phase is ±180°.

## Example 10.9 Range of Gain for Stability via Bode Plots

**PROBLEM:** Use Bode plots to determine the range of $K$ within which the unity feedback system shown in Figure 10.10 is stable. Let $G(s) = K/[(s+2)(s+4)(s+5)]$.

**SOLUTION:** Since this system has all of its open-loop poles in the left-half-plane, the open-loop system is stable. Hence, from the discussion of Section 10.5, the closed-loop system will be stable if the frequency response has a gain less than unity when the phase is 180°.

Begin by sketching the Bode magnitude and phase diagrams shown in Figure 10.36. In Section 10.2, we summed normalized plots of each factor of $G(s)$ to create the Bode plot. We saw that at each break frequency, the slope of the resultant Bode plot changed by an amount equal to the new slope that was added. Table 10.6 demonstrates this observation. In this example, we use this fact to draw the Bode plots faster by avoiding the sketching of the response of each term.

The low-frequency gain of $G(s)H(s)$ is found by setting $s$ to zero. Thus, the Bode magnitude plot starts at $K/40$. For convenience, let $K = 40$ so that the log-magnitude plot starts at 0 dB. At each break frequency, 2, 4, and 5, a 20 dB/decade increase in negative slope is drawn, yielding the log-magnitude plot shown in Figure 10.36.

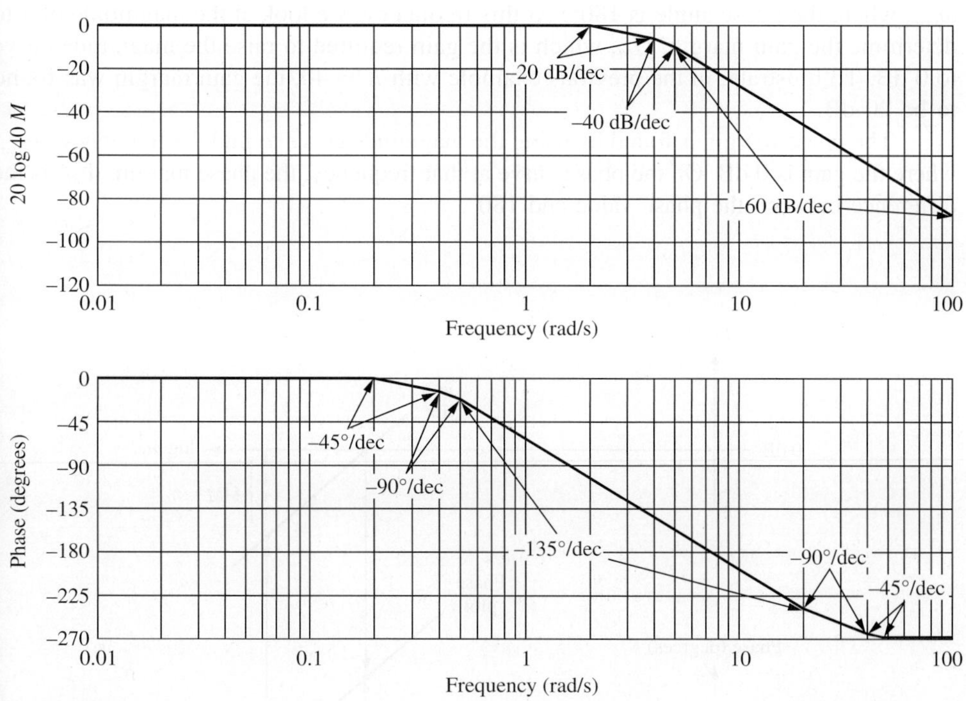


**FIGURE 10.36** Bode log-magnitude and phase diagrams for the system of Example 10.9

The phase diagram begins at 0° until a decade below the first break frequency of 2 rad/s. At 0.2 rad/s the curve decreases at a rate of –45°/decade, decreasing an additional

45°/decade at each subsequent frequency (0.4 and 0.5 rad/s) a decade below each break. At a decade above each break frequency, the slopes are reduced by 45°/decade at each frequency.

The Nyquist criterion for this example tells us that we want zero encirclements of $-1$ for stability. Thus, we recognize that the Bode log-magnitude plot must be less than unity when the Bode phase plot is 180°. Accordingly, we see that at a frequency of 7 rad/s, when the phase plot is $-180°$, the magnitude plot is $-20$ dB. Therefore, an increase in gain of $+20$ dB is possible before the system becomes unstable. Since the gain plot was scaled for a gain of 40, $+20$ dB (a gain of 10) represents the required increase in gain above 40. Hence, the gain for instability is $40 \times 10 = 400$. The final result is $0 < K < 400$ for stability.

This result, obtained by approximating the frequency response by Bode asymptotes, can be compared to the result obtained from the actual frequency response, which yields a gain of 378 at a frequency of 6.16 rad/s.

MATLAB
**ML**

`Students who are using MATLAB should now run ch10p4 in Appendix B. You will learn how to use MATLAB to find the range of gain for stability via frequency response methods. This exercise solves Example 10.10 using MATLAB.`

## Evaluating Gain and Phase Margins

Next we show how to evaluate the gain and phase margins by using Bode plots (Figure 10.37). The gain margin is found by using the phase plot to find the frequency, $\omega_{G_M}$, where the phase angle is 180°. At this frequency, we look at the magnitude plot to determine the gain margin, $G_M$, which is the gain required to raise the magnitude curve to 0 dB. To illustrate, in the previous example with $K = 40$, the gain margin was found to be 20 dB.

The phase margin is found by using the magnitude curve to find the frequency, $\omega_{\Phi_M}$, where the gain is 0 dB. On the phase curve at that frequency, the phase margin, $\Phi_M$, is the difference between the phase value and 180°.

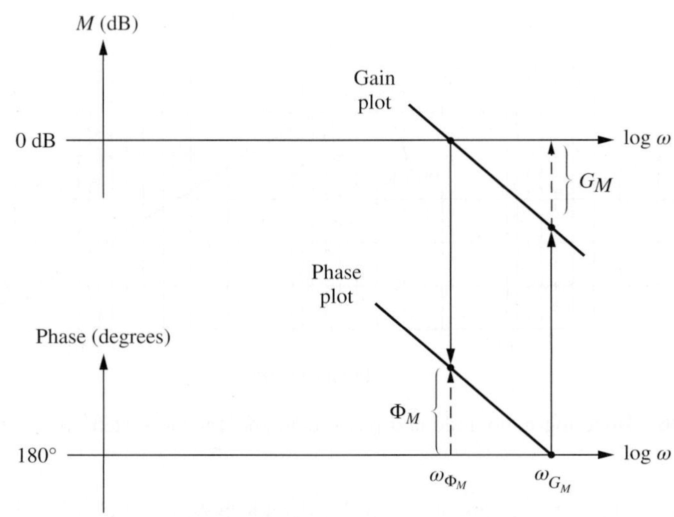


**FIGURE 10.37** Gain and phase margins on the Bode diagrams

## Example 10.10 Gain and Phase Margins from Bode Plots

**PROBLEM:** If $K = 200$ in the system of Example 10.9, find the gain margin and the phase margin.

**SOLUTION:** The Bode plot in Figure 10.36 is scaled to a gain of 40. If $K = 200$ (five times as great), the magnitude plot would be $20 \log 5 = 13.98$ dB higher.

To find the gain margin, look at the phase plot and find the frequency where the phase is 180°. At this frequency, determine from the magnitude plot how much the gain can be increased before reaching 0 dB. In Figure 10.36, the phase angle is 180° at approximately 7 rad/s. On the magnitude plot, the gain is $-20 + 13.98 = -6.02$ dB. Thus, the gain margin is 6.02 dB.

To find the phase margin, we look on the magnitude plot for the frequency where the gain is 0 dB. At this frequency, we look on the phase plot to find the difference between the phase and 180°. This difference is the phase margin. Again, remembering that the magnitude plot of Figure 10.36 is 13.98 dB lower than the actual plot, the 0 dB crossing ($-13.98$ dB for the normalized plot shown in Figure 10.36) occurs at 5.5 rad/s. At this frequency the phase angle is $-165°$. Thus, the phase margin is $-165° - (-180°) = 15°$.

GUI Tool

**GUIT**

`MATLAB's LTI Viewer, with Bode plots selected, is another method that may be used to find gain margin, phase margin, zero dB frequency, and 180° frequency. You are encouraged to study Appendix E at www.wiley.com/college/nise, which contains a tutorial on the LTI Viewer as well as some examples. Example E.3 solves Example 10.10 using the LTI Viewer.`

## Skill-Assessment Exercise 10.6

**PROBLEM:** For the system shown in Figure 10.10, where

$$G(s) = \frac{K}{(s+5)(s+20)(s+50)}$$

do the following:

**a.** Draw the Bode log-magnitude and phase plots.

**b.** Find the range of $K$ for stability from your Bode plots.

**c.** Evaluate gain margin, phase margin, zero dB frequency, and 180° frequency from your Bode plots for $K = 10{,}000$.

**ANSWERS:**

**a.** See the answer at www.wiley.com/college/nise.

**b.** $K < 96{,}270$

**c.** Gain margin = 19.67 dB, phase margin = 92.9°, zero dB frequency = 7.74 rad/s, and 180° frequency = 36.7 rad/s

The complete solution is at www.wiley.com/college/nise.

### TryIt 10.4

Use MATLAB, the Control System Toolbox, and the following statements to solve Skill-Assessment Exercise 10.6(c) using Bode plots.

```
G=zpk([],...
 [-5,-20,-50],10000)
bode(G)
grid on
```

After the Bode plot appears:

1. Right-click in the graph area.
2. Select **Characteristics**.
3. Select **All Stability Margins**.
4. Let the mouse rest on the margin points to read the gain and phase margins.

We have seen that the open-loop frequency response curves can be used not only to determine whether a system is stable but to calculate the range of loop gain that will ensure stability. We have also seen how to calculate the gain margin and the phase margin from the Bode diagrams.

Is it then possible to parallel the root locus technique and analyze and design systems for transient response using frequency response methods? We will begin to explore the answer in the next section.

# 10.8 Relation Between Closed-Loop Transient and Closed-Loop Frequency Responses

## Damping Ratio and Closed-Loop Frequency Response

In this section, we will show that a relationship exists between a system's transient response and its closed-loop frequency response. In particular, consider the second-order feedback control system of Figure 10.38, which we have been using since Chapter 4, where we derived relationships between the closed-loop transient response and the poles of the closed-loop transfer function,

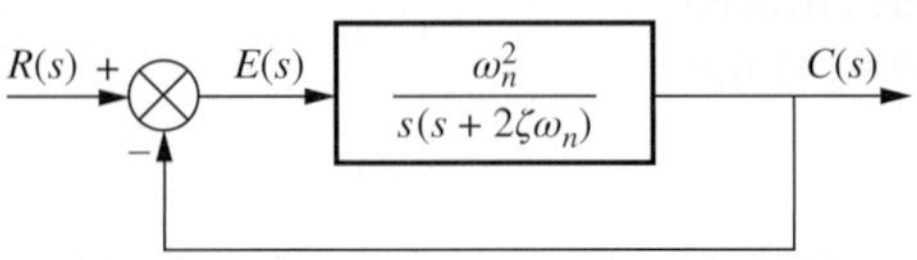


**FIGURE 10.38** Second-order closed-loop system

$$\frac{C(s)}{R(s)} = T(s) = \frac{\omega_n^2}{s^2 + 2\zeta\omega_n s + \omega_n^2} \tag{10.49}$$

We now derive relationships between the transient response of Eq. (10.49) and characteristics of its frequency response. We define these characteristics and relate them to damping ratio, natural frequency, settling time, peak time, and rise time. In Section 10.10, we will show how to use the frequency response of the open-loop transfer function

$$G(s) = \frac{\omega_n^2}{s(s + 2\zeta\omega_n)} \tag{10.50}$$

shown in Figure 10.38, to obtain the same transient response characteristics.

Let us now find the frequency response of Eq. (10.49), define characteristics of this response, and relate these characteristics to the transient response. Substituting $s = j\omega$ into Eq. (10.49), we evaluate the magnitude of the closed-loop frequency response as

$$M = |T(j\omega)| = \frac{\omega_n^2}{\sqrt{(\omega_n^2 - \omega^2)^2 + 4\zeta^2\omega_n^2\omega^2}} \tag{10.51}$$

A representative sketch of the log plot of Eq. (10.51) is shown in Figure 10.39.

We now show that a relationship exists between the peak value of the closed-loop magnitude response and the damping ratio. Squaring Eq. (10.51), differentiating with respect to $\omega^2$, and setting the derivative equal to zero yields the maximum value of $M$, $M_p$, where

$$M_p = \frac{1}{2\zeta\sqrt{1 - \zeta^2}} \tag{10.52}$$

at a frequency, $\omega_p$, of

$$\omega_p = \omega_n\sqrt{1 - 2\zeta^2} \tag{10.53}$$

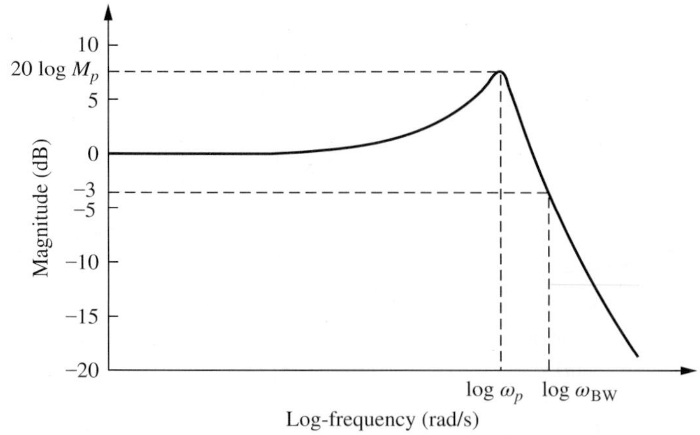


**FIGURE 10.39** Representative log-magnitude plot of Eq. (10.51)

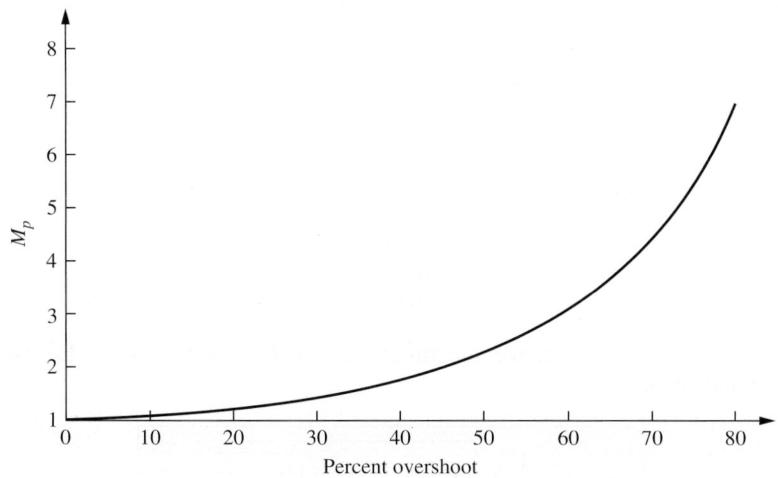


**FIGURE 10.40** Closed-loop frequency response peak vs. percent overshoot for a two-pole system

Since $\zeta$ is related to percent overshoot, we can plot $M_p$ vs. percent overshoot. The result is shown in Figure 10.40.

Equation (10.52) shows that the maximum magnitude on the frequency response curve is directly related to the damping ratio and, hence, the percent overshoot. Also notice from Eq. (10.53) that the peak frequency, $\omega_p$, is not the natural frequency. However, for low values of damping ratio, we can assume that the peak occurs at the natural frequency. Finally, notice that there will not be a peak at frequencies above zero if $\zeta > 0.707$. This limiting value of $\zeta$ for peaking on the magnitude response curve should not be confused with overshoot on the step response, where there is overshoot for $0 < \zeta < 1$.

## Response Speed and Closed-Loop Frequency Response

Another relationship between the frequency response and time response is between the speed of the time response (as measured by settling time, peak time, and rise time) and the *bandwidth* of the closed-loop frequency response. Bandwidth is defined here as the frequency, $\omega_{BW}$, at which the magnitude response curve is 3 dB down from its value at zero frequency (see Figure 10.39).

The bandwidth of a two-pole system can be found by finding that frequency for which $M = 1/\sqrt{2}$ (that is, $-3$ dB) in Eq. (10.51). The derivation is left as an exercise for the student. The result is

$$\omega_{BW} = \omega_n\sqrt{(1-2\zeta^2)+\sqrt{4\zeta^4-4\zeta^2+2}} \tag{10.54}$$

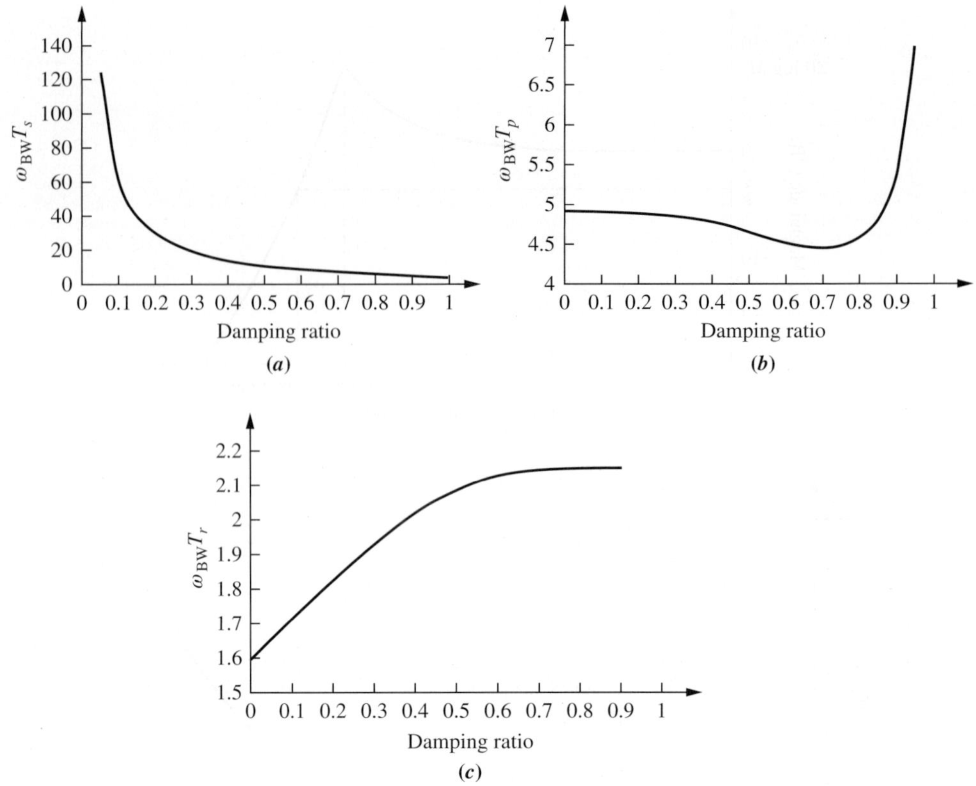


**FIGURE 10.41** Normalized bandwidth vs. damping ratio for **a.** settling time; **b.** peak time; **c.** rise time

To relate $\omega_{BW}$ to settling time, we substitute $\omega_n = 4/T_s\zeta$, derived from Eq. (4.42), into Eq. (10.54) and obtain

$$\omega_{BW} = \frac{4}{T_s\zeta}\sqrt{(1-2\zeta^2)+\sqrt{4\zeta^4-4\zeta^2+2}} \tag{10.55}$$

Similarly, since $\omega_n = \pi/(T_p\sqrt{1-\zeta^2})$,

$$\omega_{BW} = \frac{\pi}{T_p\sqrt{1-\zeta^2}}\sqrt{(1-2\zeta^2)+\sqrt{4\zeta^4-4\zeta^2+2}} \tag{10.56}$$

To relate the bandwidth to rise time, $T_r$, we use Figure 4.16, knowing the desired $\zeta$ and $T_r$. For example, assume $\zeta = 0.4$ and $T_r = 0.2$ second. Using Figure 4.16, the ordinate $T_r\omega_n = 1.463$, from which $\omega_n = 1.463/0.2 = 7.315$ rad/s. Using Eq. (10.54), $\omega_{BW} = 10.05$ rad/s. Normalized plots of Eqs. (10.55) and (10.56) and the relationship between bandwidth normalized by rise time and damping ratio are shown in Figure 10.41.

## Skill-Assessment Exercise 10.7

**PROBLEM:** Find the closed-loop bandwidth required for 20% overshoot and 2-seconds settling time.

**ANSWERS:** $\omega_{BW} = 5.79$ rad/s

The complete solution is at www.wiley.com/college/nise.

In this section, we related the closed-loop transient response to the closed-loop frequency response via bandwidth. We continue by relating the closed-loop frequency response to the open-loop frequency response and explaining the impetus.

# 10.9 Relation Between Closed- and Open-Loop Frequency Responses

At this point, we do not have an easy way of finding the closed-loop frequency response from which we could determine $M_p$ and thus the transient response.[2] As we have seen, we are equipped to rapidly sketch the open-loop frequency response but not the closed-loop frequency response. However, if the open-loop response is related to the closed-loop response, we can combine the ease of sketching the open-loop response with the transient response information contained in the closed-loop response.

## Constant *M* Circles and Constant *N* Circles

Consider a unity feedback system whose closed-loop transfer function is

$$T(s) = \frac{G(s)}{1 + G(s)} \tag{10.57}$$

The frequency response of this closed-loop function is

$$T(j\omega) = \frac{G(j\omega)}{1 + G(j\omega)} \tag{10.58}$$

Since $G(j\omega)$ is a complex number, let $G(j\omega) = P(\omega) + jQ(\omega)$ in Eq. (10.58), which yields

$$T(j\omega) = \frac{P(\omega) + jQ(\omega)}{[(P(\omega) + 1) + jQ(\omega)]} \tag{10.59}$$

Therefore,

$$M^2 = |T^2(j\omega)| = \frac{P^2(\omega) + Q^2(\omega)}{[(P(\omega) + 1)^2 + Q^2(\omega)]} \tag{10.60}$$

Equation (10.60) can be put into the form

$$\left(P + \frac{M^2}{M^2 - 1}\right)^2 + Q^2 = \frac{M^2}{(M^2 - 1)^2} \tag{10.61}$$

which is the equation of a circle of radius $M/(M^2 - 1)$ centered at $[-M^2/(M^2 - 1),\ 0]$. These circles, shown plotted in Figure 10.42 for various values of $M$, are called *constant M circles* and are the locus of the closed-loop magnitude frequency response for unity feedback systems. Thus, if the polar frequency response of an open-loop function, $G(s)$, is plotted and superimposed on top of the constant $M$ circles, the closed-loop magnitude frequency response is determined by each intersection of this polar plot with the constant $M$ circles.

[2] At the end of this subsection, we will see how to use MATLAB to obtain closed-loop frequency responses.

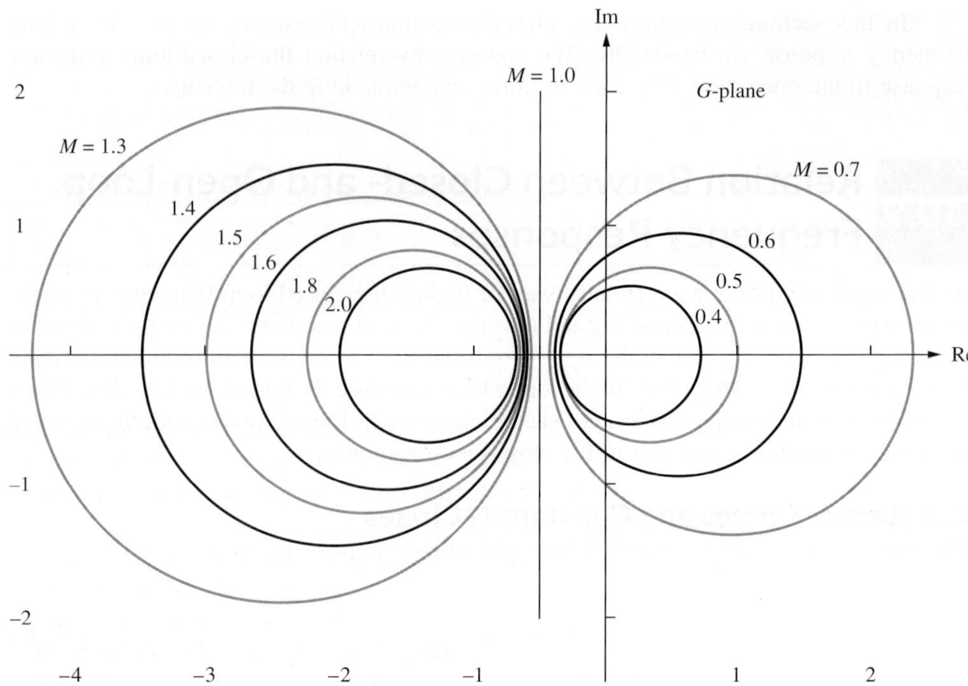


**FIGURE 10.42** Constant $M$ circles

Before demonstrating the use of the constant $M$ circles with an example, let us go through a similar development for the closed-loop phase plot, the constant $N$ circles. From Eq. (10.59), the phase angle, $\phi$, of the closed-loop response is

$$\begin{aligned}\phi &= \tan^{-1}\frac{Q(\omega)}{P(\omega)} - \tan^{-1}\frac{Q(\omega)}{P(\omega)+1} \\ &= \tan^{-1}\frac{\dfrac{Q(\omega)}{P(\omega)} - \dfrac{Q(\omega)}{P(\omega)+1}}{1+\dfrac{Q(\omega)}{P(\omega)}\left(\dfrac{Q(\omega)}{P(\omega)+1}\right)}\end{aligned} \tag{10.62}$$

after using $\tan(\alpha-\beta) = (\tan\alpha - \tan\beta)/(1+\tan\alpha\tan\beta)$. Dropping the functional notation,

$$\tan\phi = N = \frac{Q}{P^2+P+Q^2} \tag{10.63}$$

Equation (10.63) can be put into the form of a circle,

$$\left(P+\frac{1}{2}\right)^2 + \left(Q-\frac{1}{2N}\right)^2 = \frac{N^2+1}{4N^2} \tag{10.64}$$

which is plotted in Figure 10.43 for various values of $N$. The circles of this plot are called *constant N circles*. Superimposing a unity feedback, open-loop frequency response over the constant $N$ circles yields the closed-loop phase response of the system. Let us now look at an example of the use of the constant $M$ and $N$ circles.

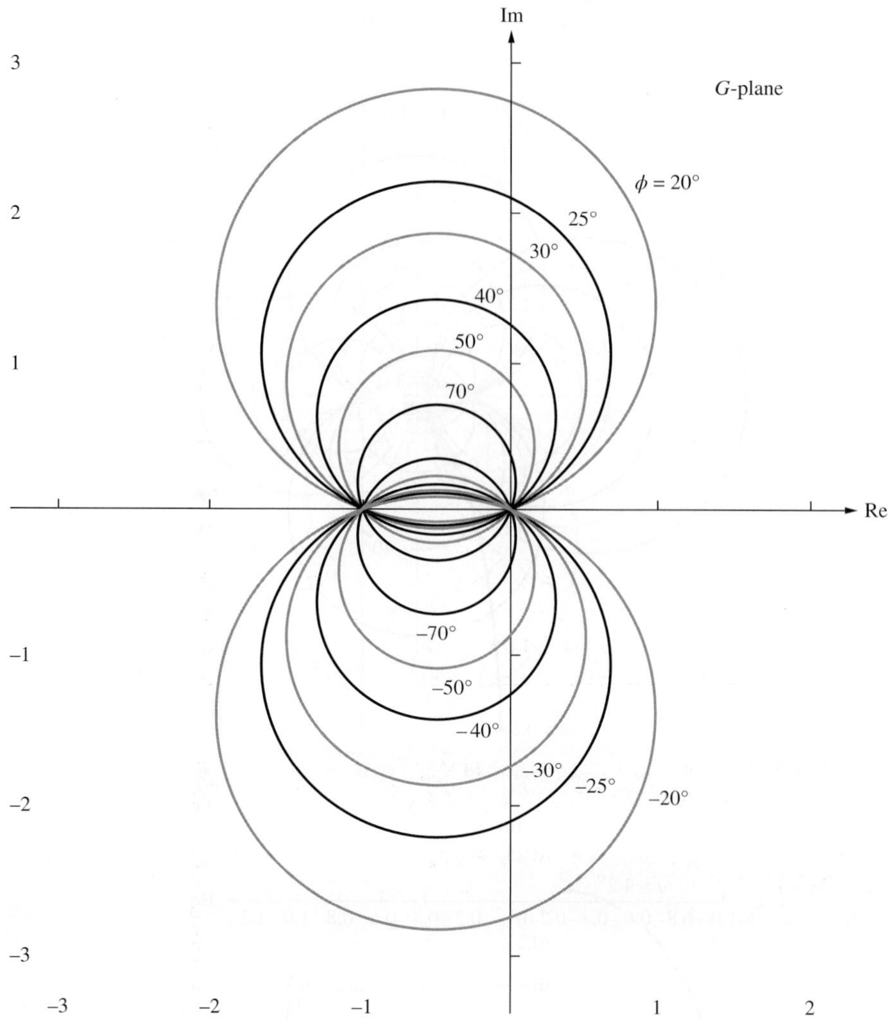


**FIGURE 10.43** Constant *N* circles

## Example 10.11 Closed-Loop Frequency Response from Open-Loop Frequency Response

**PROBLEM:** Find the closed-loop frequency response of the unity feedback system shown in Figure 10.10, where $G(s) = 50/[s(s+3)(s+6)]$, using the constant *M* circles, *N* circles, and the open-loop polar frequency response curve.

**SOLUTION:** First evaluate the open-loop frequency function and make a polar frequency response plot superimposed over the constant *M* and *N* circles. The open-loop frequency function is

$$G(j\omega) = \frac{50}{-9\omega^2 + j(18\omega - \omega^3)} \tag{10.65}$$

from which the magnitude, $|G(j\omega)|$, and phase, $\angle G(j\omega)$, can be found and plotted. The polar plot of the open-loop frequency response (Nyquist diagram) is shown superimposed over the *M* and *N* circles in Figure 10.44.

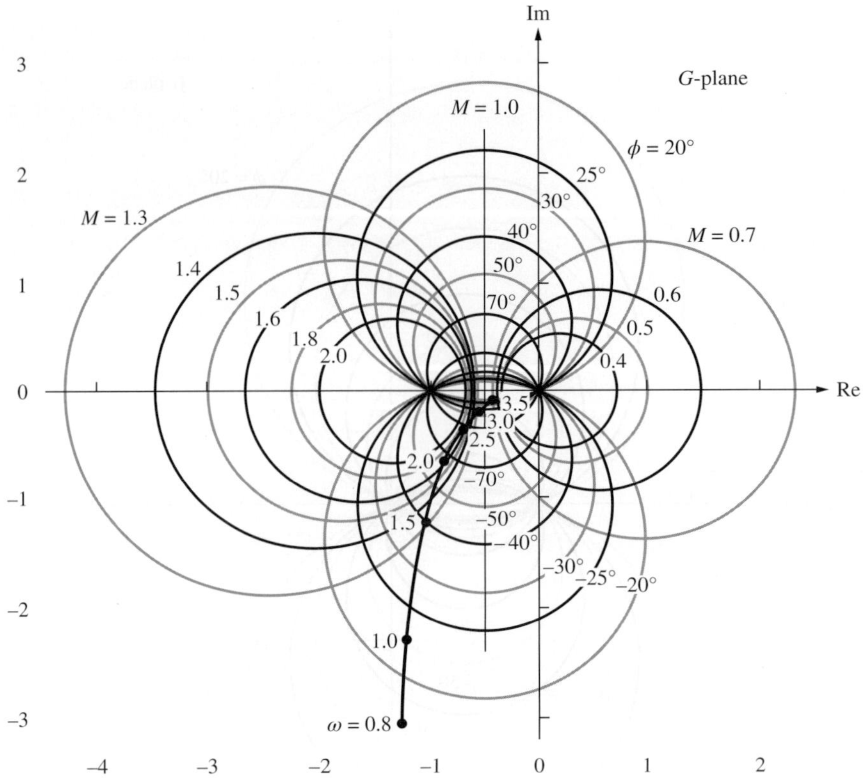


**FIGURE 10.44** Nyquist diagram for Example 10.11 and constant *M* and *N* circles

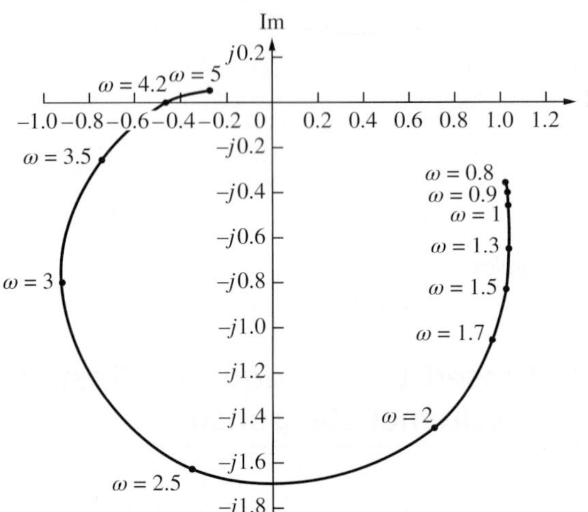


**FIGURE 10.45** Closed-loop frequency response for Example 10.11

The closed-loop magnitude frequency response can now be obtained by finding the intersection of each point of the Nyquist plot with the *M* circles. The closed-loop phase response can be obtained by finding the intersection of each point of the Nyquist plot with the *N* circles. The result is shown in Figure 10.45.[3]

MATLAB
**ML**

`Students who are using MATLAB should now run ch10p5 in Appendix B. You will learn how to use MATLAB to find the closed-loop frequency response. This exercise solves Example 10.11 using MATLAB.`

[3] You are cautioned not to use the *closed-loop* polar plot for the Nyquist criterion. The closed-loop frequency response, however, can be used to determine the closed-loop transient response, as discussed in Section 10.8.

## Nichols Charts

A disadvantage of using the $M$ and $N$ circles is that changes of gain in the open-loop transfer function, $G(s)$, cannot be handled easily. For example, in the Bode plot, a gain change is handled by moving the Bode magnitude curve up or down an amount equal to the gain change in dB. Since the $M$ and $N$ circles are not dB plots, changes in gain require each point of $G(j\omega)$ to be multiplied in length by the increase or decrease in gain.

Another presentation of the $M$ and $N$ circles, called a *Nichols chart*, displays the constant $M$ circles in dB, so that changes in gain are as simple to handle as in the Bode plot. A Nichols chart is shown in Figure 10.46. The chart is a plot of open-loop magnitude in dB vs. open-loop phase angle in degrees. Every point on the $M$ circles can be transferred to the Nichols chart. Each point on the constant $M$ circles is represented by magnitude and angle (polar coordinates). Converting the magnitude to dB, we can transfer the point to the Nichols chart, using the polar coordinates with magnitude in dB plotted as the ordinate, and the phase angle plotted as the abscissa. Similarly, the $N$ circles also can be transferred to the Nichols chart.

For example, assume the function

$$G(s) = \frac{K}{s(s+1)(s+2)} \tag{10.66}$$

Superimposing the frequency response of $G(s)$ on the Nichols chart by plotting magnitude in dB vs. phase angle for a range of frequencies from 0.1 to 1 rad/s, we obtain the plot in Figure 10.47 for $K = 1$. If the gain is increased by 10 dB, simply raise the curve for $K = 1$ by 10 dB and obtain the curve for $K = 3.16$ (10 dB). The intersection of the plots of $G(j\omega)$ with the Nichols chart yields the frequency response of the closed-loop system.

MATLAB
ML

Students who are using MATLAB should now run ch10p6 in Appendix B. You will learn how to use MATLAB to make a Nichols plot. This exercise makes a Nichols plot of $G(s) = 1/[s(s+1)(s+2)]$ using MATLAB.

GUI Tool
GUIT

MATLAB's LTI Viewer is an alternative method of obtaining the Nichols chart. You are encouraged to study Appendix E at www.wiley.com/college/nise, which contains a tutorial on the LTI Viewer as well as some examples. Example E.4 shows how to obtain Figure 10.47 using the LTI Viewer.

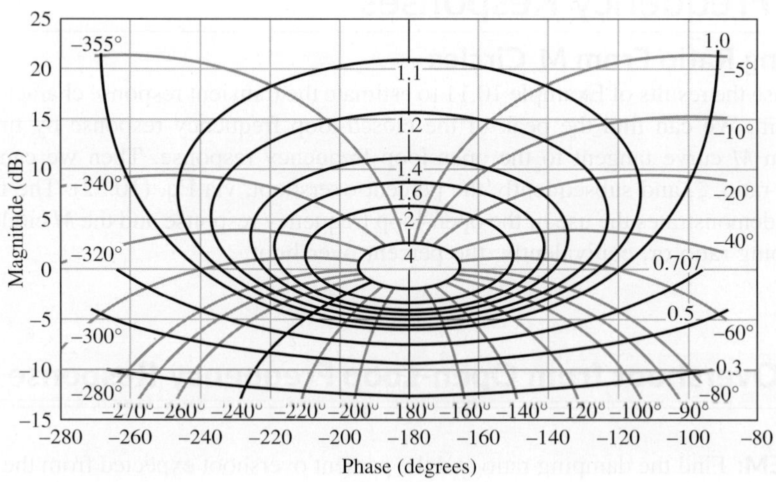


**FIGURE 10.46** Nichols chart

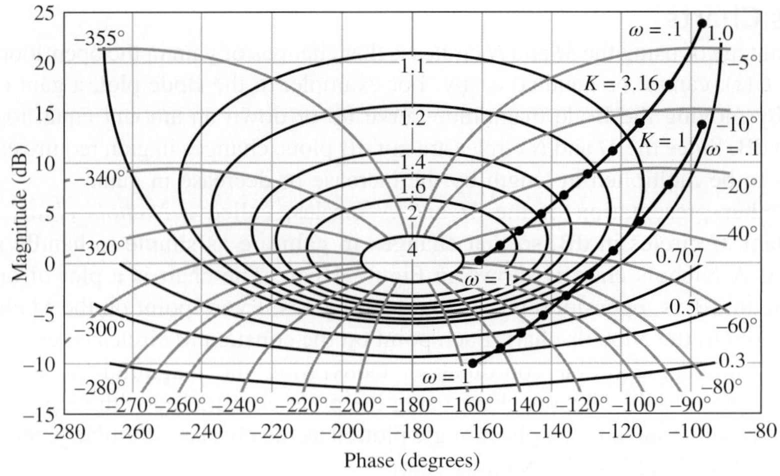


**FIGURE 10.47** Nichols chart with frequency response for $G(s) = K/[s(s+1)(s+2)]$ superimposed. Values for $K = 1$ and $K = 3.16$ are shown.

## Skill-Assessment Exercise 10.8

**TryIt 10.5**

Use MATLAB, the Control System Toolbox, and the following statements to make a Nichols chart of the system given in Skill-Assessment Exercise 10.8

```
G=zpk([],...
  [-5,-20,-50],8000)
nichols(G)
grid on
```

**PROBLEM:** Given the system shown in Figure 10.10, where

$$G(s) = \frac{8000}{(s+5)(s+20)(s+50)}$$

plot the closed-loop log-magnitude and phase frequency response plots using the following methods:

**a.** $M$ and $N$ circles

**b.** Nichols chart

**ANSWERS:** The complete solution is at www.wiley.com/college/nise.

# 10.10 Relation Between Closed-Loop Transient and Open-Loop Frequency Responses

## Damping Ratio From *M* Circles

We can use the results of Example 10.11 to estimate the transient response characteristics of the system. We can find the peak of the closed-loop frequency response by finding the maximum $M$ curve tangent to the open-loop frequency response. Then we can find the damping ratio, $\zeta$, and subsequently the percent overshoot, via Eq. (10.52). The following example demonstrates the use of the open-loop frequency response and the $M$ circles to find the damping ratio or, equivalently, the percent overshoot.

## Example 10.12 Percent Overshoot from Open-Loop Frequency Response

**PROBLEM:** Find the damping ratio and the percent overshoot expected from the system of Example 10.11, using the open-loop frequency response and the $M$ circles.

**SOLUTION:** Equation (10.52) shows that there is a unique relationship between the closed-loop system's damping ratio and the peak value, $M_P$, of the closed-loop system's magnitude frequency plot. From Figure 10.44, we see that the Nyquist diagram is tangent to the 1.8 $M$ circle. We see that this is the maximum value for the closed-loop frequency response. Thus, $M_p = 1.8$.

We can solve for $\zeta$ by rearranging Eq. (10.52) into the following form:

$$\zeta^4 - \zeta^2 + (1/4M_p^2) = 0 \quad (10.67)$$

Since $M_p = 1.8$, then $\zeta = 0.29$ and $0.96$. From Eq. (10.53), a damping ratio larger than 0.707 yields no peak above zero frequency. Thus, we select $\zeta = 0.29$, which is equivalent to 38.6% overshoot. Care must be taken, however, to be sure we can make a second-order approximation when associating the value of percent overshoot to the value of $\zeta$. A computer simulation of the step response shows 36% overshoot.

So far in this section, we have tied together the system's transient response and the peak value of the closed-loop frequency response as obtained from the open-loop frequency response. We used the Nyquist plots and the $M$ and $N$ circles to obtain the closed-loop transient response. Another association exists between the open-loop frequency response and the closed-loop transient response that is easily implemented with the Bode plots, which are easier to draw than the Nyquist plots.

## Damping Ratio from Phase Margin

Let us now derive the relationship between the phase margin and the damping ratio. This relationship will enable us to evaluate the percent overshoot from the phase margin found from the open-loop frequency response.

Consider a unity feedback system whose open-loop function

$$G(s) = \frac{\omega_n^2}{s(s + 2\zeta\omega_n)} \quad (10.68)$$

yields the typical second-order, closed-loop transfer function

$$T(s) = \frac{\omega_n^2}{s^2 + 2\zeta\omega_n s + \omega_n^2} \quad (10.69)$$

In order to evaluate the phase margin, we first find the frequency for which $|G(j\omega)| = 1$. Hence,

$$|G(j\omega)| = \frac{\omega_n^2}{|-\omega^2 + j2\zeta\omega_n\omega|} = 1 \quad (10.70)$$

The frequency, $\omega_1$, that satisfies Eq. (10.70) is

$$\omega_1 = \omega_n\sqrt{-2\zeta^2 + \sqrt{1 + 4\zeta^4}} \quad (10.71)$$

The phase angle of $G(j\omega)$ at this frequency is

$$\begin{aligned} \angle G(j\omega) &= -90 - \tan^{-1}\frac{\omega_1}{2\zeta\omega_n} \\ &= -90 - \tan^{-1}\frac{\sqrt{-2\zeta^2 + \sqrt{4\zeta^4 + 1}}}{2\zeta} \end{aligned} \quad (10.72)$$

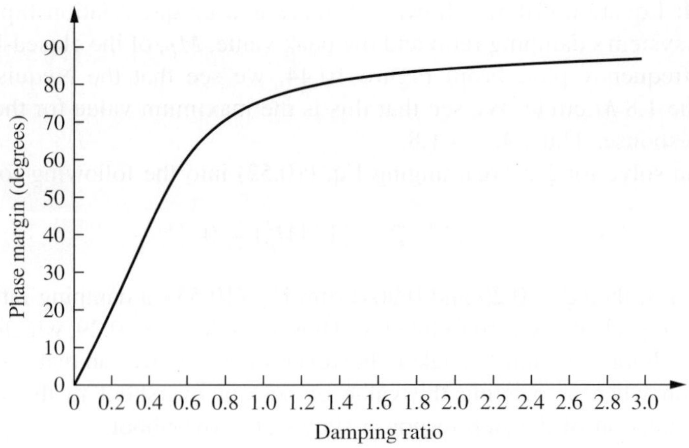


**FIGURE 10.48** Phase margin vs. damping ratio

The difference between the angle of Eq. (10.72) and $-180°$ is the phase margin, $\phi_M$. Thus,

$$\begin{aligned}\Phi_M &= 90 - \tan^{-1}\frac{\sqrt{-2\zeta^2+\sqrt{1+4\zeta^4}}}{2\zeta}\\ &= \tan^{-1}\frac{2\zeta}{\sqrt{-2\zeta^2+\sqrt{1+4\zeta^4}}}\end{aligned} \tag{10.73}$$

Equation (10.73), plotted in Figure 10.48, shows the relationship between phase margin and damping ratio.

As an example, Eq. (10.53) tells us that there is no peak frequency if $\zeta = 0.707$. Hence, there is no peak to the closed-loop magnitude frequency response curve for this value of damping ratio and larger. Thus, from Figure 10.48, a phase margin of $65.52°(\zeta = 0.707)$ or larger is required from the *open-loop* frequency response to ensure there is no peaking in the *closed-loop* frequency response.

## Response Speed from Open-Loop Frequency Response

Equations (10.55) and (10.56) relate the closed-loop bandwidth to the desired settling or peak time and the damping ratio. We now show that the closed-loop bandwidth can be estimated from the open-loop frequency response. From the Nichols chart in Figure 10.46, we see the relationship between the open-loop gain and the closed-loop gain. The $M = 0.707(-3\text{ dB})$ curve, replotted in Figure 10.49 for clarity, shows the open-loop gain when the closed-loop gain is $-3$ dB. This relationship typically occurs at $\omega_{BW}$ if the low-frequency closed-loop gain is 0 dB. We can approximate Figure 10.49 by saying that

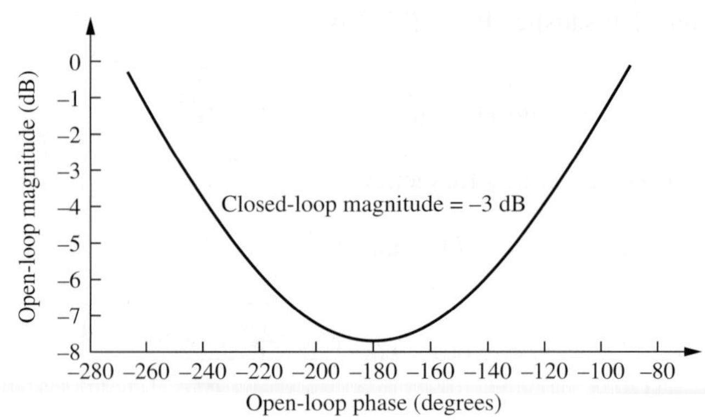


**FIGURE 10.49** Open-loop gain vs. open-loop phase angle for $-3$ dB closed-loop gain

the closed-loop bandwidth, $\omega_{BW}$ (the frequency at which the closed-loop magnitude response is $-3$ dB), equals the frequency at which the open-loop magnitude response is between $-6$ and $-7.5$ dB if the open-loop phase response is between $-135°$ and $-225°$. Then, using a second-order system approximation, Eqs. (10.55) and (10.56) can be used, along with the desired damping ratio, $\zeta$, to find settling time and peak time, respectively. Let us look at an example.

## Example 10.13 Settling and Peak Times from Open-Loop Frequency Response

**PROBLEM:** Given the system of Figure 10.50($a$) and the Bode diagrams of Figure 10.50($b$), estimate the settling time and peak time.

**SOLUTION:** Using Figure 10.50($b$), we estimate the closed-loop bandwidth by finding the frequency where the open-loop magnitude response is in the range of $-6$ to $-7.5$ dB if the phase response is in the range of $-135°$ to $-225°$. Since Figure 10.50($b$) shows $-6$ to $-7.5$ dB at approximately 3.7 rad/s with a phase response in the stated region, $\omega_{BW} \cong 3.7$ rad/s.

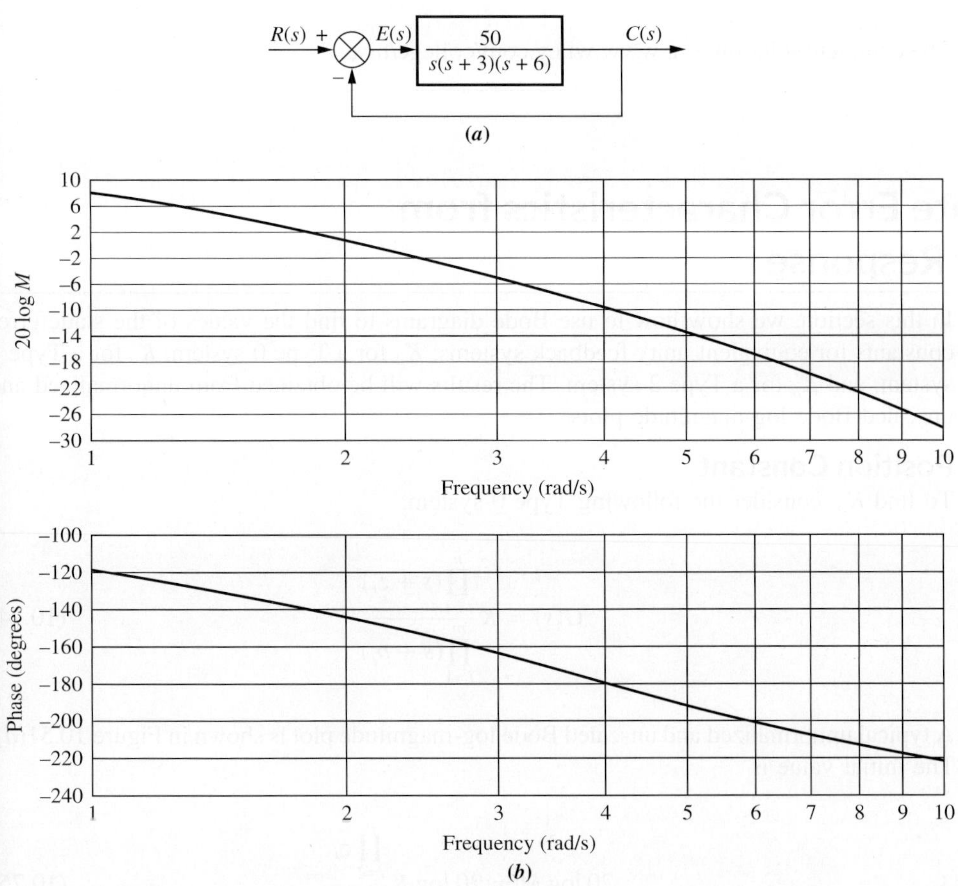


**FIGURE 10.50** **a.** Block diagram; **b.** Bode diagrams for system of Example 10.13

Next find $\zeta$ via the phase margin. From Figure 10.50(*b*), the phase margin is found by first finding the frequency at which the magnitude plot is 0 dB. At this frequency, 2.2 rad/s, the phase is about $-145^\circ$. Hence, the phase margin is approximately $(-145^\circ - (-180^\circ)) = 35^\circ$. Using Figure 10.48, $\zeta = 0.32$. Finally, using Eqs. (10.55) and (10.56), with the values of $\omega_{BW}$ and $\zeta$ just found, $T_s = 4.86$ seconds and $T_p = 129$ seconds. Checking the analysis with a computer simulation shows $T_s = 5.5$ seconds, and $T_p = 1.43$ seconds.

## Skill-Assessment Exercise 10.9

**PROBLEM:** Using the open-loop frequency response for the system in Figure 10.10, where

$$G(s) = \frac{100}{s(s+5)}$$

estimate the percent overshoot, settling time, and peak time for the closed-loop step response.

**ANSWERS:** $\%OS = 44\%$, $T_s = 1.64$ s, and $T_P = 0.33$ s

The complete solution is at www.wiley.com/college/nise.

# 10.11 Steady-State Error Characteristics from Frequency Response

In this section, we show how to use Bode diagrams to find the values of the static error constants for equivalent unity feedback systems: $K_p$ for a Type 0 system, $K_v$ for a Type 1 system, and $K_a$ for a Type 2 system. The results will be obtained from unnormalized and unscaled Bode log-magnitude plots.

### Position Constant

To find $K_p$, consider the following Type 0 system:

$$G(s) = K\frac{\prod_{i=1}^{n}(s+z_i)}{\prod_{i=1}^{m}(s+p_i)} \tag{10.74}$$

A typical unnormalized and unscaled Bode log-magnitude plot is shown in Figure 10.51(*a*). The initial value is

$$20 \log M = 20 \log K\frac{\prod_{i=1}^{n} z_i}{\prod_{i=1}^{m} p_i} \tag{10.75}$$

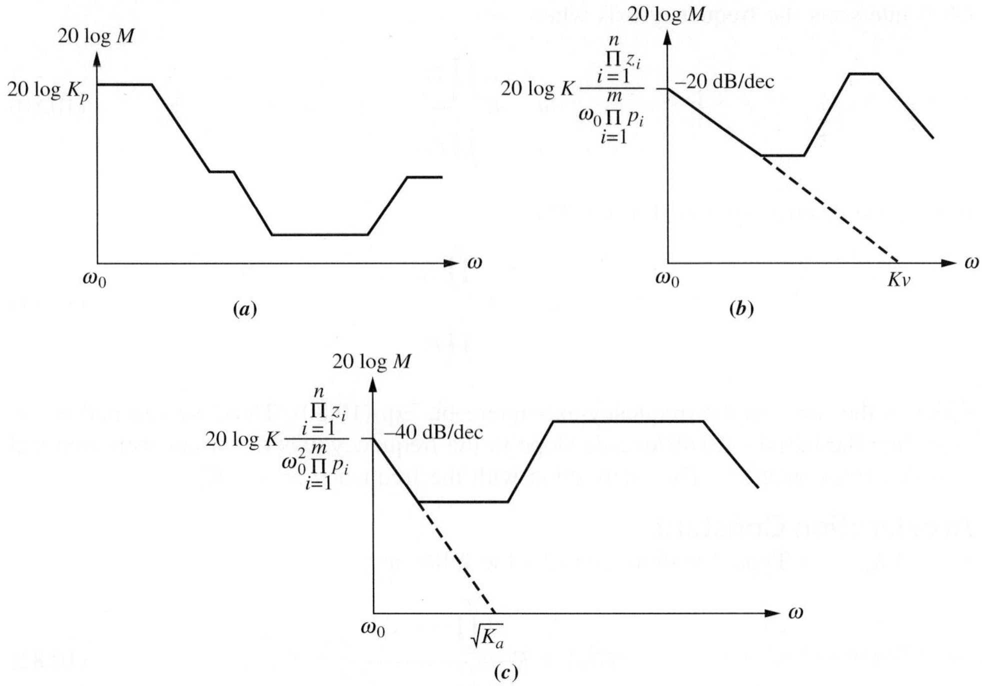


**FIGURE 10.51** Typical unnormalized and unscaled Bode log-magnitude plots showing the value of static error constants: **a.** Type 0; **b.** Type 1; **c.** Type 2

But for this system

$$K_p = K\frac{\prod_{i=1}^{n} z_i}{\prod_{i=1}^{m} p_i} \tag{10.76}$$

which is the same as the value of the low-frequency axis. Thus, for an unnormalized and unscaled Bode log-magnitude plot, the low-frequency magnitude is 20 log $K_p$ for a Type 0 system.

## Velocity Constant

To find $K_v$ for a Type 1 system, consider the following open-loop transfer function of a Type 1 system:

$$G(s) = K\frac{\prod_{i=1}^{n}(s + z_i)}{s\prod_{i=1}^{m}(s + p_i)} \tag{10.77}$$

A typical unnormalized and unscaled Bode log-magnitude diagram is shown in Figure 10.51(*b*) for this Type 1 system. The Bode plot starts at

$$20 \log M = 20 \log K\frac{\prod_{i=1}^{n} z_i}{\omega_0 \prod_{i=1}^{m} p_i} \tag{10.78}$$

The initial −20 dB/decade slope can be thought of as originating from a function,

$$G'(s) = K\frac{\prod_{i=1}^{n} z_i}{s\prod_{i=1}^{m} p_i} \tag{10.79}$$

$G'(s)$ intersects the frequency axis when

$$\omega = K\frac{\prod_{i=1}^{n} z_i}{\prod_{i=1}^{m} p_i} \tag{10.80}$$

But for the original system (Eq. (10.77)),

$$K_v = K\frac{\prod_{i=1}^{n} z_i}{\prod_{i=1}^{m} p_i} \tag{10.81}$$

which is the same as the frequency-axis intercept, Eq. (10.80). Thus, we can find $K_v$ by extending the initial $-20$ dB/decade slope to the frequency axis on an unnormalized and unscaled Bode diagram. The intersection with the frequency axis is $K_v$.

## Acceleration Constant

To find $K_a$ for a Type 2 system, consider the following:

$$G(s) = K\frac{\prod_{i=1}^{n} (s+z_i)}{s^2\prod_{i=1}^{m} (s+p_i)} \tag{10.82}$$

A typical unnormalized and unscaled Bode plot for a Type 2 system is shown in Figure 10.51($c$). The Bode plot starts at

$$20\log M = 20\log K\frac{\prod_{i=1}^{n} z_i}{\omega_0^2\prod_{i=1}^{m} p_i} \tag{10.83}$$

The initial $-40$ dB/decade slope can be thought of as coming from a function,

$$G'(s) = K\frac{\prod_{i=1}^{n} z_i}{s^2\prod_{i=1}^{m} p_i} \tag{10.84}$$

$G'(s)$ intersects the frequency axis when

$$\omega = \sqrt{K\frac{\prod_{i=1}^{n} z_i}{\prod_{i=1}^{m} p_i}} \tag{10.85}$$

But for the original system (Eq. (10.82)),

$$K_a = K\frac{\prod_{i=1}^{n} z_i}{\prod_{i=1}^{m} p_i} \tag{10.86}$$

Thus, the initial $-40$ dB/decade slope intersects the frequency axis at $\sqrt{K_a}$.

## Example 10.14 Static Error Constants from Bode Plots

**PROBLEM:** For each unnormalized and unscaled Bode log-magnitude plot shown in Figure 10.52,

**a.** Find the system type.

**b.** Find the value of the appropriate static error constant.

**SOLUTION:** Figure 10.52(*a*) is a Type 0 system, since the initial slope is zero. The value of $K_p$ is given by the low-frequency asymptote value. Thus, $20 \log K_p = 25$, or $K_p = 17.78$.

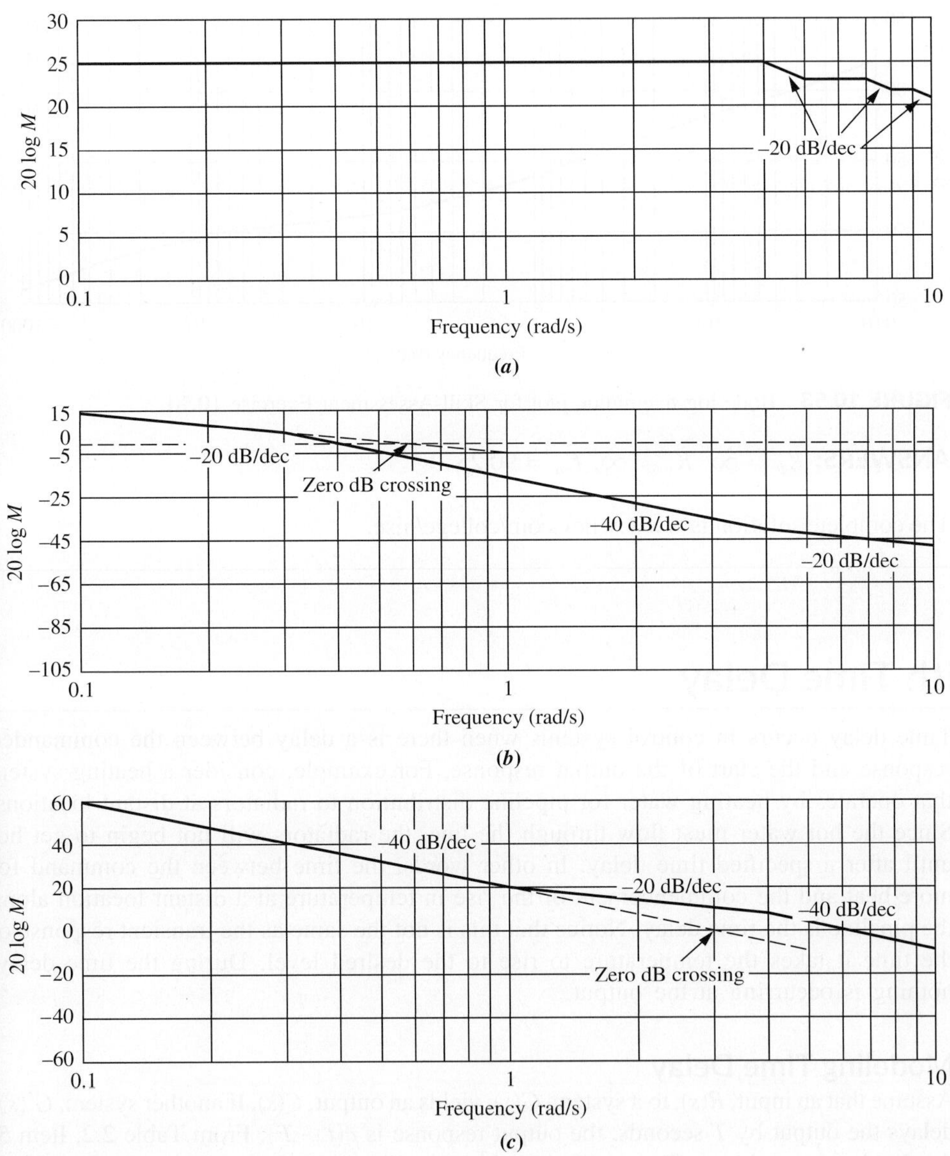


**FIGURE 10.52** Bode log-magnitude plots for Example 10.14

Figure 10.52(*b*) is a Type 1 system, since the initial slope is −20 dB/decade. The value of $K_v$ is the value of the frequency that the initial slope intersects at the zero dB crossing of the frequency axis. Hence, $K_v = 0.55$.

Figure 10.52(*c*) is a Type 2 system, since the initial slope is −40 dB/decade. The value of $\sqrt{K_a}$ is the value of the frequency that the initial slope intersects at the zero dB crossing of the frequency axis. Hence, $K_a = 3^2 = 9$.

## Skill-Assessment Exercise 10.10

**PROBLEM:** Find the static error constants for a stable unity feedback system whose open-loop transfer function has the Bode magnitude plot shown in Figure 10.53.

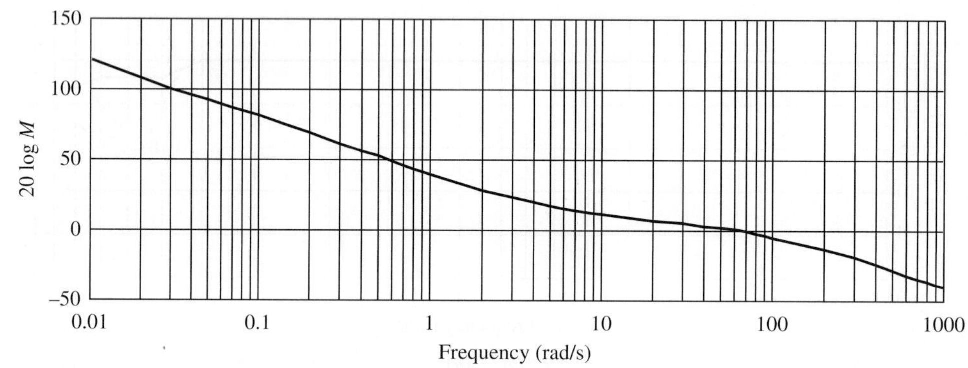


**FIGURE 10.53** Bode log-magnitude plot for Skill-Assessment Exercise 10.10

**ANSWERS:** $K_p = \infty$, $K_v = \infty$, $K_a = 90.25$

The complete solution is www.wiley.com/college/nise.

# 10.12 Systems with Time Delay

Time delay occurs in control systems when there is a delay between the commanded response and the start of the output response. For example, consider a heating system that operates by heating water for pipeline distribution to radiators at distant locations. Since the hot water must flow through the line, the radiators will not begin to get hot until after a specified time delay. In other words, the time between the command for more heat and the commencement of the rise in temperature at a distant location along the pipeline is the time delay. Notice that this is not the same as the transient response or the time it takes the temperature to rise to the desired level. During the time delay, nothing is occurring at the output.

## Modeling Time Delay

Assume that an input, $R(s)$, to a system, $G(s)$, yields an output, $C(s)$. If another system, $G'(s)$, delays the output by $T$ seconds, the output response is $c(t - T)$. From Table 2.2, Item 5, the Laplace transform of $c(t - T)$ is $e^{-sT}C(s)$. Thus, for the system without delay, $C(s) = R(s)G(s)$, and for the system with delay, $e^{-sT}C(s) = R(s)G'(s)$. Dividing these

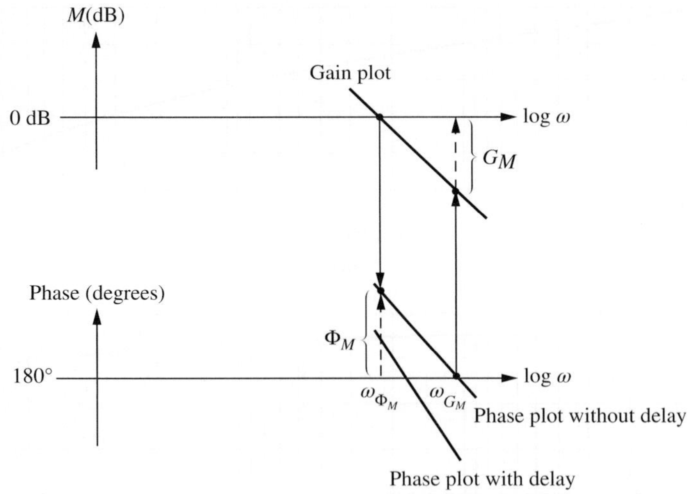


**FIGURE 10.54** Effect of delay upon frequency response

two equations, $G'(s)/G(s) = e^{-sT}$. Thus, a system with time delay $T$ can be represented in terms of an equivalent system without time delay as follows:

$$G'(s) = e^{-sT}G(s) \tag{10.87}$$

The effect of introducing time delay into a system can also be seen from the perspective of the frequency response by substituting $s = j\omega$ in Eq. (10.87). Hence,

$$G'(j\omega) = e^{-j\omega T}G(j\omega) = |G(j\omega)|\angle\{-\omega T + \angle G(j\omega)\} \tag{10.88}$$

In other words, the time delay does not affect the magnitude frequency response curve of $G(j\omega)$, but it does subtract a linearly increasing phase shift, $\omega T$, from the phase frequency response plot of $G(j\omega)$.

The typical effect of adding time delay can be seen in Figure 10.54. Assume that the gain and phase margins as well as the gain- and phase-margin frequencies shown in the figure apply to the system without delay. From the figure, we see that the reduction in phase shift caused by the delay reduces the phase margin. Using a second-order approximation, this reduction in phase margin yields a reduced damping ratio for the closed-loop system and a more oscillatory response. The reduction of phase also leads to a reduced gain-margin frequency. From the magnitude curve, we can see that a reduced gain-margin frequency leads to reduced gain margin, thus moving the system closer to instability.

An example of plotting frequency response curves for systems with delay follows.

## Example 10.15 Frequency Response Plots of a System with Time Delay

**PROBLEM:** Plot the frequency response for the system $G(s) = K/[s(s+1)(s+10)]$ if there is a time delay of 1 second through the system. Use the Bode plots.

**SOLUTION:** Since the magnitude curve is not affected by the delay, it can be plotted by the methods previously covered in the chapter and is shown in Figure 10.55($a$) for $K = 1$.

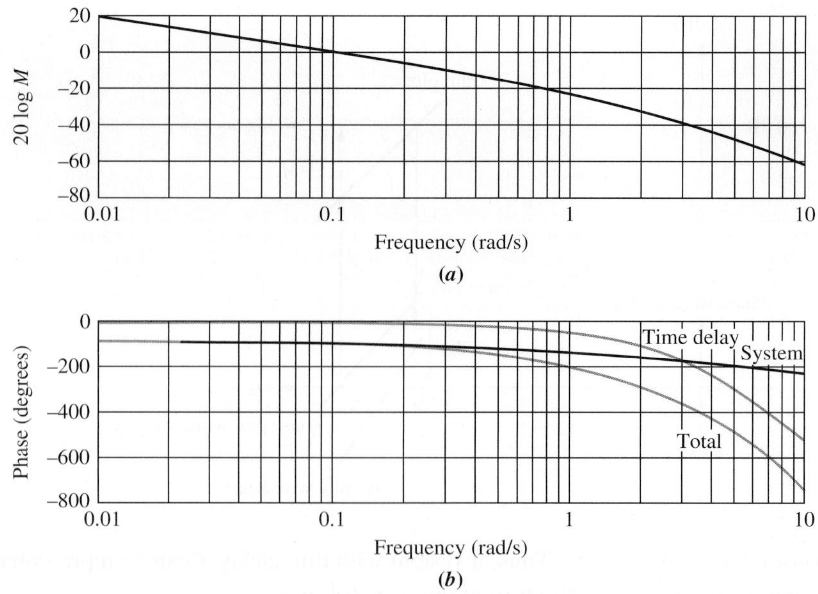


**FIGURE 10.55** Frequency response plots for $G(s) = K/[s(s+1)(s+10)]$ with a delay of 1 second and $K = 1$: **a.** magnitude plot; **b.** phase plot

The phase plot, however, is affected by the delay. Figure 10.55(*b*) shows the result. First draw the phase plot for the delay, $e^{-j\omega T} = 1\angle -\omega T = 1\angle -\omega$, since $T = 1$ from the problem statement. Next draw the phase plot of the system, $G(j\omega)$, using the methods previously covered. Finally, add the two phase curves together to obtain the total phase response for $e^{-j\omega T}G(j\omega)$. Be sure to use consistent units for the phase angles of $G(j\omega)$ and the delay; either degrees or radians.

Notice that the delay yields a decreased phase margin, since at any frequency the phase angle is more negative. Using a second-order approximation, this decrease in phase margin implies a lower damping ratio and a more oscillatory response for the closed-loop system.

Further, there is a decrease in the gain-margin frequency. On the magnitude curve, note that a reduction in the gain-margin frequency shows up as reduced gain margin, thus moving the system closer to instability.

MATLAB

ML

Students who are using MATLAB should now run ch10p7 in Appendix B. You will learn how to use MATLAB to include time delay on Bode plots. You will also use MATLAB to make multiple plots on one graph and label the plots. This exercise solves Example 10.15 using MATLAB.

Let us now use the results of Example 10.15 to design stability and analyze transient response and compare the results to the system without time delay.

## Example 10.16 Range of Gain for Stability for System with Time Delay

**PROBLEM:** The open-loop system with time delay in Example 10.15 is used in a unity feedback configuration. Do the following:

**a.** Find the range of gain, *K*, to yield stability. Use Bode plots and frequency response techniques.

**b.** Repeat Part a for the system without time delay.

**SOLUTION:**

**a.** From Figure 10.55, the phase angle is $-180°$ at a frequency of 0.81 rad/s for the system with time delay, marked "Total" on the phase plot. At this frequency, the magnitude curve is at $-20.39$ dB. Thus, $K$ can be raised from its current value of unity to $10^{20.39/20} = 10.46$. Hence, the system is stable for $0 < K \leq 10.46$.

**b.** If we use the phase curve without delay, marked "System," $-180°$ occurs at a frequency of 3.16 rad/s, and $K$ can be raised 40.84 dB or 110.2. Thus, without delay the system is stable for $0 < K \leq 110.2$, an order of magnitude larger.

## Example 10.17 Percent Overshoot for System with Time Delay

**PROBLEM:** The open-loop system with time delay in Example 10.15 is used in a unity feedback configuration. Do the following:

**a.** Estimate the percent overshoot if $K = 5$. Use Bode plots and frequency response techniques.

**b.** Repeat Part **a** for the system without time delay.

**SOLUTION:**

**a.** Since $K = 5$, the magnitude curve of Figure 10.55 is raised by 13.98 dB. The zero dB crossing then occurs at a frequency of 0.47 rad/s with a phase angle of $-145°$, as seen from the phase plot marked "Total." Therefore, the phase margin is $(-145° - (-180°)) = 35°$. Assuming a second-order approximation and using Eq. (10.73) or Figure 10.48, we find $\zeta = 0.33$. From Eq. (4.38), $\%OS = 33\%$. The time response, Figure 10.56($a$), shows a 38% overshoot instead of the predicted 33%. Notice the time delay at the start of the curve.

**b.** The zero dB crossing occurs at a frequency of 0.47 rad/s with a phase angle of $-118°$, as seen from the phase plot marked "System." Therefore, the phase margin is $(-118° - (-180°)) = 62°$. Assuming a second-order approximation and using

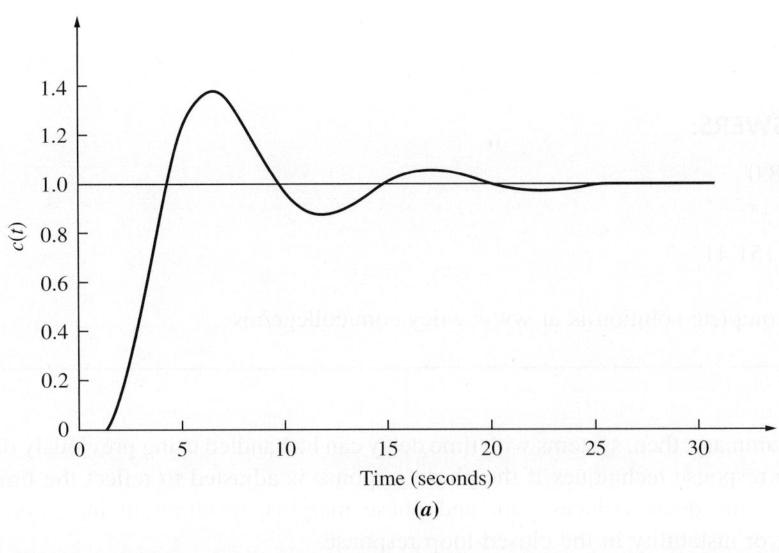


**FIGURE 10.56** Step response for closed-loop system with $G(s) = 5/[s(s+1)(s+10)]$: **a.** with a 1-second delay; *(figure continues)*

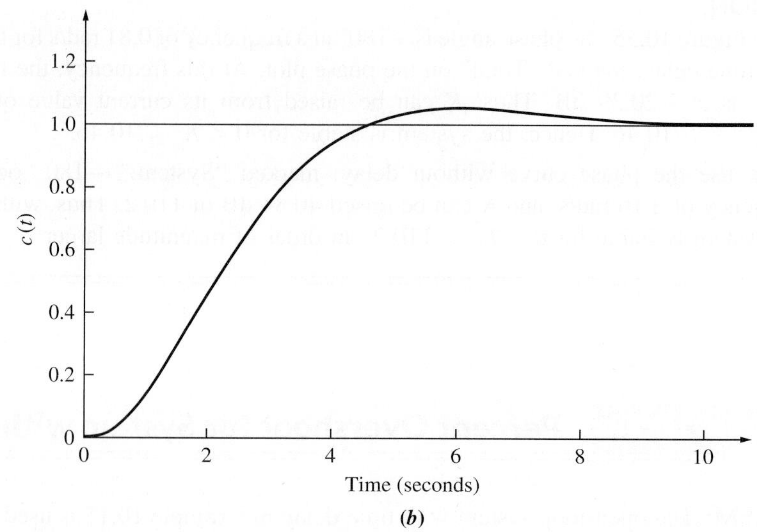


**FIGURE 10.56** *(Continued)*
**b.** without delay

Eq. (10.73) or Figure 10.48, we find $\zeta = 0.64$. From Eq. (4.38), $\%OS = 7.3\%$. The time response is shown in Figure 10.56(*b*). Notice that the system without delay has less overshoot and a smaller settling time.

## Skill-Assessment Exercise 10.11

**PROBLEM:** For the system shown in Figure 10.10, where

$$G(s) = \frac{10}{s(s+1)}$$

find the phase margin if there is a delay in the forward path of

**a.** 0 s

**b.** 0.1 s

**c.** 3 s

**ANSWERS:**

**a.** 18.0°

**b.** 0.35°

**c.** −151.41°

The complete solution is at www.wiley.com/college/nise.

### TryIt 10.6

Use MATLAB, the Control System Toolbox, and the following statements to solve Skill-Assessment Exercise 10.11. For each part of the problem let $d =$ the specified delay.

```
G=zpk([],[0,-1],10)
d=0
[numGd,denGd]=pade...
  (d,12)
Gd=tf(numGd,denGd)
Ge=G*Gd
bode(Ge)
grid on
```

After the Bode diagrams appear:

1. Right-click in the graph area.
2. Select **Characteristics.**
3. Select **All Stability Margins.**
4. Let the mouse rest on the margin point on the phase plot to read the phase margin.

In summary, then, systems with time delay can be handled using previously described frequency response techniques if the phase response is adjusted to reflect the time delay. Typically, time delay reduces gain and phase margins, resulting in increased percent overshoot or instability in the closed-loop response.

## 10.13 Obtaining Transfer Functions Experimentally

In Chapter 4, we discussed how to obtain the transfer function of a system through step-response testing. In this section, we show how to obtain the transfer function using sinusoidal frequency response data.

The analytical determination of a system's transfer function can be difficult. Individual component values may not be known, or the internal configuration of the system may not be accessible. In such cases, the frequency response of the system, from input to output, can be obtained experimentally and used to determine the transfer function. To obtain a frequency response plot experimentally, we use a sinusoidal force or signal generator at the input to the system and measure the output steady-state sinusoid amplitude and phase angle (see Figure 10.2). Repeating this process at a number of frequencies yields data for a frequency response plot. Referring to Figure 10.2(*b*), the amplitude response is $M(\omega) = M_o(\omega)/M_i(\omega)$, and the phase response is $\phi(\omega) = \phi_o(\omega) - \phi_i(\omega)$. Once the frequency response is obtained, the transfer function of the system can be estimated from the break frequencies and slopes. Frequency response methods can yield a more refined estimate of the transfer function than the transient response techniques covered in Chapter 4.

Bode plots are a convenient presentation of the frequency response data for the purpose of estimating the transfer function. These plots allow parts of the transfer function to be determined and extracted, leading the way to further refinements to find the remaining parts of the transfer function.

Although experience and intuition are invaluable in the process, the following steps are still offered as a guideline:

1. Look at the Bode magnitude and phase plots and estimate the pole-zero configuration of the system. Look at the initial slope on the magnitude plot to determine system type. Look at phase excursions to get an idea of the difference between the number of poles and the number of zeros.
2. See if portions of the magnitude and phase curves represent obvious first- or second-order pole or zero frequency response plots.
3. See if there is any telltale peaking or depressions in the magnitude response plot that indicate an underdamped second order pole or zero, respectively.
4. If any pole or zero responses can be identified, overlay appropriate $\pm 20$ or $\pm 40$dB/decade lines on the magnitude curve or $\pm 45^\circ$/decade lines on the phase curve and estimate the break frequencies. For second-order poles or zeros, estimate the damping ratio and natural frequency from the standard curves given in Section 10.2.
5. Form a transfer function of unity gain using the poles and zeros found. Obtain the frequency response of this transfer function and subtract this response from the previous frequency response (*Franklin, 1991*). You now have a frequency response of reduced complexity from which to begin the process again to extract more of the system's poles and zeros. A computer program such as MATLAB is of invaluable help for this step.

Let us demonstrate.

### Example 10.18 Transfer Function from Bode Plots

**PROBLEM:** Find the transfer function of the subsystem whose Bode plots are shown in Figure 10.57.

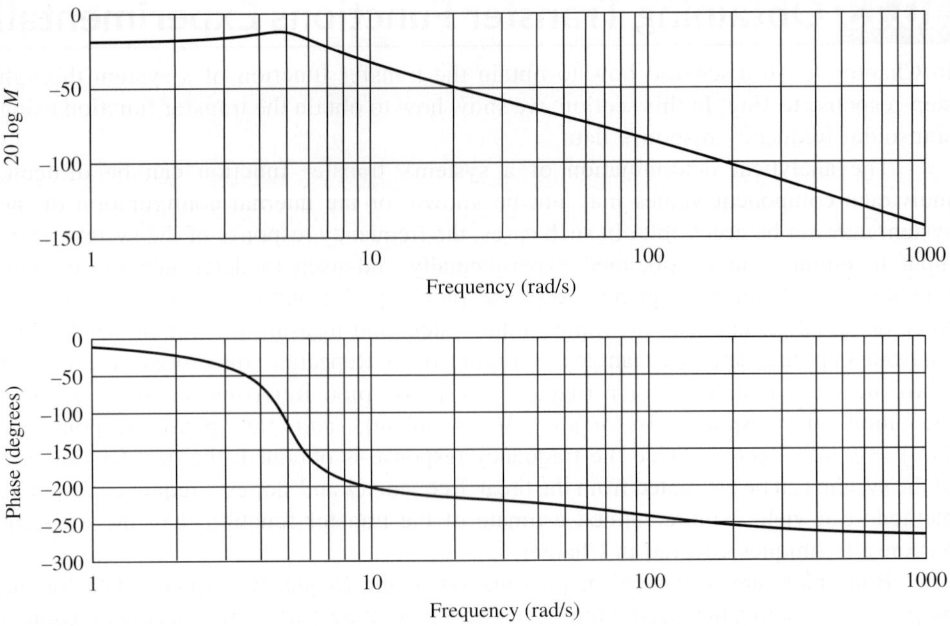


**FIGURE 10.57** Bode plots for subsystem with undetermined transfer function

**SOLUTION:** Let us first extract the underdamped poles that we suspect, based on the peaking in the magnitude curve. We estimate the natural frequency to be near the peak frequency, or approximately 5 rad/s. From Figure 10.57, we see a peak of about 6.5 dB, which translates into a damping ratio of about $\zeta = 0.24$ using Eq. (10.52). The unity gain second-order function is thus $G_1(s) = \omega_n^2/(s^2 + 2\zeta\omega_n s + \omega_n^2) = 25/(s^2 + 2.4s + 25)$. The frequency response plot of this function is made and subtracted from the previous Bode plots to yield the response in Figure 10.58.

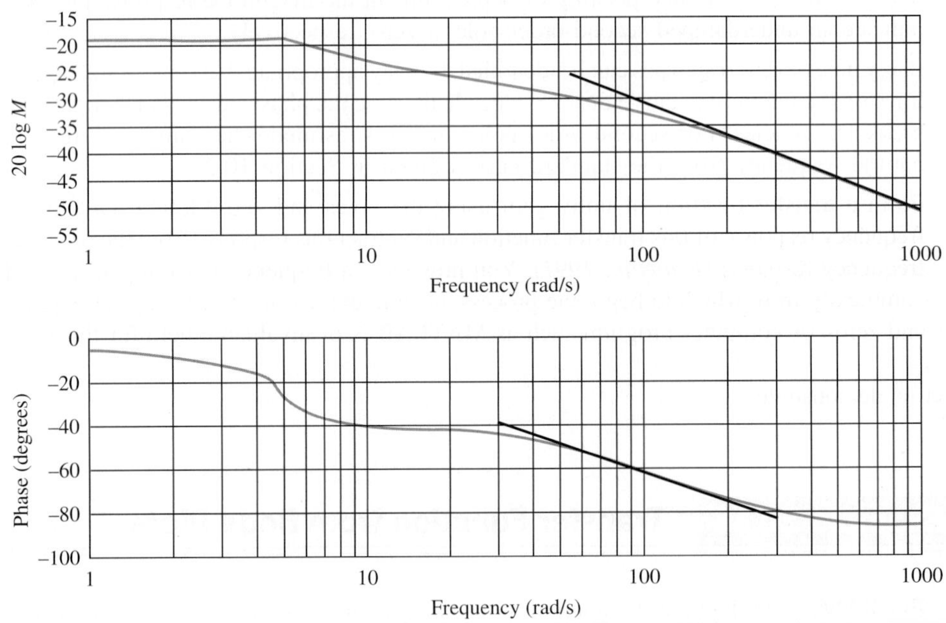


**FIGURE 10.58** Original Bode plots minus response of $G_1(s) = 25/(s^2 + 2.4s + 25)$

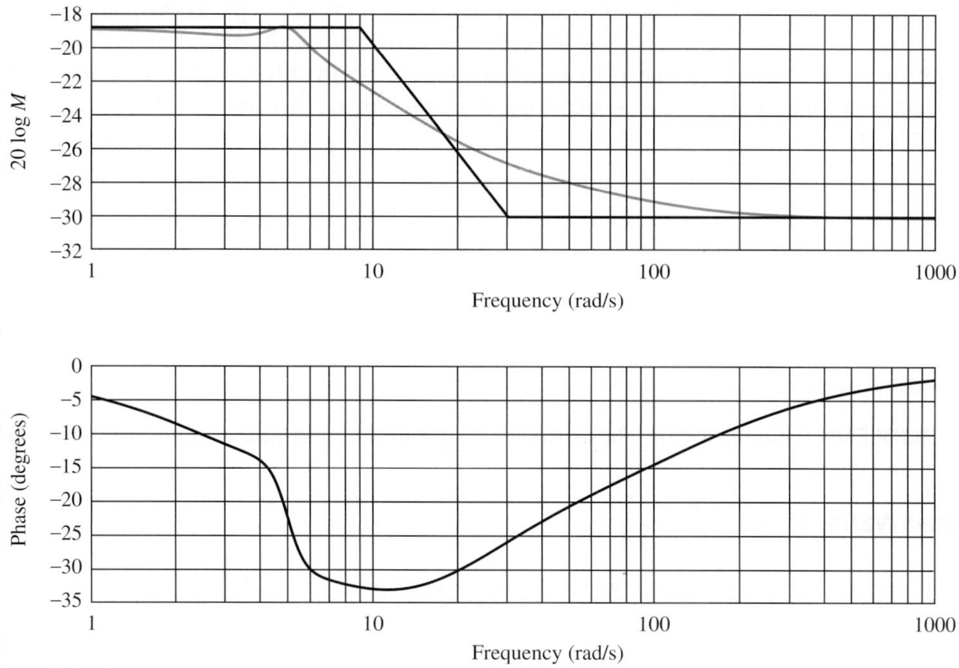


**FIGURE 10.59** Original Bode plot minus response of $G_1(s)G_2(s) = [25/(s^2 + 2.4s + 25)]\ [90/(s + 90)]$

Overlaying a $-20$ dB/decade line on the magnitude response and a $-45°$/decade line on the phase response, we detect a final pole. From the phase response, we estimate the break frequency at 90 rad/s. Subtracting the response of $G_2(s) = 90/(s + 90)$ from the previous response yields the response in Figure 10.59.

Figure 10.59 has a magnitude and phase curve similar to that generated by a lag function. We draw a $-20$ dB/decade line and fit it to the curves. The break frequencies are read from the figure as 9 and 30 rad/s. A unity gain transfer function containing a pole at $-9$ and a zero at $-30$ is $G_3(s) = 0.3(s + 30)/(s + 9)$. Upon subtraction of $G_1(s)G_2(s)G_3(s)$, we find the magnitude frequency response flat $\pm 1$ dB and the phase response flat at $-3°\pm 5°$. We thus conclude that we are finished extracting dynamic transfer functions. The low-frequency, or dc, value of the original curve is $-19$ dB, or 0.11. Our estimate of the subsystem's transfer function is $G(s) = 0.11G_1(s)G_2(s)G_3(s)$, or

$$\begin{aligned} G(s) &= 0.11\left(\frac{25}{s^2 + 2.4s + 25}\right)\left(90\frac{1}{s + 90}\right)\left(0.3\frac{s + 30}{s + 9}\right) \\ &= 74.25\frac{s + 30}{(s + 9)(s + 90)(s^2 + 2.4s + 25)} \end{aligned} \tag{10.89}$$

It is interesting to note that the original curve was obtained from the function

$$G(s) = 70\frac{s + 20}{(s + 7)(s + 70)(s^2 + 2s + 25)} \tag{10.90}$$

MATLAB

ML

Students who are using MATLAB should now run ch10 p8 in Appendix B. You will learn how to use MATLAB to subtract Bode plots for the purpose of estimating transfer functions through sinusoidal testing. This exercise solves a portion of Example 10.18 using MATLAB.

## Skill-Assessment Exercise 10.12

**PROBLEM:** Estimate $G(s)$, whose Bode log-magnitude and phase plots are shown in Figure 10.60.

**ANSWERS:** $G(s) = \dfrac{30(s+5)}{s(s+20)}$

The complete solution is at www.wiley.com/college/nise.

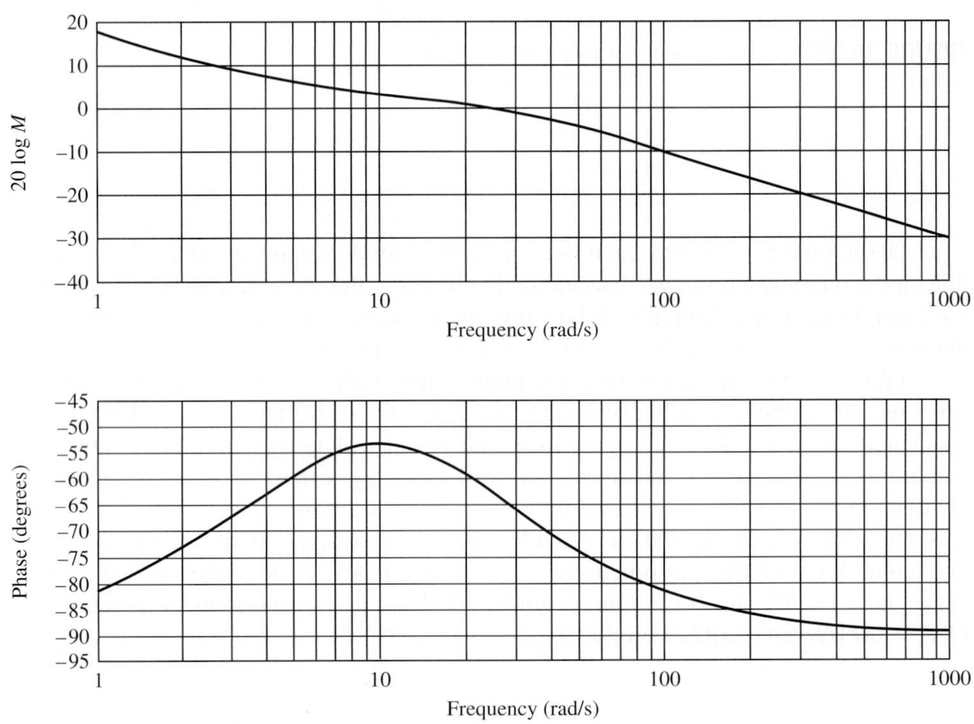


**FIGURE 10.60** Bode plots for Skill-Assessment Exercise 10.12

In this chapter, we derived the relationships between time response performance and the frequency responses of the open- and closed-loop systems. The methods derived, although yielding a different perspective, are simply alternatives to the root locus and steady-state error analyses previously covered.

## Case Study Antenna Control: Stability Design and Transient Performance

Design D

Our ongoing antenna position control system serves now as an example that summarizes the major objectives of the chapter. The case study demonstrates the use of frequency response methods to find the range of gain for stability and to design a value of gain to meet a percent overshoot requirement for the closed-loop step response.

**PROBLEM:** Given the antenna azimuth position control system shown on the front endpapers, Configuration 1, use frequency response techniques to find the following:

**a.** The range of preamplifier gain, $K$, required for stability

**b.** Percent overshoot if the preamplifier gain is set to 30

**c.** The estimated settling time

**d.** The estimated peak time

**e.** The estimated rise time

**SOLUTION:** Using the block diagram (Configuration 1) shown on the front endpapers and performing block diagram reduction yields the loop gain, $G(s)H(s)$, as

$$G(s)H(s) = \frac{6.63K}{s(s+1.71)(s+100)} = \frac{0.0388K}{s\left(\frac{s}{1.71}+1\right)\left(\frac{s}{100}+1\right)} \tag{10.91}$$

Letting $K = 1$, we have the magnitude and phase frequency response plots shown in Figure 10.61.

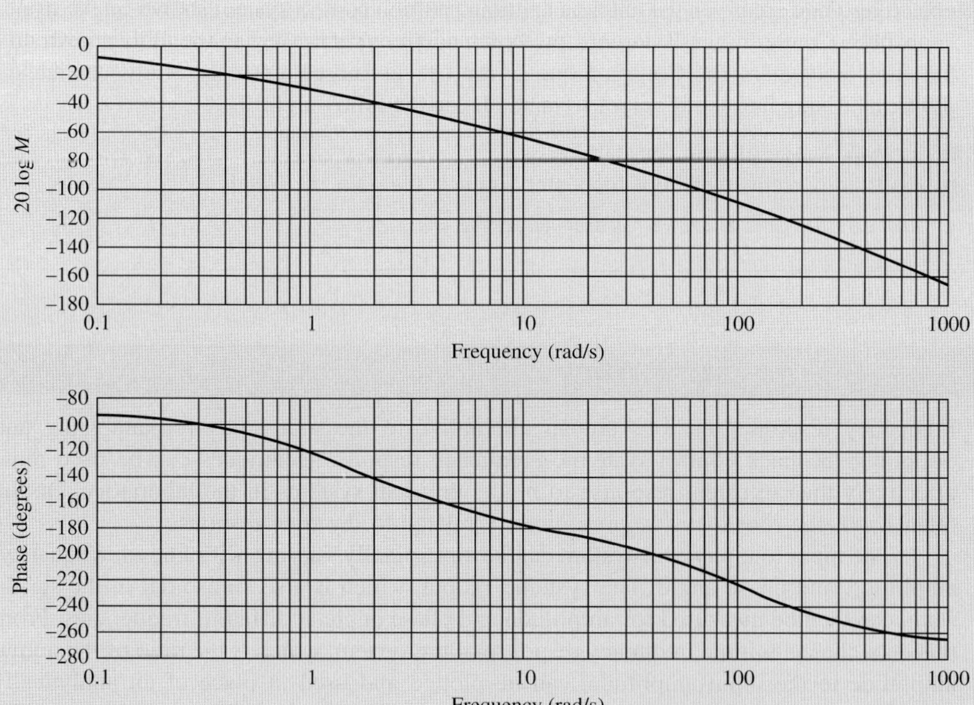


**FIGURE 10.61** Open-loop frequency response plots for the antenna control system ($K = 1$)

**a.** In order to find the range of $K$ for stability, we notice from Figure 10.61 that the phase response is $-180^\circ$ at $\omega = 13.1$ rad/s. At this frequency, the magnitude plot is $-68.41$ dB. The gain, $K$, can be raised by 68.41 dB. Thus, $K = 2633$ will cause the system to be marginally stable. Hence, the system is stable if $0 < K < 2633$.

**b.** To find the percent overshoot if $K = 30$, we first make a second-order approximation and assume that the second-order transient response equations relating percent overshoot, damping ratio, and phase margin are true for this system. In other words, we assume that Eq. (10.73), which relates damping ratio to phase margin, is valid. If $K = 30$, the magnitude curve of Figure 10.61 is moved up by $20 \log 30 = 29.54$ dB. Therefore, the adjusted magnitude curve goes through zero dB at $\omega = 1$. At this frequency, the phase angle is $-120.9^\circ$, yielding a phase margin of $59.1^\circ$. Using Eq. (10.73) or Figure 10.48, $\zeta = 0.6$, or 9.48% overshoot. A computer simulation shows 10%.

**c.** To estimate the settling time, we make a second-order approximation and use Eq. (10.55). Since $K = 30$ (29.54 dB), the open-loop magnitude response is $-7$ dB when the normalized magnitude response of Figure 10.61 is $-36.54$ dB. Thus, the estimated bandwidth is 1.8 rad/s. Using Eq. (10.55), $T_s = 4.25$ seconds. A computer simulation shows a settling time of about 4.4 seconds.

**d.** Using the estimated bandwidth found in Part **c** along with Eq. (10.56) and the damping ratio found in **a**, we estimate the peak time to be 2.5 seconds. A computer simulation shows a peak time of 2.8 seconds.

**e.** To estimate the rise time, we use Figure 4.16 and find that the normalized rise time for a damping ratio of 0.6 is 1.854. Using Eq. (10.54), the estimated bandwidth found in **c**, and $\zeta = 0.6$, we find $\omega_n = 1.57$. Using the normalized rise time and $\omega_n$, we find $T_r = 1.854/1.57 = 1.18$ seconds. A simulation shows a rise time of 1.2 seconds.

**CHALLENGE:** You are now given a problem to test your knowledge of this chapter's objectives. You are given the antenna azimuth position control system shown on the front endpapers, Configuration 3. Record the block diagram parameters in the table shown on the front endpapers for Configuration 3 for use in subsequent case study challenge problems. Using frequency response methods, do the following:

**a.** Find the range of gain for stability.

**b.** Find the percent overshoot for a step input if the gain, $K$, equals 3.

MATLAB
**ML**

**c.** `Repeat Parts` **a** `and` **b** `using MATLAB.`

## Summary

Frequency response methods are an alternative to the root locus for analyzing and designing feedback control systems. Frequency response techniques can be used more effectively than transient response to model physical systems in the laboratory. On the other hand, the root locus is more directly related to the time response.

The input to a physical system can be sinusoidally varying with known frequency, amplitude, and phase angle. The system's output, which is also sinusoidal in the steady state, can then be measured for amplitude and phase angle at different frequencies. From this data, the magnitude frequency response of the system, which is the ratio of the output amplitude to the input amplitude, can be plotted and used in place of an analytically obtained magnitude frequency response. Similarly, we can obtain the phase response by finding the difference between the output phase angle and the input phase angle at different frequencies.

The frequency response of a system can be represented either as a polar plot or as separate magnitude and phase diagrams. As a polar plot, the magnitude response is the length of a vector drawn from the origin to a point on the curve, whereas the phase response is the angle of that vector. In the polar plot, frequency is implicit and is represented by each point on the polar curve. The polar plot of $G(s)H(s)$ is known as a *Nyquist diagram.*

Separate magnitude and phase diagrams, sometimes referred to as *Bode plots*, present the data with frequency explicitly enumerated along the abscissa. The magnitude curve can be a plot of log-magnitude versus log-frequency. The other graph is a plot of phase angle versus log-frequency. An advantage of Bode plots over the Nyquist diagram is that they can easily be drawn using asymptotic approximations to the actual curve.

The Nyquist criterion sets forth the theoretical foundation from which the frequency response can be used to determine a system's stability. Using the Nyquist criterion and Nyquist diagram, or the Nyquist criterion and Bode plots, we can determine a system's stability.

Frequency response methods give us not only stability information but also transient response information. By defining such frequency response quantities as gain margin and phase margin, the transient response can be analyzed or designed. *Gain margin* is the amount that the gain of a system can be increased before instability occurs if the phase angle is constant at 180°. *Phase margin* is the amount that the phase angle can be changed before instability occurs if the gain is held at unity.

While the open-loop frequency response leads to the results for stability and transient response just described, other design tools relate the closed-loop frequency response peak and bandwidth to the transient response. Since the closed-loop response is not as easy to obtain as the open-loop response, because of the unavailability of the closed-loop poles, we use graphical aids in order to obtain the closed-loop frequency response from the open-loop frequency response. These graphical aids are the $M$ and $N$ circles and the Nichols chart. By superimposing the open-loop frequency response over the $M$ and $N$ circles or the Nichols chart, we are able to obtain the closed-loop frequency response and then analyze and design for transient response.

Today, with the availability of computers and appropriate software, frequency response plots can be obtained without relying on the graphical techniques described in this chapter. The program used for the root locus calculations and described in Appendix H.2 is one such program. MATLAB is another.

We concluded the chapter discussion by showing how to obtain a reasonable estimate of a transfer function using its frequency response, which can be obtained experimentally. Obtaining transfer functions this way yields more accuracy than transient response testing.

This chapter primarily has examined *analysis* of feedback control systems via frequency response techniques. We developed the relationships between frequency response and both stability and transient response. In the next chapter, we apply the concepts to the *design* of feedback control systems, using the Bode plots.

## Review Questions

1. Name four advantages of frequency response techniques over the root locus.
2. Define frequency response as applied to a physical system.
3. Name two ways to plot the frequency response.
4. Briefly describe how to obtain the frequency response analytically.
5. Define Bode plots.
6. Each pole of a system contributes how much of a slope to the Bode magnitude plot?

7. A system with only four poles and no zeros would exhibit what value of slope at high frequencies in a Bode magnitude plot?
8. A system with four poles and two zeros would exhibit what value of slope at high frequencies in a Bode magnitude plot?
9. Describe the asymptotic phase response of a system with a single pole at $-2$.
10. What is the major difference between Bode magnitude plots for first-order systems and for second-order systems?
11. For a system with three poles at $-4$, what is the maximum difference between the asymptotic approximation and the actual magnitude response?
12. Briefly state the Nyquist criterion.
13. What does the Nyquist criterion tell us?
14. What is a Nyquist diagram?
15. Why is the Nyquist criterion called a frequency response method?
16. When sketching a Nyquist diagram, what must be done with open-loop poles on the imaginary axis?
17. What simplification to the Nyquist criterion can we usually make for systems that are open-loop stable?
18. What simplification to the Nyquist criterion can we usually make for systems that are open-loop unstable?
19. Define gain margin.
20. Define phase margin.
21. Name two different frequency response characteristics that can be used to determine a system's transient response.
22. Name three different methods of finding the closed-loop frequency response from the open-loop transfer function.
23. Briefly explain how to find the static error constant from the Bode magnitude plot.
24. Describe the change in the open-loop frequency response magnitude plot if time delay is added to the plant.
25. If the phase response of a pure time delay were plotted on a linear phase versus linear frequency plot, what would be the shape of the curve?
26. When successively extracting component transfer functions from experimental frequency response data, how do you know when you are finished?

## Problems

1. Find analytical expressions for the magnitude and phase response for each $G(s)$ below. [Section: 10.1]

   **a.** $G(s) = \dfrac{1}{s(s+2)(s+4)}$

   **b.** $G(s) = \dfrac{(s+5)}{(s+2)(s+4)}$

   **c.** $G(s) = \dfrac{(s+3)(s+5)}{s(s+2)(s+4)}$

2. For each function in Problem 1, make a polar plot of the frequency response. [Section: 10.1]

3. For each function in Problem 1, make a plot of the log-magnitude and the phase, using log-frequency in rad/s as the ordinate. Do not use asymptotic approximations. [Section: 10.1]

4. For each function in Problem 1, sketch the Bode asymptotic magnitude and asymptotic phase plots. Compare your results with your answers to Problem 1. [Section: 10.2]

5. Sketch the Nyquist diagram for each of the systems in Figure P10.1. [Section: 10.4]

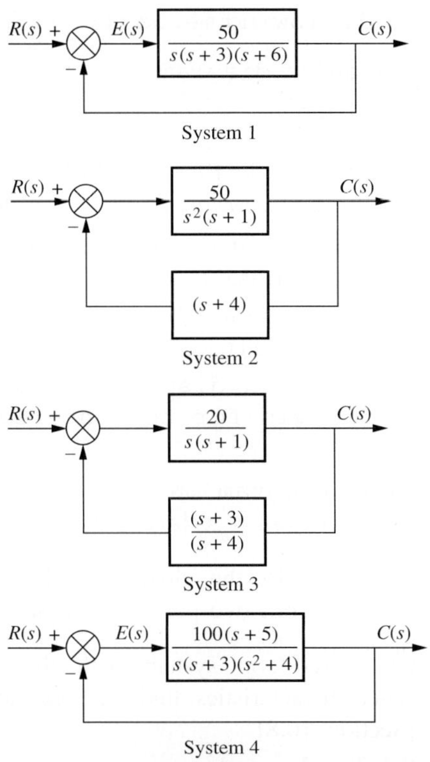


**FIGURE P10.1**

6. Draw the polar plot from the separate magnitude and phase curves shown in Figure P10.2. [Section: 10.1]

7. Draw the separate magnitude and phase curves from the polar plot shown in Figure P10.3. [Section: 10.1]

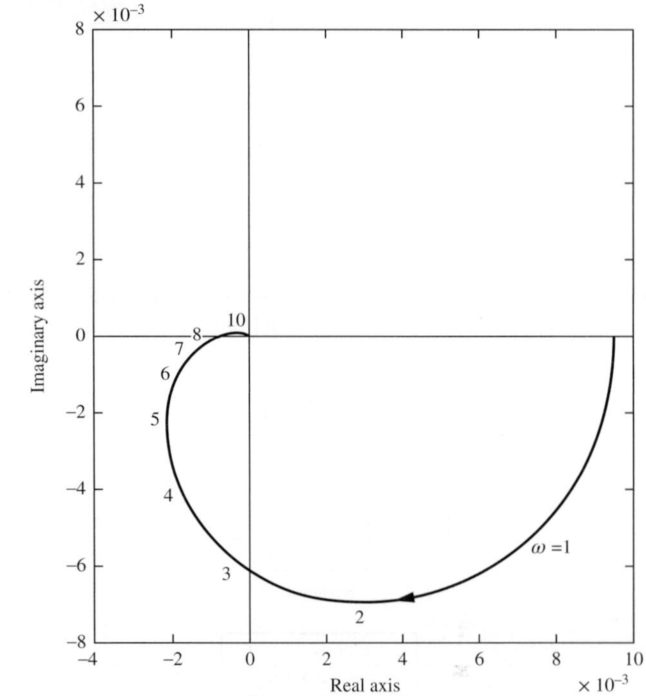


**FIGURE P10.3**

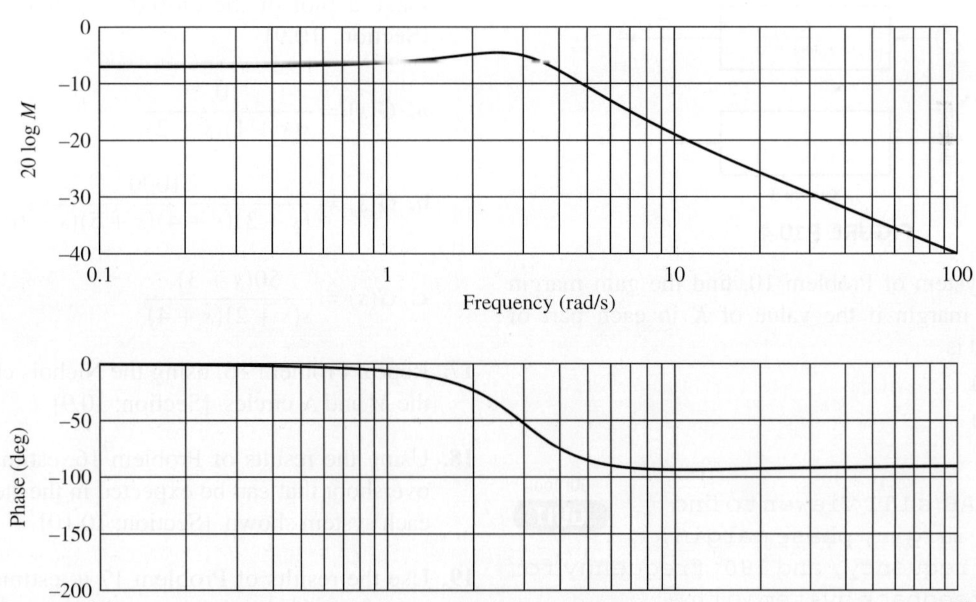


**FIGURE P10.2**

**8.** Using the Nyquist criterion, find out whether each system of Problem 5 is stable. [Section: 10.3]

**9.** `Write a program in MATLAB that will do the following:` MATLAB ML

**a.** `Plot the Nyquist diagram of a system`

**b.** `Display the real-axis crossing value and frequency`

Apply your program to a unity feedback system with

$$G(s) = \frac{K(s+5)}{(s^2+6s+100)(s^2+4s+25)}$$

**10.** Using the Nyquist criterion, find the range of $K$ for stability for each of the systems in Figure P10.4: [Section: 10.3]

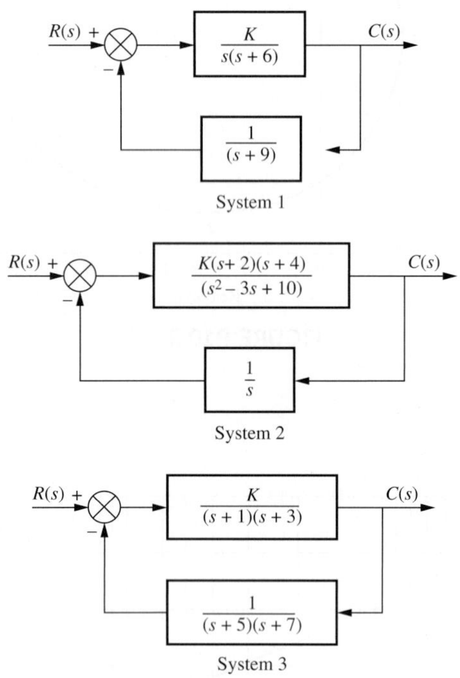


**FIGURE P10.4**

**11.** For each system of Problem 10, find the gain margin and phase margin if the value of $K$ in each part of Problem 10 is

**a.** $K = 600$

**b.** $K = 100$

**c.** $K = 10$

**12.** `Use MATLAB's LTI Viewer to find the gain margin, phase margin, zero dB frequency, and 180° frequency for a unity feedback system with` GUI Tool GUIT

$$G(s) = \frac{10{,}000}{(s+5)(s+18)(s+30)}$$

`Use the following methods:`

**a.** `The Nyquist diagram`

**b.** `Bode plots`

**13.** `Write a program in MATLAB that will do the following:` MATLAB ML

**a.** `Allow a value of gain,` $K$`, to be entered from the keyboard`

**b.** `Display the Bode plots of a system for the entered value of` $K$

**c.** `Calculate and display the gain and phase margin for the entered value of` $K$

`Test your program on a unity feedback system with` $G(s) = K/[s(s+3)(s+12)]$.

**14.** Derive Eq. (10.54), the closed-loop bandwidth in terms of $\zeta$ and $\omega_n$ of a two-pole system. [Section: 10.8]

**15.** For each closed-loop system with the following performance characteristics, find the closed-loop bandwidth: [Section: 10.8]

**a.** $\zeta = 0.3$, $T_s = 2$ seconds

**b.** $\zeta = 0.3$, $T_p = 2$ seconds

**c.** $T_s = 5$ seconds, $T_p = 3$ seconds

**d.** $\zeta = 0.38$, $T_r = 2.8$ seconds

**16.** Consider the unity feedback system of Figure 10.10. For each $G(s)$ that follows, use the $M$ and $N$ circles to make a plot of the closed-loop frequency response: [Section: 10.9]

**a.** $G(s) = \dfrac{10}{s(s+1)(s+2)}$

**b.** $G(s) = \dfrac{1000}{(s+3)(s+4)(s+5)(s+6)}$

**c.** $G(s) = \dfrac{50(s+3)}{s(s+2)(s+4)}$

**17.** Repeat Problem 16, using the Nichols chart in place of the $M$ and $N$ circles. [Section: 10.9]

**18.** Using the results of Problem 16, estimate the percent overshoot that can be expected in the step response for each system shown. [Section: 10.10]

**19.** Use the results of Problem 17 to estimate the percent overshoot if the gain term in the numerator of the forward path of each part of the problem is respectively changed as follows: [Section: 10.10]

a. From 10 to 30

b. From 1000 to 2500

c. From 50 to 75

20. Write a program in MATLAB that will do the following: MATLAB ML

a. Allow a value of gain, $K$, to be entered from the keyboard

b. Display the closed-loop magnitude and phase frequency response plots of a unity feedback system with an open-loop transfer function, $KG(s)$

c. Calculate and display the peak magnitude, frequency of the peak magnitude, and bandwidth for the closed-loop frequency response and the entered value of $K$

Test your program on the system of Figure P10.5 for $K = 40$.

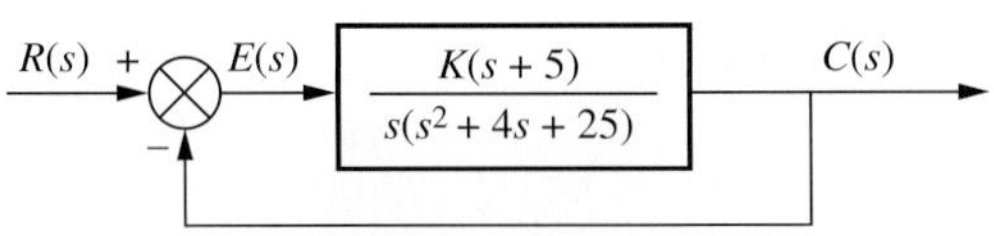


**FIGURE P10.5**

21. Use MATLAB's LTI Viewer with the Nichols plot to find the gain margin, phase margin, zero dB frequency, and 180° frequency for a unity feedback system with the forward-path transfer function GUI Tool GUIT

$$G(s) = \frac{7(s+5)}{s(s^2+4s+10)}$$

22. Write a program in MATLAB that will do the following: MATLAB ML

a. Make a Nichols plot of an open-loop transfer function

b. Allow the user to read the Nichols plot display and enter the value of $M_p$

c. Make closed-loop magnitude and phase plots

d. Display the expected values of percent overshoot, settling time, and peak time

e. Plot the closed-loop step response

Test your program on a unity feedback system with the forward-path transfer function

$$G(s) = \frac{7(s+5)}{s(s^2+4s+10)}$$

23. For the system of Figure P10.5, do the following: [Section: 10.10]

a. Plot the Bode magnitude and phase plots.

b. Assuming a second-order approximation, estimate the transient response of the system if $K = 40$.

c. Use MATLAB or any other program to check your assumptions by simulating the step response of the system. MATLAB ML

24. Using Bode plots, estimate the transient response of the systems in Figure P10.6. [Section: 10.10]

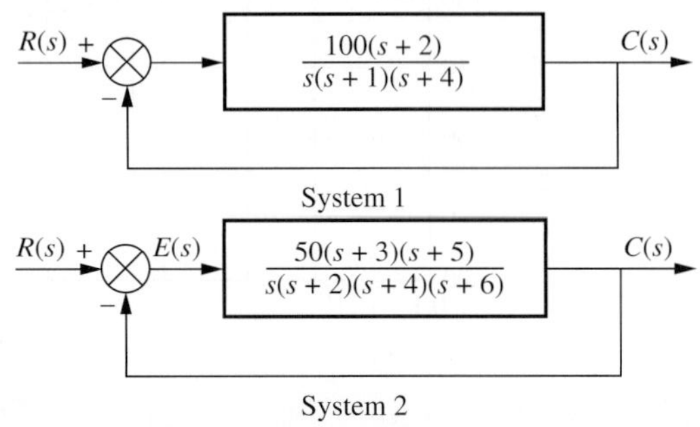


**FIGURE P10.6**

25. Write a program in MATLAB that will use an open-loop transfer function, $G(s)$, to do the following: MATLAB ML

a. Make a Bode plot

b. Use frequency response methods to estimate the percent overshoot, settling time, and peak time

c. Plot the closed-loop step response

Test your program by comparing the results to those obtained for the systems of Problem 24.

26. The Bode plots for a plant, $G(s)$, used in a unity feedback system are shown in Figure P10.7. Do the following:

a. Find the gain margin, phase margin, zero dB frequency, 180° frequency, and the closed-loop bandwidth.

b. Use your results in Part **a** to estimate the damping ratio, percent overshoot, settling time, and peak time.

27. The open-loop frequency response shown in Figure P10.8 was experimentally obtained from a unity feedback system. Estimate the percent overshoot and steady-state error of the closed-loop system. [Sections: 10.10, 10.11]

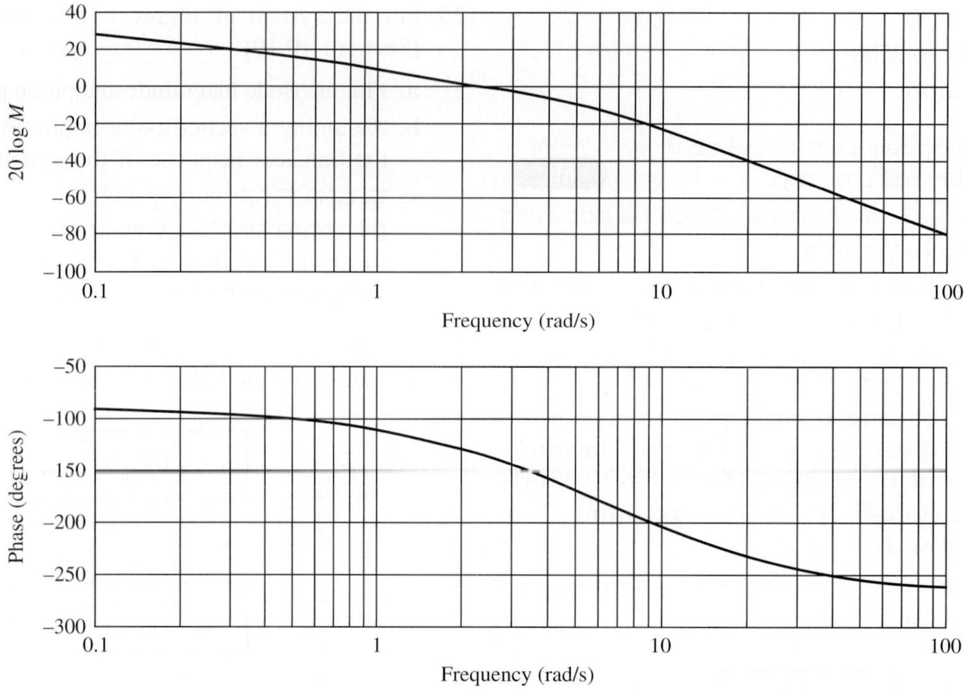


**FIGURE P10.7**

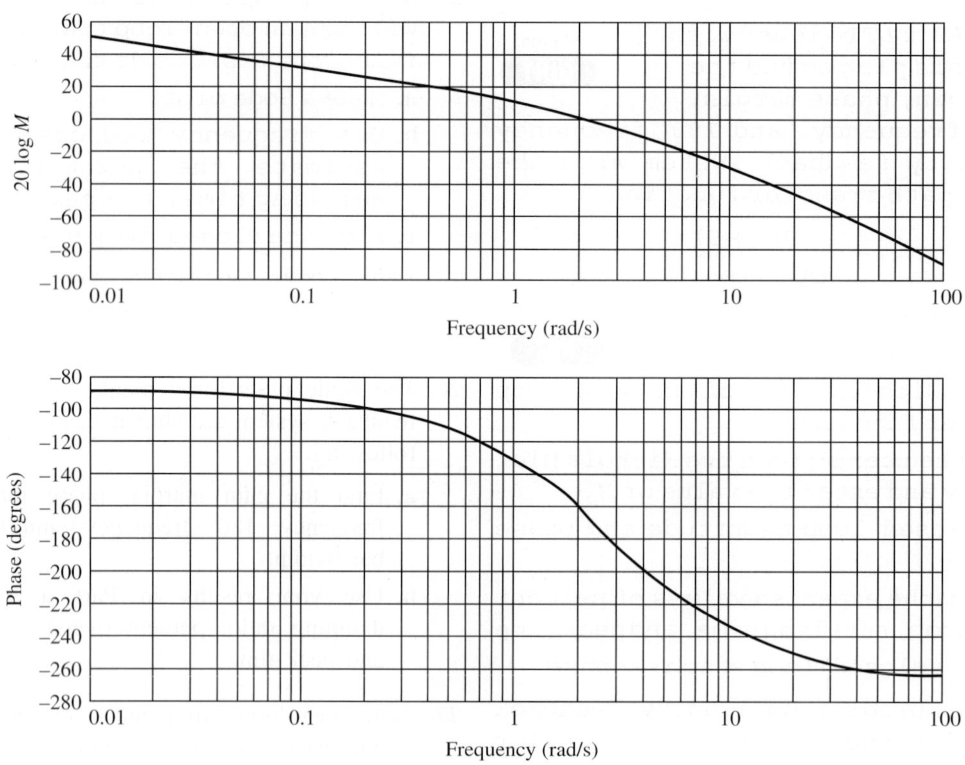


**FIGURE P10.8**

28. Consider the system in Figure P10.9. [Section: 10.12]

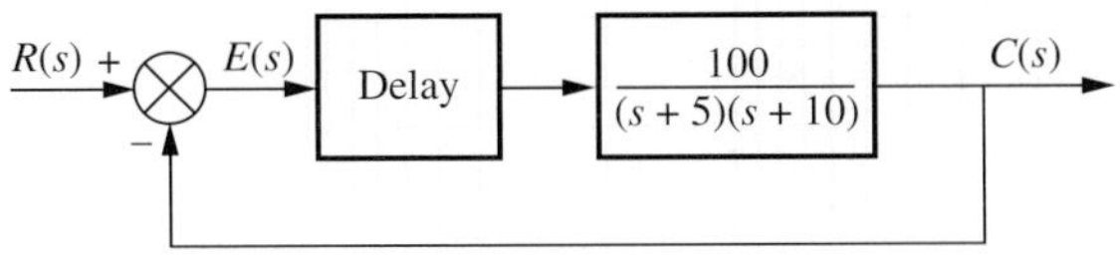


**FIGURE P10.9**

a. Find the phase margin if the system is stable for time delays of 0, 0.1, 0.2, 0.5, and 1 second.

b. Find the gain margin if the system is stable for each of the time delays given in Part **a**.

c. For what time delays mentioned in Part **a** is the system stable?

d. For each time delay that makes the system unstable, how much reduction in gain is required for the system to be stable?

29. Given a unity feedback system with the forward-path transfer function

$$G(s) = \frac{K}{(s+1)(s+3)(s+6)}$$

and a delay of 0.5 second, find the range of gain, $K$, to yield stability. Use Bode plots and frequency response techniques. [Section: 10.12]

30. Given a unity feedback system with the forward-path transfer function

$$G(s) = \frac{K}{s(s+3)(s+12)}$$

and a delay of 0.5 second, make a second-order approximation and estimate the percent overshoot if $K = 40$. Use Bode plots and frequency response techniques. [Section: 10.12]

31. Use the MATLAB function `pade(T,n)` to model the delay in Problem 30. Obtain the unit step response and evaluate your second-order approximation in Problem 30. MATLAB **ML**

32. For the Bode plots shown in Figure P10.10, determine the transfer function by hand or via MATLAB. [Section: 10.13]

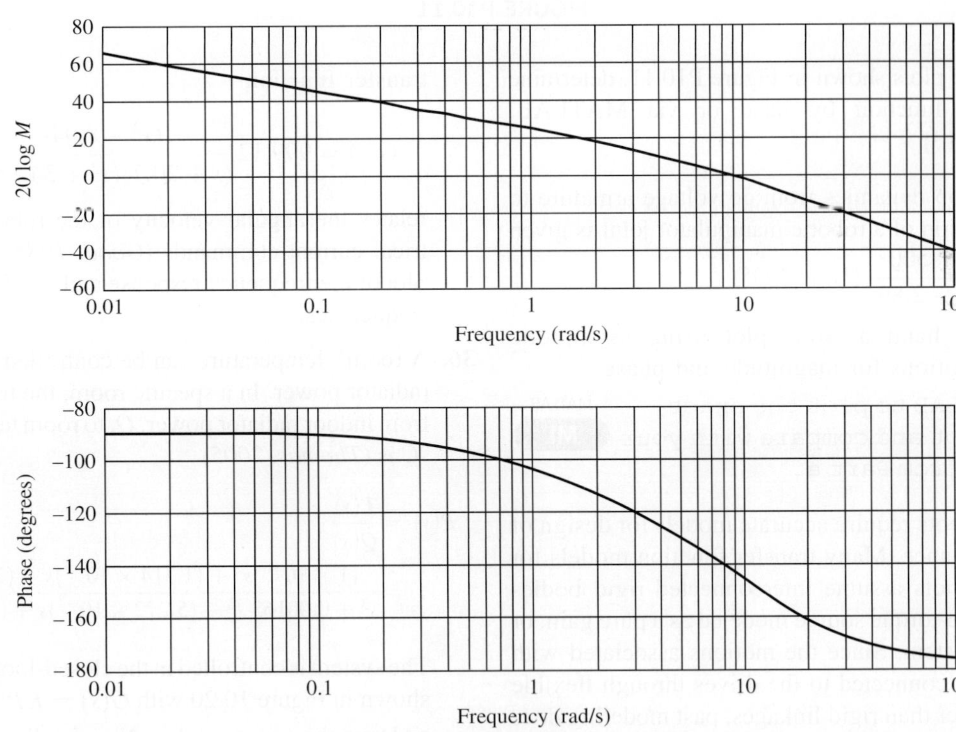


**FIGURE P10.10**

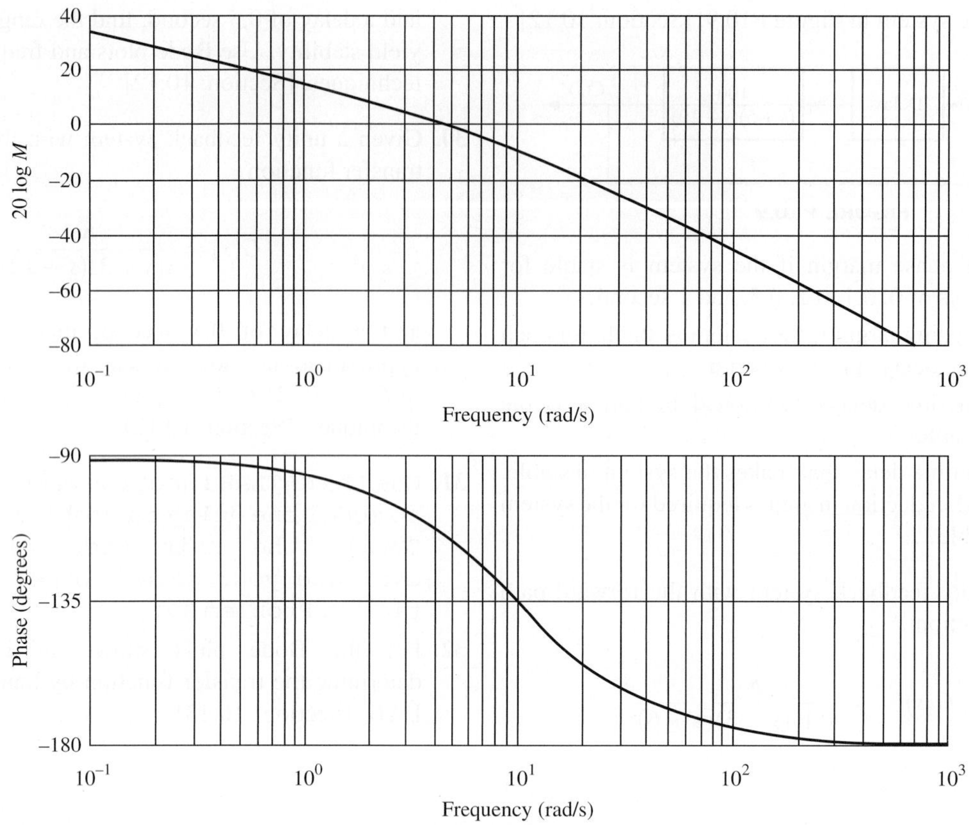


**FIGURE P10.11**

**33.** For the Bode plots shown in Figure P10.11, determine the transfer function by hand or via MATLAB. [Section: 10.13]

**34.** The open-loop dynamics from dc voltage armature to angular position of a robotic manipulator joint is given by $P(s) = \dfrac{48500}{s^2 + 2.89s}$ (*Low, 2005*).

**a.** Draw by hand a Bode plot using asymptotic approximations for magnitude and phase.

**b.** `Use MATLAB to plot the exact Bode plot and compare with your sketch from Part` **`a`**`.` MATLAB **ML**

**35.** Industrial robots require accurate models for design of high performance. Many transfer function models for industrial robots assume interconnected rigid bodies with the drive-torque source modeled as a pure gain, or first-order system. Since the motions associated with the robot are connected to the drives through flexible linkages rather than rigid linkages, past modeling does not explain the resonances observed. An accurate, small-motion, linearized model has been developed that takes into consideration the flexible drive. The transfer function

$$G(s) = 999.12\frac{(s^2 + 8.94s + 44.7^2)}{(s + 20.7)(s^2 + 34.858s + 60.1^2)}$$

relates the angular velocity of the robot base to electrical current commands (*Good, 1985*). Make a Bode plot of the frequency response and identify the resonant frequencies.

**36.** A room's temperature can be controlled by varying the radiator power. In a specific room, the transfer function from indoor radiator power, $\dot{Q}$, to room temperature, $T$ in °C is (*Thomas, 2005*)

$$P(s) = \frac{T(s)}{\dot{Q}(s)} = \frac{(1 \times 10^{-6})s^2 + (1.314 \times 10^{-9})s + (2.66 \times 10^{-13})}{s^3 + 0.00163\, s^2 + (5.272 \times 10^{-7})s + (3.538 \times 10^{-11})}$$

The system is controlled in the closed-loop configuration shown in Figure 10.20 with $G(s) = KP(s)$, $H = 1$.

**a.** Draw the corresponding Nyquist diagram for $K = 1$.

**b.** Obtain the gain and phase margins.

**c.** Find the range of $K$ for the closed-loop stability.

**37.** An overhead crane consists of a horizontally moving trolley of mass $m_T$ dragging a load of mass $m_L$, which dangles from its bottom surface at the end of a rope of fixed length, $L$. The position of the trolley is controlled in the feedback configuration shown in Figure 10.20. Here, $G(s) = KP(s)$, $H = 1$, and

$$P(s) - \frac{X_T(s)}{F_T(s)} = \frac{1}{m_T} \frac{s^2 + \omega_0^2}{s^2(s^2 + a\omega_0^2)}$$

The input is $f_T(t)$, the input force applied to the trolley. The output is $x_T(t)$, the trolley displacement. Also, $\omega_0 = \sqrt{\frac{g}{L}}$ and $a = (m_L + m_T)/m_T$ (*Marttinen, 1990*). Make a qualitative Bode plot of the system assuming $a > 1$.

**38.** The simplified and linearized model for the transfer function of a certain bicycle from steer angle ($\delta$) to roll angle ($\varphi$) is given by (*Åstrom, 2005*)

$$P(s) = \frac{\varphi(s)}{\delta(s)} = \frac{10(s + 25)}{s^2 + 25}$$

Assume the rider can be represented by a gain $K$, and that the closed-loop system is shown in Figure 10.20 with $G(s) = KP(s)$ and $H = 1$. Use the Nyquist stability criterion to find the range of $K$ for closed-loop stability.

**39.** A floppy disk drive was discussed in Problem 57 in Chapter 8. Assuming the system block diagram shown in Figure P10.12, use frequency response techniques to determine the following:

**a.** Gain margin, phase margin, zero dB frequency, 180° frequency, and closed-loop bandwidth

**b.** Percent overshoot, settling time, and peak time

**c.** `Use MATLAB to simulate the closed-loop step response and compare the results to those obtained in Part` **`b`**`.` MATLAB ML

**40.** A robot arm called ISAC (Intelligent Soft Arm Control) can be used as part of a system to feed people with disabilities. The control system guides the spoon to the food and then to a position near the person's mouth. The arm expansion and contraction in length can drive a pulley or other device. A video camera provides the sight for the robot and the tracking loop (*Kara, 1992*). Assuming the system block diagram shown in Figure P10.13, use frequency response techniques to determine the following (*Kara, 1992*):

**a.** Gain margin, phase margin, zero dB frequency, and 180° frequency

**b.** Is the system stable? Why?

**41.** The control of the radial pickup position of a digital versatile disk (DVD) was discussed in Problem 53, Chapter 9. There, the open-loop transfer function from coil input voltage to radial pickup position was given as (*Bittanti, 2002*)

$$P(s) = \frac{0.63}{\left(1 + \frac{0.36}{305.4}s + \frac{s^2}{305.4^2}\right)\left(1 + \frac{0.04}{248.2}s + \frac{s^2}{248.2^2}\right)}$$

Assume the plant is in cascade with a controller,

$$M(s) = \frac{0.5(s + 1.63)}{s(s + 0.27)}$$

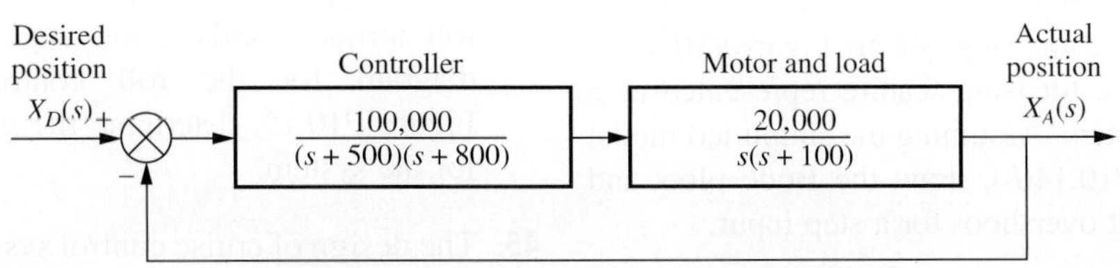


**FIGURE P10.12** Floppy disk drive block diagram

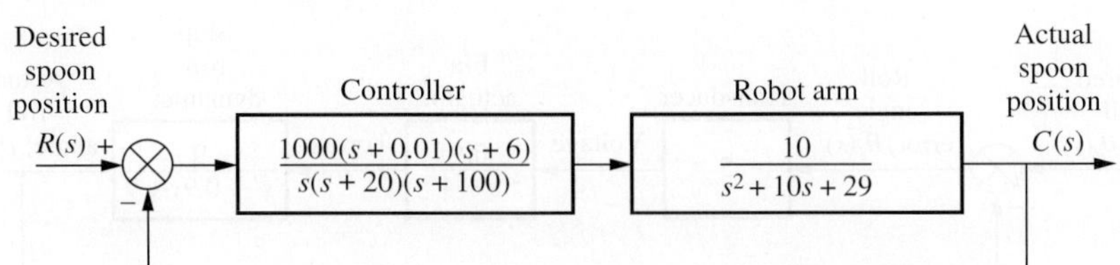


**FIGURE P10.13** Soft Arm position control system block diagram

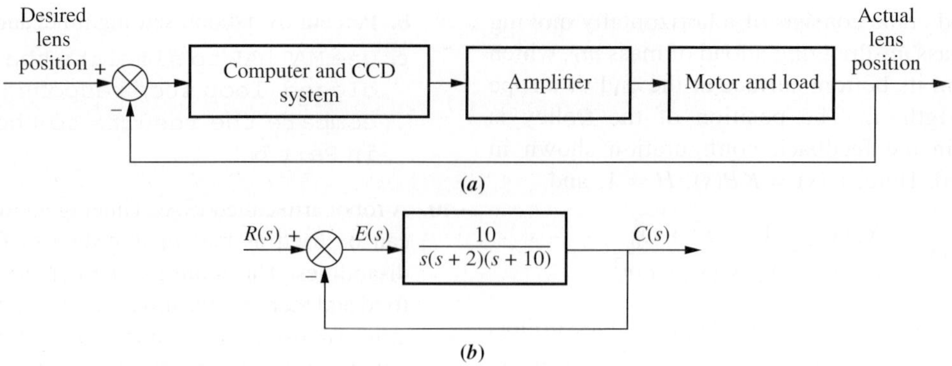


**FIGURE P10.14** a. Functional block diagram; b. block diagram

and in the closed-loop configuration shown in Figure 10.20, where $G(s) = M(s)P(s)$ and $H = 1$. Do the following:

**a.** Draw the open-loop frequency response in a Nichols chart.

**b.** Predict the system's response to a unit step input. Calculate the $\%OS$, $c_{\text{final}}$, and $T_s$.

**c.** `Verify the results of Part b using MATLAB simulations.` MATLAB ML

**42.** The charge-coupled device (CCD) that is used in video movie cameras to convert images into electrical signals can be used as part of an automatic focusing system in cameras. Automatic focusing can be implemented by focusing the center of the image on a charge-coupled device array through two lenses. The separation of the two images on the CCD is related to the focus. The camera senses the separation, and a computer drives the lens and focuses the image. The automatic focus system is a position control, where the desired position of the lens is an input selected by pointing the camera at the subject. The output is the actual position of the lens. The camera uses a CCD automatic focusing system. Figure P10.14(*a*) shows the automatic focusing feature represented as a position control system. Assuming the simplified model shown in Figure P10.14(*b*), draw the Bode plots and estimate the percent overshoot for a step input.

**43.** Fruit flies' flight dynamics are interesting to study because they provide a proof-of-concept framework and inspiration for the invention of man-made machines. In an experiment (*Roth, 2012*), flies are stimulated to follow, in flight, an oscillating vertical bar. Through frequency response measurements, the obtained open-loop transfer function from stimulus stripe position to a voltage proportional to wingbeat amplitudes is

$$G(s) = e^{-0.032s}\frac{0.181s^2 + 1.23s + 8.68}{s^3 + 20.6s^2 + 277s + 1098}$$

Assume a unit feedback system.

**a.** Make a Bode plot of the open-loop transfer function.

**b.** Find the gain and phase margins.

**c.** Use a computer program to make a plot of the corresponding Nyquist diagram.

**d.** Use the result in Part **b** to find the range of $K$ for closed-loop stability if the open-loop transfer function becomes $KG(s)$.

**44.** A ship's roll can be stabilized with a control system. A voltage applied to the fins' actuators creates a roll torque that is applied to the ship. The ship, in response to the roll torque, yields a roll angle. Assuming the block diagram for the roll control system shown in Figure P10.15, determine the gain and phase margins for the system.

**45.** The design of cruise control systems in heavy vehicles, such as big rigs, is especially challenging due to the extreme variations in payload. Assume that the

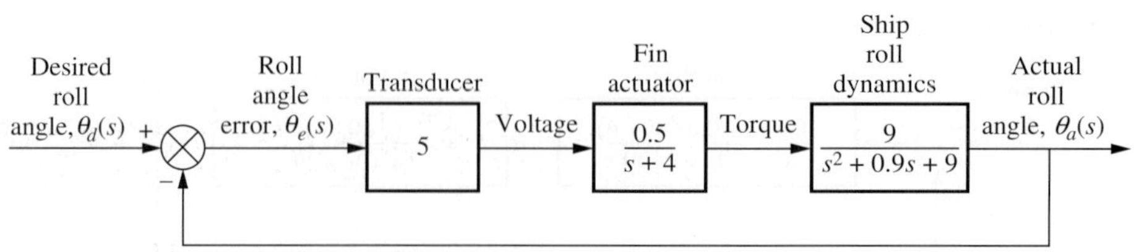


**FIGURE P10.15** Block diagram of a ship's roll-stabilizing system

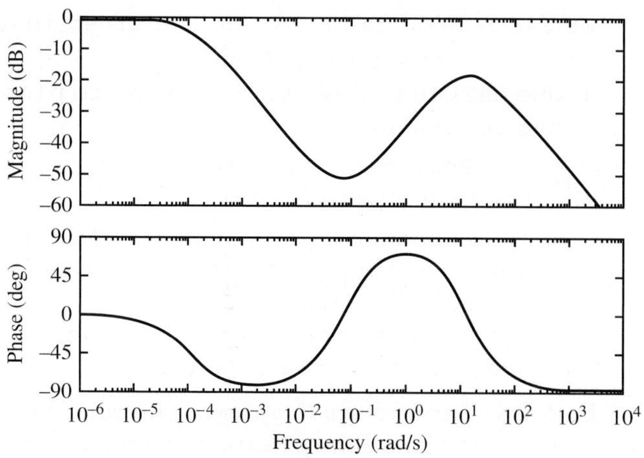


**FIGURE P10.16**

frequency response for the transfer function from fuel mass flow to vehicle speed is shown in Figure P10.16.

This response includes the dynamics of the engine, the gear box, the propulsion shaft, the differential, the drive shafts, the chassis, the payload, and tire dynamics. Assume that the system is controlled in a closed-loop, unity-feedback loop using a proportional compensator (*van der Zalm, 2008*).

**a.** Make a plot of the Nyquist diagram that corresponds to the Bode plot of Figure P10.16.

**b.** Assuming there are no open-loop poles in the right half-plane, find out if the system is closed-loop stable when the proportional gain $K = 1$.

**c.** Find the range of positive $K$ for which the system is closed-loop stable.

**46.** `Modify the MATLAB program you developed in Problem 10.20 to do the following:` MATLAB ML

**a.** `Display the closed-loop magnitude and phase frequency response plots for the drive system (`*`Thomsen, 2011`*`) presented in Problem 56, Chapter 8. Using the graph properties, specify the value of` $K$ `in the Bode plot title.`

**b.** `Calculate and display the closed-loop transfer function,` $T(s)$`, the peak magnitude, frequency of the peak magnitude, and bandwidth for the closed-loop frequency response at the following two values of the proportional controller's gain,` $K = K_P = 3.2$ `and 10.`

**47.** `Use LabVIEW with the Control Design and Simulation Module and MathScript RT Module to do the following: Modify the CDEx Nyquist Analysis.vi to obtain the range of` $K$ `for stability using the Nyquist plot for any system you enter. In addition, design a LabVIEW VI that will accept as an input the polynomial numerator and polynomial denominator of an open-loop transfer function and obtain a Nyquist plot for a value of` $K = 10{,}000$`. Your VI will also display the following as generated from the Nyquist plot: (1) gain margin, (2) phase margin, (3) zero dB frequency, and (4) 180 degrees frequency. Use the system and results of Skill-Assessment Exercise 10.6 to test your VIs.` LabVIEW LV MATLAB ML

**48.** A new measurement-based technique to design fixed-structure controllers for unknown SISO systems, which does not require system identification, has been proposed. The fourth-order transfer function shown below and modified to have a unity steady-state gain is used as an example (*Khadraoui, 2013*).

$$G(s) = \frac{0.1111(4s^2 + 5s + 1)}{s^4 + 3.1s^3 + 0.85s^2 + 0.87s + 0.1111}$$

The interested reader is referred to the reference to explore this new technique. In this problem and its companion design problem in Chapter 11, however, we take a standard approach as covered in Chapters 10 and 11.

`Assuming that a cascade-connected proportional controller,` $G_C(s) = K$`, is used, utilize MATLAB and frequency response techniques to obtain the Bode plots for this system and find:` MATLAB ML

**a.** `The range of` $K$ `for system stability`

**b.** `The gain margin, phase margin, zero dB frequency, and 180° frequency, if` $K = 0.3$`.`

**49.** `Use LabVIEW with the Control Design and Simulation Module, and MathScript RT Module to build a VI that will accept an open-loop transfer function, plot the Bode diagram, and plot the closed-loop step response. Your VI will also use the CD Parametric Time Response VI to display (1) rise time, (2) peak time, (3) settling time, (4) percent overshoot, (5) steady-state value, and` LabVIEW LV MATLAB ML

(6) peak value. Use the system in Skill-Assessment Exercise 10.9 to test your VI. Compare the results obtained from your VI with those obtained in Skill-Assessment Exercise 10.9.

**50.** A linear model of the $\alpha$-subsystem of a grid-connected voltage-source converter (VSC) with a Y-Y transformer (*Mahmood, 2012*) was presented in Problem 66, Chapter 8. In Figure P8.21(*b*), $G_C(s) = K$ and $G_P(s)$ is given in a pole zero form (with a unity gain and slightly modified parameters) as follows:

$$G_P(s) = \frac{V_\alpha(s)}{M_\alpha(s)} = \frac{(s+2200)}{(s+220)(s^2+120s+16\times10^6)}$$

Use MATLAB and frequency response techniques to obtain the Bode plots for this system and find the following:

MATLAB ML

**a.** The range of $K$ for system stability

**b.** The gain margin, phase margin, zero dB frequency, and 180° frequency, if $K = 5\times10^5$.

## PROGRESSIVE ANALYSIS AND DESIGN PROBLEMS

**51. Control of HIV/AIDS.** The linearized model for an HIV/AIDS patient treated with RTIs was obtained in Chapter 6 as (*Craig, 2004*);

$$P(s) = \frac{Y(s)}{U_1(s)} = \frac{-520s - 10.3844}{s^3 + 2.6817s^2 + 0.11s + 0.0126}$$

**a.** Consider this plant in the feedback configuration in Figure 10.20 with $G(s) = P(s)$ and $H(s) = 1$. Obtain the Nyquist diagram. Evaluate the system for closed-loop stability.

**b.** Consider this plant in the feedback configuration in Figure 10.20 with $G(s) = -P(s)$ and $H(s) = 1$. Obtain the Nyquist diagram. Evaluate the system for closed-loop stability. Obtain the gain and phase margins.

**52. Hybrid vehicle.** In Problem 71, Chapter 8, we used MATLAB to plot the root locus for the speed control of an HEV rearranged as a unity-feedback system, as shown in Figure P7.31 (*Preitl, 2007*). The plant and compensator were given by

MATLAB ML

$$G(s) = \frac{K(s+0.6)}{(s+0.5858)(s+0.0163)}$$

and we found that $K = 0.78$, resulted in a critically damped system.

**a.** Use MATLAB or any other program to plot the following:

**i.** The Bode magnitude and phase plots for that system, and

**ii.** The response of the system, $c(t)$, to a step input, $r(t) = 4\,u(t)$. Note on the $c(t)$ curve the rise time, $T_r$, and settling time, $T_s$, as well as the final value of the output.

**b.** Now add an integral gain to the controller, such that the plant and compensator transfer function becomes

$$G(s) = \frac{K_1(s+Z_C)(s+0.6)}{s(s+0.5858)(s+0.0163)}$$

where $K_1 = 0.78$ and $Z_C = \dfrac{K_2}{K_1} = 0.4$. Use MATLAB or any other program to do the following:

**i.** Plot the Bode magnitude and phase plots for this case.

**ii.** Obtain the response of the system to a step input, $r(t) = 4\,u(t)$. Plot $c(t)$ and note on it the rise time, $T_r$, percent overshoot, $\%OS$, peak time, $T_p$, and settling time, $T_s$.

**c.** Does the response obtained in Parts **a** or **b** resemble a second-order overdamped, critically damped, or underdamped response? Explain.

**53. Parabolic trough collector.** As discussed in Section 10.12, the Nyquist stability criterion can be applied to systems with pure time delay without the need for rational approximations as required in Problems 8.72 and 9.57. You will verify this by applying the Nyquist stability criterion to the parabolic trough collector by assuming a unity-feedback system and a forward-path transfer function (*Camacho, 2012*),

$$G(s) = \frac{137.2\times10^{-6}K}{s^2 + 0.0224s + 196\times10^{-6}}e^{-39s}$$

**a.** Draw the corresponding Nyquist diagram for $K = 1$.

**b.** Use the Nyquist diagram to find the range of $K$ for which the system is closed-loop stable.

**c.** Find the value of $K$ that will make the system marginally stable and the associated frequency of oscillation.

# Cyber Exploration Laboratory

## Experiment 10.1

**Objectives** To examine the relationships between open-loop frequency response and stability, open-loop frequency response and closed-loop transient response, and the effect of additional closed-loop poles and zeros upon the ability to predict closed-loop transient response

**Minimum Required Software Packages** MATLAB, and the Control System Toolbox

### Prelab

1. Sketch the Nyquist diagram for a unity negative feedback system with a forward transfer function of $G(s) = \dfrac{K}{s(s+2)(s+10)}$. From your Nyquist plot, determine the range of gain, $K$, for stability.
2. Find the phase margins required for second-order closed-loop step responses with the following percent overshoots: 5%, 10%, 20%, 30%.

### Lab

1. Using the SISO Design Tool, produce the following plots simultaneously for the system of Prelab 1: root locus, Nyquist diagram, and step response. Make plots for the following values of $K$: 50, 100, the value for marginal stability found in Prelab 1, and a value above that found for marginal stability. Use the zoom tools when required to produce an illustrative plot. Finally, change the gain by grabbing and moving the closed-loop poles along the root locus and note the changes in the Nyquist diagram and step response.
2. Using the SISO Design Tool, produce Bode plots and closed-loop step responses for a unity negative feedback system with a forward transfer function of $G(s) = \dfrac{K}{s(s+10)^2}$. Produce these plots for each value of phase margin found in Prelab 2. Adjust the gain to arrive at the desired phase margin by grabbing the Bode magnitude curve and moving it up or down. Observe the effects, if any, upon the Bode phase plot. For each case, record the value of gain and the location of the closed-loop poles.
3. Repeat Lab 2 for $G(s) = \dfrac{K}{s(s+10)}$.

### Postlab

1. Make a table showing calculated and actual values for the range of gain for stability as found in Prelab 1 and Lab 1.
2. Make a table from the data obtained in Lab 2 itemizing phase margin, percent overshoot, and the location of the closed-loop poles.
3. Make a table from the data obtained in Lab 3 itemizing phase margin, percent overshoot, and the location of the closed-loop poles.
4. For each Postlab task 1 to 3, explain any discrepancies between the actual values obtained and those expected.

## Experiment 10.2

**Objectives** To use LabVIEW and Nichols charts to determine the closed-loop time response performance

**Minimum Required Software Packages** LabVIEW, Control Design and Simulation Module, MathScript RT Module, and MATLAB

### Prelab

1. Assume a unity-feedback system with a forward-path transfer function, $G(s) = \dfrac{100}{s(s+5)}$. Use MATLAB or any method to determine gain and phase margins. In addition, find the percent overshoot, settling time, and peak time of the closed-loop step response.
2. Design a LabVIEW VI that will create a Nichols chart. Adjust the Nichols chart's scale to estimate gain and phase margins. Then, prompt the user to enter the values of gain and phase margins found from the Nichols chart. In response, your VI will produce the percent overshoot, settling time, and peak time of the closed-loop step response.

**Lab** Run your VI for the system given in the Prelab. Test your VI with other systems of your choice.

**Postlab** Compare the closed-loop performance calculated in the Prelab with those produced by your VI.

## Bibliography


Åstrom, K., Klein, R. E., and Lennartsson, A. Bicycle Dynamics and Control. *IEEE Control System*, August 2005, pp. 26–47.

Bhambhani, V., and Chen, YQ. Experimental Study of Fractional Order Proportional Integral (FOPI) Controller for Water Level Control. *47th IEEE Conference on Decision and Control*, 2008, pp. 1791–1796.

Bittanti, S., Dell'Orto, F., DiCarlo, A., and Savaresi, S. M. Notch Filtering and Multirate Control for Radial Tracking in High Speed DVD-Players. *IEEE Transactions on Consumer Electronics*, vol. 48. 2002, pp. 56–62.

Bode, H. W. *Network Analysis and Feedback Amplifier Design*. Van Nostrand, Princeton, NJ, 1945.

Camacho, E. F., Berenguel, M., Rubio, F. R., and Martinez, D. *Control of Solar Energy Systems*. Springer-Verlag, London, 2012.

Craig, I. K., Xia, X., and Venter, J. W. Introducing HIV/AIDS Education into the Electrical Engineering Curriculum at the University of Pretoria. *IEEE Transactions on Education*, vol. 47, no. 1, February 2004, pp. 65–73.

Dorf, R. C. *Modern Control Systems*, 5th ed. Addison-Wesley, Reading, MA, 1989.

Franklin, G., Powell, J. D., and Emami-Naeini, A. *Feedback Control of Dynamic Systems*, 2d ed. Addison-Wesley, Reading, MA, 1991.

Galvão, R. K. H., Yoneyama, T., and de Araújo, F. M. U. A Simple Technique for Identifying a Linearized Model for a Didactic Magnetic Levitation System. *IEEE Transactions on Education*, vol. 46, no. 1, February 2003, pp. 22–25.

Hollot, C. V., Misra, V., Towsley, D., and Gong, W. A Control Theoretic Analysis of RED. Proceedings of IEEE INFOCOM, 2001, pp. 1510–1519.

Hostetter, G. H., Savant, C. J., Jr., and Stefani, R. T. *Design of Feedback Control Systems*, 2d ed. Saunders College Publishing, New York, 1989.

Khadraoui, S., Nounou, H., Nounou, M., Datta, A., and Bhattacharyya, S. P. A measurement-based approach for tuning of reduced-order controllers. *American Control Conference (ACC)*, June 2013, pp. 3876–3881.

Kim, S.-H., Kim, J. H., Yang, J., Yang, H., Park, J.-Y., and Park, Y.-P. Tilt Detection and Servo Control Method for the Holographic Data Storage System. *Microsyst Technol*, vol. 15, 2009. pp. 1695–1700.

Kuo, B. C. *Automatic Control Systems*, 5th ed. Prentice Hall, Upper Saddle River, NJ, 1987.

Kuo, F. F. *Network Analysis and Synthesis*. Wiley, New York, 1966.

Lam, P. Y. Gyroscopic Stabilization of a Kid-Size Bicycle. *IEEE 5th International Conference on Cybernetics and Intelligent Systems*, 2011, pp. 247–252.

Low, K. H., Wang, H., Liew, K. M., and Cai, Y. Modeling and Motion Control of Robotic Hand for Telemanipulation Application. *International Journal of Software Engineering and Knowledge Engineering*, vol. 15, 2005, pp. 147–152.

Mahmood, H., and Jiang, J. Modeling and Control System Design of a Grid Connected VSC Considering the Effect of the Interface Transformer Type. *IEEE Transactions on Smart Grid*, vol. 3, no. 1, March 2012, pp. 122–134.

Marttinen, A., Virkkunen, J., and Salminen, R. T. Control Study with Pilot Crane. *IEEE Transactions on Education*, vol. 33, no. 3, August 1990, pp. 298–305.

Nilsson, J. W. *Electric Circuits*, 3d ed. Addison-Wesley, Reading, MA, 1990.

Nyquist, H. Regeneration Theory. *Bell Systems Technical Journal*, January 1932, pp. 126–147.

Ogata, K. *Modern Control Engineering*, 2d ed. Prentice Hall, Upper Saddle River, NJ, 1990.

Preitl, Z., Bauer, P., and Bokor, J. A Simple Control Solution for Traction Motor Used in Hybrid Vehicles. *Fourth International Symposium on Applied Computational Intelligence and Informatics*. IEEE, 2007, pp. 157–162.

Roth, E., Reiser, M. B., Dickinson, M. H., Noah, J., and Cowan, N. J. A Task-level Model for Optomotor Yaw Regulation in Drosophila Melanogaster: A Frequency-Domain System Identification Approach. *51st IEEE Conference on Decision and Control*, December 10–13, 2012, pp. 3721–3726.

Thomas, B., Soleimani-Mosheni, M., and Fahlén, P. Feed-forward in Temperature Control of Buildings. *Energy and Buildings*, vol. 37, 2005, pp. 755–761.

Thomsen, S., Hoffmann, N., and Fuchs, F. W. PI Control, PI-based State Space Control, and Model-Based Predictive Control for Drive Systems With Elastically Coupled Loads—A Comparative Study. *IEEE Transactions on Industrial Electronics*, vol. 58, no. 8, August 2011, pp. 3647–3657.

Van der Zalm, G., Huisman, R., Steinbuch, M., and Veldpaus, F. Frequency Domain Approach for the Design of Heavy-Duty Vehicle Speed Controllers. *Int. J. Heavy Vehicle Systems*, vol. 15, no. 1, 2008. pp. 107–123.

Wang, X.-K., Yang, X.-H., Liu, G., and Qian, H. Adaptive Neuro-Fuzzy Inference System PID Controller for Steam Generator Water Level of Nuclear Power Plant, *Proceedings of the Eighth International Conference on Machine Learning and Cybernetics*, 2009, pp. 567–572.

# 11

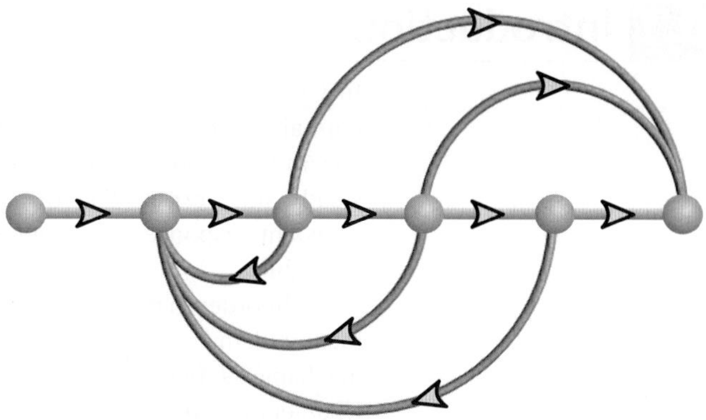

# Design via Frequency Response

## Chapter Learning Outcomes

After completing this chapter the student will be able to:

- Use frequency response techniques to adjust the gain to meet a transient response specification (Sections 11.1–11.2)
- Use frequency response techniques to design cascade compensators to improve the steady-state error (Section 11.3)
- Use frequency response techniques to design cascade compensators to improve the transient response (Section 11.4)
- Use frequency response techniques to design cascade compensators to improve both the steady-state error and the transient response (Section 11.5)

## ▶ Case Study Learning Outcomes

You will be able to demonstrate your knowledge of the chapter objectives with case studies as follows:

- Given the antenna azimuth position control system shown on the front endpapers, you will be able to use frequency response techniques to design the gain to meet a transient response specification.
- Given the antenna azimuth position control system shown on the front endpapers, you will be able to use frequency response techniques to design a cascade compensator to meet both transient and steady-state error specifications.

# 11.1 Introduction

In Chapter 8, we designed the transient response of a control system by adjusting the gain along the root locus. The design process consisted of finding the transient response specification on the root locus, setting the gain accordingly, and settling for the resulting steady-state error. The disadvantage of design by gain adjustment is that only the transient response and steady-state error represented by points along the root locus are available.

In order to meet transient response specifications represented by points not on the root locus and, independently, steady-state error requirements, we designed cascade compensators in Chapter 9. In this chapter, we use Bode plots to parallel the root locus design process from Chapters 8 and 9.

Let us begin by drawing some general comparisons between root locus and frequency response design.

*Stability and transient response design via gain adjustment.* Frequency response design methods, unlike root locus methods, can be implemented conveniently without a computer or other tool except for testing the design. We can easily draw Bode plots using asymptotic approximations and read the gain from the plots. Root locus requires repeated trials to find the desired design point from which the gain can be obtained. For example, in designing gain to meet a percent overshoot requirement, root locus requires the search of a radial line for the point where the open-loop transfer function yields an angle of 180°. To evaluate the range of gain for stability, root locus requires a search of the $j\omega$-axis for 180°. Of course, if one uses a computer program, such as MATLAB, the computational disadvantage of root locus vanishes.

*Transient response design via cascade compensation.* Frequency response methods are not as intuitive as the root locus, and it is something of an art to design cascade compensation with the methods of this chapter. With root locus, we can identify a specific point as having a desired transient response characteristic. We can then design cascade compensation to operate at that point and meet the transient response specifications. In Chapter 10, we learned that phase margin is related to percent overshoot (Eq. (10.73)) and bandwidth is related to both damping ratio and settling time or peak time (Eqs. (10.55) and (10.56)). These equations are rather complicated. When we design cascade compensation using frequency response methods to improve the transient response, we strive to reshape the open-loop transfer function's frequency response to meet both the phase-margin requirement (percent overshoot) and the bandwidth requirement (settling or peak time). There is no easy way to relate all the requirements prior to the reshaping task. Thus, the reshaping of the open-loop transfer function's frequency response can lead to several trials until all transient response requirements are met.

*Steady-state error design via cascade compensation.* An advantage of using frequency design techniques is the ability to design derivative compensation, such as lead compensation, to speed up the system and at the same time build in a desired steady-state error requirement that can be met by the lead compensator alone. Recall that in using root locus there are an infinite number of possible solutions to the design of a lead compensator. One of the differences between these solutions is the steady-state error. We must make numerous tries to arrive at the solution that yields the required steady-state error performance. With frequency response techniques, we build the steady-state error requirement right into the design of the lead compensator.

You are encouraged to reflect on the advantages and disadvantages of root locus and frequency response techniques as you progress through this chapter. Let us take a closer look at frequency response design.

When designing via frequency response methods, we use the concepts of stability, transient response, and steady-state error that we learned in Chapter 10. First, the Nyquist criterion tells us how to determine if a system is stable. Typically, an open-loop stable system is stable in closed-loop if the open-loop magnitude frequency response has a gain of less than 0 dB at the frequency where the phase frequency response is 180°. Second, percent overshoot is reduced by increasing the phase margin, and the speed of the response is increased by increasing the bandwidth. Finally, steady-state error is improved by increasing the low-frequency magnitude responses, even if the high-frequency magnitude response is attenuated.

These, then, are the basic facts underlying our design for stability, transient response, and steady-state error using frequency response methods, where the Nyquist criterion and the Nyquist diagram compose the underlying theory behind the design process. Thus, even though we use the Bode plots for ease in obtaining the frequency response, the design process can be verified with the Nyquist diagram when questions arise about interpreting the Bode plots. In particular, when the structure of the system is changed with additional compensator poles and zeros, the Nyquist diagram can offer a valuable perspective.

The emphasis in this chapter is on the design of lag, lead, and lag-lead compensation. General design concepts are presented first, followed by step-by-step procedures. These procedures are only suggestions, and you are encouraged to develop other procedures to arrive at the same goals. Although the concepts in general apply to the design of PI, PD, and PID controllers, in the interest of brevity, detailed procedures and examples will not be presented. You are encouraged to extrapolate the concepts and designs covered and apply them to problems involving PI, PD, and PID compensation presented at the end of this chapter. Finally, the compensators developed in this chapter can be implemented with the realizations discussed in Section 9.6.

## 11.2 Transient Response via Gain Adjustment

Let us begin our discussion of design via frequency response methods by discussing the link between phase margin, transient response, and gain. In Section 10.10, the relationship between damping ratio (equivalently percent overshoot) and phase margin was derived for $G(s) = \omega_n^2/s(s + 2\zeta\omega_n)$. Thus, if we can vary the phase margin, we can vary the percent overshoot. Looking at Figure 11.1, we see that if we desire a phase margin, $\Phi_M$, represented by $CD$, we would have to raise the magnitude curve by $AB$. Thus, a simple gain adjustment can be used to design phase margin and, hence, percent overshoot.

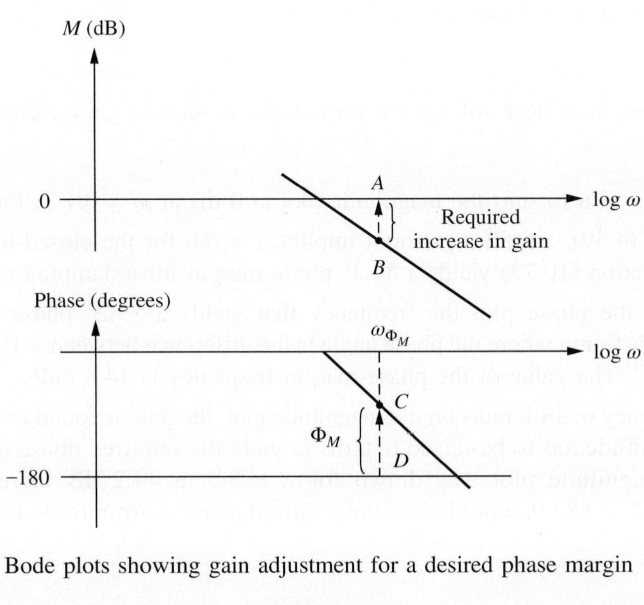


**FIGURE 11.1** Bode plots showing gain adjustment for a desired phase margin

We now outline a procedure by which we can determine the gain to meet a percent overshoot requirement using the open-loop frequency response and assuming dominant second-order closed-loop poles.

## Design Procedure

1. Draw the Bode magnitude and phase plots for a convenient value of gain.
2. Using Eqs. (4.39) and (10.73), determine the required phase margin from the percent overshoot.
3. Find the frequency, $\omega_{\Phi_M}$, on the Bode phase diagram that yields the desired phase margin, $CD$, as shown on Figure 11.1.
4. Change the gain by an amount $AB$ to force the magnitude curve to go through 0 dB at $\omega_{\Phi_M}$. The amount of gain adjustment is the additional gain needed to produce the required phase margin.

We now look at an example of designing the gain of a third-order system for percent overshoot.

## Example 11.1 Transient Response Design via Gain Adjustment

Design
D

**PROBLEM:** For the position control system shown in Figure 11.2, find the value of preamplifier gain, $K$, to yield a 9.5% overshoot in the transient response for a step input. Use only frequency response methods.

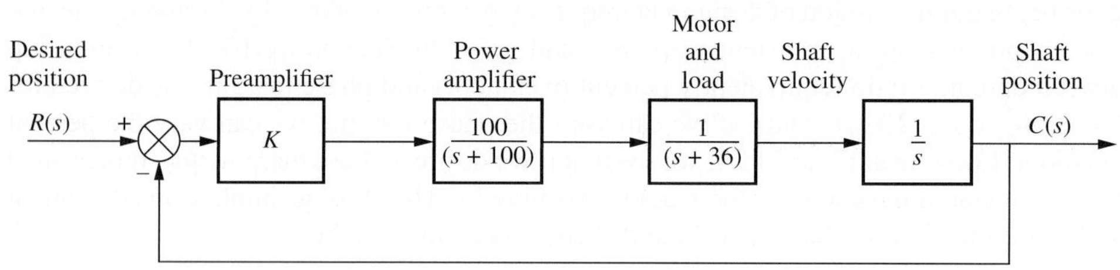


**FIGURE 11.2** System for Example 11.1

**SOLUTION:** We will now follow the previously described gain adjustment design procedure.

1. Choose $K = 3.6$ to start the magnitude plot at 0 dB at $\omega = 0.1$ in Figure 11.3.
2. Using Eq. (4.39), a 9.5% overshoot implies $\zeta = 0.6$ for the closed-loop dominant poles. Equation (10.73) yields a 59.2° phase margin for a damping ratio of 0.6.
3. Locate on the phase plot the frequency that yields a 59.2° phase margin. This frequency is found where the phase angle is the difference between $-180°$ and 59.2°, or $-120.8°$. The value of the phase-margin frequency is 14.8 rad/s.
4. At a frequency of 14.8 rad/s on the magnitude plot, the gain is found to be $-44.2$ dB. This magnitude has to be raised to 0 dB to yield the required phase margin. Since the log-magnitude plot was drawn for $K = 3.6$, a 44.2 dB increase, or $K = 3.6 \times 162.2 = 583.9$, would yield the required phase margin for 9.48% overshoot.

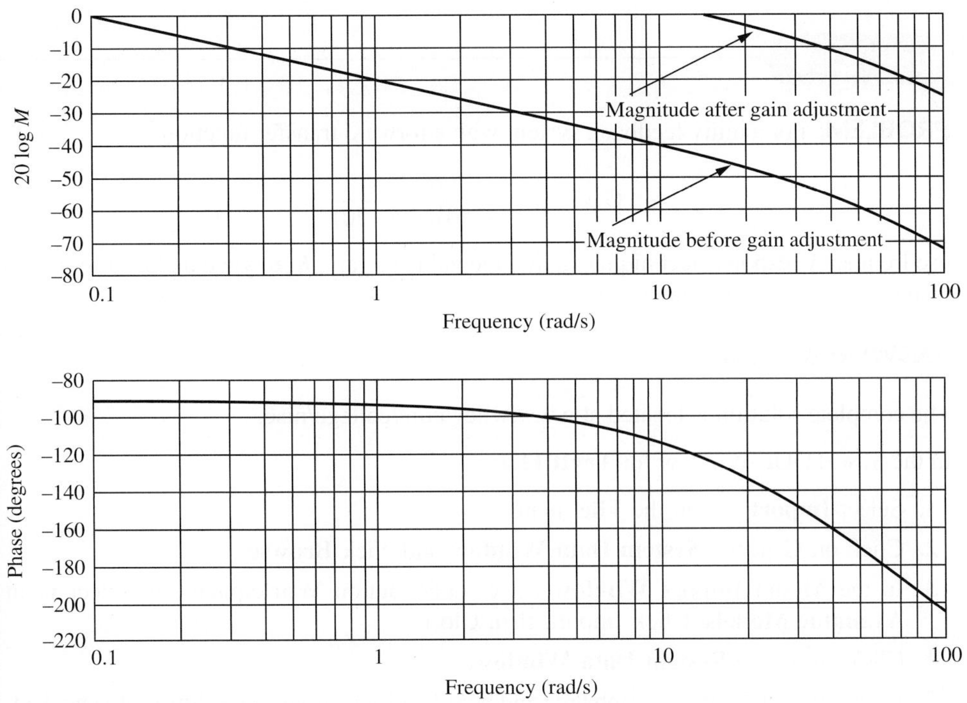


**FIGURE 11.3** Bode magnitude and phase plots for Example 11.1

The gain-adjusted open-loop transfer function is

$$G(s) = \frac{58{,}390}{s(s+36)(s+100)} \tag{11.1}$$

Table 11.1 summarizes a computer simulation of the gain-compensated system.

**TABLE 11.1** Characteristic of gain-compensated system of Example 11.1

| Parameter | Proposed specification | Actual value |
|---|---|---|
| $K_v$ | — | 16.22 |
| Phase margin | 59.2° | 59.2° |
| Phase-margin frequency | — | 14.8 rad/s |
| Percent overshoot | 9.5 | 10 |
| Peak time | — | 0.18 second |

MATLAB **ML**

Students who are using MATLAB should now run ch11p1 in Appendix B. You will learn how to use MATLAB to design a gain to meet a percent overshoot specification using Bode plots. This exercise solves Example 11.1 using MATLAB.

## Skill-Assessment Exercise 11.1

**PROBLEM:** For a unity feedback system with a forward transfer function

$$G(s) = \frac{K}{s(s+50)(s+120)}$$

use frequency response techniques to find the value of gain, $K$, to yield a closed-loop step response with 20% overshoot.

**ANSWER:** $K = 194{,}200$

The complete solution is located at www.wiley.com/college/nise.

**TryIt 11.1**

Use MATLAB, the Control System Toolbox, and the following statements to solve Skill-Assessment Exercise 11.1.

```
pos=20
z=(-log(pos/100))/...
 (sqrt(pi^2+...
 log(pos/100)^2))
Pm=atan(2*z/...
 (sqrt(-2*z^2+...
 sqrt(1+4*z^4))))*...
 (180/pi)
G=zpk([],...
 [0,-50,-120],1)
sisotool
```

In the SISOTOOL Window for **TryIt 11.1**:

1. Select **Import** . . . in the **File** menu.
2. Click on **G** in the **System Data Window** and click **Browse** . . .
3. In the **Model Import Window** select radio button **Workspace** and select **G** in **Available Models**. Click Import, then Close.
4. Click Ok in the **System Data Window.**
5. Right-click in the Bode graph area and be sure all selections under **Show** are checked.
6. Grab the stability margin point in the magnitude diagram and raise the magnitude curve until the phase curve shows the phase margin calculated by the program and shown in the MATLAB **Command Window** as Pm.
7. Right-click in the Bode plot area, select **Edit Compensator** . . . and read the gain under **Compensator** in the resulting window.

In this section, we paralleled our work in Chapter 8 with a discussion of transient response design through gain adjustment. In the next three sections, we parallel the root locus compensator design in Chapter 9 and discuss the design of lag, lead, and lag-lead compensation via Bode diagrams.

# 11.3 Lag Compensation

In Chapter 9, we used the root locus to design lag networks and PI controllers. Recall that these compensators permitted us to design for steady-state error without appreciably affecting the transient response. In this section, we provide a parallel development using the Bode diagrams.

## Visualizing Lag Compensation

The function of the lag compensator as seen on Bode diagrams is to (1) improve the static error constant by increasing only the low-frequency gain without any resulting instability, and (2) increase the phase margin of the system to yield the desired transient response. These concepts are illustrated in Figure 11.4.

The uncompensated system is unstable, since the gain at 180° is greater than 0 dB. The lag compensator, while not changing the low-frequency gain, does reduce the high-frequency gain.[1] Thus, the low-frequency gain of the system can be made high to yield

[1] The name *lag compensator* comes from the fact that the typical phase angle response for the compensator, as shown in Figure 11.4, is always negative, or *lagging* in phase angle.

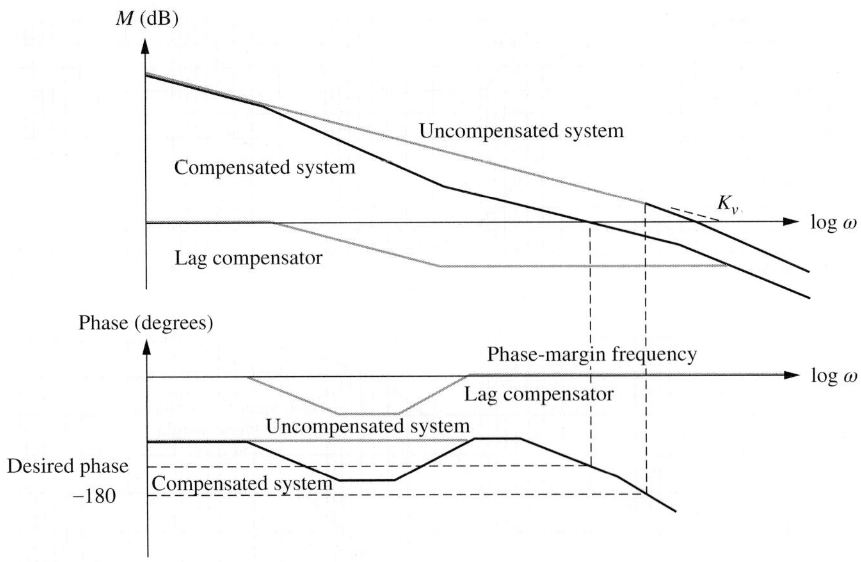


**FIGURE 11.4** Visualizing lag compensation

a large $K_v$ without creating instability. This stabilizing effect of the lag network comes about because the gain at 180° of phase is reduced below 0 dB. Through judicious design, the magnitude curve can be reshaped, as shown in Figure 11.4, to go through 0 dB at the desired phase margin. Thus, both $K_v$ and the desired transient response can be obtained. We now enumerate a design procedure.

## Design Procedure

1. Set the gain, $K$, to the value that satisfies the steady-state error specification and plot the Bode magnitude and phase diagrams for this value of gain.
2. Find the frequency where the phase margin is 5° to 12° greater than the phase margin that yields the desired transient response (*Ogata, 1990*). This step compensates for the fact that the phase of the lag compensator may still contribute anywhere from −5° to −12° of phase at the phase-margin frequency.
3. Select a lag compensator whose magnitude response yields a composite Bode magnitude diagram that goes through 0 dB at the frequency found in Step 2 as follows: Draw the compensator's high-frequency asymptote to yield 0 dB for the compensated system at the frequency found in Step 2. Thus, if the gain at the frequency found in Step 2 is $20 \log K_{PM}$, then the compensator's high-frequency asymptote will be set at $-20 \log K_{PM}$. Select the upper break frequency to be 1 decade below the frequency found in Step 2;[2] select the low-frequency asymptote to be at 0 dB. Connect the compensator's high- and low-frequency asymptotes with a −20 dB/decade line to locate the lower break frequency.
4. Reset the system gain, $K$, to compensate for any attenuation in the lag network in order to keep the static error constant the same as that found in Step 1.

From these steps, you see that we are relying upon the initial gain setting to meet the steady-state requirements. Then, we rely upon the lag compensator's −20 dB/decade slope to meet the transient response requirement by setting the 0 dB crossing of the magnitude plot.

[2] This value of break frequency ensures that there will be only −5° to −12° phase contribution from the compensator at the frequency found in Step 2.

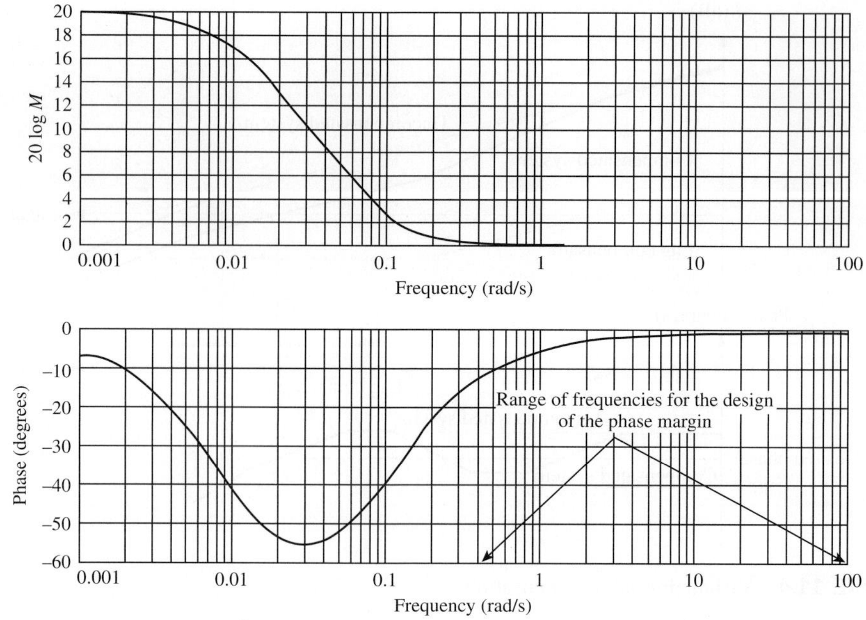


**FIGURE 11.5** Frequency response plots of a lag compensator, $G_c(s) = (s+0.1)/(s+0.01)$

The transfer function of the lag compensator is

$$G_c(s) = \frac{s + \frac{1}{T}}{s + \frac{1}{\alpha T}} \tag{11.2}$$

where $\alpha > 1$.

Figure 11.5 shows the frequency response curves for the lag compensator. The range of high frequencies shown in the phase plot is where we will design our phase margin. This region is after the second break frequency of the lag compensator, where we can rely on the attenuation characteristics of the lag network to reduce the total open-loop gain to unity at the phase-margin frequency. Further, in this region the phase response of the compensator will have minimal effect on our design of the phase margin. Since there is still some effect, approximately 5° to 12°, we will add this amount to our phase margin to compensate for the phase response of the lag compensator (see Step 2).

## Example 11.2 Lag Compensation Design

Design
D

**PROBLEM:** Given the system of Figure 11.2, use Bode diagrams to design a lag compensator to yield a tenfold improvement in steady-state error over the gain-compensated system while keeping the percent overshoot at 9.5%.

**SOLUTION:** We will follow the previously described lag compensation design procedure.

1. From Example 11.1 a gain, $K$, of 583.9 yields a 9.5% overshoot. Thus, for this system, $K_v = 16.22$. For a tenfold improvement in steady-state error, $K_v$ must increase by a factor of 10, or $K_v = 162.2$. Therefore, the value of $K$ in Figure 11.2 equals 5839, and

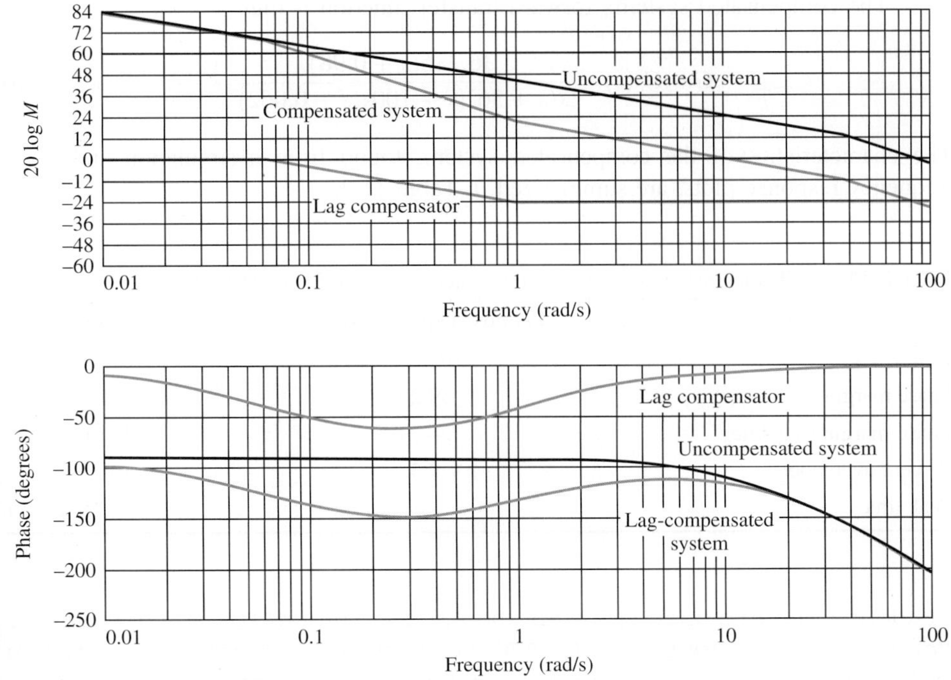


**FIGURE 11.6** Bode plots for Example 11.2

the open-loop transfer function is

$$G(s) = \frac{583{,}900}{s(s+36)(s+100)} \tag{11.3}$$

The Bode plots for $K = 5839$ are shown in Figure 11.6.

**2.** The phase margin required for a 9.5% overshoot ($\zeta = 0.6$) is found from Eq. (10.73) to be 59.2°. We increase this value of phase margin by 10° to 69.2° in order to compensate for the phase angle contribution of the lag compensator. Now find the frequency where the phase margin is 69.2°. This frequency occurs at a phase angle of $-180° + 69.2° = -110.8°$ and is 9.8 rad/s. At this frequency, the magnitude plot must go through 0 dB. The magnitude at 9.8 rad/s is now +24 dB (exact, that is, nonasymptotic). Thus, the lag compensator must provide −24 dB attenuation at 9.8 rad/s.

**3.&4.** We now design the compensator. First draw the high-frequency asymptote at −24 dB. Arbitrarily select the higher break frequency to be about one decade below the phase-margin frequency, or 0.98 rad/s. Starting at the intersection of this frequency with the lag compensator's high-frequency asymptote, draw a −20 dB/decade line until 0 dB is reached. The compensator must have a dc gain of unity to retain the value of $K_v$ that we have already designed by setting $K = 5839$. The lower break frequency is found to be 0.062 rad/s. Hence, the lag compensator's transfer function is

$$G_c(s) = \frac{0.063(s+0.98)}{(s+0.062)} \tag{11.4}$$

where the gain of the compensator is 0.063 to yield a dc gain of unity.

The compensated system's forward transfer function is thus

$$G(s)G_c(s) = \frac{36{,}786(s+0.98)}{s(s+36)(s+100)(s+0.062)} \tag{11.5}$$

The characteristics of the compensated system, found from a simulation and exact frequency response plots, are summarized in Table 11.2.

**TABLE 11.2** Characteristics of the lag-compensated system of Example 11.2

| Parameter | Proposed specification | Actual value |
|---|---|---|
| $K_v$ | 162.2 | 161.5 |
| Phase margin | 59.2° | 62° |
| Phase-margin frequency | — | 11 rad/s |
| Percent overshoot | 9.5 | 10 |
| Peak time | — | 0.25 second |

MATLAB
ML

Students who are using MATLAB should now run ch11p2 in Appendix B. You will learn how to use MATLAB to design a lag compensator. You will enter the value of gain to meet the steady-state error requirement as well as the desired percent overshoot. MATLAB then designs a lag compensator using Bode plots, evaluates $K_v$, and generates a closed-loop step response. This exercise solves Example 11.2 using MATLAB.

## Skill-Assessment Exercise 11.2

**PROBLEM:** Design a lag compensator for the system in Skill-Assessment Exercise 11.1 that will improve the steady-state error tenfold, while still operating with 20% overshoot.

**ANSWER:**

$$G_{\text{lag}}(s) = \frac{0.0691(s+2.04)}{(s+0.141)}; \quad G(s) = \frac{1{,}942{,}000}{s(s+50)(s+120)}$$

The complete solution is at www.wiley.com/college/nise.

**TryIt 11.2**

Use MATLAB, the Control System Toolbox, and the following statements to solve Skill-Assessment Exercise 11.2.

```
pos=20
Ts=0.2
z=(-log(pos/100))/(sqrt(pi^2+log(pos/100)^2))
Pm=atan(2*z/(sqrt(-2*z^2+sqrt(1+4*z^4))))*(180/pi)
Wbw=(4/(Ts*z))*sqrt((1-2*z^2)+sqrt(4*z^4-4*z^2+2))
K=1942000
G=zpk([], [0,-50,-120], K)
sisotool(G,1)
```

(TryIt continues)

*(TryIt continued)*

When the **SISO Design for SISO Design Task Window** appears:

1. Right-click on the Bode plot area and select **Grid.**
2. Note the phase margin shown in the MATLAB **Command Window.**
3. Using the Bode phase plot, estimate the frequency at which the phase margin from Step 2 occurs.
4. On the **SISO Design for SISO Design Task Window toolbar**, click on the red zero.
5. Place the zero of the compensator by clicking on the gain plot at a frequency that is 1/10 that found in Step 3.
6. On the **SISO Design for SISO Design Task Window toolbar**, click on the red pole.
7. Place the pole of the compensator by clicking on the gain plot to the left of the compensator zero.
8. Grab the pole with the mouse and move it until the phase plot shows a P.M. equal to that found in Step 2.
9. Right-click in the Bode plot area and select **Edit Compensator . . .**
10. Read the lag compensator in the **Control and Estimation Tools Manager Window.**

In this section, we showed how to design a lag compensator to improve the steady-state error while keeping the transient response relatively unaffected. We next discuss how to improve the transient response using frequency response methods.

## 11.4 Lead Compensation

For second-order systems, we derived the relationship between phase margin and percent overshoot as well as the relationship between closed-loop bandwidth and other time-domain specifications, such as settling time, peak time, and rise time. When we designed the lag network to improve the steady-state error, we wanted a minimal effect on the phase diagram in order to yield an imperceptible change in the transient response. However, in designing lead compensators via Bode plots, we want to change the phase diagram. We want to increase the phase margin to reduce the percent overshoot, and increase the gain crossover to realize a faster transient response.

### Visualizing Lead Compensation

The lead compensator increases the bandwidth by increasing the gain crossover frequency. At the same time, the phase diagram is raised at higher frequencies. The result is a larger phase margin and a higher phase-margin frequency. In the time domain, lower percent overshoots (larger phase margins) with smaller peak times (higher phase-margin frequencies) is the result. The concepts are shown in Figure 11.7.

The uncompensated system has a small phase margin ($B$) and a low phase-margin frequency ($A$). Using a phase lead compensator, the phase angle plot (compensated system) is raised for higher frequencies.[3] At the same time, the gain crossover frequency in the magnitude plot is increased from $A$ rad/s to $C$ rad/s. These effects yield a larger phase margin ($D$), a higher phase-margin frequency ($C$), and a larger bandwidth.

One advantage of the frequency response technique over the root locus is that we can implement a steady-state error requirement and then design a transient response. This specification of transient response with the constraint of a steady-state error is easier to implement with the frequency response technique than with the root locus. Notice that the initial slope, which determines the steady-state error, is not affected by the design for the transient response.

[3] The name *lead compensator* comes from the fact that the typical phase angle response shown in Figure 11.7 is always positive, or *leading* in phase angle.

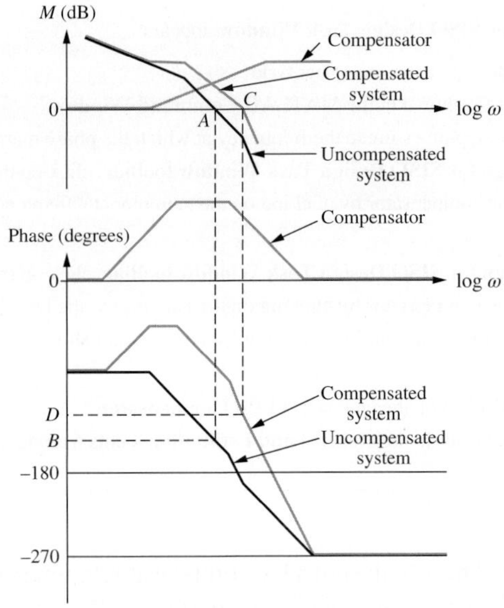


**FIGURE 11.7** Visualizing lead compensation

## Lead Compensator Frequency Response

Let us first look at the frequency response characteristics of a lead network and derive some valuable relationships that will help us in the design process. Figure 11.8 shows plots of the lead network

$$G_c(s) = \frac{1}{\beta}\frac{s+\dfrac{1}{T}}{s+\dfrac{1}{\beta T}} \tag{11.6}$$

for various values of $\beta$, where $\beta < 1$. Notice that the peaks of the phase curve vary in maximum angle and in the frequency at which the maximum occurs. The dc gain of the compensator is set to unity with the coefficient $1/\beta$, in order not to change the dc gain designed for the static error constant when the compensator is inserted into the system.

In order to design a lead compensator and change both the phase margin and phase-margin frequency, it is helpful to have an analytical expression for the maximum value of phase and the frequency at which the maximum value of phase occurs, as shown in Figure 11.8.

From Eq. (11.6) the phase angle of the lead compensator, $\phi_c$, is

$$\phi_c = \tan^{-1}\omega T - \tan^{-1}\omega\beta T \tag{11.7}$$

Differentiating with respect to $\omega$, we obtain

$$\frac{d\phi_c}{d\omega} = \frac{T}{1+(\omega T)^2} - \frac{\beta T}{1+(\omega\beta T)^2} \tag{11.8}$$

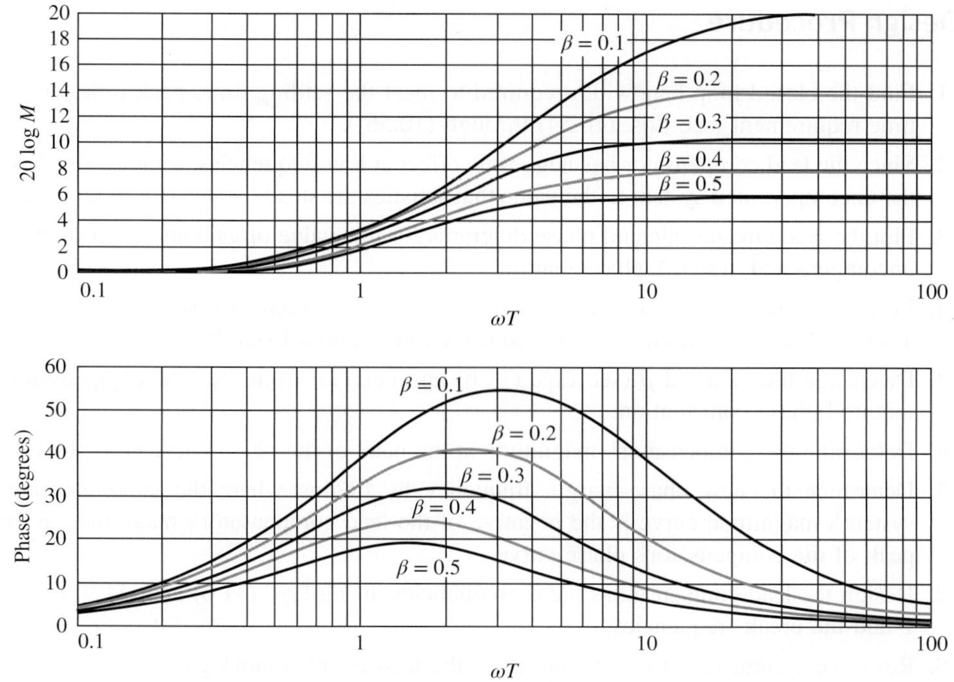


**FIGURE 11.8** Frequency response of a lead compensator, $G_c(s) = [1/\beta][(s + 1/T)/(s + 1/\beta T)]$

Setting Eq. (11.8) equal to zero, we find that the frequency, $\omega_{max}$, at which the maximum phase angle, $\phi_{max}$, occurs is

$$\omega_{max} = \frac{1}{T\sqrt{\beta}} \tag{11.9}$$

Substituting Eq. (11.9) into Eq. (11.6) with $s = j\omega_{max}$,

$$G_c(j\omega_{max}) = \frac{1}{\beta}\frac{j\omega_{max} + \frac{1}{T}}{j\omega_{max} + \frac{1}{\beta T}} = \frac{j\frac{1}{\sqrt{\beta}} + 1}{j\sqrt{\beta} + 1} \tag{11.10}$$

Making use of $\tan(\phi_1 - \phi_2) = (\tan\phi_1 - \tan\phi_2)/(1 + \tan\phi_1 \tan\phi_2)$, the maximum phase shift of the compensator, $\phi_{max}$, is

$$\phi_{max} = \tan^{-1}\frac{1-\beta}{2\sqrt{\beta}} = \sin^{-1}\frac{1-\beta}{1+\beta} \tag{11.11}$$

and the compensator's magnitude at $\omega_{max}$ is

$$|G_c(j\omega_{max})| = \frac{1}{\sqrt{\beta}} \tag{11.12}$$

We are now ready to enumerate a design procedure.

## Design Procedure

1. Find the closed-loop bandwidth required to meet the settling time, peak time, or rise time requirement (see Eqs. (10.54) through (10.56)).
2. Since the lead compensator has negligible effect at low frequencies, set the gain, $K$, of the uncompensated system to the value that satisfies the steady-state error requirement.
3. Plot the Bode magnitude and phase diagrams for this value of gain and determine the uncompensated system's phase margin.
4. Find the phase margin to meet the damping ratio or percent overshoot requirement. Then evaluate the additional phase contribution required from the compensator.[4]
5. Determine the value of $\beta$ (see Eqs. (11.6) and (11.11)) from the lead compensator's required phase contribution.
6. Determine the compensator's magnitude at the peak of the phase curve (Eq. (11.12)).
7. Determine the new phase-margin frequency by finding where the uncompensated system's magnitude curve is the negative of the lead compensator's magnitude at the peak of the compensator's phase curve.
8. Design the lead compensator's break frequencies, using Eqs. (11.6) and (11.9) to find $T$ and the break frequencies.
9. Reset the system gain to compensate for the lead compensator's gain.
10. Check the bandwidth to be sure the speed requirement in Step 1 has been met.
11. Simulate to be sure all requirements are met.
12. Redesign, if necessary, to meet requirements.

From these steps, we see that we are increasing both the amount of phase margin (improving percent overshoot) and the gain crossover frequency (increasing the speed). Now that we have enumerated a procedure with which we can design a lead compensator to improve the transient response, let us demonstrate.

### Example 11.3 Lead Compensation Design

Design D

**PROBLEM:** Given the system of Figure 11.2, design a lead compensator to yield a 20% overshoot and $K_v = 40$, with a peak time of 0.1 second.

**SOLUTION:** The uncompensated system is $G(s) = 100K/[s(s+36)(s+100)]$. We will follow the outlined procedure.

1. We first look at the closed-loop bandwidth needed to meet the speed requirement imposed by $T_p = 0.1$ second. From Eq. (10.56), with $T_p = 0.1$ second and $\zeta = 0.456$ (i.e., 20% overshoot), a closed-loop bandwidth of 46.6 rad/s is required.
2. In order to meet the specification of $K_v = 40$, $K$ must be set at 1440, yielding $G(s) = 144{,}000/[s(s+36)(s+100)]$.

[4] We know that the phase-margin frequency will be increased after the insertion of the compensator. At this new phase-margin frequency, the system's phase will be smaller than originally estimated, as seen by comparing points $B$ and $D$ in Figure 11.7. Hence, an additional phase should be added to that provided by the lead compensator to correct for the phase reduction caused by the original system.

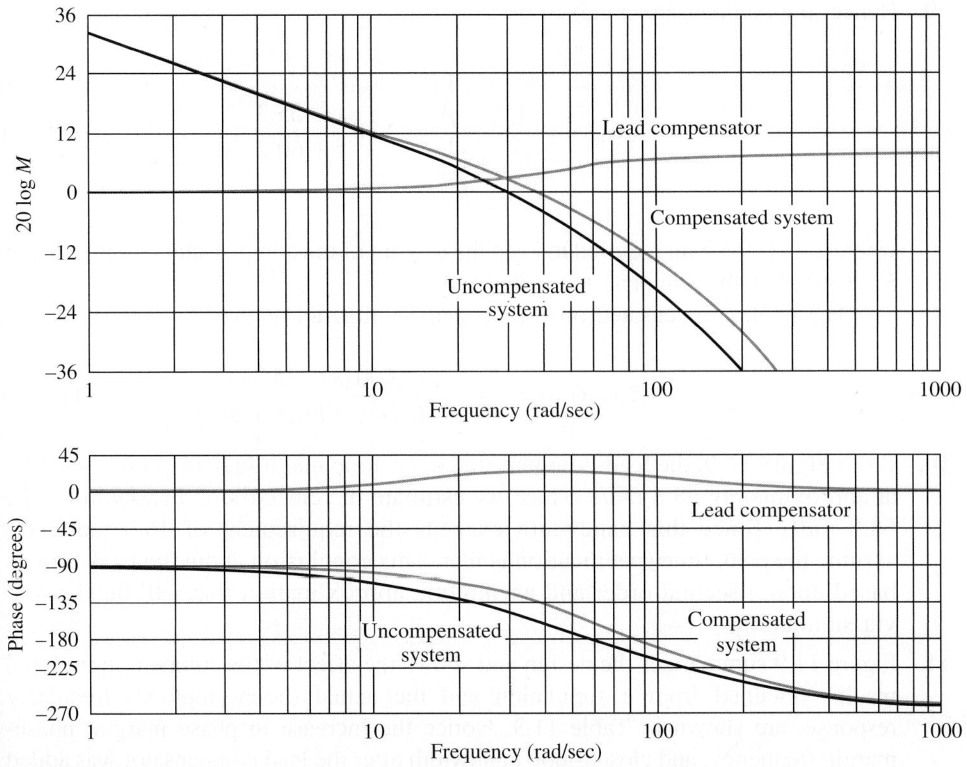


**FIGURE 11.9** Bode plots for lead compensation in Example 11.3

3. The uncompensated system's frequency response plots for $K = 1440$ are shown in Figure 11.9.
4. A 20% overshoot implies a phase margin of 48.1°. The uncompensated system with $K = 1440$ has a phase margin of 34° at a phase-margin frequency of 29.6. To increase the phase margin, we insert a lead network that adds enough phase to yield a 48.1° phase margin. Since we know that the lead network will also increase the phase-margin frequency, we add a correction factor to compensate for the lower uncompensated system's phase angle at this higher phase-margin frequency. Since we do not know the higher phase-margin frequency, we assume a correction factor of 10°. Thus, the total phase contribution required from the compensator is $48.1° - 34° + 10° = 24.1°$. In summary, our compensated system should have a phase margin of 48.1° with a bandwidth of 46.6 rad/s. If the system's characteristics are not acceptable after the design, then a redesign with a different correction factor may be necessary.
5. Using Eq. (11.11), $\beta = 0.42$ for $\phi_{\max} = 24.1°$.
6. From Eq. (11.12), the lead compensator's magnitude is 3.76 dB at $\omega_{\max}$.
7. If we select $\omega_{\max}$ to be the new phase-margin frequency, the uncompensated system's magnitude at this frequency must be −3.76 dB to yield a 0 dB crossover at $\omega_{\max}$ for the compensated system. The uncompensated system passes through −3.76 dB at $\omega_{\max} = 39$ rad/s. This frequency is thus the new phase-margin frequency.
8. We now find the lead compensator's break frequencies. From Eq. (11.9), $1/T = 25.3$ and $1/\beta T = 60.2$.

**9.** Hence, the compensator is given by

$$G_c(s) = \frac{1}{\beta}\frac{s+\dfrac{1}{T}}{s+\dfrac{1}{\beta T}} = 2.38\frac{s+25.3}{s+60.2} \tag{11.13}$$

where 2.38 is the gain required to keep the dc gain of the compensator at unity so that $K_v = 40$ after the compensator is inserted.

The final, compensated open-loop transfer function is then

$$G_c(s)G(s) = \frac{342{,}600\,(s+25.3)}{s(s+36)(s+100)(s+60.2)} \tag{11.14}$$

**10.** From Figure 11.9, the lead-compensated open-loop magnitude response is −7 dB at approximately 68.8 rad/s. Thus, we estimate the closed-loop bandwidth to be 68.8 rad/s. Since this bandwidth exceeds the requirement of 46.6 rad/s, we assume the peak time specification is met. This conclusion about the peak time is based upon a second-order and asymptotic approximation that will be checked via simulation.

**11.** Figure 11.9 summarizes the design and shows the effect of the compensation. Final results, obtained from a simulation and the actual (nonasymptotic) frequency response, are shown in Table 11.3. Notice the increase in phase margin, phase-margin frequency, and closed-loop bandwidth after the lead compensator was added to the gain-adjusted system. The peak time and the steady-state error requirements have been met, although the phase margin is less than that proposed and the percent overshoot is 2.6% larger than proposed. Finally, if the performance is not acceptable, a redesign is necessary.

**TABLE 11.3** Characteristic of the lead-compensated system of Example 11.3

| Parameter | Proposed specification | Actual gain-compensated value | Actual lead-compensated value |
|---|---|---|---|
| $K_v$ | 40 | 40 | 40 |
| Phase margin | 48.1° | 34° | 45.5° |
| Phase-margin frequency | — | 29.6 rad/s | 39 rad/s |
| Closed-loop bandwidth | 46.6 rad/s | 50 rad/s | 68.8 rad/s |
| Percent overshoot | 20 | 37 | 22.6 |
| Peak time | 0.1 second | 0.1 second | 0.075 second |

MATLAB
ML

Students who are using MATLAB should now run ch11p3 in Appendix B. You will learn how to use MATLAB to design a lead compensator. You will enter the desired percent overshoot, peak time, and $K_v$. MATLAB then designs a lead compensator using Bode plots, evaluates $K_v$, and generates a closed-loop step response. This exercise solves Example 11.3 using MATLAB.

## Skill-Assessment Exercise 11.3

**PROBLEM:** Design a lead compensator for the system in Skill-Assessment Exercise 11.1 to meet the following specifications: $\%OS = 20\%$, $T_s = 0.2$ s and $K_v = 50$.

**ANSWER:**

$$G_{\text{lead}}(s) = \frac{2.27(s+33.2)}{(s+75.4)}; \quad G(s) = \frac{300,000}{s(s+50)(s+120)}$$

The complete solution is at www.wiley.com/college/nise.

**TryIt 11.3**

Use MATLAB, the Control System Toolbox, and the following statements to solve Skill-Assessment Exercise 11.3.

```
pos=20
Ts=0.2
z=(-log(pos/100))/(sqrt(pi^2+log(pos/100)^2))
Pm=atan(2*z/(sqrt(-2*z^2+sqrt(1+4*z^4))))*(180/pi)
Wbw=(4/(Ts*z))*sqrt((1-2*z^2)+sqrt(4*z^4-4*z^2+2))
K=50*50*120
G=zpk([], [0,-50,-120],K)
sisotool(G,1)
```

When the **SISO Design for SISO Design Task Window** appears:

1. Right-click on the Bode plot area and select **Grid.**
2. Note the phase margin and bandwidth shown in the MATLAB **Command Window.**
3. On the **SISO Design for SISO Design Task Window toolbar,** click on the red pole.
4. Place the pole of the compensator by clicking on the gain plot at a frequency that is to the right of the desired bandwidth found in Step 2.
5. On the **SISO Design for SISO Design Task Window toolbar,** click on the red zero.
6. Place the zero of the compensator by clicking on the gain plot to the left of the desired bandwidth.
7. Reshape the Bode plots: alternately grab the pole and the zero with the mouse and alternately move them along the phase plot until the phase plot shows a P.M. equal to that found in Step 2 and a phase-margin frequency close to the bandwidth found in Step 2.
8. Right-click in the Bode plot area and select **Edit Compensator** . . .
9. Read the lead compensator in the **Control and Estimation Tools Manager Window.**

Keep in mind that the previous examples were designs for third-order systems and must be simulated to ensure the desired transient results. In the next section, we look at lag-lead compensation to improve steady-state error and transient response.

# 11.5 Lag-Lead Compensation

In Section 9.4, using root locus, we designed lag-lead compensation to improve the transient response and steady-state error. Figure 11.10 is an example of a system to which lag-lead compensation can be applied. In this section we repeat the design, using frequency response techniques. One method is to design the lag compensation to lower the high-frequency gain,

Katharina Bosse/laif/Redux Pictures.

(a)

Katharina Bosse/laif/Redux Pictures.

(b)

**FIGURE 11.10** **a.** The National Advanced Driving Simulator at the University of Iowa; **b.** test driving the simulator with its realistic graphics

stabilize the system, and improve the steady-state error and then design a lead compensator to meet the phase-margin requirements. Let us look at another method.

Section 9.6 describes a passive lag-lead network that can be used in place of separate lag and lead networks. It may be more economical to use a single, passive network that performs both tasks, since the buffer amplifier that separates the lag network from the lead network may be eliminated. In this section, we emphasize lag-lead design, using a single, passive lag-lead network.

The transfer function of a single, passive lag-lead network is

$$G_c(s) = G_{\text{Lead}}(s)G_{\text{Lag}}(s) = \left(\frac{s+\dfrac{1}{T_1}}{s+\dfrac{\gamma}{T_1}}\right)\left(\frac{s+\dfrac{1}{T_2}}{s+\dfrac{1}{\gamma T_2}}\right) \tag{11.15}$$

where $\gamma > 1$. The first term in parentheses produces the lead compensation, and the second term in parentheses produces the lag compensation. The constraint that we must follow here is that the single value $\gamma$ replaces the quantity $\alpha$ for the lag network in Eq. (11.2) and the quantity $\beta$ for the lead network in Eq. (11.6). For our design, $\alpha$ and $\beta$ must be reciprocals of each other. An example of the frequency response of the passive lag-lead is shown in Figure 11.11.

We are now ready to enumerate a design procedure.

## Design Procedure

1. Using a second-order approximation, find the closed-loop bandwidth required to meet the settling time, peak time, or rise time requirement (see Eqs. (10.55) and (10.56)).
2. Set the gain, $K$, to the value required by the steady-state error specification.
3. Plot the Bode magnitude and phase diagrams for this value of gain.

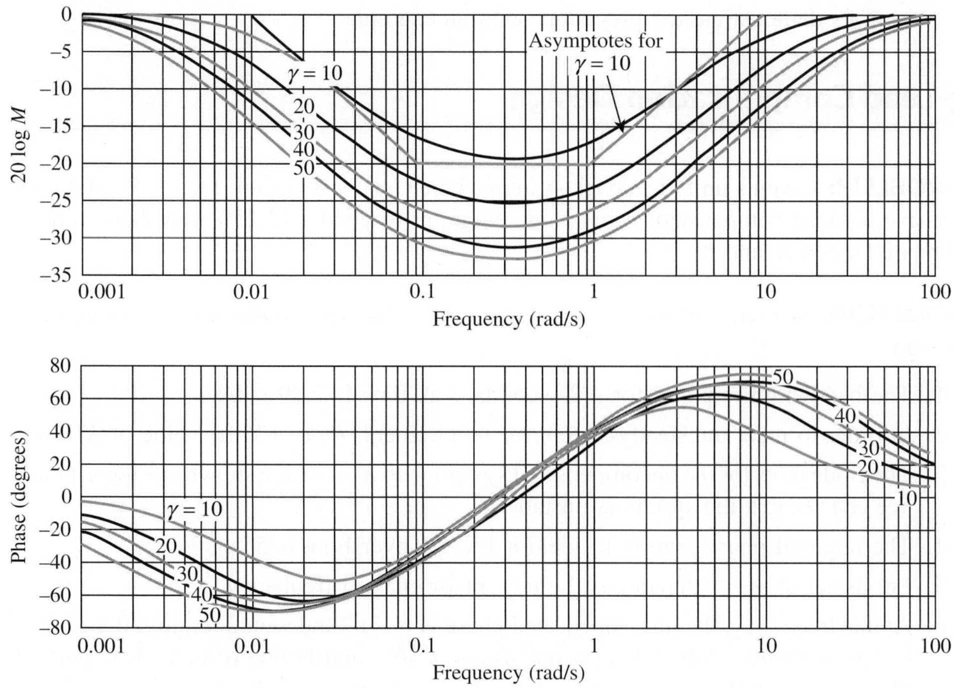


**FIGURE 11.11** Sample frequency response curves for a lag-lead compensator, $G_c(s) = [(s+1)(s+0.1)]/\left[(s+\gamma)\left(s+\frac{0.1}{\gamma}\right)\right]$

4. Using a second-order approximation, calculate the phase margin to meet the damping ratio or percent overshoot requirement, using Eq. (10.73).
5. Select a new phase-margin frequency near $\omega_{BW}$.
6. At the new phase-margin frequency, determine the additional amount of phase lead required to meet the phase-margin requirement. Add a small contribution that will be required after the addition of the lag compensator.
7. Design the lag compensator by selecting the higher break frequency one decade below the new phase-margin frequency. The design of the lag compensator is not critical, and any design for the proper phase margin will be relegated to the lead compensator. The lag compensator simply provides stabilization of the system with the gain required for the steady-state error specification. Find the value of $\gamma$ from the lead compensator's requirements. Using the phase required from the lead compensator, the phase response curve of Figure 11.8 can be used to find the value of $\gamma = 1/\beta$. This value, along with the previously found lag's upper break frequency, allows us to find the lag's lower break frequency.
8. Design the lead compensator. Using the value of $\gamma$ from the lag compensator design and the value assumed for the new phase-margin frequency, find the lower and upper break frequencies for the lead compensator, using Eq. (11.9) and solving for $T$.
9. Check the bandwidth to be sure the speed requirement in Step 1 has been met.
10. Redesign if phase-margin or transient specifications are not met, as shown by analysis or simulation.

Let us demonstrate the procedure with an example.

## Example 11.4 Lag-Lead Compensation Design

**PROBLEM:** Given a unity feedback system where $G(s) = K/[s(s+1)(s+4)]$, design a passive lag-lead compensator using Bode diagrams to yield a 13.25% overshoot, a peak time of 2 seconds, and $K_v = 12$.

**SOLUTION:** We will follow the steps previously mentioned in this section for lag-lead design.

1. The bandwidth required for a 2-seconds peak time is 2.29 rad/s.
2. In order to meet the steady-state error requirement, $K_v = 12$, the value of $K$ is 48.
3. The Bode plots for the uncompensated system with $K = 48$ are shown in Figure 11.12. We can see that the system is unstable.
4. The required phase margin to yield a 13.25% overshoot is 55°.
5. Let us select $\omega = 1.8$ rad/s as the new phase-margin frequency.
6. At this frequency, the uncompensated phase is $-176°$ and would require, if we add a $-5°$ contribution from the lag compensator, a 56° contribution from the lead portion of the compensator.
7. The design of the lag compensator is next. The lag compensator allows us to keep the gain of 48 required for $K_v = 12$ and not have to lower the gain to stabilize the system. As long as the lag compensator stabilizes the system, the design parameters are not critical, since the phase margin will be designed with the lead compensator. Thus, choose the lag compensator so that its phase response will have minimal effect at the new phase-margin frequency. Let us choose the lag compensator's higher break frequency to be 1 decade below the new phase-margin frequency, at 0.18 rad/s. Since we need to add 56° of phase shift with the lead compensator at $\omega = 1.8$ rad/s, we estimate from Figure 11.8 that, if $\gamma = 10.6$ (since $\gamma = 1/\beta$, $\beta = 0.094$), we can obtain about 56° of phase shift from the lead compensator. Thus with $\gamma = 10.6$ and a new phase-margin frequency of $\omega = 1.8$ rad/s, the transfer function of the lag compensator is

$$G_{\text{lag}}(s) = \frac{1}{\gamma}\frac{\left(s + \frac{1}{T_2}\right)}{\left(s + \frac{1}{\gamma T_2}\right)} = \frac{1}{10.6}\frac{(s+0.183)}{(s+0.0172)} \tag{11.16}$$

where the gain term, $1/\gamma$, keeps the dc gain of the lag compensator at 0 dB. The lag-compensated system's open-loop transfer function is

$$G_{\text{lag-comp}}(s) = \frac{4.53(s+0.183)}{s(s+1)(s+4)(s+0.0172)} \tag{11.17}$$

8. Now we design the lead compensator. At $\omega = 1.8$, the lag-compensated system has a phase angle of 180°. Using the values of $\omega_{\text{max}} = 1.8$ and $\beta = 0.094$, Eq. (11.9) yields the lower break, $1/T_1 = 0.56$ rad/s. The higher break is then $1/\beta T_1 = 5.96$ rad/s.

   The lead compensator is

$$G_{\text{lead}}(s) = \gamma\frac{\left(s + \frac{1}{T_1}\right)}{\left(s + \frac{\gamma}{T_1}\right)} = 10.6\frac{(s+0.56)}{(s+5.96)} \tag{11.18}$$

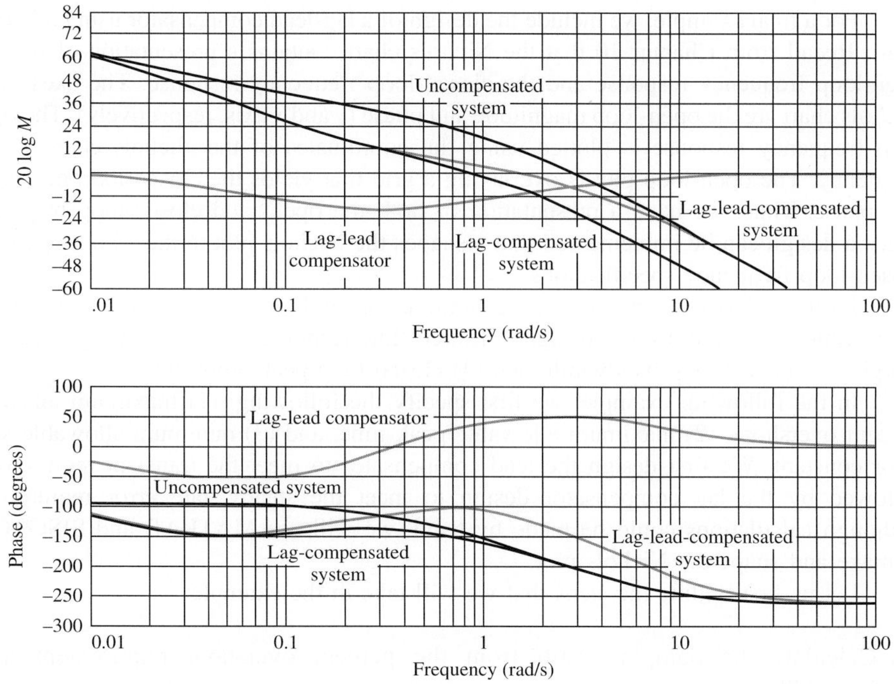


**FIGURE 11.12** Bode plots for lag-lead compensation in Example 11.4

The lag-lead-compensated system's open-loop transfer function is

$$G_{\text{lag-lead-comp}}(s) = \frac{48(s+0.183)(s+0.56)}{s(s+1)(s+4)(s+0.0172)(s+5.96)} \tag{11.19}$$

**9.** Now check the bandwidth. The closed-loop bandwidth is equal to that frequency where the open-loop magnitude response is approximately −7 dB. From Figure 11.12, the magnitude is −7 dB at approximately 3 rad/s. This bandwidth exceeds that required to meet the peak time requirement.

The design is now checked with a simulation to obtain actual performance values. Table 11.4 summarizes the system's characteristics. The peak time requirement is also met. Again, if the requirements were not met, a redesign would be necessary.

**TABLE 11.4** Characteristics of gain-compensated system of Example 11.4

| Parameter | Proposed specification | Actual value |
|---|---|---|
| $K_v$ | 12 | 12 |
| Phase margin | 55° | 59.3° |
| Phase-margin frequency | — | 1.63 rad/s |
| Closed-loop bandwidth | 2.29 rad/s | 3 rad/s |
| Percent overshoot | 13.25 | 10.2 |
| Peak time | 2.0 seconds | 1.61 seconds |

Students who are using MATLAB should now run ch11p4 in Appendix B. You will learn how to use MATLAB to design a lag-lead compensator. You will enter the desired percent overshoot, peak time, and $K_v$. MATLAB then designs a lag-lead compensator using Bode plots, evaluates $K_v$, and generates a closed-loop step response. This exercise solves Example 11.4 using MATLAB.

MATLAB

ML

For a final example, we include the design of a lag-lead compensator using a Nichols chart. Recall from Chapter 10 that the Nichols chart contains a presentation of both the open-loop frequency response and the closed-loop frequency response. The axes of the Nichols chart are the open-loop magnitude and phase ($y$ and $x$ axis, respectively). The open-loop frequency response is plotted using the coordinates of the Nichols chart at each frequency. The open-loop plot is overlaying a grid that yields the closed-loop magnitude and phase. Thus, we have a presentation of both the open- and closed-loop responses. Thus, a design can be implemented that reshapes the Nichols plot to meet both open- and closed-loop frequency specifications.

From a Nichols chart, we can see simultaneously the following frequency response specifications that are used to design a desired time response: (1) phase margin, (2) gain margin, (3) closed-loop bandwidth, and (4) closed-loop peak amplitude.

In the following example, we first specify the following: (1) maximum allowable percent overshoot, (2) maximum allowable peak time, and (3) minimum allowable static error constant. We first design the lead compensator to meet the transient requirements followed by the lag compensator design to meet the steady-state error requirement. Although calculations could be made by hand, we will use MATLAB and SISOTOOL to make and shape the Nichols plot.

Let us first outline the steps that we will take in the example:

1. Calculate the damping ratio from the percent overshoot requirement using Eq. (4.39)
2. Calculate the peak amplitude, $M_p$, of the closed-loop response using Eq. (10.52) and the damping ratio found in (1).
3. Calculate the minimum closed-loop bandwidth to meet the peak time requirement using Eq. (10.56), with peak time and the damping ratio from (1).
4. Plot the open-loop response on the Nichols chart.
5. Raise the open-loop gain until the open-loop plot is tangent to the required closed-loop magnitude curve, yielding the proper $M_p$.
6. Place the lead zero at this point of tangency and the lead pole at a higher frequency. Zeros and poles are added in SISOTOOL by clicking either one on the tool bar and then clicking the position on the open-loop frequency response curve where you desire to add the zero or pole.
7. Adjust the positions of the lead zero and pole until the open-loop frequency response plot is tangent to the same $M_p$ curve, but at the approximate frequency found in (3). This yields the proper closed-loop peak and proper bandwidth to yield the desired percent overshoot and peak time, respectively.
8. Evaluate the open-loop transfer function, which is the product of the plant and the lead compensator, and determine the static error constant.
9. If the static error constant is lower than required, a lag compensator must now be designed. Determine how much improvement in the static error constant is required.
10. Recalling that the lag pole is at a frequency below that of the lag zero, place a lag pole and zero at frequencies below the lead compensator and adjust to yield the desired improvement in static error constant. As an example, recall from Eq. (9.5) that the improvement in static error constant for a Type 1 system is equal to the ratio of the lag zero value divided by the lag pole value. Readjust the gain if necessary.

## Example 11.5 Lag-Lead Design Using the Nichols Chart, MATLAB, and SISOTOOL

MATLAB **ML**

GUI Tool **GUIT**

**PROBLEM:** Design a lag-lead compensator for the plant, $G(s) = \dfrac{K}{s(s+5)(s+10)}$, to meet the following requirements: (1) a maximum of 20% overshoot, (2) a peak time of no more than 0.5 seconds, (3) a static error constant of no less than 6.

**SOLUTION:** We follow the steps enumerated immediately above,

1. Using Eq. (4.39), $\zeta = 0.456$ for 20% overshoot.
2. Using Eq. (10.52), $M_p = 1.23 = 1.81$ dB for $\zeta = 0.456$.
3. Using Eq. (10.56), $\omega_{\mathrm{BW}} = 9.3$ r/s for $\zeta = 0.456$ and $T_p = 0.5$.
4. Plot the open-loop frequency response curve on the Nichols chart for $K = 1$.
5. Raise the open-loop frequency response curve until it is tangent to the closed-loop peak of 1.81 dB curve as shown in Figure 11.13. The frequency at the tangent point is approximately 3 r/s, which can be found by letting your mouse rest on the point of tangency. On the menu bar, select **Designs/Edit Compensator** . . . and find the gain added to the plant. Thus, the plant is now $G(s) = \dfrac{150}{s(s+5)(s+10)}$. The gain-adjusted closed-loop step response is shown in Figure 11.14. Notice that the peak time is about 1 second and must be decreased.
6. Place the lead zero at this point of tangency and the lead pole at a higher frequency.
7. Adjust the positions of the lead zero and pole until the open-loop frequency response plot is tangent to the same $M_p$ curve, but at the approximate frequency found in **3**.

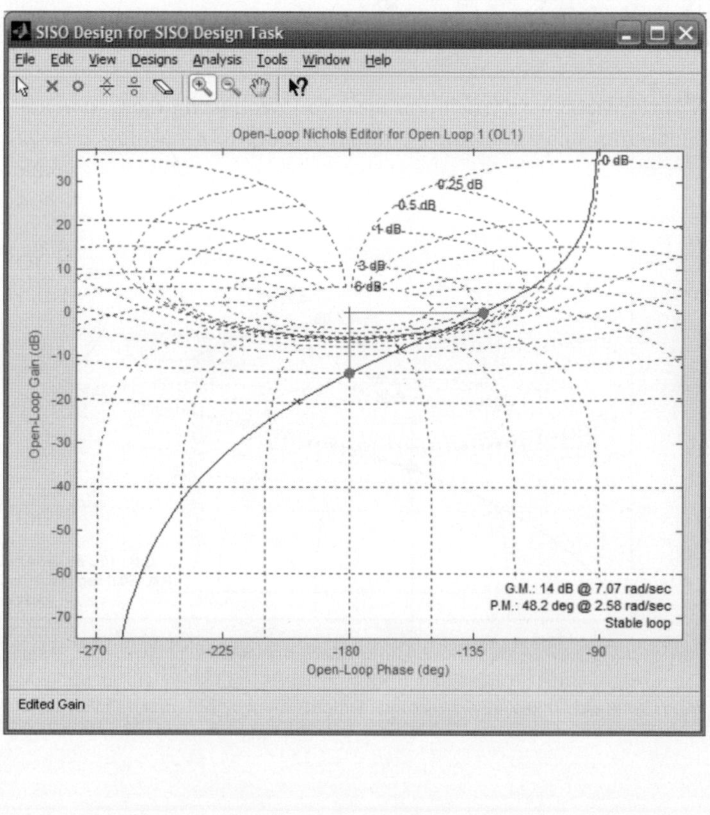


**FIGURE 11.13** Nichols chart after gain adjustment

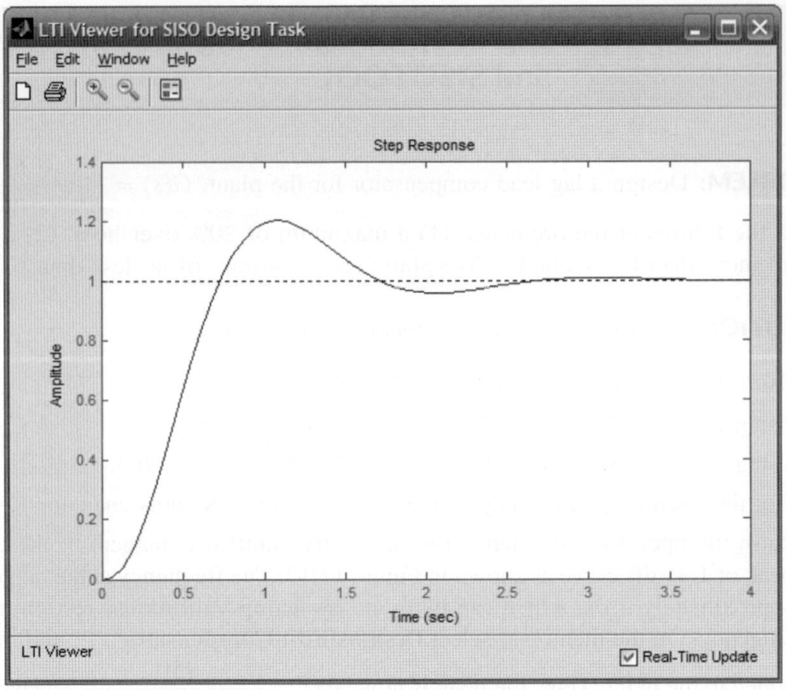


**FIGURE 11.14** Gain-adjusted closed-loop step response

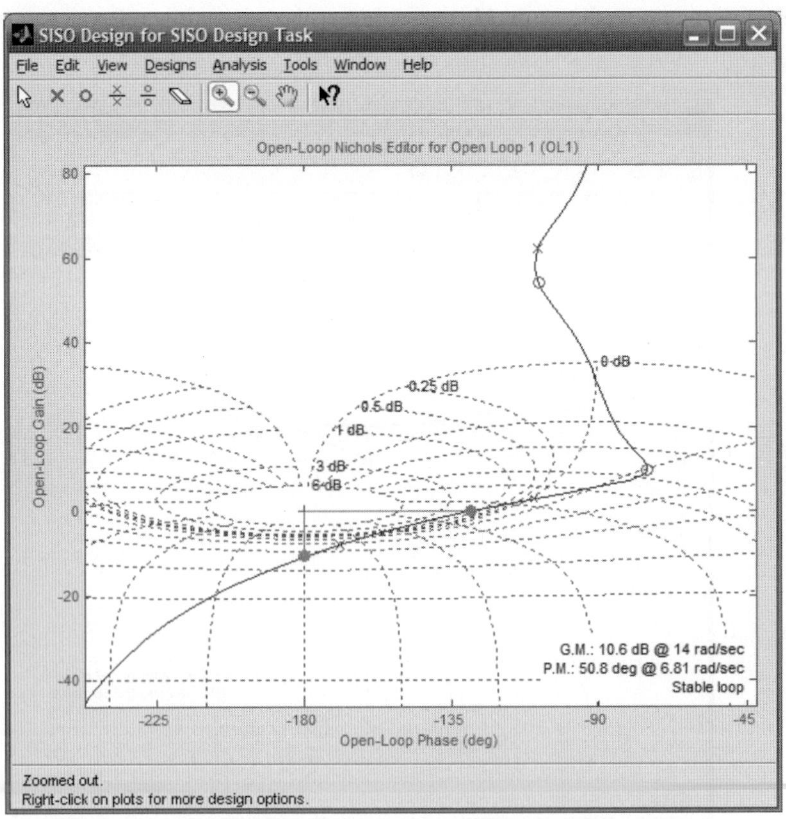


**FIGURE 11.15** Nichols chart after lag-lead compensation

8. Checking **Designs/Edit Compensator** . . . shows $G(s)G_{\text{lead}}(s) = \dfrac{1286(s+1.4)}{s(s+5)(s+10)(s+12)}$, which yields a $K_v = 3$.

9. We now add lag compensation to improve the static error constant by at least 2.

10. Now add a lag pole at $-0.004$ and a lag zero at $-0.008$. Readjust the gain to yield the same tangency as after the insertion of the lead. The final forward path is found to be $G(s)G_{\text{lead}}(s)G_{\text{lag}}(s) = \dfrac{1381(s+1.4)(s+0.008)}{s(s+5)(s+10)(s+12)(s+0.004)}$. The final Nichols chart is shown in Figure 11.15, and the compensated time response is shown in Figure 11.16. Notice that the time response has the expected slow climb to the final value that is typical of lag compensation. If your design requirements require a faster climb to the final response, then redesign the system with a larger bandwidth or attempt a design only with lead compensation. A problem at the end of the chapter provides the opportunity for practice.

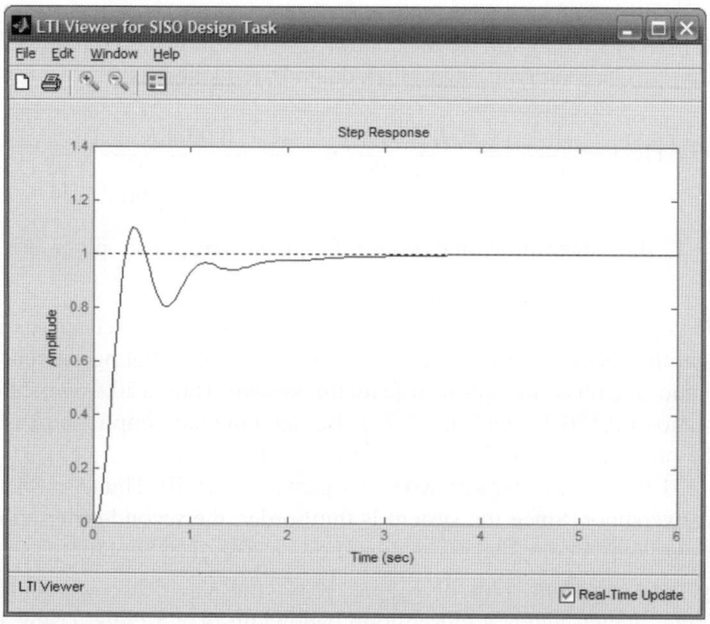


**FIGURE 11.16** Lag-lead compensated closed-loop step response

## Skill-Assessment Exercise 11.4

**PROBLEM:** Design a lag-lead compensator for a unity feedback system with the forward-path transfer function

$$G(s) = \frac{K}{s(s+8)(s+30)}$$

to meet the following specifications: $\%OS = 10\%$, $T_p = 0.6$ s, and $K_v = 10$. Use frequency response techniques.

**ANSWER:** $G_{\text{lag}}(s) = 0.456\dfrac{(s+0.602)}{(s+0.275)}$; $G_{\text{lead}}(s) = 2.19\dfrac{(s+4.07)}{(s+8.93)}$; $K = 2400$.

The complete solution is at www.wiley.com/college/nise.

## Case Studies

Our ongoing antenna azimuth position control system serves now as an example to summarize the major objectives of the chapter. The following cases demonstrate the use of frequency response methods to (1) design a value of gain to meet a percent overshoot requirement for the closed-loop step response and (2) design cascade compensation to meet both transient and steady-state error requirements.

Design
D

### Antenna Control: Gain Design

**PROBLEM:** Given the antenna azimuth position control system shown on the front endpapers, Configuration 1, use frequency response techniques to do the following:

**a.** Find the preamplifier gain required for a closed-loop response of 20% overshoot for a step input.

**b.** Estimate the settling time.

**SOLUTION:** The block diagram for the control system is shown on the inside front cover (Configuration 1). The loop gain, after block diagram reduction, is

$$G(s) = \frac{6.63K}{s(s+1.71)(s+100)} = \frac{0.0388K}{s\left(\frac{s}{1.71}+1\right)\left(\frac{s}{100}+1\right)} \tag{11.20}$$

Letting $K = 1$, the magnitude and phase frequency response plots are shown in Figure 10.61.

**a.** To find $K$ to yield a 20% overshoot, we first make a second-order approximation and assume that the second-order transient response equations relating percent overshoot, damping ratio, and phase margin are true for this system. Thus, a 20% overshoot implies a damping ratio of 0.456. Using Eq. (10.73), this damping ratio implies a phase margin of 48.1°. The phase angle should therefore be $(-180° + 48.1°) = -131.9°$. The phase angle is $-131.9°$ at $\omega = 1.49$ rad/s, where the gain is $-34.1$ dB. Thus $K = 34.1$ dB $= 50.7$ for a 20% overshoot. Since the system is third-order, the second-order approximation should be checked. A computer simulation shows a 20% overshoot for the step response.

**b.** Adjusting the magnitude plot of Figure 10.61 for $K = 50.7$, we find $-7$ dB at $\omega = 2.5$ rad/s, which yields a closed-loop bandwidth of 2.5 rad/s. Using Eq. (10.55) with $\zeta = 0.456$ and $\omega_{BW} = 2.5$, we find $T_s = 4.63$ seconds. A computer simulation shows a settling time of approximately 5 seconds.

**CHALLENGE:** We now give you a problem to test your knowledge of this chapter's objectives. You are given the antenna azimuth position control system shown on the inside front cover (Configuration 3). Using frequency response methods, do the following:

**a.** Find the value of $K$ to yield 25% overshoot for a step input.

MATLAB
ML

**b.** `Repeat Part a using MATLAB.`

Design
D

### Antenna Control: Cascade Compensation Design

**PROBLEM:** Given the antenna azimuth position control system block diagram shown on the front endpapers, Configuration 1, use frequency response techniques and design cascade compensation for a closed-loop response of 20% overshoot for a step input, a fivefold improvement in steady-state error over the gain-compensated system operating at 20% overshoot, and a settling time of 3.5 seconds.

**SOLUTION:** Following the lag-lead design procedure, we first determine the value of gain, $K$, required to meet the steady-state error requirement.

1. Using Eq. (10.55) with $\zeta = 0.456$, and $T_s = 3.5$ seconds, the required bandwidth is 3.3 rad/s.
2. From the preceding case study, the gain-compensated system's open-loop transfer function was, for $K = 50.7$,

$$G(s)H(s) = \frac{6.63K}{s(s+1.71)(s+100)} = \frac{336.14}{s(s+1.71)(s+100)} \tag{11.21}$$

   This function yields $K_v = 1.97$. If $K = 254$, then $K_v = 9.85$, a fivefold improvement.
3. The frequency response curves of Figure 10.61, which are plotted for $K = 1$, will be used for the solution.
4. Using a second-order approximation, a 20% overshoot requires a phase margin of 48.1°.
5. Select $\omega = 3$ rad/s to be the new phase-margin frequency.
6. The phase angle at the selected phase-margin frequency is $-152°$. This is a phase margin of 28°. Allowing for a 5° contribution from the lag compensator, the lead compensator must contribute $(48.1° - 28° + 5°) = 25.1°$.
7. The design of the lag compensator now follows. Choose the lag compensator upper break one decade below the new phase-margin frequency, or 0.3 rad/s. Figure 11.8 says that we can obtain 25.1° phase shift from the lead if $\beta = 0.4$ or $\gamma = 1/\beta = 2.5$. Thus, the lower break for the lag is at $1/(\gamma T) = 0.3/2.5 = 0.12$ rad/s.

   Hence,

$$G_{\text{lag}}(s) = 0.4\frac{(s+0.3)}{(s+0.12)} \tag{11.22}$$

8. Finally, design the lead compensator. Using Eq. (11.9), we have

$$T = \frac{1}{\omega_{\max}\sqrt{\beta}} = \frac{1}{3\sqrt{0.4}} = 0.527 \tag{11.23}$$

   Therefore, the lead compensator lower break frequency is $1/T = 1.9$ rad/s, and the upper break frequency is $1/(\beta T) = 4.75$ rad/s. Thus, the lag-lead-compensated forward path is

$$G_{\text{lag-lead-comp}}(s) = \frac{(6.63)(254)(s+0.3)(s+1.9)}{s(s+1.71)(s+100)(s+0.12)(s+4.75)} \tag{11.24}$$

9. A plot of the open-loop frequency response for the lag-lead-compensated system shows −7 dB at 5.3 rad/s. Thus, the bandwidth meets the design requirements for settling time. A simulation of the compensated system shows a 20% overshoot and a settling time of approximately 3.2 seconds, compared to a 20% overshoot for the uncompensated system and a settling time of approximately 5 seconds. $K_v$ for the compensated system is 9.85 compared to the uncompensated system value of 1.97.

**CHALLENGE:** We now give you a problem to test your knowledge of this chapter's objectives. You are given the antenna azimuth position control system shown on the front endpapers (Configuration 3). Using frequency response methods, do the following:

**a.** Design a lag-lead compensator to yield a 15% overshoot and $K_v = 20$. In order to speed up the system, the compensated system's phase-margin frequency will be set to 4.6 times the phase-margin frequency of the uncompensated system.

MATLAB ML **b.** `Repeat Part a using MATLAB.`

## Summary

This chapter covered the design of feedback control systems using frequency response techniques. We learned how to design by gain adjustment as well as cascaded lag, lead, and lag-lead compensation. Time response characteristics were related to the phase margin, phase-margin frequency, and bandwidth.

Design by gain adjustment consisted of adjusting the gain to meet a phase-margin specification. We located the phase-margin frequency and adjusted the gain to 0 dB.

A lag compensator is basically a low-pass filter. The low-frequency gain can be raised to improve the steady-state error, and the high-frequency gain is reduced to yield stability. Lag compensation consists of setting the gain to meet the steady-state error requirement and then reducing the high-frequency gain to create stability and meet the phase-margin requirement for the transient response.

A lead compensator is basically a high-pass filter. The lead compensator increases the high-frequency gain while keeping the low-frequency gain the same. Thus, the steady-state error can be designed first. At the same time, the lead compensator increases the phase angle at high frequencies. The effect is to produce a faster, stable system, since the uncompensated phase margin now occurs at a higher frequency.

A lag-lead compensator combines the advantages of both the lag and the lead compensator. First, the lag compensator is designed to yield the proper steady-state error with improved stability. Next, the lead compensator is designed to speed up the transient response. If a single network is used as the lag-lead, additional design considerations are applied so that the ratio of the lag zero to the lag pole is the same as the ratio of the lead pole to the lead zero.

In the next chapter, we return to state space and develop methods to design desired transient and steady-state error characteristics.

## Review Questions

1. What major advantage does compensator design by frequency response have over root locus design?
2. How is gain adjustment related to the transient response on the Bode diagrams?
3. Briefly explain how a lag network allows the low-frequency gain to be increased to improve steady-state error without having the system become unstable.
4. From the Bode plot perspective, briefly explain how the lag network does not appreciably affect the speed of the transient response.
5. Why is the phase margin increased above that desired when designing a lag compensator?
6. Compare the following for uncompensated and lag-compensated systems designed to yield the same transient response: low-frequency gain, phase-margin frequency, gain curve value around the phase-margin frequency, and phase curve values around the phase-margin frequency.

7. From the Bode diagram viewpoint, briefly explain how a lead network increases the speed of the transient response.
8. Based upon your answer to Question 7, explain why lead networks do not cause instability.
9. Why is a correction factor added to the phase margin required to meet the transient response?
10. When designing a lag-lead network, what difference is there in the design of the lag portion as compared to a separate lag compensator?

## Problems

1. Design the value of gain, $K$, for a gain margin of 10 dB in the unity feedback system of Figure P11.1 if [Section: 11.2]

   **a.** $G(s) = \dfrac{K}{(s+3)(s+9)(s+15)}$

   **b.** $G(s) = \dfrac{K}{s(s+3)(s+9)}$

   **c.** $G(s) = \dfrac{K(s+2)}{s(s+3)(s+4)(s+5)}$

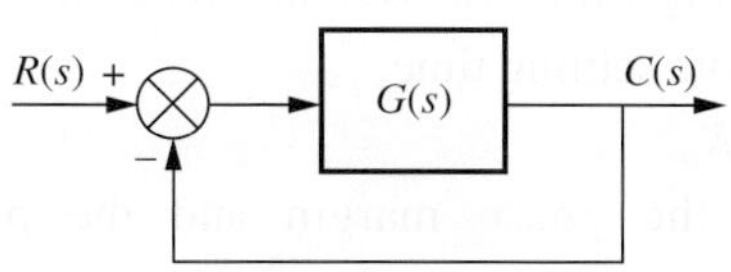


**FIGURE P11.1**

2. For each of the systems in Problem 1, design the gain, $K$, for a phase margin of 40°. [Section: 11.2]
3. Given the unity feedback system of Figure P11.1, use frequency response methods to determine the value of gain, $K$, to yield a step response with a 20% overshoot if [Section: 11.2]

   **a.** $G(s) = \dfrac{K}{s(s+6)(s+12)}$

   **b.** $G(s) = \dfrac{K(s+4)}{s(s+8)(s+10)(s+12)}$

   **c.** $G(s) = \dfrac{K(s+2)(s+7)}{s(s+4)(s+8)(s+10)(s+15)}$

4. Given the unity feedback system of Figure P11.1 with

$$G(s) = \frac{K(s+20)(s+25)}{s(s+5)(s+8)(s+14)}$$

   do the following: [Section: 11.2]

   **a.** Use frequency response methods to determine the value of gain, $K$, to yield a step response with a 15% overshoot. Make any required second-order approximations.

   **b.** Use MATLAB or any other computer program to test your second-order approximation by simulating the system for your designed value of $K$.

   MATLAB ML

5. Given the unity feedback system of Figure P11.1 with

$$G(s) = \frac{K(s+10)(s+11)}{s(s+3)(s+6)(s+9)}$$

   do the following: [Section: 11.3]

   **a.** Use frequency response methods to design a lag compensator to yield $K_v = 1000$ and 15% overshoot for the step response. Make any required second-order approximations.

   **b.** Use MATLAB or any other computer program to test your second-order approximation by simulating the system for your designed value of $K$ and lag compensator.

   MATLAB ML

6. The unity feedback system of Figure P11.1 with

$$G(s) = \frac{K}{s(s+7)}$$

   is operating with 15% overshoot. Using frequency response techniques, design a compensator to yield $K_v = 50$ with the phase-margin frequency and phase margin remaining approximately the same as in the uncompensated system. [Section: 11.3]

7. The unity feedback system shown in Figure P11.1with

$$G(s) = \frac{K}{s(s+5)(s+8)}$$

   is operating with 20% overshoot. Using frequency response methods, design a compensator to yield a five-fold improvement in steady-state error without appreciably changing the transient response.

8. Design a lag compensator so that the system of Figure P11.1 where

$$G(s) = \frac{K(s+4)}{(s+2)(s+6)(s+8)}$$

operates with a 45° phase margin and a static error constant of 100. [Section: 11.3]

9. Design a PI controller for the system of Figure 11.2 that will yield zero steady-state error for a ramp input and a 9.48% overshoot for a step input. [Section: 11.3]

10. For the system of Problem 5, do the following: [Section: 11.3]
   **a.** Use frequency response methods to find the gain, $K$, required to yield about 15% overshoot. Make any required second-order approximations.
   **b.** Use frequency response methods to design a PI compensator to yield zero steady-state error for a ramp input without appreciably changing the transient response characteristics designed in Part **a**.
   **c.** Use MATLAB or any other computer program to test your second-order approximation by simulating the system for your designed value of $K$ and PI compensator. MATLAB ML

11. Write a MATLAB program that will design a PI controller assuming a second-order approximation as follows: MATLAB ML
   **a.** Allow the user to input from the keyboard the desired percent overshoot
   **b.** Design a PI controller and gain to yield zero steady-state error for a closed-loop step response as well as meet the percent overshoot specification
   **c.** Display the compensated closed-loop step response

   Test your program on

$$G(s) = \frac{K}{(s+5)(s+10)}$$

   and 25% overshoot.

12. Design a compensator for the unity feedback system of Figure P11.1 with

$$G(s) = \frac{K}{s(s+3)(s+15)(s+20)}$$

to yield a $K_v = 4$ and a phase margin of 40°. [Section: 11.4]

13. Consider the unity feedback system of Figure P11.1 with

$$G(s) = \frac{K}{s(s+5)(s+20)}$$

The uncompensated system has about 55% overshoot and a peak time of 0.5 second when $K_v = 10$. Do the following: [Section: 11.4]
   **a.** Use frequency response methods to design a lead compensator to reduce the percent overshoot to 10%, while keeping the peak time and steady-state error about the same or less. Make any required second-order approximations.
   **b.** Use MATLAB or any other computer program to test your second-order approximation by simulating the system for your designed value of $K$. MATLAB ML

14. Repeat Problem 13 using a PD compensator. [Section: 11.4]

15. The unity feedback system of Figure P11.1 with

$$G(s) = \frac{K(s+5)}{(s+2)(s+6)(s+10)}$$

is operating with 20% overshoot. [Section: 11.4]
   **a.** Find the settling time.
   **b.** Find $K_p$.
   **c.** Find the phase margin and the phase-margin frequency.
   **d.** Using frequency response techniques, design a compensator that will yield a threefold improvement in $K_p$ and a twofold reduction in settling time while keeping the overshoot at 20%.

16. Repeat the design of Example 11.3 in the text using a PD controller. [Section: 11.4]

17. Write a MATLAB program that will design a lead compensator assuming second-order approximations as follows: MATLAB ML
   **a.** Allow the user to input from the keyboard the desired percent overshoot, peak time, and gain required to meet a steady-state error specification
   **b.** Display the gain-compensated Bode plot
   **c.** Calculate the required phase margin and bandwidth
   **d.** Display the pole, zero, and gain of the lead compensator
   **e.** Display the compensated Bode plot

f. Output the step response of the lead-compensated system to test your second-order approximation

Test your program on a unity feedback system where

$$G(s) = \frac{K(s+1)}{s(s+2)(s+6)}$$

and the following specifications are to be met: percent overshoot = 10%, peak time = 0.1 second, and $K_v = 30$.

**18.** Repeat Problem 17 for a PD controller. MATLAB ML

**19.** Use frequency response methods to design a lag-lead compensator for a unity feedback system where [Section: 11.4]

$$G(s) = \frac{K(s+7)}{s(s+5)(s+15)}$$

and the following specifications are to be met: percent overshoot = 15%, settling time = 0.1 second, and $K_v = 1000$.

**20.** Write a MATLAB program that will design a lag-lead compensator assuming second-order approximations as follows: [Section: 11.5] MATLAB ML

**a.** Allow the user to input from the keyboard the desired percent overshoot, settling time, and gain required to meet a steady-state error specification

**b.** Display the gain-compensated Bode plot

**c.** Calculate the required phase margin and bandwidth

**d.** Display the poles, zeros, and the gain of the lag-lead compensator

**e.** Display the lag-lead-compensated Bode plot

**f.** Display the step response of the lag-lead compensated system to test your second-order approximation

Use your program to do Problem 19.

**21.** Given a unity feedback system with

$$G(s) = \frac{K}{s(s+2)(s+5)}$$

design a PID controller to yield zero steady-state error for a ramp input, as well as a 20% overshoot, and a peak time less than 2 seconds for a step input. Use only frequency response methods. [Section: 11.5]

**22.** A unity feedback system has MATLAB ML

$$G(s) = \frac{K}{s(s+3)(s+6)}$$

If this system has an associated 0.5 second delay, use MATLAB to design the value of $K$ for 20% overshoot. Make any necessary second-order approximations, but test your assumptions by simulating your design. The delay can be represented by cascading the MATLAB function padé $(T, n)$ with $G(s)$, where $T$ is the delay in seconds and $n$ is the order of the Pade approximation (use 5). Write the program to do the following:

**a.** Accept your value of percent overshoot from the keyboard

**b.** Display the Bode plot for $K = 1$

**c.** Calculate the required phase margin and find the phase-margin frequency and the magnitude at the phase-margin frequency

**d.** Calculate and display the value of $K$

## DESIGN PROBLEMS

**23.** Aircraft are sometimes used to tow other vehicles. A roll control system for such an aircraft was discussed in Problem 57 in Chapter 6. If Figure P11.2 represents the roll control system, use only frequency response techniques to do the following (*Cochran, 1992*):

**a.** Find the value of gain, $K$, to yield a closed-loop step response with 10% overshoot.

**b.** Estimate peak time and settling time using the gain-compensated frequency response.

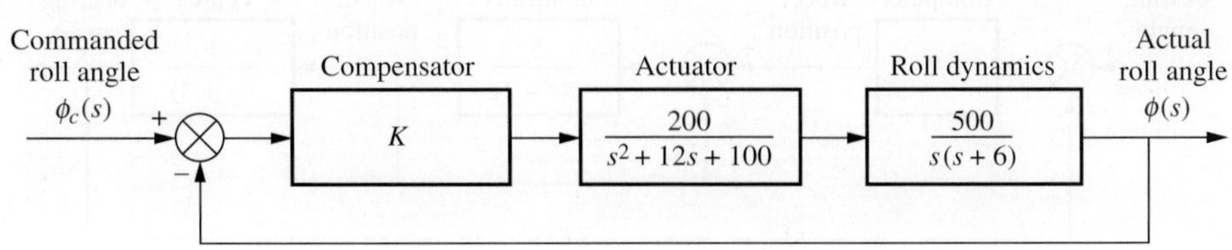


**FIGURE P11.2** Towed-vehicle roll control

c. Use MATLAB to simulate your system. Compare the results of the simulation with the requirements in Part **a** and your estimation of performance in Part **b**. MATLAB ML

24. The model for a specific linearized TCP/IP computer network queue, working under a random early detection (RED) algorithm, has been modeled using the block diagram of Figure P11.1, where $G(s) = M(s)P(s)$, with

$$M(s) = \frac{0.005L}{s + 0.005}$$

and

$$P(s) = \frac{140{,}625e^{-0.1s}}{(s + 2.67)(s + 10)}$$

Also, $L$ is a parameter to be varied (*Hollot, 2001*).

a. Adjust $L$ to obtain a 15% overshoot in the transient response for step inputs.

b. Verify Part **a** with a Simulink unit step response simulation. Simulink SL

25. An electric ventricular assist device (EVAD) that helps pump blood concurrently to a defective natural heart in sick patients can be shown to have a transfer function

$$G(s) = \frac{P_{ao}(s)}{E_m(s)} = \frac{1361}{s^2 + 69s + 70.85}$$

The input, $E_m(s)$, is the motor's armature voltage, and the output is $P_{ao}(s)$, the aortic blood pressure (*Tasch, 1990*). The EVAD will be controlled in the closed-loop configuration shown in Figure P11.1.

a. Design a phase lag compensator to achieve a tenfold improvement in the steady-state error to step inputs without appreciably affecting the transient response of the uncompensated system.

b. Use MATLAB to simulate the uncompensated and compensated systems for a unit step input. MATLAB ML

26. A Tower Trainer 60 Unmanned Aerial Vehicle has a transfer function

$$P(s) = \frac{h(s)}{\delta_e(s)} = \frac{-34.16s^3 - 144.4s^2 + 7047s + 557.2}{s^5 + 13.18s^4 + 95.93s^3 + 14.61s^2 + 31.94s}$$

where $\delta_e(s)$ is the elevator angle and $h(s)$ is the change in altitude (*Barkana, 2005*).

a. Assuming the airplane is controlled in the closed-loop configuration of Figure P11.1 with $G(s) = KP(s)$, find the value of $K$ that will result in a 30° phase margin.

b. For the value of $K$ calculated in Part **a**, obtain the corresponding gain margin.

c. Obtain estimates for the system's $\%OS$ and settling times $T_s$ for step inputs.

d. Simulate the step response of the system using MATLAB. MATLAB ML

e. Explain the simulation results and discuss any inaccuracies in the estimates obtained in Part **c**.

27. Self-guided vehicles are used in factories to transport products from station to station. One method of construction is to embed a wire in the floor to provide guidance. Another method is to use an onboard computer and a laser scanning device. Bar-coded reflective devices at known locations allow the system to determine the vehicle's angular position. This system allows the vehicle to travel anywhere, including between buildings (*Stefanides, 1987*). Figure P11.3 shows a simplified block diagram of the vehicle's bearing control system. For 11% overshoot, $K$ is set equal to 2. Design a lag compensator using frequency response techniques to improve the steady-state error by a factor of 30 over that of the uncompensated system.

28. An aircraft roll control system is shown in Figure P11.4. The torque on the aileron generates a roll rate. The

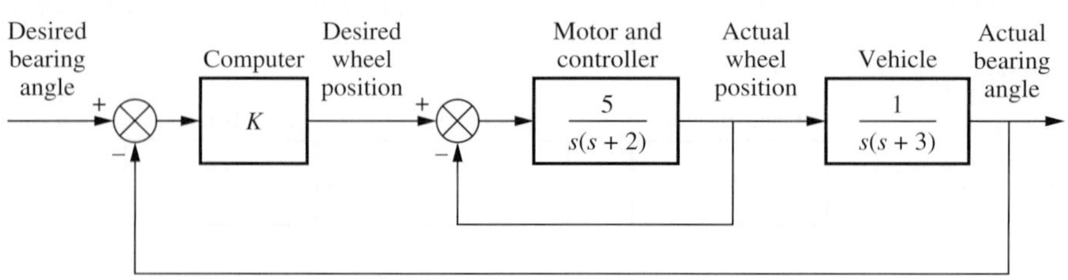


**FIGURE P11.3** Simplified block diagram of a guided cart

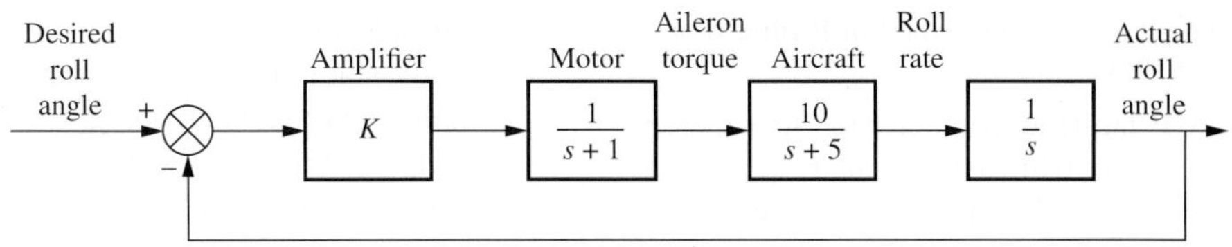


**FIGURE P11.4**

resulting roll angle is then controlled through a feedback system as shown. Design a lead compensator for a 60° phase margin and $K_v = 5$.

**29.** Figure P11.5 illustrates a set of booms used for the delivery of chemicals in agriculture (*Sun, 2011*). Each of the booms has equally spaced nozzles, the purpose of which is to maintain a constant gap between the nozzles and the soil despite car movements due to road unevenness. The booms are tethered to a vehicle (not shown in figure), and the gap is measured using an infrared sensor. This measurement is fed to a controller that drives two hydraulic cylinders to adjust the boom's positions. Under certain operating conditions, it was found that the system can be described by the unity-feedback configuration of Figure P11.1 where

$$G(s) = \frac{K}{s}\frac{2.78\times10^{-4}}{\left(\dfrac{s^2}{60^2}+\dfrac{s}{60}+1\right)}\frac{509.3}{\left(\dfrac{s^2}{213^2}+\dfrac{3s}{213}+1\right)}$$

**a.** Design a lag compensator to achieve $K_v = 30$, and $\%OS = 10\%$.

**b.** Use a computer program to obtain the step response of the closed-loop system and verify its performance.

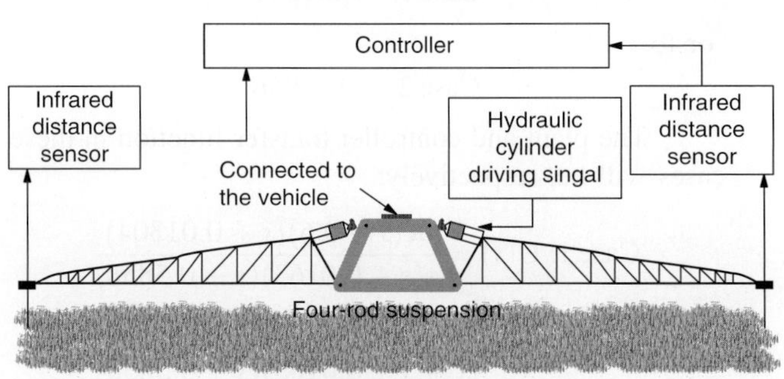


**FIGURE P11.5**[5]

**30.** The amount of leftover moisture in a grain drying conveyor process can be controlled by varying the conveyor's motor speed (*Mansor, 2011*). Although the process is highly variable, the transfer function at one of the operating points has been found to be

$$G(s) = \frac{0.0169s + 0.03558}{3.368s^3 + 3.762s^2 + 32.19s + 1}e^{-27s}$$

A unity feedback system will be built around this plant such as the one shown in Figure P11.1 to maintain grain moisture at specified levels. The specifications are: $e_{ss} = 0$ for a unit-step input; $e_{ss}$ is half that of the uncompensated system for a ramp input; settling time = 500 seconds, and $\%OS = 13\%$.

Since the system is Type 0, it must be augmented to a Type 1 using an integrator in order to achieve $e_{ss} = 0$ for a step input. In addition, a phase lead-lag compensator will be required to achieve the specifications. Simulate your design using a computer program.

**31.** Active front steering is used in front-steering four-wheel cars to control the yaw rate of the vehicle as a function of changes in wheel-steering commands. For a certain car, and under certain conditions, it has been shown that the transfer function from steering wheel angle to yaw rate is given by (*Zhang, 2008*):

$$P(s) = \frac{28.4s + 119.7}{s^2 + 7.15s + 14.7}$$

The system is controlled in a unity-feedback configuration.

**a.** Use the Nichols chart and follow the procedure of Example 11.5 to design a lag-lead compensator such that the system has zero steady-state error for a step input. The bandwidth of the closed-loop system must be $\omega_B = 10$ rad/sec. Let the open-loop magnitude response peak be less than 1 dB and the steady-state error constant $K_v = 20$.

**b.** Relax the bandwidth requirement to $\omega_B \geq 10$ rad/sec. Design the system for a steady-state error of zero for a step input. Let the open-loop magnitude response peak be less than 1 dB and $K_v = 20$ using only a lead compensator.

**c.** `Simulate the step response of both designs using MATLAB.` MATLAB ML

[5] Sun, J., and Miao, Y. Modeling and simulation of the agricultural sprayer boom leveling system. *IEEE Third International Conf. on Measuring Tech. and Mechatronics Automation, 2011*, pp. 613–618. Figure 2, p. 613. 2011 Third International Conference on Measuring Technology and Mechatronics Automation by IEEE. Reproduced with permission of IEEE in the format Republish in a book via Copyright Clearance Center.

32. For the heat exchange system described in Problem 38, Chapter 9 (*Smith, 2002*):
    a. Design a passive lag-lead compensator to achieve 5% steady-state error with a transient response of 10% overshoot and a settling time of 60 seconds for step inputs.
    b. Use MATLAB to simulate and verify your design. MATLAB ML

## PROGRESSIVE ANALYSIS AND DESIGN PROBLEMS

33. **Control of HIV/AIDS.** In Chapter 6, the model for an HIV/AIDS patient treated with RTIs was linearized and shown to be

$$P(s) = \frac{Y(s)}{U_1(s)} = \frac{-520s - 10.3844}{s^3 + 2.6817s^2 + 0.11s + 0.0126} = \frac{-520(s+0.02)}{(s+2.2644)(s^2+0.04s+0.0048)}$$

It is assumed here that the patient will be treated and monitored using the closed-loop configuration shown in Figure P11.1 Since the plant has a negative dc gain, assume for simplicity that $G(s) = G_c(s)P(s)$ and $G_c(0) < 0$. Assume also that the specifications for the design are (1) zero steady-state error for step inputs, (2) overdamped time-domain response, and (3) settling time $T_s \approx 100$ days (*Craig, 2004*).
    a. The overdamped specification requires a $\Phi_M \approx 90°$. Find the corresponding bandwidth required to satisfy the settling time requirement.
    b. The zero steady-state error specification implies that the open-loop transfer function must be augmented to Type 1. The $-0.02$ zero of the plant adds too much phase lead at low frequencies, and the complex conjugate poles, if left uncompensated within the loop, result in undesired oscillations in the time domain. Thus, as an initial approach to compensation for this system we can try

$$G_c(s) = \frac{-K(s^2+0.04s+0.0048)}{s(s+0.02)}$$

For $K = 1$, make a Bode plot of the resulting system. Obtain the value of $K$ necessary to achieve the design demands. Check for closed-loop stability.
    c. Simulate the unit step response of the system using MATLAB. Adjust $K$ to achieve the desired response. MATLAB ML

34. **Hybrid vehicle.** In Part **b** of Problem 52 in Chapter 10, we used a proportional-plus-integral (PI) speed controller that resulted in an overshoot of 20% and a settling time, $T_s = 3.92$ seconds (*Preitl, 2007*).
    a. Now assume that the system specifications require a steady-state error of zero for a step input, a ramp input steady-state error $\leq 2\%$, a $\%OS \leq 4.32\%$, and a settling time $\leq 4$ seconds. One way to achieve these requirements is to cancel the PI-controller's zero, $Z_I$, with the real pole of the uncompensated system closest to the origin (located at $-0.0163$). Assuming exact cancellation is possible, the plant and controller transfer function becomes

$$G(s) = \frac{K(s+0.6)}{s(s+0.5858)}$$

Design the system to meet the requirements. You may use the following steps:
        i. Set the gain, $K$, to the value required by the steady-state error specifications. Plot the Bode magnitude and phase diagrams.
        ii. Calculate the required phase margin to meet the damping ratio or equivalently the $\%OS$ requirement, using Eq. (10.73). If the phase margin found from the Bode plot obtained in Step **i** is greater than the required value, simulate the system to check whether the settling time is less than 4 seconds and whether the requirement of a $\%OS \leq 4.32\%$ has been met. Redesign if the simulation shows that the $\%OS$ and/or the steady-state error requirements have not been met. If all requirements are met, you have completed the design.
    b. In most cases, perfect pole-zero cancellation is not possible. Assume that you want to check what happens if the PI-controller's zero changes by ± 20%, e.g., if $Z_I$ moves to:

Case 1: $-0.01304$

or to

Case 2: $-0.01956$.

The plant and controller transfer function in these cases will be, respectively:

$$\text{Case 1: } G(s) = \frac{K(s+0.6)(s+0.01304)}{s(s+0.0163)(s+0.5858)}$$

$$\text{Case 2: } G(s) = \frac{K(s+0.6)(s+0.01956)}{s(s+0.0163)(s+0.5858)}$$

Set $K$ in each case to the value required by the steady-state error specifications and plot the Bode magnitude and phase diagrams. Simulate the closed-loop step response for each of the three locations of $Z_I$: pole/zero cancellation, Case 1, and Case 2, given in the problem.

Do the responses obtained resemble a second-order overdamped, critically damped, or underdamped response? Is there a need to add a derivative mode?

**35. Parabolic trough collector.** In order to reduce the steady-state error of the parabolic trough collector system, a PI controller is added to the open-loop transfer function so that (*Camacho, 2012*)

$$G(s) = \frac{137.2 \times 10^{-6} K(s + 0.01)}{s(s^2 + 0.0224s + 196 \times 10^{-6})} e^{-39s}$$

**a.** Draw the new resulting Nyquist diagram when $K = 1$.

**b.** Find the range of $K$ for closed-loop stability.

**c.** Use a phase margin argument to find the value of $K$ that will yield $\zeta = 0.5$ damping factor.

**d.** Using the value found in Part **a**, simulate the system for a unit-step response using a computer program.

# Cyber Exploration Laboratory

## Experiment 11.1

**Objectives** To design a PID controller using MATLAB's SISO Design Tool; To observe the effect of a PI and a PD controller on the magnitude and phase responses at each step of the design of a PID controller

**Minimum Required Software Packages** MATLAB, and the Control System Toolbox

### Prelab

1. What is the phase margin required for 12% overshoot?
2. What is the bandwidth required for 12% overshoot and a peak time of 2 seconds?
3. Given a unity feedback system with $G(s) = \dfrac{K}{s(s+1)(s+4)}$, what is the gain, $K$, required to yield the phase margin found in Prelab 1? What is the phase-margin frequency?
4. Design a PI controller to yield a phase margin 5° more than that found in Prelab 1.
5. Complete the design of a PID controller for the system of Prelab 3.

### Lab

1. Using MATLAB's SISO Design Tool, set up the system of Prelab 3 and display the open-loop Bode plots and the closed-loop step response.
2. Drag the Bode magnitude plot in a vertical direction until the phase margin found in Prelab 1 is obtained. Record the gain $K$, the phase margin, the phase-margin frequency, the percent overshoot, and the peak time. Move the magnitude curve up and down and note the effect upon the phase curve, the phase margin, and the phase-margin frequency.
3. Design the PI controller by adding a pole at the origin and a zero one decade below the phase-margin frequency found in Lab 2. Readjust the gain to yield a phase margin 5° higher than that found in Prelab 1. Record the gain $K$, the phase margin, the phase-margin frequency, the percent overshoot, and the peak time. Move the zero back and forth in the vicinity of its current location and note the effect on the magnitude and phase curve. Move the magnitude curve up and down and note its effect on the phase curve, the phase margin, and the phase-margin frequency.
4. Design the PD portion of the PID controller by first adjusting the magnitude curve to yield a phase-margin frequency slightly below the bandwidth calculated in Prelab 2. Add a zero to the system and move it until you obtain the phase margin calculated in Prelab 1. Move the zero and note its effect. Move the magnitude curve and note its effect.

### Postlab

1. Compare the Prelab PID design with that obtained via the SISO Design Tool. In particular, compare the gain $K$, the phase margin, the phase-margin frequency, the percent overshoot, and the peak time.

2. For the uncompensated system, describe the effect of changing gain on the phase curve, the phase margin, and the phase-margin frequency.
3. For the PI-compensated system, describe the effect of changing gain on the phase curve, the phase margin, and the phase-margin frequency. Repeat for changes in the zero location.
4. For the PID-compensated system, describe the effect of changing gain on the phase curve, the phase margin, and the phase-margin frequency. Repeat for changes in the PD zero location.

# Bibliography


Barkana, I. Classical and Simple Adaptive Control of Nonminimum Phase Autopilot Design. *Journal of Guidance, Control, and Dynamics*, vol. 28, 2005, pp. 631–638.

Camacho, E. F., Berenguel, M., Rubio, F. R., and Martinez, D. *Control of Solar Energy Systems*. Springer-Verlag, London, 2012.

Craig, I. K., Xia, X., and Venter, J. W. Introducing HIV/AIDS Education into the Electrical Engineering Curriculum at the University of Pretoria. *IEEE Transactions on Education*, vol. 47, no. 1, February 2004, pp. 65–73.

D'Azzo, J. J., and Houpis, C. H. *Feedback Control System Analysis and Synthesis*, 2d ed. McGraw-Hill, New York, 1966.

Dorf, R. C. *Modern Control Systems*, 5th ed. Addison-Wesley, Reading, MA, 1989.

Flower, T. L., and Son, M. Motor Drive Mechanics and Control Electronics for a High Performance Plotter. *HP Journal*, November 1981, pp. 12–15.

Hollot, C. V., Misra, V., Towsley, D., and Gong, W. A Control Theoretic Analysis of RED. Proceedings of IEEE INFOCOM, 2001, pp. 1510–1519.

Hostetter, G. H., Savant, C. J., and Stefani, R. T. *Design of Feedback Control Systems*, 2d ed. Saunders College Publishing, New York, 1989.

Khadraoui, S., Nounou, H., Nounou, M., Datta, A., and Bhattacharyya, S. P. A Measurement-based approach for tuning of reduced-order controllers. *American Control Conference (ACC)*, June 2013, pp. 3876–3881.

Kuo, B. C. *Automatic Control Systems*, 5th ed. Prentice Hall, Upper Saddle River, NJ, 1987.

Mansor, H., Noor, S. B. M., Ahmad, R. K. R., and Taip, F. S. Online Quantitative feedback theory (QFT)-based self-tuning controller for grain drying process. *Scientific Research and Essays*, vol. 6, no. 31, December 16, 2011, pp. 6520–6534.

Ogata, K. *Modern Control Engineering*, 2d ed. Prentice Hall, Upper Saddle River, NJ, 1990.

Phillips, C. L., and Harbor, R. D. *Feedback Control Systems*. Prentice Hall, Upper Saddle River, NJ, 1988.

Preitl, Z., Bauer, P., and Bokor, J. A Simple Control Solution for Traction Motor Used in Hybrid Vehicles. *Fourth International Symposium on Applied Computational Inteligence and Informatics*. IEEE, 2007, pp. 157–162.

Raven, F. H. *Automatic Control Engineering*, 4th ed. McGraw-Hill, New York, 1987.

Smith, C. A. *Automated Continuous Process Control*. Wiley, New York, 2002.

Sun, J., and Miao, Y. Modeling and simulation of the agricultural sprayer boom leveling system. *IEEE Third International Conf. on Measuring Tech. and Mechatronics Automation*, 2011, pp. 613–618.

Tasch, U., Koontz, J. W., Ignatoski, M. A., and Geselowitz, D. B. An Adaptive Aortic Pressure Observer for the Penn State Electric Ventricular Assist Device. *IEEE Transactions on Biomedical Engineering*, vol. 37, 1990, pp. 374–383.

Yan, T., and Lin, R. Experimental Modeling and Compensation of Pivot Nonlinearity in Hard Disk Drives. *IEEE Transactions on Magnetics*, vol. 39, 2003, pp. 1064–1069.

Zhang, J.-Y., Kim, J.-W., Lee, K.-B., and Kim, Y.-B. Development of an Active Front Steering (AFS) System With QFT Control. *International Journal of Automotive Technology*, vol. 9, no. 6, 2008, pp. 695–702.

# 12

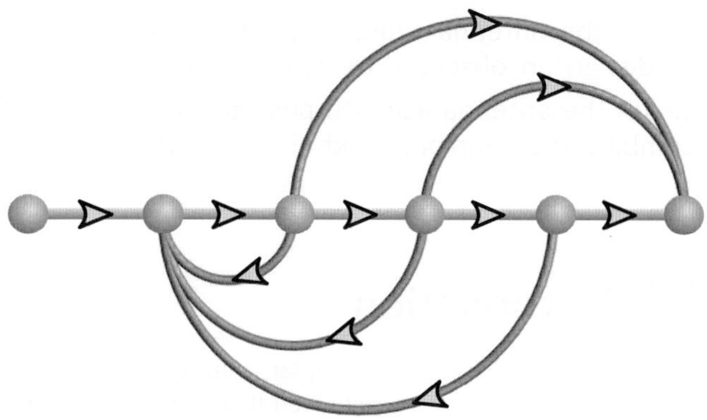

# Design via State Space

This chapter covers only state-space methods.

State Space

SS

## Chapter Learning Outcomes

After completing this chapter the student will be able to:

- Design a state-feedback controller using pole placement for systems represented in phase-variable form to meet transient response specifications (Sections 12.1–12.2)
- Determine if a system is controllable (Section 12.3)
- Design a state-feedback controller using pole placement for systems not represented in phase-variable form to meet transient response specifications (Section 12.4)
- Design a state-feedback observer using pole placement for systems represented in observer canonical form (Section 12.5)
- Determine if a system is observable (Section 12.6)
- Design a state-feedback observer using pole placement for systems not represented in observer canonical form (Section 12.7)
- Design steady-state error characteristics for systems represented in state space (Section 12.8)

## ▶ Case Study Learning Outcomes

You will be able to demonstrate your knowledge of the chapter objectives with case studies as follows:

- Given the antenna azimuth position control system shown on the front endpapers, you will be able to specify all closed-loop poles and then design a state-feedback controller to meet transient response specifications.

- Given the antenna azimuth position control system shown on the front endpapers, you will be able to design an observer to estimate the states.
- Given the antenna azimuth position control system shown on the front endpapers, you will be able to combine the controller and observer designs into a viable compensator for the system.

## 12.1 Introduction

Chapter 3 introduced the concepts of state-space analysis and system modeling. We showed that state-space methods, like transform methods, are simply tools for analyzing and designing feedback control systems. However, state-space techniques can be applied to a wider class of systems than transform methods. Systems with nonlinearities, such as that shown in Figure 12.1, and multiple-input, multiple-output systems are just two of the candidates for the state-space approach. In this book, however, we apply the approach only to linear systems.

In Chapters 9 and 11, we applied frequency domain methods to system design. The basic design technique is to create a compensator in cascade with the plant or in the feedback path that has the correct additional poles and zeros to yield a desired transient response and steady-state error.

One of the drawbacks of frequency domain methods of design, using either root locus or frequency response techniques, is that after designing the location of the dominant second-order pair of poles, we keep our fingers crossed, hoping that the higher-order poles do not affect the second-order approximation. What we would like to be able to do is specify

**FIGURE 12.1** A robot in a hospital pharmacy selects medications by bar code[1]

[1] Tadeo F., Perez, Loepez O., and Alvarez T. Control of Neutralization Processes by Robust Loopsharing. *IEEE Trans. on Cont. Syst. Tech.*, vol. 8, no. 2, 2000. Fig. 2, p. 239. IEEE Transactions on Control Systems Technology by Institute of Electrical and Electronics Engineers; IEEE Control Systems Society Reproduced with permission of Institute of Electrical and Electronics Engineers, in the format Republish in a book via Copyright Clearance Center.

*all* closed-loop poles of the higher-order system. Frequency domain methods of design do not allow us to specify all poles in systems of order higher than 2 because they do not allow for a sufficient number of unknown parameters to place all of the closed-loop poles uniquely. One gain to adjust, or compensator pole and zero to select, does not yield a sufficient number of parameters to place all the closed-loop poles at desired locations. Remember, to place $n$ unknown quantities, you need $n$ adjustable parameters. State-space methods solve this problem by introducing into the system (1) other adjustable parameters and (2) the technique for finding these parameter values, so that we can properly place all poles of the closed-loop system.[2]

On the other hand, state-space methods do not allow the specification of closed-loop zero locations, which frequency domain methods do allow through placement of the lead compensator zero. This is a disadvantage of state-space methods, since the location of the zero does affect the transient response. Also, a state-space design may prove to be very sensitive to parameter changes.

Finally, there is a wide range of computational support for state-space methods; many software packages support the matrix algebra required by the design process. However, as mentioned before, the advantages of computer support are balanced by the loss of graphic insight into a design problem that the frequency domain methods yield.

This chapter should be considered only an introduction to state-space design; we introduce one state-space design technique and apply it only to linear systems. Advanced study is required to apply state-space techniques to the design of systems beyond the scope of this textbook.

## 12.2 Controller Design

This section shows how to introduce additional parameters into a system so that we can control the location of all closed-loop poles. An $n$th-order feedback control system has an $n$th-order closed-loop characteristic equation of the form

$$s^n + a_{n-1}s^{n-1} + \cdots + a_1 s + a_0 = 0 \tag{12.1}$$

Since the coefficient of the highest power of $s$ is unity, there are $n$ coefficients whose values determine the system's closed-loop pole locations. Thus, if we can introduce $n$ adjustable parameters into the system and relate them to the coefficients in Eq. (12.1), all of the poles of the closed-loop system can be set to any desired location.

### Topology for Pole Placement

In order to lay the groundwork for the approach, consider a plant represented in state space by

$$\dot{\mathbf{x}} = \mathbf{Ax} + \mathbf{B}u \tag{12.2a}$$

$$y = \mathbf{Cx} \tag{12.2b}$$

and shown pictorially in Figure 12.2(*a*), where light lines are scalars and the heavy lines are vectors.

[2] This is an advantage as long as we know where to place the higher-order poles, which is not always the case. One course of action is to place the higher-order poles far from the dominant second-order poles or near a closed-loop zero to keep the second-order system design valid. Another approach is to use optimal control concepts, which are beyond the scope of this text.

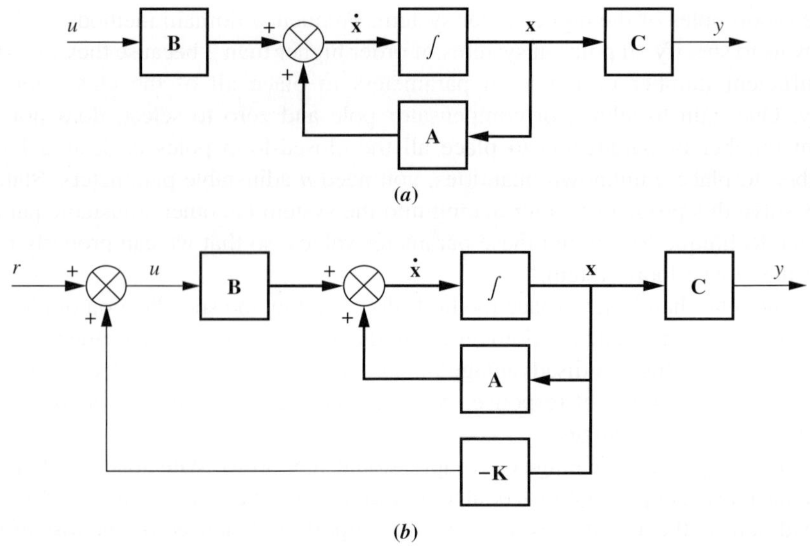


**FIGURE 12.2** **a.** State-space representation of a plant; **b.** plant with state-variable feedback

In a typical feedback control system, the output, $y$, is fed back to the summing junction. It is now that the topology of the design changes. Instead of feeding back $y$, what if we feed back all of the state variables? If each state variable is fed back to the control, $u$, through a gain, $k_i$, there would be $n$ gains, $k_i$, that could be adjusted to yield the required closed-loop pole values. The feedback through the gains, $k_i$, is represented in Figure 12.2($b$) by the feedback vector $-\mathbf{K}$.

The state equations for the closed-loop system of Figure 12.2($a$) can be written by inspection as

$$\dot{\mathbf{x}} = \mathbf{A}\mathbf{x} + \mathbf{B}u = \mathbf{A}\mathbf{x} + \mathbf{B}(-\mathbf{K}\mathbf{x} + r) = (\mathbf{A} - \mathbf{B}\mathbf{K})\mathbf{x} + \mathbf{B}r \tag{12.3a}$$

$$y = \mathbf{C}\mathbf{x} \tag{12.3b}$$

Before continuing, you should have a good idea of how the feedback system of Figure 12.2($b$) is actually implemented. As an example, assume a plant signal-flow graph in phase-variable form, as shown in Figure 12.3($a$). Each state variable is then fed back to the plant's input, $u$, through a gain, $k_i$, as shown in Figure 12.3($b$). Although we will cover other representations later in the chapter, the phase-variable form, with its typical lower companion system matrix, or the controller canonical form, with its typical upper companion system matrix, yields the simplest evaluation of the feedback gains. In the ensuing discussion, we use the phase-variable form to develop and demonstrate the concepts. End-of-chapter problems will give you an opportunity to develop and test the concepts for the controller canonical form.

The design of state-variable feedback for closed-loop pole placement consists of equating the characteristic equation of a closed-loop system, such as that shown in Figure 12.3($b$), to a desired characteristic equation and then finding the values of the feedback gains, $k_i$.

If a plant like that shown in Figure 12.3($a$) is of high order and not represented in phase-variable or controller canonical form, the solution for the $k_i$'s can be intricate. Thus, it is advisable to transform the system to either of these forms, design the $k_i$'s, and then transform the system back to its original representation. We perform

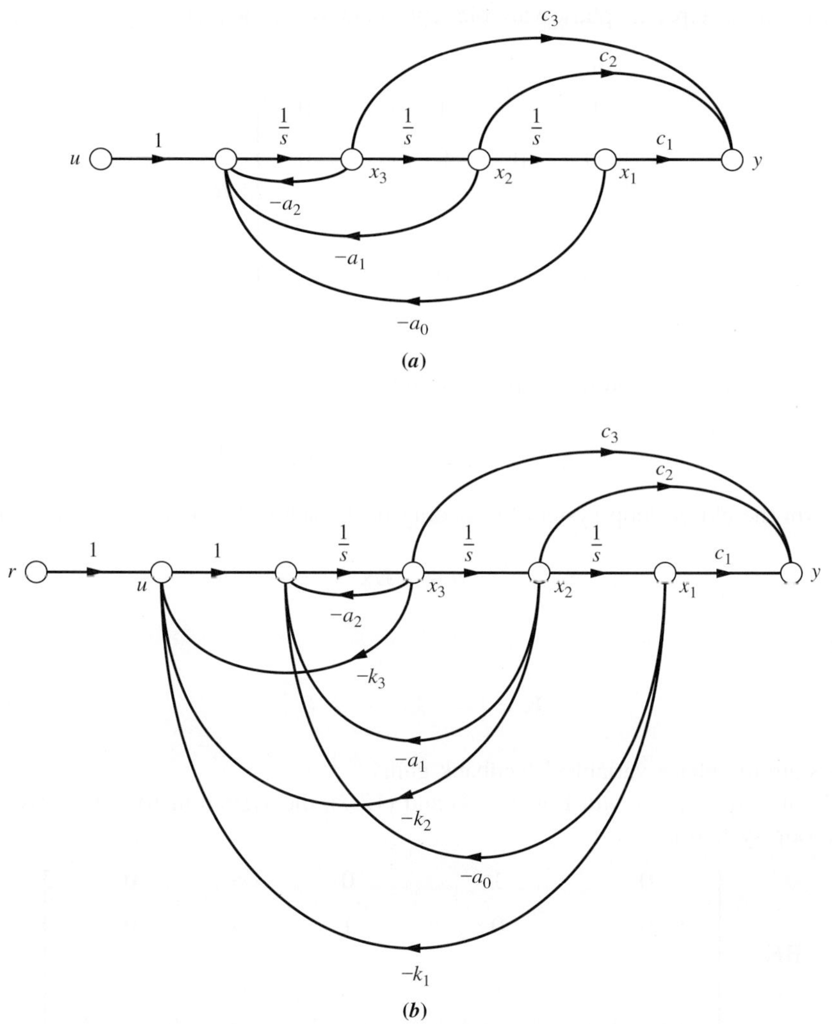


**FIGURE 12.3** **a.** Phase-variable representation for plant; **b.** plant with state-variable feedback

this conversion in Section 12.4, where we develop a method for performing the transformations. Until then, let us direct our attention to plants represented in phase-variable form.

## Pole Placement for Plants in Phase-Variable Form

To apply pole-placement methodology to plants represented in phase-variable form, we take the following steps:

1. Represent the plant in phase-variable form.
2. Feed back each phase variable to the input of the plant through a gain, $k_i$.
3. Find the characteristic equation for the closed-loop system represented in Step 2.
4. Decide upon all closed-loop pole locations and determine an equivalent characteristic equation.
5. Equate like coefficients of the characteristic equations from Steps 3 and 4 and solve for $k_i$.

Following these steps, the phase-variable representation of the plant is given by Eq. (12.2), with

$$\mathbf{A} = \begin{bmatrix} 0 & 1 & 0 & \cdots & 0 \\ 0 & 0 & 1 & \cdots & 0 \\ \vdots & \vdots & \vdots & \vdots & \vdots \\ -a_0 & -a_1 & -a_2 & \cdots & -a_{n-1} \end{bmatrix}; \; B = \begin{bmatrix} 0 \\ 0 \\ \vdots \\ 1 \end{bmatrix};$$
$$\mathbf{C} = [\, c_1 \quad c_2 \quad \cdots \quad c_n \,] \tag{12.4}$$

The characteristic equation of the plant is thus

$$s^n + a_{n-1}s^{n-1} + \cdots + a_1 s + a_0 = 0 \tag{12.5}$$

Now form the closed-loop system by feeding back each state variable to $u$, forming

$$u = -\mathbf{Kx} \tag{12.6}$$

where

$$\mathbf{K} = [\, k_1 \quad k_2 \quad \cdots \quad k_n \,] \tag{12.7}$$

The $k_i$'s are the phase variables' feedback gains.

Using Eq. (12.3*a*) with Eqs. (12.4) and (12.7), the system matrix, $\mathbf{A} - \mathbf{BK}$, for the closed-loop system is

$$\mathbf{A} - \mathbf{BK} = \begin{bmatrix} 0 & 1 & 0 & \cdots & 0 \\ 0 & 0 & 1 & \cdots & 0 \\ \vdots & \vdots & \vdots & \vdots & \vdots \\ -(a_0 + k_1) & -(a_1 + k_2) & -(a_2 + k_3) & \cdots & -(a_{n-1} + k_n) \end{bmatrix} \tag{12.8}$$

Since Eq. (12.8) is in phase-variable form, the characteristic equation of the closed-loop system can be written by inspection as

$$\begin{aligned} \det(s\mathbf{I} - (\mathbf{A} - \mathbf{BK})) = s^n + (a_{n-1} + k_n)s^{n-1} + (a_{n-2} + k_{n-1})s^{n-2} \\ + \cdots (a_1 + k_2)s + (a_0 + k_1) = 0 \end{aligned} \tag{12.9}$$

Notice the relationship between Eqs. (12.5) and (12.9). For plants represented in phase-variable form, we can write by inspection the closed-loop characteristic equation from the open-loop characteristic equation by adding the appropriate $k_i$ to each coefficient.

Now assume that the desired characteristic equation for proper pole placement is

$$s^n + d_{n-1}s^{n-1} + d_{n-2}s^{n-2} + \cdots + d_2 s^2 + d_1 s + d_0 = 0 \tag{12.10}$$

where the $d_i$'s are the desired coefficients. Equating Eqs. (12.9) and (12.10), we obtain

$$d_i = a_i + k_{i+1} \quad i = 0,\ 1,\ 2, \ldots, n-1 \tag{12.11}$$

from which

$$k_{i+1} = d_i - a_i \tag{12.12}$$

Now that we have found the denominator of the closed-loop transfer function, let us find the numerator. For systems represented in phase-variable form, we learned that the numerator polynomial is formed from the coefficients of the output coupling matrix, **C**. Since Figures 12.3(*a*) and (*b*) are both in phase-variable form and have the same output coupling matrix, we conclude that the numerators of their transfer functions are the same. Let us look at a design example.

## Example 12.1 Controller Design for Phase-Variable Form

**PROBLEM:** Given the plant

$$G(s) = \frac{20(s+5)}{s(s+1)(s+4)} \tag{12.13}$$

design the phase-variable feedback gains to yield 9.5% overshoot and a settling time of 0.74 second.

**SOLUTION:** We begin by calculating the desired closed-loop characteristic equation. Using the transient response requirements, the closed-loop poles are $-5.4 \pm j7.2$. Since the system is third-order, we must select another closed-loop pole. The closed-loop system will have a zero at $-5$, the same as the open-loop system. We could select the third closed-loop pole to cancel the closed-loop zero. However, to demonstrate the effect of the third pole and the design process, including the need for simulation, let us choose $-5.1$ as the location of the third closed-loop pole.

Now draw the signal-flow diagram for the plant. The result is shown in Figure 12.4(*a*). Next feed back all state variables to the control, $u$, through gains $k_i$, as shown in Figure 12.4(*b*).

Writing the closed-loop system's state equations from Figure 12.4(*b*), we have

$$\dot{\mathbf{x}} = \begin{bmatrix} 0 & 1 & 0 \\ 0 & 0 & 1 \\ -k_1 & -(4+k_2) & -(5+k_3) \end{bmatrix} \mathbf{x} + \begin{bmatrix} 0 \\ 0 \\ 1 \end{bmatrix} r \tag{12.14a}$$

$$y = [\,100 \quad 20 \quad 0\,]\,\mathbf{x} \tag{12.14b}$$

Comparing Eqs. (12.14) to Eqs. (12.3), we identify the closed-loop system matrix as

$$\mathbf{A} - \mathbf{BK} = \begin{bmatrix} 0 & 1 & 0 \\ 0 & 0 & 1 \\ -k_1 & -(4+k_2) & -(5+k_3) \end{bmatrix} \tag{12.15}$$

To find the closed-loop system's characteristic equation, form

$$\det(s\mathbf{I} - (\mathbf{A} - \mathbf{BK})) = s^3 + (5+k_3)s^2 + (4+k_2)s + k_1 = 0 \tag{12.16}$$

This equation must match the desired characteristic equation

$$s^3 + 15.9s^2 + 136.08s + 413.1 = 0 \tag{12.17}$$

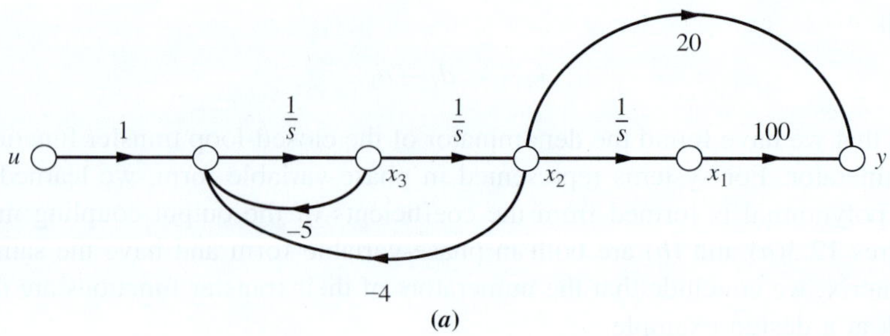


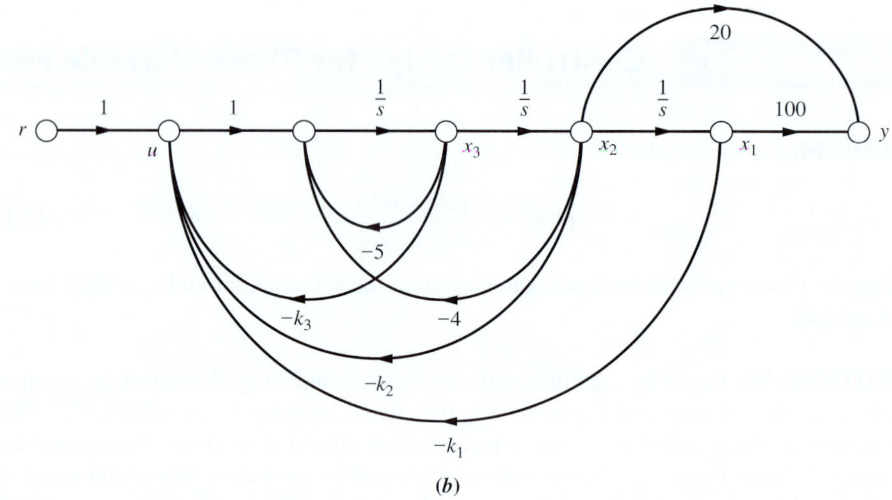


**FIGURE 12.4** **a.** Phase-variable representation for plant of Example 12.1; **b.** plant with state-variable feedback

formed from the poles $-5.4 + j7.2$, $-5.4 - j7.2$, and $-5.1$, which were previously determined.

Equating the coefficients of Eqs. (12.16) and (12.17), we obtain

$$k_1 = 413.1; \quad k_2 = 132.08; \quad k_3 = 10.9 \tag{12.18}$$

Finally, the zero term of the closed-loop transfer function is the same as the zero term of the open-loop system, or $(s + 5)$.

Using Eqs. (12.14), we obtain the following state-space representation of the closed-loop system:

$$\dot{\mathbf{x}} = \begin{bmatrix} 0 & 1 & 0 \\ 0 & 0 & 1 \\ -413.1 & -136.08 & -15.9 \end{bmatrix} \mathbf{x} + \begin{bmatrix} 0 \\ 0 \\ 1 \end{bmatrix} r \tag{12.19a}$$

$$y = [100 \quad 20 \quad 0]\,\mathbf{x} \tag{12.19b}$$

The transfer function is

$$T(s) = \frac{20(s+5)}{s^3 + 15.9s^2 + 136.08s + 413.1} \tag{12.20}$$

Figure 12.5, a simulation of the closed-loop system, shows 11.5% overshoot and a settling time of 0.8 second. A redesign with the third pole canceling the zero at $-5$ will yield performance equal to the requirements.

Since the steady-state response approaches 0.24 instead of unity, there is a large steady-state error. Design techniques to reduce this error are discussed in Section 12.8.

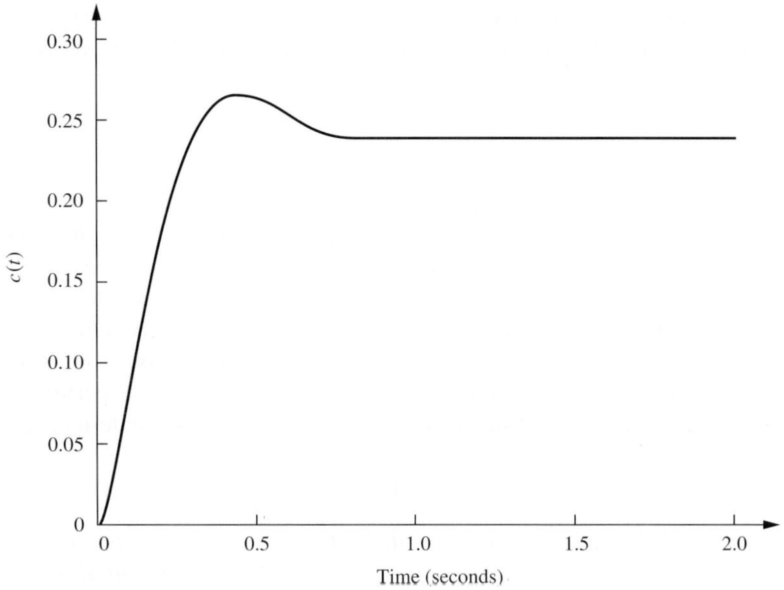


**FIGURE 12.5** Simulation of closed-loop system of Example 12.1

MATLAB

**ML**

`Students who are using MATLAB should now run ch12 p1 in Appendix B. You will learn how to use MATLAB to design a controller for phase variables using pole placement. MATLAB will plot the step response of the designed system. This exercise solves Example 12.1 using MATLAB.`

## Skill-Assessment Exercise 12.1

**PROBLEMS:** For the plant

$$G(s) = \frac{100(s+10)}{s(s+3)(s+12)}$$

represented in the state space in phase-variable form by

$$\dot{\mathbf{x}} = \mathbf{A}\mathbf{x} + \mathbf{B}u = \begin{bmatrix} 0 & 1 & 0 \\ 0 & 0 & 1 \\ 0 & -36 & -15 \end{bmatrix}\mathbf{x} + \begin{bmatrix} 0 \\ 0 \\ 1 \end{bmatrix}u$$

$$y = \mathbf{C}\mathbf{x} = [\,1000 \quad 100 \quad 0\,]\mathbf{x}$$

design the phase-variable feedback gains to yield 5% overshoot and a peak time of 0.3 second.

**ANSWER:** $\mathbf{K} = [\,2094 \quad 373.1 \quad 14.97\,]$

The complete solution is located at www.wiley.com/college/nise.

**TryIt 12.1**

Use MATLAB, the Control System Toolbox, and the following statements to solve for the phase-variable feedback gains to place the poles of the system in Skill-Assessment Exercise 12.1 at $-3+j5$, $-3-j5$, and $-10$.

```
A=[0 1 0
   0 0 1
   0 -36 -15]
B=[0;0;1]
poles=[-3+5j,...
  -3-5j,-10]
K=acker(A,B,poles)
```

In this section, we showed how to design feedback gains for plants represented in phase-variable form in order to place all of the closed-loop system's poles at desired locations on the $s$-plane. On the surface, it appears that the method should always work for any system. However, this is not the case. The conditions that must exist in order to uniquely place the closed-loop poles where we want them is the topic of the next section.

## 12.3 Controllability

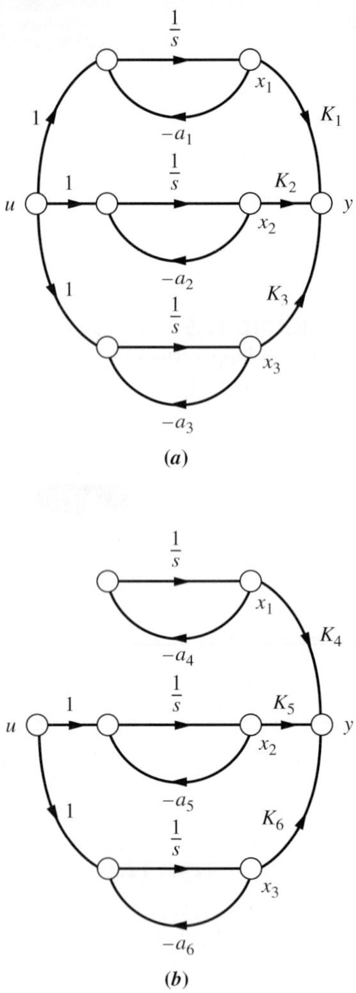


**FIGURE 12.6** Comparison of **a.** controllable and **b.** uncontrollable systems

Consider the parallel form shown in Figure 12.6($a$). To control the pole location of the closed-loop system, we are saying implicitly that the control signal, $u$, can control the behavior of each state variable in $x$. If any one of the state variables cannot be controlled by the control $u$, then we cannot place the poles of the system where we desire. For example, in Figure 12.6($b$), if $x_1$ were not controllable by the control signal and if $x_1$ also exhibited an unstable response due to a nonzero initial condition, there would be no way to effect a state-feedback design to stabilize $x_1$. State variable $x_1$ would perform in its own way regardless of the control signal, $u$. Thus, in some systems, a state-feedback design is not possible.

We now make the following definition based upon the previous discussion:

*If an input to a system can be found that takes every state variable from a desired initial state to a desired final state, the system is said to be* ***controllable****; otherwise, the system is* ***uncontrollable****.*

Pole placement is a viable design technique only for systems that are controllable. This section shows how to determine, a priori, whether pole placement is a viable design technique for a controller.

### Controllability by Inspection

We can explore controllability from another viewpoint: that of the state equation itself. When the system matrix is diagonal, as it is for the parallel form, it is apparent whether or not the system is controllable. For example, the state equation for Figure 12.6($a$) is

$$\dot{\mathbf{x}} = \begin{bmatrix} -a_1 & 0 & 0 \\ 0 & -a_2 & 0 \\ 0 & 0 & -a_3 \end{bmatrix} \mathbf{x} + \begin{bmatrix} 1 \\ 1 \\ 1 \end{bmatrix} u \tag{12.21}$$

or

$$\dot{x}_1 = -a_1 x_1 \qquad\qquad + u \tag{12.22a}$$

$$\dot{x}_2 = \qquad -a_2 x_2 \qquad + u \tag{12.22b}$$

$$\dot{x}_3 = \qquad\qquad -a_3 x_3 + u \tag{12.22c}$$

Since each of Eqs. (12.22) is independent and decoupled from the rest, the control $u$ affects each of the state variables. This is controllability from another perspective.

Now let us look at the state equations for the system of Figure 12.6($b$):

$$\dot{\mathbf{x}} = \begin{bmatrix} -a_4 & 0 & 0 \\ 0 & -a_5 & 0 \\ 0 & 0 & -a_6 \end{bmatrix} \mathbf{x} + \begin{bmatrix} 0 \\ 1 \\ 1 \end{bmatrix} u \tag{12.23}$$

or

$$\dot{x}_1 = -a_4 x_1 \tag{12.24a}$$

$$\dot{x}_2 = \qquad -a_5 x_2 \qquad + u \tag{12.24b}$$

$$\dot{x}_3 = \qquad\qquad -a_6 x_3 + u \tag{12.24c}$$

From the state equations in (12.23) or (12.24), we see that state variable $x_1$ is not controlled by the control $u$. Thus, the system is said to be uncontrollable.

In summary, a system with distinct eigenvalues and a diagonal system matrix is controllable if the input coupling matrix **B** does not have any rows that are zero.

## The Controllability Matrix

Tests for controllability that we have so far explored cannot be used for representations of the system other than the diagonal or parallel form with distinct eigenvalues. The problem of visualizing controllability gets more complicated if the system has multiple poles, even though it is represented in parallel form. Further, one cannot always determine controllability by inspection for systems that are not represented in parallel form. In other forms, the existence of paths from the input to the state variables is not a criterion for controllability since the equations are not decoupled.

In order to be able to determine controllability or, alternatively, to design state feedback for a plant under any representation or choice of state variables, a matrix can be derived that must have a particular property if all state variables are to be controlled by the plant input, $u$. We now state the requirement for controllability, including the form, property, and name of this matrix.[3]

An $n$th-order plant whose state equation is

$$\dot{\mathbf{x}} = \mathbf{A}\mathbf{x} + \mathbf{B}u \tag{12.25}$$

is completely controllable[4] if the matrix

$$\mathbf{C_M} = [\mathbf{B} \quad \mathbf{AB} \quad \mathbf{A}^2\mathbf{B} \quad \cdots \quad \mathbf{A}^{n-1}\mathbf{B}] \tag{12.26}$$

is of rank $n$, where $\mathbf{C_M}$ is called the *controllability* matrix.[5] As an example, let us choose a system represented in parallel form with multiple roots.

## Example 12.2 Controllability via the Controllability Matrix

**PROBLEM:** Given the system of Figure 12.7, represented by a signal-flow diagram, determine its controllability.

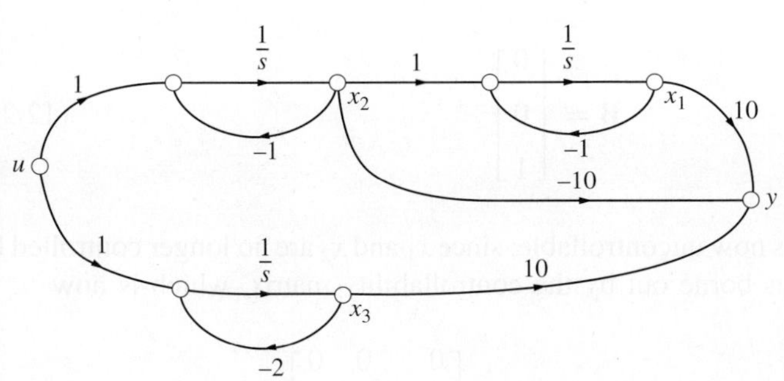


**FIGURE 12.7** System for Example 12.2

[3] See the work listed in the Bibliography by *Ogata* (*1990*: *699–702*) for the derivation.

[4] *Completely controllable* means that all state variables are controllable. This textbook uses *controllable* to mean *completely controllable*.

[5] See Appendix G at www.wiley.com/college/nise for the definition of rank. For single-input systems, instead of specifying rank $n$, we can say that $\mathbf{C_M}$ must be nonsingular, possess an inverse, or have linearly independent rows and columns.

**SOLUTION:** The state equation for the system written from the signal-flow diagram is

$$\dot{\mathbf{x}} = \mathbf{A}\mathbf{x} + \mathbf{B}u = \begin{bmatrix} -1 & 1 & 0 \\ 0 & -1 & 0 \\ 0 & 0 & -2 \end{bmatrix} \mathbf{x} + \begin{bmatrix} 0 \\ 1 \\ 1 \end{bmatrix} u \tag{12.27}$$

At first, it would appear that the system is not controllable because of the zero in the **B** matrix. Remember, though, that this configuration leads to uncontrollability only if the poles are real and distinct. In this case, we have multiple poles at $-1$.

The controllability matrix is

$$\mathbf{C_M} = \begin{bmatrix} \mathbf{B} & \mathbf{AB} & \mathbf{A}^2\mathbf{B} \end{bmatrix} = \begin{bmatrix} 0 & 1 & -2 \\ 1 & -1 & 1 \\ 1 & -2 & 4 \end{bmatrix} \tag{12.28}$$

The rank of $\mathbf{C_M}$ equals the number of linearly independent rows or columns. The rank can be found by finding the highest-order square submatrix that is nonsingular. The determinant of $\mathbf{C_M} = -1$. Since the determinant is not zero, the $3 \times 3$ matrix is nonsingular, and the rank of $\mathbf{C_M}$ is 3. We conclude that the system is controllable since the rank of $\mathbf{C_M}$ equals the system order. Thus, the poles of the system can be placed using state-variable feedback design.

MATLAB

ML

Students who are using MATLAB should now run ch12p2 in Appendix B. You will learn how to use MATLAB to test a system for controllability. This exercise solves Example 12.2 using MATLAB.

In the previous example, we found that even though an element of the input coupling matrix was zero, the system was controllable. If we look at Figure 12.7, we can see why. In this figure, all of the state variables are driven by the input $u$.

On the other hand, if we disconnect the input at either $dx_1/dt$, $dx_2/dt$, or $dx_3/dt$, at least one state variable would not be controllable. To see the effect, let us disconnect the input at $dx_2/dt$. This causes the **B** matrix to become

$$\mathbf{B} = \begin{bmatrix} 0 \\ 0 \\ 1 \end{bmatrix} \tag{12.29}$$

We can see that the system is now uncontrollable, since $x_1$ and $x_2$ are no longer controlled by the input. This conclusion is borne out by the controllability matrix, which is now

$$\mathbf{C_M} = \begin{bmatrix} \mathbf{B} & \mathbf{AB} & \mathbf{A}^2\mathbf{B} \end{bmatrix} = \begin{bmatrix} 0 & 0 & 0 \\ 0 & 0 & 0 \\ 1 & -2 & 4 \end{bmatrix} \tag{12.30}$$

Not only is the determinant of this matrix equal to zero, but so is the determinant of any $2 \times 2$ submatrix. Thus, the rank of Eq. (12.30) is 1. The system is uncontrollable because the rank of $\mathbf{C_M}$ is 1, which is less than the order, 3, of the system.

## Skill-Assessment Exercise 12.2

**PROBLEM:** Determine whether the system

$$\dot{\mathbf{x}} = \mathbf{Ax} + \mathbf{B}u = \begin{bmatrix} -1 & 1 & 2 \\ 0 & -1 & 5 \\ 0 & 3 & -4 \end{bmatrix} \mathbf{x} + \begin{bmatrix} 2 \\ 1 \\ 1 \end{bmatrix} u$$

is controllable.

**ANSWER:** Controllable

The complete solution is located at www.wiley.com/college/nise.

**TryIt 12.2**

Use MATLAB, the Control System Toolbox, and the following statements to solve Skill-Assessment Exercise 12.2.

```
A=[-1  1  2
    0 -1  5
    0  3 -4]
B=[2;1;1]
Cm=ctrb(A,B)
Rank=rank(Cm)
```

In summary, then, pole-placement design through state-variable feedback is simplified by using the phase-variable form for the plant's state equations. However, controllability, the ability for pole-placement design to succeed, can be visualized best in the parallel form, where the system matrix is diagonal with distinct roots. In any event, the controllability matrix will always tell the designer whether the implementation is viable for state-feedback design.

The next section shows how to design state-variable feedback for systems not represented in phase-variable form. We use the controllability matrix as a tool for transforming a system to phase-variable form for the design of state-variable feedback.

## 12.4 Alternative Approaches to Controller Design

Section 12.2 showed how to design state-variable feedback to yield desired closed-loop poles. We demonstrated this method using systems represented in phase-variable form and saw how simple it was to calculate the feedback gains. Many times the physics of the problem requires feedback from state variables that are not phase variables. For these systems we have some choices for a design methodology.

The first method consists of matching the coefficients of $\det(s\mathbf{I} - (\mathbf{A} - \mathbf{BK}))$ with the coefficients of the desired characteristic equation, which is the same method we used for systems represented in phase variables. This technique, in general, leads to difficult calculations of the feedback gains, especially for higher-order systems not represented with phase variables. Let us illustrate this technique with an example.

### Example 12.3 Controller Design by Matching Coefficients

**PROBLEM:** Given a plant, $Y(s)/U(s) = 10/[(s+1)(s+2)]$, design state feedback for the plant represented in cascade form to yield a 15% overshoot with a settling time of 0.5 second.

**SOLUTION:** The signal-flow diagram for the plant in cascade form is shown in Figure 12.8(*a*). Figure 12.8(*b*) shows the system with state feedback added. Writing the state equations from Figure 12.8(*b*), we have

$$\dot{\mathbf{x}} = \begin{bmatrix} -2 & 1 \\ -k_1 & -(k_2+1) \end{bmatrix} \mathbf{x} + \begin{bmatrix} 0 \\ 1 \end{bmatrix} r \tag{12.31a}$$

$$y = [\,10 \quad 0\,]\,\mathbf{x} \tag{12.31b}$$

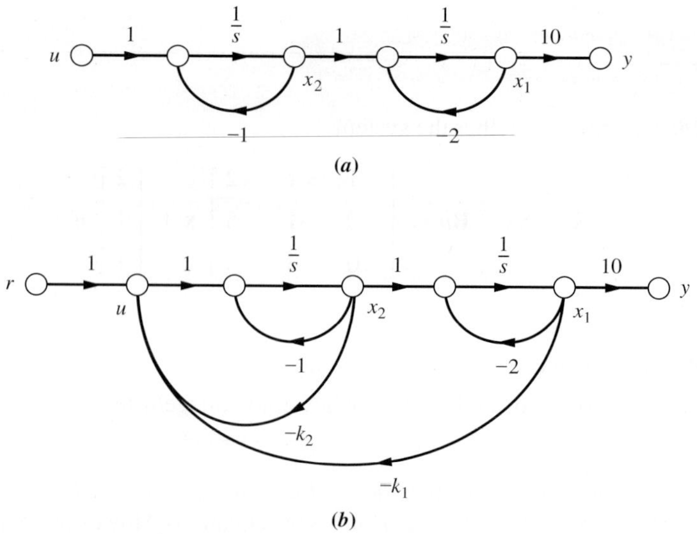


**FIGURE 12.8** **a.** Signal-flow graph in cascade form for $G(s) = 10/[(s+1)(s+2)]$; **b.** system with state feedback added

where the characteristic equation is

$$s^2 + (k_2 + 3)s + (2k_2 + k_1 + 2) = 0 \tag{12.32}$$

Using the transient response requirements stated in the problem, we obtain the desired characteristic equation

$$s^2 + 16s + 239.5 = 0 \tag{12.33}$$

Equating the middle coefficients of Eqs. (12.32) and (12.33), we find $k_2 = 13$. Equating the last coefficients of these equations along with the result for $k_2$ yields $k_1 = 211.5$.

The second method consists of transforming the system to phase variables, designing the feedback gains, and transforming the designed system back to its original state-variable representation.[6] This method requires that we first develop the transformation between a system and its representation in phase-variable form.

Assume a plant not represented in phase-variable form

$$\dot{\mathbf{z}} = \mathbf{A}\mathbf{z} + \mathbf{B}u \tag{12.34a}$$

$$y = \mathbf{C}\mathbf{z} \tag{12.34b}$$

whose controllability matrix is

$$\mathbf{C_{Mz}} = [\mathbf{B} \quad \mathbf{AB} \quad \mathbf{A}^2\mathbf{B} \cdots \mathbf{A}^{n-1}\mathbf{B}] \tag{12.35}$$

[6] See the discussions of Ackermann's formula in (*Franklin, 1994*) and (*Ogata, 1990*), listed in the Bibliography.

Assume that the system can be transformed into the phase-variable ($\mathbf{x}$) representation with the transformation

$$\boxed{\mathbf{z} = \mathbf{P}\mathbf{x}} \tag{12.36}$$

Substituting this transformation into Eqs. (12.34), we get

$$\dot{\mathbf{x}} = \mathbf{P}^{-1}\mathbf{A}\mathbf{P}\mathbf{x} + \mathbf{P}^{-1}\mathbf{B}u \tag{12.37a}$$

$$y = \mathbf{C}\mathbf{P}\mathbf{x} \tag{12.37b}$$

whose controllability matrix is

$$\begin{aligned}\mathbf{C_{Mx}} &= [\mathbf{P}^{-1}\mathbf{B} \quad (\mathbf{P}^{-1}\mathbf{A}\mathbf{P})(\mathbf{P}^{-1}\mathbf{B}) \quad (\mathbf{P}^{-1}\mathbf{A}\mathbf{P})^2(\mathbf{P}^{-1}\mathbf{B}) \quad \cdots \quad (\mathbf{P}^{-1}\mathbf{A}\mathbf{P})^{n-1}(\mathbf{P}^{-1}\mathbf{B})] \\ &= [\mathbf{P}^{-1}\mathbf{B} \quad (\mathbf{P}^{-1}\mathbf{A}\mathbf{P})(\mathbf{P}^{-1}\mathbf{B}) \quad (\mathbf{P}^{-1}\mathbf{A}\mathbf{P})(\mathbf{P}^{-1}\mathbf{A}\mathbf{P})(\mathbf{P}^{-1}\mathbf{B}) \quad \cdots \quad (\mathbf{P}^{-1}\mathbf{A}\mathbf{P}) \\ &\qquad (\mathbf{P}^{-1}\mathbf{A}\mathbf{P})(\mathbf{P}^{-1}\mathbf{A}\mathbf{P}) \quad \cdots \quad (\mathbf{P}^{-1}\mathbf{A}\mathbf{P})(\mathbf{P}^{-1}\mathbf{B})] \\ &= \mathbf{P}^{-1}[\mathbf{B} \quad \mathbf{A}\mathbf{B} \quad \mathbf{A}^2\mathbf{B} \quad \cdots \quad \mathbf{A}^{n-1}\mathbf{B}]\end{aligned} \tag{12.38}$$

Substituting Eq. (12.35) into (12.38) and solving for $\mathbf{P}$, we obtain

$$\boxed{\mathbf{P} = \mathbf{C_{Mz}}\mathbf{C_{Mx}}^{-1}} \tag{12.39}$$

Thus, the transformation matrix, $\mathbf{P}$, can be found from the two controllability matrices.

After transforming the system to phase variables, we design the feedback gains as in Section 12.2. Hence, including both feedback and input, $u = -\mathbf{K_x}\mathbf{x} + r$, Eqs. (12.37) becomes

$$\begin{aligned}\dot{\mathbf{x}} &= \mathbf{P}^{-1}\mathbf{A}\mathbf{P}\mathbf{x} - \mathbf{P}^{-1}\mathbf{B}\mathbf{K_x}\mathbf{x} + \mathbf{P}^{-1}\mathbf{B}r \\ &= (\mathbf{P}^{-1}\mathbf{A}\mathbf{P} - \mathbf{P}^{-1}\mathbf{B}\mathbf{K_x})\mathbf{x} + \mathbf{P}^{-1}\mathbf{B}r\end{aligned} \tag{12.40a}$$

$$\mathbf{y} = \mathbf{C}\mathbf{P}\mathbf{x} \tag{12.40b}$$

Since this equation is in phase-variable form, the zeros of this closed-loop system are determined from the polynomial formed from the elements of $\mathbf{CP}$, as explained in Section 12.2.

Using $\mathbf{x} = \mathbf{P}^{-1}\mathbf{z}$, we transform Eqs. (12.40) from phase variables back to the original representation and get

$$\dot{\mathbf{z}} = \mathbf{A}\mathbf{z} - \mathbf{B}\mathbf{K_x}\mathbf{P}^{-1}\mathbf{z} + \mathbf{B}r = (\mathbf{A} - \mathbf{B}\mathbf{K_x}\mathbf{P}^{-1})\mathbf{z} + \mathbf{B}r \tag{12.41a}$$

$$y = \mathbf{C}\mathbf{z} \tag{12.41b}$$

Comparing Eqs. (12.41) with (12.3), the state variable feedback gain, $\mathbf{K_z}$, for the original system is

$$\boxed{\mathbf{K_z} = \mathbf{K_x}\mathbf{P}^{-1}} \tag{12.42}$$

The transfer function of this closed-loop system is the same as the transfer function for Eqs. (12.40), since Eqs. (12.40) and (12.41) represent the same system. Thus, the zeros of the closed-loop transfer function are the same as the zeros of the uncompensated plant, based upon the development in Section 12.2. Let us demonstrate with a design example.

## Example 12.4 Controller Design by Transformation

**PROBLEM:** Design a state-variable feedback controller to yield a 20.8% overshoot and a settling time of 4 seconds for a plant,

$$G(s) = \frac{(s+4)}{(s+1)(s+2)(s+5)} \tag{12.43}$$

that is represented in cascade form as shown in Figure 12.9.

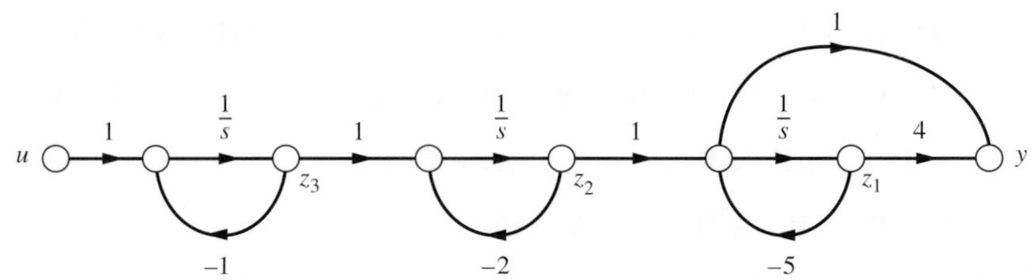


**FIGURE 12.9** Signal-flow graph for plant of Example 12.4

**SOLUTION:** First find the state equations and the controllability matrix. The state equations written from Figure 12.9 are

$$\dot{\mathbf{z}} = \mathbf{A_z z} + \mathbf{B_z}u = \begin{bmatrix} -5 & 1 & 0 \\ 0 & -2 & 1 \\ 0 & 0 & -1 \end{bmatrix}\mathbf{z} + \begin{bmatrix} 0 \\ 0 \\ 1 \end{bmatrix}u \tag{12.44a}$$

$$y = \mathbf{C_z z} = \begin{bmatrix} -1 & 1 & 0 \end{bmatrix}\mathbf{z} \tag{12.44b}$$

from which the controllability matrix is evaluated as

$$\mathbf{C_{Mz}} = \begin{bmatrix} \mathbf{B_z} & \mathbf{A_z B_z} & \mathbf{A_z^2 B_z} \end{bmatrix} = \begin{bmatrix} 0 & 0 & 1 \\ 0 & 1 & -3 \\ 1 & -1 & 1 \end{bmatrix} \tag{12.45}$$

Since the determinant of $\mathbf{C_{Mz}}$ is $-1$, the system is controllable.

We now convert the system to phase variables by first finding the characteristic equation and using this equation to write the phase-variable form. The characteristic equation, $\det(s\mathbf{I} - \mathbf{A_z})$, is

$$\det(s\mathbf{I} - \mathbf{A_z}) = s^3 + 8s^2 + 17s + 10 = 0 \tag{12.46}$$

Using the coefficients of Eq. (12.46) and our knowledge of the phase-variable form, we write the phase-variable representation of the system as

$$\dot{\mathbf{x}} = \mathbf{A_x x} + \mathbf{B_x}u = \begin{bmatrix} 0 & 1 & 0 \\ 0 & 0 & 1 \\ -10 & -17 & -8 \end{bmatrix}\mathbf{x} + \begin{bmatrix} 0 \\ 0 \\ 1 \end{bmatrix}u \tag{12.47a}$$

$$y = \begin{bmatrix} 4 & 1 & 0 \end{bmatrix}\mathbf{x} \tag{12.47b}$$

The output equation was written using the coefficients of the numerator of Eq. (12.43), since the transfer function must be the same for the two representations. The controllability matrix, $\mathbf{C_{Mx}}$, for the phase-variable system is

$$\mathbf{C_{Mx}} = \begin{bmatrix} \mathbf{B_x} & \mathbf{A_x B_x} & \mathbf{A_x^2 B_x} \end{bmatrix} = \begin{bmatrix} 0 & 0 & 1 \\ 0 & 1 & -8 \\ 1 & -8 & 47 \end{bmatrix} \quad (12.48)$$

Using Eq. (12.39), we can now calculate the transformation matrix between the two systems as

$$\mathbf{P} = \mathbf{C_{Mz} C_{Mx}^{-1}} = \begin{bmatrix} 1 & 0 & 0 \\ 5 & 1 & 0 \\ 10 & 7 & 1 \end{bmatrix} \quad (12.49)$$

We now design the controller using the phase-variable representation and then use Eq. (12.49) to transform the design back to the original representation. For a 20.8% overshoot and a settling time of 4 seconds, a factor of the characteristic equation of the designed closed-loop system is $s^2 + 2s + 5$. Since the closed-loop zero will be at $s = -4$, we choose the third closed-loop pole to cancel the closed-loop zero. Hence, the total characteristic equation of the desired closed-loop system is

$$D(s) = (s+4)(s^2+2s+5) = s^3 + 6s^2 + 13s + 20 = 0 \quad (12.50)$$

The state equations for the phase-variable form with state-variable feedback are

$$\dot{\mathbf{x}} = (\mathbf{A_x} - \mathbf{B_x K_x})\mathbf{x} = \begin{bmatrix} 0 & 1 & 0 \\ 0 & 0 & 1 \\ -(10+k_{1_x}) & -(17+k_{2_x}) & -(8+k_{3_x}) \end{bmatrix} \mathbf{x} \quad (12.51a)$$

$$y = [4 \quad 1 \quad 0]\mathbf{x} \quad (12.51b)$$

The characteristic equation for Eqs. (12.51) is

$$\det(s\mathbf{I} - (\mathbf{A_x} - \mathbf{B_x K_x})) = s^3 + (8+k_{3_x})s^2 + (17+k_{2_x})s + (10+k_{1_x}) = 0 \quad (12.52)$$

Comparing Eq. (12.50) with (12.52), we see that

$$\mathbf{K_x} = [k_{1_x} \quad k_{2_x} \quad k_{3_x}] = [10 \quad -4 \quad -2] \quad (12.53)$$

Using Eqs. (12.42) and (12.49), we can transform the controller back to the original system as

$$\mathbf{K_z} = \mathbf{K_x P^{-1}} = [-20 \quad 10 \quad -2] \quad (12.54)$$

The final closed-loop system with state-variable feedback is shown in Figure 12.10, with the input applied as shown.

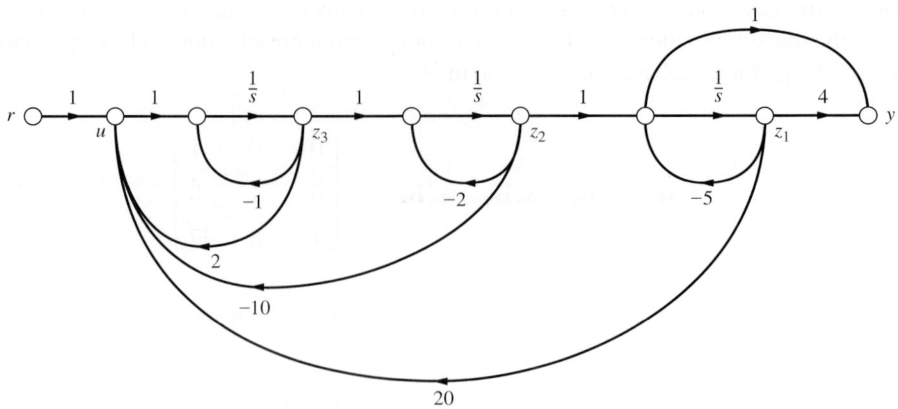


**FIGURE 12.10** Designed system with state-variable feedback for Example 12.4

Let us now verify our design. The state equations for the designed system shown in Figure 12.10 with input $r$ are

$$\dot{\mathbf{z}} = (\mathbf{A_z} - \mathbf{B_z K_z})\mathbf{z} + \mathbf{B_z} r = \begin{bmatrix} -5 & 1 & 0 \\ 0 & -2 & 1 \\ 20 & -10 & 1 \end{bmatrix}\mathbf{z} + \begin{bmatrix} 0 \\ 0 \\ 1 \end{bmatrix} r \tag{12.55a}$$

$$y = \mathbf{C_z z} = [-1 \quad 1 \quad 0]\,\mathbf{z} \tag{12.55b}$$

Using Eq. (3.73) to find the closed-loop transfer function, we obtain

$$T(s) = \frac{(s+4)}{s^3 + 6s^2 + 13s + 20} = \frac{1}{s^2 + 2s + 5} \tag{12.56}$$

The requirements for our design have been met.

MATLAB

**ML**

`Students who are using MATLAB should now run ch12p3 in Appendix B. You will learn how to use MATLAB to design a controller for a plant not represented in phase-variable form. You will see that MATLAB does not require transformation to phase-variable form. This exercise solves Example 12.4 using MATLAB.`

## Skill-Assessment Exercise 12.3

**PROBLEM:** Design a linear state-feedback controller to yield 20% overshoot and a settling time of 2 seconds for a plant

$$G(s) = \frac{(s+6)}{(s+9)(s+8)(s+7)}$$

that is represented in state space in cascade form by

$$\dot{\mathbf{z}} = \mathbf{Az} + \mathbf{B}u = \begin{bmatrix} -7 & 1 & 0 \\ 0 & -8 & 1 \\ 0 & 0 & -9 \end{bmatrix}\mathbf{z} + \begin{bmatrix} 0 \\ 0 \\ 1 \end{bmatrix} u$$

$$y = \mathbf{Cz} = [-1 \quad 1 \quad 0]\,\mathbf{z}$$

**ANSWER:** $\mathbf{K_z} = [-40.23 \quad 62.24 \quad -14]$

The complete solution is located at www.wiley.com/college/nise.

In this section, we saw how to design state-variable feedback for plants not represented in phase-variable form. Using controllability matrices, we were able to transform a plant to phase-variable form, design the controller, and finally transform the controller design back to the plant's original representation. The design of the controller relies on the availability of the states for feedback. In the next section, we discuss the design of state-variable feedback when some or all of the states are not available.

## 12.5 Observer Design

Controller design relies upon access to the state variables for feedback through adjustable gains. This access can be provided by hardware. For example, gyros can measure position and velocity on a space vehicle. Sometimes it is impractical to use this hardware for reasons of cost, accuracy, or availability. For example, in powered flight of space vehicles, inertial measuring units can be used to calculate the acceleration. However, their alignment deteriorates with time; thus, other means of measuring acceleration may be desirable (*Rockwell International, 1984*). In other applications, some of the state variables may not be available at all, or it is too costly to measure them or send them to the controller. If the state variables are not available because of system configuration or cost, it is possible to estimate the states. Estimated states, rather than actual states, are then fed to the controller. One scheme is shown in Figure 12.11(*a*). An *observer*, sometimes called an *estimator*, is used to calculate state variables that are not accessible from the plant. Here the observer is a model of the plant.

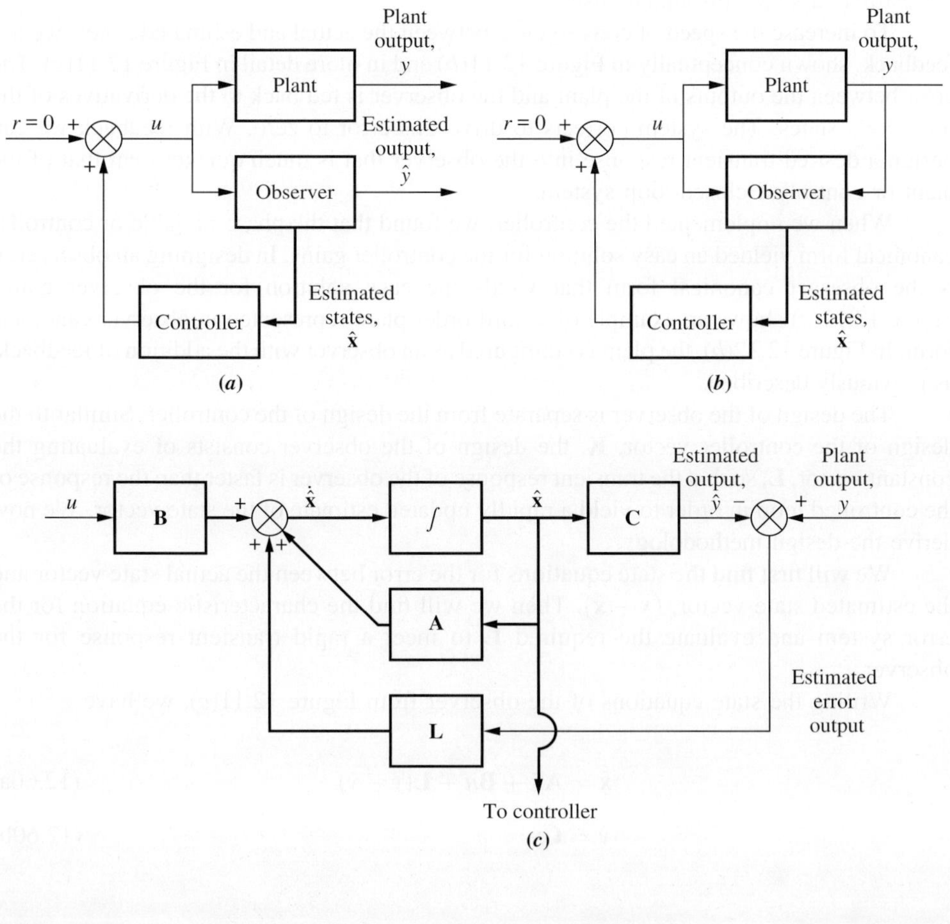


**FIGURE 12.11** State-feedback design using an observer to estimate unavailable state variables: **a.** open-loop observer; **b.** closed-loop observer; **c.** exploded view of a closed-loop observer, showing feedback arrangement to reduce state-variable estimation error

Let us look at the disadvantages of such a configuration. Assume a plant,

$$\dot{\mathbf{x}} = \mathbf{A}\mathbf{x} + \mathbf{B}u \tag{12.57a}$$

$$y = \mathbf{C}\mathbf{x} \tag{12.57b}$$

and an observer,

$$\dot{\hat{\mathbf{x}}} = \mathbf{A}\hat{\mathbf{x}} + \mathbf{B}u \tag{12.58a}$$

$$\hat{\mathbf{y}} = \mathbf{C}\hat{\mathbf{x}} \tag{12.58b}$$

Subtracting Eqs. (12.58) from (12.57), we obtain

$$\dot{\mathbf{x}} - \dot{\hat{\mathbf{x}}} = \mathbf{A}(\mathbf{x} - \hat{\mathbf{x}}) \tag{12.59a}$$

$$y - \hat{y} = \mathbf{C}(\mathbf{x} - \hat{\mathbf{x}}) \tag{12.59b}$$

Thus, the dynamics of the difference between the actual and estimated states is unforced, and if the plant is stable, this difference, due to differences in initial state vectors, approaches zero. However, the speed of convergence between the actual state and the estimated state is the same as the transient response of the plant since the characteristic equation for Eq. (12.59a) is the same as that for Eq. (12.57a). Since the convergence is too slow, we seek a way to speed up the observer and make its response time much faster than that of the controlled closed-loop system, so that, effectively, the controller will receive the estimated states instantaneously.

To increase the speed of convergence between the actual and estimated states, we use feedback, shown conceptually in Figure 12.11(*b*) and in more detail in Figure 12.11(*c*). The error between the outputs of the plant and the observer is fed back to the derivatives of the observer's states. The system corrects to drive this error to zero. With feedback we can design a desired transient response into the observer that is much quicker than that of the plant or controlled closed-loop system.

When we implemented the controller, we found that the phase-variable or controller canonical form yielded an easy solution for the controller gains. In designing an observer, it is the observer canonical form that yields the easy solution for the observer gains. Figure 12.12(*a*) shows an example of a third-order plant represented in observer canonical form. In Figure 12.12(*b*), the plant is configured as an observer with the addition of feedback, as previously described.

The design of the observer is separate from the design of the controller. Similar to the design of the controller vector, **K**, the design of the observer consists of evaluating the constant vector, **L**, so that the transient response of the observer is faster than the response of the controlled loop in order to yield a rapidly updated estimate of the state vector. We now derive the design methodology.

We will first find the state equations for the error between the actual state vector and the estimated state vector, $(\mathbf{x} - \hat{\mathbf{x}})$. Then we will find the characteristic equation for the error system and evaluate the required **L** to meet a rapid transient response for the observer.

Writing the state equations of the observer from Figure 12.11(*c*), we have

$$\dot{\hat{\mathbf{x}}} = \mathbf{A}\hat{\mathbf{x}} + \mathbf{B}u + \mathbf{L}(y - \hat{y}) \tag{12.60a}$$

$$\hat{y} = \mathbf{C}\hat{\mathbf{x}} \tag{12.60b}$$

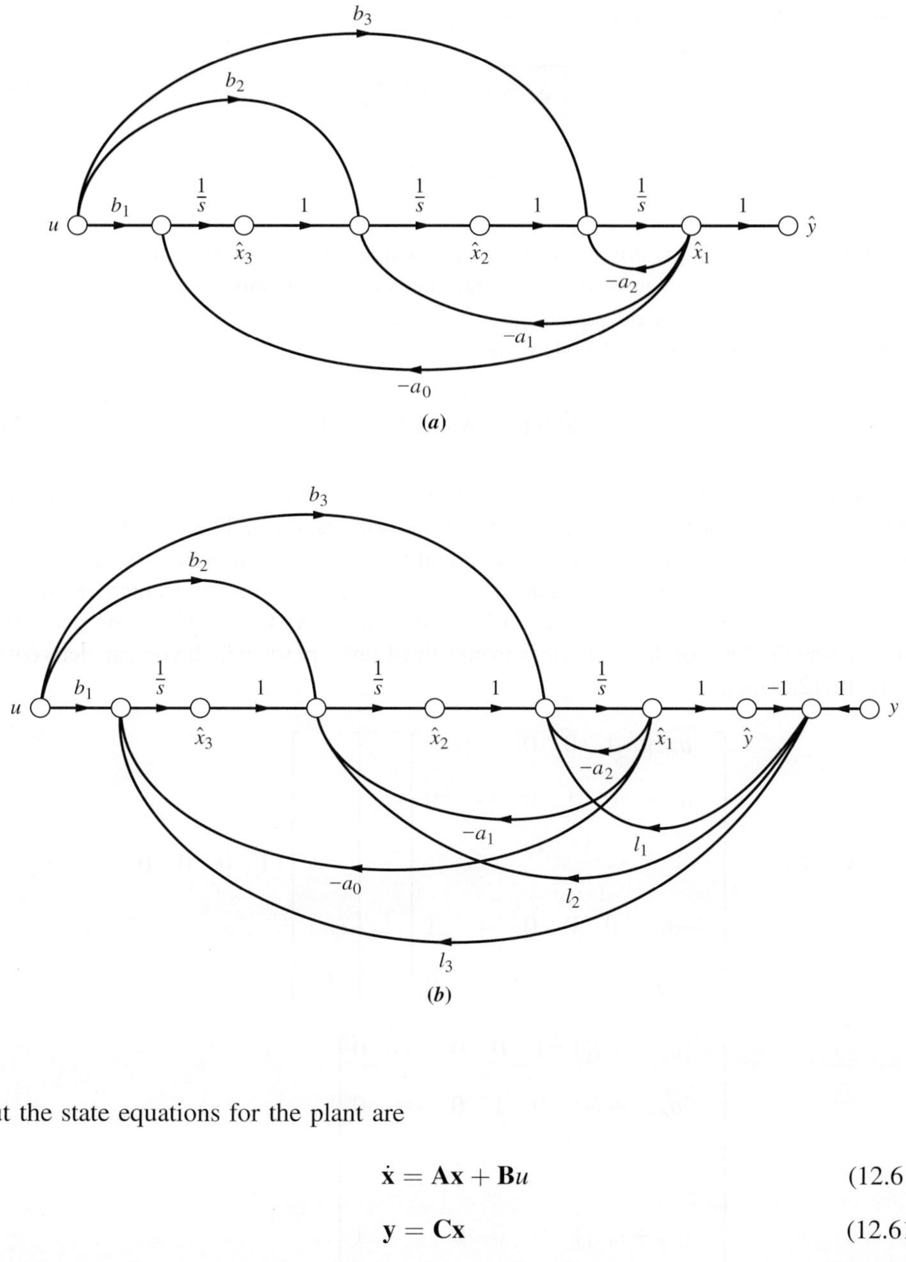


**FIGURE 12.12** Third-order observer in observer canonical form: **a.** before the addition of feedback; **b.** after the addition of feedback

But the state equations for the plant are

$$\dot{\mathbf{x}} = \mathbf{A}\mathbf{x} + \mathbf{B}u \tag{12.61a}$$

$$\mathbf{y} = \mathbf{C}\mathbf{x} \tag{12.61b}$$

Subtracting Eqs. (12.60) from (12.61), we obtain

$$(\dot{\mathbf{x}} - \dot{\hat{\mathbf{x}}}) = \mathbf{A}(\mathbf{x} - \hat{\mathbf{x}}) - \mathbf{L}(y - \hat{y}) \tag{12.62a}$$

$$(y - \hat{y}) = \mathbf{C}(\mathbf{x} - \hat{\mathbf{x}}) \tag{12.62b}$$

where $\mathbf{x} - \hat{\mathbf{x}}$ is the error between the actual state vector and the estimated state vector, and $y - \hat{y}$ is the error between the actual output and the estimated output.

Substituting the output equation into the state equation, we obtain the state equation for the error between the estimated state vector and the actual state vector:

$$(\dot{\mathbf{x}} - \dot{\hat{\mathbf{x}}}) = (\mathbf{A} - \mathbf{L}\mathbf{C})(\mathbf{x} - \hat{\mathbf{x}}) \tag{12.63a}$$

$$(y - \hat{y}) = \mathbf{C}(\mathbf{x} - \hat{\mathbf{x}}) \tag{12.63b}$$

Letting $\mathbf{e_x} = (\mathbf{x} - \hat{\mathbf{x}})$, we have

$$\dot{\mathbf{e}}_\mathbf{x} = (\mathbf{A} - \mathbf{LC})\mathbf{e_x} \tag{12.64a}$$

$$y - \hat{y} = \mathbf{C}\mathbf{e_x} \tag{12.64b}$$

Equation (12.64a) is unforced. If the eigenvalues are all negative, the estimated state vector error, $\mathbf{e_x}$, will decay to zero. The design then consists of solving for the values of $\mathbf{L}$ to yield a desired characteristic equation or response for Eqs. (12.64). The characteristic equation is found from Eqs. (12.64) to be

$$\det[\lambda\mathbf{I} - (\mathbf{A} - \mathbf{LC})] = 0 \tag{12.65}$$

Now we select the eigenvalues of the observer to yield stability and a desired transient response that is faster than the controlled closed-loop response. These eigenvalues determine a characteristic equation that we set equal to Eq. (12.65) to solve for $\mathbf{L}$.

Let us demonstrate the procedure for an $n$th-order plant represented in observer canonical form. We first evaluate $\mathbf{A} - \mathbf{LC}$. The form of $\mathbf{A}$, $\mathbf{L}$, and $\mathbf{C}$ can be derived by extrapolating the form of these matrices from a third-order plant, which you can derive from Figure 12.12. Thus,

$$\mathbf{A} - \mathbf{LC} = \begin{bmatrix} -a_{n-1} & 1 & 0 & 0 & \cdots & 0 \\ -a_{n-2} & 0 & 1 & 0 & \cdots & 0 \\ \vdots & \vdots & \vdots & \vdots & \vdots & \vdots \\ -a_1 & 0 & 0 & 0 & \cdots & 1 \\ -a_0 & 0 & 0 & 0 & \cdots & 0 \end{bmatrix} - \begin{bmatrix} l_1 \\ l_2 \\ \vdots \\ l_{n-1} \\ l_n \end{bmatrix} \begin{bmatrix} 1 & 0 & 0 & 0 & \cdots & 0 \end{bmatrix}$$

$$= \begin{bmatrix} -(a_{n-1}+l_1) & 1 & 0 & 0 & \cdots & 0 \\ -(a_{n-2}+l_2) & 0 & 1 & 0 & \cdots & 0 \\ \vdots & \vdots & \vdots & \vdots & \vdots & \vdots \\ -(a_1+l_{n-1}) & 0 & 0 & 0 & \cdots & 1 \\ -(a_0+l_n) & 0 & 0 & 0 & \cdots & 0 \end{bmatrix} \tag{12.66}$$

The characteristic equation for $\mathbf{A} - \mathbf{LC}$ is

$$s^n + (a_{n-1}+l_1)s^{n-1} + (a_{n-2}+l_2)s^{n-2} + \cdots + (a_1+l_{n-1})s + (a_0+l_n) = 0 \tag{12.67}$$

Notice the relationship between Eq. (12.67) and the characteristic equation, $\det(s\mathbf{I} - \mathbf{A}) = 0$, for the plant, which is

$$s^n + a_{n-1}s^{n-1} + a_{n-2}s^{n-2} + \cdots + a_1 s + a_0 = 0 \tag{12.68}$$

Thus, if desired, Eq. (12.67) can be written by inspection if the plant is represented in observer canonical form. We now equate Eq. (12.67) with the desired closed-loop observer characteristic equation, which is chosen on the basis of a desired transient response. Assume the desired characteristic equation is

$$s^n + d_{n-1}s^{n-1} + d_{n-2}s^{n-2} + \cdots + d_1 s + d_0 = 0 \tag{12.69}$$

We can now solve for the $l_i$'s by equating the coefficients of Eqs. (12.67) and (12.69):

$$l_i = d_{n-i} - a_{n-i} \quad i = 1, 2, \ldots, n \tag{12.70}$$

Let us demonstrate the design of an observer using the observer canonical form. In subsequent sections we will show how to design the observer for other than observer canonical form.

## Example 12.5 Observer Design for Observer Canonical Form

**PROBLEM:** Design an observer for the plant

$$G(s) = \frac{(s+4)}{(s+1)(s+2)(s+5)} = \frac{s+4}{s^3 + 8s^2 + 17s + 10} \tag{12.71}$$

which is represented in observer canonical form. The observer will respond 10 times faster than the controlled loop designed in Example 12.4.

**SOLUTION:**

1. First represent the estimated plant in observer canonical form. The result is shown in Figure 12.13(*a*).
2. Now form the difference between the plant's actual output, $y$, and the observer's estimated output, $\hat{y}$, and add the feedback paths from this difference to the derivative of each state variable. The result is shown in Figure 12.13(*b*).
3. Next find the characteristic polynomial. The state equations for the estimated plant shown in Figure 12.13(*a*) are

$$\dot{\hat{\mathbf{x}}} = \mathbf{A}\hat{\mathbf{x}} + \mathbf{B}u = \begin{bmatrix} -8 & 1 & 0 \\ -17 & 0 & 1 \\ -10 & 0 & 0 \end{bmatrix}\hat{\mathbf{x}} + \begin{bmatrix} 0 \\ 1 \\ 4 \end{bmatrix} u \tag{12.72a}$$

$$\hat{y} = \mathbf{C}\hat{\mathbf{x}} = [1 \quad 0 \quad 0]\,\hat{\mathbf{x}} \tag{12.72b}$$

From Eqs. (12.64) and (12.66), the observer error is

$$\dot{\mathbf{e}}_\mathbf{x} = (\mathbf{A} - \mathbf{LC})\mathbf{e}_\mathbf{x} = \begin{bmatrix} -(8+l_1) & 1 & 0 \\ -(17+l_2) & 0 & 1 \\ -(10+l_3) & 0 & 0 \end{bmatrix}\mathbf{e}_\mathbf{x} \tag{12.73}$$

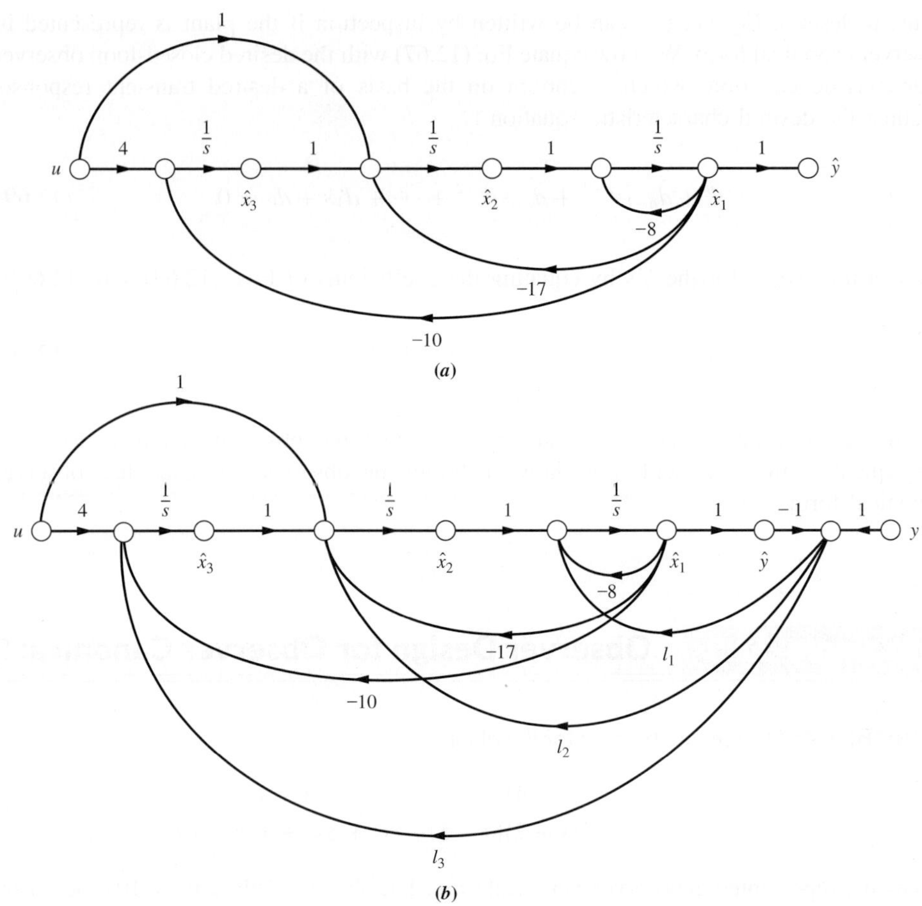


**FIGURE 12.13** **a.** Signal-flow graph of a system using observer canonical form variables; **b.** additional feedback to create observer

Using Eq. (12.65), we obtain the characteristic polynomial

$$s^3 + (8 + l_1)s^2 + (17 + l_2)s + (10 + l_3) \tag{12.74}$$

**4.** Now evaluate the desired polynomial, set the coefficients equal to those of Eq. (12.74), and solve for the gains, $l_i$. From Eq. (12.50), the closed-loop controlled system has dominant second-order poles at $-1 \pm j2$. To make our observer 10 times faster, we design the observer poles to be at $-10 \pm j20$. We select the third pole to be 10 times the real part of the dominant second-order poles, or $-100$. Hence, the desired characteristic polynomial is

$$(s + 100)(s^2 + 20s + 500) = s^3 + 120s^2 + 2500s + 50{,}000 \tag{12.75}$$

Equating Eqs. (12.74) and (12.75), we find $l_1 = 112$, $l_2 = 2483$, and $l_3 = 49{,}990$.

A simulation of the observer with an input of $r(t) = 100t$ is shown in Figure 12.14. The initial conditions of the plant were all zero, and the initial condition of $\hat{x}_1$ was 0.5.

Since the dominant pole of the observer is $-10 \pm j20$, the expected settling time should be about 0.4 second. It is interesting to note the slower response in Figure 12.14(*b*),

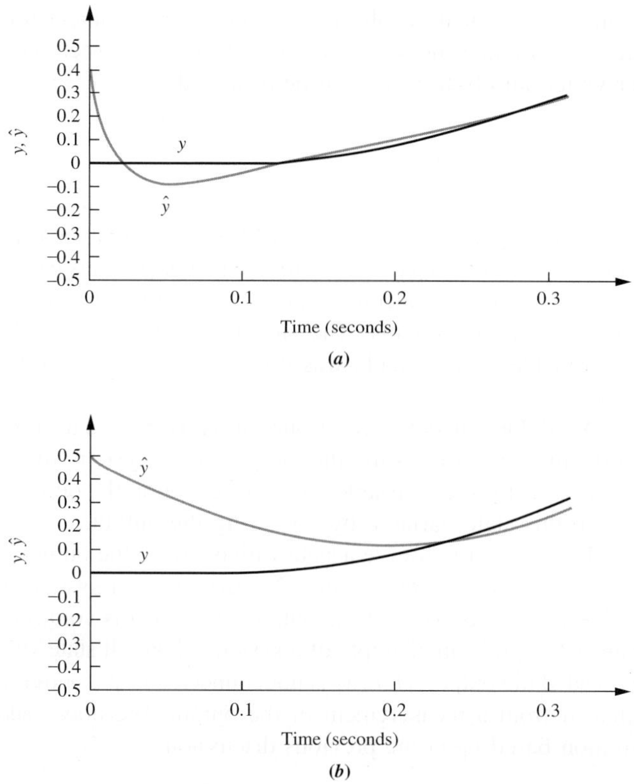


**FIGURE 12.14** Simulation showing response of observer: **a.** closed-loop; **b.** open-loop with observer gains disconnected

where the observer gains are disconnected, and the observer is simply a copy of the plant with a different initial condition.

MATLAB **ML**

`Students who are using MATLAB should now run ch12p4 in Appendix B. You will learn how to use MATLAB to design an observer using pole placement. This exercise solves Example 12.5 using MATLAB.`

## Skill-Assessment Exercise 12.4

**PROBLEM:** Design an observer for the plant

$$G(s) = \frac{(s+6)}{(s+7)(s+8)(s+9)}$$

whose estimated plant is represented in state space in observer canonical form as

$$\dot{\hat{\mathbf{x}}} = \mathbf{A}\hat{\mathbf{x}} + \mathbf{B}u = \begin{bmatrix} -24 & 1 & 0 \\ -191 & 0 & 1 \\ -504 & 0 & 0 \end{bmatrix} \hat{\mathbf{x}} + \begin{bmatrix} 0 \\ 1 \\ 6 \end{bmatrix} u$$

$$\hat{y} = \mathbf{C}\hat{\mathbf{x}} = [1 \quad 0 \quad 0]\hat{\mathbf{x}}$$

The observer will respond 10 times faster than the controlled loop designed in Skill-Assessment Exercise 12.3.

**ANSWER:** $L = [216 \quad 9730 \quad 383{,}696]^T$, where T signifies vector transpose

The complete solution is located at www.wiley.com/college/nise.

### TryIt 12.3

Use MATLAB, the Control System Toolbox, and the following statements to solve Skill-Assessment Exercise 12.4.

```
A=[-24 1 0
   -191 0 1
   -504 0 0]
C=[1 0 0]
pos=20
Ts=2
z=(-log(pos/100))/...
 (sqrt(pi^2+...
 log(pos/100)^2));
wn=4/(z*Ts);
r=roots([1,2*z*wn,...
 wn^2]);
poles=10*[r' 10*...
 real(r(1))]
l=acker(A',C',poles)'
```

In this section, we designed an observer in observer canonical form that uses the output of a system to estimate the state variables. In the next section, we examine the conditions under which an observer cannot be designed.

## 12.6 Observability

Recall that the ability to control all of the state variables is a requirement for the design of a controller. State-variable feedback gains cannot be designed if any state variable is uncontrollable. Uncontrollability can be viewed best with diagonalized systems. The signal-flow graph showed clearly that the uncontrollable state variable was not connected to the control signal of the system.

A similar concept governs our ability to create a design for an observer. Specifically, we are using the output of a system to deduce the state variables. If any state variable has no effect upon the output, then we cannot evaluate this state variable by observing the output.

The ability to observe a state variable from the output is best seen from the diagonalized system. Figure 12.15($a$) shows a system where each state variable can be observed at the output since each is connected to the output. Figure 12.15($b$) is an example of a system where all state variables cannot be observed at the output. Here $x_1$ is not connected to the output and could not be estimated from a measurement of the output. We now make the following definition based upon the previous discussion:

> If the initial-state vector, $\mathbf{x}(t_0)$, can be found from $u(t)$ and $y(t)$ measured over a finite interval of time from $t_0$, the system is said to be *observable*; otherwise the system is said to be *unobservable*.

Simply stated, observability is the ability to deduce the state variables from a knowledge of the input, $u(t)$, and the output, $y(t)$. Pole placement for an observer is a viable design technique only for systems that are observable. This section shows how to determine, a priori, whether or not pole placement is a viable design technique for an observer.

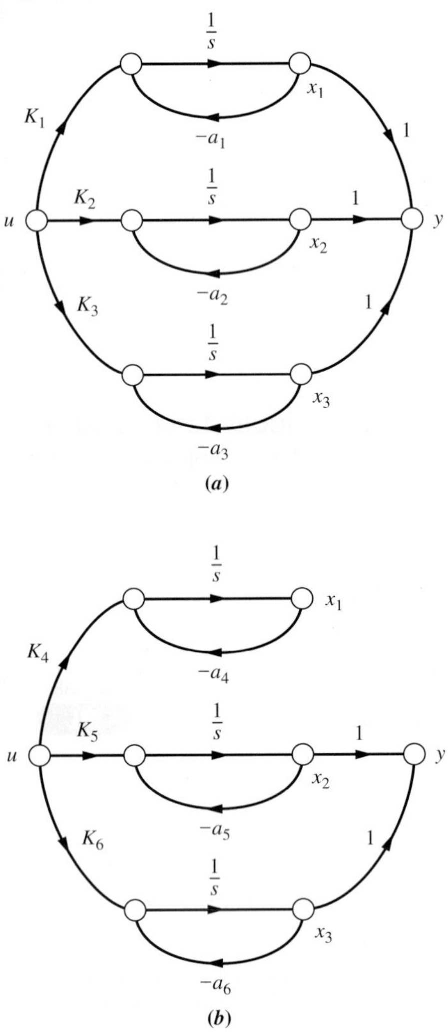


**FIGURE 12.15** Comparison of **a.** observable, and **b.** unobservable systems

### Observability by Inspection

We can also explore observability from the output equation of a diagonalized system. The output equation for the diagonalized system of Figure 12.15($a$) is

$$y = \mathbf{Cx} = [1 \quad 1 \quad 1]\,\mathbf{x} \tag{12.76}$$

On the other hand, the output equation for the unobservable system of Figure 12.15($b$) is

$$y = \mathbf{Cx} = [0 \quad 1 \quad 1]\,\mathbf{x} \tag{12.77}$$

Notice that the first column of Eq. (12.77) is zero. For systems represented in parallel form with distinct eigenvalues, if any column of the output coupling matrix is zero, the diagonal system is not observable.

### The Observability Matrix

Again, as for controllability, systems represented in other than diagonalized form cannot be reliably evaluated for observability by inspection. In order to determine observability for systems under any representation or choice of state variables, a matrix can be derived that must have a particular property if all state variables are to be observed at the output. We now state the requirements for observability, including the form, property, and name of this matrix.

An $n$th-order plant whose state and output equations are, respectively,

$$\dot{\mathbf{x}} = \mathbf{A}\mathbf{x} + \mathbf{B}\mathbf{u} \tag{12.78a}$$

$$\mathbf{y} = \mathbf{C}\mathbf{x} \tag{12.78b}$$

is completely observable[7] if the matrix

$$\mathbf{O_M} = \begin{bmatrix} \mathbf{C} \\ \mathbf{CA} \\ \vdots \\ \mathbf{CA}^{n-1} \end{bmatrix} \tag{12.79}$$

is of rank $n$, where $\mathbf{O_M}$ is called the *observability matrix*.[8]

The following two examples illustrate the use of the observability matrix.

## Example 12.6 Observability via the Observability Matrix

**PROBLEM:** Determine if the system of Figure 12.16 is observable.

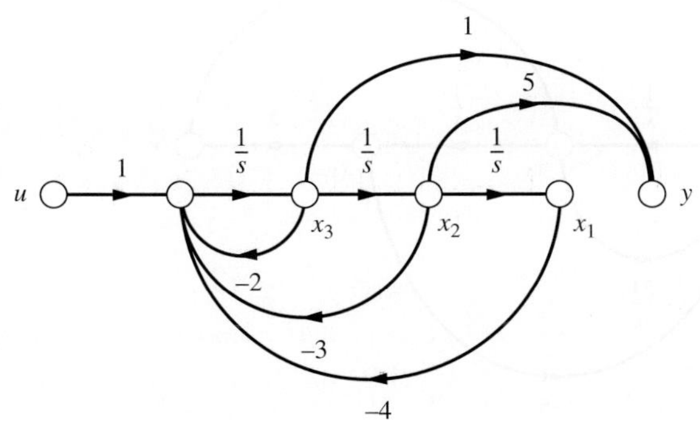


**FIGURE 12.16** System of Example 12.6

**SOLUTION:** The state and output equations for the system are

$$\dot{\mathbf{x}} = \mathbf{A}\mathbf{x} + \mathbf{B}u = \begin{bmatrix} 0 & 1 & 0 \\ 0 & 0 & 1 \\ -4 & -3 & -2 \end{bmatrix}\mathbf{x} + \begin{bmatrix} 0 \\ 0 \\ 1 \end{bmatrix}u \tag{12.80a}$$

$$y = \mathbf{C}\mathbf{x} = [0 \quad 5 \quad 1]\,\mathbf{x} \tag{12.80b}$$

[7] *Completely observable* means that all state variables are observable. This textbook uses *observable* to mean *completely observable*.

[8] See *Ogata* (*1990*: *706–708*) for a derivation.

Thus, the observability matrix, $\mathbf{O_M}$, is

$$\mathbf{O_M} = \begin{bmatrix} \mathbf{C} \\ \mathbf{CA} \\ \mathbf{CA}^2 \end{bmatrix} = \begin{bmatrix} 0 & 5 & 1 \\ -4 & -3 & 3 \\ -12 & -13 & -9 \end{bmatrix} \tag{12.81}$$

Since the determinant of $\mathbf{O_M}$ equals $-344$, $\mathbf{O_M}$ is of full rank equal to 3. The system is thus observable.

You might have been misled and concluded by inspection that the system is unobservable because the state variable $x_1$ is not fed *directly* to the output. Remember that conclusions about observability by inspection are valid only for diagonalized systems that have distinct eigenvalues.

MATLAB

ML

`Students who are using MATLAB should now run ch12p5 in Appendix B. You will learn how to use MATLAB to test a system for observability. This exercise solves Example 12.6 using MATLAB.`

## Example 12.7 Unobservability via the Observability Matrix

**PROBLEM:** Determine whether the system of Figure 12.17 is observable.

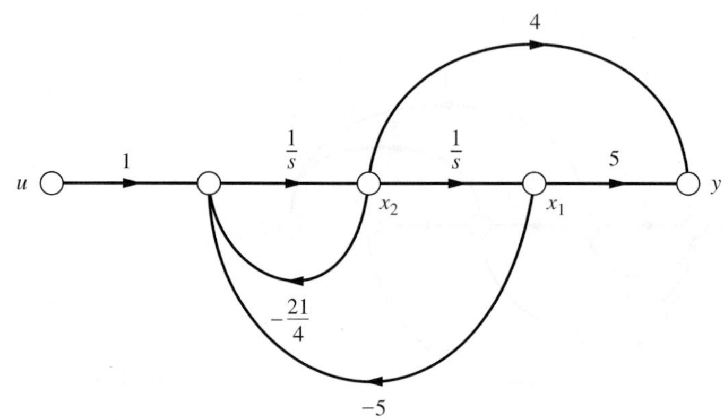


**FIGURE 12.17** System of Example 12.7

**SOLUTION:** The state and output equations for the system are

$$\dot{\mathbf{x}} = \mathbf{Ax} + \mathbf{B}u = \begin{bmatrix} 0 & 1 \\ -5 & -21/4 \end{bmatrix}\mathbf{x} + \begin{bmatrix} 0 \\ 1 \end{bmatrix} u \tag{12.82a}$$

$$y = \mathbf{Cx} = [5 \quad 4]\mathbf{x} \tag{12.82b}$$

The observability matrix, $\mathbf{O_M}$, for this system is

$$\mathbf{O_M} = \begin{bmatrix} \mathbf{C} \\ \mathbf{CA} \end{bmatrix} = \begin{bmatrix} 5 & 4 \\ -20 & -16 \end{bmatrix} \tag{12.83}$$

The determinant for this observability matrix equals 0. Thus, the observability matrix does not have full rank, and the system is not observable.

Again, you might conclude by inspection that the system is observable because all states feed the output. Remember that observability by inspection is valid only for a diagonalized representation of a system with distinct eigenvalues.

## Skill-Assessment Exercise 12.5

**PROBLEM:** Determine whether the system

$$\dot{\mathbf{x}} = \mathbf{A}\mathbf{x} + \mathbf{B}u = \begin{bmatrix} -2 & -1 & -3 \\ 0 & -2 & 1 \\ -7 & -8 & -9 \end{bmatrix}\mathbf{x} + \begin{bmatrix} 2 \\ 1 \\ 2 \end{bmatrix}u$$

$$y = \mathbf{C}\mathbf{x} = [4 \quad 6 \quad 8]\mathbf{x}$$

is observable.

**ANSWER:** Observable

The complete solution is located at www.wiley.com/college/nise.

### TryIt 12.4

Use MATLAB, the Control System Toolbox, and the following statements to solve Skill-Assessment Exercise 12.5.

```
A=[-2  1  3
    0 -2  1
   -7 -8 -9]
C=[4 6 8]
Om=obsv(A,C)
Rank=rank(Om)
```

Now that we have discussed observability and the observability matrix, we are ready to talk about the design of an observer for a plant not represented in observer canonical form.

# 12.7 Alternative Approaches to Observer Design

Earlier in the chapter, we discussed how to design controllers for systems not represented in phase-variable form. One method is to match the coefficients of $\det[s\mathbf{I} - (\mathbf{A} - \mathbf{B}\mathbf{K})]$ with the coefficients of the desired characteristic polynomial. This method can yield difficult calculations for higher-order systems. Another method is to transform the plant to phase-variable form, design the controller, and transfer the design back to its original representation. The transformations were derived from the controllability matrix.

In this section, we use a similar idea for the design of observers not represented in observer canonical form. One method is to match the coefficients of $\det[s\mathbf{I} - (\mathbf{A} - \mathbf{L}\mathbf{C})]$ with the coefficients of the desired characteristic polynomial. Again, this method can yield difficult calculations for higher-order systems. Another method is first to transform the plant to observer canonical form so that the design equations are simple, then perform the design in observer canonical form, and finally transform the design back to the original representation.

Let us pursue this second method. First we will derive the transformation between a system representation and its representation in observer canonical form. Assume a plant not represented in observer canonical form,

$$\dot{\mathbf{z}} = \mathbf{A}\mathbf{z} + \mathbf{B}u \qquad (12.84a)$$

$$y = \mathbf{C}\mathbf{z} \qquad (12.84b)$$

whose observability matrix is

$$\mathbf{O_{Mz}} = \begin{bmatrix} \mathbf{C} \\ \mathbf{CA} \\ \mathbf{CA}^2 \\ \vdots \\ \mathbf{CA}^{n-2} \\ \mathbf{CA}^{n-1} \end{bmatrix} \tag{12.85}$$

Now assume that the system can be transformed to the observer canonical form, $\mathbf{x}$, with the transformation

$$\mathbf{z} = \mathbf{Px} \tag{12.86}$$

Substituting Eq. (12.86) into Eqs. (12.84) and premultiplying the state equation by $\mathbf{P}^{-1}$, we find that the state equations in observer canonical form are

$$\dot{\mathbf{x}} = \mathbf{P}^{-1}\mathbf{APx} + \mathbf{P}^{-1}\mathbf{B}u \tag{12.87a}$$

$$y = \mathbf{CPx} \tag{12.87b}$$

whose observability matrix, $\mathbf{O_{Mx}}$, is

$$\mathbf{O_{Mx}} = \begin{bmatrix} \mathbf{CP} \\ \mathbf{CP}(\mathbf{P}^{-1}\mathbf{AP}) \\ \mathbf{CP}(\mathbf{P}^{-1}\mathbf{AP})(\mathbf{P}^{-1}\mathbf{AP}) \\ \vdots \\ \mathbf{CP}(\mathbf{P}^{-1}\mathbf{AP})(\mathbf{P}^{-1}\mathbf{AP}) \cdots (\mathbf{P}^{-1}\mathbf{AP}) \end{bmatrix} = \begin{bmatrix} \mathbf{C} \\ \mathbf{CA} \\ \mathbf{CA}^2 \\ \vdots \\ \mathbf{CA}^{n-1} \end{bmatrix} \mathbf{P} \tag{12.88}$$

Substituting Eq. (12.85) into (12.88) and solving for $\mathbf{P}$, we obtain

$$\mathbf{P} = \mathbf{O_{Mz}}^{-1}\mathbf{O_{Mx}} \tag{12.89}$$

Thus, the transformation, $\mathbf{P}$, can be found from the two observability matrices.

After transforming the plant to observer canonical form, we design the feedback gains, $\mathbf{L_x}$, as in Section 12.5. Using the matrices from Eqs. (12.87) and the form suggested by Eqs. (12.64), we have

$$\dot{\mathbf{e}}_\mathbf{x} = (\mathbf{P}^{-1}\mathbf{AP} - \mathbf{L_x CP})\,\mathbf{e_x} \tag{12.90a}$$

$$y - \hat{y} = \mathbf{CPe_x} \tag{12.90b}$$

Since $\mathbf{x} = \mathbf{P}^{-1}\mathbf{z}$, and $\hat{\mathbf{x}} = \mathbf{P}^{-1}\hat{\mathbf{z}}$, then $\mathbf{e_x} = \mathbf{x} - \hat{\mathbf{x}} = \mathbf{P}^{-1}\mathbf{e_z}$. Substituting $\mathbf{e_x} = \mathbf{P}^{-1}\mathbf{e_z}$ into Eqs. (12.90) transforms Eqs. (12.90) back to the original representation. The result is

$$\dot{\mathbf{e}}_\mathbf{z} = (\mathbf{A} - \mathbf{PL_x C})\,\mathbf{e_z} \tag{12.91a}$$

$$y - \hat{y} = \mathbf{Ce_z} \tag{12.91b}$$

Comparing Eq. (12.91a) to (12.64a), we see that the observer gain vector is

$$\mathbf{L_z} = \mathbf{L}\mathbf{P_x} \tag{12.92}$$

We now demonstrate the design of an observer for a plant not represented in observer canonical form. The first example uses transformations to and from observer canonical form. The second example matches coefficients without the transformation. This method, however, can become difficult if the system order is high.

## Example 12.8 Observer Design by Transformation

**PROBLEM:** Design an observer for the plant

$$G(s) = \frac{1}{(s+1)(s+2)(s+5)} \tag{12.93}$$

represented in cascade form. The closed-loop performance of the observer is governed by the characteristic polynomial used in Example 12.5: $s^3 + 120s^2 + 2500s + 50{,}000$.

**SOLUTION:** First represent the plant in its original cascade form.

$$\dot{\mathbf{z}} = \mathbf{Az} + \mathbf{B}u = \begin{bmatrix} -5 & 1 & 0 \\ 0 & -2 & 1 \\ 0 & 0 & -1 \end{bmatrix}\mathbf{z} + \begin{bmatrix} 0 \\ 0 \\ 1 \end{bmatrix}u \tag{12.94a}$$

$$y = \mathbf{Cz} = [1 \quad 0 \quad 0]\,\mathbf{z} \tag{12.94b}$$

The observability matrix, $\mathbf{O_{Mz}}$, is

$$\mathbf{O_{Mz}} = \begin{bmatrix} \mathbf{C} \\ \mathbf{CA} \\ \mathbf{CA}^2 \end{bmatrix} = \begin{bmatrix} 1 & 0 & 0 \\ -5 & 1 & 0 \\ 25 & -7 & 1 \end{bmatrix} \tag{12.95}$$

whose determinant equals 1. Hence, the plant is observable.

The characteristic equation for the plant is

$$\det(s\mathbf{I} - \mathbf{A}) = s^3 + 8s^2 + 17s + 10 = 0 \tag{12.96}$$

We can use the coefficients of this characteristic polynomial to form the observer canonical form

$$\dot{\mathbf{x}} = \mathbf{A_x x} + \mathbf{B_x}u \tag{12.97a}$$

$$y = \mathbf{C_x x} \tag{12.97b}$$

where

$$\mathbf{A_x} = \begin{bmatrix} -8 & 1 & 0 \\ -17 & 0 & 1 \\ -10 & 0 & 0 \end{bmatrix}; \quad \mathbf{C_x} = [1 \quad 0 \quad 0] \tag{12.98}$$

The observability matrix for the observer canonical form is

$$\mathbf{O_{Mx}} = \begin{bmatrix} \mathbf{C_x} \\ \mathbf{C_x A_x} \\ \mathbf{C_x A_x^2} \end{bmatrix} = \begin{bmatrix} 1 & 0 & 0 \\ -8 & 1 & 0 \\ 47 & -8 & 1 \end{bmatrix} \tag{12.99}$$

We now design the observer for the observer canonical form. First form $(\mathbf{A_x} - \mathbf{L_x C_x})$,

$$\mathbf{A_x} - \mathbf{L_x C_x} = \begin{bmatrix} -8 & 1 & 0 \\ -17 & 0 & 1 \\ -10 & 0 & 0 \end{bmatrix} - \begin{bmatrix} l_1 \\ l_2 \\ l_3 \end{bmatrix} [1 \quad 0 \quad 0] = \begin{bmatrix} -(8+l_1) & 1 & 0 \\ -(17+l_2) & 0 & 1 \\ -(10+l_3) & 0 & 0 \end{bmatrix} \tag{12.100}$$

whose characteristic polynomial is

$$\det[s\mathbf{I} - (\mathbf{A_x} - \mathbf{L_x C_x})] = s^3 + (8+l_1)s^2 + (17+l_2)s + (10+l_3) \tag{12.101}$$

Equating this polynomial to the desired closed-loop observer characteristic equation, $s^3 + 120s^2 + 2500s + 50{,}000$, we find

$$\mathbf{L_x} = \begin{bmatrix} 112 \\ 2483 \\ 49{,}990 \end{bmatrix} \tag{12.102}$$

Now transform the design back to the original representation. Using Eq. (12.89), the transformation matrix is

$$\mathbf{P} = \mathbf{O_{Mz}}^{-1}\mathbf{O_{Mx}} = \begin{bmatrix} 1 & 0 & 0 \\ -3 & 1 & 0 \\ 1 & -1 & 1 \end{bmatrix} \tag{12.103}$$

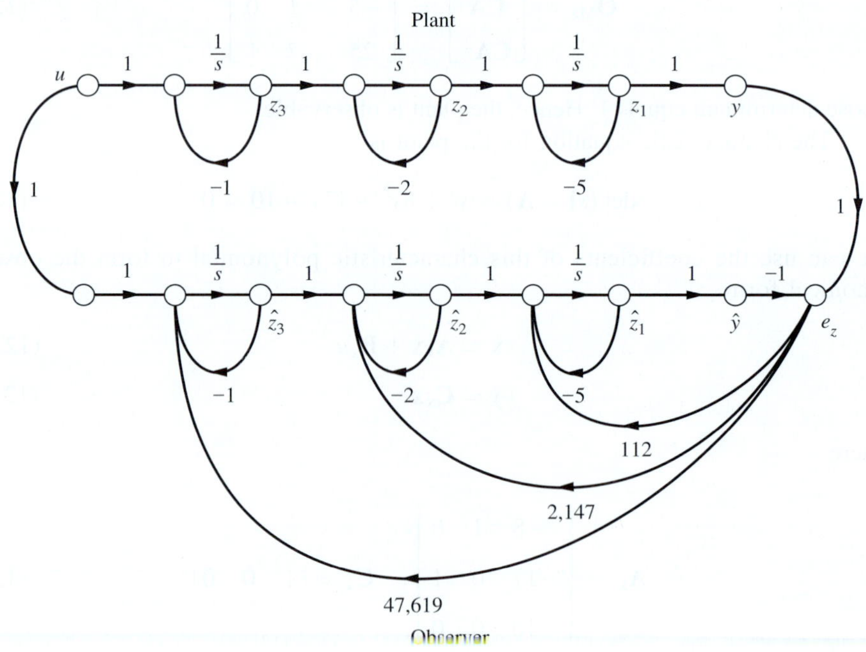


**FIGURE 12.18** Observer design

Transforming $\mathbf{L_x}$ to the original representation, we obtain

$$\mathbf{L_z} = \mathbf{PL_x} = \begin{bmatrix} 112 \\ 2147 \\ 47{,}619 \end{bmatrix} \tag{12.104}$$

The final configuration is shown in Figure 12.18.

A simulation of the observer is shown in Figure 12.19(*a*). To demonstrate the effect of the observer design, Figure 12.19(*b*) shows the reduced speed if the observer is simply a copy of the plant and all observer feedback paths are disconnected.

MATLAB ML

`Students who are using MATLAB should now run ch12p6 in Appendix B. You will learn how to use MATLAB to design an observer for a plant not represented in observer canonical form. You will see that MATLAB does not require transformation to observer canonical form. This exercise solves Example 12.8 using MATLAB.`

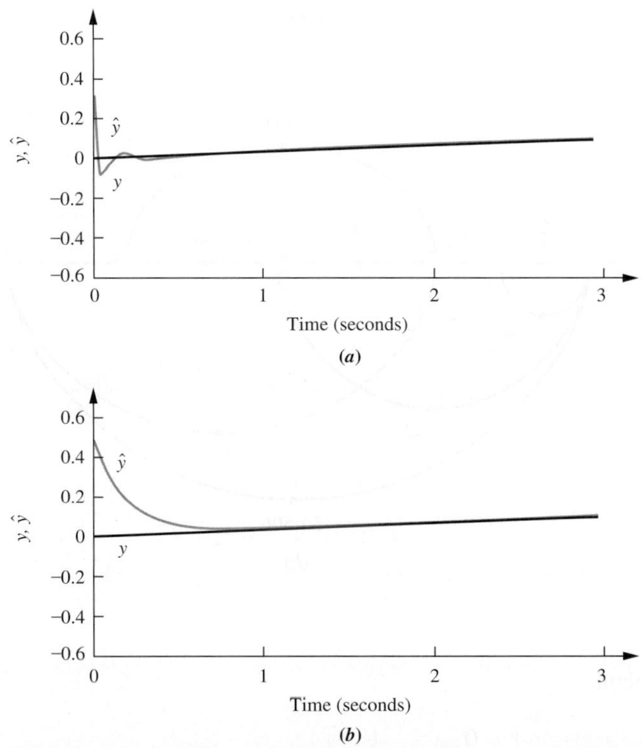


**FIGURE 12.19** Observer design step response simulation: **a.** closed-loop observer; **b.** open-loop observer with observer gains disconnected

## Example 12.9 Observer Design by Matching Coefficients

**PROBLEM:** A time-scaled model for the body's blood glucose level is shown in Eq. (12.105). The output is the deviation in glucose concentration from its mean value in mg/100 ml, and the input is the intravenous glucose injection rate in g/kg/hr (*Milhorn, 1966*).

$$G(s) = \frac{407(s+0.916)}{(s+1.27)(s+2.69)} \tag{12.105}$$

Design an observer for the phase variables with a transient response described by $\zeta = 0.7$ and $\omega_n = 100$.

**SOLUTION:** We can first model the plant in phase-variable form. The result is shown in Figure 12.20($a$).

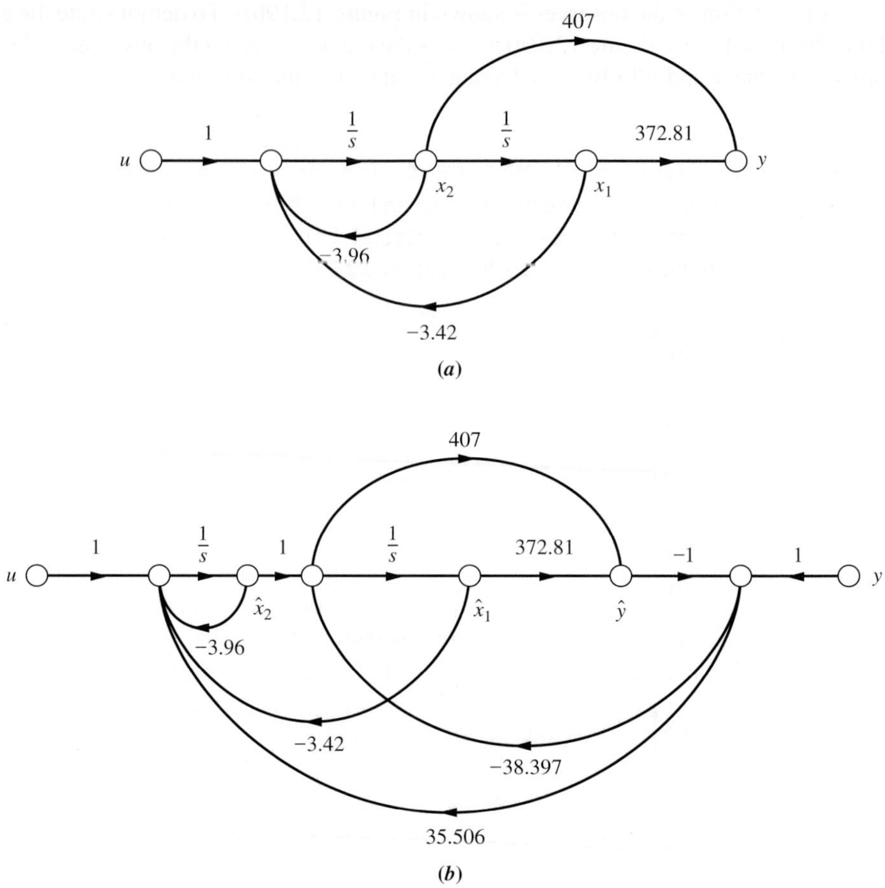


**FIGURE 12.20** **a.** Plant; **b.** designed observer for Example 12.9

For the plant,

$$\mathbf{A} = \begin{bmatrix} 0 & 1 \\ -3.42 & -3.96 \end{bmatrix}; \quad \mathbf{C} = [\,372.81 \quad 407\,] \tag{12.106}$$

Calculation of the observability matrix, $\mathbf{O_M} = [\,\mathbf{C} \quad \mathbf{CA}\,]^T$, shows that the plant is observable and we can proceed with the design. Next find the characteristic equation of the observer. First we have

$$\begin{aligned} \mathbf{A} - \mathbf{LC} &= \begin{bmatrix} 0 & 1 \\ -3.42 & -3.96 \end{bmatrix} - \begin{bmatrix} l_1 \\ l_2 \end{bmatrix} [\,372.81 \quad 407\,] \\ &= \begin{bmatrix} -372.81 l_1 & (1 - 407 l_1) \\ -(3.42 + 372.81 l_2) & -(3.96 + 407 l_2) \end{bmatrix} \end{aligned} \tag{12.107}$$

Now evaluate det $[\lambda\mathbf{I} - (\mathbf{A} - \mathbf{LC})] = 0$ in order to obtain the characteristic equation:

$$\det[\lambda\mathbf{I} - (\mathbf{A} - \mathbf{LC})] = \det\begin{bmatrix} (\lambda + 372.81l_1) & -(1 - 407l_1) \\ (3.42 + 372.81l_2) & (\lambda + 3.96 + 407l_2) \end{bmatrix}$$
$$= \lambda^2 + (3.96 + 372.81l_1 + 407l_2)\lambda + (3.42 + 84.39l_1 + 372.81l_2)$$
$$= 0 \quad (12.108)$$

From the problem statement, we want $\zeta = 0.7$ and $\omega_n = 100$. Thus,

$$\lambda^2 + 140\lambda + 10{,}000 = 0 \quad (12.109)$$

Comparing the coefficients of Eqs. (12.108) and (12.109), we find the values $l_1$ and $l_2$ to be $-38.397$ and $35.506$, respectively. Using Eq. (12.60), where

$$\mathbf{A} = \begin{bmatrix} 0 & 1 \\ -3.42 & -3.96 \end{bmatrix}; \ \mathbf{B} = \begin{bmatrix} 0 \\ 1 \end{bmatrix}; \ \mathbf{C} = [372.81 \quad 407];$$
$$\mathbf{L} = \begin{bmatrix} -38.397 \\ 35.506 \end{bmatrix} \quad (12.110)$$

the observer is implemented and shown in Figure 12.20($b$).

## Skill-Assessment Exercise 12.6

**PROBLEM:** Design an observer for the plant

$$G(s) = \frac{1}{(s+7)(s+8)(s+9)}$$

whose estimated plant is represented in state space in cascade form as

$$\dot{\hat{\mathbf{z}}} = \mathbf{A}\hat{\mathbf{z}} + \mathbf{B}u = \begin{bmatrix} -7 & 1 & 0 \\ 0 & -8 & 1 \\ 0 & 0 & -9 \end{bmatrix}\hat{\mathbf{z}} + \begin{bmatrix} 0 \\ 0 \\ 1 \end{bmatrix} u$$
$$\hat{y} = \mathbf{C}\hat{\mathbf{x}} = [1 \quad 0 \quad 0]\hat{\mathbf{z}}$$

The closed-loop step response of the observer is to have 10% overshoot with a 0.1 second settling time.

**ANSWER:**

$$\mathbf{L_z} = \begin{bmatrix} 456 \\ 28{,}640 \\ 1.54 \times 10^6 \end{bmatrix}$$

The complete solution is located at www.wiley.com/college/nise.

Now that we have explored transient response design using state-space techniques, let us turn to the design of steady-state error characteristics.

## 12.8 Steady-State Error Design via Integral Control

In Section 7.8, we discussed how to *analyze* systems represented in state space for steady-state error. In this section, we discuss how to *design* systems represented in state space for steady-state error.

Consider Figure 12.21. The previously designed controller discussed in Section 12.2 is shown inside the dashed box. A feedback path from the output has been added to form the error, $e$, which is fed forward to the controlled plant via an integrator. The integrator increases the system type and reduces the previous finite error to zero. We will now derive the form of the state equations for the system of Figure 12.21 and then use that form to design a controller. Thus, we will be able to design a system for zero steady-state error for a step input as well as design the desired transient response.

An additional state variable, $x_N$, has been added at the output of the leftmost integrator. The error is the derivative of this variable. Now, from Figure 12.21,

$$\dot{x}_N = r - \mathbf{Cx} \tag{12.111}$$

Writing the state equations from Figure 12.21, we have

$$\dot{\mathbf{x}} = \mathbf{Ax} + \mathbf{B}u \tag{12.112a}$$

$$\dot{x}_N = -\mathbf{Cx} + r \tag{12.112b}$$

$$y = \mathbf{Cx} \tag{12.112c}$$

Equations (12.112) can be written as augmented vectors and matrices. Hence,

$$\begin{bmatrix} \dot{\mathbf{x}} \\ \dot{x}_N \end{bmatrix} = \begin{bmatrix} \mathbf{A} & 0 \\ -\mathbf{C} & 0 \end{bmatrix} \begin{bmatrix} \mathbf{x} \\ x_N \end{bmatrix} + \begin{bmatrix} \mathbf{B} \\ 0 \end{bmatrix} u + \begin{bmatrix} \mathbf{0} \\ 1 \end{bmatrix} r \tag{12.113a}$$

$$y = [\mathbf{C} \quad 0] \begin{bmatrix} \mathbf{x} \\ x_N \end{bmatrix} \tag{12.113b}$$

But

$$u = -\mathbf{Kx} + K_e x_N = -[\mathbf{K} \quad -K_e] \begin{bmatrix} \mathbf{x} \\ x_N \end{bmatrix} \tag{12.114}$$

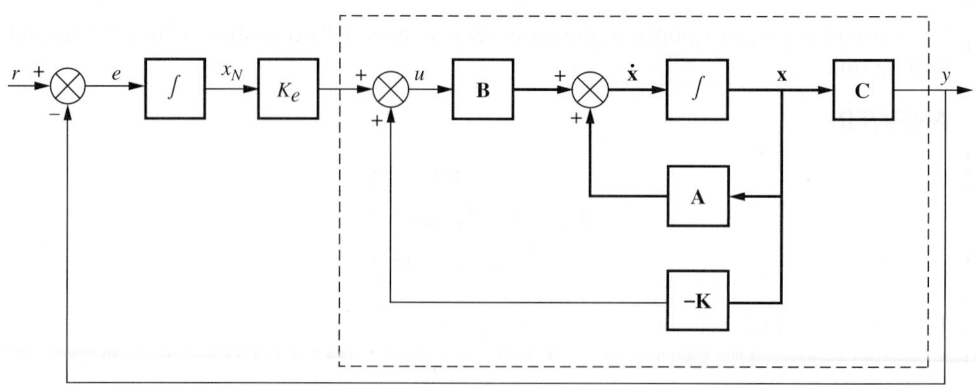


**FIGURE 12.21** Integral control for steady-state error design

Substituting Eq. (12.114) into (12.113a) and simplifying, we obtain

$$\begin{bmatrix} \dot{\mathbf{x}} \\ \dot{x}_N \end{bmatrix} = \begin{bmatrix} (\mathbf{A}-\mathbf{BK}) & \mathbf{B}K_e \\ -\mathbf{C} & 0 \end{bmatrix} \begin{bmatrix} \mathbf{x} \\ x_N \end{bmatrix} + \begin{bmatrix} \mathbf{0} \\ 1 \end{bmatrix} r \tag{12.115a}$$

$$y = [\mathbf{C} \quad 0] \begin{bmatrix} \mathbf{x} \\ x_N \end{bmatrix} \tag{12.115b}$$

Thus, the system type has been increased, and we can use the characteristic equation associated with Eq. (12.115a) to design **K** and $K_e$ to yield the desired transient response. Realize, we now have an additional pole to place. The effect on the transient response of any closed-loop zeros in the final design must also be taken into consideration. One possible assumption is that the closed-loop zeros will be the same as those of the open-loop plant. This assumption, which of course must be checked, suggests placing higher-order poles at the closed-loop zero locations. Let us demonstrate with an example.

## Example 12.10 Design of Integral Control

**PROBLEM:** Consider the plant of Eqs. (12.116):

$$\dot{\mathbf{x}} = \begin{bmatrix} 0 & 1 \\ -3 & -5 \end{bmatrix} \mathbf{x} + \begin{bmatrix} 0 \\ 1 \end{bmatrix} u \tag{12.116a}$$

$$y = [1 \quad 0]\,\mathbf{x} \tag{12.116b}$$

**a.** Design a controller without integral control to yield a 10% overshoot and a settling time of 0.5 second. Evaluate the steady-state error for a unit step input.

**b.** Repeat the design of Part **a** using integral control. Evaluate the steady-state error for a unit step input.

**SOLUTION:**

**a.** Using the requirements for settling time and percent overshoot, we find that the desired characteristic polynomial is

$$s^2 + 16s + 183.1 \tag{12.117}$$

Since the plant is represented in phase-variable form, the characteristic polynomial for the controlled plant with state-variable feedback is

$$s^2 + (5 + k_2)s + (3 + k_1) \tag{12.118}$$

Equating the coefficients of Eqs. (12.117) and (12.118), we have

$$\mathbf{K} = [k_1 \quad k_2] = [180.1 \quad 11] \tag{12.119}$$

From Eqs. (12.3), the controlled plant with state-variable feedback represented in phase-variable form is

$$\dot{\mathbf{x}} = (\mathbf{A} - \mathbf{BK})\mathbf{x} + \mathbf{B}r = \begin{bmatrix} 0 & 1 \\ -183.1 & -16 \end{bmatrix}\mathbf{x} + \begin{bmatrix} 0 \\ 1 \end{bmatrix} r \tag{12.120a}$$

$$y = \mathbf{Cx} = \begin{bmatrix} 1 & 0 \end{bmatrix}\mathbf{x} \tag{12.120b}$$

Using Eq. (7.96), we find that the steady-state error for a step input is

$$\begin{aligned} e(\infty) &= 1 + \mathbf{C}(\mathbf{A} - \mathbf{BK})^{-1}\mathbf{B} \\ &= 1 + \begin{bmatrix} 1 & 0 \end{bmatrix}\begin{bmatrix} 0 & 1 \\ -183.1 & -16 \end{bmatrix}^{-1}\begin{bmatrix} 0 \\ 1 \end{bmatrix} \\ &= 0.995 \end{aligned} \tag{12.121}$$

**b.** We now use Eqs. (12.115) to represent the integral-controlled plant as follows:

$$\begin{bmatrix} \dot{x}_1 \\ \dot{x}_2 \\ \dot{x}_N \end{bmatrix} = \begin{bmatrix} \left(\begin{bmatrix} 0 & 1 \\ -3 & -5 \end{bmatrix} - \begin{bmatrix} 0 \\ 1 \end{bmatrix}\begin{bmatrix} k_1 & k_2 \end{bmatrix}\right) & \begin{bmatrix} 0 \\ 1 \end{bmatrix} K_e \\ -\begin{bmatrix} 1 & 0 \end{bmatrix} & 0 \end{bmatrix}\begin{bmatrix} x_1 \\ x_2 \\ x_N \end{bmatrix} + \begin{bmatrix} 0 \\ 0 \\ 1 \end{bmatrix} r$$

$$= \begin{bmatrix} 0 & 1 & 0 \\ -(3 + k_1) & -(5 + k_2) & K_e \\ -1 & 0 & 0 \end{bmatrix}\begin{bmatrix} x_1 \\ x_2 \\ x_N \end{bmatrix} + \begin{bmatrix} 0 \\ 0 \\ 1 \end{bmatrix} r \tag{12.122a}$$

$$y = \begin{bmatrix} 1 & 0 & 0 \end{bmatrix}\begin{bmatrix} x_1 \\ x_2 \\ x_N \end{bmatrix} \tag{12.122b}$$

Using Eq. (3.73) and the plant of Eqs. (12.116), we find that the transfer function of the plant is $G(s) = 1/(s^2 + 5s + 3)$. The desired characteristic polynomial for the closed-loop integral-controlled system is shown in Eq. (12.117). Since the plant has no zeros, we assume no zeros for the closed-loop system and augment Eq. (12.117) with a third pole, $(s + 100)$, which has a real part greater than five times that of the desired dominant second-order poles. The desired third-order closed-loop system characteristic polynomial is

$$(s + 100)(s^2 + 16s + 183.1) = s^3 + 116s^2 + 1783.1s + 18{,}310 \tag{12.123}$$

The characteristic polynomial for the system of Eqs. (12.112) is

$$s^3 + (5 + k_2)s^2 + (3 + k_1)s + K_e \tag{12.124}$$

Matching coefficients from Eqs. (12.123) and (12.124), we obtain

$$k_1 = 1780.1 \tag{12.125a}$$

$$k_2 = 111 \tag{12.125b}$$

$$k_e = 18{,}310 \tag{12.125c}$$

Substituting these values into Eqs. (12.122) yields this closed-loop integral-controlled system:

$$\begin{bmatrix} \dot{x}_1 \\ \dot{x}_2 \\ \dot{x}_N \end{bmatrix} = \begin{bmatrix} 0 & 1 & 0 \\ -1783.1 & -116 & 18{,}310 \\ -1 & 0 & 0 \end{bmatrix} \begin{bmatrix} x_1 \\ x_2 \\ x_N \end{bmatrix} + \begin{bmatrix} 0 \\ 0 \\ 1 \end{bmatrix} r \qquad (12.126a)$$

$$y = [1 \quad 0 \quad 0] \begin{bmatrix} x_1 \\ x_2 \\ x_N \end{bmatrix} \qquad (12.126b)$$

In order to check our assumption for the zero, we now apply Eq. (3.73) to Eqs. (12.126) and find the closed-loop transfer function to be

$$T(s) = \frac{18{,}310}{s^3 + 116s^2 + 1783.1s + 18{,}310} \qquad (12.127)$$

Since the transfer function matches our design, we have the desired transient response.

Now let us find the steady-state error for a unit step input. Applying Eq. (7.96) to Eqs. (12.126), we obtain

$$e(\infty) = 1 + [1 \quad 0 \quad 0] \begin{bmatrix} 0 & 1 & 0 \\ -1783.1 & -116 & 18{,}310 \\ -1 & 0 & 0 \end{bmatrix}^{-1} \begin{bmatrix} 0 \\ 0 \\ 1 \end{bmatrix} = 0 \qquad (12.128)$$

Thus, the system behaves like a Type 1 system.

## Skill-Assessment Exercise 12.7

**PROBLEM:** Design an integral controller for the plant

$$\dot{\mathbf{x}} = \begin{bmatrix} 0 & 1 \\ -7 & -9 \end{bmatrix} \mathbf{x} + \begin{bmatrix} 0 \\ 1 \end{bmatrix} u$$

$$y = [4 \quad 1]\mathbf{x}$$

to yield a step response with 10% overshoot, a peak time of 2 seconds, and zero steady-state error.

**ANSWER:** $\mathbf{K} = [2.21 \quad -2.7]$, $K_e = 3.79$

The complete solution is located at www.wiley.com/college/nise.

Now that we have designed controllers and observers for transient response and steady-state error, we summarize the chapter with a case study demonstrating the design process.

## Case Study Antenna Control: Design of Controller and Observer

Design D

In this case study, we use our ongoing antenna azimuth position control system to demonstrate the combined design of a controller and an observer. We will assume that the states are not available and must be estimated from the output. The block diagram of the original system is shown on the front endpapers, Configuration 1. Arbitrarily setting the preamplifier gain to 200 and removing the existing feedback, the forward transfer function is simplified to that shown in Figure 12.22.

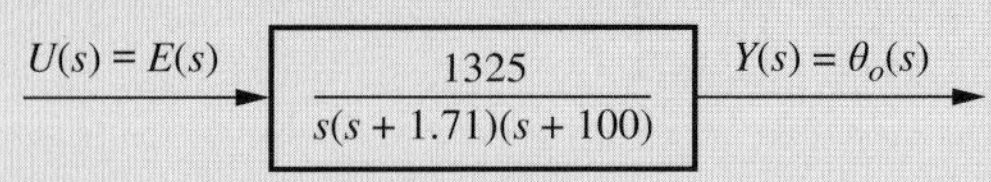


**FIGURE 12.22** Simplified block diagram of antenna control system shown on the front endpapers (Configuration 1) with $K = 200$

The case study will specify a transient response for the system and a faster transient response for the observer. The final design configuration will consist of the plant, the observer, and the controller, as shown conceptually in Figure 12.23. The design of the observer and the controller will be separate.

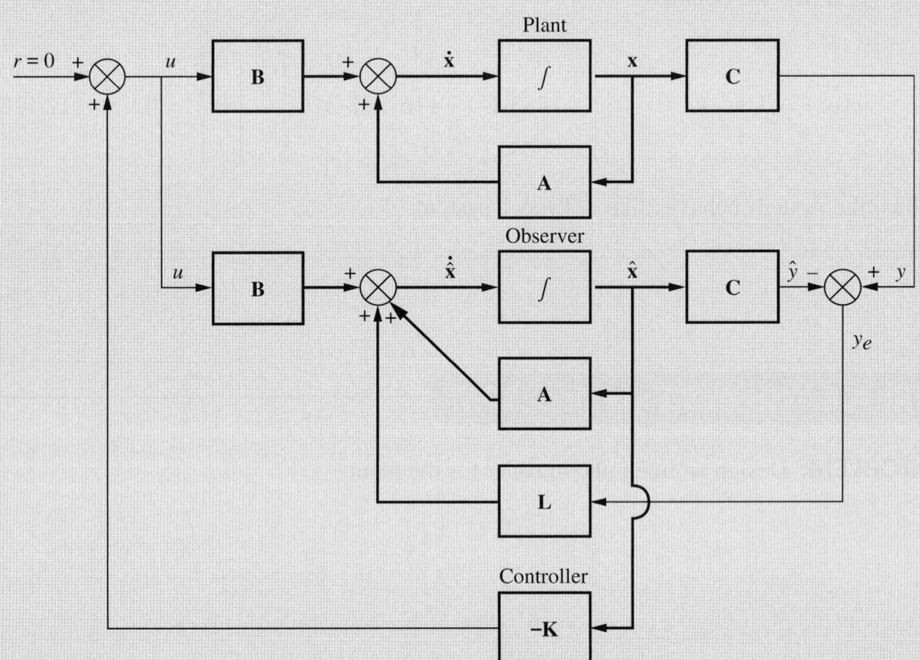


**FIGURE 12.23** Conceptual state-space design configuration, showing plant, observer, and controller

**PROBLEM:** Using the simplified block diagram of the plant for the antenna azimuth position control system shown in Figure 12.22, design a controller to yield a 10% overshoot and a settling time of 1 second. Place the third pole 10 times as far from the imaginary axis as the second-order dominant pair.

Assume that the state variables of the plant are not accessible and design an observer to estimate the states. The desired transient response for the observer is a 10% overshoot and a natural frequency 10 times as great as the system response above. As in the case of the controller, place the third pole 10 times as far from the imaginary axis as the observer's dominant second-order pair.

**SOLUTION: Controller Design:** We first design the controller by finding the desired characteristic equation. A 10% overshoot and a settling time of 1 second yield $\zeta = 0.591$ and $\omega_n = 6.77$. Thus, the characteristic equation for the dominant poles is $s^2 + 8s + 45.8 = 0$, where the dominant poles are located at $-4 \pm j5.46$. The third pole will be 10 times as far from the imaginary axis, or at $-40$. Hence, the desired characteristic equation for the closed-loop system is

$$(s^2 + 8s + 45.8)(s + 40) = s^3 + 48s^2 + 365.8s + 1832 = 0 \qquad (12.129)$$

Next we find the actual characteristic equation of the closed-loop system. The first step is to model the closed-loop system in state space and then find its characteristic equation. From Figure 12.22, the transfer function of the plant is

$$G(s) = \frac{1325}{s(s + 1.71)(s + 100)} = \frac{1325}{s(s^2 + 101.71s + 171)} \qquad (12.130)$$

Using phase variables, this transfer function is converted to the signal-flow graph shown in Figure 12.24, and the state equations are written as follows:

$$\dot{\mathbf{x}} = \begin{bmatrix} 0 & 1 & 0 \\ 0 & 0 & 1 \\ 0 & -171 & -101.71 \end{bmatrix} \mathbf{x} + \begin{bmatrix} 0 \\ 0 \\ 1 \end{bmatrix} u = \mathbf{Ax} + \mathbf{B}u \qquad (12.131a)$$

$$y = \begin{bmatrix} 1325 & 0 & 0 \end{bmatrix} \mathbf{x} = \mathbf{Cx} \qquad (12.131b)$$

We now pause in our design to evaluate the controllability of the system. The controllability matrix, $\mathbf{C_M}$, is

$$\mathbf{C_M} = \begin{bmatrix} \mathbf{B} & \mathbf{AB} & \mathbf{A}^2\mathbf{B} \end{bmatrix} \begin{bmatrix} 0 & 0 & 1 \\ 0 & 1 & -101.71 \\ 1 & -101.71 & 10{,}173.92 \end{bmatrix} \qquad (12.132)$$

The determinant of $\mathbf{C_M}$ is $-1$; thus, the system is controllable.

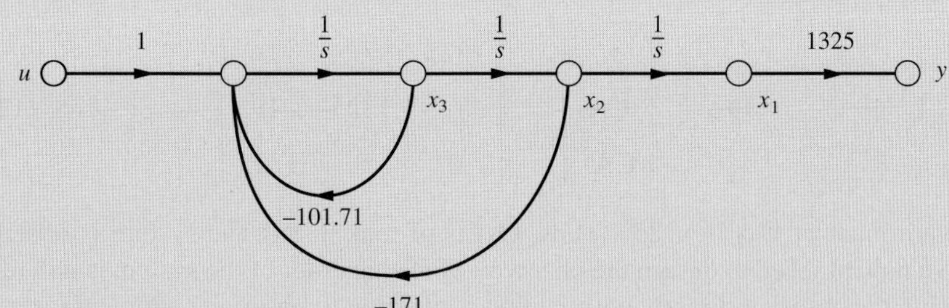


**FIGURE 12.24** Signal-flow graph for $G(s) = 1325/[s(s^2 + 101.71s + 171)]$

Continuing with the design of the controller, we show the controller's configuration with the feedback from all state variables in Figure 12.25. We now find the characteristic

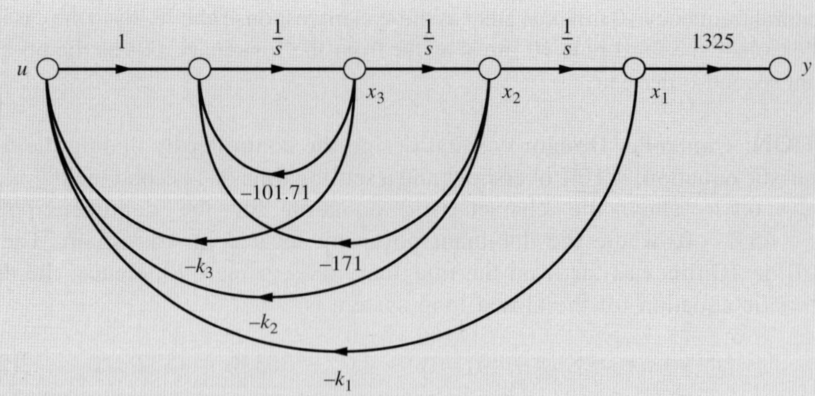


**FIGURE 12.25** Plant with state-variable feedback for controller design

equation of the system of Figure 12.25. From Eq. (12.7) and Eq. (12.131a), the system matrix, $\mathbf{A}-\mathbf{BK}$, is

$$\mathbf{A}-\mathbf{BK}=\begin{bmatrix}0 & 1 & 0\\ 0 & 0 & 1\\ -k_1 & -(171+k_2) & -(101.71+k_3)\end{bmatrix} \tag{12.133}$$

Thus, the closed-loop system's characteristic equation is

$$\det[s\mathbf{I}-(\mathbf{A}-\mathbf{BK})]=s^3+(101.71+k_3)s^2+(171+k_2)s+k_1=0 \tag{12.134}$$

Matching the coefficients of Eq. (12.129) with those of Eq. (12.134), we evaluate the $k_i$'s as follows:

$$k_1=1832 \tag{12.135a}$$

$$k_2=194.8 \tag{12.135b}$$

$$k_3=-53.71 \tag{12.135c}$$

**Observer Design:** Before designing the observer, we test the system for observability. Using the **A** and **C** matrices from Eqs. (12.131), the observability matrix, $\mathbf{O_M}$, is

$$\mathbf{O_M}=\begin{bmatrix}\mathbf{C}\\ \mathbf{CA}\\ \mathbf{CA}^2\end{bmatrix}=\begin{bmatrix}1325 & 0 & 0\\ 0 & 1325 & 0\\ 0 & 0 & 1325\end{bmatrix} \tag{12.136}$$

The determinant of $\mathbf{O_M}$ is $1325^3$. Thus, $\mathbf{O_M}$ is of rank 3, and the system is observable.

We now proceed to design the observer. Since the order of the system is not high, we will design the observer directly without first converting to observer canonical form. From Eq. (12.64a) we need first to find $\mathbf{A}-\mathbf{LC}$. **A** and **C** from Eqs. (12.131) along with

$$\mathbf{L}=\begin{bmatrix}l_1\\ l_2\\ l_3\end{bmatrix} \tag{12.137}$$

are used to evaluate $\mathbf{A}-\mathbf{LC}$ as follows:

$$\mathbf{A}-\mathbf{LC} = \begin{bmatrix} -1325l_1 & 1 & 0 \\ -1325l_2 & 0 & 1 \\ -1325l_3 & -171 & -101.71 \end{bmatrix} \quad (12.138)$$

The characteristic equation for the observer is now evaluated as

$$\begin{aligned}\det[\lambda\mathbf{I}-(\mathbf{A}-\mathbf{LC})] &= \lambda^3 + (1325l_1 + 101.71)\,\lambda^2 \\ &\quad + (134{,}800l_1 + 1325l_2 + 171)\,\lambda \\ &\quad + (226{,}600l_1 + 134{,}800l_2 + 1325l_3) \\ &= 0 \end{aligned} \quad (12.139)$$

From the problem statement, the poles of the observer are to be placed to yield a 10% overshoot and a natural frequency 10 times that of the system's dominant pair of poles. Thus, the observer's dominant poles yield $[s^2 + (2 \times 0.591 \times 67.7)s + 67.7^2] = (s^2 + 80s + 4583)$. The real part of the roots of this polynomial is $-40$. The third pole is then placed 10 times farther from the imaginary axis at $-400$. The composite characteristic equation for the observer is

$$(s^2 + 80s + 4583)(s + 400) = s^3 + 480s^2 + 36{,}580s + 1{,}833{,}000 = 0 \quad (12.140)$$

Matching coefficients from Eqs. (12.139) and (12.140), we solve for the observer gains:

$$l_1 = 0.286 \quad (12.141a)$$

$$l_2 = -1.57 \quad (12.141b)$$

$$l_3 = 1494 \quad (12.141c)$$

Figure 12.26, which follows the general configuration of Figure 12.23, shows the completed design, including the controller and the observer.

The results of the design are shown in Figure 12.27. Figure 12.27(*a*) shows the impulse response of the closed-loop system without any difference between the plant and its modeling as an observer. The undershoot and settling time approximately meet the requirements set forth in the problem statement of 10% and 1 second, respectively. In Figure 12.27(*b*), we see the response designed into the observer. An initial condition of 0.006 was given to $x_1$ in the plant to make the modeling of the plant and observer different. Notice that the observer's response follows the plant's response by the time 0.06 second is reached.

**CHALLENGE:** You are now given a case study to test your knowledge of this chapter's objectives: You are given the antenna azimuth position control system shown on the front endpapers, Configuration 3. If the preamplifier gain $K = 20$, do the following:

**a.** Design a controller to yield 15% overshoot and a settling time of 2 seconds. Place the third pole 10 times as far from the imaginary axis as the second-order dominant pole pair. Use physical variables as follows: power amplifier output, motor angular velocity, and motor displacement.

**b.** Redraw the schematic shown on the front endpapers, showing a tachometer that yields rate feedback along with any added gains or attenuators required to implement the state-variable feedback gains.

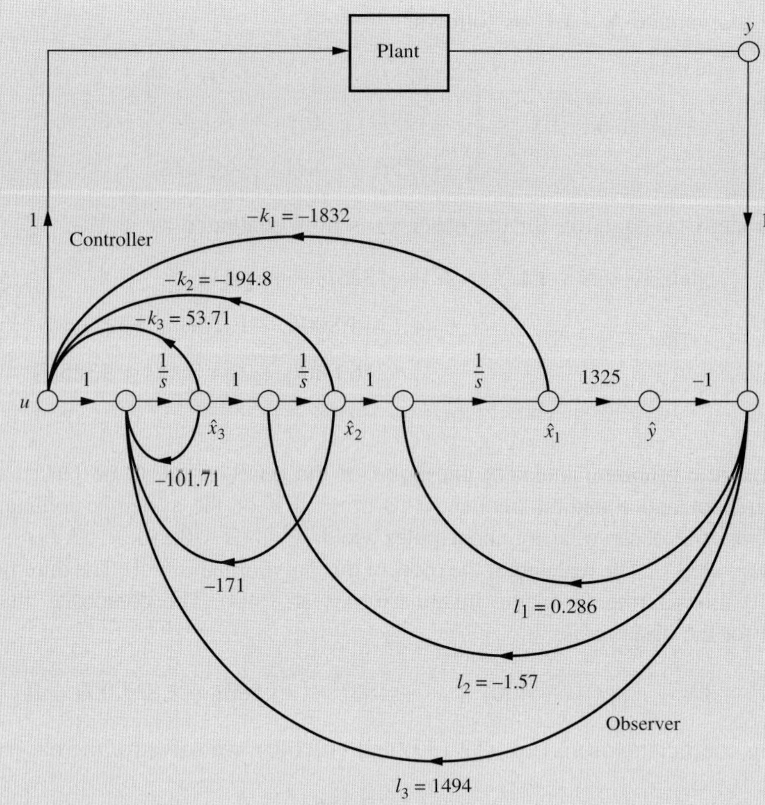


**FIGURE 12.26** Completed state-space design for the antenna azimuth position control system, showing controller and observer

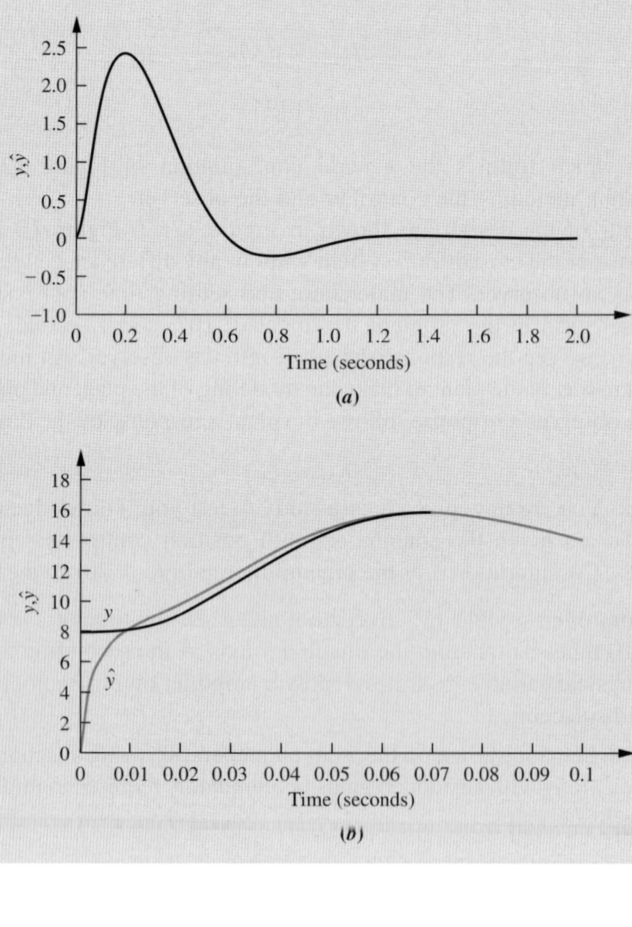


**FIGURE 12.27** Designed response of antenna azimuth position control system: **a.** impulse response—plant and observer with the same initial conditions, $x_1(0) = \hat{x}_1(0) = 0$; **b.** portion of impulse response—plant and observer with different initial conditions, $\hat{x}_1(0) = 0.006$ for the plant, $\hat{x}_1(0) = 0$ for the observer

**c.** Assume that the tachometer is not available to provide rate feedback. Design an observer to estimate the physical variables' states. The observer will respond with 10% overshoot and a natural frequency 10 times as great as the system response. Place the observer's third pole 10 times as far from the imaginary axis as the observer's dominant second-order pole pair.

**d.** Redraw the schematic on the front endpapers, showing the implementation of the controller and the observer.

**e.** `Repeat Parts a and c using MATLAB.`

MATLAB ML

## Summary

This chapter has followed the path established by Chapters 9 and 11—control system design. Chapter 9 used root locus techniques to design a control system with a desired transient response. Sinusoidal frequency response techniques for design were covered in Chapter 11, and in this chapter we used state-space design techniques.

State-space design consists of specifying the system's desired pole locations and then designing a controller consisting of state-variable feedback gains to meet these requirements. If the state variables are not available, an observer is designed to emulate the plant and provide estimated state variables.

Controller design consists of feeding back the state variables to the input, $u$, of the system through specified gains. The values of these gains are found by matching the coefficients of the system's characteristic equation with the coefficients of the desired characteristic equation. In some cases, the control signal, $u$, cannot affect one or more state variables. We call such a system *uncontrollable*. For this system, a total design is not possible. Using the controllability matrix, a designer can tell whether or not a system is controllable prior to the design.

Observer design consists of feeding back the error between the actual output and the estimated output. This error is fed back through specified gains to the derivatives of the estimated state variables. The values of these gains are also found by matching the coefficients of the observer's characteristic equation with the coefficients of the desired characteristic equation. The response of the observer is designed to be faster than that of the controller, so the estimated state variables effectively appear instantaneously at the controller. For some systems, the state variables cannot be deduced from the output of the system, as is required by the observer. We call such systems *unobservable*. Using the observability matrix, the designer can tell whether or not a system is observable. Observers can be designed only for observable systems.

Finally, we discussed ways of improving the steady-state error performance of systems represented in state space. The addition of an integration before the controlled plant yields improvement in the steady-state error. In this chapter, this additional integration was incorporated into the controller design.

Three advantages of state-space design are apparent. First, in contrast to the root locus method, all pole locations can be specified to ensure a negligible effect of the nondominant poles upon the transient response. With the root locus, we were forced to justify an assumption that the nondominant poles did not appreciably affect the transient response. We were not always able to do so. Second, with the use of an observer, we are no longer forced to acquire the actual system variables for feedback. The advantage here is that sometimes the variables cannot be physically accessed, or it may be too expensive to provide that access. Finally, the methods shown lend themselves to design automation using the digital computer.

A disadvantage of the design methods covered in this chapter is the designer's inability to design the location of open- or closed-loop zeros that may affect the transient response. In root locus or frequency response design, the zeros of the lag or lead compensator can be specified. Another disadvantage of state-space methods concerns the designer's ability to relate all pole locations to the desired response; this relationship is not always apparent. Also, once the design is completed, we may not be satisfied with the sensitivity to parameter changes.

Finally, as previously discussed, state-space techniques do not satisfy our intuition as much as root locus techniques, where the effect of parameter changes can be immediately seen as changes in closed-loop pole locations.

In the next chapter we return to the frequency domain and design digital systems using gain adjustment and cascade compensation.

## Review Questions

1. Briefly describe an advantage that state-space techniques have over root locus techniques in the placement of closed-loop poles for transient response design.
2. Briefly describe the design procedure for a controller.
3. Different signal-flow graphs can represent the same system. Which form facilitates the calculation of the variable gains during controller design?
4. In order to effect a complete controller design, a system must be controllable. Describe the physical meaning of controllability.
5. Under what conditions can inspection of the signal-flow graph of a system yield immediate determination of controllability?
6. In order to determine controllability mathematically, the controllability matrix is formed, and its rank evaluated. What is the final step in determining controllability if the controllability matrix is a square matrix?
7. What is an observer?
8. Under what conditions would you use an observer in your state-space design of a control system?
9. Briefly describe the configuration of an observer.
10. What plant representation lends itself to easier design of an observer?
11. Briefly describe the design technique for an observer, given the configuration you described in Question 9.
12. Compare the major difference in the transient response of an observer to that of a controller. Why does this difference exist?
13. From what equation do we find the characteristic equation of the controller-compensated system?
14. From what equation do we find the characteristic equation of the observer?
15. In order to effect a complete observer design, a system must be observable. Describe the physical meaning of observability.
16. Under what conditions can inspection of the signal-flow graph of a system yield immediate determination of observability?
17. In order to determine observability mathematically, the observability matrix is formed and its rank evaluated. What is the final step in determining observability if the observability matrix is a square matrix?

# Problems

1. Consider the following open-loop transfer functions, where $G(s) = Y(s)/U(s)$, $Y(s)$ is the Laplace transform of the output, and $U(s)$ is the Laplace transform of the input control signal:

   **i.** $G(s) = \dfrac{(s+1)}{s(s+2)}$

   **ii.** $G(s) = \dfrac{(s+2)}{(s+3)(s+4)}$

   **iii.** $G(s) = \dfrac{10(s+2)(s+3)}{s(s+4)(s+5)}$

   **iv.** $G(s) = \dfrac{10s}{(s+2)(s+4)(s+6)}$

   **v.** $G(s) = \dfrac{s^2+2s+10}{(s+7)(s^2+3s+100)}$

   For each of these transfer functions, do the following: [Section: 12.2]

   **a.** Draw the signal-flow graph in phase-variable form.

   **b.** Add state-variable feedback to the signal-flow graph.

   **c.** For each closed-loop signal-flow graph, write the state equations.

   **d.** Write, *by inspection*, the closed-loop transfer function, $T(s)$, for your closed-loop signal-flow graphs.

   **e.** Verify your answers for $T(s)$ by finding the closed-loop transfer functions from the state equations and Eq. (3.73).

2. The following open-loop transfer functions can be represented by signal-flow graphs in cascade form.

   **i.** $G(s) = \dfrac{20(s+5)}{(s+1)(s+4)(s+6)}$

   **ii.** $G(s) = \dfrac{5(s^2+3s+7)}{(s+2)(s^2+2s+10)}$

   For each, do the following: [Section: 12.4]

   **a.** Draw the signal-flow graph and show the state-variable feedback.

   **b.** Find the closed-loop transfer function with state-variable feedback.

3. The following open-loop transfer functions can be represented by signal-flow graphs in parallel form.

   **i.** $G(s) = \dfrac{50(s^2+7s+25)}{s(s+10)(s+20)}$

   **ii.** $G(s) = \dfrac{50(s+3)(s+4)}{(s+5)(s+6)(s+7)}$

   For each, do the following: [Section: 12.4]

   **a.** Draw the signal-flow graph and show the state-variable feedback.

   **b.** Find the closed-loop transfer function with state-variable feedback.

4. Given the following open-loop plant [Section: 12.2]

$$G(s) = \frac{20}{(s+1)(s+3)(s+7)}$$

   design a controller to yield a 15% overshoot and a settling time of 0.75 second. Place the third pole 10 times as far from the imaginary axis as the dominant pole pair. Use the phase variables for state-variable feedback.

5. Section 12.2 showed that controller design is easier to implement if the uncompensated system is represented in phase-variable form with its typical lower companion matrix. We alluded to the fact that the design can just as easily progress using the controller canonical form with its upper companion matrix. [Section: 12.2]

   **a.** Redo the general controller design covered in Section 12.2, assuming that the plant is represented in controller canonical form rather than phase-variable form.

   **b.** Apply your derivation to Example 12.1 if the uncompensated plant is represented in controller canonical form.

6. Given the following open-loop plant: [Section: 12.2]

$$G(s) = \frac{100(s+2)(s+20)}{(s+1)(s+3)(s+4)}$$

   design a controller to yield 15% overshoot with a peak time of 0.5 second. Use the controller canonical form for state-variable feedback.

7. Given the following open-loop plant: [Section: 12.2]

$$G(s) = \frac{20(s+2)}{s(s+4)(s+6)}$$

   design a controller to yield a 10% overshoot and a settling time of 2 seconds. Place the third pole 10 times as far from the imaginary axis as the dominant pole pair. Use the phase variables for state-variable feedback.

8. Repeat Problem 4 assuming that the plant is represented in the cascade form. Do not convert to phase-variable form. [Section: 12.4]

**9.** Repeat Problem 7 assuming that the plant is represented in the parallel form. Do not convert to phase-variable form. [Section: 12.4]

**10.** Given the plant shown in Figure P12.1, what relationship exists between $b_1$ and $b_2$ to make the system uncontrollable? [Section: 12.3]

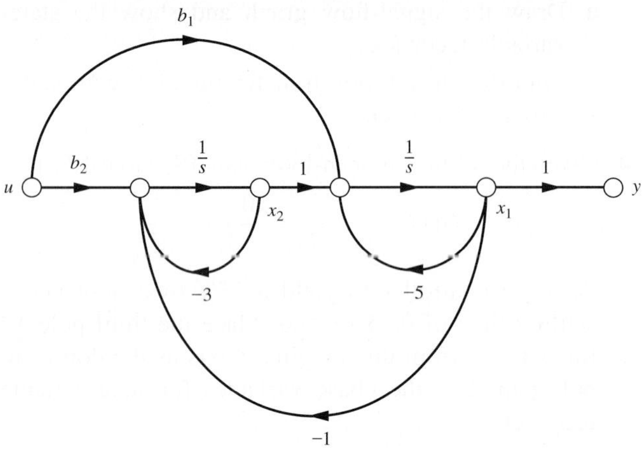


**FIGURE P12.1**

MATLAB
ML

**11.** `Use MATLAB to determine the controllability of the systems of Figure P12.2(d) and (f).`

**12.** For each of the plants represented by signal-flow graphs in Figure P12.2, determine the controllability. If the controllability can be determined by inspection, state that it can and then verify your conclusions using the controllability matrix. [Section: 12.3]

**13.** In Section 12.4, we discussed how to design a controller for systems not represented in phase-variable form with its typical lower companion matrix. We described how to convert the system to phase-variable form, design the controller, and convert back to the original representation. This technique can be applied just as easily if the original representation is converted to controller canonical form with its typical upper companion matrix. Redo Example 12.4 in the text by designing the controller after converting the uncompensated plant to controller canonical form. [Section: 12.4]

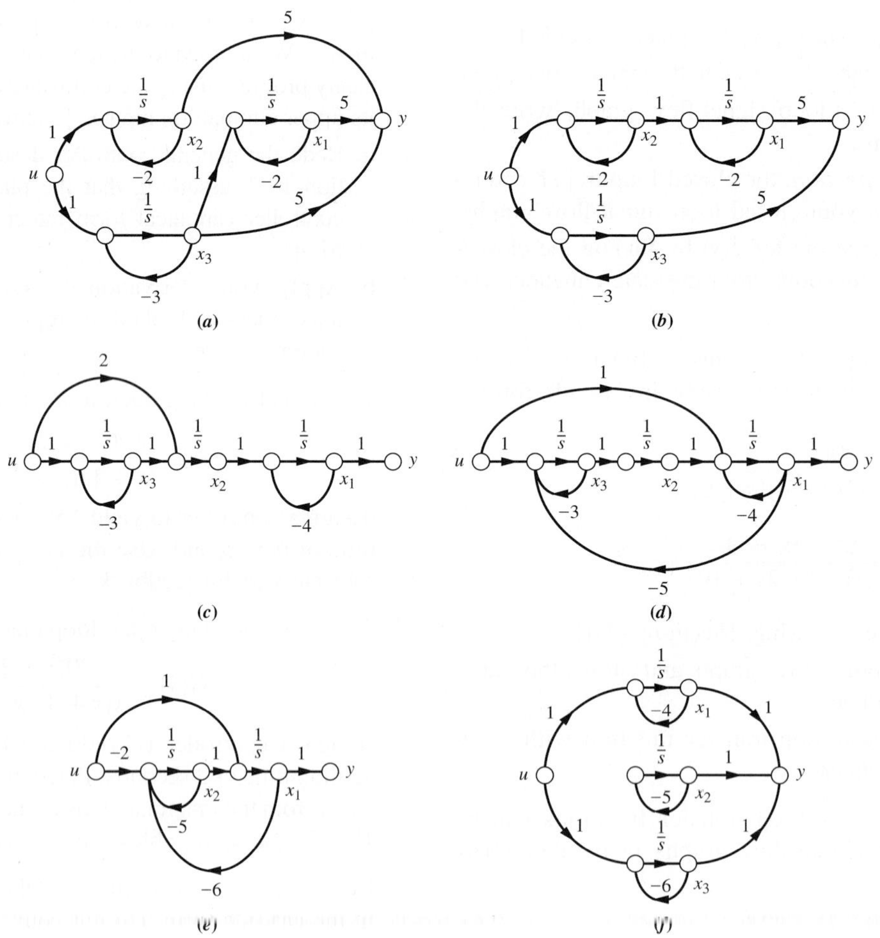


**FIGURE P12.2**

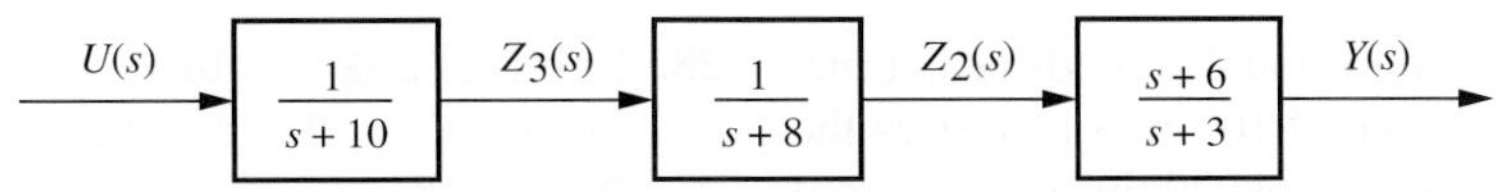


**FIGURE P12.3**

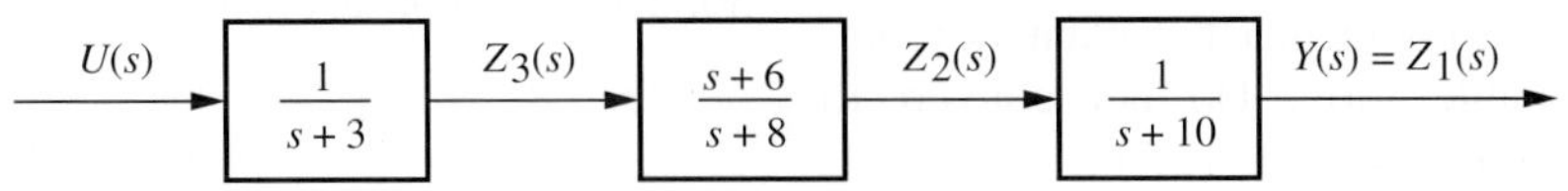


**FIGURE P12.4**

**14.** Consider the following transfer function:

$$G(s) = \frac{(s+6)}{(s+3)(s+8)(s+10)}$$

If the system is represented in cascade form, as shown in Figure P12.3, design a controller to yield a closed-loop response of 10% overshoot with a settling time of 1 second. Design the controller by first transforming the plant to phase variables. [Section: 12.4]

**15.** `Use MATLAB to design the controller gains for the system given in Problem 14.` MATLAB **ML**

**16.** Repeat Problem 14 assuming that the plant is represented in parallel form. [Section: 12.4]

**17.** The open-loop system of Problem 14 is represented as shown in Figure P12.4. If the output of each block is assigned to be a state variable, design the controller gains for feedback from these state variables. [Section: 12.4]

**18.** If an open-loop plant,

$$G(s) = \frac{100}{s(s+4)(s+8)}$$

is represented in parallel form, design a controller to yield a closed-loop response of 15% overshoot and a peak time of 0.2 second. Design the controller by first transforming the plant to controller canonical form. [Section: 12.4]

**19.** `For a specific individual, the linear time-invariant model of the hypothalamic-pituitary-adrenal axis of the endocrine system with five state variables has been found to be (Kyrylov, 2005)` MATLAB **ML**

$$\begin{bmatrix}\dot{x}_1\\ \dot{x}_2\\ \dot{x}_3\\ \dot{x}_4\\ \dot{x}_5\end{bmatrix} = \begin{bmatrix}-0.014 & 0 & -1.4 & 0 & 0\\ 0.023 & -0.023 & -0.023 & 0 & 0\\ 0.134 & 0.67 & -0.67 & 0.38 & 0.003264\\ 0 & 0 & 0.06 & -0.06 & 0\\ 0 & 0 & 0.0017 & 0 & -0.001\end{bmatrix} \times \begin{bmatrix}x_1\\ x_2\\ x_3\\ x_4\\ x_5\end{bmatrix} + \begin{bmatrix}1\\ 0\\ 0\\ 0\\ 0\end{bmatrix} d_0$$

`The state-variable definitions were given in Problem 24, Chapter 3.`

**a.** `Use MATLAB to determine if the system is controllable.`

**b.** `Use MATLAB to express the matrices A and B in phase-variable form.`

**20.** Consider the plant

$$G(s) = \frac{1}{s(s+3)(s+7)}$$

whose state variables are not available. Design an observer for the observer canonical variables to yield a transient response described by $\zeta = 0.4$ and $\omega_n = 75$. Place the third pole 10 times farther from the imaginary axis than the dominant poles. [Section: 12.5]

**21.** Design an observer for the plant

$$G(s) = \frac{10}{(s+2)(s+6)(s+12)}$$

operating with 10% overshoot and 2 seconds peak time. Design the observer to respond 10 times as fast as the plant. Place the observer third pole 20 times as far from the imaginary axis as the observer dominant poles. Assume the plant is represented in observer canonical form. [Section: 12.5]

**22.** Repeat Problem 20 assuming that the plant is represented in phase-variable form. Do not convert to observer canonical form. [Section: 12.7]

**23.** Consider the plant

$$G(s) = \frac{(s+1)}{(s+4)(s+6)}$$

whose phase variables are not available. Design an observer for the phase variables with a transient response described by $\zeta = 0.7$ and $\omega_n = 100$. Do not convert to observer canonical form. [Section: 12.7]

**24.** Determine whether or not each of the systems shown in Figure P12.2 is observable. [Section: 12.6]

**25.** `Use MATLAB to determine the observability of the systems of Figure P12.2(a) and (f).` MATLAB ML

**26.** Given the plant of Figure P12.5, what relationship must exist between $c_1$ and $c_2$ in order for the system to be unobservable? [Section: 12.6]

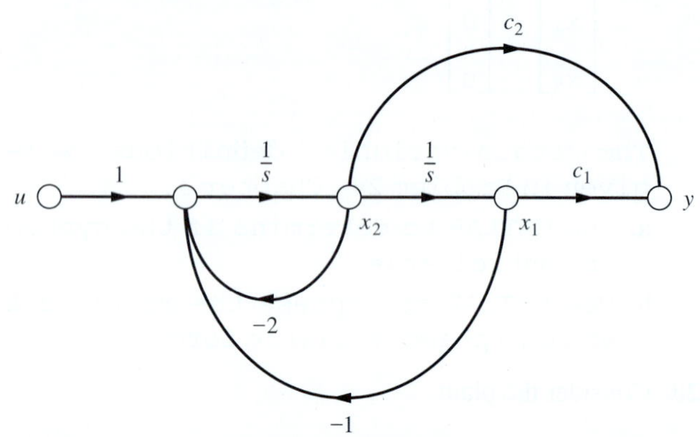


**FIGURE P12.5**

**27.** Design an observer for the plant

$$G(s) = \frac{1}{(s+5)(s+13)(s+20)}$$

represented in cascade form. Transform the plant to observer canonical form for the design. Then transform the design back to cascade form. The characteristic polynomial for the observer is to be $s^3 + 600s^2 + 40{,}000s + 1{,}500{,}000$.

**28.** `Use MATLAB to design the observer gains for the system given in Problem 27.` MATLAB ML

**29.** Repeat Problem 27 assuming that the plant is represented in parallel form. [Section: 12.7]

**30.** Design an observer for

$$G(s) = \frac{50}{(s+3)(s+6)(s+9)}$$

represented in phase-variable form with a desired performance of 10% overshoot and a settling time of 0.5 second. The observer will be 10 times as fast as the plant, and the observer's nondominant pole will be 10 times as far from the imaginary axis as the observer's dominant poles. Design the observer by first converting to observer canonical form. [Section: 12.7]

**31.** Observability and controllability properties depend on the state-space representation chosen for a given system. In general, observability and controllability are affected when pole-zero cancellations are present in the transfer function. Consider the following two systems with representations:

$$\dot{x}_i = \mathbf{A}_i\mathbf{x}_i = \mathbf{B}_i r$$
$$y = \mathbf{C}_i\mathbf{x}_i;$$

$$\mathbf{A}_1 = \begin{bmatrix} 0 & 1 \\ -2 & -3 \end{bmatrix}; \ \mathbf{B}_1 = \begin{bmatrix} 0 \\ 1 \end{bmatrix}; \ \mathbf{C}_1 = [2 \quad 0]$$

$$\mathbf{A}_2 = \begin{bmatrix} 0 & 1 & 0 \\ 0 & 0 & 1 \\ -6 & -11 & -6 \end{bmatrix}; \ \mathbf{B}_2 = \begin{bmatrix} 0 \\ 0 \\ 1 \end{bmatrix}; \ \mathbf{C}_2 = [6 \quad 2 \quad 0]$$

**a.** Show that both systems have the same transfer function $G_i(s) = \frac{Y(s)}{R(s)}$ after pole-zero cancellations.

**b.** Evaluate the observability of both systems.

**32.** Given the plant

$$\dot{\mathbf{x}} = \begin{bmatrix} -2 & 1 \\ 0 & -5 \end{bmatrix}\mathbf{x} + \begin{bmatrix} 0 \\ 1 \end{bmatrix}\mathbf{u}; \quad y = [1 \quad 1]\mathbf{x}$$

design an integral controller to yield a 15% overshoot, 0.6-second settling time, and zero steady-state error for a step input. [Section: 12.8]

**33.** Repeat Problem 32 for the following plant: [Section: 12.8]

$$\dot{\mathbf{x}} = \begin{bmatrix} -1 & 1 \\ 0 & 2 \end{bmatrix}\mathbf{x} + \begin{bmatrix} 0 \\ 1 \end{bmatrix}\mathbf{u}; \quad y = [1 \quad 1]\mathbf{x}$$

## DESIGN PROBLEMS

**34.** Figure P12.6 shows a continuous stirred tank reactor in which an aqueous solution of sodium acetate ($CH_3COONa$) is neutralized in the mixing tank with hydrochloric acid (HCl) to maintain a particular pH in the mixing tank.

The amount of acid in the mix is controlled by varying the rotational speed of a feeding peristaltic pump. A nominal linearized transfer function from HCl flowrate to pH has been shown to be (*Tadeo, 2000*)

$$G(s) = \frac{-0.9580 \times 10^{-4}s - 0.01197 \times 10^{-4}}{s^3 + 0.5250s^2 + 0.01265s + 0.000078}$$

**a.** Write the system in state-space phase-variable form.

**b.** Use state-feedback methods to design a matrix **K** that will yield an overdamped output pH response with a settling time of $T_s \approx 5$ min for a step input change in pH.

**c.** `Simulate the step response of the resulting closed-loop system using MATLAB.` MATLAB ML

**35.** In the past, Type-1 diabetes patients had to inject themselves with insulin three to four times a day. New delayed-action insulin analogues such as insulin Glargine require a single daily dose. A similar procedure is used to find a model for the concentration-time evolution of plasma for insulin Glargine (*Tarín, 2005*). The model to find the response for a specific patient to medication can be expressed in phase-variable form with

$$\mathbf{A} = \begin{bmatrix} 0 & 1 & 0 \\ 0 & 0 & 1 \\ -501.6 \times 10^{-6} & -128.8 \times 10^{-3} & -854 \times 10^{-3} \end{bmatrix};$$

$$\mathbf{B} = \begin{bmatrix} 1 \\ 0 \\ 0 \end{bmatrix}; \quad \mathbf{C} = \begin{bmatrix} 0.78 \times 10^{-4} & 41.4 \times 10^{-4} & 0.01 \end{bmatrix};$$

$$\mathbf{D} = 0$$

The state variables will take on a different significance in this expression, but the input and the output remain the same. Recall that $u$ = external insulin flow, and $y$ = plasma insulin concentration.

**a.** Obtain a state-feedback gain matrix so that the closed-loop system will have two of its poles placed at $-1/15$ and the third pole at $-1/2$.

**b.** `Use MATLAB to verify that the poles appear at the positions specified in Part a.` MATLAB ML

**36.** In the dc-dc converter of Problem 67, Chapter 4 (*Van Dijk, 1995*), with $L = 6$ mH, $C = 1$ mF, $R = 100\ \Omega$, a 50% PWM duty cycle, and assuming the system's

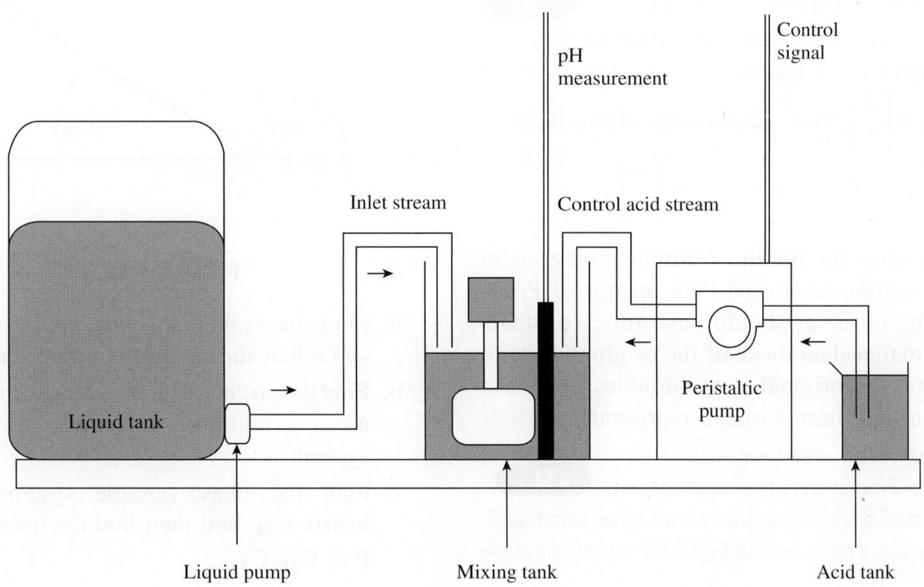


**FIGURE P12.6**[9]

[9] Tadeo F., Perez, Loepez O., and Alvarez T. Control of Neutralization Processes by Robust Loopsharing. *IEEE Trans. on Cont. Syst. Tech.*, vol. 8, no. 2, 2000. Fig. 2, p. 239. IEEE Transactions on Control Systems Technology by Institute of Electrical and Electronics Engineers; IEEE Control Systems Society. Reproduced with permission of Institute of Electrical and Electronics Engineers, in the format Republish in a book via Copyright Clearance Center.

output is the voltage across the capacitor, the model can be expressed as

$$\begin{bmatrix} \dot{i}_L \\ \dot{u}_C \end{bmatrix} = \begin{bmatrix} 0 & -83.33 \\ 500 & -10 \end{bmatrix} \begin{bmatrix} i_L \\ u_C \end{bmatrix} + \begin{bmatrix} 166.67 \\ 0 \end{bmatrix} E_s$$

$$y = [0 \quad 1] \begin{bmatrix} i_L \\ u_C \end{bmatrix}$$

**a.** Find the system's transfer function.

**b.** Express the system's state equations in phase-variable form.

**c.** Find a set of state-feedback gains to obtain 20% overshoot and a settling time of 0.5 second in the phase-variable system.

**d.** Obtain the corresponding set of state-feedback gains in the original system.

**e.** Verify that the set of gains in Part **d** places the closed-loop poles at the desired positions.

**f.** `Simulate the unit step response of the system using MATLAB.` MATLAB ML

**37. a.** Design an observer for the dc-dc converter of Problem 36. The observer should have time constants 10 times smaller than those of the original system.

**b.** `Simulate your system and observer for a unit step input using Simulink. Assume that the initial conditions for the original system are` $\mathbf{x}(0) = \begin{bmatrix} 2 \\ 1 \end{bmatrix}$. `The observer should have initial conditions` $\hat{\mathbf{x}}(0) = \begin{bmatrix} 0 \\ 0 \end{bmatrix}$. Simulink SL

**38. a.** Design an observer for the neutralization system using the continuous stirred tank reactor of Problem 34. The observer should have time constants 10 times smaller than those of the original system. Assume that the original state variables are those obtained in the phase-variable representation.

**b.** `Simulate your system and observer for a unit step input using Simulink. Assume that the initial conditions for the original system are` $\mathbf{x}(0) = \begin{bmatrix} -1 \\ -10 \\ 3 \end{bmatrix}$. `The observer should have initial conditions` $\hat{\mathbf{x}}(0) = \begin{bmatrix} 0 \\ 0 \\ 0 \end{bmatrix}$. Simulink SL

**39.** The study of the flexible links, such as the one shown in Figure P12.7, is important because of their application to the control of flexible lightweight robots (*Saini, 2012*). The flexible link angle is deflected by a servomotor. It is assumed that the base angle, $\theta(t)$, and the tip angular deflection relative to the undeformed link, $\alpha(t)$, can be measured. For a specific setup, a state-space model of the system was developed. The state vector is $\mathbf{x} = [\theta \quad \alpha \quad \omega \quad \dot{\alpha}]^T$, where $\omega(t) = \dot{\theta}(t)$ and input $u(t)$ is the voltage applied to the servomotor. Thus the system is represented as $\dot{\mathbf{x}} = \mathbf{A}\mathbf{x} + \mathbf{B}u$, $y = \mathbf{C}\mathbf{x}$ where

$$\mathbf{A} = \begin{bmatrix} 0 & 0 & 1 & 0 \\ 0 & 0 & 0 & 1 \\ 0 & 673.07 & -35.1667 & 0 \\ 0 & -1023.07 & 35.1667 & 0 \end{bmatrix}$$

$$\mathbf{B} = \begin{bmatrix} 0 \\ 0 \\ 61.7325 \\ -61.7325 \end{bmatrix}$$

$$\mathbf{C} = [1 \quad 1 \quad 0 \quad 0]$$

It is desired to build state-feedback compensation around this system so that the system's characteristic equation becomes $D(s) = (s + 10)^4$. In order to do this:

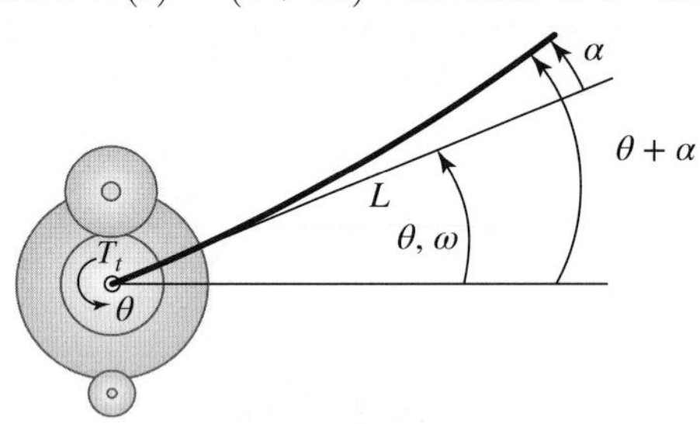


**FIGURE P12.7**[10]

**a.** Find the system's controllability matrix $\mathbf{C}_{\mathbf{M_O}}$ and show that the system is controllable.

**b.** Find the original system's characteristic equation and use it to find a phase-variable representation of the system.

**c.** Find the phase-variable system's controllability matrix $\mathbf{C}_{\mathbf{M_P}}$ and then find the transformation matrix $\mathbf{P} = \mathbf{C}_{\mathbf{M_O}} \mathbf{C}_{\mathbf{M_P}}^{-1}$.

[10] Saini, S. C., Sharma , Y., Bhandari, M., and Satija, U. Comparison of Pole Placement and LQR Applied to Single Link Flexible Manipulator, *International Conference on Communication Systems and Network Technologies*, IEEE Computer Society, 2012, pp. 843–847, Figure 3, p. 844.

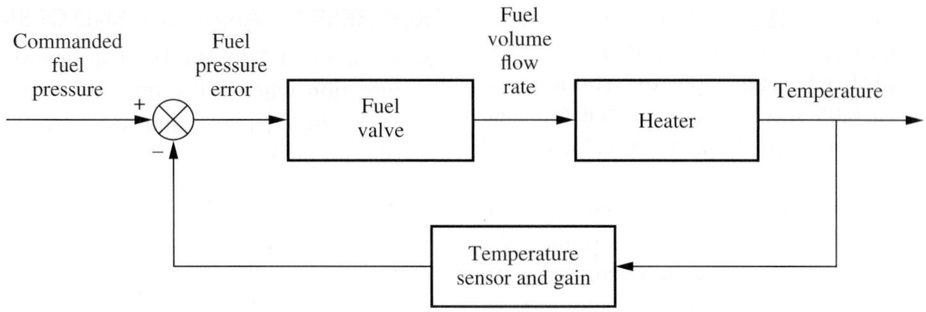


**FIGURE P12.8** Block diagram of a gas-fired heater

**d.** Use the phase-variable representation to find a feedback gain matrix $\mathbf{K_P}=[k_{1p} \quad k_{2p} \quad k_{3p} \quad k_{4p}]$ that will place the closed-loop poles in the desired positions.

**e.** Find the corresponding feedback gain matrix $\mathbf{K_O}=\mathbf{K_P}\,\mathbf{P}^{-1}$.

**40.** The conceptual block diagram of a gas-fired heater is shown in Figure P12.8. The commanded fuel pressure is proportional to the desired temperature. The difference between the commanded fuel pressure and a measured pressure related to the output temperature is used to actuate a valve and release fuel to the heater. The rate of fuel flow determines the temperature. When the output temperature equals the equivalent commanded temperature as determined by the commanded fuel pressure, the fuel flow is stopped and the heater shuts off (*Tyner, 1968*).

If the transfer function of the heater, $G_H(s)$, is

$$G_H(s)=\frac{1}{(s+0.4)(s+0.8)}\,\frac{\text{degrees F}}{\text{ft}^3/\text{min}}$$

and the transfer function of the fuel valve, $G_v(s)$, is

$$G_v(s)=\frac{5}{s+5}\,\frac{\text{ft}^3/\text{min}}{\text{psi}}$$

replace the temperature feedback path with a phase-variable controller that yields a 5% overshoot and a settling time of 10 minutes. Also, design an observer that will respond 10 times faster than the system but with the same percent overshoot.

**41.** The use of feedback control to vary the pitch angle in the blades of a variable speed wind turbine allows power generation optimization under variable wind conditions (*Liu, 2008*). At a specific operating point, it is possible to linearize turbine models. For example, the model of a three-blade turbine with a 15 m radius working in 12 m/s wind-speed and generating 220 V can be expressed as

$$\dot{\mathbf{x}}=\begin{bmatrix}-5 & 0 & 0 & 0 & 0\\ 0 & 0 & 1 & 0 & 0\\ -10.5229 & -1066.67 & -3.38028 & 23.5107 & 0\\ 0 & 993.804 & 3.125 & -23.5107 & 0\\ 0 & 0 & 0 & 10 & -10\end{bmatrix}\mathbf{x}+\begin{bmatrix}5\\0\\0\\0\\0\end{bmatrix}u$$

$$y=\begin{bmatrix}0 & 0 & 0 & 1.223x10^5 & 0\end{bmatrix}x$$

where the state variable vector is given by

$$\mathbf{x}=\begin{bmatrix}\beta & \xi & \dot{\xi} & \omega_g & \omega_{gm}\end{bmatrix}$$

Here, $\beta$ = pitch angle of the wind turbine blades, $\xi$ = relative angle of the secondary shaft, $\omega_g$ = generator speed, and $\omega_{gm}$ = generator measurement speed. The system input is $u$, the pitch angle reference, and the output is $y$, the active power generated.

**a.** Find a state feedback vector gain such that the system responds with a 10% overshoot and a settling time of 2 seconds for a step input.

**b.** Use MATLAB to verify the operation of the system under state feedback. MATLAB ML

**42.** The floppy disk drive of Problem 57 in Chapter 8 is to be redesigned using state-variable feedback. The controller is replaced by a unity dc gain amplifier,

$G_a(s) = 800/(s + 800)$. The plant, $G_p(s) = 20{,}000/[s(s + 100)]$, is in cascade with the amplifier.

**a.** Design a controller to yield 10% overshoot and a settling time of 0.05 second. Assume that the state variables are the output position, output velocity, and amplifier output.

**b.** Evaluate the steady-state error and redesign the system with an integral controller to reduce the steady-state error to zero. (Use of a program with symbolic capability is highly recommended.)

**c.** Simulate the step response for both the controller-compensated and integral controller-compensated systems. Use MATLAB or any other computer program. MATLAB **ML**

**43. a.** Redesign the dc-dc converter system of Problem 36 to include integral control.

**b.** Simulate your system for a step input using Simulink and verify that the specifications are met. In particular, verify that the system has zero steady-state error. Simulink **SL**

**44.** We want to use an observer in a textile machine to estimate the state variables. The 2-input, 1-output system's model is $\dot{\mathbf{x}} = \mathbf{Ax} + \mathbf{Bu}$; $y = \mathbf{Cx}$, where (*Cardona, 2010*)

$$\mathbf{A} = \begin{bmatrix} 0 & 1 & 0 \\ -52.6532 & -4.9353 & -2768.1557 \\ -0.001213 & 0 & -0.06106 \end{bmatrix}$$

$$\mathbf{B} = \begin{bmatrix} 0 & 0 \\ 0 & 0 \\ 0.001613 & -0.001812 \end{bmatrix}$$

$$\mathbf{C} = [1 \quad 0 \quad 0]$$

**a.** Find the system's observability matrix $\mathbf{O_{Mz}}$ and show that the system is observable.

**b.** Find the original system's characteristic equation and use it to find an observable canonical representation of the system.

**c.** Find the observable canonical system's observability matrix $\mathbf{O_{Mx}}$ and then find the transformation matrix $\mathbf{P} = \mathbf{O_{Mz}^{-1}O_{Mx}}$.

**d.** Use the observable canonical representation to find an observer gain matrix $\mathbf{L_x} = [l_{1x} \quad l_{2x} \quad l_{3x} \quad l_{4x}]^T$ so that the observer characteristic polynomial is $D(s) = s^3 + 30s^2 + 316s + 1160$.

**e.** Find the corresponding observer gain matrix $\mathbf{L_z} = \mathbf{PL_x}$.

## PROGRESSIVE ANALYSIS AND DESIGN PROBLEMS

**45. Control of HIV/AIDS.** The linearized model of HIV infection when RTIs are used for treatment was introduced in Chapter 4 and repeated here for convenience (*Craig, 2004*):

$$\begin{bmatrix} \dot{T} \\ \dot{T}^* \\ \dot{v} \end{bmatrix} = \begin{bmatrix} -0.04167 & 0 & -0.0058 \\ 0.0217 & -0.24 & 0.0058 \\ 0 & 100 & -2.4 \end{bmatrix} \begin{bmatrix} T \\ T^* \\ v \end{bmatrix} + \begin{bmatrix} 5.2 \\ -5.2 \\ 0 \end{bmatrix} u_1$$

$$y = [0 \quad 0 \quad 1] \begin{bmatrix} T \\ T^* \\ v \end{bmatrix}$$

$T$ represents the number of healthy T-cells, $T^*$ the number of infected cells, and $v$ the number of free viruses.

**a.** Design a state-feedback scheme to obtain

**(1)** zero steady-state error for step inputs

**(2)** 10% overshoot

**(3)** a settling time of approximately 100 days

(Hint: the system's transfer function has an open-loop zero at approximately $-0.02$. Use one of the poles in the desired closed-loop-pole polynomial to eliminate this zero. Place the higher-order pole 6.25 times farther than the dominant pair.)

**b.** Simulate the unit step response of your design using Simulink. Simulink **SL**

**46. Hybrid vehicle.** In Problem 31, Chapter 3, we introduced the idea that when an electric motor is the sole motive force provider for a hybrid electric vehicle (HEV), the forward paths of all HEV topologies are similar. It was noted that, in general, the forward path of an HEV cruise control system can be represented by a block diagram similar to that of Figure P3.19 (*Preitl, 2007*). The diagram is shown in Figure P12.9, with the parameters substituted by their numerical values from Problem 68, Chapter 6; the motor armature represented as a first-order system with a unity steady-state gain and a time constant of 50 ms; and the power amplifier gain set to 50. Whereas the state variables remain as the motor angular speed, $\omega(t)$, and armature current, $I_a(t)$, we assume now that we have only one input variable, $u_c(t)$, the command voltage from the electronic control unit, and one output variable, car speed, $v = r\omega/i_{tot} = 0.06154\omega$. The change in the load

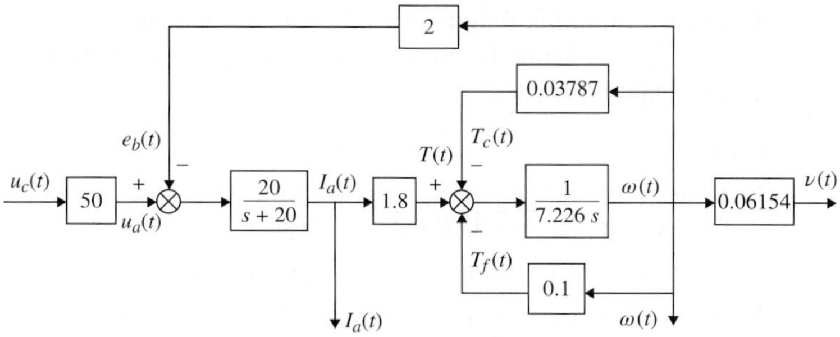


**FIGURE P12.9**

torque, $T_c(t)$, is represented as an internal feedback proportional to $\omega(t)$.

Looking at the diagram, the state equations may be written as

$$\begin{bmatrix} \dot{I}_a \\ \dot{\omega} \end{bmatrix} = \begin{bmatrix} -20 & -40 \\ 0.2491 & -0.0191 \end{bmatrix} \begin{bmatrix} I_a \\ \omega \end{bmatrix} + \begin{bmatrix} 0 \\ 1000 \end{bmatrix} u_c(t)$$

$$y(t) = v(t) = [0 \quad 0.05154] \begin{bmatrix} I_a \\ \omega \end{bmatrix}$$

**a.** Design an integral controller for $\%OS \leq 4.32\%$, a settling time, $T_s \leq 4.4$ sec, and a zero steady-state error for a step input (Hint: To account for the effect of the integral controller on the transient response, use $T_s = 4$ seconds in your calculation of the value of the natural frequency, $\omega_n$, of the required dominant poles).

**b.** `Use MATLAB to verify that the design requirements are met.` MATLAB **ML**

**47. Parabolic trough collector.** A parabolic trough collector can be designed using state-space techniques. For simplicity, pure time delay will be ignored here, although it could be handled in several different ways. Consider the open-loop transfer function (*Camacho, 2012*):

$$G(s) = \frac{137.2 \times 10^{-6} K}{s^2 + 0.0224s + 196 \times 10^{-6}}$$

Design a state feedback controller with integral control to yield zero steady-state error, such that the system transient response results in a damping factor of $\zeta = 0.5$ with a settling time $T_s = 200$ sec. Simulate the step response of your designed system using a computer program.

# Cyber Exploration Laboratory

## Experiment 12.1

**Objective** To simulate a system that has been designed for transient response via a state-space controller and observer.

**Minimum Required Software Packages** MATLAB, Simulink, and the Control System Toolbox

### Prelab

1. This experiment is based upon your design of a controller and observer as specified in the Case Study Challenge problem in Chapter 12. Once you have completed the controller and observer design in that problem, go on to Prelab 2.
2. What is the controller gain vector for your design of the system specified in the Case Study Challenge problem in Chapter 12?
3. What is the observer gain vector for your design of the system specified in the Case Study Challenge problem in Chapter 12?

4. Draw a Simulink diagram to simulate the system. Show the system, the controller, and the observer using the physical variables specified in the Case Study Challenge problem in Chapter 12.

### Lab

1. Using Simulink and your diagram from Prelab 4, produce the Simulink diagram from which you can simulate the response.
2. Produce response plots of the system and the observer for a step input.
3. Measure the percent overshoot and the settling time for both plots.

### Postlab

1. Make a table showing the design specifications and the simulation results for percent overshoot and settling time.
2. Compare the design specifications with the simulation results for both the system response and the observer response. Explain any discrepancies.
3. Describe any problems you had implementing your design.

## Experiment 12.2

**Objective** To use LabVIEW to design a controller and observer

**Minimum Required Software Packages** LabVIEW, the Control Design and Simulation Module, and the MathScript RT Module.

**Prelab** Create a LabVIEW VI that will design the controller and observer for the Antenna Control Case Study in this chapter. Your VI will have the following inputs: phase-variable form of the plant, the controller poles, and the observer poles to meet the requirements. Your indicators will display the following: the phase-variable equation of the plant, whether or not the system is controllable, the observer canonical equation of the observer, whether or not the system is observable, the gains for the controller, and the gains for the observer. Also provide the impulse response and initial response curves shown in Figure 12.27. In addition, provide similar response curves for the state variables.

**Lab** Run your VI and collect the data from which to compare the results of the case study with those found from your VI.

**Postlab** Compare and summarize the results found from your VI with those of the Chapter 12 Antenna Control Case Study.

# Bibliography

Camacho, E. F., Berenguel, M., Rubio, F. R., and Martinez, D. *Control of Solar Energy Systems*. Springer-Verlag, London, 2012.

Cardona, J. E., and Cárdenas, M. O. Design and implementation of a state observer for a textile machine, *IEEE ANDESCON*, 2010, pp. 1–6.

Craig, I. K, Xia, X., and Venter, J. W. Introducing HIV/AIDS Education into the Electrical Engineering Curriculum at the University of Pretoria. *IEEE Transactions on Education*, vol. 47, no. 1, February 2004, pp. 65–73.

D'Azzo, J. J., and Houpis, C. H. *Linear Control System Analysis and Design: Conventional and Modern*, 3d ed. McGraw-Hill, New York, 1988.

Franklin, G. F, Powell, J. D, and Emami-Naeini, A. *Feedback Control of Dynamic Systems*, 3d ed. Addison-Wesley, Reading, MA, 1994.

Hostetter, G. H, Savant, C. J. Jr., and Stefani, R. T. *Design of Feedback Control Systems*, 2d ed. Saunders College Publishing, New York, 1989.

Kailath, T. *Linear Systems*. Prentice Hall, Upper Saddle River, NJ, 1980.

Kyrylov, V., Severyanova, L. A., and Vieira, A. Modeling Robust Oscillatory Behavior of the Hypothalamic-Pituitary-Adrenal Axis. *IEEE Transactions on Biomedical Engineering*, vol. 52, no. 12, 2005, pp. 1977–1983.

Liu, J.-H., Xu, D.-P., and Yang, X.-Y. Multi-Objective Power Control of a Variable Speed Wind Turbine Based on Theory. *Proceedings of the Seventh International Conference on Machine Learning and Cybernetics*, July 2008, pp. 2036–2041.

Luenberger, D. G. Observing the State of a Linear System. *IEEE Transactions on Military Electronics*, vol. MIL-8, April 1964, pp. 74–80.

Milhorn, H. T. Jr., *The Application of Control Theory to Physiological Systems*. W. B. Saunders, Philadelphia, 1966.

Ogata, K. *Modern Control Engineering*, 2d ed. Prentice Hall, Upper Saddle River, NJ, 1990.

Ogata, K. *State Space Analysis of Control Systems*. Prentice Hall, Upper Saddle River, NJ, 1967.

Prasad, L., Tyagi, B., and Gupta, H. Modeling & Simulation for Optimal Control of Nonlinear Inverted Pendulum Dynamical System using PID Controller & LQR. *IEEE Computer Society Sixth Asia Modeling Symposium*, 2012, pp.138–143.

Preitl, Z., Bauer, P., and Bokor, J. A Simple Control Solution for Traction Motor Used in Hybrid Vehicles. *Fourth International Symposium on Applied Computational Intelligence and Informatics*. IEEE, 2007, pp. 157–162.

Rockwell International. *Space Shuttle Transportation System*. 1984 (press information).

Saini, S. C., Sharma, Y., Bhandari, M., and Satija, U. Comparison of Pole Placement and LQR Applied to Single Link Flexible Manipulator. *International Conference on Communication Systems and Network Technologies*, IEEE Computer Society, 2012, pp. 843–847.

Shinners, S. M. *Modern Control System Theory and Design*. Wiley, New York, 1992.

Sinha, N. K. *Control Systems*. Holt, Rinehart & Winston, New York, 1986.

Tadeo, F., Pérez López, O., and Alvarez, T. Control of Neutralization Processes by Robust Loop-shaping. *IEEE Transactions on Control Systems Technology*, vol. 8, no. 2, 2000, pp. 236–246.

Tarín, C., Teufel, E., Picó, J., Bondia, J., and Pfleiderer, H. J. Comprehensive Pharmacokinetic Model of Insulin Glargine and Other Insulin Formulations. *IEEE Transactions on Biomedical Engineering*, vol. 52, no. 12, 2005, pp. 1994–2005.

Thomsen, S., Hoffmann, N., and Fuchs, F. W. PI Control, PI-Based State Space Control, and Model-Based Predictive Control for Drive Systems With Elastically Coupled Loads—A Comparative Study. *IEEE Transactions on Industrial Electronics*, vol. 58, no. 8, August 2011, pp. 3647–3657.

Timothy, L. K, and Bona, B. E. *State Space Analysis: An Introduction*. McGraw-Hill, New York, 1968.

Van Dijk, E., Spruijt, J. N., O'Sullivan, D. M., and Klaasens, J. B. PWM-Switch Modeling of DC-DC Converters. *IEEE Transactions on Power Electronics*, vol. 10, 1995, pp. 659–665.

# 13

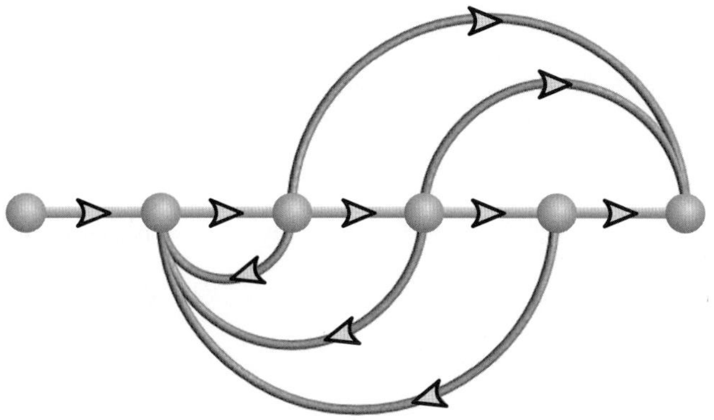

# Digital Control Systems

## Chapter Learning Outcomes

After completing this chapter the student will be able to:

- Model the digital computer in a feedback system (Sections 13.1–13.2)
- Find *z*- and inverse *z*-transforms of time and Laplace functions (Section 13.3)
- Find sampled-data transfer functions (Section 13.4)
- Reduce an interconnection of sampled-data transfer functions to a single sampled-data transfer function (Section 13.5)
- Determine whether a sampled-data system is stable and determine sampling rates for stability (Section 13.6)
- Design digital systems to meet steady-state error specification (Section 13.7)
- Design digital systems to meet transient response specifications using gain adjustment (Sections 13.8–13.9)
- Design cascade compensation for digital systems (Sections 13.10–13.11)

## ▶ Case Study Learning Outcomes

You will be able to demonstrate your knowledge of the chapter objectives with a case study as follows:

- Given the analog antenna azimuth position control system shown on the front endpapers and in Figure 13.1(*a*), you will be able to convert the system to a digital system as shown in Figure 13.1(*b*) and then design the gain to meet a transient response specification.
- Given the digital antenna azimuth position control system shown in Figure 13.1(*b*), you will be able to design a digital cascade compensator to improve the transient response.

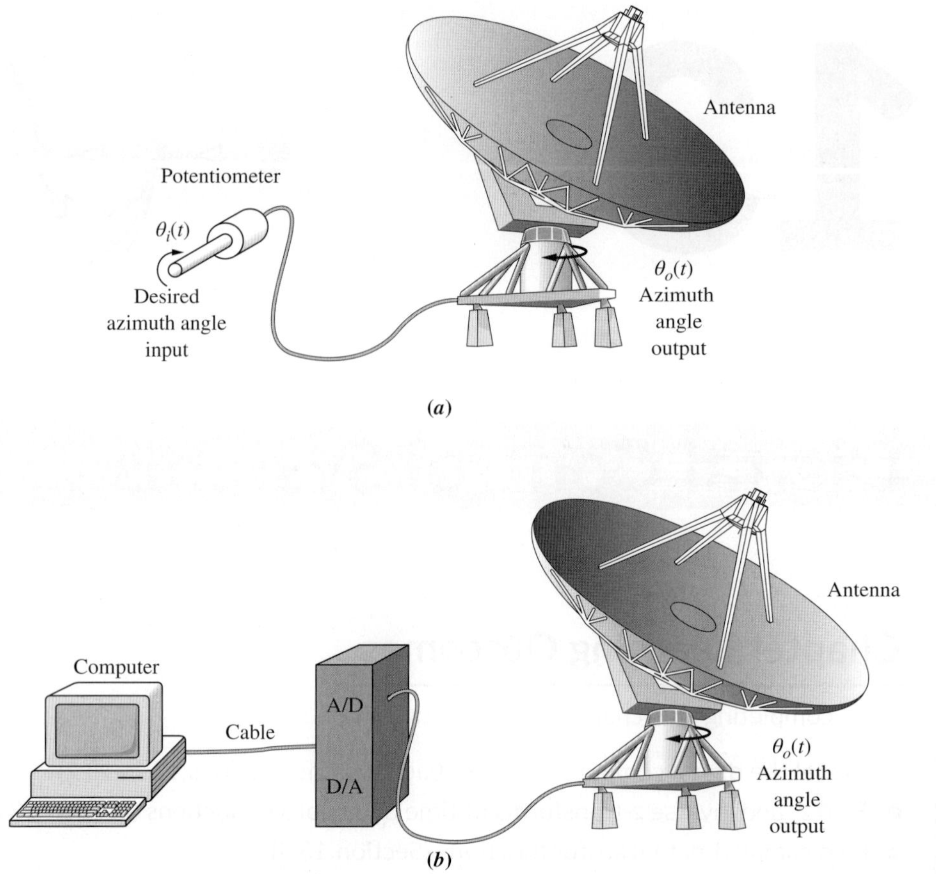


**FIGURE 13.1** Conversion of antenna azimuth position control system from **a.** analog control to **b.** digital control

## 13.1 Introduction

This chapter is an introduction to digital control systems and will cover only frequency-domain analysis and design. You are encouraged to pursue the study of state-space techniques in an advanced course in sampled-data control systems. In this chapter, we introduce analysis and design of stability, steady-state error, and transient response for computer-controlled systems.

With the development of the minicomputer in the mid-1960s and the microcomputer in the mid-1970s, physical systems need no longer be controlled by expensive mainframe computers. For example, milling operations that required mainframe computers in the past can now be controlled by a personal computer.

The digital computer can perform two functions: (1) supervisory—external to the feedback loop; and (2) control—internal to the feedback loop. Examples of supervisory functions consist of scheduling tasks, monitoring parameters and variables for out-of-range values, or initiating safety shutdown. Control functions are of primary interest to us, since a computer that performs within the feedback loop replaces the methods of compensation heretofore discussed. Examples of control functions are lead and lag compensation.

Transfer functions, representing compensators built with analog components, are now replaced with a digital computer that performs calculations that emulate the physical compensator. What advantages are there to replacing analog components with a digital computer?

## Advantages of Digital Computers

The use of digital computers in the loop yields the following advantages over analog systems: (1) reduced cost, (2) flexibility in response to design changes, and (3) noise immunity. Modern control systems require control of numerous loops at the same time—pressure, position, velocity, and tension, for example. In the steel industry, a single digital computer can replace numerous analog controllers with a subsequent reduction in cost. Where analog controllers implied numerous adjustments and resulting hardware, digital systems are now installed. Banks of equipment, meters, and knobs are replaced with computer terminals, where information about settings and performance is obtained through menus and screen displays. Digital computers in the loop can yield a degree of flexibility in response to changes in design. Any changes or modifications that are required in the future can be implemented with simple software changes rather than expensive hardware modifications. Finally, digital systems exhibit more noise immunity than analog systems by virtue of the methods of implementation.

Where then is the computer placed in the loop? Remember that the digital computer is controlling numerous loops; thus, its position in the loop depends upon the function it performs. Typically, the computer replaces the cascade compensator and is thus positioned at the place shown in Figure 13.2(*a*).

The signals $r$, $e$, $f$, and $c$ shown in Figure 13.2(*a*) can take on two forms: digital or analog. Up to this point we have used analog signals exclusively. Digital signals, which consist of a sequence of binary numbers, can be found in loops containing digital computers.

Loops containing both analog and digital signals must provide a means for conversion from one form to the other as required by each subsystem. A device that converts analog signals to digital signals is called an *analog-to-digital* (*A/D*) *converter*. Conversely, a device that converts digital signals to analog signals is called a *digital-to-analog* (*D/A*) *converter*. For example, in Figure 13.2(*b*), if the plant output, $c$, and the system input, $r$, are analog signals, then an analog-to-digital converter must be provided at the input to the digital computer.

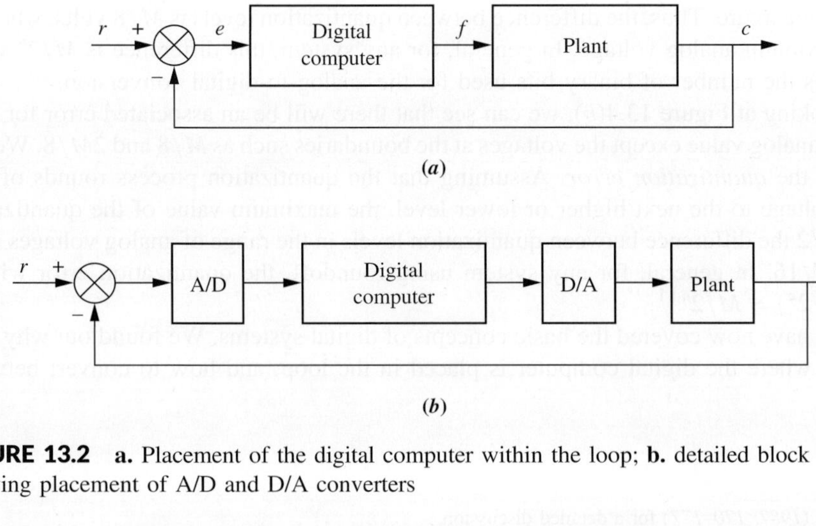


**FIGURE 13.2** **a.** Placement of the digital computer within the loop; **b.** detailed block diagram showing placement of A/D and D/A converters

Also, if the plant input, *f*, is an analog signal, then a digital-to-analog converter must be provided at the output of the digital computer.

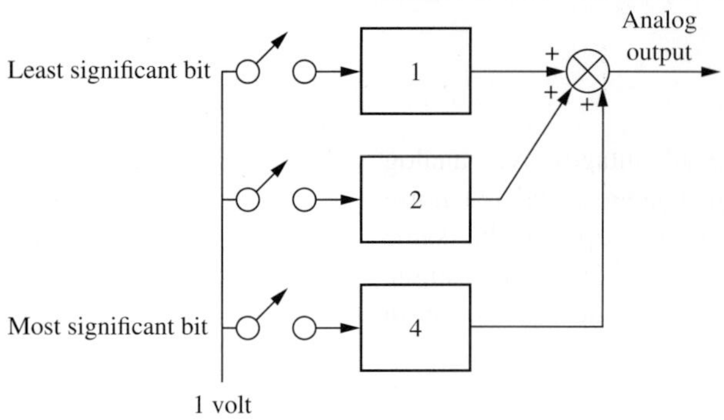


**FIGURE 13.3** Digital-to-analog converter

## Digital-to-Analog Conversion

Digital-to-analog conversion is simple and effectively instantaneous. Properly weighted voltages are summed together to yield the analog output. For example, in Figure 13.3, three weighted voltages are summed. The three-bit binary code is represented by the switches. Thus, if the binary number is $110_2$, the center and bottom switches are on, and the analog output is 6 volts. In actual use, the switches are electronic and are set by the input binary code.

## Analog-to-Digital Conversion

Analog-to-digital conversion, on the other hand, is a two-step process and is not instantaneous. There is a delay between the input analog voltage and the output digital word. In an analog-to-digital converter, the analog signal is first converted to a sampled signal and then converted to a sequence of binary numbers, the digital signal.

The sampling rate must be at least twice the bandwidth of the signal, or else there will be distortion. This minimum sampling frequency is called the *Nyquist sampling rate*.[1]

In Figure 13.4(*a*), we start with the analog signal. In Figure 13.4(*b*), we see the analog signal sampled at periodic intervals and held over the sampling interval by a device called a *zero-order sample-and-hold* (*z.o.h.*) that yields a staircase approximation to the analog signal. Higher-order holds, such as a first-order hold, generate more complex and more accurate waveshapes between samples. For example, a first-order hold generates a ramp between the samples. Samples are held before being digitized because the analog-to-digital converter converts the voltage to a digital number via a digital counter, which takes time to reach the correct digital number. Hence, the constant analog voltage must be present during the conversion process.

After sampling and holding, the analog-to-digital converter converts the sample to a digital number (as shown in Figure 13.4(*c*)), which is arrived at in the following manner. The dynamic range of the analog signal's voltage is divided into discrete levels, and each level is assigned a digital number. For example, in Figure 13.4(*b*), the analog signal is divided into eight levels. A three-bit digital number can represent each of the eight levels as shown in the figure. Thus, the difference between quantization levels is $M/8$ volts, where $M$ is the maximum analog voltage. In general, for any system, this difference is $M/2^n$ volts, where $n$ is the number of binary bits used for the analog-to-digital conversion.

Looking at Figure 13.4(*b*), we can see that there will be an associated error for each digitized analog value except the voltages at the boundaries such as $M/8$ and $2M/8$. We call this error the *quantization error*. Assuming that the quantization process rounds off the analog voltage to the next higher or lower level, the maximum value of the quantization error is $1/2$ the difference between quantization levels in the range of analog voltages from 0 to $15M/16$. In general, for any system using roundoff, the quantization error will be $(1/2)(M/2^n) = M/2^{n+1}$.

We have now covered the basic concepts of digital systems. We found out why they are used, where the digital computer is placed in the loop, and how to convert between

[1] See *Ogata* (*1987: 170–177*) for a detailed discussion.

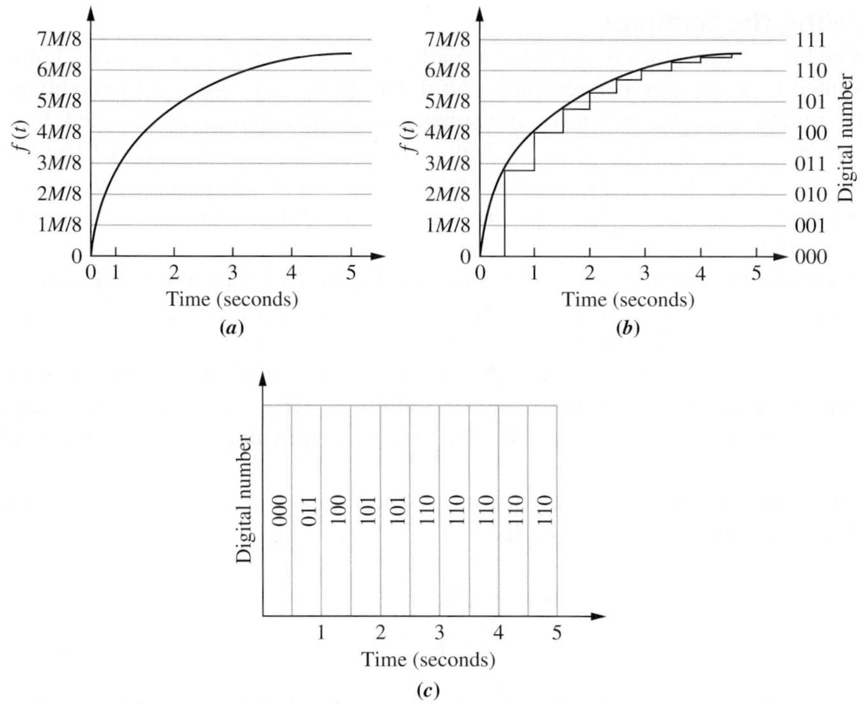


**FIGURE 13.4** Steps in analog-to-digital conversion: **a.** analog signal; **b.** analog signal after sample-and-hold; **c.** conversion of samples to digital numbers

analog and digital signals. Since the computer can replace the compensator, we have to realize that the computer is working with a quantized amplitude representation of the analog signal, formed from values of the analog signal, at discrete intervals of time. Ignoring the quantization error, we see that the computer performs just as the compensator does, except that signals pass through the computer only at the sampled intervals of time. We will find that the sampling of data has an unusual effect upon the performance of a closed-loop feedback system, since stability and transient response are now dependent upon the sampling rate. If the sampling rate is too slow, the system can be unstable, since the values are not being updated rapidly enough. If we are to analyze and design feedback control systems with digital computers in the loop, we must be able to model the digital computer and associated digital-to-analog and analog-to-digital converters. The modeling of the digital computer along with associated converters is covered in the next section.

## 13.2 Modeling the Digital Computer

If we think about it, the form of the signals in a loop is not as important as what happens to them. For example, if analog-to-digital conversion could happen instantaneously, and time samples occurred at intervals of time that approached zero, there would be no need to differentiate between the digital signals and the analog signals. Thus, previous analysis and design techniques would be valid regardless of the presence of the digital computer.

The fact that signals are sampled at specified intervals and held causes the system performance to change with changes in sampling rate. Basically, then, the computer's effect upon the signal comes from this sampling and holding. Thus, in order to model digital control systems, we must come up with a mathematical representation of this sample-and-hold process.

## Modeling the Sampler

Our objective at this point is to derive a mathematical model for the digital computer as represented by a sampler and zero-order hold. Our goal is to represent the computer as a transfer function similar to that for any subsystem. When signals are sampled, however, the Laplace transform that we have dealt with becomes a bit unwieldy. The Laplace transform can be replaced by another related transform called the *z-transform*. The z-transform will arise naturally from our development of the mathematical representation of the computer.

Consider the models for sampling shown in Figure 13.5. The model in Figure 13.5(*a*) is a switch turning on and off at a uniform sampling rate. In Figure 13.5(*b*), sampling can also be considered to be the product of the time waveform to be sampled, $f(t)$, and a sampling function, $s(t)$. If $s(t)$ is a sequence of pulses of width $T_W$, constant amplitude, and uniform rate as shown, the sampled output, $f^*_{T_W}(t)$, will consist of a sequence of sections of $f(t)$ at regular intervals. This view is equivalent to the switch model of Figure 13.5(*a*).

We can now write the time equation of the sampled waveform, $f^*_{T_W}(t)$. Using the model shown in Figure 13.5(*b*), we have

$$f^*_{T_W}(t) = f(t)s(t) = f(t)\sum_{k=-\infty}^{\infty} u(t-kT) - u(t-kT-Tw) \tag{13.1}$$

where $k$ is an integer between $-\infty$ and $+\infty$, $T$ is the period of the pulse train, and $T_W$ is the pulse width.

Since Eq. (13.1) is the product of two time functions, taking the Laplace transform in order to find a transfer function is not simple. A simplification can be made if we assume that the pulse width, $T_W$, is small in comparison to the period, $T$, such that $f(t)$ can

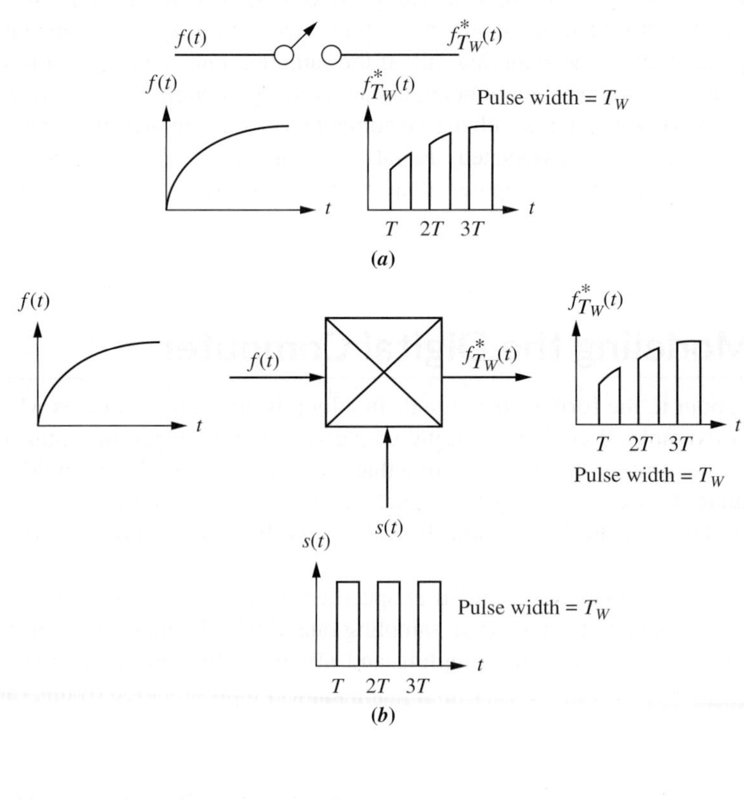


**FIGURE 13.5** Two views of uniform-rate sampling: **a.** switch opening and closing; **b.** product of time waveform and sampling waveform

be considered constant during the sampling interval. Over the sampling interval, then, $f(t) = f(kT)$. Hence,

$$f^*_{T_W}(t) = \sum_{k=-\infty}^{\infty} f(kT)[u(t - kT) - u(t - kT - T_W)] \tag{13.2}$$

for small $T_W$.

Equation (13.2) can be further simplified through insight provided by the Laplace transform. Taking the Laplace transform of Eq. (13.2), we have

$$F^*_{T_W}(s) = \sum_{k=-\infty}^{\infty} f(kT)\left[\frac{e^{-kTs}}{s} - \frac{e^{-kTs-T_Ws}}{s}\right] = \sum_{k=-\infty}^{\infty} f(kT)\left[\frac{1 - e^{-T_Ws}}{s}\right]e^{-kTs} \tag{13.3}$$

Replacing $e^{-T_Ws}$ with its series expansion, we obtain

$$F^*_{T_W}(s) = \sum_{k=-\infty}^{\infty} f(kT)\left[\frac{1 - \left\{1 - T_Ws + \frac{(T_Ws)^2}{2!} - \cdots\right\}}{s}\right]e^{-kTs} \tag{13.4}$$

For small $T_W$, Eq. (13.4) becomes

$$F^*_{T_W}(s) = \sum_{k=-\infty}^{\infty} f(kT)\left[\frac{T_Ws}{s}\right]e^{-kTs} = \sum_{k=-\infty}^{\infty} f(kT)T_We^{-kTs} \tag{13.5}$$

Finally, converting back to the time domain, we have

$$f^*_{T_W}(t) = T_W \sum_{k=-\infty}^{\infty} f(kT)\delta(t - kT) \tag{13.6}$$

where $\delta(t - kT)$ are Dirac delta functions.

Thus, the result of sampling with rectangular pulses can be thought of as a series of delta functions whose area is the product of the rectangular pulse width and the amplitude of the sampled waveform, or $T_Wf(kT)$.

Equation (13.6) is portrayed in Figure 13.6. The sampler is divided into two parts: (1) an ideal sampler described by the portion of Eq. (13.6) that is not dependent upon the sampling waveform characteristics,

$$f^*(t) \sum_{k=-\infty}^{\infty} f(kT)\delta(t - kT) \tag{13.7}$$

and (2) the portion dependent upon the sampling waveform's characteristics, $T_W$.

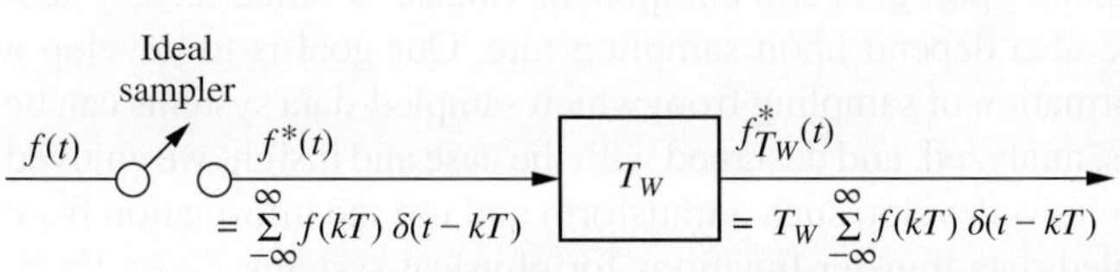


**FIGURE 13.6** Model of sampling with a uniform rectangular pulse train

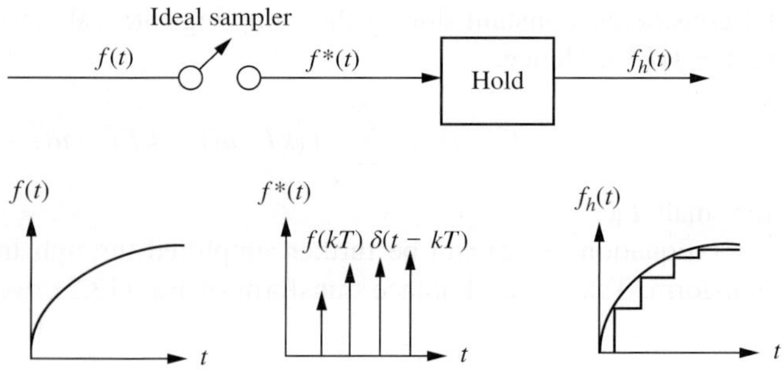


**FIGURE 13.7** Ideal sampling and the zero-order hold

## Modeling the Zero-Order Hold

The final step in modeling the digital computer is modeling the zero-order hold that follows the sampler. Figure 13.7 summarizes the function of the zero-order hold, which is to hold the last sampled value of $f(t)$. If we assume an ideal sampler (equivalent to setting $T_W = 1$), then $f^*(t)$ is represented by a sequence of delta functions. The zero-order hold yields a staircase approximation to $f(t)$. Hence, the output from the hold is a sequence of step functions whose amplitude is $f(t)$ at the sampling instant, or $f(kT)$. We have previously seen that the transfer function of any linear system is identical to the Laplace transform of the impulse response, since the Laplace transform of a unit impulse or delta function input is unity. Since a single impulse from the sampler yields a step over the sampling interval, the Laplace transform of this step, $G_h(s)$, which is the impulse response of the zero-order hold, is the transfer function of the zero-order hold. Using an impulse at zero time, the transform of the resulting step that starts at $t = 0$ and ends at $t = T$ is

$$G_h(s) = \frac{1 - e^{-Ts}}{s} \tag{13.8}$$

In a physical system, samples of the input time waveform, $f(kT)$, are held over the sampling interval. We can see from Eq. (13.8) that the hold circuit integrates the input and holds its value over the sampling interval. Since the area under the delta functions coming from the ideal sampler is $f(kT)$, we can then integrate the ideal sampled waveform and obtain the same result as for the physical system. In other words, if the ideal sampled signal, $f^*(t)$, is followed by a hold, we can use the ideal sampled waveform as the input, rather than $f^*_{T_W}(t)$.

In this section, we modeled the digital computer by cascading two elements: (1) an ideal sampler and (2) a zero-order hold. Together, the model is known as a *zero-order sample-and-hold*. The ideal sampler is modeled by Eq. (13.7), and the zero-order hold is modeled by Eq. (13.8). In the next section, we start to create a transform approach to digital systems by introducing the *z*-transform.

# 13.3 The *z*-Transform

The effect of sampling within a system is pronounced. Whereas the stability and transient response of analog systems depend upon gain and component values, sampled-data system stability and transient response also depend upon sampling rate. Our goal is to develop a transform that contains the information of sampling from which sampled-data systems can be modeled with transfer functions, analyzed, and designed with the ease and insight we enjoyed with the Laplace transform. We now develop such a transform and use the information from the last section to obtain sampled-data transfer functions for physical systems.

Equation (13.7) is the ideal sampled waveform. Taking the Laplace transform of this sampled time waveform, we obtain

$$F^*(s) = \sum_{k=0}^{\infty} f(kT)e^{-kTs} \tag{13.9}$$

Now, letting $z = e^{Ts}$, Eq. (13.9) can be written as

$$F(z) = \sum_{k=0}^{\infty} f(kT)z^{-k} \tag{13.10}$$

Equation (13.10) defines the *z-transform*. That is, an $F(z)$ can be transformed to $f(kT)$, or an $f(kT)$ can be transformed to $F(z)$. Alternately, we can write

$$f(kT) \leftrightarrows F(z) \tag{13.11}$$

Paralleling the development of the Laplace transform, we can form a table relating $f(kT)$, the value of the sampled time function at the sampling instants, to $F(z)$. Let us look at an example.

## Example 13.1 z-Transform of a Time Function

**PROBLEM:** Find the $z$-transform of a sampled unit ramp.

**SOLUTION:** For a unit ramp, $f(kT) = kT$. Hence the ideal sampled step can be written from Eq. (13.7) as

$$f^*(t) = \sum_{k=0}^{\infty} kt\delta(t - kT) \tag{13.12}$$

Taking the Laplace transform, we obtain

$$F^*(s) = \sum_{k=0}^{\infty} kTe^{-kTs} \tag{13.13}$$

Converting to the $z$-transform by letting $e^{-kTs} = z^{-k}$, we have

$$F(z) = \sum_{k=0}^{\infty} kTz^{-k} = T\sum_{k=0}^{\infty} kz^{-k} = T(z^{-1} + 2z^{-2} + 3z^{-3} + \cdots) \tag{13.14}$$

Equation (13.14) can be converted to a closed form by forming the series for $zF(z)$ and subtracting $F(z)$. Multiplying Eq. (13.14) by $z$, we get

$$zF(z) = T(1 + 2z^{-1} + 3z^{-2} + \cdots) \tag{13.15}$$

Subtracting Eq. (13.14) from Eq. (13.15), we obtain

$$zF(z) - F(z) = (z-1)F(z) = T(1 + z^{-1} + z^{-2} + \cdots) \tag{13.16}$$

But

$$\frac{1}{1 - z^{-1}} = 1 + z^{-1} + z^{-2} + z^{-3} + \cdots \tag{13.17}$$

which can be verified by performing the indicated division. Substituting Eq. (13.17) into (13.16) and solving for $F(z)$ yields

$$F(z) = T\frac{z}{(z-1)^2} \tag{13.18}$$

as the $z$-transform of $f(kT) = kT$.

Symbolic Math
**SM**

```
Students who are performing the MATLAB exercises and want to
explore the added capability of MATLAB's Symbolic Math Toolbox
should now run ch13sp1 in Appendix F located at www.wiley.com/
college/nise. You will learn how to find the z-transform of time
functions. Example 13.1 will be solved using MATLAB and the
Symbolic Math Toolbox.
```

The example demonstrates that any function of $s$, $F^*(s)$, that represents a sampled time waveform can be transformed into a function of $z$, $F(z)$. The final result, $F(z) = Tz/(z-1)^2$, is in a closed form, unlike $F^*(s)$. If this is the case for numerous other sampled time waveforms, then we have the convenient transform that we were looking for. In a similar way, $z$-transforms for other waveforms can be obtained that parallel the table of Laplace transforms in Chapter 2. A partial table of $z$-transforms is shown in Table 13.1, and a partial table of $z$-transform theorems is shown in Table 13.2. For functions not in the table, we must perform an inverse $z$-transform calculation similar

**TABLE 13.1** Partial table of $z$- and $s$-transforms

| | $f(t)$ | $F(s)$ | $F(z)$ | $f(kT)$ |
|---|---|---|---|---|
| 1. | $u(t)$ | $\frac{1}{s}$ | $\frac{z}{z-1}$ | $u(kT)$ |
| 2. | $t$ | $\frac{1}{s^2}$ | $\frac{Tz}{(z-1)^2}$ | $kT$ |
| 3. | $t^n$ | $\frac{n!}{s^{n+1}}$ | $\lim_{a\to 0}(-1)^n\frac{d^n}{da^n}\left[\frac{z}{z-e^{-aT}}\right]$ | $(kT)^n$ |
| 4. | $e^{-at}$ | $\frac{1}{s+a}$ | $\frac{z}{z-e^{-aT}}$ | $e^{-akT}$ |
| 5. | $t^n e^{-at}$ | $\frac{n!}{(s+a)^{n+1}}$ | $(-1)^n\frac{d^n}{da^n}\left[\frac{z}{z-e^{-aT}}\right]$ | $(kT)^n e^{-akT}$ |
| 6. | $\sin\omega t$ | $\frac{\omega}{s^2+\omega^2}$ | $\frac{z\sin\omega T}{z^2-2z\cos\omega T+1}$ | $\sin\omega kT$ |
| 7. | $\cos\omega t$ | $\frac{s}{s^2+\omega^2}$ | $\frac{z(z-\cos\omega T)}{z^2-2z\cos\omega T+1}$ | $\cos\omega kT$ |
| 8. | $e^{-at}\sin\omega t$ | $\frac{\omega}{(s+a)^2+\omega^2}$ | $\frac{ze^{-aT}\sin\omega T}{z^2-2ze^{-aT}\cos\omega T+e^{-2aT}}$ | $e^{-akT}\sin\omega kT$ |
| 9. | $e^{-at}\cos\omega t$ | $\frac{s+a}{(s+a)^2+\omega^2}$ | $\frac{z^2-ze^{-aT}\cos\omega T}{z^2-2ze^{-aT}\cos\omega T+e^{-2aT}}$ | $e^{-akT}\cos\omega kT$ |

**TABLE 13.2** z-transform theorems

| | Theorem | Name |
|---|---|---|
| 1. | $z\{af(t)\} = aF(z)$ | Linearity theorem |
| 2. | $z\{f_1(t) + f_2(t)\} = F_1(z) + F_2(z)$ | Linearity theorem |
| 3. | $z\{e^{-aT}f(t)\} = F(e^{aT}z)$ | Complex differentiation |
| 4. | $z\{f(t - nT)\} = z^{-n}F(z)$ | Real translation |
| 5. | $z\{tf(t)\} = -Tz\dfrac{dF(z)}{dz}$ | Complex differentiation |
| 6. | $f(0) = \lim_{z \to \infty} F(z)$ | Initial value theorem |
| 7. | $f(\infty) = \lim_{z \to 1}(1 - z^{-1})F(z)$ | Final value theorem |

Note: $kT$ may be substituted for $t$ in the table.

to the inverse Laplace transform by partial-fraction expansion. Let us now see how we can work in the reverse direction and find the time function from its $z$-transform.

## The Inverse z-Transform

Two methods for finding the inverse $z$-transform (the sampled time function from its $z$-transform) will be described: (1) partial-fraction expansion and (2) the power series method. Regardless of the method used, remember that, since the $z$-transform came from the sampled waveform, the inverse $z$-transform will yield only the values of the time function at the sampling instants. Keep this in mind as we proceed, because even as we obtain closed-form time functions as results, they are valid only at sampling instants.

**Inverse z-Transforms via Partial-Fraction Expansion** Recall that the Laplace transform consists of a partial fraction that yields a sum of terms leading to exponentials, that is, $A/(s + a)$. Taking this lead and looking at Table 13.1, we find that sampled exponential time functions are related to their $z$-transforms as follows:

$$e^{-akT} \leftrightarrows \frac{z}{z - e^{aT}} \tag{13.19}$$

We thus predict that a partial-fraction expansion should be of the following form:

$$F(z) = \frac{Az}{z - z_1} + \frac{Bz}{z - z_2} + \cdots \tag{13.20}$$

Since our partial-fraction expansion of $F(s)$ did not contain terms with $s$ in the numerator of the partial fractions, we first form $F(z)/z$ to eliminate the $z$ terms in the numerator, perform a partial-fraction expansion of $F(z)/z$, and finally multiply the result by $z$ to replace the $z$'s in the numerator. An example follows.

### Example 13.2 Inverse z-Transform via Partial-Fraction Expansion

**PROBLEM:** Given the function in Eq. (13.21), find the sampled time function.

$$F(z) = \frac{0.5z}{(z - 0.5)(z - 0.7)} \tag{13.21}$$

**SOLUTION:** Begin by dividing Eq. (13.21) by $z$ and performing a partial-fraction expansion.

$$\frac{F(z)}{z} = \frac{0.5}{(z-0.5)(z-0.7)} = \frac{A}{z-0.5} + \frac{B}{z-0.7} = \frac{-2.5}{z-0.5} + \frac{2.5}{z-0.7} \quad (13.22)$$

Next, multiply through by $z$.

$$F(z) = \frac{0.5z}{(z-0.5)(z-0.7)} = \frac{-2.5z}{z-0.5} + \frac{2.5z}{z-0.7} \quad (13.23)$$

Using Table 13.1, we find the inverse $z$-transform of each partial fraction. Hence, the value of the time function at the sampling instants is

$$f(kT) = -2.5(0.5)^k + 2.5(0.7)^k \quad (13.24)$$

Also, from Eqs. (13.7) and (13.24), the ideal sampled time function is

$$f^*(t) = \sum_{k=-\infty}^{\infty} f(kT)\delta(t-kT) = \sum_{k=-\infty}^{\infty} [-2.5(0.5)^k + 2.5(0.7)^k]\delta(t-kT) \quad (13.25)$$

If we substitute $k = 0, 1, 2,$ and 3, we can find the first four samples of the ideal sampled time waveform. Hence,

$$f^*(t) = 0\delta(t) + 0.5\delta(t-T) + 0.6\delta(t-2T) + 0.545\delta(t-3T) \quad (13.26)$$

Symbolic Math

SM

Students who are performing the MATLAB exercises and want to explore the added capability of MATLAB's Symbolic Math Toolbox should now run ch13sp2 in Appendix F located at www.wiley.com/college/nise. You will learn how to find the inverse z-transform of sampled time functions. Example 13.2 will be solved using MATLAB and the Symbolic Math Toolbox.

**Inverse *z*-Transform via the Power Series Method** The values of the sampled time waveform can also be found directly from $F(z)$. Although this method does not yield closed-form expressions for $f(kT)$, it can be used for plotting. The method consists of performing the indicated division, which yields a power series for $F(z)$. The power series can then be easily transformed into $F^*(s)$ and $f^*(t)$.

## Example 13.3 Inverse *z*-Transform via Power Series

**PROBLEM:** Given the function in Eq. (13.21), find the sampled time function.

**SOLUTION:** Begin by converting the numerator and denominator of $F(z)$ to polynomials in $z$.

$$F(z) = \frac{0.5z}{(z-0.5)(z-0.7)} = \frac{0.5z}{z^2 - 1.2z + 0.35} \quad (13.27)$$

Now perform the indicated division.

$$\begin{array}{r l} & \quad 0.5z^{-1} + 0.6z^{-2} + 0.545z^{-3} \\ z^2 - 1.2z + 0.35 & \overline{)\, 0.5z} \\ & \underline{0.5z - 0.6 + 0.175z^{-1}} \\ & \quad 0.6 - 0.175z^{-1} \\ & \quad \underline{0.6 - 0.720z^{-1} + 0.21} \\ & \qquad 0.545z^{-1} - 0.21 \end{array} \tag{13.28}$$

Using the numerator and the definition of $z$, we obtain

$$F^*(s) = 0.5e^{-Ts} + 0.6e^{-2Ts} + 0.545e^{-3Ts} + \cdots \tag{13.29}$$

from which

$$f^*(t) = 0.5\delta(t - T) + 0.6\delta(t - 2T) + 0.545\delta(t - 3T) + \cdots \tag{13.30}$$

You should compare Eq. (13.30) with Eq. (13.26), the result obtained via partial expansion.

## Skill-Assessment Exercise 13.1

**PROBLEM:** Derive the $z$-transform for $f(t) = \sin \omega t\, u(t)$.

**ANSWER:** $F(z) = \dfrac{z^{-1}\sin(\omega T)}{1 - 2z^{-1}\cos(\omega T) + z^{-2}}$

The complete solution is located at www.wiley.com/college/nise.

## Skill-Assessment Exercise 13.2

**PROBLEM:** Find $f(kT)$ if $F(z) = \dfrac{z(z+1)(z+2)}{(z-0.5)(z-0.7)(z-0.9)}$.

**ANSWER:** $f(kT) = 46.875(0.5)^k - 114.75(0.7)^k + 68.875(0.9)^k$

The complete solution is located at www.wiley.com/college/nise.

# 13.4 Transfer Functions

Now that we have established the $z$-transform, let us apply it to physical systems by finding transfer functions of sampled-data systems. Consider the continuous system shown in Figure 13.8($a$). If the input is sampled as shown in Figure 13.8($b$), the output is still a continuous signal. If, however, we are satisfied with finding the output at the sampling instants and not in between, the representation of the sampled-data system can be greatly simplified. Our assumption is visually described in Figure 13.8($c$), where the output is conceptually sampled in synchronization with the input by a phantom sampler.

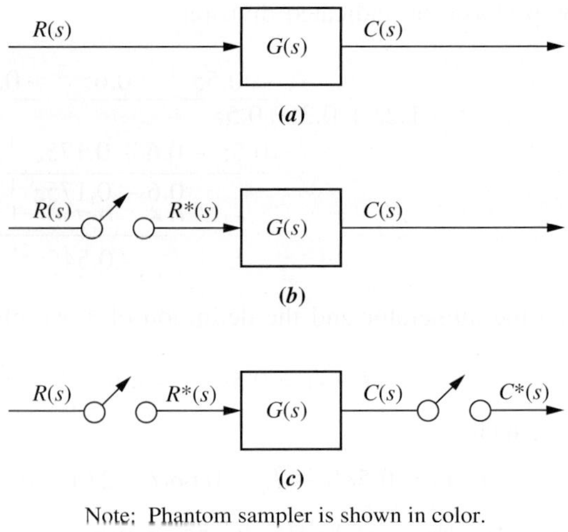


**FIGURE 13.8** Sampled-data systems: **a.** continuous; **b.** sampled input; **c.** sampled input and output

Using the concept described in Figure 13.8(*c*), we derive the pulse transfer function of $G(s)$.

## Derivation of the Pulse Transfer Function

Using Eq. (13.7), we find that the sampled input, $r^*(t)$, to the system of Figure 13.8(*c*) is

$$r^*(t) = \sum_{n=0}^{\infty} r(nT)\delta(t - nT) \tag{13.31}$$

which is a sum of impulses. Since the impulse response of a system, $G(s)$, is $g(t)$, we can write the time output of $G(s)$ as the sum of impulse responses generated by the input, Eq. (13.31). Thus,

$$c(t) = \sum_{n=0}^{\infty} r(nT)g(t - nT) \tag{13.32}$$

From Eq. (13.10),

$$C(z) = \sum_{k=0}^{\infty} c(kT)z^{-k} \tag{13.33}$$

Using Eq. (13.32) with $t = kT$, we obtain

$$c(kT) = \sum_{n=0}^{\infty} r(nT)g(kT - nT) \tag{13.34}$$

Substituting Eq. (13.34) into Eq. (13.33), we obtain

$$C(z) = \sum_{k=0}^{\infty}\sum_{n=0}^{\infty} r(nT)g[(k - n)T]z^{-k} \tag{13.35}$$

Letting $m = k - n$, we find

$$C(z) = \sum_{m+n=0}^{\infty} \sum_{n=0}^{\infty} r(nT)g(mT)z^{-(m+n)} \\ = \left\{\sum_{m=0}^{\infty} g(mT)z^{-m}\right\}\left\{\sum_{n=0}^{\infty} r(nT)z^{-n}\right\} \tag{13.36}$$

where the lower limit, $m + n$, was changed to $m$. The reasoning is that $m + n = 0$ yields negative values of $m$ for all $n > 0$. But, since $g(mT) = 0$ for all $m < 0$, $m$ is not less than zero. Alternately, $g(t) = 0$ for $t < 0$. Thus, $n = 0$ in the first sum's lower limit.

Using the definition of the $z$-transform, Eq. (13.36) becomes

$$C(z) = \sum_{m=0}^{\infty} g(mT)z^{-m} \sum_{n=0}^{\infty} r(nT)z^{-n} = G(z)R(z) \tag{13.37}$$

Equation (13.37) is a very important result, since it shows that the transform of the sampled output is the product of the transforms of the sampled input and the pulse transfer function of the system. Remember that although the output of the system is a continuous function, we had to make an assumption of a sampled output (phantom sampler) in order to arrive at the compact result of Eq. (13.37).

One way of finding the pulse transfer function, $G(z)$, is to start with $G(s)$, find $g(t)$, and then use Table 13.1 to find $G(z)$. Let us look at an example.

## Example 13.4 Converting $G_1(s)$ in Cascade with z.o.h. to $G(z)$

**PROBLEM:** Given a z.o.h. in cascade with $G_1(s) = (s + 2)/(s + 1)$ or

$$G(s) = \frac{1 - e^{-Ts}}{s}\frac{(s + 2)}{(s + 1)} \tag{13.38}$$

find the sampled-data transfer function, $G(z)$, if the sampling time, $T$, is 0.5 second.

**SOLUTION:** Equation (13.38) represents a common occurrence in digital control systems, namely a transfer function in cascade with a zero-order hold. Specifically, $G_1(s) = (s + 2)/(s + 1)$ is in cascade with a zero-order hold, $(1 - e^{-Ts})/s$. We can formulate a general solution to this type of problem by moving the $s$ in the denominator of the zero-order hold to $G_1(s)$, yielding

$$G(s) = (1 - e^{-Ts})\frac{G_1(s)}{s} \tag{13.39}$$

from which

$$G(z) = (1 - z^{-1})z\left\{\frac{G_1(s)}{s}\right\} = \frac{z - 1}{z}z\left\{\frac{G_1(s)}{s}\right\} \tag{13.40}$$

Thus, begin the solution by finding the impulse response (inverse Laplace transform) of $G_1(s)/s$. Hence,

$$G_2(s) = \frac{G_1(s)}{s} = \frac{s + 2}{s(s + 1)} = \frac{A}{s} + \frac{B}{s + 1} = \frac{2}{s} - \frac{1}{s + 1} \tag{13.41}$$

Taking the inverse Laplace transform, we get

$$g_2(t) = 2 - e^{-t} \tag{13.42}$$

from which

$$g_2(kT) = 2 - e^{-kt} \tag{13.43}$$

Using Table 13.1, we find

$$G_2(z) = \frac{2z}{z-1} - \frac{z}{z-e^{-T}} \tag{13.44}$$

Substituting $T = 0.5$ yields

$$G_2(z) = z\left\{\frac{G_1(s)}{s}\right\} = \frac{2z}{z-1} - \frac{z}{z-0.607} = \frac{z^2 - 0.213z}{(z-1)(z-0.607)} \tag{13.45}$$

From Eq. (13.40),

$$G(z) = \frac{z-1}{z}G_2(z) = \frac{z-0.213}{z-0.607} \tag{13.46}$$

**TryIt 13.1**

Use MATLAB, the Control System Toolbox, and the following statements to find $G_1(s)$ in Example 13.4 given $G(z)$ in Eq. (13.46)

```
num=0.213;
den=0.607;
K=1;
T=0.5;
Gz=zpk(num,den,K,T)
Gs=d2c(Gz,'zoh')
```

MATLAB **ML**

Students who are using MATLAB should now run ch13p1 in Appendix B. You will learn how to use MATLAB to convert $G_1(s)$ in cascade with a zero-order hold to $G(z)$. This exercise solves Example 13.4 using MATLAB.

Symbolic Math **SM**

Students who are performing the MATLAB exercises and want to explore the added capability of MATLAB's Symbolic Math Toolbox should now run ch13sp3 in Appendix F located at www.wiley.com/college/nise. MATLAB's Symbolic Math Toolbox yields an alternative method of finding the $z$-transform of a transfer function in cascade with a zero-order hold. Example 13.4 will be solved using MATLAB and the Symbolic Math Toolbox with a method that follows closely the hand calculation shown in that example.

MATLAB **ML**

Students who are using MATLAB should now run ch13p2 in Appendix B. You will learn how to use MATLAB to convert $G(s)$ to $G(z)$ when $G(s)$ is not in cascade with a zero-order hold. This is the same as finding the $z$-transform of $G(s)$.

MATLAB **ML**

Students who are using MATLAB should now run ch13p3 in Appendix B. You will learn how to create digital transfer functions directly.

MATLAB **ML**

Students who are using MATLAB should now run ch13p4 in Appendix B. You will learn how to use MATLAB to convert $G(z)$ to $G(s)$ when $G(s)$ is not in cascade with a zero-order hold. This is the same as finding the Laplace transform of $G(z)$.

## Skill-Assessment Exercise 13.3

**TryIt 13.2**

Use MATLAB, the Control System Toolbox, and the following statements to solve Skill-Assessment Exercise 13.3.

```
Gs=zpk([],-4,8)
Gz=c2d(Gs,0.25,'zoh')
```

**PROBLEM:** Find $G(z)$ for $G(s) = 8/(s+4)$ in cascade with a zero-order sample and hold. The sampling period is 0.25 second.

**ANSWER:** $G(z) = 1.264/(z - 0.3679)$

The complete solution is located at www.wiley.com/college/nise.

The major discovery in this section is that once the pulse transfer function, $G(z)$, of a system is obtained, the transform of the sampled output response, $C(z)$, for a given sampled input can be evaluated using the relationship $C(z) = R(z)G(z)$. Finally, the time function can be found by taking the inverse $z$-transform, as covered in Section 13.3. In the next section, we look at block diagram reduction for digital systems.

## 13.5 Block Diagram Reduction

Up to this point, we have defined the $z$-transform and the sampled-data system transfer function and have shown how to obtain the sampled response. Basically, we are paralleling our discussions of the Laplace transform in Chapters 2 and 4. We now draw a parallel with some of the objectives of Chapter 5, namely block diagram reduction. Our objective here is to be able to find the closed-loop sampled-data transfer function of an arrangement of subsystems that have a computer in the loop.

When manipulating block diagrams for sampled-data systems, you must be careful to remember the definition of the sampled-data system transfer function (derived in the last section) to avoid mistakes. For example, $z\{G_1(s)G_2(s)\} \neq G_1(z)G_2(z)$, where $z\{G_1(s)G_2(s)\}$ denotes the $z$-transform. The $s$-domain functions have to be multiplied together before taking the $z$-transform. In the ensuing discussion, we use the notation $G_1G_2(s)$ to denote a single function that is $G_1(s)G_2(s)$ after evaluating the product. Hence, $z\{G_1(s)G_2(s)\} = z\{G_1G_2(s)\} = G_1G_2(z) \neq G_1(z)G_2(z)$.

Let us look at the sampled-data systems shown in Figure 13.9. The sampled-data systems are shown under the column marked $s$. Their $z$-transforms are shown under the column marked $z$. The standard system that we derived earlier is shown in Figure 13.9(*a*), where the transform of the output, $C(z)$, is equal to $R(z)G(z)$. This system forms the basis for the other entries in Figure 13.9.

In Figure 13.9(*b*), there is no sampler between $G_1(s)$ and $G_2(s)$. Thus, we can think of a single function, $G_1(s)G_2(s)$, denoted $G_1G_2(s)$, existing between the two samplers and yielding a single transfer function, as shown in Figure 13.9(*a*). Hence, the pulse transfer function is $z\{G_1G_2(s)\} = G_1G_2(z)$. The transform of the output, $C(z) = R(z)G_1G_2(z)$.

In Figure 13.9(*c*), we have the cascaded two subsystems of the type shown in Figure 13.9(*a*). For this case, then, the $z$-transform is the product of the two $z$-transforms, or $G_2(z)G_1(z)$. Hence the transform of the output $C(z) = R(z)G_2(z)G_1(z)$.

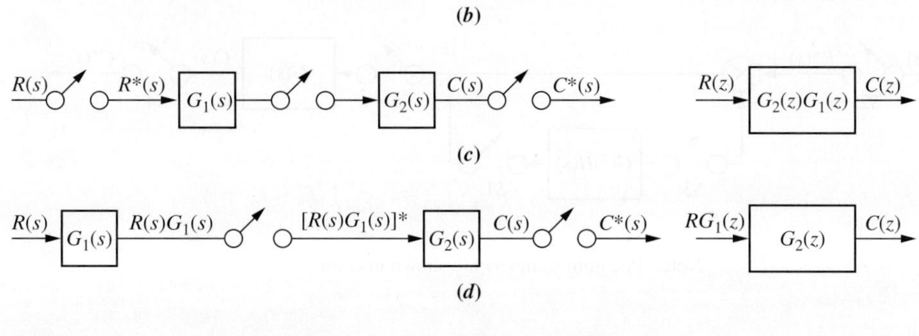


**FIGURE 13.9** Sampled-data systems and their $z$-transforms

Finally, in Figure 13.9(*d*), we see that the continuous signal entering the sampler is $R(s)G_1(s)$. Thus, the model is the same as Figure 13.9(*a*) with $R(s)$ replaced by $R(s)G_1(s)$, and $G_2(s)$ in Figure 13.9(*d*) replacing $G(s)$ in Figure 13.9(*a*). The $z$-transform of the input to $G_2(s)$ is $z\{R(s)G_1(s)\} = z\{RG_1(s)\} = RG_1(z)$. The pulse transfer function for the system $G_2(s)$ is $G_2(z)$. Hence, the output $C(z) = RG_1(z)G_2(z)$.

Using the basic forms shown in Figure 13.9, we can now find the $z$-transform of feedback control systems. We have shown that any system, $G(s)$, with sampled input and sampled output, such as that shown in Figure 13.9(*a*), can be represented as a sampled-data transfer function, $G(z)$. Thus, we want to perform block diagram manipulations that result in subsystems, as well as the entire feedback system, that have sampled inputs and sampled outputs. Then we can make the transformation to sampled-data transfer functions. An example follows.

## Example 13.5 Pulse Transfer Function of a Feedback System

**PROBLEM:** Find the $z$-transform of the system shown in Figure 13.10(*a*).

**SOLUTION:** The objective of the problem is to proceed in an orderly fashion, starting with the block diagram of Figure 13.10(*a*) and reducing it to the one shown in Figure 13.10(*f*).

One operation we can always perform is to place a phantom sampler at the output of any subsystem that has a sampled input, provided that the nature of the signal sent to any

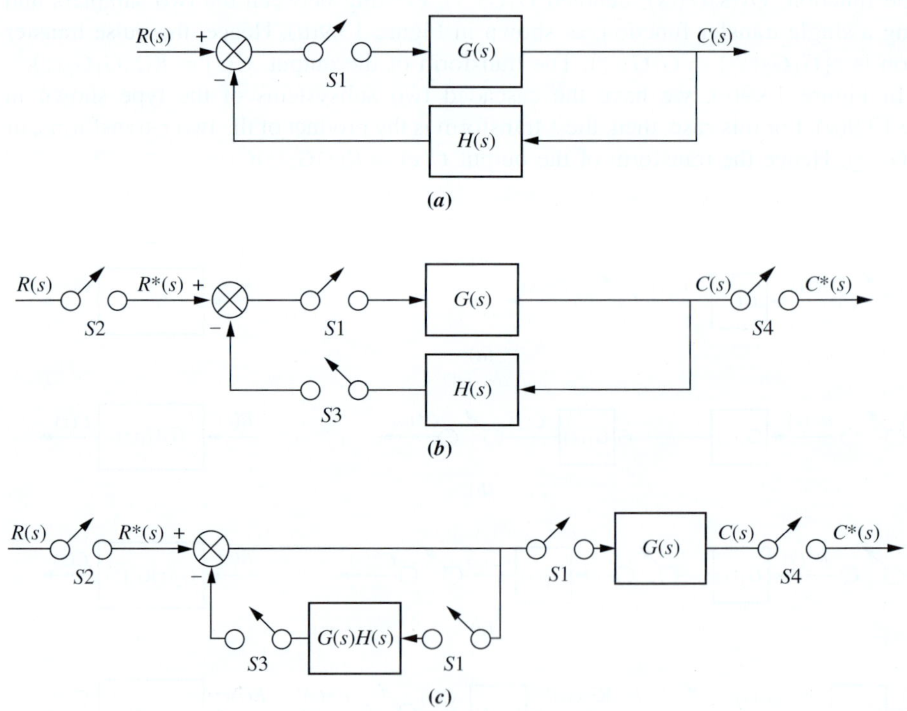


**FIGURE 13.10** Steps in block diagram reduction of a sampled-data system (*figure continues*)

Note: Phantom samplers are shown in color.

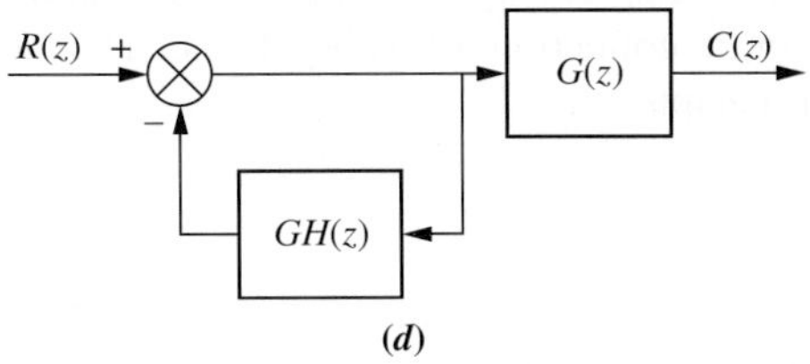


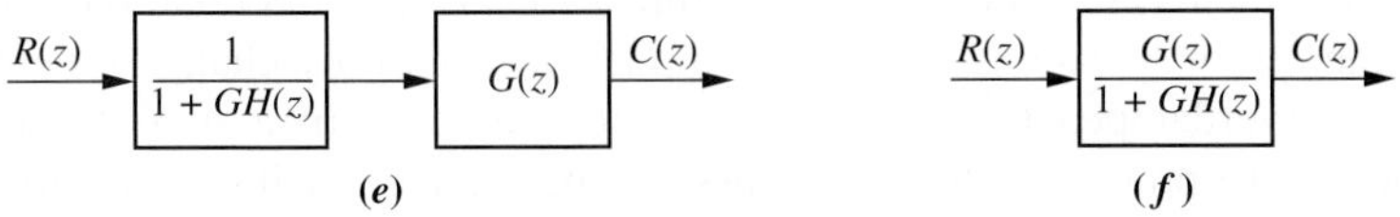


**FIGURE 13.10** *(Continued)*

other subsystem is not changed. For example in Figure 13.10(*b*), phantom sampler $S4$ can be added. The justification for this, of course, is that the output of a sampled-data system can only be found at the sampling instants anyway, and the signal is not an input to any other block.

Another operation that can be performed is to add phantom samplers $S2$ and $S3$ at the input to a summing junction whose output is sampled. The justification for this operation is that the sampled sum is equivalent to the sum of the sampled inputs, provided, of course, that all samplers are synchronized.

Next, move sampler $S1$ and $G(s)$ to the right past the pickoff point, as shown in Figure 13.10(*c*). The motivation for this move is to yield a sampler at the input of $G(s)H(s)$ to match Figure 13.9(*b*). Also, $G(s)$ with sampler $S1$ at the input and sampler $S4$ at the output matches Figure 13.9(*a*). The closed-loop system now has a sampled input and a sampled output.

$G(s)H(s)$ with samplers $S1$ and $S3$ becomes $GH(z)$, and $G(s)$ with samplers $S1$ and $S4$ becomes $G(z)$, as shown in Figure 13.10(*d*). Also, converting $R^*(s)$ to $R(z)$ and $C^*(s)$ to $C(z)$, we now have the system represented totally in the $z$-domain.

The equations derived in Chapter 5 for transfer functions represented with the Laplace transform can be used for sampled-data transfer functions with only a change in variables from $s$ to $z$. Thus, using the feedback formula, we obtain the first block of Figure 13.10(*e*). Finally, multiplication of the cascaded sampled-data systems yields the final result shown in Figure 13.10(*f*).

## Skill-Assessment Exercise 13.4

**PROBLEM:** Find $T(z) = C(z)/R(z)$ for the system shown in Figure 13.11.

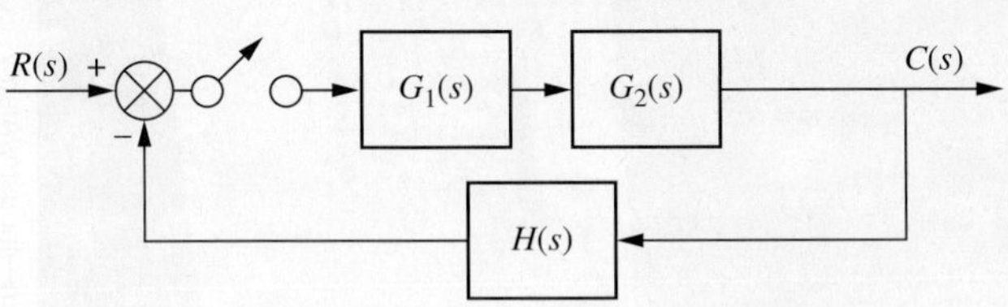


**FIGURE 13.11** Digital system for Skill-Assessment Exercise 13.4

**ANSWER:** $T(z) = \dfrac{G_1G_2(z)}{1 + HG_1G_2(z)}$

The complete solution is located at www.wiley.com/college/nise.

This section paralleled Chapter 5 by showing how to obtain the closed-loop, sampled-data transfer function for a collection of subsystems. The next section parallels the discussion of stability in Chapter 6.

## 13.6 Stability

The glaring difference between analog feedback control systems and digital feedback control systems, such as the one shown in Figure 13.12, is the effect that the sampling rate has on the transient response. Changes in sampling rate not only change the nature of the response from overdamped to underdamped, but also can turn a stable system into an unstable one. As we proceed with our discussion, these effects will become apparent. You are encouraged to be on the lookout.

We now discuss the stability of digital systems from two perspectives: (1) $z$-plane and (2) $s$-plane. We will see that the Routh-Hurwitz criterion can be used only if we perform our analysis and design on the $s$-plane.

### Digital System Stability via the z-Plane

In the $s$-plane, the region of stability is the left half-plane. If the transfer function, $G(s)$, is transformed into a sampled-data transfer function, $G(z)$, the region of stability on the $z$-plane can be evaluated from the definition, $z = e^{Ts}$. Letting $s = \alpha + j\omega$, we obtain

$$\begin{aligned} z = e^{Ts} &= e^{T(\alpha+j\omega)} = e^{\alpha T} e^{j\omega T} \\ &= e^{\alpha T}(\cos \omega T + j \sin \omega T) \\ &= e^{\alpha T} \angle \omega T \end{aligned} \tag{13.47}$$

since $(\cos \omega T + j \sin \omega T) = 1\angle\omega T$.

Each region of the $s$-plane can be mapped into a corresponding region on the $z$-plane (see Figure 13.13). Points that have positive values of $\alpha$ are in the right half of the $s$-plane,

David J. Green—Industry/Alamy

**FIGURE 13.12** A lathe using digital numerical control

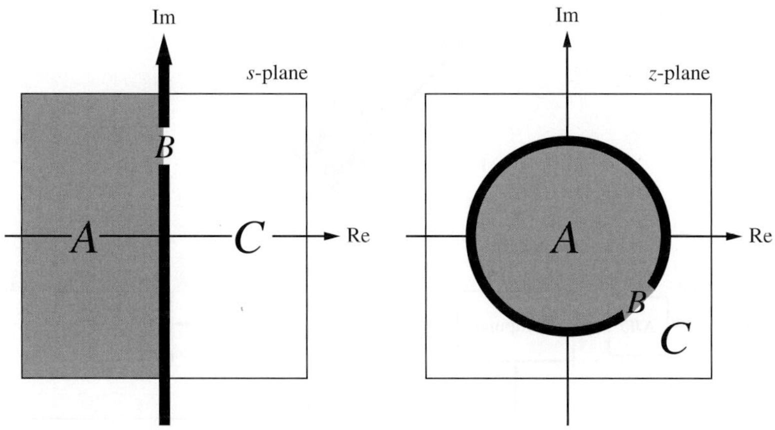


**FIGURE 13.13** Mapping regions of the $s$-plane onto the $z$-plane

region $C$. From Eq. (13.47), the magnitudes of the mapped points are $e^{\alpha T} > 1$. Thus points in the right half of the $s$-plane map into points outside the unit circle on the $z$-plane.

Points on the $j\omega$-axis, region $B$, have zero values of $\alpha$ and yield points on the $z$-plane with magnitude $= 1$, the unit circle. Hence, points on the $j\omega$-axis in the $s$-plane map into points on the unit circle on the $z$-plane.

Finally, points on the $s$-plane that yield negative values of $\alpha$(left–half-plane roots, region A) map into the inside of the unit circle on the $z$-plane.

Thus, a digital control system is (1) stable if all poles of the closed-loop transfer function, $T(z)$, are inside the unit circle on the $z$-plane; (2) unstable if any pole is outside the unit circle and/or there are poles of multiplicity greater than one on the unit circle; and (3) marginally stable if poles of multiplicity 1 are on the unit circle and all other poles are inside the unit circle. Let us look at an example.

## Example 13.6 Modeling and Stability

**PROBLEM:** The missile shown in Figure 13.14($a$) can be aerodynamically controlled by torques created by the deflection of control surfaces on the missile's body. The commands to deflect these control surfaces come from a computer that uses tracking data along with programmed guidance equations to determine whether the missile is on track. The information from the guidance equations is used to develop flight-control commands for the missile. A simplified model is shown in Figure 13.14($b$). Here the computer performs the function of controller by using tracking information to develop input commands to the missile. An accelerometer in the missile detects the actual acceleration, which is fed back to the computer. Find the closed-loop digital transfer function for this system and determine if the system is stable for $K = 20$ and $K = 100$ with a sampling interval of $T = 0.1$ second.

**SOLUTION:** The input to the control system is an acceleration command developed by the computer. The computer can be modeled by a sample-and-hold. The $s$-plane model is shown in Figure 13.14($c$). The first step in finding the $z$-plane model is to find $G(z)$, the forward-path transfer function. From Figure 13.14($c$) or ($d$),

$$G(s) = \frac{1 - e^{-Ts}}{s} \frac{Ka}{s(s+a)} \tag{13.48}$$

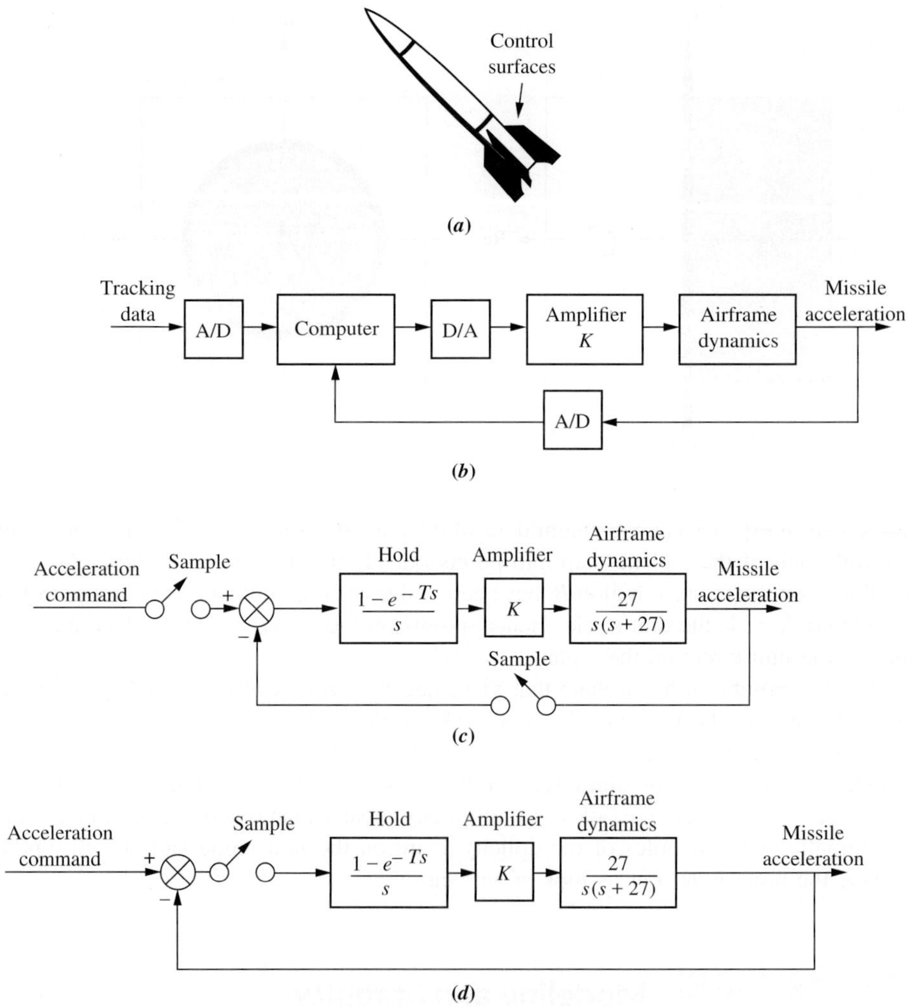


**FIGURE 13.14** Finding stability of a missile control system: **a.** missile; **b.** conceptual block diagram; **c.** block diagram; **d.** block diagram with equivalent single sampler

where $a = 27$. The $z$-transform, $G(z)$, is $(1 - z^{-1})z\{Ka/[s^2(s + a)]\}$.

The term $Ka/[s^2(s + a)]$ is first expanded by partial fractions, after which we find the $z$-transform of each term from Table 13.1. Hence,

$$
\begin{aligned}
z\left\{\frac{Ka}{s^2(s+a)}\right\} &= Kz\left\{\frac{a}{s^2(s+a)}\right\} = Kz\left\{\frac{1}{s^2} - \frac{1/a}{s} + \frac{1/a}{s+a}\right\} \\
&= K\left\{\frac{Tz}{(z-1)^2} - \frac{z/a}{z-1} + \frac{z/a}{z-e^{-aT}}\right\} \\
&= K\left\{\frac{Tz}{(z-1)^2} - \frac{(1-e^{-aT})z}{a(z-1)(z-e^{-aT})}\right\} \qquad (13.49)
\end{aligned}
$$

Thus,

$$
G(z) = K\left\{\frac{T(z-e^{-aT}) - (z-1)\left(\dfrac{1-e^{-aT}}{a}\right)}{(z-1)(z-e^{-aT})}\right\} \qquad (13.50)
$$

Letting $T = 0.1$ and $a = 27$, we have

$$G(z) = \frac{K(0.0655z + 0.02783)}{(z-1)(z-0.0672)} \quad (13.51)$$

Finally, we find the closed-loop transfer function, $T(z)$, for a unity feedback system:

$$T(z) = \frac{G(z)}{1+G(z)} = \frac{K(0.0655z + 0.02783)}{z^2 + (0.0655K - 1.0672)z + (0.02783K + 0.0672)} \quad (13.52)$$

The stability of the system is found by finding the roots of the denominator. For $K = 20$, the roots of the denominator are $0.12 \pm j0.78$. The system is thus stable for $K = 20$, since the poles are inside the unit circle. For $K = 100$, the poles are at $-0.58$ and $-4.9$. Since one of the poles is outside the unit circle, the system is unstable for $K = 100$.

Students who are using MATLAB should now run ch13p5 in Appendix B. You will learn how to use MATLAB to determine the range of $K$ for stability in a digital system. This exercise solves Example 13.6 using MATLAB.

MATLAB
ML

In the case of continuous systems, the determination of stability hinges upon our ability to determine whether the roots of the denominator of the closed-loop transfer function are in the stable region of the $s$-plane. The problem for high-order systems is complicated by the fact that the closed-loop transfer function denominator is in polynomial form, not factored form. The same problem surfaces with closed-loop sampled-data transfer functions.

Tabular methods for determining stability, such as the Routh-Hurwitz method used for higher-order continuous systems, exist for sampled-data systems. These methods, which are not covered in this introductory chapter to digital control systems, can be used to determine stability in higher-order digital systems. If you wish to go further into the area of digital system stability, you are encouraged to look at Raible's tabular method or Jury's stability test for determining the number of a sampled-data system's closed-loop poles that exist outside the unit circle and thus indicate instability.[2]

The following example demonstrates the effect of sampling rate on the stability of a closed-loop feedback control system. All parameters are constant except for the sampling interval, $T$. We will see that varying $T$ will lead us through regions of stability and instability just as though we were varying the forward-path gain, $K$.

## Example 13.7 Range of $T$ for Stability

**PROBLEM:** Determine the range of sampling interval, $T$, that will make the system shown in Figure 13.15 stable, and the range that will make it unstable.

**SOLUTION:** Since $H(s) = 1$, the $z$-transform of the closed-loop system, $T(z)$, is found from Figure 13.10 to be

$$T(z) = \frac{G(z)}{1+G(z)} \quad (13.53)$$

[2] A discussion of Raible's tabular method and Jury's stability test can be found in *Kuo* (*1980: 278–286*).

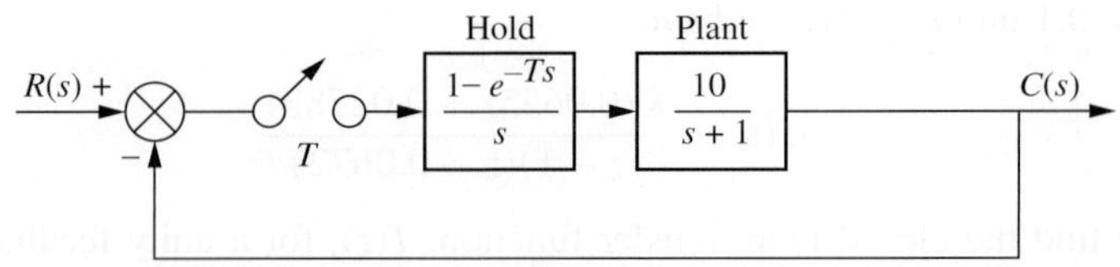


**FIGURE 13.15** Digital system for Example 13.7

To find $G(z)$, first find the partial-fraction expansion of $G(s)$.

$$G(s) = 10\frac{1 - e^{-Ts}}{s(s+1)} = 10(1 - e^{-Ts})\left(\frac{1}{s} - \frac{1}{s+1}\right) \tag{13.54}$$

Taking the $z$-transform, we obtain

$$G(z) = \frac{10(z-1)}{z}\left[\frac{z}{z-1} - \frac{z}{z-e^{-T}}\right] = 10\frac{(1-e^{-T})}{(z-e^{-T})} \tag{13.55}$$

Substituting Eq. (13.55) into (13.53) yields

$$T(z) = \frac{10(1-e^{-T})}{z - (11e^{-T} - 10)} \tag{13.56}$$

The pole of Eq. (13.56), $(11e^{-T} - 10)$, monotonically decreases from $+1$ to $-1$ for $0 < T < 0.2$. For $0.2 < T < \infty, (11e^{-T} - 10)$ monotonically decreases from $-1$ to $-10$. Thus, the pole of $T(z)$ will be inside the unit circle, and the system will be stable if $0 < T < 0.2$. In terms of frequency, where $f = 1/T$, the system will be stable as long as the sampling frequency is $1/0.2 = 5$ hertz or greater.

We now have found, via the $z$-plane, that sampled systems are stable if their poles are inside the unit circle. Unfortunately, this stability criterion precludes the use of the Routh-Hurwitz criterion, which detects roots in the right half-plane rather than outside the unit circle. However, another method exists that allows us to use the familiar $s$-plane and the Routh-Hurwitz criterion to determine the stability of a sampled system. Let us introduce this topic.

## Bilinear Transformations

*Bilinear transformations* give us the ability to apply our $s$-plane analysis and design techniques to digital systems. We can analyze and design on the $s$-plane as we have done in Chapters 8 and 9 and then, using these transformations, convert the results to a digital system that contains the same properties. Let us look further into this topic.

We can consider $z = e^{Ts}$ and its inverse, $s = (1/T)\ln z$, as the exact transformations between $z$ and $s$. Thus, if we have $G(z)$ and substitute $z = e^{Ts}$, we obtain $G(e^{Ts})$ as the result of converting to $s$. Similarly, if we have $G(s)$ and substitute $s = (1/T)\ln z$, we obtain $G((1/T)\ln z)$ as the result of converting to $z$. Unfortunately, both transformations yield transcendental functions, which we of course take care of through the rather complicated $z$-transform.

What we would like is a simple transformation that would yield linear arguments when transforming in both directions (bilinear) through direct substitution and without the complicated $z$-transform.

Bilinear transformations of the form

$$z = \frac{as + b}{cs + d} \tag{13.57}$$

and its inverse,

$$s = \frac{-dz + b}{cz - a} \tag{13.58}$$

have been derived to yield linear variables in $s$ and $z$. Different values of $a$, $b$, $c$, and $d$ have been derived for particular applications and yield various degrees of accuracy when comparing properties of the continuous and sampled functions.

For example, in the next subsection we will see that a particular choice of coefficients will take points on the unit circle and map them into points on the $j\omega$-axis. Points outside the unit circle will be mapped into the right half-plane, and points inside the unit circle will be mapped into the left half-plane. Thus, we will be able to make a simple transformation from the $z$-plane to the $s$-plane and obtain stability information about the digital system by working in the $s$-plane.

Since the transformations are not exact, only the property for which they are designed can be relied upon. For the stability transformation just discussed, we cannot expect the resulting $G(s)$ to have the same transient response as $G(z)$. Another transformation will be covered that will retain that property.

## Digital System Stability via the *s*-Plane

In this subsection, we look at a bilinear transformation that maps $j\omega$-axis points on the $s$-plane to unit-circle points on the $z$-plane. Further, the transformation maps right–half-plane points on the $s$-plane to points outside the unit circle on the $z$-plane. Finally, the transformation maps left–half-plane points on the $s$-plane to points inside the unit circle on the $z$-plane. Thus, we are able to transform the denominator of the pulsed transfer function, $D(z)$, to the denominator of a continuous transfer function, $D(s)$, and use the Routh-Hurwitz criterion to determine stability.

The bilinear transformation

$$s = \frac{z + 1}{z - 1} \tag{13.59}$$

and its inverse

$$z = \frac{s + 1}{s - 1} \tag{13.60}$$

perform the required transformation (*Kuo, 1995*). We can show this fact as follows: Letting $s = \alpha + j\omega$ and substituting into Eq. (13.60),

$$z = \frac{(\alpha + 1) + j\omega}{(\alpha - 1) + j\omega} \tag{13.61}$$

from which

$$|z| = \frac{\sqrt{(\alpha + 1)^2 + \omega^2}}{\sqrt{(\alpha - 1)^2 + \omega^2}} \tag{13.62}$$

Thus,

$$|z| < 1 \quad \text{when } \alpha < 0 \tag{13.63a}$$

$$|z| > 1 \quad \text{when } \alpha > 0 \tag{13.63b}$$

and

$$|z| = 1 \quad \text{when } \alpha = 0 \tag{13.63c}$$

Let us look at an example that shows how the stability of sampled systems can be found using this bilinear transformation and the Routh-Hurwitz criterion.

## Example 13.8 Stability via Routh-Hurwitz

**PROBLEM:** Given $T(z) = N(z)/D(z)$, where $D(z) = z^3 - z^2 - 0.2z + 0.1$, use the Routh-Hurwitz criterion to find the number of $z$-plane poles of $T(z)$ inside, outside, and on the unit circle. Is the system stable?

**SOLUTION:** Substitute Eq. (13.60) into $D(z) = 0$ and obtain[3]

$$s^3 - 19s^2 - 45s - 17 = 0 \tag{13.64}$$

The Routh table for Eq. (13.64), Table 13.3, shows one root in the right–half-plane and two roots in the left–half-plane. Hence, $T(z)$ has one pole outside the unit circle, no poles on the unit circle, and two poles inside the unit circle. The system is unstable because of the pole outside the unit circle.

**TABLE 13.3** Routh table for Example 13.8

| | | |
|---|---|---|
| $s^3$ | 1 | −45 |
| $s^2$ | 19 | −17 |
| $s^1$ | −45.89 | 0 |
| $s^0$ | −17 | 0 |

## Skill-Assessment Exercise 13.5

**PROBLEM:** Determine the range of sampling interval, $T$, that will make the system shown in Figure 13.16 stable.

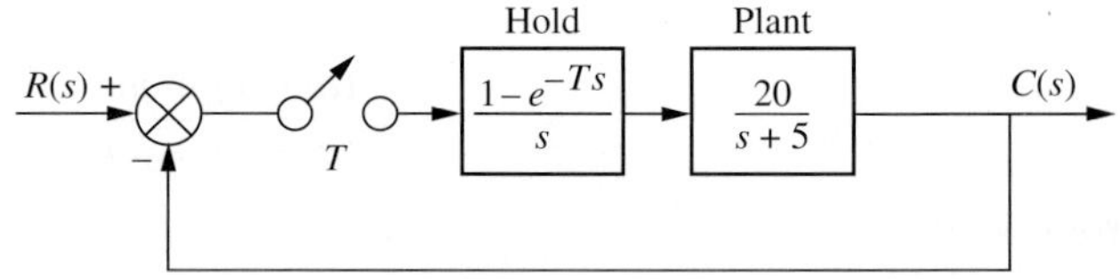


**FIGURE 13.16** Digital system for Skill-Assessment Exercise 13.5

**ANSWER:** $0 < T < 0.1022$ second

The complete solution is located at www.wiley.com/college/nise.

[3] Symbolic math software, such as MATLAB's Symbolic Math Toolbox, is recommended to reduce the labor required to perform the transformation.

## Skill-Assessment Exercise 13.6

**PROBLEM:** Given $T(z) = N(z)/D(z)$, where $D(z) = z^3 - z^2 - 0.5z + 0.3$, use the Routh-Hurwitz criterion to find the number of $z$-plane poles of $T(z)$ inside, outside, and on the unit circle. Is the system stable?

**ANSWER:** $T(z)$ has one pole outside the unit circle, no poles on the unit circle, and two poles inside the unit circle. The system is unstable.

The complete solution is located at www.wiley.com/college/nise.

In this section, we covered the concepts of stability for digital systems. Both $z$- and $s$-plane perspectives were discussed. Using a bilinear transformation, we are able to use the Routh-Hurwitz criterion to determine stability.

The highlight of the section is that sampling rate (along with system parameters, such as gain and component values) helps to determine or destroy the stability of a digital system. In general, if the sampling rate is too slow, the closed-loop digital system will be unstable. We now move from stability to steady-state errors, paralleling our previous discussion of steady-state errors in analog systems.

# 13.7 Steady-State Errors

We now examine the effect of sampling upon the steady-state error for digital systems. Any general conclusion about the steady-state error is difficult because of the dependence of those conclusions upon the placement of the sampler in the loop. Remember that the position of the sampler could change the open-loop transfer function. In the discussion of analog systems, there was only one open-loop transfer function, $G(s)$, upon which the general theory of steady-state error was based and from which came the standard definitions of static error constants. For digital systems, however, the placement of the sampler changes the open-loop transfer function and thus precludes any general conclusions. In this section, we assume the typical placement of the sampler after the error and in the position of the cascade controller, and we derive our conclusions accordingly about the steady-state error of digital systems.

Consider the digital system in Figure 13.17($a$), where the digital computer is represented by the sampler and zero-order hold. The transfer function of the plant is represented by $G_1(s)$ and the transfer function of the z.o.h. by $(1 - e^{-Ts})/s$. Letting $G(s)$ equal the product of the z.o.h. and $G_1(s)$, and using the block diagram reduction techniques for sampled-data systems, we can find the sampled error, $E^*(s) = E(z)$. Adding synchronous samplers at the input and the feedback, we obtain Figure 13.17($b$). Pushing $G(s)$ and its input sampler to the right past the pickoff point yields Figure 13.17($c$). Using Figure 13.9($a$), we can convert each block to its $z$-transform, resulting in Figure 13.17($d$).

From this figure, $E(z) = R(z) - E(z)G(z)$, or

$$E(z) = \frac{R(z)}{1 + G(z)} \tag{13.65}$$

The final value theorem for discrete signals states that

$$e^*(\infty) = \lim_{z \to 1}(1 - z^{-1})E(z) \tag{13.66}$$

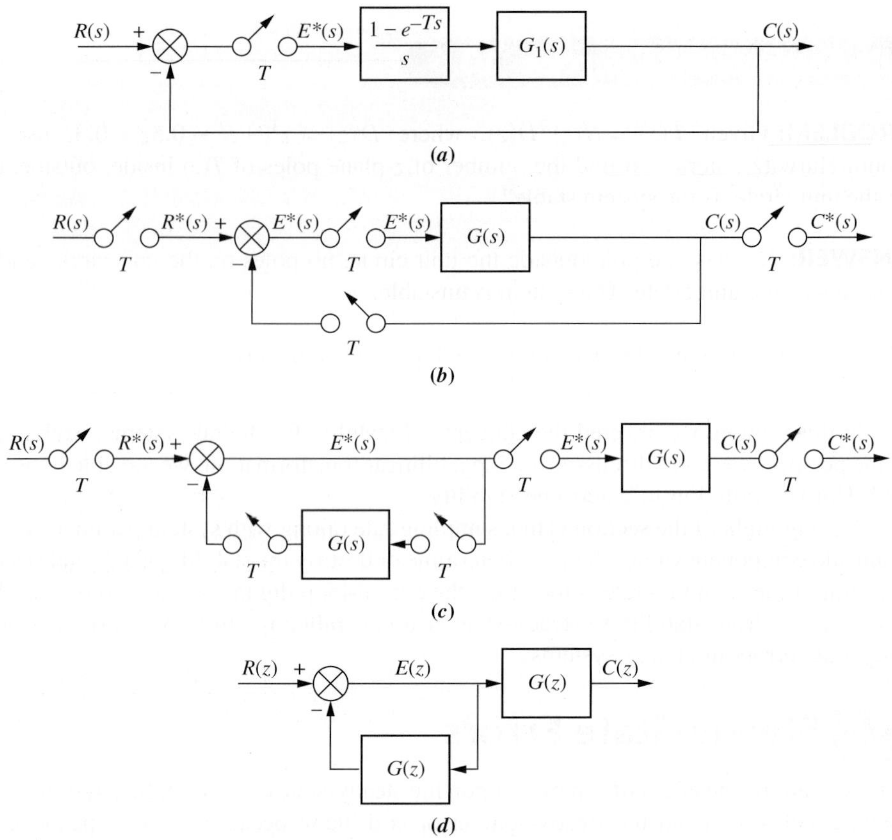


Note: Phantom samplers are shown in color.

**FIGURE 13.17** **a.** Digital feedback control system for evaluation of steady-state errors; **b.** phantom samplers added; **c.** pushing $G(s)$ and its samplers to the right past the pickoff point; **d.** $z$-transform equivalent system

where $e^*(\infty)$ is the final sampled value of $e(t)$, or (alternatively) the final value of $e(kT)$.[4]

Using the final value theorem on Eq. (13.65), we find that the sampled steady-state error, $e^*(\infty)$, for unity negative-feedback systems is

$$e^*(\infty) = \lim_{z\to 1}(1 - z^{-1})E(z) = \lim_{z\to 1}(1 - z^{-1})\frac{R(z)}{1 + G(z)} \tag{13.67}$$

Equation (13.67) must now be evaluated for each input: step, ramp, and parabola.

## Unit Step Input

For a unit step input, $R(s) = 1/s$. From Table 13.1,

$$R(z) = \frac{z}{z - 1} \tag{13.68}$$

Substituting Eq. (13.68) into Eq. (13.67), we have

$$e^*(\infty) = \frac{1}{1 + \lim_{z\to 1} G(z)} \tag{13.69}$$

[4] See *Ogata (1987: 59)* for a derivation.

Defining the static error constant, $K_p$, as

$$K_p = \lim_{z \to 1} G(z) \tag{13.70}$$

we rewrite Eq. (13.69) as

$$e^*(\infty) = \frac{1}{1 + K_p} \tag{13.71}$$

## Unit Ramp Input

For a unit ramp input, $R(z) = Tz/(z - 1)^2$. Following the procedure for the step input, you can derive the fact that

$$e^*(\infty) = \frac{1}{K_v} \tag{13.72}$$

where

$$K_v = \frac{1}{T} \lim_{z \to 1} (z - 1)G(z) \tag{13.73}$$

## Unit Parabolic Input

For a unit parabolic input, $R(z) = T^2 z/(z + 1)/[2(z - 1)^3]$. Similarly,

$$e^*(\infty) = \frac{1}{K_a} \tag{13.74}$$

where

$$K_a = \frac{1}{T^2} \lim_{z \to 1} (z - 1)^2 G(z) \tag{13.75}$$

## Summary of Steady-State Errors

The equations developed above for $e^*(\infty)$, $K_p$, $K_v$, and $K_a$ are similar to the equations developed for analog systems. Whereas multiple pole placement at the origin of the $s$-plane reduced steady-state errors to zero in the analog case, we can see that multiple pole placement at $z = 1$ reduces the steady-state error to zero for digital systems of the type discussed in this section. This conclusion makes sense when one considers that $s = 0$ maps into $z = 1$ under $z = e^{Ts}$.

For example, for a step input, we see that if $G(z)$ in Eq. (13.69) has one pole at $z = 1$, the limit will become infinite, and the steady-state error will reduce to zero.

For a ramp input, if $G(z)$ in Eq. (13.73) has two poles at $z = 1$, the limit will become infinite, and the error will reduce to zero.

Similar conclusions can be drawn for the parabolic input and Eq. (13.75). Here, $G(z)$ needs three poles at $z = 1$ in order for the steady-state error to be zero. Let us look at an example.

## Example 13.9 Finding Steady-State Error

**PROBLEM:** For step, ramp, and parabolic inputs, find the steady-state error for the feedback control system shown in Figure 13.17(*a*) if

$$G_1(s) = \frac{10}{s(s+1)} \tag{13.76}$$

**SOLUTION:** First find $G(s)$, the product of the z.o.h. and the plant.

$$G(s) = \frac{10(1-e^{-Ts})}{s^2(s+1)} = 10(1-e^{-Ts})\left[\frac{1}{s^2} - \frac{1}{s} + \frac{1}{s+1}\right] \tag{13.77}$$

The $z$-transform is then

$$\begin{aligned} G(z) &= 10(1-z^{-1})\left[\frac{Tz}{(z-1)^2} - \frac{z}{z-1} + \frac{z}{z-e^{-T}}\right] \\ &= 10\left[\frac{T}{z-1} - 1 + \frac{z-1}{z-e^{-T}}\right] \end{aligned} \tag{13.78}$$

For a step input,

$$K_p = \lim_{z\to 1} G(z) = \infty; \quad e^*(\infty) = \frac{1}{1+K_p} = 0 \tag{13.79}$$

For a ramp input,

$$K_v = \frac{1}{T}\lim_{z\to 1}(z-1)G(z) = 10; \quad e^*(\infty) = \frac{1}{K_v} = 0.1 \tag{13.80}$$

For a parabolic input,

$$K_a = \frac{1}{T^2}\lim_{z\to 1}(z-1)^2 G(z) = 0; \quad e^*(\infty) = \frac{1}{K_a} = \infty \tag{13.81}$$

You will notice that the answers obtained are the same as the results obtained for the analog system. However, since stability depends upon the sampling interval, be sure to check the stability of the system after a sampling interval is established before making steady-state error calculations.

MATLAB

ML

Students who are using MATLAB should now run `ch13p6` in Appendix B. You will learn how to use MATLAB to determine $K_p$, $K_v$, and $K_a$ in a digital system as well as check the stability. This exercise solves Example 13.9 using MATLAB.

## Skill-Assessment Exercise 13.7

**PROBLEM:** For step, ramp, and parabolic inputs, find the steady-state error for the feedback control system shown in Figure 13.17(*a*) if

$$G_1(s) = \frac{20(s+3)}{(s+4)(s+5)}$$

Let $T = 0.1$ second. Repeat for $T = 0.5$ second.

**ANSWER:** For $T = 0.1$ second, $K_p = 3$, $K_v = 0$, and $K_a = 0$; for $T = 0.5$ second, the system is unstable.

The complete solution is located at www.wiley.com/college/nise.

In this section, we discussed and evaluated the steady-state error of digital systems for step, ramp, and parabolic inputs. The equations for steady-state error parallel those for analog systems. Even the definitions of the static error constants were similar. Poles at the origin of the $s$-plane for analog systems were replaced with poles at +1 on the $z$-plane to improve the steady-state error. We continue our parallel discussion by moving into a discussion of transient response and the root locus for digital systems.

## 13.8 Transient Response on the z-Plane

Recall that for analog systems a transient response requirement was specified by selecting a closed-loop, $s$-plane pole. In Chapter 8, the closed-loop pole was on the existing root locus, and the design consisted of a simple gain adjustment. If the closed-loop pole was not on the existing root locus, then a cascade compensator was designed to reshape the original root locus to go through the desired closed-loop pole. A gain adjustment then completed the design.

In the next two sections, we want to parallel the described analog methods and apply similar techniques to digital systems. For this introductory chapter, we will parallel the discussion through design via gain adjustment. The design of compensation is left to you to pursue in an advanced course.

Chapter 4 established the relationships between transient response and the $s$-plane. We saw that vertical lines on the $s$-plane were lines of constant settling time, horizontal lines were lines of constant peak time, and radial lines were lines of constant percent overshoot. In order to draw equivalent conclusions on the $z$-plane, we now map those lines through $z = e^{sT}$.

The vertical lines on the $s$-plane are lines of constant settling time and are characterized by the equation $s = \sigma_1 + j\omega$, where the real part, $\sigma_1 = -4/T_s$, is constant and is in the left–half-plane for stability. Substituting this into $z = e^{sT}$, we obtain

$$z = e^{\sigma_1 T} e^{j\omega T} = r_1 e^{j\omega T} \tag{13.82}$$

Equation (13.82) denotes concentric circles of radius $r_1$. If $\sigma_1$ is positive, the circle has a larger radius than the unit circle. On the other hand, if $\sigma_1$ is negative, the circle has a smaller radius than the unit circle. The circles of constant settling time, normalized to the sampling interval, are shown in Figure 13.18 with radius $e^{\sigma_1 T} = e^{-4/(T_s/T)}$. Also, $T_s/T = -4/\ln(r)$, where $r$ is the radius of the circle of constant settling time.

The horizontal lines are lines of constant peak time. The lines are characterized by the equation $s = \sigma + j\omega_1$, where the imaginary part, $\omega_1 = \pi/T_p$, is constant. Substituting this into $z = e^{sT}$, we obtain

$$z = e^{\sigma T} e^{j\omega_1 T} = e^{\sigma T} e^{j\theta_1} \tag{13.83}$$

Equation (13.83) represents radial lines at an angle of $\theta_1$. If $\sigma$ is negative, that section of the radial line lies inside the unit circle. If $\sigma$ is positive, that section of the radial line lies outside the unit circle. The lines of constant peak time normalized to the sampling interval are shown in Figure 13.18. The angle of each radial line is $\omega_1 T = \theta_1 = \pi/(T_p/T)$, from which $T_p/T = \pi/\theta_1$.

Finally, we map the radial lines of the $s$-plane onto the $z$-plane. Remember, these radial lines are lines of constant percent overshoot on the $s$-plane. From Figure 13.19, these radial lines are represented by

$$\frac{\sigma}{\omega} = -\tan(\sin^{-1}\zeta) = -\frac{\zeta}{\sqrt{1-\zeta^2}} \tag{13.84}$$

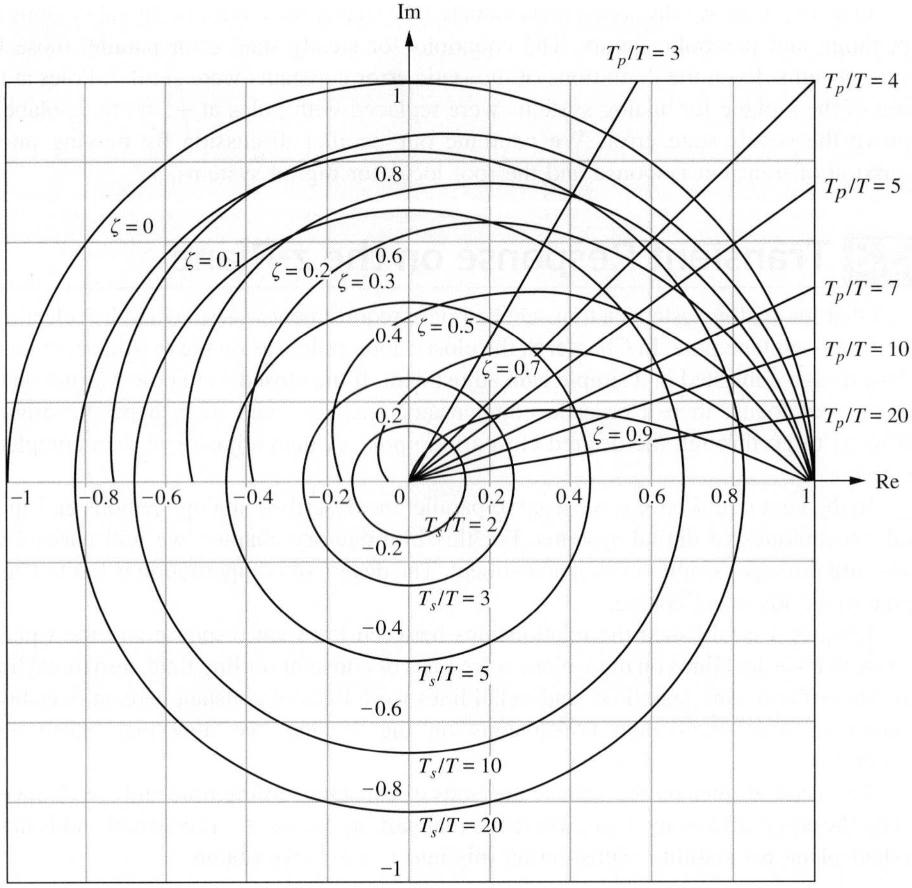


**FIGURE 13.18** Constant damping ratio, normalized settling time, and normalized peak time plots on the $z$-plane

Hence,

$$s = \sigma + j\omega = -\omega \frac{\zeta}{\sqrt{1-\zeta^2}} + j\omega \tag{13.85}$$

Transforming Eq. (13.85) to the $z$-plane yields

$$z = e^{sT} = e^{-\omega T\left(\zeta/\sqrt{1-\zeta^2}\right)} e^{j\omega T} = e^{-\omega T\left(\zeta/\sqrt{1-\zeta^2}\right)} \angle \omega T \tag{13.86}$$

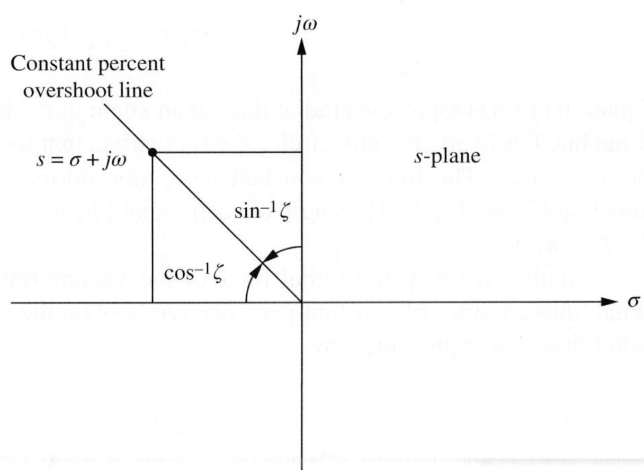


**FIGURE 13.19** The $s$-plane sketch of constant percent overshoot line

Thus, given a desired damping ratio, $\zeta$, Eq. (13.86) can be plotted on the $z$-plane through a range of $\omega T$ as shown in Figure 13.18. These curves can then be used as constant percent overshoot curves on the $z$-plane.

This section has set the stage for the analysis and design of transient response for digital systems. In the next section, we apply the results to digital systems using the root locus.

## 13.9 Gain Design on the z-Plane

In this section, we plot root loci and determine the gain required for stability as well as the gain required to meet a transient response requirement. Since the open-loop and closed-loop transfer functions for the generic digital system shown in Figure 13.20 are identical to the continuous system except for a change in variables from $s$ to $z$, we can use the same rules for plotting a root locus.

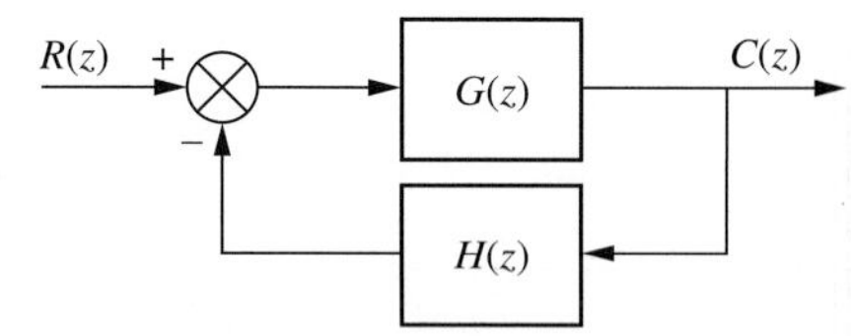


**FIGURE 13.20** Generic digital feedback control system

However, from our previous discussion, the region of stability on the $z$-plane is within the unit circle and not the left half-plane. Thus, in order to determine stability, we must search for the intersection of the root locus with the unit circle rather than the imaginary axis.

In the last section, we derived the curves of constant settling time, peak time, and damping ratio. In order to design a digital system for transient response, we find the intersection of the root locus with the appropriate curves as they appear on the $z$-plane in Figure 13.18. Let us look at the following example.

### Example 13.10 Stability Design via Root Locus

**PROBLEM:** Sketch the root locus for the system shown in Figure 13.21. Also, determine the range of gain, $K$, for stability from the root locus plot.

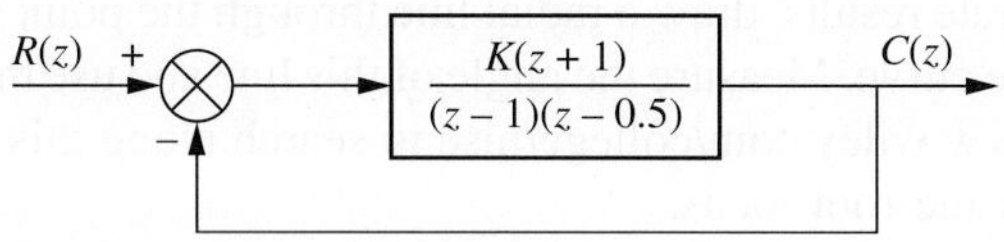


**FIGURE 13.21** Digital feedback control for Example 13.10

**SOLUTION:** Treat the system as if $z$ were $s$, and sketch the root locus. The result is shown in Figure 13.22. Using the root locus program discussed in Appendix H.2 at www.wiley.com/college/nise, search along the unit circle for 180. Identification of the gain, $K$, at this point yields the range of gain for stability. Using the program, we find that the intersection of the root locus with the unit circle is $1\angle 60$. The gain at this point is 0.5. Hence, the range of gain for stability is $0 < K < 0.5$.

MATLAB
**ML**

`Students who are using MATLAB should now run ch13p7 in Appendix B. You will learn how to use MATLAB to plot a root locus on the z-plane as well as superimpose the unit circle. You will learn how to select interactively the intersection of the root locus and the unit circle to obtain the value of gain for stability. This exercise solves Example 13.10 using MATLAB.`

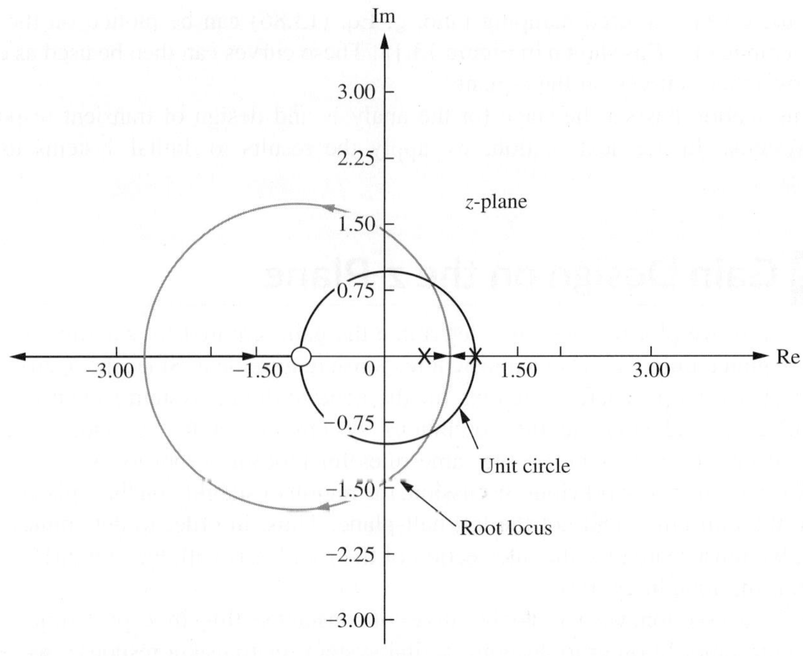


**FIGURE 13.22** Root locus for the system of Figure 13.21

In the next example, we design the value of gain, $K$, in Figure 13.21 to meet a transient response specification. The problem is handled similarly to the analog system design, where we found the gain at the point where the root locus crossed the specified damping ratio, settling time, or peak time curve. In digital systems, these curves are as shown in Figure 13.18. In summary, then, draw the root locus of the digital system and superimpose the curves of Figure 13.18. Then find out where the root locus intersects the desired damping ratio, settling time, or peak time curve and evaluate the gain at that point. In order to simplify the calculations and obtain more accurate results, draw a radial line through the point where the root locus intersects the appropriate curve. Measure the angle of this line and use the root locus program in Appendix H.2 at www.wiley.com/college/nise to search along this radial line for the point of intersection with the root locus.

## Example 13.11 Transient Response Design via Gain Adjustment

**PROBLEM:** For the system of Figure 13.21, find the value of gain, $K$, to yield a damping ratio of 0.7.

**SOLUTION:** Figure 13.23 shows the constant damping ratio curves superimposed over the root locus for the system as determined from the last example. Draw a radial line from the origin to the intersection of the root locus with the 0.7 damping ratio curve (a 16.62 line). The root locus program discussed in Appendix H.2 at www.wiley.com/college/nise can now be used to obtain the gain by searching along a 16.62 line for 180, the intersection with the root locus. The results of the program show that the gain, $K$, is 0.0627 at $0.719 + j0.215$, the point where the 0.7 damping ratio curve intersects the root locus.

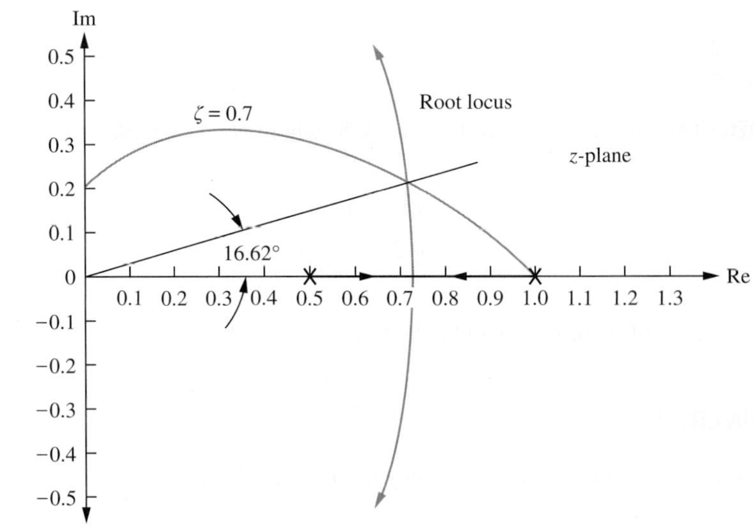


**FIGURE 13.23** Root locus for the system of Figure 13.21 with constant 0.7 damping ratio curve

We can now check our design by finding the unit sampled step response of the system of Figure 13.21. Using our design, $K = 0.0627$, along with $R(z) = z/(z-1)$, a sampled step input, we find the sampled output to be

$$C(z) = \frac{R(z)G(z)}{1+G(z)} = \frac{0.0627z^2 + 0.0627z}{z^3 - 2.4373z^2 + 2z - 0.5627} \tag{13.87}$$

Performing the indicated division, we obtain the output valid at the sampling instants, as shown in Figure 13.24. Since the overshoot is approximately 5%, the requirement of a 0.7 damping ratio has been met. You should remember, however, that the plot is valid only at integer values of the sampling instants.

MATLAB ML

`Students who are using MATLAB should now run ch13p8 in Appendix B. You will learn how to use MATLAB to plot a root locus on the z-plane as well as superimpose a grid of damping ratio curves. You will learn how to obtain the gain and a closed-loop step response of a digital system after interactively selecting the operating point on the root locus. This exercise solves Example 13.11 using MATLAB.`

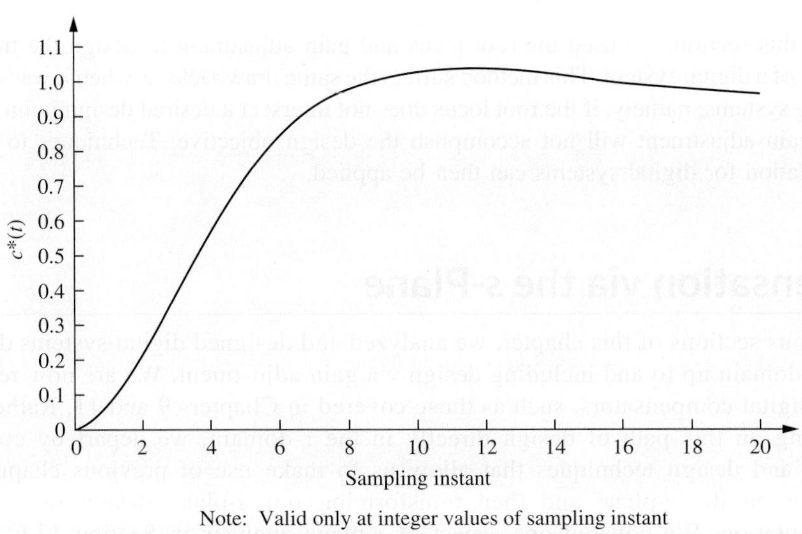


**FIGURE 13.24** Sampled step response of the system of Figure 13.21 with $K = 0.0627$

## Skill-Assessment Exercise 13.8

**TryIt 13.3**

Use MATLAB, the Control System Toolbox, and the following statements to solve Skill-Assessment Exercise 13.8.

```
Gz=zpk(-0.5,[0.25,0.75],...
1,[])
rlocus(Gz)
zgrid(0.5, [])
[K,p]=rlocfind(Gz)
```

Note: When the root locus appears, click on the intersection of the 0.5 damping ratio curve and the root locus to calculate the gain.

**PROBLEM:** For the system of Figure 13.20 where $H(z) = 1$ and

$$G(z) = \frac{K(z + 0.5)}{(z - 0.25)(z - 0.75)}$$

find the value of gain, $K$, to yield a damping ratio of 0.5.

**ANSWER:** $K = 0.31$

The complete solution is at www.wiley.com/college/nise.

Simulink
**SL**

Simulink provides an alternative method of simulating digital systems to obtain the time response. Students who are performing the MATLAB exercises and want to explore the added capability of Simulink should now consult Appendix C, Simulink Tutorial. Example C.4 in the tutorial shows how to use Simulink to simulate digital systems.

GUI Tool
**GUIT**

MATLAB's LTI Viewer provides another method of simulating digital systems to obtain the time response. Students who are performing the MATLAB exercises and want to explore the added capability of MATLAB's LTI Viewer should now consult Appendix E at www.wiley.com/college/nise, which contains a tutorial on the LTI Viewer as well as some examples. One of the illustrative examples, Example E.5, finds the closed-loop step response of a digital system using the LTI Viewer.

In this section, we used the root locus and gain adjustment to design the transient response of a digital system. This method suffers the same drawbacks as when it was applied to analog systems; namely, if the root locus does not intersect a desired design point, then a simple gain adjustment will not accomplish the design objective. Techniques to design compensation for digital systems can then be applied.

# 13.10 Cascade Compensation via the *s*-Plane

In previous sections of this chapter, we analyzed and designed digital systems directly in the $z$-domain up to and including design via gain adjustment. We are now ready to design digital compensators, such as those covered in Chapters 9 and 11. Rather than continuing on this path of design directly in the $z$-domain, we depart by covering analysis and design techniques that allow us to make use of previous chapters by designing on the $s$-plane and then transforming our $s$-plane design to a digital implementation. We covered one aspect of $s$-plane analysis in Section 13.6, where we used a bilinear transformation to analyze stability. We now continue with $s$-plane analysis and design by applying it to cascade compensator design. Direct design of compensators on the $z$-plane is left for a dedicated course in digital control systems.

## Cascade Compensation

In order to perform design in the $s$-plane and then convert the continuous compensator to a digital compensator, we need a bilinear transformation that will preserve, at the sampling instants, the response of the continuous compensator. The bilinear transformation covered in Section 13.6 will not meet that requirement. A bilinear transformation that can be performed with hand calculations and yields a digital transfer function whose output response at the sampling instants is approximately the same as the equivalent analog transfer function is called the *Tustin transformation*. This transformation is used to transform the continuous compensator, $G_c(s)$, to the digital compensator, $G_c(z)$. The Tustin transformation is given by[5]

$$s = \frac{2(z-1)}{T(z+1)} \tag{13.88}$$

and its inverse by

$$z = \frac{-\left(s+\dfrac{2}{T}\right)}{\left(s-\dfrac{2}{T}\right)} = \frac{1+\dfrac{T}{2}s}{1-\dfrac{T}{2}s} \tag{13.89}$$

As the sampling interval, $T$, gets smaller (higher sampling rate), the designed digital compensator's output yields a closer match to the analog compensator. If the sampling rate is not high enough, there is a discrepancy at higher frequencies between the digital and analog filters' frequency responses. Methods are available to correct the discrepancy, but they are beyond the scope of our discussion. The interested reader should investigate the topic of *prewarping*, covered in books dedicated to digital control and listed in the Bibliography at the end of this chapter.

*Astrom and Wittenmark* (*1984*) have developed a guideline for selecting the sampling interval, $T$. Their conclusion is that the value of $T$ in seconds should be in the range $0.15/\omega_{\Phi_M}$ to $0.5/\omega_{\Phi_M}$, where $\omega_{\Phi_M}$ is the zero dB frequency (rad/s) of the magnitude frequency response curve for the cascaded analog compensator and plant.

In the following example, we will design a compensator, $G_c(s)$, to meet the required performance specifications. We will then use the Tustin transformation to obtain the model for an equivalent digital controller. In the next section, we will show how to implement the digital controller.

[5] See *Ogata* (*1987: 315–318*) for a derivation.

## Example 13.12 Digital Cascade Compensator Design

**PROBLEM:** For the digital control system of Figure 13.25($a$), where

$$G_p(s) = \frac{1}{s(s+6)(s+10)} \tag{13.90}$$

design a digital lead compensator, $G_c(z)$, as shown in Figure 13.25($c$), so that the system will operate with 20% overshoot and a settling time of 1.1 seconds. Create your design in the $s$-domain and transform the compensator to the $z$-domain.

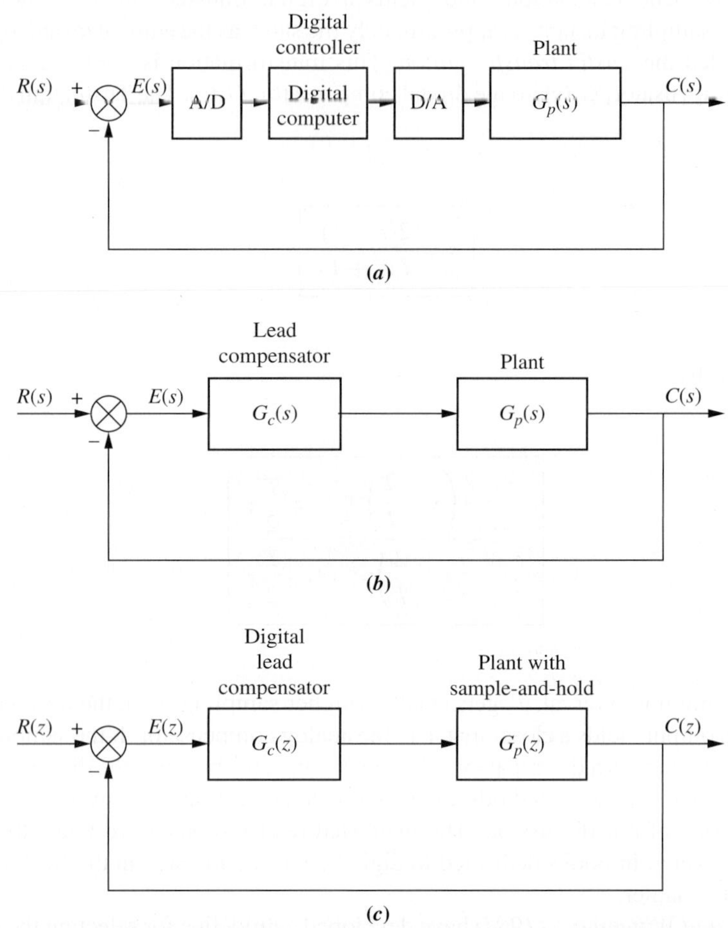


**FIGURE 13.25** **a.** Digital control system showing the digital computer performing compensation; **b.** continuous system used for design; **c.** transformed digital system

**SOLUTION:** Using Figure 13.25($b$), design a lead compensator using the techniques described in Chapter 9 or 11. The design was created as part of Example 9.6, where we found that the lead compensator was

$$G_c(s) = \frac{1977(s+6)}{(s+29.1)} \tag{13.91}$$

Using Eqs. (13.90) and (13.91), we find that the zero dB frequency, $\omega_{\Phi_M}$, for $G_p(s)G_c(s)$ is 5.8 rad/s. Using the guideline described by *Astrom and Wittenmark* (*1984*), the lowest value of $T$ should be in the range $0.15/\omega_{\Phi_M} = 0.026$ to $0.5/\omega_{\Phi_M} = 0.086$ second. Let us use $T = 0.01$ second.

Substituting Eq. (13.88) into Eq. (13.91) with $T = 0.01$ second yields

$$G_c(z) = \frac{1778z - 1674}{z - 0.746} \tag{13.92}$$

The $z$-transform of the plant and zero-order hold, found by the method discussed in Section 13.4 with $T = 0.01$ second, is

$$G_p(z) = \frac{(1.602 \times 10^{-7}z^2) + (6.156 \times 10^{-7}z) + (1.478 \times 10^{-7})}{z^3 - 2.847z^2 + 2.699z - 0.8521} \tag{13.93}$$

The time response in Figure 13.26 ($T = 0.01$ s) shows that the compensated closed-loop system meets the transient response requirements. The figure also shows the response for a compensator designed with sampling times at the extremes of Astrom and Wittenmark's guideline.

MATLAB ML

`Students who are using MATLAB should now run ch13p9 in Appendix B. You will learn how to use MATLAB to design a digital lead compensator using the Tustin transformation. This exercise solves Example 13.12 using MATLAB.`

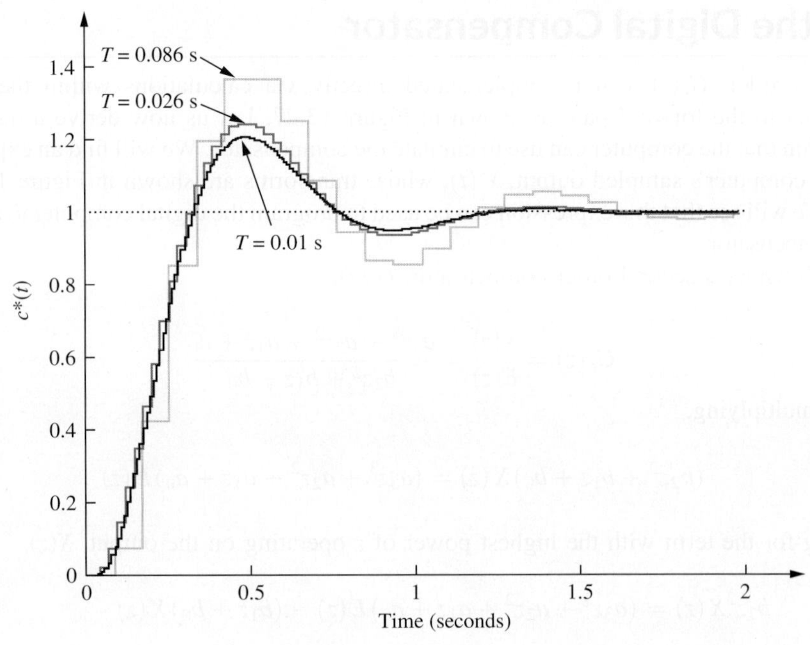


**FIGURE 13.26** Closed-loop response for the compensated system of Example 13.12 showing effect of three different sampling frequencies

## Skill-Assessment Exercise 13.9

**PROBLEM:** In Example 11.3, a lead compensator was designed for a unity feedback system whose plant was

$$G(s) = \frac{100K}{s(s+36)(s+100)}$$

The design specifications were as follows: percent overshoot = 20%, peak time = 0.1 second, and $K_v = 40$. In order to meet the requirements, the design yielded $K = 1440$ and a lead compensator,

$$G_c(s) = 2.38\frac{s+25.3}{s+60.2}$$

If the system is to be computer controlled, find the digital controller, $G_c(z)$.

**ANSWER:** $G_c(z) = 2.34\dfrac{z-0.975}{z-0.9416}$, $T = 0.001$ second

The complete solution is at www.wiley.com/college/nise.

Now that we have learned how to design a digital cascade compensator, $G_c(z)$, the next section will teach us how to use the digital computer to implement it.

## 13.11 Implementing the Digital Compensator

The controller, $G_c(z)$, can be implemented directly via calculations within the digital computer in the forward path as shown in Figure 13.27. Let us now derive a numerical algorithm that the computer can use to emulate the compensator. We will find an expression for the computer's sampled output, $x^*(t)$, whose transforms are shown in Figure 13.27 as $X(z)$. We will see that this expression can be used to program the digital computer to emulate the compensator.

Consider a second-order compensator, $G_c(z)$,

$$G_c(z) = \frac{X(z)}{E(z)} = \frac{a_3z^3 + a_2z^2 + a_1z + a_0}{b_2z^2 + b_1z + b_0} \tag{13.94}$$

Cross-multiplying,

$$(b_2z^2 + b_1z + b_0)X(z) = (a_3z^3 + a_2z^2 + a_1z + a_0)E(z) \tag{13.95}$$

Solving for the term with the highest power of $z$ operating on the output, $X(z)$,

$$b_2z^2X(z) = (a_3z^3 + a_2z^2 + a_1z + a_0)E(z) - (b_1z + b_0)X(z) \tag{13.96}$$

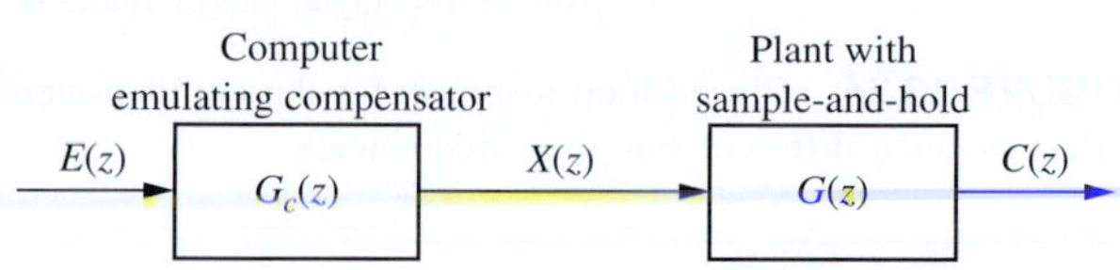


**FIGURE 13.27** Block diagram showing computer emulation of a digital compensator

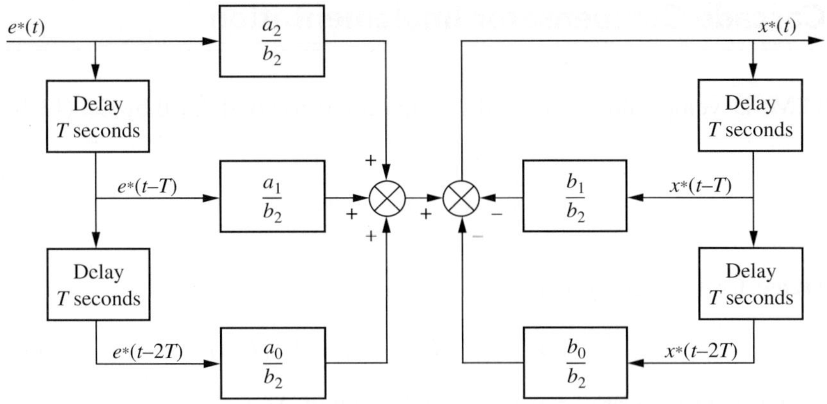


**FIGURE 13.28** Flowchart for a second-order digital compensator[6]

Dividing by the coefficient of $X(z)$ on the left-hand side of Eq. (13.96) yields

$$X(z) = \left(\frac{a_3}{b_2}z + \frac{a_2}{b_2} + \frac{a_1}{b_2}z^{-1} + \frac{a_0}{b_2}z^{-2}\right)E(z) - \left(\frac{b_1}{b_2}z^{-1} + \frac{b_0}{b_2}z^{-2}\right)X(z) \tag{13.97}$$

Finally, taking the inverse $z$-transform,

$$\begin{aligned} x^*(t) = {} & \frac{a_3}{b_2}e^*(t+T) + \frac{a_2}{b_2}e^*(t) + \frac{a_1}{b_2}e^*(t-T) + \frac{a_0}{b_2}e^*(t-2T) \\ & - \frac{b_1}{b_2}x^*(t-T) - \frac{b_0}{b_2}x^*(t-2T) \end{aligned} \tag{13.98}$$

We can see from this equation that the present sample of the compensator output, $x^*(t)$, is a function of future $(e^*(t+T))$, present $(e^*(t))$, past $(e^*(t-T))$, and $e^*(t-2T))$ samples of $e(t)$, along with past values of the output, $x^*(t-T)$ and $x^*(t-2T)$. Obviously, if we are to physically realize this compensator, the output sample cannot be dependent upon future values of the input. Hence, to be physically realizable, $a_3$ must equal zero for the future value of $e(t)$ to be zero. We conclude that the numerator of the compensator's transfer function must be of equal or lower order than the denominator in order that the compensator be physically realizable.

Now assume that $a_3$ does indeed equal zero. Equation (13.98) now becomes

$$x^*(t) = \frac{a_2}{b_2}e^*(t) + \frac{a_1}{b_2}e^*(t-T) + \frac{a_0}{b_2}e^*(t-2T) - \frac{b_1}{b_2}x^*(t-T) - \frac{b_0}{b_2}x^*(t-2T) \tag{13.99}$$

Hence, the output sample is a function of current and past input samples of the input as well as past samples of the output. Figure 13.28 shows the flowchart of the compensator from which a program can be written for the digital computer.[7] The figure shows that the compensator can be implemented by storing several successive values of the input and output. The output is then formed by a weighted linear combination of these stored variables. Let us now look at a numerical example.

[6] Adapted from Chassing, R. *Digital Signal Processing* (New York: John Wiley & Sons, Inc., 1999), p. 137. © 1999 John Wiley & Sons, Inc.

[7] For an excellent discussion on basic flowcharts to represent digital compensators, including the representation shown in Figure 13.28 and alternative flowcharts with half as many delays, see *Chassaing (1999, pp. 135–143)*.

## Example 13.13 Digital Cascade Compensator Implementation

**PROBLEM:** Develop a flowchart for the digital compensator defined by Eq. (13.100).

$$G_c(z) = \frac{X(z)}{E(z)} = \frac{z+0.5}{z^2 - 0.5z + 0.7} \tag{13.100}$$

**SOLUTION:** Cross-multiply and obtain

$$(z^2 - 0.5z + 0.7)X(z) = (z+0.5)E(z) \tag{13.101}$$

Solve for the highest power of $z$ operating on the output, $X(z)$,

$$z^2X(z) = (z+0.5)E(z) - (-0.5z + 0.7)X(z) \tag{13.102}$$

Solving for $X(z)$ on the left-hand side,

$$X(z) = (z^{-1} + 0.5z^{-2})E(z) - (-0.5z^{-1} + 0.7z^{-2})X(z) \tag{13.103}$$

Implementing Eq. (13.103) with the flowchart of Figure 13.29 completes the design.

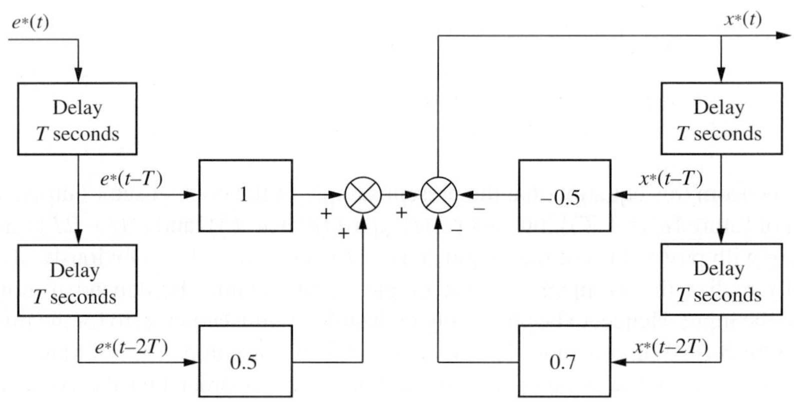


**FIGURE 13.29** Flowchart to implement[8] $G_c(z) = \dfrac{z+0.5}{z^2 - 0.5z + 0.7}$

## Skill-Assessment Exercise 13.10

**PROBLEM:** Draw a flowchart from which the compensator

$$G_c(z) = \frac{1899z^2 - 3761z + 1861}{z^2 - 1.908z + 0.9075}$$

can be programmed if the sampling interval is 0.1 second.

**ANSWER:** The complete solution is at www.wiley.com/college/nise.

[8] Adapted from Chassaing, R. *Digital Signal Processing* (New York: John Wiley & Sons, Inc., 1999), p. 137. © 1999 John Wiley & Sons, Inc.

In this section, we learned how to implement a digital compensator. The resulting flowchart can serve as the design of a digital computer program for the computer in the loop. The design consists of delays that can be thought of as storage for each sampled value of input and output. The stored values are weighted and added. The engineer then can implement the design with a computer program.

In the next section, we will put together the concepts of this chapter as we apply the principles of digital control system design to our antenna azimuth control system.

## Case Studies Antenna Control: Transient Design via Gain

Design D

We now demonstrate the objectives of this chapter by turning to our ongoing antenna azimuth position control system. We will show where the computer is inserted in the loop, model the system, and design the gain to meet a transient response requirement. Later, we will design a digital cascade compensator.

The computer will perform two functions in the loop. First, the computer will be used as the input device. It will receive digital signals from the keyboard in the form of commands, and digital signals from the output for closed-loop control. The keyboard will replace the input potentiometer, and an analog-to-digital (A/D) converter along with a unity gain feedback transducer will replace the output potentiometer.

Figure 13.30(*a*) shows the original analog system, and Figure 13.30(*b*) shows the system with the computer in the loop. Here the computer is receiving digital signals from two sources: (1) the input via the keyboard or other tracking commands and (2) the output via an A/D converter. The plant is receiving signals from the digital computer via a digital-to-analog (D/A) converter and the sample-and-hold.

Figure 13.30(*b*) shows some simplifying assumptions we have made. The power amplifier's pole is assumed to be far enough away from the motor's pole that we can represent the power amplifier as a pure gain equal to its dc gain of unity. Also, we have absorbed any preamplifier and potentiometer gain in the computer and its associated D/A converter.

**PROBLEM:** Design the gain for the antenna azimuth position control system shown in Figure 13.30(*b*) to yield a closed-loop damping ratio of 0.5. Assume a sampling interval of $T = 0.1$ second.

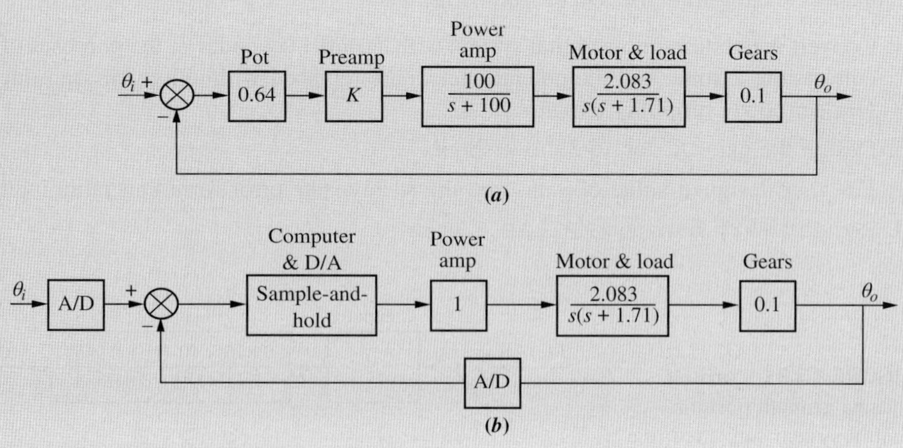


**FIGURE 13.30** Antenna control system: **a.** analog implementation; **b.** digital implementation

**SOLUTION: Modeling the System:** Our first objective is to model the system in the $z$-domain. The forward transfer function, $G(s)$, which includes the sample-and-hold, power amplifier, motor and load, and the gears, is

$$G(s) = \frac{1-e^{-Ts}}{s}\frac{0.2083}{s(s+a)} = \frac{0.2083}{a}(1-e^{Ts})\frac{a}{s^2(s+a)} \tag{13.104}$$

where $a = 1.71$, and $T = 0.1$.

Since the $z$-transform of $(1-e^{-Ts})$ is $(1-z^{-1})$ and, from Example 13.6, the $z$-transform of $a/[s^2(s+a)]$ is

$$z\left\{\frac{a}{s^2(s+a)}\right\} = \left[\frac{Tz}{(z-1)^2} - \frac{(1-e^{-aT})z}{a(z-1)(z-e^{-aT})}\right] \tag{13.105}$$

the $z$-transform of the plant, $G(z)$, is

$$\begin{aligned} G(z) &= \frac{0.2083}{a}(1-z^{-1})z\left\{\frac{a}{s^2(s+a)}\right\} \\ &= \frac{0.2083}{a^2}\left[\frac{[aT-(1-e^{-aT})]z + [(1-e^{-aT}) - aTe^{-aT}]}{(z-1)(z-e^{-aT})}\right] \end{aligned} \tag{13.106}$$

Substituting the values for $a$ and $T$, we obtain

$$G(z) = \frac{9.846 \times 10^{-4}(z+0.945)}{(z-1)(z-0.843)} \tag{13.107}$$

Figure 13.31 shows the computer and plant as part of the digital feedback control system.

**Designing for Transient Response:** Now that the modeling in the $z$-domain is complete, we can begin to design the system for the required transient response. We superimpose the root locus over the constant damping ratio curves in the $z$-plane, as shown in Figure 13.32. A line drawn from the origin to the intersection forms an 8.58 angle. Searching along this line for 180, we find the intersection to be $(0.915 + j0.138)$, with a loop gain, $9.846 \times 10^{-4}K$, of 0.0135. Hence, $K = 13.71$.

Checking the design by finding the unit sampled step response of the closed-loop system yields the plot of Figure 13.33, which exhibits 20% overshoot ($\zeta = 0.456$).

**CHALLENGE:** We now give you a case study to test your knowledge of this chapter's objectives: You are given the antenna azimuth position control system shown on the front endpapers, Configuration 2. Do the following:

**a.** Convert the system into a digital system with $T = 0.1$ second. For the purposes of the conversion, assume that the potentiometers are replaced with unity gain transducers. Neglect power amplifier dynamics.

**b.** Design the gain, $K$, for 16.3% overshoot.

**c.** For your designed value of gain, find the steady-state error for a unit ramp input.

**d.** Repeat Part **b** using MATLAB.

MATLAB
ML

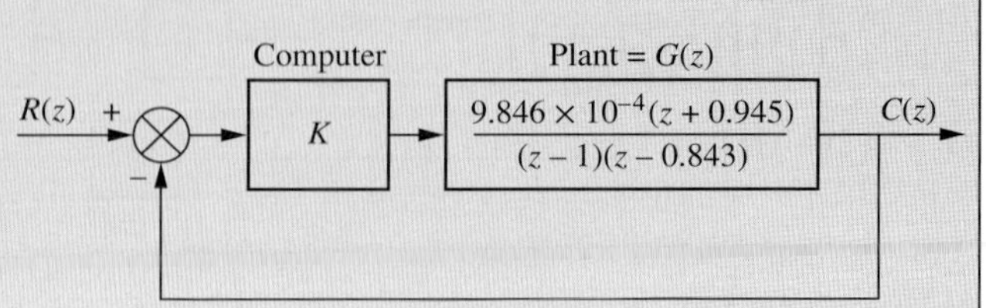


**FIGURE 13.31** Analog antenna azimuth position control system converted to a digital system

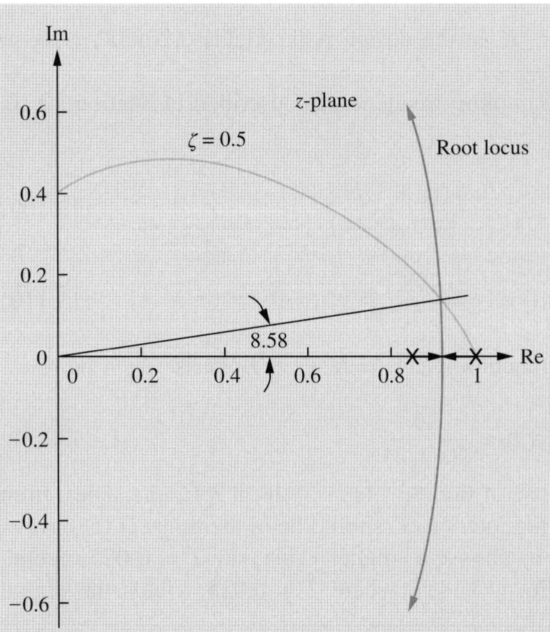


**FIGURE 13.32** Root locus superimposed over constant damping ratio curve

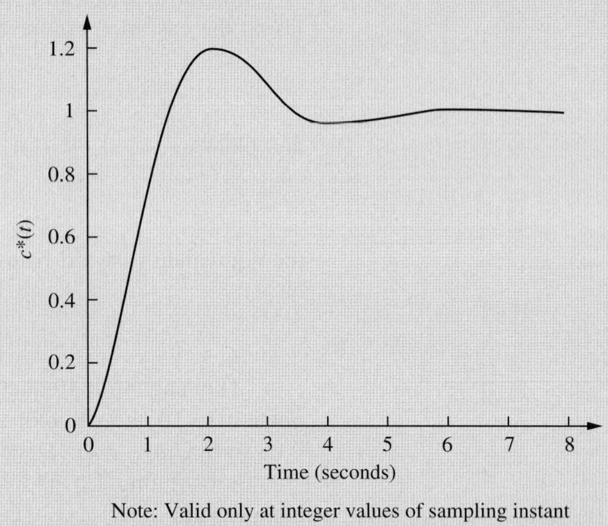


**FIGURE 13.33** Sampled step response of the antenna azimuth position control system

## Antenna Control: Digital Cascade Compensator Design

Design
D

**PROBLEM:** Design a digital lead compensator to reduce the settling time by a factor of 2.5 from that obtained for the antenna azimuth control system in the previous Case Study problem in this chapter.

**SOLUTION:** Figure 13.34 shows a simplified block diagram of the continuous system, neglecting power amplifier dynamics and assuming that the potentiometers are replaced with unity gain transducers as previously explained.

We begin with an $s$-plane design. From Figure 13.33, the settling time is about 5 seconds. Thus, our design requirements are a settling time of 2 seconds and a damping ratio of 0.5. The natural frequency is $\omega_n = 4/(\zeta T_s) = 4$ rad/s. The compensated dominant poles are located at $-\zeta\omega_n \pm j\omega_n\sqrt{1-\zeta^2} = -2 \pm j3.464$.

Designing a lead compensator zero to cancel the plant pole on the $s$-plane at $-1.71$ yields a lead compensator pole at $-4$. Hence the lead compensator is given by

$$G_c(s) = \frac{s+1.71}{s+4} \tag{13.108}$$

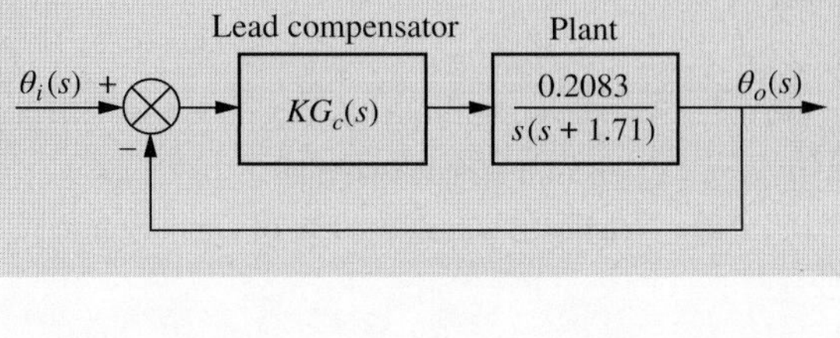


**FIGURE 13.34** Simplified block diagram of antenna azimuth control system

Using root locus to evaluate the gain, $K$, at the design point yields $0.2083K = 16$, or $K = 76.81$.

We now select an appropriate sampling frequency as described in Section 13.10. Using the cascaded compensator,

$$KG_c(s) = \frac{76.81(s+1.71)}{(s+4)} \tag{13.109}$$

and plant,

$$G_p(s) = \frac{0.2083}{s(s+1.71)} \tag{13.110}$$

the equivalent forward-path transfer function, $G_e(s) = KG_c(s)G_p(s)$, is

$$G_e(s) = \frac{16}{s(s+4)} \tag{13.111}$$

The magnitude frequency response of Eq. (13.111) is 0 dB at 3.1 rad/s. Thus, from Section 13.10, the value of the sampling interval, $T$, should be in the range $0.15/\omega_{\Phi_M} = 0.05$ to $0.5/\omega_{\Phi_M} = 0.16$ second. Let us choose a smaller value, say $T = 0.025$ second.

Substituting Eq. (13.89) into Eq. (13.111), where $T = 0.025$, yields the digital compensator

$$KG_c(z) = \frac{74.72z - 71.59}{z - 0.9048} \tag{13.112}$$

In order to simulate the digital system, we calculate the $z$-transform of the plant in Figure 13.34 in cascade with a zero-order sample-and-hold. The $z$-transform of the sampled plant is evaluated by the method discussed in Section 13.4 using $T = 0.025$. The result is

$$G_p(z) = \frac{6.418 \times 10^{-5}z + 6.327 \times 10^{-5}}{z^2 - 1.958z + 0.9582} \tag{13.113}$$

The step response in Figure 13.35 shows approximately 20% overshoot and a settling time of 2.1 seconds for the closed-loop digital system.

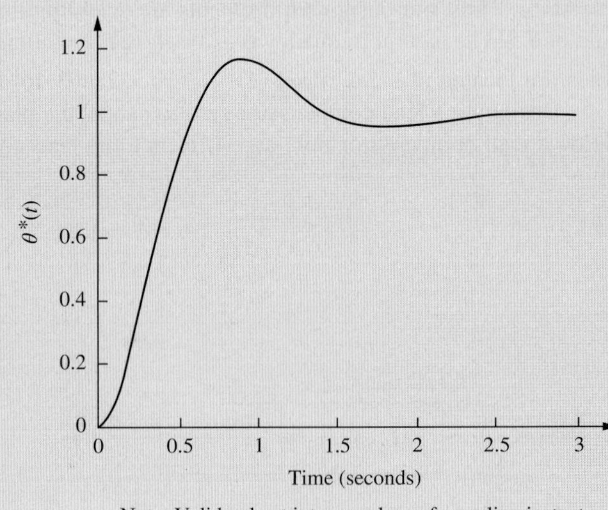


**FIGURE 13.35** Closed-loop digital step response for antenna control system with a lead compensator

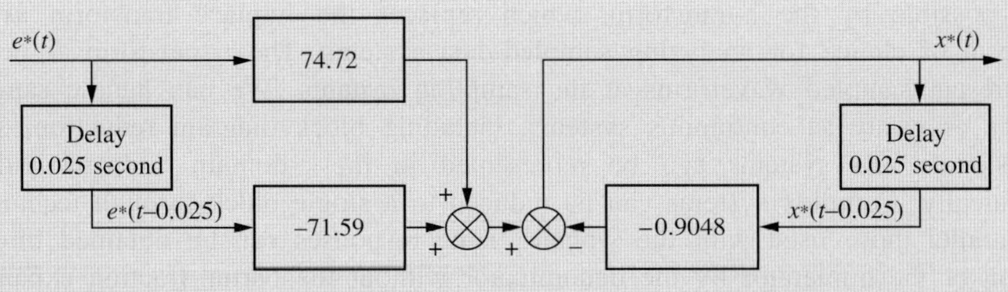


**FIGURE 13.36** Flowchart for digital lead compensator[9]

We conclude the design by obtaining a flowchart for the digital compensator. Using Eq. (13.112), where we define $KG_c(z) = X(z)/E(z)$, and cross-multiplying yields

$$(z - 0.9048)X(z) = (74.72z - 71.59)E(z) \tag{13.114}$$

Solving for the highest power of $z$ operating on $X(z)$,

$$zX(z) = (74.72z - 71.59)E(z) + 0.9048X(z) \tag{13.115}$$

Solving for $X(z)$,

$$X(z) = (74.72 - 71.59z^{-1})E(z) + 0.9048z^{-1}X(z) \tag{13.116}$$

Implementing Eq. (13.116) as a flowchart yields Figure 13.36.

**CHALLENGE:** You are now given a case study to test your knowledge of this chapter's objectives. You are given the antenna azimuth position control system shown on the front endpapers, Configuration 2. Replace the potentiometers with unity gain transducers, neglect power amplifier dynamics, and do the following:

**a.** Design a digital lead compensator to yield 10% overshoot with a 1-second peak time. Design in the $s$-plane and use the Tustin transformation to specify and implement a digital compensator. Choose an appropriate sampling interval.

**b.** Draw a flowchart for your digital lead compensator.

**c.** `Repeat Part` **a** `using MATLAB.`

MATLAB ML

# Summary

In this chapter, we covered the design of digital systems using classical methods. State-space techniques were not covered. However, you are encouraged to pursue this topic in a course dedicated to sampled-data control systems.

We looked at the advantages of digital control systems. These systems can control numerous loops at reduced cost. System modifications can be implemented with software changes rather than hardware changes.

Typically, the digital computer is placed in the forward path preceding the plant. Digital-to-analog and analog-to-digital conversion is required within the system to ensure compatibility of the analog and digital signals throughout the system. The digital computer in the loop is modeled as a sample-and-hold network along with any compensation that it performs.

Throughout the chapter, we saw direct parallels to the methods used for $s$-plane analysis of transients, steady-state errors, and the stability of analog systems. The parallel is

[9] Adapted from Chassing, R. *Digital Signal Processing* (New York: John Wiley & Sons, Inc., 1999), p. 137. © 1999 John Wiley & Sons, Inc.

made possible by the $z$-transform, which replaces the Laplace transform as the transform of choice for analyzing sampled-data systems. The $z$-transform allows us to represent sampled waveforms at the sampling instants. We can handle sampled systems as easily as continuous systems, including block diagram reduction, since both signals and systems can be represented in the $z$-domain and manipulated algebraically. Complex systems can be reduced to a single block through techniques that parallel those used with the $s$-plane. Time responses can be obtained through division of the numerator by the denominator without the partial-fraction expansion required in the $s$-domain.

Digital systems analysis parallels the $s$-plane techniques in the area of stability. The unit circle becomes the boundary of stability, replacing the imaginary axis.

We also found that the concepts of root locus and transient response are easily carried into the $z$-plane. The rules for sketching the root locus do not change. We can map points on the $s$-plane into points on the $z$-plane and attach transient response characteristics to the points. Evaluating a sampled-data system shows that the sampling rate, in addition to gain and load, determines the transient response.

Cascade compensators also can be designed for digital systems. One method is to first design the compensator on the $s$-plane or via frequency response techniques described in Chapters 9 and 11, respectively. Then the resulting design is transformed to a digital compensator using the Tustin transformation. Designing cascade compensation directly on the $z$-plane is an alternative method that can be used. However, these techniques are beyond the scope of this book.

This introductory control systems course is now complete. You have learned how to analyze and design linear control systems using frequency-domain and state-space techniques. This course is only a beginning. You may consider furthering your study of control systems by taking advanced courses in digital, nonlinear, and optimal control, where you will learn new techniques for analyzing and designing classes of systems not covered in this book. We hope we have whetted your appetite to continue your education in control systems engineering.

## Review Questions

1. Name two functions that the digital computer can perform when used with feedback control systems.
2. Name three advantages of using digital computers in the loop.
3. Name two important considerations in analog-to-digital conversion that yield errors.
4. Of what does the block diagram model for a computer consist?
5. What is the $z$-transform?
6. What does the inverse $z$-transform of a time waveform actually yield?
7. Name two methods of finding the inverse $z$-transform.
8. What method for finding the inverse $z$-transform yields a closed-form expression for the time function?
9. What method for finding the inverse $z$-transform immediately yields the values of the time waveform at the sampling instants?
10. In order to find the $z$-transform of a $G(s)$, what must be true of the input and the output?
11. If input $R(z)$ to system $G(z)$ yields output $C(z)$, what is the nature of $c(t)$?

12. If a time waveform, $c(t)$, at the output of system $G(z)$ is plotted using the inverse $z$-transform, and a typical second-order response with damping ratio $= 0.5$ results, can we say that the system is stable?
13. What must exist in order for cascaded sampled-data systems to be represented by the product of their pulse transfer functions, $G(z)$?
14. Where is the region for stability on the $z$-plane?
15. What methods for finding the stability of digital systems can replace the Routh-Hurwitz criterion for analog systems?
16. To drive steady-state errors in analog systems to zero, a pole can be placed at the origin of the $s$-plane. Where on the $z$-plane should a pole be placed to drive the steady-state error of a sampled system to zero?
17. How do the rules for sketching the root locus on the $z$-plane differ from those for sketching the root locus on the $s$-plane?
18. Given a point on the $z$-plane, how can one determine the associated percent overshoot, settling time, and peak time?
19. Given a desired percent overshoot and settling time, how can one tell which point on the $z$-plane is the design point?
20. Describe how digital compensators can be designed on the $s$-plane.
21. What characteristic is common between a cascade compensator designed on the $s$-plane and the digital compensator to which it is converted?

# Problems

1. Derive the $z$-transforms for the time functions listed below. Do not use any $z$-transform tables. Use the plan $f(t) \rightarrow f^*(t) \rightarrow F^*(s) \rightarrow F(z)$, followed by converting $F(z)$ into closed form making use of the fact that $1/(1-z^{-1}) = 1+z^{-1}+z^{-2}+z^{-3}+\cdots$. Assume ideal sampling. [Section: 13.3]

   **a.** $e^{-at}u(t)$

   **b.** $u(t)$

   **c.** $t^2e^{-at}u(t)$

   **d.** $\cos \omega t\, u(t)$

2. `Repeat all parts of Problem 1 using MATLAB and MATLAB's Symbolic Math Toolbox.` Symbolic Math SM

3. For each $F(z)$, find $f(kT)$ using partial-fraction expansion. [Section: 13.3]

   **a.** $F(z) = \dfrac{z(z+3)(z+5)}{(z-0.4)(z-0.6)(z-0.8)}$

   **b.** $F(z) = \dfrac{(z+0.2)(z+0.4)}{(z-0.1)(z-0.5)(z-0.9)}$

   **c.** $F(z) = \dfrac{(z+1)(z+0.3)(z+0.4)}{z(z-0.2)(z-0.5)(z-0.7)}$

4. `Repeat all parts of Problem 3 using MATLAB and MATLAB's Symbolic Math Toolbox.` Symbolic Math SM

5. For each $F(z)$ in Problem 3, do the following: [Section: 13.3]

   **a.** Find $f(kT)$ using the power series expansion.

   **b.** Check your results against your answers from Problem 3.

6. Using partial-fraction expansion and Table 13.1, findthe $z$-transform for each $G(s)$ shown below if $T = 0.5$ second.

   **a.** $G(s) = \dfrac{(s+4)}{(s+2)(s+5)}$

   **b.** $G(s) = \dfrac{(s+1)(s+2)}{s(s+3)(s+4)}$

   **c.** $G(s) = \dfrac{27}{(s+2)(s^2+4s+13)}$

   **d.** $G(s) = \dfrac{10}{s(s+2)(s^2+12s+61)}$

7. `Repeat all parts of Problem 6 using MATLAB and MATLAB's Symbolic Math Toolbox.`

Symbolic Math
SM

8. Find $G(z) = C(z)/R(z)$ for each of the block diagrams shown in Figure P13.1 if $T = 0.1$ second.

9. Find $T(z) = C(z)/R(z)$ for each of the systems shown in Figure P13.2. [Section: 13.5]

10. Find $C(z)$ in general terms for the digital system shown in Figure P13.3. [Section: 13.5]

11. Find the closed-loop transfer function, $T(z) = C(z)/R(z)$, for the system shown in Figure P13.4. [Section: 13.5]

12. `Write a MATLAB program that can be used to find the range of sampling time, T, for stability. The`

MATLAB
ML

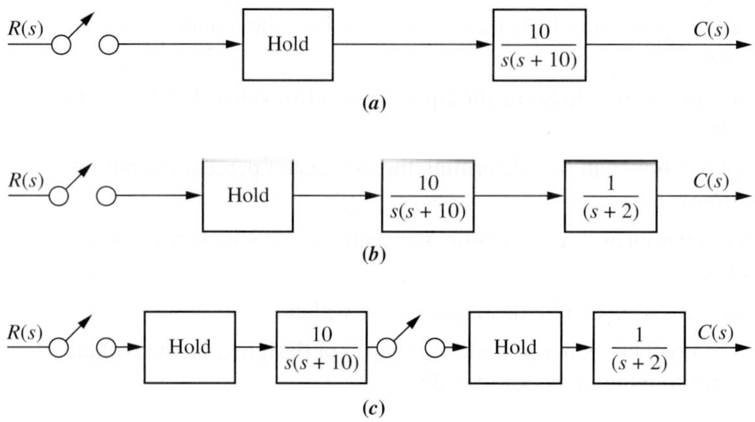


FIGURE P13.1

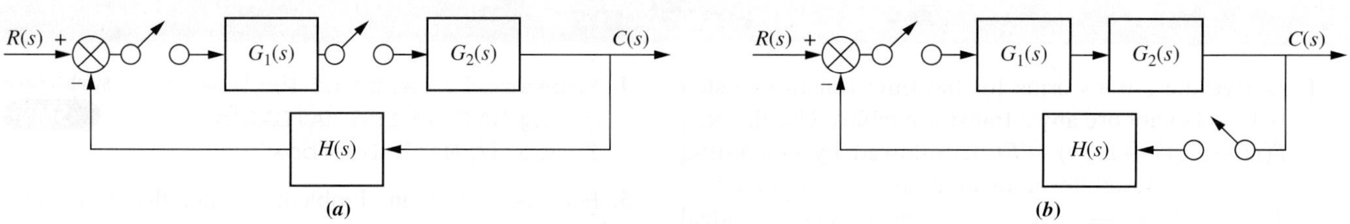


FIGURE P13.2

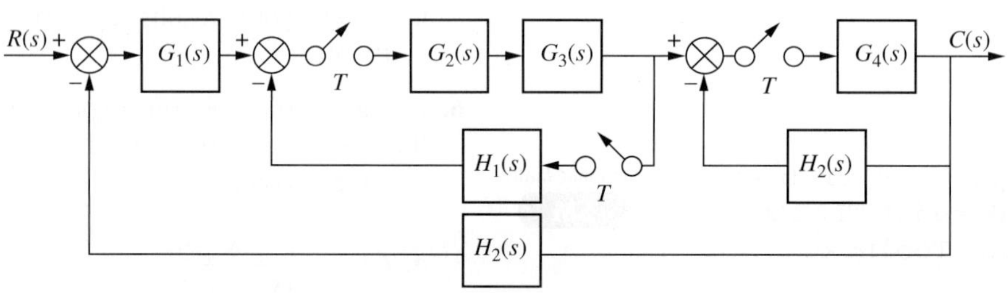


FIGURE P13.3

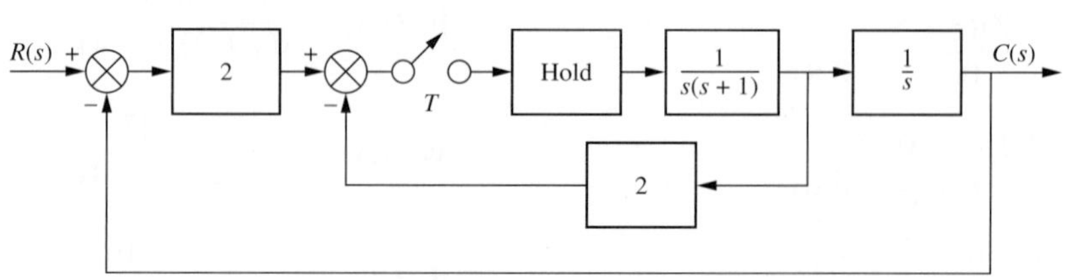


FIGURE P13.4

program will be used for systems of the type represented in Figure P13.5 and should meet the following requirements:

**a.** MATLAB will convert $G_1(s)$ cascaded with a sample-and-hold to $G(z)$.

**b.** The program will calculate the $z$-plane roots of the closed-loop system for a range of $T$ and determine the value of $T$, if any, below which the system will be stable. MATLAB will display this value of $T$ along with the $z$-plane poles of the closed-loop transfer function.

Test the program on

$$G_1(s) = \frac{10(s+7)}{(s+1)(s+3)(s+4)(s+5)}$$

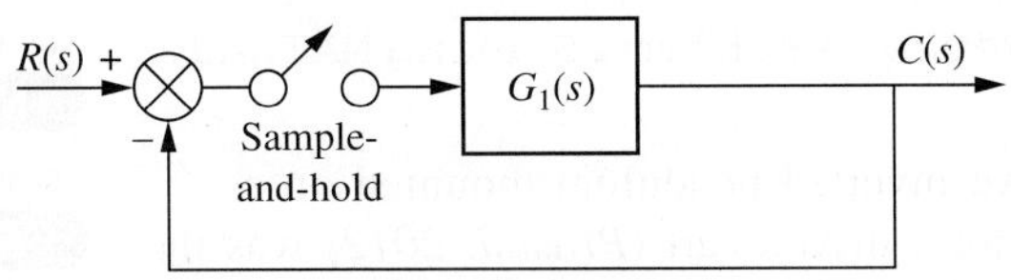


**FIGURE P13.5**

**13.** Given the system in Figure P13.6, find the range of sampling interval, $T$, that will keep the system stable. [Section: 13.6]

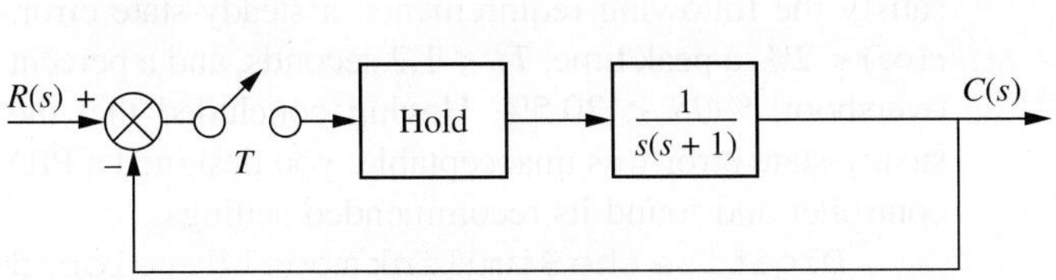


**FIGURE P13.6**

**14.** Find the range of gain, $K$, to make the system shown in Figure P13.7 stable. [Section: 13.6]

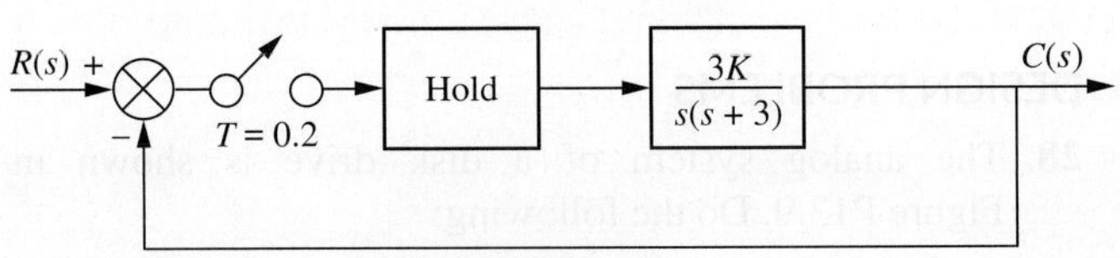


**FIGURE P13.7**

**15.** Find the static error constants and the steady-state error for each of the digital systems shown in Figure P13.8 if the inputs are [Section: 13.7]

**a.** $u(t)$

**b.** $tu(t)$

**c.** $\frac{1}{2}t^2u(t)$

**16.** Write a MATLAB program that can be used to find $K_p$, $K_v$, and $K_a$ for digital systems. The program will be used for systems of the type represented in Figure P13.5. Test your program for

MATLAB **ML**

$$G(z) = \frac{0.04406z^3 - 0.03624z^2 - 0.03284z + 0.02857}{z^4 - 3.394z^3 + 4.29z^2 - 2.393z + 0.4966}$$

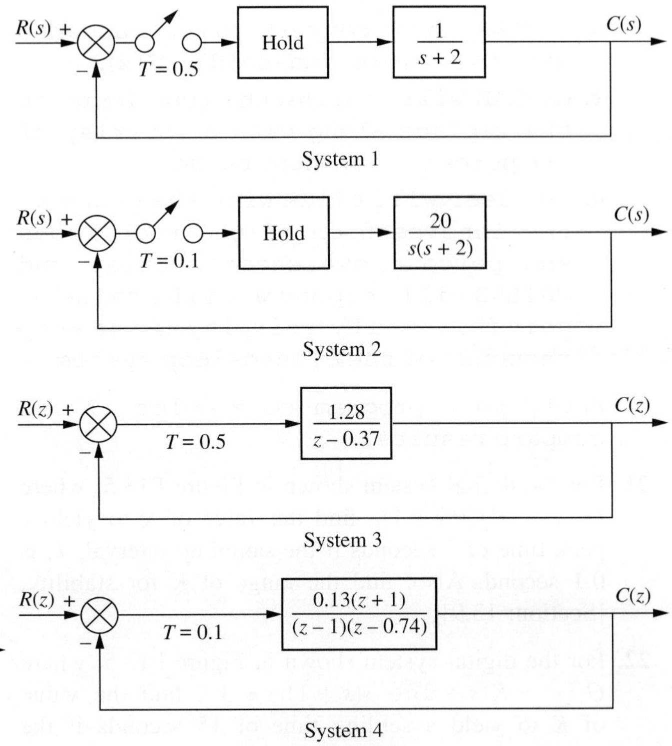


**FIGURE P13.8**

where $G(z)$ is the pulse transfer function for $G_1(s)$ in cascade with the z.o.h. and $T = 0.1$ second.

**17.** For the digital system shown in Figure P13.5, where $G_1(s) = K/[(s+1)\times(s+4)]$, find the value of $K$ to yield a 16.3% overshoot. Also find the range of $K$ for stability. Let $T = 0.1$ second. [Section: 13.9]

**18.** Use Simulink to simulate the step response for the system of Problem 17. Set the value of gain, $K$, to that designed in Problem 17 for 16.3% overshoot.

Simulink **SL**

**19.** Use MATLAB's LTI Viewer to determine the peak time and settling time of the closed-loop step response for System 4 in Figure P13.8.

GUI Tool **GUIT**

20. Write a MATLAB program that can be used to design the gain of a digital control system to meet a percent overshoot requirement. The program will be used for systems of the type represented in Figure P13.5 and meet the following requirements: MATLAB ML

**a.** The user will input the desired percent overshoot.

**b.** MATLAB will convert $G_1(s)$ cascaded with the sample-and-hold to $G(z)$.

**c.** MATLAB will display the root locus on the $z$-plane along with an overlay of the percent overshoot curve.

**d.** The user will click with the mouse at the intersection of the root locus and percent overshoot overlay and MATLAB will respond with the value of gain followed by a display of the step response of the closed-loop system.

Apply your program to Problem 17 and compare results.

21. For the digital system shown in Figure P13.5, where $G_1(s) = K/[s(s+1)]$, find the value of $K$ to yield a peak time of 2 seconds if the sampling interval, $T$, is 0.1 second. Also, find the range of $K$ for stability. [Section: 13.9]

22. For the digital system shown in Figure P13.5, where $G_1(s) = K(s+2) \div [s(s+1)(s+3)]$, find the value of $K$ to yield a settling time of 15 seconds if the sampling interval, $T$, is 1 second. Also, find the range of $K$ for stability. [Section: 13.9]

23. For the digital system shown in Figure P13.5, where $G_1(s) = K/[s(s+1)(s+3)]$, find the value of $K$ to yield a 20% overshoot if the sampling interval, $T$, is 0.1 second. Also, find the range of $K$ for stability. [Section: 13.9]

24. A PID controller was designed in Example 9.5 for a continuous system with unity feedback. The system's plant was

$$G(s) = \frac{(s+8)}{(s+3)(s+6)(s+10)}$$

The designed PID controller was

$$G_c(s) = 4.6\frac{(s+55.92)(s+0.5)}{s}$$

Find the digital transfer function, $G_c(z)$, of the PID controller in order for the system to be computer controlled if the sampling interval, $T$, is 0.01 second. [Section: 13.10]

25. A continuous unity feedback system has a forward transfer function of

$$G(s) = \frac{1}{s(s+5)(s+8)}$$

The system is to be computer controlled with the following specifications:

Percent overshoot: 10%
Settling time: 2 seconds
Sampling interval: 0.01 second

Design a lead compensator for the digital system to meet the specifications. [Section: 13.10]

26. Repeat Problem 25 using MATLAB. MATLAB ML

27. An inverted pendulum mounted on a motor-driven cart (*Prasad, 2012*) was the subject of Problem 33, Chapter 9. In that problem you were asked to develop Simulink models for two feedback systems, one of which was to control the cart position, $x(t)$. At the recommended settings, the step response of that system was expected to satisfy the following requirements: a steady-state error, $e(\infty) < 2\%$, a peak time, $T_P < 1.2$ seconds, and a percent overshoot, $\%OS < 20.5\%$. Having concluded that the steady-state error was unacceptable, you designed a PID controller and found its recommended settings. Simulink SL

Digitize the Simulink model developed for PID control of the cart position in Part **d** of Problem 33, Chapter 9, by adding a zero-order-hold set to 0.01 second. Then, run a simulation to evaluate performance.

## DESIGN PROBLEMS

28. The analog system of a disk drive is shown in Figure P13.9. Do the following:

**a.** Convert the disk drive to a digital system. Use a sampling time of 0.01 second.

**b.** Find the range of digital controller gain to keep the system stable.

**c.** Find the value of digital controller gain to yield 15% overshoot for a digital step response.

**d.** Repeat all previous parts and obtain the step response for Part **c** using MATLAB. MATLAB ML

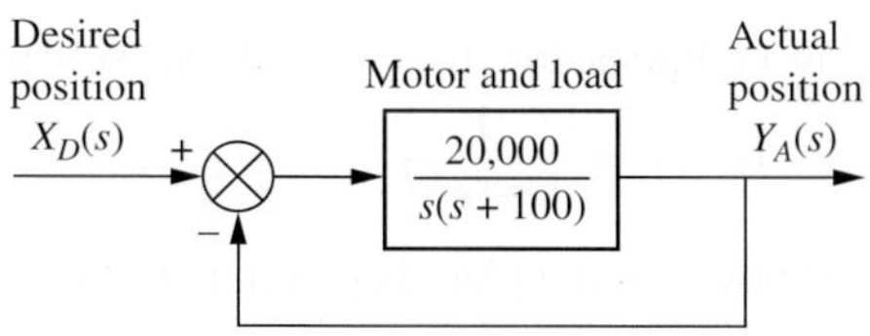


**FIGURE P13.9** Simplified block diagram of a floppy disk drive

**29. a.** Convert the heading control for the UFSS vehicle shown on the back endpapers (*Johnson, 1980*) into a digitally controlled system.

**b.** Find the closed-loop pulse transfer function, $T(z)$, if $T = 0.1$ second.

**c.** Find the range of heading gain to keep the digital system stable.

**30.** Problem 36, Chapter 9 described a two-tank system where the objective was to maintain a constant liquid level in one of the tanks via control of an inflow valve. Assume for this problem that the transfer function relating liquid-level output, $Y(s)$, to flow rate input $F_\theta(s)$, for the lower tank is (*Romagnoli, 2006*)

$$G(s) = \frac{Y(s)}{F_\theta(s)} = \frac{0.0187}{s^2 + 0.237s + 0.00908}$$

Assume that the system will be controlled in closed loop by means of a digital computer system with a sample period $T = 1$ second, as shown in Figure P13.5, with $G_1(s) = KG(s)$. Use the bilinear transformation and the Routh-Hurwitz method to find the range of $K$ that will result in a stable closed-loop system.

**31.** Assume that the two-tank system of Problem 30 is controlled by a digital computer in the configuration of Figure P13.5, where $G_1(s) = KG(s)$. If a sampling period of $T = 1$ second is used, do the following (*Romagnoli, 2006*).

**a.** `Use MATLAB to draw the root locus.` MATLAB **ML**

**b.** Find the value of $K$ that will result in a stable system with a damping factor of $\zeta = 0.7$.

**c.** Use the root locus of Part **b** to predict the step-response settling time, $T_s$, and peak time, $T_p$.

**d.** Calculate the final value of the closed-loop system to a unit step input.

**e.** `Obtain the step response of the system using Simulink. Verify the predictions you made in Parts c and d.` Simulink **SL**

**32.** In Problem 26, Chapter 11, we discussed an EVAD, a device that works in parallel with the human heart to help pump blood in patients with cardiac conditions. The device has a transfer function

$$G(s) = \frac{P_{ao}(s)}{E_m(s)} = \frac{1361}{s^2 + 69s + 70.85}$$

where $E_m(s)$ is the motor's armature voltage, and $P_{ao}(s)$ is the aortic blood pressure (*Tasch, 1990*). Using continuous techniques, a cascaded compensator is designed in a unity feedback configuration with a transfer function

$$G_c(s) = \frac{0.5(s+1)}{s + 0.05}$$

Selecting to control the device using a microcontroller, a discrete equivalent has to be found for $G_c(s)$. Do the following:

**a.** Find an appropriate sampling frequency for the discretization.

**b.** Translate the continuous compensator into a discrete compensator using the sampling frequency found in Part **a**.

**c.** `Use Simulink to simulate the continuous and discrete systems on the same graph for a unit step input. There should be little difference between the compensated continuous and discrete systems.` Simulink **SL**

**33.** In Problem 52, Chapter 9, a steam-driven turbine-governor system was implemented by a unity feedback system with a forward-path transfer function (*Khodabakhshian, 2005*)

$$G(s) = \frac{K}{(s + 0.08)(s + 2)(s + 5)}$$

**a.** Use a sampling period of $T = 0.5$ s and find a discrete equivalent for this system.

**b.** `Use MATLAB to draw the root locus.` MATLAB **ML**

**c.** Find the value of $K$ that will result in a stable system with a damping factor of $\zeta = 0.7$.

**d.** Use the root locus found in Part **a** to predict the step-response settling time, $T_s$, and peak time, $T_p$.

**e.** Calculate the final value of the closed-loop system unit step response.

**f.** `Obtain the step response of the system using Simulink. Verify the predictions you made in Parts c and d.` Simulink **SL**

**34.** Obtaining an exact shape in metalforming can be tricky because of material springback. A feedback system has been devised in which critical deviations from specifications are measured as soon as a part is formed and automatic incremental corrections to the forming tools are made before the next part is formed. Eventually the system compensates for material springback and results in parts compliant with specifications. A unity-feedback digital system with a forward path

$$G(z) = \frac{k}{z-1}$$

can be used as a simplified representation of the system (*Fu, 2013*).

**a.** Make a sketch of the system's root locus.

**b.** Find the range of $k$ for which the system is closed-loop stable.

**c.** Find the system's steady-state error for a step input.

**d.** Find the value of $k$ that will result in the fastest possible response.

**35.** Discrete time controlled systems can exhibit unique characteristics not available in continuous controllers. For example, assuming a specific input and some conditions, it is possible to design a system to achieve steady state within one single time sample without overshoot. This scheme is well known and referred to as *deadbeat control*. We illustrate deadbeat control design with a simple example. For a more comprehensive treatment see (*Ogata, 1987*).

Assume in Figure 13.25(*a*) that $G_p(s) = \frac{1}{s+1}$. The purpose of the design will be to find a compensator, $G_c(z)$, such that for a step input the system achieves steady state within one sample. We start by translating the system into the discrete domain to obtain the equivalent of Figure 13.25(*c*). The pulse transfer function, $G_p(z) = \frac{(1-e^{-T})z^{-1}}{1-e^{-T}z^{-1}}$, is found using Eq. (13.40), since it is assumed that the compensator will be followed by a zero-order hold. In Figure 13.25(*c*), the closed-loop transfer function is given by $\frac{C(z)}{R(z)} = T(z) = \frac{G_c(z)G_p(z)}{1+G_c(z)G_p(z)}$, or, solving for the compensator, we get $G_c(z) = \frac{1}{G_p(z)}\frac{T(z)}{1-T(z)}$. The desired system output is a unit step delayed by one unit sample. Thus, $C(z) = \frac{z}{z-1}z^{-1} = \frac{1}{z-1}$. Since the input is a unit step, $R(z) = \frac{z}{z-1}$; the desired closed-loop transfer function is $\frac{C(z)}{R(z)} = T(z) = z^{-1}$, and the resulting compensator, found by direct substitution, is given by $G_c(z) = \frac{1}{1-e^{-T}}\frac{(z-e^{-T})}{z-1}$.

Assume now that the plant is given by $G_p(s) = \frac{1}{s}$, and a sampling period of $T = 0.05$ second is used.

**a.** Design a deadbeat compensator to reach steady state within one time sample for a step input.

**b.** Calculate the resulting steady-state error for a unit-slope ramp input.

**c.** Simulate your system using SIMULINK. (Hint: Following Figure 13.25, the forward path will consist of the cascading of $G_c(z)$, a zero-order hold, and $G_p(s)$.) Show that the system reaches steady state after one sample. Also verify your steady-state error ramp result.

Simulink **SL**

**36.** Given

LabVIEW **LV**

$$G(z) = \frac{K(z+0.5)}{(z-0.25)(z-0.75)}$$

Use the LabVIEW Control Design and Simulation Module and the MathScript RT Module to (1) obtain the value of $K$ that will yield a damping ratio of 0.5 for the closed-loop system in Figure 13.20, where $H(z)=1$; and (2) display the step response of the closed-loop system in Figure 13.20 where $H(z) = 1$. Compare your results with those of Skill-Assessment Exercise 13.8.

MATLAB **ML**

**37.** A linear model of the $\alpha$-subsystem of a grid-connected voltage-source converter with a Y-Y transformer (*Mahmood, 2012*) was presented in Problem 69, Chapter 8, and Problem 51, Chapter 10. The system was represented with unity-feedback and a forward path consisting of the cascading of a compensator and a plant. The plant is given by

$$G_P(s) = \frac{V_\alpha(s)}{M_\alpha(s)} = \frac{(s+2200)}{(s+220)(s^2+120s+16\times10^6)}$$

This system is now to be digitally controlled with the following specifications: percent overshoot, $\%OS = 10\%$; settling time, $T_S = 0.1$ second; and sampling interval, $T = 0.001$ second. Design a lead compensator for that system to meet these specifications. [Section: 13.10]

## PROGRESSIVE ANALYSIS AND DESIGN PROBLEMS

**38. Control of HIV/AIDS.** In Chapter 11, a continuous cascaded compensator for a unity feedback system was designed for the treatment of the HIV-infected patient treated with RTIs (*Craig, 2004*). The transfer function of the designed compensator was

$$G_c(s) = \frac{-2 \times 10^{-4}(s^2 + 0.04s + 0.0048)}{s(s + 0.02)}$$

The linearized plant was given by

$$P(s) = \frac{Y(s)}{U_1(s)} = \frac{-520s - 10.3844}{s^3 + 2.6817s^2 + 0.11s + 0.0126}$$

The compensated system is overdamped with an approximate settling time of 100 seconds. This system must be discretized for practical reasons: (1) HIV patient cannot be monitored continuously and (2) medicine dosage cannot be adjusted continuously.

**a.** Show that a reasonable sampling period for this system is $T = 8$ days (medicine dosage will be updated on a weekly basis).

**b.** Use Tustin's method and $T = 8$ days to find a discrete equivalent to $G_c(s)$.

**c.** `Use Simulink to simulate the continuous and discrete compensated systems for a unit step input. Plot both responses on the same graph.` Simulink SL

**39. Hybrid vehicle.** In Problem 67, Chapter 7 (Figure P7.32), the block diagram of a cascade scheme for the speed control of an HEV (*Preitl, 2007*) was represented as a unity feedback system. In that diagram the output of the system is the speed transducer's output voltage, $C(s) = K_{ss}V(s)$. In Part **b** of Problem 34, Chapter 11, where a compensator was designed for this problem, we discussed the feasibility of achieving full pole-zero cancellation when we place a PI speed controller's zero, $Z_I$, on top of the uncompensated system's real pole, closest to the origin (located at $-0.0163$). Noting that perfect pole-zero cancellation may not be maintained, we studied a case where the PI-controller's zero changed by +20%, moving to $-0.01304$. In that case, the transfer function of the plant with a PI speed controller, which has a proportional gain = $K$, was given by MATLAB ML

$$G(s) = \frac{K(s + 0.6)(s + 0.01304)}{s(s + 0.0163)(s + 0.5858)}$$

Assuming that $G_1(s)$ in Figure P13.5 equals the transfer function, $G(s)$, given above for the vehicle with the speed controller:

**a.** `Develop a MATLAB M-file that would allow you to do the following: (Hint: Refer to the M-files you developed for Problems 12 and 20 of this chapter)`

(1) `Convert` $G_1(s)$ `cascaded with a sample-and-hold to` $G(z)$;

(2) `Search over the range` $0 < T < 5$ `seconds for the largest sampling period` $T_{max}$ `below which the system is stable. Calculate the` $z$`-plane roots of the closed-loop system for the whole range of the sampling time,` $T$. `Subsequently set` $T = 0.75T_{\max}$;

(3) `Design the gain of a digital control system to meet a percent overshoot requirement,` $\%OS$, `allowing the user to input the value of the desired` $\%OS$ `and the value of the PI speed controller's proportional gain,` $K$;

(4) `Plot the step response of that digital system (in per unit, p. u., vs. time in seconds).`

**b.** `Run the M-file you developed in Part` **a** `and enter the values of the desired percent overshoot,` $\%OS = 0$, `and the PI speed controller's proportional gain,` $K = 61$;

**c.** `Select a point in the graphics window displaying the root locus, such that all poles of the closed-loop transfer function,` $T_z$, `are inside the unit circle.`

**d.** `Write the sampled-data transfer functions obtained,` $G_z$ `and` $T_z$, `indicating the corresponding value of the sampling time,` $T$, `and all poles,` $r$, `of the closed-loop transfer function,` $T_z$;

**e.** `Plot the step response of that digital system (in per unit, p. u., vs. time in seconds) noting the following characteristics: final value, rise time, and settling time.`

**40. Parabolic trough collector.** In Problem 35, Chapter 11, a zero steady-state error for a unit-step input was achieved through the design of a lag compensator with integral control. In that problem, the open-loop transmission can be written as $L(s) = G_c(s)G(s)$, where the parabolic trough plant is given by (*Camacho, 2012*)

$$G(s) = \frac{137.2 \times 10^{-6}}{s^2 + 0.0224s + 196 \times 10^{-6}} e^{-39s}$$

and the lag compensator is given by

$$G_c(s) = 1.12 \frac{(s + 0.01)}{s}$$

We want to substitute for the continuous compensator with a digital one.

**a.** Find a suitable sampling period for the system.

**b.** Find the equivalent compensator's transfer function in $z$-domain.

**c.** `Use Simulink to simulate the digital compensator with the continuous plant. Compare the resulting response with that of the original system using the continuous compensator on the same graph.` Simulink **SL**

## Cyber Exploration Laboratory

### Experiment 13.1

**Objectives** To design the gain of a digital control system to meet a transient response requirement; to simulate a digital control system to test a design; to see the effect of sampling rate upon the time response of a digital system

**Minimum Required Software Packages** MATLAB, Simulink, and the Control System Toolbox

#### Prelab

1. Given the antenna azimuth control system shown on the front endpapers, use Configuration 2 to find the discrete transfer function of the plant. Neglect the dynamics of the power amplifier and include the preamplifier, motor, gears, and load. Assume a zero-order hold and a sampling interval of 0.01 second.
2. Using the digital plant found in Prelab 1, find the preamplifier gain required for a closed-loop digital system response with 10% overshoot and a sampling interval of 0.01 second. What is the peak time?
3. Given the antenna azimuth control system shown on the front endpapers, use Configuration 2 to find the preamplifier gain required for the continuous system to yield a closed-loop step response with 10% overshoot. Consider the open-loop system to be the preamplifier, motor, gears, and load. Neglect the dynamics of the power amplifier.

#### Lab

1. Verify your value of preamplifier gain found in Prelab 2 using the SISO Design Tool to generate the root locus for the digital open-loop transfer function found in Prelab 1. Use the Design Constraints capability to generate the 10% overshoot curve and place your closed-loop poles at this boundary. Obtain a plot of the root locus and the design boundary. Record the value of gain for 10% overshoot. Also, obtain a plot of the closed-loop step response using the LTI Viewer and record the values of percent overshoot and peak time. Use the same tool to find the range of gain for stability.
2. Using Simulink, set up the closed-loop digital system whose plant was found in Prelab 1. Make two diagrams: one with the digital transfer function for the plant and another using the continuous transfer function for the plant preceded by a zero-order sample-and-hold.

Use the same step input for both diagrams and obtain the step response of each. Measure the percent overshoot and peak time.

3. Using Simulink, set up both the digital and continuous systems calculated in Prelab 2 and Prelab 3, respectively, to yield 10% overshoot. Build the digital system with a sample-and-hold rather than the $z$-transform function. Plot the step response of each system and record the percent overshoot and the peak time.
4. For one of the digital systems built in Lab 2, vary the sampling interval and record the responses for a few values of sampling interval above 0.01 second. Record sampling interval, percent overshoot, and peak time. Also, find the value of sampling interval that makes the system unstable.

## Postlab

1. Make a table containing the percent overshoot, peak time, and gain for each of the following closed-loop responses: the digital system using the SISO Design Tool; the digital system using Simulink and the digital transfer functions; the digital system using Simulink and the continuous transfer functions with the zero-order sample-and-hold; and the continuous system using Simulink.
2. Using the data from Lab 4, make a table containing sampling interval, percent overshoot, and peak time. Also, state the sampling interval that makes the system unstable.
3. Compare the responses of all of the digital systems with a sampling interval of 0.01 second and the continuous system. Explain any discrepancies.
4. Compare the responses of the digital system at different sampling intervals with the continuous system. Explain the differences.
5. Draw some conclusions about the effect of sampling.

## Experiment 13.2

**Objective** To use the various functions from the LabVIEW Control Design and Simulation Module for the analysis of digital control systems.

**Minimum Required Software Packages** LabVIEW with the Control Design and Simulation Module and the MathScript RT Module; MATLAB with the Control Systems Toolbox.

**Prelab** You are given Figure P8.22 and the parameters listed in the Prelab of Cyber Exploration Laboratory Experiment 8.2 for the open-loop NASA eight-axis ARMII (Advanced Research Manipulator II) electromechanical shoulder joint/link, actuated by an armature-controlled dc servomotor.

1. Obtain the open-loop transfer function of the shoulder joint/link, $G(s) = \dfrac{\theta_L(s)}{V_{ref}(s)}$, or use your calculation from Cyber Exploration Laboratory Experiment 8.2.
2. Use MATLAB and design a digital compensator to yield a closed-loop response with zero steady-state error and a damping ratio of 0.7. If you already have performed Cyber Exploration Laboratory Experiment 8.2, modify your M-file from that experiment. Test your design using MATLAB.

**Lab** Simulate your Prelab design using a Simulation Loop from the LabVIEW Control Design and Simulation Module. Plot the step response of two loops as follows: (1) a unity feedback with the forward path consisting of the continuous system transfer function preceded by a zero-order hold, and (2) a unity feedback with the forward path consisting of the equivalent discrete transfer function of your compensator in cascade with the open-loop plant.

**Postlab** Compare the results obtained with those from your prelab MATLAB program. Comment on time-performance specifications.

## Bibliography


Astrom, K. J., and Wittenmark, B. *Computer Controlled Systems*. Prentice Hall, Upper Saddle River, NJ, 1984.

Bittanti, S., Dell'Orto, F., Di Carlo, A., and Savaresi, S. M. Notch Filtering and Multirate Control for Radial Tracking in High Speed DVD-Players. *IEEE Transactions on Consumer Electronics*, vol. 48, 2002, pp. 56–62.

Boyd, M., and Yingst, J. C. PC-Based Operator Control Station Simplifies Process, Saves Time. *Chilton's I & CS*, September 1988, pp. 99–101.

Camacho, E. F., Berenguel, M., Rubio, F. R., and Martinez, D. *Control of Solar Energy Systems*. Springer-Verlag, London, 2012.

Chassaing, R. *Digital Signal Processing*. Wiley, New York, 1999.

Craig, I. K., Xia, X., and Venter, J. W. Introducing HIV/AIDS Education into the Electrical Engineering Curriculum at the University of Pretoria. *IEEE Transactions on Education*, vol. 47, no. 1, February 2004, pp. 65–73.

Craig, J. J. *Introduction to Robotics. Mechanics and Control*, 3d ed. Prentice Hall, Upper Saddle River, NJ, 2005.

Fu, Z., Mo, J., Han, F., and Gong, P. Tool path correction algorithm for single-point incremental forming of sheet metal. *Int. J. Adv. Manuf. Technol.*, vol. 64, 2013, Springer, pp. 1239–1248.

Hostetter, G. H. *Digital Control System Design*. Holt, Rinehart & Winston, New York, 1988.

Johnson, H. et al. *Unmanned Free-Swimming Submersible (UFSS) System Description.* NRL Memorandum Report 4393. Naval Research Laboratory, Washington, D.C., 1980.

Katz, P. *Digital Control Using Microprocessors*. Prentice Hall, Upper Saddle River, NJ, 1981.

Khodabakhshian, A., and Golbon, N. Design of a New Load Frequency PID Controller Using QFT. *Proceedings of the 13th Mediterranean Conference on Control and Automation*, 2005, pp. 970–975.

Kuo, B. C. *Automatic Control Systems*, 7th ed. Prentice Hall, Upper Saddle River, NJ, 1995.

Kuo, B. C. *Digital Control Systems*. Holt, Rinehart & Winston, New York, 1980.

Mahmood, H., and Jiang, J. Modeling and Control System Design of a Grid Connected VSC Considering the Effect of the Interface Transformer Type. *IEEE Transactions on Smart Grid*, vol. 3, no. 1, March 2012, pp. 122–134.

Neogi, B., Ghosh, R., Tarafdar, U., and Das, A. Simulation Aspect of an Artificial Pacemaker. *International Journal of Information Technology and Knowledge Management*, vol. 3, no. 2, 2010, pp. 723–727.

Nyzen, R. J. *Analysis and Control of an Eight-Degree-of-Freedom Manipulator*, Ohio University Masters Thesis, Mechanical Engineering, Dr. Robert L. Williams II, Advisor, August 1999.

Ogata, K. *Discrete-Time Control Systems*. Prentice Hall, Upper Saddle River, NJ, 1987.

Phillips, C. L., and Nagle, H. T. Jr. *Digital Control System Analysis and Design*. Prentice Hall, Upper Saddle River, NJ, 1984.

Prasad, L., Tyagi, B., and Gupta, H. Modeling & Simulation for Optimal Control of Nonlinear Inverted Pendulum Dynamical System using PID Controller and LQR. *IEEE Computer Society Sixth Asia Modeling Symposium*, 2012, pp. 138–143.

| | |
|---|---|
| $R(s)$ | Laplace transform of the input to a system |
| $r$ | Nonlinear electrical resistance |
| $r(t)$ | Input to a system |
| $R_a$ | Motor armature resistance in ohms |
| rad | Radian—unit of angular displacement |
| rad/s | Radian/second—unit of angular velocity |
| s | Second—unit of time |
| $s$ | Complex variable for the Laplace transform |
| $S_{F:P}$ | Sensitivity of $F$ to a fractional change in $P$ |
| $T$ | Time constant; sampling interval for digital signals |
| $T(s)$ | Closed-loop transfer function; Laplace transform of mechanical torque |
| $T(t)$ | Mechanical torque in N-m |
| $T_m(t)$ | Torque at the armature developed by a motor in N-m |
| $T_m(s)$ | Laplace transform of the torque at the armature developed by a motor |
| $T_p$ | Peak time in seconds |
| $T_r$ | Rise time in seconds |
| $T_s$ | Settling time in seconds |
| $T_w$ | Pulse width in seconds |
| $\mathbf{u}$ | Input or control vector for state-space representation |
| $u$ | Input control signal for state-space representation |
| $u(t)$ | Unit step input |
| V-s/rad | Volt-seconds/radian—unit of motor back emf constant |
| $v(t)$ | Mechanical translation velocity in m/s; electrical voltage |
| $v_b(t)$ | Motor back emf in volts |
| $v_e(t)$ | Error voltage |
| $v_p(t)$ | Power amplifier input in volts |
| $\mathbf{x}$ | State vector for state-space representation |
| $x(t)$ | Mechanical translation displacement in meters; a state variable |
| $\dot{x}$ | Time derivative of a state variable |
| $\dot{\mathbf{x}}$ | Time derivative of the state vector |
| $\mathbf{y}$ | Output vector for state-space representation |
| $y(t)$ | Output scalar for state-space representation |
| $z$ | Complex variable for the z-transform |
| $z_c$ | Compensator zero |
| $\alpha$ | Pole-scaling factor for a lag compensator, where $\alpha > 1$; angle of attack |
| $\beta$ | Pole-scaling factor for a lead compensator, where $\beta < 1$ |
| $\gamma$ | Pole-scaling factor for a lag-lead compensator, where $\gamma > 1$ |
| $\delta$ | Thrust angle |
| $\zeta$ | Damping ratio |
| $\theta$ | Angle of a vector with the positive extension of the real axis |
| $\theta(t)$ | Angular displacement |
| $\theta_a$ | Angle of a root locus asymptote with the positive extension of the real axis |
| $\theta_c$ | Angular contribution of a compensator on the $s$-plane |
| $\theta_m(t)$ | Angular displacement of the armature of a motor |
| $\lambda$ | Eigenvalue of a square matrix |

| | |
|---|---|
| $\sigma$ | Real part of the Laplace transform variable, $s$ |
| $\sigma_a$ | Real-axis intercept of the root locus asymptotes |
| $\Phi_M$ | Phase margin |
| $\Phi(t)$ | State transition matrix |
| $\phi$ | Sinusoidal phase angle; body angle |
| $\phi_c$ | Sinusoidal phase angle of a compensator |
| $\phi_{max}$ | Maximum sinusoidal phase angle |
| Ω | Ohm—unit of electrical resistance |
| ℧ | Mho—unit of electrical conductance |
| $\omega$ | Imaginary part of the Laplace transform variable, $s$ |
| $\omega(t)$ | Angular velocity in rad/s |
| $\omega_{\mathrm{BW}}$ | Bandwidth in rad/s |
| $\omega_d$ | Damped frequency of oscillation in rad/s |
| $\omega_{\Phi_M}$ | Phase-margin frequency in radians |
| $\omega_{G_M}$ | Gain-margin frequency in radians |
| $\omega_n$ | Natural frequency in rad/s |
| $\omega_p$ | Peak-magnitude frequency of the magnitude frequency response in rad/s |

# Appendix B: MATLAB Tutorial

## B.1 Introduction

MATLAB is a high-level technical computing environment suitable for solving scientific and engineering problems. When used with routines from its companion software, the Control System Toolbox, MATLAB can be used to analyze and design control systems problems such as those covered in this textbook. MATLAB and the Control System Toolbox are commercial software products available from MathWorks, Inc., 3 Apple Hill Drive, Natick, MA 01760-2098. Phone: (508) 647-7000. Email: info@mathworks.com. URL: www.mathworks.com.

The MATLAB examples in this tutorial consist of solved problems that demonstrate the application of MATLAB to the analysis and design of control systems. Many problems were taken from examples in the text (identified with a **MATLAB** icon) that were solved without MATLAB. A Command Summary at the end of this appendix lists key MATLAB statements and their descriptions.

The code in this tutorial is also available in the Control Systems Engineering Toolbox folder at www.wiley.com/college/nise and at www.mathworks.com/matlabcentral/fileexchange. You should have MATLAB Version 8.3 (R2014a) and the Control System Toolbox Version 9.7 installed on your machine to execute this appendix's code in the Control Systems Engineering Toolbox Version 7.

To run the M-files, first be sure the files are either added to the search path in **Set Path** under the **HOME** tab in the **ENVIRONMENT** section or appear in the **Current Folder** window, which is part of the **MATLAB** window. To see the computer responses after installing the M-files, run each problem by typing the M-file name, such as ch2p1, after the prompt (>>) in the **Command Window.** You may also run the files by right-clicking the file name, if it appears in the **Current Folder** window, and select **Run.**

To view all or part of the M-file in the **Command Window,** enter "type <file name>" or "help <file name>," respectively, after the prompt. You may also view and make changes to the M-file by double-clicking the file in the **Current Folder** window. This action brings up the editor. After editing, be sure to save the revised file before executing.

If you do not have the Control Systems Engineering Toolbox M-files, you can create your own M-files by typing the code for each problem in this appendix into a separate M-file (there is no need to type the final pause statement or comments), and naming each M-file with a .m extension, as in ch2p1.m. You can also type the code for more than one problem into an M-file, including the pause command, and name the M-file with the .m extension. You can then call the file from the **Command Window,** and continue past the pause statements to the next problem by pressing any key.

By its nature, this appendix cannot cover all the background and details necessary for a complete understanding of MATLAB. For further details, you are referred to other sources, including MATLAB reference manuals and instructions specific to your particular computer. The bibliography at the end of this appendix provides a partial listing of references. This appendix should give you enough information to be able to apply MATLAB to the analysis and design problems covered in this book.

The code will also run on workstations that support MATLAB. Consult the MATLAB *Installation Guide* for your platform for minimum system hardware requirements.

# B.2 MATLAB Examples

## Chapter 2: Modeling in the Frequency Domain

**ch2p1** Bit strings will be used to identify parts of this tutorial on the computer output. Bit strings are represented by the text enclosed in apostrophes, such as `'ab'`. Comments begin with `%` and are ignored by MATLAB. Numbers are entered without any other characters. Arithmetic can be performed using the proper arithmetic operator. Numbers can be assigned using a left-hand argument and an equals sign. Finally, we can find the magnitude and angle of a complex number, $Q$ using `abs` ($Q$) and `angle` ($Q$), respectively.

```
'(ch2p1)'                    % Display label.
'How are you?'               % Display string.
-3.96                        % Display scalar number -3.96.
-4 + 7i                      % Display complex number -4+7i.
-5-6j                        % Display complex number -5-6j.
(-4+7i)+(-5-6i)              % Add two complex numbers and
                             % display sum.
(-4+7j)*(-5-6j)              % Multiply two complex numbers and
                             % display product.
M=5                          % Assign 5 to M and display.
N=6                          % Assign 6 to N and display.
P=M+N                        % Assign M+N to P and display.
Q=3+4j                       % Define complex number, Q.
MagQ=abs(Q)                  % Find magnitude of Q.
ThetaQ=(180/pi)*angle(Q)     % Find the angle of Q in degrees.
pause
```

**ch2p2** Polynomials in $s$ can be represented as row vectors containing the coefficients. Thus $P_1 = s^3 + 7s^2 - 3s + 23$ can be represented by the vector shown below with elements separated by a space or comma. Bit strings can be used to identify each section of this tutorial.

```
'(ch2p2)'                    % Display label.
P1=[1 7 -3 23]               % Store polynomial s^3 + 7s^2 -3s+
                             % 23 as P1 and display.
pause
```

**ch2p3** Running the previous statements causes MATLAB to display the results. Ending the command with a semicolon suppresses the display. Typing an expression without a left-hand assignment and without a semicolon causes the expression to be evaluated and the result displayed. Enter `P2` in the **MATLAB Command Window** after execution.

```
'(ch2p3)'                            % Display label.
P2=[3 5 7 8] ;                       % Assign 3s^3 + 5s^2 +7s + 8 to P2
                                     % without displaying.
3*5                                  % Evaluate 3*5 and display result.
pause
```

**ch2p4** An $F(s)$ in factored form can be represented in polynomial form. Thus $P_3 = (s+2)(s+5)(s+6)$ can be transformed into a polynomial using `poly (V)`, where `V` is a row vector containing the roots of the polynomial and `poly (V)` forms the coefficients of the polynomial.

```
'(ch2p4)'                            % Display label.
P3=poly([-2 -5 -6])                  % Store polynomial
                                     % (s+2)(s+5)(s+6) as P3 and
                                     % display the coefficients.
pause
```

**ch2p5** We can find roots of polynomials using the `roots (V)` command. The roots are returned as a column vector. For example, find the roots of $5s^4 + 7s^3 + 9s^2 - 3s + 2 = 0$.

```
'(ch2p5)'                            % Display label.
P4=[5 7 9 -3 2]                      % Form 5s^4+7s^3+9s^2-3s+2 and
                                     % display.
rootsP4=roots(P4)                    % Find roots of 5s^4+7s^3+9s^2
                                     %-3s+2,
                                     % assign to rootsP4, and display.
pause
```

**ch2p6** Polynomials can be multiplied together using the `conv(a,b)` command (standing for convolve). Thus, $P_5 = (s^3 + 7s^2 + 10s + 9)(s^4 - 3s^3 + 6s^2 + 2s + 1)$ is generated as follows:

```
'(ch2p6)'                            % Display label.
P5=conv([1 7 10 9],[1 -3 6 2 1])     % Form (s^3+7s^2+10s+9)(s^4-
                                     % 3s^3+6s^2+2s+1), assign to P5,
                                     % and display.
pause
```

**ch2p7** The partial-fraction expansion for $F(s) = b(s)/a(s)$ can be found using the `[K, p, k ]= residue (b, a)` command (`K` = residue; `p` = roots of denominator; `k` = direct quotient, which is found by dividing polynomials prior to performing a partial-fraction expansion). We expand $F(s) = (7s^2 + 9s + 12)/[s(s+7)(s^2 + 10s + 100)]$ as an example. Using the results from MATLAB yields: $F(s) = [(0.2554 - 0.3382i)/(s+5.0000-8.6603i)]+[(0.2554+0.3382i)/(s+5.0000+8.6603i)]-[0.5280/(s+7)]+[0.0171/s]$.

```
'(ch2p7)'                            % Display label.
numf=[7 9 12];                       % Define numerator of F(s).
denf=conv(poly([0 -7]),[1 10 100]);  % Define denominator of F(s).
[K,p,k]=residue(numf,denf)           % Find residues and assign to K;
                                     % find roots of denominator and
                                     % assign to p; find
                                     % constant and assign to k.
pause
```

**ch2p8 (Example 2.3)** Let us do Example 2.3 in the book using MATLAB.

```
'(ch2p8) Example 2.3'               % Display label.
numy=32;                            % Define numerator.
deny=poly([0 -4 -8]);               % Define denominator.
[r,p,k]=residue(numy,deny)          % Calculate residues, poles, and
                                    % direct quotient.
pause
```

## ch2p9 Creating Transfer Functions

**Vector Method, Polynomial Form** A transfer function can be expressed as a numerator polynomial divided by a denominator polynomial, that is, $F(s) = N(s)/D(s)$. The numerator, $N(s)$, is represented by a row vector, `numf`, that contains the coefficients of $N(s)$. Similarly, the denominator, $D(s)$, is represented by a row vector, `denf`, that contains the coefficients of $D(s)$. We form $F(s)$ with the command, `F=tf(numf,denf)`. `F` is called a linear time-invariant (LTI) object. This object, or transfer function, can be used as an entity in other operations, such as addition or multiplication. We demonstrate with $F(s) = 150(s^2 + 2s + 7)/[s(s^2 + 5s + 4)]$. Notice after executing the `tf` command, MATLAB prints the transfer function.

**Vector Method, Factored Form** We also can create LTI transfer functions if the numerator and denominator are expressed in factored form. We do this by using row vectors containing the roots of the numerator and denominator. Thus $G(s) = K^*N(s)/D(s)$ can be expressed as an LTI object using the command, `G=zpk(numg,deng,K)`, where `numg` is a row vector containing the roots of $N(s)$ and `deng` is a row vector containing the roots of $D(s)$. The expression `zpk` stands for zeros (roots of the numerator), poles (roots of the denominator), and gain, $K$. We demonstrate with $G(s) = 20(s + 2)(s + 4)/[(s + 7)(s + 8)(s + 9)]$. Notice after executing the `zpk` command, MATLAB prints the transfer function.

**Rational Expression in s Method, Polynomial Form (Requires Control System Toolbox 8.4)** This method allows you to type the transfer function as you normally would write it. The statement `s=tf('s')` must precede the transfer function if you wish to create an LTI transfer function in polynomial form equivalent to using `F=tf(numf, denf)`.

**Rational Expression in s Method, Factored Form (Requires Control System Toolbox 8.4)** This method allows you to type the transfer function as you normally would write it. The statement `s=zpk('s')` must precede the transfer function if you wish to create an LTI transfer function in factored form equivalent to using `G=zpk(numg,deng,K)`.

For both rational expression methods the transfer function can be typed in any form regardless of whether `s=tf('s')` or `s=zpk('s')` is used. The difference is in the created LTI transfer function. We use the same examples above to demonstrate the rational expression in $s$ methods.

```
'(ch2p9)'                           % Display label.
'Vector Method, Polynomial Form'    % Display label.
numf=150*[1 2 7]                    % Store 150(s^2+2s+7) in numf and
                                    % display.
denf=[1 5 4 0]                      % Store s(s+1)(s+4) in denf and
                                    % display.
'F(s)'                              % Display label.
F=tf(numf,denf)                     % Form F(s) and display.
clear                               % Clear previous variables from
                                    % workspace.
```

```
'Vector Method, Factored Form'           % Display label.
numg=[-2 -4]                             % Store (s+2)(s+4) in numg and
                                         % display.
deng=[-7 -8 -9]                          % Store (s+7)(s+8)(s+9) in deng
                                         % and display.
K=20                                     % Define K.
'G(s)'                                   % Display label.
G=zpk(numg,deng,K)                       % Form G(s) and display.
clear                                    % Clear previous variables from
                                         % workspace.
'Rational Expression Method, Polynomial Form'
                                         % Display label.
s=tf('s')                                % Define 's' as an LTI object in
                                         % polynomial form.
F=150*(s^2+2*s+7)/[s*(s^2+...            % Form F(s) as an LTI transfer
 5*s+4)]                                 % function in polynomial form.
G=20*(s+2)*(s+4)/[(s+7)*...              % Form G(s) as an LTI transfer
(s+8)*(s+9)]                             % function in polynomial form.
clear                                    % Clear previous variables from
                                         % workspace.
'Rational Expression Method, Factored Form'
                                         % Display label.
s=zpk('s')                               % Define 's' as an LTI object in
                                         % factored form.
F=150*(s^2+2*s+7)/[s*(s^2+5*s+4)]
                                         % Form F(s) as an LTI transfer
                                         % function in factored form.
G=20*(s+2)*(s+4)/[(s+7)*(s+8)*(s+9)]
                                         % Form G(s) as an LTI transfer
                                         % function in factored form.
pause
```

**ch2p10** Transfer function numerator and denominator vectors can be converted between polynomial form containing the coefficients and factored form containing the roots. The MATLAB function, `tf2zp(numtf,dentf)`, converts the numerator and denominator from coefficients to roots. The results are in the form of column vectors. We demonstrate this with $F(s) = (10s^2 + 40s + 60)/(s^3 + 4s^2 + 5s + 7)$. The MATLAB function, `zp2tf(numzp,denzp,K)`, converts the numerator and denominator from roots to coefficients. The arguments `numzp` and `denzp` must be column vectors. In the demonstration that follows, apostrophes signify transpose. We demonstrate the conversion from roots to coefficients with $G(s) = 10(s + 2)\ (s + 4)/[s(s + 3)(s + 5)]$.

```
'(ch2p10)'                               % Display label.
'Coefficients for F(s)'                  % Display label.
numftf=[10 40 60]                        % Form numerator of F(s) =
                                         % (10s^2+40s+60)/(s^3+4s^2+5s
                                         % +7).
denftf=[1 4 5 7]                         % Form denominator of F(s) =
                                         % (10s^2+40s+60)/(s^3+4s^2+5s
                                         % +7).
'Roots for F(s)'                         % Display label.
[numfzp,denfzp]=tf2zp(numftf,denftf)
                                         % Convert F(s) to factored form.
'Roots for G(s)'                         % Display label.
numgzp=[-2 -4]                           % Form numerator of
```

```
K=10                                    % G(s)=10(s+2)(s+4)/[s(s + 3)
                                        % (s+5)].
dengzp=[0 -3 -5]                        % Form denominator of
                                        % G(s)=10(s+2)(s+4)/[s(s+3)(s+5)].
'Coefficients for G(s)'                 % Display label.
[numgtf,dengtf]=zp2tf(numgzp',dengzp',K)
                                        % Convert G(s) to polynomial form.
pause
```

**ch2p11** LTI models can also be converted between polynomial and factored forms. MATLAB commands `tf` and `zpk` are also used for the conversion between LTI models. If a transfer function, *Fzpk*(*s*), is expressed as factors in the numerator and denominator, then `tf (Fzpk)` converts *Fzpk*(*s*) to a transfer function expressed as coefficients in the numerator and denominator. Similarly, if a transfer function, *Ftf*(*s*), is expressed as coefficients in the numerator and denominator, then `zpk(Ftf)` converts *Ftf*(*s*) to a transfer function expressed as factors in the numerator and denominator. The following example demonstrates the concepts.

```
'(ch2p11)'                              % Display label.
'Fzpk1(s)'                              % Display label.
Fzpk1=zpk([-2 -4],[0 -3 -5],10)         % Form Fzpk1(s)=
                                        % 10(s+2)(s+4)/[s(s+3)(s+5)].
'Ftf1'                                  % Display label.
Ftf1=tf(Fzpk1)                          % Convert Fzpk1(s) to
                                        % coefficients form.
'Ftf2'                                  % Display label.
Ftf2=tf([10 40 60],[1 4 5 7])           % Form Ftf2(s)=
                                        % (10s^2+40s+60)/(s^3+4s^2+5s
                                        % +7).
'Fzpk2'                                 % Display label.
Fzpk2=zpk(Ftf2)                         % Convert Ftf2(s) to
                                        % factored form.
pause
```

**ch2p12** Functions of time can be easily plotted using MATLAB's `plot(X,Y,S)`, where `X` is the independent variable, `Y` is the dependent variable, and `S` is a character string describing the plot's color, marker, and line characteristic. Type `HELP PLOT` in the **Command Window** to see a list of choices for `S`. Multiple plots also can be obtained using `plot (X1,Y1,S1,X2,Y2,S2,X3,Y3,S3, . . . )`. In the following example we plot on the same graph sin(5*t*) in red and cos(5*t*) in green for $t = 0$ to 10 seconds in 0.01 second increments. Time is specified as `t=start: increment: final`.

```
'(ch2p12)'                              % Display label.
t=0:0.01:10;                            % Specify time range and increment.
f1=cos(5*t);                            % Specify f1 to be cos(5t).
f2=sin(5*t);                            % Specify f2 to be sin (5t).
plot(t,f1,'r',t,f2,'g')                 % Plot f1 in red and f2 in green.
pause
```

## Chapter 3: Modeling in the Time Domain

**ch3p1** The square system matrix, $\mathbf{A} = \begin{bmatrix} 0 & 1 & 0 \\ 0 & 0 & 1 \\ -9 & -8 & -7 \end{bmatrix}$ is written with a space or comma separating the elements of each row. The next row is indicated with a semicolon or carriage return. The entire matrix is then enclosed in a pair of square brackets.

```
'(ch3p1)'                          % Display label.
A=[0 1 0;0 0 1; -9 -8 -7]          % Represent A.
'or'
A=[0 1 0                           % Represent A.
0 0 1
-9 -8 -7]
pause
```

**ch3p2** A row vector, such as the output matrix **C**, can be represented with elements separated by spaces or commas and enclosed in square brackets. A column vector, such as input matrix **B**, can be written as elements separated by semicolons or carriage returns, or as the transpose (') of a row vector.

```
'(ch3p2)'                          % Display label.
C=[2 3 4]                          % Represent row vector C.
B=[7;8;9]                          % Represent column vector B.
'or'
B=[7                               % Represent column vector B.
8
9]
'or'
B=[7 8 9]'                         % Represent column vector B.
pause
```

**ch3p3** The state-space representation consists of specifying the **A**, **B**, **C**, and **D** matrices followed by the creation of an LTI state-space object using the MATLAB command, `ss(A,B,C,D)`. Hence, for the matrices in (ch3p1) and (ch3p2), the state-space representation would be:

```
'(ch3p3)'                          % Display label.
A=[0 1 0;0 0 1;-9 -8 -7];          % Represent A.
B=[7;8;9];                         % Represent column vector B.
C=[2 3 4];                         % Represent row vector C.
D=0;                               % Represent D.
F=ss(A,B,C,D)                      % Create an LTI object and display.
```

**ch3p4 (Example 3.4)** Transfer functions represented either by numerator and denominator or an LTI object can be converted to state space. For numerator and denominator representation, the conversion can be implemented using `[A, B, C, D]=tf2ss(num,den)`. The **A** matrix is returned in a form called the controller canonical form, which will be explained in Chapter 5 in the text. To obtain the phase-variable form, `[Ap,Bp,Cp,Dp]`, we perform the following operations: `Ap=inv(P)*A*P;Bp=inv(P)*B;Cp=C*P,Dp=D`, where **P** is a matrix with 1's along the anti-diagonal and 0's elsewhere. These transformations will be explained in Chapter 5. The command `inv (X)` finds the inverse of a square matrix. The symbol * signifies multiplication. For systems represented as LTI objects, the command `ss(F)`, where `F` is an LTI transfer-function object, can be used to convert `F` to a state-space object. Let us look at Example 3.4 in the text. For the numerator-denominator representation, notice that the MATLAB response associates the gain, 24, with the vector **C** rather than the vector **B** as in the example in the text. Both representations are equivalent. For the LTI transfer-function object, the conversion to state space does not yield the phase-variable form. The result is a balanced model that improves the accuracy of calculating eigenvalues, which are covered in Chapter 4. Since `ss(F)` does not yield familiar forms of the state equations (nor is it possible to easily convert to familiar forms), we will have limited use for that transformation at this time.

```
'(ch3p4) Example 3.4'                   % Display label.
'Numerator-denominator representation conversion'
                                        % Display label.
'Controller canonical form'             % Display label.
num=24;                                 % Define numerator of
                                        % G(s)=C(s)/R(s).
den=[1 9 26 24];                        % Define denominator of G(s).
[A,B,C,D]=tf2ss(num,den)                % Convert G(s) to controller
                                        % canonical form, store matrices
                                        % A, B, C, D, and display.
'Phase-variable form'                   % Display label.
P=[0 0 1;0 1 0;1 0 0];                  % Form transformation matrix.
Ap=inv(P)*A*P                           % Form A matrix, phase-variable
                                        % form.
Bp=inv(P)*B                             % Form B vector, phase-variable
                                        % form.
Cp=C*P                                  % Form C vector, phase-variable
                                        % form.
Dp=D                                    % Form D phase-variable form.
'LTI object representation'             % Display label.
T=tf(num,den)                           % Represent T(s)=24/(s^3+9s^2 +
                                        % 26s+24) as an LTI transfer-
                                        % function object.
Tss=ss(T)                               % Convert T(s) to state space.
pause
```

**ch3p5** State-space representations can be converted to transfer functions represented by a numerator and a denominator using `[num,den]=ss2tf(A,B,C,D,iu)`, where `iu` is the input number for multiple-input systems. For single-input, single-output systems `iu=1`. For an LTI state-space system, `Tss`, the conversion can be implemented using `Ttf=tf(Tss)` to yield the transfer function in polynomial form or `Tzpk=zpk(Tss)` to yield the transfer function in factored form. For example, the transfer function represented by the matrices described in (ch3p3) can be found as follows:

```
'(ch3p5)'                               % Display label.
      'Non LTI'                         % Display label.
A=[0 1 0;0 0 1;-9 -8 -7];               % Represent A.
B=[7;8;9];                              % Represent B.
C=[2 3 4];                              % Represent C.
D=0;                                    % Represent D.
'Ttf(s)'                                % Display label.
[num,den]=ss2tf(A,B,C,D,1)              % Convert state-space
                                        % representation to a
                                        % transfer function represented as
                                        % a numerator and denominator in
                                        % polynomial form, G(s)=num/den,
                                        % and display num and den.
      'LTI'                             % Display label.
Tss=ss(A,B,C,D)                         % Form LTI state-space model.
'Polynomial form, Ttf(s)'               % Display label.
```

```
Ttf=tf(Tss)                      % Transform from state space to
                                 % transfer function in polynomial
                                 % form.
'Factored form, Tzpk(s)'         % Display label.
Tzpk=zpk(Tss)                    % Transform from state space to
                                 % transfer function in factored
                                 % form.
pause
```

## Chapter 4: Time Response

**ch4p1 (Example 4.6)** We can use MATLAB to calculate characteristics of a second-order system, such as damping ratio, $\zeta$; natural frequency, $\omega_n$; percent overshoot, $\%OS$ (pos); settling time, $T_s$; and peak time, $T_p$. Let us look at Example 4.6 in the text.

```
'(ch4p1) Example 4.6'            % Display label.
p1=[1 3+7*i];                    % Define polynomial containing
                                 % first pole.
p2=[1 3-7*i];                    % Define polynomial containing
                                 % second pole.
deng=conv(p1,p2);                % Multiply the two polynomials to
                                 % find the 2nd order polynomial,
                                 % as^2+bs+c.
omegan=sqrt(deng(3)/deng(1))     % Calculate the natural frequency,
                                 % sqrt(c/a).
zeta=(deng(2)/deng(1))/(2*omegan)
                                 % Calculate damping ratio,
                                 % ((b/a)/2*wn).
Ts=4/(zeta*omegan)               % Calculate settling time,
                                 % (4/z*wn).
Tp=pi/(omegan*sqrt(1-zeta^2))    % Calculate peak time,
                                 % pi/wn*sqrt(1-z^2).
pos=100*exp(-zeta*pi/sqrt(1-zeta^2))
                                 % Calculate percent overshoot
                                 % (100*e^(-z*pi/sqrt(1-z^2)).
pause
```

**ch4p2 (Example 4.8)** We can use MATLAB to obtain system step responses. These responses are particularly valuable when the system is not a pure two-pole system and has additional poles or zeros. We can obtain a plot of the step response of a transfer function, $T(s)$ =`num/den`, using the command `step(T)`, where `T` is an LTI transfer-function object. Multiple plots also can be obtained using `step(T1,T2,...)`.

Information about the plots obtained with `step(T)` can be found by left-clicking the mouse on the curve. You can find the curve's label as well as the coordinates of the point on which you clicked. Right-clicking away from a curve brings up a menu. From this menu you can select (1) system responses to be displayed and (2) response characteristics to be displayed, such as peak response. When selected, a dot appears on the curve at the appropriate point. Let your mouse rest on the point to read the value of the characteristic. You may also select (3) choice for grid on or off, (4) choice to normalize the curve, and (5) properties, such as labels, limits, units, style, and characteristics.

If we add the left-hand side, `[y,t]=step(T)`, we create vectors containing the plot's points, where `y` is the output vector and `t` is the time vector. For this case, a plot is not made until the `plot(t,y)` command is given, where we assume we want to plot the output `(y)` versus time `(t)`. We can label the plot, the $x$-axis, and the $y$-axis with `title('ab')`, `xlabel('ab')`, and `ylabel('ab')`, respectively. The command `clf` clears the graph prior to plotting. Finally, text can be placed anywhere on the graph using the command `text(X,Y,'text')`, where `(X,Y)` are the graph coordinates where `'text'` will be displayed. Let us look at Example 4.8 in the text.

```
'(ch4p2) Example 4.8'                 % Display label.
'Test Run'                            % Display label.
clf                                   % Clear graph.
numt1=[24.542];                       % Define numerator of T1.
dent1=[1 4 24.542];                   % Define denominator of T1.
'T1(s)'                               % Display label.
T1=tf(numt1,dent1)                    % Create and display T1(s).
step(T1)                              % Run a demonstration step response
                                      % plot
title('Test Run of T1(s)')            % Add title to graph.
pause
'Complete Run'                        % Display label.
[y1,t1]=step(T1);                     % Run step response of T1 and
                                      % collect points.
numt2=[245.42];                       % Define numerator of T2.
p1=[1 10];                            % Define (s+10) in denominator
                                      % of T2.
p2=[1 4 24.542];                      % Define (s^2+4s+24.542) in
                                      % denominator of T2.
dent2=conv(p1,p2);                    % Multiply (s + 10)(s^2+4s+24.542)
                                      % for denominator of T2.
'T2(s)'                               % Display label.
T2=tf(numt2,dent2)                    % Create and display T2.
[y2,t2]=step(T2);                     % Run step response of T2 and
                                      % collect points.
numt3=[73.626];                       % Define numerator of T3.
p3=[1 3];                             % Define (s+3) in denominator
                                      % of T3.
dent3=conv(p3,p2);                    % Multiply (s+3)(s^2+4s+24.542)
                                      % for denominator of T3.
'T3(s)'                               % Display label.
T3=tf(numt3,dent3)                    % Create and display T3.
[y3,t3]=step(T3);                     % Run step response of T3 and
                                      % collect points.
clf                                   % Clear graph.
plot(t1,y1,t2,y2,t3,y3)               % Plot acquired points with all
                                      % three plots on one graph.
title('Step Responses of T1(s),T2(s), and T3(s)')
                                      % Add title to graph.
xlabel('Time(seconds)')               % Add time axis label.
ylabel('Normalized Response')         % Add response axis label.
text(0.7,0.7,'c3(t)')                 % Label step response of T1.
text(0.7,1.1,'c2(t)')                 % Label step response of T2.
text(0.5,1.3,'c1(t)')                 % Label step response of T3.
pause
step(T1,T2,T3)                        % Use alternate method of plotting
                                      % step responses.
title('Step Responses of T1(s), T2(s), and T3(s)')
                                      % Add title to graph.
pause
```

**ch4p3** We also can plot the step response of systems represented in state space using the `step(T,t)` command. Here `T` is any LTI object and `t=a:b:c` is the range for the time axis, where a is the initial time, b is the time step size, and c is the final time. For example, `t=0:1:10` means time from 0 to 10 seconds in steps of 1 second. The `t` field is optional. Finally, in this example we introduce the command `grid on`, which superimposes a grid over the step response. Place the `grid on` command after the `step(T,t)` command.

```
'(ch4p3)'                        % Display label.
clf                              % Clear graph.
A=[0 1 0;0 0 1;-24 -26 -9];      % Generate A matrix.
B=[0;0;1];                       % Generate B vector.
C=[2 7 1];                       % Generate C vector.
D=0;                             % Generate D.
T=ss(A,B,C,D)                    % Generate LTI object, T, in state
                                 % space and display.
t=0:0.1:10;                      % Define range of time for plot.
step(T,t)                        % Plot step response for given
                                 % range of time.
grid on                          % Turn grid on for plot.
pause
```

**ch4p4 (Antenna Control Case Study)** We now use MATLAB to plot the step response requested in the Antenna Control Case Study.

```
' (ch4p4) Antenna Control Case Study'
                                 % Display label.
clf                              % Clear graph.
numg=20.83;                      % Define numerator of G(s).
deng=[1 101.71 171];             % Define denominator of G(s).
'G(s)'                           % Display label.
G=tf(numg,deng)                  % Form and display transfer
                                 % Function G(s).
step(G);                         % Generate step response.
title(' Angular Velocity Response')
                                 % Add title.
pause
```

**ch4p5 (UFSS Case Study)** As a final example, let us use MATLAB to do the UFSS Case Study in the text (*Johnson, 1980*). We introduce table lookup to find the rise time. Using the `interp1(y,t,y1)` command, we set up a table of values of amplitude, `y`, and time, `t`, from the step response and look for the value of time for which the amplitude is `y1`= 0.1 and 0.9. We also generate time response data over a defined range of time using `t=a:b:c` followed by `[y,t]=step(G,t)`. Here `G` is an LTI transfer-function object and `t` is the range for the time axis, where a is the initial time, b is the time step size, and c is the final time; y is the output.

```
' (ch4p5) UFSS Case Study'       % Display label.
clf                              % Clear graph.
'(a)'                            % Display label.
numg=0.0169;                     % Define numerator of 2nd order
                                 % approximation of G(s).
deng=[1 0.226 0.0169];           % Define 2nd order term of
                                 % denominator of G(s).
'G(s)'                           % Display label.
G=tf(numg,deng)                  % Create and display G(s).
omegan=sqrt(deng(3))             % Find natural frequency.
zeta=deng(2)/(2*omegan)          % Find damping ratio.
Ts=4/(zeta*omegan)               % Find settling time.
Tp=pi/(omegan*sqrt(1-zeta^2))    % Find peak time.
```

```
pos=exp(-zeta*pi/sqrt(1-zeta^2))*100
                                    % Find percent overshoot.
t=0:0.1:35;                         % Limit time to find rise time. t=0
                                    % to 35 in steps of 0.1.
[y,t]=step(G,t);                    % Generate and save points of step
                                    % response over defined range of t.
Tlow=interp1(y,t,0.1);              % Search table for time when
                                    % y=0.1*finalvalue.
Thi=interp1(y,t,0.9);               % Search table for
                                    % time=0.9*finalvalue.
Tr=Thi-Tlow                         % Calculate rise time.
'(b)'                               % Display label.
numc=0.125*[1 0.435];               % Define numerator of C(s).
denc=conv(poly([0 -1.23]),[1 0.226 0.0169]);
                                    % Define denominator of C(s).
[K,p,k]=residue(numc,denc)          % Find partial-fraction expansion.
'(d)'                               % Display label.
numg=0.125*[1 0.435];               % Define numerator of G(s).
deng=conv([1 1.23],[1 0.226 0.0169]);
                                    % Define denominator of G(s).
'G(s)'                              % Display label.
G=tf(numg,deng)                     % Create and display G(s).
[y,t]=step(G);                      % Generate complete step response
                                    % and collect points.
plot(t,y)                           % Plot points.
title ('Pitch Angle Response')      % Add title.
xlabel('Time (seconds)')            % label time axis.
ylabel ('Pitch Angle (radians)')    % Label y-axis.
pause
```

## Chapter 5: Reduction of Multiple Subsystems

**ch5p1 (UFSS Pitch Control System)** MATLAB can be used for block diagram reduction. Three methods are available: (1) Solution via Series, Parallel, & Feedback Commands, (2) Solution via Algebraic Operations, and (3) Solution via Append & Connect Commands. Let us look at each of these methods.

1. **Solution via Series, Parallel, & Feedback Commands**
   The closed-loop transfer function is obtained using the following commands successively, where the arguments are LTI objects: `series(G1,G2)` for a cascade connection of $G_1(s)$; and $G_2(s)$; `parallel(G1,G2)` for a parallel connection of $G_1(s)$ and $G_2(s)$; `feedback(G,H,sign)` for a closed-loop connection with $G(s)$ as the forward path, $H(s)$ as the feedback, and `sign` is – 1 for negative-feedback systems or +1 for positive-feedback systems. The sign is optional for negative-feedback systems.
2. **Solution via Algebraic Operations**
   Another approach is to use arithmetic operations successively on LTI transfer functions as follows: `G2*G1` for a cascade connection of $G_1(s)$ and $G_2(s)$; `G1+G2` for a parallel connection of $G_1(s)$ and $G_2(s)$; `G/(1+G*H)` for a closed-loop negative-feedback connection with $G(s)$ as the forward path and $H(s)$ as the feedback; `G/(1-G*H)` for positive-feedback systems. When using division we follow with the function `minreal (sys)` to cancel common terms in the numerator and denominator.
3. **Solution via Append & Connect Commands**
   The last method, which defines the topology of the system, may be used effectively for complicated systems. First, the subsystems are defined. Second, the subsystems are

appended, or gathered, into a multiple-input/multiple-output system. Think of this system as a single system with an input for each of the subsystems and an output for each of the subsystems. Next, the external inputs and outputs are specified. Finally, the subsystems are interconnected. Let us elaborate on each of these steps.

The subsystems are defined by creating LTI transfer functions for each. The subsystems are appended using the command `G=append(G1,G2,G3,G4, . . . . .Gn)`, where the `Gi` are the LTI transfer functions of the subsystems and `G` is the appended system. Each subsystem is now identified by a number based upon its position in the append argument. For example, `G3` is 3, based on the fact that it is the third subsystem in the append argument (not the fact that we write it as `G3`).

Now that we have created an appended system, we form the arguments required to interconnect their inputs and outputs to form our system. The first step identifies which subsystems have the external input signal and which subsystems have the external output signal. For example, we use `inputs=[1 5 6]` and `outputs=[3 4]` to define the external inputs to be the inputs of subsystems 1,5, and 6 and the external outputs to be the outputs of subsystems 3 and 4. For single-input/single-output systems, these definitions use scalar quantities. Thus `inputs=5,outputs=8` define the input to subsystem 5 as the external input and the output of subsystem 8 as the external output.

At this point we tell the program how all of the subsystems are interconnected. We form a $Q$ matrix that has a row for each subsystem whose input comes from another subsystem's output. The first column contains the subsystem's number. Subsequent columns contain the numbers of the subsystems from which the inputs come. Thus, a typical row might be as follows: `[3 6 −7]`, or subsystem 3's input is formed from the sum of the output of subsystem 6 and the negative of the output of subsystem 7.

Finally, all of the interconnection arguments are used in the `connect(G Q inputs, outputs)` command, where all of the arguments have been previously defined.

Let us demonstrate the three methods for finding the total transfer function by looking at the back endpapers and finding the closed-loop transfer function of the pitch control loop for the UFSS with $K_1 = K_2 = 1$ (*Johnson, 1980*). The last method using append and connect requires that all subsystems be proper (the order of the numerator cannot be greater than the order of the denominator). The pitch rate sensor violates this requirement. Thus, for the third method, we perform some block diagram maneuvers by pushing the pitch rate sensor to the left past the summing junction and combining the resulting blocks with the pitch gain and the elevator actuator. These changes are reflected in the program. You should verify all computer results with hand calculations.

```
'(ch5p1) UFSS Pitch Control System'
' Solution via Series, Parallel, & Feedback Commands'
                                        % Display labels.
numg1=[-1];                             % Define numerator of G1(s).
deng1=[1];                              % Define denominator of G1(s).
numg2=[0 2];                            % Define numerator of G2(s).
deng2=[1 2];                            % Define denominator of G2 (s).
numg3=-0.125*[1 0.435];                 % Define numerator of G3(s).
deng3=conv([1 1.23],[1 0.226 0.0169]);
                                        % Define denominator of G3(s).
numh1=[-1 0];                           % Define numerator of H1(s).
denh1=[0 1];                            % Define denominator of H1(s).
G1=tf(numg1,deng1);                     % Create LTI transfer function,
                                        % G1(s).
G2=tf(numg2,deng2);                     % Create LTI transfer function,
                                        % G2(s).
G3=tf(numg3,deng3);                     % Create LTI transfer function,
                                        % G3(s).
```

```
H1=tf(numh1,denh1);                   % Create LTI transfer function,
                                      % H1(s).
G4=series(G2,G3);                     % Calculate product of elevator
                                      % and vehicle dynamics.
G5=feedback(G4,H1);                   % Calculate closed-loop transfer
                                      % function of inner loop.
Ge=series(G1,G5);                     % Multiply inner-loop transfer
                                      % function and pitch gain.
'T(s) via Series, Parallel, & Feedback Commands'
                                      % Display label.
T=feedback(Ge,1)                      % Find closed-loop transfer
                                      % function.
'Solution via Algebraic Operations'
                                      % Display label.
clear                                 % Clear session.
numg1=[-1];                           % Define numerator of G1(s).
deng1=[1];                            % Define denominator of G1(s).
numg2=[0 2];                          % Define numerator of G2(s).
deng2=[1 2];                          % Define denominator of G2(s).
numg3=-0.125*[1 0.435];               % Define numerator of G3(s).
deng3=conv([1 1.23],[1 0.226 0.0169]);
                                      % Define denominator of G3(s).
numh1=[-1 0];                         % Define numerator of H1(s).
denh1=[0 1];                          % Define denominator of H1(s).
G1=tf(numg1,deng1);                   % Create LTI transfer function, G1(s).
G2=tf(numg2,deng2);                   % Create LTI transfer function, G2(s).
G3=tf(numg3,deng3);                   % Create LTI transfer function, G3(s).
H1=tf(numh1,denh1);                   % Create LTI transfer function, H1(s).
G4=G3*G2;                             % Calculate product of elevator and
                                      % vehicle dynamics.
G5=G4/(1+G4*H1);                      % Calculate closed-loop transfer
                                      % function of inner loop.
G5=minreal(G5);                       % Cancel common terms.
Ge=G5*G1;                             % Multiply inner-loop transfer
                                      % functions.
'T(s) via Algebraic Operations'       % Display label.
T=Ge/(1+Ge);                          % Find closed-loop transfer function.
T=minreal(T)                          % Cancel common terms.
'Solution via Append & Connect Commands'
                                      % Display label.
'G1(s)=(-K1)*(1/(-K2s))=1/s'          % Display label.
numg1=[1];                            % Define numerator of G1(s).
deng1=[1 0];                          % Define denominator of G1(s).
G1=tf(numg1,deng1)                    % Create LTI transfer function,
                                      % G1(s)=pitch gain*
                                      % 1(1/Pitch rate sensor).
'G2(s)=(-K2s)*(2/(s+2)'               % Display label.
numg2=[-2 0];                         % Define numerator of G2(s).
deng2=[1 2];                          % Define denominator of G2(s).
G2=tf(numg2,deng2)                    % Create LTI transfer function,
                                      % G2(s)=pitch rate sensor*vehicle
                                      % dynamics.
'G3(s)=-0.125(s+0.435)/((s+1.23)(s^2+0.226s+0.0169))'
                                      % Display label.
numg3=-0.125*[1 0.435];               % Define numerator of G3(s).
deng3=conv([1 1.23],[1 0.226 0.0169]);
                                      % Define denominator of G3(s).
```

```
G3=tf(numg3,deng3);                  % Create LTI transfer function,
                                     % G3(s)=vehicle dynamics.
System=append(G1,G2,G3)              % Gather all subsystems.
input=1;                             % Input is at first subsystem,
                                     % G1(s).
output=3;                            % Output is output of third
                                     % subsystem, G3(s).
Q=[1 -3 0                            % Subsystem 1, G1(s), gets its
                                     % input from the negative of the
                                     % output of subsystem 3, G3(s).
2 1 -3                               % Subsystem 2, G2(s), gets its
                                     % input from subsystem 1, G1(s),
                                     % and the negative of the output
                                     % of subsystem 3, G3(s).
3 2 0];                              % Subsystem 3, G3(s), gets its
                                     % input from subsystem 2, G2(s).
T=connect(System,Q,input,output);
                                     % Connect the subsystems.
'T(s) via Append & Connect Commands'
                                     % Display label.
T=tf(T);                             % Create LTI closed-loop transfer
                                     % function.
T=minreal(T)                         % Cancel common terms.
pause
```

**ch5p2 (Example 5.3)** We can use MATLAB to calculate the closed-loop characteristics of a second-order system, such as damping ratio, $\zeta$; natural frequency, $\omega_n$; percent overshoot, $\%OS$ (pos); settling time, $T_s$; and peak time, $T_p$. The command `[numt,dent]=tfdata(T,'v')` extracts the numerator and denominator of $T(s)$ for a single-input/single-output system from which the calculations are based. The argument `'v'` returns the numerator and denominator as simple row vectors. Omitting `'v'` would return the numerator and denominator as cell arrays requiring more steps to obtain the row vectors. We end by generating a plot of the closed-loop step response. Let us look at Example 5.3 in the text.

```
'(ch5p2) Example 5.3'                % Display label.
numg=[25];                           % Define numerator of G(s).
deng=poly([0 -5]);                   % Define denominator of G(s).
'G(s)'                               % Display label.
G=tf(numg,deng)                      % Create and display G(s).
'T(s)'                               % Display label.
T=feedback(G,1)                      % Find T(s).
[numt,dent]=tfdata(T,'v');           % Extract numerator & denominator
                                     % of T(s).
wn=sqrt(dent(3))                     % Find natural frequency.
z=dent(2)/(2*wn)                     % Find damping ratio.
Ts=4/(z*wn)                          % Find settling time.
Tp=pi/(wn*sqrt(1-z^2))               % Find peak time.
pos=exp(-z*pi/sqrt(1-z^2))*100       % Find percent overshoot.
step(T)                              % Generate step response.
pause
```

**ch5p3** MATLAB can be used to convert transfer functions to state space in a specified form. The command `[Acc Bcc Ccc Dcc]=tf2ss(num,den)` can be used to convert $T(s)$=`num/den` into controller canonical form with matrices and vectors `Acc`, `Bcc`, `Ccc`, and `Dcc`. We can then form an LTI state space object using `Scc=ss(Acc,Bcc,Ccc,Dcc)`. This object can then be converted into parallel form using `Sp=canon(Scc,'type')`, where `type=modal` yields the parallel form. Another choice, not used here, is `type=companion`, which yields a right companion system matrix. Transformation matrices can be used to convert to other representations. As an example, let us convert $C(s)/R(s) = 24/[(s+2)(s+3)(s+4)]$ into a parallel representation in state space, as is done in Section 5.7—Parallel Form. Notice that the product of values in the **B** and **C** vectors yields the same product as the results in Eqs. (5.49) and (5.50). Thus, the two solutions are the same, but the state variables are ordered differently, and the gains are split between the **B** and **C** vectors. We can also extract the system matrices from the LTI object using `[A,B,C,D]=ssdata(S)`, where `S` is a state-space LTI object and `A, B, C, D`, are its associated matrices and vectors.

```
'(ch5p3)'                          % Display label.
numt=24;                           % Define numerator of T(s).
dent=poly([-2 -3 -4]);             % Define denominator of T(s).
'T(s)'                             % Display label.
T=tf(numt,dent)                    % Create and display T(s).
[Acc Bcc Ccc Dcc]=tf2ss(numt,dent);
                                   % Convert T(s) to controller
                                   % canonical form.
Scc=ss(Acc,Bcc,Ccc,Dcc);           % Create LTI controller canonical
                                   % state-space object.
Sp=canon(Scc,'modal');             % Convert controller canonical form
                                   % to parallel form.
'Controller Canonical Form'        % Display label.
[Acc,Bcc,Ccc,Dcc]=ssdata(Scc)      % Extract and display controller
                                   % canonical form matrices.
'Parallel Form'                    % Display label.
[Ap,Bp,Cp,Dp]=ssdata(Sp)           % Extract and display parallel form
                                   % matrices.
pause
```

**ch5p4 (Example 5.9)** We can use MATLAB to perform similarity transformations to obtain other forms. Let us look at Example 5.9 in the text.

```
'(ch5p4) Example 5.9'              % Display label.
Pinv=[2 0 0; 3 2 0; 1 4 5];        % Define P inverse.
P=inv(Pinv)                        % Calculate P.
'Original'                         % Display label.
Ax=[0 1 0;0 0 1;-2 -5 -7]          % Define original A.
Bx=[0 0 1]                         % Define original B.
Cx=[1 0 0]                         % Define original C.
'Transformed'                      % Display label.
Az=Pinv*Ax*P                       % Calculate new A.
Bz=Pinv*Bx                         % Calculate new B.
Cz=Cx*P                            % Calculate new C.
pause
```

**ch5p5** Using MATLAB's `[P,d]=eig(A)` command, where the columns of `P` are the eigenvectors of `A` and the diagonal elements of `d` are the eigenvalues of `A`, we can find the eigenvectors of the system matrix and then proceed to diagonalize the system. We can also use `canon (S,'modal')` to diagonalize an LTI object, `S`, represented in state space.

```
'(ch5p5)'                          % Display label.
A=[3 1 5;4 -2 7;2 3 1];            % Define original A.
B=[1;2;3];                         % Define original B.
```

```
C=[2 4 6];                        % Define original C.
[P,d]=eig(A)                      % Generate transformation matrix,
                                  % P, and eigenvalues, d.
'Via Transformation'              % Display label.
Adt=inv(P)*A*P                    % Calculate diagonal system A.
Bdt=inv(P)*B                      % Calculate diagonal system B.
Cdt=C*P                           % Calculate diagonal system C.
'Via Canon Command'               % Display label.
S=ss(A,B,C,0)                     % Create state-space LTI object
                                  % for original system.
Sp=canon(S,'modal')               % Calculate diagonal system via
                                  % canon command.
pause
```

## Chapter 6: Stability

**ch6p1 (Example 6.7)** MATLAB can solve for the poles of a transfer function in order to determine stability. To solve for the poles of $T(s)$ use the `pole(T)` command. Let us look at Example 6.7 in the text.

```
'(ch6p1) Example 6.7'             % Display label.
numg=1;                           % Define numerator of G(s).
deng=conv([1 0],[2 3 2 3 2]);     % Define denominator of G(s).
G=tf(numg,deng);                  % Create G(s) object.
'T(s)'                            % Display label.
T=feedback(G,1)                   % Calculate closed-loop T(s)
                                  % object.
                                  % Negative feedback is default
                                  % when there is no sign parameter.
poles=pole(T)                     % Find poles of T(s).
pause
```

**ch6p2 (Example 6.9)** We can use MATLAB to find the range of gain for stability by generating a loop, changing gain, and finding at what gain we obtain right–half-plane poles.

```
'(ch6p2) Example 6.9'             % Display label.
K=[1:1:2000];                     % Define range of K from 1 to 2000
                                  % in steps of 1.
for n=1:length(K);                % Set up length of DO LOOP to equal
                                  % number of K values to be tested.
      dent=[1 18 77 K(n)];        % Define the denominator of T(s)
                                  % for the nth value of K.
      poles=roots(dent);          % Find the poles for the nth value
                                  % of K.
      r=real(poles);              % Form a vector containing the real
                                  % parts of the poles for K(n).
         if max(r) >=0,           % Test poles found for the nth
                                  % value of K for a real value ≥ 0.
          poles                   % Display first pole values where
                                  % there is a real part ≥ 0.
          K=K(n)                  % Display corresponding value of K.
          break                   % Stop loop if rhp poles are found.
      end                         % End if.
end                               % End for
pause
```

**ch6p3 (Example 6.11)** We can use MATLAB to determine the stability of a system represented in state space by using the command `eig(A)` to find the eigenvalues of the system matrix, `A`. Let us apply the concept to Example 6.11 in the text.

```
'(ch6p3) Example 6.11'                % Display label.
A=[0 3 1;2 8 1;-10 -5 -2]             % Define system matrix, A.
eigenvalues=eig(A)                    % Find eigenvalues.
pause
```

## Chapter 7: Steady-State Errors

**ch7p1 (Example 7.4, sys. b)** Static error constants are found using lim $s^nG(s)$ as $s \to 0$. Once the static error constant is found, we can evaluate the steady-state error. To evaluate the static error constant we can use the command `dcgain(G)`, which evaluates $G(s)$ at $s = 0$. Let us look at Example 7.4, system (b), in the text.

```
'(ch7p1) Example 7.4, sys.b'          % Display label.
numg=500*poly([-2 -5 -6]);            % Define numerator of G(s).
deng=poly([0 -8 -10 -12]);            % Define denominator of G(s).
G=tf(numg,deng);                      % Form G(s).
' Check Stability'                    % Display label.
T=feedback(G,1);                      % Form T(s).
poles=pole(T)                         % Display closed-loop poles.
' Step Input'                         % Display label.
Kp=dcgain(G)                          % Evaluate Kp=numg/deng for s=0.
ess=1/(1+Kp)                          % Evaluate ess for step input.
' Ramp Input'                         % Display label.
numsg=conv([1 0],numg);               % Define numerator of sG(s).
densg=poly([0 -8 -10 -12]);           % Define denominator of sG(s).
sG=tf(numsg,densg);                   % Create sG(s).
sG=minreal(sG);                       % Cancel common 's' in
                                      % numerator(numsg) and
                                      % denominator(densg).
Kv=dcgain(sG)                         % Evaluate Kv=sG(s) for s=0.
ess=1/Kv                              % Evaluate steady-state error for
                                      % ramp input.
' Parabolic Input'                    % Display label.
nums2g=conv([1 0 0],numg);            % Define numerator of s^2G(s).
dens2g=poly([0 -8 -10-12]);           % Define denominator of s^2G(s).
s2G=tf(nums2g,dens2g);                % Create s^2G(s).
s2G=minreal(s2G);                     % Cancel common 's' in
                                      % numerator(nums2g) and
                                      % denominator(dens2g).
Ka=dcgain(s2G)                        % Evaluate Ka=s^2G(s) for s=0.
ess=1/Ka                              % Evaluate steady-state error for
                                      % parabolic input.
pause
```

**ch7p2 (Example 7.6)** We can use MATLAB to evaluate the gain, $K$, required to meet a steady-state error specification. Let us look at Example 7.6 in the text.

```
'(ch7p2) Example 7.6'                 % Display label.
numgdK=[1 5];                         % Define numerator of G(s)/K.
dengdK=poly([0 -6 -7 -8]);            % Define denominator of G(s)/K.
GdK=tf(numgdK,dengdK);                % Create G(s)/K.
numgkv=conv([1 0],numgdK);            % Define numerator of sG(s)/K.
```

```
dengkv=dengdK;                    % Define denominator of sG(s)/K.
GKv=tf(numgkv,dengkv);            % Create sG(s)/K.
GKv=minreal(GKv);                 % Cancel common 's' in numerator
                                  % and denominator of sG(s)/K.
KvdK=dcgain(GKv)                  % Evaluate (Kv/K)=(numgkv/dengkv)
                                  % for s=0.
ess=0.1                           % Enumerate steady-state error.
K=1/(ess*KvdK)                    % Solve for K.
'Check Stability'                 % Display label.
T=feedback(K*GdK,1);              % Form T(s).
poles=pole(T)                     % Display closed-loop poles.
pause
```

## Chapter 8: Root Locus Techniques

**ch8p1 (Example 8.7)** MATLAB allows root loci to be plotted with the `rlocus(GH)` command, where $G(s)H(s)$=`numgh/dengh` and `GH` is an LTI transfer-function object. Points on the root locus can be selected interactively using the `[K,p]=rlocfind(GH)` command. MATLAB then yields the gain `(K)` at that point as well as all other poles `(p)` that have that gain. We can zoom in and out of the root locus by changing the range of axis values using the command `axis([xmin,xmax,ymin,ymax])`. The root locus can be drawn over a grid that shows constant damping ratio `(z)` and constant natural frequency `(wn)` curves using the `sgrid(z,wn)` command. To plot multiple $\zeta$ and $\omega_n$ curves, use `z=zmin:zstep:zmax` and `wn=wnmin:wn-step:wnmax` to specify ranges of values.

```
'(ch8p1) Example 8.7'             % Display label.
clf                               % Clear graph.
numgh=[1 -4 20];                  % Define numerator of G(s)H(s).
dengh=poly([-2 -4]);              % Define denominator of G(s)H(s).
'G(s)H(s)'                        % Display label.
GH=tf(numgh,dengh)                % Create G(s)H(s) and display.
rlocus(GH)                        % Draw root locus.
z=0.2:0.05:0.5;                   % Define damping ratio values : 0.2
                                  % to 0.5 in steps of 0.05.
wn=0:1:10;                        % Define natural frequency values:
                                  % 0 to 10 in steps of 1.
sgrid(z,wn)                       % Generate damping ratio and
                                  % natural frequency grid lines for
                                  % root locus.
title ('Root Locus')              % Define title for root locus.
pause
rlocus(GH)                        % Draw close-up root locus.
axis([-3 1 -4 4])                 % Define range on axes for root
                                  % locus close-up view.
title('Close-up')                 % Define title for close-up root
                                  % locus.
z=0.45;                           % Define damping ratio line for
                                  % overlay on close-up root locus.
wn=0;                             % Suppress natural frequency
                                  % overlay curves.
sgrid(z,wn)                       % Overlay damping ratio curve on
                                  % close-up root locus.
for k=1: 3                        % Loop allows 3 points to be
                                  % selected as per Example 8.7,
                                  % (z=0.45, jwcrossing,breakaway).
```

```
[K,p]=rlocfind(GH)                    % Generate gain, K, and closed-loop
                                      % poles, p, for point selected
                                      % interactively on the root locus.
end                                   % End loop.
pause
```

**ch8p2 (Example 8.8)** We can couple the design of gain on the root locus with a step-response simulation for the gain selected. We introduce the command `rlocus(G,K)`, which allows us to specify the range of gain, K, for plotting the root locus. This command will help us smooth the usual root locus plot by equivalently specifying more points via the argument, K. Notice that the first root locus plotted without the argument K is not smooth. We also introduce the command `x=input('prompt')`, which allows keyboard entry of a value for x in response to a prompt. We apply this command to enter the desired percent overshoot. We also add a variable's value to the title of the root locus and step-response plots by inserting another field in the title command and use `num2str(value)` to convert value from a number to a character string for display. Let us apply the concepts to Example 8.8 in the text.

```
'(ch8p2) Example 8.8'                 % Display label.
clear                                 % Clear variables from workspace.
clf                                   % Clear graph.
numg=[1 1.5];                         % Define numerator of G(s).
deng=poly([0 -1 -10]);                % Define denominator of G(s).
'G(s)'                                % Display label.
G=tf(numg,deng)                       % Create and display G(s).
rlocus(G)                             % Draw root locus (H(s)=1).
title('Original Root Locus')          % Add title.
pause
K=0:0.005:50;                         % Specify range of gain to smooth
                                      % root locus.
rlocus(G,K)                           % Draw smoothed root locus
                                      % (H(s)=1).
title('Smoothed Root Locus')          % Add title.
pos=input('Type %OS ');               % Input desired percent overshoot
                                      % from the keyboard.
z=-log(pos/100)/sqrt(pi^2+[log(pos/100)]^2)
                                      % Calculate damping ratio.
sgrid(z,0)                            % Overlay desired damping ratio
                                      % line on root locus.
title(['Root Locus with', num2str(pos),'% overshoot line'])
                                      % Define title for root locus
                                      % showing percent overshoot used.
[K,p]=rlocfind(G)                     % Generate gain, K, and closed-loop
                                      % poles, p, for point selected
                                      % interactively on the root locus.
pause
'T(s)'                                % Display label.
T=feedback(K*G,1)                     % Find closed-loop transfer
                                      % function
                                      % with selected K and display.
step(T)                               % Generate closed-loop step
                                      % response for point select on
                                      % root locus.
title(['Step Response for K=',num2str(K)])
                                      % Give step response a title which
                                      % includes the value of K.
pause
```

## Chapter 9: Design via Root Locus

**ch9p1 (Example 9.3)** We can use MATLAB to design PD controllers. The program allows us to input a desired percent overshoot via the keyboard. MATLAB then produces a root locus for the uncompensated system with an overlay of the percent overshoot line. We interactively select the intersection of the root locus and the desired percent overshoot line to set the gain. MATLAB outputs an estimate of the uncompensated system's performance specifications and a step response of the uncompensated system for us to determine the required settling time. After we input the settling time through the keyboard, MATLAB designs the PD controller and produces a root locus of the PD compensated system from which we can interactively select the gain. Finally, MATLAB produces an estimate of the PD compensated system's performance specifications and a step response of the PD compensated system.

```
'(ch9p1) Example 9.3'                   % Display label.
clf                                     % Clear graph.
'Uncompensated System'                  % Display label.
numg=1;                                 % Generate numerator of G(s).
deng=poly([0 -4 -6]);                   % Generate denominator of G(s).
'G(s)'                                  % Display label.
G=tf(numg,deng)                         % Create and display G(s).
pos=input('Type desired percent overshoot ');
                                        % Input desired percent overshoot.
z=log(pos/100)/sqrt(pi^2+[log(pos/100)]^2);
                                        % Calculate damping ratio.
rlocus(G)                               % Plot uncompensated root locus.
sgrid(z,0)                              % Overlay desired percent
                                        % overshoot line.
title (['Uncompensated Root Locus with ', num2str(pos),...
'% Overshoot Line'])                    % Title uncompensated root locus.
[K,p]=rlocfind(G);                      % Generate gain, K, and closed-loop
                                        % poles, p, for point selected
                                        % interactively on the root locus.
'Closed-loop poles='                    % Display label.
p                                       % Display closed-loop poles.
f=input('Give pole number that is operating point ');
                                        % Choose uncompensated system
                                        % dominant pole.
'Summary of estimated specifications for selected point on'
'uncompensated root locus'              % Display label.
operatingpoint=p(f)                     % Display uncompensated dominant
                                        % pole.
gain=K                                  % Display uncompensated gain.
estimated_settling_time=4/abs(real(p(f)))
                                        % Display uncompensated settling
                                        % time.
estimated_peak_time=pi/abs(imag(p(f)))
                                        % Display uncompensated peak time.
estimated_percent_overshoot=pos
                                        % Display uncompensated percent
                                        % overshoot.
estimated_damping_ratio=z               % Display uncompensated damping
                                        % ratio.
estimated_natural_frequency=sqrt(real(p(f))^2+imag(p(f))^2)
                                        % Display uncompensated natural
                                        % frequency.
numkv=conv([1 0],numg);                 % Set up numerator to evaluate Kv.
```

```
denkv=deng;                              % Set up denominator to evaluate Kv.
sG=tf(numkv,denkv);                      % Create sG(s).
sG=minreal(sG);                          % Cancel common poles and zeros.
Kv=dcgain(K*sG)                          % Display uncompensated Kv.
ess=1/Kv                                 % Display uncompensated
                                         % steady-state
                                         % error for unit ramp input.
'T(s)'                                   % Display label.
T=feedback(K*G,1)                        % Find uncompensated T(s).
step(T)                                  % Plot step response of
                                         % uncompensated system.
title(['Uncompensated System Step Response with ',num2str(pos),...
'% Overshoot'])                          % Add title to uncompensated step
                                         % response.
'Press any key to go to PD compensation'
                                         % Display label.
pause
'Compensated system'                     % Display label.
Ts=input('Type Desired Settling Time ');
                                         % Input desired settling time from
                                         % the keyboard.
wn=4/(Ts*z);                             % Calculate natural frequency.
desired_pole=(-z*wn)+(wn*sqrt(1-z^2)*i);
                                         % Calculate desired dominant pole
                                         % location.
angle_at_desired_pole=(180/pi)*...
angle(polyval(numg,desired_pole)/polyval(deng,desired_pole));
                                         % Calculate angular contribution
                                         % to desired pole without PD
                                         % compensator.
PD_angle=180-angle_at_desired_pole;
                                         % Calculate required angular
                                         % contribution from PD
                                         % compensator.
zc=((imag(desired_pole)/tan(PD_angle*pi/180))...
-real(desired_pole));                    % Calculate PD zero location.
'PD Compensator'                         % Display label.
numc=[1 zc];                             % Calculate numerator of Gc(s).
denc=[0 1];                              % Calculate denominator of Gc(s).
'Gc(s)'                                  % Display label.
Gc=tf(numc,denc)                         % Create and display Gc(s).
'G(s)Gc(s)'                              % Display label.
Ge=G*Gc                                  % Cascade G(s) and Gc(s).
rlocus(Ge,0:0.005:100)                   % Plot root locus of PD compensated
                                         % system.
sgrid(z,0)                               % Overlay desired percent
                                         % overshoot line.
title(['PD Compensated Root Locus with ', num2str(pos),...
'% Overshoot Line'])                     % Add title to PD compensated root
                                         % locus.
[K,p]=rlocfind(Ge);                      % Generate gain, K, and closed-loop
                                         % poles, p, for point selected
                                         % interactively on the root locus.
'Closed-loop poles='                     % Display label.
p                                        % Display PD compensated systems'
                                         % closed-loop poles.
f=input('Give pole number that is operating point ');
                                         % Choose PD compensated system
                                         % dominant pole.
```

```
'Summary of estimated specifications for selected point on PD'
'compensated root locus'            % Display label.
operatingpoint=p(f)                 % Display PD compensated dominant
                                    % pole.
gain=K                              % Display PD compensated gain.
estimated_settling_time=4/abs(real(p(f)))
                                    % Display PD compensated settling
                                    % time.
estimated_peak_time=pi/abs(imag(p(f)))
                                    % Display PD compensated peak time.
estimated_percent_overshoot=pos     % Display PD compensated percent
                                    % overshoot.
estimated_damping_ratio=z           % Display PD compensated damping
                                    % ratio.
estimated_natural_frequency=sqrt(real(p(f))^2+imag(p(f))^2)
                                    % Display PD compensated natural
                                    % frequency.
s=tf([1 0],1);                      % Create transfer function, 's'.
sGe=s*Ge;                           % Create sGe(s).
sGe=minreal(sGe);                   % Cancel common poles and zeros.
Kv=dcgain(K*sGe)                    % Display compensated Kv.
ess=1/Kv                            % Display compensated
                                    % steady-state error for
                                    % unit ramp input.
'T(s)'                              % Display label.
T=feedback(K*Ge,1)                  % Create and display PD compensated
                                    %T(s).
'Press any key to continue and obtain the PD compensated step'
'response'                          % Display label.
pause
step(T)                             % Plot step response for PD
                                    % compensated system.
title(['PD Compensated System Step Response with '...
num2str(pos),'% Overshoot'])        % Add title to step response
                                    % of PD compensated system.
pause
```

**ch9p2 (Example 9.4)** We can use MATLAB to design a lead compensator. The program allows us to input a desired percent overshoot via the keyboard. MATLAB then produces a root locus for the uncompensated system with an overlay of the percent overshoot line. We interactively select the intersection of the root locus and the desired percent overshoot line to set the gain. MATLAB outputs an estimate of the uncompensated system's performance specifications and a step response of the uncompensated system for us to determine the required settling time. Next we input the settling time and the lead compensator zero through the keyboard. At this point we take a different approach from that of the previous example. Rather than letting MATLAB calculate the lead compensator pole directly, MATLAB produces a root locus for every interactive guess of a lead compensator pole. Each root locus contains the desired damping ratio and natural frequency curves. When our guess is correct, the root locus, the damping ratio line, and the natural frequency curve will intersect. We then interactively select this point of intersection to input the gain. Finally, MATLAB produces an estimate of the lead-compensated system's performance specifications and a step response of the lead-compensated system.

```
'(ch9p2) Example 9.4'               % Display label.
clf                                 % Clear graph.
```

```
'Uncompensated System'                   % Display label.
numg=1;                                  % Generate numerator of G(s).
deng=poly([0 -4 -6]);                    % Generate denominator of G(s).
'G(s)'                                   % Display label.
G=tf(numg,deng)                          % Create and display G(s).
pos=input('Type desired percent overshoot ');
                                         % Input desired percent overshoot.
z=-log(pos/100)/sqrt(pi^2+[log(pos/100)]^2);
                                         % Calculate damping ratio.
rlocus(G)                                % Plot uncompensated root locus.
sgrid(z,0)                               % Overlay desired percent
                                         % overshoot line.
title(['Uncompensated Root Locus with ', num2str(pos),...
'% Overshoot Line'])                     % Title uncompensated root locus.
[K,p]=rlocfind(G);                       % Generate gain, K, and closed-loop
                                         % poles, p, for point selected
                                         % interactively on the root locus.
'Closed-loop poles='                     % Display label.
p                                        % Display closed-loop poles.
f=input('Give pole number that is operating point ');
                                         % Choose uncompensated system
                                         % dominant pole.
'Summary of estimated specifications for selected point on'
'uncompensated root locus'               % Display label.
operatingpoint=p(f)                      % Display uncompensated dominant
                                         % pole.
gain=K                                   % Display uncompensated gain.
estimated_settling_time=4/abs(real(p(f)))
                                         % Display uncompensated settling
                                         % time.
estimated_peak_time=pi/abs(imag(p(f)))
                                         % Display uncompensated peak time.
estimated_percent_overshoot=pos          % Display uncompensated percent
                                         % overshoot.
estimated_damping_ratio=z                % Display uncompensated damping
                                         % ratio.
estimated_natural_frequency=sqrt(real(p(f))^2+imag(p(f))^2)
                                         % Display uncompensated natural
                                         % frequency.
numkv=conv([1 0],numg);                  % Set up numerator to evaluate Kv.
denkv=deng;                              % Set up denominator to evaluate Kv.
sG=tf(numkv,denkv);                      % Create sG(s).
sG=minreal(sG);                          % Cancel common poles and zeros.
Kv=dcgain(K*sG)                          % Display uncompensated Kv.
ess=1/Kv                                 % Display uncompensated
                                         % steady-state error for
                                         % unit ramp input.
'T(s)'                                   % Display label.
T=feedback(K*G,1)                        % Create and display T(s).
step(T)                                  % Plot step response of
                                         % uncompensated system.
title(['Uncompensated System Step Response with ',...
num2str(pos),'% Overshoot'])             % Add title to uncompensated step
                                         % response.
'Press any key to go to lead compensation'
                                         % Display label.
pause
```

```
Ts=input('Type Desired Settling Time');
                                      % Input desired settling time.
b=input('Type Lead Compensator Zero, (s+b). b= ');
                                      % Input lead compensator zero.
done=1;                               % Set loop flag.
while done==1                         % Start loop for trying lead
                                      % compensator pole.
a=input ('Enter a Test Lead Compensator Pole, (s+a). a= ');
                                      % Enter test lead compensator pole.
numge=conv(numg,[1 b]);               % Generate numerator of Gc(s)G(s).
denge=conv([1 a],deng);               % Generate denominator
                                      % of Gc(s)G(s).
Ge=tf(numge,denge);                   % Create Ge(s)=Gc(s)G(s).
wn=4/(Ts*z);                          % Evaluate desired natural
                                      % frequency.
clf                                   % Clear graph.
rlocus(Ge)                            % Plot compensated root locus with
                                      % test lead compensator pole.
axis([-10,10,-10,10])                 % Change lead-compensated
                                      % root locus axes.
sgrid(z,wn)                           % Overlay grid on lead-compensated
                                      % root locus.
title (['Lead-Compensated Root Locus with ', num2str(pos),...
'% Overshoot Line, Lead Pole at ',...
num2str(-a),' and Required Wn'])      % Add title to lead-compensated
                                      % root locus.
done=input (' Are you done? (y=0,n=1) ');
                                      % Set loop flag.
end                                   % End loop for trying compensator
                                      % pole.
[K,p]=rlocfind (Ge);                  % Generate gain, K, and closed-loop
                                      % poles, p, for point selected
                                      % interactively on the root locus.
'Gc(s)'                               % Display label.
Gc=tf ([1 b],[1 a])                   % Display lead compensator.
'Gc(s)G(s)'                           % Display label.
Ge                                    % Display Gc(s)G(s).
'Closed-loop poles='                  % Display label.
p                                     % Display lead-compensated
                                      % system's
                                      % closed-loop poles.
f=input('Give pole number that is operating point ');
                                      % Choose lead-compensated system
                                      % dominant pole.
'Summary of estimated specifications for selected point on lead'
'compensated root locus'              % Display label.
operatingpoint=p(f)                   % Display lead-compensated
                                      % dominant pole.
gain=K                                % Display lead-compensated gain.
estimated_settling_time=4/abs(real(p(f)))
                                      % Display lead-compensated
                                      % settling time.
estimated_peak_time=pi/abs(imag(p(f)))
                                      % Display lead-compensated
                                      % peak time.
estimated_percent_overshoot=pos       % Display lead-compensated
                                      % percent overshoot.
```

```
estimated_damping_ratio=z          % Display lead-compensated
                                   % damping ratio.
estimated_natural_frequency=sqrt(real(p(f))^2+imag(p(f))^2)
                                   % Display lead-compensated
                                   % natural frequency.
s=tf([1 0],1);                     % Create transfer Function, 's'.
sGe=s*Ge;                          % Create sGe(s) to evaluate Kv.
sGe=minreal(sGe);                  % Cancel common poles and zeros.
Kv=dcgain(K*sGe)                   % Display lead-compensated Kv.
ess=1/Kv                           % Display lead-compensated steady-
                                   % state error for unit ramp input.
'T(s)'                             % Display label.
T=feedback(K*Ge,1)                 % Create and display lead-
                                   % compensated T(s).
'Press any key to continue and obtain the lead-compensated step'
'response'                         % Display label.
pause
step(T)                            % Plot step response for lead-
                                   % compensated system.
title(['Lead-Compensated System Step Response with ',...
num2str(pos),'% Overshoot'])       % Add title to step response
                                   % of lead-compensated system.
pause
```

## Chapter 10: Frequency Response Techniques

**ch10p1 (Example 10.3)** We can use MATLAB to make Bode plots using `bode(G)`, where `G/(s)=numg/deng` and `G` is an LTI transfer-function object. Information about the plots obtained with `bode(G)` can be found by left-clicking the mouse on the curve. You can find the curve's label, as well as the coordinates of the point on which you clicked. Right-clicking away from a curve brings up a menu if the icons on the menu bar are deselected. From this menu you can select (1) system responses to be displayed and (2) characteristics, such as peak response. When selected, a dot appears on the curve at the appropriate point. Let your mouse rest on the point to read the value of the characteristic. You may also select (3) which curves to view, (4) choice for grid on or off, (5) returning to full view after zooming, and (6) properties, such as labels, limits, units, style, and characteristics. We can obtain points on the plot using `[mag, phase, w]=bode(G)`, where magnitude, phase, and frequency are stored in `mag, phase,` and `w`, respectively. Magnitude and phase are stored as 3-D arrays. We use `mag(:,:)', phase(:,:)'` to convert the arrays to column vectors, where the apostrophe signifies matrix transpose. Let us look at Example 10.3 in the text.

```
'(ch10p1) Example 10.3'            % Display label.
clf                                % Clear graph.
numg=[1 3];                        % Define numerator of G(s).
deng=conv([1 2],[1 2 25]);         % Define denominator of G(s).
'G(s)'                             % Display label.
G=tf(numg,deng)                    % Create and display G(s).
bode(G)                            % Make a Bode plot.
grid on                            % Turn on grid for Bode plot.
title('Open-Loop Frequency Response')
                                   % Add a title to the Bode plot.
[mag,phase,w]=bode(G);             % Store points on the Bode plot.
points=[20*log10(mag(:,:))',phase(:,:)',w]
                                   % List points on Bode plot with
                                   % magnitude in dB.
pause
```

**ch10p2 (Example 10.5)** We can use MATLAB to make Nyquist diagrams using `nyquist(G)`, where $G(s)$=numg/deng and G is an LTI transfer-function object. Information about the plots obtained with `nyquist(G)` can be found by left-clicking the mouse on the curve. You can find the curve's label, as well as the coordinates of the point on which you clicked and the frequency. Right-clicking away from a curve brings up a menu if the icons on the menu bar are deselected. From this menu you can select (1) system responses to be displayed and (2) characteristics, such as peak response. When selected, a dot appears on the curve at the appropriate point. Let your mouse rest on the point to read the value of the characteristic. You may also select (3) whether or not to show negative frequencies, (4) choice for grid on or off, (5) choice for zooming to (–1,0), (6) returning to full view after zooming, and (7) properties, such as labels, limits, units, style, and characteristics. We can obtain points on the plot by using `[re,im,w]=nyquist(G)`, where the real part, imaginary part, and frequency are stored in `re`, `im`, and `w`, respectively, and `re` and `im` are 3-D arrays. We can specify a range of `w` by using `[re,im]=nyquist(G,w)`. We use `re(:,:)'`, and `im(:,:)'` to convert the arrays to column vectors. Let us look at Example 10.5 in the text.

```
'(ch10p2) Example 10.5'                % Display label.
clf                                    % Clear graph.
numg=[1 2];                            % Define numerator of G(s).
deng=[1 0 0];                          % Define denominator of G(s).
'G(s)'                                 % Display label.
G=tf(numg,deng)                        % Create and display G(s).
nyquist(G)                             % Make a Nyquist diagram.
grid on                                % Turn on grid for Nyquist diagram.
title('Open-Loop Frequency Response')
                                       % Add a title to the Nyquist
                                       % diagram.
w=0:0.5:10;                            % Let 0 <w< 10 in steps of 0.5.
[re,im]=nyquist(G,w);                  % Get Nyquist diagram points for a
                                       % range of w.
points=[re(:,:)',im(:,:)',w']          % List specified range of points
                                       % in Nyquist diagram.
pause
```

**ch10p3 (Example 10.8)** We can use MATLAB to find gain margin (`Gm`), phase margin (`Pm`), the gain-margin frequency, where the phase plot goes through 180 degrees (`Wcg`), and the phase-margin frequency, where the magnitude plot goes through zero dB (`Wcp`). To find these quantities we use $[G_m, P_m, W_{cg}, W_{cp}]$=margin(G), where $G(s)$=numg/deng and G is an LTI transfer-function object. Let us look at Example 10.8 in the text.

```
'(ch10p3) Example 10.8'                % Display label.
clf                                    % Clear graph.
numg=6;                                % Define numerator of G(s).
deng=conv([1 2],[1 2 2]);              % Define denominator of G(s).
'G(s)'                                 % Display label.
G=tf(numg,deng)                        % Create and display G(s).
nyquist(G)                             % Make a Nyquist diagram.
grid on                                % Turn on grid for the Nyquist
                                       % diagram.
title ('Open-Loop Frequency Response')
                                       % Add a title to the Nyquist
                                       % diagram.
[Gm,Pm,Wcg,Wcp]=margin(G);             % Find margins and margin
                                       % frequencies.
```

```
'Gm(dB); Pm(deg.); 180 deg. freq.(r/s); 0 dB freq. (r/s)'
                                     % Display label.
margins=[20*log10(Gm),Pm,Wcg,Wcp]
                                     % Display margin data.
pause
```

**ch10p4 (Example 10.9)** We can use MATLAB to determine the range of *K* for stability using frequency response methods. Let us look at Example 10.9 in the text.

```
'(ch10p4) Example 10.9'              % Display label.
numg=1;                              % Define numerator of G(s).
deng=poly([-2 -4 -5]);               % Define denominator of G(s).
'G(s)'                               % Display label.
G=tf(numg,deng)                      % Create and display G(s).
[Gm,Pm,Wcg,Wcp]=margin(G);           % Find margins and margin
                                     % frequencies.
K=Gm                                 % Display K for stability.
pause
```

**ch10p5 (Example 10.11)** We can use MATLAB to find the closed-loop frequency response. Let us look at Example 10.11 in the text.

```
'(ch10p5) Example 10.11'             % Display label.
clf                                  % Clear graph.
numg=50;                             % Define numerator of G(s).
deng=poly([0 -3 -6]);                % Define denominator of G(s).
'G(s)'                               % Display label.
G=tf(numg,deng)                      % Create and display G(s).
'T(s)'                               % Display label.
T=feedback(G,1)                      % Find and display closed-loop
                                     % transfer function.
bode(T)                              % Make a Bode plot.
grid on                              % Turn on the grid for the plots.
title('Closed-Loop Frequency Response')
                                     % Add a title to the Bode plot.
pause
nyquist(T)                           % Make a Nyquist diagram.
title('Closed-Loop Frequency Response')
                                     % Add a title to the Nyquist
                                     % diagram.
pause
```

**ch10p6** We can use MATLAB to plot Nichols charts using `nichols(G)`, where *G(s)*=`numg/deng` and `G` is an LTI transfer-function object. The Nichols grid can be added using the `ngrid` command after the `nichols(G)` command. Information about the plots obtained with `nichols(G)` can be found by left-clicking the mouse on the curve. You can find the curve's label, as well as the coordinates of the point on which you clicked and the frequency. Right-clicking away from a curve brings up a menu if the icons on the menu bar are deselected. From this menu you can select (1) system responses to be displayed and (2) characteristics, such as peak response. When selected, a dot appears on the curve at the appropriate point. Let your mouse rest on the point to read the value of the characteristic. You may also select (3) choice for grid on or off, (4) returning to full view after zooming, and (5) properties, such as labels, limits, units, style, and characteristics. Let us make a Nichols chart of $G(s) = 1/[s(s+1)(s+2)]$.

```
'(ch10p6)'                          % Display label.
clf                                 % Clear graph.
numg=1;                             % Define numerator of G(s).
deng=poly([0 -1 -2]);               % Define denominator of G(s).
'G(s)'                              % Display label.
G=tf(numg,deng)                     % Create and display G(s).
nichols(G)                          % Make a Nichols plot.
ngrid                               % Add Nichols grid.
pause
```

**ch10p7 (Example 10.15)** We can use MATLAB and frequency response methods to include time delay in the loop. Time delay is represented by `[numd,dend]=pade(T,n)`, where `T` is the delay time in seconds and `n` is the order. Larger values of `n` give better approximations to the delay, $G_d(s)$=`numd/dend`. Since we are plotting multiple plots, we first collect the data for the Bode plots by using `[mag,phase]=bode(G,w)`, where `w` is specified as a range of frequencies. We then use the generic plotting command. Also notice the commands used to label the axes and the plots on the Bode plot (see the MATLAB instruction manual for details). Let us look at Example 10.15 in the text.

```
'(ch10p7) Example 10.15'            % Display label.
clf                                 % Clear graph.
hold off                            % Turn graph hold off.
numg=1;                             % Define numerator of G(s).
deng=poly([0 -1 -10]);              % Define denominator of G(s).
'G(s)'                              % Display label.
G=tf(numg,deng)                     % Create and display G(s).
w=0.01:0.1:10;                      % Let 0.01<w<10 in steps of 0.1.
[magg,phaseg]=bode(G,w);            % Collect Bode data for G(s).
[numd,dend]=pade(1,6);              % Represent the delay.
Gd=tf(numd,dend);                   % Create and display the delay,
                                    % Gd(s).
[magd,phased]=bode(Gd,w);           % Collect Bode data for Gd(s).
Ge=Gd*G;                            % Form Gd(s)G(s).
[mage,phasee]=bode(Ge,w);           % Collect Bode data for Gd(s)G(s).
subplot(2,1,1)                      % Subdivide plot area for plot 1.
semilogx(w,20*log10(mage(:,:)))     % Plot magnitude response.
grid on                             % Turn on grid for magnitude plot.
axis([0.01,10,-80,20]);             % Limit Bode plot axes.
title('Magnitude Response with Delay')
                                    % Add title to magnitude response.
xlabel('Frequency (rad/s)')         % Label x-axis of magnitude
                                    % response.
ylabel('20log M')                   % Label y-axis of magnitude
                                    % response.
subplot(2,1,2)                      % Subdivide plot area for plot 2.
phased=phased-1080;                 % Adjust phase offset to compensate
                                    % for modulo 360.
phasee=phasee-1080;                 % Adjust phase offset to compensate
                                    % for modulo 360.
semilogx(w,phaseg(:,:),w,phased(:,:),w,phasee(:,:))
                                    % Plot phase response for G(s),
                                    % Gd(s), and G(s)Gd(s) on one
                                    % graph.
```

```
grid on                                   % Turn on grid for phase plot.
axis([0.01,10,-900,0]);                   % Limit Bode plot axes.
title ('Phase Response with Delay')
                                          % Add title to phase response.
xlabel('Frequency (rad/s)')               % Label x-axis of phase response.
ylabel('Phase (degrees)')                 % Label y-axis of phase response.
text(1.5,-50,'Time Delay')                % Label time delay curve.
text(4,-150,'System')                     % Label system curve.
text(2.7,-300,'Total')                    % Label total curve.
pause
```

**ch10p8 (Example 10.18)** We can use MATLAB and frequency response methods to determine experimentally a transfer function from frequency response data. By determining simple component transfer functions and then successively subtracting their frequency response, we can approximate the complete transfer function. Let us look at Example 10.18 in the text and use MATLAB for a portion of the problem. You can complete the program for practice. For this problem we generate the original frequency response plot via a transfer function. Normally, the data for the original frequency response plot would be tabular, and the program would begin at the step `[M0,P0]=bode(G0,w)` where the tabular data is generated. In other words, in a real application, the data would consist of column vectors `M0, P0, and w'`.

```
'(ch10p8) Example 10.18'                  % Display label.
clf                                       % Clear graph.
hold off                                  % Turn graph hold off.
% Generate the experimental Bode plots for G0(s)=numg0/deng0, that
% is, M0,P0.
numg0=70*[1 20];                          % Define numerator of G0(s).
deng0=conv([1 7],[1 2 25]);               % Partially define denominator of
                                          % G0(s).
deng0=conv(deng0,[1 70]);                 % Complete the denominator of
                                          % G0(s).
G0=tf(numg0,deng0);                       % Create G0(s).
w=1:0.5:1000;                             % Let 1<w<1000 in steps of 0.5.
[M0,P0]=bode(G0,w);                       % Generate the tabular data.
[20*log10(M0(:,:))',P0(:,:)',w'];
                                          % Convert magnitude data to dB.
bode(G0,w)                                % Generate a Bode plot.
grid on                                   % Turn on grid for Bode plot.
title('Experimental')                     % Add title.
pause
clf                                       % Clear graph.
% Estimate a component part of the transfer function as
% G1 (s)=25/(s^2+2*0.22*5s+5^2) and subtract it from the experimental
% frequency response
numg1=5^2;                                % Define numerator of G1(s).
deng1=[1 2*0.22*5 5^2];                   % Define denominator of G1(s).
'First estimate'                          % Display label.
G1=tf(numg1,deng1)                        % Create and display G1(s).
[M1,P1]=bode(G1,w);                       % Generate Bode data for G1(s).
M2=20*log10(M0(:,:))-20*log10(M1(:,:));
                                          % Subtract Bode magnitude data of
                                          % G1 from original magnitude data.
P2=P0(:,:)-P1(:,:);                       % Subtract Bode phase data of G1
                                          % from original phase data.
subplot(2,1,1)                            % Divide plot area in two for
                                          % magnitude plot.
semilogx(w(:,:),M2)                       % Plot magnitude response after
                                          % subtracting.
```

```
grid on                              % Turn on grid for magnitude plot.
xlabel('Frequency (rad/sec)')        % Add x-axis label.
ylabel('Gain dB')                    % Add y-axis label.
subplot(2,1,2)                       % Divide plot area in two for phase
                                     % plot.
semilogx(w,P2)                       % Plot the phase response after
                                     % subtracting.
grid on                              % Turn on grid for phase plot.
title('Experimental Minus 25/(s^2+2*0.22*5s+5^2)')
                                     % Add title.
xlabel('Frequency (rad/sec)')        % Add x-axis label.
ylabel('Phase deg')                  % Add y-axis label.
'This completes a portion of Example 10.18.'
'The student should continue the program for practice.'
pause
```

## Chapter 11: Design via Frequency Response

**ch11p1 (Example 11.1)** We can design via gain adjustment on the Bode plot using MATLAB. You will input the desired percent overshoot from the keyboard. MATLAB will calculate the required phase margin and then search the Bode plot for that phase margin. The magnitude at the phase-margin frequency is the reciprocal of the required gain. MATLAB will then plot a step response for that gain. Let us look at Example 11.1 in the text.

```
'(ch11p1) Example 11.1'              % Display label.
clf                                  % Clear graph.
numg=[100];                          % Define numerator of G(s).
deng=poly([0 -36 -100]);             % Define denominator of G(s).
G=tf(numg,deng)                      % Create and display G(s).
pos=input('Type %OS ');              % Input desired percent overshoot.
z=(-log(pos/100))/(sqrt(pi^2+log(pos/100)^2));
                                     % Calculate required damping ratio.
Pm=atan(2*z/(sqrt(-2*z^2+sqrt(1+4*z^4))))*(180/pi);
                                     % Calculate required phase margin.
w=0.01:0.01:1000;                    % Set range of frequency from 0.01
                                     % to 1000 in steps of 0.01.
[M,P]=bode(G,w);                     % Get Bode data.
Ph=-180+Pm;                          % Calculate required phase angle.
for k=1: 1: length(P);               % Search Bode data for required
                                     % phase angle.
if P(k)-Ph <=0;                      % If required phase angle is found,
                                     % find the value of
M=M(k);                              % magnitude at the same frequency.
'Required K'                         % Display label.
K=1/M                                % Calculate the required gain.
break                                % Stop the loop.
end                                  % End if.
end                                  % End for.
T=feedback(K*G,1);                   % Find T(s) using the calculated K.
step(T)                              % Generate a step response.
title(['Closed-Loop Step Response for K= ',num2str(K)])
                                     % Add title to step response.
pause
```

**ch11p2 (Example 11.2)** Let us use MATLAB to design a lag compensator. The program solves Example 11.2 in the text and follows the same design technique demonstrated in that example. You will input the value of gain to meet the steady-state error requirement followed by the desired percent overshoot. MATLAB then designs a lag compensator, evaluates $K_v$, and generates a closed-loop step response.

```
'(ch11p2) Example 11.2'                   % Display label.
clf                                       % Clear graph.
K=input('Type value of K to meet steady-state error requirement ');
                                          % Input K.
pos=input ('Type %OS ');                  % Input desired percent overshoot.
numg=[100*K];                             % Define numerator of G(s).
deng=poly([0 -36 -100]);                  % Define denominator of G(s).
'G(s)'                                    % Display label.
G=tf(numg,deng)                           % Create and display G(s).
z=(-log(pos/100))/(sqrt(pi^2+log(pos/100)^2));
                                          % Calculate required damping
                                          % ratio.
Pm=atan(2*z/(sqrt(-2*z^2+sqrt(1+4*z^4))))*(180/pi)+10;
                                          % Calculate required phase margin.
w=0.01:0.01:100;                          % Set range of frequency from 0.01
                                          % to 1000 in steps of 0.01.
[M,P]=bode(G,w);                          % Get Bode data.
Ph=-180+Pm;                               % Calculate required phase angle.
for k=1:1:length(P);                      % Search Bode data for required
                                          % phase angle.
if P(k)-Ph <=0;                           % If required phase angle is found,
                                          % find the value of
M=M(k);                                   % magnitude at the same frequency.
wf=w(k);                                  % At this frequency the magnitude
                                          % plot must go through 0 dB.
break                                     % Stop the loop.
end                                       % End if.
end                                       % End for.
wh=wf/10;                                 % Calculate the high-frequency
                                          % break of the lag compensator.
wl=(wh/M);                                % Calculate the low-frequency
                                          % break of the lag compensator;
                                          % found from lag compensator,
                                          % Gc(s)=Kc(s+wh)/(s+wl), high & low
                                          % frequency gain requirements.
                                          % At low w, gain=1. Thus,
                                          % Kc*wh/wl=1. At high w, gain=1/M.
                                          % Thus Kc=1/M. Hence
                                          % Kc=wl/wh=1/M, or wl=wh/M.
numc=[1 wh];                              % Generate numerator of lag
                                          % compensator, Gc(s).
denc=[1 wl];                              % Generate denominator of lag
                                          % compensator, Gc(s).
Kc=wl/wh;                                 % Generate K for Gc(s).
'Lag compensator'                         % Display label.
Kc                                        % Display lag compensator K.
'Gc(s)'                                   % Display label.
Gc=tf(Kc*numc,denc)                       % Create and display Gc(s).
' Gc(s)G(s)'                              % Display label.
GcG=Gc*G                                  % Create and display Gc(s)G(s).
s=tf([1 0],1);                            % Create transfer function,'s'.
sGcG=s*GcG;                               % Create sGc(s)G(s).
sGcG=minreal(sGcG);                       % Cancel common terms.
```

```
Kv=dcgain(sGcG)                    % Evaluate Kv.
T=feedback(GcG,1);                 % Create T(s).
step(T)                            % Generate a closed-loop, lag-
                                   % compensated step response.
title ('Closed-Loop Step Response for Lag-Compensated System')
                                   % Add title to step response.
pause
```

**ch11p3 (Example 11.3)** Let us use MATLAB to design a lead compensator. The program solves Example 11.3 in the text and follows the same design technique demonstrated in that example. You will enter desired percent overshoot, peak time, and $K_v$. MATLAB then designs the lead compensator using Bode plots, calculates $K_v$, and plots a closed-loop step response.

```
'(ch11p3) Example 11.3'            % Display label.
clf                                % Clear graph.
pos=input ('Type %OS ');           % Input desired percent overshoot.
Tp=input (' Type peak time ');     % Input desired peak time.
Kv=input (' Type value of Kv ');   % Input Kv.
numg=[100];                        % Define numerator of G(s).
deng=poly ([0 -36 -100]);          % Define denominator of G(s).
G=tf (numg,deng);                  % Create G(s).
s=tf ([1 0],1);                    % Create transfer function,'s'.
sG=s*G;                            % Create sG(s).
sG=minreal (sG);                   % Cancel common factors.
K=dcgain (Kv/sG);                  % Solve for K.
'G(s)'                             % Display label.
G=zpk (K*G)                        % Put K into G (s), convert to
                                   % factored form, and display.
z=(-log (pos/100))/(sqrt (pi^2+log (pos/100)^2));
                                   % Calculate required damping
                                   % ratio.
Pm=atan (2*z/(sqrt (-2*z^2+sqrt (1+4*z^4))))*(180/pi);
                                   % Calculate required phase margin.
wn=pi/(Tp*sqrt (1-z^2));           % Calculate required natural
                                   % frequency.
wBW=wn*sqrt ((1-2*z^2)+sqrt (4*z^4-4*z^2+2));
                                   % Determine required bandwidth.
w=0.01:0.5:1000;                   % Set range of frequency from 0.01
                                   % to 1000 in steps of 0.5
[M,P]=bode (G,w);                  % Get Bode data.
[Gm,Pm,Wcg,Wcp]=margin (G);        % Find current phase margin.
Pmreq=atan (2*z/(sqrt (-2*z^2+sqrt (1+4*z^4))))*(180/pi);
                                   % Calculate required phase margin.
Pmreqc=Pmreq+10;                   % Add a correction factor of 10
                                   % degrees.
Pc=Pmreqc-Pm;                      % Calculate phase contribution
                                   % required from lead compensator.
% Design lead compensator
beta=(1-sin (Pc*pi/180))/(1+sin (Pc*pi/180));
                                   % Find compensator beta.
magpc=1/sqrt (beta);               % Find compensator peak magnitude.
for k=1:1:length (M);              % Find frequency at which
                                   % uncompensated system has a
                                   % magnitude of 1/magpc.
                                   % This frequency will be the new
                                   % phase margin frequency.
if M(k)-(1/magpc) <=0;             % Look for peak magnitude.
```

```
wmax=w(k);                              % This is the frequency at the
                                        % peak magnitude.
break                                   % Stop the loop,
end                                     % End if.
end                                     % End for.
% Calculate lead compensator zero, pole, and gain.
zc=wmax*sqrt(beta);                     % Calculate the lead compensators'
                                        % low break frequency.
pc=zc/beta;                             % Calculate the lead compensators'
                                        % high break frequency.
Kc=1/beta;                              % Calculate the lead compensators'
                                        % gain.
'Gc(s)'                                 % Display label.
Gc=tf(Kc*[1 zc],[1 pc]);                % Create Gc(s).
Gc=zpk(Gc)                              % Convert Gc(s) to factored form
                                        % and display.
'Ge(s)=G(s)Gc(s)'                       % Display label.
Ge=G*Gc                                 % Form Ge(s)=Gc(s)G(s).
sGe=s*Ge;                               % Create sGe(s).
sGe=minreal(sGe);                       % Cancel common factors.
Kv=dcgain(sGe)                          % Calculate Kv.
T=feedback(Ge,1);                       % Find T(s).
step(T)                                 % Generate closed-loop, lead-
                                        % compensated step response.
title('Lead-Compensated Step Response')
                                        % Add title to lead-compensated
                                        % step response.
pause
```

**ch11p4 (Example 11.4)** Let us use MATLAB to design a lag-lead compensator. The program solves Example 11.4 in the text and follows the same design technique demonstrated in that example. You will enter desired percent overshoot, peak time, and $K_v$. MATLAB then designs the lag-lead compensator using Bode plots, calculates $K_v$, and plots a closed-loop step response.

```
'(ch11p4) Example 11.4'                 % Display label.
clf                                     % Clear graph.
pos=input('Type %OS ');                 % Input desired percent overshoot.
Tp=input('Type peak time ');            % Input desired peak time.
Kv=input('Type value of Kv ');          % Input desired Kv.
numg=[1];                               % Define numerator of G(s).
deng=poly([0 -1 -4]);                   % Define denominator of G(s).
G=tf(numg,deng);                        % Create G(s) without K.
s=tf([1 0],1);                          % Create transfer function,'s'.
sG=s*G;                                 % Create sG(s).
sG=minreal(sG);                         % Cancel common factors.
K=dcgain(Kv/sG);                        % Solve for K.
'G(s)'                                  % Display label.
G=tf(K*numg,deng);                      % Put K into G(s).
G=zpk(G)                                % Convert G(s) to factored form and
                                        % display.
z=(-log(pos/100))/(sqrt(pi^2+log(pos/100)^2));
                                        % Calculate required damping ratio.
Pmreq=atan(2*z/(sqrt(-2*z^2+sqrt(1+4*z^4))))*(180/pi);
                                        % Calculate required phase margin.
wn=pi/(Tp*sqrt(1-z^2));                 % Calculate required natural
                                        % frequency.
```

```
wBW=wn*sqrt((1-2*z^2)+sqrt(4*z^4-4*z^2+2));
                                    % Determine required bandwidth.
wpm=0.8*wBW;                        % Choose new phase-margin
                                    % frequency.
[M,P]=bode(G,wpm);                  % Get Bode data.
Pmreqc=Pmreq-(180+P)+5;             % Find phase contribution required
                                    % from lead compensator
                                    % with additional 5 degrees.
beta=(1-sin(Pmreqc*pi/180))/(1+sin(Pmreqc*pi/180));
                                    % Find beta.
                                    % Design lag compensator zero, pole,
                                    % and gain.
zclag=wpm/10;                       % Calculate zero of lag compensator.
pclag=zclag*beta;                   % Calculate pole of lag compensator.
Kclag=beta;                         % Calculate gain of lag compensator.
'Lag compensator, Glag(s)'          % Display label.
Glag=tf(Kclag*[1 zclag],[1 pclag]); % Create lag compensator.
Glag=zpk(Glag)                      % Convert Glag(s) to factored form
                                    % and display.
                                    % Design lead compensator zero,
                                    % pole, and gain.
zclead=wpm*sqrt(beta);              % Calculate zero of lead
                                    % compensator.
pclead=zclead/beta;                 % Calculate pole of lead
                                    % compensator.
Kclead=1/beta;                      % Calculate gain of lead
                                    % compensator.
'Lead compensator'                  % Display label.
Glead=tf(Kclead*[1 zclead],[1 pclead]);
                                    % Create lead compensator.
Glead=zpk(Glead)                    % Convert Glead(s) to factored form
                                    % and display.
'Lag-Lead Compensated Ge(s)'        % Display label.
Ge=G*Glag*Glead                     % Create compensated system,
                                    % Ge(s)=G(s) Glag(s) Glead(s).
sGe=s*Ge;                           % Create sGe(s).
sGe=minreal(sGe);                   % Cancel common factors.
Kv=dcgain(sGe)                      % Calculate Kv.
T=feedback(Ge,1);                   % Find T(s).
step(T)                             % Generate closed-loop, lag-lead-
                                    % compensated step response.
title('Lag-Lead-Compensated Step Response')
                                    % Add title to lag-lead-
                                    % compensated
                                    % step response.
pause
```

## Chapter 12: Design via State Space

**ch12p1 (Example 12.1)** We can use MATLAB to design controller gains using pole placement. You will enter the desired percent overshoot and settling time. We introduce the following commands: `[num,den]=ord2(wn,z)`, which produces a second-order system, given the natural frequency `(wn)` and the damping ratio `(z)`. Then we use the denominator `(den)` to specify the dominant poles; and `K=acker (A,B,-poles)`, which calculates controller gains from the system matrix `(A)`, the input matrix `(B)`, the desired poles `(poles)`. Let us look at Example 12.1 in the text.

```
'(ch12p1) Example 12.1'                 % Display label.
clf                                     % Clear graph.
numg=20*[1 5];                          % Define numerator of G(s).
deng=poly([0 -1 -4]);                   % Define denominator of G(s).
'Uncompensated G(s)'                    % Display label.
G=tf(numg,deng)                         % Create and display G(s).
pos=input('Type desired %OS ');         % Input desired percent overshoot.
Ts=input('Type desired settling time ');
                                        % Input desired settling time.
z=(-log(pos/100))/(sqrt(pi^2+log(pos/100)^2));
                                        % Calculate required damping ratio.
wn=4/(z*Ts);                            % Calculate required natural
                                        % frequency.
[num,den]=ord2(wn,z);                   % Produce a second-order system
                                        % that meets the transient response
                                        % requirements.
r=roots(den);                           % Use denominator to specify
                                        % dominant poles.
poles=[r(1) r(2) -5.1];                 % Specify pole placement for all
                                        % poles.
characteristiceqdesired=poly(poles)
                                        % Form desired characteristic
                                        % polynomial for display.
[Ac Bc Cc Dc]=tf2ss(numg,deng);         % Find controller canonical form
                                        % of state-space representation
                                        % of G(s).
P=[0 0 1;0 1 0;1 0 0];                  % Transformation matrix for
                                        % controller canonical to phase-
                                        % variable form.
Ap=inv(P)*Ac*P;                         % Transform Ac to phase-variable
                                        % form.
Bp=inv(P)*Bc;                           % Transform Bc to phase-variable
                                        % form.
Cp=Cc*P;                                % Transform Cc to phase-variable
                                        % form.
Dp=Dc;                                  % Transform Dc to phase-variable
                                        % form.
Kp=acker(Ap,Bp,poles)                   % Calculate controller gains in
                                        % phase-variable form.
Apnew=Ap-Bp*Kp;                         % Form compensated A matrix.
Bpnew=Bp;                               % Form compensated B matrix.
Cpnew=Cp;                               % Form compensated C matrix.
Dpnew=Dp;                               % Form compensated D matrix.
[numt,dent]=ss2tf(Apnew,Bpnew,Cpnew,Dpnew);
                                        % Form T(s) numerator and
                                        % denominator.
'T(s)'                                  % Display label.
T=tf(numt,dent)                         % Create and display T(s).
poles=roots(dent)                       % Display poles of T(s).
Tss=ss(Apnew,Bpnew,Cpnew,Dpnew)         % Create and display Tss, an LTI
                                        % state-space object.
step(Tss)                               % Produce compensated step
                                        % response.
title('Compensated Step Response')
                                        % Add title to compensated step
                                        % response.
pause
```

**ch12p2 (Example 12.2)** We can test controllability by using the MATLAB command `Cm=ctrb(A,B)` to find the controllability matrix given the system matrix `(A)` and the input matrix `(B)`. This command is followed by `rank(Cm)` to test the rank of the controllability matrix `(Cm)`. Let us apply the commands to Example 12.2.

```
'(ch12p2) Example 12.2'          % Display label.
A=[-1 1 0;0 -1 0;0 0 -2]         % Define compensated A matrix.
B=[0;1;1]                        % Define compensated B matrix.
Cm=ctrb(A,B)                     % Calculate controllability
                                 % matrix.
Rank=rank(Cm)                    % Find rank of controllability
                                 % matrix.
pause
```

**ch12p3 (Example 12.4)** If we design controller gains using MATLAB, we do not have to convert to phase-variable form. MATLAB will give us the controller gains for any state-space representation we input. Let us look at Example 12.4 in the text.

```
'(ch12p3) Example 12.4'          % Display label.
clf                              % Clear graph.
A=[-5 1 0;0 -2 1;0 0 -1];        % Define system matrix A.
B=[0;0;1];                       % Define input matrix B.
C=[-1 1 0];                      % Define output matrix C.
D=0;                             % Define matrix D.
pos=input('Type desired %OS ');  % Input desired percent overshoot.
Ts=input('Type desired settling time ')
                                 % Input desired settling time.
z=(-log(pos/100))/(sqrt(pi^2+log(pos/100)^2));
                                 % Calculate required damping ratio.
wn=4/(z*Ts);                     % Calculate required natural
                                 % frequency.
[num,den]=ord2(wn,z);            % Produce a second-order system
                                 % that meets the transient
                                 % requirements.
r=roots(den);                    % Use denominator to specify
                                 % dominant poles.
poles=[r(1) r(2) -4];            % Specify pole placement for all
                                 % poles.
K=acker(A,B,poles)               % Calculate controller gains.
Anew=A-B*K;                      % Form compensated A matrix.
Bnew=B;                          % Form compensated B matrix.
Cnew=C;                          % Form compensated C matrix.
Dnew=D;                          % Form compensated D matrix.
Tss=ss(Anew,Bnew,Cnew,Dnew);     % Form LTI state-space object.
'T(s)'                           % Display label.
T=tf(Tss);                       % Create T(s).
T=minreal(T)                     % Cancel common terms and display
                                 % T(s).
poles=pole(T)                    % Display poles of T(s).
step(Tss)                        % Produce compensated step
                                 % response.
title('Compensated Step Response')
                                 % Add title to compensated step
                                 % response.
pause
```

**ch12p4 (Example 12.5)** We can design observer gains by using the command `l=acker(A',C',poles)'`. Notice we use the transpose of the system matrix `(A)` and output matrix `(C)` along with the desired poles `(poles)`. Let us look at Example 12.5 in the text.

```
'(ch12p4) Example 12.5'                % Display label.
numg=[1 4];                            % Define numerator of G(s).
deng=poly([-1 -2 -5]);                 % Define denominator of G(s).
'G(s)'                                 % Display label.
G=tf(numg,deng)                        % Create and display G(s).
[Ac,Bc,Cc,Dc]=tf2ss(numg,deng);        % Transform G(s) to controller
                                       % canonical form in state space.
Ao=Ac';                                % Transform Ac to observer
                                       % canonical form.
Bo=Cc';                                % Transform Bc to observer
                                       % canonical form.
Co=Bc';                                % Transform Cc to observer
                                       % canonical form.
Do=Dc;                                 % Transform Dc to observer
                                       % canonical form.
r=roots([1 2 5])                       % Find the controller-compensated
                                       % system poles.
poles=10*[r' 10*real(r(1))]            % Make observer poles 10x bigger.
lp=acker(Ao',Co',poles)'               % Find the observer gains in
                                       % observer canonical form.
pause
```

**ch12p5 (Example 12.6)** We can test observability using the MATLAB command `Om=obsv(A,C)` to find the observability matrix given the system matrix `(A)` and the output matrix `(C)`. This command is followed by `rank(Om)` to test the rank of the observability `matrix (Om)`. Let us apply the commands to Example 12.6.

```
'(ch12p5) Example 12.6'                % Display label.
A=[0 1 0;0 0 1;-4 -3 -2]               % Define compensated A matrix.
C=[0 5 1]                              % Define compensated C matrix.
Om=obsv(A,C)                           % Form observability matrix.
Rank=rank(Om)                          % Find rank of observability
                                       % matrix.
pause
```

**ch12p6 (Example 12.8)** We can design observer gains using the command `l=acker (A',C',poles)'` without transforming to observer canonical form. Let us look at Example 12.8 in the text.

```
'(ch12p6) Example 12.8'                % Display label.
A=[-5 1 0;0 -2 1;0 0 -1];              % Define system matrix A.
B=[0;0;1];                             % Define input matrix B.
C=[1 0 0];                             % Define output matrix C.
D=0;                                   % Define matrix D.
poles=roots([1 120 2500 50000])        % Specify pole placement for all
                                       % poles.
l=acker(A',C',poles)'                  % Calculate observer gains.
pause
```

## Chapter 13: Digital Control Systems

**ch13p1 (Example 13.4)** We can convert $G_1(s)$ in cascade with a zero-order hold (z.o.h.) to $G(z)$ using MATLAB's `G=c2d (G1,T,' zoh')` command, where `G1` is an LTI continuous-system object and `G` is an LTI sampled-system object. `T` is the sampling interval and `'zoh'` is a method of transformation that assumes $G_1(s)$ in cascade with a z.o.h. We simply put $G_1(s)$ into the command (the z.o.h. is automatically taken care of) and the command returns $G(z)$. Let us apply the concept to Example 13.4. You will enter `T` through the keyboard.

```
'(ch13p1) Example 13.4'          % Display label.
T=input('Type T ');              % Input sampling interval.
numg1s=[1 2];                    % Define numerator of G1(s).
deng1s=[1 1];                    % Define denominator of G1(s).
'G1(s)'                          % Display label.
G1=tf(numg1s,deng1s)             % Create G1(s) and display.
'G(z)'                           % Display label.
G=c2d(G1,T,'zoh')                % Convert G1(s) in cascade with
                                 % z.o.h. to G(z) and display.
pause
```

**ch13p2** We also can use MATLAB to convert $G(s)$ to $G(z)$ when $G(s)$ is not in cascade with a z.o.h. The command `H=c2d(F,T,'zoh')` transforms $F(s)$ in cascade with a z.o.h. to $H(z)$, where $H(z) = ((z-1)/z)*z\{F(s)/s\}$. If we let $F(s) = sG(s)$, the command solves for $H(z)$, where $H(z) = ((z-1)/z)*z\{G(s)\}$. Hence, $z\{G(s)\} = (z/[z-1])*H(z)$. In summary, input $F(s) = sG(s)$, and multiply the result of `H=c2d(F,T,'zoh')` by $(z/[z-1])$ . This process is equivalent to finding the $z$-transform. We convert $G(s) = (s+3)/(s^2+6s+13)$ into $G(z)$. You will enter `T`, the sampling interval, through the keyboard. `T` is used to form $H(z)$. We use an unspecified sampling interval, `T=[ ]`, to form $z/(z-1)$.

```
'(ch13p2)'                       % Display label.
T=input('Type T ');              % Input sampling interval.
numgs=[1 3];                     % Define numerator of G(s).
dengs=[1 6 13];                  % Define denominator of G(s).
'G(s)'                           % Display label.
Gs=tf(numgs,dengs)               % Create and display G(s).
Fs=Gs*tf([1 0],1);               % Create F(s)=sG(s).
Fs=minreal(Fs);                  % Cancel common poles and zeros.
Hz=c2d(Fs,T,'zoh');              % Convert F(s) to H(z) assuming
                                 % z.o.h.
Gz=Hz*tf([1 0],[1 -1],[]);       % Form G(z)=H(z)*z/(z-1).
'G(z)'                           % Display label.
Gz=minreal(Gz)                   % Cancel common poles and zeros.
pause
```

### ch13p3 Creating Digital Transfer Functions Directly

#### Vector Method, Polynomial Form

A digital transfer function can be expressed as a numerator polynomial divided by a denominator polynomial, that is, $F(z) = N(z)/D(z)$. The numerator, $N(z)$, is represented by a vector, `numf`, that contains the coefficients of $N(z)$. Similarly, the denominator, $D(z)$, is represented by a vector, `denf`, that contains the coefficients of $D(z)$. We form $F(z)$ with the command, `F=tf(numf,denf,T)`, where `T` is the sampling interval. `F` is called a linear time-invariant (LTI) object. This object, or transfer function, can be used as an entity in other operations, such as addition or multiplication. We demonstrate with

$F(z) = 150(z^2 + 2z + 7)/(z^2 - 0.3z + 0.02)$. We use an unspecified sampling interval, `T=[]`. Notice after executing the `tf` command, MATLAB prints the transfer function.

## Vector Method, Factored Form

We also can create digital LTI transfer functions if the numerator and denominator are expressed in factored form. We do this by using vectors containing the roots of the numerator and denominator. Thus, $G(s) = K*N(z)/D(z)$ can be expressed as an LTI object using the command, `G=zpk(numg,deng,K,T)`, where `numg` is a vector containing the roots of $N(z)$, `deng` is a vector containing the roots of $D(z)$, `K` is the gain, and `T` is the sampling interval. The expression `zpk` stands for zeros (roots of the numerator), poles (roots of the denominator), and gain, `K`. We demonstrate with $G(z) = 20(z+2)(z+4)/[(z-0.5)(z-0.7)(z-0.8)]$ and an unspecified sampling interval. Notice after executing the `zpk` command, MATLAB prints the transfer function.

## Rational Expression in *z* Method, Polynomial Form (Requires Control System Toolbox 9.7)

This method allows you to type the transfer function as you normally would write it. The statement `z=tf('z')` must precede the transfer function if you wish to create a digital LTI transfer function in polynomial form equivalent to using `G=tf(numg, deng,T)`.

## Rational Expression in *z* Method, Factored Form (Requires Control System Toolbox 9.7)

This method allows you to type the transfer function as you normally would write it. The statement `z=zpk('z')` must precede the transfer function if you wish to create a digital LTI transfer function in factored form equivalent to using `G=zpk(numg,-deng,K,T)`.

For both rational expression methods the transfer function can be typed in any form regardless of whether `z=tf('z')` or `z=zpk('z')` is used. The difference is in the created digital LTI transfer function. We use the same examples above to demonstrate the rational expression in $z$ methods.

```
'(ch13p3)'                           % Display label.
'Vector Method, Polynomial Form'     % Display label.
numf=150*[1 2 7]                     % Store 150(z^2+2z+7) in numf and
                                     % display.
denf=[1 -0.3 0.02]                   % Store (z^2-0.3z+0.02) in denf and
                                     % display.
'F(z)'                               % Display label.
F=tf(numf,denf,[ ])                  % Form F(z) and display.
clear                                % Clear previous variables from
                                     % workspace.
'Vector Method, Factored Form'       % Display label.
numg=[-2 -4]                         % Store (s+2)(s+4) in numg and
                                     % display.
deng=[0.5 0.7 0.8]                   % Store (s-0.5)(s-0.7)(s-0.8) in
                                     % deng and display.
K=20                                 % Define K.
'G(z)'                               % Display label.
G=zpk(numg,deng,K,[ ])               % Form G(z) and display,
clear                                % Clear previous variables from
                                     % workspace.
'Rational Expression Method, Polynomial Form'
                                     % Display label.
z=tf('z')                            % Define z as an LTI object in
                                     % polynomial form.
```

```
F=150*(z^2+2*z+7)/(z^2-0.3*z+0.02)
                                        % Form F(z) as an LTI transfer
                                        % function in polynomial form.
G=20*(z+2)*(z+4)/[(z-0.5)*(z-0.7)*(z-0.8)]
                                        % Form G(z) as an LTI transfer
                                        % function in polynomial form.
clear                                   % Clear previous variables from
                                        % workspace.
'Rational Expression Method, Factored Form'
                                        % Display label.
z=zpk('z')                              % Define z as an LTI object in
                                        % factored form.
F=150*(z^2+2*z+7)/(z^2-0.3*z+0.02)
                                        % Form F(z) as an LTI transfer
                                        % function in factored form.
G=20*(z+2)*(z+4)/[(z-0.5)*(z-0.7)*(z-0.8)]
                                        % Form G(z) as an LTI transfer
                                        % function in factored form.
pause
```

**ch13p4** We also can use MATLAB to convert $G(z)$ to $G(s)$ when $G(s)$ is not in cascade with a z.o.h. First, we create a sampled LTI transfer function, as discussed in ch13p3. The command `F=d2c(H,'zoh')` transforms $H(z)$ to $F(s)$ in cascade with a z. o.h., where $H(z) = ((z-1)/z)z\{F(s)/s\}$. If we consider $F(s) = sG(s)$, the command solves for $sG(s)$ given $H(z)$. Finally, $sG(s)/s = G(s)$ yields the final result. In summary, form $H(z)$, where $H(z) = ((z-1)/z)G(z)$. Use `F=d2c(H,'zoh')` to find $F(s) = sG(s)$. Divide the result by $s$ and obtain $G(s)$. We convert $G(z) = z/(z-0.3)$ into $G(s)$. You will enter `T`, the sampling interval, through the keyboard.

```
'(ch13p4)'                              % Display label.
T=input('Type T ');                     % Input sampling interval.
numgz=[1 0];                            % Define numerator of G(z).
dengz=[1 -.3];                          % Define denominator of G(z).
'G(z)'                                  % Display label.
Gz=tf(numgz,dengz,T)                    % Create and display G(z).
Hz=Gz*tf([1 -1],[1 0],T);               % Create H(z)=((z-1)/z)*G(z).
Hz=minreal(Hz);                         % Cancel common poles and zeros.
Fs=d2c(Hz,'zoh');                       % Convert from H(z) to F(s)=sG(s).
Gs=Fs*tf(1,[1 0]);                      % Create G(s)=F(s)(1/s).
'G(s)'                                  % Display label.
Gs=minreal(Gs)                          % Cancel common poles and zeros.
pause
```

**ch13p5 (Example 13.6)** We can use MATLAB to find the gain for stability. Let us look at Example 13.6 in the text.

```
'(ch13p5) Example 13.6'                 % Display label.
numgas=27;                              % Define numerator of Ga(s).
dengas=[1 27 0];                        % Define denominator of Ga(s).
'Ga(s)'                                 % Display label.
Ga=tf(numgas,dengas)                    % Create and display Ga(s).
'G(z)'                                  % Display label.
Gz=c2d(Ga,0.1,'zoh')                    % Find G(z) assuming Ga(s) in
                                        % cascade with z.o.h. and display.
```

```
for K=1:0.1:50;                         % Set range of K to look for
                                        % stability.
Tz=feedback(K*Gz,1);                    % Find T(z).
r=pole(Tz);                             % Get poles for this value of K.
rm=max(abs(r));                         % Find pole with maximum absolute
                                        % value for this value of K.
if rm>=1,                               % See if pole is outside unit
                                        % circle.
break;                                  % Stop if pole is found outside
                                        % unit circle.
end;                                    % End if.
end;                                    % End for.
K                                       % Display K value.
r                                       % Display closed-loop poles for
                                        % this value of K.
rm                                      % Display absolute value of pole.
pause
```

**ch13p6 (Example 13.9)** We can use MATLAB's command `dcgain(Gz)` to find steady-state errors. The command evaluates the dc gain of Gz, a digital LTI transfer function object, by evaluating Gz at $z = 1$. We use the dc gain to evaluate, $K_p$, $K_v$, and $K_a$. Let us look at Example 13.9 in the text. You will input `T`, the sampling interval, through the keyboard to test stability.

```
'(ch13p6) Example 13.9'                 % Display label.
T=input('Type T ');                     % Input sampling interval.
numg1s=[10];                            % Define numerator of G1(s).
deng1s=poly([0 -1]);                    % Define denominator of G1(s).
'G1(s)'                                 % Display label.
G1s=tf(numg1s,deng1s)                   % Create and display G1(s).
'G(z)'                                  % Display label.
Gz=c2d(G1s,T,'zoh')                     % Convert G1(s) and z.o.h. to G(z)
                                        % and display.
'T(z)'                                  % Display label.
Tz=feedback(Gz,1)                       % Create and display T(z).
'Closed-Loop z-Plane Poles'             % Display label.
r=pole(Tz)                              % Check stability.
M=abs(r)                                % Display magnitude of roots.
pause
Kp=dcgain(Gz)                           % Calculate Kp.
GzKv=Gz*(1/T)*tf([1 -1],[1 0],T);       % Multiply G(z) by(1/T)*(z-1).
                                        % Also, divide G(z) by z , which
                                        % makes transfer function proper
                                        % and yields same Kv.
GzKv=minreal(GzKv,0.00001);             % Cancel common poles and zeros.
Kv=dcgain(GzKv)                         % Calculate Kv.
GzKa=Gz*(1/T^2)*tf([1 -2 1],[1 0 0],T); % Multiply G(z) by (1/T^2)(z-1)^2.
                                        % Also, divide G(z)by z^2 , which
                                        % makes the transfer function
                                        % proper and yields the same Ka.
GzKa=minreal(GzKa,0.00001);             % Cancel common poles and zeros.
Ka=dcgain(GzKa)                         % Calculate Ka.
pause
```

**ch13p7 (Example 13.10)** We now use the root locus to find the gain for stability. First, we create a digital LTI transfer-function object for $G(z) = N(z)/D(z)$, with an unspecified sampling interval. The LTI object is created using `tf(numgz,dengz,[])`, where `numgz` represents

*N*(*z*), `dengz` represents *D*(*z*), and `[]` indicates an unspecified sampling interval. MATLAB produces a *z*-plane root locus along with the unit circle superimposed using the command, `zgrid ([],[])`. We then interactively select the intersection of the root locus and the unit circle. MATLAB responds with the value of gain and the closed-loop poles. Let us look at Example 13.10.

```
'(ch13p7) Example 13.10'           % Display label.
clf                                % Clear graph.
numgz=[1 1];                       % Define numerator of G(z).
dengz=poly([1 0.5]);               % Define denominator of G(z).
'G(z)'                             % Display label.
Gz=tf(numgz,dengz,[])              % Create and display G(z).
rlocus(Gz)                         % Plot root locus.
zgrid([],[])                       % Add unit circle to root locus.
title (['z-Plane Root Locus'])     % Add title to root locus.
[K,p]=rlocfind(Gz)                 % Allows input of K by selecting
                                   % point on graphic.
pause
```

**ch13p8 (Example 13.11)** We now use the root locus to find the gain to meet a transient response requirement. After MATLAB produces a *z*-plane root locus, along with damping ratio curves superimposed using the command `zgrid`, we interactively select the desired operating point at a damping ratio of 0.7, thus determining the gain. MATLAB responds with a gain value as well as the step response of the closed-loop sampled system using `step(Tz)`, where `Tz` is a digital LTI transfer-function object. Let us look at Example 13.11.

```
'(ch13p8) Example 13.11'           % Display label.
clf                                % Clear graph.
numgz=[1 1];                       % Define numerator of G(z).
dengz=poly([1 0.5]);               % Define denominator of G(z).
'G(z)'                             % Display label.
Gz=tf(numgz,dengz,[])              % Create and display G(z).
rlocus(Gz)                         % Plot root locus.
axis([0,1,-1,1])                   % Create close-up view.
zgrid                              % Add damping ratio curves to root
                                   % locus.
title(['z-Plane Root Locus'])      % Add title to root locus.
[K,p]=rlocfind(Gz)                 % Allows input of K by selecting
                                   % point on graphic.
'T(z)'                             % Display label.
Tz=feedback(K*Gz,1)                % Find T(z).
step(Tz)                           % Find step response of gain-
                                   % compensated system.
title (['Gain Compensated Step Response'])
                                   % Add title to step response of
                                   % gain-compensated system.
pause
```

**ch13p9 (Example 13.12)** Let us now use MATLAB to design a digital lead compensator. The *s*-plane design was performed in Example 9.6. Here we convert the design to the *z*-plane and run a digital simulation of the step response. Conversion of the *s*-plane lead compensator, *Gc*(*s*)=`numgcs/dengcs`, to the *z*-plane compensator, *Gc*(*z*)=`numgcz/dengcz`, is accomplished using the `Gcz=c2d(numgcs,dengcs,T,'tustin')` command to perform a Tustin transformation, where `T`=sampling interval, which for this example is 1/300. This exercise solves Example 13.12 using MATLAB.

```
'(ch13p9) Example 13.12'                  % Display label.
clf                                       % Clear graph.
T=0.01                                    % Define sampling interval.
numgcs=1977*[1 6];                        % Define numerator of Gc(s).
dengcs=[1 29.1];                          % Define denominator of Gc(s).
'Gc(s) in polynomial form'                % Print label.
Gcs=tf(numgcs,dengcs)                     % Create Gc(s) in polynomial form
                                          % and display.
'Gc(s) in polynomial form'                % Display label.
Gcszpk=zpk(Gcs)                           % Create Gc(s) in factored form
                                          % and display.
'Gc(z) in polynomial form via Tustin Transformation'
                                          % Display label.
Gcz=c2d(Gcs,T,'tustin')                   % Form Gc(z) via Tustin
                                          % transformation.
'Gc(z) in factored form via Tustin Transformation'
                                          % Display label.
Gczzpk=zpk(Gcz)                           % Show Gc(z) in factored form.
numgps=1;                                 % Define numerator of Gp(s).
dengps=poly([0 -6 -10]);                  % Define denominator of Gp(s).
'Gp(s) in polynomial form'                % Display label.
Gps=tf(numgps,dengps)                     % Create Gp(s) in polynomial form
                                          % and display.
'Gp(s) in factored form'                  % Display label.
Gpszpk=zpk(Gps)                           % Create Gp(s) in factored form
                                          % and display.
'Gp(z) in polynomial form'                % Display label.
Gpz=c2d(Gps,T,'zoh')                      % Form Gp(z) via zoh transformation.
'Gp(z) in factored form'                  % Display label.
Gpzzpk=zpk(Gpz)                           % Form Gp(z) in factored form.
Gez=Gcz*Gpz;                              % Form Ge(z) = Gc(z)Gp(z).
'Ge(z)=Gc(z)Gp(z) in factored form'       % Display label.
Gezzpk=zpk(Gez)                           % Form Ge(z) in factored form
                                          % and display.
'z-1'                                     % Display label.
zm1=tf([1 -1],1,T)                        % Form z-1.
zm1Gez=minreal(zm1*Gez,0.00001);          % Cancel common factors.
'(z-1)Ge(z) for finding steady-state error'
                                          % Display label.
zm1Gezzpk=zpk(zm1Gez)                     %Form & display(z-1)Ge(z) in
                                          % factored form.
Kv=(1/T)*dcgain(zm1Gez)                   % Find Kv.
'T(z)=Ge(z)/(1+Ge(z))'                    % Display label.
Tz=feedback(Gez,1)                        % Find closed-loop
                                          % transfer function, T(z)
step(Tz,0:T:2)                            % Find step reponse.
title('Closed-Loop Digital Lead Compensated Step Response')
                                          % Add title to step response.
```

## B.3 Command Summary

| | |
|---|---|
| `abs(x)` | Obtain absolute value of *x*. |
| `acker(A,B,poles)` | Find gains for pole placement. |
| `angle(x)` | Compute the angle of *x* in radians. |
| `atan(x)` | Compute arctan(*x*). |
| `axis([xmin,xmax,ymin,ymax])` | Define range on axes of a plot. |

| | |
|---|---|
| `bode(G,w)` | Make a Bode plot of transfer function $G(s)$ over a range of frequencies, $\omega$.<br>Field $\omega$ is optional. |
| `break` | Exit loop. |
| `c2d(G,T,'tustin')` | Convert $G(s)$ to $G(z)$ using the Tustin transformation.<br>$T$ is the sampling interval. |
| `c2d(G,T,'zoh')` | Convert $G(s)$ in cascade with a zero-order hold to $G(z)$. $T$ is the sampling interval. |
| `canon(S,'modal')` | Convert an LTI state-space object, $S$, to parallel form. |
| `clear` | Clear variables from workspace. |
| `clf` | Clear current figure. |
| `conv([a b c d],[e f g h])` | Multiply $(as^3 + bs^2 + cs + d)$ by $(es^3 + fs^2 + gs + h)$. |
| `ctrb(A,B)` | Find controllability matrix. |
| `d2c(G,'zoh')` | Convert $G(z)$ to $G(s)$ in cascade with a zero-order hold. |
| `dcgain(G)` | Find dc gain for $G(s)$ (that is, $s = 0$), or $G(z)$ (that is, $z = 1$). |
| `eig(A)` | Find eigenvalues of matrix **A**. |
| `end` | End the loop. |
| `exp(a)` | Obtain $e^a$. |
| `feedback(G,H,sign)` | Find $T(s) = G(s)/[1 \pm G(s)H(s)]$. Sign $= -1$ or is optional for negative feedback systems.<br>Sign $= +1$ for positive feedback systems. |
| `grid on` | Put grid lines on a graph. |
| `hold off` | Turn off graph hold; start new graph. |
| `imag(P)` | Form a matrix of the imaginary parts of the components of matrix **P**. |
| `input('str')` | Permit variable values to be entered from the keyboard with prompt `str`. |
| `interp1(x,y,x1)` | Perform table lookup by finding the value of $y$ at the value of $x = x_1$. |
| `inv(P)` | Find the inverse of matrix **P**. |
| `length(P)` | Obtain dimension of vector **P**. |
| `log(x)` | Compute natural log of $x$. |
| `log10(x)` | Compute log to the base 10 of $x$. |
| `margin(G)` | Find gain and phase margins, and gain and phase margin frequencies of transfer function, $G(s)$.<br>Return [Gain margin, Phase margin, 180° frequency, 0 dB frequency]. |
| `max(P)` | Find the maximum component of **P**. |
| `minreal(G,tol)` | Cancel common factors from transfer function $G(s)$ within tolerance, tol.<br>If `'tol'` field is blank, a default value is used. |
| `ngrid` | Superimpose grid over a Nichols plot. |
| `nichols(G,w)` | Make a Nichols plot of transfer function $G(s)$ over a range of frequencies, $\omega$.<br>Field $\omega$ is optional. |

| | |
|---|---|
| `nyquist(G,w)` | Make a Nyquist diagram of transfer function $G(s)$ over a range of frequencies, $\omega$. Field $\omega$ is optional. |
| `obsv(A,C)` | Find observability matrix. |
| `ord2(wn,z)` | Create a second-order system, $G(s) = 1/[s^2 + 2\zeta\omega_n s + \omega_n^2]$. |
| `pade(T,n)` | Obtain *n*th order Padé approximation for delay, $T$. |
| `pause` | Pause program until any key is pressed. |
| `plot(t1,y1,t2,y2,t3,y3)` | Plot $y1$ versus $t1$, $y2$ versus $t2$, and $y3$ versus $t3$ on the same graph. |
| `pole(G)` | Find poles of LTI transfer function object, $G(s)$. |
| `poly([-a -b -c])` | Form polynomial $(s + a)(s + b)(s + c)$. |
| `polyval(P,a)` | Find polynomial $P(s)$ evaluated at $a$, that is, $P(a)$. |
| `rank(A)` | Find rank of matrix $\mathbf{A}$. |
| `real(P)` | Form a matrix of the real parts of the components of matrix $\mathbf{P}$. |
| `residue(numf,denf)` | Find residues of $F(s)$ = numf/denf. |
| `rlocfind(GH)` | Allow interactive selection of points on a root locus plot for loop gain, $G(s)H(s)$. Return value for $K$ and all closed-loop poles at that $K$. |
| `rlocus(GH,K)` | Plot root locus for loop gain, $G(s)H(s)$, over a range of gain, $K$. The $K$ field is optional. |
| `roots(P)` | Find roots of polynomial, $P$. |
| `semilogx(w,P1)` | Make a semilog plot of $P_1$ versus $\log_{10}(\omega)$. |
| `series(G1,G2)` | Find $G_1(s)G_2(s)$. |
| `sgrid(z,wn)` | Overlay $z(\zeta)$ and wn($\omega_n$) grid lines on a root locus. |
| `sin(x)` | Find $\sin(x)$. |
| `sqrt(a)` | Compute $\sqrt{a}$. |
| `ss2tf(A,B,C,D,1)` | Convert a state-space representation to a transfer function. Return `[num,den]`. |
| `ss(A,B,C,D)` | Create an LTI state-space object, $S$. |
| `ss(G)` | Convert an LTI transfer function object, $G(s)$, to an LTI state-space object. |
| `ssdata(S)` | Extract $\mathbf{A}$, $\mathbf{B}$, $\mathbf{C}$, and $\mathbf{D}$ matrices from LTI state-space object, $S$. |
| `step(G1,G2,.. Gn,t)` | Plot step responses of $G_1(s)$ through $G_n(s)$ on one graph over a range of time, $t$. Field $t$ is optional as are fields $G_2$ through $G_n$. |
| `subplot(xyz)` | Divide plotting area into an $x$ by $y$ grid with $z$ as the window number for the current plot. |
| `tan(x)` | Find tangent of $x$ radians. |
| `text(a,b,'str')` | Put `str` on graph at graph coordinates, $x = a$, $y = b$. |
| `tf2ss(numg,deng)` | Convert $G(s)$ = `numg/deng` to state space in controller canonical form. Return `[A,B,C,D]`. |
| `vpa(a,D)` | Calculate $a$ with $D$ digits and convert to a symbolic with $D$ digits. |

| | |
|---|---|
| `tf2zp(numg,deng)` | Convert $G(s)$ = `numg/deng` in polynomial form to factored form.<br>Return `[zeros, poles, gains]`. |
| `tf(numg,deng,T)` | Create an LTI transfer function, $G(s)$ = `numg/deng`, in polynomial form.<br>$T$ is the sampling interval and should be used only if $G$ is a sampled transfer function. |
| `tf(G)` | Convert an LTI transfer function, $G(s)$, to polynomial form. |
| `tfdata(G,'v')` | Extract numerator and denominator of an LTI transfer function, $G(s)$, and convert values to a vector. Return `[num, den]`. |
| `title('str')` | Put title `str` on graph. |
| `xlabel('str')` | Put label `str` on $x$ axis of graph. |
| `ylabel('str')` | Put label `str` on $y$ axis of graph. |
| `zgrid` | Superimpose `z` ($\zeta$) and `wn` ($\omega_n$) grid curves on a $z$-plane root locus. |
| `zgrid([ ],[ ])` | Superimpose the unit circle on a $z$-plane root locus. |
| `zp2tf ([-a -b]',[-c -d]', K)` | Convert $F(s) = K(s+a)(s+b)/(s+c)(s+d)$ to polynomial form.<br>Return `[num,den]`. |
| `zpk(numg,deng,K,T)` | Create an LTI transfer function, $G(s)$ = `numg/deng`, in factored form.<br>$T$ is the sampling interval and should be used only if $G$ is a sampled transfer function. |
| `zpk(G)` | Convert an LTI transfer function, $G(s)$, to factored form. |

# Bibliography


Johnson, H. et al. *Unmanned Free-Swimming Submersible (UFFS) System Description*. NRL Memorandum Report 4393. Naval Research Laboratory, Washington, D.C., 1980.

MathWorks. *Control System Toolbox™ Getting Started Guide R2014a*. MathWorks, Natick, MA, 2000–2014.

MathWorks. *Control System Toolbox™ User's Guide R2014a*. MathWorks, Natick, MA, 2001–2014.

MathWorks. MATLAB® *Primer R2014a*. MathWorks, Natick, MA, 1984–2014.

MathWorks. *MATLAB® Graphics R2014a*. MathWorks, Natick, MA, 1984–2014.

MathWorks. *MATLAB® Mathematics R2014a*. MathWorks, Natick, MA, 1984–2014.

MathWorks. *MATLAB® Programming Fundamentals R2014a*. MathWorks, Natick, MA, 1984–2014.

MathWorks. *Simulink® Getting Started Guide R2014a*. MathWorks, Natick, MA, 1990–2014.

MathWorks. *Simulink® User's Guide R2014a*. MathWorks, Natick, MA, 1990–2014.

MathWorks *MATLAB® & Simulink® Installation Guide R2014a*. MathWorks, Natick, MA, 1996–2014.

# Appendix C: Simulink Tutorial

## C.1 Introduction

Readers who are studying MATLAB may want to explore the functionality and convenience of Simulink. Before proceeding, the reader should have studied Appendix B, the MATLAB Tutorial, including Section B.1, which is applicable to this appendix.

Simulink Version 8.3 and MATLAB Version 8.3(R2014a) are required in order to use Simulink. In addition, if you wish to pursue the design of PID controllers discussed in Section C.4, you will need the Simulink Control Design Version 4(R2014a) add-on.

The models described in this appendix are available at www.wiley.com/college/nise. The code will also run on workstations that support MATLAB. Consult the MATLAB Installation Guide for your platform for minimum system hardware requirements.

Simulink is used to simulate systems. It uses a graphical user interface (GUI) for you to interact with blocks that represent subsystems. You can position the blocks, resize the blocks, label the blocks, specify block parameters, and interconnect blocks to form complete systems from which simulations can be run.

Simulink has block libraries from which subsystems, sources (that is, function generators), and sinks (that is, scopes) can be copied. Subsystem blocks are available for representing linear, nonlinear, and discrete systems. LTI objects can be generated if the Control System Toolbox is installed.

Help is available on the **Quick Access Toolbar** at the top of the **MATLAB R2014a** window. Click the circled question mark and select **Simulink**. Help is also available for each block in the block library and is accessed either by right-clicking a block's icon in the **Simulink Library Browser** and selecting **Help for** or by double-clicking the block's icon and then clicking the **Help** button. Finally, screen tips are available for some toolbar buttons. Let your mouse's pointer rest on the button for a few seconds to see the explanation.

## C.2 Using Simulink

The following summarize the steps to take to use Simulink. Section C.3 will present four examples that demonstrate and clarify these steps.

1. **Access Simulink** The **Simulink Library Browser**, from where we begin Simulink, is accessed by typing *simulink* in the **MATLAB Command Window** or by clicking on the **Simulink Library** button on the toolstrip, shown circled in Figure C.1.

   In response, MATLAB displays the **Simulink Library Browser** shown in Figure C.2(*a*). We now create an **untitled** window, Figure C.2(*b*), by clicking on the **New model** button (shown circled in Figure C.2(*a*)) on the toolbar of the **Simulink**

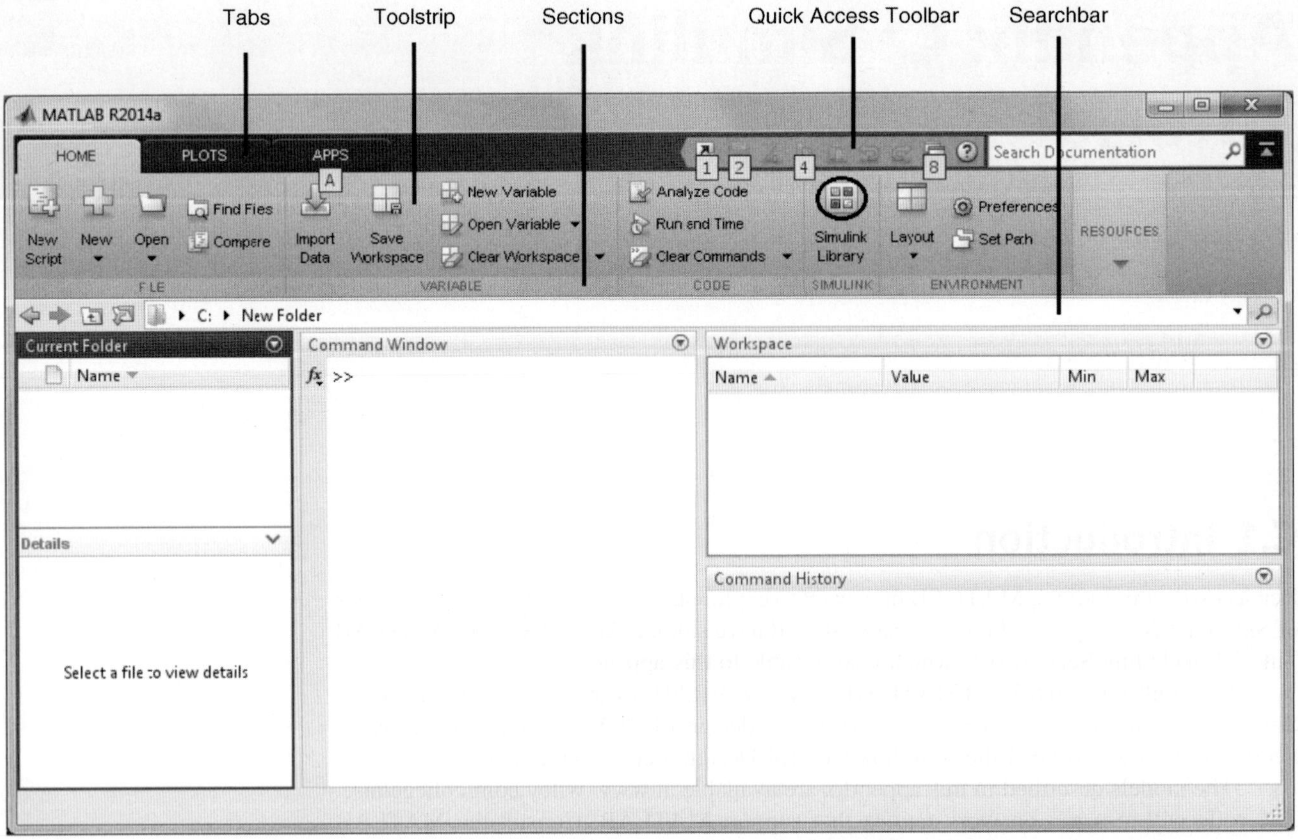


**FIGURE C.1** **MATLAB** Window showing how to access Simulink. The **Simulink Library Browser** button is shown circled

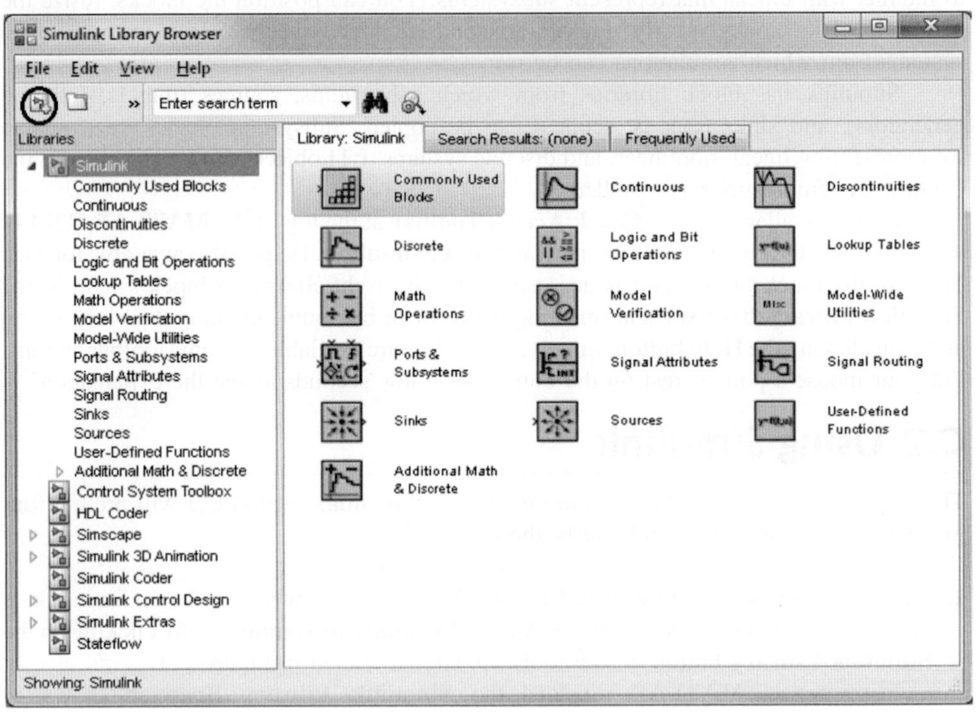


(*a*)

**FIGURE C.2** a. **Simulink Library Browser** window showing the **New model** button encircled; (*figure continues*)

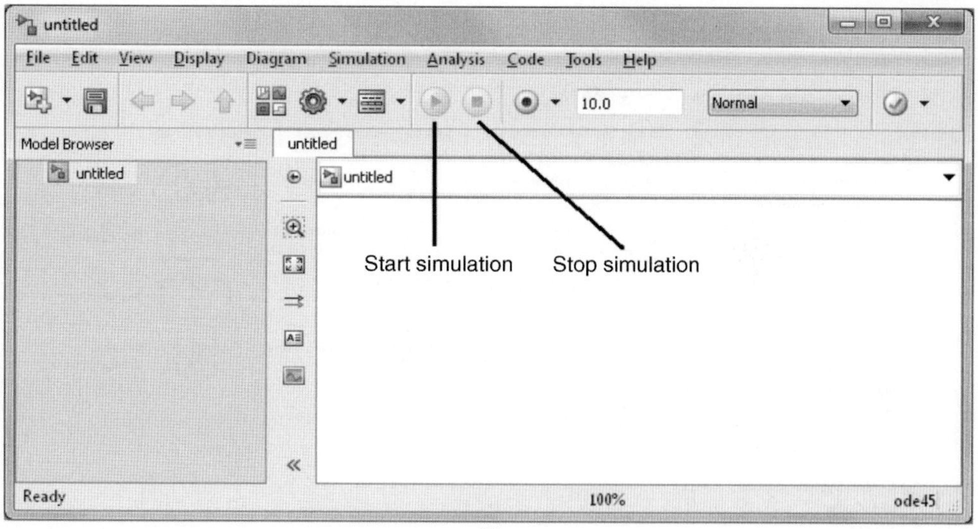


(*b*)

**FIGURE C.2** (*Continued*) **b.** resulting **untitled** model window

**Library Browser**. You will build your system in this window. Existing models may be opened by clicking on the **Open model** button on the **Simulink Library Browser** toolbar. This button is immediately to the right of the **New model** button. Existing models may also be opened from the **File** section of the **Home** tab on the **Toolstrip** or from the **Searchbar** below the **Toolstrip** on the **MATLAB** window.

2. **Select blocks** Figure C.2(*a*) shows the **Simulink Library Browser** from which all blocks can be accessed. The left-hand side of the browser shows major libraries, such as **Simulink**, as well as underlying block libraries, such as **Continuous**. The right-hand side of Figure C.2(*a*) also shows the underlying block libraries. To reveal a block library's underlying blocks, select the block library on the left-hand side or double-click the block library on the right-hand side. As an example, the **Continuous** library blocks under the **Simulink** major library are shown exposed in Figure C.3(*a*). Figures C.3(*b*) and C.3(*c*) show some of the **Sources** and **Sinks** library blocks, respectively.

   Another approach to revealing the Simulink block library is to type *open_system* (*'simulink'*) in the **MATLAB Command Window**. The window shown in Figure C.4 is the result. Double-clicking any of the libraries in Figure C.4 reveals an individual window containing that library's blocks, equivalent to the right-hand side of the **Simulink Library Browser** as shown in the examples of Figure C.3.

3. **Assemble and label subsystems** Drag required subsystems (blocks) to your model window from the browser, such as those shown in Figure C.3. Also, you may access the blocks by double-clicking the libraries shown in Figure C.4. You can position, resize, and rename the blocks. To position, drag with the mouse; to resize, click on the subsystem and drag the handles; to rename, click on the existing name, select the existing text, and type the new name.

4. **Interconnect subsystems and label signals** Position the pointer on the small arrow on the side of a subsystem, press the mouse button, and drag the resulting cross-hair pointer to the small arrow of the next subsystem. A line will be drawn between the two subsystems. Blocks may also be interconnected by single-clicking the first block followed by single-clicking the second block while holding down the control key. You can move line segments by positioning the pointer on the line, pressing the mouse button, and dragging. Branches to line segments can be drawn by positioning the pointer where you want to create a line segment, holding down the mouse's right button, and

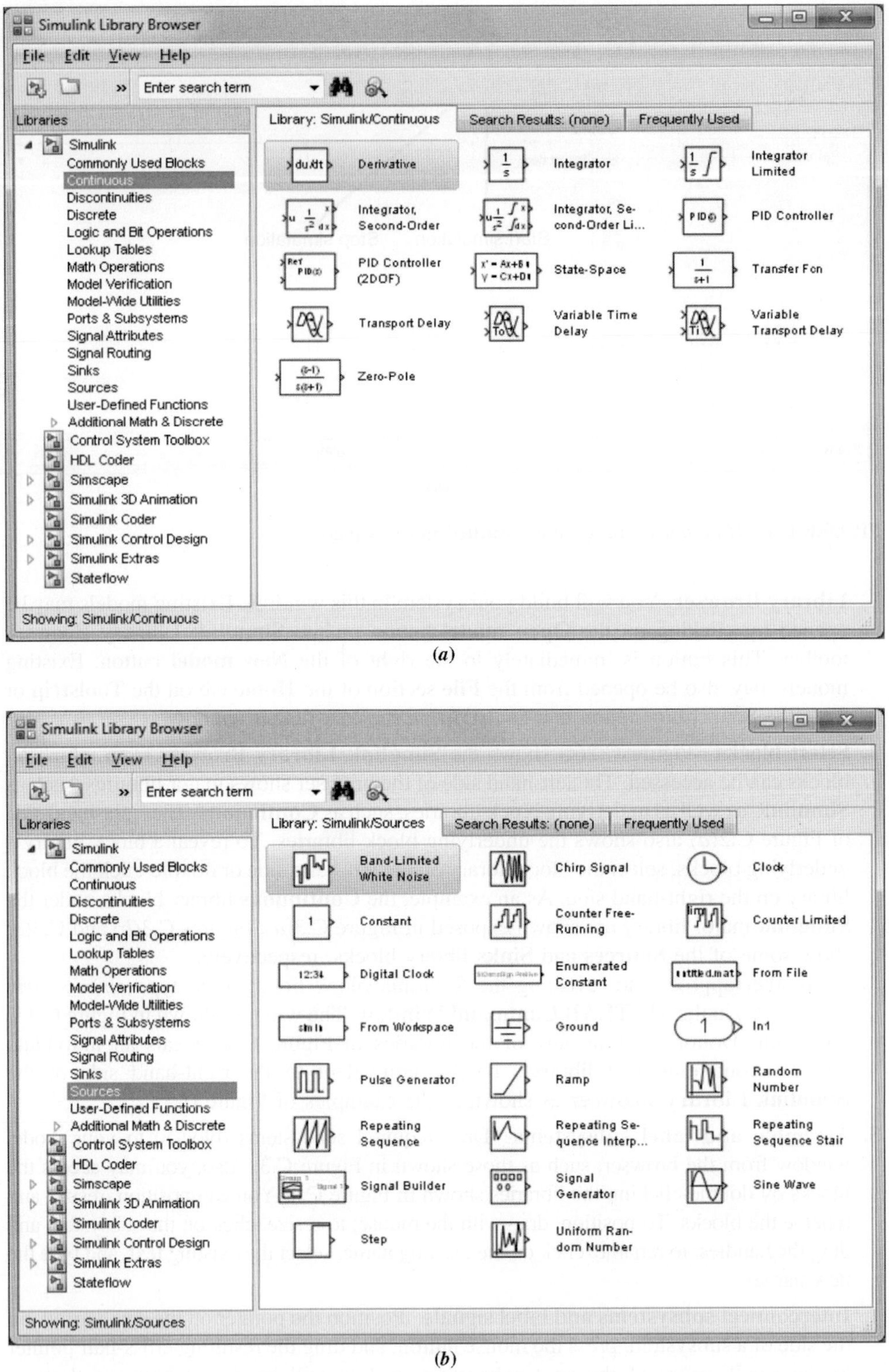


**FIGURE C.3** Simulink block libraries: **a.** Continuous systems; **b.** Sources (*figure continues*)

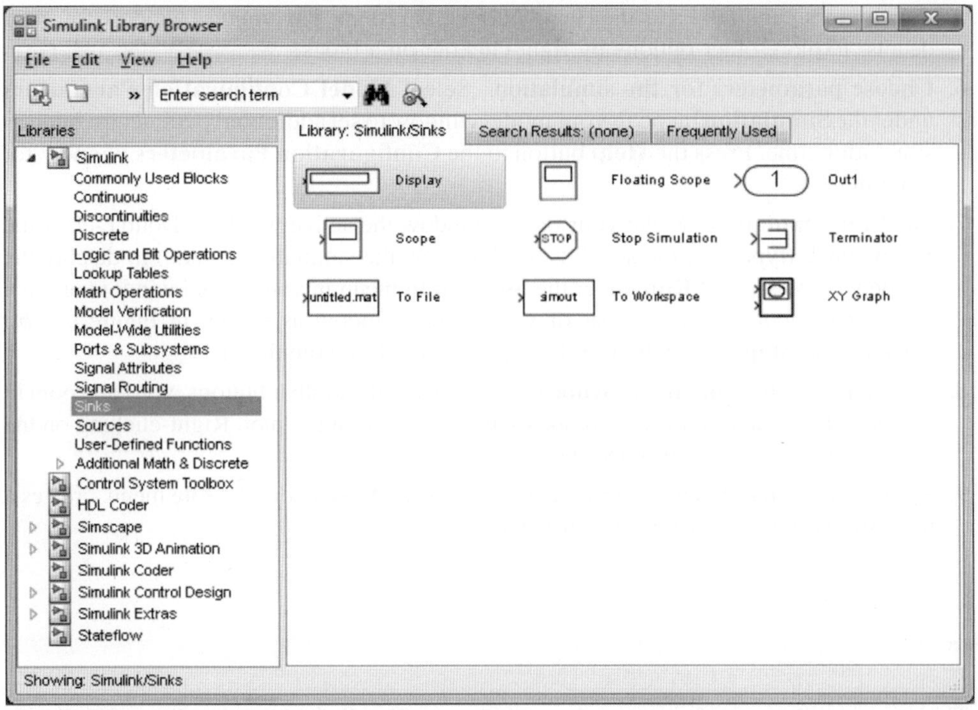


(*c*)

**FIGURE C.3** (*Continued*) **c.** Sinks

dragging the resulting cross hairs. A new line segment will form. Signals can be labeled by double-clicking the line and typing. Finally, labels can be placed anywhere by double-clicking and typing into the resulting box.

5. **Choose parameters for the subsystems** Double-click a subsystem in your model window and type in the desired parameters. Some explanations are provided in the **Block**

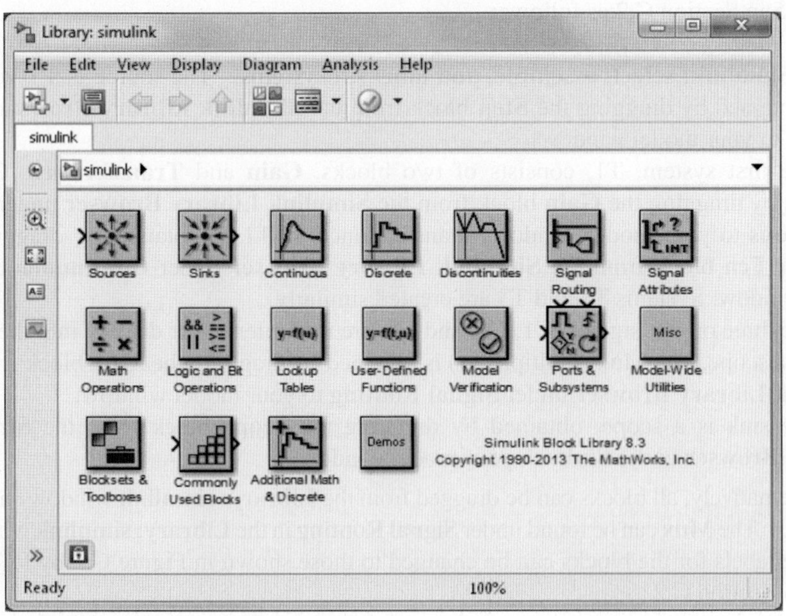


**FIGURE C.4** Simulink Block Library window

**Parameters** window. Press the Help button in the **Block Parameters** window for more details. Explore other options by right-clicking on a block.

6. **Choose parameters for the simulation** Select **Model Configuration Parameters** under the **Simulation** menu in your model window to set additional parameters, such as simulation time. Press the **Help** button in the **Configuration Parameters** window for more details.
7. **Start the simulation** Make your model window the active window. Double-click the **Scope** block (typically, the scope is used to view the simulation results) to display the **Scope** window. Select **Run** under the **Simulation** menu in your model window or click on the **Run** icon on the toolbar of your model window as shown in Figure C.2(*b*). Clicking the **Stop** icon will stop the simulation before completion.
8. **Interact with the plot** In the **Scope** window, using the toolbar buttons, you can zoom in and out, change axes ranges, save axis settings, and print the plot. Right-clicking on the **Scope** window brings up other choices.
9. **Save your model** Saving your model, by choosing **Save** under the **File** menu, creates a file with an .mdl extension, which is required.

## C.3 Examples

This section will present four examples of the use of Simulink to simulate linear, nonlinear, and digital systems. Examples will show the Simulink block diagrams as well as explain the settings of parameters for the blocks. Finally, the results of the simulations will be shown.

### Example C.1 Simulation of Linear Systems

Our first example develops a simulation of three linear systems to compare their step responses. In particular, we solve Example 4.8 and reproduce the responses shown in Figure 4.24. Figure C.5 shows a Simulink block diagram formed by following Steps 1 through 5 in Section C.2 as follows:

**Access Simulink; select, assemble, and label subsystems** The source is a 1-volt step input, obtained by dragging the **Step** block from the **Simulink Library Browser** under **Sources** to your model window.

The first system, T1, consists of two blocks, **Gain** and **Transfer Fcn**. Gain is obtained by dragging the **Gain** block from the **Simulink Library Browser** under **Math Operations** to your model window. Transfer function, T1, is obtained by dragging the **Transfer Fcn** block from the **Simulink Library Browser** under **Continuous** to your model window. Systems T2 and T3 are created similarly.

The three output signals, C1, C2, and C3, are multiplexed for display into the single input of a scope. The Mux (multiplexer) is obtained by dragging the **Mux** block from the **Simulink Library Browser** under **Signal Routing** to your model window.

The sink is a scope, obtained by dragging the **Scope** block from the **Simulink Library Browser** under **Sinks** to your model window.

Alternatively, all blocks can be dragged from the **Library: simulink** window shown in Figure C.4. The **Mux** can be found under **Signal Routing** in the **Library: simulink** window.

The labels for the blocks can be changed to those shown in Figure C.5 by following Step 3 in Section C.2.

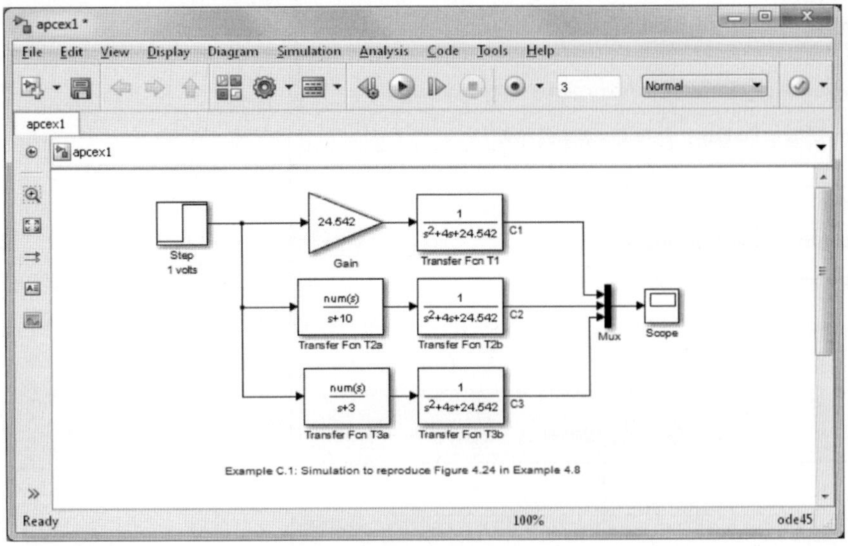


**FIGURE C.5** Simulink block diagram for Example C.1

**Interconnect subsystems and label signals** Follow Step 4 to interconnect the subsystems and label the signals. You must set the mux's parameters before the wiring can be completed. See the next paragraph.

**Choose parameters for the subsystems** Let us now set the parameters of each block using Step 5. The **Block Parameters** window for each block is accessed by double-clicking the block on your model window. Figure C.6 shows the **Block Parameters** windows for the 1-Volt step input, gain, transfer function 1, and mux. Set the parameters to the required values as shown.

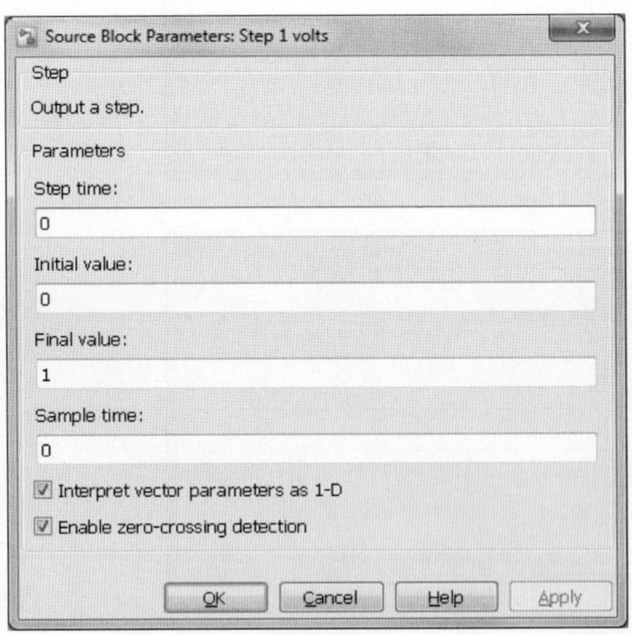


(*a*)

**FIGURE C.6** **Block parameters** windows for **a.** 1-Volt step source; (*figure continues*)

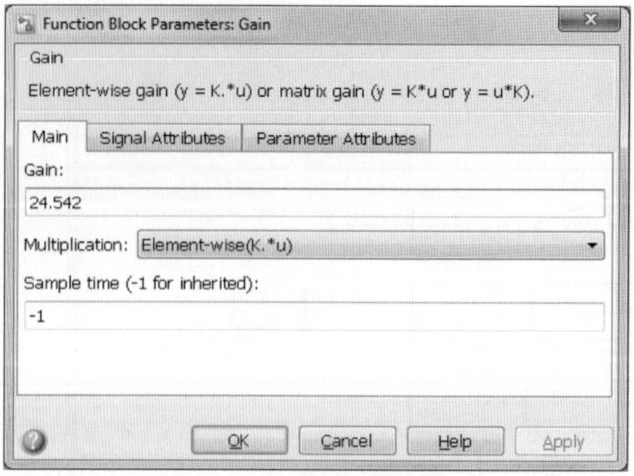


(*b*)

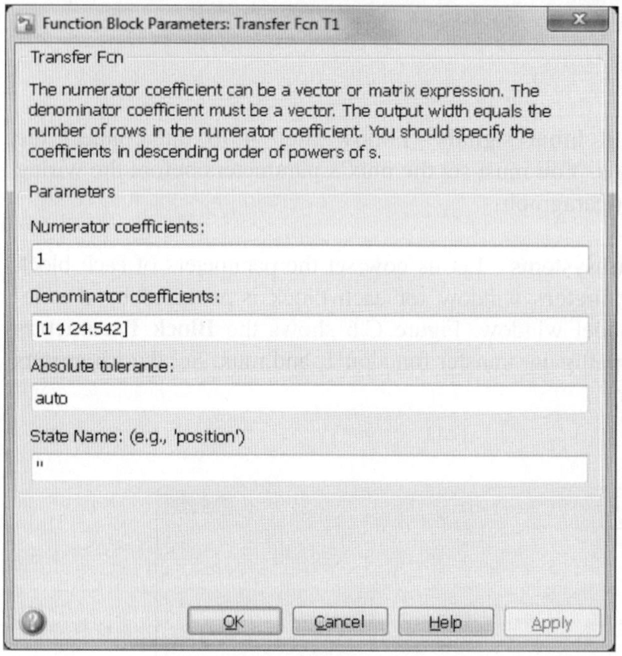


(*c*)

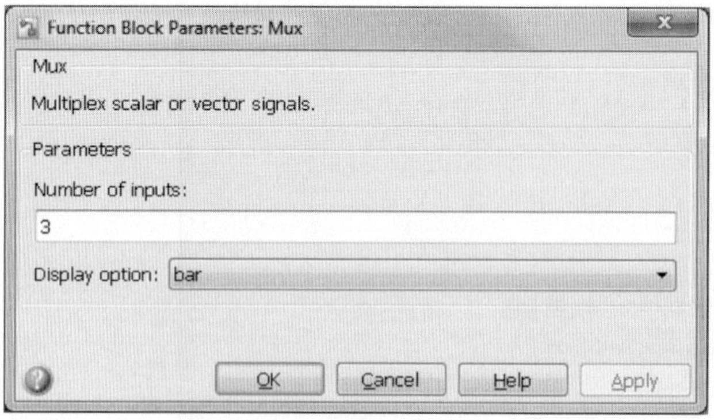


(*d*)

**FIGURE C.6** (*Continued*) **b.** gain; **c.** transfer function 1; **d.** mux

The scope requires further explanation. Double-clicking the **Scope** block in your model window accesses the scope's display, Figure C.7(*a*).

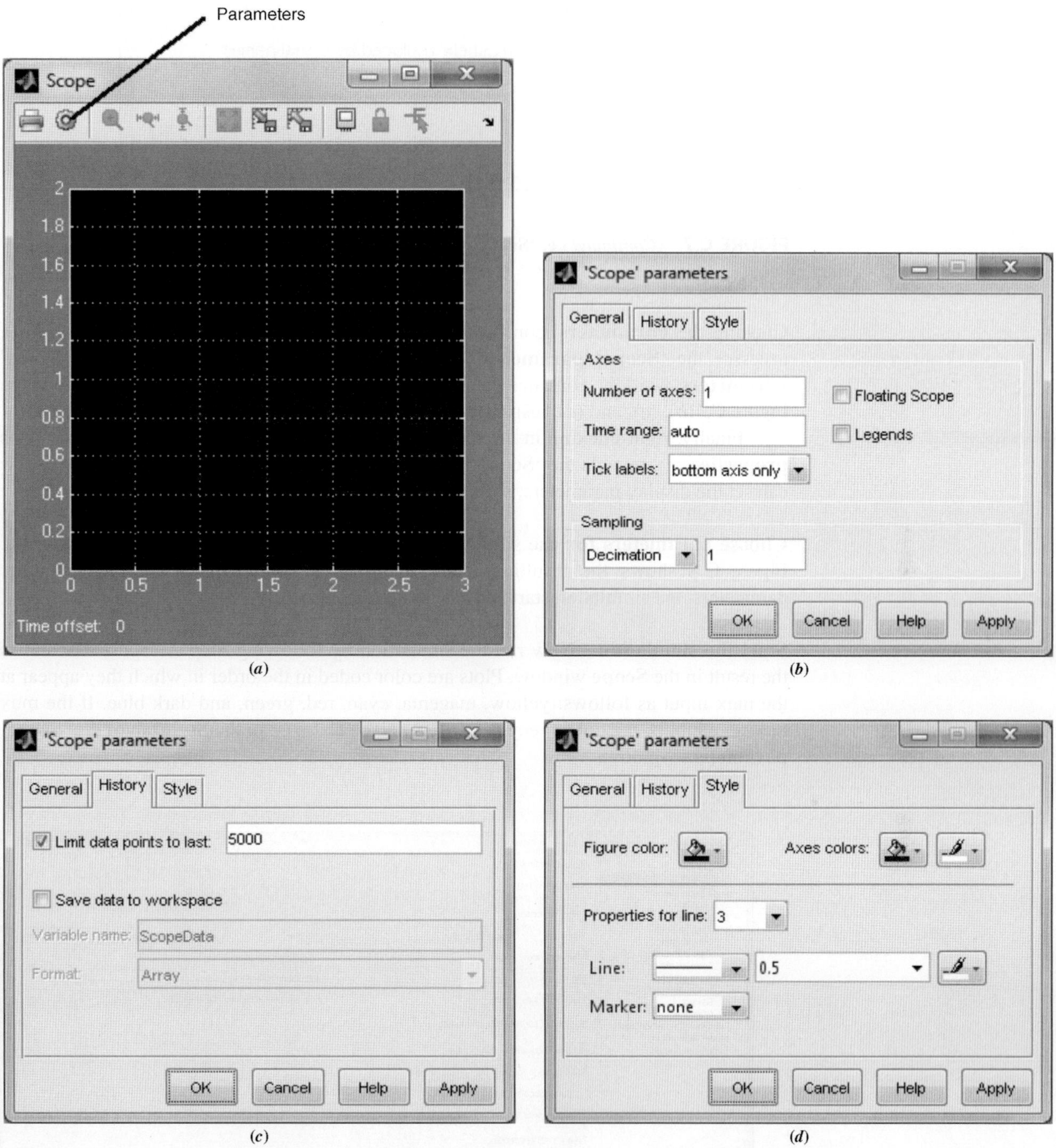


**FIGURE C.7** Windows for the scope: **a. Scope**; **b. 'Scope' parameters, General** tab; **c. 'Scope' parameters, History** tab; **d. 'Scope' parameters, Style** tab; (*figure continues*)

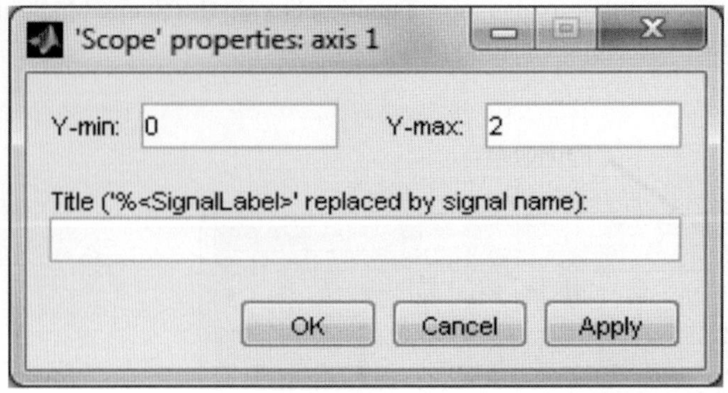


(*e*)

**FIGURE C.7** (*Continued*) **e. 'Scope' properties axis 1**

Clicking the **Parameters** icon on the **Scope** window toolbar, shown in Figure C.7(*a*), accesses the **'Scope' parameters** window as shown in Figure C.7(*b*). The **'Scope' parameters** window contains three tabs, **General, History,** and **Style**, as shown in Figure C.7(*b*), (*c*), and (*d*), respectively.

Finally, right-clicking in the plotting area in the Scope window and selecting **Axis properties** . . . reveals the **'Scope' properties: axis 1** window, Figure C.7(*e*). We now can set the display parameters, such as amplitude range.

**Choose parameters for the simulation** Follow Step 6 to set simulation parameters. Figure C.8 shows the resulting **Configuration Parameters** window. Among other parameters, the simulation start and stop times can be set.

**Start the simulation** Now run the simulation by following Step 7. Figure C.9 shows the result in the **Scope** window. Plots are color coded in the order in which they appear at the mux input as follows: yellow, magenta, cyan, red, green, and dark blue. If the mux has more inputs, the colors recycle. Colors can be changed in the **Style** tab of the '**Scope**' **parameters** window.

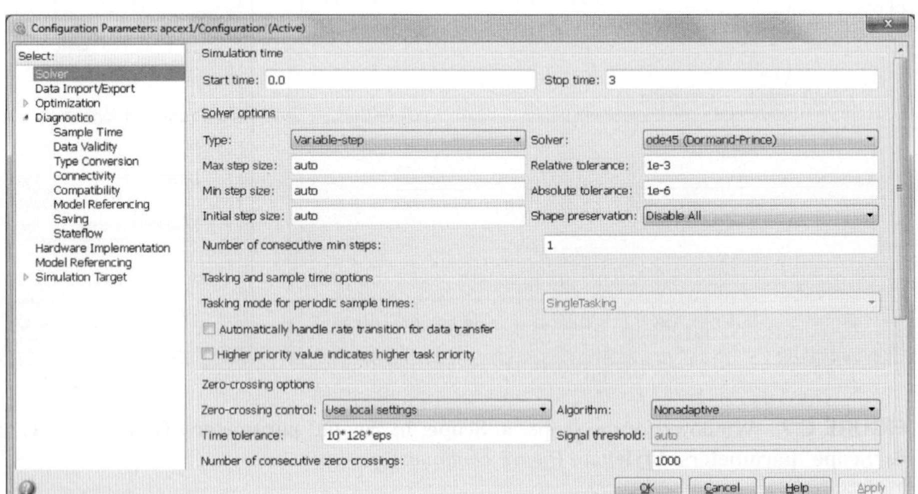


**FIGURE C.8** **Configuration Parameters** window for **Solver** tab

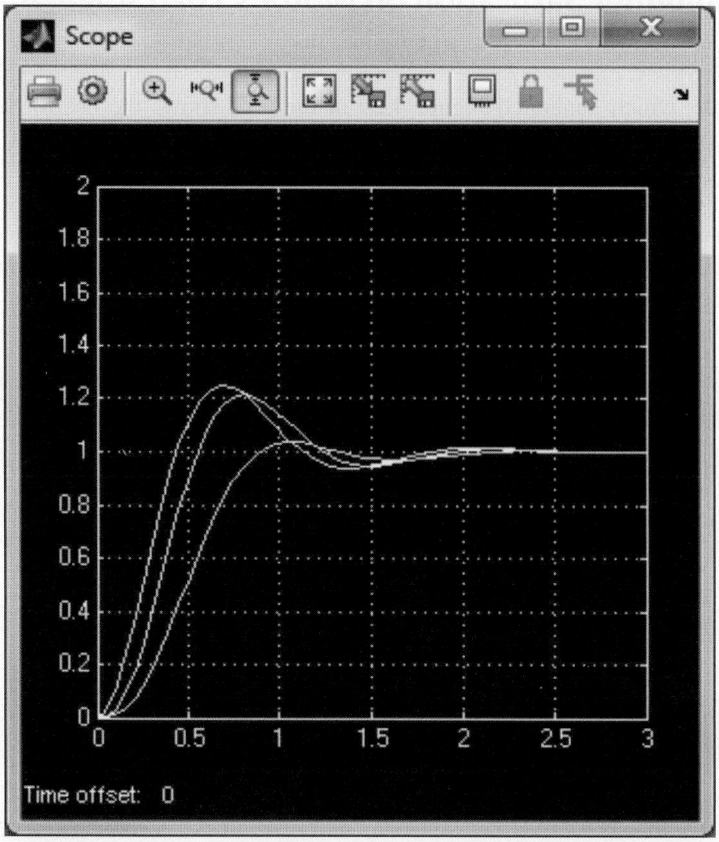


**FIGURE C.9** **Scope** window after Example C.1 simulation stops

**Interact with the plot** The toolbar of the **Scope** window shown in Figure C.9 has several buttons that can be used to interact with the plot. Let us summarize the function and operation of each, starting with the left-most button:

**Button 1** executes a plot print.

**Button 2** has already been explained and is used to set scope parameters.

**Button 3** permits zooming into the plot in both the $x$ and $y$ directions. Press the button and drag a rectangle over the portion of the curve you want to expand.

**Button 4** allows zooming in the $x$ direction only. Drag a horizontal line over the plot covering the extent of $x$ you want to expand.

**Button 5** allows zooming in the $y$ direction only. Drag a vertical line over the plot covering the range of $y$ you want to expand.

**Button 6** autoscales axis for use after zooming.

**Button 7** saves current axis settings.

**Button 8** restores saved axis settings.

**Button 9** toggles floating scope. It must be turned off to use zooming. See documentation for use of floating scopes.

**Button 10** toggles lock for current axis selection.

**Button 11** allows selection of signals to view when using floating scope.

## Example C.2 Effect of Amplifier Saturation on Motor's Load Angular Velocity

This example, which generated Figure 4.29 in the text, shows the use of Simulink to simulate the effect of saturation nonlinearity on an open-loop system. Figure C.10 shows a Simulink block diagram formed by following Steps 1 through 5 in Section C.2 above.

Saturation nonlinearity is an additional block that we have not used before. Saturation is obtained by dragging to your model window the **Saturation** block in the **Simulink Library Browser** window under **Discontinuities** as shown in Figure C.11(*a*) and setting its parameters to those shown in Figure C.11(*b*).

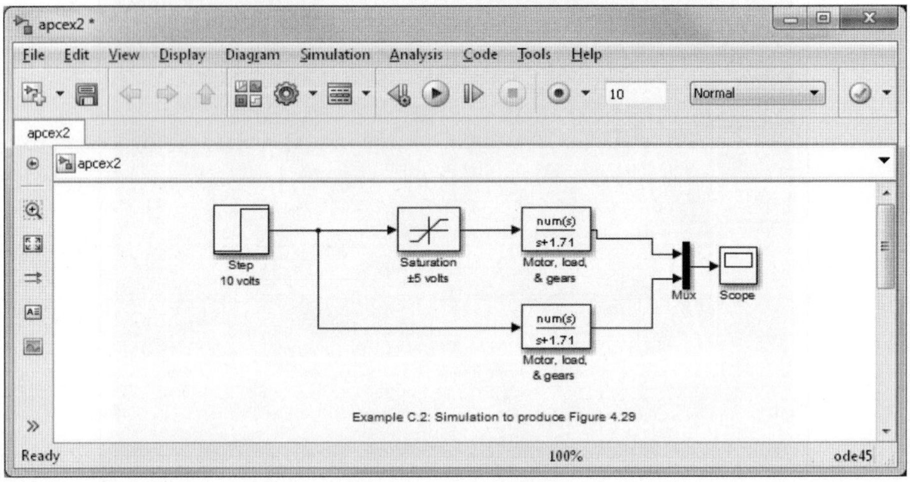


**FIGURE C.10** Simulink block diagram for Example C.2

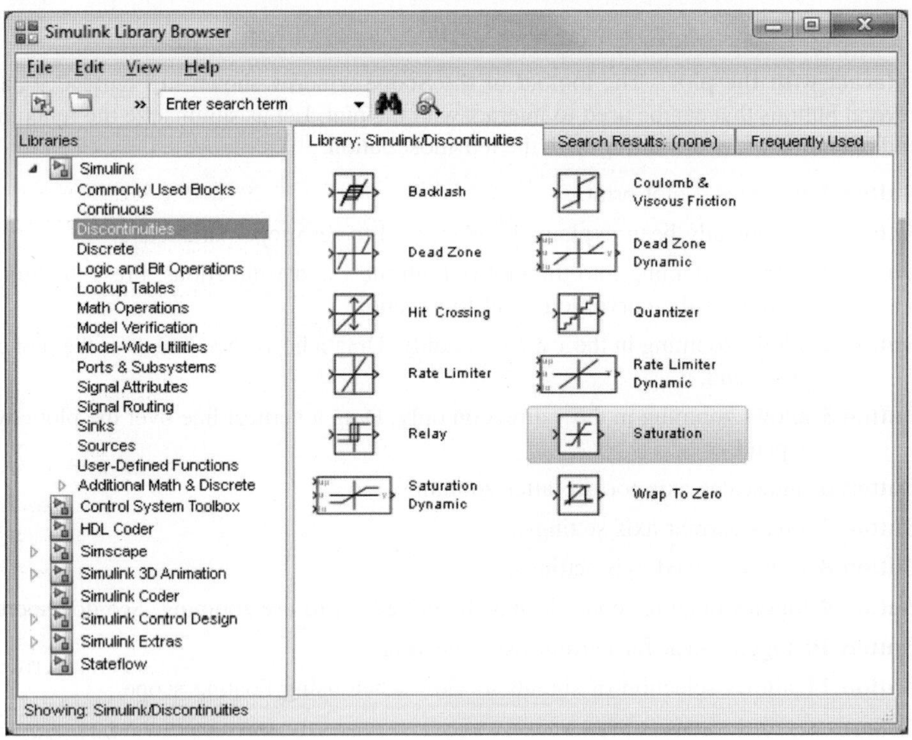


(*a*)

**FIGURE C.11** **a.** Simulink library for nonlinearities; (*figure continues*)

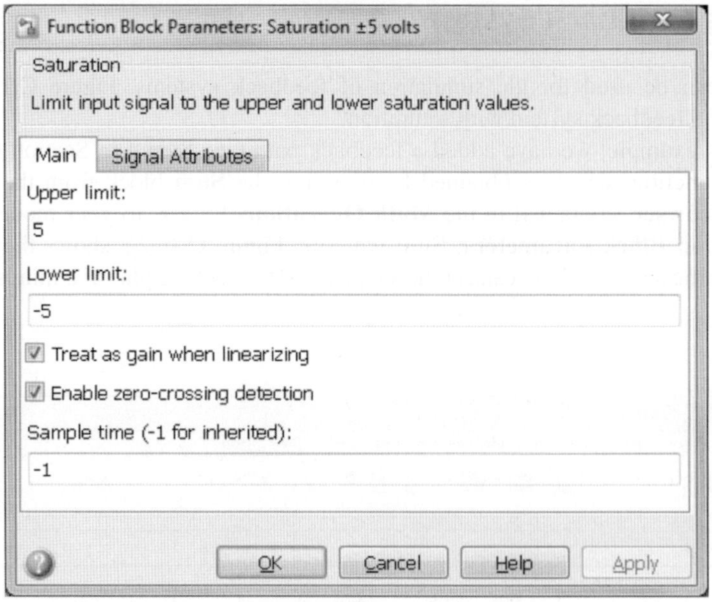


(*b*)

**FIGURE C.11** (*Continued*) **b.** parameter settings for saturation

Now run the simulation by making your model window active and selecting **Run** under the **Simulation** menu of your model window or clicking on the **Run** button on your model window toolbar. Figure C.12 shows the result in the **Scope** window.

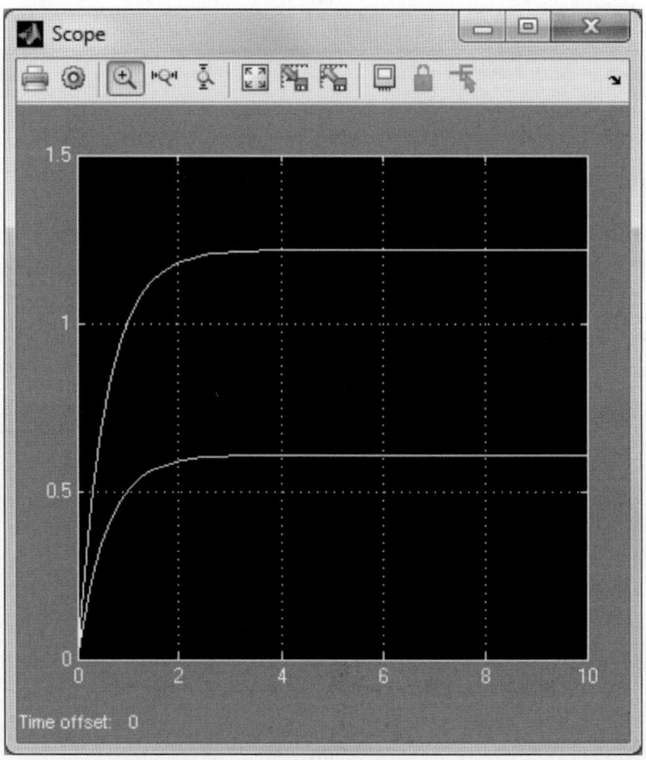


**FIGURE C.12** **Scope** window after simulation of Example C.2 stops. The lower curve is the output with saturation

## Example C.3 Simulating Feedback Systems

Simulink can be used for the simulation of feedback systems. Figure C.13(*a*) is an example of a feedback system with saturation.

In this example, we have added a feedback path (see Step 4 in Section C.2) and a summing junction, which is obtained by dragging the **Sum** block from the **Simulink Library Browser**, contained in the **Math Operations** library, to your model window. The **Function Block Parameters: Sum** window, Figure C.13(*b*), shows the parameter settings for the summer. You can set the shape as well as set the plus and minus inputs. In

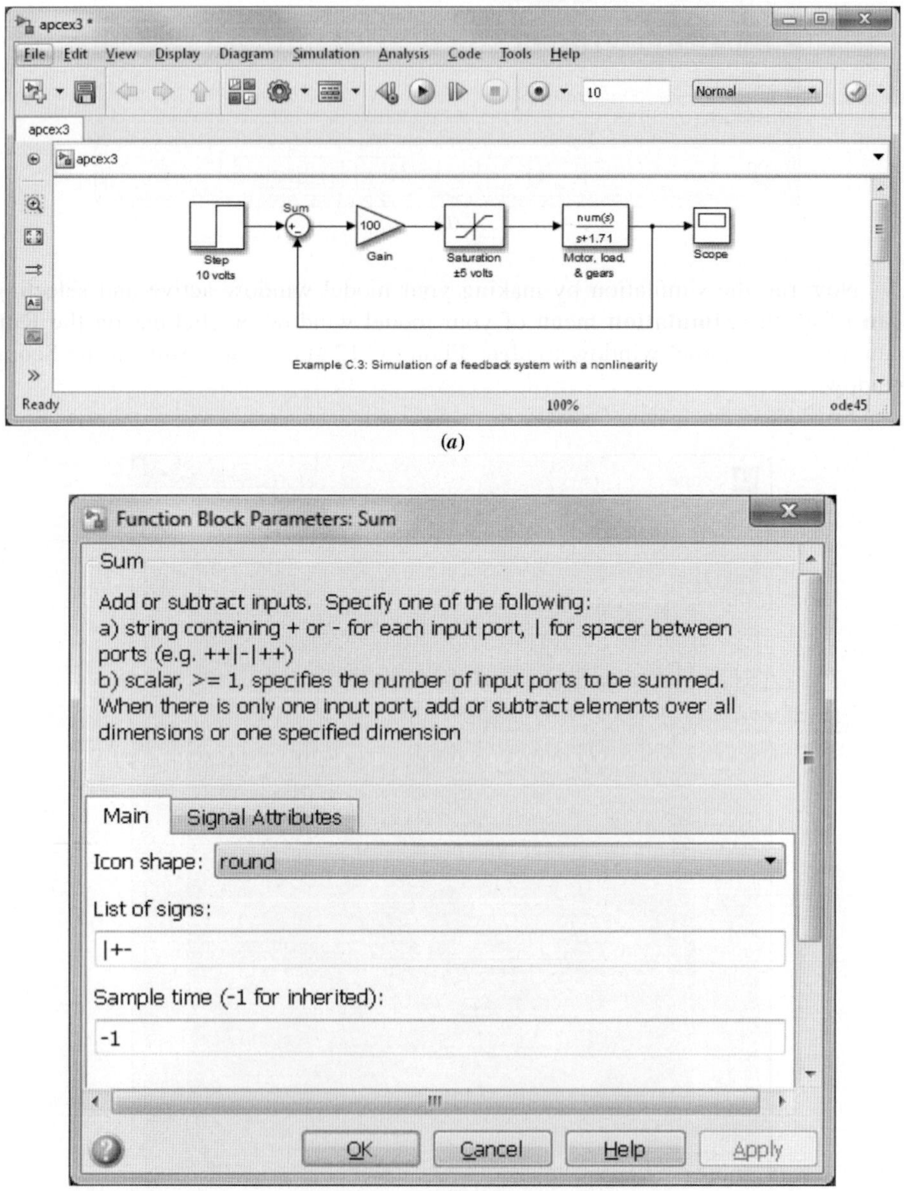


(*b*)

**FIGURE C.13** **a.** Simulation block diagram for a feedback system with saturation; **b.** block parameter window for the summer

the list of signs, the "|"symbol signifies a space. We place it at the beginning to start the signs at "nine o'clock," conforming to our standard symbol, rather than at "12 o'clock." The result of the simulation is shown in Figure C.14.

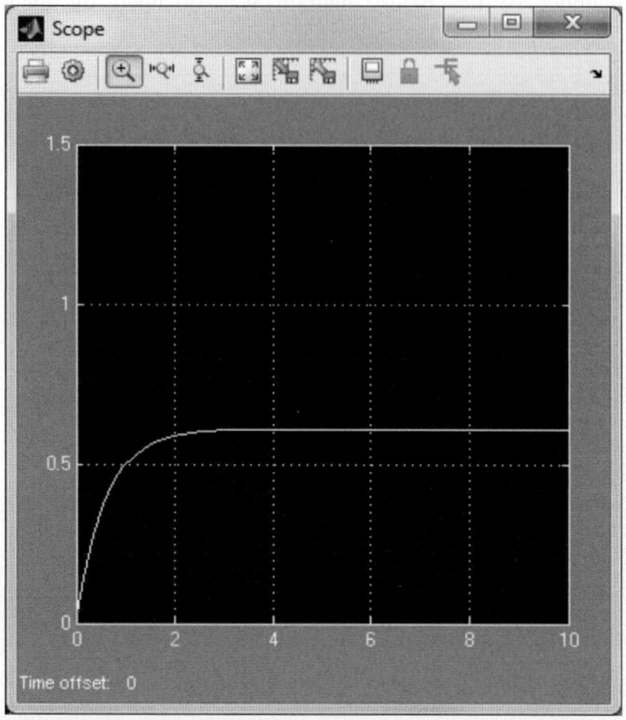


**FIGURE C.14** Simulation output for Example C.3

## Example C.4 Simulating Digital Systems

This example demonstrates two methods of generating digital systems via Simulink for the purpose of simulation, as shown in Figure C.15.

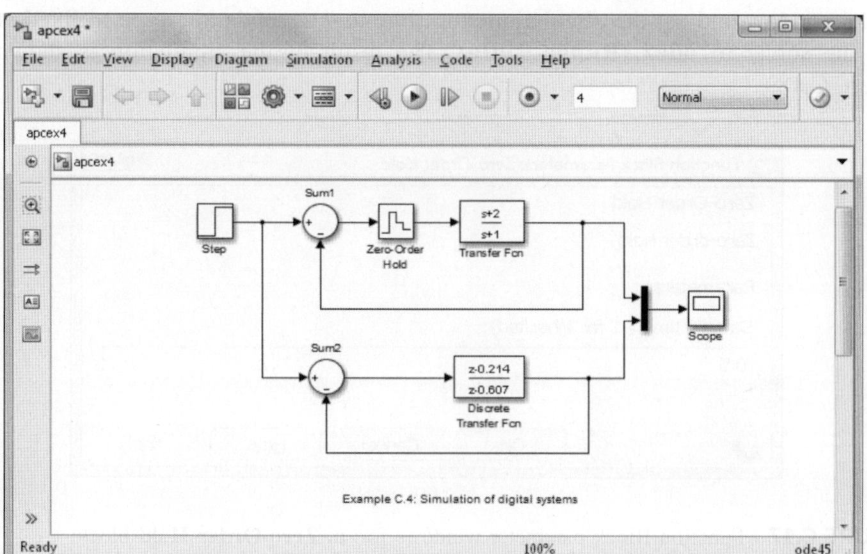


**FIGURE C.15** Simulink block diagram for simulating digital systems two ways

The first approach uses a linear transfer function cascaded with a **Zero-Order Hold** block obtained from the **Simulink Library Browser** under the **Discrete** block library, shown on the right-hand side of Figure C.16. The second method uses a discrete transfer function also obtained from the **Simulink Library Browser** under the **Discrete** block library. The remainder of the block diagram was obtained by methods previously described.

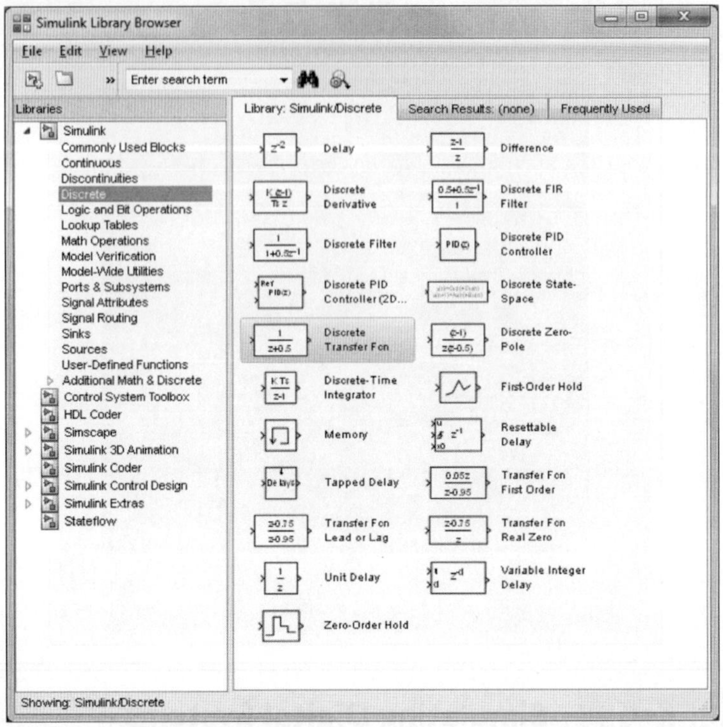


**FIGURE C.16** Simulink library of discrete blocks

The block parameters for the **Zero-Order Hold** and **Discrete Transfer Fcn** blocks are set as shown in Figures C.17(*a*) and (*b*), respectively.

Select **Model Configuration Parameters** under the **Simulation** menu in your model window and set the simulation stop time to 4 seconds, the type to **fixed-step**, and the solver to **ode4 (Runge-Kutta)**. The result of the simulation is shown in Figure C.18.

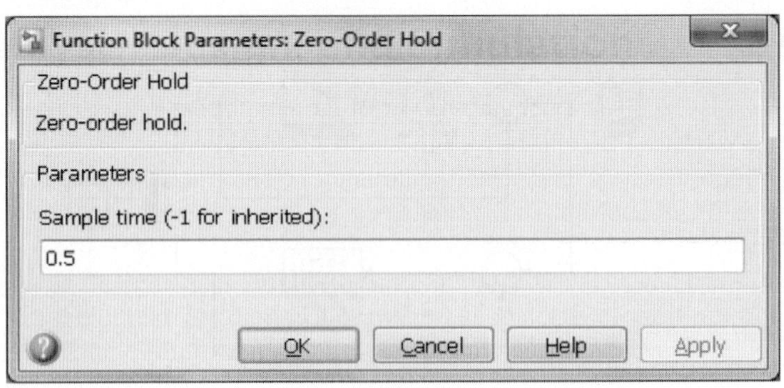


(*a*)

**FIGURE C.17** Function Block parameter windows for: **a. Zero-Order Hold** block; (*figure continues*)

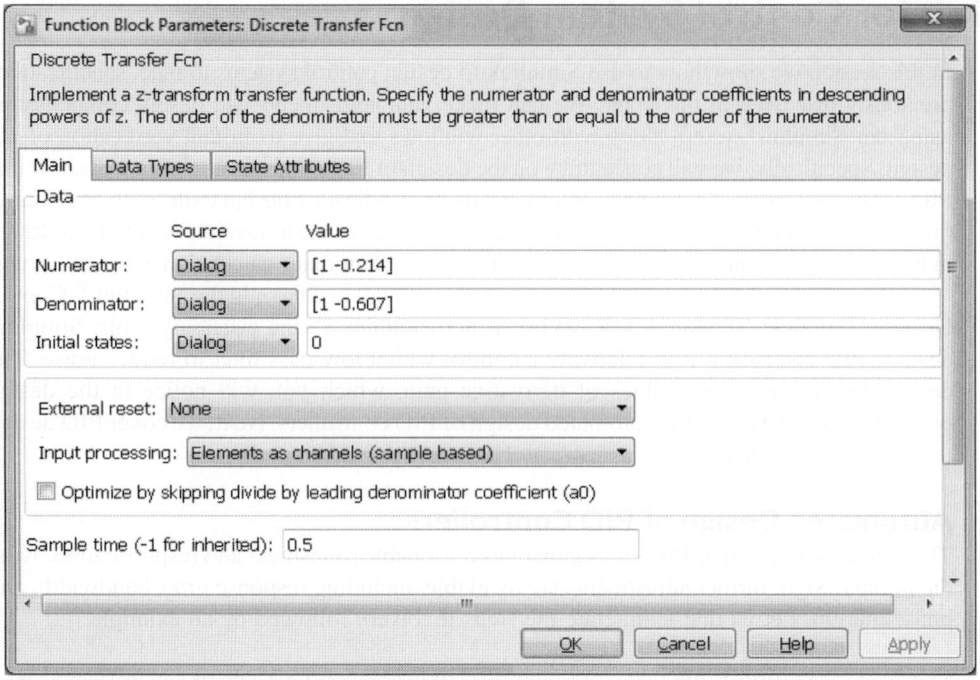


(*b*)

**FIGURE C.17** (*Continued*) **b. Discrete Transfer Fcn** block

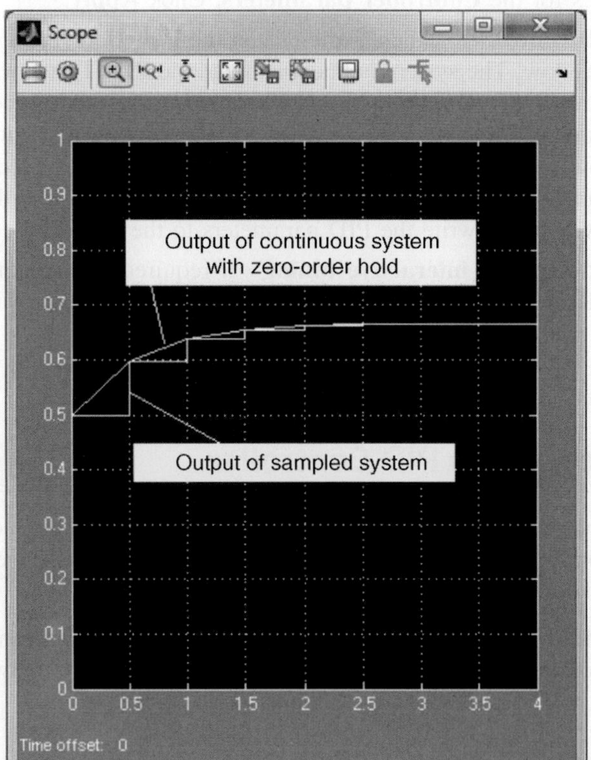


**FIGURE C.18** Outputs of the digital systems

# C.4 Using Simulink for Control System Design

In this section we show how to use Simulink to design control systems to meet specifications previously discussed in this book. We can make gain adjustments and design compensators using our Simulink system along with other windows that give us instant verification of our design. Specifically, we will concentrate on the design of PID compensators. We will show that PID compensators can be designed automatically or by adjusting design tools, such as response time and transient behavior. As we make adjustments, we see the immediate result of our design in lists of specifications or time responses along with the automatic calculation of the PID gains.

In order to perform control system design, you will need to add the Simulink® Control Design™ module Version 4.0 (R2014a), which contains all the necessary tools. Simulink designs PID controllers using derivative control with a low-pass filter to reduce noise. The design requires negative values of derivative gain, which you will notice in the design examples. We first cover the automated design of PID controllers. Next, we cover PID design using graphical methods.

## Automated Design of PID Controllers

The automated design of PID gains generates reasonable robustness and response time. After the initial design, further adjustments are available, including response time, bandwidth, and phase margin. Let us first enumerate the steps involved, followed by an example.

1. **Create a Simulink diagram** Begin with a linear or nonlinear feedback control system containing a PID controller.
2. **Set initial values for the PID controller** Double-click the PID controller and launch the **Function Block Parameters: PID Controller** window. On the **MAIN** tab, input nominal values for the **Controller parameters.** Click **Apply**.
3. **Tune the PID controller** Click **Tune . . .** in the **Function Block Parameters: PID Controller** window. The system is linearized and the **PID Tuner** window is launched showing the nominal values response (**Block response**) and the designed response (**Tuned response**). Click the **Show parameters** button on the **Toolstrip** of the **PID Tuner** window to expose performance data, including the designed PID gains. If the response meets requirements, then click the **Update Block** button on the **Toolstrip** of the **PID Tuner** window to write the PID parameters to the controller.
4. **Modify the design via interactive tuning** If required, change the performance by moving the **Response Time** and **Transient Behavior** sliders in the middle of the **Toolstrip** of the **PID Tuner** window. Click the **Update Block** to write the PID parameters to the controller. If the system is nonlinear, you should run the simulation to see the effect of the nonlinearity on the designed response.

## Example C.5 Automated Design of a PID Controller

In this example we follow the previously enumerated steps to automate the design of a PID controller for the system of Figure C.19. The requirements are (1) less than 1 second settling time; and (2) less than 5% overshoot.

**Create a Simulink diagram** We create the Simulink block diagram of Figure C.19 where the PID controller block is found in the **Simulink Library Browser** as shown in Figure C.3(*a*).

**Set initial values for the PID controller** Double-clicking the PID controller results in Figure C.20. For initial values we have chosen unity and clicked **Apply**.

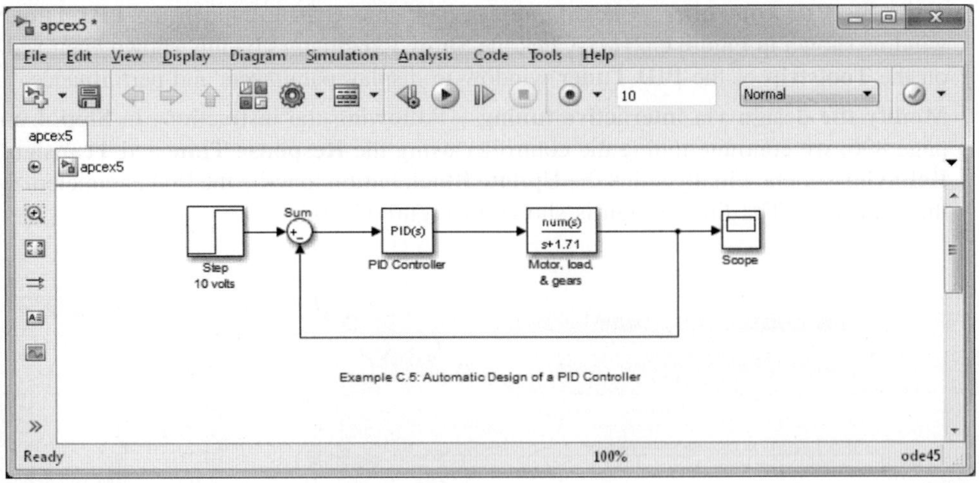


**FIGURE C.19** Simulink block diagram for automated design of a PID controller

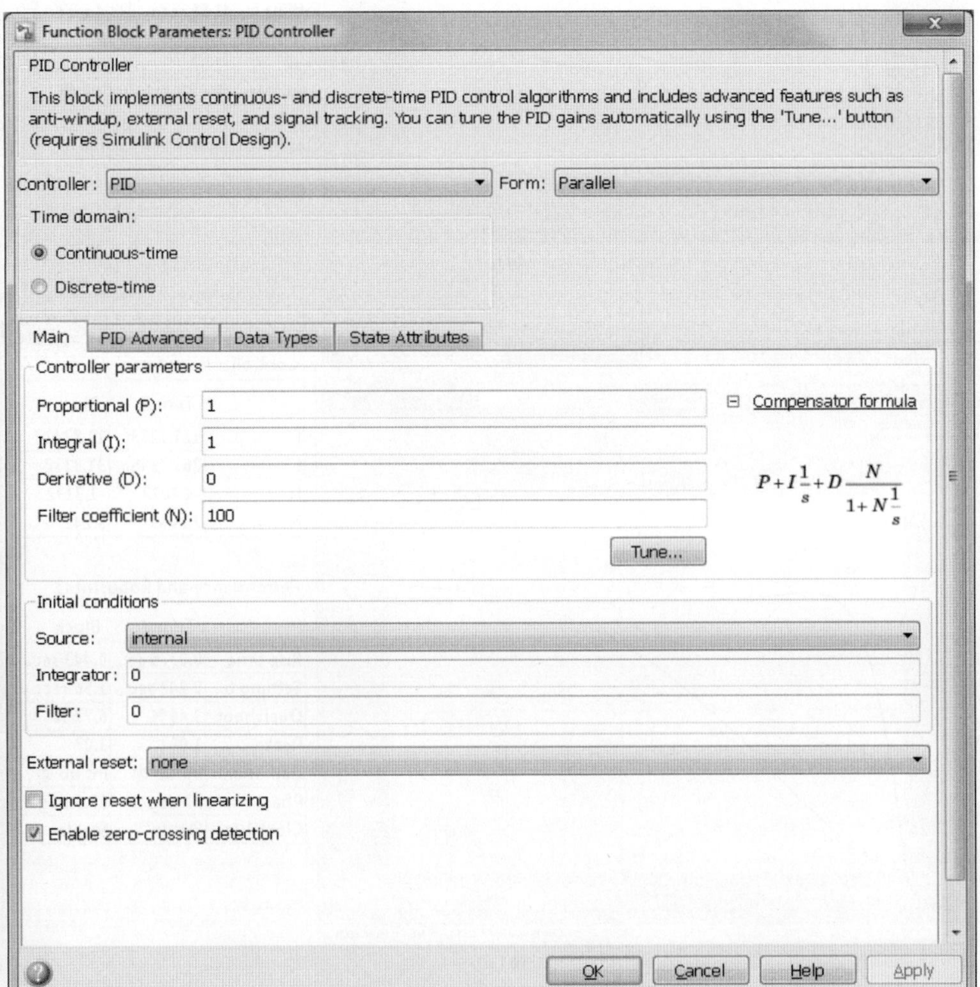


**FIGURE C.20** **Function Block Parameters** window for the PID controller

**Tune the PID controller** Click **Tune . . .** in Figure C.20 to launch the **PID Tuner** window shown in Figure C.21(*a*). Click the **Show Parameters** button, shown encircled, on the **Toolstrip** of the **PID Tuner** window to display parameters and performance.

**Modify the design via interactive tuning** Following the instructions in **Step 4** on page 836, we continue tuning the controller using the **Response Time** and **Transient Bahavior** sliders. Finally, click the **Update Block** button to write the PID parameters to the controller. The final design is shown in Figure C.21(*b*).

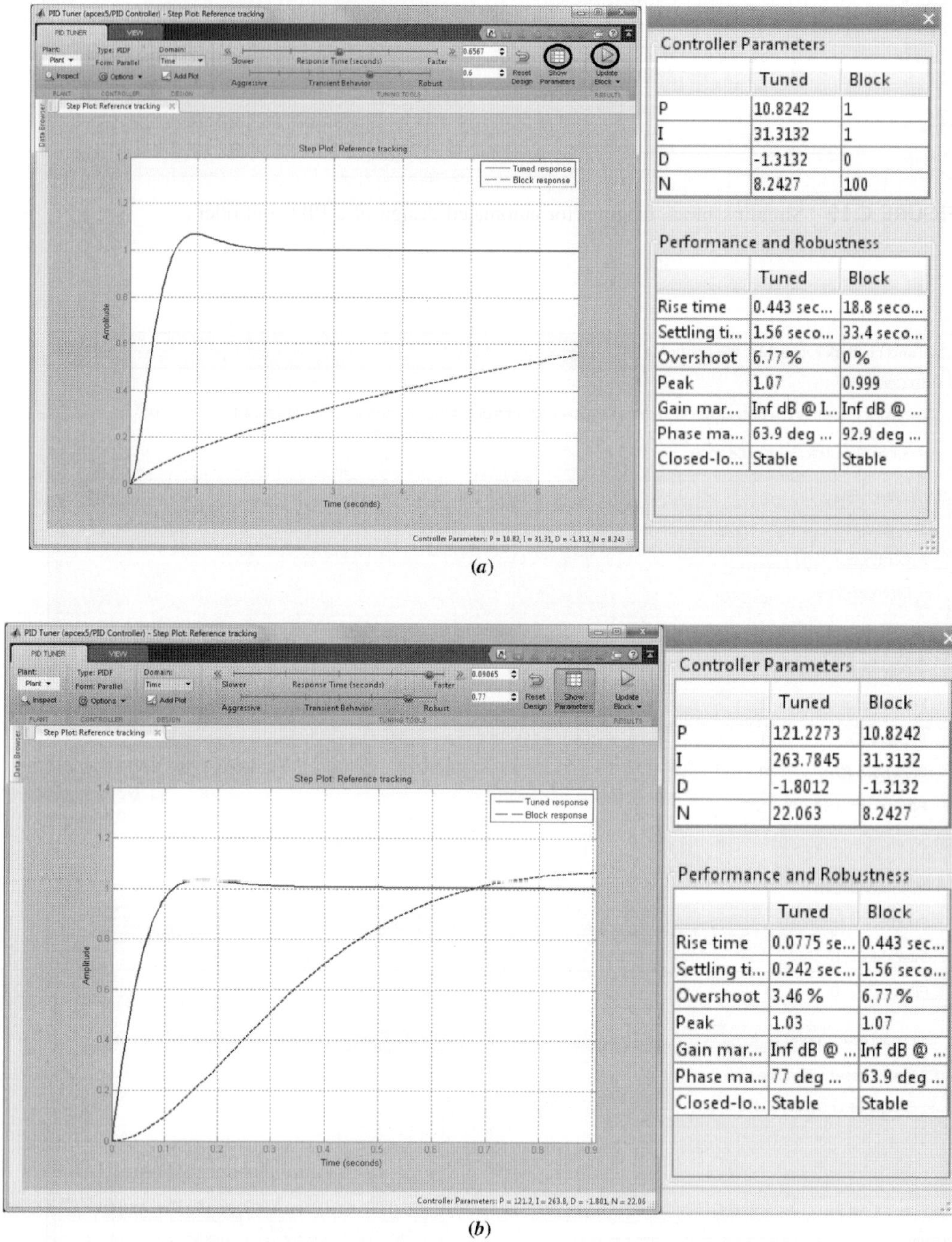


**FIGURE C.21** **PID Tuner** window: **a.** before additional tuning; **b.** after additional tuning

## Automated Tuning of PID Controllers and Graphical Design

In this subsection, we begin PID design with automated tuning followed by graphical design of our choice using Bode plots, root locus, etc. Let us first enumerate the steps involved followed by an example.

1. **Create a Simulink diagram** Begin with a linear or nonlinear feedback control system containing a PID controller.
2. **Begin compensator design** From the menu bar of your Simulink block diagram select **Analysis/Control Design/Control System Designer** . . . launching the **Control and Estimation Tools Manager** window.
3. **Select block to tune** In the **Control and Estimation Tools Manager** window, click **Select Blocks**. . . . In the resulting **Select Blocks to Tune** window, choose **PID Controller** and click **OK**.
4. **Choose the closed-loop response for which response will be analyzed** On the Simulink block diagram, choose the input and output for the closed-loop. Right-click at the input and choose **Linear Analysis Points/Input Perturbation**. Right-click at the output point and choose **Linear Analysis Points/Output Measurement**. Small vertical arrows result at these points to indicate your selection.
5. **Tune blocks and launch the Design Configuration Wizard** Click **Tune Blocks**. . . and launch the **Design Configuration Wizard**. Click **Next** and select the graphical plots you want to use for design. Click **Next** and choose the **Analysis Plots** you want to use to analyze the design. Check the **Contents of Plots** boxes that correspond to the **Responses** you want your **Analysis Plots** to display. Click **Finish** and launch the **LTI Viewer for SISO Design Task** window that contains the **Analysis Plots** you selected. Also launched is the **SISO Design for SISO Design Task** window that contains the graphical design plots that you previously selected.
6. **Perform automatic tuning of the PID controller** In the **Control and Estimation Tools Manager** window select the **Automated Tuning** tab at the top right. Select **PID Tuning** in the **Design method:** drop-down menu. Select **Tuning method** and **Controller Type**. Click **Update Compensator**, which selects compensator gains and updates the response and graphical design curves. Click **Update Simulink Block Parameters**.
7. **Evaluate performance** Right-click the response curve and select **Characteristics**. Select each desired choice under **Characteristics**, adding a plot marker to the curve at the appropriate measurement point. Click on each marker to read the performance data.
8. **Improve performance by interacting with the graphical design plots** Change PID poles and gains until your response meets the requirements. You may add performance boundaries by right-clicking the design graphic and selecting **Design Requirements/New. . . .** From the pull-down menu select the desired requirements. Upon closing the window the performance boundary is placed on the graphical design plot.

### Example C.6 Automated Tuning of a PID Controller and Graphical Design

In this example we follow the previously enumerated steps to automatically design a PID controller for the system of Figure C.19 and follow with further improvement in performance using root locus. The requirements are: (1) less than 1 second settling time; and (2) less than 4% overshoot.

**Create a Simulink diagram** We create the Simulink block diagram of Figure C.19 where the PID controller block is found in the **Simulink Browser Library** as shown in Figure C.3(*a*).

**Begin Compensator design** In Figure C.19, we select **Analysis/Control Design/ Compensator Design. . . ,** launching the **Control and Estimation Tools Manager** window shown in Figure C.22(*a*).

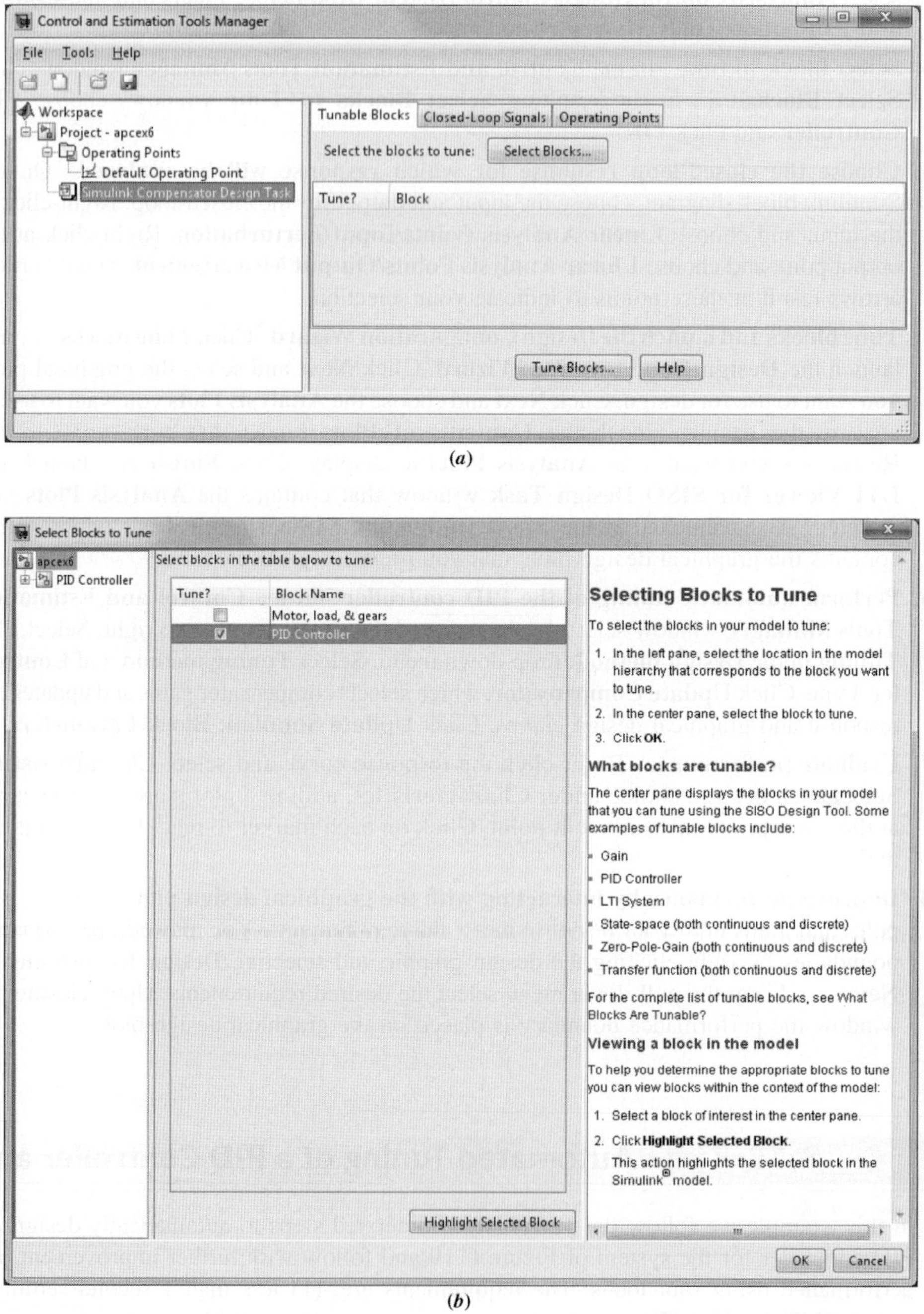


**FIGURE C.22** **a. Control and Estimation Tools Manager** window before selecting block to tune; **b. Select blocks to Tune** window; (*figure continues*)

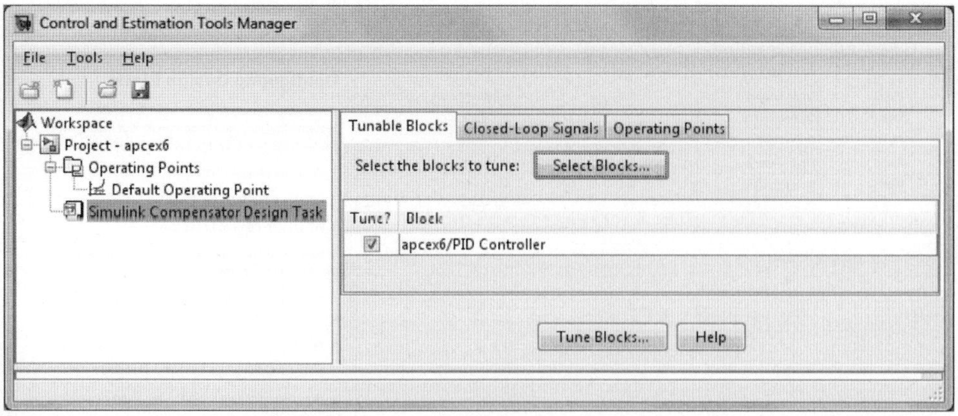


(*c*)

**FIGURE C.22** (*Continued*) **c. Control and Estimation Tools Manager** window after selecting block to tune

**Select blocks to tune** Click **Select Blocks . . .** in Figure C.22(*a*) and launch the **Select Blocks to Tune** window shown in Figure C.22(*b*). Select **PID Controller** and click **OK**. The resulting **Control and Estimation Tools Manager** is shown in Figure C.22(*c*).

**Choose the closed-loop response for which response will be analyzed** On the Simulink block diagram, right-click the output of the step input and select **Linear Analysis Points/Input Perturbation**, placing a small down-arrow with a summing junction at that point, signifying the input of the closed-loop response. Similarly, the closed-loop output is chosen by right-clicking the output of the plant and selecting **Linear Analysis Points/Output Measurement**. A small up-arrow results, signifying the output of the closed-loop system. The resulting Simulink block diagram is shown in Figure C.23.

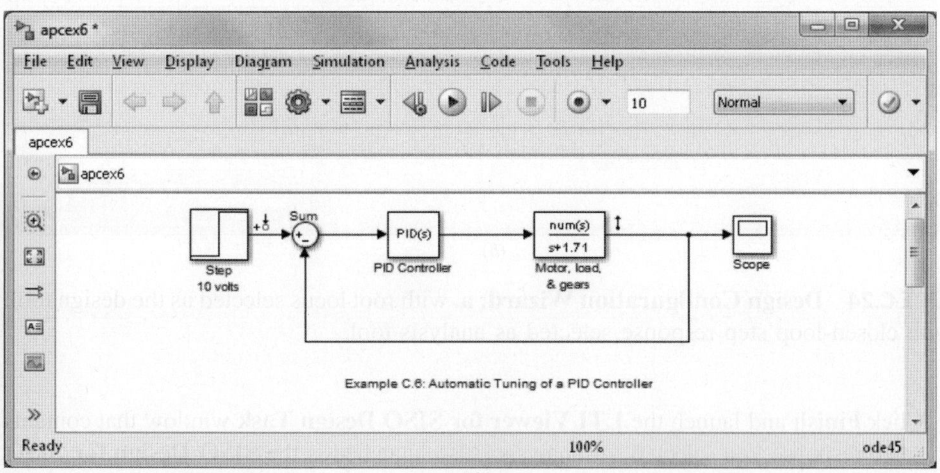


**FIGURE C.23** Simulink block diagram showing closed-loop input and output points

**Tune blocks and launch the Design Configuration Wizard** Click **Tune Blocks . . .** in Figure C.22(*c*) and launch the **Design Configuration Wizard**. Click **Next**. The resulting window is shown in Figure C24(*a*). Select **Root Locus** in the **Plot Type** pull-down menu for **Plot 1**. Click **Next** to select the plot to be used for analysis as shown in Figure C.24(*b*). For this case we choose **Step** for **Plot 1**, **Plot Type**. Also check the box for **Plot 1** below to indicate the **Responses** as shown.

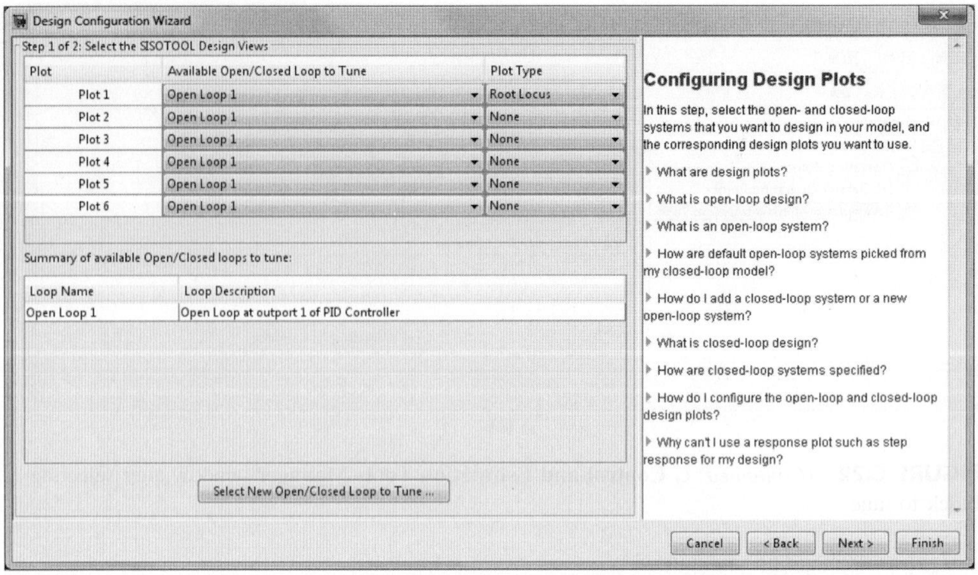


(*a*)

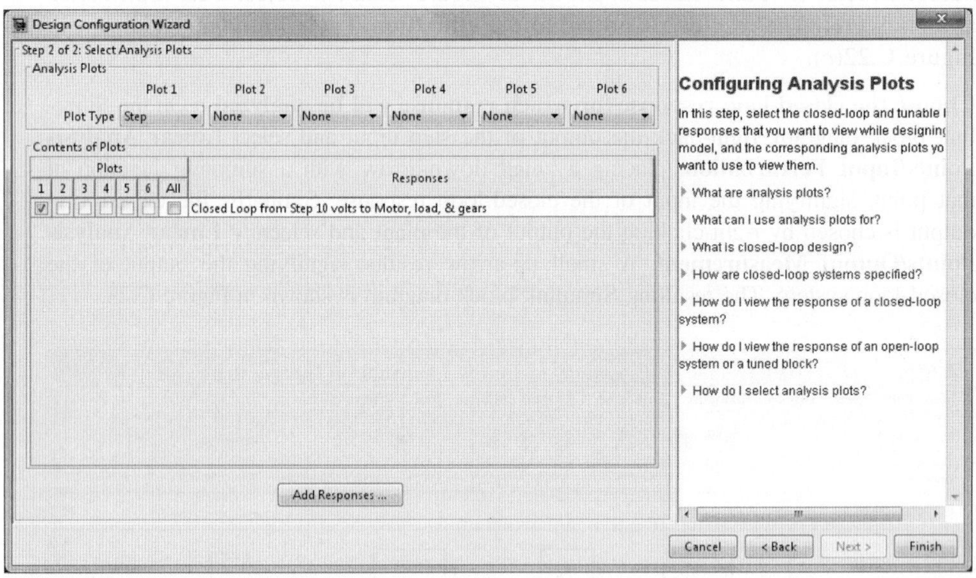


(*b*)

**FIGURE C.24 Design Configuration Wizard: a.** with root locus selected as the design tool; **b.** with closed-loop step response selected as analysis tool

Click **Finish** and launch the **LTI Viewer for SISO Design Task** window that contains the step response you selected for analysis. Also launched is the **SISO Design for SISO Design Task** window that contains the root locus plot that you previously selected for design. Before we discuss these two plots, we first perform automatic PID tuning.

**Perform automated tuning of the PID controller** The **Control and Estimation Tools Manager** now looks like that shown in Figure C.25 after selecting the **Automated Tuning** tab at the top and making the selections shown in the figure. Click **Update Compensator** and **Update Simulink Block Parameters**. You should notice a marked improvement in the step response as shown in Figure C.26.

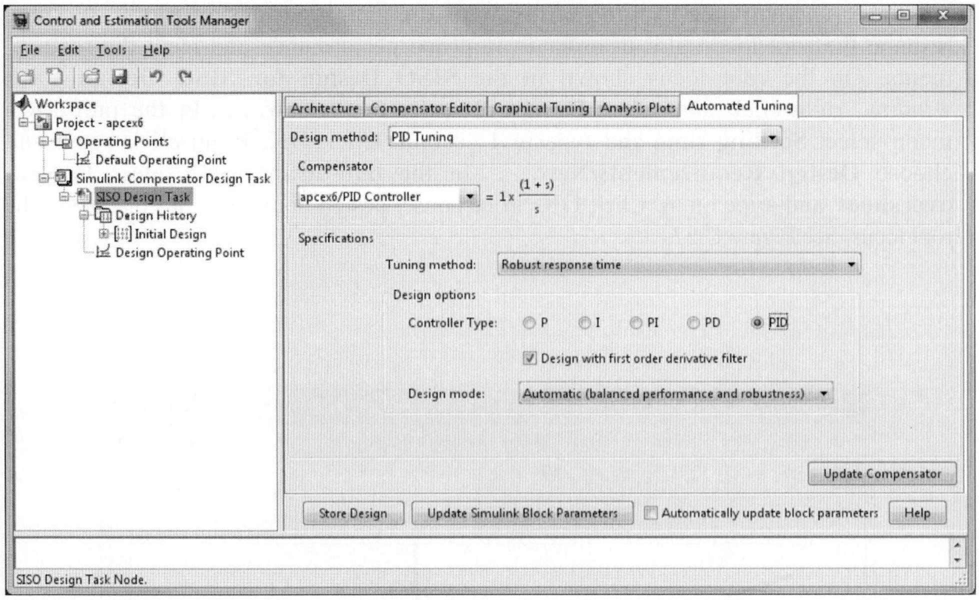


**FIGURE C.25** **Control and Estimation Tools Manager** prepared for automatic PID tuning

**Evaluate performance** Right-click on the response curve shown in Figure C.26 and choose **Characteristics**. Select **Peak Response** and **Settling Time**. Blue dots are placed on the response curve at the appropriate points. Click on each dot to reveal the performance. In this case we see that automatic tuning yields 6.77% overshoot and 1.56 settling time. The requirements are not met.

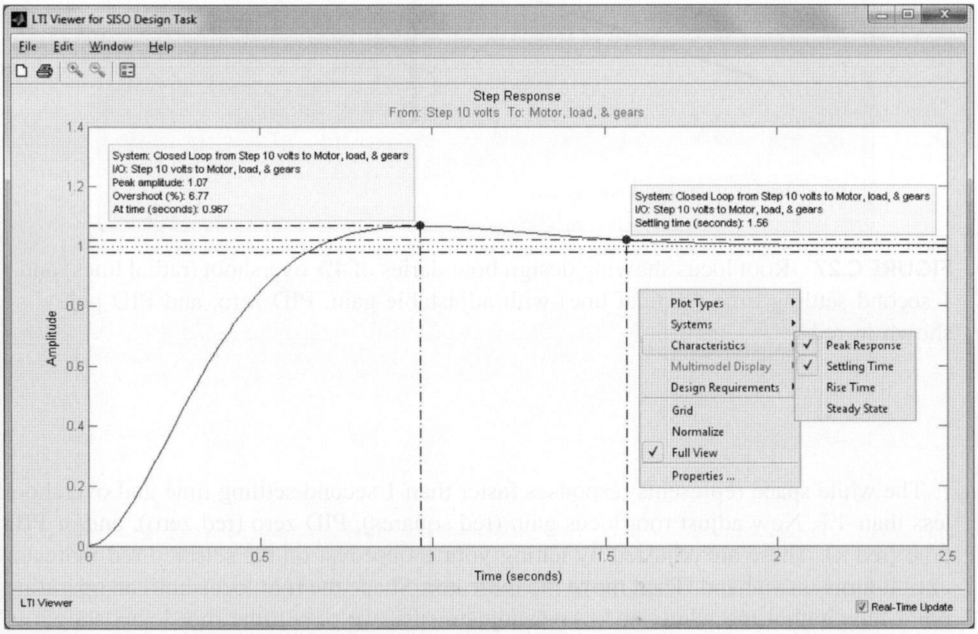


**FIGURE C.26** **LTI Viewer for SISO Design Task** window showing closed-loop step response after automatic PID tuning

**Improve performance by interacting with the graphical design plots** We now reshape the root locus plot in order to improve performance and meet the requirements. On the root locus shown in the **SISO Design for SISO Design Task** window, right-click and select **Design Requirements/New. . . .** In the pull-down menu select **Settling time** and type in 1 second. Click **OK**. Right-click again and choose **Design Requirements/New. . . .** In the pull-down menu select **Percent overshoot** and type in 4. Click **OK**. In response, design boundaries are placed on the root locus. See Figure C.27.

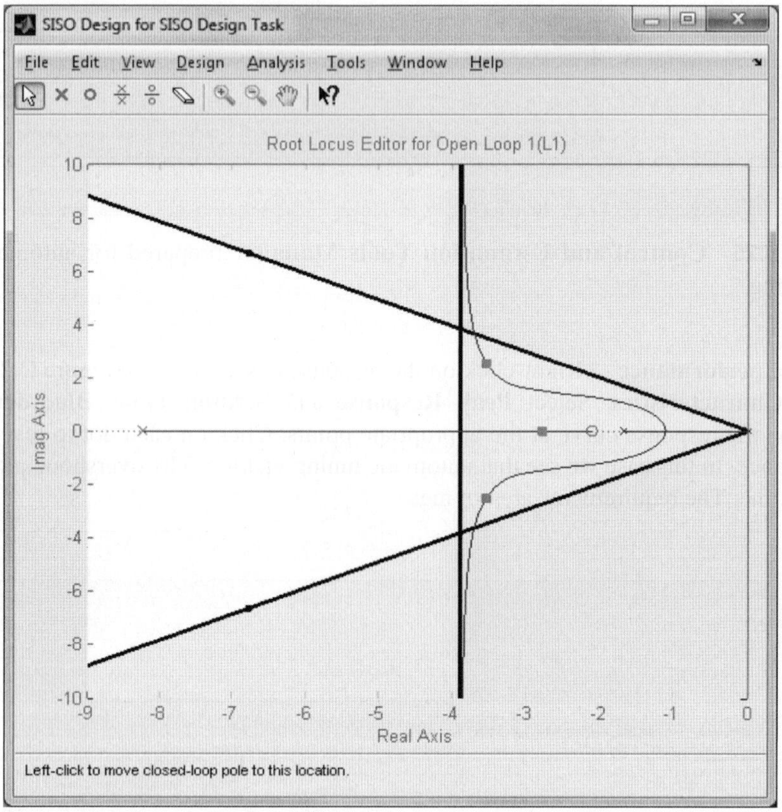


**FIGURE C.27** Root locus showing design boundaries of 4% overshoot (radial lines) and 1 second settling time (vertical line) with adjustable gain, PID zero, and PID pole shown in red

The white space represents responses faster than 1 second settling time and overshoot less than 4%. Now adjust root locus gain (red squares), PID zero (red zero), and/or PID pole (red x). These are adjusted by letting your mouse rest on the selected red indicator until it turns into a hand. Then move the indicator. Shape the root locus until an operating point falls within the white area. After adjustments, we obtain the root locus shown in Figure C.28($a$) and the time response shown in Figure C.28($b$) . We conclude that the design requirements are met.

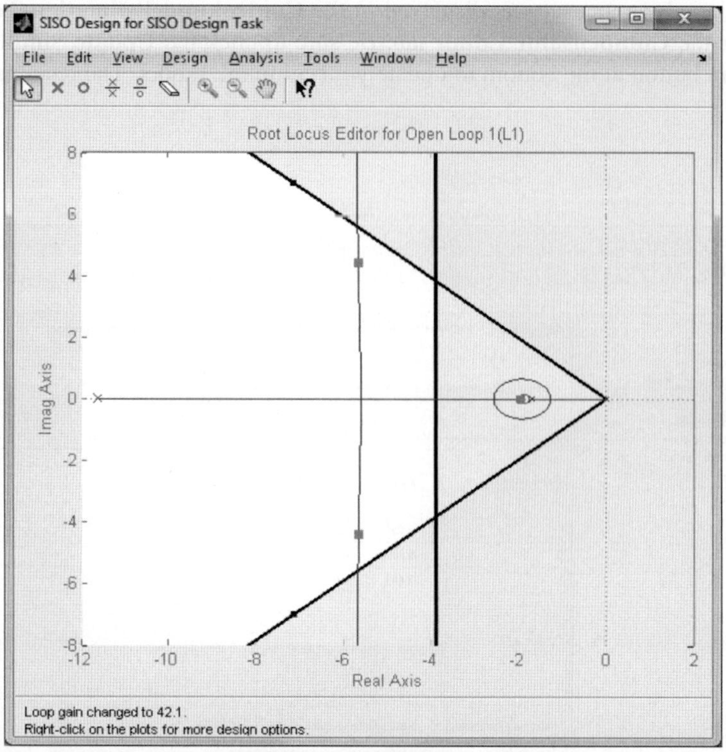


(*a*)

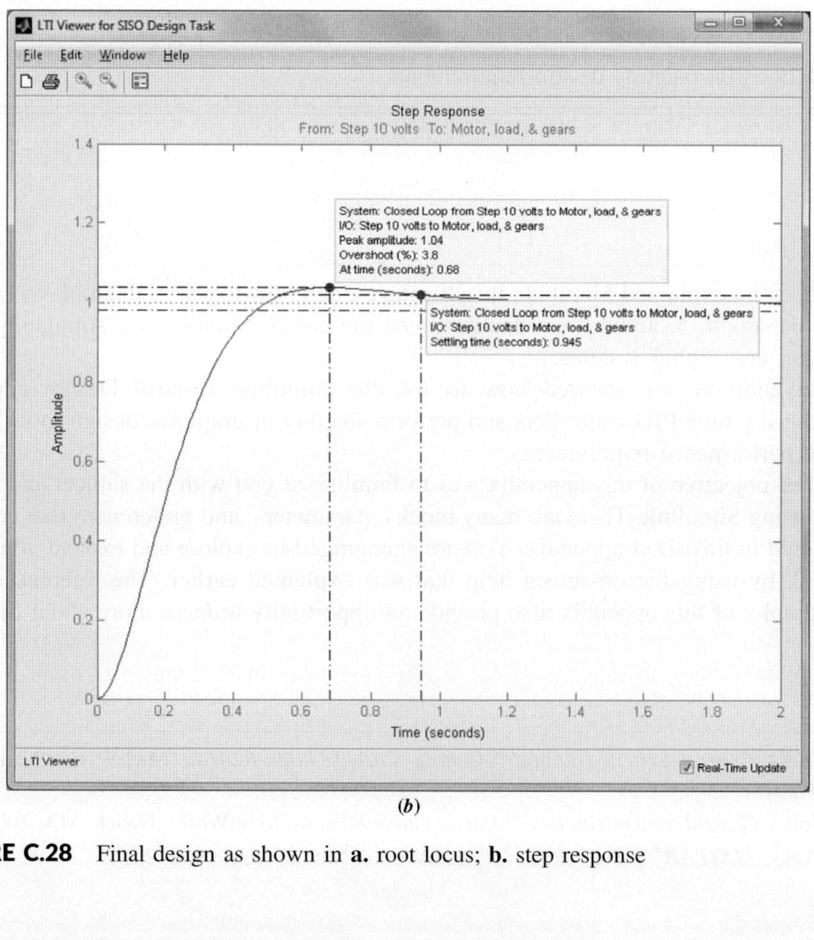


(*b*)

**FIGURE C.28** Final design as shown in **a.** root locus; **b.** step response

The designed compensator is displayed in the **Control and Estimation Tools Manager** under the **Compensator Editor** tab. Be sure to click the **Update Simulink Block Parameters** button at the bottom to update your Simulink model. The result is shown in Figure C.29.

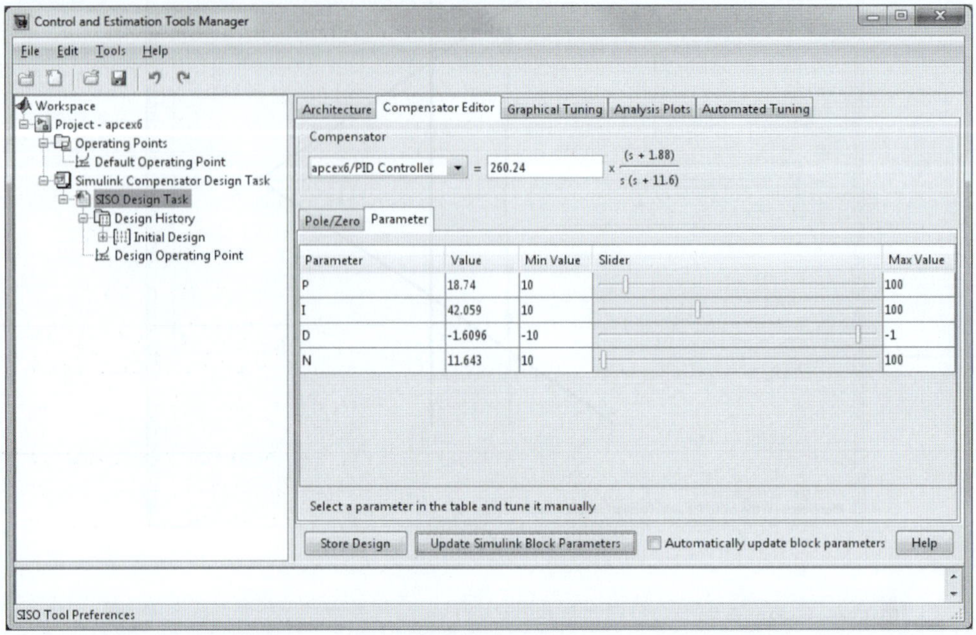


**FIGURE C.29** PID meeting design requirements

# Summary

This appendix explained Simulink, its advantages, and how to use it. Examples were taken from Chapters 4, 5, and 13 and demonstrated the use of Simulink for simulating linear, nonlinear, and digital systems.

In addition, we showed how to use the Simulink Control Design add-on to automatically tune PID controllers and perform shaping of graphical design tools in order to meet performance requirements.

The objective of this appendix was to familiarize you with the subject and get you started using Simulink. There are many blocks, parameters, and preferences that could not be covered in this short appendix. You are encouraged to explore and expand your use of Simulink by using the on-screen help that was explained earlier. The references in the Bibliography of this appendix also provide an opportunity to learn more about Simulink.

# Bibliography

MathWorks. *Control System Toolbox™ Getting Started Guide R2014a*. MathWorks, Natick, MA, 2000–2014.

MathWorks. *Control System Toolbox™ User's Guide R2014a*. MathWorks, Natick, MA, 2001–2014.

MathWorks. *MATLAB® Primer R2014a*. MathWorks, Natick, MA, 1984–2014.

MathWorks. *MATLAB® Graphics R2014a*. MathWorks, Natick, MA, 1984–2014.

MathWorks. *MATLAB® Mathematics R2014a*. MathWorks, Natick, MA, 1984–2014.

MathWorks. *MATLAB® Programming Fundamentals R2014a* . MathWorks, Natick, MA, 1984–2014.

MathWorks. *Simulink® Getting Started Guide R2014a*. MathWorks, Natick, MA, 1990–2014.

MathWorks. *Simulink® User's Guide R2014a*. MathWorks, Natick, MA, 1990–2014.

MathWorks. *Simulink® Control Design™ Getting Started Guide R2014a*. MathWorks, Natick, MA, 2004–2014.

MathWorks. *Simulink® Control Design™ User's Guide R2014a*. MathWorks, Natick, MA, 2004–2014.

*All MATLAB and Simulink screenshots in this book are reprinted with permission from The MathWorks, Inc.

# Appendix D: LabVIEW Tutorial

## D.1 Introduction

LabVIEW is a programming environment that is presented here as an alternative to MATLAB. Although not necessary, the reader is encouraged to become acquainted with MATLAB before proceeding, since familiarity with MATLAB can enhance the understanding of the relationship between textual (MATLAB) and graphical (LabVIEW) programming languages and extend the functionality of LabVIEW. In this tutorial, we will show how to use LabVIEW to **(1)** analyze and design control systems, and **(2)** simulate control systems. This appendix was developed using LabVIEW 2013.

LabVIEW is a graphical programming environment that produces virtual instruments (VI's). A VI is a pictorial reproduction of a hardware instrument on your computer screen, such as an oscilloscope or waveform generator. The VI can consist of various controls and indicators, which become inputs and outputs, respectively, to your program. Underlying each control and indicator is an associated block of code that defines its operation. The LabVIEW model thus consists of two windows: **(1) Front Panel**, which is a replica of the hardware front panel showing the controls and indicators, and **(2) Block Diagram**, which contains the underlying code for the controls and indicators on the **Front Panel**.

Associated with the **Front Panel** window is a **Controls** palette window containing numerous icons representing controls and indicators. The icons can be dragged onto a **Front Panel** window to create that control or indicator. Simultaneously, the associated code block is formed on the **Block Diagram** window.

Alternately, the block diagram can be formed first, and then the front panel is created from the block diagram. Associated with the **Block Diagram** window is a **Functions** palette window containing numerous icons representing a wide range of functions. Icons can be dragged onto a **Block Diagram** window to create that code block.

For example, Figure D.1(*a*) is the front panel of a signal generator. The generator consists of a control to select the signal type and a waveform graph that shows the output waveform. Figure D.1(*b*) shows the underlying code, which is contained in the code blocks. Here, the signal type selector is a control, while the waveform graph is an indicator. Later we will show how to make connections to other VI's. The palette windows for the front panel and block diagram are shown respectively in Figures D.1(*c*) and (*d*).

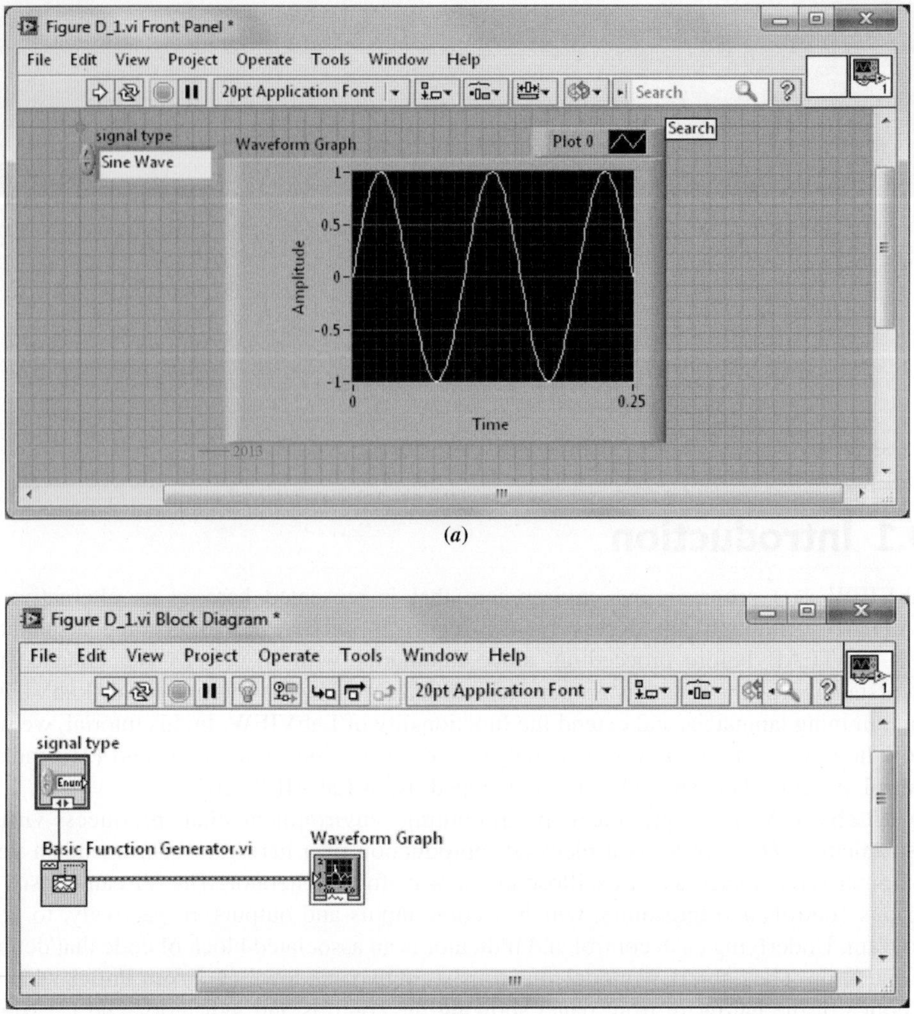


(*a*)

(*b*)

**FIGURE D.1** A LabVIEW function generator VI: **a. Front Panel** window; **b. Block Diagram** window; (*figure continues*)

# D.2 Control Systems Analysis, Design, and Simulation

LabVIEW can be used as an alternative to or in conjunction with MATLAB to analyze, design, simulate, build, and deploy control systems. In addition to LabVIEW, you will need the LabVIEW Control Design and Simulation Module. Finally, as an option that will be explained later, you may want to install the MathScript RT Module.

Analysis and design can be thought of as similar to writing MATLAB code, while simulation can be thought of as similar to Simulink. In LabVIEW, analysis and design, as opposed to simulation, are handled from different subpalettes of the **Functions** window's **Control Design and Simulation** palette. See Figure D.1(*d*). Analysis and design, and simulation will typically begin with the **Block Diagram** window, where icons representing code blocks will be interconnected. Parameters used by the code can be conveniently selected, changed, and passed to the code through VI controls on the **Front Panel** window created from the code icons. Any results, such as time response, can be displayed through VI indicators on the **Front Panel** window created from the code icons.

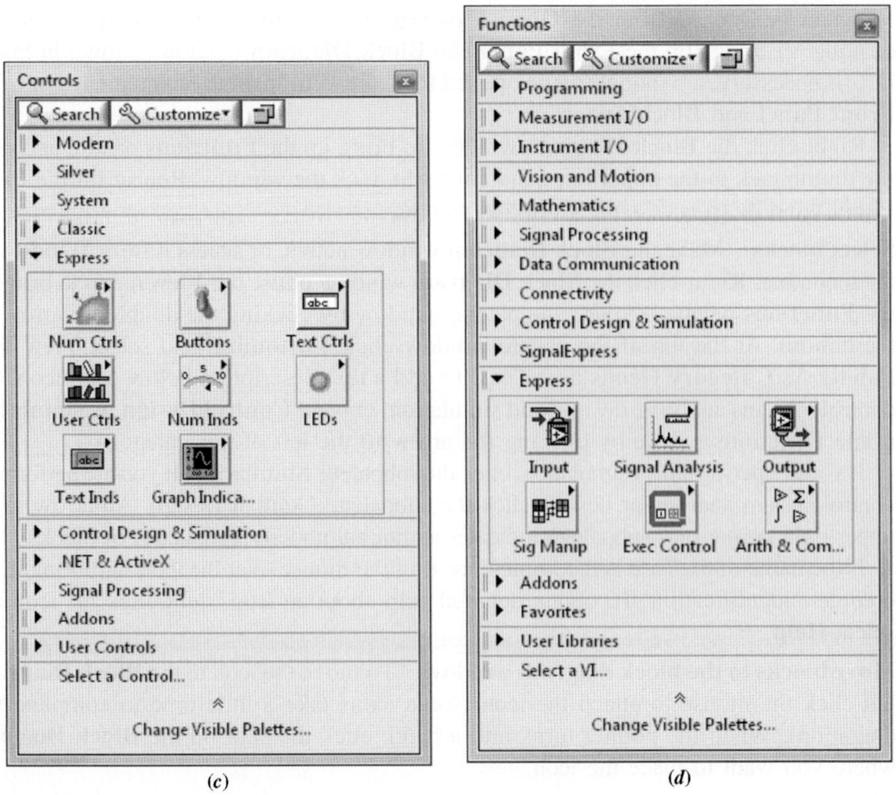


**FIGURE D.1** (*Continued*) **c. Controls** palette; **d. Functions** palette

# D.3 Using LabVIEW

The following steps start you on your way to using LabVIEW for control systems analysis, design, and simulation. These steps will be illustrated in the examples that follow.

1. **Start LabVIEW** LabVIEW starts with the window shown in Figure D.2, where you can select a **New VI** or **Open** an existing VI from the **File** menu. Alternatively, existing

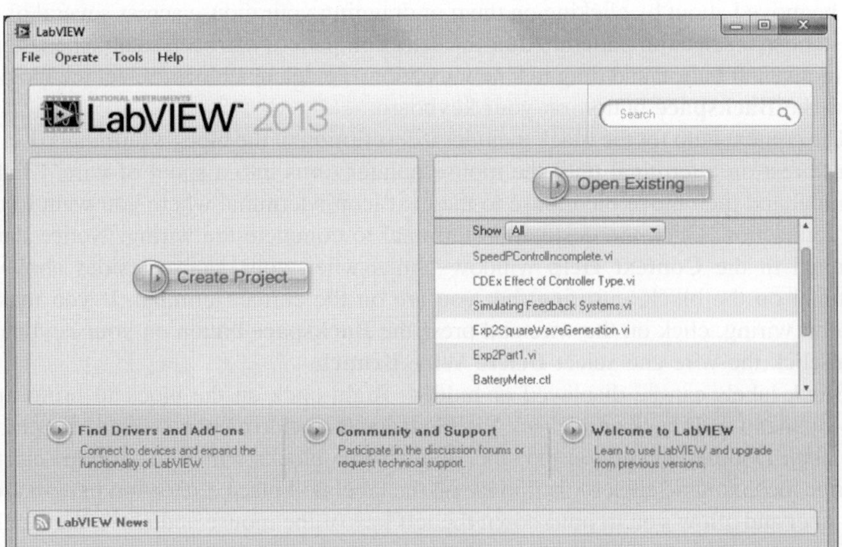




**FIGURE D.2** LabVIEW window

VI's can be opened from the **Open Existing** table on the right. Selecting a new or existing VI brings up the **Front Panel** and **Block Diagram** windows shown in Figure D.1. If necessary, a window can be opened from the **Window** tab on the menu bar of the **Front Panel** and **Block Diagram**.

Right-click the **Block Diagram** window to bring up the **Functions** palette and click the thumb tack in the upper left-hand corner to dock the window. Repeat for the **Front Panel** window to access the **Controls** palette.

2. **Select blocks** Make the **Block Diagram** window active, or access it from **Window** on the menu bar. Right-click the **Block Diagram** window or use the **View** menu to bring up the **Functions** palette. Expand the palette window by clicking the double-up arrows at the bottom. At the top of the palette window click **Customize**, and select **View This Palette As/Category (Icons and Text)** to add a text description below each icon. For control systems analysis, design, and simulation, expand **Control Design & Simulation** in the **Functions** palette by clicking the arrow to the left of this category.

   If you are performing a simulation, click the subpalette **Simulation**. If you are performing control system analysis or design, click the subpalette **Control Design**. An arrow in the upper-right corner of a subpalette indicates additional underlying palettes or blocks.

   If the name of the icon is incomplete, resting the mouse over the icon will bring up its complete identification. To obtain detailed help about an icon, right-click the icon and select **Help**.

3. **Move blocks to the block diagram window** To move the icon to the **Block Diagram**, left-click the mouse to attach the icon (some icons take a little time to complete this operation). When the pointer turns into a hand, click the spot on the **Block Diagram** where you want to place the icon.

4. **Obtain information about the block** You will now want to obtain information about how to interconnect the block to other blocks and pass parameters to the block as well as other characteristics about the block. Select the yellow question mark at the right of the **Block Diagram** toolbar to turn on the **Context Help window**. This window will provide help about a particular icon if you rest your mouse over that icon. Additional help is available under the **Help** menu on the **Block Diagram** menu bar. Finally, right-click the icon to bring up a menu with additional choices, such as **Properties**, if any. In particular, you will use this menu to create the block's front panel's controls and indicators. This front panel will be your interface with the block to choose parameters and see responses.

5. **Interconnect and label blocks** Once blocks are placed on the **Block Diagram** they can be moved about by clicking on them or dragging your mouse across several of them to establish a selection pattern. After the selection pattern has been established, depress the mouse left button and drag to a new location. To delete a block, select the block and press the **Backspace** button on your keyboard.

   The context help for the block includes a description of the block's terminals. Let your mouse rest on a terminal until the mouse pointer turns into a spool of wire. Click the terminal and then move the mouse to the next icon's terminal where you want to make the connection. Click the destination terminal to complete the wiring. Notice that the terminal in the **Context Help** window blinks when your mouse resides above that terminal on the block, ensuring that you are on the correct terminal. If you make an error in wiring, click on the wire and press the **Backspace** button on your keyboard or right-click the wire and select **Delete Wire Branch**.

   Block labels can be displayed or hidden. Right-click on the block to bring up the pop-up menu and check or uncheck **Visible Items/Label** to display or hide, respectively, the label. Double-clicking on the label above some blocks will allow you to select and change the text. One click of the mouse on the label will place a selection pattern around the label and allow you to hold down the left key of the mouse and move the label to a different location.

6. **Create the interface to your block** You will now want to create the interface to your block in order to control or select functions, specify parameters, or view responses. This interface will be accessed via the **Front Panel** window. Right-click a terminal on a block for which you want to create an interface. On the pop-up menu, choose **Create/Control** to be able to interact with the block or **Create/Indicator** to view a response or setting.
7. **Set the controls** Switch to the **Front Panel** window and set your controls. For example, enter parameter values, select functions, etc. If you want to change values and at some future time return to the current values, click on **Edit** on the **Block Diagram** menu bar and select **Make Current Values Default**. To return to the default values in the future, click on **Edit** on the **Block Diagram** menu bar and select **Reinitialize Values to Default**.
8. **Run the program** Click on the arrow at the left of the toolbar on either the **Block Diagram** or **Front Panel** window to run the program. The program can be run continuously by clicking the curved arrows button on the toolbar second from the left. Continuously running your program permits changing functions and parameter values during execution.

   In order to identify the buttons, let your mouse rest on a button to bring up a context menu. Stop your simulation by pressing the red-dot button, third from the left. If you are performing control systems analysis and design, another way to continuously run the program is to place a **While Loop** around your block diagram. The loop is available in the **Functions** palette at **Express/Execution Control/While Loop**. This loop also places a **Stop** button on the **Front Panel**. The program executes until you press the stop button. In lieu of the **Stop** button, any true/false Boolean can be wired to the condition block (red dot) created inside the **While Loop**.

   If you are performing simulation, you can use a **Simulation Loop** available in the **Functions** palette at **Control Design and Simulation/Simulation/Control & Simulation Loop**. Place the **Control & Simulation Loop** around your simulation block diagram by dragging the mouse. Right-click on the **Control & Simulation Loop** outline and choose **Configure Simulation Parameters** . . . to determine the parameters for executing the simulation. The **Front Panel** indicators and controls are also configurable. Right-click on the indicator or control and select **Properties**.

# D.4 Analysis and Design Examples

In this section, we will present some examples showing the use of LabVIEW for the analysis and design of control systems. In the next section, examples of the use of LabVIEW for simulation will be presented.

Analysis and design examples use icons selected from the **Control Design** subpalette under the **Control Design and Simulation** palette. In the next section showing examples of simulation, we will use icons taken from the **Simulation** subpalette under the **Control Design and Simulation** palette.

## Example D.1 Open-Loop Step Response

Analysis and design usually begins by selecting icons from the **Control Design** subpalette and dragging them to the **Block Diagram** window. The icons represent blocks of code and the cascading of code blocks can be thought of as a sequence of lines of code. Thus, an advantage of LabVIEW over MATLAB is that the programmer does not need to memorize coding language. For example, consider the MATLAB code shown in TryIt D.1 that produces the step response of $G(s) = 100/(s^2 + 2s + 100)$.

**TryIt D.1**

```
numg=100;
deng=[1 2 100];
'G(s)'
G=tf(numg,deng)
step(G);
title('Angular Velocity')
```

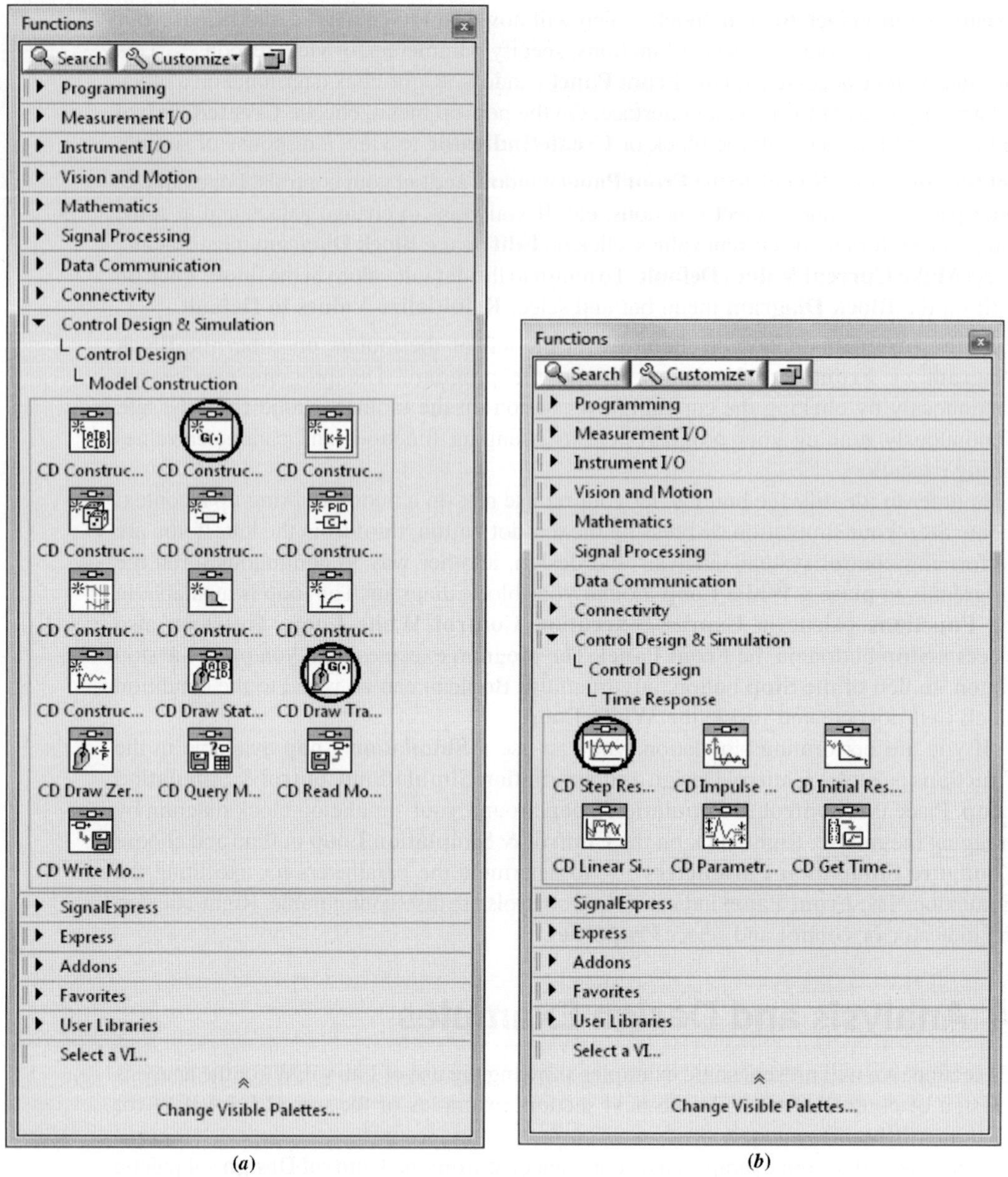


**FIGURE D.3** Selecting **a. CD Construct . . .** and **CD Draw . . .**; **b. CD Step Response . . .**

This step response can be produced in LabVIEW without knowing any coding language. We now demonstrate by following each step of Section D.3:

1. **Start LabVIEW** Start LabVIEW and select **New VI** from the **File** menu shown in Figure D.2.
2. **Select blocks** From the **Functions** palette, select the blocks shown in Figure D.3(*a*) and (*b*).
3. **Move/blocks to the Block Diagram window** Drag your icons one at a time to the **Block Diagram** window, Figure D.4.
4. **Obtain information about the block** Right-click each of the blocks and be sure the first two items under **Visible Items** are checked. Look at the **CD Construct Transfer Function Model.vi.** A **Polymorphic VI Selector** is shown at the bottom of the icon.

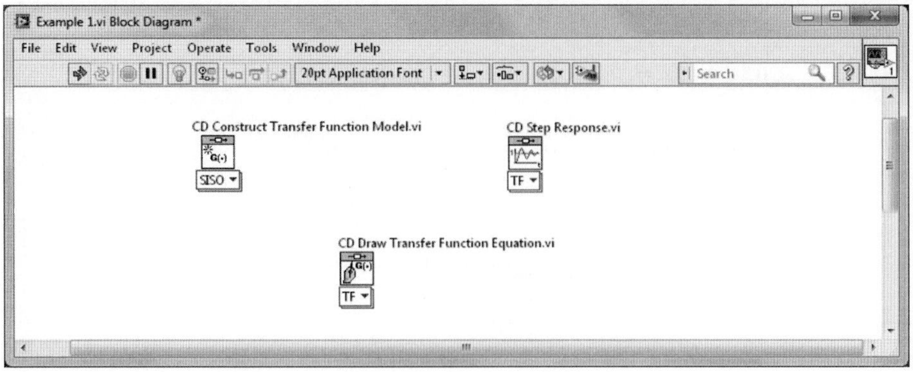


**FIGURE D.4** **Block Diagram** window

Click the selector to bring up the menu. Select **SISO**. This block effectively creates the transfer function shown in the first four steps of the MATLAB code in TryIt D.1.

Repeat for the **CD Draw Transfer Function Equation.vi** and select **TF** from the **Polymorphic VI Selector**. This block will write the transfer function symbolically in the display. Your selection from the polymorphic vi selector should match the format of the transfer function created by the **CD Construct Transfer Function Model.vi**.

Repeat for the **CD Step Response.vi.** and select **TF** from the **Polymorphic VI Selector**. This block will collect the data for the step response and permit plotting the data. This block effectively creates the last two commands of the MATLAB code shown in TryIt D.1.

5. **Interconnect and label blocks** You should now have the **Block Diagram** window shown in Figure D.4. Interconnect the code blocks. Click on the question mark on the right side of the toolbar to bring up the context menu. As your mouse passes above an icon, its context menu appears, showing the terminals. See Figure D.5. Interconnect the terminals by letting the mouse rest on a terminal until it becomes a spool of wire.

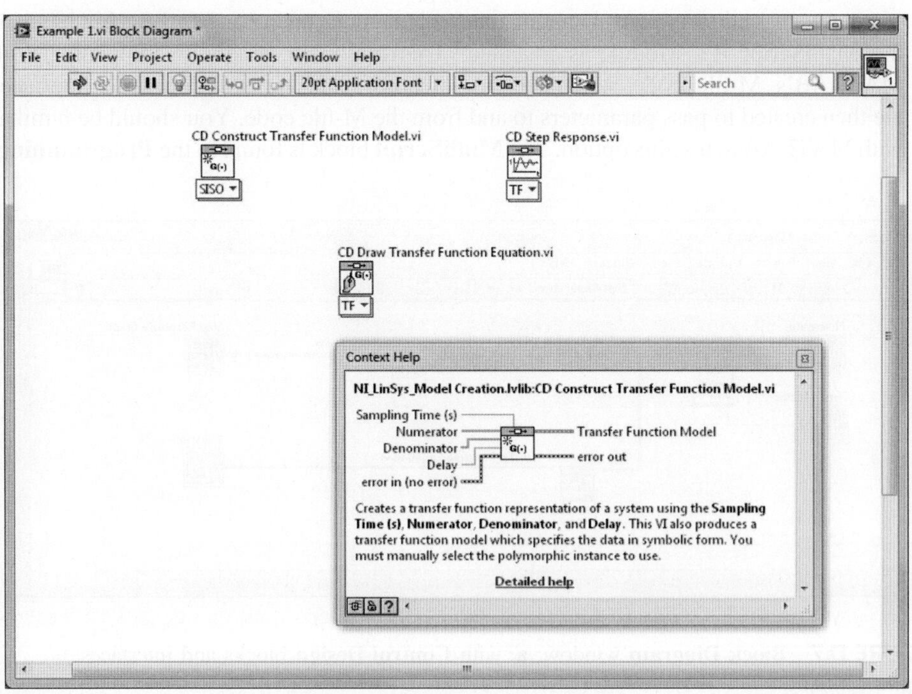


**FIGURE D.5** **Context Help** for **CD Construct Transfer Function Model.vi**

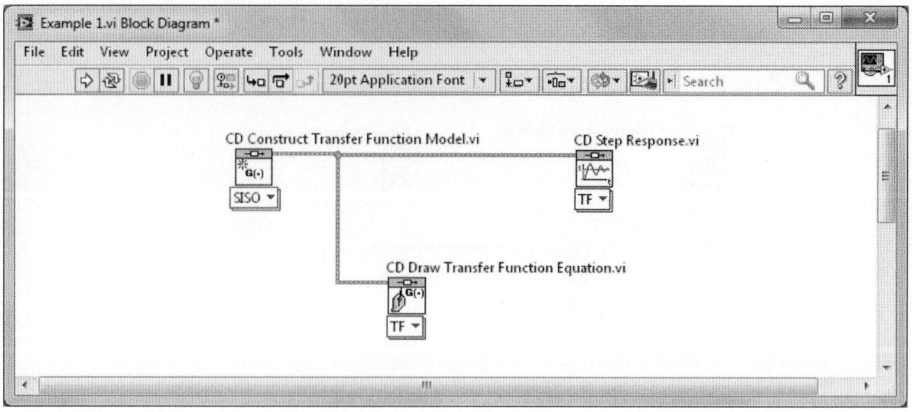


**FIGURE D.6** Interconnected blocks

Click on the terminal and then click on the destination terminal. The two terminals will appear as wired together. Continue wiring terminals until you have the **Block Diagram** window shown in Figure D.6. Mid-wire connections as shown can be made by letting your mouse rest at the connection point until it becomes a spool of wire.

6. **Create the interface to your block** You will now want to create the interface to specify parameters and view responses. This step will create the interface that will be accessed on the **Front Panel** window. The interfaces we will create are:
   - **CD Construct Transfer Function Model.vi** input parameter controls. Right-click on the numerator terminal shown in Figure D.5 and select **Create/Control**. Repeat for the denominator.
   - **CD Step Response.vi** response plot indicator. Right-click on the **Step Response Graph** terminal and select **Create/Indicator**.
   - **CD Draw Transfer Function Equation.vi** symbolic transfer function indicator. Right-click on the **Equation** terminal and select **Create/Indicator**. Your **Block Diagram** should now look similar to Figure D.7(*a*).

   As an option, you can create transfer functions using a **MathScript** block if the MathScript RT Module is installed. This option is generally compatible with MATLAB's M-file code statements for creating your transfer function. Interfaces are then created to pass parameters to and from the M-file code. You should be familiar with MATLAB to use this option. The **MathScript** block is found in the **Programming/**

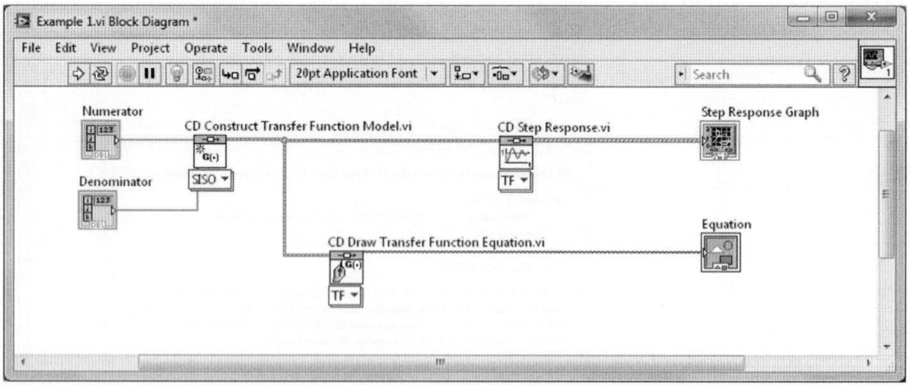


(*a*)

**FIGURE D.7** **Block Diagram** window, **a.** with **Control Design** blocks and interfaces. (*figure continues*)

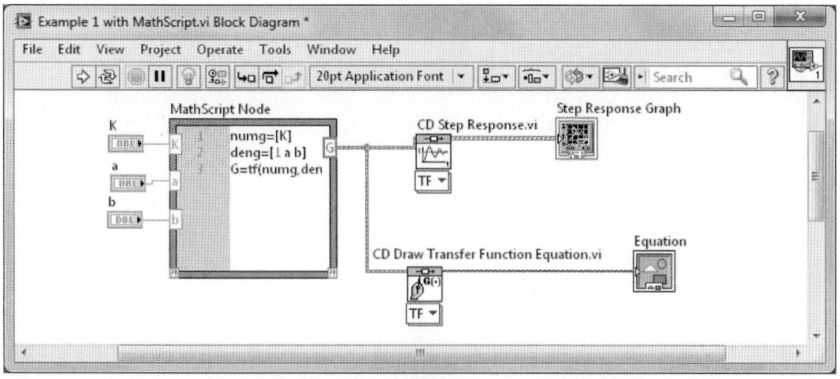


(*b*)

**FIGURE D.7** (*Continued*) **b.** with **MathScript** block

**Structures/MathScript** palette. You create M-file code inside the **MathScript** block. Inputs, outputs, and controls are created as follows. Right-click on the left side of the **MathScript Node** and select **Add Input**. Name the input **K**. Right-click your terminal **K** and select **Create/Control**. A control is formed on both the **Block Diagram** and **Front Panel**. Repeat the same process to create inputs and controls for parameters **a** and **b**. Now create the output to the **MathScript Node**. Right-click the right-hand side of the **MathScript Node** and select **Add Output/G**. After wiring your inputs and outputs, your Block Diagram will be that shown in Figure D.7(*b*).

On the **Block Diagram** window menu bar, select **Window/Show Front Panel**. You will see the **Front Panel** shown in Figure D.8 created by your interfaces. You can double-click the labels above your interfaces either in the **Front Panel** window or the **Block Diagram** window to change the label to be more descriptive of your project.

7. **Set the controls** Using the **Front Panel** window, enter polynomial coefficients for the numerator and denominator in ascending order—lowest to highest. The selector to the left of the numerator and denominator shows the power of $s$ for the left-most coefficient. Increasing the counter allows entry of higher-order coefficients not visible originally. To make all coefficients of a polynomial visible, let the mouse move on the right-hand edge of the polynomial indicator until the pointer becomes a double arrow

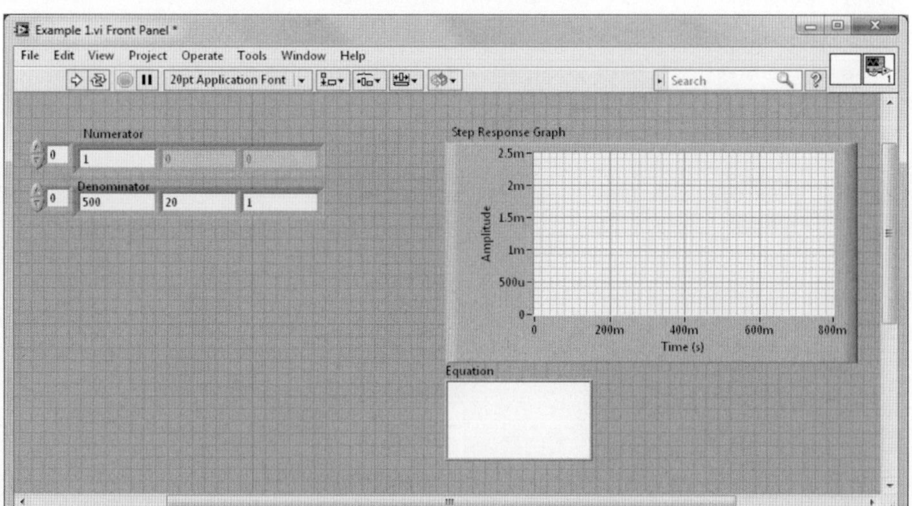


(*a*)

**FIGURE D.8** **Front Panel: a.** for **Block Diagram** shown in Figure D.7(*a*); (*figure continues*)

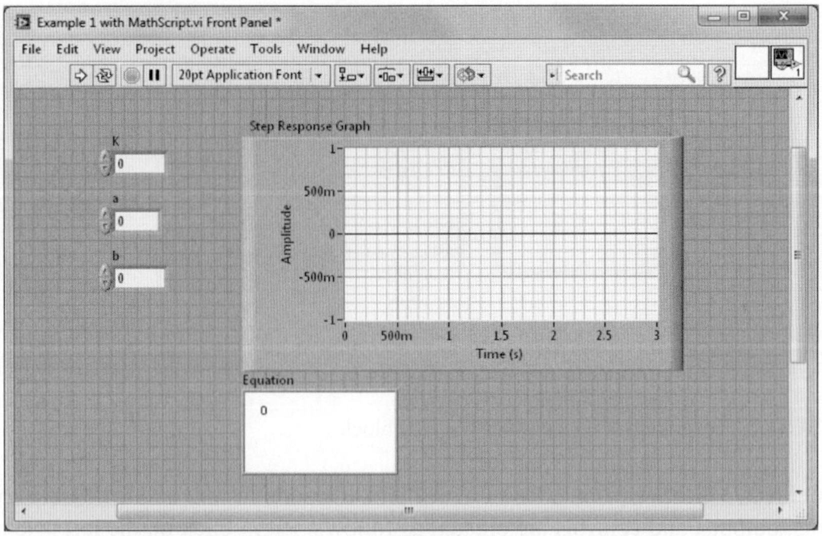


(*b*)

**FIGURE D.8** (*Continued*) **b.** for **Block Diagram** shown in Figure D.7(*b*)

and blue dots appear at the left and right edges of the entire polynomial indicator. You can then drag the right blue dot to expose more cells.

Familiarize yourself with the choices on the menu bar as well as those on the pop-up menus created when you right-click on any indicator or control. For example, under the **Edit** menu, among other choices, you can **Make Current Values Default** or **Reinitialize Values to Default**. Right-clicking the indicators or controls brings up a menu from which, among other choices, **Properties** can be selected to configure the indicator or control as desired.

8. **Run the program** Figure D.9 shows Example D.1 after execution. The figure shows the values entered, the equation, and the step response. Execution was initiated by clicking the arrow at the left of the toolbar.

   The program can run continuously by clicking the curved arrows on the toolbar. Now, change values; hit the **Enter** key and see the results immediately displayed. Stop the program execution by clicking on the red hexagon on the toolbar. Another way of

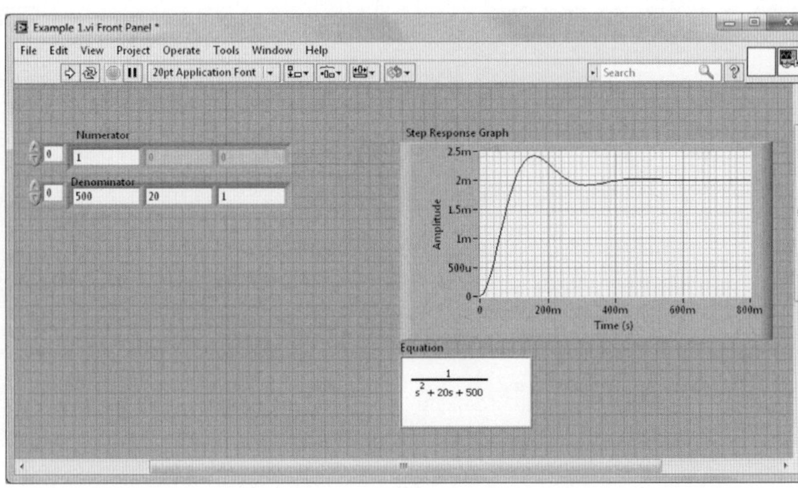


(*a*)

**FIGURE D.9** **Front Panel** after execution: **a.** for block diagram in Figure D.7(*a*); (*figure continues*)

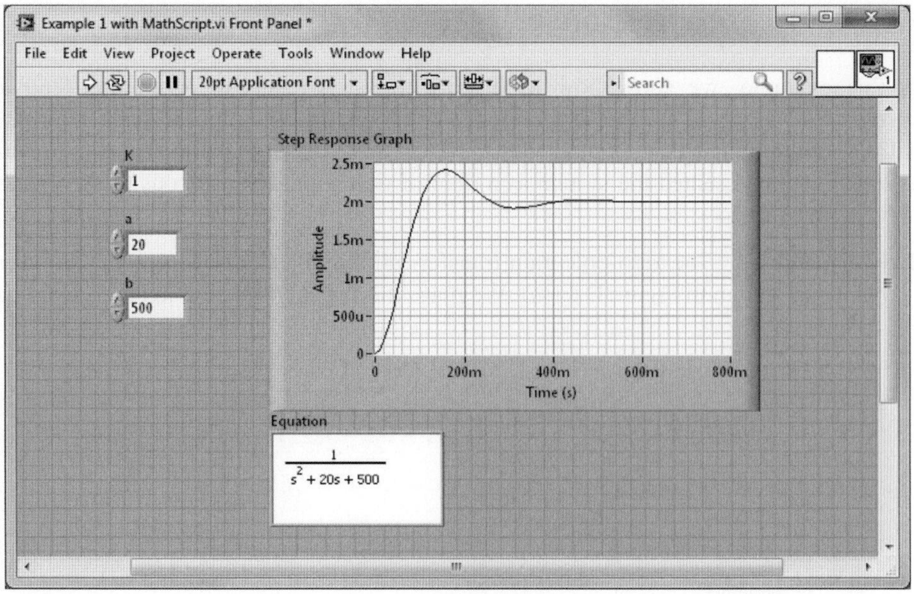


(*b*)

**FIGURE D.9** (*Continued*) **b.** for block diagram in Figure D.7(*b*)

continuously running the program is to place a **While Loop** around the block diagram as shown in Figure D.10(*a*). The loop is accessed from **Functions/Express/Execution Control** as shown in Figure D.10(*b*). After selecting the **While Loop**, drag the cursor across the block diagram to create the continuous loop. A **stop** button will appear on the block diagram as well as on the **Front Panel**. At the lower right is a **Conditional Terminal** icon, which can be used to control the **While Loop**. The reader should consult the on-line documentation for further information.

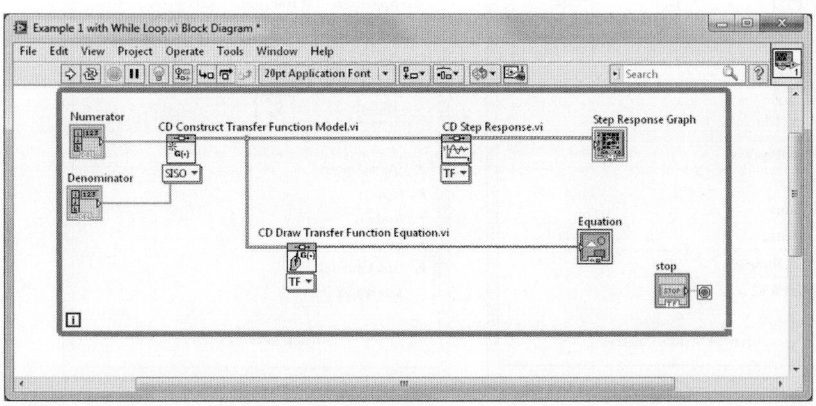


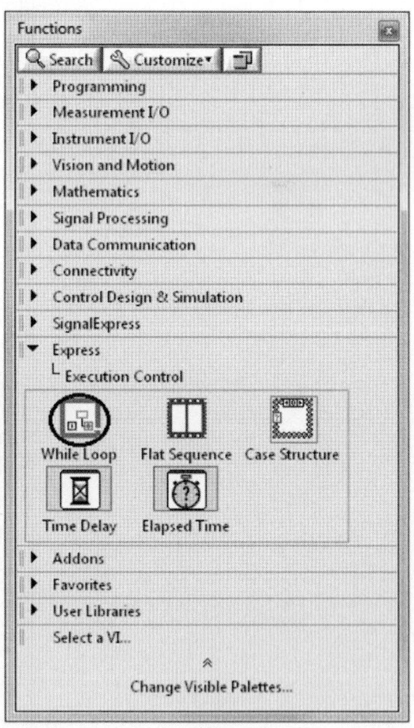


(*a*) (*b*)

**FIGURE D.10** **a.** Block diagram with **While Loop**; **b. Functions** palette showing **While Loop** location

## Example D.2 Closed-Loop Step Response

In this example, we show how to display the step response of a unity-feedback system. For variety, we represent the open-loop system as a ratio of zeros over poles with a multiplying gain, analogous to MATLAB's **zpk** function. In the previous example, we represented the system as a ratio of polynomials, analogous to MATLAB's **tf** function.

1. **Select blocks** The zero-pole-gain transfer function is obtained from the **Functions** palette as shown in Figure D.11(*a*). We place this transfer function in the forward path

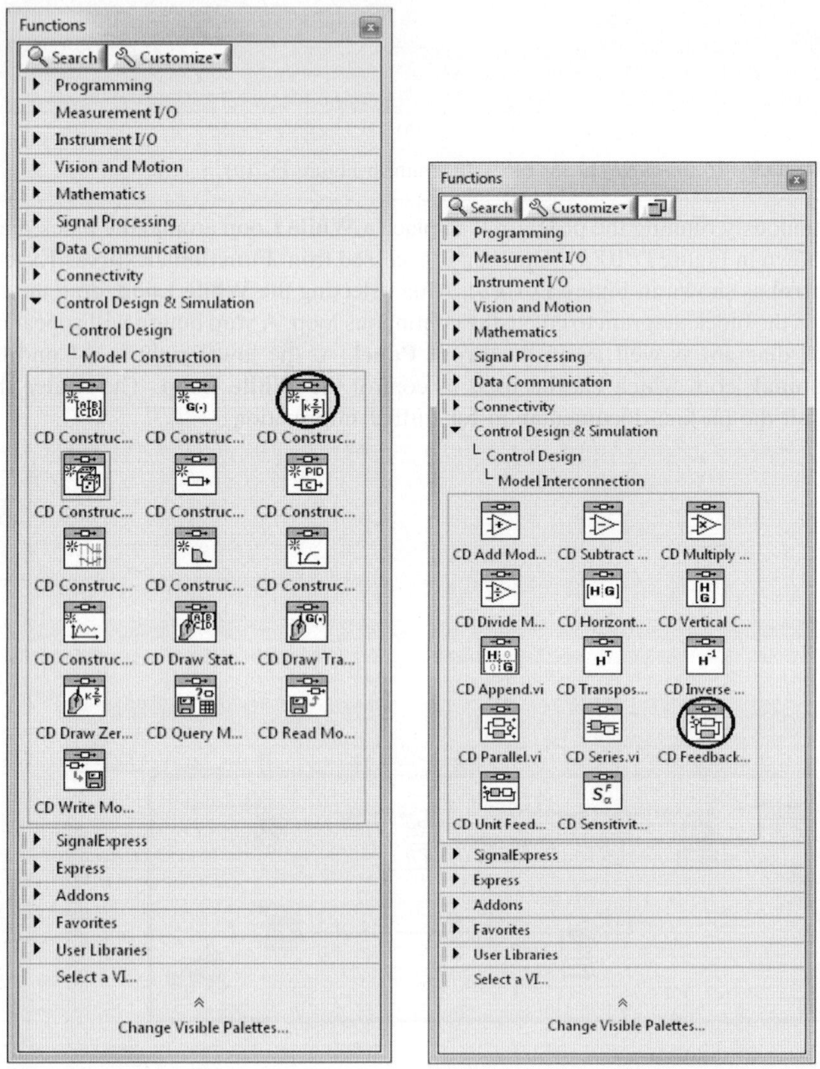


(*a*) (*b*)

**FIGURE D.11** **a.** Obtaining zero-pole-gain transfer function from the **Functions** palette; **b.** Obtaining **Feedback** interconnection from **Functions** palette

of a unity-feedback system by following its block with a **Feedback** block obtained from the **Functions** palette as shown in Figure D.11(*b*). If the **Model 2** input to the **Feedback** block is left unconnected, then a unity-feedback interconnection is assumed. Other options for interconnection, such as parallel and series, are shown on the palette of Figure D.11(*b*).

2. **Interconnect and label blocks** Producing the closed-loop step response is similar to Example D.1, except the step-response blocks are placed at the output of the **Feedback** block. The equation writer is wired to the system output as in Example D.1. All data types must be compatible and are shown selected with the pull-down menu at the base of the blocks. If you select **Automatic** in the pull-down menu, LabVIEW will select the correct form for you as you connect the blocks.

 The final **Block Diagram** and **Front Panel** for this example are shown in Figure D.12 (*a*) and (*b*), respectively. Notice that you enter open-loop poles, zeros, and gain on the **Front Panel** in place of polynomial open-loop numerator and denominator coefficients.

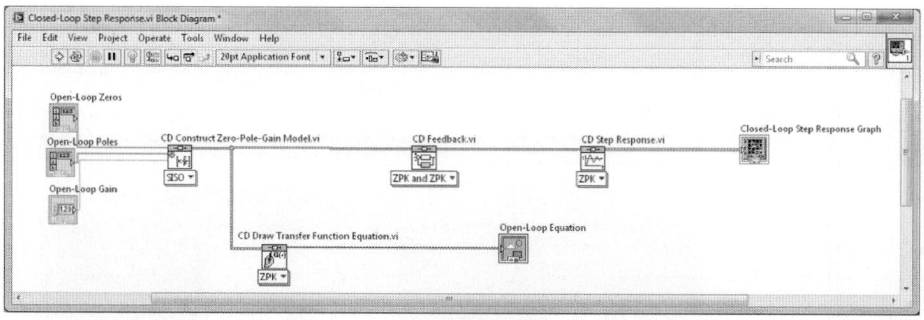


(*a*)

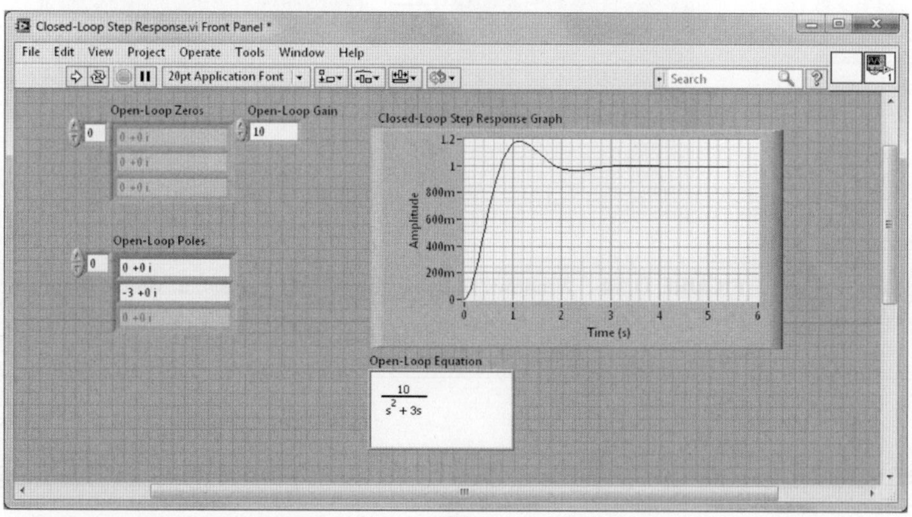


(*b*)

**FIGURE D.12** a. **Block Diagram** for Example D.2; b. **Front Panel** for Example D.2

## Example D.3 Root Locus Analysis and Design

We can obtain root locus plots by adding the **Root Locus** block obtained from the **Functions** palette as shown in Figure D.13. The **Root Locus** block is connected to the output of the open-loop system and a **Root Locus Graph** indicator is formed at the output of the **Root Locus** block. The resultant **Block Diagram** and **Front Panel** are shown in Figure D.14(*a*) and (*b*) respectively.

Figure D.13 shows other characteristic blocks that can be added. For example, closed-loop poles and zeros, as well as damping ratio and natural frequency, can be displayed.

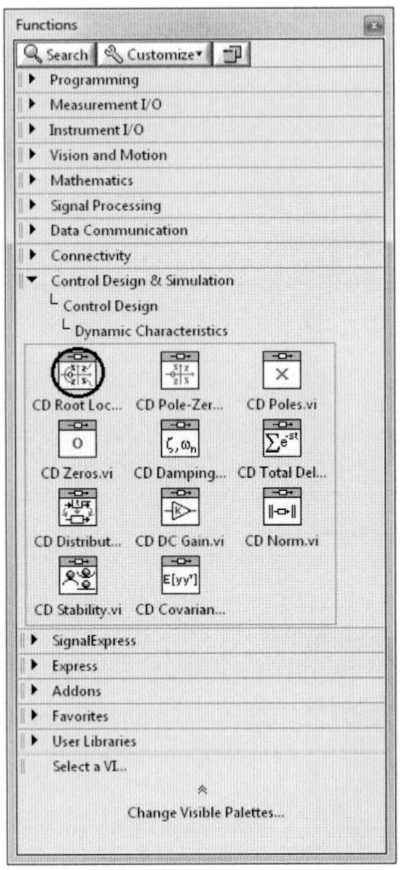


**FIGURE D.13** **Functions** palette showing location of **Root Locus** block

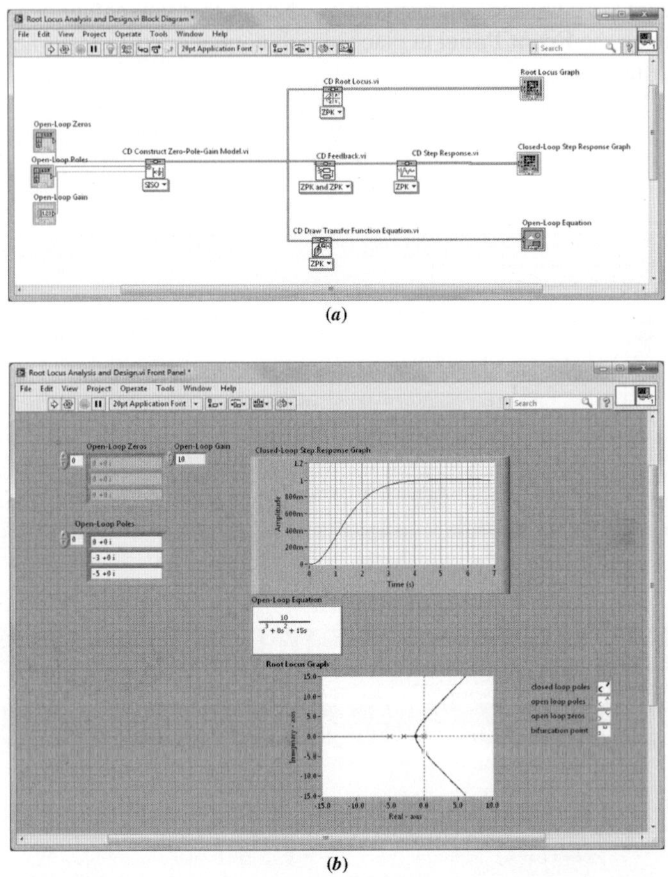


**FIGURE D.14** Windows showing root locus analysis: **a. Block Diagram**; **b. Front Panel**

## Example D.4 Open- and Closed-Loop Sinusoidal Frequency Analysis and Design

We can obtain open- and closed-loop sinusoidal frequency response curves by replacing the **Root Locus** block with the **Bode** block to yield the open-loop frequency response. A copy of the **Bode** block can be added at the output of the **Feedback** block to obtain the closed loop frequency response. Figure D.15 shows where to obtain the **Bode** block.

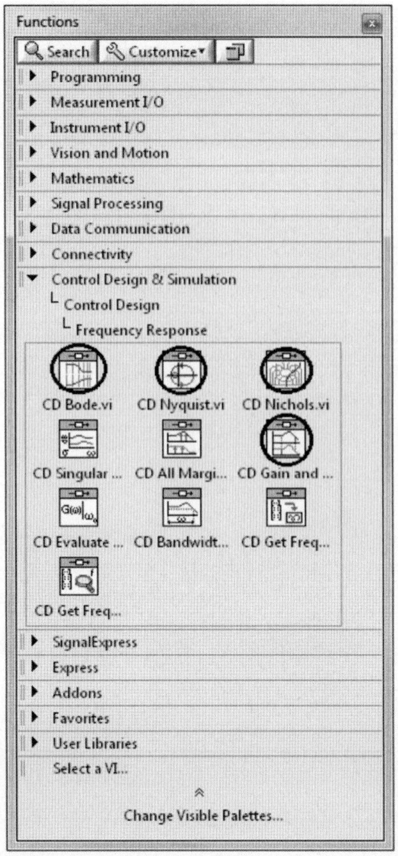


**FIGURE D.15** **Functions** window showing frequency response blocks, such as **Bode**, **Nyquist**, **Nichols**, and **Gain and Phase Margin** blocks

Figure D.16 shows the **Block Diagram** and **Front Panel** with open- and closed-loop Bode analysis. In order to display the plots, the indicators shown at the outputs of the **Bode** blocks were created.

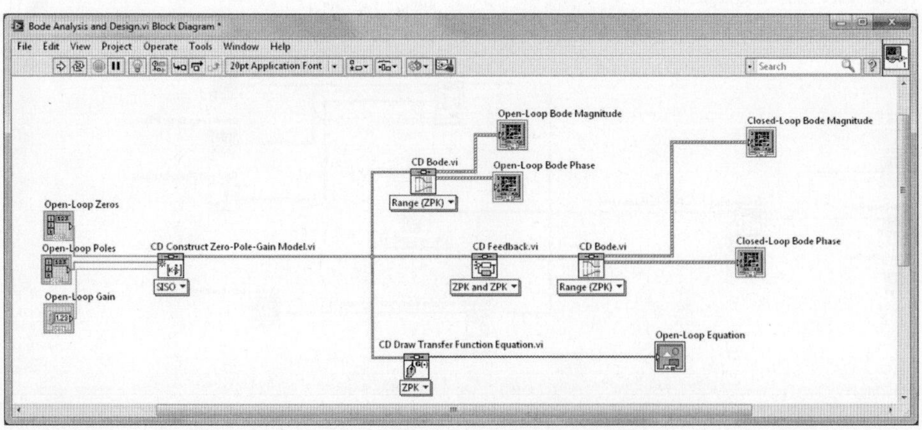


(*a*)

**FIGURE D.16** Bode analysis via LabVIEW: **a. Block Diagram**; (*figure continues*)

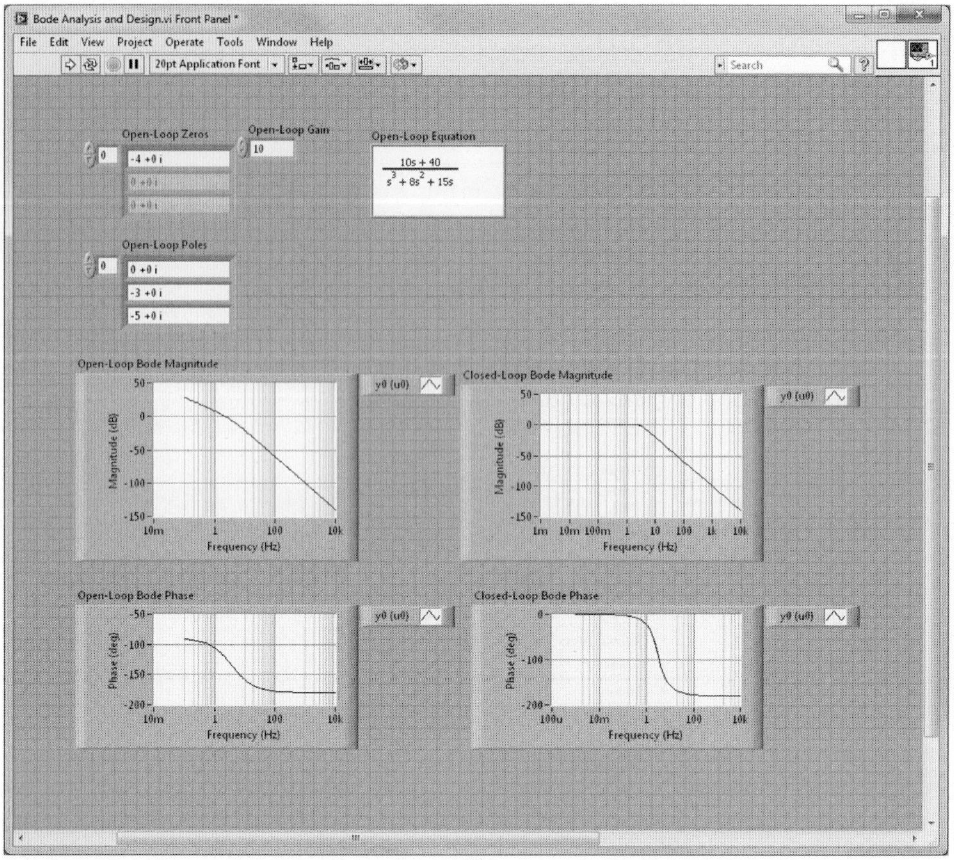


(*b*)

**FIGURE D.16** (*Continued*) **b. Front Panel**

Figure D.15 shows other alternatives for frequency response analysis. For example, in addition to the Bode plots, you can create an indicator telling you the gain and phase margins by using the **Gain and Phase Margin** block. Figure D.17 shows that result.

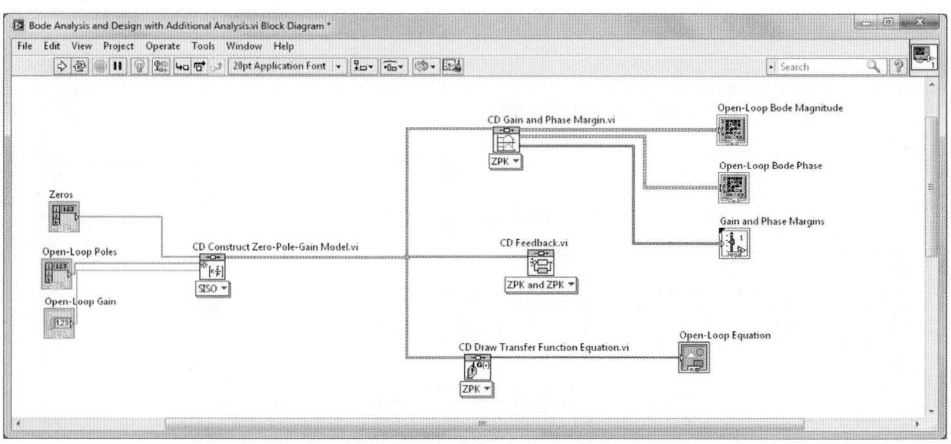


(*a*)

**FIGURE D.17** Bode analysis with gain and phase margin: **a. Block Diagram**; (*figure continues*)

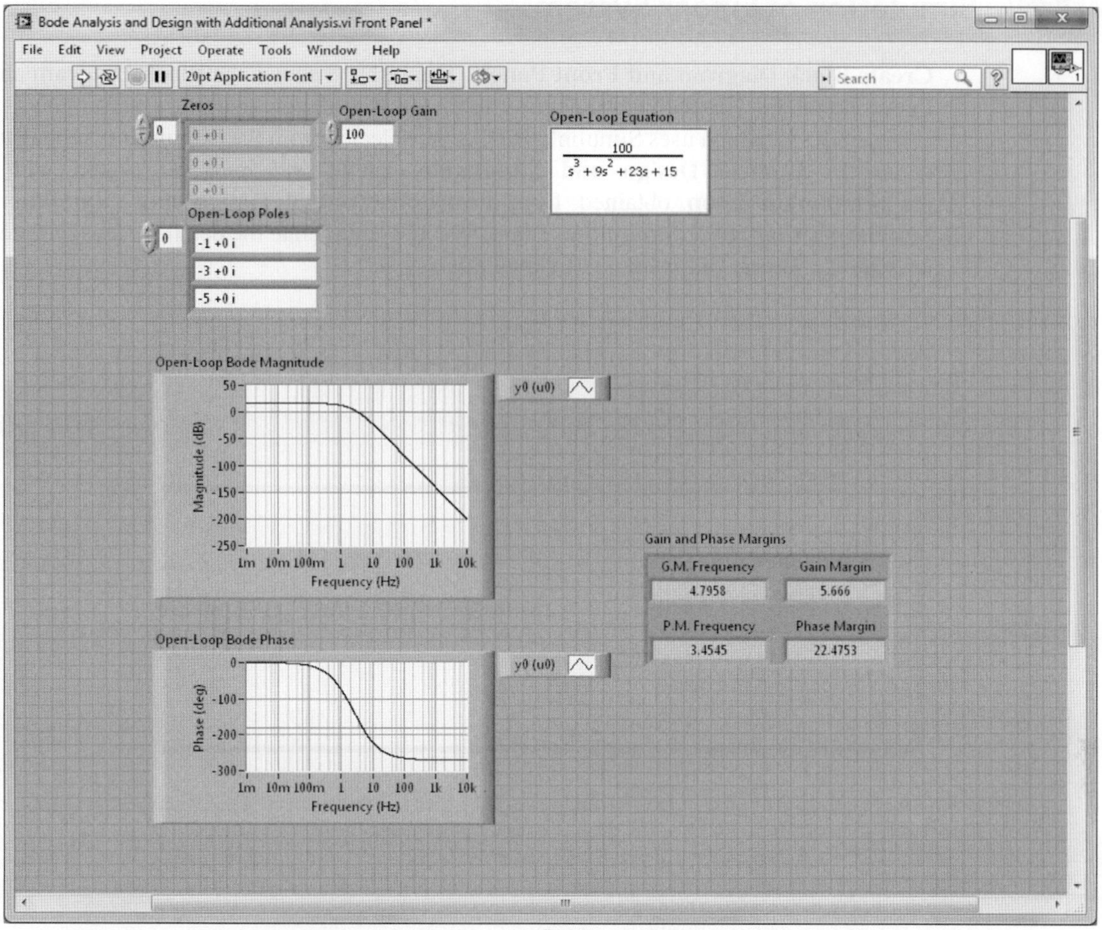


(*b*)

**FIGURE D.17** (*Continued*) **b. Front Panel**

Finally, if you need to use Nyquist or Nichols charts, the associated blocks are shown in Figure D.15 and can replace the **Bode** blocks.

# D.5 Simulation Examples

Whereas the LabVIEW block sequence for design and analysis is analogous to following the code statement sequence in a MATLAB M-file, the LabVIEW block sequence for simulation is analogous to following the block sequence of a Simulink diagram.

In this section, we show examples of simulation using LabVIEW. For control system simulation, icons for the block diagram are taken from the **Simulation** subpalette under the **Control Design and Simulation** palette. Our examples will parallel the examples shown in Appendix C which uses Simulink.

## Example D.5 Simulation of Linear Systems

**Create Block Diagram and Front Panel** Figure D.18 shows the **Block Diagram** and **Front Panel** for simulating a linear system. The simulation reproduces Example C.1 in Appendix C, which uses Simulink. Blocks are selected from the **Simulation** subpalette under the **Control Design and Simulation** palette and must be placed within the **Simulation Loop** obtained from **Functions/Control Design and Simulation/ Simulation/Control & Simulation Loop**. We now enumerate the detailed steps required to create the **Block Diagram** and **Front Panel:**

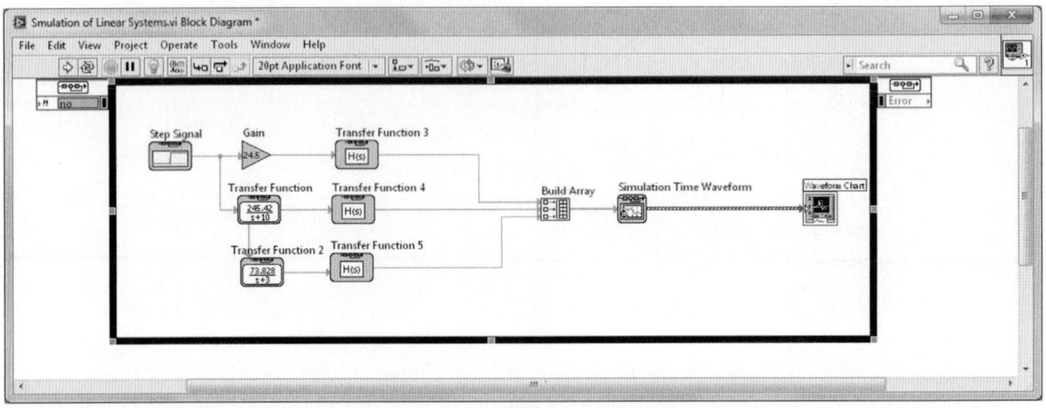


(*a*)

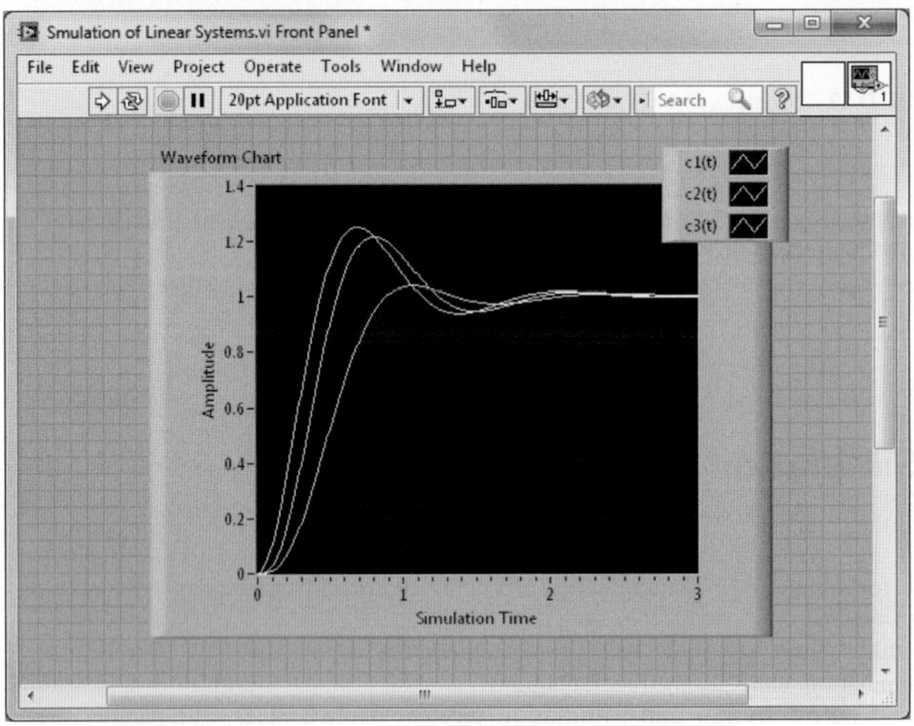


(*b*)

**FIGURE D.18** Simulation of linear systems: **a. Block Diagram**; **b. Front Panel**

1. Transfer functions are obtained from **Functions/Control Design and Simulation/Simulation/Continuous Linear Systems/Transfer Function**. Right-click on each transfer function and select **Configuration** to enter the parameter values shown in Figure D.18(*a*) or equivalently in Figure C.5.
2. The gain block is obtained from **Functions/Control Design and Simulation/Simulation/Signal Arithmetic/Gain**. Right-click on the gain block and select **Configuration** to enter the parameter value.
3. The step-input block is obtained from **Functions/Control Design and Simulation/Simulation/Signal Generation/Step Signal**. Right-click on the gain block and select **Configuration** to enter the parameter value.
4. In order to display the three step-response curves simultaneously, we use a **Build Array** block obtained from **Functions/Programming/Array/Build Array**. Drag the bottom of the icon to expose the correct number of inputs (three for this case).
5. To create the display, we use the **Simulation Time Waveform** block obtained from **Functions/Control Design and Simulation/Simulation/Graph Utilities/Simtime Waveform**. Right-click the output of the **Simtime Waveform** block and select **Create/Indicator** to produce the **Waveform Chart** icon and the **Front Panel** display.

**Configure simulation loop** Finally, set the simulation parameters by right-clicking the **Simulation Loop** and selecting **Configure Simulation Parameters**. . . . Set the parameters as shown in Figure D.19.

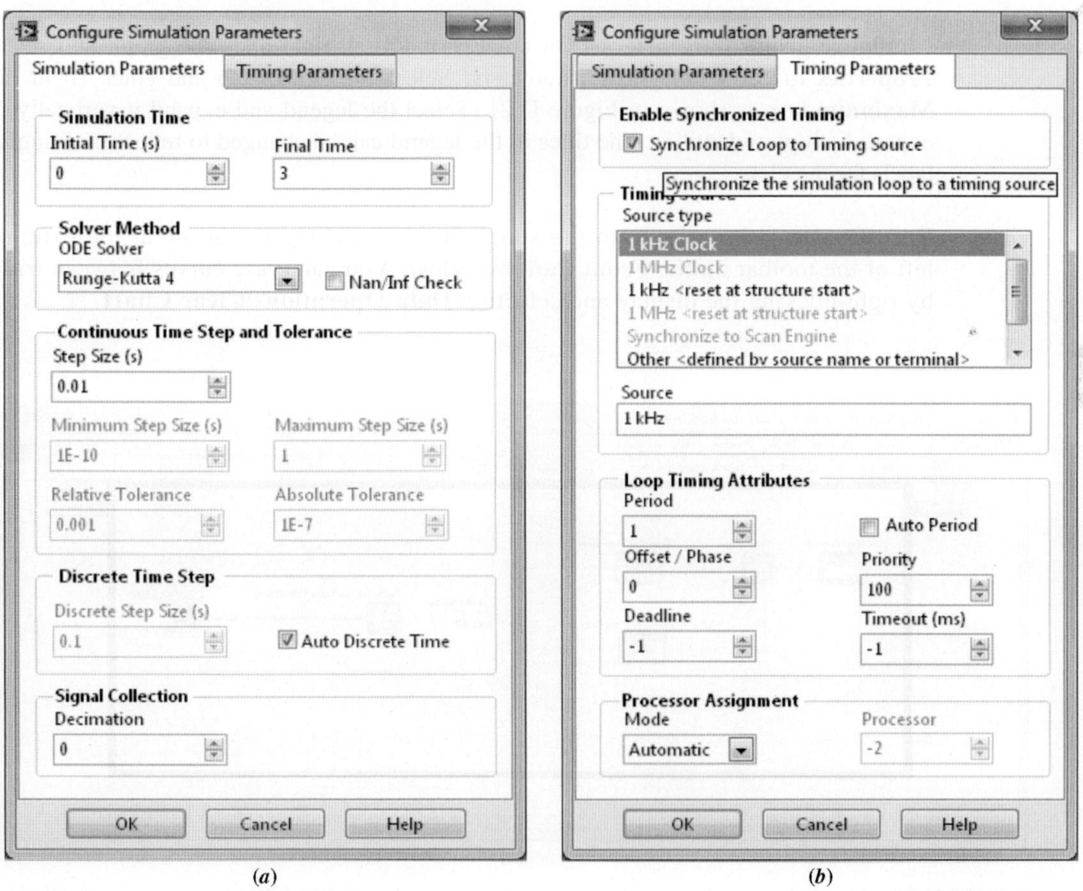


**FIGURE D.19** Configuring the **Simulation Loop** parameters: **a.** Simulation parameters; **b.** Timing parameters

**Configure graph parameters** On the **Front Panel**, right-click the graph and select **Properties** to configure graph parameters if required. Select the legend and expand it vertically to expose all three plot identities. The titles in the legend can be changed to reflect meaningful labels for the plots.

**Run the simulation** Perform the simulation by clicking the arrow at the extreme left of the toolbar on the **Front Panel** window. You can erase curves between trials by right-clicking the display and selecting **Data Operations/Clear Chart**.

## Example D.6 Effect of Amplifier Saturation on Motor's Load Angular Velocity

**Create Block Diagram and Front Panel** The **Block Diagram** and **Front Panel** for simulating a dc motor with and without saturation are shown in Figure D.20. The **Saturation** block is obtained from **Control Design & Simulation/Simulation/Nonlinear Systems/Saturation.**

**Configure simulation loop** Configure the simulation loop as shown in figure D.19, except change the **Final Time (s)** in Figure D.19(*a*) to 10.

**Configure graph parameters** On the **Front Panel**, right-click the graph and select **Properties** to configure graph parameters. Select the **Scales** tab and enter 10 in the **Maximum** box as shown in Figure D.21. Select the legend and expand it vertically to expose both plot identities. The titles in the legend can be changed to reflect meaningful labels for the plots.

**Run the simulation** Perform the simulation by clicking the arrow at the extreme left of the toolbar on the **Front Panel** window. You can erase curves between trials by right-clicking the display and selecting **Data Operations/Clear Chart**.

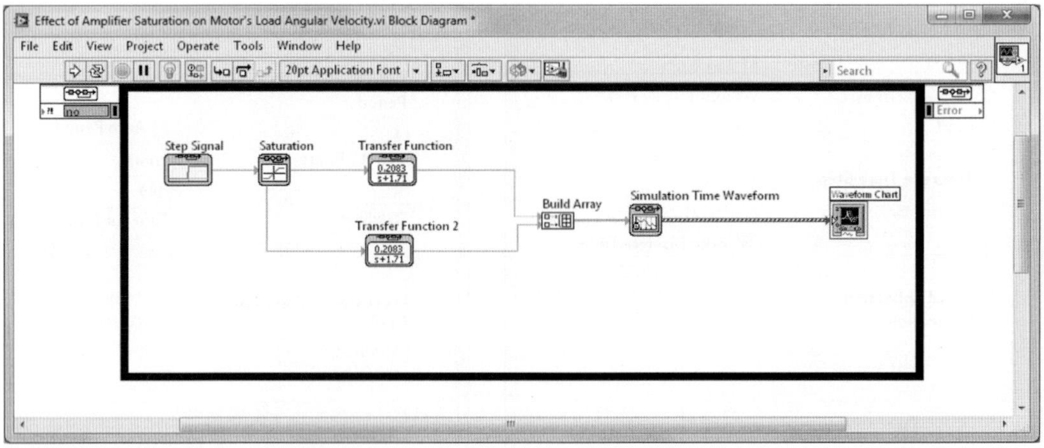


(*a*)

**FIGURE D.20** Simulation of a dc motor with and without saturation: **a. Block Diagram**, (*figure continues*)

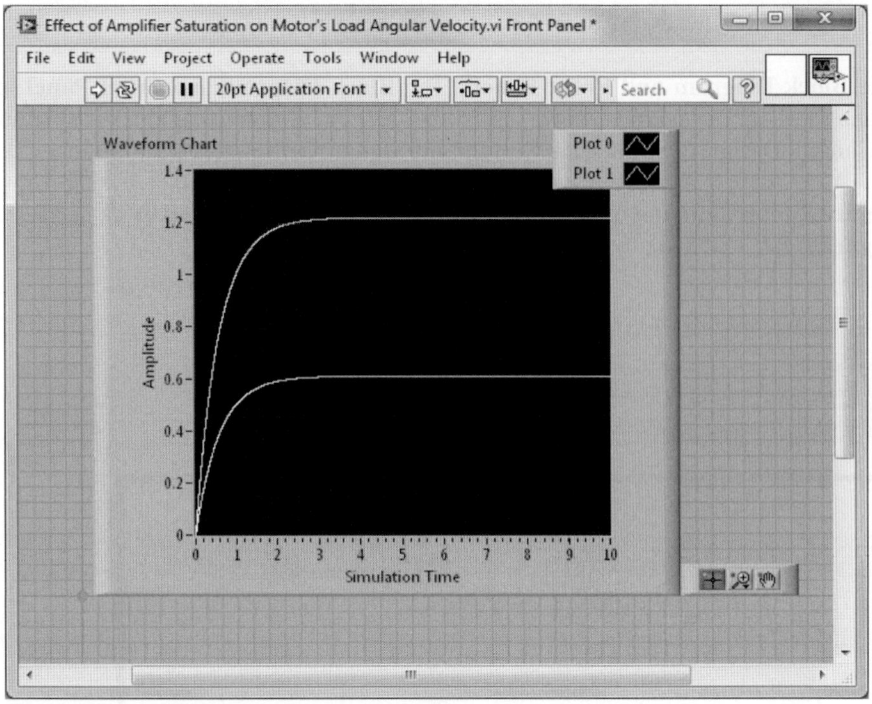


(*b*)

**FIGURE D.20** (*Continued*) **b. Front Panel**

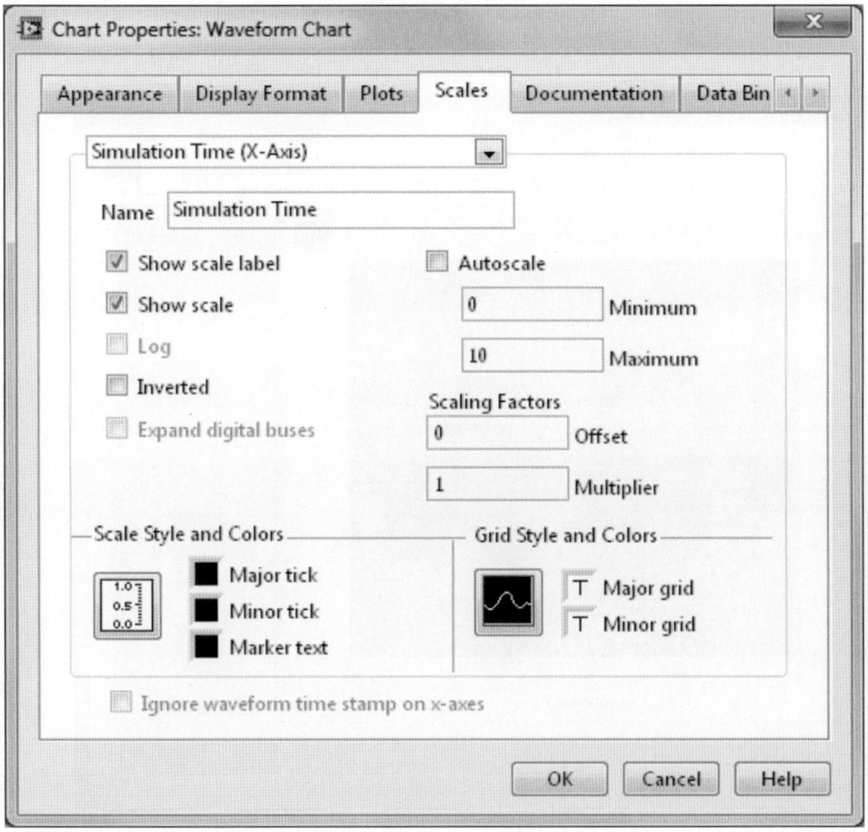


**FIGURE D.21** **Chart Properties: Waveform Chart** Window

## Example D.7 Simulating Feedback Systems

**Create Block Diagram and Front Panel** The **Block Diagram** and **Front Panel** for simulating feedback systems is shown in Figure D.22. The **Summation** block is obtained from **Control Design & Simulation/Simulation/Signal Arithmetic/Summation**.

**Configure Summation and other blocks** Right-click the **Summation** block and select **Configuration** . . . Repeat for other blocks.

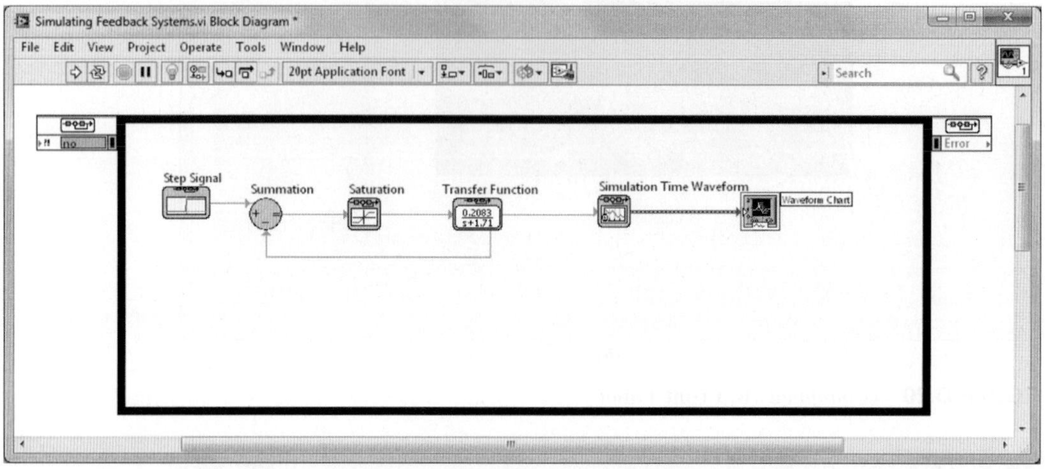


(*a*)

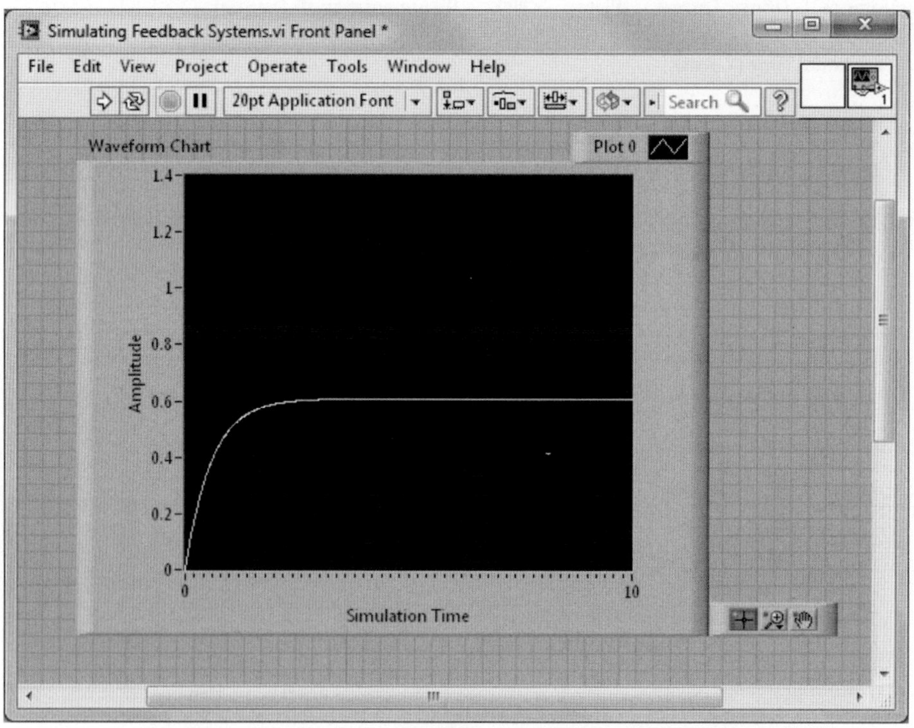


(*b*)

**FIGURE D.22** Simulation of feedback systems: **a. Block Diagram**; **b. Front Panel**

**Configure simulation loop** Configure the simulation loop as shown in Figure D.19, except change the **Final Time (s)** in Figure D.19(*a*) to 10.

**Configure graph parameters** On the **Front Panel**, right-click the graph and select **Properties** to configure graph parameters. Select the **Scales** tab and enter 10 in the **Maximum** box as shown in Figure D.21.

**Run the simulation** Perform the simulation by clicking the arrow at the extreme left of the toolbar on the **Front Panel** window. You can erase curves between trials by right-clicking the display and selecting **Data Operations/Clear Chart**.

## Example D.8 Simulating Digital Systems with the Simulation Palette

Digital systems, such as Example C.4 in Appendix C, can be simulated using LabVIEW. However, there are restrictions on the transfer functions used in the simulation. LabVIEW requires that all inputs to the transfer functions be present at the beginning of the simulation or else a cycle error will result. Unfortunately, this requirement limits the use of transfer functions to those with a denominator of higher order than the numerator. Under these conditions, the reader is advised to use either MATLAB or the **Control Design** palette rather than the **Simulation** palette of the **Control Design & Simulation** function.

Our first digital example will simulate a digital feedback system using the **Simulation** palette with proper transfer functions. The next example will simulate Example C.4 in Appendix C, which does not have proper transfer functions, using LabVIEW's **Control Design** palette.

**Create Block Diagram and Front Panel** The **Block Diagram** and **Front Panel** for simulating digital systems is shown in Figure D.23. The **Discrete Zero-Order Hold** block is obtained from **Control Design & Simulation/Simulation/Discrete Linear Systems/**

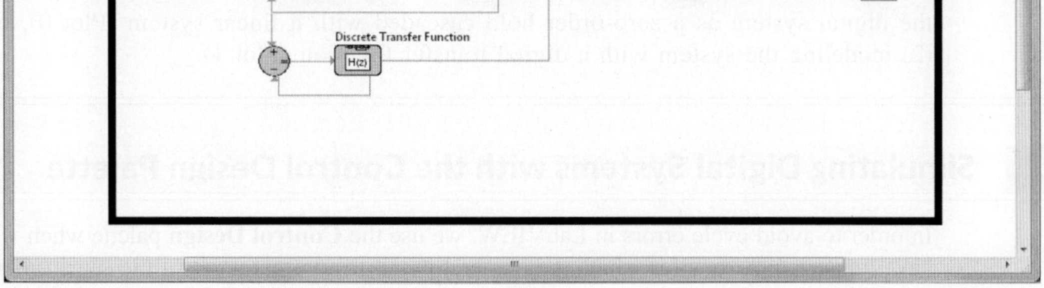


(*a*)

**FIGURE D.23** Simulation of digital systems with **Simulation** palette: **a. Block Diagram**; (*figure continues*)

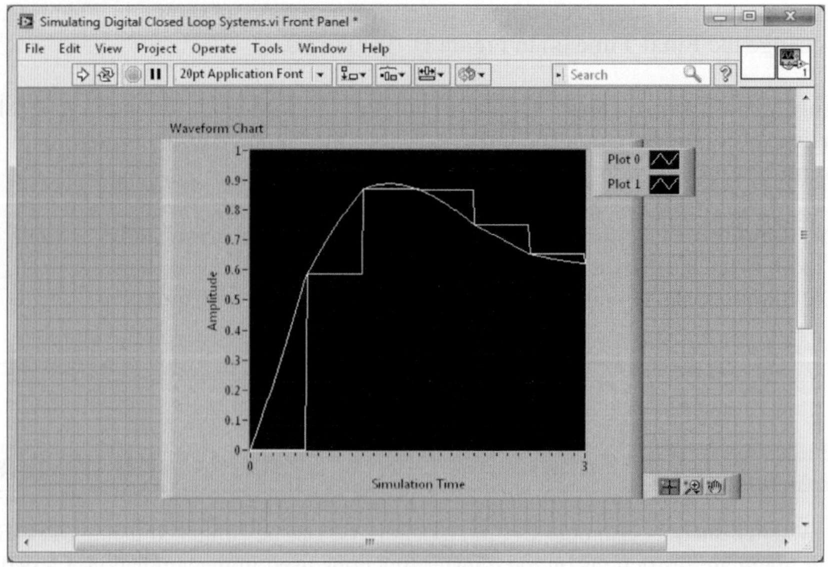


(*b*)

**FIGURE D.23** (*Continued*) **b. Front Panel**

**Discrete Zero-Order Hold**. The **Discrete Transfer Function** is obtained from **Control Design & Simulation/Simulation/Discrete Linear Systems/Discrete Transfer Function**.

**Configure Discrete Zero-Order Hold and other blocks** Right click the **Discrete Zero-Order Hold** block and select **Configuration . . .** Set the sample period to 0.5 second. Configure the transfer functions as shown on the **Block Diagram**. Configure the **Step Signal** to be a unit step.

**Configure simulation loop** Configure the simulation loop as shown in Figure D.19.

**Configure graph parameters** On the **Front Panel**, right-click the graph and select **Properties** to configure graph parameters. Select the **Scales** tab and enter 3 in the **Maximum** box for the *x* axis and 1 for the *y* axis. Select the legend and expand it vertically to expose both plot identities. The titles in the legend can be changed to reflect meaningful labels for the plots.

**Run the simulation** Perform the simulation by clicking the arrow at the extreme left of the toolbar on the **Front Panel** window. You can erase curves between trials by right-clicking the display and selecting **Data Operations/Clear Chart**.

The simulation shows the difference in responses obtained by (1) modeling the digital system as a zero-order hold cascaded with a linear system (Plot 0), or (2) modeling the system with a digital transfer function (Plot 1).

## Example D.9 Simulating Digital Systems with the Control Design Palette

In order to avoid cycle errors in LabVIEW, we use the **Control Design** palette when we have transfer functions for which the numerator and denominator are of the same order. This example reproduces Simulink Example C.4.

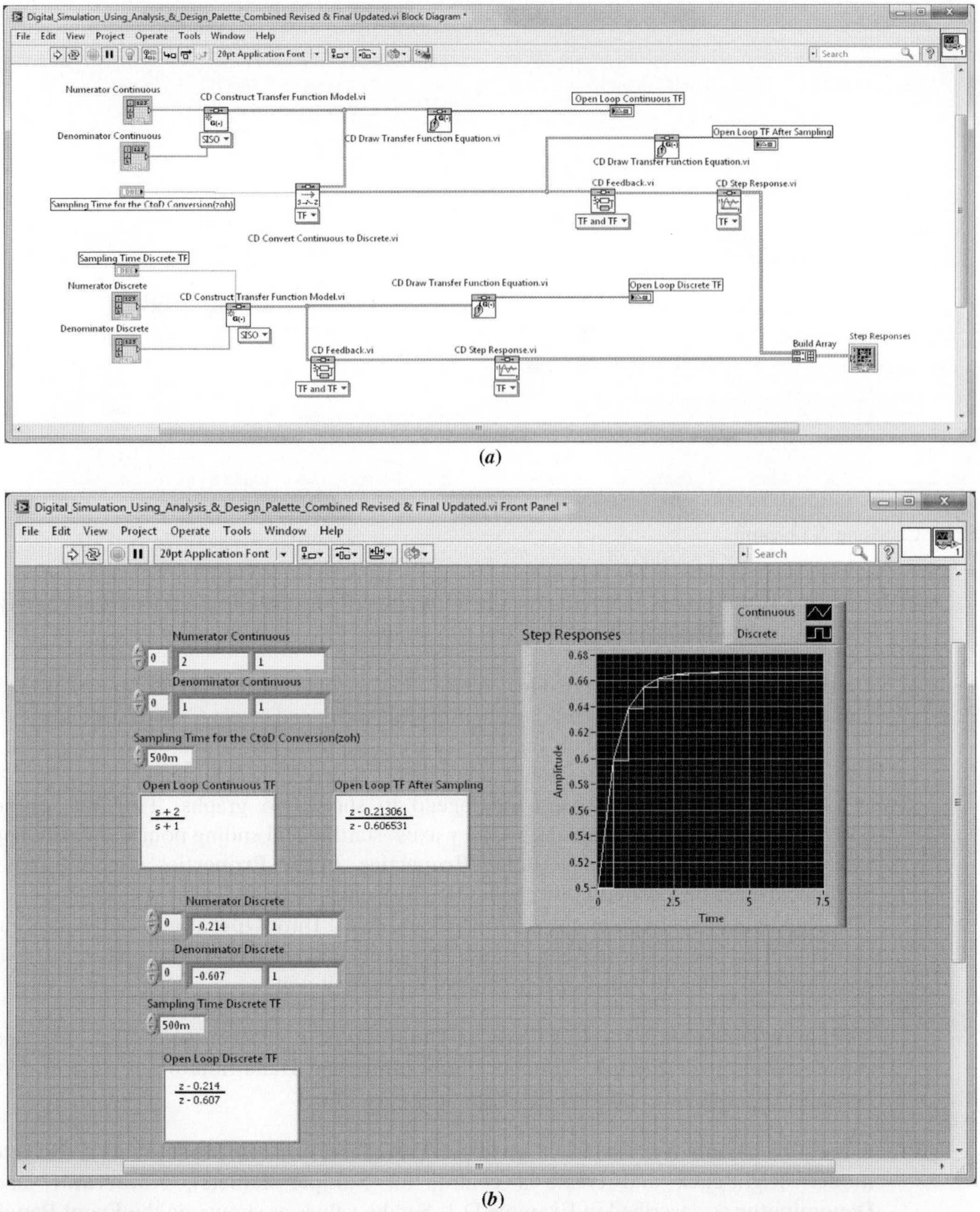


**FIGURE D.24** Simulation of digital systems with **the Control Design** palette: **a. Block Diagram**; **b. Front Panel**

**Create Block Diagram and Front Panel** The **Block Diagram** and **Front Panel** for this example are shown in Figure D.24. Wire the blocks as shown.

Most of the blocks were previously discussed in Examples D.1 and D.2. Digital transfer functions are created using the same blocks as continuous systems, but with a nonzero **Sampling Time(s)** input.

The **CD Convert Continuous to Discrete.vi**, is obtained from **Functions/ Control Design & Simulation/Control Design/Model Conversion/CD Convert Continuous to Discrete.vi**.

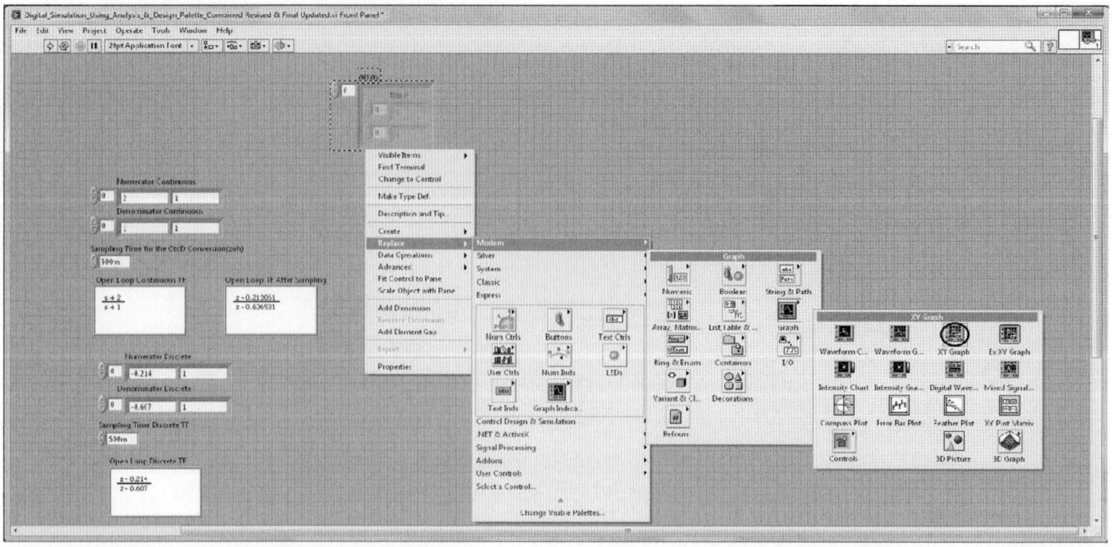

**FIGURE D.25** Choosing **XYGraph**

The **Build Array** is obtained from **Functions/Programming/Array/Build Array**. Expand the **Build Array** block to show two inputs.

**Configure parameters for Build Array** Right-click on **Build Array** and select **Concatenate Inputs**. Right-click again on **Build Array** and select **Create/Indicator**.

Select and then right-click the indicator on the front panel and choose **Replace**.[1] Using the resulting palettes as shown in Figure D.25, select the **XY Graph**.[2]

On the front panel expand the legend to show two graphs. Title the legend components as shown. Change the *x*- and *y*-axes' starting and ending points as desired by right-clicking the graph and selecting **Properties**. In the **Properties** window, select **Scales** and enter the desired information. Finally, select **Plots** and enter your choices.

Right-click the graph on the front panel and select **Data Operations** and make your current values the default. Also, right-click again and choose to reinitialize to your default values. You may also choose to clear the current plot.

**Configure parameters for CD Convert Continuous to Discrete.vi** Right-click and create a control for **Sample Time(s), Numerator**, and **Denominator** as described in Example D.1. Set the values as shown on the **Front Panel**.

**Configure parameters for CD Construct Transfer Function Model.vi as a discrete model** Right-click and create a control for **Sample Time(s)**, **Numerator**, and **Denominator** as described in Example D.1. Set the values as shown on the **Front Panel**.

**Configure parameters for all CD Draw Transfer Function Equation.vi** Right-click and create a control for **Equation** as described in Example D.1. Set the values as shown on the **Front Panel**.

**Run simulation** See Example D.1 for a description. The results are shown in Figure D.24(*b*).

[1] Be sure to select the entire indicator. Then place your mouse on the outer edge and be sure it is a pointer before selecting **Replace**.

[2] Right click the **XY Graph** on the front panel and open **Properties**. On the **Appearance** tab, select 2 for **Plots shown**.

## D.6 Interfacing with External Hardware

This section provides an introduction to the use of LabVIEW virtual instruments to control external hardware with the **NI myDAQ** for students. Specifically, we concentrate on using the **NI myDAQ** to analyze and design an actual feedback system used in the Hardware Exploration Laboratory sections at the end of select chapters.

1. **Required Hardware** **NI myDAQ** is a data acquisition module that is ideal for student experimentation, since it is portable and of low cost. In order to perform the experiments in this book, you will need a motor control chip and a gearmotor. Although other alternatives exist, we use the following: (1) ROHM motor control chip, BA6886N; and (2) 9.7:1 metal gearmotor 25Dx48L mm HP with 48 CPR encoder, which can be obtained from www.pololu.com.
2. **Required Software** The software required to support the experiments in this book are: (1) LabVIEW or the Student Version of LabVIEW and (2) **NI ELVISmx Soft Front Panel (SFP) Instruments**, which comes bundled with the **NI myDAQ. NI ELVISmx** provides virtual instruments that will generate input signals to and acquire output signals from your external control system.
3. **Basic Configuration** Figure D.26 shows the **NI myDAQ** and the basic configuration that will be used to perform control system design and analysis. Detailed wiring diagrams will accompany specific experiments.
4. **Launching NI myDAQ** The **NI myDAQ** kit comes with the following cables and connectors: (1) USB cable; (2) 20-position screw terminal connector; (3) audio cable; and (4) DMM banana cable. The following steps will launch the **NI myDAQ** and **NI ELVISmx**, which provides convenient virtual instruments for control and data acquisition.

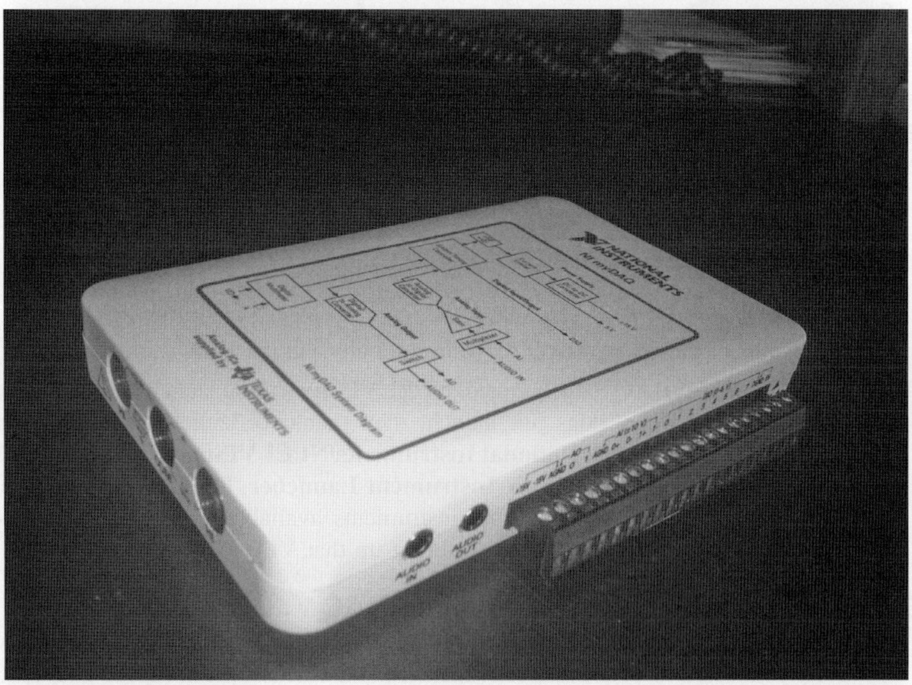


*(a)*

**FIGURE D.26** a. NI myDAQ; (*figure continues*)

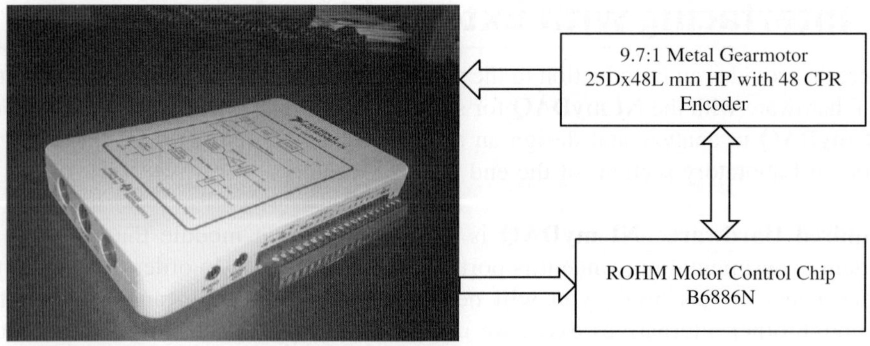


(*b*)

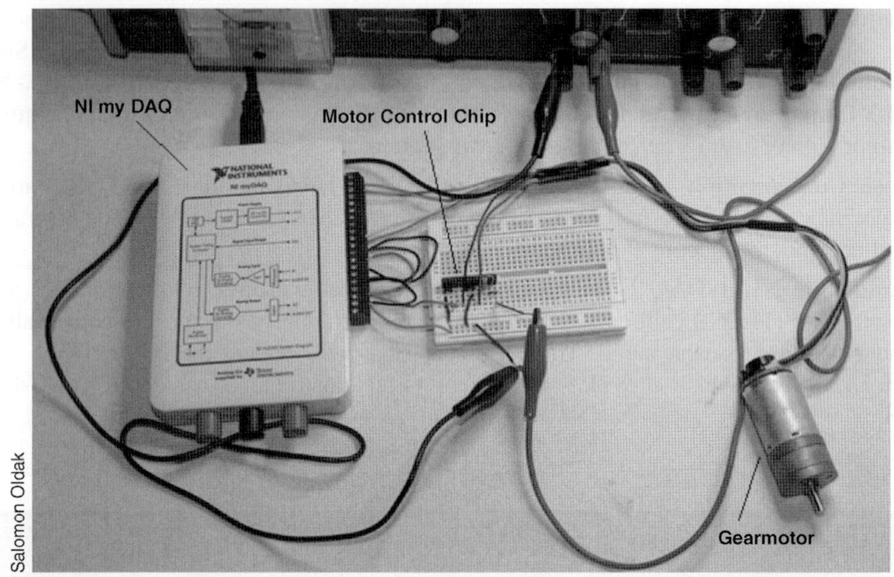


(*c*)

**FIGURE D.26** (*Continued*) **b.** basic configuration showing **NI myDAQ** interfaced with motor control chip and gearmotor; **c.** interconnected hardware

**Step 1:** Be sure LabVIEW is installed on your computer.

**Step 2:** Install the **NI myDAQ** software.

**Step 3:** Connect the **NI myDAQ** to your computer via the USB cable.

**Step 4:** After **NI myDAQ** is recognized, **NI ELVISmx** should launch automatically. If **NI ELVISmx** does not launch, then manually launch from the **Start** menu **All Programs/National Instruments/NI ELVISmx for NI ELVIS & NI myDAQ/ NI ELVISmx Instrument Launcher**. Figure D.27 shows the window containing the virtual instruments available from **NI ELVISmx**. Clicking on any instrument will bring up that VI.

## Simple Experiments Using the NI myDAQ

1. **Measuring battery voltage using the NI myDAQ Digital Multimeter (DMM)** With myDAQ connected to your computer, launch the DMM shown in Figure D.27. Attach the DMM probes between the myDAQ and a battery as shown in Figure D.28(*a*). Press the green **Run** arrow.

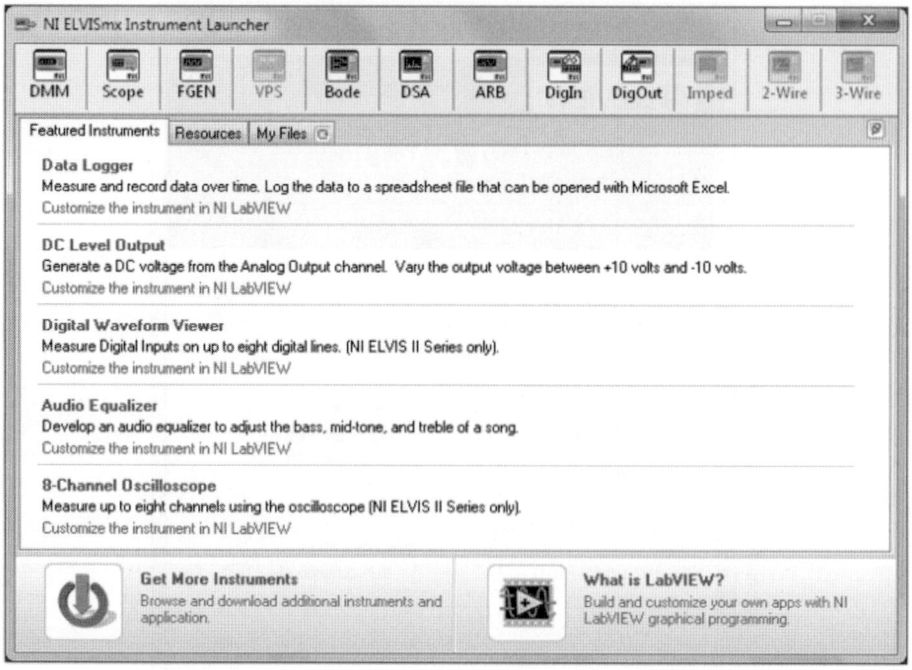




**FIGURE D.27** The **NI ELVISmx Instrument Launcher** window

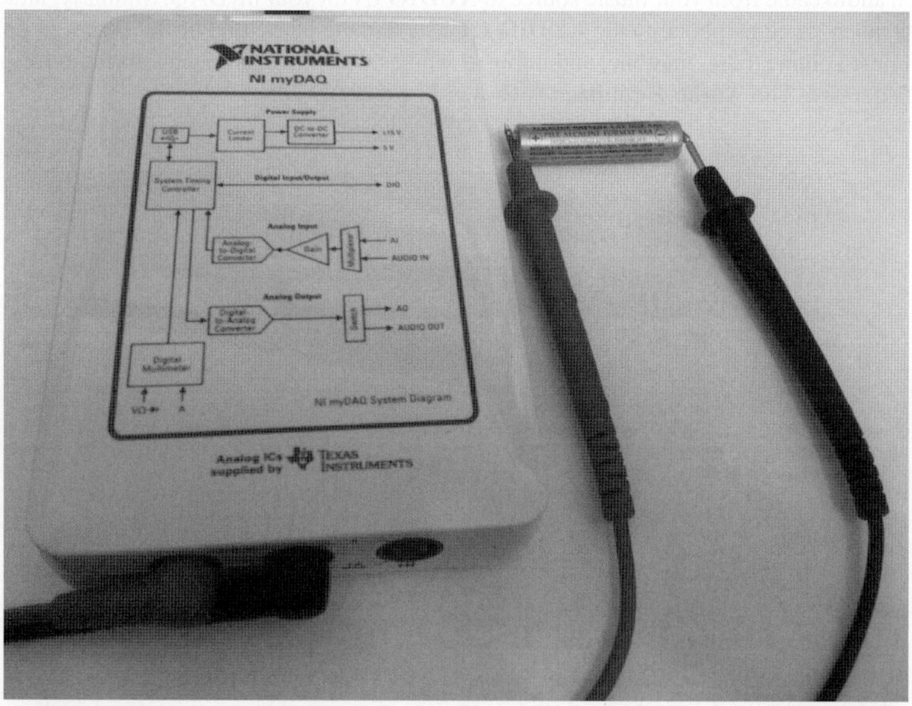


*(a)*

**FIGURE D.28** Battery voltage measurement: **a.** Connections to myDAQ; (*figure continues*)

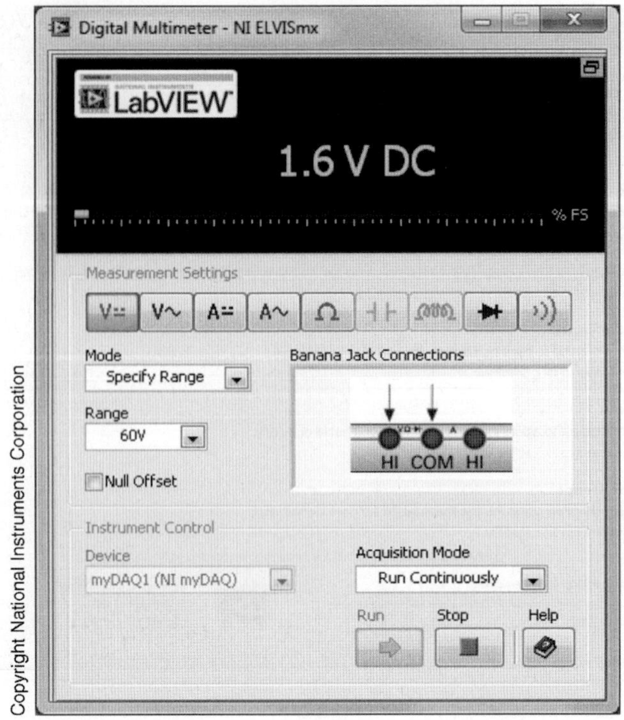


(*b*)

**FIGURE D.28** (*Continued*) **b.** DMM reading

2. **The NI myDAQ audio equalizer** With myDAQ connected to your computer, launch the **Audio Equalizer** in the **Featured Instruments** menu shown in Figure D.27. Attach an audio cable from your music source to **AUDIO IN** on your myDAQ. Similarly, attach speakers or an earphone to the **AUDIO OUT**. Press the blue **Start** arrow. You can now adjust volume, bass, midtone, and treble as well as watch the audio waveform in time or frequency. The **Audio Equalizer** is shown in Figure D.29.

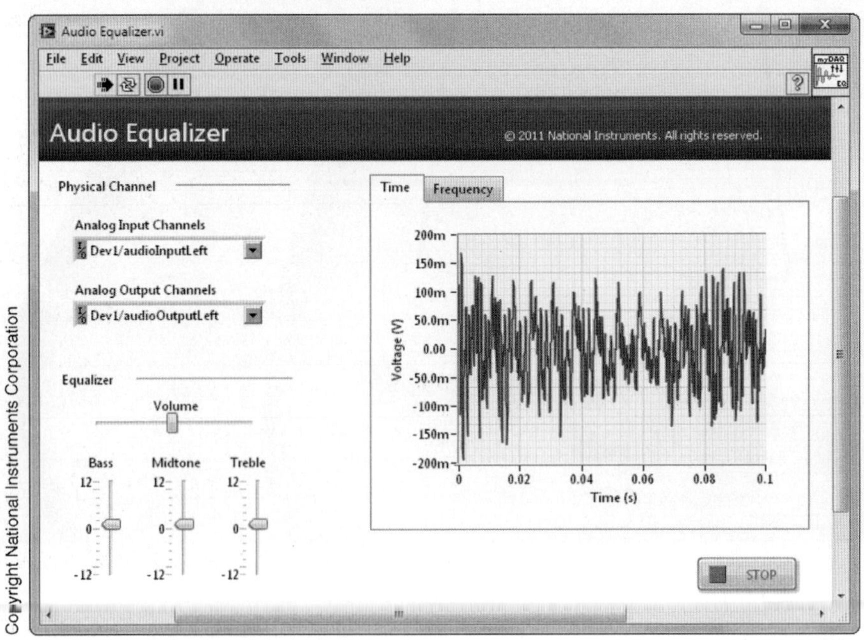


**FIGURE D.29** The NI myDAQ Audio Equalizer

## Summary

This appendix presented LabVIEW as an alternative to MATLAB for analysis, design, and simulation. Our discussion was divided into analysis and design, and simulation.

Analysis and design is performed by interconnecting code blocks, which is analogous to writing in-line code for MATLAB M-files. Since the LabVIEW code blocks are represented by icons, an advantage of using LabVIEW is that you do not have to know specific code statements.

Simulation is performed by interconnecting code blocks and is analogous to Simulink flow diagrams.

LabVIEW has more extensive applications. You can create virtual instruments on your computer monitor that can operate external hardware and transmit and receive telemetric data. We covered a few of these applications in this appendix using the NI myDAQ. It is left to the interested reader to pursue more of these applications.

## Bibliography

National Instruments. *Getting Started with LabVIEW*. National Instruments, Austin, TX. 2003–2013.

National Instruments. *LabVIEW Fundamentals*. National Instruments, Austin, TX. 2005.

National Instruments. *LabVIEW™ Control Design User Manual*. National Instruments, Austin, TX. 2004–2009.

National Instruments. *Introduction to LabVIEW in 3 Hours for Control Design and Simulation*. National Instruments Course Notes, Austin, TX. 2009.

National Instruments. *NI myDAQ User Guide and Specifications*. National Instruments, Austin, TX. 2010–2011.

# Glossary

**Acceleration constant** $\lim_{s \to 0} s^2 G(s)$

**Actuating signal** The signal that drives the controller. If this signal is the difference between the input and output, it is called the *error*.

**Analog-to-digital converter** A device that converts analog signals to digital signals.

**Armature** The rotating member of a dc motor through which a current flows.

**Back emf** The voltage across the armature of a motor.

**Bandwidth** The frequency at which the magnitude frequency response is −3 dB below the magnitude at zero frequency.

**Basis** Linearly independent vectors that define a space.

**Bilinear transformation** A mapping of the complex plane where one point, $s$, is mapped into another point, $z$, through $z = (as + b)/(cs + d)$.

**Block diagram** A representation of the interconnection of subsystems that form a system. In a linear system, the block diagram consists of blocks representing subsystems, arrows representing signals, summing junctions, and pickoff points.

**Bode diagram (plot)** A sinusoidal frequency response plot where the magnitude response is plotted separately from the phase response. The magnitude plot is dB versus log $\omega$, and the phase plot is phase versus log $\omega$. In control systems, the Bode plot is usually made for the open-loop transfer function. Bode plots can also be drawn as straight-line approximations.

**Branches** Lines that represent subsystems in a signal-flow graph.

**Break frequency** A frequency where the Bode magnitude plot changes slope.

**Breakaway point** A point on the real axis of the $s$-plane where the root locus leaves the real axis and enters the complex plane.

**Break-in point** A point on the real axis of the $s$-plane where the root locus enters the real axis from the complex plane.

**Characteristic equation** The equation formed by setting the characteristic polynomial to zero.

**Characteristic polynomial** The denominator of a transfer function. Equivalently, the unforced differential equation, where the differential operators are replaced by $s$ or $\lambda$.

**Classical approach to control systems** *See* **frequency domain techniques**.

**Closed-loop system** A system that monitors its output and corrects for disturbances. It is characterized by feedback paths from the output.

**Closed-loop transfer function** For a generic feedback system with $G(s)$ in the forward path and $H(s)$ in the feedback path, the closed-loop transfer function, $T(s)$, is $G(s)/[1 \pm G(s)H(s)]$, where the + is for negative feedback, and the – is for positive feedback.

**Compensation** The addition of a transfer function in the forward path or feedback path for the purpose of improving the transient or steady-state performance of a control system.

**Compensator** A subsystem inserted into the forward or feedback path for the purpose of improving the transient response or steady-state error.

**Constant *M* circles** The locus of constant, closed-loop magnitude frequency response for unity feedback systems. It allows the closed-loop magnitude frequency response to be determined from the open-loop magnitude frequency response.

**Constant *N* circles** The locus of constant, closed-loop phase frequency response for unity feedback systems. It allows the closed-loop phase frequency response to be determined from the open-loop phase frequency response.

**Controllability** A property of a system by which an input can be found that takes every state variable from a desired initial state to a desired final state in finite time.

**Controlled variable** The output of a plant or process that the system is controlling for the purpose of desired transient response, stability, and steady-state error characteristics.

**Controller** The subsystem that generates the input to the plant or process.

**Critically damped response** The step response of a second-order system with a given natural frequency that is characterized by no overshoot and a rise time that is faster than any possible overdamped response with the same natural frequency.

**Damped frequency of oscillation** The sinusoidal frequency of oscillation of an underdamped response.

**Damping ratio** The ratio of the exponential decay frequency to the natural frequency.

**Decade** Frequencies that are separated by a factor of 10.

**Decibel (dB)** The decibel is defined as $10 \log P_G$, where $P_G$ is the power gain of a signal. Equivalently, the decibel is also $20 \log V_G$, where $V_G$ is the voltage gain of a signal.

**Decoupled system** A state-space representation in which each state equation is a function of only one state variable. Hence, each differential equation can be solved independently of the other equations.

**Digital compensator** A sampled transfer function used to improve the response of computer-controlled feedback systems. The transfer function can be emulated by a digital computer in the loop.

**Digital-to-analog converter** A device that converts digital signals to analog signals.

**Disturbance** An unwanted signal that corrupts the input or output of a plant or process.

**Dominant poles** The poles that predominantly generate the transient response.

**Eigenvalues** Any value, $\lambda_i$, that satisfies $\mathbf{AX_i} = \lambda_i \mathbf{x_i}$ for $\mathbf{x_i} \neq 0$. Hence, any value, $\lambda_i$, that makes $\mathbf{x_i}$ an eigenvector under the transformation $\mathbf{A}$.

**Eigenvector** Any vector that is collinear with a new basis vector after a similarity transformation to a diagonal system.

**Electric circuit analog** An electrical network whose variables and parameters are analogous to another physical system. The electric circuit analog can be used to solve for variables of the other physical system.

**Electrical admittance** The inverse of electrical impedance. The ratio of the Laplace transform of the current to the Laplace transform of the voltage.

**Electrical impedance** The ratio of the Laplace transform of the voltage to the Laplace transform of the current.

**Equilibrium** The steady-state solution characterized by a constant position or oscillation.

**Error** The difference between the input and the output of a system.

**Euler's approximation** A method of integration where the area to be integrated is approximated as a sequence of rectangles.

**Feedback** A path through which a signal flows back to a previous signal in the forward path in order to be added or subtracted.

**Feedback compensator** A subsystem placed in a feedback path for the purpose of improving the performance of a closed-loop system.

**Forced response** For linear systems, that part of the total response function due to the input. It is typically of the same form as the input and its derivatives.

**Forward-path gain** The product of gains found by traversing a path that follows the direction of signal flow from the input node to the output node of a signal-flow graph.

**Frequency domain techniques** A method of analyzing and designing linear control systems by using transfer functions and the Laplace transform as well as frequency response techniques.

**Frequency response techniques** A method of analyzing and designing control systems by using the sinusoidal frequency response characteristics of a system.

**Gain** The ratio of output to input; usually used to describe the amplification in the steady state of the magnitude of sinusoidal inputs, including dc.

**Gain margin** The amount of additional open-loop gain, expressed in decibels (dB), required at 180° of phase shift to make the closed-loop system unstable.

**Gain-margin frequency** The frequency at which the phase frequency response plot equals 180°. It is the frequency at which the gain margin is measured.

**Homogeneous solution** *See* **natural response.**

**Ideal derivative compensator** *See* **proportional-plus-derivative controller**.

**Ideal integral compensator** *See* **proportional-plus-integral controller**.

**Instability** The characteristic of a system defined by a natural response that grows without bounds as time approaches infinity.

**Kirchhoff's law** The sum of voltages around a closed loop equals zero. Also, the sum of currents at a node equals zero.

**Lag compensator** A transfer function, characterized by a pole on the negative real axis close to the origin and a zero close and to the left of the pole, that is used for the purpose of improving the steady-state error of a closed-loop system.

**Lag-lead compensator** A transfer function, characterized by a pole-zero configuration that is the combination of a lag and a lead compensator, that is used for the purpose of improving both the transient response and the steady-state error of a closed-loop system.

**Laplace transformation** A transformation that transforms linear differential equations into algebraic expressions. The transformation is especially useful for modeling, analyzing, and designing control systems as well as solving linear differential equations.

**Lead compensator** A transfer function, characterized by a zero on the negative real axis and a pole to the left of the zero, that is used for the purpose of improving the transient response of a closed-loop system.

**Linear combination** A linear combination of $n$ variables, $x_i$, for $i = 1$ to $n$, given by the following sum, $S$:

$$S = K_n X_n + K_{n-1} X_{n-1} + \cdots + K_1 X_1$$

where each $K_i$ is a constant.

**Linear independence** The variables $x_i$, for $i = 1$ to $n$, are said to be linearly independent if their linear combination, $S$, equals zero *only* if every $K_i = 0$ and *no* $x_i = 0$. Alternatively, if the $x_i$'s are linearly independent, then $K_n x_n + K_{n-1} x_{n-1} + \cdots + K_1 x_1 = 0$ cannot be solved for any $x_k$. Thus, no $x_k$ can be expressed as a linear combination of the other $x_i$'s.

**Linear system** A system possessing the properties of superposition and homogeneity.

**Linearization** The process of approximating a nonlinear differential equation with a linear differential equation valid for small excursions about equilibrium.

**Loop gain** For a signal-flow graph, the product of branch gains found by traversing a path that starts at a node and ends at the same node without passing through any other node more than once, and following the direction of the signal flow.

**Major-loop compensation** A method of feedback compensation that adds a compensating zero to the open-loop transfer function for the purpose of improving the transient response of the closed-loop system.

**Marginal stability** The characteristic of a system defined by a natural response that neither decays nor grows, but remains constant or oscillates as time approaches infinity as long as the input is not of the same form as the system's natural response.

**Mason's rule** A formula from which the transfer function of a system consisting of the interconnection of multiple subsystems can be found.

**Mechanical rotational impedance** The ratio of the Laplace transform of the torque to the Laplace transform of the angular displacement.

**Mechanical translational impedance** The ratio of the Laplace transform of the force to the Laplace transform of the linear displacement.

**Minor-loop compensation** A method of feedback compensation that changes the poles of a forward-path transfer function for the purpose of improving the transient response of the closed-loop system.

**Modern approach to control systems** *See* **state-space representation.**

**Natural frequency** The frequency of oscillation of a system if all the damping is removed.

**Natural response** That part of the total response function due to the system and the way the system acquires or dissipates energy.

**Negative feedback** The case where a feedback signal is subtracted from a previous signal in the forward path.

**Newton's law** The sum of forces equals zero. Alternatively, after bringing the *ma* force to the other side of the equality, the sum of forces equals the product of mass and acceleration.

**Nichols chart** The locus of constant closed-loop magnitude and closed-loop phase frequency responses for unity feedback systems plotted on the open-loop dB versus phase-angle plane. It allows the closed-loop frequency response to be determined from the open-loop frequency response.

**Nodes** Points in a signal-flow diagram that represent signals.

**No-load speed** The speed produced by a motor with constant input voltage when the torque at the armature is reduced to zero.

**Nonminimum-phase system** A system whose transfer function has zeros in the right half-plane. The step response is characterized by an initial reversal in direction.

**Nontouching-loop gain** The product of loop gains from nontouching loops taken two, three, four, and so on at a time.

**Nontouching loops** Loops that do not have any nodes in common.

**Notch filter** A filter whose magnitude frequency response dips at a particular sinusoidal frequency. On the $s$-plane, it is characterized by a pair of complex zeros near the imaginary axis.

**Nyquist criterion** If a contour, A, that encircles the entire right half-plane is mapped through $G(s)H(s)$, then the number of closed-loop poles, $Z$, in the right half-plane equals the number of open-loop poles, $P$, that are in the right half-plane minus the number of counterclockwise revolutions, $N$, around $-1$, of the mapping; that is, $Z = P - N$. The mapping is called the *Nyquist diagram* of $G(s)H(s)$.

**Nyquist diagram (plot)** A polar frequency response plot made for the open-loop transfer function.

**Nyquist sampling rate** The minimum frequency at which an analog signal should be sampled for correct reconstruction. This frequency is twice the bandwidth of the analog signal.

**Observability** A property of a system by which an initial state vector, $x(t_0)$, can be found from $u(t)$ and $y(t)$ measured over a finite interval of time from $t_0$. Simply stated, observability is the property by which the state variables can be estimated from a knowledge of the input, $u(t)$, and output, $y(t)$.

**Observer** A system configuration from which inaccessible states can be estimated.

**Octave** Frequencies that are separated by a factor of two.

**Ohm's law** For dc circuits the ratio of voltage to current is a constant called resistance.

**Open-loop system** A system that does not monitor its output nor correct for disturbances.

**Open-loop transfer function** For a generic feedback system with $G(s)$ in the forward path and $H(s)$ in the feedback path, the open-loop transfer function is the product of the forward-path transfer function and the feedback transfer function, or $G(s)H(s)$.

**Operational amplifier** An amplifier—characterized by a very high input impedance, a very low output impedance, and a high gain—that can be used to implement the transfer function of a compensator.

**Output equation** For linear systems, the equation that expresses the output variables of a system as linear combinations of the state variables.

**Overdamped response** A step response of a second-order system that is characterized by no overshoot.

**Partial-fraction expansion** A mathematical equation where a fraction with $n$ factors in its denominator is represented as the sum of simpler fractions.

**Particular solution** *See* **forced response.**

**Passive network** A physical network that only stores or dissipates energy. No energy is produced by the network.

**Peak time, $T_p$** The time required for the underdamped step response to reach the first, or maximum, peak.

**Percent overshoot, $\%OS$** The amount that the underdamped step response overshoots the steady-state, or final, value at the peak time, expressed as a percentage of the steady-state value.

**Phase margin** The amount of additional open-loop phase shift required at unity gain to make the closed-loop system unstable.

**Phase-margin frequency** The frequency at which the magnitude frequency response plot equals zero dB. It is the frequency at which the phase margin is measured.

**Phase variables** State variables such that each subsequent state variable is the derivative of the previous state variable.

**Phasor** A rotating vector that represents a sinusoid of the form $A\cos(\omega t + \phi)$.

**Pickoff point** A block diagram symbol that shows the distribution of one signal to multiple subsystems.

**Plant or process** The subsystem whose output is being controlled by the system.

**Poles** (1) The values of the Laplace transform variable, $s$, that cause the transfer function to become infinite; and (2) any roots of factors of the characteristic equation in the denominator that are common to the numerator of the transfer function.

**Position constant** $\lim_{s\to 0} G(s)$

**Positive feedback** The case where a feedback signal is added to a previous signal in the forward path.

**Proportional-plus-derivative (PD) controller** A controller that feeds forward to the plant a proportion of the actuating signal plus its derivative for the purpose of improving the transient response of a closed-loop system.

**Proportional-plus-integral (PI) controller** A controller that feeds forward to the plant a proportion of the actuating signal plus its integral for the purpose of improving the steady-state error of a closed-loop system.

**Proportional-plus-integral-plus-derivative (PID) controller** A controller that feeds forward to the plant a proportion of the actuating signal plus its integral plus its derivative for the purpose of improving the transient response and steady-state error of a closed-loop system.

**Quantization error** For linear systems, the error associated with the digitizing of signals as a result of the finite difference between quantization levels.

**Raible's tabular method** A tabular method for determining the stability of digital systems that parallels the Routh-Hurwitz method for analog signals.

**Rate gyro** A device that responds to an angular position input with an output voltage proportional to angular velocity.

**Residue** The constants in the numerators of the terms in a partial-fraction expansion.

**Rise time, $T_r$** The time required for the step response to go from 0.1 of the final value to 0.9 of the final value.

**Root locus** The locus of closed-loop poles as a system parameter is varied. Typically, the parameter is gain. The locus is obtained from the open-loop poles and zeros.

**Routh-Hurwitz criterion** A method for determining how many roots of a polynomial in $s$ are in the right half of the $s$-plane, the left half of the $s$-plane, and on the imaginary axis. Except in some special cases, the Routh-Hurwitz criterion does not yield the coordinates of the roots.

**Sensitivity** The fractional change in a system characteristic for a fractional change in a system parameter.

**Settling time, $T_s$** The amount of time required for the step response to reach and stay within $\pm 2\%$ of the steady-state value. Strictly speaking, this is the definition of the 2% settling time. Other percentages, for example 5%, also can be used. This book uses the 2% settling time.

**Signal-flow graph** A representation of the interconnection of subsystems that form a system. It consists of nodes representing signals and lines representing subsystems.

**Similarity transformation** A transformation from one state-space representation to another state-space representation. Although the state variables are different, each representation is a valid description of the same system and the relationship between the input and the output.

**Stability** That characteristic of a system defined by a natural response that decays to zero as time approaches infinity.

**Stall torque** The torque produced at the armature when a motor's speed is reduced to zero under a condition of constant input voltage.

**State equations** A set of $n$ simultaneous, first-order differential equations with $n$ variables, where the $n$ variables to be solved are the state variables.

**State space** The $n$-dimensional space whose axes are the state variables.

**State-space representation** A mathematical model for a system that consists of simultaneous, first-order differential equations and an output equation.

**State-transition matrix** The matrix that performs a transformation on $\mathbf{x}(0)$, taking $\mathbf{x}$ from the initial state, $\mathbf{x}(0)$, to the state $\mathbf{x}(t)$ at any time, $t \geq 0$.

**State variables** The smallest set of linearly independent system variables such that the values of the members of the set at time $t_0$ along with known forcing functions completely determine the value of all system variables for all $t \geq t_0$.

**State vector** A vector whose elements are the state variables.

**Static error constants** The collection of position constant, velocity constant, and acceleration constant.

**Steady-state error** The difference between the input and the output of a system after the natural response has decayed to zero.

**Steady-state response** *See* **forced response**.

**Subsystem** A system that is a portion of a larger system.

**Summing junction** A block diagram symbol that shows the algebraic summation of two or more signals.

**System type** The number of pure integrations in the forward path of a unity feedback system.

**System variables** Any variable that responds to an input or initial conditions in a system.

**Tachometer** A voltage generator that yields a voltage output proportional to rotational input speed.

**Time constant** The time for $e^{-at}$ to decay to 37% of its original value at $t = 0$.

**Time-domain representation** *See* **state-space representation**.

**Torque-speed curve** The plot that relates a motor's torque to its speed at a constant input voltage.

**Transducer** A device that converts a signal from one form to another, for example from a mechanical displacement to an electrical voltage.

**Transfer function** The ratio of the Laplace transform of the output of a system to the Laplace transform of the input.

**Transient response** That part of the response curve due to the system and the way the system acquires or dissipates energy. In stable systems it is the part of the response plot prior to the steady-state response.

**Tustin transformation** A bilinear transformation that converts transfer functions from continuous to sampled and vice versa. The important characteristic of the Tustin transformation is that both transfer functions yield the same output response at the sampling instants.

**Type** *See* **system type**.

**Undamped response** The step response of a second-order system that is characterized by a pure oscillation.

**Underdamped response** The step response of a second-order system that is characterized by overshoot.

**Velocity constant** $\lim_{s \to 0} sG(s)$

**$z$-transformation** A transformation related to the Laplace transformation that is used for the representation, analysis, and design of sampled signals and systems.

**Zero-input response** That part of the response that depends upon only the initial state vector and not the input.

**Zero-order sample-and-hold (z.o.h.)** A device that yields a staircase approximation to the analog signal.

**Zeros** (1) Those values of the Laplace transform variable, $s$, that cause the transfer function to become zero; and (2) any roots of factors of the numerator that are common to the characteristic equation in the denominator of the transfer function.

**Zero-state response** That part of the response that depends upon only the input and not the initial state vector.

# Answers to Selected Problems

## Chapter 1

**20. b.** $x(t) = -e^{-t} + 9te^{-t} + 5e^{-2t} + t - 2$

**21. c.** $x(t) = \frac{2}{5} - e^{-4t}\left(\frac{2}{5}\cos 3t + \frac{8}{15}\sin 3t\right)$

## Chapter 2

**4. b.** $x(t) = \frac{15}{26}e^{-2t} - \frac{3}{10}e^{-4t} - \frac{18}{65}\cos 3t - \frac{1}{65}\sin 3t$

**7.** $\frac{Y(s)}{X(s)} = \frac{s^3 + 2s^2 + 3s + 7}{s^3 + 5s^2 + 7s + 1}$

**8. c.** $\frac{d^3x}{dt^3} + 10\frac{d^2x}{dt^2} + 11\frac{dx}{dt} + 18x = \frac{df}{dt} + 2f(t)$

**16. a.** $\frac{V_o(s)}{V_i(s)} = \frac{1}{s+2}$

**18. b.** $\frac{V_o(s)}{V_i(s)} = \frac{s^2 + 2s + 2}{s^4 + 2s^3 + 3s^2 + 3s + 2}$

**34.** $\frac{\theta_2(s)}{T(s)} = \frac{3}{20s^2 + 13s + 4}$

**35.** $\frac{\theta_2(s)}{T(s)} = \frac{1}{130s^2 + 400s + 55}$

**44.** $\frac{\theta_2(s)}{E_a(s)} = \frac{0.0833}{s(s + 0.75)}$

## Chapter 3

**1.** $$\dot{\mathbf{x}} = \begin{bmatrix} -\frac{2}{3} & -\frac{1}{3} & \frac{1}{3} \\ -\frac{1}{3} & -\frac{2}{3} & \frac{2}{3} \\ -\frac{1}{3} & -\frac{2}{3} & -\frac{1}{3} \end{bmatrix} \begin{bmatrix} i_{L_1} \\ i_{L_2} \\ v_o \end{bmatrix} + \begin{bmatrix} \frac{2}{3} \\ \frac{1}{3} \\ \frac{1}{3} \end{bmatrix} v_i$$

$$y = \begin{bmatrix} 0 & 0 & 1 \end{bmatrix} \begin{bmatrix} i_{L_1} \\ i_{L_2} \\ v_o \end{bmatrix}$$

Note: $L_1$ is left-most inductor in Figure P3.1 in the text.

**11. a.** $$\dot{\mathbf{x}} = \begin{bmatrix} 0 & 1 & 0 & 0 \\ 0 & 0 & 1 & 0 \\ 0 & 0 & 0 & 1 \\ -10 & -5 & -1 & -2 \end{bmatrix} \mathbf{x} + \begin{bmatrix} 0 \\ 0 \\ 0 \\ 1 \end{bmatrix} r(t)$$

$$c(t) = \begin{bmatrix} 10 & 5 & 0 & 0 \end{bmatrix} \mathbf{x}$$

**14. a.** $$\frac{Y(s)}{R(s)} = \frac{10}{s^3 + 5s^2 + 2s + 3}$$

**17.** $$\dot{\mathbf{x}} = \begin{bmatrix} -\frac{D_{eq}}{J_{eq}} & 0 & \frac{K_t N_1}{J_{eq} N_2} \\ 1 & 0 & 0 \\ -\frac{K_b N_2}{L_a N_1} & 0 & -\frac{R_a}{L_a} \end{bmatrix} \begin{bmatrix} \omega_L \\ \theta_L \\ i_a \end{bmatrix} + \begin{bmatrix} 0 \\ 0 \\ \frac{1}{L_a} \end{bmatrix} e_a$$

$$y = \begin{bmatrix} 0 & \frac{N_2}{N_1} & 0 \end{bmatrix} \begin{bmatrix} \omega_L \\ \theta_L \\ i_a \end{bmatrix}$$

## Chapter 4

**14.** $$x(t) = \frac{1}{5}\left[1 - \sqrt{\frac{20}{19}} e^{-0.5t} \cos\left(\frac{\sqrt{19}}{2} t - \arctan \frac{1}{\sqrt{19}}\right)\right]$$

**20. a.** $\zeta = 0.375$; $\omega_n = 4$ rad/s; $T_s = 2.67$ s; $T_p = 0.847$ s; $\%OS = 28.06$

**23. a.** $s = -20 \pm j29.59$

**37.** $s = -5.79, -1.21$

**38. a.** $s^3 - 5s^2 - 8s + 2 = 0$ **b.** $s = 1.453, 0.22086, 6.2322$

**42.** $y(t) = \frac{2}{5} - \frac{2}{5} e^{-5t}$

**45.** $$\mathbf{\Phi}(t) = \begin{bmatrix} 1.0455e^{-0.20871t} - 0.045545e^{-4.7913t} & 0.21822e^{-0.20871t} - 0.21822e^{-4.7913t} \\ -0.21822e^{-0.20871t} + 0.21822e^{-4.7913t} & -0.045545e^{-0.20871t} + 1.0455e^{-4.7913t} \end{bmatrix}$$

$$\mathbf{x}(t) = \begin{bmatrix} 1.0455e^{-0.20871t} - 0.045545e^{-4.7913t} \\ -0.21822e^{-0.20871t} + 0.21822e^{-4.7913t} \end{bmatrix}$$

$$y(t) = 0.60911e^{-0.20871t} + 0.39089e^{-4.7913t}$$

**75.** $D = 0.1824$ N-m-s/rad

**78.** $R = 456\ \Omega$

## Chapter 5

**3.** $$\frac{C(s)}{R(s)} = \frac{G_3(G_1G_2 + 1)}{1 + G_1H_1}$$

**4.** $$\frac{C(s)}{R(s)} = \frac{G_1G_2 + G_3}{1 + G_3H + G_1G_2H + G_2G_4}$$

**6.** $$\frac{C(s)}{R(s)} = \frac{G_1G_5}{1 + G_1G_2 + G_1G_3G_4G_5 + G_1G_3G_5G_6G_7 + G_1G_5G_8}$$

**10.** $$\frac{C(s)}{R(s)} = \frac{G_4G_6 + G_2G_5G_6 + G_3G_5G_6}{1 + G_6 + G_1G_2 + G_1G_3 + G_1G_2G_6+}$$

$$G_1G_3G_6 + G_4G_6G_7 + G_2G_5G_6G_7 + G_3G_5G_6G_7$$

**26.** $$\frac{C(s)}{R(s)} = \frac{s^3 + 1}{2s^4 + s^2 + 2s}$$

**30. b.** $$\dot{\mathbf{x}} = \begin{bmatrix} -5 & 1 & 0 & 0 \\ 0 & -5 & 0 & 0 \\ 0 & 0 & -7 & 1 \\ 0 & 0 & 0 & -7 \end{bmatrix}\mathbf{x} + \begin{bmatrix} 0 \\ 1 \\ 0 \\ 1 \end{bmatrix} r(t)$$

$$\mathbf{y} = \begin{bmatrix} -\frac{3}{4} & 1 & -\frac{5}{4} & -1 \end{bmatrix}\mathbf{x}$$

**37.** $$\dot{\mathbf{x}} = \begin{bmatrix} 0 & 1 & 0 & 0 \\ -1 & 0 & 1 & 0 \\ 0 & 0 & 0 & 1 \\ 1 & -1 & 0 & 0 \end{bmatrix}\mathbf{x} + \begin{bmatrix} 0 \\ 0 \\ 0 \\ 1 \end{bmatrix} r(t)$$

$$c = \begin{bmatrix} -1 & 1 & 0 & 0 \end{bmatrix}\mathbf{x}$$

**69.** $D_L = 3560$ N-m-s/rad

**71.** $D_L = 4192.11$ N-m-s/rad

## Chapter 6

**3.** 3 rhp, 2 lhp, 0 $j\omega$

**4.** 1 rhp, 0 lhp, 4 $j\omega$

**5.** 0 rhp, 2 lhp, 2 $j\omega$

**9.** Stable (0 rhp, 4 lhp)

**17.** 1 rhp, 2 lhp, 4 $j\omega$

**23.** Stable for $0 < K < 140{:}8$

**39.** **a.** $-3 < K < 14.22$; **b.** 1.291 rad/s

**41.** **a.** $0 < K < 19.69$; **b.** $K = 19.69$; **c.** $s = \pm j1.118, -4.5, -3.5$

**42.** $-\frac{2}{3} < K < 0$

## Chapter 7

**4.** $e_{\text{step}}(\infty) = 0$; $e_{\text{ramp}}(\infty) = 127.5$; $e_{\text{parabola}}(\infty) = \infty$

**7.** $\dot{e}(\infty) = 0.9$

**10.** **a.** $\%OS = 4.32$; **b.** $T_s = 0.16$ sec; **c.** $e_{ss} = 0$;
**d.** $e_{ss} = 0.2$; **e.** $e_{ss} = \infty$

**13.** **a.** $K_p = 0.25$, $K_v = 0$, $K_a = 0$; **b.** $e(\infty) = 40, \infty, \infty$, respectively;
**c.** Type 0

**20.** $K = 110,000$

**27.** $\beta = 1$, $K = 1.16$, $\alpha = 7.76$, or $\beta = -1$, $K = 5.16$, $\alpha = 1.74$

**32.** **a.** $K = 207,936$; $a = 415.872$; **b.** $K = 3,492.8$; $a = 69.86$

**40.** **a.** Step: $e(\infty) = 0.1622$; ramp: $e(\infty) = \infty$

## Chapter 8

**16.** Breakaway point $= -2.333$; asymptotes: $\sigma_a = -5$; $j\omega$-axis crossing $= \pm j7.35$

**19.** **b.** Asymptotes: $\sigma_a = -\frac{8}{3}$; **c.** $K = 140.8$; **d.** $K = 13.12$

**20.** $K = 9997$; $\alpha = 7$

**23.** **a.** $\sigma_a = -\frac{5}{2}$; **b.** $s = -1.38, -3.62$; **c.** $0 < K < 126$; **d.** $K = 10.3$

**26.** **b.** $K = 14.0$; **c.** $T_s = 4.23$ s, $T_p = 1.698$ s; **d.** $\sigma = -7.0$; **e.** $0 < K < 89.7$

**30.** $\alpha = 9$

**39.** **a.** $0 < K < 4366$; **b.** $K = 827.2$; **c.** $K = 527.6$

**42.** **a.** $K = 170.1$; **b.** $K = 16.95$

## Chapter 9

**1.** $G_c(s) = \frac{s+0.1}{s}$; $K \simeq 72.23$ for both cases; $K_{p_O} = 2.44$; $K_{p_N} = \infty$;
$\%OS_O = \%OS_N = 16.3$; $T_{s_O} = T_{s_N} = 2.65$ s

**9.** **a.** $s = -2.5 \pm j5.67$; **b.** Angle $= -59.27°$; **c.** $s = -5.87$
**d.** $K = 225.79$; **e.** $s = -11.59, -1.36$

**10.** **a.** $s = -2.2 \pm j3.81$; **b.** $s = -5.694$; **c.** $K = 25.1$;
**d.** $s = -1.29$; **f.** $K_a = 4.4$

**15.** **a.** $G_c(s) = \frac{s+7}{s+37.42}$, $K = 5452$; dominant poles $= -4.13 \pm j10.78$

**24. a.** $K_{uc} = 10$; $K_c = 9.95$; **b.** $K_{p_{uc}} = 1.25$; $K_{p_c} = 6.22$;

**c.** $\%OS_{uc} = \%OS_c = 4.32$;

**d.** Uncompensated: exact second-order system, approximation OK; compensated: closed-loop pole at $-0.3$, closed-loop zero at $-0.5$, simulate

**e.** Approach to final value longer than settling time of uncompensated system

**f.** $G_{\text{LLC}}(s) = \dfrac{404.1(s+0.5)(s+4)}{(s+2)(s+4)(s+0.1)(s+28.36)}$ yields approximately a 5 times improvement in speed.

**25.** $G_c(s) = \dfrac{(s+6.93)(s+0.1)}{s}$, $K = 3.08$

**28.** Poles $= -0.747 \pm j1.237,\ -2.51$; zeros–none

## Chapter 10

**10.** System a: $0 < K < 810$; System 2: If $0 < K < 3.94$, the system is unstable. If $K > 3.94$, the system is stable; System 3: If $0 < K < 720$, the system is stable. If $K > 720$, the system is unstable. (Answers are from exact frequency response)

**11. a.** System 1: Gain margin = 4 dB; Phase margin = 15°.

**15. c.** $\omega_{\text{BW}} = 1.5$ rad/s

**24.** System 2: $T_s = 2.23$ sec, $T_p = 0.476$ s, $\%OS = 42.62$ (Answers are from exact frequency response)

**44.** $G_M = 1.17$ dB, $\boldsymbol{\Phi}_M = 6.01°$ (Answers are from exact frequency response)

## Chapter 11

**1. a.** $K = 1639$ (Answer is from exact frequency response)

**2. a.** $K = 1782$ (Answer is from exact frequency response)

**12.** $G_c(s) = \dfrac{s+2.701}{s+5.954)}$, $K = 7936$
(Answer is from exact frequency response)

**21.** $G_c(s) = \dfrac{(s+0.097)(s+2.16)}{s}$, $K = 25.189$
(Answer is from exact frequency response)

## Chapter 12

**1. d.** For function **i**: $T(s) = \dfrac{s+1}{s^2 + (2+k_2)s + k_1}$

**3. b.** For function **i**: $G(s) = \dfrac{6.25}{s} - \dfrac{27.5}{s+10} + \dfrac{71.25}{s+20}$, $T(s) = \dfrac{200(s^2+7s+25)}{4s^3 + as^2 + bs + c}$

where
$$a = (25k_3 - 110k_2 + 285k_1 + 120)$$
$$b = (750k_3 - 2200k_2 + 2850k_1 + 800)$$
$$c = 5000k_3$$

and $\mathbf{C} = [1 \quad 1 \quad 1]$; $\mathbf{B} = [71.25 \quad -27.5 \quad 6.25]^T$ was used

**12. a.** Uncontrollable; **b.** Controllable; **c.** Controllable

**14.** $\mathbf{K} = [92.35 \quad 36.78 \quad -7]$ for a characteristic polynomial of $(s+6)(s^2+8s+45.78) = s^3 + 14s^2 + 93.78s + 274.7$

**23.** $\mathbf{l}_1 = -656.4$; $\mathbf{l}_2 = -786.4$

## Chapter 13

**3. a.** $f(kT) = 229.5(0.4)^k - 504(0.6)^k + 275.5(0.8)^k$

**6. c.** $G(z) = 1.02557\dfrac{z(z+0.36788)}{(z-0.36788)(z^2-0.052046z+0.13534)}$

**8. b.** $G(z) = 0.0012549\,\dfrac{(z+0.19485)(z+2.821)}{(z-1)(z-0.36788)(z-0.81873)}$

**9. a.** $T(z) = \dfrac{G_1(z)G_2(z)}{1+G_1(z)G_2H(z)}$

**14.** $0 < K < 11.104$

**15. a.** $K_p = \dfrac{1}{2}$, $e^*(\infty) = \dfrac{2}{3}$; $K_v = 0$, $e^*(\infty) = \infty$; $K_a = 0$, $e^*(\infty) = \infty$

**17.** $K = 14.18$ for 16.3% of overshoot; $0 < K < 109.28$ for stability

# Index

**D**

**E**

**N**

## Q



## R

**T**

**Z**

# Key Equations

## Modeling

$$\frac{V_o(s)}{V_i(s)} = -\frac{Z_2(s)}{Z_1(s)} \quad (2.97); \qquad \frac{V_o(s)}{V_i(s)} = \frac{Z_1(s)+Z_2(s)}{Z_1(s)} \quad (2.104)$$

$$\frac{\theta_2}{\theta_1} = \frac{r_1}{r_2} = \frac{N_1}{N_2} \quad (2.133); \qquad \frac{T_2}{T_1} = \frac{\theta_1}{\theta_2} = \frac{N_2}{N_1} \quad (2.135)$$

$$\left(\frac{\text{Number of teeth of gear on } \textit{destination} \text{ shaft}}{\text{Number of teeth of gear on } \textit{source} \text{ shaft}}\right)^2 \quad \text{(see after 2.138)}$$

$$\frac{\theta_m(s)}{E_a(s)} = \frac{K_t/(R_aJ_m)}{s\left[s+\frac{1}{J_m}\left(D_m+\frac{K_tK_b}{R_a}\right)\right]} \quad (2.153)$$

$$\frac{K_t}{R_a} = \frac{T_{\text{stall}}}{e_a} \quad (2.162); \quad K_b = \frac{e_a}{\omega_{\text{no-load}}} \quad (2.163)$$

$$T(s) = \frac{Y(s)}{U(s)} = \mathbf{C}(s\mathbf{I}-\mathbf{A})^{-1}\mathbf{B}+\mathbf{D} \quad (3.73)$$

## Time Response

$$T_r = \frac{2.2}{a} \quad (4.9); \quad T_s = \frac{4}{a} \quad (4.10)$$

$$G(s) = \frac{\omega_n^2}{s^2+2\zeta\omega_n s+\omega_n^2} \quad (4.22)$$

$$\%OS = e^{-(\zeta\pi/\sqrt{1-\zeta^2})} \times 100 \quad (4.38)$$

$$\zeta = \frac{-\ln(\%OS/100)}{\sqrt{\pi^2+\ln^2(\%OS/100)}} \quad (4.39)$$

$$T_p = \frac{\pi}{\omega_n\sqrt{1-\zeta^2}} \quad (4.34); \quad T_s = \frac{4}{\zeta\omega_n} \quad (4.42)$$

## Steady-State Error

$$e(\infty) = e_{\text{step}}(\infty) = \frac{1}{1+\lim_{s\to 0} G(s)} \quad (7.30); \quad K_p = \lim_{s\to 0} G(s) \quad (7.33)$$

$$e(\infty) = e_{\text{ramp}}(\infty) = \frac{1}{\lim_{s\to 0} sG(s)} \quad (7.31); \quad K_v = \lim_{s\to 0} sG(s) \quad (7.34)$$

$$e(\infty) = e_{\text{parabola}}(\infty) = \frac{1}{\lim_{s\to 0} s^2G(s)} \quad (7.32); \quad K_a = \lim_{s\to 0} s^2G(s) \quad (7.35)$$

## Root Locus

$$\angle KG(s)H(s) = -1 = 1\angle(2k+1)180^\circ \quad (8.13)$$

$$\sigma_a = \frac{\sum \text{finite poles} - \sum \text{finite zeros}}{\#\text{ finite poles} - \#\text{ finite zeros}} \quad (8.27)$$

$$\theta_a = \frac{(2k+1)\pi}{\#\text{ finite poles} - \#\text{ finite zeros}} \quad (8.28)$$

$$\theta = \sum \text{finite zero angles} - \sum \text{finite pole angles}$$

$$K = \frac{1}{|G(s)H(s)|} = \frac{1}{M} = \frac{\prod \text{finite pole lengths}}{\prod \text{finite zero lengths}} \quad (8.51)$$

## Frequency Response

$$M_p = \frac{1}{2\zeta\sqrt{1-\zeta^2}} \quad (10.52); \quad \omega_p = \omega_n\sqrt{1-2\zeta^2} \quad (10.53)$$

$$\omega_{\text{BW}} = \omega_n\sqrt{(1-2\zeta^2)+\sqrt{4\zeta^4-4\zeta^2+2}} \quad (10.54)$$

$$\Phi_M = \tan^{-1}\frac{2\zeta}{\sqrt{-2\zeta^2+\sqrt{1+4\zeta^4}}} \quad (10.73)$$

$$\phi_{\max} = \tan^{-1}\frac{1-\beta}{2\sqrt{\beta}} = \sin^{-1}\frac{1-\beta}{1+\beta} \quad (11.11)$$

$$\omega_{\max} = \frac{1}{T\sqrt{\beta}} \quad (11.9); \quad |G_c(j\omega_{\max})| = \frac{1}{\sqrt{\beta}} \quad (11.12)$$

## State Space

$$\mathbf{C_M} = [\mathbf{B} \quad \mathbf{AB} \quad \mathbf{A}^2\mathbf{B} \quad \cdots \quad \mathbf{A}^{n-1}\mathbf{B}] \quad (12.26)$$

$$\dot{\mathbf{x}} = (\mathbf{A}-\mathbf{BK})\mathbf{x}+\mathbf{B}r; \quad y = \mathbf{Cx} \quad (12.3); \quad \mathbf{O_M} = \begin{bmatrix} \mathbf{C} \\ \mathbf{CA} \\ \vdots \\ \mathbf{CA}^{n-1} \end{bmatrix} \quad (12.79)$$

$$\dot{\mathbf{e}}_\mathbf{x} = (\mathbf{A}-\mathbf{LC})\mathbf{e}_\mathbf{x}; \quad y-\hat{y} = \mathbf{Ce}_\mathbf{x} \quad (12.64)$$

## Digital Control

$$e^*(\infty) = \lim_{z\to 1}(1-z^{-1})E(z) \quad (13.66)$$

$$K_p = \lim_{z\to 1} G(z) \quad (13.70); \quad K_v = \frac{1}{T}\lim_{z\to 1}(z-1)G(z) \quad (13.73)$$

$$K_a = \frac{1}{T^2}\lim_{z\to 1}(z-1)^2G(z) \quad (13.75)$$

# Unmanned Free-Swimming Submersible Vehicle

## Pitch Control System

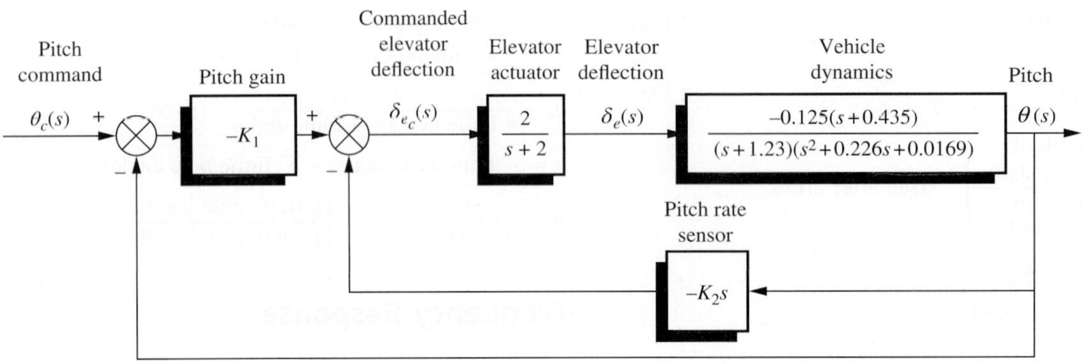


## Heading Control System

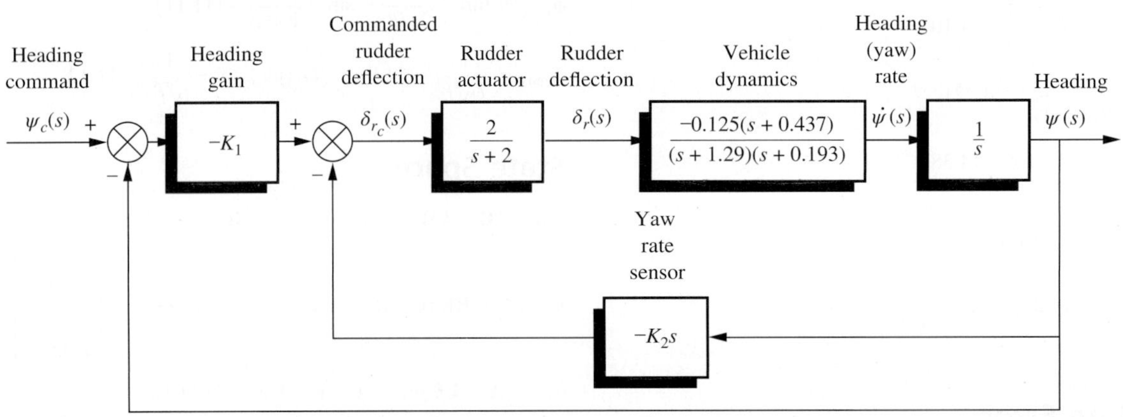